著名病毒学家曾毅院士论文集 IV（外文部分）

Selected Workers of Zeng Yi
Volume 4（In English）

邵一鸣　周　玲　主编
Shao Yi-Ming　Zhou Ling

中国科学技术出版社
·北　京·

图书在版编目（CIP）数据

著名病毒学家曾毅院士论文集. 4，1957 - 2009：英文/邵一鸣，周玲主编.
—北京：中国科学技术出版社，2010. 9
ISBN 978 - 7 - 5046 - 5662 - 9

Ⅰ. ①著… Ⅱ. ①邵… ②周… Ⅲ. ①病毒学 - 文集 - 英文 Ⅳ. ①Q939. 4 - 53

中国版本图书馆 CIP 数据核字（2010）第 129715 号

本社图书贴有防伪标志，未贴为盗版

责任编辑：张　楠
责任校对：林　华　刘红岩
责任印制：安利平

About the author

Zeng Yi, graduated from Shanghai first medical college in 1952, is a distinguished virologist, an Academician of Chinese Academy of Sciences, a foreign member of France National Academy of Medical Sciences and Russian Academy of Medical Sciences, a tutor of postgraduate candidates and graduate candidates.

He used to work as research fellow in MRC Virology Centre in Glasgow UK from 1974 to 1975 and his main concern was cancer basic research. He worked as research fellow in France National Research Centre from 1986 to 1987 and studied on HIV. He used to be the director of Institute for Virology, the president of Chinese Academy of Preventive Medicine, a member of Academic Degrees Committee of the State Council and WHO Expert Advisory Panel on Cancer and Executive Board of International Union of Microbiology, the president of Chinese Foundation for Prevention of STD and AIDS, the president of Chinese Preventive Medicine Association, an executive member of steering committee of Asia Pacific Leadership Forum on HIV/AIDS and Development.

He is the chief scientist of National Center for AIDS/STD Control and Prevention, the chief of Academician's Laboratory in Institutes of Viral Disease Control and Prevention, the dean of Life Science and Bioengineering College of Beijing University of Technology, a senior counselor of China National Accreditation Service, an executive member of Chinese Medical Association, and the honor president of China Preventive Medicine Association.

Academician Zeng Yi is the first scientist studying tumor virus and HIV in China. Since 1961 he has been working on oncogenic virus and mainly focuses on animal leukaemia virus, polyoma virus, human papilloma virus, adenovirus and other associated cancer viruses. Since 1973 he has been mainly working on the relationship between EB virus and nasopharyngeal carcinoma as well as the relationship between other tumor viruses and cancer, including HPV, cervical cancer and esophageal cancer, HBV and liver cancer, HTLV-1 and adult T cell leukaemia, HHV-8 and kaposi sarcoma.

Academician Zeng Yi has more than 500 papers and 6 books published in Chinese and English and has been honored 20 awards.

Introduction

Professor Zeng Yi, a member of the Chinese Academy of Sciences, is an internationally renowned virologist and oncologist. He graduated from Shanghai First Medical College soon after the People's Republic of China was established. In the early years of his career, he worked in medical schools as a researcher and teacher in microbiology and virology. Later he transferred to the Chinese Academy of Medical Sciences, and then the Chinese Academy of Preventive Medicine and the Chinese Center for Disease Control, which were founded later, and has devoted to his beloved medical virology research to this day. Even during the Cultural Revolution, he managed to overcome all difficulties and continued his research, traveling between home, research institute, and cancer hospital by bicycle for 4 hours everyday. It is his strong curiosity towards scientific truths and strong sense of responsibility on developing the preventive medicine in our country that has inspired Professor Zeng Yi to go through nearly 60 years of ups and downs and become a great scientist in tumor virology and HIV/AIDS preventive medicine.

Professor Zeng Yi has made a lot of achievements in virology, oncology and HIV/AIDS prevention and treatment, which is clearly reflected by the 509 scientific publications in Chinese or English collected in this book. These research results have won a number of science and technology progress awards from the state and the ministries, and many other science and technology awards such as Tan Kah Kee award. Because of his profound scientific attainments, Professor Zeng Yi was elected member of the Chinese Academy of Sciences in 1993, and was elected foreign member of the French National Academy of Medical Sciences and the Russian Academy of Medical Sciences by international scholars. As Professor Zeng Yi's students, we have learned from him the outstanding characters uniquely owned by the old generation of scientists of China. His scientific spirit and methodologies will benefit us for life. Among Professor Zeng Yi's numerous outstanding scientific characters, what impressed us the most is his sharp insight on the direction of scientific development, his courage to practice and his strong sense of responsibility on disease prevention in the country.

At the end of 1950s, Professor Zeng Yi proposed the hypothesis that since many animal tumors are caused by viruses, human tumors might also be caused by viruses.

He began to study tumor viruses in 1961, such as adenovirus, chicken leukemia virus, and polyoma virus. He began to study human tumor viruses in 1973, including EB virus and HPV. He systematically studied the role of EB virus in the occurrence and development of nasopharyngeal carcinoma (NPC), and by creatively combining the molecular virology techniques developed by international researchers with field epidemiological investigation, he established a NPC early diagnosis technology system employing EB virus and EA/IgA, VCA/IgA antibody screening supplemented by clinical diagnosis and biopsy, which significantly improved the early diagnosis rate of NPC and has saved the lives of many patients. Application of serological markers in the diagnosis of tumor was an innovation in tumor virology and cancer diagnostics. It was a successful practice of applying the results of basic research to clinical medicine, an ideal design called "from bench to bedside". In the research area of viral etiology of NPC, Professor Zeng Yi also carried out large – scale field etiological investigation through multi – disciplinary collaboration. Combining laboratory research results, he proposed a multiple etiologies theory on NPC with EB virus as the cause, environmental carcinogens playing a synergistic role, and the genetic susceptibility as the underlying factor. This theory has an important position in the NPC etiology field and promoted the advance of tumor virology research.

In the early 1980s, a ferocious new infectious disease, AIDS, was discovered in the United States. As a tumor virologist, Professor Zeng Yi immediately set up a laboratory and related techniques to study the virus and closely followed the international research progress in this field. In 1983, the French scientists first reported the discovery of AIDS virus (HIV), and in 1984, Professor Zeng Yi carried out HIV screening in China and was the first one to perform epidemiological study on sera from AIDS patients in our country. In 1985, Professor Zeng Yi reported the first 4 HIV – infected cases in China. Later, Professor Zeng Yi's laboratory undertook the diagnosis, training, technical support on AIDS, and the research and development of AIDS diagnostic reagents in the early stage of China's AIDS control, and strongly supported the AIDS diagnosis work and serological study in the early stage. Professor Zeng Yi also collaborated with his wife, Professor Li Zelin of the China Academy of Traditional Chinese Medicine, in the research of screening anti – HIV ingredient from traditional Chinese medicine. After more than a decade's unremitting efforts, the study has advanced to clinical trials. These studies have also been recorded in this collection of publications.

As a member of the Chinese Academy of Sciences and a leader of China's disease control agency and academic society, Professor Zeng Yi has been continuously calling the government to strengthen the fight against AIDS and increase input on AIDS research in various historical periods, he also constantly lectured around the country, and personally involved in organizing AIDS control knowledge tour, specific AIDS control campaigns, and AIDS control fund – raising work. The rapid development of China's AIDS control work today can not be separated from the powerful promotion by the scientist and social activist Professor Zeng Yi and a number of other well – known scientists.

Professor Zeng Yi has published a wealth of research articles in 56 years of medical virology study. We have tried our best to collect and organize 509 articles published by Mr. Zeng from 1957 to early 2010 (397 articles in Chinese and 112 in English) and put them into this collection. Because the articles spanned a period of more than half a century and were published in dozens of journals or materials, the original articles were presented in all sorts of layout styles and formats. In order to keep the original style, we decided to respect for history and keep the major contents of the original articles. However, for today's readers, we have taken current editing norm into consideration, and generally unified the format of all articles. Because of the passage of time, a number of articles are hard to find, so we have to leave them out; the quality of some original figures and tables is now too poor to be copied, so they have to be omitted. As time went by, the name of the author's institute has been changed several times. The names in the original articles will be used in this book, which will not be explained separately. This book is divided into Volume I to IV in chronological order. The fist three volumes are articles in Chinese and the fourth volume contains articles in English. We received a few early articles from time to time in the late stage of compilation, which can only be appended at the end of the book as "Addendum". The whole book contains approximately 4.30 million words. We truly believe that the publication of this book will definitely contribute to the development of China's medical virology. We would express our sincere gratitude to all colleagues at the Institute for Viral Disease Control and Prevention of China CDC, National Center for AIDS/STD Control and Prevention of China CDC, and everyone involved in the compilation of this book.

As this is a large amount of work finished in a relatively short period of time, it is inevitable that the book may have some errors. Your criticism and correction are welcome!

Shao Yiming and Zhou Ling

June 1, 2009

CONTENT

1978—1985

1986—2000

2001—2005

2006—2009

Addendum

1. Influence of Placental Globulin on Clinical and Immunologic Effects of the Vaccine

HUANG Chenh-siang[1], CHU Fu-tang[2], CHIA Ping-yi[3], LIN Chuan-chia[2], ZENG Yi[1]

1. Department of Virology, Chinese Academy of Medical Sciences;

2. Peking Children's Hospital, Peking; 3. Academy of Medical Sciences of Hepei Province

This paper reports our studies on the influence of placental globulin on the clinical and immunologic response of 130 children previously inoculated with 6 $TCID_{50}$ virus vaccine L_4 mentioned in the foregoing communication.

Material and Methods

1. Placental globulin: The placental globulin (lot no. 6097), prepared by the National Vaccine and Serum Institute, has a gamma globulin content of approximately 3.5%. Its titer of measles antibody as determined by tissue culture neutralization test, using 100 $TCID_{50}$ virus for neutralization, and hemagglutination inhibition test was both 1:50.

2. Methods: 130 healthy normally developed children, aged 1/2 – 6 years (only 2 below 1 year of age), with no history of measles and no recent history of contact with measles patients were chosen and studied in 4 kindergartens in the winter of 1960. Children having a history of convulsion during high fever and those whose parents objected were not vaccinated. They lived together with the vaccinated children and served as indicators to determine the communicability of the vaccinated and as controls on the effectiveness of the vaccine if measles broke out.

3. Method of immunization: The 130 children were all vaccinated subcutaneously with 0.5 ml of 6 $TCID_{50}$ virus vaccine L_4. They were divided into 12 groups and given different amounts (1 – 3 ml) of placental globulin at intervals of 1 – 4 days after inoculation of the vaccine (Tab. 1). One group of vaccinated children who were not inoculated with placental globulin served as controls. In another kindergarten, 3 children were given 0.6 $TCID_{50}$ virus vaccine L_4 and one of them was given 3 ml of placental globulin 6 days later.

Observations of the clinical reactions to the vaccine were made by the same method as stated in the previous paper.

4. Immunologic evaluation: Evaluation of the effectiveness of the immunization was based on both serologic and epidemiologic studies. The hemagglutination inhibition method was used to study the antibody content of the sera. Blood was taken from the ear lobe and the test was carried out in a plastic plate. The antigens were prepared by propagating the virus in a stable line of either human amnion cells or human kidney cells. Two fold dilutions of the serum were made 0.025 ml of antigen containing 4 hemagglutination units was added to an equal volume of different dilutions of the serum. They were mixed, kept at room temperature for 1 hour and then 0.025 ml of 1% monkey red blood cells was added to each. The results were and after 1 hour of incubation at 37 ℃. The highest dilution of the serum that gave complete inhibition of hemagglutination was taken as the antibody titer. A comparative study of the hemagglutination inhibition and tissue culture neutralization tests (using 100 TCID$_{50}$ of the virus for neutralization) carried out in our laboratory showed that the titer was approximately the same[16].

Tab. 1 Influence of placental globulin administration at different intervals after inoculation of attenuated measles vaccine on the clinical reactions

Group No.	Placental globulin Amount (ml)	Interval after vaccination (days)	Kindergarten	Children inoculated	Age (yr)	Febrile reaction Children	%	Incubation (days)	Average incubation (days)	Range of maximum temperature (℃)	Average maximum temperature (℃)	Range of duration (days)	Average duration (days)	No. with rash	No. with catarrh
1	0	—	A,B	18	2–6	13	72.2	5-10	8.5	37.2–40.4	39.1	1–6	3.9	7	8
2		1	B	4	2–4	3	75.0	7–10	8.7	38.8–39.5	38.9	2–5	3.7	1	2
3	1.0	2	C,D	18	1–5	13	72.2	8–12	9.8	37.8–40.2	38.7	1–5	2.8	0	11
4		3	A	15	1–6	13	80.0	7–14	10.7	37.5–39.5	38.6	1–5	3.3	5	6
5	1.5	2	D	9	2/3–5	5	55.6	8–9	9.0	37.8–40.0	38.7	1–5	2.8	2	4
6		1	B	4	2–4	3	75.0	11	11.0	37.4–49.0	38.4	1–3	2.0	0	0
7	2.0	2	B,C	12	1/2–6	4	33.4	7–10	9.5	33.0–39.6	38.8	2–3	2.5	1	3
8		3	A	14	1–4	9	64.3	9–13	10.8	38.2–39.7	38.9	1–4	2.8	3	5
9		1	B	4	3–4	2	50.0	9–12	10.5	37.5–38.0	37.8	1–2	1.5	0	1
10	3.0	2	B,C	13	1–6	4	30.8	5–10	8.0	37.2–39.1	38.2	1–3	2.2	1	0
11		3	A	13	1–6	6	46.1	8–13	10.5	37.5–38.4	37.8	1–3	2.0	0	3
12		4	D	6	1–2	1	16.7	11	11.0	39.4	39.4	3	3.0	0	1

Experimental Results

1. Influence of placental globulin on clinical reactions to the vaccine: Successful application of gamma globulin to decrease or abolish the clinical reactions in natural measles infection depends on 3 variable factors: the infective dose, the amount of gamma globulin used and the interval from the time of exposure to the administration of gamma blobulin. Our results are recorded in Tab. 1, 2 and 3. From Tab. 1, in can be seen that the period between inoculation of the vaccine and administration of placental globulin did not seem to affect the clinical reactions but that different doses of placental globulin had considerable influence on them. As shown in Tab. 2, with increase of placental globulin the percentage of children showing febrile reaction decreased, 75.7%, 55.6%, 53.3%, 36.1% for 1 ml, 1.5 ml, 2 ml, 3 ml respectively. While, as has been stated in the foregoing paper 6 TCID$_{50}$ of the vaccine gave rise to 100% febrile reaction, the results recorded in Tab. 2 show that 27.8% of the control group and 24.3% of the group given 1 ml placental globulin had no febrile reaction. This probably indicates that these children had inapparent infection before vaccination. The degree and duration of febrile reaction also decreased with increase in the amount of placental globulin. The average highest temperature of the control group was 39.1 ℃, the average duration of fever was 3.9 days and the percentage of those having a higher temperature that 39 ℃ was 61.5%, while in the groups given placental globulin the corresponding figures were 38.8 ℃, 2.9 days and 37.5% in the 2 ml group and 38 ℃, 1.8 days and 15.4% in the 3 ml group.

Tab. 2　Influence of dosage of placental globulin on the clinical reactions to the vaccine

| Group | Children inoculated | Febrlle reaction | | | | | | | | | No. with rash | No. with catarrh |
		Children	%	Incubation (days)	Average incubation (days)	Range of maximum temperature (℃)	Average maximum temperature (℃)	Range of duration (days)	Average duration (days)	Temperature above 39 ℃ (%)		
Vaccine alone (control)	18	13	72.2	5 – 10	8.5	37.2 – 40.4	39.1	1 – 6	3.9	61.5	7	8
Vaccine followed by 1 ml placental globulin	37	28	75.7	6 – 14	10.0	37.5 – 40.2	38.9	1 – 5	2.9	50.0	6	9
Vaccine followed by 1.5 ml placental globulin	9	5	55.6	8 – 11	9.0	37.8 – 40.0	38.7	1 – 5	2.8	40.0	2	4
Vaccine followed by 2 ml placental globulin	30	16	53.3	7 – 13	10.0	37.4 – 39.7	38.8	1 – 7	2.9	37.5	4	8
Vaccine followed by 3 ml placental globulin	36	13	36.1	5 – 11	9.8	37.1 – 39.4	38.0	1 – 3	1.8	15.4	1	5

Since the amount of gamma globulin used to modify the clinical eoures of natural infection of measles was calculated according to the body weight, we tried to ascertain if the percentage of febrile reaction in the younger age group (1/2 – 2 years) should be lower than that of the older age group (3 – 6 years). Tab. 3 shows that such difference was observed only when 3 ml of placental globulin were used.

Tab. 3　Influence of dose of placental globulin and age of vaccinated children on clinical reactions

| Amount of placental globulin given after vaccination (ml) | 1/2 – 2 years | | | 3 – 6 years | | |
	No. inoculated	No. without clinical reactions	% without clinical reactions	No. inoculated	No. without clinical reactions	% without clinical reactions
1 – 1.5	25	8	32.0	20	5	25.0
2	11	4	36.4	19	9	47.4
3	15	12	80.0	21	11	52.4
1 – 3	51	24	47.1	60	25	41.6
Vaccine only	4	1	25.0	15	4	26.7

2. Antibody response after cobined use of vaccine and placental globulin: Blood specimens were taken from the ear lobe 1 day before and 2 and 6 – 7 months after inoculation of the vaccine. While the tissue culture neutralization test of the sera collected 1 day before and 2 months after inoculation of the vaccing was made, the temperature in the incubator reached 65℃. Because of this accident only a few paired sera that had been tested by the hemagglutination inhibition method conld be recorded. As shown in Tab. 4, the combined use of vaccine and placental globulin induced inapparent infection. 3 children inoculated with 6 TCID$_{50}$, and 1 child inoculated with 0.6 TCID$_{50}$, with no detectable antibody before vaccination, showed antibody when tested 2 or 6 months later even though the clinical reactions had completely abated after the use of placental globulin.

Tab. 4 Hemagglutination – inhibiting antibody titer before and after immunization with vaccine followed by placental globulin

Vacclin	Placental globulin		Febrile reaction	Amtibody titer			
Infective doses TCID$_{50}$	Day given after inoculation of vaccine	Amout (ml)		Before vaccination	1 month after vaccination	2 months after vaccination	6 months after vaccination
*6	—	—	+	<1:5	1:80		
	—	—	+	<1:5	1:40		
	—	—	+	<1:5	1:80		
	—	—	+	<1:5	1:40		
6	3	2	+	<1:4			1:16
	3	2	—	<1:4		1:64	
	2	3	—	<1:4			1:16
	3	3	—	<1:4		1:32	
*0.6	—	—	+	<1:5	1:40		
	—	—	+	<1:5	1:80		
0.6	6	3	—	<1:4		>1:128	

Note: * Data form previous paper[3] using neutralization test for comparison

103 children were available for antibody study 6 – 7 months after the combined use of vaccine and placental globulin. The results show that all except 2 had detectable antibody 6 – 7 months after inoculation (Tab. 5). The titer varied from 1:40 to 1:32. The use of various doses of placental globulin after vaccination did not seem to influence the antibody level as compared with the control group who received vaccine alone. The average antibody titer of the children showing no clinical reactions was not significantly different from those showing cliical reactions.

Tab. 5 Hemagglutination inhibiting antibody titer 6 – 7 months after administration of vaccine alone or vaccine followed by placental globulin

Amount of placental globulin given after vaccination (ml)	Febrle reaction		Amtibody titer 6 – 7 months after inoculaltion					
	No. with reaction	No. without reaction	<1:4	1:4	1:8	1:16	1:32	Average
1.0	20				13	4	3	1:13.2
		7		3	2	1	1	1:10.9
1.5 – 2.0	15			1	5	7	2	1:14.7
		13		1	7	4	1	1:12.0
3.0	13		2	3	5	2	1	1:8.9
		13		7	3	6	2	1:11.8
1.0 – 3.0	48		2	4	23	13	6	1:12.5
		38		11	12	11	4	1:11.7
Vaccine only	13			1	2	10		1:13.8
		4		1	2	1		1:9.0

3. *Epidemiologic evaluation of the combined use of vaccine and placental globulin*: Outbreaks of measles among the unvaccinated susceptible children occurred at different intervals after inoculation in 5 classes as 2 kindergartes. In order to obtains a better evaluation of the effectiveness of the combined use of vaccine and placental golbulin, measles patients and vaccinated children were not isolated. The epidemiologic observations are summarized in Tab. 6.

Tab. 6   Epidemiologic evaluation of the immunization

Kin-der-garten	Class No.	Unvaccinated susceptible group			Vaccinated group					
		Chil-dren	No. with measles	Development of measles after inoculation (days)	Inoculated with vaccine alone		Inoculated with vaccine followed by placental globulin		To-tal	No. with mea-sles
					No. with fever	No. without fever	No. with fever	No. without fever		
A	1	3	3	2,18,31	1	1			2	0
	2	2	2	47,59	1	1		3	5	
D	2	12	12	13,28,29,29,30,43,43,44 44,55,56			7	2	9	0
	1	9	7	27,35,36,38,38,39,39			2	2	4	0
	3	15	15	127,139,139,139,139,141,141,141 141,141,143,145,150,150,150			2	6	8	0
Total		41	39		2	2	11	13	28	0

In Class 1 of kindergarten A, there were 3 unvaccinated susceptible children, and they had typical measles 2, 18 and 31 days after the other children of the same class had been vaccinated. In this calss, 2 children who received the vaccine alone (1 had febrile reaction 9 days after vaccination and the other had no reaction) had intimate contact with the 3 measles children for a period of 30 days but none caught the disease. In Class 2 of the same kindergarten, the only 2 unvaccinated susceptible children had typical measles, one 47 days and the other 59 days after 5 children of the same class had revieved the vaccine. All these 5 vaccinated children did not have measles in spite of close contact with the 2 measles chidren for at least 12 days.

In kindergarten D, 3 of the classes had measles outbreaks among the unvaccinated susceptible chidren. In Class 2, there were 12 unvaccinated susceptible children who had typical measles within a period of 43 days. The first case occurred 13 days and the last case 56 days after the other 9 children had received the vaccine. In spite of close contact with the measles children for so long a period, none of the 9 children inoculated with the vaccine and later with 1 − 1.5 ml of placental globulin showed any signs of measles. In Class 1 of the same kindergarten, 7 out of 9 unvaccinated susceptible children had typical measles 27 − 39 days after 4 other children had received inoculation. None of the latter had measles. These 4 children were given first the vaccine and then 1, 1.5, 3 and 3 ml placental globulin respectively. It should be pointed out that the children of Class 1 lived together with those of Class 2 for 4 days, and during these 4 days the first case of measles occurred in Class 2, so actually the 4 vaccinated children of Class 1 had repeated contact with measles patients for a period of 26 days (Tab. 6). In Class 3 of the same kindergarten, there were 15 unvaccinated susceptible children and 8 who received both the vaccine and 1. 5 – 3 ml of placental globulin. All the 15 unvaccinated susceptible children came down with measles 127 – 150 days after the vaccinated group had received the vaccine, but none of the 8 vaccinated children had measles. Among these 8, 2 had and 6 had no febrile reaction. Of the 6 children with no febrile reaction, 3 each had received 3 ml and 1. 5 ml of placental globulin.

The above epidemiologic observations indicate that after inoculation of the vaccine, the administration of placental globulin did not influence the development of immunity to measles but it did reduce the degree and percentage of febrile reaction. They also show that the vaccine virus was probably not communicable by mere contact of susceptible children with the vaccinated.

Discussion

From the above studies, it can be seen clearly that during the first four days after inoculation of 6 TCID$_{50}$ virus vaccine L$_4$, the administration of placental globulin reduced both the degree and percentage of febrile reaction without influencing the development of immunity. This is different from the old concept that immunity afforded by the administration of placental globulin after natural infection of measles is only transient if on signs of the disease appeared[17]. Black[18], who has recently studied the clinical and serologic response of susceptible children after receiving gamma globulin during an outbreak of measles in the same cottage reports that all the children who had modified measles had neu tralizing antibody in their sera, and that of those showing no apparent clinical illness only some had neutralizing antibody. While inapparent infection can be acquired in natural infection through the use of gamma globulin, it is difficult to bring about inapparent infection with certainty by the administration of gamma globulin alone in natural infection, as it is not possible to determine the infective dose and the time required. But with the combined use of attenuated vaccine and gamma globulin, the infective dose, the dosage of gamma globulin and the time interval between inoculation of virus and gamma globulin can be accuragely ascertained. Hence, further studies on the relation of larger and smaller infective doses of the vaccine to the dose of piacental globulin, the relation of the neutralizing antibody titer in different lots of placental globulin to the same infective dose of vaccine, and the relation of body weight to the dosage fo placental globulin should be made.

It is important to know whether or not the immunity acquired through the combined use of vaccine and placental globulin is of long duration. Enders and his co – workers[6], using chick embryo measles attenuated vaccine alone, found that antibody reached the peak titer 3 weeks after vaccination, and that the titer, though greatly decreased after 6 months, remained a low level a year later. The change in titer corresponds closely to that seen after natural infection[19]. Hence they suggested that the immunity acquired by the use of chick embryo attenuated vaccine is probably of long duration. In our studies, we found no difference in the antibody titer of subjects inoculated with the vaccine alone and that of those inoculated with the vaccine followed by placental globulin. There was also no significant difference between the average titer of those showing febrile reaction to the combined use of the vaccine and placental globulin and that of those showing no febrile reaction, they being 1∶12. 5 and 1∶11. 7 respectively. The titer observed 6 – 7 months after vaccination was lower than that obtained by other workers[6]. However, we feel that the titer is strong enough to resist the natural infection. As pointed out in the foregoing paper, the titer of neutralizing antibody in the serum as measured by the tissue culture neutralization test is actually neutralization of at least 10000 virus units with clinical reactions in man as criteria, since susceptible children were found to be at least 100 times more sensitive to the virus of vaccine L_4 than to that of the tissue culture system. According to Stokes[13], children with a serum neutralizing antibody titer of 1∶1 could still resist the infection of the attenuated virus. Thus, we believe that the immunity acquired after the combined use of vaccine and placental globulin, with or without clinical reactions, is probably of long duration.

The rate of multiplication of measles virus in tissue culture varies with the type of tissue culture system used. The shortest interval from the time of infection to the appearance of detectable virus in the culture fluid, found in a stable cell line, was 27 – 30 hours in Hep_2 cells[20]. If the incubation period of virus multiplication in susceptible children after inoculation of 6 $TCID_{50}$ virus vaccine L_4 is not less that 24 hours, then the administration of placental globulin given 24 hours after inoculation of the vaccine should prevent the spread of infection from the initial site or sites of infection. That immunity can be acquired through such a localized primary infection is of interest.

The advantages of using the vaccine followed by placental globulin are many. This method can be used in a pre – epidemic period when the respiratory infections are minimal, and, probably, also during an epidemic. We found a vaccinated dhild who did not show increased clinical reactions after close contact with a measles patient just before or immediately after vaccination, probably owing to the shorter incubation period after inoculation of attenuated measles vaccine than after the natural infection. There is no transmission of clinical infection from vaccinated to susceptible subjects. This facilitates immunization on a large scale. The noncommunicability of the vaccine is in accord with that observed with chick embryo vaccine[6,7]. Finally, since so small a dose of virus, 6 $TCID_{50}$, is sufficient to induce immunity, it is possible to prepare a vaccine from one human amnion for 10 million persons a very economical product for mass vaccination.

The drawbacks of the combined use of vaccine and placental globulin are that 2 inoculations are needed and the supply of placental globulin is limited. In regard to the second point, although in a previous study of the content of measles antibody in different lost of placental globulin in our laboratory[16], it was found that the majority had a titer of 1∶150 – 1∶200, yet in the present study the titer of the placental globulin was only 1∶50. If further studies should prove that the influence of placental globulin on febrile reaction to the vaccine is directly proportional to the amount of antibody content, then the actual amount of placental globulin can be further reduced.

Since the different measles vaccines so far prepared give rise to febrile reaction, some as high as 41 ℃, and since convincing data are still lacking to show that prologed passages in tissue culture could reduce febrile reaction, it seems desirable that the combined use of vaccine and placental globulin should be given further clinical trias.

Summary

1. The influence of placental globulin on the clinical and immunologic effects of 6 $TCID_{50}$ of measles virus attenusted by passages in human amnion cells was studied among 130 children.

2. Experimental results indicate that the degree and percentage of febrile reaction to the vaccine decreased with increased dose of placental globulin. The average maximum temperature and the duration of fever also showed regular decrease with increase in the dose of placental globulin. There was, however, no difference in the degree and extent of reactions when the same dose of placental globulin was given 1 – 4 days after inoculation of the vaccine.

3. The antibody level 6 – 7 months after immunization was found to be about the same as when the vaccine was given alone or was followed by placental globulin. The average antibody titer of children with clinical reactions to the combined use of vaccine and placental globulin was found to be slightly higher than that of those showing no reactions, but the difference in not significant.

4. 28 immunized children had intimate contact with measles children for 1 day to more than 4 months after vaccination. They lived and played together in the same room for a period of 2 – 5 weeks, but none of them had messles. Among these 28 immunized children, 4 received the vaccine alone and 24 received both the vaccine and placental globulin. By contrast, living in the same environment, 39 out of 41 unvaccinated susceptibles came down with the disease.

〔In 《The Chinese Medical Journal》 1978 (1): 15 –22〕

References

1 Enders JF, Peebles TC, Propagation in tissue cultures of cytopathogenic agents from patients with measles. Proc Soc Exp Biol Med, 1954, 86: 277

2 Maris EP, et al, Vaccination of children with various chorioallantoic passages of measles virus; follow – up study, Pediatrics, 1949, 4: 1

3 Сергиев ПТ, et al, Разработка способа активной иммунизации против кори на обезьянах. Воп Вирусол, 1959, (5): 558

4 Chang C, et al, Personal communication, 1958.

5 Chang CT, on the use of Arakawa vaccine in the prevention of measles. Yokohama Med Bull, 1960, 11: 21

6 Enders JF, et al, Development and preparation of the vaccine; technics for assay of effects of vaccination. Nwe Engl J Med, 1960, 263: 153

a Enders JF, et al, Development and preparation of the vaccine; thchnics for assay of effects of vaccination. New Engl J Med, 1960, 263: 153

b Katz SL, et al, Clinical, virologic and immunologic effffects of vaccine in institutionalized children. New Eangl J Med, 1960, 263: 159

c Kempe CH, et al, Clinical and antigenic effects of vaccine in institutionalized children. New Engl J Med, 1960, 263: 161

d Black FL, Sheridan SR, Administration of vaccine by several routes. New Engl J Med, 1960, 263: 165

e Lepow ML, et al, Clinical, antigenic and prophycaltic effects of vaccine in institutionalized ang home – dwelling children. New Engl J Med, 1960, 263: 170

f Krugman S, et al, Clinical, antigenic and urophylactic effects of vaccine in institutionalized children. New Engl J Med, 1960, 263: 174

g Haggerty RJ, et al, Clinical, antigenic and prophylactic effects of vaccine in homedwelling children. New Engl J Med, 1960, 263: 178

h Katz SL, et al, General summary and evaluation of the results of vaccine. New Engl J Med,

1960, 263: 160

7　Smorodintsev AA, et al, Clinical and immunological response to live tissue culture vaccine
against measles. Acta Virol (Eng), 1960, 4:
201

8　Жданов ВМ, Фадеева ЛЛ, Проблеме изыска-
ния коревой вакцины Воп. Вирусол, 1959,
(5): 551

9　Kress S, et al, Studies with living attenuated
measles virus vaccine. AMA J Dis Child, 1960,
100: 536

10　Huang CH, et al, Isolation of measles virus from
typical and gamma globulin modified measles children. Zhong Yixue Z, 1961, 47: 352

11　Dolgin J, et al, Imunizing properties of live attenuated measles virus. J Pediat, 1960, 57: 36

12　Enders JF, et al, Recent aduances in knowledge
of the measles virus, in "Perspectives in Virology, a Symposium" edited by Pollard M, p. 103,
Chapman & Hall, London, 1959

13　Stokes J, et al, Use of living attenuated measles
virus vaccine in early infancy. New Engl J Med
263: 230, 1960.

14　Bech V, Studies on the development of complement – fixing antibodies in measles patients; observations during a measles epidemic in Greenland. J Immun, 1959, 83: 267

15　Black FL, Measles antibodies in the population
on New Haven Connecticut. J Immun, 1959,
83: 74

16　Tseng Y, Teng YM, Application of hemagglutination inhibition test to examine heasles antibody.
Zhong Yixue Z, 1961, 47: 355

17　Karelitz S, Measles, in "Brenneman's Practice
of Pediatrics" edited by McQuarie I, Kelley VC,
vol. 2, Chap. 11, p. 18, Prior Co, Hagerstown,
Md., 1959

18　Black FL, Yannet H, Inapparent measles after
gamma globulin administration. JAMA, 1960,
173: 1183

19　Bech V, Titers of complement fixing measles antibodies in human sera collected from 1 – 5 years
after illness. Acta Path Microbiol Scand, 1960,
50: 81

20　Black FL, Growth and stability of measles virus.
Virology, 1959, 7: 184

2. Establishment of an Epithelioid Cell Line and a Fusiform Cell Line from a Patient with Nasopharyngeal Carcinoma[①]

Laboratory of Tumor Viruses of Cancer Institute, Laboratory of Tumor Virusce of
Institute of Epidemiology, Department of Radiotherapy of Cancer Institute,
and Laboratory of Cell Biology of Cancer Institute.
Chinese Academy of Medical Sciences
Laboratory of Electron Microscope, Department of Microbiology, and
Laboratory of Pathogenesis, Chuny Shan Medical Colloge
Chinese Academy of Medical Sciences

Summary

An epithelioid cell line and a fusiform eell liue were established from a tumor biopsy from a patient with nasopharyngeal carcinoma whieh was histologieally diagnosed as a well differentiated squamous eell carcinoma. Based on studies of the cell growth pattern, ehromosome analysis, heterotransplantation, and electron microscopy, these two cell lines were considered to be squamous carcinoma cells, and the fusiform cells might have originated from the epithelioid cells. There were many round cells on top of the epithelioid and fusiform cell sheets, many of which became continuously detached into the medium. Cultures initiated from these floating round cells grew into their original epithelioid or fusiform forms. No lymphoblastoid cell line could be established after cultivating these two cell lines for more than one year.

No EB virus particle or early antigen could be detected in these two cell lines by means of electron microscopic examination and indirect immunofluoresccnce test.

Introduction

The serological relationship between nasopharyngeal carcinoma (NPC) and EB virus was first demonstrated by Old et al. in 1966 using immunodiffusion test[1]. It was shown subsequently by indirect immunofluorescence test that NPC patients had various antibodies to EB virus, and the antibody spectra and titers were clearly referable to total tumor burden[2-8]; furthermore, EB virus DNA and nuclear antigen (EBNA) have been demonstrated regularly in epithelial tumor cells of NPC[9-16]. All the above rusults indicate a close association of EBV with NPC.

Extensive attempts have been made in establishing a permanent epithelial cell line from NPC

---

① First appeared in Chinese in the note form in Kexue Tongbao, p. 143, No. 3, 1977.

patients in order to investigate further the relationship between EBV and NPC but no successful results have been reported[16-18]. In our laboratory, we have succeeded in establishing an epithelioid cell line and a fusiform cell line from a patient with NPC. These cell lines have been maintained in culture for 19 months and subcultured for more than 80 times. This paper deseribes the establishment and the characteristics of these cell lines.

Materials and Methods

The tissue used for cultures was obtained from a tumor biopsy from a 58 – year – old woman with NPC on August 13, 1975. The patient had severe headache, tinnitus, and epistaxis. Clinical examination showed the presence of a tumor in the nasopharynx whice had invaded the base of skull, with compression signs of cranial nerves (Ⅲ, Ⅳ, Ⅵ, Ⅸ, Ⅻ), and with metastases to the lymph nodes on both sides of the neck. The soft tissue tumor mass on the posterior wall of nasopharynx was also confirmed on orentgeuography, X – ray film of the base of the skull showed suspected bony destruction of the left side of the external plate of pterygoidal process of sphenoid. Tumor biopsy revealed a well – differentiated squamous cell carcinoma (Fig. 1).

The tumor specimen was cut into approximately 0. 5 – 1. 0 mm pieces. Minced tumor fragments were placed on the surface of flasks which had been pretreated with rat tail collagen to aid the attachment of explants. RPMI 1640 medium supplemented with 40% of calf serum, 100 units of penicillin, and 100 μg of streptomycin per ml were added to the flasks at the opposite side of the explants. The cultures were incubated at 37℃

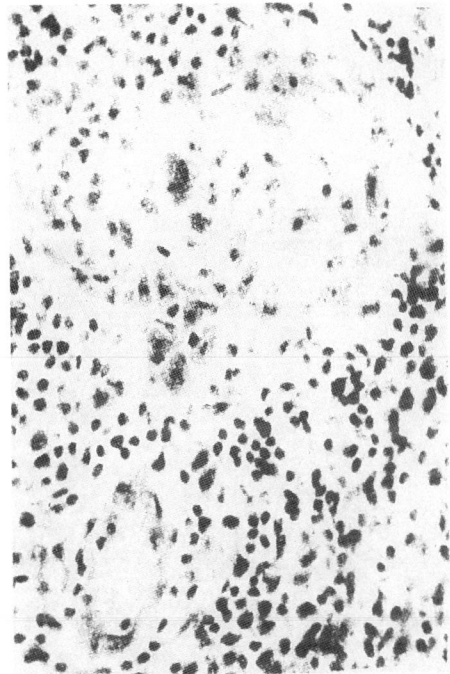

Fig. 1　Patient with NPC, histologic
section from nasopharyngeal
carcinome. HE stain (×100)

for 3 hr in an incubator with 5% CO_2 in air, and the flasks were turned over to allow the medium to cover the explants. The medium was changed twice a week. Subcultures were made by dispersing the cells with 25% trypsin: 0. 2% versene solution.

Course of Establishment of Permanent Cell Lines

Epithelioid cells began to outgrow around the tissue fragments in two of five flasks on the 10 th day of cultivation, after which the cell sheet increased in size gradually. Attempts to transfer the epithelioid cells by dispersing part of the cell sheet with trypsin. versene (0. 25% : 0. 02%) or by

· 11 ·

scraping cells with eapillary pipette failed. The first successful subculture was made by trypsin: versene 11 weeks after cultivation. Thereafter the cells were successively transferred once every week. Thus an epithelioid cell line designated as CNE was established. The epithelioid cells were polygonal in shape with the nuclei varying in size. Some multi – nucleated giant cells and cytoplasmic vacuoles were present (Fig. 2). In old cultures some cells frequently had part of the cytoplasm protruded and finally detached from the cell sheet as dead cells. At the 12th passage a few fusiform cells appeared among the epithelioid cells and gradually increased in number (Fig. 3). The fusiform cells were isolated as an independent line and designated as CNF. There were many round cells on the top of both cell sheets, especially on the fusiform ones. Many of them became continuously detached into the medium. Cultures initiated from the floating round cells reeovered their original epithelioid or fusiform cell morphology. No lymphoblastoid cell line could be established after long – term cutivation without transfer up to the 7th month, and no fibroblast cells appeared during the course of establishment of cell lines.

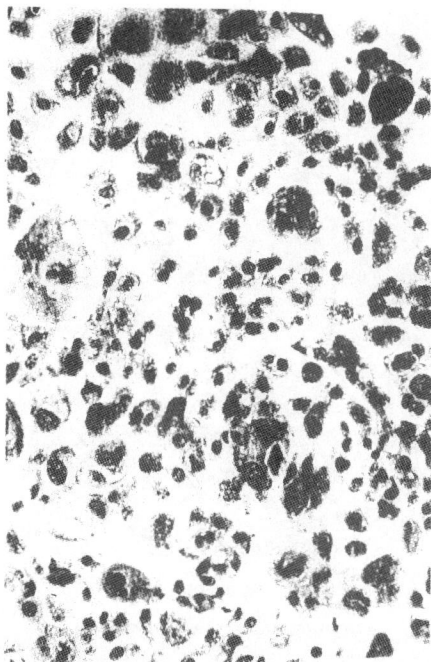

Fig. 2 Epithelioid cells, 7 th passage,
Giemsa stain (×200)

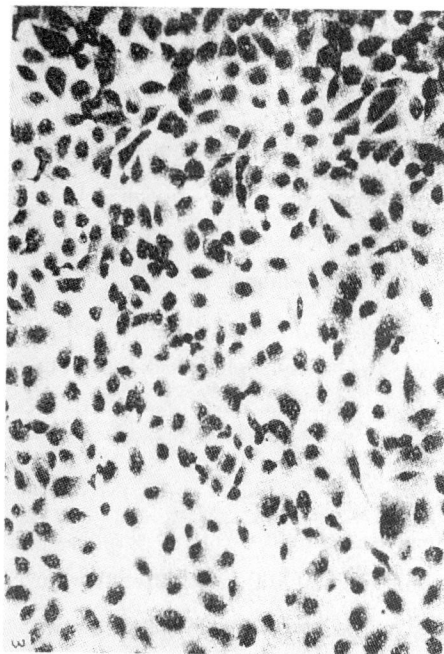

Fig. 3 Epithelioid cells, 12 th passage.
Giemsa stain (×100)

The CNE and CNF cells were dispersed by trypsin: versene (0. 25%: 0. 2%) and inoculated into separated dishes with 5 ml of medium containing 40, 20 and 10 cells respectively. Ten days after cultivation at 37℃ in 5% CO_2, the typical clones of the epithelioid and fusiform cells were isolated from dishes at terminal dilutions (Figs. 4&5). This procedure was repeated twice. The clonal epithelioid and fusiform cell lines thus obtained were designated as CNEC and CNFC respectively

(Figs. 6&7). The epithelioid cell sheet was not as readily dispersed by trypsin: versene (0.25%: 0.02%) as fusiform cell sheet unless the concentration of versene was increased to 0.2%.

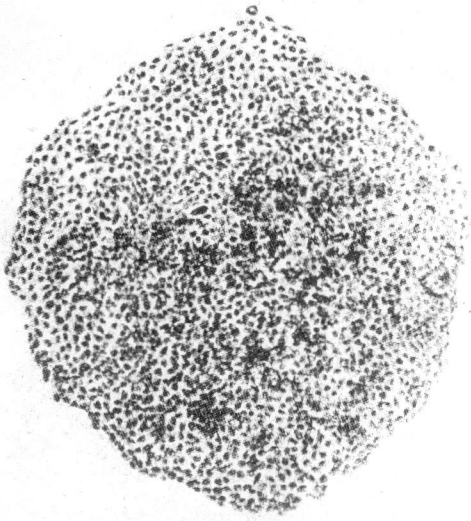

Fig. 4　Clone of ONE cell line.
Giemsa stain（×40）

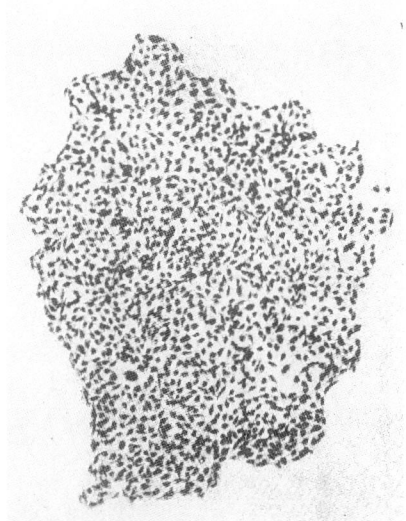

Fig. 5　Clone of CNF cell line.
Giemsa stain（×40）

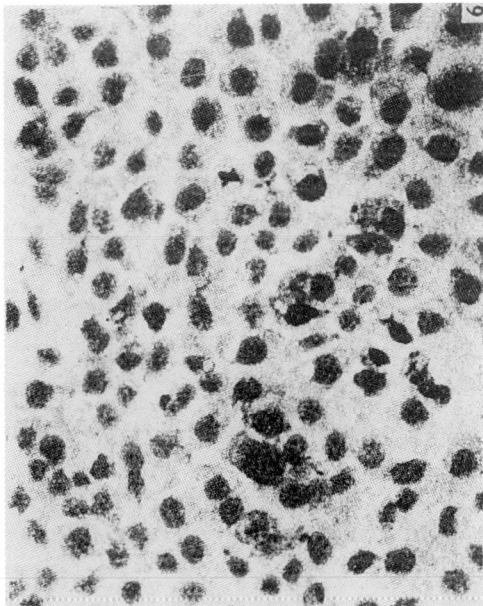

Fig. 6　Clonal cell line of epithelioid cells
（CNEC）. Giemsa stainn（×200）

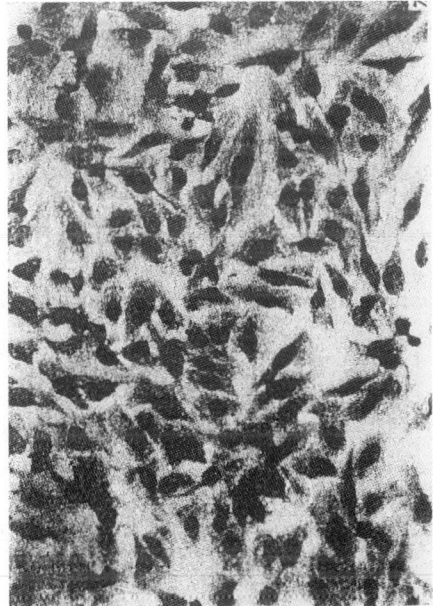

Fig. 7　Clonal cell line of fusiform
cells（CNFC）. Giemsa stain（×200）

Growth Curve, Saturation Density, and Plating Efficiency

Monolayer cells of CNE and CNF cell lines were dispersed with trypsin and versene (0.25% : 0.02%), 30 ml flasks were plated with 1×10^5 cells in 3.5 ml of complete RPMI 1640 medium. Cell cultures were kept at 37℃. The medium was changed and the cell counting from 2 flasks was performed every 3 days for 12 days. The effect of different concentrations of calf serum on the growth of both cell lines was studied. As shown in Fig. 8, the growth curves of CNE and CNF cell lines were similar in media with 20% and 40% of calf serum. The cell number increased logarithmically on the $6^{th} - 9^{th}$ day, reaching more than 20 times of that originally seeded. The growth rate of both cell lines in medium containing 5% of calf serum was slower. The cell number on the 12th day was 14 – 15 times of that originally seeded.

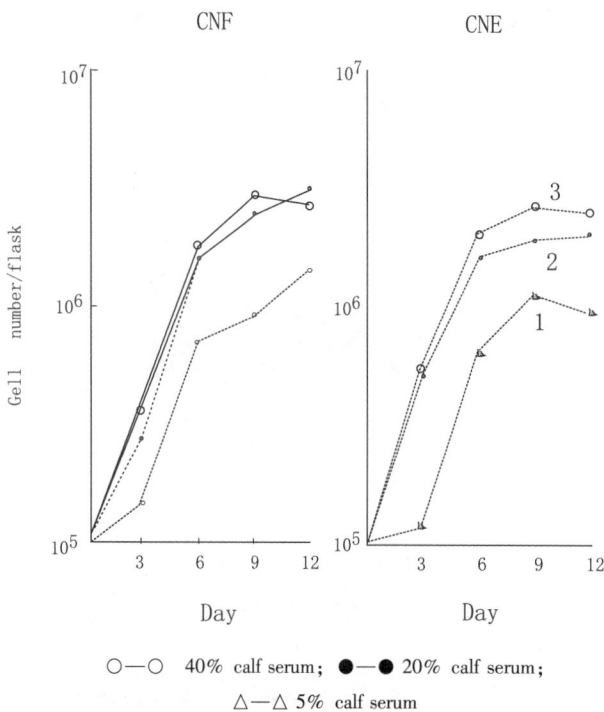

○—○ 40% calf serum; ●—● 20% calf serum;
△—△ 5% calf serum

Fig. 8　Growth curves of CNF and CNE cell lines

The value where two successive harvest showed no increase in cell number was taken as saturation density. The saturation densities of CNE and CNF cell lines were $2.13 \times 10^5/cm^2$ and $2.26 \times 10^5/cm^2$ respectively.

All suspension was diluted in RPMI 1640 medium and 200 cells in 5 ml of complete RPMI 1640 medium were plated in 6 cm dishes. After 10 days of incubation at 37℃ in 5% CO_2, the colonies were stained with the Giemsa stain and counted. The number of colonies was calculated from the average of 5 dishes. The plating efficiencies of CNE and CNF cell lines were 22% and 52% respectively. The colonies of CNE cell consisting of epithelioid cells were uniform in size with regular margin, whereas the colonies of CNF consisting of fusiform cells showed variable size with irregular margin.

Assay of Agglutination by Concanavalin – A

2×10^5 cells of CNE, CNF, CNEC, and CNFC cell lines in 3 ml of complete RPMI 1640 medium were plated into separate flasks and incubated at 37℃ for 24 hours. After washing twice with PBS lacking calcium and magnesium, the cells were dispersed with versenc (0.02%), washed again with PBS, and then suspensed in PBS containing calcium and magnesium at a concentration of 4×10^5 cells per ml. 0.1 ml of concanavalin – A at different concentrations in PBS was mixed with 0.1 ml of the cell suspension in 100mm × 12 mm tubes at room temperature for 30 minutes. The aggregates were scored under inverse microscope in a scale from − to ++++. As shown in Tab. 1, cells of all four lines could be agglutinated by 4 μg of concanavalin − A, and the size of the aggregates increased with the increasing concentrations of concanavalin − A.

Tab. 1   Characteristics of Nasopharyngeal Carcinoma Cell Lines

Cell Line	Morphology	Typsin: Versene (0.25% : 0.02%)	Saturation Density (10⁵ cells/cm²)	Plating Effieiency (%)	Concana − valin − A Agglutination	Heterotrans − plantation	Chromosome Aberration	Electron Microscopic Examination		Immunological Test	
								Cell	EBV	CF Antigen	EA
CNE	Epithelioid	Rather difficult to disperse	2.13	22	+	Poorly dif − fereutiated squamous carcinoma	Aneuploidy	Squamous carcinoma	−	−	−
CNEC	Epithelioid	Rather difficult to disperse			+		Aneuploidy				−
CNF	Fusiform	Rather easy to disperse	2.26	58	+	Poorly dif − ferentiated squamous carcinoma	Ancuploidy	Squamous carcinoma	−	−	−
CNFC	Fusiform	Rather easy to disperse			+		Aneuploidy				−

Chromosome Analysis

Cell liues of the CNE (51st passage), CNEC (20th passage), CNF (34th & 38th passages), and CNFC (16th passage) were treated with colchicine (final concentration 0.02 μg/ml) for 2 − 4 hours during the logarithmie growing phase. The chromosome preparation was made according to an air − dried technique. 100 − 200 metaphase plates of ench line were counted and chromosome aberrations also recorded. The detailed data are summarized in Tab. 2 and Fig. 9.

Tab. 2   Chromosome Aberrations of Nasopharyngeal Carcinoma Cell Lines

	Type and %						
Cell Line	Dicentric	Tricentric	Tetracentric	Fragmental	Minute	Superfrag − mental	Different Types of Aberration in Some Cells
CNE	5	0	0	2	3	0	0
CNEC	2	0	0	2	3	1	1
CNF	22	0.5	0.5	7	2	2	2
CNFC	54	0	0	11	4	0	0

Chromosome numbers of these cell lines showed a wide distribution with a mode between hypotriploid and hypotetraploid (Fig. 10). Although the stemline of the CNEC and CNF cell lines was not formed, yet cells with chromosomes numbering over 100 appeared frequently. In the CNE cell line the mode accumulated between hypertriploid and hypotetraploid with a stemline of 80 chromosomes. For the CNFC cell line, a stemline of 70 chromosomes was noted at the 16 th passage.

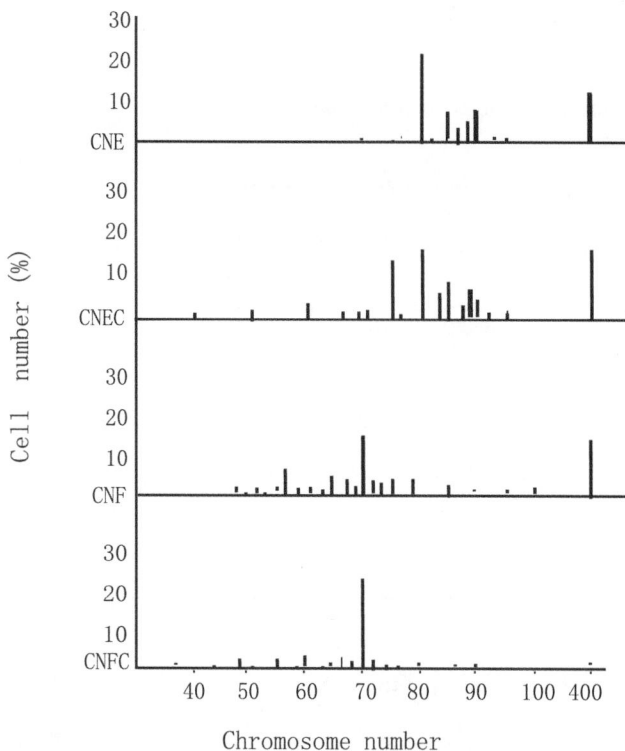

Fig. 9　Comparison of the distribution
of chromosome number of 4 cell
lines of nasopharyngeal carcinona

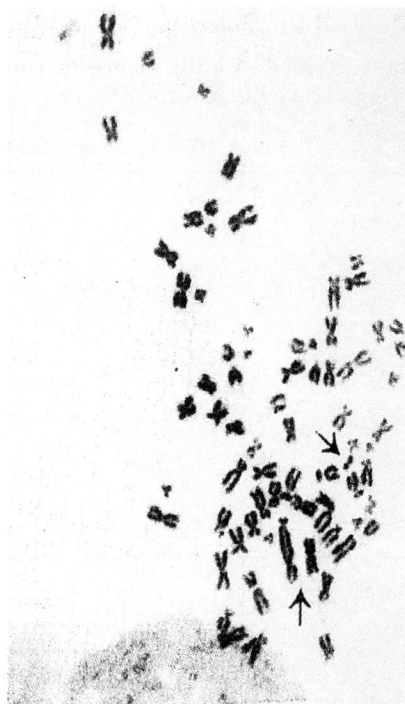

Fig. 10　Chromosome aberration
in the CNF cell line. Dicentric
and fragmental chromosome

Various types of chromosome aberration, such as the dicentric, mulitcentric, fragmental, minute, and superfragmental, were observed in different cell lines. Furthermore, different types of aberration may be noted in the same cells (Fig. 10). About 2% − 5% of chromosome aberrations occurred in the CNE and CNFC cell lines. Besides other types of chromosome aberration, dicentric chromosome was most frequently encountered in the CNF cell line, accounting for 22% of all types. The dicentric, fragmental, and minute types of chromosome aberration were frequently observed in the CNFC cell line, 54% of them were classified as the dicentric type. The reason why such a high incidence of unstable chromosome aberrations appeared in the CNF and CNFC cell lines remains to be studied.

Heterotransplantation

0. 1 ml each of the CNE (33rd passage) and CNF (19th passage) cell suspension were transplanted subcutaneously into newborn rats. Anti − thymocyte serum (0. 4 ml) was given on the day of transplantation and subsequently on the 3rd, 5th, and 8th days. 11 days after implantation the animals were sacrificed. The tumors measured 0. 5 − 0. 7 cm in diameter and were examined histologically. Anti − thymocyte rabbit serum with titer of 1∶640 was prepared in this laboratory. As shown in Tab. 1, the transplantability of CNE and CNF was 77% and 100% respectively. Histologically, these

tumors were poorly differentiated squamous carcinoma (Figs. 11&12).

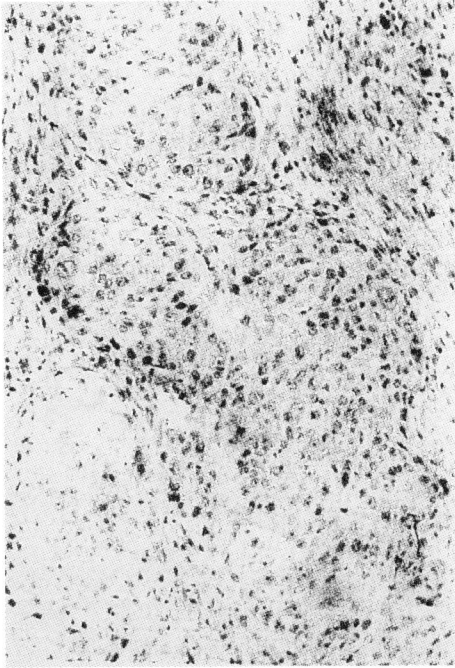

Fig. 11 Histologic section of transplanted
tumor in newborn rat, CNE, 33rd
passage. HE stain (×125)

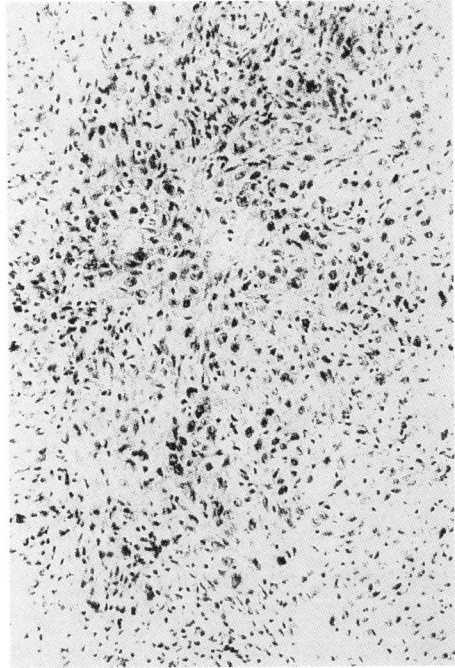

Fig. 12 Histologic section of transplanted
tumor in newborn rat, CNF, 19th
passage. HE stain (×125)

Electron Microscopic Examination

CNE and CNF cells were scraped off with a rubber policemand and centrifuged at 1000 r/min for 1 – 2 minutes. Cell pellets were fixed in 5% gluteraldehyde followed by osmie acid, then embedded in butyl methacrylate. Thin – section specimens were stained with uranyl acetate and lead citrate, and examined under electronmicroscope.

Electronmicroscopy showed that the CNE cells maintained typical features of epithelial cells, including desmosomes, tonofibrils, keratohyalin granules and membrane coating granules (Fig. 13). The CNF cells were polymorphic; some of them contained minute desmosomes and tonofilaments, but not as typical as those seen in CNE cells. The ratio of nucleus to cytoplasm of CNF cells was larger, and more endoplasmic reticula, free ribosomes, and mitochondria were seen in the cytoplasm (Fig. 14). It is rather difficult to identify the nature of some fusiform cells, so CNF cells are temporarily classfied as poorly differentiated carcinoma cells.

It is of particular interest that some round cells without the characteristic features of epithelial cells were seen in the CNE cell line. These round cells contained large nuclei and some vacuoles, and were rich in the ribosome in cytoplasm. The nature of these cells remains to be identified (Fig. 15).

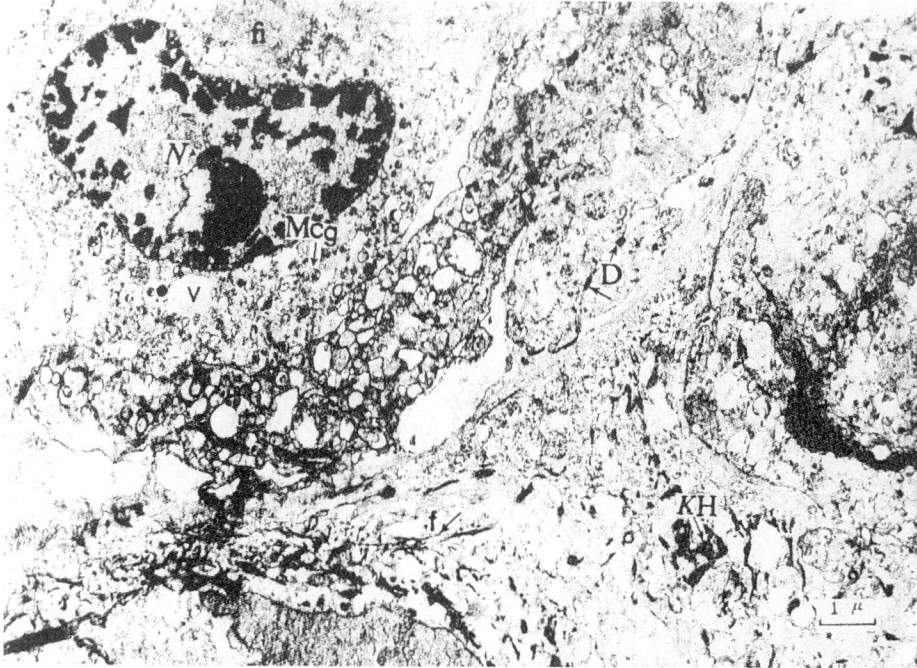

Fig. 13　Ultrastructure of CNE cell line, 38th passage: D – desmosome, f – tonofibril,
f$_1$ – tonofilament, KH – keratohyalin granule, Mcg – membrane coating grmule, V – vesicle (×5000)

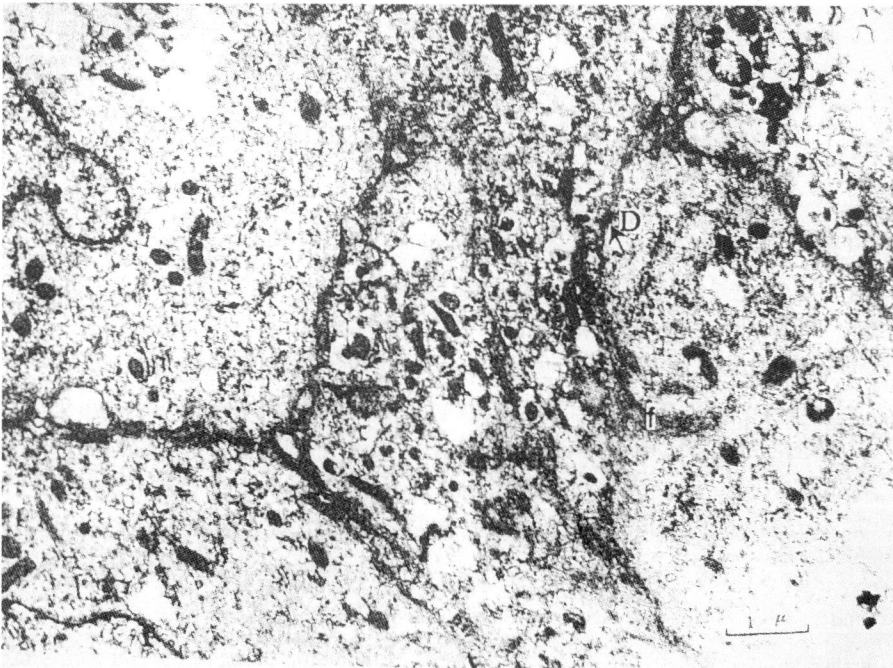

Fig. 14　Ultrastructure of CNF cell line, 24th passage:
D – desmosome, f – tonofibril (×14 000)

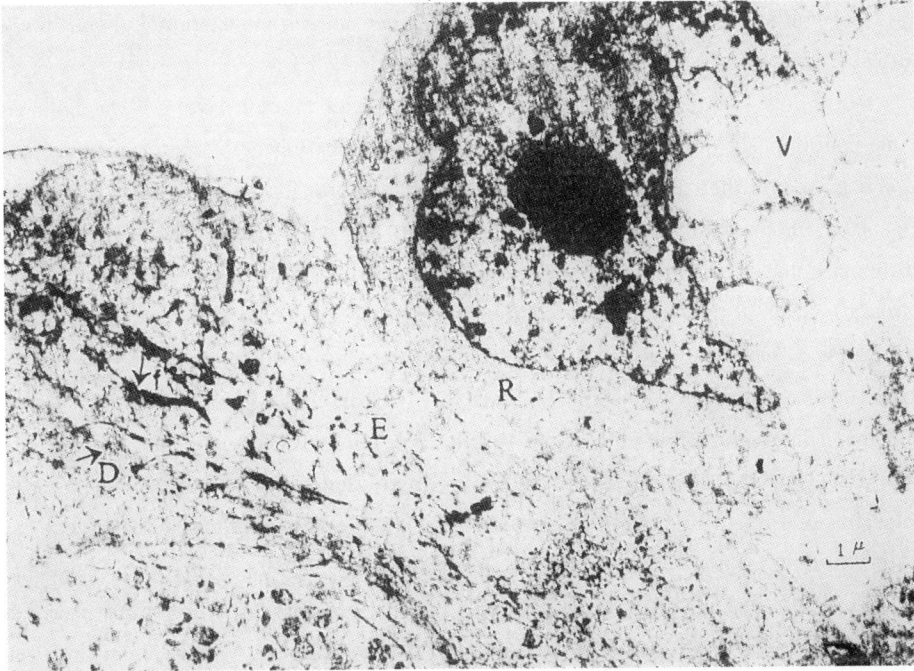

Fig. 15　Ultrastructure of CNE cell line 38th passage, a round cell with epithelial cell characteristics: D – desmosome, f – tonofibril, R – ribosome, V – vesicle (×7 800)

No virus could be found in CNE and CNF cells treated or untreated with INDR (30 μg/ml).

Detection of EBV Specific Antigen

1. Complement Fixation Test

A 5% suspension of CNE and CNF cells in veronal buffered saline was sonicated and centrifuged at 27 000 g for 30 minutes. The supernatant was used as antigen and tested with strong EBV positive serum from an NPC patient by microcomplement fixation test. The results were negative.

2. Indirect Immunofluoresccnce Test

The CNE, CNEC, CNF, and CNFC cells growing on coverslips were treated with IUDR (50 μg/ml) in complete RPMI 1640 medium for 6 days. They were examined for EBVsecific early antigen (EA) by indirect immunofluorescence test. No early antigen could be found in these lines.

Discussion

Based on studies of the cell growth pattern, chromosome analysis, heterotransplantation, and electronmicroscopy, the CNE cell line had the characteristics of epithelial cells and was confirmed to be squamous carcinoma cells. The fusiform cells could not be seen in cultures of the epithelioid cell line until the 12th passage. The CNF cell line also had some characteristics of epithelial cells and formed poorly differentiated squamous cell carcinoma in animals treated with antithymocyte serum. Therefore, the fusiform cells probably originated from the epithelioid cell line. Contamination by other malignant epithelial cell line could be ruled out, because no other epithelial cell line was present in our laboratory.

Many round cells were observed on the top of monolayer or multilayers of the both cell lines, and became continuously detached into the medium. These floating cells, after seeding into another flask, Could grow into their original epithelioid cell or fusiform forms. Since many patients with nasopharyngeal carcinoma had lymph node metastases in the neck region in the absence of notable tumor in the nasopharynx when they first came to the outpatient clinic, the easy detachment of cancer cells from the original tumor might be similar to the phemomenon as observed in tissue culture. No EB virus particle or early antigen could be detected in these two cell lines either treated or untreated with IUDR. These cell lines were established from an NPC patient with well differentiated squamous cell carcinoma. This might be similar to the results as reported by Klein et al. [14], who could not demonstrate EBV, DNA, and EBNA in the well-differentiated squamous carcinoma cells of NPC transplanted in nude mice. But Liang Po-chiang et al. [19] reported that nasopharyngeal carcinomas showed, during their course of development, a definite tendency to change their histological pattern, and did so in a definite sequence, i. e. from highly differentiated type toward poorly differentiated type and from poorly differentiated type toward undifferentiated type. Even in the same biopsy specimen of the primary growth, different parts of the tumor revealed different histologic patterus. Our two cell lines formed poorly differentiated squamous cell carcinoma in immunosuppressive animals. Whether there are viral genome and its expression in these two cell lines or not needs further study by other methods.

[In 《Scientia Sinica》 1978, 21 (1): 127－134]

References

1　Old L T, et al. Proc Nat Acad Sci (USA), 1966, 56: 1699.

2　de Schryvei A, et al. Clin Exp Immunol, 1969, 5: 443.

3　Henle W, et al. In Comparative Leukemia Research (ed. Dutcher R M), Basel: Karger, 1969, 706.

4　Henle G, et al. Int J Cancer, 1971, 8: 272

5　Henle W, et al. Cancer Res, 1973, 33: 1419

6　Henle W, et al. J Nat Cancer Inst, 1973, 51: 361

7　Henderson B E, et al. Cancer Res, 1974, 34: 1207

8　de-The G, et al. Int J Cancer, 1975, 16: 713

9　Zur Hausen H, et al. Nature (London), 1970, 228: 1056

10　Zur Hausen H, et al. Int J Cancer, 1974, 13: 657

11　Nonoyama M, et al. Proc Nat Acad Sci (USA), 1973, 70: 3267

12　Wolf H, et al. Nature, 1973, 244: 245

13　Huang D P, et al. Int J Cancer, 1974, 14: 580

14　Klein G, et al. Proc Nat Acad Sci. (USA), 1974, 71: 4737

15　Degrange C, et al. Int J Cancer, 1975, 16: 7

16　Trumper P A, et al Int J Cancer, 1976, 17: 578

17　de-The G, et al. Int J Cancer, 1970, 6: 189

18　de-The G, In Oncogenesis and Herpes Virus, eds. Biggs P M, et al. Lyon. 1972, p275

19　Liang P C, et al. Chinese M J, 1962, 81: 629

3. Application of an Immunoenzymatic Method and an Immunoautoradiograhic Method for a Mass Survey of Nasopharyngeal Carcinoma

ZENG Yi[1], LIU Yu – xi[2], LIU Chun – ren[2], CHEN San – wen[3],

WEI Jih – neng[3], ZHU Ji – song[4], ZAI Hui – jong[3]

1. Institute of Virology, Chinese Academy of Medical Sciences; 2. Cancer Institute,
Chinese Academy of Medical Sciences; 3. People's Hospital of Guangxi Zhuang
Autonomous Region; 4. Cancer Control Office of Zangwu County; Institute of Atomic Energy,
Academia Sinica – Beijing, People's Republic of China

Summary

The frequency of IgA antibody to virus capsid antigen (VCA) of EB virus was tested by an immunoenzymatic method among adults in a mass serological screening of the general population of 6 communes in South China, as a guide toward early detection of nasopharyngeal carcinoma. Sera from 56 patients already recognized as having nasopharyngeal carcinoma (NPC) also were tested. The screening was conducted with sera fom 56 584 persons age 30 years and older in Zangwu County in the Guangxi Zhuang Autonomous Region. IgA antibody to VAC was found in 96% of the NPC patients tested; the geometric mean antibody titer for this group was 1:41. The antibody also was detected in 117 of the persons surveyed in the mass screening, and among these, the geometric mean antibody titer was 1:21. 20 of the antibody – positive persons were diagnosed by clinical and pathological examination as having NPC. 18 of these cases were diagnosed soon after the initial tests in June – September, 1978, and an additional 2 were diagnosed only after follow – up examinations 10 months after the first blood sample was taken. All of the persons who are in the antibody – positive group but exhibit no detectable tumors are to be reexamined periodically.

addition to the immunoenzymatic method, and immunoautoradiographic method was also used in tests with sera from 12 328 persons in one of the communes. IgA antibody to VCA of EB virus was detected in 25 subjects by the immunoenzymatic test and in 69 by immunoautoradiogaphy. Thus the immunoautoradiographic method appears to be more sensitive. However, sera from the 18 NPC patients detected by serological screening were positive by both methods. The results indicate that both methods are simple and sensitive, and that serological screening is valuable in the early detection of NPC.

[**Key words**]   Nasopharyngeal carcinoma; EB virus; VCA antibody; IgA antibody; Immunoenzymatic method; Immunoautorediographic method

Introduction

Nasopharyngeal carcinoma (NPC) is one of the common maligancies in South China. A close association of EB Virus with NPC has been demonstrated by serological, virological and molecular biological studies. It is of interest that IgA antibody to EBV – related antigen is greatly elevated in sera of NPC patients[1]. Detection of IgA antibody has also been carried out in our laboratories by immunofluorescence, immunoenzymaticand immunoautoradiographic methods[2-5]. The frequency of antibody detection in sera from patients with NPC was over 90%, that from patients with other malignancies was less than 4%, whereas sera examined from normal persons have given negative results. Therefore, detection of IgA antibody is useful in the diagnosis of NPC and might be valuable for mass surveys to detect potential NPC patients among the population in areas at high risk for NPC. This paper reports the application of immunoenzymatic and immunoautoradiographic methods for the detection of IgA antibody to viral capsid antigen (VCA) of EB virus, in sera from the general adult population in a high – risk area for NPC.

Zungwu county is located in the eastern part of Guangxi. The population is 474 705 persons (male 244 747, female 229 958), of whom 99.6% are of Han ethnic background, while the others are of Zhuang, Yao, and other minority ethnic groups. The county is divided into 14 rural communes and one urban commune. From 1971 to 1977, the yearly mortality rate from NPC was 10.18/100 000. A clinical mass survey of NPC had been conducted in persons aged 15 years and older from December 1977 to March 1978, with a detection rate of 51.5/100 000. If all ages were included, the prevalence rate would be 29.96/100 000. In the 6 communes included in this report, the yearly incidence rate of NPC for 1975, 1976, and 1977 was 15.6, 14.8, and 15.8, respectively, per 100 000 of the total NPC patients, 91.4% were age 30 years and over. The population of these 6 communes totals 177 022 persons. Sera were collected from 56 584 persons age 30 years and over (90.4% of the age group). The sex and age distribution of the persons examined was similar to that of the general population.

Materials and Methods

1. Sera

The sera, collected by pricking the ear lobe, were stored at $-15°$. Venous blood specimens were then obtained from persons who were antibody – positive in the initial screening, and also from patients suspected or proven to have NPC.

2. Immunoenzymatic Method[4]

Cell smears were prepared from B95 – 8 cultures, fixed in acetone and used in the indirect immunoenzymatic method with peroxidase – conjugated anti – human IgA antibody. Sera diluted to 1.25 and 1:5 were added to separate wells of slides. The slides were incubated at 37℃ for 30 min in a humid atmosphere, and washed 3 times with phosphate – buffered saline (PBS). Peroxidase – conjugated anti-human IgA antibody in appropriate dilution was added to the slides. The slides were incubated for 30 min, washed 3 times with PBS, and flooded with diaminobenzidine and H_2O_2 for 10 min. Postive and negative serum controls were included in each experiment. Slides were examined under the light microscope. A serum was considered positive if the cells in the well that contained the 1:2.5 dilution showed brown color characteristic of this test. Venous blood specimens from suspected NPC patients and from persons antibody – positive in the initial screening were tested in further dilutions. The highest dilution of serum still positive for IgA antibody to VAC was considered as the antibody titer of that serum.

3. Immunoautoradiographic Method[5]

Sera diluted 1:160 and 1:640 were added to separate wells. After incubation at 37℃ in a humid atmosphere for 30 min, slides were washed 3 times with PBS containing 1% calf serum. 0.7 ml of ^{125}I – labeled antihuman IgA antibody in appropriate dilution was added on each slide. The slides were incubated at 37℃ for 30 min, washed 3 times with PBS containing 1% calf serum and left to air-dry. They were then coated with nuclear emulsion, slowly dried, and kept in the dark at room temperature for 24 h. Slides were developed in D – 19 for 15 min, placed in fixing reagent for 10 min, washed and left to air – dry, and then were examined under the light microscope. A serum was considered positive if the cells in the well that contained the 1:640 dilution showed black granules typical of this test. Venous blood was further tested in selected instances as described under Immunoenzymatic method, above.

4. Clinical and Pathological Examination

The antibody – positive persons were examined by nesopharyngoscope. Biopsies were taken from the following antibody – positive persons: (1) those diagnosed clinically as NPC patients or suspected of having NPC; (2) those having some lesions in the nasopharynx, such as hyperplasia or a residue of adenoidal tissue, rough mucosa, local congestion or inflammation; (3) those without lesions in the nasopharynx but with high antibody titer.

Results

1. Immunoenzymatic Method

i. Detection of IgA Antibody to VCA in Sera from Patients with NPC. Among patients with NPC that had developed and had been identified before the serological mass survey, 96% have IgA antibody to VCA, with a geometric mean titer (GMT) of 1:40.5.

ii. Detection of IgA Anitbody to VCA in Sera from the General Population. Among 56 584 persons age 30 years and older in the general population studied, 117 were found to have IgA antibody to VCA. The prevalence rates of antibody in different communes varied from 125/100000 to 290/100 000, with an average rate of 207/100 000 (Tab. 1).

The antibody titers among the antibodyspoitive indviduals ranged from 1:2.5 to 1:1 280 (Tab. 2). In this group of 117 persons, 91 (78%) had antibody titers of 1:10 or more;

Tab. 1   Results of IgA – VCA antibody screening

Commune No.	Number of persons examined	Number positive	Antibody prevalence (per 100 000)
1	7 430	11	148
2	7 780	15	193
3	11 380	33	290
4	12 075	26	215
5	12 328	25	203
6	5 588	7	125
Total	56 584	117	207

63 （54%）were at levels of 1:20 or above; and 38 （32%）had titers of 1:40 or higher. Among 52 patients with NPC already diagnosed before the serum was taken, 47 （90%）had antibody titers of 1:10 or higher; for 38 （73%）the antibody titers were 1:20 or more; and for 31 （60%）, the antibody levels were 1:40 or more. The GMT for the previously recognized patients was twice that calculated for the 117 antibody-positive persons who were discovered in this survey, 1:46 vs 1:21.

2. Clinical and Pathological Examination, and Antibody Titers by Two Methods

Tab. 2 Distribution of anti – VAC antibody in NPC patients previously diagnosed and in persons found antibody – positive in serological screening

Antibody titier （recip – rocal）	Patients with NPC previously diagnosed		Persons found antibody – positive in sero – logical screeing	
	number posi – tive	% posi – tive	number posi – tive	% posi – tive
2. 5	0	0	10	9
5	5	10	16	14
10	9	17	28	24
20	7	13	25	21
40	6	12	11	9
80	9	17	9	8
160	8	15	9	8
320	6	12	4	4
640	2	4	3	3
1 280	0	0	2	2
Total	52	100	117	100
Geomatric mean titer 1:46			1:21	

Biopsies were taken from 74 of the 117 perwow who were antibody – positive. 19 of these subjects were diagnosed clinically and pathologically as having malignant tumors; these included 18 NPCs and 1 skin basal – cell carcinoma of the face. Of 18 patients with NPC, 7 were in stage I, 4 in stage II, 5 in stage III, and 2 in stage IV （Tab. 3）. The 7 patients considered to be in stage I had no subjective symptoms, and the early pathological changes such as rough and nodular mucosa – could only be found after careful examination. All of the NPC patients, except 1 with undifferentiated carcinoma, showed poorly differentiated squamous carcinoma. All of these sera were antibody – positive by both the immunoenzymatic method and the immunoautoradiographic method. The GMT for these newly discovered patients was rather high – 1:109 by the immunoenzymatic method, as compared with 1:46 for the persons previously recognized as having NPC （Tab. 2）.

3. Prevalence Rate and Incidence Rate of NPC

There were 74 NPC patients in the 6 communes studied, among them the 18 patients who were first detected by this serological mass survey; this constitutes an NPC prevalence rate of 42/100 000-higher than that detected by the clinical mass survey conducted from November 1977 to March 1978 （32. 2/100 000）. The comparative data are shown in Tab 4. Cases detected in 1978 totaled 37, including the 18 first identified as a result of this serological survey, yielding an incidence rate for 1978 of 20. 9/100 000 – higher than that reported for 1975 – 1977 （which was 14. 8 to 15. 8 per 100 000）.

4. Immunoautoradiographic Method

Detection of IgA antibody to VCA in sera form the general population incommune No. 5 was done by immunoautoradiography in addition to the immunoenzymatic testing used in the overall study （Tab. 5）. Of 12 328 persons age 30 years and older in this commune, 69 were positive for IgA antibody to VAC when tested by the immunoautoradiographic method. This constituted an antibody prevalence rate of 560/100 000 – considerably higher than that detected in this ame population by use of the immunoenzymatic method, which revealed only 25 antibody – positive persons – a rate of 203/100 000. However, for 2 patients in this commune who already had NPC Classed in Stage I and confirmed by pathological examination, the antibody was detected by both methods. The immunoautoradiographic antibody titers ranged from 1:640 to 1:40 960 with a GMT of 1:748.

Tab. 3   NPC patients found in serological mass survey

Case number	Sex	Age years	Clinical stage	Pathological diagnoses	Antibody titer[1]	
					IE	IR
23	male	31	I	NPC: poorly differentiated carcinoma	10	640
30	male	47	I		160	2 560
38	male	44	I		20	2 560
41	male	58	I		40	640
75	male	39	I		320	10 240
67	female	68	I	NPC: undifferentiated carcinoma	1 208	40 960
32 218	male	36	I		80	640
70	female	54	II		160	2 560
33	female	52	II		80	10 240
31	male	59	II		80	640
63	female	33	II		160	2 560
36	male	50	III	NPC: poorly differentiated carcinoma	640	40 960
28	female	59	III		1 280	10 240
72	male	40	III		80	640
73	male	51	III		40	10 240
74	female	83	III		160	10 240
29	male	59	IV		160	640
71	male	38	IV		10	640
42	male	77		skin basal cell carcinoma	20	2 560

[1] HE = Immunoenzymatic method; IR = immunoautoradiographic method

Tab. 4   Comparison of NPC prevalence rates detected as a result of clinical and serological mass surveys

Commune No.	Number of patients with NPC	New NPC patients	Total	Prevalence rate by serological mass survey (per 100 000)	Prevalence rate byclinical mass survey[1] (per 100 000)
1	12	9	21	57. 7	30. 2
2	15	2	17	48. 9	43. 1
3	8	5	13	46. 8	28. 8
4	11	1	12	31. 0	28. 5
5	5	0	5	19. 6	31. 5
6	5	1	6	43. 0	50. 2
	56	18	74	42. 0	32. 2

Notes: [1] Clinical mass survey carried out from December 1977 to March 1978

Tab. 5　Detection of VCA – IgA antibody in persons of age 30 years and older in commune No. 5

Method	Number tested	Number positive	Antibody preva – lence（per 100 000）	Patients with stage I NPC
Immunoautoradiographic mathod	12 328	69	560	2
Immunoenzymatic method		25	203	2

5. Resurvey of the IgA Antibody – Positive individuals

Aside from the 18 NPC patients and 1 patient with skin basal cell carinoma detected by pathological examination among the 117 IgA antibody – positive persons as reported above, the rest – 98 persons – were reexamined in April 1979. 29 biopsies were obtained from the persons suspected to have NPC and from persons having high titers of IgA antibody to VCA. Among them, 2 NPC patients in stages I and II were diagnosed pathologically in this follow – up, 10 months after the first blood samples were taken. The antibody titer in the patient in stage I maintained the same level, 1∶10, but in the patient in stage II the titer increased from 1∶20 to 1∶1280. Taken together with the above, 20 NPC patients have been diagnosed among the 117 persons who were discovered in the 1978 screening to be positive for IgA antibody to VAC of EB virus.

Discussion

The results obtained for patients with NPC in the field study are in accordance with our previous reports[2-5]. Of 56 584 persons of age 30 years and over in the general population, 117 had IgA antibody to VCA. Among these 117 persons, 74 were biopsied in the first survey and 18（24%）were diagnosed by clinical and pathological examination as having NPC. 7 cases without subjective symptoms were in stage I, and 4 were in stage II. In the second surey which took place 10 months later, 2 additional NPC patients were diagnosed. These results indicate that NPC patients – including early cases not recognized clinically – could be identified by the immunoenzymatic method. The early diagnosis of NPC means a more favorable outcome of radiotherapy. It is expected that periodic follow up for the remaining antibody – positive persons without detectable tumors will reveal more NPC cases. Such studies not only are important for early detection of NPC, but also should help to clarify the relationship between EB virus and NPC.

Tests were conducted by both immunoenzymatic and immunoautoradiographic methods for detection of IgA antibody to VCA, in the sera from 12 328 persons age 30 years and older in one of the communes studied. Positive results were obtained in 25 and 69 cases, respectively. So it appears that the immunoautoradiographic method is more sensitive than the immunoenzymatic method. However, sera from the 18 NPC patients detected by this serological mass survey gave positive results by both methods. The results indicate that both methods are simple, sensitive and specific, and that antibody screening is valuable in the early detection of NPC.

〔In《Intervirology》1980, 13: 162 – 168〕

References

1 Henle G, and Henle W. Int J Cancer, 1976, 17: 1 – 7
2 Laboratory of Tumor Viruses of Cancer Institute, et al. Acta microbiol sin, 1978, 18, 253
3 Zeng Y, et al. Chinese J Oncol, 1979, 1: 2
4 Liu Yyx, et al. Chinese J Oncol, 1979, 1: 8
5 Liu CR, et al. Kexue Tongbao, 1979, 24: 715

4. Investigation of Epstein – Barr Virus Complement – fixing Antibody Levels in Sera of Patients with Nasopharyngeal Carcinoma and Nasopharyngeal Mucosal Hyperplasia

Department of Microbiology and Cancer Hospital; Zhongshan Medical College, Guangzhou; Department of Virology, Cancer Institute and Department of Tumor Viruses, Institute of Virology, Chinese Academy of Medical Sciences, Beijing; Cancer Institute of Zhongshan County, Guangdong

Summary

Anti – Epstein Barr virus (Anti – EBV) complement – fixing antibodies were measured in 118 sera samples of nasopharyngeal carcinoma cases (NPC) and 109 patients with other kinds of cancer using the micro – complement fixation method. The incidence of "high antibody" and NPC antibody geometric mean titer (GMT) were 61.8% and 1: 331.4, both being higher than in patients with cancers other than NPC. The antibody level at NPC stage I, though slightly lower than at more advanced stages, was significantly higher than in normal controls. NPC patients' complement – fixing antibodies after radiotherapy remained at a rather high titer for a relatively long period of time. The antibody level in NPC patients before radiotherapy and within 3 years after it remained about the same. Antibody levels decreased slightly in the NPC group surviving 3 – 6 years after radiotherapy and declined markedly in the NPC group surviving 7 – 15 years. This indicates that NPC is colsely related to EBV infections.

228 sera samples of patients with nasopharyngeal hyperplastic lesions and 504 sera samples of normal controls were also tested for anti – EBV antibodies. The difference in CF antibody titers in these 2 groups was significant, although it was very high (1: 640 – 1: 1 280) in 2.3% of both groups. Whether NPC would ultimately develop in this 2.3% should be followed up.

Introduction

It is well known that sera from patients with nasopharyngeal carcinoma (NPC) and Burkitt's lymphoma contain antibodies in high titers against various EBV – determined antigens, including capsid antigen (VCA), early antigen (EA), membrane antigen (MA), complement – fixing antigen (CF), and nuclear antigen (EBNA)[1]. Since 1973, investigations carried out by us on NPC etiology have resulted in the identification and establishment of more than 10 lymphoblastoid cell lines containing EBV antigens from NPC biopsy materials. From 1974 to 1975, antibody levels were determined by micro complement fixing reaction in our laboratory with the sera of NPC, non – NPC neoplasms and nasopharyngeal mucosal hyperplasia patients in order to elucidate the relationship between nasopharyngeal carcinoma and EBV infection in our country.

Material and Methods

1. Preparation of antigen: Complement fixing antigen was prepared from the lymphoblastoid cell line, CNL – 8, which was established in our laboratory from an NPC biopsy. The cells were cultured at 37℃ in medium RPMI – 1640 supplemented with 20% neonatal calf serum. As a rule, the cultures were fed every 3 to 4 days with fresh medium. After about 7 days of cultivation, the cells were harvested and washed twice in veronal buffer saline (VBS) at pH 7.2 by centrifugation at 2 000 r/min for 15 minutes. The cell pack was resuspended in VBS and the concentration adjusted to 4×10^7 cell/ml. The cell suspension was treated by ultrasonic disintegration or 4 cycles of freezing and thawing and then centrifuged at 10 000 r/min for 30 minutes at low temperature. The clear supernatant fluid was used as complement – fixing antigen with a titer of about 1:4 to 1:6.

2. Sera: A total of 477 sera specimens were tested. They came from: a. 118 cases of NPC before treatment; b. 109 cases of non – NPC neoplasms taken at Beijing and Guangzhou and c. 250 cases of NPC in Zhongshan county, Guangdong province. All patients in the latter group received radiotherapy and had survived for varying periods before the samples were taken and most showed no obvious signs of tumor burden at that time.

228 serum samples were collected from patients with hyperplastic lesions of the nasopharynx in Zhongshan county and tested to see if this type of nasopharyngeal lesions is associated with EBV infection.

128 sera specimens were collected from normal adults in Beijing or Guangzhou, to serve as control for the NPC specimens and 504 specimens were taken from normal adults of Zhongshan county to serve as control for nasopharyngeal hyperplastic lesions.

3. Microcomplement fixation test (MET): Briefly, the method[2] adopted was as follows: MFT was performed on microtitration plastic plates, using veronal buffered saline containig 5% chicken egg albumin, pH 7.2 as the diluent and guinea pig sera as sources of complement. 1.7 units of complement and 2 units of antigen were employed throughout the experiment. The indicating system of hemolysis was 4% sheep erythrocytes sensitized with an equal volume of 2 units hemolysin. For titrtion, the sera were serially 2 – fold diluted from 1: 10 to 1: 2 560 and the titers expressed as reciprocals. If the antibody titer of a serum sample was below 10, it was regarded as negative, if it was 10 or more it was designated as positive. Furthermore, if the antibody titer reached 320, it was defined as a "high titer antibody" (HTA).

Experimental Results

1. Comparison of EBV CF antibody titers in patients with NPC, non – NPC neoplasms and normal controls: As shown in Tab. 1, all the 118 cases of NPC had EBV CF antibodies and 61. 9% of them had HTA before treatment. The antibody geomtric mean titer (GMT) in this group was 331. 4. In contrast, the incidence of HTA in 109 sera samples of non – NPC neoplasm patients was only 20. 2% and the antibody GMT which was 121 was also significantly lower than that of the NPC group. It could be seen that the percentage of HTA in breast cancer and wonen's cancer (9. 5%) was the lowest in the non – NPC neoplasm group. In the normal control group, only 14% of the subjects had serum HTA and the antibody GMT was 76. 6. That is, the antibody level of NPC patients was the highest, being about 4 times greater than that of the normal adult control group and about 3 times higher than that of the non – NPC neoplasm group. Statistical analysis showed significant difference in the anti – EBV antibody levels of the 3 groups.

The sera of NPC patients were then classified into 4 subgroups according to the stages of the disease in order to determine if there was any correlation between anti – EBV antibody titers and advancement of the neoplasm (Tab. 2). It can be seen that the incidence of HTA and the antibody GMT of stage I patients was higher than in the normal group but lower than in the stage II group. The antibody GMT of the stage III group was somewhat lower than in the stage II and IV groups. The antibody GMTs of the stage I, II, III and IV groups were compared with those of the normal group, the GMT ratios were 2. 5, 5. 1, 3. 7 and 5. 2.

2. Persistence of anti – EBV CF antibodies in NPC patients after radiotherapy: Another series of sera samples was obtained from 250 NPC patients in Zhongshan county. All of them had received radiotherapy and had survived for varying periods of time afterwards and the antibody levels were determined to measure the persistence of EBV antibodies after this therapy. The patients were divided into 3 subgroups according to the length of survival after radiotherapy. As shown in Tab. 3, the antibody GMT and NPC patients in subgroup 1 who had survived 3 years, these 2 figures being similar to those in NPC patients before radiotherapy (Tab. 1). In the subgroup 2 NPC patients who had survived 3 – 7 years the antibody levels showed a moderate decrease, with a Gmt of 261. 7 but a marked decrease in antibody titer (GMT: 142. 6) and HTA was seen in subgroup 3 patients who had survived 7 – 15 years after radiotherapy. Comparing the EBV antibody GMT levels of the 3 patient subgroups with the normal group, the GMT ratios were 5. 2, 3. 9 and 2. 1, the difference being statistically significant.

Tab. 1 Comparison of sera EBV CF antibody titers in patients with NPC, non – NPC neoplasms and normal individuals

Source	Patient's sera with following titers										Total	High titer antibody (≥1:320)%	GMT
	<10	10	20	40	80	160	320	640	1 280	2 560			
NPC other cancers	–	–	–	2	12	31	24	35	10	4	118	61.9	331.4
Head and neck	–	–	–	6	6	10	4	3	–	–	29	24.1	132.2
Digestive tube and respiration organs	–	–	2	4	6	7	3	3	–	–	25	24.0	118.0
Breast and gynecologic	–	–	–	3	24	12	3	1	–	–	43	9.3	106.9
Lymph and blood	–	–	–	1	2	2	2	1	–	–	8	37.5	160.0
Miscellaneous	–	–	–	1	1	–	1	1	–	–	4	50.0	160.0
Total	–	–	2	15	39	31	13	9	–	–	109	20.2	121.0
Normal controls	5	4	4	32	37	28	18	–	–	–	128	14.1	76.6

Notes: Statistical analysis of antibody GMT in NPC, non – NPC neoplasms and normal controls: NPC and normal t = 11.3, $P <$ 0.01; NPC and othre cancers t = 8.3, $P < 0.01$; Other cancers and normal t = 3.7, $P < 0.01$

Tab. 2 Anti – EBV CF antibody titers in NPC patients at different stages of the disease

Source	Patient's sera with following titers										Total	High titer antibody (≥1:320)%	GMT
	<10	10	20	40	80	160	320	640	1 280	2 560			
NPC patients													
Stage I	–	–	–	–	3	2	1	2	–	–	8	37.5	190.3
Stage II	–	–	–	–	2	10	3	13	2	2	32	62.5	388.9
Stage III	–	–	–	2	5	10	13	14	1	1	46	63.0	287.9
Stage IV	–	–	–	–	2	9	7	6	7	1	32	65.6	397.3
Total	–	–	–	2	12	31	24	35	10	4	118	61.8	331.4
Normal controls	5	4	4	32	37	28	18	–	–	–	128	14.0	76.6

Tab. 3 Sera levels of anti – EBV CF antibody in NPC patients at different periods after radiotherapy

Years after radiotherapy	Patient's sera with following titers										Total	High titer antibody (≥1:320)%	GMT
	<10	10	20	40	80	160	320	640	1 280	2 560			
NPC													
0 – 2	–	–	–	2	12	16	29	31	10	2	102	70.6	344.8
3 – 6	–	–	–	5	8	24	39	22	2	–	100	63.0	261.7
7 – 15	–	–	–	7	14	14	6	7	–	–	48	27.1	142.6
Normal controls	33	19	48	112	116	109	55	7	5	–	504	13.3	66.2

3. Anti – EBV antibody levels of individuals with hyperplastic lesions of the nasopharyngeal mucosa: Three kinds of nasopharyngeal mucosa lesions were observed during broad scale mass survey for NPC in Zhongshan county starting from 1973, they were adenoids, hyperplastic nodules and severe inflammation with erosion. All these were designated as "hyperplastic lesions" and were suspected of having a possible relation to the subsequent development of NPC. Therefore the serum EBV antibody levels of these patients were compared with those of NPC patients and normal persons. 228 serum samples of nasopharyngeal hyper – plastic persons of Zhongshan county were tested as were 504 samples from normal healthy persons of the same county as controls. The results are presented in Tab. 4. The incidence of antibody HTA in the individuals with hyperplastic lesions, except the group with severe inflammation, was only slightly higher than that of the normal group, whereas the antibody GMTs of those with the 3 forms of hyperplastic lesions were significantly higher than those of the normal group. It is interesting to note that although the average anti – EBV antibody titers in the groups with hyperplasia and the normal group were in general significantly lower than those of NPC patients, a small number of cases, 17 sera samples, 2.3% had a high antibody titer of ⩾640.

Tab. 4   Comparison of anti – EBV CF antibody titers in patients with hyperplastic lesions of nasopharyngeal mucosa and normal adults in Zhongshan county, Guangdong province

Source	Patient's sera with following titers											High titer antibody (⩾1:320)%	GMT
	<10	10	20	40	80	160	320	640	1 280	2 560	Total		
Hyperplastic lesions:													
Adenoids	–	–	5	11	13	8	7	2	–	–	46	19.6	88.9
Nodules	–	–	9	24	28	20	17	2	–	–	100	19.0	90.6
Inflammation	–	–	5	20	24	23	9	1	–	–	82	12.2	90.1
Total	–	–	19	55	65	51	33	5	–	–	228	16.7	90.1
Normal controls	33	19	48	112	116	109	55	7	5		504	13.3	66.2

Notes: Statistical analysis: $t = 3.8$, $P < 0.01$

Discussion

Soluble extracts of CNL – 8 line lympho – blastoid cells were used as the antigen throughout our experiments. As the cell line was derived from a NPC patient who lived in Guangdong province, a high risk NPC area, this antigen seems to be suitable for examination of NPC patients' anti – EBV CF antibodies. Our results are comparable with those reported by others. The fact that anti – EBV antibody levels of NPC patient are markedly higher than those of normal individuals and patients with other cancers suggests that the development of NPC has a close serologic association with EBV infec – tions. As the serum anti – EBV antibody levels of patients with other cancers of the head and neck or other regions are substantially higher than those of the normal controls, it is possible that latent EBV infections may be activated to a certain extent by immunologic suppression in cancer patients[3]. As a result, the antibody levels are elevated although they remain definitely lower than in NPC patients.

As to the relationship between clinical status of NPC patients and anti – EBV antibody levels, it was reported by de – The[4] that anti – EBV CF antibody GMT increases gradually from stage I to stage V with the clinical deterioration. According to our study, CF – antibody GMT is relatively low at stag I of the disease, though it is apparently higher than that of normal invividuals. At stage Ⅱ, antibody level rises sharply and attains a level as high as in stage Ⅳ whereas stage Ⅲ antibody level is definitely lower than those at stages Ⅱ and Ⅳ. This feature does not coincide with that of de The. These results show that serum anti – EBV CF antibodies can be maintained at a high level over long periods of time in NPC patients. For example, the antibody titers of NPC patients surviving 3 years after radiotherapy were just as high as those of NPC patients prior to radiotherapy. At the same time, it is also apparent that the longer the survival period the lower the antibody HTA and GMT in NPC patients' serum. It is noteworthy that although the percentage of antibody HTA and GMT showed significant decline in NPC patients with long – term survivals of 7 – 15 years both figures were still higher than in normal controls. Thus, it is apparent that although the cancer may be effectively controlled by radiotherapy, even with complete regression, the elevated EBV CF antibody titers can not be brought down rapidly during rehabilitation. Instead, the antibodies may persist in the patients at a comparatively high level for a relatively long period of time. Whether the persistence of anti – EBV CF antibody represents a continuous activation of EBV infection in the host is a problem deserving further investigation. When analysing the results of nasopharyngoscopy of 430 000 people in Guangdong, the Zhongshan Medical College Tumor Hospital pointed out that hyperplastic lesions of the nesopharynx may be associated with the subsequent development of NPC. In addition, based on pathologic examination of these hyperplastic lesions, atypical hyperplssia and metaplasia are thought to have a certain relation with malignant transformation. These suspicions prompted us to survey the correlation between mucosal hyperplastic lesions and the exten of EBV infection of the nasopharynx by determining the anti – EBV antibody count. Although the anti – EBV antibody GMT of individuals with hyperplastic lesions was significantly higher than that of normal controls, only a few cases with excessively high antibody levels ($\geqslant 1: 640$) were found in each group. Whether this marked difference in antibody levels and the appearance of HTA are associated with possible NPC development need further investigation.

It was reported by de – The[4] that no substantial difference was observed in EBV CF antibody titration of normal sera samples when CF – antigens were prepared from a productive or nonproductive lymphoblastoid cell line. But when titrations were done with NPC sera, the CF – antibody titers were apparently higher with the productive lymphoblastoid cell antigens than with the nonproductive ones. He also claimed that, in addition to the soluble CF antigen, EA or VCA could be present in the supernatant of the productive cell line. Therefore, the antibody titers reflect not only the CF antibodies but also some VCA or EA antibodies. In our experiments, CF antigens were prepared from CNL – 8 cell line containing 0. 5% – 1% VCA positive cells. The GMTs of NPC anti – EBV CF antibodies and normal controls were several times higher than de – The's. The cause of this difference in antibody titers is not yet clear.

〔In 《Chines Medical Journal》1980, 93 (6): 359 – 364〕

References

1 Epstein MA, Achong BG. The EB Virus. Ann Rev Microbiol, 1973, 27: 413

2 北京协和医院检验科主编: 病毒实验诊断手册. 第一版. 北京: 人民卫生出版社, 1960: 106

3 Henle W, et al. Antibodies to Epstein – Barr virus in nasopharyngeal carcinoma, other head and neck neoplasms and control groups. J Natl Cancer Inst, 1970, 44: 225

4 de – The G, et al. Nasopharyngeal carcinoma IX. Antibodies to EBNA and correlation with response to other EBV antigen in Chinese patients. Int J Cancer, 1975, 16: 713

5. Study of Giant Group a Marker Chromosome in Several Burkitt's Lymphoma and Iymphoblastoid Cell Lines with Epstein – Barr Virus from Different Origins

Wu Bing[1], Wu Yu – qing[2], Li Yi – wan[1]

Zeng Yi[3], Wu Min[1], Zhao Zhi – hui[1], Gong Chui – hong[3]

1. Laboratory of Tumor Viruses, Cancer Institute, Chinese Academy of Medical Sciences;
2. Laboratory of Cell Biology, Cancer Institute, Chinese Academy of Medical Sciences;
3. Laboratory of Tumor Viruses, Institute of Virology, Chinese Academy of Medical Sciences

Summary

Several Burkitt's lymphoma and lymphoblestoid cell lines with Epstein – Barr virus derived from different origins have been investigated cytogenetically.

Giant group A marker chromosome was detected only in three lymphoblastoid cell lines from nasopharyngeal carcinoma (NPC). With the exception of the HS2 – IB lymphoblastoid cell line derived from a normal donor, the giant group A marker chromosome could not be found in the cell lines from the other sources, including P3HR – 1, B95 – 8, Raji and lymphoblasttoid cell lines from tonsil and so on. The results suggest that the giant group A marker chromosome might be associated with NPC.

It has been proved by G – banding technic that this giant submetacentric group A chromo – some was formed by the translocation of the short arm of chromosome 3, breaking at the point near to or even involving its centromere, to the distal light band region of the long arm of chromosome 1, namely t (1; 3) (1p ter→1q 44 :: 3 p11→3p ter) and t (1; 3) (1p ter→1q 44 :: 3q21→3p ter).

The occurrence rate of the marker chromosome was only 6.6% half a year after CNL8 cell line was established; 82.0% after two years; and 100% after five years. The occurrence rate of the marker chromosome was 32% half a year after the NPC80 cell line had been established, and 72% after three years. Increase in the occurrence rate of the marker chromosome in these cell lines was clearly paralleled to the time of the cell lines culture in vitro.

In addition, the occurrence rate of the marker chromosome was not the same in different cell lines. It was only 6.6% half a year after CNL8 cell line was established, while is was 32% for NPC 80 and 63.3% for CNL5.

Introduction

Epstein – Garr (EB) virus, the causative agent of infectious mononucleosis (IM), is closely associated with African Burkitt's lymphoma (BL) and nasopharyngeal carcinoma (NPC)[1-3]. The cells of African BL are known to carry the EB viral genome[4,5]. Both the tumor cells and cultured cell lines derived from them show a No. 14 chromosome abnormality[6], but the abnormality in the No. 14 Chromosome of BL cells appears to be unrelated to EB virus, since it has been shown to be lacking in EB virus – carrying lymphoid cells from other sources, such as blood of patients with infectious mononucleosis and normal seropositive individuals or after in vitro transformation of lymphocytes from seronegative donors. It might be an important event in the development of human lymphocytic malignancy[7-9].

The epithelial tumor cells of undifferentiated nasopharyngeal carcinoma likewise carry the EB viral genome[10,11]. In addition, biopsy samples of this tumor can give rise to EB virus – containing lymphoblastoid cell lines in vitro[12,13]. Such a material is quite suitable for cytogenetic analysis. Recently, Finerty et al[14] reported that the chromosome spreads of lymphoblastoid cell lines established from seven different NPC biopsy specimens, and the epithelial tumor cells from another five biopsy specimens freed from nonmalignant infiltration cells by transplanting into nude mice were examined with banding technique. No marker chromosome has been detected.

In 1973, eleven NPC lymphoblastoid cell lines were established in our laboratory[15]. Two of them, CNL5 and CNL8, were examined cytogenetically within half a year after their establishment, and a giant submeta – centric group A marker chromosome was seen in them[16]. In 1978, Xia et al[17] reported a giant group A marker chromosome present in all three NPC lymphoblastoid cell lines, CSN3, CSN7 and CNL8. As this marker chromosome is larger than chromosome No. 1, it was therefore designated as giant group A marker chromosome.

In this paper, we report our studies on the relationship between giant group A marker chromosome and NPC.

Materials and Methods

1. BL and lymphoblastoid cell lines: CNL5 and CNL8 lymphoblastoid cell lines: they were established from biopsy materials of NPC patients, within 198 and 159 days respectively in our laboratory in 1973.

NPC80 lymphoblastoid cell line was established from NPC18 in the Department of Microbiology and the Department of Etiology of the Cancer Institute of Zhongshan Medical College.

B95 – 8 lymphoblastoid cell line was a marmoset cell line transformed by EB virus from a patient with infectious mononucleosis. It was kindly supplied by Professor Epstein of the Department of Pathology, University of Bristol Medical School, Great Britain.

Raji Lymphoma cell line was derived from BL. It was also kindly supplied by Professor Epstein.

P3HR – 1 cell line was derived from BL cell line Jijoye.

Ton – 11 lymphoblastoid cell line was derived from the tonsil of a normal individual. It was established in this laboratory in 1978. H26 and HS2 – 1B lymphoblastoid cell lines were derived from the peripheral blood lymphocytes of a normal donor. They were kindly supplied by the Departent of Microbiology of the Shanghai Second Medical College.

2. *Chromosome analysis*: The cells of lymphoblastoid cell lines were incubated in an incubator with 5% CO_2 in air at 37℃ for 36 – 48 hrs. A drop of colchicine (final concentration in medium: 0. 02 – 0. 04 μg/ml) was added to the cultures 4 – 10 hrs before harvest. Dry metaphase spreads were made and stained by Giemsa. 25 metaphase cells were counted for each line, some of them were photographed by means of microphotography, and karyotypes were arranged according to Denver's system.

A part of the slides were processed by Seabright's trypsin banding method for G – banding[19,20].

The slides with clear bands were selected for microphotography and karyotypes were arranged according to the criteria set at Paris Conference on Standardization of Terms in Human Cytogenetics.

Results

Nine lymphoma and lymphoblastoid cell lines with EB virus from different origins were examined. Three of them, CNL5, CNL8 and NPC80, were derived from NPC. One of them, B95 – 8, was a transformed cell line by EB virus from IM. Two of them, Raji and P3HR – 1, were derived from Burkitt's lymphoma. Three of them, Ton – 11, H26 and HS2 – 1B, were derived from normal donors. Twenty – five metaphase cells were examined from each line. The mode numbers of CNL5, CNL8, NPC80 cell lines were 46, 47, and 44 respectively (Tab. 1). The mode number of B95 – 8 cell line was 44. The mode numbers of Raji and P3HR – 1 cell lines were 47 and 46 respectively. The mode numbers of Ton – 11, H26 and HS2 – 1B cell lines were 46, 46 and 43 respectively.

There were many giant group A marker chromosomes in all three NPC lymphoblastoid cell ines, CNL5, CNL8 and NPC80 (Figs. 1 – 3). The occurrence rates of this marker chromosome in these three cell lines were 63. 3%, 100%, and 72% respectively. No giant group A marker chromosome could be found in the cell lines of B95 – 8, Raji and P3HR – 1. No giant group A marker chromosome could be encountered in the cell lines Ton – 11 and H26, which were derived from normal donors, but it was detected in the HS2 – 1B cell line which was derived from normal individual (Fig. 4). The occurrence rate of this marker chromosome was 80%. On the other hand, the occurrence rate of the marker chromosome was only 6. 6% half a year after establishment of CNL8 cell line, then it became 82.0% at two years, and 100% at five years (Tab. 2).

Tab. 1 Origins of the lymphoma and lymphoblastoid cell lines and their giant group A chromosomes

Cell line	Origins of cell lines	Cells counted	Mode of chromosome	Giant group A chromosome	% of cells with giant A chromosome
CNL5	Nasopharyngeal	30	46	+	63.3
CNL8	Carclnoma	25	47	+	100.0
NPC80		25	44	+	72.0
B95 – 8	Infectious mononucleosis	25	44	–	
Raji	Burkitt's lymphoma	25	47	–	
P3HR – 1		25	46	–	
Ton – 11	Normal tissue	25	46	–	
H26		9	46	–	
HS2 – 1B		25	43	+	80.0

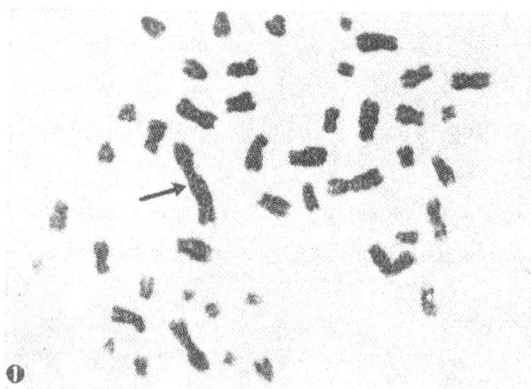

Fig. 1 The chromosome in a cell from CNL5.
The arrow shows the giant group A chromosome

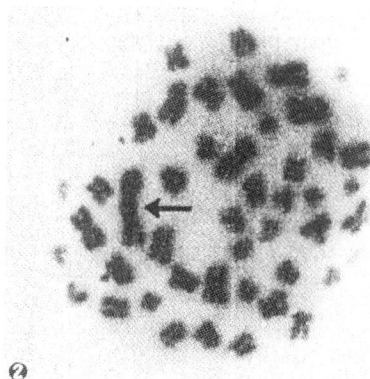

Fig. 2 The chromosome in a cell from CNL8.
The arrow shows the giant group A chromosome.

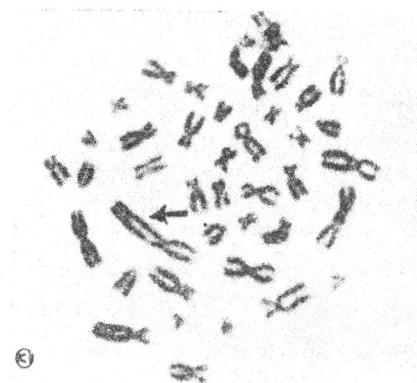

Fig. 3 The chromosome in a cell from NPC80.
The arrow shows the giant group A chromosome

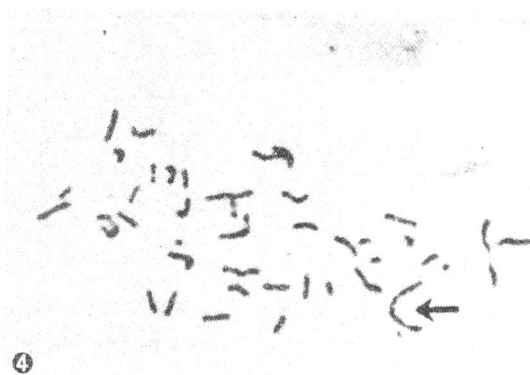

Fig. 4 The chromosome in a cell from HS2 – 1 B.
The arrow shows the giant group A chromosome

Tab. 2 Relationship between occurrence rate of giant group A chromosome and different durations after cell lines were established

Cell lines	Date of cell lines established	Chromosome examination after cell lines established (year)	% of cells with giant A chromosome
CNL_5	November, 1973	0.5	63.3
CNL_8	November, 1973	0.5	6.6
		2.0	82.0 *
		5.0	100.0
NPC_{80}	September, 1974	0.5	32.0 * *
		3.0	72.0

Notes: * From reference 17

      * * From reference 18

The occurrence rate of the marker chro – mosome was 32% [18], half a year after the NPC80 cell line was established, then 72% at the end of the 3rd year.

It seems that the occurrence rate of the marker chromosome rose with the lenght of time after these cell lines were established. Twenty – five metaphase plates from three passages of CNL8 cell line were analysed by G – banding technique. It was found that the short arm of this marker chromosome had two dark bands over the proximal part (Fig. 5) and some lighter bands over the distal part, identical with that of the shot arm of chromosome 1. In the well

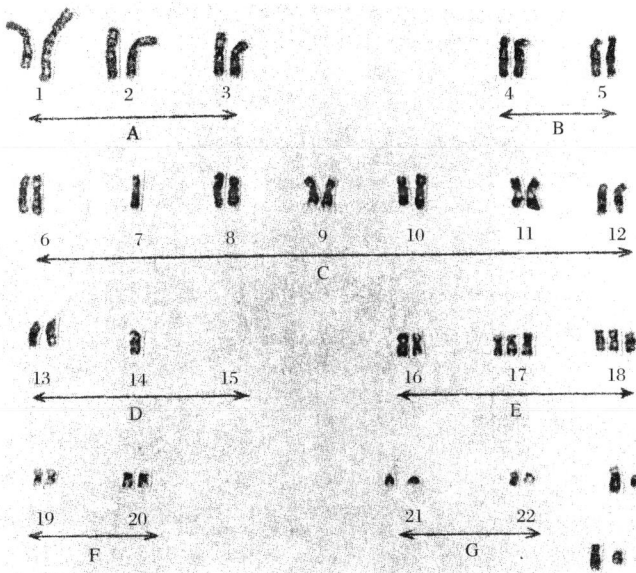

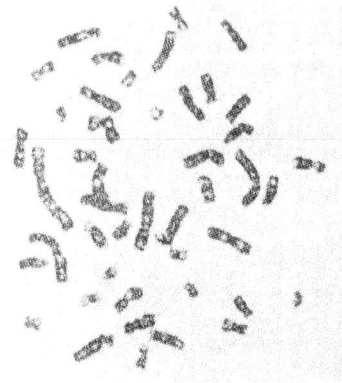

Fig. 5 **The banded chromosome in a cell from CNL 8. The arrow indicates the giant group A chromosome.**

treated preparations it could be found clearly that there were 8 deeply stained bands on the long arm of this marker chromosome, the proximal 4 darker bands were close to each other in pairs showing the characteristics of the long arm of chromosome 1. The 4 dark bands in the distal part were also close to each other in pairs and often fused into 2 broad dark bands separated by a broad light band, simulating the characteristics of the short arm of chromosome 3. The 2 arms of this giant group A chromosome were compared with the corresponding parts of chromosome 1 and chromosome 3 in the same cell. It was shown that their band patterns were completely identical (Fig. 6). It might be con – cluded that this giant submetacentric group A chromosome is formed by the translocation of short arm of chromosome 3, breaking at a point near its centromere, to distal light band region of the long

Fig. 6 Comparison of the 2 arms of this giant chromosome with the corresponding parts of chromosome 1 and chromosome 3 in the same cell

arm of chromosome 1, namely t (1；3) (1p ter→1p 44：3p 11→3p ter). But it was found these markers were dicentric in a few of CNL8 cells (Fig. 7). This kind of giant submetacentric group A chromosome appears to be formed by the translocation of the short arm of chromosome 3 including its centromere, to distal light band region of the long arm of chromosome 1, namely t (1；3) (1p ter→1q 44：3q 21→3p ter).

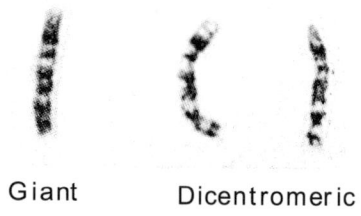

Giant   Dicentromeric

Fig. 7 A giant group A chromosome and dicentric giant group A chromosome

In view of the presence of the two homologous chromosomes 3 by karyotyping (Fig. 5), the problem of the composition of this marker chromosome awaits further elucidation.

Discussion

Among several BL and lymphoblastoid cell lines with EB virus from different origins, a giant submetacentric group A marker chromosome was detected in the cells of all three NPC lymphoblastoid cell lines. with the exception of the HS2 – 1B, the giant A marker chromosome could not be found in the cell lines from other sources, including P3HR – 1, B95 – 8, Raji and lymphoblastoid cell lines derived from tonsil and so on. Xia et al. [17] found the giant group A maker chromosome in three NPC lymphoblastoid cell lines, CSN3, CSN7 and CNL8. Ou et al. [18] examined the chromosomes of 79 NPC biopsy samples, and 45 marker chromosomes were detected in 201 cells. The occurrence oate of the giant group A marker chromosome was 22. 3%. These results suggest that this giant group A chromosome might be associated with NPC. To date there has been no suggestions regarding the presence of the giant marker chromosome in the NPC lymphoblastoid cell lines in reports. But we have observed that there was a submetacentric chromosome larger than group A chromosome 1 in the HW cells of NPC in Fig. 5 of Finerty's report[14]. It remains to be confirmed whether this is similar to our findings.

There was a giant group A marker chromosome in lymphoblastoid cell lines, CNL5, CNL8, NPC80 and HS2 – 1B. The occurrence rate of the marker chromosome was 63. 3%, 100%, 72%

and 80% respectively (Tab. 1). The occurrence rate of marker chromosome was only 6.6% half a year after CNL8 cell line had been established; 82.0% after two years, and 100% after five years (Tab. 2). The occurrence rate of the marker chromosome was 32% half a year after the NPC80 cell line had been established[18] and 72% after three years. Increase in the occurrence rate of the marker chromosome in these cell lines clearly paralleled to the culturing time of the cell lines in vitro. Whether this increase in due to increase of cells carrying the giant group A marker chromosome during the process of the cell line culture in vitro or is due to the effect of EB virus on the lymphoblastoid cell lines awaits further study. In addition, the rate of the marker chromosome was not the same in different cell lines. For example, it was only 6.6% half a year after CNL8 cell line was established, while now it was 32% for NPC80 and 63.3% for CNL5 (Tab. 2).

Recently Kovasc[21] reported that marker chromosomes involving chromosome No. 1 were studied with banding techniques in 10 human primary solid tumors. Structural or numerical aberrations of chromosome No. 1 were found in nine of them, including breast cancer and cancer of colon. The aberration of chromosome No. 1 suggests that a "weak point" exists on the long arm of chromosome No. 1 near the centromere and region q 21. The detected chromosome segments are often translocated to other chromosomes. It is therefore possible that aberrations of chromosome No. 1, particularly trisomy of region q 21 – 32, are importat for the development of certain tumor types. Nasopharyngeal carcinoma is a solid tumor too, and has also aberrations of chromosome No. 1, But the aberration of chromosome No. 1 found by Kovacs is not similar to our finding.

The giant group A marker chromosome is formed by the translocation of the short arm of chromosome 3, breaking at the point near its centromeree, to the distal light band region of the long arm of chromosome 1. And a few of the giant group A marker chromosomes were found to be formed by the translocation of the short arm of chromosome 3 involving its centromere, to the distal light band region of the long arm of chromosome 1. The morphology of this giant group A chromosome is similar to that of group A chromosome 1, but it is larger than group A chromosome 1. The largest marker chromosome is 1/5 to 2/5 longer than chromosome No. 1. It is always present singly. Sometimes two of these chromosomes are seen in a few of hte cells carrying 90 or more chromosomes. A part of our observation is similar to that of Xia et al.[17] As the two homologous chromosomes of chromosome 3 were present by karyotyping, the problem of composition of the marker chromosome should be further studied.

EB virus is the causative agent of infectious mononucleosis, and it also has a remarkable association with African (endemic) BL and with the poorly differentiated carcinoma of the nasopharynx. With EB virus playing such a variety of roles in different conditions, the question has naturally been raised as to whether there is presence of different strains of EB virus or not. But up to now, EB virus strains having different biologic properties have not been found in nature.

The marker chromosomes carried by cells of BL and lymphoblastoid cell lines with EB virus derived from different origins were different. Whether this is due to the cell lines being derived from different origins or due to the biologic properties of different EB viral strains remains to be investigated.

[In 《Chinese Medical Journal》 1980, 93 (6): 4500 – 406]

References

1 Zur Hausen H. Biochemical approaches to detection of Epatein – Barr virus in human tumors. Cancer Res, 1976, 36: 678

2 Old LJ, et al. Precipitating antibody in human serum to an antigen present in cultured Burkitt's lymphoma cells. Proc Nutl Acad Sci USA, 1996, 56: 1699

3 Henle W, et al. Antibodies to Epstein – Barr virus in nasopharyngeal carcinoma, other head and neck neoplasm, and control groups. J Natl Cancer Intst, 1970, 44: 225

4 Zur Hausen H, et al. EBV DNA in biopsies of Burkitt tumors and anaplastic carcinomas of the nasopharynx. Nature, 1970, 228: 1056

5 Nonoyama M, et al. DNA of Epstein – Barr virus detected in tissue of Burkitt's lymphoma and nasopharyngeal carcinoma. Proc Natl Acad Sci USA, 1973, 70: 3265

6 Manolov G, et al. Marker band in one chromosome 14 from Burkitt lymphomas. Nature, 1972, 237: 33

7 Kaiser – McCaw B, et al. Chromosome 14 translocation in African and North American Burkitt's lymphoma. Int J Cancer, 1977, 19: 482

8 Jarvis JE, et al. Cytogenetic studies on human lymphoblastoid cell lines from Burkitt's lymphomas and other sources. Int J Cancer, 1974, 14: 716

9 Zech L, et al. Characteristic chrommosomal abnormalites in biopsies and lymphoid cell lines from patients with Burkitt and non – Burkitt lymphomas. Int J Cancer, 1976, 17: 47

10 Wolf H, et al. EB viral genomes in epitelial nasopharyngeal carcinoma cells. Nature（New Biol）, 1973, 244: 245

11 Kiein G, et al. Direct evidence for the presence of Epstein – Barr virus DNA and nuclear antigen in malignan t epithelial cells from patients with poorly differentiated carcinoma of the nasopharynx. Proc Natl Acad Sci USA, 1974, 71: 4737

12 DE – The G, et al. Lymphoblastoid transformation and presence of herpes – type viral particles in a Chinese nasopharyngeal tumor cultured in vitro. Nature, 1969, 221: 770

13 DE – The G, et al. Nasopharyngeal carcinoma. I Types of cultures derived from tumour biopsies and non – tumorous tissues of Chinese patients with special reference to lymphoblastoid transformation. Int J Cancer, 1970, 6: 189

14 Finerty S, et al. Cytogenetics of malignant epithelial cells and lymphoblastoid cell lines from nasopharyngeal carcinoma. J Cancer, 1978, 37: 231

15 Department of Virus Cancer Institute Chinese Academy of Science Establishment of lymphoblastiod cell line and isolation of CMV Chinese J. ENT and Laryns

16 Unpublished Data Department of Cell Biology, Chinese Academy of Medical Science

17 XIA Jiahui et. al. Maker Chromosome Related to Nasopharyngeal Carcinoma J. Genetics

18 OU Baoxiang Analysis of Chromosome in Nasopharyngeal Biopsies and the Clinical Value

19 Seabright M. A rapid banding technique for human chromosomes. Lancet, 1971, 2: 971

20 Institute of Cancer Chinese Academy of Medical Science Bands of Normal Human Chromosomes J. Genetics

21 Kovacs, G. Abnormalities of chromosome No. 1 in human solid malignant tumours. Int J Cancer, 1978, 21: 688

6. Anticomplement Immunoenzymatic Method of Detecting Epstein – Barr Nuclear Antigen in Nasopharyngeal Carcinoma Celisand cells and Normal Epithelial Cells

Zeng Yi[1], Shen Shu – jing[2], Pi Guo – hua[*], Ma Jiao – lian[2],

Zhang Qin[1], Zhao Ming – lun[2], Dong Han – ji[2]

1. Institute of Virology, Chinese Academy of Medical Sciences, Beijing;

2. Zhanjiang Medical College, Zhanjiang

Summary

Exfoliated cells collected by negative gressure suction from the nasopharynx of patients with poorly differentiated and undifferentiated cell carcinoma and of suspect nasopharyngeal carcinoma (NPC) patients were examined for Epstein – Barr nuclear antigen (EBNA) by anticomplement immunoenzymatic method (ACIE). All 79 NPC patients had EBNA – positive carcinoma cells, while the positive rate of cytologic and histologic examination was only 87. 3% and 91. 1%. No EBNA was found in tumor cells from patients with head and neck malignant and benign tumors other than NPC, nor was it found in nasopharyngeal epithelial cells of dead fetuses. 6 persons with EBNA – positive cells detected by this method, with no carcinoma detected by cytology or histology, were reexamined regularly. Carcinoma cells were detected from 1 week to 3 months later in 4. This shows that this ACIE method is specific and sensitive and can be used for early NPC detection, especially in combination with serologic detection of Epstein – Barr virus (EBV) antibody.

Introduction

Besides finding EBNA in nasopharyngeal carcinoma cells, it can also be found in normal ciliated columnar epithelial cells and hyperplastic cells of the nasopharynx.

The close association of Epstein – Barr virus (EBV) with nasopharyngeal carcinmoa (NPC) is demonstrated by the regular finding of Epstein – Barr nuclear antigen (EBNA) and EBV – DNA in NPC cells by anticomplement immunofluorescence (ACIF) and nucleic acid hybridzation[1-5]. Early stage NPC can be detected by IgA serologic mass surveys[6,7]. Follow up study of VCA – IgA antibody positive persons showed that some new NPC cases could be detected[8] but that a large proportion of the group showed no evidence of NPC. Therefore, it was necessary to improve the technic of detecting the EBV marker in NPC cells. We developed an anticomplement immunoenzymatic (ACIE) method for EBNA detection[9] which is a very sensitive test. Exfoliated cells collected from the nasopharynx by negative pressure suction are used for cytologic examination. The positive carcinoma cell rate is 92. 7%[10]. This method is simple and convenient and numerous cells can be obtained. The exfoliated nasopharyngeal cells obtained by this method are examined by ACIE to detect

EBNA to diagnose NPC and study the relationship of the EBV and NPC. In this study, EBNA was found not only in NPC cells, but also in normal epithelial cells and hyperplastic cells.

Material and Methods

1. Specimens: Exfoliated nasopharyngeal cells from 145 NPC patients and NPC suspects were collected. Touch smears were made from 18 head and neck tumors other than NPC and from 21 nasopharyngeal mucosa of dead fetuses aged 3 – 8 months.

2. Method of collecting cells from nasopharynx by negative pressure suction: 1% dicaine was sprayed into the oropharyngeal and nasopharyngeal cvity. Exfoliated cells were collected by electric suction apparatus, the negative pressure being no more than – 30 mmHg. Smears were made on slides from the head of the suction apparatus. All smears were fixed in cold acetone for 10 minutes and examined by ACIE or stained with HE for cytologic examination.

3. Histologic examination: Biopsies taken from suspect nasopharyngeal lesions or tissue fragments obtained by negative pressure suction were fixed with 10% formalin stained with HE and examined.

4. ACIE test[9]: Normal human sera were used as sources of complement, some of the EBV antibody negative sera were donated by Drs de – The and Desgranges. There was no nonspecific antinuclear antibody in the sera. Balance salt solution (BSS) was used as diluent. The reference NPC sera containing EBNA antibody and normal human serum with complement at a final 1 : 10 dilution were added to the smears and the slides placed in a humidified chamber at 37℃ 1 hour. Afer washing with BSS 3 times, the anti – C_3 antibody conjugated with horse – radish peroxidase at 1 : 10 was added and the slides stored at 37℃ 30 minuts. The smears were again washed with BSS 3 times, stained with diaminobenzene and H_2O_2 and examined.

5. ACIF test: EBNA was also detected by ACIF test as described by Reedman and Klein[11].

Results

1. Comparison of ACIE and ACIF tests in EBNA detection: Tab. 1 shows that EBNA – positive carcinoma cells were found in exfoliated cells from all 8 cases by both ACIE and ACIF tests. Usually more EBNA positive carcinoma cells were detected by the ACIE test than by the ACIF test. Carcinoma cells were detected in 7 cases by both cytologic and histologic examinations. In Case 73, the histologic and cytologic examinations showed no tumor, but 1 month later, reexamination showed a poorly differentiated squamous cell carcinoma. All 8 NPC cases had VCA – IgA antibody.

Tab. 1 Comparison of ACIE and ACIF tests in EBNA detection

Case	EBNA		Cytolotic examination	Pathologic examination	VCA/IgA antibody
	ACIE	ACIF			
99	+	+	+	+	1 : 20
101	+	+	+	+	1 : 20
102	+	+	+	+	1 : 1 280
103	+	+	+	+	1 : 20
112	+	+	+	+	1 : 1 280
113	+	+	+	+	1 : 80
119	+	+	+	+	1 : 1 280
73 – 1	+	+	–	–	1 : 80
73 – 2	+	+	–	+	

2. ACIE test in EBNA detection: Exfoliated cells were collected from 145 NPC patients and NPC suspects, 18 malignant and benign head and neck tumors other than NPC and 21 nesopharyngeal mucosa of dead fetuses. All specimens were examined by ACIE test, cytologic and histologic examinations. If either cytologic or histologic examination showed carcinoma cells, the patient was diagnosed as NPC. Tab. 2. shows the comparison of the ACIE test with the cytologic and pathologic examinations for NPC diagnosis. 62 were positive by all 3 methods (Figs. 1 – 3), 10 were positive by ACIE and histologic examination and 7 were positive by ACIE and cytologic examination. All 79 patients were diagnosed as poorly differentiated or undifferentiated carcinomas. 9 with EBNA – positive cells were negative by cytologic and histologic examination and therefore could not be diagnosed as NPC. No EBNA – positive cells could be found in exfoliated cells from 57 in whom NPC was excluded, in touch smears from 18 patients with head and neck tumors other than NPC or 21 dead fetuses. As shown in Tab. 3, the number of NPC cases in clinical stages Ⅰ, Ⅱ, Ⅲ and Ⅳ was 15, 29, 31 and 4. The positive rate of carcinoma cell detection by ACIE, cytologic and histologic examinations was 100%, 87.3% and 91.1%. ACIE is more sensitive than the 2 other methods, especially in stage Ⅰ, the positive rate by the latter 2 methods was only 86.4% and 80.0%.

Tab. 2   Comparison of ACIE method, cytologic and pathologic examinations in NPC detection

Group	Cases	ACIE	Cytologic examination	Histologic examination
NPC patients	79	79	69 (87.3%)	72 (91.1%)
	62	+	+	+
	10	+	−	+
	7	+	+	−
NPC suspects	9	+	−	−
	57	−	−	−
Head and neck tumors other than NPC	18	−	−	−
Fetal nasopharynx	21	−	−	−

3. EBNA in different types of nasopharyngeal cells: In NPC patients EBNA could be found not only in carcinoma cells, but also in hyperplastic cells and normal cilated colnumar epithelial cells in the same specimens, it was scarcer in the normal squamous epithelial cells of these specimens. EBNV was present in either normal epithelial cells or hyperplastic cells in only 9 cases in whom the cytologic and histologic examinations were negative (Figs. 4 – 6).

Tab. 3   Comparison of 3 methods in different clinical stage NPC detection

Clinical stage	Cases	Positive (%)		
		ACIE	Cytologic examination	Histologic examination
Ⅰ	15	15 (100)	13 (86.4)	12 (80.0)
Ⅱ	29	29 (100)	23 (79.3)	27 (93.1)
Ⅲ	31	31 (100)	29 (93.5)	29 (93.5)
Ⅳ	4	4 (100)	4 (100)	4 (100)
Total	79	79 (100)	69 (87.3)	72 (91.1)

4. Reexamination of EBNA – positive mucosae: 6 cases who had EBNA it the carcinoma cells or hyperplastic cells, with no carcinoma cells by cytologic and hitologic examination, were reexamined from 1 week to 3 months later, carcinoma cells were detected by cytologic and histologic methods in 4 but were still absent in the otuer 2 (Tab. 4).

Tab. 4 Follow up study on the EBNA positive individuals

Patient	Data	ACIE	Cytologic examination	Histologic examination	Clinical stage	VCA – IgA antibody
Cheng	11. 01. 80	+	–	–		1:20
	18. 01. 80	+	+	+	I	
Wu	26. 01. 80	+ (HC)	–			1:20
	03. 05. 80	+	+	+	II	
Lin	02. 01. 80	+ (HC)	–			1:80
	31. 03. 80	+	+	–	II	
Le	07. 04. 80	+ (HC)	–	–		
	15. 05. 80	+	–	+	II	1:80
Sha	26. 03. 80	+ (NE)	–	–		1:80
	13. 04. 80	+ (NE)	–	–		
Ho	13. 11. 79	+ (HC)	–	–		
	11. 04. 80	+ (HC)	–	–		

Notes: + = carcinoma cell, + (HC) = hyperplastic cell, + (NE) = normal epithelial cell

Discussion

Our results indicate that EBNA can be detected in NPC patients' carcinoma cells by the ACIE test. This test is specific, because carcinoma cell nuclear staining was obtained only with sera containing EBV antibodies and not in negative sera. Compared with the ACIE test, more carcinoma cells collected by negative pressure suction could be stained by the ACIE test and the different types of EBNA – positive cells could be more easily identified. Furthermore, the ACIE test is more sensitive than cytologic or histologic detection of NPC, especially when patients are in clinical stage I. As mentioned, 6 normal individuals with EBNA – positive cells, but without carcinoma cells cytologically and histologically were reexamined from 1 week to 3 months later with positive carcinoma cell results in 4 by either of the 3 methods. Therefore, besides routine cytologic and histologic examination, detection of EBNA by ACIE and of VCA – IgA antibody is highly valuable for early NPC diagnosis.

EBNA was found in carcinoma cells, hyperplastic cells and ciliated columnar epithelial cells of NPC patients and was also found in hyperplastic epithelial cells and ciliated columnar epithelial cells from normal individual, but was scarcer in squamous epithelial cells. These data suggest that the EBV may first infect normal epithelial cells, especially ciliated columnar epithelial cells, integrating its DNA into the cellular NDA. How the malignant transfromation of epithelial cells occurs under these conditions is not known. It is of paramount interest to follow up the normal individuals having EBNA – positive cells. Up to the present, the EBV is only known to infect Blymphocytes. How the EBV enters the epithelial cells from which NPC arises, remains to be determined.

Acknowledgements: We would like to thank Dr Gcy de – The and Dr C Desgranges for donating EBV antibody negative human serum and florescence labelled antihuman C_3 antibody.

We also thank Prof CH Huang, Beijing, and Dr G de – The, France for revision of the manuscript.

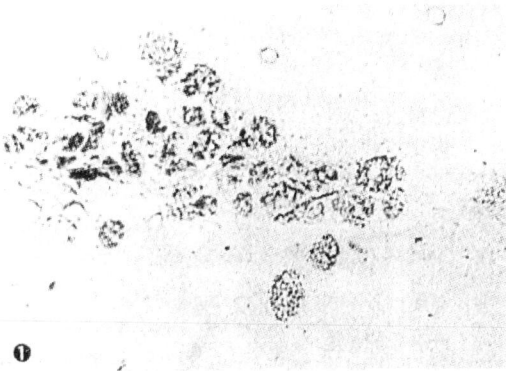

Fig. 1 EBNA positive carcinoma cells detected by ACIE test, Case 60. ×132

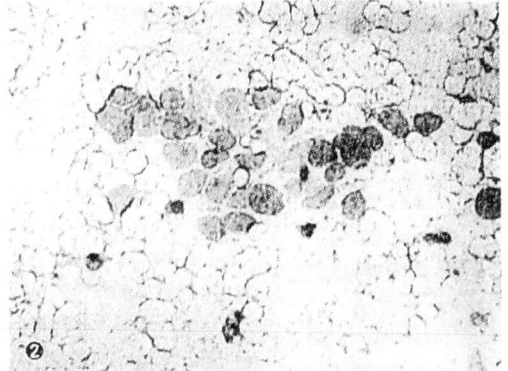

Fig. 2 Carcinoma cells detected by cytologic examination, Case 60. HE ×132

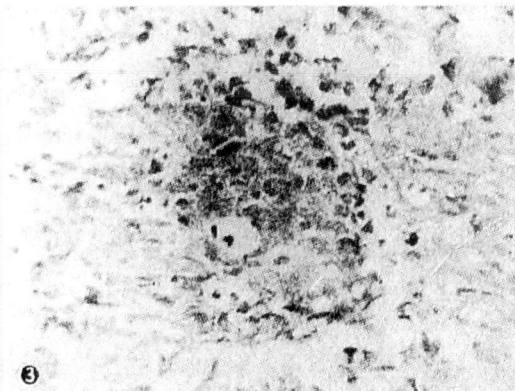

Fig. 3 Carcinoma cells detected by histologic examination, Case 60. HE ×132

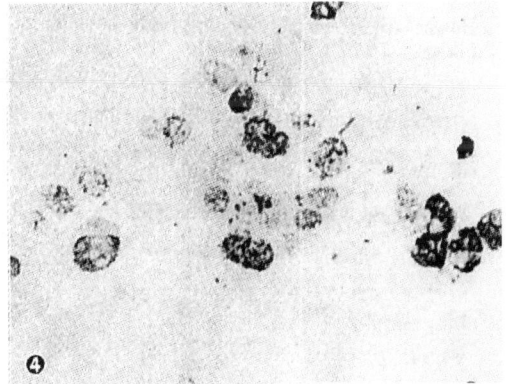

Fig. 4 EBNA – positive carcinoma cell and ciliated columnar epithelian cells from NPC patient 63. HE ×132

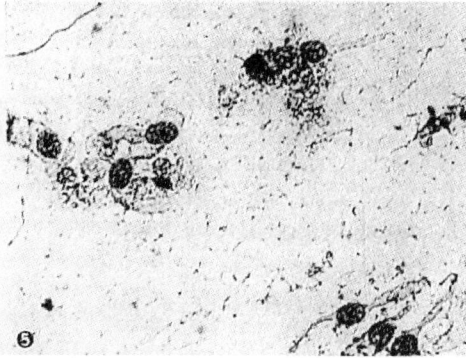

Fig. 5 EBNA – positive hyperplastic columnar epithelial cells and ciliated columnar epithelial cells from NPC Case 64. HE × 132

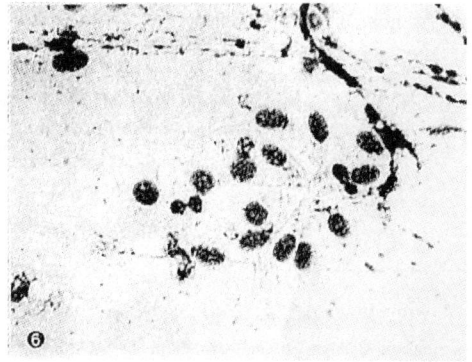

Fig. 6 EBNA – positive ciliated columnar epithelial cells from a normal individual. × 132

[In 《Chinese Medical Journal》 1981, 94 (10): 663 – 668]

References

1　Wolf H, et al. EB viral genomes in epithelial nasopharyngeal carcinoms cells. Nature, 1973, 244: 245

2　Huang DP, et al. Demonstration of Epstein – Barr Virus associated nuclear antigen in nasopharyngeal carcinoma cells from fresh biopsies. Int J Cancer, 1974, 14: 580

3　Huang DP, et al. Presence of EBNA in nasopharyngeal carcinoma and control patient tissue related to EBV serology. Int J Cancer, 1978, 22: 226

4　Klein G, et al. Direct evidence for the presence of Epstein – Barr virus DNA and nuclear antigen in malignant epithelial cells from patients with anaplastic carcinoma of the nasopharynx. Proc Natl Acad Sci USA, 1974, 71: 4737

5　Desgranges C, et al. Nasopharyngeal carcinoma: X. Presence of Epstein – Barr genomes in epithelial cell tumors from high and medium risk areas. Int J Cancer, 1975, 16: 7

6　Zeng Y, et al. Application of immunoenzymatic method and immunoautoradiogrphic method for the mass survey of nasopharyngeal carcinoma. Chin J Oncol, 1979, 1: 2

7　Zeng Y, et al. Serologic mass survey of nasopharyngeal carcinoma. Acta Acad Med Sin, 1979, 1: 123

8　Zeng Y, et al. Application of an immunoenzymatic and an immunoautoradiographic method for a mass survey of nasopharyngeal carcinoma. Intervirology, 1980, 13: 162

9　Zent Y, et al. Detection of EB virus nuclear antigen (EBNA) by anticomplement immunoenzymatic method. Acta Actd Med Sinica, 1980, 2: 134

10　Zhanjiang Medical College: Diagnosis of nasopharyngeal carcinoma by negative pressure suction and cytologic examination. Chin Med J, 1976, 89: 45

11　Reedman BM, Klein G. Cellular localization of an Epstein – Barr virus associated complement fixing antigen in producer and nonproducer lymphoblastoid cell lines. Int J Cancer, 1973, 11: 499

7. Application of Anticomplement Immunoenzymatic Method for the Detection of EBNA in Carcinoma Cells and Normal Epithelial Cells from the Nasopharynx

Zheng Yi[1], Shen Shu – jing[2], Pi Guo – hua[1], MA Jiao – lian[2]

Zhang Quin[1], Zhao Ming – lun[2], Dong Han – ji[2]

1. Institute of Virology, Chinese Academy of Medical
Sciences, Beijing, People's Republic of China;
2. Zhanjiang Medical College, Zhanjiang, People's Republic of China

Summary

Exfoliacred cells collected by negative pressure suction from the nasopharynx of patients with poorly differentiated and undifferentiated cell carcinoma, and of suspected NPC patients were examined form EBNA by means of anticomplement immunoenzymatic method. Among the 79 NPC patients all had ENBA – positive carcinoma cells, while the positive rate of cytological and histological examination was only 87.3% and 91.1% respectively. No EBNA was found in tumor cells from patients with malignant and benign tumors of head and neck other than NPC, nor was it found in nasopharyngeal epithelial cells of dead fetus. Six individuals with EBNA – positive cells detected by the present method, but where no carcinoma was detected by cytology and histologyc, were reexamined regularly. Carcinoma cells were detected from one tree weeks to three months later in four out of the six individuals. These findings show that the anticomplement immunoenzymatic method is specific and sensitive, and can be used for the early detection of NPC, especially in combination with serological methods for the detection of antibody to EB virus.

Besides the finding of EBNA in nasopharyngeal carcinoma cells, EBNA can also be found in normal ciliated columnar epithelial cells and hyperplastic cells of the nasopharynx.

Introduction

There is a close association of Epstein – Barr virus with nasopharyngeal carcinoma. EBNA and EBV – DNA are regularly found in the nasopharyngeal carcinoma cells by means of anticomplement immunofluorescent technique (ACIF) and nucleic acid hybridization (Wolf et al., 1973, Juang et al., 1974, 1978; Klein et al., 1974, Desgranges et al., 1975). NPC in early stage could be detected by IgA serological mass surveys (Zeng et al., 1979a, 1979b). Follow – up study among the VCA – IgA antibody positive persons showed that some new NPC cases could be detected in later years (Zeng et al., 1980a). There remains a large number of VCA – IgA antibody positive persons with no evidence of NPC. Therefore, it is necessary ot improve the technique for detecting the EBV marker inNPC cells. We have established an anticomplement immunoenzymatic method (ACIE) for detection of EBNA (Zeng et al. 1980b) which proved to be a very sensitive test. Exfoliated cells collected from the nasopharynx by negative pressure suction were used for cytological examination. The positive carcinoma cell rate was 92.7% (Zhanjiang Medical College, 1976). This method is simple and convienet, and a lot of cells can be obtained. The exfoliated cells from nasopharynx obtained by this method were examined by ACIE for detection of EBNA as diagnosis of NPC and for studying the relationship of EB virus and NPC. In the present study, EBNA was found not only in NPC cells, but also in normal epithelial cells and hyperplastic cells in same specimens.

Material and Methods

1. Specimens

Exfoliated cells of the nasopharynx were obtained from 145 NPC patients and suspected NPC patients. Touch smears were made from 18 tumors of head and neck other than NPC and from 21 nasopharyngeal mucosa of dead ferus aged 3 – 8 months.

2. Method for collecting cells from nasopharynx by negative pressure suction

1% Dicaine was sprayed into the oropharyngeal and nasopharyngeal cavity. Exfoliated cells were collected by an electric suction appartur (the negative pressure was not more than – 30 mmHg). Smears were prepared on slides from the head of the suction apparatus. All smears were fixed in cold acetone for 10 minutes, examined by ACIE or stained with H. E. for cytological examination.

3. Histological examination

Biopstes taken from the suspected lesions of nasopharynx or tissue fragments obtained by negative pressure suction were fixed with 10% formalin, stained with H. E. and examined.

4. Anticomplement immunoenzymatic test (Zheng et al. , 1980b)

The normal human sera were used as source of complement, some of these sera negative for EBV antibody were given by Drs. De The and Desgranges. There was no nonspecific antinuclear antibody in these sera. Balanced salt solution (BSS) was used as diluent. The reference NPC sera contagining antibody to EBNA and the normal human serum with complement in final dilution of 1 : 10 were added on the smears, and the slides were placed in a humidified chamber at 37℃ for one hour. After washing with BSS for three times, the anti – C3 antibody conjugated with horsradish peroxidase in 1 : 10 was added and the slides were placed at 37℃ for 30 minutes. The smears were washed with BSS again for three times, stained with diaminobenzene and H_2O_2 and examined.

5. Anticomplement immunofluorescence test (ACIF)

EBNA was also detected by ACIF test as described by Reedman and Klein (1973)

Results

1. Comparison of the ACIE and ACIF tests for detecion of EBNA

Tab. 1. shows that EBNA – positive carcinoma cells were found in exfoliated cells from the all 8 cases by both ACIE and ACIF tests. Usually there were more EBNA positive carcinoma cells detected by the ACIE test than by the ACIF test. Carcinoma cells were detected in seven cases by both cytological and histological examinations. In case No. 73, the histological and cytological examinations showed no tumor, but one month later, reexamination showed a poorly differentiated squamous cell carcinoma. All of 8 NPC cases had VCA – IgA antibody.

2. Application of ACIE test for detection of EBNA

Exfoliated cells were collected from 145 NPC patients and suspected NPC patients, 18 maligant and benign tumors of head and neck other than NPC and 21 nasopharyngeal mucosa of dead fetus. All specimens were examined by ACIE test, cytological and histological examinations. If either cytology or histology showed carcinoma cells, the patient was diagnosed as NPC. Tab. 2. shows the comparison of the ACIE test with cytology and pathology for the diagnosis of NPC. 62 cases were positive by all three methods (Figs. 1, 2, 3, brief), 10 cases were shown positive by ACIE and histological examination and 7 cases were positive by ACIE and

Tab. 1   Comparison of ACIE and ACIF method for detection of EBNA

Patients	EBNA ACIE	ACIF	Cytological examination	Pathological examination	VCA/IgA antibody
99	+	+	+	+	1:20
101	+	+	+	+	1:20
102	+	+	+	+	1:1 280
102	+	+	+	+	1:20
112	+	+	+	+	1:1 280
113	+	+	+	+	1:80
119	+	+	+	+	1:1 280
73 – 1	+	+	–	–	1:80
73 – 2	+	+	–	+	

cytological examination. All of those 79 patients were diagnosed as poorly differentiated or undifferentiated carcinomas. 9 cases having EBNA – positive cells were negative by cytology and histology and therefore could not be diagnosed as NPC. No EBNA – positive cells could found in exfoliated cells from 57 cases in whom NPC was exculded, as well as in touch smears from 18 patients with tumors of head and neck other than NPC and from 21 dead fetus. As shown in Tab. 3, the number of NPC cases in clinical stage I, II, III and IV was 15, 29, 31 and 4 respectively. The positive rate of carcinoma cells detected by ACIE test, cytological and histological examinations was respectively 100%, 87.3% and 91.1%. ACIE test is thus more sensitive than the two other methods, especially in stage I, the positive rate by the latter two methods was only 86.4% and 80.0%.

Tab. 2   Comparison of anticomplement immunoenzymatic method, cytological and pathological examination for detection of NPC

Group	Number of cases	ANIE	Cytological Examination	Histological Examination
NPC patients	62	+	+	+
	10	+	–	+
	7	+	｜	–
Total	79	79	69 (87.3%)	72 (91.1%)
Suspected NPC patients	9	+	–	–
	57	–	–	–
Other tumors in head and neck other than NPC	18	–	–	–
Fetal nasopharynx	21	–	–	–

Tab. 3　Comparison of three methods for detection of NPC in different clinical stage

Clinical stage	Number of cases	Positive number (%) ACIE	Cytological examination	Histological examination
I	15	15 (100)	13 (86. 4)	12 (80. 0)
II	29	29 (100)	23 (79. 3)	27 (93. 1)
III	31	31 (100)	29 (93. 5)	29 (93. 5)
IV	4	4 (100)	4 (100)	4 (100)
Total	79	79 (100)	69 (87. 3)	72 (91. 1)

3. EBNA in different types of nasopharyngeal cells

In NPC patients the EBNA could be found not only in carcinoma cells, btu also in hyperplastic cells and normal ciliared columnar epithelial cells in same specimens, whereas it was less in normal squamous epithelial cells. There was only presence of EBNA in either normal epithelial cells or in hyperplastic cells from 9 cases, in whom the cytological an dhistological examinations were negative (Figs. 4, 5, 6, Brief).

4. Reexamination of EBNA – positive mucosae

6 cases who had EBNA in the carcinoma cells or in the hyperplastic cells, but revealed no carcinoma cells by cytological and histological examinations, were reexamined from one week to three months later, carcinoma cells were detected by cytological and histological methods in 4 of them, but still absent in another two cases (Tab. 4).

Tab. 4　Follow – up study on the EBNA positive individuals

Patient	Data	ACIE	Cytologic examination	Histologic examination	Clinical stage	VCA – IgA antibody
Cheng	11. 01. 80	+	−	−		1:20
	18. 01. 80	+	+	+	I	
Wu	26. 01. 80	+ (HC)	−			1:20
	03. 05. 80	+	+	+	II	
Lin	02. 01. 80	+ (HC)	−			1:80
	31. 03. 80	+	+	−	II	
Le	07. 04. 80	+ (HC)	−	−		
	15. 05. 80	+	−	+	II	1:80
Sha	26. 03. 80	+ (NE)	−	−		1:80
	13. 04. 80	+ (NE)	−	−		
Ho	13. 11. 79	+ (HC)	−	−		
	11. 04. 80	+ (HC)	−	−		

Notes: + = carcinoma cell, + (HC) = hyperplastic cell, + (NE) = normal epithelial cell

Discussion

The results presented here indicate that EBNA can be detected in carcinoma cells from NPC patients by the ACIE test. This test is specific, because the nuclear staining of carcinoma cells was obtained only with sera containing antibodies to EBV and not with negative sera. As compared to ACIF test, more carcinoma cells collected by negative pressure suction could be stained by the ACIE test and the different types of EBNA – positive cells could be more easily identified after the ACIE test. Furthermore, the ACIE test is more sensitive than the cytology or histology for detection of NPC, especially when patients are in clinical stage I. As mentioned above, 6 normal individuals with EBNA – positive cells, but without carcinoma cells as detected by cytology and histology were reexamined from one weeks to three months later. Carcinoma cells were detected in 4 of them by either of the three methods. Therefore besides the routine cytological and histological examination, detection of EBNA by ACE and of VCA – IgA antibody is highly valuable for early diagnosis of NPC.

EBNA was found in carcinoma cells, hyperplastic cells and ciliated columnar epithelial cells from NPC patients, and also found in hyperplastic epithelial cells and ciliated columnar epithelial cells from normal individual, but it was less in squamous epithelial cells. These data suggested that the EB virus might first infect the normal epithelial cells, especially the ciliated columnar epithelial cells, and integrate its DNA into the cellular DNA. How the malignant transformation of epithelial cells occur under these conditions is not known. It is of paramount interest to follow – up the normal individuals having EBNA – positive cells. Up to now, the EB virus is only known to infect B lymphocytes. Then how does the EB virus enter into the epithelial cells from which NPC arises, remains to be determined.

Acknowledgements

We would like to thank Dr. Guy de – The and Dr. C. Desgranges for giving the human serum negative for EBV antibody and the fluorescence labelled antihuman C3 antibody.

We thank also Pr. C. H. Huang, Institute of Virology, Beijing, for critical reading and revision of the manuscript.

[In 《Cancer Campaign》 1981, 5 – sp: 139 – 143]

References

1　Desgranges C, Wolf H, de – The G, Shanmugaratnam K, Ellouz R, Cammoun N, Klein G, Zur Hausen Nasopharyngeal Carcinoma. X. Presence of Epstein – Barr genomes in epithelial cell tumors from high and medium risk areas. Int J Cancer, 1976, 16: 7 – 15

2　Huang DP, Ho J H C, Henle W, Henle G. Demonstration of Epstein – Barr virus associated nuclar antigen in nasopharyngeal carcinoma cells from fresh biopsies. Int J Cancer, 1974, 14: 580 – 588.

3　Huang DP, Ho JHC, Henle W, Henle G, Saw D, Lui M. Presence of EBNA in nasopharyngeal carcinoma and control patient tissue related to EBV serology. Int J Cancer, 1978, 22: 266 – 274

4　Klein G, Giovannella BC, Lindahl, T, Flalkow PJ, Singh S, Stehlin J. Direct evidence for the presence of Epstein – Barr virus DNA and nuclear antigen in malignant epithelial cells from patients with anaplastic carcinoma of the nasopharynx. Proc Natl Acad Sci UAS, 1974, 71: 4737 –4741

5　Reedman BM, Klein G: Cellular localization of an Epstein – Barr virus associated complement fixing antigen in producer and non – producer lymphoblastoid cell lines. Int J Cancer, 1973, 11: 499 – 520

6　Wolf H, Zur Hausen H, Becker Y. EB viral genomes in epithelial nasopharyngeal carcinoma cells. Narure, 1973, 244: 245 – 257

7　Zeng Y, Liu yx, Liu CN, Chen SW Wei JN, Zhu JS, Zai HG, Application of immunoenzymatic method and immunoautradiographic method for the mass survey of nasopharyngeal carcinoma. Chinese J Oncology, 1979a, 1: 2 – 8.

8　Zeng Y, Liu yx, Wei JN, Zhu JS, Cai SI, Wang PZ, Zhong JM, Li RC, Pan WJ, Li EJ, Tan BF. Serological mass survey of nasopharyngeal carcinoma. Acta Acad Med Sin, 1979b: 1: 123 – 126

9　Zeng Y, Liu, yx, Liu CN, Chen SW, Wei, JN, Zhu JS, Zai HJ. Application of an immunoenzymatic and an immunoautoradiographic method for a mass survey of nasopharyngeal carcinoma. Intervirology, 1980a, 13: 162 – 168

10　Zeng Y, Pi GH, Zhao WP. Detection of EB virus nuclear antigen (EBNA) by anticomplement immunoenzymaric method, Acta Acad Med Sin, 1980b, 2: 134 – 135

11　Zhangjiang Medical College: Diagnosis of nasopharyngeal carcinoma by negative pressure suction and cytological examination. Chinese Med J, 1976, 1: 45 – 48.

8.　Development of an Anticomplement Immunoenzyme Test for Detection of EB Virus Nuclear Antigen (EBNA) and Antibody to EBNA

Pi Guo – hua, Zeng Yi, Zhao Win – ping, Zhang Qin

Department of Tumor Virus. Institiute of Virology

Chinese Academy of Medical Sciences, Peking, Rep. of China

Summary

An anticomplement immunoenzyme test was developed by conjugating anti – human C3 antibody with horeseradish peroxidase. EBNA could be detected by this test in all cell lines related to EB virus, as well as in nasopharyngeal carcinoma cells, but not in unrelated cell lines. Antibody to EBNA could also be detected. The test is sensitive and does not require a fluorescence microscope which makes it particularly suitable for mass survey.

Introduction

Nasopharyngeal carcinoma (NPC) may be detected at an early stage by serological mass surveys and follow – up studies (Zeng et al. 1979a, 1979b, 1980), but there remains a large number of VCA – IgA antibody – positive persons with no evidence of NPC. EBNA and EBV DNA are regularly found in NPC cells (Wolf et al. 1973, 1975; Huang et al. 1974, 1978; Klein et al. 1974; Desgranges et al. 1975).

Detection of EBNA in nasopharyngeal cells from VCA – IgA antibody – positive individuals and NPC – suspected persons might be of help for early diagnosis of NPC. Reedman and Klein (1973) developed on anticomplement immunofluorescence test for the detection of EBNA. This test requires fluorescence microscopy and hence is not convenient for field surveys. We have, therefore, developed an anticomplement immunoenzymetic test.

Materials and Methods

1. Cell lines
Raji, Namalwa, B95 – 8, CNE, CFN, S – H and Vero cell lines were used.

2. NPC specimens
Exfoliated cells were obtained from the nasopharynx of NPC patients by negative pressure suction.

3. Sera
Sera were obtained from 5 NPC patients and normal individuals.

4. Anticomplement immunoenzyme test (ACIE)

Anticomplement (anti – C3) serum was prepared by immunizing rabbits with human complement C3 absorbed on inulin. Briefly, 1 g of inulin was washed 3 times with barbital buffer, pH 7. 6. Forty ml of fresh human serum from 6 persons were added to the inulin drop by drop. After mixing, the mixture was incubated in a waterbath at 37℃ for 1 h, with shaking every 5 min. The inulin – C3 complex was then washed 7 times with barbital buffer. Barbital buffer (19 ml) was added to the pellet of inulin – C3 complex and the preparation was stored at – 30℃. Five ml of inulin – C3 complex was emulsified with an equal amount of complete Freund's adjuvant and injected subcutaneously int 10 sites in the back and into the enlarged lymph nodes of rabbits, which had received BCG in the footpads 10 days previously. Two further injections of inulin – C3 complex in incomplete Freund's adjuvant were given at 10 – day intervals. Bleedings were taken 10 days after the last injection. The titer of anticomplement C3 antibody was 1:16 – 1:64 by double immunodiffusion.

The method used for labeling the anticomplement antibody was the horseradish peroxidease technique described by Avrameas (1969). Briefly, 5 mg of lyophilized horseradish peroxidase (HRP, Boehringer, Grade I), was dissolved in 1 ml of 0. 3 mol/L $NaHCO_3$ and treated for 1 h at room temperature with 0. 1 ml of flurodintrobenzene in ethanol. One ml of 0. 06 mol/L sodium periodate was added and kept at room temperature for 30 min. The reaction was stopped with 1 ml of 0. 16 mol/L ethylene glycol and the solution was dialyzed overnight at 4℃ in carbonate/bicarbonate buffer, 0. 01 mol/L pH 9. 5. Five mg of anti – human C3 IgG in 1 ml carbonate/bicarbonate buffer, pH 9. 5 was mixed with the activated HRP for 3 h at room temperature. Then 5 mg of $NaBH_4$ was added to the mixture overnight at 4℃. The solution was dialyzed in PBS. Horseradish peroxidase antibody conjugate was stored at – 30℃.

Some of the normal human sera were negative for EBV antibody. These were given by Drs. de –

The and Desgranges to be used as sources of complement. There was no antinuclear antibody in these sera. The balanced salt solution (BSS) was used as diluent. The cells were treated with 0.4% KCl at 4℃ for 15 min, and smears were made. After drying in air the cells were fixed with cold acetone.

The reference NPC sera containing antibody to EBNA together with normal human serum with complement (final dilution 1:10) were then added to the smears, and the slides placed in a humidified chamber at 37℃ for 1 h. After washing 3 times with BSS, the anti – C3 antibody conjugate (1:10) was added and the slides were kept at 37℃ for 30 min. The smears were again washed 3 times with BSS, stained with diaminobenzene and H_2O_2, and examined under a light microscope. Nuclei with brown coloration were considered positive for EBNA.

5. Anticomplement immunoflrorescence test (ACIF)

EBNA was also detected by ACIF as described by Reedman and Klein (1973).

Results

Detection of EBNA in different cell lines

Raji cells were examined by both ACIE and ACIF. As shown in Figs. 1 and 2, the EBNA was detected in as many Raji cells by ACIE as by ACIF. After inactivation of complement in the NPC reference sera and normal human sera at 56℃ for 30 min, no EBNA could be detected in Raji cells (Fig. 3, Tab. 1).

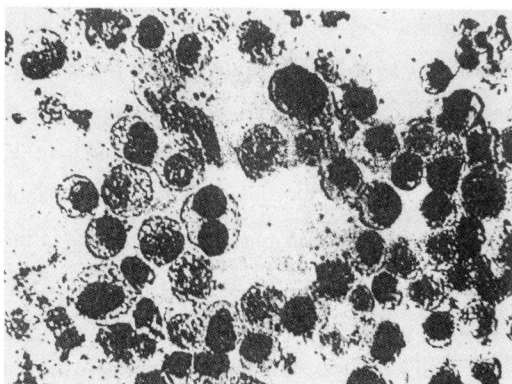

Fig. 1   Raji cells, EBNA – positive
(ACIE) ×66

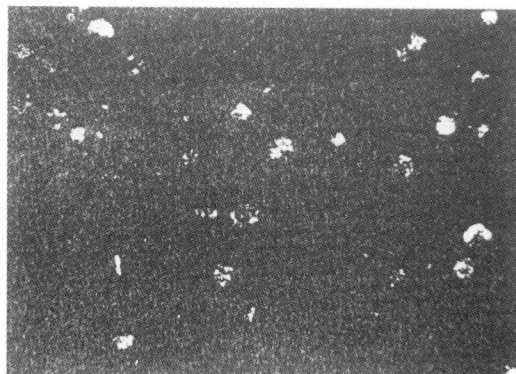

Fig. 2   Raji cells, EBNA – positive
(ACIF) ×66

EBNA was demonstrated in Namalwa and B95 – 8 cells which are related to EB virus, but not in unrelated cells, i. e. CNE, CNF, S – H and Vero cell lines.

2. Detection of EBNA in nasopharyngeal carcinoma cells from NPC patients

EBNA – positive carcinoma cells were found in exfoliated cells from 5 NPC patients by both ACIE (Fig. 4, Tab. 1) and ACIF. All these patients were diagnosed histologically as poorly or undifferentiated NPC.

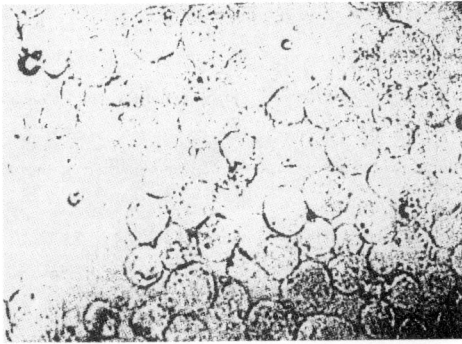

Fig. 3　Raji cells. After inactivation of
complement EBNA – negative ×66

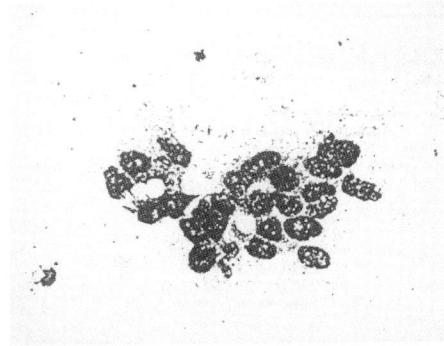

Fig. 4　EBNA – positive nasopharyngeal
carcinoma cells（ACIF）×66

Tab. 1　Detection of EBNA in different cells

	Cells	Sources	EBNA	VCA – IgA antibody
Cell lines	Raji	Burkitt's lymphoma	+	
	Namalwa	Burkitt's lymphoma	+	
	B95 – 8	EBV transformed marmoset lymphoblastoid cells	+	
	CNE	Well differentiated NPC cells	−	
	CNF	Well differentiated NPC cells	−	
	Vero	Green monkey kidney cells	−	
	S – H	Peripheral blood cells, epithelial cells	−	
Nasopharyngeal carcinoma cells	No. 1	Poorly differentiated NPC cells	+	1: 20
	No. 2	Poorly differentiated NPC cells	+	1: 80
	No. 3	Poorly differentiated NPC cells	+	1: 80
	No. 4	Poorly differentiated NPC cells	+	1: 1 280
	No. 5	Undifferentiated NPC cells	+	

3. Detection of antibody to EBNA

Antibody to EBNA was detected in sera from 5 NPC patients with titers of 1: 1280, 1: 1280, 1: 1280, 1: 320 and 1: 20, and from 5 normal indivduals with titers of 1: 1280, 1: 1280, 1: 320, 1: 80 and 1: 20, respectively.

Discussion

The results of the present study show that EBNA could be detected in all cell lines related to EB virus, but not in unrelated lines, and that antibody to EBNA could also detected by the ACIE test. This test is as sensitive as the ACIF test and much more convenient for field studies. EBNA was also found in nasopharyngeal carcinoma cells. The significance of this finding requires further study.

Acknowledgements

We would like to thank Drs. de – The and Desgranges for the human serum negative for EBV antibody and the fluorescence – labeled anti – human C3 antibody, and Dr. S. C. Shen for the exfoliated cells from NPC patients.

We are also grateful to Prof. C. H. Huang, Institute of Virology Beijing, China for critical reading and revision of the manuscript.

〔In 《Journal of Immunologycal Methods》 1981, 44: 73 –78〕

References

1 Avrameas S, Immunochemistry, 1969, 6, 43

2 Desgranges C, Wolf, H. de – The, G. Shanmugaratman, K. Ellouz, R. Cammoun, N. Klein and G. Zur Hausen, H. Int J Cancer, 1975, 16, 7

3 Huang D P Ho, J H C, Henle W. and Henle, G. Int J Cancer 1974, 14, 580

4 Huang, D P Ho, J H C, Henle, W. Henle, G. Saw D. and Lui, M. Int J Cancer, 1978, 22, 266

5 Klein G, Giovannella, B C, Lindahl T, Fialkow PJ, SinS. Natl Acad Sci. USA, 71, 4737

6 Reedman B M, Klein G, Int J Cancer, 1973, 11, 499

7 Wolf H, Zur Hausen H, Becker Y, Nature, 1973, 244, 245

8 Wolf, H, Zur Hausen H, Klein G, Becker y, Henle G hemle W. Med Microbiol Immunol, 1975, 161, 15

9 Zeng Y, Liu Yx, Liu CM, Chen SW, Wei JN, Zhu JS, Zei HG, Chin J Oncol, 1979a, 1, 2

10 Zeng Y, Liu YX, Wei JN, Zhu JS, Cai SL, Wang PZ, Zhong JN, Li RC, Pan WJ, Li EJ, Tan BF, Acta Acad Med Sin, 1979b, 1, 123

11 Zeng Y, Liu YX, Liu CN, Chen SN, Wei JN, Zhu JS, Zai HJ, Intervirology 1980, 13, 162

9. Inhibitory Effect of Retinoids on Epstein – Barr Virus Induction in Raji Cells

ZENG Y, ZHOU H M, XU SP

Institutes of Virology and Materia Medica, Chinese Academy of Medical Sciences, Beijing, China

Summary

Induction of Epstein – Barr virus (EBV) early antigen after treatment with various combinations of croton oli and n – butyrate was markedly inhibited by retinoids 7901, 7902 Ro 10 – 9359 AND Ro 11 – 1430. Possible administration of retinnoids to virus capsid antigen IgA antibody – positive individuals in high – risk areas for masopharyngeal carcinoma to prevent EBV activation and development of this cancer is discussed.

Introduction

Yamamoto et al. [1] first reported that retinoic acid interferes with Epstein – Barr virus (EBV) early antigen (EA) induction in Raji cells treated with 12 – O – tertradecanoyl – phorbol – 13 – acetate. Ito et al[2,3] also reported that increases in EBV EA and virus capsid antigen (VCA) in Raji and P3HR – 1 cells treated with croton oil and n – butyrate were markedly inhibited by retinoic acid and retinoids. Although treatment of experimental animals with vitamin A and retinoids inhibits tumors[4-8], vitamin A is highly toxic after long – term administration and cannot be used clinically. However, synthesis of vitamin A derivatives with low toxicity has been accomplished. Such retinoids, made in China and abroad, were tested for inhibitory effect on EBV induction in our laboratory for possible use in the prevention of nasopharyngeal carcinoma (NPC).

[Key words]　　Retinoid; Croton oil; n – Butyrate; Epstein – Barr virus; Early antigen

Materials and Methods

1. Retinoids, croton oil and n – butyrate

Retinoids 7901 and 7902 were made at the Institute of Materia Medica of the Chinese Academy of Medical Sciences, Beijing, and retinoids Ro 10 – 9359 and Ro 11 – 1430 were gifts from Dr. Scott. Hoffmann – La Roche Inc. Nutley, N. J. All were dissolved in dimethylsulfoxide and stored at – 30℃. Croton oil was dissovled in ethanol and stored at – 30℃, whereas n – butyrate was dissolved in RPMI 1640 medium and kept at 4℃.

2. EA Induction and Inhibition

The method used has been described by Ito et al. [2,3]. Briefly, Raji cells were cultivated in RPMI 1640 medium containing 20% calf serum, crolon oil (50 – 500 ng/ml), and n – butyrate (4 mmol/L), and were then incubated at 37℃ for 2 days. For inhibition tests, retinoids were simultaneously added to the medium as mentioned above, and the Raji cells were cultivated for 2 days at 37℃. EA – positive cells were detected in smears by immunoenzymatic means[9].

Results

1. Induction of EA in raji cells by croton oil and n – butyrate

Varying concentrations of croton oit and 4 mmol/L n – butyrate were used to examine EA induction in Raji cells. As shown in fig. 1, 47% of the cells were EA – positive when 500 ng/ml croton oil and 4 m mol/L n – butyrate were used; however, activity decreased with decreasing conocentrations of the oil. Only 3% of the cells appeared to be EA – positive when treated with 500 ng/ml croton oil alone, and only 1% of the cells were EA – positive when 4 mmol/L n – butyrate was used alone.

2. Effect of Various Concentrations of Ratinoid 7901 on EA Induction in Raji Cells

Treatment of Raji cells with retinoid 7901 effectively inhibited EA induction by croton oil (250 ng/ml) and n – butyrate (4 mmol/L). At a 10 μmol/L concentration of retinoid, EA induction was inhibited by 84%. The percent inhibition at 1, 0. 1 and 0. 01 μmol/L was 63, 60, and 15%, respectively; no suppression of EA induction was observed at 0. 001 μmol/L (Fig. 2).

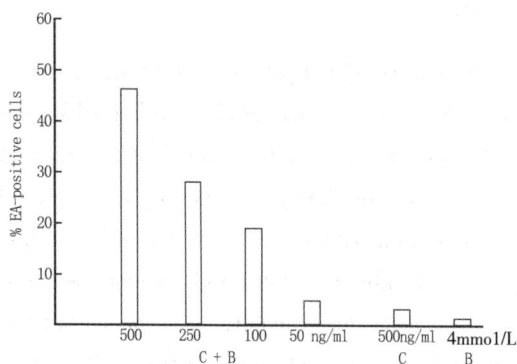

Raji cells were treated with varying concentrations of croton oil (50 – 500 ng/ml) and n – butyrate (4 mmol/L) for 48 h. C = Croton oil; B = n = butyrate.

Fig. 1　Induction of EA by croton oil and n – buty-rate in Raji cells.

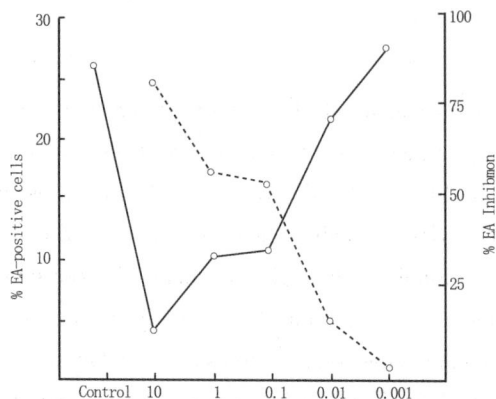

Raji ceels. Raji cells were treated with croton oil (250 ng/ml) and n – butyrate(4 mmol/L) for 48 h. ○ – ○ = % EA – positive cells; ○ ------ ○ = % EA inhibition

Fig. 2　Inhibitory effect of retinoid 7901 (0. 001 – 10 μmol/L) on EA induction in Raji cells.

3. Inhibitory effect of different fetinoids on EA induction

Retinoids 7901, Ro 10 – 9359 and Ro 11 – 1430 produced similar inhibitory effects on EA induction, but had no inhibitory effect at a concentration of 0. 01 μmol/L. Retinoid 7902, however, was more effective and exhibited some inhibitory effect at concentrations as low as 0. 01 μmol/L (Tab. 1).

Tab. 1　Comparison of the inhibitory effect of different retinoids on EA induction

Retinoid concentration μmol/L	% EA – positive cells			
	Ro 10 – 9359	Ro 11 – 1430	7901	7902
10	2. 6	4. 2	4. 3	0
1	10. 4	11. 3	10	1. 2
0. 1	11	13. 4	11	6. 7
0. 01	30	26	23	15. 3
0. 001	31	32	27. 3	22
Control	27	27	27	27

4. Effect of Retinoid 7901 on Epstein – Barr Nuclear Antigen (EBNA) in Raji Cells

Raji cells were cultivated in growth medium containing retinoid 7901 (10 μmol/L). After 5 – 10passages, EBNA was still present in the cells and EA could be induced by treatment of the cells with croton oil and n – butyrate.

Discussion

EA induction in Raji cells treated with croton oil and n – butyrate, as described by Ito et al[2,3]. provides a good model for examining and identifying inducers of EBV antigens. The present study further confirmed that EA induction by croton oil and n – butyrate could be inhibited markedly by retinoids 7901, 7902, Ro 10 – 9359 and Ro 11 – 1430 and that retinoid 7902 was more effective than the others. Since retionid 7901 has a lower toxicity than Ro 10 – 9359 [unpublis hed data]. which has been used in clinical trials, it appears that retionid 7901 could also be used clinically. The toxicity of retinoid 7902 is now under study.

Mass serological surveys carried out since 978 in Zangwu County in the Guangsi Zhuang Auto nomous Region of China[9,10] , have shown that the frequency of IgA antibo dy to EBV VCA increased with increasing age, and have enabled detection of NPC at an early stage. Follow – up studies detected additional new NPC cases[11]. These data indicate that the presence of VCA IgA antibo dy is closely related to the development of NPC; however, a large number of normal individuals also are antibody – positive. Our previous work[12] demonstrated that the level of complement fixing antibody to EBV in sera from patients over 20 years of age in areas of high risk for NPC was significantly higher than that in indivduals in low – risk areas. Thus, it appears that EBV infection is more active in persons in high – risk NPC areas. Because vitamin A and its derivatives have shown effective inhibition of EBV induction and antitumor effect , administration of retinoids to VAC IgA antibody positive individuals might prevent EBV activation and possibly NPC.

Addendum

Since submission of this paper, Lin et al. 〔Virology 111: 294 – 298, 1981〕 . have reported inhibition by retinoic acid of EBV DNA induced in P3HR – 1 cells by phorbol ester. It should be noted that the P3HR – 1 cells are producers, whereas the Raji cells used in this report are nonproducers.

Acknowledgments

We thank Prof. Y. Ito (Kyoto, Japan) for providing the unpublished data concerning the methods for induction and inhibition of EBV antigens and Prof. C. H. Huang (Beijing) for critically reading the manuscript.

〔In 《Intervirology》 1981, 16: 29 – 32〕

References

1 Yamamoto N, Bister K, zur Hausen H. Retinoic acid inhibition of Epstein – Barr virus induction. Nature, Lond, 1979, 278: 553 – 554

2 Ito Y, Kishishita M, Morigaki T, Yanase S, Hirayama T. Induction and intrevtion of EB virus antigens in human lymphoblastoid cell lines: a simulation model for study of cause and prevention of nasopharyngeal carcinoma and Burkitt lymphoma. Dusseldorf NPC Symp, 1980

3 Ito Y, Kishishita M, Yanase S. Induction of Epstein – Barr virus antigens in human lymphoblastoid P3HR 1 cells with culture fluid of Fusobacterium nucleatum. Cancer Res, 1980, 40: 4329 – 7330

4 Bollag W. Therapeutic effects of an aromatic retinoic acid analog on chemically induced skin papillomas and carcinomas of mice. Eur J Cancer, 1974, 10: 731 – 737

5 Bollag W. Prophylaxis of chemically induced epithelial tumors with an aromatic retinoic acid analog (Ro 10 – 9359). Eur J Cancer, 1975, 11: 721 – 724

6 Felix E L, Loyd B, Cohen M H. Inhibition of growh and developmant of a transplantable murine melanoma by vitamin A. Science, 1975, 189: 886 – 888

7 Trown P W, Buck M J, Hansen R. Inhibition of growth and regression of a transplantable rat chondrosarcoma by three retionids. Cancer Treatm Rep, 1976, 60: 1647 – 1653

8 Ito Y. Effect of an aromatic retinoic acid analo (Ro 10 – 9359) on growth of virus – induced papilloma (Shope) and related neoplasia of rabbits. Eur J Cancer, 1981, 17: 35 – 42.

9 Zeng Y, Liu YX, Wei JN, Zhu JS, Cai S L, Wang PH, Zhong JM, Li RH, Pan WJ. Serological mass survey of NPC. Acta Acad Med,

Sin, 1979, 1: 123 – 126.

10 Zeng Y, Liu YX, Liu CR, Chen SW, We JN, Zhu JS, Zai HJ. Application of immunoenzymatic method and immunoautoradiographicmethod for the mass survey of nasopharyngeal carcinoma. Chin J Oncol, 1979, 1: 2 – 7

11 Zeng Y, Liu Y, Liu C, Chen S, Wei J, Zhu J, Zai H. Application of an immunoenzymatic method and an immunoautoradiographic method for a mass survey of nasopharyngeal carcinoma. Intervirology, 1980, 13: 162 – 168

12 Tumor Control Team of Zhong – shan County (Guangdong et al). A study on the serum level of complement – fixing antibody to EB virus in groups of indivduals of Guangdong Province and Beijing. Chin J Otorhinolar, 1978, 13: 23 – 25

10. Serological Studies on Nasopharyngeal Carcinoma and Cerical Cancer in China

ZENG Yi

Institute of Virology, Chinese Academy of Medical
Sciences, Beijing, People's Republic of China

Nasopharyngeal Carcinoma

Nasopharyngeal carcinoma (NPC) is a very common disease in Sothern China and in South – East Asia. The distribution of NPC showed very significant geoqraphical variations[1,2]. The age – adjusted mortality rates of NPC were 2. 49 per 100 000 for men and 1. 27 for women. Among the 29 provinces and cities surveyed, all except 6 (Guangdong, Guangxi, Fujian, Hunan, Jiangxi and Zhejiang) were below the average. The central part of Guangdong and the eastern part of Guangxi formed geographically an unique high risk area with a decreasing rate tendency toward their periphery, and the mortality rates were up to 10 per 100 000. The proportion of NPC mortality to all tumor mortality as a whole in the 29 provinces and cities was 3. 11% for males and 2. 34% for females, ranking seventh and ninth in prevalence for males and females, respectively. Nevertheless, in Guangdong and Guangxi the NPC mortality ranked third for males and fourth for females. In the hyghest NPC risk area such as Szehui county it ranked first among all tumor mortality, being 31. 03% and 21. 12% for males and females, respectively. The mortality or incidence of NPC has been quite stable. Old et al. [3] first demonstrated the serological relationship between EBV and NPC by means of immunodiffusion test. studies on this problem were started in China in 1973.

1. Seroepidemiological study on EB virus infection[4]

A total of 2300 normal sera from normal individuals of different age groups in NPC low risk areas (Beijing, Lufeng and Wuhua counties of Guangdong) and in high risk areas (Zhungshan county and Guangzhow city) were tested for antibody to EB virus by means of micro – complement fixation test. The positive rate (≥1: 10) of total sera was 90. 4% with a GMT of 1: 52. 90 – 100% of children in the age group of 3 – 5 years had antibody to EB VIRUS (Fig. 1). This indicates that the age of primary infection in different areas is similar. As NPC occurs more frequently in persons above 20 years of age, it can be seen from Fig. 2 that the GMT of antibody titer from the healthy population over 20 in the high risk areas were significantly higher than those in the NPC low risk areas (Fig. 2). This kind of variation may reflect the reactivation of the EB virus in the people living in the NPC high risk area and this might be related to the occurrence of NPC.

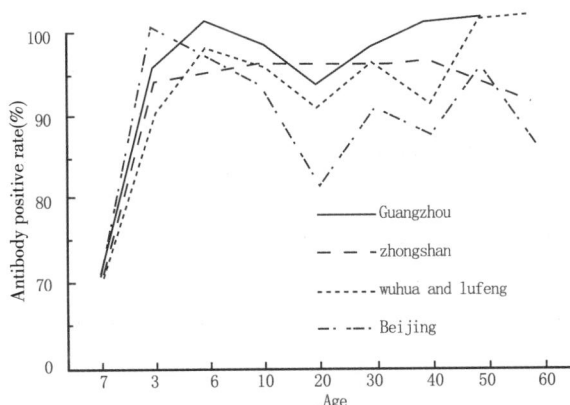

Fig. 1   Age – specific incidence of antibody to EBV

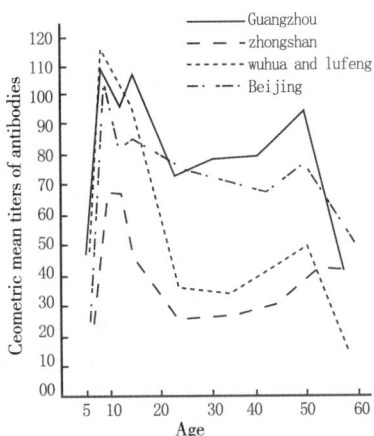

Fig. 2   Geoemtric mean titers
of antibodies to EBV

2. Serological diagnosis of NPC

Henle and Henle[5] reported the IgA/VCA and IgA/EN antibody to EB virus as an outstanding feature of NPC by immunofluorescence test. These antibodies were also detected in China[6]. 96% of NPC patients had IgA/VCA antibody with a GMT of 13. 6, whereas less than 4% of sera from patients with malignant tumor other than NPC and from normal subjects had this antibody with a GMT of 1: 1. 25 – 1. 32. Similar results were obtained in Beijing and 7 provinces and cities including NPC high risk areas (Guangdong, Guangxi, Huang and Fujian). There was no marked difference in IgA/VCA positive rate and GMT in different clinical stages, which makes it helpful for the early detection of NPC. The level of IgA/VCA antibody titier from NPC patients declined gradually with increase in survival time after radiotherapy. When these patients had recurrence or distant metastases the antibody increased again and reached its original level. The serological follow – up study might provide prognostic information[7].

The immunofluorescence technique meeds a fluorescence microscope and is not convenient for NPC field study. We therefore established a simpler and more sensitive test – immuno – exzymatic test for serological diagnosis and mass Servey ot NPC[8]. This test is now widely used in China. As shown in Tab. 1. The positive rate of IgA/VCA antibody in NPC patients, patients with maligant tumors other than NPC and normal individual groups were 92. 5% – 98%, 0 – 37% and 0 – 6%, respectively[8 – 12]. These results indicate that the frequency and GMT of IgA/VCA antibody in NPC patients differ markedly from those in patients with malignant tumor other than NPC and in normal subjects. Thus, detection of IgA/VCA antibody is valuable in the diagnosis of NPC, especially in its early stage or in NPC patients without notable tumor in the nasopharynx superficially, but with invasion beneath the mucosa or early metastasis in the neck region.

3. Serological mass survey of NPC

Serological mass surveys were carried out in 1978 – 1980 in Zangwu county of Guangxi Autonomous Region by the immunoenzymetic test[13-15]. Sera from 150 000 persons aged 30 and over were tested; 3539 sera were found to have IgA/VCA antibody to EB virus. Among IgA/VCA antiobody – positive persons, 54 cases were confirmed histologically as NPC, 12 were in Stage I, 19 in Stage II, 17 in Stage III and 6 in Stage IV. After 1 – 3 years follow – up studies of IgA/VCA antibody – positive cases 33 additional NPC cases were diagnosed: 1 was a carcinoma in situ, 10 in Stage I, 9 in Stage II, 11 in Stage III, and 2 in Stage IV[16]. There was no difference in GMT of IgA/VCA antibody in sera between the period before the noset of the disease and Stage I, but it was higher in Stage II – IV. Altogether 87 NPC cases were detected form 3539 IgA/VCA ant ibody – positive persons, the detection rate was 2458. 3 per 100 000. It was 82 times higher than the annual incidence in the general population of the same age group. These results indicate that antibody screening is valuable in the early detection of NPC and that the EB virus may play an important role in the development of NPC.

Tab. 1 Detection of IgA/VCA antibody to EB virus in sera from NPC patients and control groups

Authora	NPC patients			Patients with tumor other than NPC			Normal individuals			Refer – ence
	No. of cases	Positive rate (%)	GMT	No. of cases	Positive rate (%)	GMT	No. of persons	Positive rate (%)	GMT	
Liu et al. (1979)	80	92. 5	1:35. 7	107	0	1:1. 25	91	0	1:1. 25	8
Wei et al. (1980)	628	98. 1	1:38. 7	92	5. 4	1:1. 25	210	0. 5	1:1. 27	9
Tang et al. (1981)	156	95. 0	1:84. 4	45	37. 8	1:7. 13	179	2. 2	1:2. 6	10
Jian et al. (1981)	78	92. 3	1:78. 5				166	6	1:5. 4	11
Tang et al. (1981)	282	95. 4	1:82. 2	34	17	1:1. 8	325	4. 6	1:1. 3	12

Cervical cancer

Cervical cancer is also a very common disease in Chinese women. The age – adjusted mortality rate per 100 000 of the whole country was 9. 98. The high mortality area included Shanxi (20. 47/ 100 000), Inner Mongolia (17. 23), Shanxi (16. 64), Hubei (15. 89), Jiangxi (11. 82), with particularly high rates in West Hubei, South Shanxi, Southeast Shanxi, and the adjoining areas of Hunan and Jiangxi. It ranked second among all tumor mortality rates of Chinese women[1]. A munber of case – control seroepidemiologic studies have been reported in which antibodies to HSV – 2 were found more frequently among cases than among controls[17-19]. However, the percentages of cases and of controls with antibodies to HSV – 2 have not been consistent. The antibodies to HSV – 1 and HSV – 2 were also detected in China by immunoenzymatic test[20,21].

1. Detection of antibody to HSV – 1 and HSV – 2 from patients with cervical cancer and control women

The prevalence rate of cervical cancer among women aged over 30 in the Jienan county and Beijing was 769/100 000 and 44. 7/100 000 respectively[1]. A total of 300 sera from normal women in the same age group in cervical cancer high risk area (Jienan county) and low risk area (Beijing) were tested for IgG antibody to HSV – 1 and HSV – 2 by immunoenzymatic test[20,21]. As can be seen in Tab. 2, 80% of cervical cancer patients from the urban area of Beijing had IgG antibody to HSV – 2 with GMT of 1:887, whereas only 38% of the control women had such antibody with GMT of 1:282. The difference in frequency and GMT between these 2 groups was statistically significant. Similar results were obtained from rural area of Beijing and from the Jienan county, but in both groups of the Jienan county the positive rate was higher than in Beijing . Study on HSV – 2 isolation also showed that the positive rate of HSV – 2 isolation (39%) from cervical cancer cases was significantly higher than that (10%) from normal controls[22]. These data indicate that there is clearly an association between HSV – 2 and cervical cancer. No marked difference in frequency of antibody to HSV – 1 in patients with cervical cancer and control groups from high and low risk areas, but in cervical cancer group the GMT of antibody to HSV – 1 was a little bit higher than that in the control group; this may be due to the cross – reaction of antigen between HSV – 1 and HSV – 2.

Tab. 2 Occurrence of IgG antibody to HSV – 1 and HSV – 2 among cervical cancer cases and control

Group	Antibody to HSV											
	Urban area of Beijing				Rural area of Beijing				Jienan county			
	Cervical cancer		Control		Cervical cancer		Control		Cervical cancer		Control	
	+ %	GMT	+ %	GMT	+ %	GMT	+ %	GMT	+ %	GMT	+ %	GMT
HSV – 1	92	1:1132	94	1:640	98	1:898	98	1:631	100	1:1237	100	1:826
HSV – 2	80	1:887	38	1:282	73	1:525	24	1:134	95	1:735	62	1:404

2. Age – specific Incidence of HSV infection from cervical cancer high and low risx area

A total of 386 sera from normal women of different age groups in cervical cancer high risk area (Jienan county) and low risk area (Beijing) were tested for IgG antibody to HSV – 1 and HSV – 2[21]. As shown in Figure 3, the positive rate of antibody to HSV – 2 in the age group 5 – 9 was very low and there was no difference between the high and low risk area. Afterwards if started to increase from the age group 10 – 19 and reached its peak at age group 30 – 39. It is shown that the age – specific incidence of HSV – 2 infection in cervical cancer high risk area is significantly higher than that in the low risk area. The GMT of antibody to HSV – 2 in the group of over 20 was also higher in the high risk area. More than 70% of young girls under 19 years of age had antibody to HSV – 1 in cervical cancer high risk area and it is higher than that in low risk area, but there was no difference in frequency and GMT in different age groups aged over 20 years in both areas (Fig. 4). These data suggested that there is a relationship between the early and severe infection of HSV – 2 and the high risk of cervical cancer. The infection of genital tract by HSV – 2 is usually considered to be venereally acquired, the greatest age – specific incidence is found among young adults who are sexually active, especially those who have multiple partners. The age – specific incidence of HSV – 2 infection in China is similar to that in some western countries, but the social system and the lifestyle is quite different, so besides the venereal transmission of HSV – 2, there may be some other way for infection of the genital tract by HSV – 2.

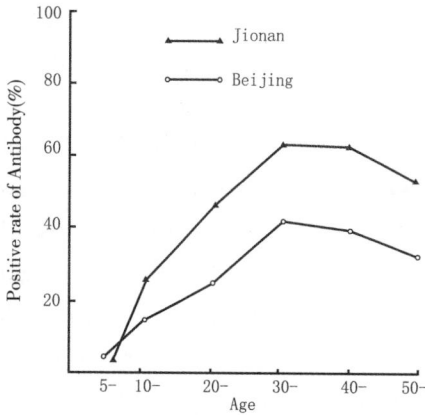

Fig. 3   Age – specific incidence of
IgG antibody to HSV – 2

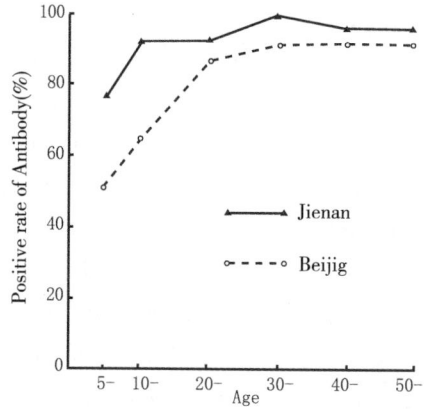

Fig. 4   Age – specific incidence of
IgG antibody to HSV – 1

Conclusion

Nasopharyngeal carcinoma is the first human cancer which can be diagnosed in the early stages by yiro – immunologi test in combination with clinical and histopathological examination. Therefore, it is possible to reduce the mortality rate of NPC through early detection and early treatment by radiotherapy. The detection rate of NPC from IgA/VCA antibody – positive persons is much higher than the annual incidence of persons in the same age group suggesting that the EB virus plays an important role in the development of NPC.

There is association between HSV – 2 and cervical cancer. The serological and virological data obtained in China also support this concept. However, the percentage of cases and of controls with antibody to HSV – 2 virus in the literature has not been consistent. Much more work is required before we can decide whether there is a direct link between the virus and the cancer.

〔In 《Herpesvirus》 1981, 380 – 388〕

References

1   Tumor Control office of the Ministry of Public Health, Survey and Investigation of Cancer Mortality in the People's Republic of China, People's Health Press. Nasopharyngeal Carcinoma, p224

– 229, Cervical Carcinoma, P, 130 – 198

2   Min HG, Wu YT, Pan QC, Huang XL, Hu MS, Zeng Yi. Preliminary Investigation of the Epidemiology of NPC in Five Provinces in South

China. Cancer Research Reports, Cancer Institute of Zhongshan Medical College, 1980, 1: 78 –89

3　Old, LJ. Boyes EA, Oettgen HF, de Harven E, Greeing G, Williamson B, Clifford P. Precipitating Antibody in Human Serum to an Antigen Present in Cultured Burkitt's Lymphoma Cells. Proc Natl Acad Sci USA, 1966, 56: 1699

4　Tumor Control Team of Zhongshan County, Department of Microbiology of Zhongshan Medical College, Department of Virology of Cancer Institute, Department of Tumor Viruses of Institute of Epidemiology, Chinese Academy of Medical Sciences, A Study on the Complement – Fixing Antibody to EB Virus in Groups of Normal Individuals in Guangdong Province and Beijing. Chinese J ENT, 1978, 1: 23

5　Henle G, Henle W. Epstein – Barr Virus – Specific IgA Serum Antibodies as an Outstanding Feature of Nasopharyngeal Carcinoma. Int J Cancer, 1976, 17: 1.

6　Laboratory of Tumor Viruses of Cancer Institute, Labortory of Tumor Viruses of Institute of Epidemiology, Department of Radiotherapy of Cancer Institute, Department of Otolaryngology of Beijing Worker – Peasant – Soldier Hospital. Detection of EB Virus – Specific Serum IgG and IgA Antibodies from Patients with Nasopharyngeal Carcinoma. Acta Microbiologica Sinica, 1978, 18: 253

7　Zeng Yi, Shan M, Liu GR, Cheng YH, Du R S, Li X H, Gan B W, Hu M J, Chen M, He S Q, Mu G P. Detection of IgA antibody to EB virus VCA from Patients with Nasopharyngeal Carcinoma by Immunofluorescence Test. Chinese J Oncology, 1979, 1: 81.

8　Liu Y. X. Zeng Y, Dong WP, Gao GR. Detection of EB Virus Specific IgA Antibody from Patients with Nasopharyngeal Carcinoma by Immunoenzymatic Method. Chinese J. Oncology, 1979, 1: 8

9　Wei JN, Zhang S, Tung SZ, Huang ZL. Detection of EB Virus IgA/VCA Antibody in Sera from patients with Nasopharyngeal Carcinoma. J Guangxi Medicine, 1980, 6: 5

10　Tang ZH, Wu YT, Yang J, Rong BP, Xy QG, Zhao CN, Meng QC. Determinationof Serum VCA IgA Antibody by Immunoenzymatic Method and Diagnosis of Nasopharyngeal Carcinoma. J Shanghai Medicine, 1981, 4: 28.

11　Li WJ, Ho PY, Li CC, Liang YR, Chen AM, Zhang SO. Ye YZ, Rong BP, Xu QG. Detection of EBV IgA/VCA Antibody in the Clinical Research of Nasopharyngeal Carcinoma, Cancer Research Report of Cancer Institute of Zhongshan Medical College, 1981, 2: 95

12　Jian SW, Li ZW, Luo WL, Li MC, Pan WZ, Zhang XH. The Application of Immunoenzymatic Method for the Diagnosis of Naspharyngeal Carcinoma. Cancer Research Reports, Cancer Institute of Zhongshan Medical College, 1981, 2: 23

13　Zeng Y, Liu YX, Liu ZY, Zhen SW, Wei JN, Zhu JS, Zai HJ. Application of Immunoenzymatic Method and Immuno – autoradiographic Method for the Mass Survey of Nasopharyngeal Carcinoma. Chinese J Oncology, 1979, 1: 2

14　Zeng Yi, Liu YX, Wei JN, Zhu JS, Cai SL, Wang PZ, Zhong JM, Li RC, Pan WJ, Li EJ, Tan BF. Serological Mass Survey of Nasopharyngeal Carcinoma. Acta Acad Med Sin, 1979, 1: 123

15　Zeng Y, Liu YX, Liu ZR, Zheng SW, Wei JN, Zhu JS, Zai HS. Application of an Immunoenzymatic Method and an Immunoautoradiographic Method for a Mass Survey of Nasopharyngeal Cancinoma. Intervirology, 1980, 13: 162.

16　Zeng Y, Wang PZ, Zhu JS, Zhong JM, Wei JN, Li EJ, Tan QF. Follow – up Studies on NPC Among IgA Antibody – positive Persons. Unpublished data.

17　Nahmies AJ, Josey WE, Naib ZM, Luce CF, Guest BA. Antibodies to Herpes Virus Hominis Type 1 and 2 in Humans. II. Women with Cervical Cancer. Amer J Epidemiol, 1970, 91: 547

18　Rawls WE, Tomkins WAF, Melnick JL. The Association of Herpes Virus Type 2 and Carcinoma of the Uterione Cervix. Amer J Epidemiol, 1969, 89: 547.

19 Royston I, Aurelian L, Devies HJ. Genital Herpes Virus Findings in Relation to Cervical Neoplasia. J Reprod Med, 1970, 4: 109

20 Fan C, Zeng Y, Liu YF. Detection of IgA and IgG Antibodies to HSV – 1, HSV – 2 in Sera from Patients with Cervical Cancer by Immunoenzymatic Method. Acta Acad Med Sin, in press, 1982

21 Fan J, Zeng Y, Zhang SJ, Xiu JL., Guong MD, Tu LZ. Detection of IgG Antibodies to HSV – 1 and HSV – 2 in Sera from Normal Women and Patients with Cervical Cancer in High and Low Risk Areas. Chinese J Microbiol Immunol, in press, 1982

22 Tan Y, Zeng Y, Lin YF. Isolation and Identification of Herpes Virus from Patients with Simplex Cervical Cancer and Controls. Acta Acad Med Sin, in press, 1982

11. Serological Mass Survey for Early Detection of Nasopharyngeal Carcinoma in Wuzhou City, China

ZENG Y[1], ZHANG L G[2], LI H Y[2], JAN M G[1], ZHANG Q[1],
Wu Y C[2], WANG Y S[2], SU G R[2]

1. Institute of Virology, Chinese Academy of Medical Sciences, Beijing; 2. Wuzhou Red Cross Hospital, Wuzhou, Guangxi Autonomous Region, People's Republic of China

Summary

A serological mass survey was carried out in Wuzhou City of the Guangxi Autonomous Region, China. Sera were collected from 12 932 persons between the ages of 40 and 59. The positive rate of VCA/IgA antibody – positive persons was 5.3%, but no EA/IgA antibody was found in sera from VCA/IgA – negative persons. Thirteen and nine nasopharyngeal carcinoma (NPC) patients were detected from the VCA/IgA and EA/IgA antibody – positive persons, respectively. With the present combination method the detection rate of NPC for 12 932 persons was 100.5/100 000 nad for 680 VCA/IgA antibody – positive persons it was 1900/100 000. Thus, the rate was twice and 37 times higher, respectively, than the annual incidence rate of NPC in persons of the same age group from 1975 – 1978 in Wuzhou CIty. Of 13 NPC patients, 9 were in stage1 (70%) and 4 stage 11 (30%). Therefore, it is possible to reduce the mortality rate of NPC in Wuzhou City by radiotherapy of NPC patients in the early stage of the disease. The present results further suggest that EB virus is closely associated with NPC.

Introduction

Serological mass surveys were carried out in 1978 – 1980 in Zangwu County of Guangxi Autonomous Region (Zeng et al. 1979a, b, 1980a). The results indicate that VCA/IgA antibody screening is valuable in the early detection of NPC.

Our previous data also showed that the EA/IgA antibody is more specific for NPC than the VCA/

IgA antibody (Laboratory of Tumor Viruses et al. 1978), and that the anticomplement immunoenzymatic method for the detection of EBNA in nasopharyngeal mucosa of VCA/IgA antibody – positive persons in also specific and sensitive for the detection of NPC (Pi et al. 1981; Zeng et al. 1980b).

In order to detect NPC early and to reduce the mortality rate, a serological mass survey was carried out in Wuzhou City using all the methods mentioned above. Located in the center of Zangwu County, Wuzhou City (population 170 000) has high risk for NPC. The mean annual incidence of NPC was 17/100 000 in 1975 – 1978. The present first – stage study reports the results of examination of 12 932 persons in the 40 – to 59 – year group.

Material and Methods

1. Sera

Sera were obtained from venous blood of 12 932 persons and stored at $-20℃$ in Wuzhou City, sent to the Institute of Virology in Beijing by air and stored at $-20℃$.

2. Immunoenzymatic method

B958 cells were used for the detection of VCA/IgA antibody, and Raji cells, with EBV early antigen induced by croton oil (200 ng/ml) and n – butyrate (4 mmol/L), were used for the detection of EA/IgA antibody. Sera diluted 1 : 5 and 1 : 10 were added to cells in separate wells of slides. The slides were incubated at 37℃ for 30 min in a humid atmosphere and washed three times with PBS. Horseradish peroxidase (Grade I) – conjugated antihuman IgA antibody in appropriate dilution was added to the slides. The slides were incubated for 30 min, washed three times with PBS and flooded with diaminobenzene solution and H_2O_2 for 10 min. Positive and negative controls were included in each experiment. A serum dilution of 1 : 10 showing brown staining characteristic of this test was considered positive and tested with further dilutions.

3. Anticomplement immunoenzymatic method (ACIE)

Reference NPC serum containing antibody to EBNA and normal human serum with complement in a final dilution of 1 : 10 were added to the smears of desquamated nasopharyngeal cells and the slides placed in a humidifed chamber at 37℃ for 1 h. After three washes, anti – C3 antibody conjugate with horseradish peroxidase in 1 : 10 was added, and the slides were placed at 37℃ for 30 min. The slides were washed again three times, stained with diaminobenzene and H_2O_2 and then examined.

4. Method for collecting cells from the nasopharynx by negative pressure suction (Zhangjiang Medical College, 1976)

Dicaine (1%) was sprayed into the oropharyngeal and nasopharyngeal cavity. Exfoliated cells were collected by an electric suction apparatus. Smears from the silk inside the head of the suction apparatus were prepared on slides. All smears were fixed in cold acetone for 10 min, examined by the ACIE method or stained with hematoxyclin and eosin (H. and E.) for cytological examination.

5. Histological examination

Biopsies were taken from suspected lesions of the nasopharynx, fixed with 10% formalin, stained with H. and E. and then examined.

Results

1. Positive rate of VCA/IgA antibody

As shown in Tab. 1 among the 12 932 person in the age group of 40 – 59 years, 680 were found to have VCA/IgA antibody with a GMT of 1: 39, the positive rate being 5.3% (680/12 932). About 70% of these people had VCA/IgA antibody titers from 1: 10 to 1: 40. The is a tendency for an increase in the antibody – positive rate and in the GMT with increasing age (Fig. 1). In the different age groups the positive rates of VCA/IgA antibody were 4.7%, 5.3%, 5.6% and 6.2%, respectively and the GMT of the antibody 1: 33.5, 1: 34.6, 1: 40 and 1: 50, respectively. The frequency of VCA/IgA antibody for males and females was similar, and the rate 1.2: 1 (Tab. 2). EA/IgA antibody among VCA/IgA antibody – positive and – negative individuals.

Tab. 1   Distribution of antibody titer and gmt of VCA/IgA antibody

| Group | Positive number and positive rate | Distribution of antibody | | | | | | | | | GMT of VCA/IgA antibody |
		10	20	40	80	160	320	640	2 560	Total	
Persons examined	Positive No. (%)	198(29)	92(14)	191(28)	46(7)	107(16)	2(0.2)	35(5)	9(1)	680[1](100)	1: 39
NPC patients	Positive No. (%)	1(7.7)		3(23.0)		5(38.5)		4(30.8)		13(100)	1: 116

Notes: [1] VCA/IgA antibody – positive rate = 5.3% (680/12 932)

Tab. 2   comparison of positive rate of VCA/IgA antibody and detection rate of NPC in males and females

Group	No. of person examined	Positive No.	Positive rate	No. of NPC cases	Detection rate of NPC/100 000
Male	7 222	414	5.7	9	124.6
Female	5 710	266	4.7	4	70.0
Total	12 932	680	5.3	13	100.5
M: F	1.3: 1	1.6: 1	1.2: 1	2.3: 1	1.8: 1

Tab. 3   comparison of vca/IgA and EA/IgA antibody

Group	No. of person examined	Positive No. of EA/IgA	Positive rae of EA/IgA(%)
VCA/IgA +	680[1]	30[1]	4.4
VCA/IgA –	507	0	0

Notes: [1] With a GMT of 1: 41.9

No EA/IgA antibody could be detected in sera from 507 VCA/IgA antibody – negative persons, but 30 out of 680 VCA/IgA antibody – positive individuals had EA/IgA antibody. The positive rate was 4.4% (Tab. 3). The antibody titers ranged from 1: 10 to 1: 640 with a GMT of 1: 41.9.

2. Detection of NPC from the VCA/IgA and EA/IgA antibody – positive persons

Clinical, cytological and histological examinations and the EBNA test were carried out in combination to detected NPC patients from the VCA/IgA and EA/IgA antibody – positive persons. Thirteen NPC cases were finally confirmed histologically (Tab. 4). The frequency of NPC among 680 VCA/IgA antibody – positive persons was 1. 9%. Nine out of 13 NPC cases also had EA/IgA antibody. The frequency of NPC among EA/IgA antibody – positive persons was 30% (Tab. 5).

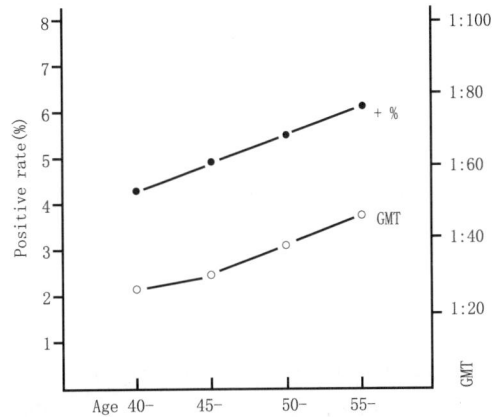

Fig. 1　VCA/IgA antibody – positive rate and GMT in different age groups

Tab. 4　NPC patients detected by serological mass survey

Case No.	Sex	Age	Histological examination	Clinical stage	Antibody	
					VCA/IgA	EA/IgA
12 660	F	50	Undiff. ca	I	1:10	–
1 171	F	46	Poorly diff. ca	I	1:40	–
11 689	M	43	Poorly diff. ca	I	1:40	1:40
361	M	47	Poorly diff. ca	I	1:40	1:40
13 684	F	46	Undiff. ca	I	1:160	1:40
1 649	F	42	Poorly diff. ca	I	1:160	–
309	M	50	Poorly diff. ca	I	1:160	1:40
7 873	M	57	Poorly diff. ca	I	1:640	1:640
12 735	M	46	Poorly diff. ca	I	1:640	1:40
68	M	50	Poorly diff. ca	II	1:160	–
3 433	M	48	Poorly diff. ca	II	1:160	1:40
23	M	50	Poorly diff. ca	II	1:640	1:10
52	M	54	Poorly diff. ca	II	1:640	1:160
GMT					1:144	1:26

Tab. 5　NPC Patients detected from VCA/IgA and EA/IgA antibody – positive person

Group	No. of person examined	No. of NPC	Positive rate of NPC(%)
VCA/IgA +	680	13	1. 9
EA/IgA +	30	9	30

The range of VCA/IgA antibody titers for NPC patients was from 1 : 10 – 1 : 640 with a GMT of 1 : 144 , and that of EA/IaA antibody titer for NPC was from 1 : 10 to 1 : 640 with a GMT of 1 : 26 (Tab. 4). All NPC patients , except for two with undifferentiated carcinoma , showed poorly

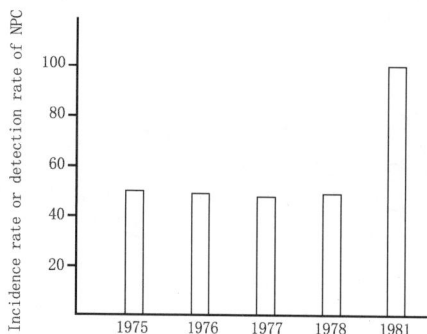

Fig. 2 Comparison of detection rate of NPC for 1981 and incidence rate of NPC for 1975 – 1978

was 1900/100 000.

differentiated carcinoma. Of 13 patients with NPC, 9 were in stage I and 4 in stage II. The majority of the stage I patients showed no notable tumor, but only some non – specific lesions, such as rough, pale or nodular mucosa.

The detection rate of NPC for 12 932 persons examined was 100. 5/100 000, and for males and females it was 124. 6/100 000, and 70/100 000, respectively(Tab. 2. Fig. 2). The detection rate of NPC for 680 VCA/IgA antibody – positive persons

Discussion

The positive rate of VCA/IgA antibody was 5. 3% , and the detection rate of NPC among the VCA/IgA antibody – positive persons was 1. 9% . The positive rate of EA/IgA antibody among VCA/IgA antibody – positive persons was 4. 4% , and the detection rate of NPC among the EA/IgA antibody – positive persons was 30% . The difference in the detection rate of NPC between these two groups was 15. 8 – fold. These data indicate that only a few VCA/IgA – positive persons eventually become EA/IgA – positive, and that EA/IgA antibody is more specific but not so sensitive for the detection of NPC as VCA/IgA antibody. The test for EA/IgA antibody in serological screening is therefore also valuable for the early diagnosis of NPC, BUT it is necessary to improve the sensitivity of the method. More attention should also be paid to careful and periodic follow – up of the EA/IgA antibody – positive persons.

The detection rate of NPC was 1900/100 000 for 680 VCA/IgA positive persons and 100. 5/ 100 000 for 12 932 persons(age 40 – 59) examined. This is 37 times and two times higher, respectively, than the incidence rate(47. 7/100 – 50/100) of NPC in the same age group in Wuzhou City (1975 – 1978) in the retrospective study (Fig. 2). These satisfactory results were obtained by combining all the clinical cytologic – histological data for individuals who were preliminarily screened out by viro – immunological methods. Thus, for example, carcinoma cells with EBNA were found in one case in stage I 2 months before the diagnosis of NPC by cytological and histological examination.

The early diagnosis of NPC means a favorable outcome of radiotherapy. Of 13 NPC patients, 9 were in stage I(70%)and 4 in stage II(30%). Therefore, it is possible to reduce the mortality rate of NPC in Wuzhou City by using the methods mentioned above. The results of the present study further suggest that EB virus in closely associated with NPC. A prospective follow – up study of the whole population is being carried out in Wuzhou City to detect the NPC patients in an early stage and to learn the time sequence: IgA – negative, IgA – positive, NPC and the proportion of NPC which does not get through the IgA – positive step. The positive rate of VCA/IgA antibody is similar for males and females, but the detection rate of NPC for males is higher than that for females. These results agree with the past annual incidence rate for males and females in Wuzhou City, and indicate that there might be some other factors favoring the development of NPC in males.

Acknowledgements

We thank Prof. C. H. Huang, Prof C. M. Chu and Dr. de – The for critically reading the manuscript

[In《Int j Cancer》1982,29:139 – 141]

References

1 Laboratory of Tumor Viruses of Cancer Institute, Laboratory of Tumor Viruses of Institute of Epidemiology; Department of Radiotherapy of Cancer Institute, Department of Otolaryngology of Beijing Worker – Peasant – Soldier Hospital. Detection of EB virus – specific serum IgA and IgA antibodies from patients with nasopharyngeal carcinoma. Acta microbiol Sin,1978,18:253 – 258

2 Pi GH,Zeng Y,Zhao WP,Zhang,QIN. Development of an anticomplement immunoenzyme test for detection of EB virus nuclear antigen (EBNA) and antibody to EBNA. J Immunol Meth,1981,44:73 – 76

3 Zeng Y,Liu YX,Lin ZY,Zhen SW,Wei JN,Zhu JS,Zai HJ. Application of immunoenzymatic method and immuno – autoradiographic method for the mass survey of nasopharyngeal carcinoma. Chinese J Oncol,1979a,1:2 – 7

4 Zeng Y,Liu YX,Lin ZR,Zhen SW,Wei JN,Zhu JS, Zai HS. Application of an immunenzymatic method and an immunoautoradigraphic method for a mass survey of nasopharyngeal carcinoma. Intervirology,1980A,13:162 – 168

5 Zeng Y,Liu YX,Wei JN,Zhu JS,Lai SL,Wang PZ,Zhong JM,Li RC,Pan WJ,Li EJ,Tan BF. Serological mass survey of nasopharyngeal carcinoma. Acta Acad Med Sin,1979b,1:123 – 126

6 Zeng Y,Pi GH,Zhang Q,Shen SJ,Zhao ML,Ma JL,Dong HJ. Application of anticomplement immunoenzymatic method for the detection of EBNA in carcinoma cells and normal epithelial cells from nasopharynx. Cancer Campaign, Vol. 5. Nasopharyngeal carcinoma. Grundmann et al. (eds). Gustav Fischer Verlag. Stuttgart. New York. 1981, 237 – 245

7 Zhangjiang Medical College,Diagnosis of nasopharyngeal carcinoma by cytological examination of exfoliated cells taken by negative pressure suction. Chinese med J,1976,1:45 – 47

12. Detection of Epstein – Barr Viral DNA Internal Repeats in the Nasopharyngeal Mucosa of Chinese with IgA/EBV – Specific Antibodies

DESGRANGES C[1] ,BORNKAMM G W[2] ,ZENG Y[3] ,WANG P C[4] ,ZHU J S[5] ,SHANG M[6] ,de – The G[1]

1. CNRS Laboratory Faculty of Medicine Alexis Carrel,69372 Lyon Cedex 2,France

2. Institute fur Virologie,Zentrum fur Hygiene,Freiburg,Germany

3. Institute of virology,Chinese Academy of Medical Sciences,Beijing,People's Republic of China

4. People's Hospital of Guang – Xi Autonomous Region,People's Republic of China

5. Cancer Control Office of Zangwu County,People's Republic of China

6. Cancer Institute,Chinese Academy of Medical Sciences,Beijing,People's Republic of China

Summary

Fifty – six South Chinese individuals exhibiting IgA antibodies to EBV for 18 months and presenting nasopharyngeal abnormalites were biopsied. Four nasopharyngeal carcinomas, two at a very early stage, were detected. In 14 futher individuals, without clinical or histopathological evidence of tumor, EBV/DNA internal repeats and/or EBNA were detected in the biopsied mucosa. The presence of IgA/EBV antibodies and/or EBV markers in the nasopharyngeal mucosa may characterize pre – cancerous conditions.

Introduction

The high level of serum IgA noted by Wara et al. (1975) in nasopharyngeal carcinoma(NPC) patients was shown by Henle and Henle(1976) to be specific for Epstein – Barr virus(EBV). It was further demonstrated that saliva of NPC patients contained IgA to viral capsid antigens(VCA) which are locally produced by the plasmocytes infiltrating the tumor itself(Desgranges et al. 1977a,b). These IgA, mostly restricted to NPC, became useful for the diagnosis and prognosis of this tumor (Coates et al. 1978a,b;Henle et al. 1977;Desgranges and de – The,1978). Ho et al. (1978) were the first to demonstrate in a retrospective study that IgA to VCA was detectable prior to the clinical diagnosis of NPC. More recently,Lanier et al. (1980) found IgA toVCA in a serum sample taken 22 months before the clinical diagnosis of NPC was made. Zeng et al. (1979,1980) in mass surveys in the Guang – Xi Autonomous Region of the People's Republic of China, assessed the value of IgA to VCA for the early detection of this cancer. In two serological surveys involving a total of 148 000 individuals, aged 30 or more, they detected, by means of an immunoenzymatic test,1267 IgA/VCA – positive individuals who were clinically examined. Of these,203 were biopsied;NPC diagnosis was established early in 46 cases, and 18 months later in 12 more cases(Zeng et al. 1979,1980,and in preparation). The present study was initiated to search for the presence of EBV markers in the nasopharyngeal mucosa of IgA/VCA – positive individuals exhibiting abnormalities in their nasopharynx(de – The et al. 1981). It enabled EBV/DNA and/or EBNA to be detected in the mucosa of 18 individuals and nasopharyngeal carcinomas to be diagnosed at early stages of the disease in four of these.

Material and Methods

1. Nasopharyngeal biopsies

The nasopharynges of 56 individuals who had been IgA/VCA – positive for at least 18 months and who exhibited abnormalities such as hypertrophic mucosa, leukoplasia or petechia, were biopsied at the site of the suspect lesions. Seven individuals who were IgA/VCA – negative but presented the same abnormalities in their nasopharynges were also biopsied.

The small biopsies were immediately divided into two parts: one was used for EBNA detection in touchsmears and for DNA extraction. The other part was fixed for histopathology. The EBNA detection was done by the ACIF test(anticomplement immunofluorescence test; Reedman and Klein, 1973), using EBNA – positive and – negative reference sera after 10 min fixation in acetone methanol. DNA was extracted from the remaining part of the biopsies(see below). The other part, fixed in 10% formalin, was embedded, sectioned and stained with hemalum – eosin and examined independently by Ms BF. Tan, Zangwu Cancer Center, Guang – Xi Autonomous Region, Dr. EJ. Li, Nanning People's Hospital, Guang – Xi Autonomous Regino, People's Republic of China, and Dr. C. Micheau, Institute Gustave – Roussy, Villejuif, France.

2. Serology

Serum and salva were collected from the 56 IgA – positive individuals and from 60 IgA – negative age/sex matched controls. IgG and IgA antibodies to VCA and EA in sera and saliva were titered by the immunofluorescence test according to Henle and Henle(1966)and Henle et al. (1970). In parallel, IgA to VCA were determined by the immunoenzymatic test described by Liu et al. (1979). Antibodies to ENBA were determined by the ACIF test, as described by Reedman and Klein(1973).

3. DNA extraction

The biopsies were ground in sand, resuspended in 0. 05 mol/L Tris pH 8. 4 10 mmol/L EDTA, 1% sarkosyl(W/V)and digested with pronase(1 mg/ml)for at least 2 h at 37℃. After two phenol extractions, the DNA was precipitated by two volumes of ethanol, redissolved in 10 mmol/L Tris pH 7. 4, 1 mmol/L EDTA, and digested with RNase(20 μg/ml)for 2 h at 37℃. After two other phenol extractions, DNA was again precipitated by ethanol and dissolved in 10 mmol/L Tris pH 7. 4, 1 mmol/L EDTA.

4. Detection of EBV/DNA sequences in biopsies

A total of 16 μg DNA from each biopsy was digested with the restriction endonucleases Bam HI, Pst I and Bgl I which cleave with the internal EBV repeat, one, twice and four times respectively (Hayward et al. 1980). After extraction with chloroform/isoamylalcohol(24: 1)DNA was concentrated by ethanol precipitation, loaded onto 0. 8% horizontal agarose gels and run for 4 – 5 h at 45 – 50 V in Trisacetate buffer(40 mmol/L Trisacetate, pH 7. 8, 1mmol/L EDTA). Bands were visualized on a

254 nm transilluminator(UR products San Gabriel) and photographed with a Polaroid camera with a Kodak wratter 23 A filter. Fragments were transferred to nitrocellulose by the method of Southern (1975). DNA from the EBV − negative cell line JM(Schneider et al. 1977) was included on each gel as a negative control. As positive controls. 500 pg or 50 pg of CC 34 −5 EBV/DNA(1 or 0. 1 viral genome copy per cell) and 8 pg internal viral repeat(1 repeat per cell) were added to 16 μg JM DNA.

The intrnal repeat of B95 − 8 virus DNA cloned in pBR 322 (kindly provided by S. D. Hayward) was labelled by nick − translation with[^{32} P] dCTP(Amersham, 400 Ci/mmol) as described by Maniatis et al. (1975).

Hybridization was carried out for days at 42℃ in polyethylene bags in a buffered solution containing 50% formamid, 5 × SSC(1 × SSC is 0. 15% sodium chloride plus 0. 015 mol/L sodium citrate) 0. 02% bovine serum albumin, 0. 02% polyvinylpyrolidone, 0. 02% Ficoll(Denhardt, 1966) 20 mmol/L sodium phosphate, 200 μg/ml of sheared denatured salmon sperm DNA and 10^6 cpm/ml heat − denatured labelled probe. The volume of the hybridization reaction was about 0. 1 ml/ cm^2. Before hybridization, the filters were incubated 1 −2 days at the same temperatute with a slightly modified buffer(0. 05 mol/L sodium phosphate, 500 μg/ml salmon sperm DNA and 5 × Denhardt solution) without the labelled probe. After hybridization, blots were washed twice in a 2 × SSC 0. 1% sodium dodecyl sulfate(SDS) at room temperature, followed by four washes in 0. 1 SSC 0. 1% SDS, each for 30 min at 50℃. Blots were exposed to Kodak Royal X − omat film at −70℃ by means of an intensifying screen(Du Pont Cronex lightning − Plus).

Results

1. Four nasopharyngeal carcinomas detected

Four patients with undifferentiated carcinomas of the nasopharynx were detected among the 56 IgA/VCA − positive individuals. Table 1 shows their clinical stages. serological profiles and the presence of viral markers in the tumors. Two of these were carcinomas at stage I. All four showed a characteristic serological pattern with high IgG/VCA and EA and IgA/VCA antibody titers. IgA antibodies directed against VCA and EA were detectable in the saliva of one patient in stage I and of those in stage Ⅱ and Ⅲ. EBNA − positive cells were observed in all four tumors. but in case 12, only a few disseminated EBNA − positive cells were found. EBV − DNA was detected in three cases, but not in case 12. This is probably due to the fact that histopathology diagnosis and search for EBV markers were made from different parts of the biopsy. The DNA analysis of two of the tumors is shown in Figure 1. The left part of the Figure shows DNAs digested by Bam HI; the right part shows the same tumor − DNAs digested by Big. 1.

The first four slots to the left of Figure 1 represent positive and negative controls, the slot nearest the left referring to one internal EBV repeat per cell, detectable in this autoradiogram. Tumors 35 and 37 were clearly positive, case 37 containing a larger amount of EBV/DNA repeat than case 35, although tumor 37 was clinically less advanced than tumor 35. Case 29 (hybridization data not shown here) was of special interest: a 38 − year − old male examined in June 1979 exhibited slight congestion of the nasopharyngeal mucosa, histologically characterized as hyperplasia of the lymphoid tissue. In May 1980, he was reexamined and a biopsy done; the clinical and histopathological findings were the same, but a few EBNA – positive cells were found in the nasopharyngeal smears and the blot analysis of biopsy DNA exhibited a small but significant presence of EBV/DNA repeats. Another clinical examination 4 months later revealed the presence of a carcinoma in situ.

Tab. 1　NPC detected by histopathology in 56 IgA – positive individuals in Zangwu county

Tm No.	Sex Age	Stage	Serology (IF)					Saliva IgA		Biopsy	
			IgG		IgA						
			VCA	EA	VCA	EA	EBNA	VCA	EA	EBNA	DNA
29	M/38	I	1 280	80	160	< 5	160	+	+	+	++
37	M/36	I	1 280	640	40	< 5	640	–	–	+	+++
35	F/47	II	≥1 280	1 280	1 280	160	1 280	++	++	+	+++
12	M/48	III	≥1 280	320	40	< 5	1 280	–	+ / –	+	–

2. Detection of EBV markers in nasopharyngeal mucosa of IgA/VCA – positive individuals without clinical or histopathological evidence of tumor

Tab. 2 summarizes the data from 14 individuals without clinically or histopathologically detectable carcinoma of the nasopharynx, btu containing a few EBNA – positive cells and /or EBV/DNA in their biopsies (invariably in lower amounts than in naso- pharyngeal carcinomas). In all of these cases, EBNA was detected only in isolated cells and not in clumps. The EBNA – positive cells usual- ly did not exceed one cell in about a thousand. The hy- bridization test with labelled EBV/DNA internal repeats revealed a weak presence of EBV/DNA in 10 cases, but at a level regularly lower than one repeat per cell. The results obtained with eight of these biopsies are shown in both Fig. 1 (for case No. 20) and Fig. 2 (for cases Nos. 10, 30, 3946, 53, 56 and 67). Cases 39 and 16

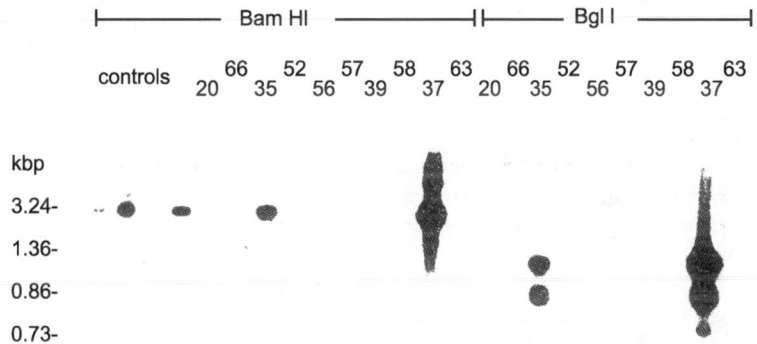

EBV DNA in nasopharyngeal mucosa

Sixteen μg DNA form each biopsy was applied to each slot. The four slots at the left – hand side are controls, the two slots nearest the left containing 16 μg JM DNA with 8 pg internal EBV repeat(one copy of repeat per cell) and 500 pg CC 34 – 5 viral DNA (one EBV genome copy per cell) respectively. The third slot contains only 16 μg JM DNA and the fourth 16 μg of JM + 1 μg Raji DNA. The autoradiogram was exposed overnight at – 70℃. 3. 2 kb is the size of the internat repeat after digestion with Bam HI. The off – size bands in DNA Nps. 35 and 37 after cleavage with Bam III represent partial digestion prod- ucts and the fragments adjacent to the intrnal repeat, which contain part of the internal re- peat sequences. 1. 36, 0. 86 and 0. 73 correspond to three of the four internal ropeat frag- ments after cleavage with Bgi II. The fourth fragment is too small to be detected.

Fig. 1　Blot hybridzation of 10 biopsies of IgA/VCA – positive Chineseondividuals after digestion with Bam HI(left) and Bgi II (right)

had faint bands at the position of the EBV internal repeats. The bands of case 16 were, however, so weak that they were regarded as doubtful. Cases 56 and 57, apparently negative after 12 h of exposure (Fig. 1). showed positive hybridization after 4 days' exposure (Fig. 2). For six biopsies, there was a good correlation between the presence of EBNA and of EBV/DNA. In one further case (No. 39) the EBV/DNA was at the borderline of significance but EBNA was present. Furthermore, in three cases, EBV/DNA could be detected but not EBNA, and in four cases the reverse was observed. EBNA – positive cells were found but no EBV/DNA could be detected. In 38 individuals with positive IgA VCA antibodies, no EBV markers were detected in the biopsies. All individuals with EBNA and/or EBV DNA in their biopsies has, besides IgA/VCA. significant IgG/EA antibody titers and seven out of 14

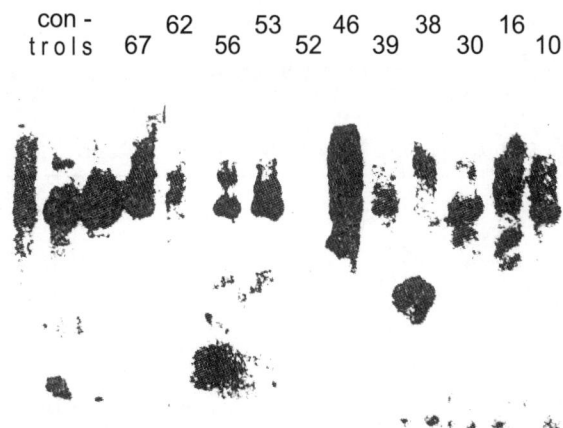

con-
trols 67 62 56 53 52 46 39 38 30 16 10

Sixteen µg of each biopsy were applied to each slot. The controls (from left to right) contained 16 µg JM DNA, 16 µg JM DNA plus 50 pg CC 34 – 5 (EBV) DNA (0. 1 EBV genome copy per cell) and 16 µg JM DNA plus 8 pg repeat (one copy of repeat per cell). The autoradiogram was exposed for 5 days at −70℃

Fig. 2   Blot hybrdization of 11 biopsies of IgA/VCA – positive individual, atter digestion with Bam HI.

exhibited IgA/VCA in their saliva.

Tab. 2   EBV markers in biopsies of IgA – positive individuals without evidence of tumor

| No. | Sex/Age | Serology | | | | | Saliva IgA | | Biopsy | |
| | | IgG | | IgA | | | | | | |
		VCA	EA	VCA	EA	EBNA	VCA	EA	EBNA	DNA
10	M/48	≥1 280	10	40	<5	160	+	−	+	+
46	F/58	1 280	20	40	<5	160	−	−	+	+
48	M/35	640	80	1 280	<5	160	+	−	+	+
56	M/37	1 280	160	20	<5	5	+	−	+	+
57	M/36	640	40	40	<5	160	−	−	+	+
67	M/30	1 280	80	40	<5	10	+	−	+	+
20	M/50	640	320	40	5	80	+	−	−	+
30	M/38	≥1 280	40	40	<5	160	−	−	−	+
53	M/36	2 560	40	10	<5	640	−	−	−	+
39	F/47	≥1 280	10	10	<5	≥1 280	−	−	+	±
61	M/66	2 560	320	160	10	160	+	+	+	−
63	F/46	1 280	10	40	<5	640	+	−	+	−
65	F/42	1 280	40	20	<5	80	−	−	+	−
66	F/43	1 640	40	80	<5	80	−	−	+	−

No EBV information was detectable in seven biopsies for IgA/VCA – negative individuals who were biopsied because of abnormalities in their nasopharyngeal mucosa similar in aspect to those observed in IgA/VCA – positive individuals.

Detailed and repeated histopathological investigations of all biopsies, by three histopathologists (Ms. Tan, Dr Li, Dr. Micheau) failed to detect any characterized dysplasia or pre – cancerous lesions. No differences between biopsies from IgA – positive or IgA – negative individuals, nor between biopsies containing or not containing EBV markers (viral DNA sequences and/or EBNA) were observed.

Discussion

The results of the seroepidemiological study of Zeng et al. (1980) have demonstrated the practical value of the IgG/VCA test in the early detection of NPC in endemic areas. In the present study, four NPC cases were detected in 56 IgA/VCA – positive but symptomless individuals. This confirms the diagnostic value of the IgA/VCA test and further indicates that long – term (12 – 18 months) IgA/VCA – positive individuals form the highest risk group for this cancer(de – The et al. 1981).

EBV/DNA inernal repeats and/or EBNA were found in the biopsies of 14 out of 52 symptomless IgA – positive individuals and in none of the seven biopsies from IgA – negative individuals. Can the presence of EBV markers in the nasopharyngeal mucosa of IgA/VCA – positive individuals reflect precancerous lesions?

This can only be answered by a careful follow – up of IgA/VCA – positive individuals with EBV markers in their nasopharyngeal mucosa. Some observations tend, however, to support such hypothesis ; in two cases reported by Lanier et al. (1981) EBV markers(EBNA or EBV/DNA) were found in nasopharyngeal biopsies histologically defined as not tumorous, but showing some abnormalitis (hyperplasia). In both cases, NPC was diagnosed a short time later. In case No. 29 reported here, the diagnosis of NPC was established after very careful histopathological reexamination of the biopsy done because of the presence of both EBNA and EBV/DNA in the specimen. It is interesing to note note here that among the IgA/VCA – positive individuals, the GMT of IgA/VCA and EA antibodies of those having detectable EBV markers(DNA or EBNA) in their nasopharynx(VCA/GMT = 1140 ; EA/GMT = 58), were higher than of those who had no detectable viral markers in the nasopharyngeal mucosa(VCA/GMT = 534 ; EA/GMT = 17). The numbers involved are however small and further field studies are required to establish if such differences are significant. To that effect, comparative evaluation of the presence of EBV/DNA and EBNA in the nasopharyngeal mucosae of larger groups of both IgA – positive and IgA – negative individuals is being implemented. The negative pressure suction apparatus used for cytological examination by Zeng et al. (in press) can collect enough cells to allow for EBNA and EBV/DNA detection from a large mumber of individuals. The blot hybidization method used here allowed cellular DNA corresponding to 3×10^{6} cells to be tested. As five to 12 internal repeats are usually present per viral genome(Hayward et al. 1980), and since a nasopharyngeal tumor cell contains about 10 to 50 EBV genomes, it should theoretically be possible to detect one EBV – positive cell among 1 000 EBV – negative cells. Furthermore, the proportion of individuals harboring viral markers in their nasopharyngeal mucosa who effectively develop clinical NPC should be assessed together with the incubation period elapsing between the first sign of viral reactivation and onset of the tumor. The answere to these critical questions necessitate the implementation of prospective epidemiological studies. As the detection of EBV/DNA by the transfer Southern technigue remains laborious and difficult to carry out in the field, the spot hybridization technique as described by Brandsma and Miller(1980) is being assayed to analyze the large numbers of samples necessary in such epidemiological studies.

Acknowledgements

The work reported here was made possible through the courtesy and help of the Medical Authorities of the Guang – Xi Autonomous Region of the People's Republic of China and especially of Dr. Li Young – Fu, Dr. Huang Li – Kua and Dr. Che.

This study was financially supported by : Institute of Virology, Chinese Academy of Medical Sciences, Bei – jing, People's Republic of China; Deutsche For schungsgemeinschaft (SFB 31), Germany; Centre National de la Recherche Scientifique (GIS 122003), France; the French Ministry of Foreign Affairs; Association pour le Developement de la Recherrche surle Cancer, Villejuif, France, and Fondation pour la Recherche Medicale, Paris, France.

[In《Int J Cancer》1982, 29 : 87 – 91]

References

1 Brandsma J, Miller G. Nucleic acid and spot hybridzation; rapid quantitative screening of lymphoid cell lines for Epstain – Barr viral DNA. Proc Nat Acad Sci(Wash), 1980, 77 : 6851 – 6855

2 Coates HL, Pearson GR, Neel HB, Weiland LH. Epstein – Barr associated antigens in nasopharyngeal carcinoma : relation to clinical course of American patients. Arch ophthal, 1978a, 104 : 427 – 430

3 Coates HL, Pearson GR, Neel HB, Ⅲ, Weiland LH, Devine KD. An immunologic basis for detection of occult primary malingnantes of the head and neck. Cancer, 1978b, 41 : 912 – 918

4 Denhardt DT. A membrane filter technique for the detection of complementary DNA. Biochem Biochem Biophys Res Commun, 1966, 23 : 641 – 646

5 Desaganges C, de – The G. IgA and nasopharyngeal carcinoma. In : G de – The, W Henle, F Rapp (ed). Oncogenesis and herpesviruses Ⅲ. IARC Scientific publication, Lyon, 1978, 24 : 883 – 891

6 Desgranges C, de – The G, Ho JHC, Ellouz R. Neutralizing EBV – specific IgA in throat washings of nasopharyngeal(NPC)patients. Int J Cancer, 1977a, 19 : 627 – 633

7 Desgranges C, Li JY, de – The G. EBV specific secretory IgA in saliva of NPC patients. Presence of secratory piece in epithelial malignant cells. Int J Cancer, 1977b, 20 : 881 – 886

8 de – The G, Desgranges C, Zeng Y, Wang PC, Bornkamm. G W, Zhu J S, Shang M. In E. Grundmann G R F, Krueger, D V Ablashi(ed). Search for pre – cancerous lesions and EBV markers in the nasopharynx of IgA – positive indivduals. Cancer campaign, Vol. 5, Gustav Fischer Verlag Stuttgart, 1981, 111 – 118

9 Hayward S D, Nogee L, Hayward G S. Organization of repeated regions within the Epstein – Barr virus DNA, molecule. J Virol 1980, 33 : 507 – 521.

10 Henle G, Henle W. Immunofluorescence in cell derived from Burkitt's lymphoma. J Bact, 1966, 91 : 1248 – 1256

11 Henle G, Henle W. Epstein – Barr virus – specific IgA serum antibodies as an outstanding feature of nasopharyngeal carcinoma. Int J Cancer, 1976, 17 : 1 – 7

12 Henle W, Henle G, Zajac B A. Pearson G, Waubke R, Scriba M. Differential reactivity of human ser-

ums with early antigens induced by Epstein – Barr virus. Science,1970,169:188 – 190

13 Henle W,Ho J H C,Henle G,Chau J C W,Kwan H C. Nasopharyngeal carcinoma: significance of changes in Epstein – Barr virus – related antibody patterns following therapy. Int J Cancer,1977,20: 663 – 672

14 Ho J H C,Kwan H C,Ng M H,de – The G. Serum IgA antibodies to EBV capsid antigens preceding symptoms of nasopharyngeal carcinoma. Lancet, 1978,1:436 – 437

15 Lanier A P,Bornkamm G W,Henle W,Henle G, Bender TR,Talbot ML,Dohan P H. Association of Epstein – Barr virus with nasopharyngeal carcinoma in Alaskan mative patients: serum antibodies and tissue EBNA and DNA. Int J Cancer,1981, 28:301 – 305.

16 Lanier AP,Henle W,Benoer TR,Henle G,Talbot ML. Epstein – Barr virus – specific antibody titers in seven Alaskan natives before and after diagnosis of nasopharyngeal carcinoma. Int J Cancer, 1980,26:133 – 137

17 Liu Y X,Zeng Y. Detection of Epstein – Barr virus IgA antibody from patients with nasopharyngeal carcinoma by immunoenzymatic method. Clin J Oncol,1979,1:8

18 Maniatis T,Jeffrey A,Kleid D G. Nucleotide sequence of the rightward operator of phage λ. Proc Nat Acad Sci(Wash),1975,72:1184 – 1188

19 Reedman BM,Kiein G. Cellular localization of an Epstein – Barr virus – associated complement fixing antigen in producer and non – producer lym-

phoblastoid cell lines. Int J Cancer, 1973, 11: 499 – 500

20 Schneider U,Schwenk H U,Bornkamm. G W. Characterization of EBV genome negative " null " and " T " cell lines derived from children with acute lymphoblastoid leukemia and leukemic transformed non – Hodgkin lymphoma. Int J Cancer, 1977,19:621 – 626

21 Southern E M. Detection of specific sequences among DNA fragments separated by gel electrophoresis. J mol Biol,1975,98:503 – 517

22 Wara W M,Wara D W,Philips T L,Ammann A J. Elevated IgA in carcinoma of the nasopharynx. Cancer,1975,35:1313 – 1315

23 Zeng Y,Liu Y X,Liu C R,Chen S W,Wei J N, Zhu J S,Zai HJ. Application of immunoenzymatic method and immunoautoradiographic method for the mass survey of nasopharyngeal carcinoma. Intervirology,1980,13:162 – 168

24 Zeng Y,Liu YX,Wei J N,Zhu J S,Cai S L,Wang P H,Zhong J M,Li R C,Ran WJ,Li E J,Tan B F. Serological mass survey of nasopharyngeal carcinoma. Acta Acad Med Sin,1979,1,123 – 126

25 Zeng Y,Shen S,Pi G H,Ma J,Zhang Q,Zhao M, Dong H. Application of anticomplement immunoenzymatic method for detection of EBNA in carcinoma cell and normal epithelial cells from the nasopharynx. In:E. Grundmann. G R F Krueger, D V Ablashi (eds). Nasopharyngeal carcinoma. Cancer campaign. Vol 5. Gustav Fischer Verlag. Stuttgart,1981:237 – 246

13. Enhancement of Spontaneous VCA and EA Induction in B95 – 8 Cells and EA Induction in Raji Cells Treated with Human Leukocyte Interferon

ZENG Y[1], ZHONG J M[2], de – The G[3], WU S H[1], HOU Y T[1], MiAO X Q[1]

1. Institute of Virology, Chinese Academy of Medical Sciences, Beijing, China

2. Cancer Control office of Zangwu County, China

3. CNRS, Faculte de Medecine Alexis Carrel, Lyon, France

Summary

The purpose of this study was to determine the antiviral effect of human leukocyte interferon on Epstein – Barr virus (EBV) viral capsid antigen (VCA) and early antigen (EA) induction in B95 – 8 and Raji cells. Interferon made at the Institute of Virology, Beijing, and that provided by Dr. Cantell gave unexpected results. Both interferon preparations markedly enhance spontaneous VCA and EA induction in B95 – 8 cells and EA inductionin Raji cells simultaneousl treated with croton oil and n – butyrate. Interferon treatment alone had no effect on EA induction in Raji cells. Thus, the effect on interferon on EA and VCA induction was related to the type of EBV infection, i. e, productive or latent. The enhancing activity of interferon could only be partially inhibited by retinoid 7901. It is suggested that mechanism for enhancement of EA induction by interferon is different from that of EA induction in Raji cells by croton oil and n – butyrate.

[**Key words**]   Human leukocyte interferon; Viral capsid antigen; Early antigen; Epstein – Barr virus

Interferon has been shown to have antiviral and antitumor activities. Treuner et al. [1] reported that a nasopharyngeal carcinoma (NPC) case in the advanced stage showed complete regression of the tumor after treatment with fibroblast interferon. However, other NPC patients did not respond to this treatment[2]. Hence, interferon may be more effective for treatment of patients in the early stages of NPC or for the prevention of NPC development in individuals positive for IgA antibody to Epstein – Barr virus (EBV) viral capsid antigen (VCA). An early – stage NPC patient has been treated with human leukocyte interferon by our research group, and temporal regression of the tumor occurred[3].

In a clinical trial currently in progress, 15 randomly selected NPC patients at stages I and II are being treated with leukocyte interferon provided by Dr. K. Cantell. At the same time, the antiviral activity of interferon in vitron, using B95 – 8 and Raji cells, has been studied. Those results are reported here.

Materials and Methods

1. Interferon

Human leukocyte interferon preparation A was made at the Institute of Virology, Beijing, and human interferon lecocyte preparation B was provided by Dr. K. Cantell(Central Public Health Laboratory, Helsinki, Finland). The specific activities of both interferon preparations were approximately 10^6 U/mg protein.

2. Cells

Raji and B95 – 8 cells, which carry the EBV genome, were cultivated in RPMI 1640 medium containing 20% calf serum.

3. Procedure for treatment of cells with interferon

Interferon at varying concentrations(200, 1 000 and 5 000 U/ml)was added to B95 – 8 cell suspensions(5×10^5 cells/ml). After cultivation at 37℃ for 48 h, smears were made from the cell suspensions, and the VCA – and EA – positive cells were counted by the immunoenzymatic method[4]. In each assay, 500 cells were counted, and the percentage of EBV – positive cells was recorded.

Raji cells were cultivated in medium[5] containing croton oil (500 ng/ml) , n – butyrate (4 mmol/L)and interferon at varying concentrations(200, 1000 and 5000 U/ml). EA – positive cells were counted as described above.

4. Inhibition of the effect of interferon by retinoid 7901

The same procedure as described above was followed except that retionid 7901(10 μmol/L)[6] was simultaneously added to the B95 – 8 and Raji cell suspensions containing interferon at varying concentrations.

Results

1. Enhancement of spontaneous VCA and EA induction in B95 – 8 cells treated with interferon

As shown in Fig. 1, both preparations of interferon clearly enhanced spontaneous VCA and EA induction in B95 – 8 cells, the activity decreasing with lower concentrations of interferon. The percentages of VCA – and EA – positive B95 – 8 cells treated with interferon preparation A at concentrations of 200, 1000, and 5000 U/ml were 5.6%, 7.2%, and 8.0%, respectively; those in B95 – 8 cells treacted with interferon preparation B were 4.8%, 5.6%, and 7.6%, respectively, while that for the control(spontaneous VCA and EA

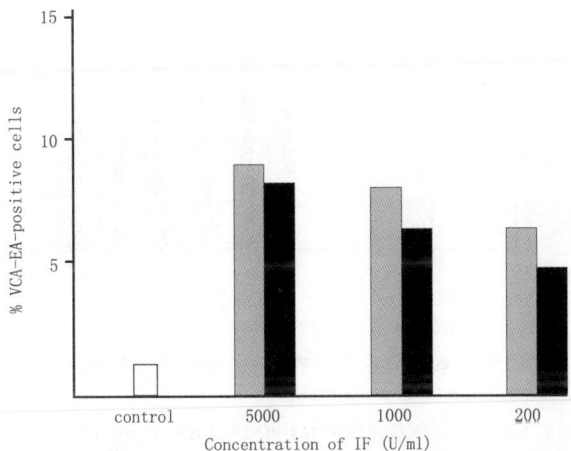

Stippled areas = IF preparation A; darkened areas = IF preparation B; open area = control

Fig. 1   Effect of interferon(IF)on VCA and EA induction in B95 – 8 cells.

induction) was only 1%. There were no significant differences in the percentages of VCA – and EA – positive B95 – 8 cells treated with interferon preparations A or B.

2. Enhancement of EA induction in Raji cells treated with croton oil, n – butyrate and interferon

Croton oil(500 ng/ml) and n – butyrate(4 mmol/L) together markedly induced EA(12. 4%) in Raji cells, while croton oil or n – but vratealonehad a lesse effect(0. 4% or 3. 2%, respectively). The percentages of EA – positive Raji cells treated simultaneously with croton oil, n – butyrate and interferon preparation A at concentrations of 1000 and 5000 U/ml were 21. 6% and 23. 2%, respectively, and those in Raji cells treated simultaneously with croton oil, n – butyrate and interferon preparation B were 18. 4% and 22. 8%, respectively. Thus, interferon preparations A and B (1000 or 5000 U/ml) enhanced EA induction in Raji cells simultaneously treated with croton oil and n – butyrate, but no significant enhancement was found when interferon preparation A or B was used at a concentration of 200 U/ml, when inferferon was used alone, or when interferon was used in combination with either croton oil or n – butyrate only(Fig. 2).

3. Effect of retinoid 7901 on VCA and EA induction in B95 – 8 and Raji cells

The results in figure 2 also showed that the enhancement of EA induction in Raji cells by interferon preparation A or B could be partially inhibited by retinoid 7901 to the level of EA – positive cells induced by croton oil plus n – butyrate only. When interferon preparation A or B at a concentration of 5000 U/ml was used, the inhibition rates of EA – positive Raji cells were 34. 5% for interferon preparation A and 40. 4% for interferon preparation B. The inhibition rates in B95 – 8 cells were less marked, being 13. 3% for interferon preparation A and 17. 4% for interferon preparation B (Fig. 3).

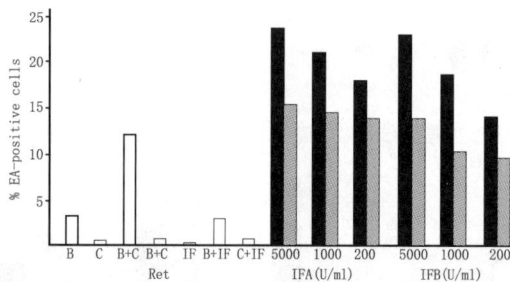

B = n – Butyrate; C = croton oil; IFA = IF preparation A; IFB = IF preparation B. Darkened areas = B + C + IF; stippled areas = B + C + IF + Ret

Fig. 2　Effect of interferon(IF) and retionoid 7901 (Ret) on EA induction in Raji cells.

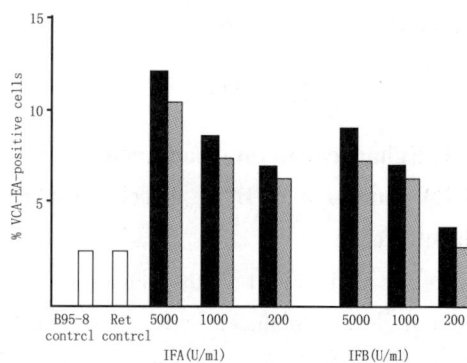

B95 – 8 cells were treated with varying concentrations of interferon preparation A(IFA) or B(IFB) (200, 1000 or 5000 U/ml) and with 10 μmol/L retionid 7901. Darkened areas = B95 – 8 cessl + IF; stippled areas = B95 – 8 cells + IF + Ret

Fig. 3　Effect of retinoid 7901(Ret) on VCA and EA induction in B95 – 8 cells.

Discussion

EBV EA in Raji cells can be induced by iodoeoxyuridine(IUdR)[7], IUdR plus phytohemagglutionin[8]. IUdR plus anti - IgM[9], or infection of cells with P3HR - 1 virus[10], but there has been no report of EA and VCA induction by human leukocyte interferon. The present study shows that EA induction in Raji cells, when simultaneously treated with croton oil and n - butyrate, and spontaneous EA induction in B95 - 8 cells can be enhanced 3 to 8 - fold by interferon. Both preparations of interferon showed similar effects on EA and VCA induction. The enhancing activity was proportional to the concentration of interferon. Interferon enhance spontaneous VCA and EA inductionin B95 - 8 cells, but showed no effect on Raji cells with out simultaneous treatment with croton oil and n - butyrate. Thus, the effect of interferorn on EA and VCA induction is related to the type of EBV infection, i. e. productive infection(B95 - 8) cells or latent infection (Raji cells). The mechanism for enhancement of EA induction by interferon may be different from that for EA induction in Raji cells by croton oil and n - butyrate. This needs to be further studied.

We do not know whether or not the enhancing activity of interferon for EA and VCA induction is favorable for the treatment of NPC patients. In our experience, there was no change of EBV - specific IgG and IgA anti body levels in the sera of NPC patients before and after treatment with interferon for 10 weeks, and the tumors showed either temporary or medium term regression[unpublished data]. Thus, the enhancing activity of interferon may not be harmful in the treatment of NPC patients. The enhancing activity may cause the activation of EBV in NPC cells and lead to a lytic infection which might be favorable for the NPC patient.

Acknowledgments

We thank Prof C H. Huang and Prof. C M. Chu for critical reading of the manuscript

[In《IntervirologY》1982,18:33 - 37]

References

1 Treuner J, Niethamme D, Dannecker G, Hagmann R, Neef V, Hofschneider P H. Successful treatment of nasopharyngeal carcinoma with interferon. Lancet, 1980, i:817 - 818

2 Treuner J. Treatment of nasopharyngeal carcinoma with interferon. Nasopharyngeal Carcinoma Symposium, Dusseldorf, 1980

3 Wang P C, Deng H, Wu S H, Wang I Z, Zeng Y Hu Y T. Preliminary treatment of nasopharyngeal carcinoma with human interferon. Interferon Scientific Memoranda, 1981, A1071

4 Zeng Y, Liu Y X, Liu Z Y, Zhen S W, Wer J N, Zhu J S, Zai H J. Application of immunoezymatic method and immunoautoradiographic method for the mass survey of nasopharyngeal carcinoma. Chin J Oncol, 1979, 1:2 - 7

5 Ito Y, Kishishita M, Morigak T, Yanase S, Hirayama T. Induction and intervention of EB virus antigens in human lymphoblastoid cell lines: a simulation model for study of cause and prevention of nasopharyngeal carcinoma and Burkitt lymphoma. Nasopharyngeal Carcinoma Symposium, Dusseldorf, 1980

6 Zeng Y, Zhou H M, Xu S P. Inhibitory effect of retinoids on Epstein - Barr virus induction in Raji cells. Intervirology, 1981, 16:29 - 32

7   Gerber P. Activation of Epstein – Barr virus by
    5 – b romodeoxyuridine in virus – free human
    cells. Proc natn Acad Sci USA,1972,69:83 – 85

8   Lenoir G, Tovey MG, Lavoue M F. Induction of
    Epstein – Barr virus early antigen by phytohemag-
    glutionin in the presence of 5 – iodo – 2 – de-
    oxyuridine:application to EBV serology. J immu-
    nol Methods,1980,34:23 – 29

9   Tovey M G,Lenoir G,Begon – Lours,J. Activation
    of latent Epstein – Barr virus by antibody to hu-
    man IgM. Nature Lond,1978,276:270 – 272

10 Henle G,Henle W,Klein G. Demonstration of two
    distinct components in the early antigen complex
    of Epstein – Barr virus – infected cells. Int J
    Cancer,1971,8:272 – 282

14.   Cytogenetic Studies on an Epithelioid
Cell Line Derived from Nasopharyngeal Carcinoma

ZHANG Si – zhong[1] ,WU Yin – tang[2] ,ZENG Yi[3] ,ZECH Lore[1] ,KLEIN George[2]

1. Department of Medical Cell Geneitcs,Medical Nobel
Institute,Karolinska Institute,Stockholm Sweden

2. Department of Tumor Biology,Karolinska Institutet,Stockholm,Sweden

3. Institute of Virology,Chinese Academy of Medical Sciences

Summary

The cytogenetics of the first epithelioid cell line(CNE) derived from the biopsy of a nasopharyngeal carcinoma has been investigated by chromosome banding techniques. The cell line was pseudo
neartriploid,and the modal number was 67 – 68 without major variation. All cells contained a series
of characteristic marker chromosomes. Most of them were either consistent or found repeatedly. The
origin of the marker was traced and many markers were identified. Two of the consistent acrocentric
markers were identified as 2p – and 14q + ,which may be identical with the large acrocentric mark-
ers(LAM) previously found in cells of the biopsies of nasopharyngeal carcinomas.

Introduction

Nasopharyngeal carcinoma(NPC) is characterized by its close association with Epstein – Batt virus
(EBV) (Old et al. 1966;Desgranges et al. 1975;Klein 1979) and great variation in incidence among
different ethnic populations(Muir 1971;Ho 1972;Shanmugaratnam 1973). The latter means that some
genetic or immunogenetic factors may play a role in the development of the tumor(Simon et al. 1975;
Willams and de – The 1974). However,the experimental studies on NPC cells have been hampered by
the fact that,although NPC is a tumor of epithelial origin,nearly all the cell lines derived so far from the
tumors were lymphoblastoid. In 1975,an EBV – negative epithelioid cell line(CNE) was established from a
well – differentiated carcinoma of an NPC patient in China(LTVCI et al. 1978). Some years later,another

epithelioid line(NPC/HK1)was also established form a differentiated squamous carcinoma of the naso-pharynx(Huang et al. 1980). Although there are some data concerning the variance of chromosome num-ber,and a general statement on the presence of many structural rearrangements,the cytogenetics of both cell lines has not been fully elaborated.

As an established epithelioid cell line,CNE is being used increasingly for immunological, viro-logical and many other studies. Therefore,it would be of interest to obtain more information on its cy-togenetics ingreater details. By using chromosome banding techniques,the present work reports the chromosome status of the CNE line and provides a description of its multiple characteristic chromo-some markers.

Material and Methods

1. The CNE line

The cell line was established from a primary tumor biopsy of a 58 – year – old female with NPC. The EBV – negative character of the cell line was confirmed by EBV specific complement fixa-tion antigen and viral capsid antigen measurement in the early generations. Recently it was also shown that the line was EBNA negative. Other biological and morphological aspects were reported previously(see LTVCI et al. 1978). Cells of the CNE line were kept in Leibovitz medium(L – 15) supplemented with 10% fetal calf serum,cultured at 37℃ with 5% CO_2 humidified air.

2. Chromosome analysis

Aliquots of cells in exponential growth were treated with Colcemid in concentration of 0. 05 μg/ml for 10,30,60 and 120 min. After trypsinization, the cells were collected and hypotonized in 0. 56% KCl solution for 10 min. Following fixation in methanol and glacial acetic acid at a ratio of 3: 1,chromosome preparations were made and airdried. Q – C – and – G – banding were carried out according to the methods of Caspersson et al. (1971) , Sumner et al. (1971) and Zhang et al. (1979) ,respectively. Banded metaphases wer photographed in a Leitz microscope with Kodak Tri – X pan films. Chromosome analysis was carried out on negatives,using av TV – screen for contrast en-hancement(Caspersson et al. 1970) , or on photocopies from the negatives. Four experiments were performed in two months with identical results. In total,more than 60 metaphases have been photo-graphed,30 of which were analyzed in detail and karyotyped.

Results

The CNE line has a pseudo near – triploid chromosome constitution,with the modal number around 67 –68(Tab. 1). Metaphases with higher ploidies were only occasionally seen. The triploid character of the cell line is only simulated(Fig. 1) ,since merely the chromosome Nos. 1 ,9 ,13 ,16 , 20 were present in trisomic condition in most metaphases. The other chromosomes occurred in two copies,and a few chromosomes were always monosomic(Nos. 3 ,13 and 18) ,or even missing in some cases(No. 19). Nos. 11 and 22 were usually present in tetrasomic or trisomic state.

The CNE line was characterized by multiple structural chromosomal rearrangements resulting in formation of a series of marker chromosomes. It is very remarkable that most of the markers were consistent and could be seen in nearly all the metaphases analyzed. A few other markers were quite frequent, although not consistent. Some markers were present only in individual cells. The following is a short description of the chromosome markers of the first two categorites.

1. Markers present in all metaphases, $M_1 - M_{13}$ (Fig. 1 and 2)

M_1, t(8q;8q +) : Large submetacentric chromosome two 8q take part in its formation.

M_2, iso8q: Large metacentric chromosome similar to M_1 but without the distal dark part on the long arm.

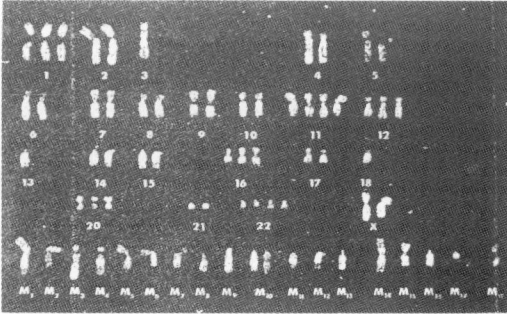

Fig. 1 A Q – banded karyotype representiative of the CNE cell line, showing 68 chromosomes and including all consistent and most of the frequently seen marker chromosomes (M_1 – M_{13}, M_{14} – M_{17}, respectively). M_{18} was an occasional finding in this karyotype

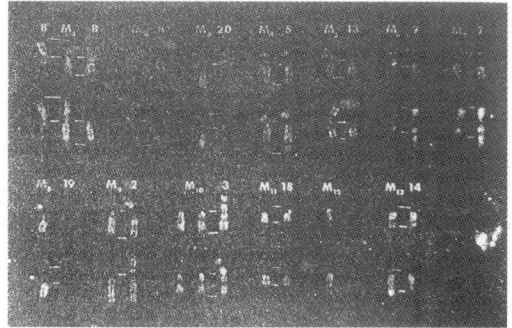

Fig. 2 Analysis of consistent marker chromosomes (M_1 – M_{13}). For each marker, the upper partial karyotype is taken from Fig. 1. and the lower, from other karyotypes. The white lines indicate the identical or similar parts of the marker and normal chromosomes. Q – banding

M_3, t(20;?) : Large submetacentric chromosome. No. 20 seems to take part in its formation.

M_4, t(?;5q) : Large submetacentric chromosome, its long arm is identical with that of No. 5.

Tab. 1 Chromosome numbers of the CNE cell line in 30 cells counted

Item	Chromosome number						
	64	65	66	67	68	69	Total
No. of cells	1	5	3	9	11	1	30

M_5, t(?;13q) : Metacentric chromosome, one of its arms seems to be identical with the long arm of No. 13.

M_6, t(7;?) : Submetacentric chromosome, its short arm and proximal part of the long arm are identical with those of No. 7.

M_7, t(7;?) : Submetacentric chromosome which is similar to marker 6 but differs from it in the banding pattern of the distal part of the long arm.

M_8, t(19;?) : Submetacentric chromosome, the short arm and proximal part of the long arm are similar to those of No. 19. In C – banded chromosomes (Fig. 4). there is a C – positive segment in the long arm.

M_9, 2p − : Large acrocentric chromosome, identical with the long arm of No. 2.

M_{10}, t(? ;3q) : Subacrocentric chromosome which is always present in duplicate. The long arm is identical with that of No. 3.

M_{11}, t(18 ; ?) : Subacrocentric chromosome. No. 18 takes part in its formation. Therefore, it can also be designated as 81q +.

M_{12}, Subacrocentric chromosome of medium size.

M_{13}, t(14 ; ?) : Medium − sized acrocentric chromosome which can be designated as 14q + according to its banding pattern. It is quie similar to the marker seen in Burkitt lymphomas (Manolov and Manolova 1971), although the origin of the extra segment needs further clarification.

Markers present in some but no all cells, M_{14} − M_{19} (Fig. 1 and 3).

M_{14}, t(2q ;6q) : Large metaccenttric chromosome, its short arm is identical with that of No. 2, and the long arm with that of No. 6.

M_{15}, 7q − : Medium − sized submetacentric chromosome formed by deletion of the distal part of the long arm of No. 7.

M_{16} : Medium − sized acrocentric chromosome, which has some resmblance to 3p.

M_{17} : Small acrocentric chromosome which is about half the size of G group chromosome.

M_{19}, t(3p + ;3q +) (see Fig. 3) : Large submetacentric chromosome which is similar to No. 3 but has an extra segment on both arms. It is interesting to notice that, once this marker is present, the single normal No. 3 is always missing.

Other markers which were present only occasionally will not be described in detail. However, in C − banded metaphases some marker chromosomes, both consistent and occasional, show an extra C − positive region (Fig. 4). In some cases these segments are extra centromeres (dicentric markers), but in others they look more like C − positive bands. The latter phenomenon is quite interesting, but further studies are necessary to clarify their origin and implication.

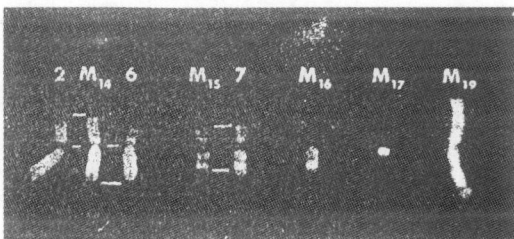

Fig. 3　Analysis of the frequently seen markers of the CNE line. All except the last are taken from Fig. 1. The white lines indicate the identical or similar parts of the marker and normal chromosomes. Q − banding

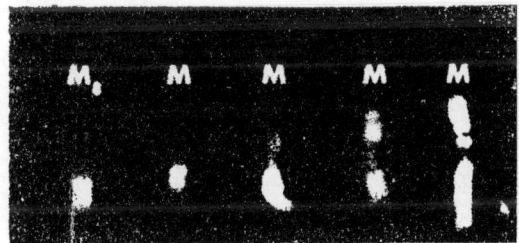

Fig. 4　Marker chromosomes with double centromeres or an extra C − positive band. Mis a consistent marker. others are found occasionally. Q − and C − banding

Discussion

There are two difficulties in studying cytogenetics of NPC. The first is, that, being an epithelial solid tumor, NPC seldom offers any metaphases of good quality in direct preparations from biopsy materials or nude – mouse – grown tumors. Secondly, because of the huge number of infiltrating lymphocytes in the tumor and the inactivity of tumor cells growing in vitro, almost all the cell lines established from tumor explants are of lymphoblastoid character. It is still unclear to what extent these lines will reflect the true picture of the tumor. In this respect, the CNE cell line has some advantages and therefore deserves special attention, since it has been derived from a NPC patient and the epithelioid nature has been confirmed by histologic and electron microscopic studies (LTVCI et al. 1978).

The present work demonstrated that, during the time of investigation, the CNE line had a fairly stable chromosome constitution. The modal number was 67 – 68. The variation in chromosome number was minor and, to a certain extent, might be due to the chromosome loss during the preparation of the slides. In early studies the chromosome number was found to be 80 (LTVCI et al. 1978). It was also somewhat less than the modal number (74) of the NPC/HKl line (Huang et al. 1980). This, naturally, may be explained by the evolution of the cell line during the culture period since its establishment, and the different origin, respectively.

The NCE cell line was remarkable for the presence of series of marker chromosomes. By using banding techniques, we were able to trace these markers in many cells and reveal that most of them were consistent or could be seen frequently. The non – randomness in marker formation, therefore, was striking, although it may to some extent account for the clonal selection in the process of establishment and long – term culture.

It was reported that, in 10 of 47 tumor biopsies, large acrocentric markers (LAM) were found. The size of them varied, and the largest ones were about the size of chromosomes Nos. 8 – 10. (Ou et al. 1979). Such abnormal long acrocentric chromosomes were also constantly found in near-triploid cells of a nude – mouse – grown epithelioid tumor (Finerty et al. 1978). Since the structure of the reported markers needs further clarification by banding pattern analysis, it is uncertain whether tyey represent the same as the acrocentric markers M_9 (2p –) and M_{13} (14q +) of the CNE line. The same can be said about the identity between our M_3 and the "large subacroccentric marker" repeatedly found in some biopsy materials (Ou et al. 1979).

The so – called giant group A chromosome, is a submetacentric one, larger than chromosome 1. It has been found in some NPC biopsies (Ou et al. 1979), in nearly all lymphoblastoid cell lines derived rom NPC (Ou et al. 1979; Wu et al. 1980; Hsia and Lu 1978) and in peripheral blood lymphocyte cultures of a few NPC patients (Zhang et al. 1980). Whether this giant chromosome has a lymphoblastoid origin and results from progressive passages is obscure (Macek et al. 1971). No such chromosome was found in CNE line.

The question whether there does exist a chromosome marker which is specific for or closely associated with NPC remains unanswered. To solve this problem, it is necessary to study the cytogenetics of NPC directly on biopsy materials or after a short – term culture. However, with the exception of tumors of hemopoietic tissue, it is very difficult to obtain well – banded metaphases from solid epithelial tumors. As a compromise, the second possibility could be to establish more epithelial cell lines and to trace their cytogenetic development from the very beginnning. This approach has some problems because of the selective growth of cultured cells but, nevertheless, the comparative study of many epithelioid cell lines may provide valuable data for the better understanding of the cytogenetics of the nasopharyngeal carcinoma.

Acknowledgments

We thank professor Yao Zhen for supplying the cell line used in this study. We also thank C. Bjorklund for skilful technical assistance.

This work was supported by the Swedish Cancer Society. Karolinska Institutests Forskningsfonder. and by Grant 5 R 01 CA 14054 – 08 from N I H to G. Klein.

〔In《Hereditas》1982, 97:23 – 28〕

Literature cited

1 Caspersson T, Lindsten J, Lomakka G, Wallman H, Zech L. Rapid identification of human chromosomes by TV – techniques. – Exp. Cell Res, 1970, 63:477 – 479

2 Caspersson T. Lomakka G, Zech L. The 24 fluorescence patterns of the human metaphase chromosomes – distinguishing characters and variability. Hereditas, 1971. 67:89 – 102

3 Desgranges C, Wolf H, de – The G, Shanmugaratnam K, Ellouz R, Cammoun N, Klein G, Zur Hausen M. Nasopharyngeal carcinome X. Presence of Epstein – Barr genomes in epithelial cells of tumors from high and medium risk areas. Int J Cancer, 1975:16:7 – 15.

4 Finerty S, Jarvis J E, Epstein M A, Trumper P A, Ball G, Giovanella B C. Cytogenetics of malignant epithelial cells and lymphoblastoid cell lines from nasopharyngeal carcinoma. Br J Cancer, 1978, 37:231 – 239

5 Ho J H C. Nasopharyngeal carcinoma (NPC). Adv. Cancer Res, 1972, 16:57 – 92

6 Hsia C, Lu H. Giant group A marker chromosome in three human lymphoblastoid cell strains from nasopharyngeal carcinoma. Chin Med J, 1978, 91:130 – 134

7 Huang D P, Ho J H C, Poon Y E, Chew E C. Saw D, Liu M, Li C L, Mak L S, Lau S H, Lau W

H. Establishment of a cell line(NPC/HKI)from a differentiated squamous carcinoma of the nasopharynx. Int J Cancer, 1980, 26:127 – 132

8 Klein G. The relationship of the virus to nasopharyngeal carcinoma. In The Epstein – Barr Virus (Eds. M . A. Epstein and B G. Achong). Springer Verlag, 1979, 339 – 350

9 Ltvci, et al. (Laboratory of Tumor Viruses of Cancer Institute and other Institutes in China). Establishment of an epithelioid cell line and a fusiform cell line from a patient with nasopharyngeal carcinoma. Sci Sin, 1978, 21:127 – 134

10 Macek M, Seidel E H, Lewis R T, Brunschwig J P, Wimberly 1, Benyesh – melnik M. Cytogenetic studies of EB virus – positive and virus – negative lymphobalastoid cell lines. Cancer Res, 1971, 31:308 – 321

11 Manolov G, Manolova Y. A marker band in one chromosome No. 14 in Burkitt lymphomas. Hereditas, 1971, 69:300

12 Muir C S, Nasopharyngeal carcinoma in non – Chinese populations with special reference to South – East Asia and Africa. Int J Cancer, 1971, 8:351 – 363

13 Old L J, Boyse E A, Oettgen H F, de Harven E, Williamson B, Clifford P. Precipitation antibody in human serum to an antigen present in cultured

Bur kitt's lymphoma cells: Proc Natl Acad Sci, 1966,56:1699 – 1704

14　Ou B,Fang,Y,Lai H. Chromosome analysis of nasopharyngeal carcinoma and its clinical significance. Zonghua Med J,1979,59:333 – 358

15　Shanmugaratnam K. Cancer in Singapore, ethnic and dialect group variations in cancer incidence. Singapore Med J,1973,14:68 – 81

16　Simon M J,Day N E,Wee G B,Chan S H,Shanmugaratnam K, de – The G. Immunogenetic aspects of nasopharyngeal carcinoa(NPC) , Ⅲ. HLA type as a genetic marker of NPC predisposition to test the hypothesis that EBV is an aetiologic factor in NPC. In Oncogeneses and Herpesviruses H. Part 2 (Eds. G. de – The. M A . Epstein and H. Zur Hausen) ,Lyon:IARC,1975,249 – 258

17　Sumner A T,Evans H J,Buckland R A. New techniques for distinguishing between human chromo-

somes. Nature New Biol,1971,232:31 – 32

18　Williams E H,de – The. G. Familiar aggregation in nasopharyngeal carcinoma. Lancet,1974,2:295 – 296

19　Wu B,Wu Y,Li Y,Zeng Y,Wu M,Zhao Z,Gong C. Study of giant group A marker chromosome in several Burkitt's lymphoma and lymphoblastiod cell lines with Epstein – Barr virus from different origins. Chin Med J,1980,93:400 – 406

20　Zhang S, Qiu J, Gao X. G – banding of human chromosomes. Zoghuma Med J, 1979,59:210 – 213

21　Zhang S, Qiu J, Gao X, Lin D, Liu Y, Xu L. Cytogenetic studies by chromosome banding techniques onperipheral blood cells from nasopharyngeal carcinoma patients. Chin Med J, 1980,93:251 – 259

15.　Follow – up Studies on Epstein – Barr Virus IgA/VCA Antibody – Positive Persons in Zangwu County, China

ZANG Y[1],ZHONG J M[2],LI L Y[2],WANG P Z[3],TANG H[2],MA Y R[3],ZHU J S[2],

PAN W J[2],LIU Y X[4],WEI Z N[3],CHEN J C[2],MO Y K[2],LI E J[3],TAN B F[2]

1. Institute of Virology,Chinese Academy of Medical Sciences,Beijing
2. Cancer Control Office,Zangwu County
3. People's Hospital of Guangxi Autonomous Region
4. Cancer Institute,Chinese Academy of Medical Sciences,Beijing,China

Summary

Serological mass surveys were carried out in Zangwu County,China,using an immunoenzymatic test. 3533 persons were found to have Epstein – Barr virus(EBV)IgA/VCA antibody among 148 029 persons age 30 years and older who were tested during 1978 – 1980. Among the IgA/VCA antibody – positive persons,55 nasopharyngeal carcinoma (NPC) caser were detected. Follow – up studies were carried out yearly on the IgA/VCA antibody – positive persons for 1 – 3 years,and 32 additional NPC patients were diagnosed. IgA/VCA antibody was detected 8 – 30 months(average,13 months) prior to the clinical diagnosis of stage I NPC. There was no marked difference in geometric mean titers of IgA/VCA antibody between the period before onset of NPC and after diagnosis at stage I, but antibody titers were higher during stages Ⅱ − Ⅳ. The NPC detection rates for all persons tested serologically and for IgA/VCA antibody − positive persons, respectively, was 2 − and 82 − fold the annual incidence of NPC in the general population of the same age group. These data further indicate that serological testing is valuable for the diagnosis of NPC, especially in its early stages, and that EBV may play and important role in the development of NPC.

〔**Key words**〕 Epstein − Barr virus; IgA/VCA antibody; Nasopharyngedl carcinoma

Introduction

148 029 persons aged 30 years or older were screened during 1978 − 1980 in Zangwu County, Guangxi Autonomous Region, by animmunoenzymatic test[1,2]. 3533 persons were found to have Epstein − Barr virus(EBV) IgA/VCA antibody(≥1:5). Among these, 55 cases of nasopharyngeal carcinoma(NPC), especially in its early stages, were detected. To study the relationship betwween EBV and NPC, follow − up studies on IgA/VCA antibody − positive person were carried out yearly from 1979 to 1981, anol 32 new NPC cases were detected. The result of this prospective seroepidemiological study are reported here.

Materials and Methods

1. Sera

Sera were collected in plastic capillary tubes ty pricking the ear lobe and were stored at − 15℃.

2. Immunoenzymatic test

This test was described previously[1]. B95 − 8 cells and horseradish peroxidase − conjugated antihuman IgA antibody were used for the detection of IgA/VCA antibody.

3. Clinical and histological examination

EBV IgA/VCA antibody − positive persons were examined clinically once a year. Biopsies were taken from suspected NPC cases and from individuals with high IgA/VCA antibody titers(≥1:80). Sera were again collected and examined.

Results

1. Distribution of EBV IgA/VCA antibody

As shown in Tab. 1, among 148 029 persons tested 3533 persons had IgA/VCA antibody(positive rate, 2.4%). The antibody titers ranged from 1:5 to 1:2560 with a geometric mean titer(GMT) of 1: 16. 87% had antibody titers of 1:5 − 1:40, and only 13% had antibody at higher titers(≥1:80).

2. NPC Patients Detected by Serological Screening

All IgA/VCA antibody − positive persons(3533) were examined clinically and histologically, and biopsies were taken from suspected cases and from persons with high antibody titers(≥1:80). Among these, 55 cases were diagnosed histologically as NPC(Tab. 2), and 31(57%) of these were in early stages(I and II). The range of IgA/VCA antibody titers for these NPC patients was from 1:10 to 1:2 560, with a GMT of 1:99. 3 cases(62%) had antibody titers of ≥1:80(Tab. 1).

Tab. 1 **Distribution of EBV IgA/VCA antibody among NPC patients**
and normal individuals age 30 years and older in Zangwu County

Item	Antibody titer										Total	GMT
	5	10	20	40	80	160	320	640	1280	2560		
Individuals age 30 years and older												
Number positiver for IgA/VCA antibody	837	967	808	474	335	55	37	10	8	2	3533	1:16
Percent positive(%)	23.7	27.3	22.9	13.4	9.5	1.6	1.0	0.3	0.2	0.06	100	
Serological mass survey												
Number of NPC cases	0	4	8	9	12	8	4	4	4	2	55	1:99
Percent positive(%)	0	7.3	14.5	16.4	21.8	14.5	7.3	7.3	7.3	3.6	100	
Follow – up study												
Number of NPC cases	0	2	5	4	8	8	2	1	2	0	32	1:85
Percent positive(%)	0	6.3	15.6	12.5	25	25	6.3	3.1	6.3	0	100	

3. NPC patients detected by follow – up studies

The IgA/VCA antibody – positive persons were reexamined clinically once a year, and 32 new NPC patients were detected over a 3 – year period(Tab. 2). All of the new cases were diagnosed clinically as poorly differentiated squamous cell carcinoma or undifferentiated carcinoma. 21 cases (66%) had IgA/VCA antibody titers of ≥1:80. The GMT was 1:85(Tab. 1).

As shown in figure 1, IgA/VCA antibody was detected 8 – 30 months(average, 13 months)prior to the clinical diagnosis of NPC at stage I, 10 – 18 months(average, 14.4 months)prior to stage II, 9 – 34 months(average, 17.3 months)prior to stage III, and 12 – 24 months(average, 18 months)prior to stage IV.

There was no marked difference in GMT of IgA/VCA antibody in sera between the period before onset of NPC and the diagnosis made at stage I, but antibody titers were higher at stages II – IV (Fig. 2), i. e, when the carcinoma cells metastasized to the cervical lymph nodes.

Tab. 2 **Detection of NPC cases in Zangwu County during**
1978 – 1980 by serological mass survey and follow – up study

Group	Number of NPC cases					
	carcinoma in situ	stage I	stage II	stage III	stage IV	total
1978 – 1980						
Serological mass survey	1	12	19	17	6	55
Follo – up study	0	10	9	11	2	32
Total	1	22	28	28	8	87

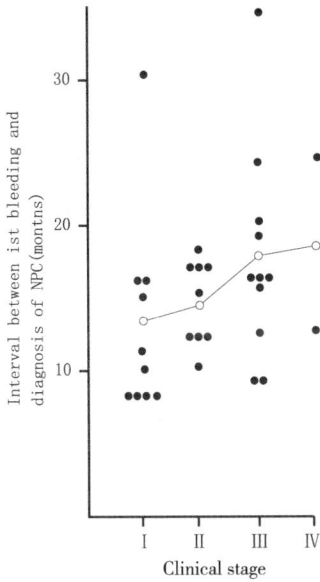

Fig. 1 Follow – up studies on IgA/VCA antibody positive persons

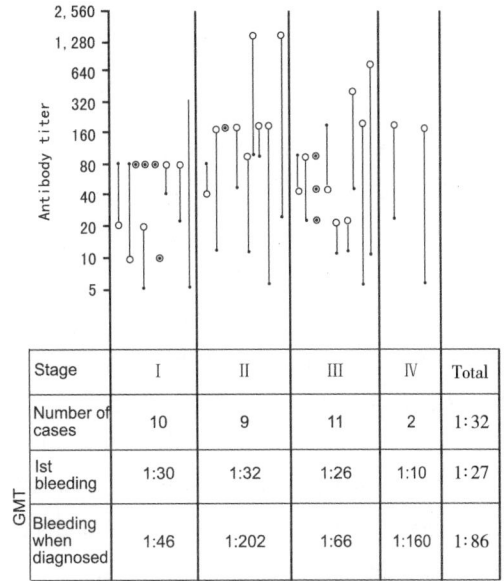

Stage	I	II	III	IV	Total
Number of cases	10	9	11	2	1:32
Ist bleeding	1:30	1:32	1:26	1:10	1:27
Bleeding when diagnosed	1:46	1:202	1:66	1:160	1:86

● = First bleeding; ○ = bleeding when diagnosis was confirmed

Fig. 2 Change of titer and GMT of IgA/VCA and body from first bleeding to diagnosis of NPC.

4. Detection rate of NPC

87 NPC cases were detected among 3533 IgA/VCA antibody – positive persons by serological screening and follow – up studies. The NPC detection rate was 59/100 000 for 148 029 persons tested and 2462/100 000 for 3533 IgA/VCA antibody – positive persons. This is 2 – and 82 – fold, respectively, the annual incidence of NPC(30/100 000) in the general population of the same age group in Zangwu County. Similar results were obtained in the city of Wuzhow[3] (Tab. 3). These rusults further indicate the close relationship of EBV to NPC.

Tab. 3 Comparison of NPC detection and incidence rates

Areas	Number of persons examined	Number positive for IgA/VCA antibody	Number of NPC cases	NPC detection rate among all persons examined	NPC detection rate among IgA/VCA antibody – positive persons	NPC incdence rate per annum
Zangwu County(1978 –1980)	148 029	3533	87	59[a](1. 96)[b]	2 462(82)	30
Wuzhow city(1981)	12 930	680	13	100. 5[a](2)	1900(38)	50

Notes:[a] Per 100 000

[b] Number in parentheses is the detection rate/incidence rate

Discussion

Serological studies indicate that there is a close relationship between EBV and NPC. The detection of IgA/VCA antibody is helpful for the diagnosis of NPC, especially in its early stages[1,2].

The percentage(25.3%)of NPC cases at stage I detected by serological mass surveys was much higher than that(6%)diagnosed in outpatient clinics[4]. 59% of NPC patients diagnosed during follow-up studies were at stages I and II and responded favorably to radiotherapy. 41% of NPC cases were diagnosed at advanced stages(III and IV), because many of the antibody-positive subjects lived in remote areas and clinical examinations were not carried out periodically or the diagnosis was missed at early stages.

IgA/VCA antibody was detected 8-30 months prior to clinical diagnosis of NPC in stage I cases. These results are similar to those reported by Ho et al. [5] and Lanier et al. [6] in their retrospective studies. They found that IgA/VCA antibody was detectable 22-72 months before clinical diagnosis was made in the advanced stages.

How long does IgA/VCA or IgA/EA antibody persist and how many antibody-negative persons become positive? What is the exact proportion of IgA/VCA or IgA/EA-positive and -negative individuals who develop NPC? These are questions that remain to be answered. More than 25 000 serum samples from normal individuals of age 40 years and older from the city of Wuchow were collected and stored at the Institute of Virology, Chinese Academy of Medical Sciences, Beijing. Careful and periodical follow-up studies on IgA/VCA- or IgA/EA-positive and -negative individuals should answer these questions. Successful intervention against EBV may provide further evidence for the relationship between EBV and NPC. A study of the inhibition of EBV by retinoids is in progress.

Serological diagnosis of NPC is now widely used in China. It is possible to reduce the mortality of NPC through early detection and treatment even though the exact etiological causes have not been elucidated.

The IgA/VCA antibody-positive rate from persons examined in 1978 was low, perhaps due to the grade III, Koch-light horseradish peroxidase(Rz =0.62)used. Reexamination of sera from these persons was sone in 1980, using grade I horseradish peroxidase, and a high positive rate was obtained. An additional explanation might be that we were not as skilled at the technique during the first serological screening in 1978.

Acknowledgments

We thank Prof C H. Huang and Prof. C M. Chu for critical reading of the manuscript.

〔In《Intervirology》1983,20:190-194〕

References

1　Zeng Y, Liu Y, Liu C, Chen S, Wei J, Zhu J, Zai H. Application of an immunoenzymatic method and an immunoautoradiographic method for a mass sur-

vey of nasopharyngeal carcinoma. Intervirology, 1980,13:162-168

2　Zeng Y, Liu Y X, Wei J N, Zhu J S, Cai S L, Wang

P Z,Zhong J M,Li R C,Pan W J,Li E J. Tan B
F. Serological mass survey of nasopharyngeal carcinoma. Acta Acad Med Sin I,1979,1:123 – 126

3　Zeng Y,Zhang L G,Li H Y,Jan M C,Zhang Q,Wa
　　Y C,Wang Y S,Su G R. Serological mass survey
　　for early detection of nasopharyngeal carcinoma in
　　Wuzhow city,china. Int J Cancer,1982,29:139 –
　　141

4　Li C C,Chen J J,Li B J. Study on precancer ous
　　changes 1. Screening for early detection of nasopha-
　　ryngeal carcinoma and hyperplastic lesions in naso-

pharynx. Cancer Chin,1982,2:81 – 83

5　Ho J H C,Kwan H C,Ng M H,de – The G. Serum
　　IgA antibodies to EBV capsid antigens preceding
　　symptoms of nasopharyngeal carcinoma. Lancet I,
　　1978,:436 – 437

6　Lanier A P,Henle W,Benden T R,Henle G,Talbot
　　M L. Epstein – Barr virus – specific antibody titers
　　in seven Alaskan natives before and after diangosis
　　of nasopharyngeal carcinoma. Int J Cancer,1980,
　　26:133 – 137

16.　Epstein – Barr Virus Early Antigen Induction in Raji Cells by Chinese Medicinal Herbs

ZENG Y[1],ZHONG J M[2],MO Y K[2],MIAO X C[1]

1. Institute of Virology,Chinese Academy of Medical Sciences,Beijing
2. Cancer Control Office of Zangwu County,Guangxi Autonomous Region,China

Summary

Ether extracts of 495 Chinese medicinal herbs from 106 families were studied for Epstein – Barr virus(EBV) early antigen (EA) induction in the Raji cell system. 15 herbs from 10 families were found to have inducing activity. Water extracts of the same herbs also had inducing activity,but it was not as strong. The significance of these herbs in the activation of EBV in vivo and their relation to the development of nasopharyngeal carcinoma are disussed. No EA – inducing activity was found in 73 samples of 14 different foods tested.

〔**Key words**〕　Epstein – Barr virus;Early antigen;Raji cells;Chinese medicinal herbs

Introduction

Ito and coworkers[1,2] reported that the Raji cell inducing(n – butyrate) system detects most,if not all,of the known promoters derived from Euphorbiaceae plants and their active components,including TPA and related compounds(mezerein and teleocidin). This test system is simple to perform and reproducible. Our previous data[3,4] showed that the geometric mean titer of complement – fixing antibody to Epstein – Barr virus(EBV) in sera from persons aged 20 years and older in highrisk areas for nasopharyngeal carcinoma(NPC) was significantly higher than that from persons of the same age group in NPC low – risk areas,and that the positive rate of IgA/VCA antibody to EBV from normal individuals in NPC high – risk areas increased with increasing age. These data indicate that EBV is more active in persons living in NPC high – risk areas, as a result of the activation of EBV by internal or environmental factors. Therefore, it is necessary to determine whether factors causing activation of EBV are present in foods, Chinese medicinal herbs, etc. As reported here, by using the Raji cell induction system, EBV activtors were found in some of the Chinese medicinal herbs tested.

Materials and Methods

1. Cells

Raji cells carrynig the EBV genome were cultivated in RPMI 1640 medium containing 20% calf serum. No spontaneous induction of early antigen(EA) in Raji cells was found. Viability of the cells was checked before and after treatment.

2. Chinese medicinal herbs

495 Chinese medicinal herbs were obtained from pharmacies in Beijing and Zangwu County of Guangxi Autonomous Region.

3. Foods

Peanut oil, pork oil. salted fish, pickles, honey and some dry salt vegetables were tested.

4. Plant Extraction

The method for ether extraction was as described by Ito et al. [1] Briefly, 10 g of test material was cutinto small pieces and extracted with 100 ml of ether for 72 h. The ether solution was then evaporated, and the residual oily extract was used as test substance. Finally, the extract was dissolved in ethanol as a 10 mg/ml stock solution and was stored at – 10℃. The water extract was made by adding 50 ml of distilled water to 5 g of test material. After boilling for 10 min and centrifugation at low speed, the supernatant was collected and stored at – 10℃ as a stock solution(100 mg ml)

5. Experimental Procedure

Raji cells were cultivated in RPMI 1640 medium containing 20% calf serum and 4 mmol/L n – butyrate; the test substance was added at varying concentrations. After cultivation at 37℃ for 48 h, smears were mede from the cell suspensions and EA – positive cells were detected by the immunoenzymatic test[4]. The test substance was also added to Raji cell suspensions in medium without n – butyrate. Croton oil, Euphorbia lathyris, Euphorbia kansur and tung oil, which were fund to be positive by Ito et al. [1,5] were used as controls(with and without n – butyrate). In each assay 500 cells were counted, and the ratio of EA – positive cells was recorded.

Results

1. EA induction in Raji cells by ether extracts of test materials

495 Chinese medicinal herbs belonging to 106 families and 73 samples of 14 different foods were tested for EBV EA – inducing activity. The results are shown in Tab. 1. Among 495 Chinese medicinal herbs, Daphne genkwa, Stellera chamaejasme, Wikstroemia chamaedaphne, Wisktroemia indica, Edgeworthia chrysantha, Premna fulva and Datura stramonium showed strong activity in inducing EA – positive cells, with the highest positive rates ranging from 43% to 53%. The other 8 herbs

(Caesalpinia sappan, Desmodium styracifolium, Sparganium stoloniferum, Tinospora sp, Aleuritopteris argentea, Clematis intricata, Knoxia valerianoides and Angelica pubescens) showed a weaker positive reaction, with the highest positive rate ranging from 1% to 17%. The EA cell – positive rate of 4 mmol/L n – butyrate alone was only 1%. These 15 herbs belong to 10 plant families (Tab. 2). The controls(Euphorbia kansui, Euphorbia lathyris, croton oil and tung oil) belong to the Euphorbiaceae family and showed strong EA – inducing activity, with the highest positive rates ranging from 42% to 53%. Eleven herbs as well as the 4 controls (Tab. 1) also induced EA – positive cells in medium without n – butyrate, but percentage positive was not as high. The other 4 herbs were negative.

Tab. 1 Effect of Chinese medicinal herbs on EBV EA induction in Raji cells

Medicinal herbs	Ether extraction[a]				Water extraction[b]			
	butyrate in assay medium(μg/ml)		without butyrate (μg/ml)		butyrate in assay medium(mg/ml)		without butyrate (mg/ml)	
	10	10	10	10	10	10	10	10
Daphne genkwa	46[c]	45	1	2	15	34	2	1
Stellera chamaejasme	36	46	2	2	2	1	2	1
Wikstroemia chamaedaphne	32	53	1	2	2	20	0.4	1
Wikstroemia indica	42	50	4	2	25	17	1	0
Edgeworihia chrysantha	43	28	1	0	0	0	0	0
Caesalpinia sappan	8	4	0	0	0	0	0	0
Desmodium styracifolium	6	4	0	0	0	0	0	0
Premna fulva	52	42	2	0.4	0.2	0.2	0	0
Sparganium stoloniferum	16	2	0.4	0	0	0	0	0
Datura stramonium	50	25	1	0	1	0	0	0
Tinospora sp	7	3	0.2	0	0	0	0	0
Aleuritopteris argentea	6	17	1	0	0	0	0	0
Clematis intricata	3	2	0.4	0	0.2	0.2	0	0
Knoxia valerianoides	3	1	0	0	0	0	0	0
Angelica pubescens	1	0	0	0	0	0	0	0
Controls								
Euphorbia kansui	28	37	1	1	5	4	1	0
Euphorbia lathyris	30	53	2	2	38	33	4	3
Croton oil	18	40	1	0.4	39	28	2	4
Tung oil	38	42	1	1				

Notes: [a] Stock ether extraction = 10 mg/ml; [b] Stock water extraction = 100 mg/ml; [c] = Percent EA – positive cells. 4 mmol/L n – butyrate alone gave a 1% positive rate

No EA – inducing activity was found in 73 samples from 14 different foods.

2. EA induction in Raji cells by water extracts of test materials

As shown in Tab. 1, EA – inducing activity was also found in water extracts of Daphne genkwa, Stallera chamaejasme, Wisktroemia chamaedaphne, Wikstroemia indica, Premnafulve, Datura stramonium and Clematis intricata, as well as 3 control herbs when the medium contained n – butyrate, but their activity was not as strong as that of ether extracts under the same experimental conditions. The EA cell – positive rate ranged from 0. 2% to 25%. Water extracts of the other 8 herbs were negative. Water extracts of 4 herbs from the Tyhmelacaceae family in medium without n – butyrate also showed weak activity.

Discussion

Immunological studies strongly suggest that EBV plays a causative role in the development of NPC. Environmental and genetic factors, and their relationship to EBV, are also considered to be involved in the development of NPC. The present results showed that among 495 Chinese medicinal herbs from 106 families, 15 herbs from 10 families were found to have EA – inducing activity. Daphne genkwa also was found by Ito[5] to have EA – inducing activity. These data extended the work of Ito et al. [1] and confirmed that the Raji cell – inducing system is a useful model for the detection of EBV EA – inducing activity indifferent materials. Some herbs having EA – inducing activity belong to the same family, such as the Thymelacaceae family where 5 species have EA – inducing activity and the Euphorbiaceae family where at least 10 species were found by Ito et al[1]., to have such activity, or they belong to the same group with similar pharmacological action. For example, 5 herbs (Daphne genkwa, Knoxia valerianoides, Euphorbia kansui, Euphorbia lathyris, and croton oil) are all purgative drugs.

Tab. 2　Chinese medicinal herbs having EA – inducing activity from different families

Family	Species
Thymelacaceae	Daphne genkwa
	Stellera chamaejasme
	Wikstroemia chamaedaphne
	Wikstroemia indica
	Edgeworthia chrysanthea
Leguminosae	Caesalpinia sappan
	Desmodium styracifolium
Verbenaceae	Premna fulva
Sparganiaceae	Sparganium stoloniferum
Solanaceae	Datura stramonium
Menispermaceae	Tinospora sp.
Sinopteridaceae	Aleuritopteris argentea
Ranunculaceae	Clematis intricata
Rubiaceae	Knoxia valerianoides
Umbelliferae	Angelica pubescens

Chinese medicinal herbs are usually administered in solution by boiling the herbs in water. Our findings showed that water extracts of herbs also have EA – inducing activity, but not as strong as that of ether extracts. Int et al[6]. reported that n – butyric acid in the culture medium of Fusobacterium nucleatum isolated from the oral cavity and upper respiratory tract was able to induce EBV. n – Butyrate plus extracts of the herbs mentioned above markedly enhanced the induction of EA antigen in Raji cells. For example, the EA cell – positive rate induced by n – butyrate plus water extract of Wikstroemia indica was 25%. This herb is produced in Guangdong, Guangxi, Fujian provinces and others. It is the main component for making tablets for detoxication and antiphlogosis in Guangdong province. The Premna fulva extract is used for treatment of rheumatism. Tung oil is a strong inducer, and people living in NPC high – risk areas come in close contact with tung oil trees. The relationship between activation of EBV by these herbs and the development of NPC needs to be further studied.

Acknowledgments

We thank Prof. Z W. Xie for identifiction of the Chinese medicinal herbs and Prof. C H . Huang, Prof. C M. Chu, and Prof Y. Ito for critical reding of the manuscript.

[In 《Intervirology》1983, 19:201 - 204]

References

1　Ito Y, Yanase S, Fujita J, Harayama T, Takashima M, Imanaka H. Ashort - term in vitro assay for promoter substances using human lymphoblastoid cells latently infected with Epstein - Barr virus. Cancer Lett, 1981, 13:29 - 37

2　Kawanishi M, Sugawara K, Ito Y. Epstein - Barr virus - induced early peptides in Raji and NC37 cells activated by diterpene ester TPA in combination with n - butyrate. Virology, 1981, 115:406 - 409

3　Tumor Control Team of Zhongshan County, Department of Microbiology of Zhongshan Medical College, Laboratory of Tumor Viruses of Cancer Institute, and Institute of Epidemiology of the Chinese Academy of Medical Sciences: A study on the complement fixing antibody to EB virus in groups of normal individuals in Guangdong province and Beijing. Chinese J. ENT, 1978, 1:23 - 25

4　Zeng Y, Liu Y X, Liu Z Y, Zhen S W, Wei J N, Zhu J S, Zai H J. Application of immunoenzymatic method and immunoautoradiographic method for the mass survey of nasopharyngeal carcinoma. Chinese J Oncol, 1979, 1:2 - 7

5　Ito Y. Personal commun, 1981

6　Ito Y, Kishishita M, Yanase S. Induction of Epstein - Barr virus antigens in human lymphoblastoid P3HR - 1 cells with culture fluid of Fusobacterium nucleatum. Cancer Res, 1980, 40:4329 - 4330

17.　Nasopharyngeal Carcinoma: Can Antiviral Interventions be Contemplated to Prevent this Cancer?

de - The G[1], ZENG Yi[2]

1. Laboratory of Epidemiology and Immunovirology of Tumors, Faculty of Medicine A. Carrel, Lyon, France; 2. Institute of Virology, Chinese Academy of Medical Sciences, Beijing, People's Rep. of China

Introduction

Undifferentiated nasopharyngeal carcinoma (NPC) is a leading cause of cancer death for large populations in Southeast Asia, Africa and around the Mediterranea. The question therefore arises as to whether the association with the Epstein - Barr virus (EBV) can be utilized for the control of this disease through early diagnosis and ultimately in preventing the disease.

The existence of pre - neoplastic conditions characterized by increasing IgA/VCA antibody titers and cytological changes in the nasopharyngeal mucosa gives an opportunity to evaluate the feasibility of conducting intervention trials aimed at reducting or eliminating EBV reactivation and cyto-logical abnormalities with the ultimate goal of reducing NPC incidence. The purpose of the present review is not to describe the epidmiology of NPC as this has been repeatedly and recently done (de – The, Ho, Muir, 1976; Hirayama, 1978), but rather to assess the feasibility of implementing preventive and antiviral interventions as a mean to try and prevent NPC and to clarify the role of EBV in NPC development.

The situation of EBV vaccination in relation to NPC will be quite different when EBV vaccine becomes available. The ethical problem of EBV vaccination will be quite critical as there is no known deadly disease associated with primary EBV infection worth preventing in the large populations at high risk for NPC.

These difficulties should not prevent considering the use of an EBV vaccine when available, but points out the need of evaluating other avenues to try and prevent NPC.

1. EBV/IgA antibodies allow early diagnosis of NPC and represent a critical risk factor for NPC development

Sero – epidemiological surveys of large population groups in the People's Republic of China by Zeng and collaborators have demonstrated the practical value of the IgA/VCA test in early detection of NPC in the highly endemic areas of South China (Zeng et al. , 1980, 1982). This was achieved by the development of an immunoenzymatic test (Pi et al. , 1981) more adapted to the field conditions than the classic immunofluorescence test as developed by Henle and Henle (1966).

The routine use of this test in certain areas of the Guang – Xi Autonomous Region of the People's Republic of China allowed to substantially improve the detection rate of NPC (Zeng et al. , 1980) and to uncover most of the NPC cases at a very early stage of the disease. In the first study (Zeng et al. , 1980), about 50% of the diagnosed NPC were at stages Ⅰ and Ⅱ, while in a more recent survey, all cases were discovered at stage Ⅰ and Ⅱ (Zeng et al. , 1982).

The impact of early diagnosis of NPC by a simple serological test is in itself a major achievement for better control of the disease, as up to 85% of NPC at stage I can be cured (five years survival disease free in Hongkong (Ho, 1972).

To try and estimate the Incidence Rate (IR) of NPC among IgA positive individuals, two pieces of information were available:

a) out of 99 persons found to have IgA antibodies in 1977 and reexamined ten months later, two were found to have developed NPC in the meantime (Zeng et al. , 1980).

b) the follow – up of the 3350 individuals, found to have IgA/VCA antibodies in the Zangwu county survey, led to the uncovering of 33 new cases within 8 – 30 months of follow – up (estimated mean: 24 months) (Zeng et al, unpublished data). From these data, a yearly rate of 10/1 00 persons can be estimated for males aged 40 to 50 years and having IgA/VCA antibodies.

The Relative Risk (RR) to develop NPC for IgA/VCA positive individuals was found to be 50 to 80 times higher than that of the general population. Furthermore, the NPC prevalence among individuals having IgA/VCA antibodies was found to be related to the antibody titers: 1% for individuals with IgA/VCA titers between 20 and 40, 3% for individuals with IgA/VCA titers 80/160 and

9% when IgA/VCA titers reached 320/640 (calculated from Zeng et al. , 1982).

2. Epidemiological characteristics of this IgA/VCA subpopulation group

The first question coming to mind is : How long before clinical onset of NPC do IgA/VCA antibodies develop? From the data of Ho et al (1978), Lanier et al. (1980) and Zeng et al. , (1980, 1982, unpublished data), it can be estimated that IgA/VCA antibodies can be detected up to five years prior to diagnosis of late NPC cases and up to 15 to 18 months prior to the diagnosis of early NPC cases. These estimations are based on small numbers and only a prospective population study will enable to establish the mean duration of the IgA/VCA positive period preceeding NPC development. This point is critical since the clinical onset of the disease may be much delayed from the time of appearance of epithelial tumorous cells in nasopharyngeal mucosa. Subclinical disease may last for years for certain carcinomas.

The second question relates to the proportion of persons entering or leaving the IgA/VCA positive subpopulation group. The IgA/VCA positive individuals as tested by IE represent 2. 5% to 5. 5% of the population aged 35 years and above. This proportion was shown to increase with age (Pi et al. , 1981) and the age – specific prevalence curve of IgA antibodies paralleled that of the NPC age – specific incidence. The population based study, as implemented in Wuchow (Zeng et al. , 1982) should give precise data on the rate of seroconversion from IgA negative to IgA positive. The type of test used to detect IgA/VCA antibodies will indeed influence directly such rate. For all practical purposes, and overly sensitive test is to be avoided as this increases the proportion of false positive, as compared to a less sensitive test. In this context, it is of interest to note that Ho in Hongkong detected only 1% of the population above 30 years of age as IgA/VCA positive, by immunofluorescence, versus 5% for Zeng by the immunoenzymatic test.

The term "retroversion" is proposed to describe the loss of detectable IgA/VCA antibodies in an individual. The retroversion rate has been estimated to be around 9% yearly by Zeng in his survey in the zangwu county (1980 and unpublished data). It is of critical importance to establish firmly such retroversion rates in the study population where intervention trials are being conteplated.

In summary, the on – going sero – epidemiological survey of Wuchow city with its yearly follow – up of 1000 IgA positive and 500 IgA negative individuals aged 40 to 50 years should establish firmly the annual rates of sero "conversion" and of "retroversion" .

3. Pre – neoplastic cytological changes in the nasopharyngeal mucosa

In the sero – upidemiological surveys referred to above, the IgA/VCA positive individuals were clinically examined and a series of macrosopical abnormalities were noted in their nasopharynx: mostly hyperplasia, metaplasia, but also chronic nasopharyngitis, enlarged adenoids, polyps, etc. In the Zangwu survey of Zeng et al. , (1980), 207 individuals with the above described macroscopic lesions were biopsied: besides 47 NPC, histopathology showed metaplasia in 32% of the non – tumorous biopsies and hyperplasia in 7%. Up to 22% of persons in the latter group noted as having "chronic inflammation" had IgA/EA antibodies. In the Wuchow survey (Zeng et al. , 1982), biopsies were taken from 38 persons with such abnormalities and 21 were diagnosed histopathologically as NPC, 13 being at stage I.

When 17 persons exhibiting severe hyperplsia or metaplasia in the nasopharynx and 38 indivi duals with minor hyperplasia or normal nasopharynx were followed up for 8 to 33 months, it was found that 5 of the 17 (28%) metaplastic or hyperplastic cases developed NPC (and were detected either at stage I or II), whereas only 1 of the 38 controls was detected at stage I, 8 months after the first biopsy was taken (Li et al. , submitted for publication).

Serological and cytological anomalies can increase the risk for NPC from 1 to 60 times, or even up to 1500 times as compared to that of the general population. It is of paramount importance that the Wuchow on going population study verifies these estimations since tyey were made on relatively small numbers. These data will form the basis on which intervention trials will be designed.

4. Detection of viral markers in nasopharyngeal mucosa

In order to see whether the secretion of EBV/IgA antibodies corresponded to an EBV activity in the nasopharyngeal mucosa, 56 symptomless individuals having serum IgA/VCA antibodies for 15 to 18 months were carefully clinically examined and biopsied (de – The et al. , 1981). Four early NPC were discovered and 14 further individuals, for whom no histopathological nor clinical evidence of NPC could be noted, were found to have detectable EBV/DNA sequences and/or EBNA in their nasopharyngeal mucosa (Desgranges et al. , 1982). In the remaining 38 biopsies of IgA positive individuals, no EBV markers were detected. 7 biopsies done on IgA/VCA negative individuals at the same time showed no detectable viral markers in their NP mucosa.

With these results in hand, it seemed that another critical and immediate NPC risk marker could be the presence of EB viral markers in the nasopharyngeal mucosa. As it was not possible to biopsy NP mucosa of normal, IgA/VCA negative individuals, it was difficult to establish the absence of EBV infection in normal NP mucosa. This possibility came about with the use of a negative pressure apparatus allowing to collect exfoliated cells from the nasopharynx. Enough cells could be obtained to make smears for cytological examination and EBNA studies, and to extract DNA for search of EBV/DNA sequences. Exfoliated cells, collected from the nasopharynx of 62 IgA/VCA positive and of 39 IgA/VCA negative individuals in Wuchow city were investigated. DNA was extracted from these cells and tested by spot and then by blot hybridization techniques using the cloned internal repeat of B95 – 8 virus DNA as probe. Unexpected findings were observed. Both IgA positive and IgA negative groups exhibited a similar proportion of EBV/DNA positive specimens (Desgranges et al, in preparation). Spot hybridization is known to yield some false positive results and thus all the spot positive specimens were reevaluated by blot hybridization halving the proportion of positive. In this context, it is to be noted that the 3 individuals positive ty spot and negative by blot also lack IgG antibodies to EA. Is EBV present in non – tumorous NP epithelial cells? The anticomplement immunoenzymatic test (ACIE) as developed by Pi et al. (1981) was used to detect EBNA in exfoliated cells. The study of 79 NPC patients exhibited EBNA in tumor cells but also in inflammatory or hyperplastic part of their nasopharynx. Furthermore, 4% of patients with ENT diseases other than NPC similarly exhibited EBNA in their NP cells (Zeng et al. , in preparation). Simlar observations were obtained by Chen et al. (1982) and Ho (personal communication).

The above data, which need confirmation by other technics, suggest that EBV might be present in non – tumorous nasopharynx but the extent of such a presence, in acute, chronic or latent EBV infection, the cell type involved and the variations to be observed between geographical areas and ethnic groups, remain to be better determined. The relatively easy access to exfoliated NP cells should promote the implementation of such studies aimed at identifying the risk factor associated with the presence of EBV markers either in epithelial cells columnar or in the basal layer or in the submucosal lymphoid cells.

5. Preliminary steps to be implemented before intervention trials can be contemplated

As mentioned above, it is critical to clarify certain characteristics of the natural history of EBV reactivation in relation to risk of NPC before implementing any intervention trial. Within the ongoing population study of Wuchow, the following will have to be characterized.

— the age/sex specific prevalence of IgA/VCA positive and IgA/EA positive individuals,

— the annual age/sex specific rate of retroversion (from IgA + to IgA –),

— the annual age/sex specific rate of conversion (from IgA – to IgA +),

— the annual age/sex specific rate of significant increases and decreases in IgA/VCA and IgA/EA antibodies,

— the annual NPC incidence among IgA + and IgA – subpopulation groups.

The registration, initial bleejing, proper follow – up and cancer registration for ideally 40 000 persons 35 years of age and above in Wuchow, should give the necessary demographic and epidemiological base lines. The ongoing serological surveillance of 1 000 IgA + and 500 IgA – individuals is minimal to obtain reliable figures concerning the rate of conversion, retroversion and antibody changes. The doubling of these figures would be desirable.

a) a better serological and cytological characterization of pre – NPC conditions are needed before contemplating interventions. Serologically, the parameters to involve will be the ADCC titers, anti – EBV/DNAse antibodies and the IgG/EA antibodies of individuals with high or low NPC risk.

b) the characteristic atypic, metaplastic cytological changes will have to be agreed upon by an international group.

c) which individuals will benefit most from intervention? Epidemilogically, individuals at the borderline of IgA positivity would be eligible for studying the rate of negative connection from IgA + to IgA – . But this would be difficult to achieve ethically and technically because of the variations at the laboratory tests. Motivation will be best for individuals with higher IgA/VCA titers and at higher risk for NPC (5% prevalence of NPC in individuals with IgA/VCA titers around 160/320 as tested by the immunoenzymatic test in China). This group should yield a large proportion of cooperative individuals after proper information on NPC is given.

6. What type of intervention can be contemplated?

This question remains wide open as there is yet no EBV specific proposal to make. However a few possibilities merit discussion

a) non – specific anti – promoting agent such as betacaroten

Peto et al. , (1981) recently reviewed the epidemiological data suggesting a two fold decrease in carcinoma risk for individuals having a high beta – caroten diet. A large intervention study involving 20 000 American physicians is being implemented by the Harvard School of Medicine (Hennekens, personal communication) to assess the role of beta – caroten intake and blood retinol in prevention of common carcinomas. In the on – going prospective serological survey in Wuchow, one could contemplate to intervene by distributing beta – caroten or placebo to randomized individuals at high risk for NPC (IgA + with cytological atypia or metaplasia).

b) immunological modifiers, such as interferon and gammaglobulin therapy

Experimentally, interferon was shown to be most effective when the tumor burden was low but to be relatively inefficient on large tumors (Gresser and tovey, 1978). A similar situation might prevail for human tumors and up to now mostly terminal cancers were treated by interferon with deceiving results.

Alpha interferon treatment (3.10^6 IU dialy for 5 weeks, followed by 3.10^6 IU thrice weekly, for 5 weeks) was given to 15 Chinese patients at early stages of NPC (stage Ⅱ of Ho). The results were encouraging but no regression of more than 50% was observed (Wang et al. , submitted for publication).

ADCC titers of NPC patients prior to radiotherapy appeared to be of high pronostic value, according to Pearson et al (1978) and Mathew et al (1981). These authors have initiated at the mayo Clinic serotherapy of terminal NPC cases, using semi – purified gamma – globulins having high ADCC activity. They observed clinical and serological effects of such a serotherapy on IgG/EA and IgA/VCA titers (Pearson, personal communication). If this is confirmed, gammaglobulin therapy possibly together with interferon therapy might be of use to try, as complementary to radiotherapy. The source of immunoglobulins remains an open question, as one must have standardized batches of sera with high ADCC activity. Preliminary investigations have shown that gamma globulins from blood donors were a better source for high ADCC activity than plancental extracts.

c) antiviral drugs

Very few compounds have yet shown major clinical effect on herpesviruses in general and on EBV in particular. It is hoped however that both pharmaceutical and public research laboratories will progressively come up with clinically useful compounds and it would be important to be epidemiologically ready to test them when needed. The aim would be to try and inhibit EBV reactivation and to prevent NPC clinical onset.

It is too early to judge whether Acyclovir might be useful in this context, since its effectiveness on EBV – related diseases, such as severe infectious mononucleosis (IM) has not yet been completed in vivo. A clinical trial is actually in progress in the USA and once the results are available, consideration will be given in the present context.

The preliminary studies described above should imply small groups of patients (for example 15 to 20 males aged 40 to 50 years with IgA/VCA titers ranging from 80 to 320) to test the feasibility and possible effects of such interventions.

Population Interventions

To test the validity of the preliminary studies, to select the best drug combination to assess the feasibility of a major population intervention, several hundred persons with appropriate premalignant conditions will have to be identified with relatively homogenous risk for the disease.

It appears that randomized intervention groups (40 to 50 persons each) would be appropriate with a 100 person placebo group, to detect an effect on EBV (significant change in IgA titera i. e. 2 dilutions) and on exfoliated NP cytology, but not on NPC incidence. The number of group depends indeed on the results of the preliminary studies and on the decision to combine certain drugs.

If the pilot study were to be successful, an intervention at a population level might be decided with the aim to reduce the incidence of NPC. The study design will be very similar to the pilot study involving randomization of participants into intervention or placebo group. A vigorous follow-up for 3 to 5 years of all individuals concerned would be required.

Acknowledgements

The substantial help of Dr. N. Gutensohn, Harvard School of Public Health, Boston, for reanalyzing and assessing the data from China, was greatly appreciated during Dr. de The s stay at the Harvard School of Public Health.

[In 《13th International Cancer Congress, Part B, Biology of Cancer》 1983: 37 – 47]

References

1 Shen SK, Tan SW, Liu B, Lu TS, Li WJ, Zhong F, Lo HL, Li SW, Li MJ, Siao JD. Detection of EBNA ant ibody and localizat: on of EBNA from patients with NPC. Cancer, 1982, 1: 40 – 44

2 Desgranges C, Borkamm GM, Zeng Y, Wang PC, Zhu JS, Shang M, de – The G. Detection of Epstein – Barr viral DNA internal repeats in the nasopharyngeal mucosa of Chinese with IgA/EBV – specific antibodies. Int J Cancer, 1982, 29: 87 – 91

3 De – The G, Desgranges C, Zeng Y, Wang PC, Bornkamm GW, Zhu JS, Shang M. Serach for precancerous lesions and EBV markers in the nasopharynx of IgA positive individuals. In: Cancer Campaign, vol 5, Nasopharyngeal Carcinoma, Grudmann, Krueger and Ablashi (eds), Gustav Fischer Verlag, Stuttgart, 1981, 5: 111 – 117

4 De – The G, Ho JHC, Muir CS (1996). Nasopharyngeal Carcinoma. In: Epidmiology and Control, AS Evans (ed), Plenum Press, New – York, PP: 539 – 563, 2nd edition, 1981, 621 – 652

5 Gresser I, Tovey G. Antitumor effects of interferon. Bioch. Bioph. Acta, 1978, 516: 231 – 247

6 Henle G and Henle W. Immunofluorescence in cells derived from Burkitt's lymphoma. J Bacteriol, 1966, 91: 1248 – 1256

7 Hirayama T. Descriptive and analytical epidemiology of nasopharyngeal cancer. In: Nasopharyngeal Carcinoma Etiology and Control, de – The G and Ito Y (eds). IARC scientific publication, 1978, 20: 167 – 189

8 Ho JHC. Nasopharyngeal Carcinoma (NPC). In: Advances in Cancer Research, vol. 15, Klein G and Weinhouse S (eds), Academic Press, New – York, 1972, 15: 57 – 92

9 Ho JHC, Kwan HC, Ng MH, de – The G. Serum Iga antibodies to EBV capsid antigens preceeding symptons of nasopharyngeal carcinoma. Lancet, 1978, 1: 436 – 437

10 Lanier AP, Henle W, Bender TR, Henle G, Talbot ML. Epstein – Barr virus – specific antibody titers in seven Alaskan natives before and after diagnosis of nasopharyngeal carcinoma. Int. J Cancer, 1980, 26: 133 – 137

11 Li EJ, Tan BF, Zeng Y, Wang PZ, Zhong JM, Deng H, Zhu CS, Wei JN, Pan WJ (submitted for publication). Some changes observed on nasopharyngeal mucosa of persons showing EBV – VCA IgA antibodies

12 Maupas P, Goudeau A, coursaget P, Chiron JP, Drucker J, Barin F, Perrin J, Denis F, Diop Mar I, Summers J. Hepatitis B virus infection and primary hepato cellular carcinoma: Epi-

demiological, clinical and virological studies in Senegal from the perspective of prevention by active immunization. In: Cold Spring Harbor Conferences on Cell Proliferation, 1980, 7: 481 – 506

13 Methew GD, Qualtiere LF, Bryan Neel Ⅲ H and Pearson GR. IgA antibody, antibody – dependent cellular cytotoxicity and prognosis in patients with nasopharyngeal carcinoma. Int J. Cancer, 1981, 27: 175 – 180

14 Pearson GR, Johansson B, Klein G. Antibody – dependent cellular cytotoxicity against Epstein – Barr virus – associated antigens in African patients with nasopharyngeal carcinoma. Int J Cancer, 1978, 22: 120 – 125

15 Peto R, Doll R, Buckley JD and Sporn MB. Can dietary beta – carotene materially reduce human cancer rates? Nature, 1981, 290

16 Pi GH, Zeng Y, Zhao WP, Zhang Q, Development of an anticomplement immunoenzyme test for detection of EBNA and antibody to EBNA. J Immunol Method, 1981, 44: 73 – 78

17 Wang PZ, Zeng Y, Dan H, Ma ER, Li EJ. Gresser I Cantell K, de – The G (submitted for publication). Progress report of alpha interferon therapeutic trial of early cases of nasopharyngeal carcinoma in south China. Interferon Scientific Memoranda (USA)

18 Zeng Y, Liu UX, Liu CR, Chen SW, Wei JN, Zhu JS, Zai HJ. Application of an immunoenzymaticm method and immuno – autoradiographic method for the mass survey of nasopharyngeal carcinoma. Intervirology, 1980, 13: 162 – 168

19 Zeng Y, Zhang LG, Li HY, Jan MG, Zhang Q, Wu YC, Wang YS, Su GR. Serological mass survey for early detection of nasopharyngeal carcinoma in Wuzhou city, China. Int J Cancer, 1982, 29: 139 – 141.

18. Epstein – Barr Virus – Activating Principle in the Ether Extracts of Soils Collected from under Plants which Contain Active Diterpene Esters

ITO Yohei[1], OHIGASHI Hajme[2], KOSHIMIZU Koichi[2], ZENG Yi[3]

1. Department of Micro biology, Faculty of Medicine; 2. Department of Food Science and Technology, Faculty of Agriculture, Kyoto University, Kyoto 606 (Japan) and; 3. Institute of Virology, Chinese Acodemy of Medical Sciences, Beijing (China)

Summary

Soil samples were collected from the ground under the plants of Euphorbiaceae and Thymelaeaceae known to possess Epstein – Barr virus – activating diterpene esters. In a test system, the ether extracts of such soil samples at a concentration of 20 μg/ml induced Epstein – Barr virus early antigen in approximately 5% – 25% of the non – producer Raji cells. These findings suggest a possible interaction between plant – derived diterpene esters and the human system, and provide a new aspect in considering the cause of Epstein – Barr virus – associated diseases, particularly nasopharyngeal carcinoma.

Introduction

The epstein – Barr virus (EBV) – activating diterpene esters are widely distributed among plants of the Euphorbiaceae and Thymelaeaceae families, many of which are currently used as folk remedies in areas where the EBV – associated diseases, Burkitt's lymphoma (BL) and nasopharyngeal carcinoma (NPC), are endemic[2-4]. The Chinese tung oil tree (Aleurites fordii), a member of Euphorbiaceae, is a popular plant in the southen provinces of China and is cultivated chiefly for an industrial purpose. The plant perse is also used as a herbal drug[6].

While investigating possible routes through which active diterpene esters might gain access to the human system, we came across an idea that such plant constituents could be released into the soil where they are growing and subsequently adsorbed on to the soil particles. These particles may later enter the nasal cavity in the form of soil dust.

Here we report that highly active substances capable of inducing EBV early antigen (EA) in the non – producer Raji cell system[5] cand be extracted from soil samples collected under the Aleurites fordii and other Euphorbiaceae and Thymelaeaceae phants with such activities.

Materials and methods

The EBV genome – carrying human lymphoblastoid Raji cells were cultivated in RPMI 1640 medium containing 10% fetal calf serum, 100 units of penicllin and 250 µg/ml of streptomycin. Under these culture conditions, our Raji subline exerted a spontaneous induction rate of EBV EA of less than 0.01%. The cells were adjusted to a density of 1×10^6 cells/ml and incubated with 4 mmol/L n – butyrate and various test extracts. 12 – O – Tetradecanoylphorbol – 13 – acetate (TPA) served as a positive control for assessing the activity of the extracts. Assays containing only n – butyrate were used as negative controls.

The soil samples (20 g) were collected from the ground under the plants within a distance of 0.5 m from the base of the stem and to a depth of 0.2 m from the soil surface. The samples were extracted with an equal volume of ether for 20 min at room temperature (20℃). After evaporating the solvent, the crude extracts were weighed and redissolved in dimethylsulfoxide (DMSO) as a stock solution (10 mg/ml). The extracts were prepared to a final concentration of 100, 29, 4 µg/ml and tested in the assay system. Cell smears from a 48 h culture were prepared on glass slides, airdried, and fixed in acetone at room temperature for 10 min. The activated Raji cells expressing EBV EA were stained with EA (+) VCA (+) serum (titer 1:1280) from a patient with NPC using an indirect immunofluorescent method[1]. The NPC serum was used at a dilution of 1:40. Untreated cultures served as the controls. In each assay, at least 500 cells were counted randomly and the results were read. The number of viable cells in the culture were determined by the methylene – blue exclusion test.

The n – butyrate and other chemical reagents were purchased from Nararai Chemicals Ltd. Kyoto, Japan. The TPA was obtained from the Chemicals for Cancer Research, Inc. Minnesota, USA. The high – titer NPC serum was a gift of Prof. H. Hattori, Department of Otorhinolaryngology, Kobe University School of Medicine.

Results and discussion

An interaction between higher plants involving chemicals which are shed by one species and affect the growth and other biological conditions of the other plant is termed allelopathy[7]. Such chemicals are released from the living plants by rain – wash, root excretion or exudation, decomposition of fallen leaves, flowers, fruits and dead remains of roots. Thus, it is feasible that EBV – activating diterpene esters of Euphorbiaceae and Thymelaeaceae plants may appear in the soil surrounding the organism.

As expected, the ether extracts of soil samples collected from under the plants of the 2 families exerted EBV – activating capacity. Such results are shown in Tab. 1. On the contrary, the extracts of soil samples taken from under the plants without such activity and from randomly selected areas of our university campus and other places showed little or no EBV EA – inducing effect.

In order to carry out a more precise survey of the EBV – activating substance (s) contained in the soil around the plant, we selected a Chinese tung oil tree, growing in the botanical garden of Kyoto College of Pharmacy as a model (Fig. 1). The soil samples were collected in a southerly direction at distances of 0. 5, 1. 0 and 2. 0 m from the base of the tree trunk. The data show that the EBV – activationg effect of the soil extracts is highest in the sample from 0. 5 m and decreases as the distance from the base increases (Tab. 2). The activity of soil samples from the nearest point (0. 5 m) is greater than or equal to the positive TPA control (10 ng/ml).

Tab. 1   Induction of Epstein – Barr virus EA in human lymphoblastoid Raji cells with extracts of soil samples from under various plants

Soil samples from	Inducing extracts (μg/ml)[a]	EA positive cells (%)	Cell viability (%)	Soil samples from	Inducing extracts (μg/ml)[a]	EA positive cells (%)	Cell viability (%)
Under specines (family) of plants							
Ficus elastica	20	0. 1[b]	49. 1	soil	4	0. 1	74. 4
Sapium japonica	20	26. 5	81. 4	(Moraceae)	4	0. 1	42. 5
(Euphorbiaceae)	4	26. 0	86. 4	Vinca rosea	20	0. 1	61. 5
Sapium sebiferum (Euphorbiaceae)	20	23. 8	77. 6	(Apocynaceae)	4	0. 1	42. 4
Codiaeum uariegatum	20	12. 0	55. 8	Diospyros kaki (Ebenaceae)	20	0. 1	64. 7
(Euphorbiaceae)	4	9. 8	41. 7	Castanea crenata (Fagaceae)	20	0. 1	67. 9
Euphorbia lathyris	20	3. 6	45. 4	Control area and material			
(Euphorbiaceae)	4	7. 5	51. 1	Campus ground	20	0. 4	65. 9
Daphne odora	20	15. 6	49. 3	(Kyoto U)	4	0. 3	59. 3
(Thymelaeaceae)	4	10. 5	65. 3	Personal garden (Prof T)	20	0. 1	86. 7
Edgeworthia papyrifera (Thymelaeaceae)	20	5. 9	73. 9	Commercial gardeing	20	0. 1	65. 2

Notes: [a] Dissolved in DMSO and used with 4 mmol/L n – butyrate in the assay[5]; [b] Represents figure less than or equal to 0. 1%.

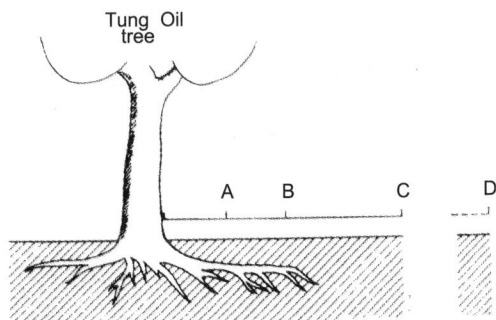

Tung Oil tree

A = 0. 5 m, B = 1. 0 m, C = 2. 0 m, D = distant point

Fig. 1 Scheme for collecting soil samples from under Chinese tung oil tree (aleurites fordii).

Tab. 2 Epstein-barr virus activation with ether extracts of soil samples from under the active diterpene esters-containing plant (aleurites fordii)

Soil sample[a]	Concentration of extract (μg/ml)[b]	EA-positive cells (%)	Cell viability (%)
A	100	0. 1[c]	9. 8
	20	23. 0	74. 5
	4	3. 8	75. 5
B	100	0. 1	8. 1
	20	11. 3	79. 6
	4	4. 3	84. 9
C	100	0. 1	23. 7
	20	6. 7	74. 5
	4	0. 2	68. 2
D	20	0. 1	80. 5
Controls TPA	0. 01[b]	21. 0	86. 7
n-Butyrate	4 mmol/L[d]	0. 1	70. 0

[a] Refer to the codes given in Fig. 1; [b] Dissolved in DMSO and used with 4 mmol/L n-butyrate[5]; [c] Represents figure less than or equal to 0. 1%; [d] Used alone

Recently we have observed that the EBV-inducing activity of tung oil can withstand heating at 100℃ for over 6 h (data not shown). Such durability may accord with the notion that the EBV-activating principle (s) released into the soil from the plants may persist in an active form for a considerable length of time. Soil particles carrying such principles may be blown up as dust and gain access to the mucous membrane of the nasopharynx, eventually leading to the emergence of malignancy.

Acknowledgements

We thank Dr. Katsumi Goto, Kyoto College of Pharmacy for his kind cooperation in collecting soil samples and Mr. Peadar Mac-Gabhan for his assistnce in editing the manuscripth. We also with to thank H. Tokuda and S. Yanase for performing the immunofluorescent tests. This work was supported in part by a grant-in-aid for Cancer Research from the Ministry of Education, Science and Culture of Japan and, by a grant from Japanese Human Cancer Virus Task Force.

〔In 《Cancer Letters》 1983, 19: 113-117〕

References

1 Henle G, Henle W. Immunofluorescence in cells derived from Burkitt's lymphoma. J. Bacteriol, 1966, 91: 1248-1256

2 Hirayama T, Ito Y. A new view of the etiology of nasopharyngeal carcinoma. Prev. Med, 1981, 10: 614-622

3 Ito Y, Kawanishi M, Harayama T, Takabayashi S Combined effect of the extracts from Croton tiglium, Euphorbia lathyris or Euphorbia tirucalli and n-butyrate on Epstein-Barr virus expression in Human lymphoblastoid P3HR-1 and Raji cells. Cancer Letters, 1981, 12: 175-180

4 Ito Y, Kishishita M, Morigaki T, Yanase S, Hirayama, T, Induction and intervention of Epstein-Barr virus expression in human lymphoblastoid cells: A siulation model for study of cause and prevention of nasopharyngeal carcinoma and Burkitt's lymphoma. Cancer Campaign, 1981, 5: 255-262.

5 Ito Y, Yanase S, Fujita J, Harayama T,

Takashima M, Imanaka H. A short – term in vitro assay for promoter substances using human lymphoblastoid cells latently infected with Epstein – Barr virus. Cancer Letters, 1981, 13: 29 – 37

6 Ito Y, Yanase S, Tokuda H, Kishishita M, Ohigashi H, Hirota M, Koshimizu K. Epstein – Barr virus activation by tung oil. extracts of Aleurites fordii and its diterpene ester HHPA. Cancer Letters, 1983, 18: 87 – 95

7 Muller C, H, Chou CH. Phytotoxins: an ecological phase of phytochemistry. In: Phytochemical Ecology, 1972, 201 – 216. Editor J B Harborne. Academic Press, London.

19. Cytogenetic Studies on An Epithelial Cell Line Derived from Poorly Differentiated Nasopharyngeal Carcinoma

ZHANG Si – zhong[1], GAO Xiu – kung[1], ZENG Yi[2]

1. Sichuan Medical College, Chengdu, Sichuan;

2. Institute of Virology, Chinese Academy of Medical Sciences, Peking, China

Summary

An epithelial cell line, CNE – 2, has been recently established from a poorly differentiated nasophryngeal carcinoma, and it represents the first of its kind. Using chromosome banding techniques, cytogenetic analysis of the cell line was carried out. It was demonstrated that the chromosome numbers of the CNE – 2 cells varied from 87 to 107 and the modal nunber was 104 – 103. All cells contained a serles of structurally abnormal chromosomes, and most of them were either consistent or frequently found. Among these chromosomes there were two giant markers which, by banding pattern analysis, proved to be distinct from the so – called glant group A marker chromosomes previously found in many lymphoblastoid cell lines from NPC. Comparison between the CNE – 2 and CNE, another epithellal cell line, which was established from well – differentiated squamous NPC, showed that while they were quite different in many cytogenetic aspects, they had three marker chromosomes in common, namely, an $\text{iso}^8 \text{q}$, 2 t (?; 3q) and a small acrocentric one. The question of whether chromosome markers specific for NPC exist is discussed in the light of the data presented.

Introduction

Nasopharyngeal carcinoma is of special interest for its close association with EB virus (Desgranges et al. , 1975) and for the possible genetic background of its origin (Simons et al. , 1975). Cytogenetic studies of NPC usually aim either to clear the question whether there is a general chromosomal instability in patients with NPC (Zhang et al. , 1980), or to reveal chromosomal abnormalities in NPC cells per se (Ou et al. , 1979). Due to the difficulty of obtaining good chromosome preparations directly from solid tumors, most investigations of the latter type were carried out on cell lines derived from patients with NPC.

It is well known that NPC is an epithelial tumor, but nearly all cell lines derived from it are lymphoblastoid. Therefore, it is uncertain to what extent cytogenetic data obtained from these lines reflect the reality of the tumor. Only two epithelial cell lines have been so far established (LTVCI et al. 1978; Huang et al. 1980). The first one, CNE by name, was reported to have a modal number of 67 – 68 with a series of consistent or frequently seen marker chromosomes (Zhang et al., 1982). The second (NPC/HK – 1) was said to have a modal number of 74 and its detailed cytogenetic study is still in progress. Both cell lines are EBV – negative and derived from well – differentiated squamous NPCs.

Recently, a new epithelial cell line (CNE – 2) has been established from a poorly differentiated NPC, and represents the first of its kind. In view of the fact that most NPCs are poorly differentiated, here we present the results of cytogenetic studies on the CNE – 2 cell line by chromosome banding techniques. It is hoped that the data reported here will allow better understanding of the chromosomal pathology of NPC and throw light on the problem of whether there exist any chromosome marker (s) specific to nasopharyngeal carcinoma.

Material and Methods

1. The cell line

CNE – 2 was established from a tumor biopsy from a 68 – year – old male patient with NPC who had typical clinical manifestations and some positive immunologic findings. The diagnosis was nasopharyngeal carcinoma, stage Ⅲ. Cytological examination and anticomplement immunoenzymatic test of exfoliated cells from the nasopharynx showed poorly differentiated carcinoma and EB nuclear antigen (EBNA) respectivley. Detailed characteristics of the cell line have been published elsewehere (Gu et al., 1982).

The cell line was maintained in culture medium RPMI 1640 supplemented with 20% calf serum at 37℃. The medium was changed twice a week. Subcultures were made at regular 6 – to 7 – day intervals by trypsinzing the cells with 0. 06% trypsin and 0. 02% EDTA. When the studies started the cell line was at its 62 nd passage.

2. Cytogenetic techniques

Cells in exponential growth were treated with Colcemid in a concentration of 0. 025 – 0. 05 μg/ml for 2 – 6 h. Then they were collected by trypsinizing and hypotonized in 0. 075 mol/L KCl for 8 min. Thereafter the cells were fixed in a 3∶1 mixture of methanol and glacial acetic acid with three changes of fixatives. Chromosome preparations were made using the conventional air – drying method.

G – and C – chromosome banding was carried out according to the method of Seabright modified as described previously (Zhang et al., 1979) and the method of Sumner (1971) respectively. Banded metaphases, after preliminary scanning in a Leitz microscope, were selected for photography. Chromosome counting and karyotypic analysis were performed on enlarged photocopies. Since the cell line appeared to be a hyper – aneuploid one, this was more reliable than direct microscopical analysis. In total, more than 50 wellspread and banded mitoses were photographed and analyzed.

Results

1. Number of chromosomes

The chromosome number distribution in CNE – 2 cells is given in Table 1. Among 50 mitoses 11 had 104 and 7 had 103 chromosomes. Thus the modal number was 104 – 103. Cells with many more chromosomes (for instance, 204 etc). were occasionally found (13/500, e. g. 2. 6%). A few cells (0. 4%) with endomitotic endoreduplication were also observed.

Tab. 1 Chromosome counts in 50 metaphases of the CNE – 2 cell line

No. of chromosomes	87	88	89	90	91	92	93	94	95	96	97	98	99	100	101	102	103	104	105	106	107	Total
No. of metaphases	2	0	1	0	0	0	0	1	1	2	3	5	1	3	5	5	7	11	1	1	1	50

Although the chromosome count seemed to be near – tetraploid this was not in fact the case because all cells contained many abnormal chromosomes and the normal chromosome types were also present in variable numbers (Fig. 1). For instance, 5 chromosomes No. 1. 3 No 2, 2 No. 3, 2 No. 4 and 3 No. 5 were usually found. Other chromosome types variced from 1 to 4 but chromosome 8 was nearly always monosomic and chromosome 13 was duplicated. Therefore, in is spite of the munerical variability of each chromosome type, there seemed to be some relative consatancy in karyotypic constituion of all CNE – 2 cells during the period of examination.

2. Marker chromosomes and their analysis

All CNE – 2 cells contained many abnormal chromosomes, usually more than 30 (Fig 1). Most of them were either consistent e. g. present in all cells, such as M1, M2, M4 – 8, M11, M12, M14, M16 – 17, M19, M21 – 23, M31 – 37; or could be seen frequently, such as M9, M13, M18, M20, M24 – 27 and M39. The description and analysis of some consistent or frequent marker chromosomes are given below.

M1: Giant submetacentric chromosome, which was 1/3 larger than normal chromosome No. 1 and represented a product of manifold chromosomal rearrangement (Figs. 1. and 3).

M2: 3q + : Giant submetacentric chromosome, slightly larger then chromosome No. 1. Its banding pattern was completely identical with that of chromosome No. 3, but there was an extra segment at the distal end of its long arm. Therefore it could be designated as 3q + marker (Figs. 1, 2).

M4, 1p – : Large submetacentric chromosome, identical with chromosome No. 1 but without the distal part of the short arm. The breakpoint was close to 1p32.

M5, 4p + : Large submetacentric chromosome, identical with chromosome No. 4, but with an extra segment at the end of the short arm.

M6, iso8q: Large metacentric chromosome. The banding pattern of both its arms was identical with that of the long arm of chromosome No. 8, so it was an isochromosome of 8q.

M7, t (X;?): Large submetacentric chromosome with the short arm and proximal part of the long arm similar to those of X chromosome. The distal part of the long arm was translocated from other unidentified chromosome (s).

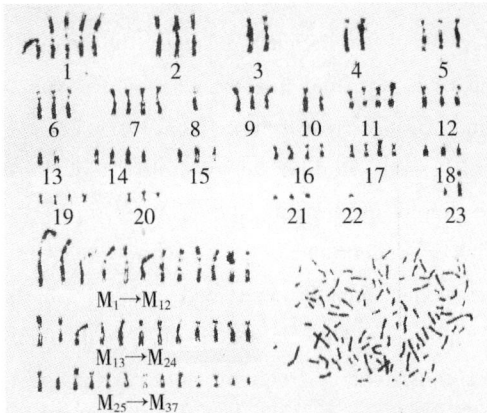

Fig. 1 A representative karyotype of a CNE – 2 cell, showing 104 chromosomes including many marker chromosomes (M1 – M37). G – banding

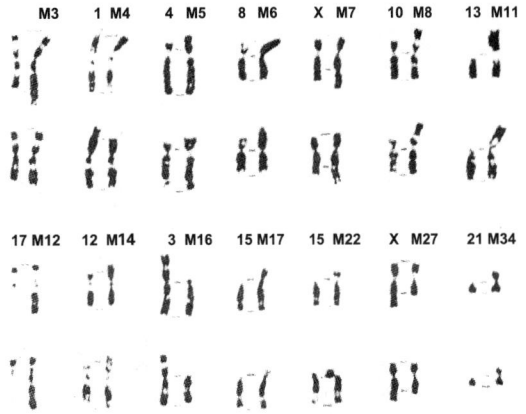

Fig. 2 Analysis of some consistent and frequent markers. For each marker, the upper partial karyotype is taken from Fig. 1 and the lower, from other karyotypes. The black bars indicate identical or similar parts of the marker and the normal chromosomes. G – banding

M8, iso10q: Large metacentric chromosome. Both its arms were identical with the long arm of chromosome No. 10 Therefore, it was an isochromosome of 10q.

M11, iso13q: Metacentric chromosome. Both its arms were similar to the long arm of chromosome 13, so it should be designated as iso13q.

M12, t (17;?): Submetacentric chromosome. Chromosome No. 17 took part in its formation.

M14, 12p +: Medium – sized submetacentric chromosome, which was identical with chromosome No. 12 but with an extra segment at the end of its short arm.

M16, t (?; 3q): Subacrocentric chromosome, the long arm of which was indentical whith that of chromosome No. 3.

M17, 15p +: Submetacentric chromosome which was identical with chromosome No. 15 but had a much longer short arm.

M22, 15P +: Another submetacentric chromosome identical with chromosome No. 15, but the short arm was different from that of M17.

M27, t (X;?): Metacentric chromosome; the short arm and the proximal part of the long arm were similar to corresponding parts of X chromosome.

M32, t (22q?; 21q): Small metacentric chromosome probably formed by fusion of long arms of chromosomes Nos. 22 and 21.

It must be added that not all marker chomosomes were easily to be identified. Due to breakage and reunion the structure and origin of some chromosomes or chromosome segments remained obscure, even through their banding patterns were clearly presented.

Apart from the markers shown in Fig. 1, some other markers such as 7q –6p – , t (8q; 8q +) etc. were occasionally found in a few cells.

Discussion

It has been shown that more than 90% of nasopharyngeal carcinomas are poorly differentiated. Therefore, the establishment of the epithelial cell line CNE – 2 from this pathohistologic type deserves attention and may play a role in experimental and clinincal research on NPC.

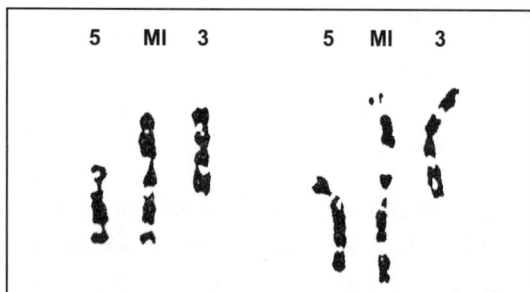

Fig. 3　Analysis of the giant submetacentric marker M1. Detailed explanation in text.

Present studies demonstrated that during the time of examination the CNE – 2 represented a hyper – aneuploid cell line with 104 – 103 as the modal number and had multiple marker chromosomes. The presence of modal number and consistent or frequent markers may be explained by clonal origin and clonal selection in the development of the tumor or the cell line from it.

In comparison with CNE, another epithelial NPC cell line reported previously (Zhang et al., 1982), the CNE – 2 has many more chromosomes. While the modal number of CNE is 68, the modal number of CNE – 2 IS 104 – 103. Both cell lines have many marker chromosomes and some of them are in common. They are iso8q or M6, t (?; 3q) or M16 and M32, corresponding to M2, M10 and M16 in CNE respectively. The finding of common markers attracts attention, since it is well known that there is a 14q + marker common to all cell lines derived from Burkitt lymphomas. However, as the epithelial cell linees of NPC are still few, it seems too early to draw any conclusion about their significance or specificity for NPC cell lines.

It has been demonstrated that the chromosome aberrations in human neoplasia are non – random (reviewed by Mitelman and Levan, 1981). The Ph[1] chromosome in chronic myeloid leukemia and the 14q + marker chromosome in Burkitt lymphoma are examples of this non – randomness (Nowell and Hungerford, 1960; Manolov and Manolova, 1972). For this reason it is not surprising that much attention has been paid to the problem of specific marker chromosomes in NPC. Thus, a large submetacentric marker chromosome ("giant group A" according to some authors) was observed in a few lymphoblastoid cell lines derived from NPC and was reported to represent a transiocation between chromosomes No. 1 and No 3 (Hsia and Lu, 1978; Wu et al., 1980). A similar, but not banded "giant group A" was also found in some biopsies from nasopharyngeal carcinomas (OU et al., 1979). However, it was not observed in other lymphoblastoid cell lines (Finerty et al., 1978). It should be pointed out that the sturcture of the "giant group A" reported by different authors may not be the same, and it is not excluded that the giant submetacentric marker is the result of long – term culturing of lymphoblastoid cell lines (Macek et al., 1971) and not necessarily associated with NPC.

With regrad to the epithelial cell lines, the CNE. which was derived from differentiated squamous NPC, did not sow any consistent abnormal chromosome larger than Chromosome No. 1. In contrast with this, the CNE – 2 revealed two giant chromosome. M1 and M2. As mentioned above, the former was a giant submetacentric chromosome formed by complex rearrangement. The banding pattern of the distal part of the short arm and the central part of the long arm were similar to corresponding segments of chromosome No. 3 and chromosome No 5. respectively. The origin of other segments remained unidentified (Fig. 3). The structure of M2 has been described earlier. While both markers involved chromosome No. 3 in their formation, they were unlike the "giant group A" described in the literature. As M1 and M2 are found in the first epithelial cell line from poorly differentiated NPC and their banding pattern is clearly presented, they may stimulate further interest in establishment and comparative cytogenetic studies of more NPC cell lines. In spite of the disadvantages of selective pressure in long – term culture, before we can obtain well – banded chromosomes in quantity directly from tumor tissues, such studies may be the only practicable approach to the perplexed problem of specific marker chromosomes in NPC.

[In 《Int J Cancer》 1983, 31: 587 – 590]

References

1 Desgranges C, Wolf H, De – The G, Shanmugaratnam K, Ellounz R, Cammoun N, Klein G, zur Hausen, M. Nasopharyngeal carcinoma X. Presence of Epstein – Barr genomes in epithelial cells of tumors from hith and medium risk areas. Int J Cancer, 1975, 16: 7 – 15

2 Finerty S, Jarvas J E, Epstein MA, Trumper P A. Ball G. and Giovanella B C. Cytogenetics of malignant epithelial cells and lymphoblastoid cell lines from nasopharyngeal carcinoma. Brit J. Cancer, 1978, 37: 231 – 239

3 Gu S Y, Tan W P, Zeng Y, Zhao WP. Li K, Deng H H. and Zhao M L. Establishment of an epithelial cell line from patient with poorly differentiated nasopharyngeal carcinoma. IV International Symposium on nasopharyngeal carcinoma, September, 1982, Kuala Lumpur, Malaysia

4 Hais C H, Lu H L. Giant group A marker chromosome in three human lymphoblastoid cell strains from nasopharyngeal carcinoma. Chin Med J, 1978, 4: 130 – 134

5 Huang D P, Ho J H C, Poon Y F, Chew E C, Saw D, Lui M, Li C E, Mak L SMLai S AH, Lau, W H. Establishment of a cell line (NPC/HK1) from a differential squamous carcinoma of the nasopharynx. Int J Cancer, 1980, 26, 127 – 132

6 Ltvci et al. (Laboratory of Tumor Viruses of Cancer Institute and other Institutes in China). Establishment of an epithelioid cell line and a fusiform cell line from a patient with nasopharyngeal carcinoma. Sci Sinica, 1978, 21: 127 – 134

7 Macek M, Seidel EH, Lewis R T, Brunschwig J P, Wimberly I, Benyesh – Melnick M. Cytogenetic studies of EB virus – positive and EB virus – negative lymphobtestoid cell lines. Cancer Res, 1971, 31: 308 – 321

8 Manolov G, Manolova Y. A marker boan in one chromosome No. 14 in Burkitt lymphomas. Hereditas, 1972, 69, 300

9 Mitelman F, Levan G. Clustering of aberrations to specific chromosomes in human neoplasms. IV. A survey of 1 871 cases. Hereditas, 1981, 95: 79 – 139

10 Nowell P C, Hungerford D A. A minute chromosome in human granulocytic leukemia. Seience, 1960, 132, 1497

11 Ou B, Fang Y, Lai H. Chromosmome analysis of nasopharyngeal carcinoma and its clinical significance. Nat Med J. China, 1979, 59: 333 – 338

12 Simons MJ, Day N E, Wee G B, Chan S H, Shan mugaratnam K, de – The G. Immunogenetic aspects of nasopharyngeal carcinoma (NPC). III. HLA type as a genctic marker of NPC predisposition to test the hypothesis that EBV is an aetiologic

afctor in NPC. In. G. de – The M A. Epstein and H. zur Hausen (ed). Oncogenesis and herpesviruses Ⅱ. Part 2, 1975 pp. 249 – 258. IARC Lyon

13 Sumner A T, Evans H J, Buckland R A. New techniques for distinguishing between human chromosomes. Nature New Biol, 1971, 23: 231 – 232

14 Wu B, Wu Y, Li Y, Zeng Y, Wu M, Zhao Z, Gong C. Study of giant group A marker chromosome in several Burkitt lymphoma and lymphoblastoid cell lines with Epstein-Barr virus from diffecrent origins. Chin. Med J, 1980, 93: 400 – 406

15 Zhang S, Qiu J, Gao X. G – banding of human chromosomes. Nat Med J. China, 1979, 59: 210 – 213.

16 Zhang S, Qiu J, Gao X. Cytogenetic studies by chromosome banding techniques on peripheral blood cells from nasopharyngeal carcinoma patients. Chin Med J, 1980, 93: 251 – 259

17 Zhang S, Wu Y, Zeng Y, Zech L, Klein G. Cytogenetics studies on an epithelioid cell line derived from nasopharyngeal carcinoma. Hereilias, 1982, 9: 723 – 728

20. Comparative Evaluation of Various Techniques to Detect EBV DNA in Exfoliated Nasopharyngeal Cells

Pi G H[1,2], Desgranges C[2], Bornkamm G W[3], Shen S J[1], Zeng Y[1], de – The G[2]

1. Institute of Virology, Chinese Academy of Medical Sciences, Beijing

(People's Republic of China); 2. Laboraloire d Epidemiologie el d.

Immunouirologic des Tmmerurs, Faculte de Medecine Alexis Carrel. 69372 Lyon Crder 2 (France)

3. Institute fur Virologie, Zentrum fur Hggiene, Freiburg (Federal Republic of Germang), and

4. Zhanjiang Medical College, Zhanjiang (People's Republic of China)

Summary

Exfoliated epithelial cells from nasopharynx of 27 individuals were collected with a low – negative – pressure suction apparatus. These cells were examined cytologically and stained for Epstein – Barr virus (EBV) nuclear antigen on touch smears by immunoperoxidase. DNA was extraeled from the remaining cells and studied for the presence of EBV sequences by blot and spot hybridization. In the 22 positive nasopharyngeal carcinoma cases (confirmed by clinical, histological and cytological examination) EBV DNA was found with both techniques in 21 cases. while in the 5 cases of chronic inflammation without tumorous cells. no EBV DNA was detected.

[**Key Words**]   Epstein Barr virus; Nasopharyngeal carcinoma; DNA hybridization; Exfoliated cells; China

Introduction

The presence of the Epstein – Barr virus (EBV) genome in the epithelial tumorous cells of nasopharyngeal carcinoma (NPC) is now well docnmented[16,11,4]. An 《outstanding feature》 of NPC is also the presence of EBV – specific IgA antibodies in the sera and saliva of the patients[8,9,5].

The presence of these EBV – specific IgA antibodies appears before the tumour is clinically detectable[10]. Recent serological mass surveys in China[17,19] have shown that it is possible to detect early NPC cases. using anti – virus capsid antigen IgA antibodies. A proportion of these IgA – positive individuals exhibit EBV DNA and Epstein – Barr nuclear antigen (EBNA) in their nasopharynx[6], believed to represent precancerous conditions.

In the present study, we compared two sensitive techniques for DNA detection (blot versus spot hybridization) on exfoliated cells collected from the nasopharynx by negative – pressure suction[20]. This procedure of cell collection does not traumatize the nasopharyngeal mucosa and allows to have sufficient celluar DNA for spot hvbridization, which is much easier to apply to large numbers of specimens than is blot hybrdization.

Materials and Methods

Twenty – seven individuals 22 histologically confirmed NPC cases at different stages of the disease and 5 clinically suspected but histologically unconfirmed NPC. subjects from the Zhanjiang eity of the People's Republic of China were investigated.

1. Specimens collected by negative – pressure suction technique

Exfoliated cells from the nasopharynx were aspira led by negative – pressure suction (30 nn Hg) with an S – Shaped aspiration device after 10% dicaine had been sprayed into the oropharyngeal cavity, as previously described be the Zhanjiang Medical College[20]. The cells, including small clumps of tiseue eollerted on a silk net in the head of the suction apparatus, could be resuspended in buffer. centrifuged and divided into 3 parts: one part was fixed in 10% formalin, stained with haemalun – eosin and examined cytologically, whilst the remaining two parts were used for DNA extraction and EBNA detection, respectively.

2. Detection of EBV DNA by molecular hybridization

After washing in Hank's solution, exfoliated cells from the second of the three parts collected (see above) were resuspended in 0.05 mol/L Tris. pH 8.1, 10 mmol/L EDTA and 1% sarkosyl (W/V), and digested with pronase (1 mg/ml pre – incubated for 1 h at 37℃) for at least 2 h at 37℃. After two phenol extractions, the DNA was precipitated by two volumes of ethanol, redissolved in 10 mmol/L Tris, pH 7.4 and 1 mmol/L EDTA, and digested with RNase (20 µg/ml) for 1 h at 37℃. After two further phenol extractions, DNA was again precipitated by ethanol and dissolved in 10 mmol/L Tris, pH 7.4 and 1 mmol/L EDTA.

AGIE = anti – complement immunoenzymatic (test); acif = anti – complement immunofluorescence (test); EBNA = epstein – Barr nuclear antigen; EBV = Epstein – Barr virus; NPC = nasopharyngeal carcinoma.

3. Blot transler hybridizalion technique

A total of 16 µg of DNA from each specimen was digested with the restriction endonuclease Pst I (Providencia stuarti, Boehringer Mannheim, Gn. bH) which cleaves four times within the internal EBV repeat[7]. After extraction with cloroform/isoamylalcohol (24/1), the DNA was concentrated by ethanol precipitation and loaded onto 0.8% horizontal agarose gels and then run for 4 – 5 h at 15 – 50 V in Tris – acetate buffer (40 mmol/L Tris – acetate, pH 7.8 mmol/L EDTA). Bands were visualized on a 254 nm transilluminator (UV prducts, San Gabriel) and photographed with a 《Polaroid》 camera with a 《Kodak》 wratten 23A filler. Fragments were transferred to nitrocellulose by the Southern's method[15]. DNA from an EBV – negative cell line. JM[14] was included on each gel as a negative control. As positive controls we used 500 or 50 pg B95 – 8 (EBV – transforming virus). EBV DNA (1 or 0.1 viral genome copy per cell) and an 8 – pg internal viral repeat (1 repeat per cell) added to 16 μg JM DNA. The internal repeat (IR) of B95 – 8 virus DNA cloned in pBR 322[7] (IR + pBR 322 = pSL9) was labelled by nick – translation with ^{32}P – dCTP (Amersham; 400 Ci/mmol) as previously described[6].

Hybridization was carried out for 4 days at 42℃ in polyethylene bags in buffer containing 50% formamid, 5 × SSC (1 × SSC = 0.15 mol/L sodium chloride + 0.015 mol/L sodium citrate), 0.02% bovine serum albumin, 0.02% polyvinylpyrolidone, 0.02% Ficoll[3], 20 mmol/L sodium phosphate, 200 μg/ml of sheared denatured salmon sperm DNA and 10^6 cpm/ml heat – denatured labelled probe.

The volume of the hybridization reaction was about 0.1 ml/cm^2. Before hybridization, the filters were incubated for 1 – 2 days at the same temperature with a slightly modified buffer (0.05 mol/L sodium phosphate, 500 μg/ml salmon sperm DNA and 5 × Denhardt solution) without the labelled prbe. After hvbridzation. blots were washed twice in 2 × SSC and 0.1% sodium dodecyl sulphate (SDS) at room temperature, followed by 4 washes in 0.1 SSC and 0.1% SDS each wash lasting 30 min at 50℃. Blots were exposed to 《Kodak Boyal X – omat》 film at – 70℃ by using an intensifying screen (Du Pent Cronex lightning – Plus).

4. Spot hybridization

The detection of EBV DNA by the spol technique was first described by Brandsma and Miller[2], by placing the cells directly onto nitrocellulose lillers. With tumour specimens, it is difficult to accurately determine the mumber of cells; so we placed on the filter amounts of 0.1, 0.5 and 1 μg of DNA from each specimen and let them dry. After 7 min in 0.5 mol/L NaOH, the lixation was carried out for 10 min in 0.1 mol/L NaOH and 1.5 mol/L NaCl and neutralization was allowed for 2 min in 0.2 mol/L Tris HCl, pH 7.5 and 2 mmol/L EDTA. After two washings in 2 × SSC, the filter was dried for 1 h at room temperature and for 2 h at 80℃. Hybridization and prebridization were then carried out as for the blot technique, but we observed that a temperature of 50℃ was better as it produced less nonspecific background.

As controls, 7 well – known EBV DNA were used: K3 (NPC tumour, positive for EBV), A, B and C (3 tumours for the head and neck, negative for EBV). Raji (a lymphoblastoid cell line, positive for EBV) and JM and Molt (two lymphoblastoid cell lines, negative for EBV).

5. Anticomplement immunoenzymatic (ACIE) test for EBNA detection

The third part of the collected (see above) was treated for EBNA detection. The anti – complement immunoenzymatic (ACIE) test, previouslv described by Zeng et al[8] and Pi et al[12]. (a simple modification of the ACIE test described by Reedman and Klein[13], was carried out in the ficld. The anti – complement C3 antiserum was prepared in China by the immunization of rabbits with human C3 adsorbed on insulin (titre of 1/16 to 1/64 by double immunodiffusion).

Labelling of the anti – C3 with horseradish peroxidase was performed as described by Avrameas[1]. EBV – negative French human sera were used as a source of complement, and balanced salt solution (BSS) as diluent. The positive or negative EBNA reference sera and the complement (diluted 1/10) were added onto the smears for 1 h at 37℃. After three washings in BSS, the anti – C3 horseradish peroxidase conjugate was added for 30 min at 37℃, washed three times again in BSS, stained with diaminobenzene and H_2O_2 and then examined.

Results

1. Cytological and histological examination

Among the 27 specimens, 22 were confirmed histologically and 24 cytologically as NPC cases (see Tab. 1). The five individuals clinically suspected of NPC (cases 11, 12, 14, 20 and 27) were histopathologically labelled as hyperplasia (cases 11, 14 and 27) or metaplasia (cases 12 and 20).

Tab. 1   Clinical and cytological data compared to EBV markers in exfoliated cells from nasopharyngeal mucosa of 27 Chinese subjects.

Case n°	Clinical stage of NPC	Tumorous cells by cytology	Tumorous eells with EBNA by peroxidase test (ACIE)	EBV DNA by blot	EBV DNA by spot
8	I	+	+	+	+
15		+	+	+	+
17		+	+	+	−
18		+	+	+	+
3	II	+	+	+	+
6		+	+	+	+
19		+	+	+	−
21		+	+	+	+
22		+	+	+	ND
23		+	+	+	ND
25		+	+	+	+
26		+	+	+	+
1	III	+	+	+	+
2		+	+	+	+
4		+	+	+	+
5		+	+	+	+
7		+	+	+	+
16		+	+	+	+
24		+	+	+	+
9	IV	+	+	+	+
10		+	+	+	+
13		+	+	+	+
11	Non – PC	—	— (∗)	—	—
12	(hyperplasia)	—	— (∗)	—	—
14		—	— (∗)	—	ND
20		—	— (∗)	—	ND
27		—	— (∗)	—	—

Notes: (∗) in these 《chronic inflammation》 cases, some epithelial and lymphoid eells were found positive for EBNA + = only positive after 5 days of autoradiography. ND = not done

2. Detection of EBV markers in cell smears

（1）DNA detection

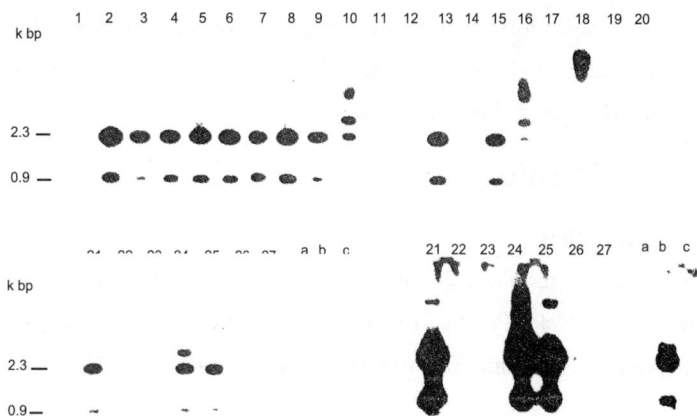

The upper part and the left of the lower part were exposed for 20 h and the right of the lower part for five days, respectively, al – 70℃ with an intensifying screen. The controls contained 16 μg DNA of the EBV – negative cell line JM:

（a）without viral DNA;

（b）with one copy of B95 – 8 DNA per cell;

（c）with one copy of pSL9 per cell（from left to right）

Fig. 1   Blot hybridization of 16 μg of cell DNA collected by aspiration from the nasopharynx of 27 different individuals

Sufficient cells were obtained by aspiration to yield about 50 μg of cellular DNA, i. e. enough to carry out experiments with different restriction enzymes. By blot transfer hybridization with the laberlled internal EBV repeat（after digestion with Pst I）, we obtained 2 major positive bands in 21 of the 22 NPC cases. The quantity of delected EBV DNA was too low in 5 of them（cases 1, 17, 19, 22 and 26）to be detected by blot hybridzation within 20 h exposure, but was seen after 5 days（Fig 1.）. No EBV DNA was detected in the 5 non – NPC specimens（cases 11, 12, 14,

20 and 27）. For some tumours there were more than 2 major bands, due to partial digestion of both the products and the ragment adjaccnl to the internal repeat which contains part of the internal sequeuces（Fig 1）. On the right – hand side of Fig. 1 can be seen the blots of seven cellular DNA exposed for 5 days, and some of the samples which appeared negative after 20 h of exposure（see left – had side of Fig. 1）were now clearly positive.

As seen in Fig. 2, the results obtained by spot hybridization usually correlated well with those obtained by blot hybridzation. The use of amounts of 0. 1, 0. 5 and 1 μg of cellular DNA enabled us to obtain a gradient of intensity in spot hybridization which proved useful to distinguish between clearly positive and doubtful cases. For 4 samples（cases 14, 22, 23 and 27）we did not have sufficient cellular DNA to do spot hybridization because we repeated blot hybridization 2 or 3 times in order to be sure of the results.

（2）EBNA detection.

EBNA was detected in the exfoliated cells in all 22 NPC cases（Tab. 1）. Positive cells were either isolated or in clumps. Whereas no carcinoma cells were observed cytologically in the 5 cases of hyperplasia and metaplasia, a few lymphocytes in two cases（11 and 14）exhibited EBNA, while in case 27 it appeared as if normal epithelial cells exhibited EBNA.

In this series of specimens, we observed good correlation between the detection of EBNA by the ACIE test and the detection of EBV DNA by spot and blot techniques. Yet in 3 out of the 5 《non –

NPC》 patients where no EBV DNA was detected by blot or spot hybridzation. EBNA was detected in some lymphocytes and possibly in normal epithelial cells.

Discussion

Negative – pressure aspiration of exfolated cells of the nasopharynx provides enough cells to detect both tumorous cells, by cytology, as well as the presence of EBNA and EBV DNA sequences. This simple technique preventing traumatic biopsies, is ideal for epidemiological surveys aimed at screening individuals with anti – virus capsid antigen IgA antibodies who represent the population at highest risk for NPC[19]. The ACIE technique described by Pi et al[12]. is more adapted to the field studies than the ACIE test[13] to detect EBNA in exfolidated nasopharyngeal cells. However, being more sensitive than the ACIE test, the possibility of false positive results needs to be carefully assessed.

The blot hybridization technique is difficult to apply to large numbers of DNA specimens orginating from lield studies, as it is expensive inlime and money. Spot hybridization is much better adapted to explore large numbers of samples, as needed in lield studies. The best approach seems to be to screen samples first by spot hybridzation and then to test the spot – positive specimens by the blot technique. As there are false positive results by spot hybridzation, the second step (blot hybridization) is crtical.

As a probe, we used the cloned internal repeat of B95 – 8 DNA (pSI, 9 a gift from Dr Hayward), as we assumed that this internal repeal was invariably present within the viral genome. We have shown that 0. 1 copy of the repeat per cell could readily be detected (Bornkamm et al. in press) since a tumour cell usually contains about ten copies or more of the viral genome, and since one viral genome usually carries 5 to 12 repeals it should be possible to detect one such EBV – carrying cell in about 500 –

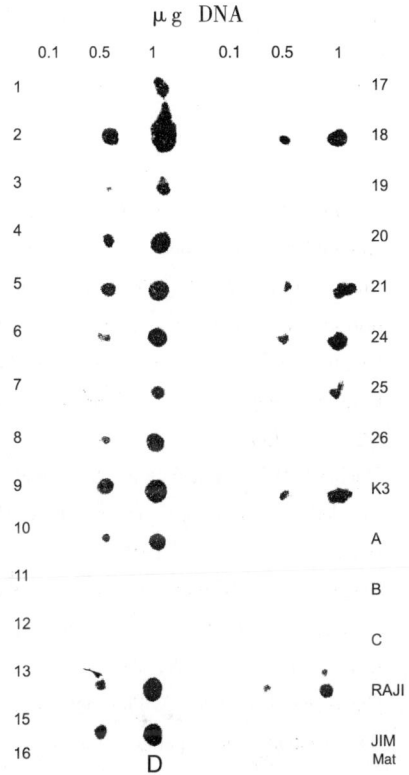

μg DNA

Three different amounts of cellular DNA were spotted onto nitrocclulose (0. 1, 0. 5, 1μg), The filter was exposed for 20 h at – 70℃ with an intensifying screen. Cellular DNA of 1 different tumour tissues and of 3 lymphoblastoid cell lines (LCL) was used as control.

K3 = EBV – positive NPC, A, B and C = EBV – negative head and meck tumours Raji EBA positive LCL. JM and Molt = EBV – negative LCL

Fig. 2 Spot hybridization of the same DNA samples as those in figure 4

2000 cells. Yet by these techniques (spot or blot), it is not possible to distinguish whether the viral genome is in tumorous or normal cells from epithelial or lymphoid origin. As it is difficult to use a short – lived isotope in the lield for epidemiology, this problem might be overcome in the future by labelling probes with precursors copuled covalently with biotin, and the hybrids formed could be identified with avidin or anti – avidin antibodies linked to fluorescein or enzymes.

Acknowledgments

We thank S. D. Hayward for a gift of pSL9.

Dr G H. Pi was supported by the 《Fondation pour la Recherche Medirale》 (Paris) and the 《Foundation Merieux》 (Lyon, France).

This work was financed by 《die Deutsche Krebshilfe》 (Fruerkennung und I) ifferenzierung Epstein – Barr virus – assozierter Tumoran). the CNBS (GIS 122003) and be National Cancer Institute (contact NO1CP – 91035).

[In 《Ann Virol》 1983, 134: 21 – 32]

References

1 Arbameas S, Coupling of enzymes to protein with gluteraldehyde. Immnnochem, 1969, 6: 43 – 52

2 Brandsma J, Miller G. Nucleic acid and spot hybridization: rapid quantitative screening of lymphoid cell lines for Epstein – Barr viral DNA. Proc nat Acad Sci (Wash), 1980, 77: 6851 – 6855

3 Denhardt D T, A membrane filter technique for the detection of complementary DNA. Biochem biophys Res Commun, 1966, 23: 611 – 616

4 Desgranges C, Wolf H, de – The G, Shanmcgaratsam, K, Cammoun, N. Ellouz, R. Klein G. Lennert K. Mcnoz N, Zeb Haesen H. Nasopharyngeal carcinoma. X. Presence of Epstein – Barr genomes in separated epithelial cells of tumors in patients from Singapore. Tunisia and Kenya. Int. J Cancer, 1975, 16: 7 – 15

5 Desgranges C, de – The G, Ho J H C, Ellouz R. Neutralizing EBV – specific IgA in throat washings of nasopharyngeal carcinoma (NPC) patients. Int J Cancer, 1977, 19: 627 – 633

6 Desgranges C, Bornkamm G W, Zeng Y, Wang P C, Zhu J S, Shang M, de – The G. Detection of Epstein – Barr viral DNA internal repeats in the nasopharyngeal mucosa of Chinese with IgA/EB-Vspecific antibodies. Int J Cancer, 1982, 29: 87 – 91

7 Hayward S D, Nogee L, Hayward G. S. Oranization of repeated regions within the Epstein – Barr virus DNA molecule. J Virol, 1980, 33: 507 – 521

8 Henle G, Henle W. Epstein – Barr virus – specitic IgA serum antibodies as an outstanding feature of nasopharyngeal carcinoma. Int J Cancer, 1976, 17: 1 – 7

9 Ho J H C, Ng M H. Kwan H C, Chan J C W, Epstein – Barr virusspecific IgA and IgG serum antibodies in nasopharyngeal carcinoma. Brit J Cancer, 1976, 34: 655 – 660

10 Ho J H C, Kwan H C, Ng M H, de – The G. Sernm IgA antibodies to Epstein – Barr virus capsid antigen preceding symptoms of nasopharvngeal carcinoma. Lancel, 1978, I: 136 – 137

11 Klein G, Giovanella B, C. Lindahi T, Flalrow P J, Singh S, Stehlin J. Direct evidemce for the presence of Epslein – Barr virns DNA and nuclear antigen in malignant epithelial crlls from palienls with anaplaslic carcinoma of the nasopharynx. Proe nat Aced Sci (Wash), 1974, 71: 1737 – 1741

12 Pi G H, Zeng Y, Zhao W P, Zhang Q I X. Developmend of an anticomplement immunoenzyme test for detection of EB virus nuclear antigen (EBNA) and antibody to EBNA. J imunol Methods, 1981, 44: 73 – 76

13 Beedman B M. Klein G. Cellular localization of an Epscin – Barr virus – associated complement – lixing antigen in producer and non – producer lymphoblastoid cell lines. Int J Cancer, 1973, 11: 199 – 520

14 Schneider U, Sghwenk H C, Bornramm G W. Charaeterization of EBV genome negative 《null》 and T – cell lines derived from children with acute lymphoblastoid leukemia and leukemic –

transformed non Hodgkin lymphoma. Int J Cancer, 1977, 19: 621 – 626

15 Southern E M. Detection of specific sequences among DNA Iragments separated by gel electrophoresis J mol Biol, 1975, 98: 503 – 517

16 Wolf H, Zur Hausen H, Beckenr Y. EB viral genomes in epithelial nasopharyngeal carcinoma cells. Nature (Lond), 1978, 244: 245 – 257

17 Zeng Y, Liu Y X, Lin Z B, Zhen S W, Wei J N, Zhe J S, Zai H S. Application of an immunoenzymatic method and an immunoautoradiogaphic method for a mass survey of nasopharyngeal carcinoma. Intervirology, 1980, 13: 162 – 168

18 Zeng Y, Shen S J, Pi G H, Ma J I, Zhang Q, Zho M L, Dong H J. Application of an anticomplement mmunoenzvmatic method for the detection of EBNA in carcinoma cells and normal epithelial cells form the nasopharynx, in: Nasopharvngeal earcinome cancet campaing (Grundmann, et al) vol, 5 (pp: 237 – 245). Guslav Fischer. Stuttgart, 1981

19 Zeng Y, Zhan L G, Li H Y, Jan M G, Zhang Q, Wu Y G, Wang Y S, Su G R. Serological mass survey for early detection of nasopharyngeal carcinoma in Wuzhou city, China. Int J Caancer, 1982, 29: 139 – 144

20 Zhangjiang Medical College. Diagnosis of nasopharyngeal carcinome by cytological exmination of exfoliated cells taken by negative pressure suction. Chin med J, 1976, 1: 45 – 47

21. Detection of Epstein – Barr Virus IgA/EA Antibody for Diagnosis of Nasopharyngeal Carcinoma by immunoautoradiography

ZENG Y[1], GONG C H[1], JAN M G[1], FUN Z[1], ZHANG LG[2], LI H Y[2]

1. Institute of Virology, Chinese Academy of Medical Sciences, Beijing

2. Wuzhou Cancer Research Unit, Wuzhou, Guangxi Autonomous Region, People's Republic of China

Summary

An immunoautoradiographlc method was used for the detection of EB virus IgA/EA antibody in sera from NPC patients and other control groups. Ninety – six percent of NPC patients had IgA/EA antibody with a high titer of GMT. The positive rates of IgA/EA antibody in patients with malignant tumours other than NPC and in normal individuals were only 4% and 0%, respectively. Eleven patients histologically diagnosed as having a chronic inflammation and who showed positive for IgA/EA antibody by immunoautoradiography were rebiopsied; six of them were discovered to have squamous cell carcinoma. Fourteen NPC patients had no IgA/EA antibody detected by immunofluorescence and immunoenzymatic testing, but II and six of them had IgA/VCA and IgA/EA antibodies detected by immunoautoradiography, respectively. These data indicate that the immunoautoradiographic method is more sensitive than either the immunofluorescence or immunoenzymatic test for the detection of IgA/EA antibody, and can be used for the detection of NPC in the early stages of development.

Introduction

The detection of Epstein – Barr (EB) virus IgA/EA antibody cannot used alone for the routine diagnosis of nasopharyngeal carcinoma (NPC), because immunofluorescence testing only reveals this antibody in 50% – 70% of NPC patients (Henle and Henle, 1981; Laboratory of Tumor Viruses of Cancer Institute and Institute of Epidemiology, Chinese Academy of Medical Sciences, 1978). IgA/EA antibody can only be detected in IgA/VCA antibody – positive individuals, but not in antibody – negative individuals. The detection rate of NPC among IgA/VCA antibody – positive individuals is 1.9%, and that among IgA/EA antibody – positive individuals is 30%. The difference between these two groups is 15.8 fold (Zeng et al. 1982b). These data indicate that only a few IgA/VCA antibody – positive individuals eventually become IgA/EA antibody – positive and that IgA/EA antibody is more specific, but not so sensitive for the detection of NPC as IgA/VCA antibody. It is therefore necessary to improve the sensitivity of this test for the detection of IgA/EA antibody.

An immunoautoradiographic method was established in our laboratory using ^{125}I – labelled anti – human IgA antibody for the detection of EB virus IgA/VCA antibody in sera and saliva (Liu et al. 1979a; Liu et al, 1979b). This test is much more sensitive than the immunoenzymatic test, but rather complicated. The immunoenzymatic test is thus widely used as a routine test for the detection of IgA/VCA antibody in China, but immunoautoradiography which is more sensitive may be useful for the detection of IgA/EA antibody.

Material and Methods

1. Sera

Sera were taken from NPC patients, NPC susected patients, patients with tumors other than NPC and normal individuals, and stored at −20℃.

2. Immunoautoradiography

Smears of Raji cells treated with croton oil (500 ng/ml) and n – butyrate (4 mmol/L) were fixed on slide wells with cold acetone. Sear diluted from 1:10 to 1:5120 in four – fold dilutions were added to separate wells. After incubation at 37℃ in a humid atmosphere for 30 min, slides were washed three times with PBS containing 1% calf serum. 0.7 ml of ^{125}I – labelled anti – human IgA antibody in an appropriate dilution was added to each slide. The slides were incubated at 37℃ for 30 min, washed three times with PBS containing 1% calf serum and left to air – dry. They were then coated with nuclear emulsion, slowly dried, and kept in the dark at room temperature for 24 h. Slides were developed in D – 19 for 15 min, placed in the fixed reagent for 10 min, washed and left to air – dry, and then were examined under the light microscope. Serum was considered positive if the cells in the well that contained the diluted serum showed black granules typical of this test.

3. Immunofluorescance test and immunoenzymatic test

IgA/VCA and IgA/EA antibodies were also detected by the immunofluorescence and immunoenzymatic tests as described by Henle and Henle (1966), and Zeng et al (1982a).

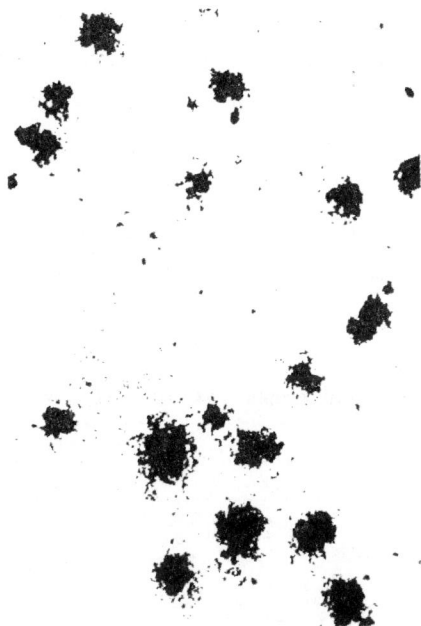

Fig. 1 Immunoautoradiography. EB virus
IgA/EA antibody positive serum (1000 ×)

Fig. 2 Immunoautroadiography. EB virus
IgA/EA antibody negative serum (1000 ×)

Results

1. Immunoautoradiographic method for detection of EBV IgA/EA antibody

As shown in Fig. 1, there were black granules in Raji cells containing early antigen reacted with IgA/EA antibody positive serum, but no such black granules were found in Raji cells reacted with IgA/EA antibody – negative serum (Fig. 2). The titers of IgA/EA antibody detected by the immunoautoradiographic method were higher than those detected by the immunoenzymatic test (Tab. 1).

Detection of EBV IgA/EA antibody from NPC patients and control by different methods

Sera from different groups were detected by immunofluorescence, and immunoenzymatic testing and by the immunoautoradiographic method. The results are shown in Tab. 2. The positive rate and GMT of IgA/EA antibody detected by immunoautroadiography in the sera from

Tab. 1 comparison of the sensitivity of immunoenzymatic testing and the immunoautoradiographic method

Serum from NPC patients	IgA/EA antibody titer	
	Immunoen – zymatic test	Immunoautorad – iography
7	–	1:40
12	–	1:40
78	–	1:10
3	1:10	1:40
5	1:10	1:160
22	1:10	1:160
13	1:20	1:640
37	1:20	1:160
39	1:20	1:160
10	1:80	1:2 560
56	1:320	1:2 560
51	1:640	1:10 240

NPC patients were 96% and 1:97, respectively. Patients with clinically suspected NP and diagnosed histologically as having a chronic inflammation in the nasopharynx had a positive rate 22% and a GMT of 1:9. Seven out of 170 patients with malignant tumours other than NPC were positive (4%) to low titer 1:5.2. All normal individuals were negative. The positive rate and GMT detected by the immunoenzymatic test in these four groups were 80% and 1:25.6, 18% and 1:7.3, 0.6% and 0% and 1:5, respectively. In the clinically suspected NPC group (chronic inflammation) biopsies were taken again from 11 IgA/EA antibody – positive individuals and the histological examination showed poorly differentiated squamous cells carcinoma in six of them.

Tab. 2 comparison between the immunoenzymatic and immunoautoradiographic test for detection of IgA/EA

Sera	Number of cases	IE[1]						IR[2]		
		IgA/VCA			IgA/EA			IgA/EA		
		(+)	(%)	GMT	(+)	(%)	GMT	(+)	(%)	GMT
NPC	56	52	(93)	1:210	45	(80)	1:25.6	54	(96)	1:97
Nasopharyngeal chronic inflammation	50	22	(93)	1:12.5	9	(18)	1:7.3	11	(22)	1:9.0
Tumours other than NPC	170	10	(5.9)	1:5.3	1	(0.6)	1:5	7	(4)	1:5.2
Normal individual	100	3	(3)	1:5	0	(0)	1:5	0	(0)	1:5

Notes:1. Immunoenzymatic test;2. Immunoautoradiographic test

Tab. 3 Immunofluorescence test and immunoautoradiography por detection of IgA/VCA AND IgA/EA antibodies in 14 NPC patients negative for these antibodies by immunoenzymatic test

Serum sample	Clinical stage	IF		IE				IR	
		IgA		IgG		IgA		IgA	
		VCA	EA	VCA	EA	VCA	EA	VCA	EA
5	I	(−)[2]	(−)	1:40	1:10	(−)	(−)	1:10	(−)
7	I	(−)	(−)	1:40	(−)	(−)	(−)	1:640	1:40
13	I	(−)	(−)	1:10	1:20	(−)	(−)	1:640	1:40
15	I	(−)	(−)	1:160	(−)	(−)	(−)	(−)	(−)
3	II	(−)	(−)	1:640	(−)	(−)	(−)	1:40	(−)
12	II	(−)	(−)	1:160	1:10	(−)	(−)	1:640	1:40
11	II	(−)	(−)	1:10	(−)	(−)	(−)	1:2 560	1:40
14	II	(−)	(−)	1:40	(−)	(−)	(−)	1:40	(−)
2	III	(−)	(−)	1:640	1:10	(−)	(−)	1:40	1:40
8	III	(−)	(−)	1:40	(−)	(−)	(−)	1:40	(−)
1		(−)	(−)	1:40	(−)	(−)	(−)	(−)	(−)
4		(−)	(−)	1:40	(−)	(−)	(−)	(−)	(−)
6		(−)	(−)	1:40	(−)	(−)	(−)	1:160	(−)
9		(−)	(−)	1:160	(−)	(−)	(−)	1:2 560	1:40
Positive rate (%)		0/14 (0%)	0/14 (0%)	14/14 (100%)	14/14 (29%)	0/14 (0%)	0/14 (0%)	11/14 (79%)	6/14 (43%)
GMT		1:5	1:5	1:59.4	1:6.4	1:5	1:5	1:65.6	1:24.4

Notes:[1] <1:10 = Negative

Sera from 14 NPC patients diagnosed histologically but showing negative IgA/VCA and IgA/EA antibody (<1:10) by the immunoenzymatic test were reexamined by immunoautoradiography, as well as by immunoenzymatic and immunofluorescence testing. As shown in Tab. 3 all of these 14 sera were still IgA/VCA and IgA/EA antibody − negative by the immunoenzymatic and immunofluorescence test, but 11/14 and 6/14 of them had IgA/VCA and IgA/EA antibodies detected by immunautoradiography, respectively.

Discussion

The EBV IgA/EA antibody is more specific for NPC than the IgA/VCA antibody[3], but its positive rate in NPC ptients is not high enough for it to be used as a diagnosis of NPC. By using the immunoautoradiographic method, satisfactory results were obtained. Ninety − six percent of NPC patients had a IgA/EA antibody with a high titer of GMT. The positive rates of IgA/EA atibody in patients with malignant tumors other than NPC and in normal individuals groups were only 4% and 0%, respectively. Out of 11 patients diagnosed histologically as having a chronic inflammation in the nasopharynx and who had IgA/EA antibody detected by immunoautoradiography, six were finally confirmed as NPC after histological examination of the second biopsies. Fourteen NPC patients had no IgA/VCA and Ig/EA antibodies detected by the immunofluorescence and immunoenzymatic test; however, among these 14 patients, 11 had IgA/VCA and six had IgA/EA antibody detected by immunoautoradiography. These data indicate that immunoautoradiography is more sensitive than either the immunoenzymatic test or the immunofluorescence test in the detection of IgA/EA antibody, and that IgA/EA antibody may serve as a specific marker for the detection of NPC. Therefore, for a serological diagnosis or serological mass survey, it is better to proceed first by using the immunoenzymatic or immunofluorescence test for the detection of IgA/VCA antibody. If this antibody is positive, the detection of IgA/EA antibody in the same serum by immunoautoradiography should then be carried out . This will mean that more NPC patients will be diagnosed in the early stages of their disease.

Acknowledgements

We thank Professor C. H. Huang for his critical appreciation of the manuscript.

[In 《Int J Cancer》 1983, 31: 599 −601]

References

1 Henle G, Henle W. Immunofluorescence in cells derived from Burkitt's lymphoma. J Bacteriol, 1966, 91: 1248 −1256

2 Henle G, Henle W. Epstein − Barr virus − specific IgA serum antibodies as an outstanding feature of nasopharyngeal carcinoma. Int, J Cancer, 1981, 17: 1 −7

3 Laboratory of tumor viruses of Cancer Institute and Institute of Epidemiology, Chinese Academy of Medical Sciences, Detection of EB virus − specific serum IgG and IgA antibodies from patients with nasopharyngeal carcinoma. Act Microbiologica Sinica, 1978, 18: 253 −258

4 Liu Z R, Shan M, Zeng Y. DaI H J Du R S. Application of immunoautoradiography for detection of IgA/VCA antibody in saliva from pa-

tients with NPC. Chinese J Medical Examination, 1979a, 2: 197 – 198

5 Liu Z R, Shan M, Zeng Y, Han Z S, Dai H J, Hu Y L, Cao G R, Dong W P. Immunoautora-diography and its application in the detection of EBV IgA antibody in NPC patients. Kexue Tong-bao, 1979b, 24: 715 – 720

6 Zeng Y, Zhang L G, Li H Y, Tan M C, Zhang Q, Wang Y S, Su G R. Serological mass survey for early detection of nasopharyngeal carcinoma in Wuzhou city, China. Int J Cancer, 1982b, 29: 139 – 141

7 Zeng Y, Liu Y X, Liu Z R, Chen S W, Wei J N, Zhu J S, Zai H J. Application of an immu-noenzymatic method and an immunoautoradio-graphic method for a mass survey of nasopharynge-al carcinoma. Intervirology, 1982a, 13: 162 – 168

22. Brief Communication: Detection of EB Virus IgA/EA Antibody for Diagnosis of Nasopharyngeal Carcinoma by Immunoautoradiography

ZENG Y[1], GONG C H[1], JAN M G[1], ZHANG L G[2], FUN Z[1]

1. Institute of Virology, Chinese Academy of Medical Sciences, Beijing
2. Wuzhou Cancer Institute, Wuzhou, Guangxi Autonomous Region, People's Republic of China

The detection of EB virus IgA/EA antibody cannot be used alone for the routine diagnosis fo na-soppharyngeal carcinoma (NPC), because only 50% – 75% of NPC patients have such antibody detectable by immunofluorescence test. IgA/EA antibody can only be detected in IgA/VCA antibody positive persons, but not in negative persons. The detection rate of NPC among IgA/VCA antibody positive persons in 1.9%, and that among IgA/EA antibody positive persons in 30%. The differ-ence between these two groups is 15.8 fold. These data indicate that only a few IgA/VCA antibody positive persons eventually become IgA/EA antibody positive and that IgA/EA antibody is more spe-cific, but not so sensitive for the detection of NPC as IgA/VCA antibody. Therefore it is necessary to improve the sensitivity of the test.

An immunoautoradiographic method was established in our laboratory using ^{125}I labelled antihu-man IgA antibody for the detection of EB virus IgA/VCA antibody in sera and saliva. This test is rather complicated and too sensitive as compared to the immunoenzymatic test, so the immu-noenzymatic test is widely used as routine test for detection of IgA/VCA antibody in China, but this very sensitive test may also be useful for the detection of IgA/EA antibody.

Sera were taken from NPC patients, NPC suspected patients, patients with tumors other than NPC and normal individuals. The procedure of the immunoautoradiography, which we have de-scribed previously (Zeng et al. 1980) is as follows: Briefly, smears of Raji cells treated with cro-ton oil (500 ng/ml) and n – butyrate (4 mmol/L) were fixed on slided wells with cold acetone. Sera diluted from 1:10 to 1:5 120 in 4 – fold dilution were added to separate wells. After incubation at 37℃ in a humid atmosphere for 30 min, slides were washed 3 times with PBS containing 1% calf serum. 0. 7 ml of ^{125}I – labelled antihuman IgA antibody in appropriate dilution was added to each slide. The slides were incubated at 37℃ for 30 min, washed 3 times with PBS containing 1% calf serum and left to air – dry. They were then coated with nuclear emulsion, slowly dried, and kept in the dark at room temperature for 24 hours. Slides were developed in D – 19 for 15 min, placed in fixing reagent for 10 min, washed and left to air – dry, and then were examined under, the light microscope. Serum was considered positive if the cells in the well that contained the diluted serum showed black granules typical of this test.

The results are shown in Tab. 1. The positive rate and GMT of IgA/EA antibody detected by immunoautoradiography in sera from NPC patients were 96% and 1:97 respectively. Patients with clinically suspected NPC and diagnosed histologically as chronic inflammation in the nasopharynx had a positive rate of 22% and a GMT of 1:9. 0. Seven patients with malignant tumor other than NPC were positive (4%) to low titer 1:5. 2. All normal individuals were negative. The positive rate and GMT detected by immunoenzymatic tests in test 4 groups were 80% and 1:25. 6, 15% and 1:73, 0. 6% and 1:5, and 0% and 1:5 respectively. In the clinically suspected NPC group (chronic inflammation), biopsies were taken again and the histological examination showed poorly differentiated squamous carcinoma cells in 6 out of 11 patients (Tab. 1).

Tab. 1 Comparison of the immunoenzymatic
and immunoautoradiographic test for detection of IgA/EA

Sera	Number of Cases	IE						IR		
		IgA/VCA			IgA/EA			IgA/EA		
		(+)	(%)	GMT	(+)	(%)	GMT	(+)	(%)	GMT
NPC	56	52	(93)	1:210	45	(80)	1:25. 6	54	(96)	1:97
Nasopharyngeal chronic inflammation	50	22	(44)	1:12. 5	9	(18)	1:7. 3	11	(22)	1:9. 0
Tumour other than NPC	170	10	(5. 9)	1:5. 3	1	(0. 6)	1:5	7	(4)	5. 2
Normal individual	100	3	(3)	1:5	0	(0)	1:5	0	(0)	1:5

IE = Immunoenzymatic test; IR = Immunoautoradiographic test

Sera from 14 NPC patients diagnosed histologically but showing negative IgA/VCA and IgA/EA antibody (<1:10) by immunoenzymatic test were reexamined by immunoautoradiography, immunoenzymatic test and immunofluorescence test. All of these 14 sera were still IgA/VCA and IgA/EA antibody negative by immunoenzymatic test and immunofluorescence test, but 11/14 and 6/14 of them had IgA/VCA and IgA/EA antibody detected by immunoautoradiography, respectively.

These data indicate that immunoautoradiography is more sensitive than the immunoenzymatic test for the detection of IgA/EA antibody and that the IgA/EA antibody may serve as a specific marker for the detection of NPC, especially for early stage NPC. Therefore, for serological diagnosis or serological mass survey, it is better to use at first the immunoenzymatic test for IgA/VCA antibody. If this antibody is positive, then the detection of IgA/EA antibody in the same serum should be carried out. With this approach, more NPC patients will be diagnosed in their early stages.

〔In 《Nasopharyngeal Carcinoma: Current Concepts》 1983, 137 – 140〕

References

1   Zeng Y, Liu U X, Liu C R, Chen S W, WEi J N. Application of immunoenzymatic method and immuno – autoradiographic method for the mass survey of nasopharyngeal carinoma. Interviorology, 1980, 13: 162 – 168

23.   Brief Communication: Establishment of an Epithelial Cell Line from Patient with Poorly Differentiated Nasopharyngeal Carcinoma

GU S Y[1], TAN W P[2], Zeng Y[1], ZHAO W P[1], LI K[3], DENG H H[2], ZHAO M L[2]

1. Institute of Virology, Chinese Academy of Medical Sciences, Beijing

2. Zhangjiang Medical College, Zhangjiang

3. Institute of Basic Medical Sciences, Beijing, Chinese Academy of Medical Sciences. Pcople's Republic of China

An epithelial cell line (CNE – 1) and a fusiform (CNF – 1) cell line were established in our laboratory from a patient with nasopharyngeal carcinoma (NPC) which was histologically diagnosed as a well differentiated squamous cell carcinoma[1]. Another cell line from a well differentiated squamous cell carcinoma of the nasopharynx was also established in Hong Kong[2]. There were no EB virus markers in these cell lines.

In order to study the reationship between EBV and NPC, it is very important to establish epithelial cell lines from poorly differentiated squamous cell carcinoma or from undifferentiated carcinoma of nasopharynx which contain EB virus markers (EBNA and DNA). Here we report the establishment of an epithelial cell line (CNE – 2) from poorly differentiated squamous cell carcinoma of a NPC patient.

The biopsy for culture was obtained from a 63 – year old male patient on December 16, 1980. The patient had headache, tinnitus and epistaxis for more than one year and was diagnosed clinically as NPC at stage Ⅲ with a tumor (4 – 5 cm) in the nasopharynx which had invaded the base of the skull. Cytological examination of the exfoliated cells from the nasopharynx showed poorly differentiated squamous cell carcinoma (Fig. 1). The EBNA in exfoliated carcinoma cells was found by anticomplement immunoenzymatic test (Fig. 2). The patient had IgG/VCA (1:320), IgG/EA

(1:160), IgA/VCA (1:40) and EBNA (1:160) antibodies. The IgA/EA antibody was negative (<1:5). All these antibodies were detected by immunoenzymatic test. These serological and viral markers in the patient was inconformity with the diagnosis of poorly differentiated squamous cells carcinoma.

Biopsy was taken from the patient for cultivation. RPMI 1640 with 20% calf serum was used as culture medium. Epithelial cells began to outgrow around the tissue fragments one week later. The fibroblasts were repeatedly eliminated by treating the cell culture with trypsin (0.1%) and versene (0.01%) or scraped out by capillary pipette. The first successful subculture was made 12 weeks after cultivation and the cell culture was then transferred once a week. The epithelial cells were polygonal in shape with nuclei varying in size. This cell line was designated as CNE – 2. EBNA could be detected at the 3rd and the 10th passage by anticomplement immunofluorescence test and anticomplement immunoenzymatic test (Fig. 3). Then the EBNA stain became weaker and finally disappeared, but on occasions a few cells still showed bright stain in the nuclei. No EB virus early antigen (EA) and viral capsid antigen (VCA) could be induced in cells after the 10 th passage by treating the cells with croton oil (250 ng/ml) and n – butyrate (4 mmol/L).

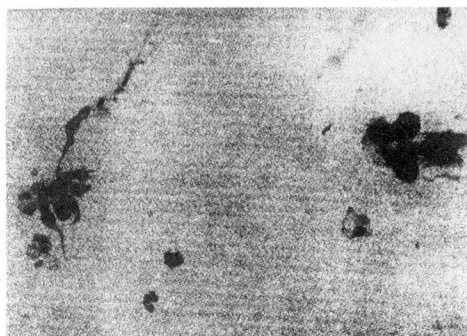

Fig. 1   Cytological examination of exfoliated
cells showed poorly differentiated squamous
cell carcinoma H. E.   (×200)

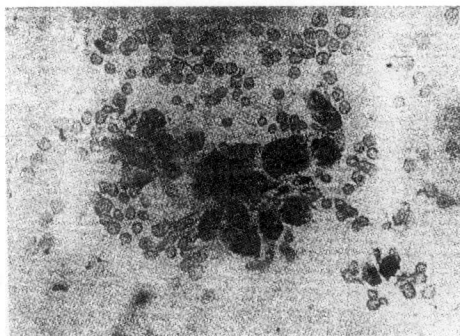

Fig. 2   EBNA in exfoliated carcinoma cells
by anticomplement immunoenzymatic
test   (×200)

The growth curve of CNE – 2 cell line is shown in Fig. 4. After seeding with 1×10^4 cells/ml, the cell numbers at the 2nd, 4th and 6th day was 1.2×10^4, 1.8×10^4 and 3.5×10^4 cells per ml respectively. The plating efficiency of CNE – 2 cell line at the 5th passage was 16%. Cells could be agglutinated by 1 μg of concanavalin – A and the size of the aggregates increased with the increasing concentration of concanavalin – A.

The chromosomes of CNE – 2 cell line showed a wide distribution with a stemline of 74 chromosomes and various types of chromosome aberration.

Under the electron microscope, the CNE – 2 cells were rich in microvilli with large irregular nuclei and 1 to 2 prominent nucleoli. Abundant euchromatin occupied most parts of the nuclei. The amount of heterochromatin was small, some of the small masses were attached beneath the nuclear membrane, and some were scattered in the central parts of the nuclei. There was a large quantity of mitochondria of different sizes and shapes in the cytoplasm. Abundant ribosomes and polyribosomes were scatrered in the cytoplasm, and only scarce rough and smooth surfaced endoplasmic reticulum could be seen. These illustrated that cell line is poorly differentiated and in active proliferation.

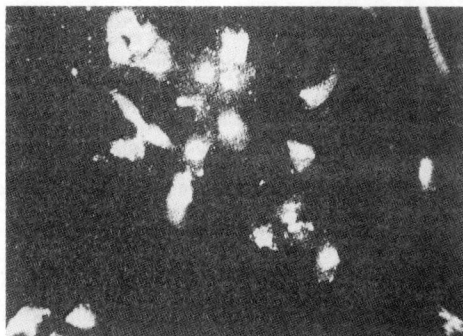

Fig. 3   EBNA in CNE − 2 cells at the
3rd passage by anticomplement
immunofluorescence test (× 200)

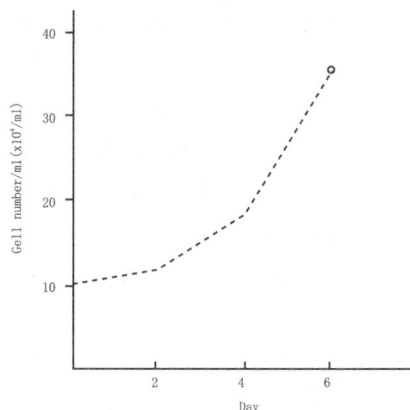

Fig. 4   Growth Curve of CNE 2 cell line

Between the cells which were in close contact with each other, some desmosomes coul be observed. Most of them were small in size and their layers and fine structures were not clear. This also illustrates that the differentiation of this cell line is poor sometimes, however, a few large sized, structurally typical desmosomes were formed between the adjacent plasma membranes. In addition to the desmosomes, a large amount of tonofibrils could be observed in the cytoplasm. Tonofilaments of these tonofibrils continuous with filaments of desmosomes could sometimes be observed. All these clearly demonstrate that this human tumor cell line is derived from epithelial cells.

CNE − 2 cells were transplanted subcutaneously into nude mice. The tumors measured 7 mm in diameter days after tansplantation.

Based on the cell growth pattern, chromosome analysis, hetrotransplantation, and electron microscope, the CNE − 2 cell line has the characteristics of epithelial cells derived from poorly differentiated squamous cell carcinomas.

This cell line has been maintained in vitro for 19 months and might be useful for the study of the relationship between EB virus and NPC.

〔In 《Nasopharyngeal Carcinoma: Current Concepts》 1983, 273 − 276〕

References

1   Laboratory of Tumor Viruses of Cancer Institute, Chinese Academy of Medical Sciences. Establishment of an epitheloid cell line and a fusiform cell line from a patient with nasopharyngeal carcinoma. Scientia Sinica, 1978, 21: 127

2   Huang D P Ho J H C Poon Y F, et al. Establishment of a cell line (NPC/HK 1) from a well differentiated squamous carcinoma of the nasopharynx. Int J Cancer, 1980, 26: 127 − 132.

24. The Existence of Pre – Nasopharyngeal Carcinoma Conditions Should Allow Preventive Interventions

de – The G[1], ZENG Y[2], DESGRANGES C[1], PI G H[1,2]

1. Laboratory of Epidemiology and Immunovirology of Tumors,
Faculty of Medicine Alexis Carrel, Lyon, France
2. Institute of Virology, Chinese Academy of Medical Science,
Beijing, People's Republic of China

Summary

That EBV fingerprints are present in undifferentiated carcinoma of the nasopharynx (NPC) in every geographical area, regardless of NPC incidence level, strongly favors, but does not prove a causal role of EBV in NPC development. How then to establish the nature of such a relationship? Seroepidemiological surveys conducted in China by Zeng and colleagues have shown that IgA/VCA antibodies were most instrumental in eary detection of the tumor, leading to lowering mortality and treatment cost. Furthermore, low levels of IgA/VCA were found in 1% to 5% of the general population aged 30 and above. The follow – up of these IgA/VCA positive individuals indicated that they were at highest risk for NPC, with an estimated annual incidence of approximately 10 per 1000 for males IgA + aged 35 to 45. The presence of IgA/VCA antibodies was accompanied by rising IgG antibodies to VCA and EA and by the presence of EBV/DNA sequences in the nasopharyngeal mucosa, suggesting that a reactivation of EBV latent infection preceded clinical onset. The origin and significance of such reactivation for tumor causation remains to be clarified. Antiviral interventions in pre – NPC conditions should enlighten NPC pathogenesis.

Introduction

Udifferentiated nasopharyngeal carcinoma (NPC) is a leading cause of cancer death for large populations in Southeast Asia, Africa and around the Mediteranean. The ouestion therefore arises as to whether the association with the Epstein – Barr virus (EBV) can be utilized for the control of this disease through earl, diagnosis and ultimately in preventing the disease.

The existence of preneoplastic conditions characterized by increasing IgA/VCA antibody titers and cytological changes in the nasopharyngeal mucosa give an opportunity to evalate the feasibility of conducting intervention trials aimed at reducing or eliminating EBV reactivation and cytologycal abnormalities with the ultimate goal of reducing NPC incidence. The purpose of the present review is not to describe the epidemiology of NPC (see de – The, Ho, Muir, 1976; Hirayama, 1978), but rather to assess the feasibility of implementing preventive and antiviral interventions as a means to try and prevent NPC and to clarify the role of EBV in NPC development.

The situation of NPC and EBV can be compared to that of emdemic primary liver carcinoma (PLC) associated with hepatitis B virus (HBV), where the viral infection long precedes tumor growth and helps to characterize individuals at hightest risk for PLC. The development of HBV vaccines allowed the implementation of field trials aimed at eliminating acute viral hepatitis, HBV carrier state and viral cirrhosis, known as critical risk factors for liver carcinoma developement (Maupas et al. , 1980). But the justification for such HBV vaccine trials was not the eradication of liver tumors a distant consequence of the viral infection, but the prevention of hepatitis, a disease with high morbidity and mortality affecting large proportions of the exposed populations.

The situation of EBV vaccination in relation to NPC will be quite different when an EBV vaccine becomes available (see Epstein, 1983). The ethical problem of EBV vaccination will be quite critical as there is no known deadly disease associated with primary EBV infection worth preventing in the large populations at high risk for NPC. These difficulties should not prevent considering the use of an EBV vaccine when available, but points out the need of evaluating other avenues to try and prevent NPC.

1. EBV/IgA antibodies allow early diagnosis of NPC and represent a critical risk factor for NPC development

Sero – epidemiological surveys of large population groups in the People's Republic of China by Zeng and collaborators have demonstrated the practical value of the IgA/VCA test in early detection of NPC in the highly endemic areas of South China (Zent et al. , 1980, 1982). This was achieved by the development of an immunoenzymatic test (Pi et al. , 1981) more adapted to field conditions than the classic immunofluorescence test as developed by Henle and Henle (1966).

The routine use of this test in certain areas of the Guangxi Autonomous Region of the People's Republic of China allowed substantial improvement of the detection rate of NPC (Zeng et al. , 1980) and to uncover most of the NPC cases at a very early stage of the disease. In the first study (Zeng et al. , 1980), about 50% of the diagnosed NPC were at stages I and II, while in a more recent survey, all cases were discovered at stage I or II (Zeng et al. , 1982).

The impact of early diagnosis of NPC by a simple serological test is in itself a major achievement for better control of the disease, as up to 85% of NPC at stage I can achieve a five years survival disease free (Ho, 1972).

To try and estimate the Incidence Rate (IR) of NPC among IgA positive individuals, two pieces of information were available:

a) out of 99 persons found to have IgA antibodies (in 1977) and reexamined ten months later, two were found to have developed NPC in the meantime (Zeng et al. , 1980).

b) the follow – up of the 3350 individuals found to have IgA/VCA antibodies in the Zangwu county survey led to the uncovering of 33 new cases within 8 – 30 months of follow – up (estimated mean: 20 months) (Zeng et al. , unpublished data). From these data, a yearly rate of 10/1000 persons can be estimated for males aged 40 to 50 years and having IgA/VCA antibodies.

The Relative Risk (RR) to develop NPC for IgA/VCA positive individuals was found to be 50 to 80 times higher than that of the general population. Furthermore, the NPC prevalence among individuals having IgA/VCA antibodies was found to be related to the antibody titers: 1% for individuals with IgA/VCA titers between 20 to 40, 3% for individuals with IgA/VCA titers 80/160 and 9% when IgA/VCA titers reached 320/640 (calculated from Zeng et al. , 1982).

2. Epidemiological characteristics of this IgA/VCA subpopulation group

The first question coming to mind is: How long before clinical onset of NPC do IgA/VCA antibodies develop? From the data of Ho et al. (1978), Lanier et al. (1980) and Zeng et al. (1980, 1982 and unpublished data), it can be estimated that IgA/VCA antibodies can be detected up to five years prior to diagnosis of late NPC cases and up to 15 to 18 months prior to the diagnosis of early NPC cases.

These estimations are based on small numbers and only a prospective population study will enable to establish the mean duration of the IgA/VCA positive period preceeding NPC development.

This point is critical since the clinical onset of the disease may be much delayed from the time of appearance of epithelial tumorous cells in nasopharyngeal mucosa. Subclinical disease may last for years for certain carcinomas.

The second question relates to the proportion of persons entering or leaving the IgA/VCA positive subpopulation group. The IgA/VCA positive individuals as tested by the Immuno – enzymatic test (IE) represent 2.5% to 5.5% for the population aged 35 and above. This proportion was shown to increase with age (Pi et al. , 1981) and the age – specific prevalence curve of IgA antibodies paralleled that of the NPC age – specific incidence. The population – based study, as implemented in Wuchow (Zeng et al. , 1982) should give precise data on the rate of sero – conversion from IgA negative to IgA positive. The type of test used to detect IgA/VCA antibodies will indeed influence directly such a rate. For all practical purpose, an over sensitive test is to be avoided as this increases the proportion of false positive, as compared to a less sensitive test. In this context, it is of interest to note that Ho in Hongkong detected only 1% of the population above 30 years of age as IgA/VCA positive, by immunofluorescence, versus 5% for Zeng by the immunoenzymatic test.

The term "retroversion" is proposed here to describe the loss of detectable IgA/VCA antibodies in an individual. The retroversion rate has been estimated to be around 9% yearly by Zeng in his survey in the Zangwu county (1980 and unpublished data). It is of critical importance to establish firmly such retroversion rates in the study poulation where intervention trials are being contemplated.

In summary, the on – going sero – epidemiological survey of Wuchow city with its yearly follow – up of 1000 IgA positive and 500 IgA negative individuals aged 40 to 50 years should establish firmly the annual rates of sero "conversion" and of "retroversion". The on – going registration of all cancers detected in Wuchow will establish the annual incidence rates of NPC in the general surveyed population and in IgA positive as well as IgA negative groups.

The study of the natural history of the reactivation of EBV latent infection preceeding NPC (as reflected by IgA/VCA antibodies) should also include the comparison between males and females; although the NPC incidence is 2.7 times higher in males than in females, the prevalence rate of IgA/VCA antibodies is similar (5.7% in males versus 4.7% in females).

The detection of IgA antibodies to EA as a risk marker for NPC has not been used up to now be-causeonly 60% to 75% of NPC patients do have IgA/EA antibodies as tested by immunofluoresce-nce. However, it is interesting to note that in the Wuchow survey (Zeng et al. , 1982) , among 680 IgA positive persons, 30 were found to have IgA/EA antibodies and that 9 of these 30 had NPC.

As discussed by Zeng (1983) , the detection of NPC using the IgA/EA test in the IgA/VCA positive individuals should be most useful. An immunoautoradiographical (IARG) test was recently developed in China (Zeng, 1983) , to detect IgA/EA antibodies, Using this test, 96% of NPC patients were found positive but no normal controls. In the Wuchow prospective study, the detection of IgA/EA by the IARG test in the IgA/VCA positive individuals could be implemented and serve as a decision point to examine very carefully those individuals and possibly to biopsy their nasopharyn-geal mucosa.

The evaluation of IgG/EA antibodies as a test to detect NPC candidates has not been epidemio-logically investigated. In the on–going prospective population based study in Wuchow, such a test should be incorporated and evaluated for NPC risk.

3. Preneoplastic cytological changes in the nasophar yngeal mucosa

In the sero–epidemiological surveys referred to above, the IgA/VCA positive individuals were clinically examined and a series of macroscopical abnormalities were noted in their nasopharynx: mostly hyperplasia, metaplasia, but also chronic nasopharyngitis, enlarged adenoids, polyps, etc. In the Zangwu survey of Zeng et al (1980) , 207 individuals with the above described macro-scopic lesions were biopsied: besides 47 NPC, histopathology showed metaplasia in 32% of the non–tumorous biopsies and hyperplasia in 7%. Up to 22% of persons in the latter group noted as hav-ing "chronic inflammation" had IgA/EA antibodies. In the Wuchow survey (Zeng et al, 1982) , biopsies were taken from 38 persons with such abnormalitises and 21 were diagnosed histopathologi-cally as NPC, 13 being at stage 1.

As seen in Tab. 1, there seems to be a progression in the frequency of metaplasia and hyperplasia from IgA/VCA negative persons, IgA/VCA positive individuals to NPC patients. When 17 persons exhibiting severe hyperplasia or meta-plasia in the nasopharynx and 38 individuals with minor hyperplasia or normal nasopharynx were followed up for 8 to 33 months, it was found that 5 of the 17 (28%) metaplasie or hyperplastic cases developed NPC (and were detected either at stage I or II) , whereas ony 1 of the 38 controls was detected at stage 1, 8 months after the first biopsy was taken (Li et al. submitted for publication) .

Tab. 1　Changes observed on nasopharyngeal mucosa of persons showing EBV – VCA IgA antibodies. Data obtained from Li (personal communication)

	NPC patients	IgA positive persons	IgA negative persons
N	75	207	75
Metaplasia	29 (39%)	48 (23%)	11 (15%)
Hyperplasia	8 (11%)	15 (7%)	3 (4%)

Serological and cytological amomalies can increase the risk for NPC from 1 to 60 times, or even up to 1500 times, as compared to that of the general population (Tab. 2). It is of paramount importance that the Wuchow on – going population study verifies these estimations since they were made on relatively small numbers. These data will form the basis on which intervention trals will be designed.

Tab. 2　Subpopulation groups and increasing relative risk (RR) for NPC development

Population groups	Estimated annual incidence of NPC	Increased relative risk (RR)
General population in the Zangwu County	17/100 000	×1
Males aged 40 – 54 years	98/100 000	×57
IgA/VCA positive males aged 40 – 54	1000/100 000	×5.7
IgA/VCA and IgA/EA positive individuals *	up to 30%	×1 700 *
IgA positive persons showing metaplasia or hyperplasia in the NP *	up to 28%	×1500 *

＊ based on small munbers and therefore subject to large error; X = multiplied by

4. Detection of viral markers in nasopharyngeal mucosa

In order to see whether the secretion of EBV/IgA antibodies correspond to an EBV activity in the nasopharyngeal mucosa, 56 asymptomatic individuals having serum IgA/VCA antibodies for 15 to 18 months were carefully examined clinically and biopsied (de – The et al. , 1981). Four early NPCs were discovered and 14 further individuals, for whom no histopathological nor clinical evidence of NPC could be noted, were found to have detectable EBV/DNA sequences and /or EBNA in their nasopharyngeal mucosa (Desgranges et al. , 1982). In the remaining 38 biopsies of IgA positive individuals, no EBV markers were detected. 7 biopsies done on IgA/VCA negative individuals at the same time showed no detectable viral markers in their NP mucosa.

With these results, it seemed that another critical and immediate NPC risk marker could be the presence of EB viral markers in the nasopharyngeal mucosa. As it was not possible to biopsy NP mucosa of normal, IgA/VCA negative individuals, it was difficult to establish the absence of EBV infection in normal NP mucosa. This possibility came about with the use of a negative pressure apparatus allowing to collect exfoliated cells from the nasopharynx. Enough cells could be obtained to make smears for cytological examination and EBNA studies, and to extract DNA for search of EBV/DNA sequences. Exfoliated cells, collected from the nasopharynx of 62 IgA/VCA positive and of 39 IgA/VCA negative individuals in Wuchow city were investigated. DNA was extracted from these cells and tested by spot and then by blot hybridization techniques using the cloned internal repeat of B95 – 8 virus DNA as probe. Unexpected findings were obsereved as shown in table 3. Both IgA positive and IgA negative groups exhibited a similar proportion of EBV/DNA positive specimens (Desgranges et al. unpublished data).

Spot hybridization is known to yield some false positive results and thus all the spot positive specimens were reevaluated by blot hybridization halving the proportion of positive.

In this context, it is to be noted that the individuals positive by spot and negative by blot also had lack of IgG antibodies to EA. Is EBV present in non – tumorous NP epithelial cells? The anticomplement immunoenzymatic test （ACIE） as developed by Pi et al. （1981）, was used to detect EBNA in exfoliated cells. The study of 79 NPC patients exhibited EBNA not only in tumor cells but also in inflammatory or hyperplastic part of their nasopharynx. Furthermore, 4% of patients with ENT disease other than NPC similarly exhibited EBNA in their NP cells （Zeng et al. , in preparation）. Similar observations were obtained by Chen et al. （1982） and by Ho （personal communication）.

Tab. 3 EBV/DNA sequences detected in NP ex – foliated cells of IgA positive or negative individuals

IgA/VCA individuals	N	Spot hybrid	Blot hybrid
Positive	62	13	7
Negative	39	7	4

Notes: （from Desgranges et al. , in preparation）

The above data, which need confirmation by other technics, suggest that EBV might be present in non – tumorous nasopharynx but the extent of such a presence, in acute, chronic or latent EBV infection, the cell type involved and the variations to be observed between geographical areas and ethnic groups, remain to be better determined. The relatively easy access to exfoliated NP cells should promote the implementation of such studies aimed at identifying the risk factor associated with the presence of EBV markers either in coulmnar epithelial cells or in the basal layer or in the submucosal lymphoid cells.

5. Preliminary steps to be implemented before intervention trials can be contemplated

As mentioned above, it is critical to clarify certain characteristics of the natural history of EBV reactivation in relation to risk of NPC before implementing any intervention trial. Within the on – going population study of Wuchow, the following will have to characterized:

—the age/sex specific prevalence of IgA/VCA positive and IgG/EA positive individuals;

—the annual age/sex specific rate of retroversion （from IgA + to IgA – ）;

—the annual age/sex specific rate of conversion （from IgA – to IgA + ）;

—the annual age/sex specific rate of significant increases and decreases in IgA/VCA and IgA/EA antibodies;

—the annual NPC incidence among IgA + and IgA – subpoulation groups.

The registration, initial bleeding, proper follow – up and cancer registration for ideally 40 000 persons 35 years of age and above in Wuchow, should give the necessary demographic and epidemiological base lines. The ongoing serological surveillance of 1000 IgA + and 500 IgA – individuals is minimal to obtain reliable figures concerning the rate of conversion, retroversion and antibody changes. The doubling of these figures would be desirable. In order to develop an intervantion program, the following considerations should be necessary:

a) a better serological and cytological characterization of pre – NPC conditions are needed before contemplating interventions. Serologically, the parameters to involve will be the ADCC titers, anti – EBV/DNAse antibodies and the IgG/EA antibodies of individuals with high or low NPC risk.

b) the characteristic atypic, metaplastic cytological changes will have to be agreed upon by an international group.

c) which individuals will benefit most from infervention?

Epidmiologically, individuals at the borderline of IgA positivity would be eligible for studying the rate of negative connection from IgA + to IgA. But this would be difficult to achieve ethically and technically because of the variations at the laboratory tests. Motivation will be best for individuals with higher IgA/VCA titers and at higher risk for NPC. Tab. 4 indicates a 5% prevalence of NPC in individuals with IgA/VCA titers around 160/320 as tested by the immunoenzymatic test in China and this group should yield a large proportion of cooperative individuals after propet, information on NPC is given.

Tab. 4　Prevalence of NPC by distribution of IgA/VCA titers

Item	Reciprocal IgA/VCA titers by IE tests			
	10 – 20	40 – 80	160 – 320	640
No. sera tested	290	237	109	44
No. NPC detected	0	4	5	4
Prevalence (%)	0	2	5	9

Notes: (from Zeng et al. , 1982)

6. What type of intervention can be contemplated?

This question remains wide open as there is yet no EBV specific proposal to make. However, a few possibilities merit discussion:

a) non – specific anti – promoting agent such as beta – carotene

Peto et al (1981) recently reviewed the epidemiological data suggesting a two fold decrease in carcinoma risk for individuals having a high beta – carotene diet. A large intervention study involving 20 000 American physicians is being implemented by the Harvard School of Medicine (Hennekens, personal communication) to assess the role of beta – carotene intake and blood retinol in prevention of common carcinomas. In the ongoing prospective serological survey in Wuchow, one could contemplate to intervene by distributing beta – carotene or placebo to randomized individuals at high risk for NPC (IgA + with cytological atypia or metaplasia).

b) immunological modifiers, such as interferon and gammaglobulin therapy

Experimentally, interferon was shown to be most effective when the tumor burden was low but to be relatively inefficient on large tumors (Gresser and Tovey, 1978). A similar situation might prevail for human tumors and up to now mostly terminal cancers were treated by interferon with deceiving results.

Alpha interferon treatment (3×10^6 IU daily for 5 weeks, followed by 3×10^6 IU thrice weekly, for 5 weeks) was given to 15 Chinese patients at early stages of NPC (stage II of Ho). As seen in Tab. 5, the results were encouraging but no regression of more than 50% was observed (Wang et al. , in press).

Tab. 5　Clinical effect of low doses of interferon on early NPC

Completed treatment by IF 1. 10^6 IU. daily for 5 weeks thrice a week for five weeks	10 NPC stage II	Cervical lymphnodes
REGRESSED	3	2
UNCHANGED	3	1
PROGRESSED	3	2
REGRESSED FOR 5 WEEKS THEN PROGRESSED	1	0

Notes: (from Wang et al. , in press)

ADCC titers of NPC patients prior to radiotherapy appeared to be of high prognostic value, according to Pearson et al. (1978) and Mathew et al. (1981). These authorshave initiated at the Mayo Clinic serotherapy of terminal NPC cases, using semipurified gamma – globulins having high ADCC activity. They observed clinical and serological effects of such as serotherapy on IgG/EA and IgA/VCA titers (Pearson, personal communication). If this is confirmed, a therapeutic trial of gamma – globulin therapy, possibly together with IF therapy, might be of value in association with radiotherapy. The source of immunoglobulins remains an open question, as one must have standardized batches of sera with high ADCC activity. Premilinary investigations have shown that gamma – globulin from blood donors was a better source for high ADCC activity than placental extracts.

c) Antiviral drugs

Very few compounds have yet shown major clinical effect on herpesviruses in general and on EBV in particular. It is hoped however that both pharmaceutical and public research laboratories will progressively come up with clinically useful compounds and it would be important to be epidemiologically ready to test them when needed. The aim would be to try and inhibit EBV reactivation and to prevent NPC clinical onset.

It is too early to judge whether Acyclovir^R might be useful in this context, since its effectiveness on EBV related diseases, such as severe in fectious mononucleosis (IM) has not yet been completed in vivo. A clinical trial is actually in progress in the USA and once the results are available, consideration will be given in the present context.

The preliminary studies described above should employ small groups of patients (for example 15 to 20 males aged 40 to 50 years with IgA/VCA titers ranging from 80 to 320) to test the feasibility and possible effects of such interventions.

7. Population in terventions

To test the validity of the preliminary studies, to select the best drug combination and to assess the feasibility of a major population intervention, several hundred persons with appropriate premalignant condition will have to be identifed with relatively homogenous risk for the disease.

It appears that randomized intervention groups (40 to 50 persons each) would be appropriate with a 100 person placebo group, to detect an effect on EBV (significant change in IgA titers i. e. 2 dilutions) and no exfoliated NP cytology, but not on NPC incidence. The number of group depends indeed on the results of the preliminary studies and on the decision to combine certain drugs.

If the pilot study were to be successful, an intervention at a population level might be decided with the amin to reduce the incidence of NPC. The study design will be very similar to the pilot study involving randomization of participants into intervention or placebo group. A vigorous follow – up for 3 to 5 years of all individuals concerned would be required.

Acknowledgments

The help of Dr. N. Gutensohn, Harvard School of Public Health, Boston, for reanalyzing and assessing the data from China, was greatly appreciated during Dr. G. de The's visiting professorship at the Harvard School of Public Health.

〔In 《Nasopharyngeal Carcinoma: Current Concepts》 1983, 365 – 374〕

References

1 Chen S K, Tan S W, Liu B, Lu T S, Li W J, Zhong F, Lo H L, Li S W, Li M J, Siao J D. Detection of EBNA antibody and localization of EBNA from patients with NPC. Cancer, 1982, 1: 40 – 42 (in Chinese)

2 Desgranges C Bornkamm G W, Zeng Y, Wang P C, Zhu J S, Shang M, de – The G. Detection of Epstein – Barr viral DNA internal repeats in the

nasopharyngeal mucosa of Chinese with IgA/EBV – specific antibodies Int J Cancer, 1982, 29: 87 – 91

3　de – The G, Desgranges C, Zeng Y, Wang P C, Bornkamm G W, Zhu J S, Shang M. Search for pre – cancerous lesions and EBV markers in the nasophrynx of IgA positive individuals. In: Nasopharyngeal Carcinoma: Cancer Campaign, vol. 5, Grundmann, E; Krueger G. R. F. and Ablashi (eds), Stutgart, New York, 1981, 111 – 117

4　de – The G, Ho J H C, Muir C S (1976): Nasopharyngeal carcinoma. In: Epidemiology and Control, A. S. Evans (ed) Plenum Press, New York, pp: 539 – 563 2 nd edition, 1981, 621 – 652

5　Epstein M A, North J R, Morgan A J. Possibilities for antiviral vaccine intervention in nasopharyngeal carcinoma. 1983, 375 – 386

6　Gresser I, Tovey G: Antitumor effects of inierferon Bioch Bioph Acta, 1978, 516: 231 – 247

7　Henle G, Henle W. Immunofluorescence in cells derived from Burkitt´s lymphoma. J Bacterial, 1966, 91: 1248 – 1256

8　Hirayama T. Descriptive and analytical epidemiology of nasopharyngeal cancer. In: Nasopharyngeal Carcinoma: Etiology and Control, de The, G, and Ito, Y. (eds) IARC, Lyon, 1978, 167 – 189

9　Ho J H C. Nasopharyngeal carcinoma (NPC). In: Advances in Cancer Resarch, vol, 15, Klein. G. and – Weinhouse, S. (eds), Academic Press, New, York, 1972, 57 – 92

10　Ho J H C, Kwan H C, Ng M H, de – The G. Serum IgA antibodies to EBV capsied antigens preceeding symptoms of nasopharyngeal carcinoma. Lancet, 1978, 1: 436 – 437

11　Lanier A P, Henle W, Bender T R, Henle G, Talbot M L. Epstein – Barr virusspecific antibody titers in seven Alaskan natives before and after diagnosis of nasopharyngeal carcinoma. In J Cancer, 1980, 26: 133 – 137

12　Maupas P, Goudeau A, Coursaget P, Chiron J P, Drucker, Barin F, Perrin J, Denis. F, Diop Mar I, Summers J. Hepatitis B virus infection and primary hepato cellular carcinoma: Epidemiological, clinical and virological studies in Senegal from the perspective of prevention by active immunization. In: Cold Spring Harbor Conferences on Cell Proliferation, 1980, 7: 481 – 506

13　Mathew G D Qualtiere L F, Bryan Neel Ⅲ H, Pearson G R. IgA antibody, antibody – dependent cellular cytotoxicity and prognosis in patients with nasopharyngeal carcinoma Int J Cancer, 1981, 27: 175 – 180

14　Pearson, G R. Johansson, B. and Klein, G(1978): Antibody – dependent cellular cytotoxicity against Epstein – Barr virus – addociated antigens in African patients with nasopharyngeal carcinoma Int. J. Cancer, 1981, 22: 120 – 125

15　Peto R, Doll R, Buckley J D, Sporn M B. Can dietary beta – carotene materially reduce human cancer rates? Nature, 1981, 290 (March 19)

16　Pi G H Zeng Y, Zhao W P, Zhang Q. Development of an anticomplement immunoenzyme test for detection of EBNA and antibody to EBNA J Immunol Method, 1981, 44: 73 – 78

17　Wang P Z, Zeng Y, DNA H, Ma E R, Li E J, Gresser I Cantell K, de – The G. Progress report of alpha interferon therapeutic trial of early cases of nasopharyngeal carcinoma in south China Interferon Scientific Memoranda (USA)

18　Zeng Y, Liu U X, Liu C R. Chen S W, Wei J N, Zhu J S, Zai H J. Application of immunoenzymatic method and immuno – autoradiographic method for the mass survey of nasopharyngeal carcinoma. Intervirology, 1980, 13: 162 – 168

19　Zeng Y, Gong C H, Jan M G, Zhang L G. Fun Z. Detection of EB virus IgA/EA antibody for diagnosis of nasopharyngeal carcinoma by immuno-autorediography, 1983, 137: 138

20　Zeng Y, Zhang L G, Li H Y, Jan M G, Zhang Q, Wu Y C, Wang Y S, Su G R. Serological mass survey for early detection of nasopharyngeal carcinoma in Wuzhou city, China. Int J Cancer, 1982, 29: 139 – 141

25. Presence of EBV – DNA Sequences in Nasopharyngeal Cells of Individuals without IgA – VCA Antibodies

DESGRANGES C[1], PI G H[2], BORNKAMN G W[3], LEGRAND G W[1], ZENG Y[2], de – The G[1]

1. Laboratory of Epidemiology and Immunovirology of Tumors, Faculty of Medicine Alexis Carrel, Lyon, France   2. Institute of Virology, Chinese Academy of Medical Sciences, Beijing, People's Republic of China   3. Zentrum fur Hygiene, Freiburg in Breisgau, Fed Rep. Germany   4. To whom reprint requests should be sent

Summary

Exfoliated nasopharyngeal (NP) cells from 62 normal Cantonese Chinese having IgA/VCA antibodies for more than a year and from 39 similar persons without IgA/VCA antibodies, were tested for the presence of EBV/DNA sequences by spot followed by blot hybridization tests, using the cloned internal repate of B95 – 8 viral DNA as probe. Thirteen out of 62 specimens from IgA/VCA – positive (21%) and six out of 39 specimens (15.4%) from IgA/VCA – negative individuals were found to contain EBV/DNA sequences. Forty – six cases (20 IgA/VCA – positive and 26 IgA/VCA – negative) were followed a year later for EBV/DNA sequences and EBV serology. Half of the individuals having EBV/DNA sequences in their exfoliated NP cells in 1981 did not have detectable EBV sequences a year later, and to out of 15 negative individuals became EBV/DNA – positive. There was no obvious correlation between EBV/DNA detectability and EBV serology. (We conclude that the best marker for NPC risk remains the increasing IgA/VCA and/or EA antibody titers).

Introduction

EBV fingerprints are present in undifferentiated carcinoma of the nasopharynx (NPC) in every geographical area, regardless of the NPC incidence level (zur Hausen et al. , 1970; Desgranges et al. , 1975a, b). Sero – epidemiological surveys in the People's Republic of China by Zeng et al. (1980, 1982) have clearly demonstrated the practical value of the presence of IgA antibodies directed to viral structural antigens (VCA) for early diagnosis of NPC in the highly endemic areas of South China. The detection of IgA/VCA – positive individuals allows most of the NPC cases to be diagnosed at a very early stage of the disease (70% in stage I in the latest survey of Zeng et al. , 1982), increasing the cure rate by cobaltotherapy.

In order see whether the presence of serum EBV/IgA antibodies corresponded to an EBV activity in the nasopharyngeal mucosa, 56 symptomless individuals having serum IgA/VCA antibodies for 15 to 18 months were clinically examined and biopsied (de – The et al. , 1981). Four early NPC were discovered and 14 further individuals, for whom no histopathological or clinical evidence of

NPC was noted, were found to have detectable EBV/DNA sequences and/or EBNA in their naso-pharyngeal mucosa (Desgranges et al. , 1982). It seemed therefore that the presence of EB viral markers in the nasopharyngeal mucosa could be another critical marker for an immediate risk of NPC.

As it was not possible to biopsy NP mucosa of normal and IgA/VCA – negative individuals. we did not know whether the normal mucosa of these subjects did or did not contain EBV markers. Using a negative pressure apparatus to collect exfoliated nasopharyngeal cells, we compared the presence of EBV markers in exfoliated NP cells from 62 IgA/VCA – positive and 39 IgA – negative individuals from the town of Wuchow, a high – incidence area for NPC in the Guangxi Autonomous Region of the People's Republic of China.

Fifty of these 101 individuals were investigated after 1 year for EBV serology as well as for EBV markers in their nasopharynx.

Material and Methods

1. Selection of individuals, collection of sera and nasopharyngeal samples

Sera from 62 individuals with IgA/VCA antibodies and from 39 individuals without IgA/VCA antibodies for a year, were selected from the on – going seroepidemiological mass survey in Wu-chow, aimed at early NPC detection.

IgG and IgA antibodies to VCA and EA in sera were titrated both by the immunoperoxidase test described by Liu and Zeng (1979) in China and in Lyon by the immunofluorescence test according to Henle and Henle (1966) and Henle et al. (1970).

Nasopharyngeal cells from individuals without apparent abnormalities in their nasopharynx were collected with a simple negative pressure suction appartus, as already described by Zeng et al. (1980) and by the Zhanginag Medical College (1976). The cells were collected on a silk net and an aliquot was examined cytologically after Giemsa staining, while the rest was used to ex-tract cellular DNA. The number of cells was sufficient to obtain 50 to 400 μg of DNA (Pi et al. , 1983).

2. Detection of EBV/DNA sequences in NP cells

DNA extiraction was carried out also already described (Desgranges et al. , 1982). In a first step, detection of EBV/DNA sequences was done by "spot gyridization". This technique, introduced by Brandsma and Miller (1980) is an adaptation to eukaryotic cells of the bacterial colony hybridization test of Grunstein and Hogness (1975). As it was difficult to dissociate and evaluate cells from the aspirates, instead of putting a known suspension of cells in dots onto ni-trocelulose, we spotted quantified aliquots of extracted cellular DNA (1.0, 0.5 and 1 μg) of each sample.

Only the positive and doubtful positive specimens were then reanalyzed by "blot hybridization" as described by Bomkamm et al. (1980, 1983) and Desgranges et al. (1982). As a probe, we used the internal repeat of B95 – 8 virus DNA cloned in pBR322 and labelled by nick – translation with 32p dCTP and dGTP, as described by Maniatis et al (1975). Hybridization was carried out for 4 days at 45℃ for both the spot and blot hybridizations. as described by Pi et al（1983）.

Results

Tab. 1. gives the list and the EBV serological profile of individuals（with or without IgA/VCA antibodies）from whom viral DNA sequences were detected in the corresponding exfoliated cells. Thirteen IgA/VCA－positive individuals out of the 62 tested were positive for viral DNA（21%）while 6 out of the 39 IgA/VCA－negative were also positive for EBV/DNA（15.4%）（Tab. 2）. The GMT of the various serological reactivities of both IgA－positive and IgA－negative groups are given in Tab. 2. There was no apparent relationship between the EBV serology and the presence of EBV markers in the NP mucosa. Strangely enough, as seen in Tab. 2, GMT of IgA－positive of negative subgroups IgA/VCA were higher in the individuals with no viral DNA in their mucosa than GMT of IgA/VCA in individuals with detectable viral DNA. These differences were at the border of significance.

All the IgA/VCA－negative samples, positive by spot hybridization, were tested again by blot hybridization（to ensure the specificity）and were found positive by this technique. Fifty individuals（23 IgA/VCA－positive and 27 IgA/VCA－negative）were analyzed a year later（1982）. Among the 23 IgA/VCA－positive, 6 were EBV/DNA－positive（26%）, while only 2 of the 27 IgA/VCA negative individuals were found to be EBV/DNA－positive（7.4%）.

Among these 50 individuals, 46 were tested 2 years consecutively（26 IgA/VCA－negative, 20 IgA/VCA－positive as observed in 1981）. Tab. 3 indicates the changes observed in the detectability of viral DNA sequences among IgA/VCA－positive or－negative individuals during the 2－year period. Among the viral DNA－positive, more than half（3/5 IgA/VCA－positive and 1/2 IgA/VCA－negative）become viral DNA－negative. Among the 15 viral DNA negative－IgA/VCA－positive individuals, 2 became viral DNA－positive, but none of the 24 viral DNA－negative－IgA/VCA－negative individuals did so. Tab. 4 gives the changes observed in the detectability of IgA/VCA antibodies in these two groups. None of the individuals who modified their viral DNA content in exfoliated NP cells exhibited IgA/VCA antibody modifications.

Tab. 1 EBV serology of individuals with EBV DNA－positive nasopharyngeal exfoliated cells

| Case number | EBV antibodies | | | | |
| | IgG | | IgA | | EBNA |
	VCA	EA	VCA	EA	
WA 11	640	40	10	<5	160
WA 24	320	80	80	20	80
WA 31	1 280	160	320	160	320
WA 44	640	40	80	<5	640
WA 57	640	<5	40	<5	320
WA 59	320	<5	10	<5	80
WB 1	320	<5	10	<5	320
WB2	160	20	10	<5	80
WB 25	160	<5	10	<5	80
WB 28	640	40	20	<5	320
WB 34	160	20	40	<5	160
WB 58	160	<5	10	<5	40
WB 59	80	<5	10	<5	40
WA 3	320	<5	<5	<5	320
WB 17	80	<5	<5	<5	160
WB 19	80	<5	<5	<5	80
WB 26	160	40	≤5	<5	160
WB 33	160	≤5	<5	<5	80
WB 43	320	10	≤5	<5	160

Tab. 2　Percentage and geometric mean titers of EBV antibodies of the various groups

		IgG VCA		IgG/EA		IgA/VCA		IgA/EA		EBNA	
		% +	GMT	% +	GMT	% +	GMT	% +	GMT	% +	GMT
IgA/VCA – positive 62/101	DNA – positive 13/62 （21%）	100	304	59	17	100	21	14	<5	100	136
	DNA – negative 49/62	100	413. 5	61	16. 8	100	25. 3	15	<5	100	120. 3
	Total 62	100	385	60	16. 9	100	24	14. 5	<5	100	124
IgA/VCA – negative 39 – 101	DNA – positive 6/39 （15. 4%）	100	139. 3	33	2. 7	0	<5	0	<5	100	142. 6
	DNA – negative 33/39	100	261. 7	18	7. 3	0	<5	0	<2. 5	100	139. 3
	Total 39	100	260	20. 5	7	0	<5	0	<2. 5	100	140

Tab. 3　EBV DNA changes after one year

1981 IgA/VCA	1981 Viral DNA	1981 No. of cases	1982 Viral DNA	1982 No. of cases
20 indiv. IgA/VCA – positive	+	5	+	2
			–	3
	–	15	–	3
			–	13
26 indiv. IgA/VCA – negative	+	2	+	1
			–	1
	–	24	+	0
			–	24

Tab. 4　IgA antibody change after one year

	1981 No. of cases	1982	No. of cases %
IgA/VCA – positive	20	+	20	100
		–	0	0
IgA/VCA – negative	20	+	6	23
		–	20	77

Discussion

It was unexpected to observe that the groups of individuals with or without IgA/VCA antibodies both had a number of cases in which EBV DNA sequences could be detected （21% versus 15. 4% in 1981; 26% versus 7. 4% in 1982）. These results suggest that these two markers （presence of EBV/DNA – sequences and IgA anti – VCA antibodies in the serum） are not directly related. We do not yet know if the individuals with detectable EBV DNA – sequences in the nasopharynx have an increased risk of developing NPC, or if this merely reflects a latent stage of EBV in the nasopharynx. To answer this question, the IgA/VCA – negative – EBV DNA – positive individuals will be followed for both the serological changes and the clinical NP abnormalities. These data, however, do not alter the value of IgA VCA for the detection of NPC at an early stage and of individuals at immediate risk of developing the disease （Zeng, 1980, 1982）.

From an epidemiological viewpoint, it will be important to assess the rate of sero – conversion to IgA VCA positivity as well as that of retroversion （from IgA/VCA positivity to negativity） in the Cantonese population. The present study shows that the IgA positivity represents a stable situation after an interval of 1 year. The variations observed in the detectability of viral DNA after a year might reflect a fluctuation in the EBV activity of the NP mucosa or a technical difficulty in detecting viral DNA when only a few copies are present in few cells.

Concerning NPC pathogenesis, these results cannot be properly evaluated until the detected EBV/DNA sequences are localized either in lymphoid or in epithelial cells, intimately mixed in the

NP mucosa. Such a study in progress and preliminary results indicate that viral DNA is present in epithelial cells (H. Wolf, personal communication). Pearson (1980), Qualtiere et al. (1982). showed that ADCC activity appears to be the best prognostic marker in NPC patients. Individuals with pre – NPC conditions (characterized by IgA/VCA antibodies, hyperplasia or metaplasia detected by cytology, with or without detectable EBV/DNA, de – The et al. , 1982). as well as the individuals with detectable EBV/DNA but without IgA/VCA antibodies, will be tested for ADCC activity. Their clinical follow – up will answer the question of the value of ADCC for charcterizing the individuals at immediate risk for NPC among the above described pre – NPC conditions. The ADCC might have a protective effect by eliminating EBV – infected malignant or pre – malignant epithelial cells.

Acknowledgements

The work reported here was made possible through the courtesy and help of the Medical Authorities of the Guang – Xi Autonomous Region of the People's Republic of China. This study was financially supported by: Institute of Virology, Chinese Academy of Medical Sciences. Beijing, People's Republic of China; Deutsche Forschungsgemeinschaft (SFB 31), Germany; Centre National de la Recherche Scientifiqus (GIS 410017), France.

〔In 《Int J Cancer》 1983, 32: 543 – 545〕

References

1　Bornkamm G W, Delius H, Zimber U, Hudewentz J, Epstein M A. Comparison of Epstein – Barr virus strains of different origin by analysis of the viral DNAs. Virol, 1980, 35: 603 – 618

2　Bornkamm G W Desgranges C, Gissmann L. Nucleic acid hybridization for the detection of viral genomes. In P A. Bachmann (ed). Current topics in microbiology and immunology, pp. 287 – 298, Springer – Verlag, Berlin, Heidelberg, New York, 1983

3　Branosma J, Hiller G. Nucleic acid and spot hybridization: rapid quantitative screening of lymphoid cell lines for Epstein – Barr viral DNA. Proc nat Acad Sci (Wash), 1980, 77: 6851 – 6855

4　Desgranges C Bornkamm G W, Zeng Y, Wang P C, Zhu J S, Shang M, de The G. Detection of Epstein – Barr viral DNA internal repeats in the nasopharyngeal mucosa of Chinese with IgA/EVA – specific antibodies. Int J Cancer, 1982, 87 – 91

5　Desgranges C, Wolf H, de – The G. Shanmugaratnamk, Ellouz R, Cammoun N, Klein G, Zur

Hausen H. Nasopharyngeal carcinoma. X. Presence of Epstein – Barr genomes in epithelial cells of tumors from high – and medium – risk areas. Int J Cancer, 1975a, 16: 7 – 15

6　Desgranges C, Wolf H, Zur Hausen H, de The G. Further studies on the detection of the Epstein – Barr viral DNA on nasopharngeal carcinoma biopsies from different parts of the world. In G. de – The M. A. Epstein and H. zur Hausen (ed). Oncogenesis and herpesviruses Ⅱ. Vol. 2, pp. 191 – 193. IARC Scientific Publiction No. 1975b, 11, Lyon

7　de – The G, Desgranges C. , Zeng Y, Wang P C, Bornkamm. G W, Zhu J S, Shang M. Search for pre – cancerous lesions and EBV markers in the nasopharynx of IgA positive individuals. In: Grundmann et al (ed). Cancer campaign. Vol. 5, Nasopharyngeal carcinoma. pp. 111 – 117, Gustav Fischer Verlag Stuttgart. New York, 1981

8　de – The G, Zeng Y, Desgranges C, P$_i$ G H. The existence of pre – nasopharyngeal carcinoma conditions should allow preventive interventions. In M J. Simons and K. Shanmugaratnam (ed). The Biology and Nasopharyngeal Carcinoma Report No. 16. Vice technical reports. Series – Volume, 1982, 71, Geneva

9　Grunstein M, Hogness D. Colony hybridization. A method for the isolation of cloned NDAs that contain a specific gene. Proc nat Acad Sci (Wash), 1975, 72, 3961

10　Henle G, Henle W. Immunofluorescence in cells derived from Burkitt´s lymphoma. J Bact, 1966, 91: 1248 – 1256

11　Henle W, HEnle G, Zajac B A, Pearson G. Waubke R, Scriba M. Differential reactivity of human serums with early antigens induced by Epstein – Barr virus. Science, 1970, 169 – 188 – 190

12　Liu Y X, Zeng Y. Detection of Epstein – Barr virus IgA antibody from patients with nasopharyngeal carcinoma by immunoenzymatic method. Chin J Oncol, 1979, 1: 8

13　Maniatis T, Jeffrey A, Kleid D G. Nucleotide sequence of the rightward operator of phage. Proc nat Acad Sci (Wash), 1975, 72: 1184 – 1188

14　Pearson G R. Epstein – Barr rivus immunolgy. In G. Klein (cd). Viral oncology p. 739, Raven Press. New York, 1980

15　Pi G H, Desgranges C, Bornkamm G W, Shen S J, Zeng Y, de – The G. Comparative evalua- tion of various techniques to detect EBV DNA in exfoliated nasopharyngeal cells. Ann Vir Inst Pasteur, 1983, 134E: 21 – 32

16　Qualtiers L F, Chase R, Pearson G R. Purification and biologic characterization of a major EBV induced membrane glycoprotein J Immunol, 1982, 129: 814 – 818

17　Zeng Y, Liu U X, Liu C R, Chen S W, Wei J N, Zhc J S, Zai H J. Application of immunoenzymatic method and immuno – autoradiographic method for the mass survey of nasopharyngeal carcinoma. Intervirology, 1980, 13: 162 – 168

18　Zeng Y, Zhang L G, Li H Y, Jan M G. Zhang Q, Wc Y C, Wang Y S, Su G R. Serological mass survey for early detection of nasopharyngeal carcinoma in Wuzhou city. China. Int J Cancer, 1982, 29: 139 – 141

19　Zhangjiang Medical College. Diagnosis of nasopharyngeal carcinoma by cytological examination of exfoliated cells taken by negative pressure suction Chin Med J, 1976, 1: 45 – 47

20　Zur Hausen H, Schulte – Holthausen H, Klein G, Henle W, Henle G, Clifford P, Santesson L. EBV – DNA in biopsies of Burkitt's tumors and anaplastic carcinomas of the nasopharynx. Natur (Lond), 1970, 228: 1056 – 1058

26. Early Nasopharyngeal Carcinoma Among IgA/VCA Antibody Positive Individuals Detected by Anticomplement Immunoenzymatic Method

ZENG Yi[1], SHEN Shu-jing[2], DENG Hong[3], MA Jiao-lian[2],

ZHANG Qin[1], ZHU Ji-song[3], CHENG Ji-ru[3]

1. Institute of Virology, Chinese Academy of Medical Sciences, Beijing 2. Zhanjiang Medical College, Guangedong province 3. Cancer Control Office, Cangwu county, Guangxi Autonmous Region

Summary

The anticomplement immunoenzymatic method (ACIE) is used to detect nasopharyngeal carcinoma in EBV IgA/VCA antibody positive individuals in a high risk area. Carcinoma cells with EBNA were found in 4 of the 64 antibody positive individuals. Cytological and histological examinations showed that stage I poorly differentiated carcinomas were in 4 cases. The intervals between the first bleeding for serology and nasopharyngeal carcinoma (NPC) diagnosis were 8 – 9 months. No further evelation of IgA/VCA antibody appeared during this period.

The result of this study confirm the value of anticomplement immunoenzyme for detecting early stage NPC.

Introduction

Serological mass survey was carried out during 1978 – 1980[1-3] in Cangwu county, Guangxi Autonomous Region. Early stage nasopharyngeal carcinoma (NPC) patients can bedetected in IgA/VCA antibody positive individuals, but a large number of antibody positive individuals have no evidence of NPC. In order to detect more NPC patients in the early stage, we have perfected a new method using anti – complement immunoenzyme (ACIE) to detect EB virus nuclear antigen (EBNA) in carcinoma cells[4]. Since it is difficult to take biopsies in the early stage, exfoliated cells of the nasopharyngeal mucosaare are collected by negative pressure suction apparatus[5] and examined by ACIE for EBNA and cytologically for carcinoma cells. Satisfactory results were obtained in the outpatient clinic[6]. This paper reports use of the ACIE method for detecting early stage NPC from IgA/VCA antibody positive individuals in a high risk area of NPC.

Material and Methods

1. IgA/VCA antibody positive individuals: 64 individuals, found to have IgA/VCA antibodies in 1978 an 1979, but lacking clinical signs of NPC 4 months previously, were included in the study. The IgA/VCA antibody was detected by the immunoenzymatic method described by Liu et al.

2. Cell smear: 1% dicaine was sprayed into the oropharyngeal and nasopharyngeal cavities. Exfoliated cells were collected by negative suction apparatus.

Smears were prepared from the silk ball inside the head of the suction apparatus, fixed in cold acetone 10 minutes, examined by ACIE method and HE staining for cytological examination.

3. Histological examination: When positive results were shown by the ACIE method or cytological examination, biopsies were taken to confirm the diagnosis.

4. ACIE test: Reference NPC serum containing EBNA antibody and normal human serum with complement at a final dilution of 1 : 10 were added to the smears and the slides placed in a humidified chamber at 37℃ for 1 hour. After washing with balanced salt solution (BSS) 3 times, anti – C_3 antibody conjugated with horseradish peroxidase at 1 : 10 was added and the slides kept at 37℃ for 30 minutes. The smears were again washed with BSS 3 times, stained with diaminobenzene and H_2O_2 and examined.

Results

As shown in Tab. 1, carcinoma cells with EBNA were found in 4 of the 64 IgA/VCA antibody positive cases. All the 4 cases were also diagnosed cytologically and histologically as having poorly differentiated nasopharyngeal carcinoma (Figs. 1 – 3).

Tab. 1　NPC patients detected by ACIE method

Case No.	Sex	Age	Clinical examination		Carcinoma cells with EBNA	Cytological examination	Histological examination	IgA/VCA antibody	
			Subjective symtoms	Stage				First bleeding	Diagnosis confirmed
10	M	45	–	I	+	+	+	80	80
12	M	53	–	I	+	+	+	80	20
43	F	38	–	I	+	+	+	80	80
58	F	47	–	I	+	+	+	40	80

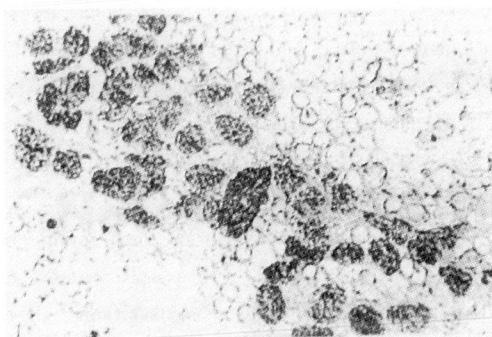

Fig. 1　EBNA positive cells detected by ACIE method (No. 43). ×200

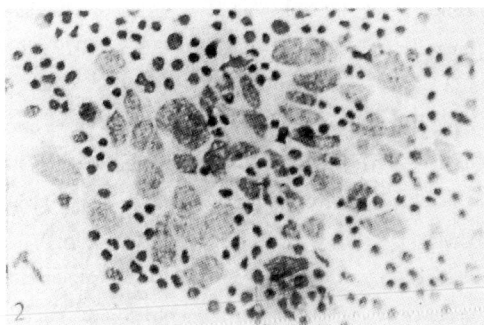

Fig. 2　Carcinoma cells detected by cytological examination (No. 43). ×200

Clinical examinations showed that these 4 were Stage I NPC patients without any subjective symptoms. Only rough mucosa on the back roof the nasopharynxes was found in 2 and 0. 5 – 0. 8 cm nodules on the nasopharynxes of the other 2. The intervals between the first serological bleeding and NPC diagnosis were 8 – 9 months. No further elevation of IgA antibody occurred in any of the 4 cases during this period.

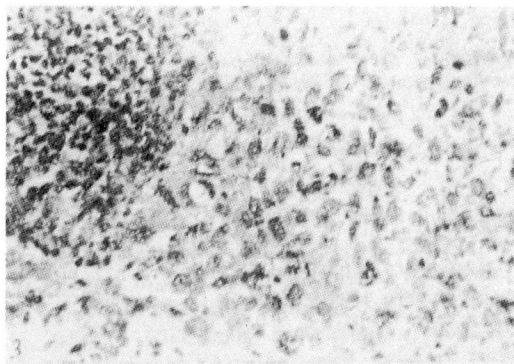

Fig. 3　Carcinoma cells detected by histological examination（No. 43）×100

Comment

Our previous results[6] indicated that ACIE method is a sensitive test which successfully detects EBNA in carcinoma cells and can be used in the outpatient clinic. This method was used to detect early NPC in a NPC high risk area. Four Stage I NPC cases without subjective symptoms were detected among 64 IgA/VCA positive individuals, a 6. 2% detection rate. The data confirm ACIE as a valuable means of diagnosing NPC and especially for follow – up study of IgA/VCA positive individuals.

Acknowledgment

We thank Prof Huang CH , Prof Chu CM and Dr G de – the for checking this manuscript.

〔In《Chinese Medical Journal》1984, 97（3）: 155 – 157〕

References

1　Zeng Yi, et al. Application of immunoenzymatic method and immunoautoradiographic method for the mass survey of nasopharyngeal carcinoma. Chin J Oncol, 1979, 1: 2

2　Zeng Yi, et al. Serological mass survey of nasopharyngeal carcinoma. Acta Acad Med Sin, 1979, 1: 123

3　Zeng Yi, et al. Application of an immunoenzymatic method and an immunoautorediographic method in a mass survey of nasopharyngeal carcinoma. Intervirology, 1980, 13: 162

4　Pi G H, et al. Development of an anticomplement immunoenzyme test for detection of EB virus nuclear antigen（EBNA）and antibody to EBNA. J Immunol Methods, 1981, 44: 73

5　Zhanjiang Medical College. Diagnosis of nasopharyngeal carcinoma by cytological examination of exfoliated cells taken by negative pressure suction apparatus. Chin Med J, 1976, 89: 45

6　Zeng Yi, et al. Application of anticomplement immunoenzymatic method for the detection of EBNA in carcinoma cells and normal epithelial cells from nasopharynx. XII International Symposium on Nasopharyngeal Carcinoma. Dusseldorf, Fed Rep Germany, Oet, 1980, 23 – 25

7　Liu Y, et al. Detection of EB virus specific IgA antibody from patients with nasopharyngeal carcinoma by immunoenzymatic method. Chin J Oncol, 1979, 1: 9

27.　HTLV Antibody in China

ZENG Y[1], LAN X Y[1], FANG J[1], WANG P Z[2], WANG Y R[3], SUI Y E[4]
WANG Z T[5], HU R J[6], HINUMA Y[7]

1. Institute of Virology, China National Centre for Preventive Medicine

2. Zhejiang Medical University   3. Shanxi Tumour Hospital;

4. No. 4 Millitary Medical School   5. Institute of Haematology Chines Academy;

6. Suzhou Medical College   7. Institute for Virus Research Kyoto university, Kyoto Japan

Sir, Antibody to HTLV (buman T – cell leukarmia – lymphoma virus) has been found in most patients with adult T – cell leukaRmia (ATL) and in 5% – 30% of healthy adults in ATL endemic areas in Japam. We have looked for HTLV antibody in China. We tested 6884 aera from normal indivduals aged 20 years of more in the cities of Beijing and Tianjin and in twenty provinces. 20 sera were from adults from Taiwan province and 25 sera from Japanese living in Beijing 510 sera were from patients with different kinds of leukaemia, including T – cell leukacmia and lymphoma (2 were suspected on cytological grounds of being ATL by Dr M, Hanaoka). Sera known to be positive or negative for HTLV antibody were used as controls. Antibody was detected by indirect immunofluorescence test with MT – 1 cells target cells.

Of the sera from normal, Chinese adults only I, from a 63-year-old woman living in Nanjing, was anti-HTLV positive (titre 160). 2 years earlier noe of us (Y. H) had found HTLV antibody in her husband's serum. The husband was Japanese, from Kagoshima, in the southern part of Japan, and had lived in China for 46 years, he died of stomach cancer in 1982. Of the 25 Japanese living in Beijing, only I had antibody to HTLV (titre 80). He is 73 years old from Chibe, Japan; his 70-year-old Japanese wife is antibody negative. All sera from patients with leukaemia were anti-HTLV negative.

850 sera, including the 2 positive sera from normal adulia and sera from leukaemia patients, were simultaneously tested with. MT – 1 and Hu T 102 cells; with identical results.

Thus the only evidence for HTLV in China was in I woman whose Japanese husband had been an HTLV carrier. HTLV antibody was also found in I Japanese, even though he has lived in China for a long time.

This work was partly suppound by a grant from the Japan Society for the Promotion of Science.

〔In 《Lancet》 1984, 799 – 800〕

28.　Epstein – Barr Virus Activation in Raji Cells with Ether Extracts of Soil from Different Areas in China

ZENG Y[1], MIAO X C[1], JAIO B[2], LI H Y[3], NI HY[4], ITO Yohei[5]

1. Institute of Virology, Chinese Academy of Medical Sciences, Beijing

2. Guangxi Autonomusregional Hospital, Nanning　3. Wuzhou Cancer Unit, Wuzhou

4. Guangxi Botanical Garden of Medicinal Plants, Nanning (China)

5. Department of Microbiology, Faculty of Medicine, Kyoto University, Kyoto (Japan)

Summary

Epstein-Barr virus(EBV)inducers were found in soil samples collected from the ground under tung oil trees and other plant species of Euphorbiaceae family growing in southern provinces of China where incidence of an EBV-associated malignancy, nasopharyngeal carcinoma(NPC)is prevalent. In such NPC high risk areas, the positive rate of EBV inducers in soil samples was up to 59.5%. Since many tung oil trees are planted along the roads and rivers of the high risk area, the possible significance of EBV inducers in soil under the trees in the development of malignancy among inhabitants of the area is discussed.

Introduction

Ito et al.[5] reported that the ether extract of soil samples collected from the ground under plants of Euphorbiaceae and Thymeleaceae families in Japan, which possess Epstein-Barr virus (EBV) - activating diterpene esters, efficiently induced EBV early antigen (EA) in Raji cells. Our previous data[6] also showed that the ether extracts of tung oil and leaves and flowers of the tung oil tree, are all strong EBV inducers. Such trees are planted in abundance along the roads (Fig. 1) and rivers in a high risk area fro NPC (Fig. 2, briefg). Here we report the detection of EBV inducers in soil samples collected from under the tung oil trees[4] growing in such areas and also in soil samples from under Euphorbiaceae plants in other different areas of the country.

Materials and Methods

1. Cells

The human lymphoblastoid Raji cells derived from Burkitt's lymphoma were cultivated in RPMI 1640 medium supplemented with 20% calf serum.

2. Soil samples

The samples were collected from the ground under the plants in different places, both from NPC endemic and non – endemic areas, to a depth of 5 cm from the surface. Five grams of such soil was extracted with 50 ml of ether at room temperature for 7 days. After removing the solution, the ether was evaporated, and the residue was dissolved in ethanol as 10 mg/ml stock soulution.

3. Experimental procedure

The synergistic assay method[2] was used. The cells were cultivated with 4 mmol/L n – butyrate and the test substance was added at varying concentration After cultivation at 37℃ for 48 h, smears were made and the detection of EBV EA – expressing cells was carried out by the immunoenzymic test[7], each assay, 500 cells were counted and the ratio of EBV EA – positive cells was recorded.

Fig. 1　Tung oil trees (Aleurites fordii) along the main road to Wuzhou City. Indicated by arrows. Photographed, December 5, 1982

Results and Discussion

As liosted in Tab. 1. The ether extracts of soil samples collected fron under Chinese tung oil trees along the Western River and Kui River in NP high risk areas, i. e. Wuzhou City and Zangwu County of Guangxi Zhan Autonomous Region, showed positive rates of 59. 5% and 40% , respectively One of 4 samples from the dock of Western River and one of 3 samples for the bank of Kuiriver in Wuzhou City showed weak activity.

The EBV-activating substance was also found in the ether extract of soil samples collected from under Euphorbiaceae and other plant families in Guangxi Botanical Garden of Medicinal Plants in Zanning (5/20 positive) (Tab. 2), parks in Beijing (3/24 positive) (Tab. 3) and in Rauzhan County (8/44 positive) (Tab. 4). Ho et al. [1] and our previous work showed that the marine population living in boat has a much higher incidence rate of NPC than the land dwellers. There are a lot of tung oil trees along the roads and rivers of the NPC high incidence area, and much rain in southern part of China. It may be assumed that the soil particles adsorbing EBV inducer substances can easily go to the river after raining. The boat people drink water from those rivers throughout their life. Thus, the high incidence of NPC among the boat people could be related to the drinking water possibly containing EBV inducers. However, more direct evidence based on extensive epidemiological studies needs to be found. Such studies are now being carried out.

Alternatively, some Chinese medicinal herbs can enhance the transformation of lymphocytes by EBV [Hu and Zeng, unpublished data] and the EBV genomes are present in normal epithelial cells. The soil dust containing EBV inducer may be inhaled by the inhabitants of the district and reach the nasopharyngeal mucosa membrane to act on the EBV geonme as inducer or on such genome – harboring epithelial cells as promoter, eventually leding to the malinant transfromation of the cells. Since many of the EBV inducers are tumor promoter[3], such an assumption is valid not only for NPC but also for other malignan tumors. Further studies are necessary.

Tab. 1 Induction of EBV EA in Raji cells by extracts of soil samples from Wuzhou city and Zangwu county

Soil samples	Positive sample No.	EBV EA positive cells (%) Ether extract (μg/ml)		
		10	2	0.4
Under tung	15	25.0	14.0	2.6
oil trees along	3	14	11.4	0.6
the road and the	5	10.6	0	0
Western River	8	4.0	3.0	2.0
(11/19 +)	11	2.6	0.6	6.0
	2	1.4	2.2	1.4
	1	2.0	0.6	1.0
	14	0	1.8	1.0
	17	0.8	1.2	1.6
	18	1.4	0.6	0.8
	4	1.0	0	0
Under tung oil	35	6.6	9.6	4.6
trees along the	21	8.0	2.8	2.0
road and the	31	2.0	8.0	1.0
Kui River[a]	22	0	4.0	0
(8/20 +)	23	8.0	1.6	0.4
	36	2.0	1.0	2.0
	29	2.0	0.4	0.8
	39	0.6	0.8	1.4
From the dock of	41	2.0	0.6	0.8
Western River[a]				
(1/4 +)				
From vegetable	54	1.0	1.0	0.6
garden				
(1/10 +)				
From the bank	56	0	1.0	0
of Kui River[*]				
(1/3 +)				
Under				
Codiacum variegatus	76	2.8	8.0	6.0
Codiacum uariegatus	77	8.0	6.0	1.6
Codiacum variegatus	78	2.0	8.0	12.0
Euphorbia milli	80	–	4.0	30.0
(4/4 +)				

Notes: Controls: B(n–butyrate) + C(eroton oil) =45.7%; B: only = 0.4%; C: only =0.8%; Untreated Raji cells =0%. –, not tested.

[a] Rivers in Guangxi Autonomous Region (NPC high risk area).

Tab. 2 Induction of EBV EA in Raji cells by extracts of soil samples from Beijing

Soil samples under	Positive sample No.	EBV EA positive cells (%) Ether extract (μg/ml)		
		10	2	0.4
Euphorbia milli	1718	0	3.0	10.6
Stellera chamaejasme	1717	7.6	1.0	0.4
Angelica pubes	1716	0.8	0.2	0
(3/24 +)				

Notes: Controls: B + C = 15.4%; B = 0.5%; C = 0.6%; untreated Raji cells =0%

Tab. 3 Induction of EBV EA in Raji cells by extracts of soil samples from Nanning

Soil samples under	Positive sample No.	EBV EA positive cells (%) Ether extract (μg/ml)		
		10	2	0.4
Euphorbia heterophylla	2117	15.6	9.6	11.7
Sapium sebiferum	2129	0.9	5.5	5.0
Euphorbia milli	2131	0	3.8	7.6
Euphorbia cochinchinensis	2120	0.8	32.4	28.2
Croton tiglium(5/20 +)	2125	30.0	36.2	31.0

Notes: Controls: B + C = 28.9%; B = 1.4%; C = 1.4%; untreated Raji cells =0%

Tab. 4 Induction of EBV EA in Raji cells by extracts of soil samples from Kauzhan county

Soil samples	Positive sample No.	EBV EA positive cells (%) Ether extracts (μg/ml)		
		10	2	0
Under				
Tung oil tree	2279	11.2	7.0	3
Tung oil tree	2273	15.0	12.7	5
Tung oil tree	2285	21.2	11.1	3
Tung oil tree	2278	15.4	5.3	–
From[a]				
Zhenan	2264	18.7	8.9	2
Tianho	2261	14.1	21.4	11
Siba	2260	13.1	6.4	1
Tianho (8/44)	2255	23.0	6.0	1

Notes: Controls: B = C = 47.6%; B = 3.1%; C = 3.4%; untreated Raji cells =0%. –, not test

[a] Local towns in Rauzhan County

[In 《Cancer Letters》 1984, 23: 53–59]

References

1 Ho J H C. An epidemiologic and clinical study of nasopharyngeal carcine Intl Radio Oncol Biol. Phys, 1978, 4: 181

2 Ito Y, Kawanishi M, Harayama T, Takabayashis. Combined effe of the extracts from Croton tiglum, Euphorbia lathyris or Euphorbia tirucalli ar n – butyrate on Epstein – Barr virus expression in human lymphoblastoid P3HR – 1 al Raji cells. Cancer Letters, 1981, 12: 175 – 180

3 Ito Y Yanase, S Fujita J, Harayama T, Takashim M, Imanaka, H. A short – term in vtiro assay for promoter substances using human lymphoblasoicl cell latently infected with Epstein – Barr virus. Cancer Letters, 1981, 13: 29 – 37

4 Ito Y, Yanase S, Tokuda H, Kishihita M, Ohigashi H, Hirota M, Koshi K. Epstein – Barr virus activation by tung oil, extracts of Aleurites fordii a diterpene ester 12 – O – hexadecanoyl – 16 – hydroxyphorbol – 13 – acetate. Cancer Let , 1982, 18: 87 – 95

5 Ito Y, Ohigashi H. Koshimizu K, Zeng Y. Epstein – Barr virus – acti principle in the ether extracts of soil collected from under plants which cont active diterpene esters. Cancer Letters, 1983, 19: 113 – 117

6 Mizuno F. Koizumi S, Osato T, Kokwaro J O, Ito, Y. Chinese African Euphorbiaceae plant extracts: markedly enhancing effect on Epstein virus – induced transformation. Cancer Letters, 1983, 19: 199 – 205

7 Zeng Y, Liu Y X, Liu Z R, Zhen S W, Wei J N, Zhu J S, Zei, H S. Application of an immunoenzymatic method and an immunoautoradiographic non for a mass survey of nasopharyngeal carcinoma. Intervirology, 1980, 13: 162 – 168

29. Epstcin – Barr Virus Activation by Human Semen Principle: Synergistic Effect of Culture Fluids of Bacteria Isolated from Patients with Carcinoma of Uterine Cervix

ZENG Y[1], GI Z W[1], ITO Y[2]

1. Institute of Virology, China National Centre for Preventive Medicine, Beijing (China)
2. Department of Microbiology, Faculty of Medicine, Kyoto University, Kyoto (Japan)

Summary

Rpsein – Barr virus early antigens (EBV EA) were induced in a Raji cell system in order to assay the activity of the EA inducer in human semen. In semen from 53 Chinese, 45. 3% induced EBV EA in Raji cells. Such positive semen and EBV – inducing positive culture fluids of bacteria isolated from the uterine cervix of patients with cervical carcinoma had a synergistic effect on the induction of EBV EA. This synergistic effect as related to the cause of cervical carcinoma is discussed.

Introduction

EBV inducers have been detected in plants of Euphorbiaceae, Thymelaeaceae, Leguminosae, etc[1,2]. and 12 – O – tetradecanoyl – phorbol – 13 – acetate （TPA）of croton oil，12 – O – hexade-canoyl – hydroxyphorbol – 13 – acetate （HHPA）of tung oil[5]，Daphne genka and Wikstroemia Chamaedaphne extracts （Y. Zeng ，unpublished data）proved to be potent tumor inducers. Some human semen also contains an EBV activating substance[6]. Here we report the detection of EBV in-ducer in human semen of Chinese and the synergistic effect of human semen and bacterial culture fluid on the induction of EBV EA.

Materials and Methods

1. Cells

Human lymphoblastoid Raji cells derived from Burkitt lymphoma were cultivated in RPMI 1640 medium supplemented with 20% calf serum，penicillin （100 units/ml）and streptomycin （100 μg/ml）.

2. Semen sample

Samples were obtained from the fertility clinic of Beijing Friendship hospital and preserved at – 20℃ until assay. After thawing，the samples were centrifuged at 18 000 r/min for 60 min and super-natant was used for EBV EA induction test.

3. Anaerobic bacterial culture fluid

Swabs were taken from the uterine cervix of patients with cervical carcinoma and were inocula-ted on the surface of plates of FM modified medium and cultured under anaerobic conditions （steel wool method，10% CO_2/90% N_2）[3]. The pure culture of gram – negative bacteria and grampositive cocci were cultivated anaerobically in both medium. The culture fluid was passed through a Seitz bacterial filter. Positive and negative samples from such bacterial cultures were both used for EB vi-rus induction.

4. Experimental procedure

The synergistic assay method[4] was used. The cells were cultivated with 4 mmol/L n – butyrate and varying concentrations of human semen were added. After cultivation at 37℃ for 48 h，smears were made and the EBV EA expressing cells were identified using an immunoenzymatic test[7]. The positive controls comprised of croton oil （500 ng/ml）plus n – butyrate （4 mmol/L）and a nega-tive control using n – butyrate alone was included. In each assay，500 cells were counted and the ra-tio of EA – positive cells recorded. For examination of the synergistic effect of semen and bacterial culture fluid，2μl or 10 μl of semen and 0. 05 ml or 0. 1 ml of bacterial culture fluid were added to 1 ml of Raji cell culture.

Results

1. Induction of EBV EA by human semen

Semen samples from 53 Chinese were assayed and the results are listed in Tab. 1. Twenty-four samples （45. 3%）induced EBV EA in Raji cells，the positive rate of smen samples was 56%，and the positive rate of cells with EA ranged from 2. 2% to 12. 2%. Three samples （nos 9，15，20）showed a higher positive rate of cells with EA （9. 8% – 12. 2%），such exceeding the 50%

induced by croton oil and n-butyrate. The remaining samples showed a lower activity. There was no marked difference in the number of positive cells with EA between concentrations of semen varying from 0. 4 μl to 10 μl in 1 ml medium.

Tab. 1　Induction by human semen of EBV EA in Raji cells[a]

No. of semen specimens	Semen	Semen + n – butyrate (%)		
	10 μl	10μl	2 μl	0. 4μl
15	1. 2	10. 8	12. 2	9. 0
9	1. 2	9. 4	10. 4	8. 0
20	1. 2	9. 2	9. 8	8. 2
36	0. 6	7. 6	6. 0	8. 2
41	0. 6	7. 4	6. 2	6. 0
33	0. 6	7. 2	6. 0	6. 2
12	0. 6	6. 8	6. 0	5. 2
17	1. 2	6. 2	5. 4	6. 6
18	1. 2	5. 8	6. 4	5. 4
23	1. 0	5. 8	6. 2	4. 8
45	0. 6	5. 4	4. 2	3. 4
13	0. 8	4. 2	5. 4	3. 4
40	0. 4	5. 2	4. 2	5. 8
48	0. 4	5. 0	4. 0	3. 4
2	0. 4	4. 6	3. 4	4. 0
3	1. 0	4. 0	4. 4	3. 0
1	0. 8	4. 2	3. 6	3. 4
14	0. 8	4. 0	2. 6	3. 4
31	0. 4	3. 2	3. 8	2. 6
27	0. 4	3. 2	3. 8	3. 8
22	0. 8	3. 4	3. 0	2. 8
25	0. 4	2. 8	3. 2	2. 2
50	0. 4	5. 2	4. 2	3. 4
58	0. 4	4. 2	3. 8	3. 0

Notes: [a] Controls: n – butyrate (4 mmol/L) EA positive rate, 0. 6% ; croton oil (500 ng/ml) EA positive rate, 1. 4% ; croton oil + n – butyrate EA positive rate, 17. 6%

2. Synergistic effect of semen and bacterial culture fluid

The synergistic effect of semen and bacterial culture fluid was assayed in Raji cells and the results are listed in Tab. 2. A positive reaction was seen only in groups with a positive semen and a positive bacterial culture fluid. The positive rate of cells with EA was from 5. 4% to 18. 6%.

Tab. 2　Synergistic effect of semen and bacterial culture fluid[a]

No. of bacterial culture fluid specimens	No. of semen specimens	Bacterial culture fluid (EA positive rate%)			
		0. 1 μl		0. 05 μl	
		10μl	2 μl	10μl	2 μl
13 (+)[b]	15 (+)	16. 2	17. 2	18. 6	14. 4
25 (+)	15 (+)	9. 2	8. 4	16. 8	17. 4
39 (+)	36 (+)	4. 65	5. 4	4. 8	4. 2
41 (+)	41 (+)	4. 8	6. 2	5. 4	4. 0
15 (+)	20 (+)	4. 8	5. 0	5. 8	4. 4
25 (+)	57 (–)	1. 2	1. 4	0. 8	0. 2
27 (+)	56 (–)	0	0	0	0
23 (–)	58 (+)	0	0	0	0
2 (–)	9 (+)	0	0	0	0
11 (–)	52 (–)	0	0	0	0
7 (–)	49 (–)	0	0	0	0

Notes: [a] Controls: n – butyrate (4 mmol/L) EA positive rate, 0. 6% ; croton oil (500 ng/ml) EA positive rate, 1. 0% ; croton oil + n – butyrate EA positive rate, 21. 4%.

[b] Activity of samples when assayed alone

The potency of the sample (bacteria no. 13 plus semen no. 15, 18. 6%) was comparable to that of the croton oil and n – butyrate (21. 4%). Induction of EA was nil in negative semen and negative bacterial culture fluid, used alone or in combination.

Discussion

The present data confirmed our previous findings[6], that the semen from 45. 3% of Japanese males induced EBV EA in Raji cells. Thus, we extended the study to determine whether the EBV EA inducer in semen possesses the activity of a tumor promoter. The EBV EA inducing substance was also found in the culture fluid of bacteria isolated from the uterine cervix of Chinese patients with cervical carcinoma. This substance was identified by gas chromatography to be butyric acid (Y. Zeng, unpublished data). For elucidating the etiology of cervical carcinoma, the possible synergistic effect of the semen and bacterial culture fluid on the induction of EBV EA was investigated. We found that these 2 natural components do have a synergistic effect, similar to the case of croton oil and n – butyrate[1]. Thus, further study on the synergistic effect of these 2 natural products in animal systems in warranted to determine if this effect plays a role as tumor promoter in the neoplastic processes of uterine carcinogenesis.

Acknowledgements

We thank M. Ohara for expert editing of the manuscript.

〔In 《Cancer Letters》 1985, 28: 311 – 315〕

References

1 Ito Y, Kawanishi M, Harayama T, Takahayashi S. Combined effect of the extracts from Croton tiglium, Euphorbia lathyris or Euphorbia tirucalli and n-butyrate on Epstein Barr virus expression in Raji cells. Cancer Letters, 1981, 12: 175 – 180

2 Zeng Y, Zhong J M, Mo Y K, Miao M C. Epstein-Barr virus early antigen induction in Raji cells by Chinese medicinal herbs. Intervirology, 1983, 19: 201 – 204

3 Ito Y, Yanase S. Induction of Epstein-Barr virus antigens in human lymphoblastoid p3HR – 1 cells with culture fluid of Fusobacterium nucleatum. Cancer Res, 1980, 40: 4329 – 4330

4 Ito Y, Yanase S, Fujita J, Harayama T, Kashima M, Imanka H. A short – term in vitro assay for promoter substances using human lymphoblastoid cells latently, with Epstein – Barr virus. Cancer Letters, 1981, 13: 29 – 37

5 Ito Y, Tokuda H, Ohigashi H, Koshimizu K. Distribution and characterization of environmental promoter sustances as assayed by synergistic Epstein – Barr – activating system. In: Cellular Interactions by Environmental Tumor Promoters, 1984, 125 – 127. Editors: Fujiki et al. Japan Sci Press Tokyo

6 Ito Y, Tokuda H, Morigaki T, Shimizu K, Kawana T, Sanada S, Yoshida O. Epstein – Barr virus – activating principle in human semen. Cetters, 1984, 23: 129 – 134

7 Zeng Y, Liu Y X, Liu Z Y, Zhen S W, Wei J N, Shu J S, Zai H J. Application of immunoanti-radiographic method for the mass survey of nasopharyngeal carcinoma. J Clin Oncol, 1980, 1: 2 – 8

30. Prospective Studies on Nasopharyngeal Carcinoma in Epstein – Barr Virus IgA/VCA Antibody – Positive Persons in Wuzhou City, China

ZENG Y[1], ZHANG L G[2], WU Y C[2], HUANG Y S[2], HUANG N Q[2], LI J Y[2],
WANG Y B[2] JIANG M K[1], FANG Z[1], MENG N N[2]

1. Institute of Virology, China National Center for Preventive Medicine, Beijing
2. Wuzhou Cancer Unit Wuzhou City, Guangxi Autonomous Region, People's Republic of China

Summary

A serological mass survey was carried out in Wuzhou City in 1980, 1136 IgA/VCA – positive persons being followed up for 4 years. Altogether 35 NPC cases were detected, of which 15 (43%) were in stage I and 17 (48.5%) in stage II, early cases (I + II) thus amounting to 91.5%. The detection rate of early cases was 2.9 times higher than in our outpatient clinic. IgA/VCA antibody could be detected 16 – 41 months prior to clinical diagnosis of NPC. We conclude that, if IgA/VCA – positive individuals are examined routinely once a year, NPC can be detected in the early stages of evolution. The annual detection rate of NPC in IgA/VCA antibody – positive individuals was 31.7 times higher than that of the annual incidence of NPC in the general population in the same age group, while during the 4 – year follow – up period the incidence was 7.5 times higher than in the general population for the same age group. These results further indicate that EB virus plays an important role in the development of NPC, and that serological screening and follow – up studies are valuable for the early detection of NPC.

Introduction

Serological mass surveys of NPC and follo – up studies on EB virus IgA/VCA antibody – positive individuals were carried out in the Zanwu County of Guangxi Autonomous Region using an immunoenzymatic test (Zeng et al., 1979, 1980, 1983). NPC patients could be detected in an early stage, but it was rather difficult to carry out follow – up studies because many of the IgA/VCA antibody – positive cases came from remote areas, and clinical examination was not carried out periodically for some patients, while in others the diagnosis was not established at an early stage. Therefore, another serological mass survey was carried out in Wuzhou City, starting in 1980 (Zeng et al., 1982). Altogether 20 726 individuals aged 40 years and more were examined and the antibody – positive cases were followed up for 4 years. The results of the follow – up studies are reported here.

Material and Methods

1. Sera

Sera obtained from venous blood were stored at $-20^{\circ}C$ in Wuzhou City and forwarded by air to the Institute of Virology, China National Center for Preventive Medicine, in Beijing.

2. Immunoenzymatic test

This was described previously (Zeng et al. , 1980). B95 – 8 and horseradish – conjugated anti – human IgA/VCA antibody were used for the detectionof IgA/VCA antibody.

3. Clinical and histological examination

EB virus IgA/VCA antibody-positive cases were examined clinically once a year for 4 years. Biopsies were taken from the suspect NPC cases and individuals with high IgA/VCA antibody titers.

Results

1. NPC patients detected by serological screening and follow – up studies

Among 20 726 persons examined 1 136 presented an IgA/VCA antibody, the positivity rate thus being 5. 5%. From this number, 18 NPC patients were diagnosed clinically and histologically. Ten were in stage I, 6 in stage II and 2 in stage III. Fourteen patients (77. 8%) had IgA/VCA antibody titers of 1:40 and over, with a geometric mean titer (GMT) of 1:93. 3 (Tab. 1).

The IgA/VCA antibody – positive cases were clinically and histologically reexamined once a year for 4 years. During this time, 17 additional NPC cases were diagnosed, 5 in stage I, 11 in stage II and one in stage III. Sixteen cases (94. 1%) were thus in early stages with a GMT of 1: 62. 6. The unique stage III case had an antibody titer of 1:40 during the first bleeding episode. Some suspect lesions in nasopharyngeal mucosa were found, but the patient refused biopsy. The NPC developed into stage III 2 years and 8 months later.

Altogether 35 NPC cases were detected by serological screening and follow – up studies within the 4 – year period. Of these, 32 cases (91. 5%) were in an early stage. As a comparison, 1036 NPC patients were diagnosed during this period in our outpatient clinic, most of them (68. 1%) being in advanced stages (III + IV), while only 330 cases (31. 9%) were in an early stage. The percentage of outpatients with NPC in stage I was 1. 7%, but it increased to 43% when serological screening and follow – up studies were applied (Tab. 2). As shown in Figure 1, IgA/VCA antibodies were detected 16 – 41 months (average, 30 months) prior to the diagnosis of NPC for stage I, and 16 – 40 months (average, 27 months) prior to diagnosis for stage II.

There was no difference in GMT of IgA/VCA antibody in sera of patients with a first bleeding and of those bleeding when diagnosis was confirmed at stage I. The GMT of IgA/VCA antibody in patients at stage II was 1: 96. 1, while it was twice as high in patients bleeding for the first time (Fig. 2). No NPC patients were found among 19 590 individuals without detectable levels of IgA/VCA antibodies, during a period of 4 years.

Tab. 1 Antibody titer of NPC patients detected during screening and follow – up studies

	Total	Titer							GMT
		10	20	40	80	160	320	640	
NPC found at									
Screening	18	4		3		7		4	1:93.3
Follow – up	17	2	1	7		6		1	1:62.6
Total	35	6	1	10		13		5	1:76.9

Tab. 2 Clinical stage of NPC diagnosed at screening and in outpatient clinic

	Clinical stages				Total
	I	II	III	IV	
Screening					
Number of cases	15	17	3	0	35
%	43.0	48.5	8.5		100.0
Outpatient clinic					
Number of cases	18	312	526	180	1 036
%	1.7	30.1	50.8	17.4	100.0

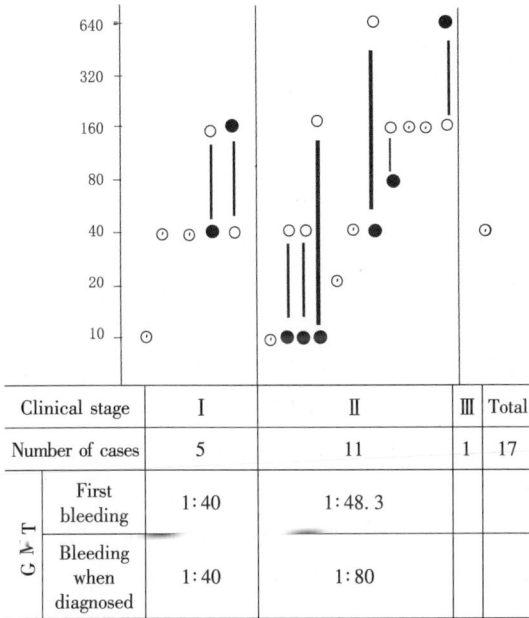

Clinical stage	I	II	III	Total
Number of cases	5	11	1	17
G M T — First bleeding	1:40	1:48.3		
G M T — Bleeding when diagnosed	1:40	1:80		

● first bleeding; ○ bleeding when diagnosis was confirmed;

⊙ two bleedings had same IgA/VCA titer

Fig. 2 Change in IgA/VCA titer and GMT from first bleeding to diagnosis of NPC.

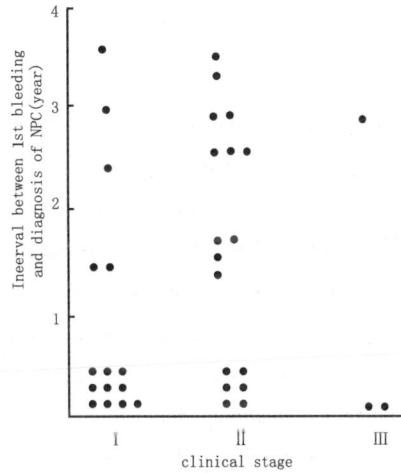

Fig. 1 Follow – up studies on IgA/VCA antibody positive individuals

2. Detection rate of NPC

Eighteen NPC patients were detected among 1136 IgA/VCA antibody – positive individuals by serological screening, the NPC detection rate being 86. 8/100 000 for 20 726 persons examined and 1584/100 000 for 1136 IgA/VCA antibody – positive persons. These rates are respectively 1. 7 and 31. 7 times higher than the annual incidence of NPC (50/100 000) in the general population at a similar age. Twenty – two IgA/VCA antibody – positive cases were found among 216 persons examined within one chemical factory. From these, 3 NPC cases (2 in stage I and 1 in stage II) were diagnosed histologically. The NPC detection rate for individuals examined and IgA/VCA antibody – positive cases are 1380/100 000 and 13 636/100 000 respectively, and these are 27. 6 and 272. 7 times higher respectively than the annual incidence in the general population of similar age. Seventeen cases were detected during the 4 – year follow – up studies, the incidence rate of NPC being 374/100 000/year. This is 7. 5 times the annual incidence in the general population of similar age (Tab. 3).

Tab. 3 Comparison of detection rate and incidence rate of NPC

	Number of persons examined	IgA/VCA positive persons	Number of NPC cases	NPC detection rate(1/100 000)		Incidence rate by follow-up studies (1/100 000 per year)	Incidence rate in general population per year	Reference
				Persons examined	IgA/VCA positive person			
Zangwu county	148 029	3 533	55	37.0(1.2)[1]	1 566.8(57.9)[1]		30	Zeng et al. 1983(screening)
Boat people in Zangwu	518	18	2	386.0(12.8)	11 111.0(370)		30	Zhu et al. 1983(screening)
Wuzhou City	12 932	680	13	100.5(2)	1 911.8(38)		50	Zeng et al. 1982(screening)
Wuzhou City	20 726	1 136	17	86.8(1.7)	1,584.5(31.7)		50	(Screening)
	20 726	1 136	17			374(7.5)[2]		(follow-up studies)
A chemical factory in Wuzhou	216	22	3	1 380(27.6)	13 636.0(272.7)		50	(screening)

[1] Detection rate/incidence rate; [2] Incidence rate(follow-up studies)/incidence rate.

Discussion

Thirty-five NPC cases were detected among 20 726 individuals by serological screening and follow-up studies during a period of 4 years, 91.5% of them being at an early stage. This was 2.9 times that (31.8%) found in our outpatient clinic. Seventeen NPC patients except one, detected by follow-up studies, were in an early stage and IgA/VCA antibodies could be detected in their serum 16—41 months prior to diagnosis of NPC. These data support the notion that serological screening and follow-up studies are valuable for early detection of NPC.

The NPC detection rate for individuals examined and for the IgA/VCA antibody-positive cases detected by serological screening were 1.7 and 31.7 times higher respectively than the annual incidence in the general population of similar age. The annual incidence rate NPC among IgA/VCA antibody-positive individuals, follow-up was 7.5 times higher than the annual incidence of NPC in the general population. These date similar to those concerning Zangwu county (Zo et al. , 1983) and indicate that EB virus plays an important role in the development of NPC.

The NPC detection rates for the individuals examined and for the IgA/VCA antibody-positive cases one chemical factory were respectively 27.6 and 272 times higher than the annual incidence in the generat population of similar age. This may indicate a relationship between environmental factors, activation of virus and development of NPC.

No NPC patients were recorded within the IgA/VCA antibody-negative group during a period of years. This fact should be useful in making decision concerning serological reexamination of IgA/VCA antibody-negative individuals.

〔In 《Int J Cancer》 1985, 36: 545-547〕

References

1 Zeng Y, Liu Y X, Liu C R, Chen S W, Wei J N, Zhu J S, Zai H J. Application of immuno autoradiographic method for the mass survey of nasopharyngeal carcinoma. Intervirology, 1980, 13: 162-168

2 Zeng Y, Liu Y X, Wei J N, Zhu J S, Cai S L, Wang P Z, Zhong J M, Li R C, Pan W J, Li E J, Tan B F. Serological screening of nasopharyngeal carcinoma. Acta Acad med Sin, 1979, 1: 123-126

3 Zeng Y, Zhang L G, Li H Y, Jan M G, Zhang Q, Wu Y C, Wang Y S, Su G R. Serological mass survey for early detection of nasopharyngeal carcinoma in Wuzhou City, China. Int J Cancer, 1982, 29: 139 – 141

4 Zeng Y, Zhong J M, Li L Y, Wang P Z, Tang H. Ma Y R, Zhu J S, Pan W J, Liu Y X, Wei Z N, Chen J Y, Mo Y K, Li E J, Tan B F. Follow – up studies on Epstein – Barr virus IgA/VCA anti-body – positive persons in Zangwu county China. Intervirology, 1983, 20: 190 – 194

5 Zhu J S, Pan W J, Li L Y, Zeng Y, Jan M K, Fang Z. Serological suvey of nasopharyngeal cancer of boat peoplc in Zangwu county. Study on preven-tion and treatment of cancer (in Chinese), 1983, 10: 189 – 190

31. Seroepidemiological studies on Nasopharyngeal Carcinoma in China

ZENG Y

Department of Tumor Viruses, Institute of Virologyc,

China National Centre for Reventive Medicing, Beijing China

Introduction

There is great variation in the incidence of nasopharyngeal carcinoma (NPC)[1] in different eth-nic groups. lt is very common in Southern China and Southeast Asia (Atlas of Cancer Mortality, Editorial Committee, 1979). The serological relationship between Epstein – Barr (EB) virus and NPC was first demonstrated by Old et al. (1966) by means of immunodiffusion. It was subsequently shown by indirect immunofluorescence that NPC patients had various antibodies to EB virus (De Schryver et al. , 1969; Henle et al. , 1970), and by anticomplement immunofluorescence (Klein et al. , 1974; Huang et al. , 1974) and nucleic acid hybridization (Zur Hausen et al. , 1970; Wolf et al. , 1973) that EB virus markers (EBNA and DNA) are also regularly found in NPC cells. Henle and Henle (1976) first reported the presence of IgA antibody to EB virus as an out-standing feature of NPC. Other authors (Ho et al. , 1976; Desgranges and de The. 1978) further confirmed that NPC patients from various geographic areas regularly exhibited IgA/VCA antibody and that detection of IgA/VCA antibody is useful for the diagnosis of NPC. The purpose of our stud-ies is first to find simple and sensitive techniques that can be conveniently applied in the high – NPC risk field for the early detection of NPC, as early treatment gives a good diagnosis; and second to find the roles of EB virus, environmental factors, and genetic factors in the development of NPC. Here we summarize our serological studies on NPC.

Distribution and Spectra of EB Virus
Antibodies in NPC Patients and Nomal Individuals

EB virus infection is prevalent all over the world, but the specific incidence of primary infection varies with different social and economic groups. In order to study the prevalence of EB virus infection in China and its relation to NPC, complement – fixing antibody to EB virus was tested in sera from different age groups of normal individuals in high NPC risk areas (Guangzhou City and Zhangshan County of Guangdong Province) as well as in low NPC risk areas (Beijing City, Lufeng and Wuhua Counties of Guangdong Province). Of a total of 2300 normal sera from different age groups, 2080 were positive (≥1:10). The positive rate was 90.4% with a geometric mean titer (GMT) of 1:52.8. Seropositivity was relatively common in both the high and low NPC risk areas of the above groups, varying from 87% to 94%, and showing no great difference. The positive rate reached 90% – 100% in the 3 – to 5 – year age group from different areas. These results indicated that EB virus spreads widely in China and people are infected with EB virus in early life. The age of primary infection in high and low NPC risk areas was similar (Tumor Control Team of Zhangshan County et al., 1978).

As NPC occurs more frequently in persons above 20 years of age in China it can be seen in Fig. 1 that the mean antibody titers of a healthy population over 20 in the high NPC risk areas were significantly higher than those in the low NPC risk areas (Tumor Control Team of Zhangshan County et al., 1978). This variation may indicate that the EB virus is more active in people living in high NPC risk areas and that this might be related to the development of NPC. The activation of EB virus may be due to internal or external factors. Recent data (Ito et al., 1980; Zeng et al., unpublished data) showed that the culture fluid of G – anerobic bacteria from the nasopharynx of patients with

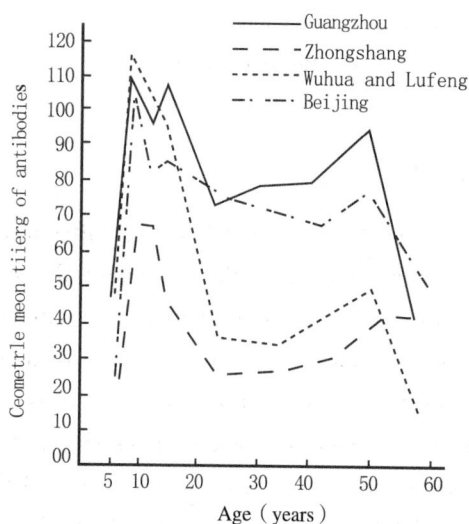

Fig. 1 Geometric mean titers of antibodies to EBV tested by complement fixation method in the population in high – and low – risk areas of NPC

NPC or with ear – nose – throat disease could induce EB virus, EA, or VCA in Raji and P3HR – 1 cells. Our data (Hu and Zeng, unpublished) also showed that n – butyrate and other extracts from the Euphorbiaceae and Thymeleaceae families could enhance the transformation of lymphocytes by EB virus. Therefore, it is necessary to study the possible role of butyric acid or these indigenous microbial products, which might activate the EB virus in vivo in the development of NPC. Ito et al (1983) reported that the ether extracts of soil samples from the ground under the plants of the Euphorbiaceae and Thymeleaceae families in Japan known to possess EB virusactivating diterpene esters induced EB virus EA in Raji cells. Our previous data (Ito et al. , 1983; Zeng et al. , 1983b) also reported that the extract from tung oil and the leaves and flowers of tung oil trees (Aleurites fordii) are strong inducers. There are plenty of such trees along the roads and rivers in NPC high – risk areas. In Wuzhou City and Zangwu County, the dust containing EB virus inducer from tung oil trees and other plants may be inhaled by the nasopharyngeal mucosa and thus acts as a promotor to the EB virus. Its genome is present in the normal epithelial cells, eventually leading to the malignant transformation of epithelial cells in the nasopharynx. The role of these environmental factors in the development of NPC needs to be further studied.

Various EB virus – specific antigens, including EA, VCA, membrane antigen (MA), and nuclear antigen (EBNA), are present in cells infected with EB virus; the corresponding antibody to these antigens are found in sera from persons after EB virus infection. IgA and IgG antibodies to EBV, VCA, or EA in sera from normal individuals were studied by immunoenzymatic test (no IgA/EA detected) (Fig. 2) (Zeng et al. , 1983c). The positive rate of IgG/VCA antibody reached 90% – 100% in the 3 – to 5 – year age group and maintained a high level for many years. The positive rate of IgG/EA antibody reached the peak (37%) at 3 – 5 years of age, then gradually declined to 17% at 20 – 29 years of age. It slightly increased at 50 – 59 years of age. The positive rate of IgA/VCA antibody also reached the

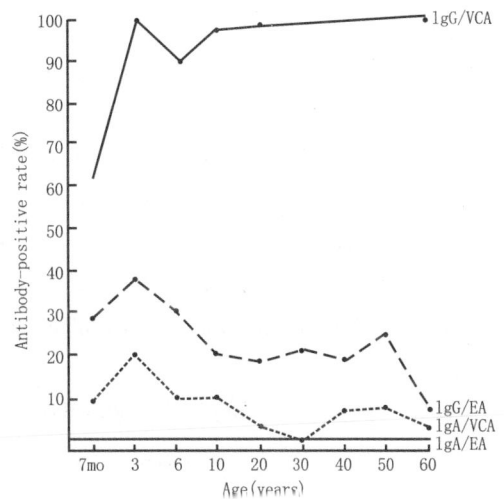

Fig. 2   **Age – specific incidence of antibodies to EBV**

peak (20%) at 3 – 5 years of age, and gradually declined. It rose slightly after 30 – 39 years of age. No IgA/EA antibody was found in any age group, even with the use of a very sensitive test: immunoantoradiography. This indicates that no such antibody, is produced or only very few people have such antibody after primary EB virus infection.

The IgG/MA and IgA/MA antibodies in sera from NPC patients and from normal individuals were detected by immunofluorescence (Zhao and Zeng, unpublished). The positive rate and GMT of IgG/MA and IgA/MA from NPC patients was 97% and 1: 210, 51% and 1: 16, respectively, and those from normal individuals was 85% and 1: 50. 0% and 1: 5, respectively. The positive rate of IgA/MA in sera from NPC patients and normal individuals is similar to that of IgA/EA from both these groups.

IgG/EA, IgA/VCA, and IgA/EA antibodies in sera from patients with NPC, with malignant tumors in head and neck other than NPC, with other malignan tumors and normal individuals were detected by immunofluorescence (Laboratory of Tumor Virus, Cancer Inst, et al. , 1978, S J . Li et al. , 1980; Zeng et al. , 1979c). Totals of 96% and 81. 5% of NPC patients had IgG/EA and IgA/VCA antibody with high GMT, respectively, whereas the positive rate of these two antibodies in three other groups was less than 6%. Of NPC patients, 50% had IgA/EA antibody, but no such antibody was found in other groups. These data indicate that IgG/EA and IgA/VCA antibodies are specific, with a high positive rate for NPC patients, and can be used for the diagnosis of NPC. But further studies showed that the positive rate of IgG/EA antibody in normal individuals was higher than that of IgA/VCA antibody, so the IgG/EA antibody was not as specific as IgA/VCA antibody for the diagnosis of NPC. The immunofluorescence test needs a fluorescence microscope and is not convenient for NPC detection in counties. An immunoenzymatic test was established in our laboratory for detection of IgA/VCA antibody (Liu et al., 1979a). As compared to immunofluorescence, the immunoenzymatic test is simpler and more sensitive. This test has been widely used in China for the diagnosis of NPC (Y. X. Liu et al., 1979; Wei et al., 1980; Jian et al., 1981; W J. Li et al., 1982). As shown in Tab. 1, the positive arte of IgA/VCA antibody (≥1:5) from NPC patients, patients with malignant tumors other than NPC, and normal persons was 92.5% – 98.1%, 0 – 5.7% and 0 – 6%, respectively, and the GMT was 1:35.7 to 1:78.5, 1:1.25 to 1:2.7, and 1:1.25 to 1:5.4, respectively. The positive rate and GMT of IgA/VCA antibody was much higher than that in other control groups. Therefore, the detection of IgA/VCA antibody is valuable for the diagnosis of NPC, especially when there was no evidence of tumor in the nasopharynx, but when it had already invaded beneath the mucosa or metastasized to the lymph node of the neck rgeion.

Tab. 1　Detection of IgA/VCA Antibody to EB Virus in Sera from NPC patients and control groups

NPC patients			Patients with tumors other than NPC			Normal individuals			Reference
No. of cases	Positive rate (%)	GMT	No. of cases	Positive rate (%)	GMT	No. of Persons	Positive rate (%)	GMT	
80	92.5	1:35.7	107	0	1:1.25	91	0	1:1.25	Liu et al. (1979)
628	98.1	1:38.7	92	5.4	1:1.25	210	0.5	1:1.27	Wei et al. (1980)
78	92.3	1:78.5				166	6	1:5.40	Jian et al. (1981)
1006	93.8	1:76.0	768	5.7	1:2.70	756	1.9	1:2.80	Li et al. (1982)

Serological Mass Surveys and Prospective Studies on NPC

Detection of IgA/VCA antibody is helpful for the diagnosis of NPC, especially in its early stage. It is very important to apply the serological test in high NPC risk areas for the detection of early NPC cases. Therefore, serological mass surveys and follow – up studies were carried out in Zangwu County, Wuzhou City, and Laucheng County of the Guangxi Autonomous Region.

1. Zangwu County

A total of 148 029 persons aged 30 and over were screened by the immunoenzymatic test in 1978 – 1980 in Zangwu County (Zeng et al., 1979a, b, 1980a, 1983c). The incidence of NPC in this county was 11/100 000, but in some communes it was up to 17/100 000. Of total NPC, 91.4% occurs at age 30 and over. Sera were collected in plastic capillary tues by pricking the ear lobe, and 3533 persons were found to have EBV IgA/VCA antibody (≥1:5). The positive rate was 2.4%. The antibody titer of the positive persons ranged from 1:5 to 1:2560, with a GMT of 1:16, and 87% had antibody titers of 1:5 to 1:40. Only 13% had antibody with higher titer (≥1:80). The positive rate of IgA/VCA antibody was shown to increase with age, and the age – specific prevalence curve of IgA/VCA antibody paralleled that of the NPC age – specific incidence.

All IgA/VCA antibody – positive persons were examined clinically and histologically. Biopsies were taken from suspected patients or persons with high antibody titer (≥1:80). Among them, 55 cases were diagnosed histoologically as NPC. One was carcinoma in situ, 12 in stage I, 19 in stage II, 17 in stage III, 6 in stage IV, and 31 cases (57%) were in early stages (I and II) (Tab. 2). The range of IgA/VCA antibody titer for these NPC patients was from 1:10 to 1:2560, with a GMT of 1:99, and 34 cases (62%) had antibody titers of ≥1:80.

The IgA/VCA antibody – positive cases were reexamined clinically and serologically, and biopsies were taken from the suspected patients once a year; 32 new NPC patients were detected after follow – up studies for 1 – 3 years. All their tumors were diagnosed histologically as poorly differentiated squamous cell carcinoma or undifferentiated carcinoma. There were 10 cases in stage I, 9 in stage II, 11 in stage III, and 2 in stage IV (Tab. 2); 21 cases (66%) had IgA/VCA antibody titer of ≥1:80.. The GMT of antibody was 1:85 (Tab. 2).

Tab. 2   Detection of NPC cases in Zangwu County during 1978—1980 by serological mass surveys and follow – up study

	Number of NPC cases					Total
	Carcinoma in situ	Stage I	Stage II	Stage III	Stage IV	
1978 – 1980						
serological mass survey	1	12	19	17	6	55
Follow. up study	0	10	9	11	2	32
Total	1	22	28	28	8	87

There was no marked difference in GMT of IgA/VCA antibody in sera of patients in stage I between the period before the onset of NPC and the diagnosis made at stage I, but antibody titer was higher in stages II – IV; e. g. when the carcinoma cells metastasized to the cervical lymph nodes, the antibody titer increased. The percentage (25.3%) of NPC at stage I detected by serological mass surveys was much higher than that (3%) diagnosed in an outpatient clinic. Of NPC patients diagnosed, 59% on follow – up studies were at stage I and II, with favorable outcome with radiotherapy. A total of 41% of NPC cases were diagnosed at advanced stages (III and IV) because many IgA/VCA antibody – positive persons live in the countryside far from town, and clinical reexaminations were not carried out periodically of diagnosis was missed in the early stages. In some, carcinomas developed more rapidly. The detection rate of NPC from IgA/VCA antibody – positive persons was 82 times higher than the incidence rate of NPC in the general population of the same age group (Tab. 3).

Tab. 3　Comparison of NPC detection rates and incidence rates

Location	No. of persons examined	No. of IgA/VCA antibody – positive persons (5%)[a]	No. of NPC	NPC detection rate(1/100 000) among all persons examined (relative incidence rate)	NPC detection rate(1/100 000) among IgA/VCA antibody – positive persons (relative incidence rate)	NPC incidence rate per annum	Reference
Zangwu							
County(1978~1980)	148 029	3 533(2.4)	87	59(1.96)[b]	2 462(82)[b]	30	Zeng et al. (1983a)
Wuzhou City	12 930	680(5.3)	13	100.5(2)[b]	1 900(38)[b]	50	Zeng et al. (1982a)
Chemical factory	216	22(10.2)	3	1380(27.6)[b]	13 636(272.7)[b]	50	Zeng et al.
Zangwu							
County boat people	518	18(3.5)	12	386(12.8)[c]	11 111(370)[c]		Zhu et al. (1983)
Laucheng							
County(Molaos)	15 324	151(0.96)	7	45.7(1.5)	4 636(154.4)[c]		Wei et al. (unpublished data)

[a] No. of IgA/VCA – positive persons/no. of persons examined × 100; [b] NPC detection rate/NPC incidence rate per annum; [c] NPC detection rate/NPC incidence rate per annum in Zangwu Conunty

How long before clinical onset does IgA antibody develop? The IgA/VCA antibody was detected 8 – 30 months (average 13 months) prior to the clinical diagnosis of NPC in stage I cases, 10 – 18 months (average 14.4 months) prior in stage II, 9 – 34 months (average 17.3 months) prior in stage III, and 12 – 24 months (average 18 months) prior stage IV (Zeng et al., 1983c). These results are similar to those reported by Ho et al. (1978) and Lanier et al. (1980) in their retrospective studies. They found that IgA/VCA antibody was detectable 22 – 72 months before the clinical diagnosis was made in the advanced stages.

How long does IgA/VCA antibody persist? In follow – up studies of sera from normal individuals, 10% converted from IgA antibody positive to negative within 1—3 years.

2. Boat people

Ho (1978) reported that the marine population living in boats in Hong Kong has a much Higher incidence rate than the land dwellers living in congested apartments. Zangwu County is a high NPC risk area and there are still some people living and working on the boats. The incidence rate of NPC for boat people from four communes of Zangwu County was 20.91/100 000 in 1978 and is higher than that (15.3100 000) for people living on the land in the same communes. A serological mass survey was carried out in poeple aged 30 and over living on the boats (Zhu et al, 1983). A total of 518 sera were taken for examination of IgA/VCA and IgA/EA antibodies, of which 18 had IgA/VCA antibody (3.47%), 5 had IgA/EA antibody (0.97%), and 2 NPC patients (stage I and II) had both antibodies. The detection rate of NPC from persons examined from IgA/VCA antibody – positive boat people was 386/100 000 and 11 111/100 000, respectively; they were 12.8 and 370 – fold higher than the NPC incidence rate (30/100 000) of the general population of the same age groups in Zangwu County (Tab.3). These data further confirmed that the boat people have a higher incidence rate of NPC than the people living on the land. There are many tung oil trees along the roads and rivers in Zangwu County and much rain in Southern China; thus, the soil particles with EB virus inducer easily reach the river after rain (Ito et al., 1983, Zeng et al., unpublished data). The boat people drink river water throughout their lives; therefore, the high incidence of NPC in boat people might be related to the drinking water containing EBV inducer from tung oil trees or from other sources. This needs to be confirmed.

3. Wuzhou city

In order to carry out a follow-up study on the EBV IgA/VCA antibody-positive persons and to reduce the mortality rate of NPC through early diagnosis and early treatment, another mass serological survey was carried out in Wuzhou City located in the center of Zangwu County. Wuzhou City (population 170 000) was a high risk for NPC; the mean annual incidence was 17/100 000. In the first atage, sera from 12 932 persons of the 40 – 59 age group were examined (Zeng et al., 1982a). IgA/VCA antibody (≥1:10) was found in 680 persons with a GMT of 1:39, the positive rate being 5.3%. About 70% of the people had VCA/IgA antibody titers from 1:10 to 1:40. There is a tendency for an increase with increasing age in the antibody-positive rate and in GMT. The frequency of VCA/IgA antibody for males and females was similar, and the rate was 1.2:1. Clinical, cytological, and histological examinations and the EBNA test were carried out in combination to detect NPC patients from IgA antibody-positive persons. Thirteen NPC cases were finally confirmed histologically. The frequency of NPC among 680 IgA/VCA antibody-positive persons was 1.9%; 9 out of 13 NPC cases also had IgA/EA antibody. The frequency among IgA/EA antibody positive was 30%. The detection rate of NPC was 1900/100 000 for 680 IgA/VCA-positive persons. This is 38 times the NPC annual incidence (50/100 000) in the general population of the same age group (Tab. 3). The early diagnosis of NPC means a favorable outcome of radiotherapy (Chang et al., 1980). Of 13 NPC patients, 9 were in stage 1 (70%) and 4 in stage II (30%). According to the data obtained from Zangwu County and Wuzhou City, the NPC prevalence among individuals having IgA/VCA antibody was found to be related to the antibody titers: 0.9% for invididuals with IgA/VCA antibody titers between 1:10 to 1:20, 2.3% for individuals with IgA/VCA antibody titers between 1:10 to 1:20, 2.3% for individuals with IgA/VCA antibody titers between 1:40 to 1:80, 5.6% for individuals with IgA/VCA antibody titers between 1:160 and 1:320, and 18.6% when IgA/VCA antibody titers reached 1:640 to 1:2560 (Tab. 4).

Tab. 4 Prevalence of NPC by distribution of IgA/VCA antibody titers

	Reciprocal IgA/VCA titers				
	10—20	40—80	160—320	640—2 560	Total
No. IgA/VCA +	850	520	178	59	1 607
No. NPC	8	12	10	11	41
Prevalence (%)	0.9	2.3	5.6	18.6	2.6

To date, 20 726 persons over 40 years of age have been examined, and 1136 persons were found to have IgA/VCA antibody. Among them 31 NPC cases were detected, 13 in stage I (41.9%), 15 in stage II (48.4%), and 3 in stage III (9.7%). The frequency of early cases

(stage Ⅰ and Ⅱ) was up to 90.3%. It was much higher than that in the outpatient clinic (3% NPC in stage I, 39.3% of NPC in stage Ⅱ). Therefore, it is possible to reduce the mortality rate of NPC in Wuzhou City through early detection and early treatment.

A serological survey was carried out in a chemical factory (Zeng et al. unpublished data). Samples of 216 sera from persons of age over 40 were tested, and 22 had IgA/VCA antibody. The positive rate was 10.2%, which was higher than that (5.3%) in Wuzhou City. Among them, 3 NPC cases in early stages (2 in stage I, 1 in stage Ⅱ) were detected. The detection rate of NPC from persons examined and from IgA/VCA – positive persons was 1380/100 000 and 13 636 /100 000, respectively. It was 27.6 and 272.7 times the incidence rate of NPC (50/100000) in the general population of the same age group (Tab. 3). The retrospective study showed that 6 NPC cases were found in this factory from 1973 to 1982, and all of them were working or had worked in the oil – refining workshop. The IgA/VCA antibody – positive and detection rate of NPC was higher than that in Wuzhou City (Tab. 3), which possibly could be related to some chemicals found in the workshops that could activate the EB virus. These data further indicate that the EB virus plays an important role in the development of NPC and that relationships between the activation of EB virus and the environmental factors should be studied. A prospective follow – up study of the whole population over 30 years of age is being carried out in Wuzhou City to detect NPC patients in the early stages and to study the causative role of EB virus in the development of NPC.

4. Laucheng county

In order to determine the prevalence of IgA antibody and the NPC detection rate in the Molaos minority, sera frol 15 324 persons age 30 and over were tested for EB virus IgA/VCA and IgA/EA antibodies by means of the imunoenzymatic test (Wei et al. unpublished data). A total of 151 persons had IgA/VCA antibody and 10 from IgA/VCA antibody – positive persons had IgA/EA antibody. The positive rate of IgA/VCA antibody was 0.98% which was lower than that in Zangwu County and Wuzhou City. The positive rate of IgA/EA from IgA/VCA – positive persons was 6.6%. We detected 7 NPC (2 in stage Ⅱ, 2 in stage Ⅲ, and 3 in stage Ⅳ) in IgA/VCA antibody – positive persons; 4 of these in stage Ⅲ and Ⅳ also had IgA/EA antibody. The detection rate of NPC from IgA/VCA and IgA/EA antibody – positive persons was 4636/100 000 and 40 000/100 000, respectively, which was similar to that in Wuzhou City (Tab. 3). The detection rate of NPC from IgA/VCA antibody – positive persons was 154.4 times higher than the incidence rate of NPC in the general population of the same age group in high – risk areas in Zangwu County. It seems that a few more NPC patients in the early stages (stages Ⅰ and Ⅱ) should have been detected, since only 2 in stage Ⅱ out of 7 NPC patients were found. They might have been missed clinically or histologically and should be followed up. More sera from other national minorities are being studied.

IgA/EA Antibody as a Specific Marker for NPC

The detection of EB virus IgA/EA antibody cannot be used alone for the routine diagnosis of NPC because other immunofluorescence test or immunoenzymatic test only reveals this antibody in

50% —75% of NPC patients (Henle and Henle, 1976; Lab. of Cancer Institute et al. , 1978).
IgA/EA antibody can only be detected in IgA/VCA antibody – positive but not in antibody – negative
individuals. The detection rate of NPC among IgA/VCA antibody – positive individuals is 1. 9% and
that among IgA/EA antibody – positive individuals is 30%. The difference between these two groups
is 15. 8% fold. These data indicate that only a few IgA/VCA antibody – positive individuals eventually become IgA/EA antibody positive, and that IgA/EA antibody is more specific, but not so sensitive for the detection of NPC as IgA/VCA antibody. It is therefore necessary to improve the sensitivity of this test for the detection of IgA/EA antibody. An immunoautoradiographic method was established in our laboratory using ^{125}I – labeled anti – human IgA antibody for the detection of EB virus
IgA/VCA antibody in sera and saliva (Liu et al. , 1979a, b). This test is much more sensitive
than the immunoenzymatic test, but it is too sensitive for detection of IgA/VCA antibody, so the
positive percentage of IgA/VCA antibody in normal individuals in too high however, this test is useful for the detection of IgA/EA antibody (Zeng et al. , 1982a). Sera from different groups were
tested by immunoflrorescence, immunoenzymatic, and immunoautoradiographic methods. The positive rate and GMT of IgA/EA antibody by immunoautoradiography in sera from NPC patients were
96% and 1:97, respectively. Those patients with clinically suspected NPC and who had been diagnosed histologically as having a chronic inflammation of the nasopharynx had a positive rate of 22%
and a GMT of 1:9 . Of 170 patients, 7 with malignant tumors other than NPC were positive (4%)
to low titer 1:5. 2. All normal individuals were negative. The positive rate and GMT detected by the
immunoenzymatic test in these four groups were 80% and 1:25. 6, 18% and 1:7. 3, 0. 6% and
1:5, and 0% and 1:5, respectively. In the suspected NPC group (chronic inflammation) biopsies
were taken again from 11 IgA/EA antibody – positive individuals, and histological examination
showed poorly differentiated squamous cell carcinoma in 6 of them. Fourteen NPC patients had no
IgA/VCA and IgA/EA antibodies detected by immunofluorescence and immunoenzymatic tests:
however, among these 14 patients, 11 had IgA/VCA and 6 had IgA/EA antibody decected by im-
munoactoradiography. These data indicate that immunoautoradiography is more sensitive than either
the immunofluorescence of the immunoenzymatic test in the dectection of IgA/EA antibody, and that
IgA/EA antibody can serve as a specific marker for the detection of NPC. Therefore, for seroepidemiological study, it is better to proceed first using the immunoenzymatic or immunofluorescence test
for the detection of IgA/VCA antibody. If this antibody is positive, the detection of IgA/EA antibody
in the same serum by immunoautoradiography should then be carried out , and more NPC patients in
the early stage will be detected. Recently, the immunoautoradiographic method was further simpli-
tied by using X – ray film instead of nuclear emulsion in our laboratory (Pi et al. 1981). Similar results were obtained with these two methods, but the use of X – ray film is more convenient for large
– scale screening in the NPC Held.

Relationship between IgA/VCA Antibody and
EB Virus Markers (DNA and EBNA)

In order to determine whether the presence of serum EBV/IgA antibody corresponded to the EB virus activity in the nasopharyngeal mucosa, 56 individuals having serum IgA/VCA antibody for 15 – 18 months were clinically examined and biopsied (de The et al. , 1981). A total of 4 NPC cases were found and 14 additional individuals, for whom no histopathological or clinical evidence of NPC was noted, were found to have detectable EBV/DNA sequences and/or EBNA in their nasopharyngeal mucosa (Desgranges et al. , 1982). It seems therefore that the presence of EB virus markers in the nasopharyngeal mucosa could be another critical marker for the immediate risk of NPC. But it is important to know whether the EB virus DNA is also present in the nasopharyngeal mucosa from IgA/VCA antibody – negative individuals. As it was not posible to biopsy nasopharyngeal mucosa of normal and IgA/VCA antibody – negative individuals, a negative – pressure apparatus was used to collect the exfoliated cells from the nasopharynx of 62 IgA/VCA antibody – positive and 39 IgA/VCA antibody – negative individuals from Wuzhou City. The exfoliated nasopharyngeal cells were tested for the presence of EBV/DNA sequences by spot followed by blot hybridization tests (Desgranges et al. , 1983). Of 62 specimens, 13 (21%) from IgA/VCA antibody – positive and 6 of 39 specimens (15. 4%) from IgA/VCA antibody – negative individuals were found to contain EBV/DNA sequences (Tab. 5). We followed 46 persons (20 IgA/VCA antibody positive and 26 IgA/VCA antibody negative) for a year. Half of the individuals having EBV/DNA sequences in their exfoliated nasopharyngeal cells in 1981 did not have detectable EBV sequences a year later, and 2 of the 15 negative individuals became EBV/DNA positive. These results suggest that the presence of EBV/DNA sequences and IgA/VCA antibody in the serum are not directly related.

EBNA is regularly found in nasopharyngeal carcinoma cells by means of the anticomplement technique (Klein et al. , 1974; Huang et al. , 1974). We have established an anticomplement immunoenzymatic method (ACIE) for the detection of EBNA (Zeng et al. , 1980b, 1981; Pi et al. , 1981) which thus proved to be a sensitive test. The exfoliated cells from the nasopharynx obtained by negative pressure were examined by ACIE for the detection of EBNA, for the diagnosis of NPC, and for the study of the relationship of EB virus and NPC. EBNA was found not only in NPC cells, but also in normal epithelial and hyperplastic cells (Zeng et al. , 1981, 1982b). Further study of EBNA in the exfoliated cells of the nasopharynx of persons with nasopharyngeal chronic inflammation showed that 34. 6% of IgA/VCA antibody – positive persons and 19. 6% of IgA/VCA antibody – negative persons had EBNA in their epithelial cells (Tab. 5) (Shen et al. , 1983). These data were similar to that of EBV DNA, but there was no obvious correlation between EB virus markers (DNA and EBNA) and EB virus serology.

Tab. 5 Detection of EBV markers in nasopharyngeal mucosa from IgA/VCA – positive and negative Individuals

EBV markers	IgA/VCA antibody positive	IgA/VCA antibody negative	Reference
DNA +	13/62(21%)	6/39(15. 4%)	Desgranges et al(1983)
DNA –	49/62(79%)	33/39(84. 6%)	
EBNA +	9/26(34. 6%)[a]	9/46(19.6%0)[a]	Shen et al (1983)
EBNA –	7/26(65. 4%)[a]	37/46(80. 4%)[a]	

Notes: [a] Chronic inflammation of nasopharynx

Relationship Between EB Virus IgA/VCA Antibody and Clinical Changes in Nasopharynx

The correlation between mucosal hyperplastic lesion and the level of antibody to EB virus was studied by complement fixtion (Dept. of Microbiology, Zhangsham Medical College et al, 1980). It was interesting to note that the GMT (1:90) of anticomplement – fixing antibody to EB virus in persons with hyperplastic lesions was significantly higher than that (1:66) in normal individuals.

A serological mass survey was carried out in Wuzhou City in persons 40 to 59 years of age (Zeng et al. , 1983a; Zhang et al. , unpublished data). Among 20 726 sera, 867 persons having IgA/VCA antibody were examined with indirect nasopharyngoscopy. The GMT of IgA/VCA antibody in persons with normal mucosa, polyps, residual adenoid, chronic inflammation, asymmetry of nasopharyngeal cavity, atropic changes of nasopharynx, and hypertrophic nodule or tumor – like nodule was 1 : 32. 6, 1 : 20, 1 : 28. 7, 1 : 28. 7, 1 : 40, 1 : 47, 1 : 52. 2, and 1 : 78. 6, respectively. According to their severity or clinical changes these seven types of conditions were classified into group 1 (normal mucosa, polyps, and residual adenoid), group 2 (chronic inflammation, asymmetry of nasopharyngeal cavity, and atrophic change of nasopharynx), and group 3 (hypertrophic nodule or tumor – like nodule). There was a marked difference in GMT between these three groups. Among 38 cases with hypertropic nodules in group 3, 24 NPC (63. 2%) were detected. A group of 248 IgA/VCA antibody – negative persons was also examined for lesions in the nasopharynx as a control. Most ('74. 6%) belonged to group 1, and only a few were in group 2 (25. 4%). NPC was not found in the antibody – negative persons. These data indicate that the clinical changes in the nasopharynx and the development of NPC were closely related to the presence of IgA/VCA antibody.

Relationship between EB Virus IgA/VCA Antibody and Histological Changes in Nasopharynx

From 1978 to 1981 serological mass surveys for NPC were carried out in Zangwu County; the antibody – positive persons were examined by nasopharyngoscope. Biopsies were taken from the following atibody – positive persons: (1) those diagnosed clinically as NPC patients or suspected of having NPC; (2) those having some lesions in the nasopharynx, such as hyperplasia or a residue of adenoidal tissue, rough mucosa, local congestion, or inflammation; and; (3) those without lesions. in the nasopharynx but with high antibody titer (E. J. Li et al. , 1983). The GMT of IgA/VCA antibody from group 1 (persons with normal mucosa, simple hyperplasia, and simple metaplasia) and group 2 (atypical hyperplasia and atypical metaplasia) was 1:25. 2, 1:30. 4, 1:22. 4; 1:105. 4, and 1:80, respectively. There was a marked difference in GMT of IgA/VCA antibody between these two groups (1:25. 3 and 1:86. 7).

A group of 45 IgA/VCA antibody – positive persons persons without carcinoma were fwollowed for 8 – 33 months with serological and histological examinations. Nasopharynx changes were of the following three types: (1) simple hyperplasia and simple metaplasia evolved into atypical hyperplasia, atypical met aplasia, or into carcinoma; (2) atypical hyperplasia and atypical metaplasia evolved into carcinoma or remained the same form of lesion; (3) carcinoma in situ with little invasion found in the first and final biopsies. When the pathological changes were more advanced, their antibody levels were elevated. The difference was significant. In 32 persons without any lesions in the nasopharynx, there was no remarkable alteration shown in the initial and final clinical examinations; their antibody levels did not show any obvious difference. When antibody levels were found to be elevated in some persons, it may be assumed that the pathological changes in the nasopharynx were more severe. Therefore, IgA/VCA antibody may be regarded as a reference index of the development of the pathological lesions in the nasopharynx. Among 45 IgA/VCA antibody – positive persons, 5 of 17 with atypical hyperplasia or atypical metaplasia developed NPC (29.4%), whereas in 38 persons with simple hyperplasia, simple metaplasia, or normal mucosa. only 1 developed NPC (2.6%). The difference in the development of NPC in these two groups was significant (Li, 1985). The frequency of atypical lesions in the antibody – positive cancerous group was highest, but it was lower in those in the antibody – positive noncancerous group, and least in the antibody – negative noncanerous group. All these results suggest that EB virus plays an important role in the development of NPG.

Conclusion

Serological studies showed that the GMT of complement – fixing antibody to EB virus in healthy individuals over 20 years of age in the high NPC risk areas was significantly higher than that in the low NPC risk areas; that the positive rate of IgA/VCA antibody in sera from the general population of 30 years of age and over in high NPC risk areas increased with increasing age; that the presence of IgA/VCA antibody was closely related to clinical and pathological changes in the nasopharynx; that IgA/VCA antibody to EB virus is specific for NPC; that the detection rate of NPC from IgA/VCA antibody – positive persons was much higher than that from the general population of the same age group; and that the IgA/VCA antibody can be detected 8 – 30 months prior to the clinical and histological diagnosis of NPC in stage I. The detection rate of NPC from IgA/EA – positive persons was up to 30% – 40%. These data suggested that EB virus plays an important role in the development of NPC and that the mortality rate of NPC can be reduced through early detection and treatment. But EB virus is not the unique factor. Others, such as environmental and genetic factors and their synergistie effects with EB vires, might also play an important role. This needs further study.

〔In 《Advance in Cancer Research》 1985, 44: 121 – 139〕

References

1 Atlas of Cancer Mortality in the People's Republic of China, Editorial Committee. China Map Press. 1979

2 Chang C P, Liu T F. Chang Y W, Cao S L. Acta Radiol. Oncol, 1980, 19: 433 – 438

3 Dept of Microbiology and Cancer, Hospital of Zhongshan Medical College, Dept. of Virology, Cancer Institute; Dept. of Tumor Viruses, Institute of Virology; Cancer Inst. of Zhongshan County Chin. Med J, 1980, 93: 359 – 364

4 De Schryver A, Friberg S Jr, Klein G, Henle G, Henle W, de – The G, Clifford. P Ho H C. Clin. Exp. Immunol, 1969, 5: 443 – 459.

5 Desgranges, de The G (1978). IARC Sci. Publ 1978, 24: 883 – 891

6 Desgranges C, Bornkamm G W, Zeng Y, Wang P C, Zhu J S, Shang M, med de The G. . Int J Cancer, 1982, 29: 87 – 91

7 Desgranges C, Pi G H, Bornkamm G W, Lagrand. C Zeng Y. and de The G. Int J. Cancer, 1983, 32: 543 – 545.

8 De The G, Desgranges C, Zong Y, Wang P C, Bornkamm G W, Zhu J S, Shang M. Int Symp Nasopharyngeal Carcinoma, 11 th, Duesseldorf, 1981, 5: 111 – 118

9 Henle G, Henle W. Int J Cancer, (1976), 17: 1 – 7

10 Henle W, Henle G, Ho H C, Burtin P, Cachin Y, Clifford P De, Schryyver A, de The G, Diehl V, Klein G. Cancer Inst, 1970, 44: 225 – 231

11 Ho J H C, Int J Radiol Oncol Biol. Phys, 1978, 4: 181

12 Ho J H E, Ng M H, Kwan H C, Chan J C W, Br J Cancer, 1976, 34: 655 – 660

13 Ho J H C, Kwan H C, Ng M H, de The G, Lancet, 1978, 1: 436 – 437

14 Huang D P, Ho J H C, Henle W, Henle G. Int J Cancer, 1974, 14: 580 – 588

15 Ito Y, Kishishita M, Yanase S. Cancer Res, 1980, 40: 4329 – 4330

16 Ito Y, Ohigashi H, Koshimizu K, Zeng Y. Cancer Lett, 1983, 19: 113 – 117

17 Jian S W, Li Z W, Luo W L, Li M S, Pan W Z, Zhang X H. Cancer Res Rep Cancer Inst. Zhongshan Med College, 1981, 2: 23 – 25

18 Klein G, Giovanella B C, Lindahl T, Fialkow P J, Singh S, Stehlin J. Proc Natl Acad Sci U S A, 1974, 71: 4737 – 4741

19 Lab of Tumor Viruses, Cancer Inst: Lab of Tumor Viruses, Inst. of Epidemiology; Dept. of Radiotherapy, Cancer Inst: Dept. of Otolaryngology. Beijing Worker – Peasant – Soldier Hospital. Acta Microbiol Sin, 1978, 18: 253 – 258

20 Lanier A P, Henle W, Bender T R, Henle G, Talbot M L. Int J Cancer, 1980, 26: 133 – 137

21 Li E J, Tan B F, Zeng Y, Wang P C, Zhong J M, Tang H, Zhu J S, Wei J N, Pan W T. Chin J Pathol, 1983, 12: 9 – 11

22 Li E J, Tan B F, Zeng Y I, Wang P Z, Zhong J M, Tang H, Zhu J S, Liu Y X, Wei J N, Pan W J. Chin Med J In Press 1985

23 Li S J, Zhon Y B, Hu X T, Zeng Yi, Chin J ENT, 1980, 15: 71 – 74

24 Li W J, Li C C, Liang Y R, Chen A M, Zhang F, Ho P Y. Cancer (Chinese), 1982, 1: 43 – 48

25 Liu Y X, Zeng Y, Dong W P, Gao G R. Chin J Oncol, 1979, 1: 8 – 12

26 Liu Z R, Shan M, Zeng Y, Han Z S, Dai H J, Hu Y L, Cao G R, Dong W P. Kexne Tongbao, 1979a, 24: 715 – 720

27 Liu Z R, Shan M, Zeng Y D, ai H J, Du R S. Chin J Med Exam, 1979b. 2: 197 – 198

28 Old L J, Boyes E A, Oettgen H F, De Harven E, Geering G, Williamson B, Clifford, P. Proc, Natl Acad Sci U S A, 1966, 56: 1699 – 1704

29 Pi G H, Zeng Y, Zhao W P, Zhang Q. J Immunol Methods, 1981, 4473 – 78

30 Shen S J, Chen C P, Ma. J L, Zhong W, and Zeng Y, Acta Zhanjiang Med, College, 1983, 1: 34 – 37

31 Tumor Control Team, Zhongshan County, Dept. Microbiology, Zhongshan Med. College; Lab of Tumor Viruses. Cancer Inst and Inst of Epidemiology, Chinese Academy of Medical Sciences. Chin J Ear Nose, Throat, 1978, 1: 23 – 25

32 Wei J N, Zhang S, Tung S Z, Huang Z L. Guangxi Yi Xue, 1980, 6: 5 – 6

33 Wolf H, Zur Hansen H, Becker V. Nature (London). New Biol, 1973, 244: 245 – 247

34 Zeng Y, Liu Y X, Liu C R, Chen S W, Wei J N, Zhu J S, Zai H G. Chin J Oncol, 1979A, 1: 2 – 7

35 Zeng Y, Liu Y X, Wei J N, Zhu J S, Cai S L, Wang P Z, Zhong J M, Li R C, Pan W J, Li E J, Tan BF. Acta Acad. Med. Sin, 1979b, 1: 123 – 126

36 Zeng Y, Zhang M, Liu Z R, Zheng Y H, Du R S, Li X H, Gan B W, Hu M G, Zhen M, He S A, Mu G P, Chin J Oncol, 1979c, 1: 81 – 83

37 Zeng Y, Liu Y X, Liu Z R, Zhen S W, Wei J N, Zhu J S, Zei H S. Intercirology, 1980a, 13: 162 – 168

38 Zeng Y, Pi C H, Zho W P. Acta Acad. Med. Sin, 1980b, 2: 134 – 135

39 Zeng Y, Pi G H, Zhang Q, Shen S J, Zhao M L, Ma J L, Dong H J. Int Symp Nasopharyngeal Carcinoma, 11 th Duesseldorf, 1981, 5: 237 – 244

40 Zeng Y, Zhang L G, Li H Y, Jan M C, Zhang Q, Wu Y C, Wang Y S, Su G R. Int J Cancer, 1982a, 29: 139 – 141

41 Zeng Y, Zhen S J, Dan H, Ma T L, Zhang Q, Zhu J S, Zheng T R, Tan B F. Acta Acad Med Sin, 1982b, 4: 254 – 255

42 Zeng Y, Cong C H, Jan M G, Fan C, Zhong L K. Int J Cancer, 1983a, 31: 599 – 601

43 Zeng Y, Zhong J M, Mo Y K, Miao X C, Intercirology, 1983b, 19: 201 – 204

44 Zeng Y, Zhong J M, Li L Y, Wang P Z, Tang H, Ma Y R, Zhu T S, Pan W J, Liu Y X, Wei J N, Chen J Y, Mo Y K, Li. E J, Tan. B F. Interoirology, 1983, 20: 190 – 194

45 Zhu Z S, Pan W J, Zhong J M, Li L Y, Zeng Y, Jan M G, Fan C. J Tumor Precent Treat Study, 1983, 10: 189 – 190

46 Zur Hausen H, Schulte Holthausen H, Klein G, Henle W, Henle G, Clifford P, Santessoni L. Nature (London), 1970, 228: 1056 – 1058

32. Nasopharyngeal Carcinoma: Early Detection and IgA – Related pre – NPC Condition, Achievements and Prospectives

ZENG Yi, de The Guy

Institute of virology – Beijing PRC and CNRS laboratory, FAC of med a carrel, lyon france

Introduction

Undifferentiated carcinomas of the nasopharynx (or NPC) represent a major cancer killer for more than 200 million people in South China, as well as in large areas of the southeast Asia, North and East Africa and in Eskimo populations. This cancer is closely associated with the ubiquitous Epstein Barr herpes virus. In contrast with the situation observed in Burkitt's Lymphoma, the association between EBV and NPC is constant in every part of the world where NPC is observed, and unrelated to its level of incidence. Such an association, most probably causative in nature, has recently been reviewed (de – The, 1982, 1984).

Following the observation of Wara, W M. et al. in 1975, that NPC patients had high level of IgA antibodies, Henle and Henle (1976) and Desgranges and de – The (1978) showed that such IgA were directed against VCA and EA and were regularly observed in NPC patients from Chinese, Arabic and Caucasian origins, but absent in Patients with other ENT tumor. These data urged us to use the IgA/VCA test for early detection of this tumor in the endemic areas of South China (Zeng et al. , 1979, 1980, 1982). We shall review first these population surveys, then discuss the pre – NPC conditions associated with rising titers of IgA.

The interplay between an ubiquitous EB Virus and nasopharyngeal carcinoma stresses the need for other environmental factors, possibly related to life – style, and to the reactivation of EBV.

1. Early detection of NPC in high risk populations of south china

(1) 1978—1980: Survey in Zang – Wu County

A major survey was implemented in 1978, in a rural area of the Eastern part of the Guang – Xi autonomous region (Zeng et al, 1979, 1980). The County of Zeng – Wu. comprises 15 communes with a total population of 450 000. Starting in 1978, individuals aged 30 and above, were registered and a small amount of blood was collected. The sera were tested for IgA/VCA antibodies by the immunoenzymatic test (Zeng et al, 1980). As seen in Tab. 1, a total of 148 029 persons were thus screened and 3 533 were found to have IgA/VCA antibodies. The positive sera were then titered and 13% showed titer superior or equal to 80, representing 460 individuals. All the 3 533 IgA/VCA positive persons, were then clinically examined, and 55 NPC patients uncovered. The clinical stages at which the patients were recognized are given in Tab. 1. where it can be seen that the majority were at stage Ⅱ and Ⅲ of the disease.

The clinical follow – up of the IgA positive individuals for 1 to 3 years led to the diagnosis of another 32 cases. The distribution of these cases according to stages, was not dramatically different from that of the main survey although there was a shift to early ages of detection (Tab. 1).

Tab. 1　Detection of NPC Cases in Zang – Wu County during
1978 – 1980 by IgA/VCA Test and during 1 to 3 years follow – up

	No of surveyed person	Ca insitute	Stages of NPC detected								Total NPC detected
			I	%	Ⅱ	%	Ⅲ	%	Ⅳ	%	
1978 – 1980 Serological mass survey	148 029	1	12	22	19	34	17	31	6	11	55
Follow – up	3 478	0	10	31	9	28	11	34	2	6	32
for 1 – 3 years of IgA/ VCA + Total		1	22		28		28		8		87

Tab. 2   Survey in Wu – Zhou City

	No persons examined	No IgA/VCA positive	%	No NPC	Stages of NPC							NPC prev. rate in survey	NPC prev in IgA/VCA + indiv.
					I	%	II	%	III	%	IV		
General Populat	20 726	1 136	5.5	35	15	43	17	48.5	3	8.5	0	1.5	27
Chemical Factory	216	22	10	3								14	136

（2） Wu – Zhou City

The City of Wu – Zhou, located in the centre of the Zang – Wu County, with a population of 170 000, was an ideal place to try and implement an early detection of NPC with a systematic survey and follow – up of the town dwellers.

Previous cancer registration in the town of Wu – Zhou, showed that the mean annual incidence of NPC was around 17 to 20 per 100 000. As seen in Tab. 2, survey of nearly 21 000 individuals aged 40 and above in Wu – Zhou, showed that 1 136 （5.5%） persons had IgA/VCA antibodies. Clinical examination of this later group allowed the detection of 31 cases of NPC, the clinical stages being given in Tab. 2. It is remarkable to see that in this survey of the Wu – Zhou City, the proportion of stage I and II represented 90% of the tumor detected. This was due to the fact that clinical detection of the tumor had been efficient since a few years in this town. The prevalence rate of NPC in the surveyed population thus reached 150/100 000, a very high figure if one considers tuat patients at stage III and IV of the disease present in hopitals were not included.

A serological survey, carried out in a chemical factory, detected more NPC than expected. As seen in Tab. 2. 216 individuals were tested in this factory and 22 or （10%） were found IgA/VCA positive. Among those, 3 were discovered having NPC, Although this may be due to chance an experimental investigation of the chemicals handled in this factory has been implemented.

（3） Laucheng County

Situated in a Northern part of the Guang – Xi autonomous region, Laucheng County is inhabited by two sub – 1 anguage groups, namely the Molaos and the Hans. As seen in table 3, the survey of 15 324 individuals from the Molaos minority gave a prevalence of 1% IgA/VCA positive （151 persons） and 7 cases of NPC were detected. In the Han majority, 0.6% of the surveyed population had IgA/VCA antibodies and 6 cases of NPC were detected （Tao et al. , in press）.

（4） IgA/EA antibodies represent a better test for early detection of NPC

Whereas immunoflorescent （Desgranges and de The, 1978） and immunoenzymatic tests （Laboratory of Cancer Institute, 1978） detected IgA/EA antibodies in about 70% to 75% of NPC patients, the immuno – autoradiographic test developed by Zeng et al, 1983 （using [125]I – labelled antihuman IgA ant ibodies） detected IgA/EA antibodies in 96% of NPC patients with a GMT titer of 1:97. Using this later test, patients with chronic inflammation of the nasopharynx, had IgA/EA antibodies in 22% of the cases with a GMT titer of 1:9. Patients with malignant tumors other than NPC were positive in 4% of the cases with a GMT titer of 1:5. All normal individuals were found negative.

Tab. 4. gives the comparative results of both the IgA/VCA and IgA/EA tests in detecting NPC in Wu – Zhou City and Launcheng County. It can be seen that 30% to 43% of individuals with IgA EA antibodies have a detectable NPC.

Tab. 3 Survey of Laucheng County

Sub – language groups	No. persons examined	No. IgA/VCA positive	%	NPC detected	NPC prev- rate in survey	NPC prev. in IgA/VCA + indiv
Molaos	15 324	151	1	7	45/100 000	4. 6%
Han	11 117	76	0. 6	6	54/100 000	7. 9%

Tab. 4 Comparison of IgA/VCA and IgA/EA in detecting NPC*

	No. persons surveyed	IgA VCA +	IgA/EA +	NPC detected	% of NPC among IgA +
Wu – Zhou City	12 930	68013		2	
			30	9	30
Laucheng County	26 441	227		13	5. 7
			14	6	43
Total NPC detected					28

* by immunoenzymatic test

Such test, sensitive and specific, could replase the IgA/VCA test for the early detection of NPC, since the background noise of IgA/EA in non NPC individuals is very low. Furthermor, IgA/EA test Will be most instrumental for detecting and investigating pre – NPC conditions (see below).

(5) an elisa test using monoclonal antibodies has been recently developed (Pi et al. submitted for publication)

Such a test, which has nearly the same sensibility and a better specificity for IgA/VCA and IgA/EA than the immunoenzymatic and immunoautoradiographical tests and which is best adapted to field conditions, should ease the implementation of population ser-oepidemiological surveys, in large areas of Souty – East Asia, where this cancer is a main killer.

2. IgA related pre-NPC conditions

It appears as if the presence of IgA antibody to VCA and to EA represents pre NPC coNditions (de – The, Zeng et al. , 1983). In order to see whether the presence of IgA antibodies correspon-ded to a specific viral activity in nasopharyngeal mucosa, 56 individuals with IgA/VCA antibodies for more than 18 months, were clinically examined, and biopsied. Four NPC cases were found (two at early stages of the disease) and further 14 individuals had detectable EBV/DNA sequences and/or EBNA antigen in their nasopharynegal mucosa without histopathological nor clinical evidence of NPC (Desgranges et al. , 1982). As it was not possible to take nasopharyngeal biopsies from normal ivdividuals lacking IgA/VCA antibodies, exfoliated cells collected from the nasopharynx (using a negative pressure apparatus developed by Zhangjiang Medical College in 1976) in 62 IgA/ VCA antibody positive and 39 IgA/VCA antibody negative individuals were tested for the presence of EBV/DNA sequences by spot followed by blot – hybridization (Desgranges et al. , 1983). As seen in Tab. 5, 13 of the 62 IgA/VCA positive specimen (21%), and 6 out of the 39 IgA negative specimen (15.4%), were found to contain EBV/DNA sequences. Among those, 20 IgA/VCA antibody positive and 26 IgA/VCA antibody negative individuals were followed a year later. Their exfoliated cells from the nasopharynx were again collected, and tested for the presence of EBV/DNA sequences. Three out of seven individuals who showed a year previously EBV/DNA sequence in exfoliated nasopharyngeal cells, failed to do so a year later. In parallel, 2 out of 15 EBV/DNA negative exfoliated cells became EBV/DNA positive a year later. Such results suggest that the presence of EBV/DNA sequences in the nasopharynx, and the presence of IgA/VCA or IgA/EA antibodies in the serum, are not directly related. Unfortunatly, the cell type harbouring the EBV/DNA could not be characterized in these studies. In situ, hybridizations made by A. Wolf et al. (unpublished) on a few samples from IgA/VCA positive individuals suggested that the EBV/DNA positive cells were of epithelial nature.

Tab. 5   Comparative detection of EBV/DNA and EBNA in IgA/VCA positive and negative individuals

	No. tested	EBV/DNA* positive	No. tested	EBNA** positive
IgA/VCA positive individuals	62	13 21%	26	9 34%
IgA/VCA negative individuals	39	6 15%	46	9 20%

Notes: * Desgranges et al, 1983; * * Shen et al, 1983

Tab. 6   Stability and fluctuations of IgA/VCA antibody over 3 years in relation to risk of NPC

Item	Perons IgA/VCA +	stability (no change in IgA/VCA Ab)	Loss of IgA/VCA Ab (retroversion)	Fluctuation in IgA/VCA Antibodies		
				Increase[a]	Decline[b]	Variations of IgA/VCA Ab[a,b]
n	1138	455	398	81	162	42
%	100%	40%	35%	7.1%	14.2%	3.9%
	NPC patients detected	6		15		

Notes: a, b: 4 fold increase (a) or decrease (b)

Using an anticomplement immunoenzymatic method (ACIE), for the detection of EBNA (Shen et al. , 1983), exfoliated cells from the nasopharynx obtained from positive and negative IgA/VCA individuals were tested by this anticomplement immunoenzymatic test. As seen in Tab. 5, 34% of IgA/VCA positive and 20% of IgA/VCA negative individuals has detectable EBNA in cells which were considered as epithelial. Thus the virus seems to be present in normal conditions in nasopharyngeal mucosa. The development of IgA/VCA antibodies must reflect a critical difference in the local immune response against the EB viral infection. In fact, IgA/EA antibodies are usually present in IgA/VCA positive individuals, thus reflecting a reactivation of the virus. That such reactivation takes place in the nasopharynx in highly probable, but not yet established, nor the fact that it precedes and not succeeds subclinical development of NPC.

In the Couny of Zang Wu 1 138 individuals with IgA/VCA antibodies were followed for 3 years from both the serological and clinical view – points. Table 6 shows that 40% of them (455 individuals) exhibited stable IgA/VCA titers. Among those, 6 developed NPC within 3 years of follow – up (1.3%). 398 individuals (35%) lost their IgA/VCA antibodies within this period and no NPC was discovered among them. IgA/VCA antibodies increased by 4 dilutions of more in 81 individuals and 15 of them developed NPC (18.5%). These results (Zeng et al. , in preparation) strongly support the hypothesis that EBV reactivation, reflected by a specific serological profile (increasing titers of IgG EA, IgA VCA, IgA EA) represents a pre – NPC condition. Whether or not such conditions reflect our inablility to detect sub – clinical tumorous growth in the submucosa of the nasopharynx remains to be determined.

3. Perspective and priorities

Early detection of NPC by the IgA/VCA or probably better by the IgA/EA test is feasible today, and should therefore be applied for the benefit of large populations at risk for this tumour which represent approximately 230 millions persons around the world. Tab. 7 gives the difference in the clinical stages of NPC patients diagnosed in out – patients clinicas, and of the patients detected during the above described early detection schemes. The shift towards early stages is obvious (43% versus 1.7% detected at stage I). Such a shift should have a critical impact on mortality by NPC, if one considers the 5 year survival rates after radiotherapy according to clinical stages. In Shanghai, for example (Zeng personal communication), more than 90% of NPC patients treated at stage I of the disease exhibited a 5 year disease free survival. In contrast, NPC patients diagnosed at stage IV and V, which represent the majority in out – patients clinics of endemic areas, have less than one year survival in 70% of the cases.

It is therefore of great interest to see that EBV serology has such a critical and practical impact for patients′ care before the nature of the relationship between the virus and this cancer was uncovered. It is a clear example where important applications for public health can be implemented prior to the understanding of the mechanism involved. If the final proof that EBV is cansually related to NPC is not yet at hand, the results shown in Tab. 1, 4, 6 and 7 strongly favour an etiolgical role of the virus in the development of undifferentiated carcinomas of the nasopharynx.

The priorities in Prevention Research concerning NPC should focus on the understanding of the virological, molecular and immunological events taking place in the nasopharyngeal mucosa during the pre – NPC events. Such an understanding will in turn permit the implementation of primary prevention, either by anti EBV interventions or by eliminating co – factors. Such co – factors may be present in the immediate environment of individuals at risk (Ito et al. , 1983, Zeng et al. , 1983). or possibly accociated with bacterial flora in the nasopharynx (Zeng et al. , in press).

Tab. 7　Comparison of NPC stage from outpatient clinic and from serological screening

	No of cases		Stages			
			I	II	III	IV
Outpatient Clinic	N	1 066	18	312	556	180
Clinic	%	100%	1.7%	31.3%	51.3%	17.2%
Serological +	N	35	15	17	3	
screening	%	100%	43%	48.5%	8.5%	

Acknowledgments:

Some aspects of these studiees were supported by the CNRS, and the Fondation pour la Recherche Medicale, Paris.

〔In《Epstein-Barr Virus and Associated Disease》1985, 151 – 163〕

References

1　Desgranges C, de The G. IgA and nasopharyngeal carcinoma, in: Oncogenesis and Herpesviruses Ⅲ (G. de The, Y. Ito, and F. Rapp eds). pp. 883 – 891, IARC Scientific Publications No, 1978, 25, Lyon

2　Desgranges C, Born kamm G W, Zeng Y, Wang P C, Zhu J S, Shang H, de The G. Detection of Epstein – Barr viral DNA in the nasopharyngeal mucosa of Chinese with IgA/EBV – specific antibodies, Int J Cancer, 1982, 29: 187 – 191

3　Desgranges C, Pi G H, Bornkamm G H. Legrand C, Zeng Y, de The G. Presence of EBV – DNA sequences in nasopharyngeal cells of individuals without IgA/VCA antibodies. Int J Cancer, 1983, 32: 543 – 545

4　De The G, Epidemiology of Epstein – Barr Virus and Associated Diseases in Man. in Man. in: The Herpesviruses, vol 1. Roizman, B (ed). pp 25 – 103, Plenum Publishing Corporation, New York, 1982

5　De The G, The role of the Epstein – Barr Virus (EBV) in the etiology and control of nasopharyngeal carcinma (NPC). In: Cancer of the Head and Neck, Williams and Wilkins Publishers, 1984

6　De The G, Zeng Y. Desgranges C, Pi G H. The Existence of Pre – Nasopharyngeal Carcinoma Conditions Should Allow Preventive Interventions. In: Nasopharyngeal Carcinoma: Gurrent Concepts, Prasad et al, (eds). University of

Malaya, Kuala Lumpur, 1983, 365 – 374

7　Ito Y, Ohigashi H. Koshimizu K, Zeng Y. Epstein – Barr Virus – activating principle in the ether extracts of soils collected from under plants which contain active diterpene esters. Cancer Letters, 19, pp. 113 – 117, Elsevier Scientific Publishers Ireland Ltd, 1983

8　Hene G, Henle W. Epstein – Barr virus – specific IgA serum antibodies as an outstanding feature of nasopharyngeal carcinoma, Int J Cancer, 1976, 17: 1 – 7

9　Laboratory of Tumor Viruses of Cancer Institute, Laboratory of Tumor Viruses of Institute of Epidemiology, Department of Radiotherapy of Cancer Institute, Department of Otolaryngology of Beijing Worker – Peasant – Soldier Hospital, Detection of EB virus – specific serum IgG and IgA antibodies from patients with nasopharyngeal carcinoma, Acta Microbiol Sin, 1978, 18: 253 – 258

10　Pi G H, Zeng Y, de The G. Enzyme – linked immunosorbent Assay for the detection of Epstein – Barr Virus IgA/EA antibody (submitted for publication).

11　Shen S J, Chen C P, Ma J L, Zhong W, Zeng Y. Further study on detection of EBNA from nasopharyngeal exfoliated cells of patients with nasopharyngeal carcinoma. Acta Zhanjiang Medical College, 1983, 1: 34 – 37

12　Tao E G, Wang P C, Wei J N, Li E J, Wei R F, Too C M, Gu S T, Tan S M, Tang H,

Zeng Y, Pi G H. Serological Mass Survey of Nasopharyngeal Carcinoma in Laucheng County, (submitted for publication)

13 Wara W M, Wara D W, Phillips T L, Ammahh A. Elevated IgA in Carcinoma of the Nasopharynx Cancer, 1975, 35: 1313 – 1315

14 Zeng Y, Liu Y X, Wei J N, Zhu J S, Cai S L, Wang P Z, Zhong J M, Li R C, Pan W J, Li E J, Tan B F. Serological mass survey of nasopharyngeal carcinoma. Acta Acad Med Sin, 1979, 1: 123 – 126

15 Zeng Y, Liu Y X, Liu Z R, Zhen S W, Wei J N, Zhu J S, Zei H S. Application of an immunoenzymatic method and an immunoautoradiographic method for a mass survey of nasopharyngeal carcinoma, Interviology, 1980, 13: 162 – 168

16 Zeng Y, Zhang L G, Li H Y, Jan M C, Zhang Q, Wu Y C, Wang Y S, Su G R, Serological mass survey for early detection of nasopharyngeal carcinoma in Wuzhou City, China. Int J Cancer, 1982, 29: 139 – 141

17 Zeng Y, Zhong J M, Li L Y, Wang P Z, Tang H, Ma Y R, Zhu T S, Pan W J, Liu Y X, Wei J N, Chen J Y, M Y K, Li E J, Tan B F. Follow – up studies on Epstein – Barr, virus IgA/VCA Antibody Positive Persons in Zangwu County, China. Intervirology, 1983, 20: 190 – 194

18 Zeng Y, Gi Z W, Wang P C, Tan H Z, Ito Y. Induction of Epstein – Barr Virus Antigen in Raji cells and P3HR – 1 cells by culture fluids of anaerobes from nasopharynx of patients with nasopharyngeal carcinoma and with other ear – nose – throat diseases. (submitted for publication)

19 Zeng Y, Miao X C, Jaio B, Li H Y, Ni H Y, Ito Y. Epstein – Barr Virus activation in Raji cells with ether extracts of soil from different areas in China. Cancer Letters, in press

20 Zeng Y, Gong M G, Jan M G, Zeng L G, Li H Y. Detection of Epstein – Barr Virus IgA/EA antibody for diagnosis of nasopharyngeal carcinoma by immunoautoradiolgraphy. Int J Cancer, 1983, 31: 599 – 601

21 Zeng Y, Zhong J M, Mo Y K, Miao X C. Epstein – Barr virus early antigen induction in Raji cells by Chinese medical herbs. Intervirology, 1983, 19: 201 – 205

22 Zhangjiang Medical College, Diagnosis of nasopharyngeal carcinoma by cytological examination of exfoliated cells taken by negative pressure suction. Chin Med J, 1976, 1: 45 – 47

33. Epstein – Barr Virus – Activating Substance (s) from soil

ITO Yohei[1], TOKUDA Harukuni[1], OHIGASHI Hajime[2], KOSHIMIZU Koichi[2] and Zeng Yi[3]

1. Department of Microbiology, Faculty of Medicine; 2. Department of Food
Sciences and Technology, Faculty of Agriculture, Kyoto University, Kyoto 606 Japan
3. Institute of Virology, National Centre for Preventive Medicine, China

Summary

The soil samples collected from under the plants Euphoribiaceae and Thymeleaeceae, which are known to contain Epstein – Barr virus (EBV) – activating and tumor – promoting diterpene ester compounds, yielded ether extracts which induced EBV early antigen (EA). Such finding were extanded to field studies in southern areas of China where one of the EBV – associated diseases, nasopharyngeal carcinoma (NPC), is endemic. The soil samples collected from such areas, particularly those from under the Chinese tung oil trees (Aleurites fordii), which contain active diterpene ester HHPA (12 – 0 – hexadecanoyl – 16 hydroxyphorbol – 13 – acetate), also exerted positive reactivity. When plants which do not contain diterpene esters were transplanted into a petridish with such active substances or were grown in soil in which they had accumulated, even those protions of the plants which were relatively free of such substances yielded extracts which activated EBV early antigen (EA) in nonproducer human lymphoblastoid cells (Raji).

Introduction

Many, if not all, species of plants belonging to the families of Euphorbiaceae and Thymeleaeceae contain irritating compounds. Such irritants are known to be diterpene esters based on the skeleton structures of tigliane, daphnane and ingenane (Evans and Schmidt, 1980). The chemical extracts from these plants have long been recognized to possess tumor – promoting capacity (Diamond et al. , 1980). The linkage between such tumor – promoting substances and EBV – EA induction was shown in recent studies (zur Hausen et al. , 1979). It was also Terealed that many species of such plants are currently used as folk remedies in areas where two EBV – associated diseases, Burkitt's lymphoma (BL) and NPC, are endemic (Hirayama and Ito, 1981; Ito et al. , 1981a). The Chinese tung oil tree, a member of Euphorbiaceae, is popular all over the southern provinces of China, where it is cultivated chiefly for industrial purposes. The plant itself is also used as a source of herbal drugs. The essential diterpene ester of this plant with EBV – activating and tumor – promoting capacity is HHPA (Ito et al. , 1983).

To determine how the EBV – activating substances gain access to the human system, we decided to test the hypothesis that such substances were in the soil in which the plants were grown. This phenomenon of higher plants affecting each other through the release of chemicals through their plant bodies, known to botanists and agriculturists, is termed allelopathy (Muller and Chou, 1972). After assaying the soil extracts, it became evident that such was the case (Ito et al. , 1983). Extracts of soil samples obtained from under the plants with EBV – activating diterpenes exerted similar activites and their potency was comparable to those of the plant extracts. The uptake of EBV – activating substances by plants which primarily lack such substances was also observed. The implications of these findings to the possible etiology of BL and NPC is discussed.

Materials and methods

1. Cells and cell culture

The Raji cell line, an EBV – nonproducer cell containing multiple copies of EBV genomes, was cultivated in RPMI 1640 medium containing 10% fetal calf serum, 100 units of penicillin and 250 μg/ml of streptomycin. Under these conditions, Raji cells showed a spontaneous rate of induction of EBV EA of less than 0.01%.

2. Assay method for EA induction

The synergistic assay for EA induction in Raji cells was employed (Ito et al. , 1981a). The cells were adjusted to a density of 1×10^6 cells – ml and were incubated with 4 mol/L n – butyrate and various test extracts at varying concentrations. 12 – 0 – tetradecanoyl phorbol – 13 – acetate (TPA), at a concentration of 10—20 ng/ml, served as a positive control, and cultures treated only with n – butyrate as a negative control. The results were read after 48 hrs incubation of the cells at 37℃, using the indirect immunofluorescence (IF) technique (Henle and Henle, 1966). Cell smears were prepared on glass slides, air – dried, and fixed with acetone at room temperatur for 10 min. Activated Raji cells expressing EBV EA were stained with EA + – virus capsid antigen (VCA +) high titer serum from an NPC patient, kindly provided by prof. H Hattori, Kobe University School of Medicine. The untreated cultures served as controls. In each assay, at least 500 cells were counted randomly and the EA + cells were determined. The number of viable cells in the culture was determined by the methyeneblue exclusion test (Ito et al. , 1981b).

3. Soil extracts

Soil samples (20 g each) were collected from under plants of Euphorbiaceae and Thymeleaeceae, and also from plants of other species (controls). The collections were made at about 0.5 cm from the base of the plant stems at a soil depth of 0.2 cm. The samples were extracted with an equal volume of ether for 20 min at room temperature (20℃). After evaporating the solvent, the crude extracts were weighed and redissolved in dimethylsulfoxide (DMSO) as a stock solution of 10 mg/ml. The extracts were prepared in final concentractions of 100, 20 and 4 ng/ml and tested for induction of EA in Raji cells.

Tab. 1　EBV – activating principles in soil extracts

Samples taken from soil under plants		Soil Extracts ($\mu g/ml$)[#]	EBV EA – positive cells(%)
Species	Family		
Sapium Sebiferum	Euphorbiaceae	20	23. 8
Codiaeum variegatum	Euphorbiaceae	20	12. 0
		4	9. 8
Euphorbia lathyris	Euphorbiaceae	20	3. 6
		4	7. 5
Daphne odora	Thymeleaceae	20	15. 6
		4	10. 5
Edgeworthia papyrifera	Thymeleaceae	20	5. 9
Vinca rosea	Apocynaceae	20	0. 1 [*]
		4	0. 1
Castanea crenata	Fagaceae	20	0. 1
Control Ground		20	0. 1
Commercial soil		20	0. 1

Notes:[#] Dissolved in DMSO and used with n – butyrate(4 mmol/L).

[*] Represents figure less than or equal to 0. 1%.

Tab. 2　Comparison of EBV – activatiiong potency of extracts from plants and from soil under the plants

Plant species	Concen – tration	EBV EA – positive cells (%)
Codiaeum variegatum		
Flowers	10	11. 4
	2	7. 5
Stems	10	13. 4
	2	10. 2
soil	20	12. 0
	4	9. 8
Daphne odora		
Leaves	10	31. 0
	2	29. 7
Soil	20	15. 6
	4	10. 5
Sapium sebiferum		
Leares	2	24. 5
	0. 4	28. 7
Soil	20	19. 3
	10	10. 3
Control		
TPA [*]		31. 1
n – butyrate		0. 1

[*] used with n – butyrate (4 mol/L).

Results and Discussion

1. Detection of EBV – activating potency in extracts from soil samples under plants containing active diterpene esters.

The phenomenon of allellopathy, described above, is accomplished by chemicals being released from plants by rain – wash, root excretion and decay; by pollens, decomposition of fallen leaves, flowers and fruit, bark. etc. In the case of the Euphorbiaceae and Thymeleaeceae plant, it seemed probable that the EBV – activating substances might also be found in the soil surrounding those plants. This was found to be the case. Table 1 shows EBV – EA induction (5. 9% – 23. 8%) in Raji cells by ether extracts of soil under plants containing diterpene esters, suggesting excretion of EBV – activating compounds; while those from under plants free of such substances, selected randomly from areas of our university campus, did not induce EA.

2. Comparison of EBV – activating potency of soil extracts with those of "parental" plants

It is of interest to determine whether the active substances detectable in the soil around the affected plants are related to the active chemical (s) in the plant bodies perse. We are currently in the process of isolating and purifying the compounds from the soil. The chromatographic data, although still too preliminary to be definitive, indicate that the compounds are at least of the diterpene ester – type. In Tab. 2, data of the EBV EA – inducing capacity of extracts from various portions of the plant bodies, as compared with those of the soil extracts beneath them, are shown. It may be noteworthy that the EBV – induction activity of extracts derived from the soil is comparable to, if not exceeding that of the plant – derived extracts, suggesting that such compounds are actively released in large quantities in the soil.

3. EBV – activating factors extracted from plants growing in media or soil containing diterpene esters and possibly related active substances.

The next question posed was whether the active substances accumulated in the soil could be absorbed by plants of other species which normally do not possess such complounds. A model experimental system was designed by culturing germinating plants in medium containing EBV – activating compounds. Extracts were prepared from the upper portion of the plant were assayed for EBV – EA induction after three days.

Bean sprouts and other plants, which are all common vegetables sold at the market, were the plants chosen to be grown in medium containing EBV – EA – inducing compounds. When 50 μg of TPA and a 5% acetone solution were added to the medium (cotton bed on which the germinating plants were transplanted and kept for 3 days), the EBV – EA induction by the plant extract was 7. 3% and 10. 2%, respectively; whereas the positive control, with 20 ng/ml of TPA in the synergistic assay system, showed 22. 6% positive cells.

FIGURE 1

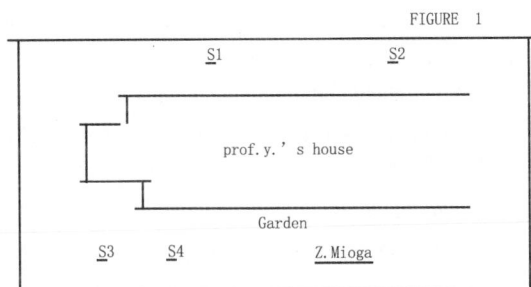

S: Sapium sebiferum (4 trees); Z: mioga: where vegetables are cultivated (2. 0 m from S_4).

Fig. 1　The location where the vegetable was grown is illustrated

The next set of experiments were carried out on a species of vegetable, Zingider mioga, a popular Japanese delicacy. We selected a location in a private garden (Dr Y S), where the vegetable was cultivated under a Sapium sebferum tree, a parental plant producing active diterpene ester HH-PA. The location where the vegetable was grown is illustrated in Fig 1., and the results of the assay of the extracts for EBV – activation are shown in Tab. 3. The portion of Z. mioga eaten grows underground (root). The results show that considerable EA – inducing activity (1% –24%) was found in the vegetable, normally free of such diterpene es-

Tab. 3　EBV – activating potency of extracts of soils and vegetables (Zingider mioga) from prof. Y S. garden

Samples	Concentration (μg/mL)	EBV EA + Cells (%)
Soil		
between S_1 & S_2	50	1. 0
	10	21. 0
	2	2. 0
between S_3 & S_4	50	24. 3
	10	11. 0
	2	7. 2
Control soil	50	0. 1
	10	0. 1
	2	0. 1
Plant (Z. mioga)		
Leaves	50	0. 1
	10	0. 1
	2	0. 1
Roots #1 [a]	50	0. 1
	10	2. 3
	2	1. 6
#2	50	1. 1
	10	5. 6
	2	6. 1
#3 (pickled)	50	2. 1
	10	4. 3
	2	2. 1
Control		
TPA		27. 8
n – butyrate		0. 1

[a] Roots are the underground portion of the vegetable eaten as delicacy in Japanese cooking

ters.

The implications of the uptake of active substances by nonactive plants may have profound significance. However, the evidence is not sufficient at the present time to suggest any link to human neoplasia. Further careful studies must be carried out to draw any conclusions.

4. Field study for EBV – activating substances from soil under trees of Aleurtes fordii (Chinese tung oil tree) and other plants in NPC endemic areas of southern China

A survey of soil samples collected from under Chinese tung oil trees and other Euphorbiaceases was carried out (Z. Y) in the southern provinces of China to confirm our findings. The rate of EBV EA induction by such samples was as high as 59.5%. Although the data are still qualitative rather than quantitative, it appears EBV – activating substances do exist in the soils of endemic NPC areas (Zeng et al. , 1984) and may be a contributing factor in the complex etiology of NPC.

〔In 《Epstein-Barr Virus and Associated Diseases》 1985, 383 – 391〕

References

1 Diamond L, O'Brien T G, Barid W M. Tumor promoters and mechanism of tumor promotion. Adv. Cancer Res, 1980, 32: 1 – 74

2 Evans F J, Schmide R J. Plants and plant products that induce contact dermatitis. J. Med. Plant Res, 1980, 38: 291 – 316

3 Henle G, Henle W. Immunofluorescence in cells derived form Burkitt's lymphoma. J Bacteriol, 1966, 91: 1248 – 1256

4 Hirayama T, Ito Y. A new view of the etiology of nasopharyngeal carcinoma. Rev Med, 1981, 10: 614 – 622

5 Ito Y, Kishishita M, Morigaki T, Yanase S, Hirayama T. Induction and intervention of Epstein – Barr virus expression in human lymphoblastoid cells: A simulation model for study of casue and prevention of nasopharyngeal carcinoma and Burkitt's lymphoma. In: E Grundmann, G R F. Krueger and D V. Ablashi (eds). Nasopharyngeal Carcinoma, Cancer Campaing, 5 (1981a), pp: 255 – 262

6 Ito Y, Yanase S, Fujita J, Hirayama T, Takashima M, Imanaka H. A short – term in vitro assay for promoter substances using human lymphoblastoid cells latently infected with Epstein – Barr virus. Cancer Letters, 1981b, 13: 29 – 37

7 Ito Y, Yanase S, Tokuda H, Kishishita M, Ohigashi H, Hirota M, Koshimizu K. Epstein – Barr virus activation by tung oil, extracts of Aleurites fordii and its diterpene ester HHPA. Cancer Letters, 1983, 18: 87 – 95

8 Muller C H, Chou C H. Phytotoxins: an ecological phase of phytochemistry. In: J B. Harborne (ed). Phytochemical Ecology, Academic Press, London (1972), pp: 201 – 206

9 Zeng Y, Miao X C, Jaio B, Li H Y, Ni H Y, Ito, Y. Epstein – Barr virus activation in Raji cells with ether extracts of soil from different areas in China. Cancer Letters, 1984, 23: 53 – 59

10 zur Hausen H, Bornkamm G W, Schmidt R, Hecker H. Tumor initiators and promoters in induction of Epstein – Barr virus Proc Natl Acad Sci USA, 1979, 76: 782 – 785

34. Nasopharyngeal Mucosal Changes in EB Virus VCA – IgA Antibody Positive Persons

LI Er – koe[1] ,TAN Bi – fang[2] ,ZENG Yi[3] ,WANG Pei – zhong[1] ,ZHONG Jian – ming[2] ,
DENG Hoog[2] ,ZHU Chi – song ,WEI Ji – neng[1] ,PAN Wen – Jun[2]

1. People's Hospital ,Guangxi Zhuang Autonomous Region ;2. Antitumor Office ,
Zangwu County ,Guangxi ;3. Institute of Virology ,Chinese Academy of Medical Sciences

Summary

The immuno – enzymatic method was used and serum VCA – IgA antibody was detected in mass surveys. The relationship between the antibody level and nasopharyngeal carcinoma(NPC) and non – NPC lesions were studied. The results showed that in the NPC or non – NPC antibody positive group , the occurrence of atypical hypereplasis or metaplasia was higher than in the antibody negative group of non – NPC patients and the antibody level in NPC atypical hyperplasia or methaplasia was higher than in simple hyperplasia or metaplasia ,the increase being related to the mucous lesion stage.

Thus we presume that NPC is closely related to atypical hyperplasia or metaplasia and that EB virus may play an important role in nasopharyngeal mucosal transformation or the development of malignancy.

Introduction

In Zangwu County of Guangxi Province ,a NPC high incidence district ,we found that positive serum titer of VCA – IgA antibodies are characteristic in diagnosing the disease. However ,the nasopharyngeal changes have not been reported so far. Our preliminary investigation of this is presented.

Materials and Methods

In July 1978 ,January 1979 and January 1980 ,we went to Zangwu County to detect VCA – IgA antibody in people over 30 years old using immunoenzymatic method. The antibody titer over 1∶2. 5 was considered to be positive. Persons showing positive VCA – IgA antibody continuously were examined by specialists and had biopsies taken to determine whether they had NPC or were NPC suspects. It was discovered that they had some nasopharyngeal lesions such as adenoid gland hyperplasia or nondevolution ,asymmetrical lymphoid tissue in both recesses or some congestion and inflammatory changes. Biopsy was perfoimed in 201 cases ;it was repeated in a few cases. Specimens were routinely made into paraffin scetions and HE stained ,and some were arhynophil stained. 75 cases in whom NPC was excluded by antibody test and biopsy served as controls.

Results

1. Distribution of antibody titer in NPC patients and non-NPC patients∶This is shown in Tab. 1.

The geometric mean titer(GMT)in NPC patients was 1:76.4 and antibody titer was over 1:40 in 57(74.7%). The GMT in non-NPC was 1:32.6 with antibody titer over 1:40 in 100(49%). The difference is obvious,the former being twice that of the latter.

Tab. 1 The EB virus VCA–IgA antibody titer distribution of NPC patients and noncancerous antibody positive persons

		Antibody titer											GMT
		2.5	5	10	20	40	80	160	320	640	1280	Total	
NPC	Cases	0	1	3	15	14	20	9	4	3	4	2	75	1:76.4
patients	%	0	1.3	4.0	20	18.7	26.7	12.0	5.3	4.0	5.3	2.7	100.0	
Antibody positive	Cases	4	18	24	60	34	37	14	9	5	1	0	206	1:32.6
noncancer persons	%	1.9	8.8	11.7	29.1	16.5	18.0	6.8	4.4	2.4	0.5	0	100.0	

There were 75 NPC cases in the series. According to the NPC clinical stage classification adopted by the National Conference on Tumors in 1965,there were 27 Stage I cases(36.0%),27 Stage II.(36.0%),14 Stage III(22.7%) and 4 Stage IV(5.3%). Of these patients 21 showed no clinical signs at examination,of whom were Stage I($T_1 N_0 M_0$)and 2 Stage II($T_2 N_0 M_0$).

2. Comparison of antibody titer in different nasopharyngeal lesions:In most cases of NPC,atypical hyperplasia and atypical metaplasia,the antibody titer was over 1:20,being 94.7%,93.8% and 100%. Frequent titer was over 1:80,being 56.0%,62.5% and 66.7% in each category. The antibody titer in simple hyperplasia, simple mataplasia and normal mucosa was rarely over 1:20,being 69.6%,82.1 and 55.6%,and titers over 1:80 were fewer,being 17.9%, 29.5% and 33.3% respectively. The difference in antibody distribution curves was statistically significant(Fig. 1).

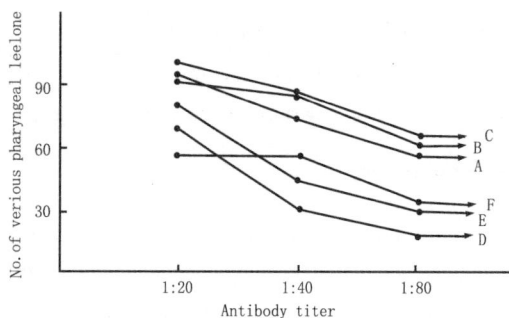

A. NPC; B. Atypical metaplasia; C. Atypical hyperplasia; D. Simple metaplasia;E. Simple hyperplasia;F. Normal mucosa

Fig. 1 Antibody titer distribution curves of NPC patients and those with various NP mucosal lesions.

3. The relation between pharyngeal lesions and VCA–IgA antibody titer in 206 non-NPC patients:No lesion was found in 9 cases (4.4%).206 cases and the geometric mean titer was 1:25.20.

Simple hyperplasia was seen in 78(37%), the GMT was 1:30.4. This included hyperplasia of the columnal,basal or goblet epithelial cells(Figs. 2-4).

Simple metaplasia was seen in 112(50.4%) and the GMT was 1:22.3. At the beginning of metaplasia,some columnal cells transformed into layers of squamous cells. When mature,keratinization appeared on the superficial platiform epithelium(Figs. 5,6).

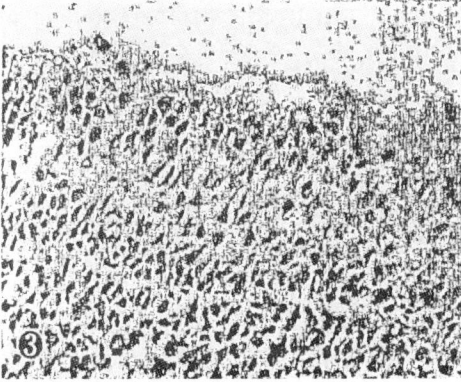

Fig. 2　Simple hyperplasia of columnar
epithelium and basal cells. HE 10 × 20

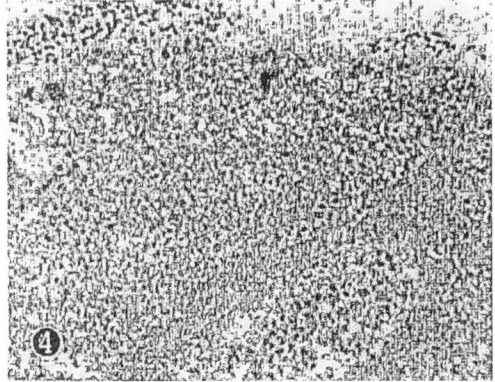

Fig. 3　Simple hyperplasia of
columnar epithelium. HE 10 × 20

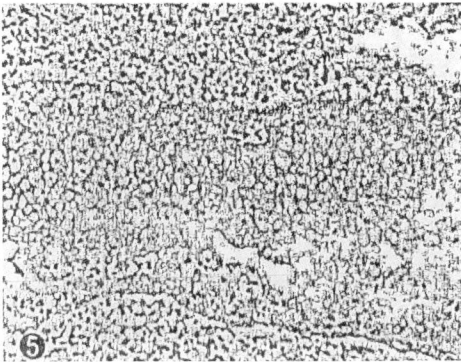

Fig. 5　Simple hyperplasia
of Globic cell. HE 10 × 20

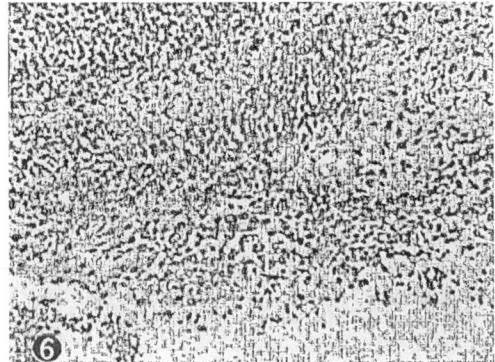

Fig. 6　Simple metaplasia
of columnar epithelium. HE 10 × 20

Atypical mucosal hyperplasia and metaplasia. Atypical hyperplasia was seen in 15(7.3%) and the GMT was 1∶105.40. There were atyplcal hyperplasia of the columnal cells or squamous cells(Figs.8,9).

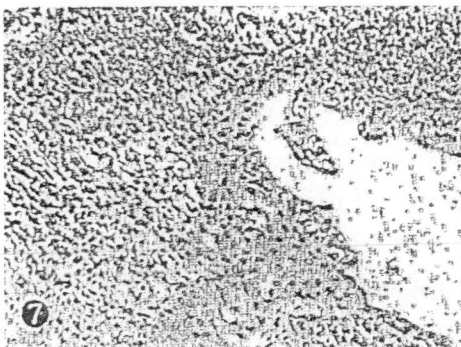

Fig. 6　Simple metaplasia of
columnar epithelium. HE 10 × 20

Fig. 7　Atypical hyperplasia of
columnar epithelium. HE 10 × 40

Atypical hyperplasia was seen in 15(7. 3%) and the GMT was 1 : 105. 40. There were atypical hyperplasia of the columnal cells or squamous cells(Figs. 7 ,8).

Atypical metaplasia was seen in 48(23. 2%)and the GMT was 1: 80. When the columnal cells transformed into squamous cells ,atypical hyperplasia also appeared(Fig. 9).

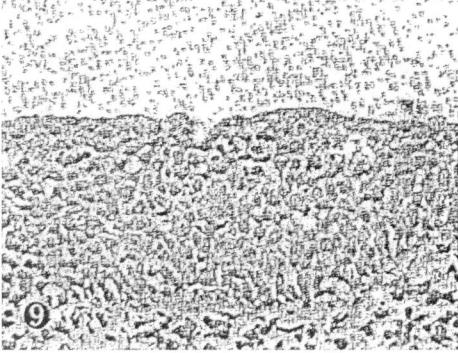

Fig. 8   Atypical hyperplasia squamous epithelium. HE 10 × 40

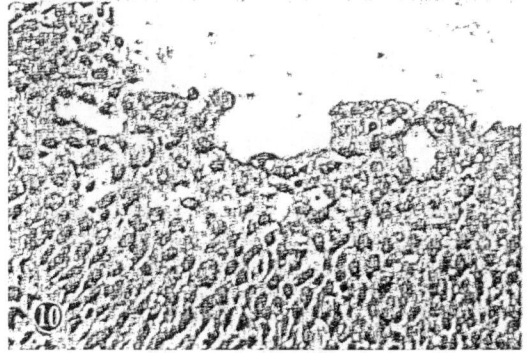

Fig. 9   Atypical metaplasia of columnar epithelium. HE 10 × 40

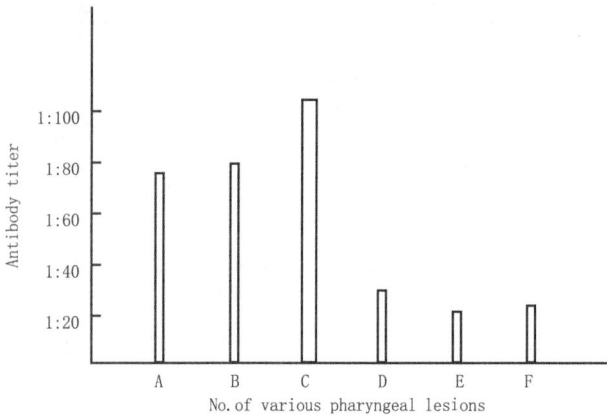

A. NPC ; B. Atypical metaplasia ; C. Atypical hyperplasia ; D. Slimple metaplasia ; E. Simple hyperplasia ; F. Normal mucosa

Fig. 10   Antibody GMT distribution of NPC patients and other non – NPC lesions

In normal mucosal simple hyperplasia or simple metaplasia the GMT was lower(under 1: 30. 4). But the GMT of atypical hyperplasia or atypical metaplasia in NPC elevated markedly (over 1 : 76. 4)as shown in Fig. 10.

Comparisons of pathologic changes of nasopharyngeal mucosa in VCA – IgA antibody positive and negative persons are shown in Tab. 2.

In tissue surrounding the cancer in 75 patients ,we found simple hyperplasia in 3 (4%), simple metaplasia in 14 (18. 5%), atypical hyperplasia in 8 (12%)and atypical metaplasia in 29(38. 7%).

Some patients with antibody were followed up after the first biopsy. The interval between the 2 biopsies was 8 – 33 months. The pathologic examination results follow.

In 17 cases who had been initially diagnosed as having atypical hyperplasia or atypical metaplasia , 5 had NPC (29. 4%), while in 38 with simple hyperplasia, simple metaplasia or normal mucosa, only in 1 (2. 6%), was NPC found (Tab. 3). One case with atypical hyperplasia of some columnar cells was found on biopsy to have NPC 33 months later. This suggests that atypical hyperplasia and atypical metaplasia are closely related with NPC.

Tab. 2　Comparison of pathological changes of nasopharyngeal mucosa in VCA – IgA antibody – positive and negative persons

Groups (NP leslon)	Antibody – positive NPC patients		Antibody – positive non – NPC patients		Antibody – negative non – cancer persons	
	Cases	%	Cases	%	Cases	%
Atypical metaplasla	29	38. 7	48	32. 2	11	14. 5
Atypical hyperplasla	8	12. 0	15	7. 3	3	4. 0
Simple metaplasla	14	18. 5	112	50. 4	44	58. 7
Simple hyperplasia	3	4. 0	76	37. 0	24	32. 0

Tab. 3　Follow – up observations of various NP region lesions

Initial biopsy	Follow – up biopsy			% of NPC proved later
	Cases	Noneancer	NPC	
Atypical hyperplasia Atypicla metaplasia	17	12	5 *	29. 4
Simple hyperplasia Simple metaplasia Normal mucosa	38	37	1 **	2. 6
Total	55	49	6	12. 2

Notes: * One case developed NPC (Stage I) after 33 months. Two had NPC (Stage II) after 12 months.

** One case developed NPC (Stage I) after 8months

Discussion

In 75 NPC patients with positive EB virus VCA – IgA antibody titration, there were 8 (12%) cases of atypical hyperplasia of the surrounding tissue and 29 (36.7%) cases of atypical metaplasia. In 75 non – NPC cases without antibody, 3 (4%) had atypical hyperplasia of the surrounding tissue and 11 (14.5%) had atypical metaplasia. There was a statistically significant difference between the 2 conditions, i. e. the frequency of atypical hyperplasia or metaplasia in the NPC and non – NPC group with antibody was higher than that in the non – NPC group without antibody. This shows that these changes are correlated with the VCA – IgA antibody level.

In the antibody positive group the GMT was higher (1 : 76. 4) in NPC patients and lower (1 : 32. 6) in non – NPC patients. It was also found that the level of patibody in simple hyperplasia and metaplasia was lower, while the level in atypical hyperplasia and metaplsia was higher.

Follow – up study showed that in some cases if the antibody level was elevated, the lesions increased 80 VCA – IgA titrations were closely related to the pathologic mucosal changes and may be taken as a reference index of epithelial lesion development. The level of VCA – IgA antibody titration was higher in NPC, atypical hyperplasia and metaplasia, the GMT being 1 : 76. 4, 1 : 105 and 1 : 80 respectively.

Some authors reported that there was a close relationship between atypical hyperplasia or atypical metaplasia and NPC. The observed the presence of some transitory morphologic changes in simple hyperplasia and metaplasia, atypical hyperplasia and metaplasia, cancer in situ and tiny infiltrative cancers.

During follow – ups of 8 – 33 months, 5 NPC cases were disovered in 17 diagnosed as having atypical hyperplasia or metapiasia. Only 1 NPC case was found in 38 patients with biopsy proved simple hyperplasia, simple metapiasia or normal mucosa. Evidently, regardless of the morphologic changes or disease stage, there is a close relationship between atypical hyperplasia or metaplasia and NPC. Therefore, we may assume that they are not only related closely to VCA – IgA antibody level, but also to NPC occurrence. Therefore EB virus activation and existence of VCA – IgA antibody are intimately concerned with NPC. This suggests that EB virus may be one of the NPC causative agents.

VCA – IgA antibody titer in patients with atypical hyperplasia is higher than in NPC patients, but the difference is not statistically significant.

Acknowledgement: We thank Prof Zong Yong – sheng for his guidance.

[In 《Chinese Medical Journal》 1985, 98: 25 – 30]

References

1 Zeng Yi, et al. Application of an immunoenzymetic method and an immunoautoradiographic method for a mass survey of nasopharyngeal carcinoma. Inter-virology, 1980, 13: 162

2 Immunopathological Department, Institute of Tumor, Zhongshand Medical College, Morphological observation on the process of carcinogencsis of nasopharyngeal epithelial cells. Selected paper, 1978, 1: 44

35. ELISA for the Detection of Nasopharyngeal Carcinoma Using IgA Antibodies to EBV Early Antigen

Pi G H[1], Zeng Y[1], de – The G[2]

1. Institute of Virology, China National Center for Preventive Medicine, Beijing (People's Republic of China); 2. Laboratoire d Epidemiologie et d. Immunovirologie des Tumeurs du CNRS, Faculte de Medecine Alexis Carrel, 69372 Lyon Cedex 08 (France)

Summary

An enzyme – linked immunosorbent assay (ELISA) was established for the detection of IgA antibodies to Epstein – Barr virus early antigens. Crude extracts from P3HRI cells treated with phorbolesters, n – butyrate and Ara – C or from Raji cells treated with phorbolesters and n – butyrate were used as antigens. The ELISA assay consisted in three steps involving test serum, mouse monoclonal antiserum to human IgA and rabbit antimouse IgG. A good correlation was obtained between this simple ELISA and the previously described immunoenzymatic test[5], but ELISA was found to be 8 times more sensitive. This ELISA is a sensitive and rapid test which has been applied to field studies for early detection of nasopharyngeal carcinoma.

[**Key Words**] EBV; Nasopharyngeal carcinoma; IgA; ELISA; Early antigen; Detection

Introduction

In our previous studies, we showed that the detection of IgA antibodies to Epstein – Barr virus (EBV) virion capsid antigen (VCA) by an immunoenzymatic test was useful for the early diagnosis of nasopharyngeal carcinoma (NPC) in high – risk populations[4,5]. We later showed that IgA antibodies against the EBV early antigen (EA) which were detected by the immunoenzymatic test were more specific than, but not as sensitive as, the IgA antibodies to VCA for the detection of NPC[6]. A very sensitive immunoautoradiographic test was then established to detect EA – IgA antibodies[7]. With such a test, 96% of NPC patients were positive for EA – IgA antibodies, but it was complicated and inconvenient for the screening of NPC in large populations. As described here, a sensitive, specific and rapid ELISA for detectionof EBV – IgA antibody was developed and tested in field conditions.

Materials and Methods

1. Cells

The Burkitt's – lymphoma – derived and EBV – producing P3HR – 1 and nonproducing Raji cell lines grown at 37℃ in RPMI – 1640 medium supplemented with 20% newborn calf serum were used to prepare the antigens.

2. Preparation of EBV EA

To prepare crude EBV EA not containing late structural antigens (VCA), two techniques were used. The flrst consisted in using P3HRI cells cultured at a concentration of 5×10^5 cells per ml in RPMI tissue culture medium supplemented with 20% newborn calf serum, containing croton oil (500 ng/ml), n – butyrate (4 mmol/L) and cytosine arabinosied (Ara – C) (0. 25 μg/ml). After 48 h, the cells were studied for the proportion of EA – positive cells (from 20% to 45%) and for the lack of VCA – positive cells. Only batches with a proportion of 30% EA – positive cells with no VCA were selected. After two washings with PBS, cells were stored at – 70℃.

The second technique used was to cultivate Raji cells at a concentration of 10^6 cells per ml for 72 h in the presence of TPA (5 ng/ml) and of BA (0. 4 ml 2 M/200 ml). After checking the proportion of EA – positive cells and the lack of VCA, batches with at least 30% EA – positive cells were washed and stored at – 70℃ as above. The proportion of EA – positive cells was determined as usual by indirect immunofluorescence (IF) or by immunoenzymatic (IE) metod[5] comparing the results of two different sera containing IgG antibodies to VCA and VCA + EA, respectively. These two sera were used at dilutions giving the same final VCA antibody titres. As stated above, only batches of cells with 30% EA – positive cells were kept for preparing crude EA batches.

Ara – C = cytosine arabinoside	IF = immunofluorescence
BA = butyric acid	IgA = immunoglobulin A
BSA = bovine serum albumin	NPC = nasopharyngeal carcinoma
EA = early antigen	OD (A) = optical density
EB = Epstein – Barr	OPD = orthophenylenediamine
EBNA = EB nuclear an tigen	PBS = phosphate – buffer saline
EBV = EB virus	PMSF = phenylmethylsulphonyl
ELISA = enzyme – linked immunosorbent assay	TPA = 12 – O – tetradecenoyl – phorbol – 13 – acetate (croton oil phorbolester)
GMT = geometric mean titre	VCA = virion capsid antigen
IE = immunoenzymatic test	

The panel of sera used for evaluation of antigenic batches consised of three normal Chinese individuals possessing anti – VCA IgG at a titre of 1/320 and anti – EBNA antibodies at 1/160, and lacking EA – IgG antibodies, as tested IF. They were used at two dilutions of 1/5 and 1/20. Three Chinese sera from NPC patients were used with the following titres: VCA – IgG at 1/1 280, VCA – IgA at 1/160, EA/IgG at 1/320, EA – IgA at 1/40 and anti – EBNA antibodies at 1/1 280, as tested by IF. These sera were used at two dilutions of 1/20 and 1/80.

To prepare crude antigenic batches, frozen pellets of 10^8 cells were suspended in 1 ml of 150 mmol/L NaCl, 20 mmol/L Tris buffer, pH 7.5, 1 mmol/L EDTA and 0.5 mmol/L phenylmethylsulphonyl fluoried (PMSF). They were sonicated 8 times for 10 seach on ice and then centrifuged at 18 000 r/min for 45 min. The supernates to be used as crude antigen were tested for protein concentration and then stored at $-70\,^{\circ}\text{C}$ until use.

3. Sera

Sera were obtained from patients with NPC, from patients with tumours other than NPC and from normal individuals, and were stored at $-20\,^{\circ}\text{C}$ until use.

4. Mouse monoclonal antibody to human IgA

A mouse monoclonal anti – human IgA antibody cell line was established in Beijing (Xia and Zeng, sFubmitted for publication). The original titre of the IgA antibody in ascitic fluid for mouse was 1/30 000.

5. Horseradish – peroxidase – conjugated rabbit anti – mouse IgG antibody

The appropriate dilrtion of this conjuagate was 1/30 000.

6. Preparation of ELISA plates

The antigen preparation was diluted in 0.05 mol/L sodium carbonate buffer at pH 9.6, and 10 μg/100 ml of protein were added to each well of polystyrene microtitre plates. The plaes were incubated overnight at $4\,^{\circ}\text{C}$, then washed twice with 0.05 mol/L sodium carbonate buffer at pH 9.6, supplemented with 0.1% bovine serum albumin (BSA) and dried for 20 min at room temperature. Unattached sites were saturated with 0.05 mol/L sodium carbonate buffer at pH 9.6, supplemented with 1% BSA for 3 h in a humid chamber at room temperature. The plates were then washed twice in PBS containing 0.05% Tween – 20, dried for 1 h at room temperature and stored at $4\,^{\circ}\text{C}$ until use.

7. ELISA method

A three – step technique was used. One – hundred microlitres of each test serum diluted 1/40 in washing buffer were added in duplicate wells, and the plates were incubated in a humid chamber at room temperature for 1 h. The plates were then washed 5 times in washing buffer. One – hundred microlitres of mouse monoclonal antibody against human IgA (Xia and Zeng, submitted for publication), diluted 1/500 in washing buffer, were added to each well, and the plates were again put in a humid chamber at room temperature for 1 h. After 5 washings with buffer, 100 μl of horseradish – peroxidase – conjugated anti – mouse IgG antibody, diluted 1/30 000 in washing buffer was added and the plates put in a humid chamber at room temperature for 1 h. After five washing with buffer, 100 μl of the substrate solution (prepared by adding 40 mg of orthophenylenediamine (OPD) and 30 μl of 30% H_2O_2 in 100 ml of phosphoric acid/citric acid buffer at pH 5.0 in 0.2 mol/L Na_2HPO_4 and 0.1 mol/L citric acid) were added to each microtitre well and the reaction was allowed to take place for 20 min at room temperature before it was stopped by addition of 50 μg of 4 mol/L H_2SO_4 The intensity of the colour reaction at 492 nm was recorded with a 《Titertek Multiskan》 .

8. Determination of results

P/N value was calculated according to the following formula:

$$\text{P/N value} = \frac{\text{serum sample } A \text{ figure} - \text{PBS control } A \text{ value}}{\text{reference negative serum } A \text{ figure} - \text{PBS control } A \text{ value}}$$

P/N value $\geqslant 2$ was condidered as positive, and P/N value < 2 as negative.

9. IE and IF tests

VCA – IgA and EA – IgA antibodies of the same sera were also detected by the IE test[5]. in Beijing, and by indirect IF in Lyon.

Results

1. Titration of antigen

In order to determine the optimal concentration of antigen for ELISA, microtitre plates were coated with serial 2 – fold dilution of cell extracts containing protein from 0.09 μg to 100 μg per well, whereas the known positive serum was also diluted from 1/40 to 1/2560. Tab. 1 gives the P/N values of a positive reference serum corresponding to the different concentrations of antigen and serum dilutions. Similar P/N value (2.4 – 3.0) and the same antibody titre were obtained when the concentration of antigen used varied from 6.25 to 100 μg per well. Thus, 10 μg of antigen per well was used as routine.

Tab. 1　Titration of EBV – EA antigen using IgA

| Serum dilutions | P/N value | | | | | | | | | | | |
| | Crude preparation of EA (μg) from P3HR1 cells | | | | | | | | | | | |
	100	50	25	12.5	6.25	3.125	1.563	0.781	0.391	0.195	0.098	0.049
1/40	7.9	6.7	6.7	5.4	4.4	3.3	2.4	2.2	2.1	2.0	1.9	2.0
1/80	7.7	7.6	7.5	5.3	4.3	2.6	1.7	1.7	1.7	1.6	2.0	2.0
1/160	5.1	7.3	6.2	4.2	3.9	2.1	1.5	1.5	1.3	1.4	1.1	1.6
1/320	5.1	4.7	5.6	4.5	3.0	2.1	1.7	1.2	1.0	1.1	1.2	1.6
1/640	3.1	3.2	3.7	2.8	2.4	1.5	1.2	0.9	1.1	1.1	0.9	1.1
1/1 280	2.9	2.4	3.0	2.6	2.4	1.1	0.7	0.8	2.0	0.8	1.0	1.5
1 /2 560	1.3	1.1	1.3	0.9	1.0	0.9	0.9	1.1	1.5	0.8	0.6	0.7

2. Titration of the monoclonal anti – human IgA antibody

To determine the optimal dilution of the mouse monoclonal anti – human IgA for the ELISA, a serial dilution from 1/125 to 1/4000 of anti – human IgA antibody was tested. As shown in Fig. 1, the P/N value of sera from NPC patients decreased with the dilution of anti – human IgA antibody. The highest dilution of anti – human IgA antibody giving positive results (P/N value $\geqslant 2$) for se rum A and B was 1/500 and 1/2 000 respectively, so 1/500 dilution of anti – human IgA antibody for ELISA was used as routine.

3. Specificity and sensitivity of the ELISA

Two known human positive sera and one negative serum with serial 2 – fold dilution were tested

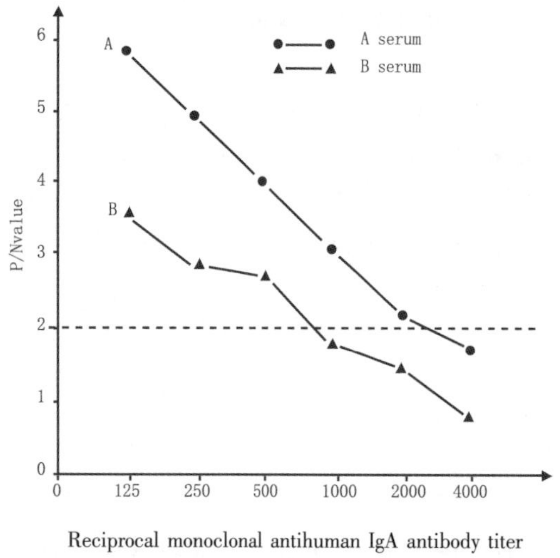

Reciprocal monoclonal antihuman IgA antibody titer

Fig. 1 Titration of the mouse monoclonal anti – human IgA antibody with two NPC sera having EA – IgA titers, as tested by IE, of 640 and 320, respectively, using a P3HRI EA preparation. IgA/EA = EA – IgA (see text)

with an optimal concentration of EA (10 μg/well). As seen in Fig. 2, the curves of the A data and P/N values of the positive and negative sera were markedly different. Then 19 known positive and 24 known negative sera were simultanously tested with this ELISA assay and the IE test. As seen in Fig. 3, a good correlation between antibody titres detected by the ELISA and the IE test was obtained using 19 sera from Chinese NPC patients and 24 controls lacking IgA antibodies to VCA and EA. The geometric mean titre (GMT) of the positive sera detected by ELISA was 1/710, or 8 times higher than the GMT (1/88.8) obtained with IE titres.

The use of 《control》 antigen (s) in the ELISA test using crude antigen extracts from an EBV – negative cell line such as BJAB or Ramos was considered. This was discarded for the following reasons: sera from patients with NPC are notoriously known to contain antinuclear factors, auto

and iso – antibodies[3], and to easily give false – positive ELISA results with non – EBV – related antigen preparations. In view of this difficulty, we preferred to assess the above ELISA in a pragmatic way by comparing the results of the present ELISA technique with the reproducible results as obtained by IE in Beijing or by indirect IF in Lyon. Experiments in Lyon comparing ELISA with IF tests gave a similar correlation to those comparing ELISA and the IE test, but with titres lower in IF than in IE.

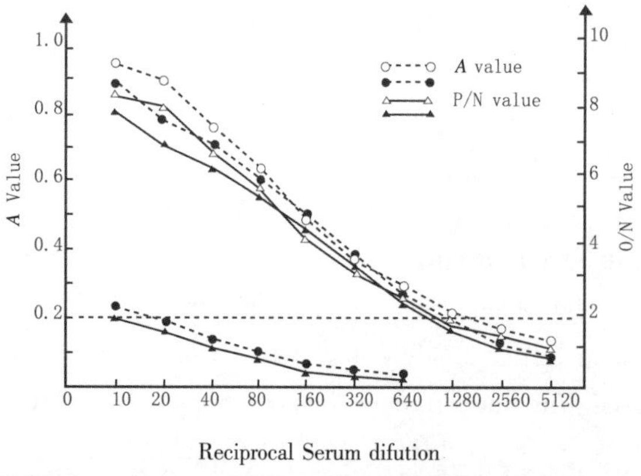

Reciprocal Serum difution

Fig. 2 Detection of EA – IgA antibodies by ELISA using 2 NPC Chinese sera with EA – IgA titres of 1/320 as tested by IE, and 1 EBV – negative serum. The antigen batch was used P3HRI – derived

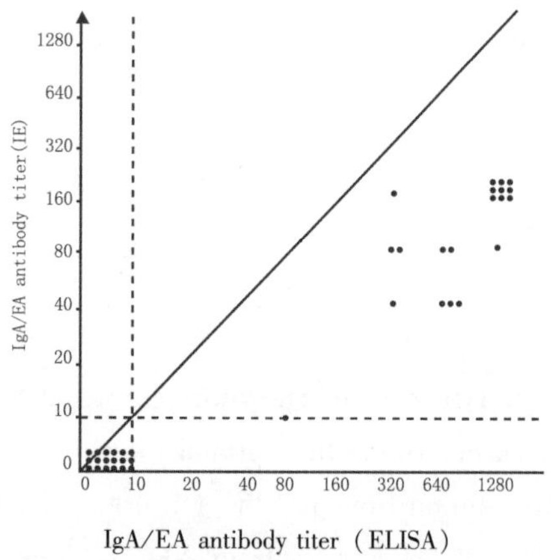

IgA/EA antibody titer (ELISA)

Fig. 3 Comparative EA – IgA antibody titres determined by ELISA and by IE[4], of 19 NPC sera and 24 controls lacking EA – IgA and VCA – IgA antibodies by the IE test. The antigenic batch was the same as that of figures 1 and 2

· 198 ·

4. Reproducibility of the ELISA

In order to determine the reproducibility of the ELISA, 10 sera from NPC patients and 8 sera from normal individuals were tested in two separate experiments. The results seen in Tab. 2. were very similar, with a variation in P/N value of each serum within a 0.1 – 0.4 range (except for 2 positive sera). The shows that the ELISA assay is reproducible.

5. Detection of EA – IgA antibody from NPC patients, patients with tumours other than NPC and normal individuals by ELISA and IE (immunoenzymatic) test

Ninety – one sera from NPC patients, 59 sera from patients with tumours other than NPC and ninety normal individuals sera were tested by both the ELISA assay and by the IE test. As shown in Tab. 3, the percentage of EA – IgA antibody detected by the IE test was 60%, 0% and 0%, respectively and that tested by ELISA using the monoclonal anti – IgA antibody was 97%, 3.4% and 2.2%, respectively. The marked difference in the sensitivity of the ELISA as compared to the IE test should hepl to better differentiate the NPC sera from the two other groups.

Tab. 2　Reproducibility of the ELISA assay

| Serum | P/N value of IgA/EA antibody | | Difference |
	Exp. 1	Exp. 2	
1) 357	5.0	4.7	0.3
397	4.6	4.4	0.2
359	4.3	5.5	1.2
361	4.2	5.0	0.8
376	3.0	2.9	0.1
360	2.7	2.4	0.3
375	2.4	3.6	1.2
365	2.4	2.2	0.2
353	2.2	2.4	0.2
366	1.2	1.5	0.3
2) 120	1.4	1.1	0.3
109	1.3	1.3	0
106	1.3	0.9	0.4
098	1.3	1.6	0.3
118	1.2	0.9	0.3
111	1.2	1.3	0.1
107	1.1	0.9	0.2
112	0.4	0.4	0

1 = NPC; 2 = normal individuals

Discussion

The present results indicate that crude antigen preparations from P3HR – 1 cells treated with croton oil, n – butyrate and Ara – C, or from Raji treated with TPA and BA can be used in ELISA for the detection of EA – IgA antibody by a simple three – step technique. This was made possible through the use of a monoclonal antibody prepared from a mouse immunized against the IgA prepared from an NPC patient (Xia and Zeng, in preparation). The fact that the results were not good when using commercially availabe anti – human IgA monoclonal antibodies suggests that our monoclonal antibody was directed against some idiotypic structure borne by IgA molecules with specificity for some epitope of EA. The use of mouse monoclonal anti – human IgA antibody probably enhances the sensitivity as well as the specificity of the ELISA assay, since EA/IgA antibodies were detected by ELISA in 97% of the NPC sera versus 60% by the IE test, while at the same time, the other two groups of sera showed only 0 ~ 3.4% of positivity. The sensitivity of the ELISA was found to be similar to that of the immunoautoradigraphy[7]. But immunoautoradiography is more complicated than the ELISA, which, being sensitive, specific and rapid was successfully applied to field studies. In the city of Wuzhou, a rate of EA – IgA antibody, as detected by IE test and by ELISA on 12 154 normal individuals was found to be 0.3% and 1.2%, respectively (Zeng, unpublished). Low – cost, standardized, industrially prepared EBV – EA – IgA ELISA kits would be of great help for promoting early detection of NPC in Southeast Asia, where this tumour represents the number one cancer killer in very large populations of approximately 250 millions inhabitants[1]. When one realizes that NPC at an early stage of the disease is a highly curable tumour, one feels the urgency from a public health viewpoint to stress the use of such a simple test.

Tab. 3　Comparison of ELISA and immunoenzymatic test (＊) for detection of IgA/EA antibody

Sera	No. of cases	IgA/VCA IE +	IgA/VCA IE %	IgA/EA IE +	IgA/EA IE %	IgA/EA ELISA +	IgA/EA ELISA %
NPC	91	91	100	55	60	88	97
Tumour other than NPC	59	2	3.4	0	0	2	3.4
Normal individuals	90	2	2.2	0	0	2	2.2

IE = immunoenzymatic test.
(＊) Using P3HR1 – derived EA.

[In 《Ann Inst Pasteur Virol》 1985, 136E: 131 – 140]

References

1　De – The G, Ho J H C, Muir C S. Nasopharyngeal carcinoma, in 《Viral infections of humans, epidemiology and control》, (A S Evans), (pp. 621 – 652). John Wiley and Sons, Chichester, 1982

2　De – The G, Zeng Y, Desgranges, C, Pi G H. The existence of prenasopharyngeal carcinoma conditions should allow preventive interventions, in 《Nasopharyngeal carcinoma: Current concepts》, (Prased et al). (p, 365 – 374). University of Malaya, Kuala Lumpur, 1983

3　Lamelis J P, De – The G, Revillard J P, Gabbiani G. Autoantibodies (cold lymphocytotxins and anti – actin and antinuclear factors) in nasopharyngeal carcinoma patients, in 《Nasopharyngeal carcinoma: Etiology and control》, CIRC Publication Scientifique N 20, (G, De – The & Y. Ito), (pp. 523 – 436). Centre International de Recherches sur le Cancer, Lyon, 1978

4　Zeng Y, Liu Y X, Wei J N, Zhu J S, Cai S L, Wang P H, Zhong, J M, Li R G, Pan W J, Li E J, Tan B F. Serological mass survey of nasophryngeal carcinoma. Acta Acad. Med. Sin, 1979, 1: 123 – 126

5　Zeng Y, Liu Y X, Lin C R, Chen S W, Wei J N, Zhu J S. Application of immunoenzymatic method for the mass survey of nasopharyngeal carcinoma. Intervirology, 1980, 13: 162 – 168

6　Zeng Y, Zhang L G, Li H Y, Jang M G, Zhang Q, Wu Y C, Wang Y S, Su G R. Serological mass survey for early detection of nasopharyngeal carcinoma in Wuzhou city, China Int J Cancer, 1982, 29: 139 – 141

7　Zeng Y, Gong C H, Jang M G, Fun Z, Zhang L G, Li H Y. Detection of Epstein – Barr virus IgA/EA antibody for diagnosis of nasophyaryngeal carcinoma by immunoautoradiography. Int J Cancer, 1983, 31: 599 – 601

36.   Prospective Studies on Nasopharyngeal Carcinoma and Epstein – Barr Virus Inducers

Zeng Yi

Institute of Virology, Chinese Academy of preventive Medicine, Beijing 100052 China

Nasopharyngeal carcinoma (NPC) is common in southern China. The purpose of our studies was first to find simple and sensitive techniques that can be conveniently applied in the high – risk NPC areas fro the early detection of NPC, and second to clarify the role of Epstein – Barr (EB) virus, environmental, and genetic factors in the development of NPC.

Prospective Studies

Serological mass surveys of NPC and follow – up studies on the EB virus IgA/VCA antbody – positive persons were carried out in Zangwu county of Guangxi Autonomous Region by an immunoenzymatic test[1]. NPC patients could be detected in their early stage, but it is rather difficult to carry out follow – up studies because many IgA/VCA antibody – positive persons live in remote areas, and clinical examinations are not carried out periodically for some persons. Therefore, another serological mass survey has been carried out in Wuzhou city since 1980[2]. Altogether 20 726 persons aged 40 years and older were examined and the antibody – positive persons were followed up for 4 years. As shown in Tab. 1, among 20 726 persons examined 1136 persons had the IgA/VCA antibody; the positive rate was 5.5%. From them 18 NPC patients were diagnosed clinically and histologically. Ten were at stage I, 6 at stage II, and 2 at stage III. The IgA/VCA antibody – positive persons were reexamined clinically and histologically once a year for 4 years; during this period 17 additional NPC cases were diagnosed, 5 at stage I 11 at stage II, and 1 at stage III. Thus 16 of the 17 cases (94.1%) were in the early stages. The only stage III case had an antibody titer of 1:40 in the first bleeding, and some suspected lesions were found in the nasopharyngeal mucosa, but this patient refused to have a biopsy taken, and the NPC developed into stage III 2 years and 8 months later. Altogether 35 NPC patients were detected, among them 15 cases (43%) at stage I and 17 (48.5%) at stage II; the early cases (I + II) amounted to 91.5%. The detection rate of early cases is 2.9 times higher than that of the outpatient clinic. IgA/VCA antibodies could be detected 16—41 months prior the clinical diagnosis of NPC patients (Tab. 1). No NPC patients were found among 19 590 persons without IgA/VCA antibody for a period of 4 years. If Iga/VCA – positive persons are examined routinely once a year, NPC patients can be detected at an early stage.

The detection rate of NPC in IgA/VCA antibody – positive persons is 61. 6 times the annual incidence of NPC in the general population of the same age group. These results further indicate that the EB virus plays an important role in the development of NPC and that serological screening and follow – up studies are valuable for the early detection of NPC.

Tab. 1 Clinical stage of NPC patients found in a mass survey and in the outpatient clinc

Group		Clinical stages				Total
		I	II	III	IV	
Screening action	Number of cases	15	17	3	0	35·
	%	43. 0	48. 5	8. 5	–	100. 0
Outpatient	Number of cases	18	312	526	180	1036
clinic	%	1. 7	30. 1	50. 8	17. 4	100. 0

Epstein – Barr Virus Inducers

Our previous data showed that the geometric mean titer of complement – fixing antibodies to EB virus in sera from persons aged 20 years and older in high – risk areas for nasopharyngeal carcinoma (NPC) was significantly higher than that from persons of the same age group in NPC low – risk areas[3] and that the positive rate of IgA/VCA antibody to EBV from normal individuals in NPC high – risk areas increased with increasing age[1]. The detection rate of NPC from persons examined serologically and from IgA antibody – positive persons in some factories was much higher than that from the general population of the same age group in Wuzhou city. Therefore, it is necessary to determine whether factors (inducers) causing an activation of the EB virus are present in foods, Chinese medical herbs, and other plant products.

1. Epstein – Barr Virus Early Antigen Induction in Raji Cells by Extracts from Plants

Ether extracts from more than 1693 plants including Chinese medicinal herbs were studied for the induction of EB virus EA in the Raji cell system. Fifity – three herbs were found to have inducing activity[4].

Many such plants such as Aleurites fordii, Sapium sebiferum, Wikstroemia indica, and Euphorbia antiquorum grow in the NPC high – risk areas (Tab. 2).

2. Epstein – Barr virus activation in Raji cells with Ether extracts of soil

The EB virus inducers were found in samples collected from the soil under tung oil trees (Aleurites fordii), Wikstroemia indica; Sapium sebiferum, Euphorbia antiquorum and other plants growing in southern China where NPC is prevalent[5] (Tab. 3). Some vegetables grown on these soils also contained EB virus inducers[6].

3. Enhancement of lymphocyte transformation In Vitro

The enhancing effect of extracts from plants on the transformation of lymphocytes by EB virus was tested in a soft agar system. Daphne genkwa. W chamaedaphne W. indica, Stellera chamaejasme, and Sparganium stoloniferum were found to induce EA of EB virus in Raji cells and also to enhance the transformation of lymphocytes by EB virus; but some plants such as Abrus fruticulosus and Bulbophyllum inconspicuum had no inducing activity of EA and also no enhancing effect on transformation (Tab. 4)[7].

Tab. 2 Effect of extracts from plants on EB virus induction in Raji cells. Four millimolar n – butyrate alone gave a 0. 4%—1% positive rate

Extract from plants (1—12. 5 μg/ml)	EA – positive cells (%)
Euphorbia humifusa	40
Euphorbia helioscopia	36. 4
Euphorbia antiquorum	36
Sapium sebiferum	24. 8
Croton lachnocarpus	22. 4
Euphorbia lumulata	17. 2
Aleurites fordii (tung oil tree)	42
Daphne genkwa	46
Stellera chamaejasme	46
Wikstroemia chamaedaphne	53
Wikstroemia indica	50
Edgeworthia chrysantha	43
Caesalpinia sppan	8
Desmodium styracifolium	6
Premna fulva	42
Sparganium stoloniferum	16
Daturamonium	50
Tinospora sp	7
Aleuritopleris argentea	17
Clematis intricata	3
Belamcanda chinesis	36
Ixeris debilis	9. 2

Tab. 3 Induction of EB in Raji cells by extracts of soil samples from Wuzhou city and Zangwu county Controls: B(n – butyrate) + C(crotonoil) =45. 7% ; B only =0. 4% ; C only =0. 8% ; untreated Raji cells =0% ; NT , not tested

Soil samples	Positive sample No.	EBV EA – positive cells (%) Ether extract (μg/ml)		
		10	2	0. 4
Under tung oil trees along the road and the Western River (11/19 +)	15	25. 0	14. 0	2. 6
	3	14. 0	11. 4	0. 6
	5	10. 6	0. 0	0. 0
	8	4. 0	3. 0	2. 0
	11	2. 6	0. 6	6. 0
	2	1. 4	2. 2	1. 4
	1	2. 0	0. 6	1. 0
	14	0. 0	1. 8	1. 0
	17	0. 8	1. 2	1. 6
	18	1. 4	0. 6	0. 8
	4	1. 0	0. 0	0. 0
Under tung oil trees along the road and the Kui River (8/20 +)	35	6. 6	9. 6	4. 6
	21	8. 0	2. 8	2. 0
	31	2. 0	8. 0	1. 0
	22	0. 0	4. 0	0. 0
	23	8. 0	1. 6	0. 4
	36	2. 0	1. 0	2. 0
	29	2. 0	0. 4	0. 8
	39	0. 6	0. 8	1. 4
From the dock of Western River (1/4 +)	41	2. 0	0. 6	0. 8
Under Codiacum variegatur	76	2. 8	8. 0	6. 0
Codiacum variegatus	77	8. 0	6. 0	1. 6
Codiacum variegatus	78	2. 0	8. 0	12. 0
Euphorbia milli (4/4 +)	80	NT	4. 0	30. 0

4. Enhancement of the growth of chicken sarcoma and papilloma by extracts from some plans

The enhancing effect of extracts of W. chamaedaphne and W. indica on the growth of sarocm was studied in chicken. The weight of tumors in these two groups is 2. 5 – 3. 0 and 2. 6 – 4. 1 times higher than that in virus control groups, respectively (Tab. 5). The effect of the extracts on tumor promotion was also studied in mice. There was no papilloma in the control group (DMBA), but the percentage of papilloma in mice in the croton oil, W. chamaedaphne, and W. indica groups was 77%, 35% and 27%, respectively (Fig. 1) These results indicate that both W. chamaedaphne and W. indica act as croton oil (TPA) and are considered to be promotors. Prospective Studies on Nasopharyngeal Carcinoma.

Tab. 4　Enhancing effect of extracts from some plants on the transformation of human lymphocytes by EB virus. TPA (5 ng/ml), 186; $P < 0.05 - 0.01$

Plant extract	Concentration (r/ml)			
	0	0.2	2	20
Stellera chamaejasme	68	131 (1.9)[a]	187 (2.8)	254 (3.7)
Wikwtroemia chamaedaphne	68	158 (2.3)	210 (3.1)	190 (2.9)
Daphne genkwa	68	160 (2.4)	210 (3.1)	229 (3.4)
Wikstroemia indica	68	111 (1.6)	200 (2.9)	280 (4.1)
Sparganium stoloniferum	63	137 (2.2)	161 (2.6)	284 (4.5)
Bulbophyllum inconspicuum	55	40 (0.7)	40 (0.7)	42 (0.7)
Abrus fruticulosus	55	44 (0.8)	42 (0.7)	48 (0.8)

Notes: [a] Ratio of colonies in experimental group to coloniese in control group

Tab. 5　Enhancing effect of extracts of Wikstroemia chamaedaphne and W. indica on the growth of chicken sarcoma by Rous sarcoma virus. Virus control, 346 mg; ratio = experimental group/virus control group

	Control (ethyl alconol)	Wikstroemia		Wikstroemia	
		chamaedaphne	(mg)	indica	(mg)
Concentration	10%	0.5	2.5	0.02	0.1
Weight of tumor (mg)	458	850	1050	900	1444
Ratio	1	2.5	3.0	2.6	4.1

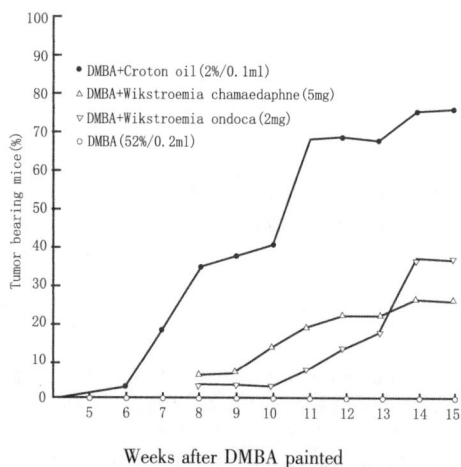

Fig. 1　Effect of plant extracts on tumor promotion

Tab. 6　Enhancement of cervical cancer in mice by Wikstroemia chemaedaphne (WC) and HHPA. Comparing data from groups, C, D. and E. the difference is significant ($P < 0.05$)

Group	No. of mice	Total cancer	%	Invasive cancer	%
WC	38	0	0	0	0
HHPA	36	0	0	0	0
MCA	23	13	56.5	4	30.7
WC + MCA	29	24	82.8	21	87.5
HHPA + MCA	45	38	84.4	30	78.5

5. Enhancement of cervical carcinoma in mice by wikstroemia chamaedaphne and HHPA

The enhancing effect of W. chamaedaphne (WC) and of HHPA on cervical cancer induced by methylcholanthrene (MCA) was studied in mice (Tab. 6). The ratio of cancer induced by WC extract and HHPA with MCA was 82.5% and 84.4%, and prospective Studies on Nasopharyngeal Carcinoma.

The rate for invasive cancer (later stage) was 87.5% and 78.5% respectively. The incidence of cancer in control groups (MCA alone) was significantly lower, with a ratio of 56.5% (noninvasive) and 30.7% (invasive). These data indicate that WC extracts and HHPA enhance the development of carvical carcinoma in mice[14].

Some extracts from plants including Chinese medicinal herbs were found to have an inducing activity of early antigen in Raji cells, to enhance the transformation of lymphocytes by EB virus in vitro, and to enhance tumor growth by Rous sarcoma virus, DMBA, and MCA in vivo. These data indicate that some extracts from plants acting as TPA could activate EB virus and were considered to be tumor promotors. There might be a relationship between the activation of EB virus by certain environmental factors and the development of NPC, but this requires confirmation.

[In 《Cancer of the liver. Esophagus and Nasopharyngeal Carcinoma》 1986, 164 – 169]

References

1　Zeng Y, Liu Y X. Liu CR, et al. Application of Immunoenzymatic Test and Immunoautoradiographic Method for the Mass Survey of Nasopharyngeal Carcinoma. Chin J, Oncol 1 (1979) 2　7 and Intervirology, 13 (1980), 162 – 168

2　Zeng Y, Zhang LG, Li HY, et al. Serological Mass Survey for Early Detection of Nasopharyngeal Cancinoma in Wuzhou City, China. Int J Cancer, 1982, 29: 139 – 141

3　Tumor Control Team, Zhoshan County; Department of Micobiology, Zhongshan Medical College; Laboratory of Tumor Viruses, Institute of Cancer, Institute of Epidemiology, Chinese Academy of Medical Sciences: A Study on the Complement Fixting Antibody to EB Virus from Normal Individuals in Guangdong Province and Beijing. Chin J ENT, 1978, 1: 23 – 25

4　Zeng Y, Zhong TM, Mo Y K, et al. Epstein – Barr Virus Early Antigen Induction in Raji Cells by Chinese Medicinal Herbs. Intervirology, 1983, 19: 201 – 204

5　Zeng Y, Miao XC, Jiao B, et al. Epstein – Barr Virus Activation in Raji Cells with Ether Extracts of Soil from Different Areas in China. Cancer Lett, 1984, 23: 53 – 59

6　Zhong JM, Zeng Y. Epstein – Barr Virus Inducers in Vegetable Grown From the Soil Containing Inducers. (Unpublished Data)

7　Hu YL, Zeng Y. The Extract from some Chinese Herbs Enhanced the Transformation of Human Lymphocytes by EB Virus. Chinese J Oncol, 1985, 1: 417 – 419

（此文原文献14条，由于版面限制，压缩成7条，请谅解——编者）

37.   Detection of Antibody to LAV/HTLV – III in Sera from Hemophiliacs in China

ZENG Yi[1], FAN Jian[1], ZHANG Qin[1], WANG Pi – chang[2], TANG De – ji[2],

ZHON Sao – cong[2], ZHENG Xi – wen[3], LIU Do – pei[4]

1. Institute of Virology, Chinese Academy of Preventive Medicine, Beijing 100052 China

2. Zhe Jiang Medical University; 3. Institute of Epidemiology and Microbiology,

Chinese Academy of Preventive Medicine; 4. Beijing Medical University, Beijing, China

Summary

Using ELISA, Western blots and immunofluorescence techniques, we identified seropositivity for lymphadenopathy – associated virus/human lymphotropic virus – III (LAV/HTLV – III) in 4 of 18 hemphiliacs and 1 AIDS patient. The four seropositive patients had received factor VIII prepared by Armour Company. The hemphiliacs are all asymptomatic. Given this documentation of introduction of LAV/HTLV – III into China, a national surveillance program is underway.

Introduction

The acquired immunodeficiency syndrome (AIDS) is a highly lethal epidemic that was first reported in 1981. There are LAV/HTLV – III antibodies in sera from AIDS patients and individuals infected with the virus. It is very important to know whether there is LAV/HTLV – III in china or not. Thus, 310 sera from normal individuals and leukemia patients in 8 provinces were tested for LAV/HTLV – III antibody by an immunofluorescence test and no seropositives were found. Godert et al. (1) reported that the positive rate of LAV/HTLV – III antibody was from 74% to 90% in hemophiliacs treated with factor VIII. Here we reported that LAV/HTLV – III antibody was found in 4 sera from hemophiliacs treated with factor VIII.

Materials and Methods

1. Sera samples: Twenty – eight sera from hemophiliacs treated with factor VIII were collected and stored at – 20℃. Among them 18 sera from hemophiliacs living in Zhejiang province who were treated with Armour Company produced factor VIII, 8 hemophiliacs in Beijing treated with Alpha Company produced factor VIII and 2 hemophiliacs in Zhejiang treated with factor VIII from 1982 to 1985 locally produced.

2. ELISA: Abbott HTLA – III EIA kits were used for detection of LAB/HTLV – III antibody.

Western blot stripa and antigen slides for immunofluorescence test were obtained from Max Von

Pettenkofer Institute in Munich, West Germany.

The procedures of the three testes were performed consequentially as required. First they were tested by ELISA and the positive sera were confirmed by Western blot assay and an immunofluorescence test.

Results

Twenty – eight sera from hemophiliacs and one from AIDS patient were tested by ELISA. Four of 18 sera from hemophiliacs in Zhejiang treated with factor Ⅷ produced by Armour Company and one sera from AIDS patients showed LAV/HTLV – Ⅲ antibody. The optical density (C. D) value for these 4 sera was 1. 831, 1. 449, 1. 854, and 0. 688, respectively. The A sera from AIDS patients was 1. 60 (Tab. 1). Eight sera from hemophiliacs in Beijing treated with Alpha Company produced factor Ⅷ and 2 treated with locally produced factor Ⅷ were negative.

Tab. 1   LAV/HTLV – Ⅲ Antibody Detection in Chinese

Method	Negative control	Positive control	Seropositive Hemophiliac Patients				AIDS Patient
			No. 5[1]	No. 9[1]	No. 11[1]	No. 55[1]	
ELISA (A value)	0. 435	1. 103	1. 831	1. 449	1. 854	0. 685	1. 600
IF	–	+	+	+	+	+	+
Western blot	–	p76	p76	p76	p76	p76	p76
		p64	p64	p64	p64	p64	p64
		p55	p55				p55
		p41	p41	p41	p41	p41	p41
		p39					p39
		p30	p30	p30	p30	p30	p30
		p24	p24	p24		p24	p24
		p16			p16		

Notes: [1] Hemophiliacs treated with factor Ⅷ from Armour Company

Abbreviations: A = optical density; IF = immunofluorescence

The 5 antibody – positive sera were tested by an immunofluorescence test and Western blot assay. They were all antibody positive (Tab. 1), but the hemophiliacs did not have symptoms related to AIDS.

Discussion

Four of 18 sera from hemophiliacs treated with Armour Company produced factor Ⅷ were LAV/HTLV – Ⅲ antibody – positive. Hence, the aera positive rate was 22. 2%. This data indicates that LAV/HTLV – Ⅲ was transmitted into China via blood products made in USA at an apparent frequency of 22. 2% which is lower than that (74% – 90%) reported by Goedert et al[1]. This difference could be due to the patients receiving fewer injections and lower dose.

To date the 4 LAV/HTLV-Ⅲ antibody – positive hemophiliacs have no symptoms related to AIDS. A follow-up study is being carried out. A nationwide serological survey of LAV/HTLV-Ⅲ antibody has been started, because the virus has been transmitted into China via the factor Ⅷ blood product.

〔In 《AIDS Research》 1986, 2: 147 – 149〕

References

1　Coedert J M, et al. Ann Int Med, 1985, 65: 493　|　2　WANG P C, et al. Chinese J Virology, 1985, 1: 391

38.　Detection of IgG and IgA Antibodies to Epstein – Barr Virus Membrane Antigen in Sera from Patients With Nasopharyngeal Carcinoma and from Normal Individuals

ZHU X X[1], ZENG Yi[1], Wolf H[2]

1. Institute of Virology, China National Centre for Preventive Medicine,
100 Ying Xing Jie, Beijing. People's Republic of China
2. Max von Pettenkofer – Institute, University of Munich,
Pettenkofer Str. 9a, D – 8000 Munich 2, Fed. Rep. of Germany

Summary

IgG and IgA antibodies to Epstein – Barr virus (EBV) membrane antigen (MA) were detected in sera from 96 NPC patients and normal individuals by the indirect immunofluorescence test. For MA/IgG antibody, 100% of NPC patients were positive with a GMT of 1:439.7 and 97.9% of normal individuals were positive with a GMT of 1:94.7. In contrast, for MA/IgA antibody, 58.3% of NPC patients were positive with a GMT of 1:7.3 and none of the normal individuals were positive. There was no difference in the detection of antibodies to EBV MA when other P3HR – 1 or B95 – 8 cell lines, differing in their major membrane antigen, were used.

Klein et al. (1966) first demonstrated EBV MA in cells from Burkitt lymphoma by the indirect immunofluorescence test, and then proved that the MA was specific to EBV by a direct blocking test (Klein et al. 1969). Other studies have shown that EBV MA exists on both the EBV envelope and the membrane of cells which carry EBV genomes and produce intact EBV particles (Sugawara and Osato, 1970; Silvestre et al. 1971). These data indicate that the antibody titer to MA correlates well with that of neutralizing antibody to EBV (Pearson et al. 1970). There are no published reports concerning IgA antibody to EBV MA. A hypothesis links blocking of ADCC with the appearance of IgA antibodies mainly directed to VCA or EA (Mathew et al. 1981). However, no data have been presented on the reactivity of IgG and particularly IgA antibodies in sera from NPC patients and normal individuals to the membrane antigen by indirect immunofluorescence tests.

Material and Methods

1. Sera

Sera were obtained from 48 NPC patients and 48 normal individuals, and stored at -20℃.

2. Indirect immunofluorescence test

The target cells used for detection of MA/IgG and MA/IgA antibodies were P3HR − 1 or B95 − 8 cells. They were cultured in RPMI 1640 medium with 20% newborn calf serum.

P3HR − 1 or B95 − 8 cells were activated for 48 hr by 4 mmol/L n − butyrate and 500 ng/m of croton oil. The activated cells were washed 3 times with Hanks´ solution and adjusted to 1×10^6 cells/ml. Then 1×10^5 cells in 100 μl were added to each well of 96 − well U − shaped hemagglutination plates. The sera were diluted from 1∶10 to 1∶640 in 2 − fold dilution and then placed in a humidified chamber at 37℃ for 45 min. After 3 washes with Hanks´solution, cell smears were prepared on slides, air − dried and fixed with cold acetone. FITC − conjugated sheep antibodies diluted 1∶10 and directed to human IgG or IgA were added and the slides were kept at 37℃ for 30 min. The smears were again washed 3 times with 0.01 mol/L PBS, pH 7.6 After counter − staining with 0.006% Evansblue for 10 min, they were examined under an Olympus fluorescence microscope. Cell membranes stained with a specific green color were considered to be positive. The number of cells positive for MA was measured with the test described above, a mixture of several sera being used as first antibody.

3. Immunoenzymatic test

The test was performed as described by Zeng et al (1979).

Results

1. Comparison of the EBV − MA − positivity in P3HR − 1 and B95 − 8 cell lines

The positivity of the EBV MA in untreated P3HR − 1 and B95 − 8 cells was 9.1% and 11.2% respectively. The numbers increased to 62.1% and 63.4% respectively after activation with croton oil and n − butyrate for 48 hr. There was no further increase in MA positivity after activation for 72 hr and more frgmnted cells were found (Fig. 1).

2. Comparison of the prevalence rate of EBV MA/IgG, MA/IgA, VCA/IgA and EA/IgA antibodies in sera from NPC patients and normal individuals.

Sera from 48 NPC patients and from 48 normal individuals were tested for EBV MA/IgG and MA/IgA antibodies by the immunofluorescence test, and for VCA/IgA and EA/IgA antibodies by the immunoenzymatic test. The positivity of the above 4 antibodies was 100%, 58.3%, 100% and 64.6% respectively, in NPC patients, and 97.9%, 0% and 0% respectively, in normal individuals (Tab. 1).

Tab. 1　Comparison of positivity rate of IgA and IgG
antibodies to VCA and MA from NPC patients and normal individuals

	Cases	MA/IgG[1]		MA/IgA[1]		VCA/IgA[2]		EA/IgA[2]	
		Number	% positivity	Number	% positivity	Number	% positivity	Number	% positivity
NPC patients	48	48	100	28	58.3	48	100	31	64.6
Normal individuals	48	47	97.9	0	0	0	0	0	0

[1] MA/IgG and MA/IgA detected by immunofluorescence test; [2] VCA/IgA and EA/IgA detected by immunoenzymatic test

3. Comparison of the distribution of EBV MA/IgG and MA/IgA antibody titers in NPC patients and normal individuals

As shown in Fig. 2, the range of the MA/IgG antibody titers for NPC patients was from 1:40 to 1:1280 with a GMT of 1:439.7, and the range for normal individuals was from 1:10 to 1: 640 with a GMT of 1:94.7. The range of MA/IgA antibody titers for NPC patients was from 1:10 to 1:160 with a GMT of 1:7.3; only 52% of the patients had antibody titers higher than 1:20, but to such antibodies could be found in normal individuals. In most NPC cases the VCA/IgA antibody titer was higher than that of MA/IgA antibodies (Fig. 3). This difference was less significant in a comparison with EA/IgA antibodies (Fig. 4).

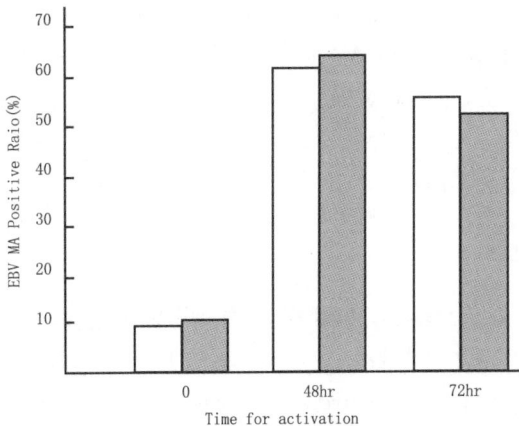

Fig. 1　Comparison of the EBV – MA positive rate in P₃HR –1 (□) and B95 –8 (▨) cells.

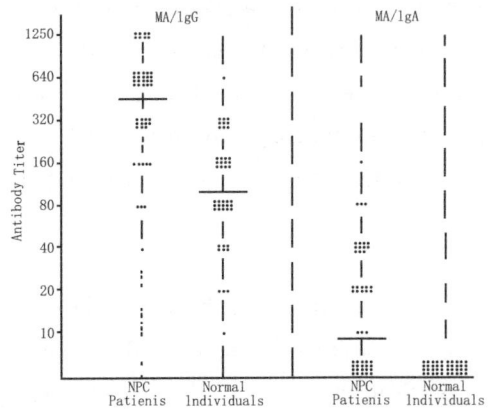

Fig. 2　Comparison of MA/IgG and MA/IgA antibodies from NPC patients and normal individuals

Discussion

It has been shown that the detection of IgA antibody to VCA and EA of EBV is of value for the diagnosis of NPC (Henal and Henle, 1976; Zeng et al, 1979a, b; 1980, 1983b). The positivity of MA/IgG antibody was very high both in NPC patients and in normal individuals. Although the GMT of MA/IgG antibody is much higher in NPC patients than in normal individuals, this test is of no value for the diagnosis of individuals cases. Of NPC patients, 58.3% had MA/IgA antibody, while all normal individuals lacked IgA antibodies to this antigen. The situation is similar to seen

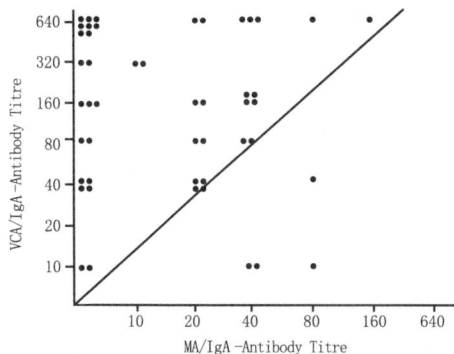

Fig. 3　Relationship between MA/IgA and VCA/IgA antibodies in sera from NPC patients.

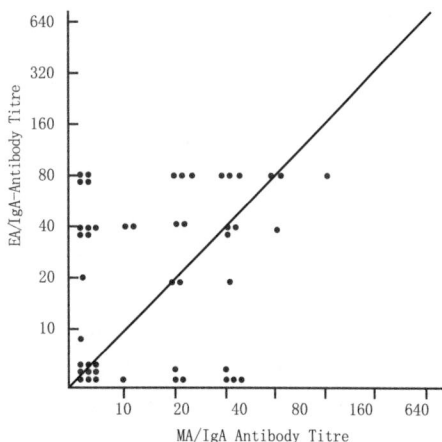

Fig. 4　Relationship between MA/IgA and EA/IgA antibodies in sera from NPC patients

with EA/IgA antibodies in NPC patients and in normal individuals (Zeng et al, 1983b), hence detection of MA/IgA antibody can be used as a marker for the diagnosis of NPC.

The positivity of MA in P3HR – 1 and B95 – 8 cells was similar, although the major membrane glycoprotein differs in both cell lines (Edson and Thorley – Lawon, 1983). suggesting that both lines can be used as targets for the detection of EBV MA/IgA antibodies. For a higher expression of MA, cells could be activated with croton oil and n – butyrate for 48 hr before use.

Detection of EBV MA/IgA antibody is more specific, but not as sensitive for the diagnosis of NPC as detection of VCA – IgA antibodies. However, a more sensitive technique for the detection of EBV MA/IgA antibodies should considerably reduce the false negatives and give an even better diagnostic value (Jilg and Wolf, 1985).

The predictive value of MA/IgA antibodies for the prognosis of patients is under investigation. An economical production of MA, using genetic engineering technology, should prove helpful for the development of simpler tests which would allow screening of large quantities of serum.

Acknowledgement

Part of this work was supported by Stiftung Volkswagenwerk.

[In 《Int J Cancer》 1986, 37: 689 – 691]

References

1　Edsdn P, Thorley – Lawson D. Synthesis and processing of the three major enaelope glycoproteins of Epstein – Barr virus. J Virol, 1983, 46: 547 – 556

2　Henle G, Henle W. Serum IgA antibodies to Epstein – Barr virus (EBV) – related antigens, a new feature for nasopharyngeal carcinoma. Bibl Haemat, 1976, 43: 322 – 325

3　Jilg W, Wolf H. Diagnostic significance of antibodies to the Epstein – Barr virus – specific membrane antigen gp 250. J infect. Dis, 1985, 152: 222 – 225

4　Klein G, Clifford P, Klen E, Stjernsward J. Search for tumor – specific immune reaction in

Burkitt lymphoma patients by the membrane immunofluorescence reaction. Proc, nat, Acad, Sci. (Wash), 1966, 55: 1628 – 1635

5　Klein G, Pearson G, Henle G, Henle W, Goldtein G, Clifford P. Relation between Epstein – Barr viral and cell membrane immunofluorescence in Burkitt tumor cells. J exp Med, 1969, 129: 679 – 705

6　Mathew G D, Qualtiere L F, Nell H B Ⅲ, Pearson G R. IgA antibody, antibody – dependent cellular cytotoxicity and prognosis in patients with nasophryngeal carcinoma. Int J Cancer, 1981, 27: 175 – 180

7　Pearson G, Dewey F, Klein G, Henle G, Henle, W. Relation between neutralization of Epstein – Barr virus and antibodies to cell – membrane antigens induced by the virus. J. nat. Cancer Inst, 1970, 45: 989

8　Silvestre D, Kourilsky F M, Klein G, Yata Y, Neauport Sautes C, levy J P. Relationship between the EBV – associated membrane antigen on Burkitt lymphoma cells and the viral envelope, demonstrated by immunoferritin labelling. Int J. Cancer, 1971, 8: 222 – 233

9　Sugawara K, Osato T. An immunoferritin study of a Burkitt lymphoma cell line harboring EB virus particles. Gann, 1970, 61: 279 – 281

10　Zeng Y, Gong C, Jan M, Fun Z, Zhang L G, Li H Y. Detection of Epstein – Barr virus IgA/EA antibody for diagnosis of nasopharyngeal carcinoma by immunoautoradiography. Int J Cancer, 1983a, 31: 599 – 601

11　Zeng Y, Liu Y, Liu C, Chen S, Wei J, Zhu J, Zai H. Application of an immunoenzymatic method and an immunoautoradiographic method for a mass survey of nasopharyngeal carcinoma. Intervirology, 1980, 13: 162 – 168

12　Zeng Y, Liu Y X, Wei J N, Zhu J S, Cai S L, Wang P Z, Zhong J M, Li R C, Pan W C, Li E J, Tan B F. Serological mass survey of nasopharyngeal carcinoma. Acta Acad Med Sin, 1979a, 1: 123 – 126

13　Zeng Y, Yuxi L, Chunren L, Sanwen C, Jineng W, Jisong Z, Hunong Z. Application of an immunoenzymatic method and an immunoautoradiographic method for a mass survey of nasopharyngeal carcinoma. Chin J Cncol, 1979b, 1: 2

14　Zeng Y, Zhong L, Li P, Wang P, Tang Y R, Ma Y R, Zhu J S, Pan W J, Liu Y X, Wei Z N, Chen J Y, Mo Y K, Li E J, Tan B F. Follow – up studies on Epstein – Barr virus IgA/VCA antibody – positive persnons in Zangwu county, China. Intervirology, 1983b, 20: 190 – 194

39. A Preliminary Analysis of HLA Studies on Multiple NPC Cases Among Siblings from the People's Republic of China, Singapore and Malaysia

Mcknight B[1], Lu S T[2], Ju L[2], Day N E[1], Degos L[3], Lepage V[3],
Chan S H[4], Prased U[5], Ho J H C[6], Simons M J[7], Zeng Y[8], de – The G[9]

1. Unit of Biostatistics IARC, Lyon, France; 2. People's Hospital Nanning, Guang – Xi Aut. Region PRC; 3. Inserm U93, Hopital St. Louis, Paris, France; 4. Microbiology Dep, National University, Singapore; 5. University of Malaya; 6. Institute of Radiology and Oncology, Queen Elizabeth Hospital, Kowloon, Hongkong; 7. Immunogene Typing Labs, Immunoresearch, Melbourne, Australia; 8. Institute of Virology, Academy Preventive Medicine, Beijing PRC; 9. Faculty of Medicine A. Carrel, Lyon, France

Earlier studies[1,2] have demonstrated an association between nasopharyngeal carcinoma (NPC) and the occurrence of HLA atigens BW46, B17 and A2. In[2], relative risks of NPC for the phenotypic expression of these antigens were estimated as 1.5 for A2, 2.14 for B17 and 1.88 for BW 46. These associations suggest that either the presence of these HLA genes themselves or the occrrence of a non – HLA disease suceptibility gene (DSG) closely linked to the HLA A and B loci is associated with an increased risk of NPC.

In order to study further the possibility of a DSG in the HLA region, investigations of families with multiple cases of NPC within a sibship were initiated in Singapore, and more recently in China. HLA A and B typing were performed in Nanning, Guang – Xi Autonomous Region, People's Republic of China on available members of Chinese families, in Singapore on families from Singapore and in Malaysia on one family from Malaysia.

The HLA phenotypes of the affected siblings and other available family members were than used to determine the affected siblings' HLA haplotypes and how many haplotypes the pairs or triples had inherited in common. Parents were often not available for HLA typing and haplotype inheritance had to be inferred from sibling and offspring phenotypes. An unequivocal determination of the number of haplotypes inherited in common was not always possible, so the families were divided into two groups: Group A, for which an unequivocal determination was possible, and group B for which an unequivocal determination was not possible. For Group B families, the haplotype sharing configuration of the siblings was inferred by assuming that any apparently unobserved haplotypes were not identical to those observed among the available siblings. Tab. 1 presents the munbers of sibships in group A alone and in groups A and B combined. Possibilites of bias introduced by either of these choices of a sample have been noted, but they are not taken into account in what is presented, here.

Tab. 1 The members of sibships in group A alone and in groups A and B combined

	Group A	Groups A and B
Total Sibling Pairs	24	31
Singapore	5	5
Malaysia	1	1
Hong Kong of China	2	6
China	16	19
Total Sibling Triples	3	4
Hong Kong of China	1	1
China	2	3

Tab. 2 The Share of the haplotypes in group A alone and in groups A and B combined

	Group A		Groups A and B	
	obs	exp	obs	exp
Sibling Pairs:				
Share 2 haplotypes	13	6	17	7.75
Share 1 haplotype	7	12	10	15.50
Share 0 heplotypes	4	6	4	7.75
Sibling Triples:				
3 pairs share 2 haplotypes	0	0.1875	0	0.25
1 pairs share 2 haplotype 2 pairs shares 1 haplotypes	1	1.1250	2	1.50
1 pairs shares 2 haplotypes 2 pairs share 0 haplotypes	1	0.5625	1	0.75
1 pair shares 0 haplotypes 2 pairs share 1 haplotype	1	1.1250	1	1.50

Sibling pairs can share zero, one or two haplotypes by descent. Sibling triples can exhibit one of four possible haplotype inheritance patterns. The observed numbers of our sibling pairs and triples exhibiting each of the possible inheritance patterns are given in Tab. 2. along with the number that would be expected if there were no DSG in the HLA region.

If a DSG in the HLA region confers an increased risk of NPC, it would be more likely to be the siblings who had inherited this gens who developed NPC. One would therefore expect a higher proportion of NPC affected sibling pairs to share two HLA haplotypes by descent than the one fourth that would be expected if there were no DSG.

To test this hypothesis, sibling triples were treated as if they contributed only two independent pairs, since in any triple the number of the first two pairs sharing two ahplotypes completely determines whether the third pair shares two haplotypes. Using this convention, the proportion of Group A pairs sharing two haplotypes is $15/30 = 0.5$ and the proportion of Group A and B pairs sharing two haplotypes is $18/39 = 0.4615$. In both cases the one sided exact test indicates that the proballity an affected sibling pair shares two haplotypes is greater than one fourth ($P = 0.003$).

Under either an autosomal dominant or recessive model for the inheritance of disease susceptibility, approximate methods of obtaining lower confidence limits for the relative risk associated with the putative DSG can be based on the proportion of affected sibling pairs that share two haplotypes by descent[3]. In our data the 95% lower bounds for the relative risk based on the pairs and triples for groups A and B combined are 6.7 for a dominant inheritance model and 5.1 for a recessive inheritance model. Performing the same calculations with data from Group. A alone or after excluding sibling triples from Group A or Groups A and B combined yielded even higher bounds.

Retrospective and more recent prospective studies of Chinese NPC patients have repeatedly demonstrated associations involving the HLA B – locus alleles[1,2]. The probability of these associations occurring by chance is small. However, population association studies depend, by their nature, on the demonstration of differences in individual allele frequencies between patients and controls. They provide conservative estimates of relative risks associated with an HLA region DSG because the marker alleles may not be completely associated with the putative DS locus. By contrast, multiple case family analyses provide a direct estimate of full genetic risk associated with a gene region.

Some of the multiple case families analyzed here were HLA typed for A and B locus alleles when HLA serology in Chinese was at a relatively early stage of development. In some families not enough family members were available for HLA typing to make unambiguous haplotype assignments. In addition, the statistical treatment of the sibling triples is imperfect. For these reasons, a conservative position has been taken and only lower bound estimates for the relative risks are reported. The lower bound estimates both dominant and recessive models are nonetheless higher than the reported relative risk associated with detectable alleles. This finding provides further support for the conclusion that risk for NPC in Chinese involves gene (s) in the HLA region.

[In 《Epstein – Barr Virus and Human Disease》 1987, 25 – 29]

References

1 Simons, et al. Natl Cancer Inst Monogr, 1977, 47: 147 – 151

2 Chan, et al. Int J Cancer, 1983, 32: 171 – 176

3 Day and Simons. Tissue Antigens, 1976, 8: 109 – 119

40. Evaluation of Monoclonal Antibodies for the Detection of Exfoliative Nasopharyngeal Carcinoma Cells

Chan K H[1], Yip T C[1], CHOY Damon [2], Chan C W[3], Zeng Y[4], Ng M H[1]

1. Departments of Microbiology, and 3. Pathology University of Hong Kong, Pokfulam Road, Hong Kong
2. Department of Radiotherapy and Oncology, Queen Mary Hospital, Pokfulam Road, Hong Kong
4. Chinese Academy of Preventive Medicine, Beijing, People's Republic of China

Summary

Exfoliative cells were aspirated from 15 patients suspected of having nasopharyngeal carcinoma (NPC) and showing the presence of lesions or other abnormalities in the nasopharynx. They were tested for binding with a ^{125}I monoclonal antibody (MAb) (MA6) which is selectively reactive against human B lymphocytes and a varitey of carcinomas. A positive result was obtained from 6/9 patients with, and from 0/5 patients without histologically confirmed disease. One patient with eskimoma also gave a negative binding result. Cytology was specific but less sensitive, tumour cells being detected in 3 of the patients with confirmed disease. Immunocytology using MA6 was limited, like cytology, by poor recovery of the tumour cells and the results were in complete concordance with cytology. The other MAbs used were raised against carcinoembryonic antigen (CEA) and a carcinoma cell line (Ca2), respectively. The latter was not reactive against the NPC tumour cells while the CEA antibody was not sufficiently selective to be useful.

Introduction

Since 1980, an ongoing serological study of healthy subjects 40 years or older for IgA antibodies to the viral capsid antigen (VCA) and other antigens of Epstein – Barr virus (EBV) has been carried out in Wuzhou City, People's Republic of China (population, 170 000) (Zeng et al, 1985). It initially involved 20 726 individuals, with others in the community being recruited to the study as they reached the age of 40. The result showed that 5. 5% of this population has an elevated level of IgA/VCA antibody and, on subsequent follow – up over 4 years, 3. 1% of the serologically positive individuals had developed NPC. This represents a substantially increased risk of the disease even when compared with the high NPC incidence of 50 per 100 000 reported fro that community at large. Earlier retrospective studies (Ho et al, 1978; Lanier et al, 1980) in other locations also suggested an association between elevated level of these EBV antibodies and increased risk of NPC. These studies showed that, for the large majority of patients, elevated serum levels of these antibodies preceded clinical onset of NPC, sometimes by up to 70 months.

Clinical onset of NPC is often insidious and, consequently, the disease may only be diagnosed at a late stage with a poor prognosis following radiation therapy (Ho, 1982). The study in Wuzhou City is especially significant in view of the finding that all but one of the 38 patients were diagnosed at an early stage when the disease was circumscribed with few symptoms or none at all. It was also noteworthy that the disease occurring in that community since the beginning of this investigation has been confined to those patients who had been identified earlier as having elevated IgA scrum EBV antibodies. With the occurrence of better prognosis following treatment of early cases, this effort has thus achieved control of one of the most prevalent malignant diseases in that community.

For early detection of the disease to be more generally possible, however, a covenient and non-aggressive investigation is necessary for screening high – risk subjects identified by serology. Nasopharyngoscopy is a common practice in the investigation of suspected NPC. Totake full advantage of this practice, a sensitive and specific method for detecting exfoliative tumour cells is required to facilitate early detection of NPC. Certain investigatiors reported detection of exfoliative tumour cells from the nasopharynx of 80% or more of patients who had histologically confirmed NPC (Dong et al, 1983). In general, however, cytology of NPC has suffered from a lack of sensitivity because of poor yield of tumour cells and interference of other exfoliative cells in their detection. The subjective element inherent in this method introduces further uncertainties which require assessment by an experienced professional staff.

We have raised an MAb against the immunoprecipitate obtained by counter – immunoelectrophoresis of an NPC patient's serum against an extract of Raji cells, a Burkitt lymphoma cell line of B-lymphocyte lineage (Chan et al, 1985). Apart from human B lymphoid cell lines and tissues, the antibody (MA6) was also reactive against different tumour cells especially carcinomas, including NPC. Although the antibody was also reactive against other cell types, notably smooth muscle and some endothelial cells, the reaction was less intense so that these cells were readily distinguished from the intensely reactive carcinoma cells. The antigen recognized by MA6, referred to as B – lymphocyte carcinoma antigen (BLCa) because of its restricted tissue distribution, was stable to a variety of tissue fixation procedures; it was preserved for 8 years of more in paraffin blocks of formaldehyde – fixed tumour tissues without deterioration. It differs from antigens identified by other MAbs with specificity for carcinomas (Bramwell et al. 1985; Jothy et al, 1986) in the BLCa is not represented on tumour cell lines other than those of B – lymphocyte origin.

We have evaluated immunocytology and direct binding of radiolabelled MA6 with exfoliative cells aspirated from the nasopharynx of 15 patients suspected of having NPC who, on nasopharyngoscopy, exhibited lesions and/or other abnormalities in their nasopharynx. The results were compared with those obtained by cytology and histopathology and by immunocytology using MAbs raised against a carcinoma cell line (Ca2) and CEA.

Material and Methods

Fifteen patients were referred to the Department of Radiation Therapy and Oncology, Queen Mary Hospital. Hong Kong, 8 for suspected NPC and 7 for suspected recurrence of the disease (Tab. 1). The latter group of patients had received radiation therapy for NPC between 11 and 187 months previously (median, 27 months), and had since been in clinical remission. Nasopharyngoscopy revealed that these 15 patients all had lesions or other abnormalities of the nasopharynx. Aspiranon was performed by instilling about 15 ml of saline through one channel of the nasopharyngoscopo. The washing was collected by a Zeng – de – The type aspirator placed in the nasopharynx through the oropharyngeal isthmus. This was followed by biopsy of the abnormal area. Disease was confirmed histologically in 9 patients – 7 new cases and 2 with recurrent disease. One patient was diagnosed as having eskimoma. There was no histological confirmation of disease for the remaining 5 patients.

The wad of silk lodged in the tip of the aspirator to entrap the exfoliative cells was removed and directly smeared onto several glass slides for cytology and immunocytology. The remaining cells (0. 8 to 8. 6×10^6) were recovered by washing the silk in Hanks'buffered salt saline (HBSS). On some occasions the cells were fixed in 10% buffered formalin and tested by the whole – cell binding assay using ^{125}I – MA6. On other occasions the cells were lysed and the extract was allowed to react with ^{125}I – MA6 by the immunodot binding assay. The specificity of binding in the case of the whole – cell binding assay was evaluated by autoradiography and, in the case of dot immunobinding assay, by Western blotting. (Yip et al, 1987a).

Radiolabelling of MA6, immunodot binding assay, wholecell radioimmunobinding assay. Western blotting and immunocytochemical staining are described in the companion report (Yip, et al. 1987a, b).

Tab. 1   Detection of exfoliative tumour cells from symptomatic nasopharyngeal carcinoma patients

Patients	Sex/Age	Histologically confirmed NPC (clinical staging)[1]	Exfoliative cells			
			Cells recovered $\times 10^{-6}$	^{125}I – MA6 binding per 3.5×10^5 cells[2]	Cytology (+ /total)[5]	Immuno – cytology (+ /total)[5]
KWP[3,7]	M/46	+ (Ⅳ)	3.7	1 070 ± 102 (+)[4]	1/3	1/2
KLS[3,7]	M/37	+ (Ⅲ)	1.3	3 237 ± 50 (+)	2/3	2/2
KK[6]	M/60	+ (Ⅳ)	8.6	1 862 ± 199 (+)	2/3	2/2
LC[6]	M/41	+ (Ⅳ)	0.8	1 013 (+)	0/3	0/2
LYM[7]	F/65	+ (Ⅳ)	0.7	1 020 (+)	0/3	0/2
CKS[7]	M/55	+ (Ⅰ)	4.5	702 ± 52 (+)	0/3	0/2
LKS[7]	F/Ad	+ (Ⅱ)	0.8	82 ± 30 (–)	0/3	0/2
CKH[7]	F/39	+ (Ⅱ)	5.5	238 ± 51 (–)	0/3	0/2
MS[7]	M/70	+ Ⅲ	0.8	302/41 (–)	0/3	0/2
YHC[7]	M/29	Eskimoma	0.9	340 (–)	0/3	0/2
Mean SD			3.2 ± 2.7	987 ± 905		
CPL[6]	M/49	– (Ⅰ)	1.6	450 ± 82 (–)	0/3	0/2
WKT[6]	F/47	– Ⅲ	7.8	232 ± 52 (–)	0/3	0/2
SLC[3,6]	M/36	– Ⅱ	1.3	143 (–)	0/3	0/2
CK[6]	F/46	– (Ⅳ)	0.9	211 ± 35 (–)	0/3	0/2
HH[6]	F/56	– Ⅳ	4.0	97 ± 18 (–)	0/3	0/2
Mean SD			2.5 ± 2.4	227 ± 123		
Raji				12 411 ± 1 932		
Molt 4				173 ± 49		

Notes: [1] Clinical staging at diagnosis according to Ho's classification (1982). – [2] Replicate testing done where amount of specimens permitted; results shown as mean ± SD, – [3] Reactivity of these exfoliative cells with MA6 was determined by whole – cell binding. and that of the other specimens by dot immunobinding. – [4] (+) A positive result was scored when net binding exceeded the mean binding obtained with exfoliative cells from patients with no histologically confirmed NPC by 2 SD more. – [5] (+) Number of smears containing tumour cells identified by cytology or immunocytology using MA6. – [6] Patients with suspected recurrent NPC. – [7] New cases.

Results

Exfoliative cells aspirated from the nasopharynx of 15 patients with symptomatic NPC were tested for binding with ^{125}I – MA6. The results were compared to those obtained by cytology and immunocytology sing MA6, Ca2 and CEA MAb. Histological confirmation of NPC was subsequently established in 9 of these patients, but not in 5 others. One patient was diagnosed as having eskimoma. For total binding. 3.6×10^5 formalin – fixed exfoliative cells (patients KWP, KLS, SLC) or extracts of the equivalent number of cells (all other patients) were employed. Replicate tests were performed where the specimens were adequate. Mean binding obtained from 5 patients who did not have confirmed disease was similar to that obtained concurrently with the non – reactive molt – 4 cells. The binding observed with individuals specimens which exceeded this mean value by 2 SD or more was considered to be significant. A positive binding result was thus obtained for 6/9 patients who had, and for 0/5 patients who had not , confirmed disease. A negative binding result was also obtained for 1 patient with eskimoa (YHC).

To evaluate the specificity of the whole – cell binding assay, the formalin – fixed exfolicative cells from patients KLS and SLC were subjected to autoradiography after they had been reacted with ^{125}I – MA6 (Fig. 1, brief). As summarized in Tab. 2. the composition of the exfoliative cell population from one patient (SLC) who did not have NPC exhibited a large preponderance of epithelial cellw which were not reactive with the antibody and only about 0. 5% of epithelial cells (columnar epithelial cells) which showed increased accumulation of silver grains. The exfoliative cell population of the patient (KLS) who had confirmed NPC was more varied. It comprised about 15% of normal epithelial cells and 12. 7% of PMN. The reactive cells included about 5. 1% of columnar epithelial cells (Fig. 1b, c, e, brief) and 13. 9% of lymphocytes (Fig. 1d, brief). About 25% of the total cell population was intensely reactive with MA6 (Fig 1a, brief). Their detailed morphology was partly obscured by the silver grains. Their tendency to aggregate, probably mediated by MA6, also made their enumeration less reliable. Nevertheless, general morphological features indicated that the reactive cells were probably tumoural in nature. The latter contention was suppoted by the stained smear of that specimen showing a preponderance of tumour cells (see Fig. 3, brief). Grain counts showed these cells to be at least 3 times more reactive with MA6 than the columnar cells or the lymphocytes (Tab. 2).

Tab. 2 Distribution of MA6 – reactive exfoliative cells from nasopharynx of suspected NPC patients[1]

Cells types	Patient KLS			Patient SLC		
	Cell count		Average grain count per cell	Cell count		Average grain count per cell
	MA6 reactive (%)	Total		MA6 reactive (%)	Total	
Epithelial cells	2 (5. 1)	39	15	5 (0. 5)	1. 015	12
Polymorphonuclear cells	0	33	0	0	0	0
Lymphocytes	17 (13. 9)	122	10	0	0	0
Suspected tumour cells	65 (100)	65	49	0	0	0

Notes: [1] Formalin – fixed exfoliative cells were reacted with ^{125}I – MA6. Smears of the reacted cells were overlaid with nuclear track emulsion (NTB – 2), exposed for 24 hr. developed and counterstained. Silver grains accumulated over different cell types were enumerated. The average grain density was calculated from the results obtained with all the reactive cells of each cell type. KLS had and SLC had not . histologically confirmed NPC.

The specimens from only 4 patients were adequate for further investigation by Western blotting (Fig. 2, brief). In the specimens from one of the 2 patients (KK) with confirmed disease, ^{125}I – MA6 identified a major antigen (arrow) which migrated in a similar way to BLCa (M_T 55 × 10^3) identified concurrently in the Raji cell extract. The antigen was not detected in the specimens from the 2 othre patients who did not have confirmed NPC (WKT and SLC), nor the specimen from the other NPC patients (LC).

Tumour cells were detected by cytology in one of the triplicate smears from one patient (KWP), and 2 of the triplicate smears from 2 patients (KLS, KK), all of whom had histologically confirmed disease (Tab. 1). Negative cytology was obtained for the remaining NPC patients and all the patients who did not have histologically confirmed disease. To directly compare the cytological and immunocytological results, the locations of the tumour cells on some of the cytology slides were marked (Fig. 3a and c, brief). The smears were treated with running tap water to remove eosin and then allowed to react with MA6. The reaction developed by the immunoalkaline phosphatase method confirmed that the tumour cells were indeed reactive with the antibody (Fig. 3b and Fig. 3d, brief).

The results obtained by immunocytology with separate smears were in complete concordance with cytological findings (Tab. 1). MA6 – reactive tumour cells were only seen in those specimens which also gave a positive cytology. The results confirmed those of earlier studies (Chan et al. 1985) using fixed tissue section, indicating that MA6 is broadly reactive against NPC tumour cells. Immunocytology was also performed using an MAb against CEA and another, Ca2, which was raised against a tumour cell line (Bramwell et al. 1985). CEA antibody was reactive against a variety of cell types, while Ca2 was not reactive against the NPC tumour cells (results not shown).

Discussion

We compared detection of tumour cells in nasopharyngeal aspirates by cytology, immunocytology and by direct binding of radiolabelled MA6 to the exfoliative cells, Whereas cytology afforded specific detection, the method suffered from a lack of sensitivity. Sensitive detection by this method depends on the number of exfoliative cells which can be practicably examined and on the tumour – cell composition of the exfoliative cell population. It was evident that sensitivity was improved by examination of replicate smears. In spite of this, however, tumour cells were only detected in 3 of the 9 patients who had histologically confirmed NPC.

The results obtained by immunocytology using MA6 were in complete concordance with cytology. They confirmed the results obtained by immunohistological examination of fixed sections of NPC (Chan et al. 1985) suggesting that the antibody is broadly reactive against NPC cells. Being subjected to the same limitations, however, immunocytology did not improve on the sensitivity afforded by cytology. The chief advantage was that the use of MA6 may minimize the subjective element inherent in cytology. The other MAbs tested concurrently did not appear to be useful for detection of exfoliative NPC tumour cells; the CEA antibody was similarly reactive against NPC cells as with a variety of other cell types, while Ca2 was not reactive against the tumour cells.

Histopathology overcomes the difficulties encountered by cytology and immunocytology. By selecting the abnormal nasopharyngeal mucosa for biopsy, it increases the chance of detecting tumour cells, whose recognition is also facilitated by the pattern assumed by tumour cells in relation to other tissues and by the greater number of cells contained in the fixed tissue sections than in average cytological smears. Where there is no obvious abnormality, however, the advantage of such selection is eliminated. Under these circumstances, an investigative approach based on random sampling of the nasopharynx by aspiration would probably be more appropriate.

The direct binding assay affords an objective assessment of the exfoliation cells in total. It is not subjected to the same limitations as those imposed on methods requiring examination of individuals cells. The method is capable of detecting as few as 5000 MA6 – reactive Raji cells (Yip et al. 1987a). By reacting 3.6×10^5 exfoliative cells, this method is therefore potentially capable of detecting about 1.4% tumour cells present in the cell population. The sensitivity of the binding assay can be further increased. limited only by the number of cells that can be recovered by aspiration. Thus, it takes full advantage of the investigative approach by nasopharyngoscopy.

The usefulness of this method, however, depends on the specificity of the reagent used. Although MA6 was reactive against a variety of cell types, the following evidence suggests that, of these, only the tumour cells contributed significantly to total binding of the antibody to the exfoliative cells or their extracts. By autoradiography of the reacted cells it was found that the tumour cells were at least 3 times more reactive than columnar epithelial cells or lymphocytes. This was supported by immunoenzymatic studies showing that the tumour cells were indeed more intensely stained by the antibody than were the other cell types. The antibody identified one or a group of similar 55×10^3 glycoproteins. Referred to as B – lymphocyte carcinoma cross – reacting antigen (BLCa). it occurs in all human B – lymphoid cell lines except one (Chan et al, 1985; Yip et al. 1987a). Extracts of biopsies from symptomatic NPC patients and of lung resections from lung cancer patients were analysed by Western blotting. BLCa was identified in specimens from all the NPC patients but not from the patients who did not have confirmed NPC (Yip et al. 1987b) or in tumour resections of lung cancer patients (not shown). The same or a similar antigen on BLCa was also identified by MA6 in exfoliative cells from one of the 2 patients who had confirmed NPC but not in 2 other patients who did not have confirmed disease. Thus, despite its cross – reaction with different cell types. the binding assay employing [125]I – MA6 and a single specimen obtained from each patient gave a positive result for 6/9 (67%) patients with, and 0/5 patients without, histologically confirmed disease. The negative results obtained with the other 3 NPC patients may be partly due to poor recovery of tumour cells and/or submucosal localization of the tumour. By repeated aspirations and by testing a greater number of exfoliative cells, the sensitivity of this method might be further increased. The specificity of the assay might be further improved in the future when other MAbs become available. Although the number of patients examined in the present study was small, the results suggested, nevertheless, that the objective binding assay using MA6 may be suitable for screening subjects at high risk of NPC.

Acknowledgement

This work was supported in part by the Croucher Foundation. Hong Kong. Part of the results were included in a thesis submitted by K. H. C. to the Institute of Medical Laboratory Sciences for andmission as a fellow to the Institute.

〔In 《Int J Cancer》 1987, 39: 455 –458〕

References

1 Bramwell M E, Ghosh A K, Smith W D. Wiseman G. Spriggs A I, Harris H. Ca2 and Ca3. New monoclonal antibodies evaluated as tumour markers in serous effusions. Cancer, 1985, 56: 105 – 110

2 Chan K H, Yip T C, Ng W L, Ng M H. A shared antigenic determinant between human B lymphocytes and carcinomas (BLCa). Int J Cancer, 1985, 36: 329 – 336

3 Dong H, Shen S, Huang S, Lo G. The cytological diagnosis of nasopharyngeal carcinoma from exfolated cells collected by suction method. J Laryng, 1983, 97: 727 – 734

4 Ho H C, Nasopharynx. In K E. Halnan (ed). Treatment of cancer, pp250 – 267. Chapman and Hall. London. 1982

5 Ho H C, Kwan H C, Ng M H, de – The G. Serum IgA antibodies to Epstein – Barr virus capsid antigens preceding symptoms of nasopharyngeal carcinoma. Lancet, 1978, 1: 436

6 Jothy S, Brazinsky S A, Chin A Y M, Haggarty A, Krantz M J, Chenung M, Fuks A. Characterization of monoclonal antibodies to carcinoembryonic antigen with increased tumour specificity. Lab Invest, 1986, 54: 108 – 117

7 Lanier A P, Henle W, Benden T R, Henle G, Talbot M L. Epstein – Barr virus – specific antibody titers in seven Alaskan natives before and after diagnosis of nasopharyngeal carcinoma. Int J Cancer, 1980, 26: 133 – 137

8 Yip T C, Chan K H, Choy D, Chan C W, Ng M H. Chatacterization of a murine monoclonal – antibody – defined B – lymphocyted carcinoma cross – reacting antigen (BLCa) from nasopharyngeal carcinoma tissues. Int J Cancer, 1987a, 39: 449 – 451

9 Yip T C, Chan K H, Ng M H, Characterization of a human B lymphocyte carcinoma cross – reacting antigen (BLCa) in B lymphocytc identified by two murine monoclonal antibodies. Int J Cancer, 1987b, 39: 452 – 458.

10 Zeng Y, Zhang LG, Wu Y C, Huang Y S, Huang N Q, Li J Y, Ng Y B, Jiang M K, Fang A, Meng N N. Prospective studies on nasopharyngeal carcinoma in Epstein – Barr virus IgA/VCA antibody – positive persons in Wuzhou City, China. Int J Cancer, 1985, 36: 545 – 547

41. Development of a Set of EBV – Specific Antigens with Recombinant Gene Technology for Diagnosis of EBV – Related Malignant of Nonmalignant Diseases

Wolf H[1], Motz M[1], Kuhbeck R[1], Jilg W[1], Fan J[1,2], Pi G H[2], Zeng Y[2]

1. Max von Pettenkofer Institute, Pettenkoferstr, 9a 8000 Munich 2, FRG

2. Chinese National Academy for Preventive Medicine, Institute of Virology, Beijing, China

Summary

Epstein – Barr virus (EBV) serology can be consideraly improved when pure defined antigens are used and antibody class specific tests are applied[1,2]. The identification of antigens is possible by series of Western blots and immunoprecipitations[3]. In combination with gene mapping data from hybrid – selected translation[3,4], recombinant DNA technology was used to produce a panel of viral antigens. These antigens were evaluated in antibody class specific tests in their diagnostic significance for infectious mononucleosis, chronic active EBV infection and nasopharyngeal carcinoma (NPC).

Introduction

EBV – related diseases have been diagnosed mainly by immunofluorescence tests using various cell lines with or without pre-treatment with inducers of viral antigen synthesis as antigens. Although considerable improvements have been made, especially by the introduction of antibody class specific tests, frequent nonspecific reactions, insufficient correlation to the state of the disease, the difficulty of quantitative and automated evaluation and above all in the case of antibodies to MA the preparation and evaluation of the test strongly demanded improved test systems.

Tab. 1   Detection rate of NPC in IgA – specific tests using sera pre – treated with Staph. aureus in comparison to untreated sera. Comparison of positive rate of IgA antibodies to VCA, EA and MA in untrealed and treated sera with SPA from NPC patients and control groups

Sera	No. of cases	IgA/VCA[a]						IgA/EA[a]						IgA/MA[b]					
		Treated by SPA			Untreated			Treated by SPA			Untreated			Treated by SPA			Untreated		
		(+)	(%)	GMT	(+)	(%)	GMT	(+)	(%)	GMT	(+)	(%)	GMT	(+)	(%)	GMT	(+)	(%)	GMT
NPC patients	48	48	100	1:562	48	100	1:247	48	100	1:196	39	77	1:34	48	100	1:141	26	54	1:14
Tumors other than NPC	40	2	5	<1:5	0	0	<1:5	0	0	<1:5	0	0	<1:5	0	0	<1:5	0	0	<1:5
Normal individuals	46	4	8.7	1:5.6	1	2	<1:5	0	0	<1:5	0	0	<1:5	0	0	<1:5	0	0	<1:5

Notes: [a] IgA/VCA and IgA/EA detected by immunoenzymatic test;

[b] IgA/MA detected by immunofluorescence test.

In an attempt to improve further the correlation of serologic tests with NPC where EBV antigen – specific IgA antiboies have been found to be of highest value, preadsorption of sera with extracts from *Staph. aureus* for removal of competing excess IgG antibodies was found to significantly increase the detection rate to almost 100% without increasing positivity in control populations (Tab. 1)[1,2].

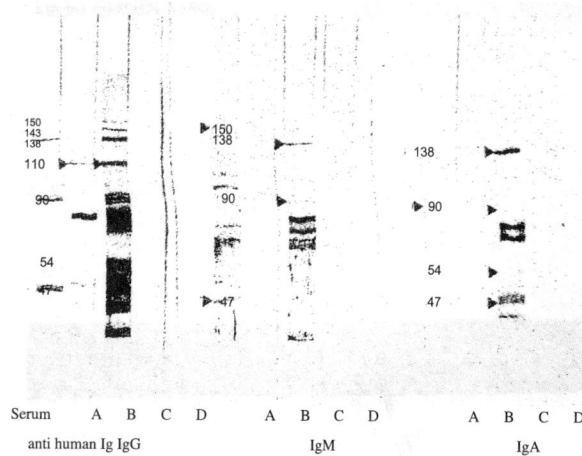

Serum   A B C D   A B C D   A B C D

anti human Ig IgG   IgM   IgA

Fig. 1   Western blot with sera

A fresh EBV infection; B: past EBV infection; C: NPC; D: negative control. As antigen source P3HR1 cells induced for two days with 3 mmol/L butyric and 40 ng/ml TPA were used. Sera were not separated by antibody classes. Antibody classspecificity was measured by using the respective second antibodies labeled with peroxidase

Albeit useful, these tests have to be replaced in the near future with machine – readable ELISA or similar tests. A prerequisite for such tests is the availability of pure viral antigens which in the absence of an effective lytic cell system have to be produced by recombinant DNA technology.

Previously we evaluated EBV – encoded polypeptides as diagnostic antigens by using series of sera in immunoprecipitation assays with radioactive lystes from labeled cells induced to produce EBV proteins[3]. The results of these tests have been extended to denatured antigens by Western blot experiments (Fig. 1). Using mapping data previously published[4] and partially extended by us (not shown) a panel of recombinant antigens given in Tab. 2 has been cloned as almost intact genes (e. g. EA – BMRF1) or as fractions of genes (EBNA1)

or autologous fusion proteins[5]. The selection of gene fragments was based on computer – aided selection of putative antigenic epitopes and deletion of major areas of the proteins which were responsible for the biological activites adverse to high – level expression in host cells.

Tab. 2   Expression products of diagnostically relevant EBV – specific antigens

Antigen class	Reading frame	EBV proteia	Classification	Expression plasmid	expressed aa/of the EBV protein	foreign aa or C – terminal	β – Gal 116 D_2 N – terminal
EBNA	BKRF1	EBNA – 1 p72	ratently expressed in infected cells	pUC8KSH1, 2	367/641 (42.7%)	623 autologous fusion protein	–
EA – D	BMRF1	p54	early protein	pUC9MBcE3. 2	404/404 (100%)	14aa	–
EA	BALF2	p138	early protein DNA binding	pUCARG1140	112/375 (33.2%)	6aa	5aa
VCA	BcLF1	p150	part of the viral capsid	pUCDc1200	386/1381 (26.6%)	9aa autologus fusion protein	9aa
MA	BLLF1	gp250/350	major late membrane protein	pUC19LP1. 9	634/907 (69.9%)	9aa	23aa
MA	BLLF1	gp250/350	major late membrane protein	pMDIIIGPTR expression in CHO cells	865/907 (95.3%) + glycosyl residues	–	5aa

The antigens have been partly purified with standard procedures or used in Western blots to evaluate their significance for the various EBV – related disorders. These antigen tests were combined with antibody class – specific reactions and the results are given in Tab. 3.

Tab. 3 Evaluation of EBV – specific recombinant antigens listed in Tab. 2 in antibody class – specific test

		IgM				IgG				IgA			
		A	P	C	N	A	P	C	N	A	P	C	N
EBNA	BKRF1	–	–	nt	–	–	+	D	+	–	–	nt	+
EA – D	BMRF1	+ D	–	nt	–	+	–	nt	++	+	–	nt	+
EA	BALF2	?	–	nt	–	+	–	nt	+	–	–	nt	+ D
VCA	BcLF1	+ (D)	–	nt	–	+	D	+	+	(+)	–	nt	+
MA	BLLFI	+ (D)	–	nt	–	+ / –	+	D	+	(+)	–	nt	+ D

Notes：A = acute infection；P = past infection；C = chronic active infection；N = NPC；D = significant for diagnosis；nt = not tested

In connection with preliminary data reported earlier for the diagnosis of chronic active EBV infection[6] and expanded with ELISA tests using recombinant MA – BLLF1, a scheme for diagnosis of EBV – related disorders is suggested (Tab. 4).

Early diagnosis of NPC has been shown to be of highest value for control of this neoplasia (e. g. 3). Similarly, possible treatment of chronic active infectious mononucleosis[7] would require diagnosis of this disease independent from clinical symptoms, preferentially by measuring serological parameters. In addition, differential early serodiagnosis of infectious mononuceosis may be greatly facilitated using test systems based on pure recombinant antigens.

Tab. 4 Suggested combination of antigens with antibody class – specific tests for diagnosis of EBV – related disease

Diagnosis	Antibody class	Antigen	Alternative/ Confirming
Acute Infection	IgM	EA (BMRF1)	IgG：EA(BMRF1) IgM：VCA(BcLF1) or IgM：MA(BLLF1)
Post – EBV Infection	IgG	VCA (PcLF1)	
Chronic Infection	IgG	VCA (BcLF1) and neg. EBNA1 (BKRF1) or neg. MA (BLLF1) and (for exclusion of earlyacute phase)：IgM： MA (BLLF1) or IgM：EA(BMRF1)
Na sopha – ryngeal carcinoma	IgA	EA(BALF2) or MA(BLLF1)	

〔In 《Epstein – Barr Virus and Human Disease》1987, 179 – 182〕

References

1 Zhu XX, Zeng Y, Wolf H. Int J Cancer, 1986, 37：689 – 691

2 Pi, G H Zeng Y, Wolf H J, Virol Methods (1986) in press

3 Wolf H, et al. Contr, Oncol. 24. (Karger Verlag, Basel, 1986) in press

4 Seibl R, Wolf H. Virology, 1985, 141：1 – 13

5 Motz M, Fan J, Seibl R, Jilg W, Wolf H. Gene, 1986, 42：303 – 312

6 Jilg W, Wolf H J, Infect Diseases, 1985, 152：222 – 225

7 Straus S E, et al. Ann Inter Med, 1985, 102：7 – 16

42. Epidemiology of Nasopharyngeal Carcinoma with Special Reference to Early Detection and Pre – NPC Conditions

de – The G[1] , ZENG Yi[2]

1. Institute Pasteur; 2. Institute of Virology chinese Academy of Preventive Medicine

The relationships between the Epstein – Barr virus (EBV) and nasopharyngeal carcinoma (NPC) are very strong and probably of a causal nature, but yet poorly understood in their molecular mechanisms. While the EBV may or may not infect the epithelial cells of the nasopharynx during primary infection, its presence in the epithelial tumor cells is constant in the undifferentiated carcinoma of the nasopharynx. Furthermore, reactivation of EBV latency in the nasopharynx represents a late step prior to clinical onset. Such a reactivation is accompanied by the synthesis of specific IgA, which was found to be most instrumental for early detection of NPC.

That NPC patients had a high level of IgA antibody was first observed by Wara et al. in 1975[7]. Soon after, Henle and Henle[5] and Desgranges and de The[2] showed that such IgA antibodies were directed against EBV/VCA and EA. The regular presence of such antibodies in patients with nasopharyngeal caracinoma originating from different geographical areas and their absence in patients with other ENT tumors suggested that such an antibody could represent a useful marker for the diagnosis of the disease.

Population Screening and Early Detection of NPC

Soon after the observations mentioned above, Zeng et al[10]. initiated a major population survey in a rural area of the eastern part of the Guang – Xi Autonomous Region of the People's Republic of China. The county of Zang – Wu, with a population of 450 000, was selected, and individuals aged 30 years and over were registered and asked to give blood to search for IgA/VCA antibodies. As seen in Tab. 1, 2.38% of the 148 029 persons screened were found to have IgA/VCA antibodies[9,12]. The 460 individuals found to have IgA/VCA antibodies at a titer equal or superior to 1:80 were clinically examined, and 55 cases of nasopharyngeal carcinoma were diagnosed at different clinical stages.

The city of Wu – Zhou, with a population of 170 000, was chosen for both cancer registration and further seroepidemiological studies. The NPC annual incidence was found to be 20 per 100 000 individuals. The survey of 20 726 workers aged 40 years and over showed that 1136 individuals had IgA/VCA antibodies[11,13]. Clinical examination of these persons led to the detection of 35 cases of NPC, the clinical stages being given in Tab. 1. The prevalence rate of NPC as detected in Wu – Zhou reached 150/100 000 in the 40 – to – 69 – year – old age group, a high figure if one considers that the hospitalized patients were not included in that figure. When one compares the results of the serological surveys among rural and urban populations, one sees that the proportion of the IgA – positive individuals was lower in Zang – Wu (2.38%) than in Wu – Zhou (5.5%). This may reflect differences in the immunoenzymatic test used. The detection rate of NPC and the proportion of NPC at early clinical stages was, however, greater in the city of Wu – Zhou than in the rural area of Zang – Wu. This might be related to a better awareness of NPC and better health delivery systems in towns than in rural areas.

Tab. 1 IgA/VCA serological population surveys in rural and urban population of the Guang – Xi autonomous region

Group	No. persons surveyed	No. with IgA/VCA	% IGA/ VCA + in surveyed population	No. NPCs detected	Ca In situ	Clinical stages				NPC detection in poputation	NPC among IgA/VCA + individuals
						I	II	III	IV		
Rural Zangwu	148 029	3533	2.38%	55	1	12 (22%)	19 (34%)	17 (31%)	6 (11%)	37/100 000	1.55%
							56%				
Urban Wuzhou	20 726	1136	5.50%	35		15 (43%)	17 (48.5)	3 (9.5%)	0	168/100 000	2.70%
							91.5%				

The use of IgA/EA antibodies instead of IgA/VCA antibodies to detect early NPC allowed us to decrease substantially the number of screened persons to be clinically examined, since only 0.05% of the general population exhibited IgA/EA antibodies, 43% of those already having clinical NPC.

Follow – up of IgA/VCA – Positive Individuals

In Zangwu county, 1138 individuals with IgA/VCA antibodies were followed up yearly for 3 years, from both serological and clinical viewpoints. It can be observed in Tab. 2 that two groups were at high risk for developing NPC: (a) individuals who increased their IgA/VCA titers fourfold or more had an 18.5% chance of developing NPC within 3 years; (b) individuals who had stable IgA/VCA antibody titers had a 1.32% chance of developing NPC within this period. The first group was of great interest, since it represented only 7.1% of the total IgA/VCA – positive population but led to 71.5% of the detected NPC. Therefore, if interventions are to be contemplated to prevent NPC, they should be performed on individuals with increasing IgA/VCA antibody titers, since one out of five will eventually develop the disease within 3 years.

If one tries to evaluate the time which elapsed between the first bleeding and subsequent tumor diagnosis in this series, one sees that, for stage I, such intervals varied from 8 to 30 months, for stage II between 10 and 48 months, and for stage III between 9 and 34 months. These surprising results may reflect important individual variations in the rate of NPC development. The survey of IgA – positive individuals in the town of Wu – Zhou is being pursued in order to better characterize the yearly rate of seroconversion (from IgA/VCA negative to positive) and of retroversion (from IgA/ VCA positive to negative) and the conditions of increasing titers.

Tab. 2　Stability and fluctuations of IgA/VCA antibody titers

over 3 years in relation to risk of developing NPC in Zangwu county

No. persons IgA/VCA +	Stability (no change in IgA/VCA)	Loss of IgA/VCA (Retroversion)	Increase[a]	Decline[a]	Up and down
1138(100%)	455(40%)	398(35%)	81(7.1%)	162(14.2%)	42(3.7%)
NPC patients detected	6[b]	0	15[c]	0	0
Total = 21 cases					

Notes: [a] By foudrold in titers

　　[b] These 6 cases represent 1.32% of the 455 individuals with stable titers and 28% of the 21 NPCs detected

　　[c] These 15 cases represent 18% of the 81 subjects with increasing IgA/VCA titers and 71% of the 21 NPCs detected

Pre – NPC Conditions

In parallel with these population surveys, we were interested in investigating what the presence of IgA antibodies to VCA and EA ment from an immunovirological viewpoint in the nasopharyngeal mucosa. In order to see whether the presence of IgA/VCA antibodies reflected a specific viral activity in the NP mucosa, 56 individuals exhibiting IgA/VCA antibodies for more than 18 months were biopsied. Four NPC cases were detected among these 56 individuals (two at an early stage of the disease and two at stage Ⅲ). A further 14 individuals without detectable abnormalities in their nasopharynx exhibited EBV – DNA sequences and/or EBNA in their nasopharyngeal mucosa[1]. In a further study, exfoliated cells from the nasopharynx of both IgA – positive and IgA – negative individuals (obtained with a negative pressure apparatus developed by Zhanjiang Medical College in 1976[14]) were investigated comparatively for the presence of EBNA and EBV – DNA sequences by spot followed by blot hybridizations. As seen in Tab. 3[3,6], both groups differed, but less than expected. One – third to one – fifth of IgA/VCA – positive individuals had detectable EBNA of EBV – DNA, respectively, in their exfoliated cells, as compared with one – fifth and one – seventh of the IgA/VCA – negative individuals. The EBNA test was regularly found to be a more sensitive test than blot hybridization. Forty – six of these IgA – positive and – negative individuals were again studied a year later, and it was noted that four out of seven individuals, who a year previously had detectable EBV – DNA sequences in their nasopharyngeal cells, failed to do so a year later. Conversely, 2 out of 15 individuals lacking detectable EBV – DNA in their exfoliated cells were positive a year later. Thus, the presence of EBV – DNA sequences in the nasopharyngeal mucosa appears not to be directly related to the presence of antibodies to IgA/VCA. This observation was surprising to us, and we are in the process of investigating whether the presence of viral DNA in nasopharyngeal cells among both groups of IgA – positive and IgA – negative individuals, might not reflect the difference in the cell population involved (lymphoid cells in IgA – negative and epithelial cells in IgA – positive individuals).

From an immunovirological viewpoint, the development of IgA/VCA and IgA/EA antibodies together with IgG/EA antibodies reflects a reactivation of the EBV/latency and characterizes pre – NPC conditions[4]. What is happening in the virus – cell relationship in the nasopharyngeal mucosa just prior to tumor development remains unknown. Further investigation of these events is critical from both the fundamental and applied viewpoints.

Tab. 3 Comparative detection of EBV/DNA and EBNA in exfoliated nasopharyngeal cells of IgA/VCA – positive and – negative individuals

Group	No. tested	EBV/DNA positive[3]	No. tested	EBNA positive[6]
IgA/VCA + individuals	62	13 (21%)	26	9 (34%)
IgA/VCA – individuals	39	6 (15%)	46	9 (20%)

Prospect of Better Control of NPC

Tab. 4 gives the comparative distribution of clinical staging of 1066 NPC cases from the outpatient clinics of hospitals outside the surveyed area and of the 35 cases detected during the serological screening in Wu – Zhou city (see also Tab. 1 and references[8,13]). Although the two groups were different by the number involved, the differences are striking: early – stages (I + II) represent 30% of the cases in outpatient clinics, whereas they represent more than 90% when systematic screening is implemented. If one considers the tumorous process limited to the nasopharyngeal mucosa (stage I), the difference is even more striking: 1.7% versus 43%. Such a shift has a tremendous impact on NPC control, since NPC patients at stage I exhibit a 90% five years disease – free survival as compared with less than 30% in patients at stage III. Today, most patients in NPC – endemic areas arrive so late that 70% of them die within 1 year following diagnosis (at stages IV. V). If early diagnosis is implemented, the situation can be completely reversed, with an 80% cure rate at 5 years for the early stages (I , II). Such an effect can be achieved before the mode of action of the EBV in the molecular pathogenesis of NPC is understood.

Tab. 4 Comparative distribution of clinical stages of NPC cases diagnosed in outpatient clinics or detected by serological screening

Group	No. cases	Stages			
		I	II	III	IV
Outpatient clinic	1066 (100%)	18 (1.7%)	312 (29.2%)	556 (52.2%)	180 (16.9%)
		(30.9%)			
Serological screening	35 (100%)	15 (43%)	17 (48.5%)	3 (8.5%)	
		(91.5)			

[In 《Cancer of the liver. Esophagus and Nasopharyngeal Carcinoma》 1987, 147 – 151]

References

1 Desgranges C, Bornkamm GW, Zeng Y, et al.　　Detection of Epstein – Barr viral DNA in the naso-

pharyngeal mucosa of Chinese with IgA/EBV – specific antibodies. Int J Cancer, 1982, 29: 187 – 191

2 Desgranges C, de The G. IgA and nasopharyngeal carcinoma. In G. de The Y. Ito , F. Rapp (Eds): Oncogenesis and Herpes – viruses III, pp, 883 – 891. Lyon: IARC, 1978: (IARC Scientific Publications No. 25)

3 Desgranges C, Pi GH, Bornkamm G et al: Presence of EBV – DNA sequences in nasopharyngeal cells of individuals without IgA/VCA antibodies. Int J Cancer, 1983, 32: 543 – 545

4 de The G, Zeng Y, Desgranges C, et al. The existence of pre – nasopharyngeal carcinoma conditions should allow preventive interventions. In Prasad et al. (Eds): Nasopharyngeal Carcinoma: Current Concepts, pp, 365 – 374. Kuala Lumpur: University of Malaya, 1983

5 Henle G, Henle W. Epstein – Barr virus – specific IgA serum antibodies as an outstanding feature of nasopharyngeal carcinoma. Int J Cancer, 1976, 17: 1 – 7

6 Shen SJ, Chen CP, Ma JL, et al. Further study on detection of EBNA from nasopharyngeal exfoliated cells of patients with nasopharyngeal carcinoma. Acta Zhanjiang med Coll, 1983, 1: 34 – 37

7 Wara WM, Wara DW, Phillips TL, et al. Elevated IgA in carcinoma of the nasopharynx. Cancer, 1975, 35: 1313 – 1315

8 Zeng Y, Gi ZW, Wang PC, et, al. Induction of Epstein – Barr virus antigen in Raji cells and P3HR – 1 cells by culture fluids of anaerobes from nasopharynx of patients with nasopharyngeal carcinoma and with other ear – nose – throat diseases. Cancer Letters (in pres)

9 Zeng Y, Liu YX, Liu ZR et al: Applcation of an immunoenzymatic method and an immunoauroradiographic method for a mass survey of nasopharyngeal carcinoma. Intervrology, 1980, 13: 162 – 168

10 Zeng Y, Liu YX, Wei J N, et al. Serological mass survey of nasopharyngeal carcinoma. Acta Acad Med Sin, 1979, 1: 123 – 126

11 Zeng Y, Zhang LG, Li HY, et al. Serological mass survey for early detection of nasopharyngeal carcinoma in Wuzhou City, China Int J. Cancer, 1982, 29: 139 – 141

12 Zeng Y, Zhong JM, Li LY, et al. Follow – up studies on Epstein – Barr virus IgA/VCA antibody positive persons in Zangwu County, China. Intervirology, 1983, 20: 190 – 194

13 Zeng Y, Zhang LG, Wu YC, et al. Follow – up studies on NPC in EB virus IgA/VCA antibody positive persons in Wuzhou City, China. Chin J Virol, 1985, 1: 7 – 11

14 Zhangjiang Medical College. Diagnosis of nasopharyngeal carcinoma by cytological examination of exfoliated cells by negative pressure suction. Chin Med J, 1976, 1: 45 – 47

43.　EB Virus and Nasopharyngeal Carcinoma

ZENG Yi

Institute of Virology, Chinese Academy of Preventive medicine

Introduction

There is a great variation of incidence of nasopharyngeal carcinoma (NPC) in different ethnic groups[1]. The incidence of NPC in Europe, America and Oceania is very low, around (0. 1 –

0. 2) /100 000/year. It is (1. 5 – 9. 0) /100 000/year in many Mediterranean countries, including North and East Africa, but it is very common is Southern China and Southeast Asia. Since old et al[2], first demonstrated the serological relationship between Epstein – Barr Virus (EBV) and NPC by means of immunodifusion test, a lot of data in the literature have further indicated the close association if EBV with NPC[3~6]. In China, studies on this problem were started in 1973. We put our emphasis on the studies and application of viro – immunological techniques in combination with clinical, cytological and histological examimation in a mass scale for the early detection of NPC. The purpose of our studies is first to find out a laboratory method for early detection of NPC patient, as early treatment gives a good prognosis and second to find out the relationship between EBV and NPC.

Epidemiology of EB Virus Infection in China

EB Virus infection is prevalent all over the world, but the age specific incidence of primary infection varies different socialeconomic group. In the low social – economic area 70% ~90% children aged 2 ~5 had already been infected with EB Virus without clinical manifestation, while in the high social – economic area primary infection is often delayed to adolescence or young adulthood, and infections mononucleosis is prevalent in these ages.

The relationship between EB Virus infection and NPC is a subject of study in many countries. In order to see the prevalence of EB Virus infection in China and its relation with NPC, complement fixing antibody to EB Virus was tested in sera from different age groups of normal individuals in high risk areas (Guangzhou city and Zhongshan county of Guangdong Province) as well as low risk areas (Beijing city, Lufeng and Wuhua counties of Guangdong province)[7]. Of a total of 2 300 normal sera, 2080 were positive (≥1:10). The positive rate was 90.4% with a GMT of 1:52.8. It was relatively high in both the high and low risk NPC areas of the above various groups, varying from 87% to 94% and showing no great difference. The positive rate (Fig. 1) reached 90% – 100% in the age group of 3 –5 years. This indicates that EB Virus spreads widely in China and people are infected with EB Virus in their early life. Similar condition prevails in the low NPC risk areas also. EB Virus IgG/VCA antibody was also tested by immunoenzymatic test[8] in the high and low NPC risk areas including Beijing, Changchun, Nanning and Cangwu county[9]. Similar results were obtained and 90% – 100% of children aged 3 –5 have already had IgG/VCA antibody (Fig. 2). de The[10] reported that in the 2 –3 year age group, 97% of Ugandans were EBV antibody positive in comparison with only 20% of Singapore Chinese and 30% of Indians. Such diffenences tapered off at around 10 years of age, when 100% of Ugandans, 75% of Chinese and 85% of Indians were found to be EBV antibody positive. So the Chinese in China have earlier EBV infection than Singapore Chinese, but later than Ugandans.

As NPC occurs more frequently in persons who are over 20 years of age in China, it can be seen from Fig. 3 (brief) that the antibody titers of healthy population who are over 20 in the high NPC risk areas were significantly higher than those in the low NPC risk areas[7]. This variation may

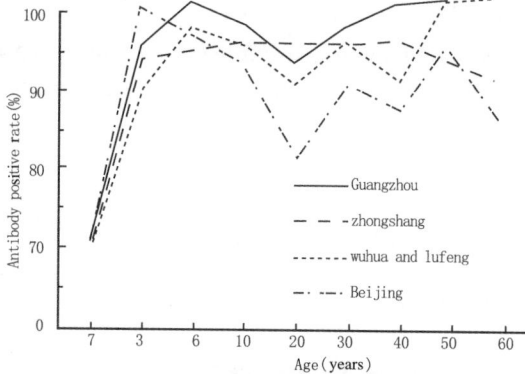

Fig. 1   Age – specific incidence of CF and to EBV

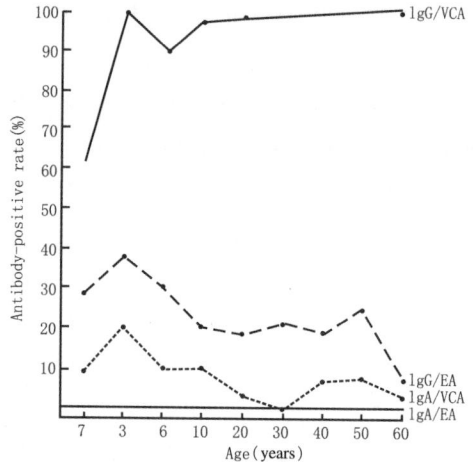

Fig. 2   Age – specific incidence of antibodies to EBV

indicate that the EB Virus is more active in people living in high NPC risk area and this might be related to the development of NPC. About 92% of NPC occurs in an age of over 30 years[11] and the peak of NPC incidence is somewhere around 50 years old, while the peak of EB Virus infection is at the age of 3 – 5 years. The interval between EB Virus infection and the development of NPC is quite long, so besides the EB Virus, other factors such as environmental carcinogen or promotor, and genetic factor may also play certain role in the development of NPC.

EB Virus Antibody Spectra in Sera from Normal Individuals and Patients with NPC

Various EB Virus specific – antigens including early antigen (EA), viral capsid antigen (VCA), membrane antigen (MA), nuclear antigen (EBNA), etc. are present in cells infected with EB Virus and the corresponding antibodies to these antigens are found in the sera from persons after having contracted infection with EB Virus. IgA and IgG antibody to EBV VCA or EA were detected by immunoenzymatic test (Fig. 2)[9]. The positive rate of IgG/VCA antibody reached 100% in the age group of 3 – 5 years and remained at high level for many years. The positive rate of IgG/ EA antibody reached the peak (37%) at the age of 3 – 5 years, then gradually declined to 17% at the age of 20 – 29 years. It slightly increased at the age of 50 – 59 years. The positive rate of IgA/ VCA antibody also reached the peak (20%) at the age of 3 – 5 years, and gradually declined. It rose slightly after the age of 30 – 39 years. No EBV IgA/EA antibody was found in all age groups even by using a very sensitive tset – immunoautoradiography. This indicates that no such antibody is produced or only very few people have such antibody after primary infection with EB Virus. In the serological mass screening carried out in Wuzhou city[12], it was found that IgA/VAC antibody negative persons had no IgA/EA antibody, and the positive rate of IgA/EA antibody among IgA/VCA positive persons was 4. 4% as compared with 0. 23% among those with IgA/VCA antibody positive and negative combined.

Although the GMT of IgG/VCA antibody or complement fixing antibody to EBV in NPC patient group was higher than that in the normal individual group, the antibody level varies from individuals to individuals, even some NPC patients had antibody with lower titer than that in the normal indivduals[13]. Thus the detection of IgG/VCA antibody or complement fixing antibody to EB Virus has no value for the diagnosis of NPC. The detection of IgG/EA, IgA/VCA and IgA/EA antibodies in the sera from patient with NPC, malignat tumor in the head and neck other than NPC, other malignant tumors and the sera of normal individuals were made by immunofluorescence test[14-15]. 96% and 81. 5% of NPC patients had IgG/EA and IgA/VCA antibodies with high titer or GMT respectively , whereas the positive rate of these two antibodies in another three groups was less than 6%. 50% of NPC patients had IgA/EA antibody, but no such antibody was found in the other groups. These data indicate that IgG/EA and IgA/VCA antibodies are more specific to high positive rate for NPC patients and can be used for the diagnosis of NPC (Fig. 4, 5). But further studies showed that the positive rate of IgG/EA antibody in normal individuals is higher than that of IgA/VCA antibody, so the IgA/EA antibody is not as specific as IgA/VCA antibody for the diagnosis of NPC.

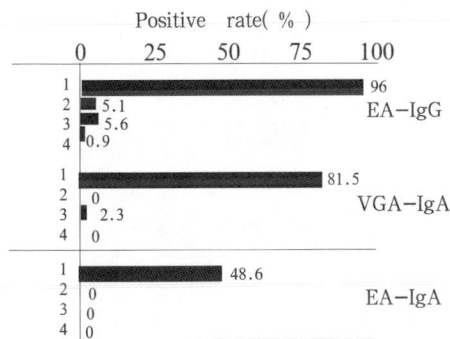

Fig. 4 Comparison of positive rate of antibody to EBV in different groups*

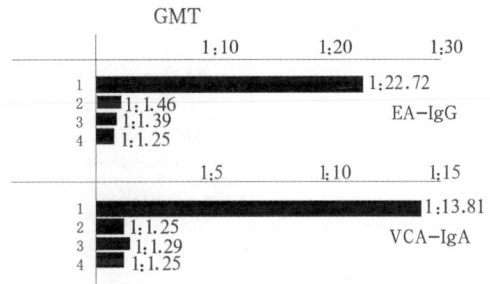

Fig. 5 Comparison of GMT to EBV in different groups*

* 1. Patients with NPC (76 cases)

2. Patients with tumor of head and neck other than NPC (58 cases)

3. Patients with other tumor (265 cases)

4. Normal individual (118 person0s)

Desgranges et al [16]. reported that 54% of NPC patients had IgA/VCA antibody in the saliva. The positive IgA/VCA antibody in the saliva detected with immunoenzymatic test[17] and immunoautoradiography[18] was 71. 1% and 85. 7% respectively in our laboratory. Only 32% of normal individuals with IgA/VCA antibody in their serum had IgA/VCA antibody in their saliva and no such antibody was found from serum IgA/VCA antibody – negative persons. The positive rate of IgA/VCA antibody in the saliva is much lower than that in the serum. Therefore, detection of IgA/VCA antibody in the sera is used as a routine test for the diagnosis of NPC.

Serological Diagnosis and Prognosis of NPC

The antibodies to EB Virus were mainly detected by immunofluorescence test as reported in the literature. This technique needs fluorescence microscope and is not convenient for NPC field study and for the diagnosis of NPC in the counties. An immunoenzymatic test was established in our laboratory[8]. As compared to the indirect immunofluorescence test, the immunoenzymatic test is simpler and more sensitive. This test has been widely used in China for the diagnosis of NPC. As shown in Tab. 1[8,19-21,63], the positive rates of IgA/VCA antibody ($\geqslant 1:5$) of NPC patients, patients with malignant tumor other than NPC and normal persons were 92.5% – 98%, 0 – 5.7% and 0 – 6% respectively, and the GMT was 1:35.7 – 1:78.5, 1:1.25 – 1:2.7 and 1:1.25 – 1:1.54 respectively. There was no marked difference in the positive rate and GMT of IgA/VCA antibody detected in the same laboratory in sera from 8 different provinces and cities including high NPC risk areas[22] (Tab. 2). The positive rate and GMT of IgA/VCA antibody was much higher than that in other control groups. Therefore, detection of IgA/VCA antibody is valuable for the diagnosis of NPC, especially when there was no evidence of tumor in the nasopharynx, but it had invaded beneath the mucosa or metastasized to the neck region.

Tab. 1　Detection of IgA/VCA antibody to EB virus in sera from NPC patients and conrtol groups

Authors	NPC patients			Patients with tumor other than NPC			Normal individuals			Reference
	No. of persons	Positive rate (%)	GMT	No. of cases	Positive rate (%)	GMT	No. of cases	Positive rate (%)	GMT	
Liu en al. (1979)	80	92.5	1:35.7	107	0	1:1.25	91	0	1:1.25	8
Wei et al. (1980)	628	98.1	1:38.7	92	5.4	1:1.25	210	0.5	1:1.27	19
Jian et al. (1981)	78	92.3	1:78.5				166	6	1:5.4	20
Li et al. (1981)	1006	93.8	1:76	768	5.7	1:2.7	756	1.9	1:2.8	21

Regarding the relationship between the positive rate and GMT, and the clinical stages, there was no marked difference in the positive percentage of IgA/VCA antibody in the sera from different stages, it is helpful to the early diagnosis of NPC. The GMT of antibody is a little lower in stage I than in stage II – IV, (Tab. 3) but Li et al[21] reported that the level of IgA antibody increased with the increase in total tumor burden and the metastasis of lymph node. It is 3 – 4 fold higher in stage IV and N3 than in stage I and N0, and there was significant difference in IgA/VCA antibody level between different clinical stages and N stages. This result was similar to the report by Henle and Henle[3]. The IgA/VCA antibody titer is related to the size of lymph node metastasis, but not the size of tumor. It was only slightly increased in the advanced NPC patient with cranial involvement, but without lymph node matastasis[21]. The positive rate of IgA antibody was very low in other control groups （below 6%）[14]. Thus, the detection of IgA/VCA antibody is valuable for differentiation of NPC from other diseases.

Tab. 2　Comparison of the Immunoenzymatic and Immunoautoradiographic test for detection of IgA/EA

Sera	Number of Cases	IE						IR		
		IgA/VCA			IgA/VCA			IgA/VCA		
		(+)	(%)	GMT	(+)	(%)	GMT	(+)	(%)	GMT
NPC	56	52	(93)	1:210	45	(80)	1:25.6	54	(96)	1:97
Nasopharyngeal chronic inflamation	50	22	(44)	1:12.5	9	(18)	1:7.3	11	(22)	1:9.0
Tumour other than NPC	170	10	(5.9)	1:5.3	1	(0.6)	1:5	7	(4)	1:5.2
Normal individual	100	3	(3)	1:5	0	(0)	1:5	0	(0)	1:5

Note: IE – Immunoenzymatic test; IR – Immunoautoradiographic test

Tab. 3　Distribution of EBV IgA/VCA antibody among NPC patients and normal individuals ago 30 years older in Zangwu County

	Antibody titer										Total	GMT
	5	10	20	40	80	160	320	640	1280	2560		
Individuals age 30 years and older												
Number positive for IgA/VCA antibody	837	967	808	474	335	55	37	10	8	2	3533	1 : 16
Percent positive	23.7	27.3	22.9	13.4	9.5	1.6	1.0	0.3	0.2	0.06	100	
Serological mass survey												
Number of NPC cases	0	4	8	9	12	8	4	4	4	2	55	1 : 99
Percent positive	0	7.3	14.5	16.4	21.8	14.5	7.3	7.3	7.3	3.6	100	
Follow-up study												
Number of NPC cases	0	2	5	4	8	8	2	1	2	0	32	1 : 85
Precent positive	0	6.3	15.6	12.5	25	25	6.3	3.1	6.3	0	100	

Although no IgA/EA antibody could be found in the sera from patients with malignant tumors other than NPC and normal individuals, the detection of IgA/EA antibody can not be used alone for the routine diagnosis of NPC, because only 50% of NPC patients have such antibody detected with immunofluorescence test[14]. The detection rate of NPC among IgA/VCA antibody positive persons is 1.9% and that among IgA/EA antibody positive persons is 30%. The difference between these two groups is 15.8 fold[12]. These data indicate that IgA/EA antibody is more specific, but not so sensitive for the detection of NPC as IgA/VCA antibody. So it is necessary to improve the sensitivity of the test.

An immunoautoradiographic method was established in our laboratory by using $^{125}2$ labelled antibuman IgA antibody for the detection of EB Virus IgA/VCA antibody in the sera[23]. This test is rather complicated as compare to the immunoenzymatic test, but this very sensitive test may be useful for the detection of IgA/EA antibody.

Sera were tested from NPC patients, NPC suspected patients, patients with tumors other than NPC and normal individuals with immunoautoradiography[24]. The positive rate and GMT of IgA/EA antibody from NPC patients were 96% and 1:97 respectively. Patients with clinically suspected NPC and diagnosed histologically as chronic inflammation if the nasopharynx had a positive rate of 22% and a GMT of 1:9. In this group, biopsies were taken and the histological examination showed carcinoma in 6 out of 11 patients. Out of 170 patiests with malignat tumor other than NPC, seven were positive (4%) to low titer 1:5.2. All normal individuals were negative (Tab. 2). Sera from 14 NPC patients diagnosed histologically, showing negative IgA/VCA and IgA/EA antibody with immunoenzymatic test and immunofluoeescence test, were reexamined with immunoautoradiograpy, and 11 and 6 of them had IgA/VCA and IgA/EA antibody respectively. These data indicate that immunoautoradiography is more sensitive than immunoenzymatic test or immunofluorescence test for the detection of IgAEA antibody and that IgA/EA abtibody may serve as specific markder for the detection of NPC especially for early stage NPC.

Prior to radiotherapy, during the course of treatment, and 1year, 1 – 4 years or 4 – 18 years after radiotherapy, the positive rate of EBV IgA/EA antibody from NPC patients was 96%, 88.8%, 89%, 75% and 73.3% respectively as compared with 91.8% from patients with recurrent of distant metastasis. The above data showed that EBV IgG/EA antibody remained unchanged for a long time after radiotherapy. However, the lever of IgA/VCA antibody declined gradually with increase in survival time among NPC patients after radiotherapy. Only 30% of NPC patients had this antibody at very low titer (GMT 1:2.8) 4 – 18 years after treatment[14]. Therefore, serological follow – up of IgA/VCA antibody, but not IgG/EA antibody, may provide prognostic information for NPC patients after radiotherapy (Fig. 6, 7).

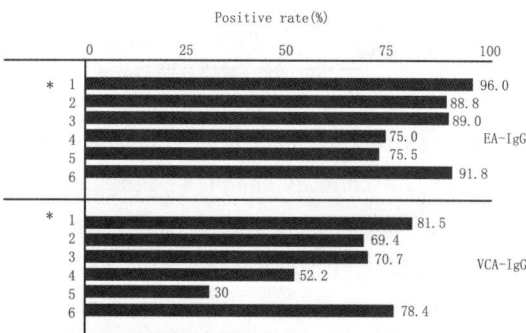

Fig. 6 Comparison of positive rate
of antibody to EBV in NPC patients
before and after radiotherapy

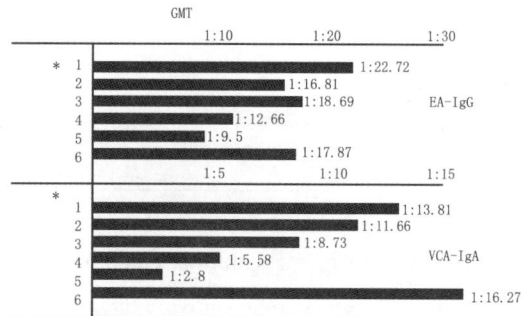

* 1. NPC patient before radiotherapy
2. NPC patients during the course of rediotherapy
3. NPC patients within one year after starting of radiotherapy
4. NPC patients 1 – 4 years after rodiotherapy
5. NPC patients 4 – 18 years after radiotherapy
6. NPC patients with recurrence of metastases

Fig. 7 Comparison of GMT to EBV in NPC
patients before and after radiotherapy

Serological Mass Surveys and Follow – up Studies on NPC

The detection of IgA antibody is valuable for the diagnosis of NPC. It is very interesting to apply the serological test to the detection of more NPC patients in their early stage and to the study on the relationship between EB Virus and the development of NPC in high NPC risk ares. A total of 148 029 persons aged 30 and over was screened for the detection of IgA/VCA antibody with immunoenzymatic test in Cangwu county of Guangxi Zhuang Autonomous Region from 1978 – 1980[11,25 - 27]. The incidence of NPC in this county was 11/100 000. A total of 91. 4% of NPC patients was aged 30 and over. The antibody – positive persons were examined with nasopharyngoscope, and biopsies were performed on the following antibody – positive persons: (1) those diagnosed clinically as NPC patients or suspected of having NPC, (2) those having some lesions in the nasopharynx, such as rough mucosa, local congestion, nodule, residue of adenoid tissue and asymmetry of nasopharyngeal cavity, and (3) those without lesion in the nasopharynx but with high antibody titer. A total of 3539 persons were found to have IgA/VCA antibody ranged from 1:5 to 1:2560, 96. 8% of antibody titer was lower than 1 : 80 (Tab. 3), the positive rate of IgA antibody increased with the increase in age. Among these 54 cases were diagnosed clinically and histologically as NPC, one was carcinoma in situ, 12 in stage I, 19 in stage II, 17 in stage III, and 6 in stage IV. Thus 31 cases (57%) were in the early stages (I and II) (Tab. 4). The IgA/VCA antibody – positive persons were examined clinically once a year, 32 new NPC patients were detected after follow – up studies for 1 – 3 years (Tab. 5). There were 10 cases in stage I, 9 in stage II, 11 in stage III, and 2 in stage IV. The IgG/VCA antibody was detected 8 – 30 months (averaging 13 months) prior to the diagnosis of NPC in stage I (Fig. 8). These results are similar to those reported by Ho et al[28] . and Lanier et al[29]. In their retrospective study. They found that IgA/VCA antibody was detectable 22 – 72 months before the diagnosis was made in the advanced stages. Thus the data indicate that the IgA/VCA antibody appeared prior to the development of NPC, and the presence of IgA/VCA antibody is closely related to the development of NPC. There was no marked difference in GMT of IgA/VCA antibody in the sera between the period before the onset of NPC and after the diagnosis was made at stage I, but the antibody was higher in stage II – IV (Fig. 9), e, g. when the carcinoma cells metastasized to the cervical lymph nodes . Altogether 87 NPC cases were detected from 3533 IgA/VCA antibody – positive persons with serological screening and follow – up studies. The detection rate of NPC was 2462/100 000 in 3533 IgA/VCA antibody – positive persons. This is 82 times the annual incidence of NPC (30/100 000) in the general population of the same age group in Gangwu county (Tab. 5). These results indicate that IgA/VCA antibody in normal individuals represents a risk ractor for the development of NPC.

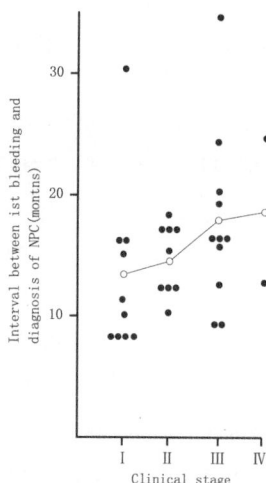

Fig. 8　Change of titer and GMT
of IgA/VCA antibody from first
bleeding to diagnosis of NPC

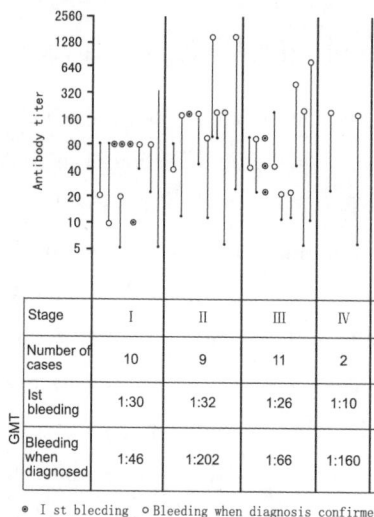

Stage	I	II	III	IV
Number of cases	10	9	11	2
1st bleeding	1:30	1:32	1:26	1:10
Bleeding when diagnosed	1:46	1:202	1:66	1:160

● I st bleeding ○Bleeding when diagnosis confirmed

Fig. 9　Follow – up studies on
IgA/VCA antibody – positive Persons

Tab. 4　Detection of NPC cases in Zangwu County during 1978 – 1980
by serlolgical mass surveys and follow – up study

Group	Number of NPC cases					total
	carcinoma in situ	stage I	stage II	stage III	stage IV	
1978 – 1980 Serological mass survey	1	12	19	17	6	55
Follow – up study	0	10	9	11	2	32
Total	1	22	28	28	8	87

Tab. 5　Comparison of NPC detection and incidence rates

Group	Number of persons examined	Number of positive for IgA/VCA antibody	Number of NPC cases	NPC detection rate among all persons examined	NPC detection rate among IgA/VCA antibody positive persons	NPC incidence rate per annum
Zangwu County (1978 – 1980)	148 029	3 533	87	59 (1. 96)	2 42 (82)	30
Wuzhou city	12 930	680	13	100. 5 (2)	1900 (38)	50

　　Five hundred and eighteen sera taken from the boat people in Cangwu county were examined for IgA/VCA and IgA/EA antibody[30]. 20 of whom had IgA/VCA antibody and 5 had IgA/EA antibody. Two NPC cases (stage I and II) detected had both IgA/VCA and IgA/EA antibody. The detection rate of NPC from IgA/VCA antibody positive boat people was 10% and was higher than that from the general population of Cangwu county. It was reported by Ho et al[31]. that these data further confirmed that the marine population living in boats has a much higher incidence rate than the land dwellers living in congested appartments.

　　In order to carry out the follow – up study on the EBV IgA/VCA antibody – positive persons and to reduce the mortality rate of NPC through early diagnosis and early treatment, another mass serological survey was carried out in Wuzhou city located in the center of Cangwu county which is also a high risk area. The incidence of NPC is 17/100 000[12]. Sera from 12 932 persons of the 40 –59 age group were examined and IgA/VCA antibody (≥1∶10) was found from 680 persons, among whom 13 NPC cases were detected, 9 (70%) in stage I, 4 (30%) in stage II (Tab. 5 – 6). The detection rate of NPC from IgA/VCA antibody – positive persons was 1900/1 000 000 which was 38 times the annual incidence (50/100 000) in the general population of the same age group. These data further confirmed that NPC in their early stage can be detected by using the serological test for the detection of IgA/VCA antibody.

Tab. 6   NPC patients detected by serological mass survey

Case No.	Sex	Age	Histological examination	Clinical stage	Antibody	
					VCA/IgA	EA/IgA
12 660	F	50	Undiff. ca	I	1∶10	—
1 171	F	46	Poorly diff. ca	I	1∶40	—
11 689	M	43	Poorly diff. ca	I	1∶40	1∶40
361	M	47	Poorly diff. ca	I	1∶40	1∶40
13 684	F	46	Undiff. ca	I	1∶160	1∶40
1 649	F	42	Poorly diff. ca	I	1∶160	—
309	M	50	Poorly diff. ca	I	1∶160	1∶40
7 873	M	57	Poorly diff. ca	I	1∶640	1∶640
12 735	M	46	Poorly diff. ca	I	1∶640	1∶40
68	M	50	Poorly diff. ca	II	1∶160	—
3 433	M	48	Poorly diff. ca	II	1∶160	1∶40
23	M	50	Poorly diff. ca	II	1∶640	1∶10
52	M	54	Poorly diff. ca	II	1∶640	1∶160
GMT					1∶144	1∶26

How long does the IgA/VCA or IgA/EA antibody persists and how many antibody negative persons will become postive? What is the exact proportion of IgA/VCA or IgA/EA antibody positive and negative individuals who will develop NPC? These are problems which remain to be further clarified. Some 25 000 serum samples from normal individuals aged 40 and over were collected and stored in the Institute of Virology, Chinese Academy of Medical Sciences, Beijing. Careful and periodic follow – up studies on IgA/VCA and IgA/EA positive and negative individuals should clarify these questions

EB Virus markers in nasopharynx

EB Virus markers (EBNA and DNA) were found in nasopharyngeal carcinoma cells with anti-complement immunofluorescence test and nucleic acid hybridization[6,32 – 34]. NPC patients can be detected through serological mass surveys and follow – up study on the IgA/VCA antibody positive persons, but a large proportion of the IgA antibody positive group showed no evidence of NPC. Detection of EB Virus markers in different cell types of the nasopharynx may be helpful to the early detection of NPC and to understand the role of EB Virus in the development of NPC. It is difficult to identify the different cell types of the nasopharynx with anticomplement immunofluorescence test, and this technique requires fluorescence microscope. We have, therefore, developed an anticomplement immunoenzymatic test (ACIE) which is simple and sensitive, and the different cell types including carcinoma cells, epithelial cells and lymphocytes can be identified[35,36]. It is helpful to localize EBNA in different cell types. These was no notable NPC in the very early and precancerous stage in the nasopharynx. Thus, it is difficult to take biopsy from the right site of nasopharyngeal mucosa, and to overcome this difficulty negative pressure suction technique was applied to collect the exfoliated cells from the nasopharynx[37]. The amount of cells is enough for EBNA and cytological examination, and for DNA detection. The cells were fixed on a slide and examined for EBNA with ACIE test. All the 79 NPC patients had EBNA – positive carcinoma cells (Fig. 10). The positive rate of carcinoma cells detected with cytological and histological examination was 87. 3% and 91. 1% respectively (Tab. 7)[38]. ACIE test is thus more sensitive than the other two methods, especially in stage I. A few cases had carcinoma cells with EBNA serveral months before confirmation of NPC with cytological and histological examination. No EBNA – positive carcinoma cells could be found in exfoliated cells from 57 persons in whom NPC was excluded, from 18 patients with head and neck tumors other than NPC in touch smears and from 21 dead fetuses. Then this test was applied to detect early stage of NPC from IgA/VCA antibody positive individuals in NPC high risk area[39]. Satisfactory results were obtained. Carcinoma cells with EBNA were found in 4 cases out of 64 IgA/VCA antibody – positive individuals. All the 4 cases were also diagnosed cytologically and histologically as poorly differentiated nasopharynaeal carcinoma. Clinical examination showed that the 4 NPC patients were in stage I without any subjective symptom. There was only rough mucosa on the back roof of nasopharynx in each of the 2 cases, and nodules 0. 5 – 0. 8 cm in size in another two cases. The interval between the first bleeding for serology and the diagnosis of NPC was 8 – 9 months. No further clevation of IgA antibody has occurred in all of the 4 cases since first detected 8 – 9 months ago. The detection rate of NPC was 6. 2%. These data further confirmed the fact that ACIE test is valuable for the diagnosis of early stage NPC, especially for the follow – up study on IgA/VCA antibody positive individuals.

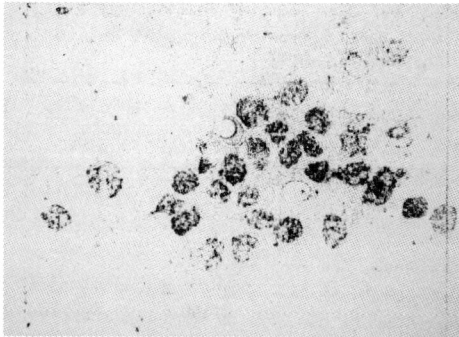

Fig. 10 EBNA Positive Carcinoma Cells (×132)

Tab. 7 Comparison of 3 methods in different clinical stage NPC detection

Clinical stage	Cases	Positive (%)		
		Cytologic	Histologic	
		ACIE	examination	examination
I	15	15 (100)	13 (86. 4)	12 (80. 0)
II	29	29 (100)	23 (79. 3)	27 (93. 1)
III	31	31 (100)	29 (93. 5)	29 (93. 5)
IV	4	4 (100)	4 (100)	4 (100)
Total	79	79 (100)	69 (87. 3)	72 (91. 1)

Tab. 8　The effect of Chinese Medicinal Herbs on EBV EA induction of Raji cells

Medicinal herbs	Etherial extraction（Cone, μg/ml）				Water extraction（Cone, μg/ml）			
	Butyrate in assay medium		Without butyrate		Butyrate in assay medium		Without butyrate	
	10	1	10	1	10	5	10	5
Daphne genkwa	46 +	45	1	2	15	31	2	1
Stellera chamae jasme	36	46	2	2	2	1	2	1
Wikstroemia chamaedaphne	32	53	1	2	2	20	0. 4	1
Wikstroemia indica	42	50	4	2	25	17	1	0
Edgeworthia chrysantha	43	28	1	0	0	0	0	0
Caesalpinia sappan	8	4	0	0	0	0	0	0
Desmodium styracifolium	6	4	0	0	0	0	0	0
Premna fulva	52	42	2	0. 4	0. 2	0. 2	0	0
Sparganium stoloniferum	16	2	0. 4	0	0	0	0	0
Datura stramonium	50	25	1	0	1	0	0	0
Tinospora sp.	7	3	0. 2	0	0	0	0	0
Aleuritopteris argentea	6	17	1	0	0	0	0	0
Clematis intricata	3	2	0. 4	0	0. 2	0. 2	0	0
Knoxia valerianoides	3	1	0	0	0	0	0	0
Angelica pubescens	1	0	0	0	0	0	0	0
Euphorbia kansui	28	37	1	1	5	4	1	0
Euphorbai lathyris	30	53	2	2	38	33	4	3
Croton oil	18	40	1	0. 4	39	28	2	4
Tung oik	38	42	1	1				

Notes： + = EA cell positive rate% ； Butyrate （4mmol/L） alone EA cell positive rate = 1% ； Stock etherial extraction （10 mg/ml）； Stock water extraction （10 mg/ml）

Tab. 9　Comparison of the inhibitory effect of different retinoids on EA induction

Retinoid concentration μmol/L	% EA – positive cells			
	Ro 10 – 9359	Ro 11 – 1430	7901	7902
10	2. 6	4. 2	4. 3	0
1	10. 4	11. 3	10	1. 2
0. 1	11	13. 4	11	6. 7
0. 01	30	26	23	15. 3
0. 001	31	32	27. 3	22
Control	27	27	27	27

EBNA could be found not only in carcinoma cells but also in the hyperplastic cells and normal ciliated columnar epithelial cells from NPC patients and normal individuals. It was found much scarcer in the normal squamous epithelial cells[38] （Fig. 11, 13）. The frequency of EBNA in ciliated columnar epithelial cells from NPC patients, patients with chronic inflammation in the nasopharynx and patients with disease of the head and neck other than NPC was 24. 8%, 17. 5% and 4%

respectively[40]. Chen et al[41]. also reported that EBNA was detected in the nasopharyngeal carcinoma cells, atypical hyperplastic cells, epithelial cells near the tumor and interstitial lymphocytes from NPC patients with anticomplement immunofluo rescence test and histological examination. These data suggest that the EB Virus may first infect normal epithelial cells especially the ciliated columnar epithelial cells, integrating ist DNA into cellular DNA, followed in certain circumstances by cell transfromation and development of NPC. This means ruling out the possibility that the presence of EB

Virus markers in NPC cells is the results of infection of epithelial cells by EB Virus subsequent to cell transformation.

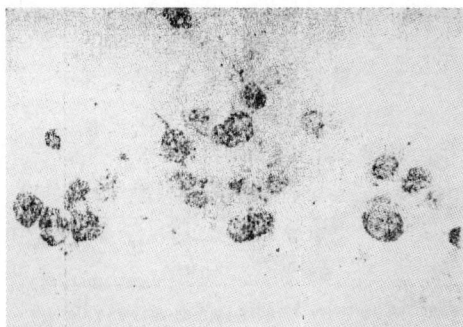

Fig. 11　EBNA – Positive Carcinoma
Cells and ciliated columnar epithelial
cells from NPC patients （×132）

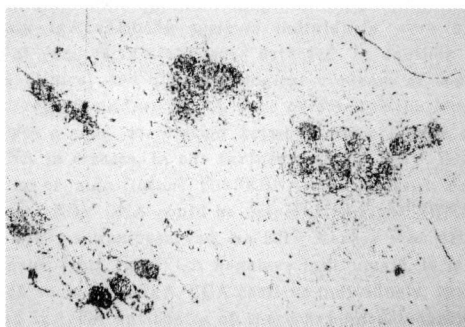

Fig. 12　EBNA – Positive hyperplastic
columnar epithelial cells and ciliated columnar
epithelial cells from NPC patients （×132）

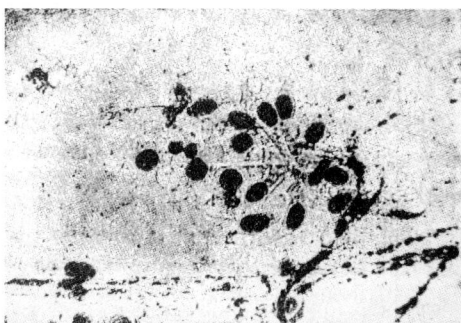

Fig. 13　EBNA – positive
ciliated columnar epithelial
cells from normal individual （×100）

The detection of EB Virus DNA and EBNA in the nasopharyngeal mucosa with blot hybridization and anticomplement immunofluorescence test was carried out in Cangwu county by de The et al.[42,43]. Fifty six IgA antibody positive individuals were biopsied. Four NPC, 2 at very early stage, were detected. In another 14 individuals without clinical and histopathological evidence of tumor, EB Virus internal repeats and/or EBNA were detected in the biopsied mucosa. There was a good correlation between the presence of EBNA and EBV DNA in 6 cases, In one further case the EBV DNA was at the borderline of significance, but EBNA was present. Furthermore, in three cases EBV DNA could be detected, but not EBNA, and in 4 cases the reverse was observed. No EBV marker was detectable in seven biopsies from IgA/VCA negative individuals. It is interesting to note that among the IgA/VCA positive individuals, the GMT of IgA/VCA and IgA/EA antibodies of those having detectable EBV marker （DNA and EBNA） in their nasopharynx was higher than those who had no detectable viral markers in their nasopharyngeal mucosa. But the possibility that EBV DNA positive in the nasopharyngeal mucosa of IgA/VCA antibody – positive persons may be present in lymphocytes, can not be completely ruled out. It is necessary to apply the technique of DNA hybridization in situ to make sure the localization of the small amoumt of EBV DNA is in epithelial cells or in lymphocytes.

Relationship between Environmental Factors, Activation of EB Virus and Development of NPC.

There is a close association between EB Virus and nasopharyngeal carcinoma, but EB Virus infection prevailed the world over and the development of NPC showed marked geographical difference. Thus, besides EB Virus, the enviromental and genetic factors may also play a role in NPC development. Huang et al[44]. reported that salted fish contained appreciable quantities of nitrosodimethylamines which can induce tumor in rats, and suggested that the eating of salted fish might be an important risk factor. Food habit of NPC patients in Tunisia[45], Cangwu county and Wuzhou city in China[46] was surveyed. They do not eat much fish and no clear relationship between salted fish and NPC was established. EB Virus can be induced and activated from EBV genome – carrying human lymphoblastoid cells by treatment with chemicals such as halogenated pirimidines, short – chain fatty acids, tumor promoting phorbol diesters (TPA) and also by antibodies to human IgM. When the cells were treated with n – butyrate and TPA or corton oil, a dramatic increase in EB Vrius antigen – producing cells was found. Ito et al. [47–49] found more than 10 Enphorbiaceae plant extracts containing inducing activity, such as Euphorbia lathyris, EKansui, Tung oil, etc. by using Raji cells and n – butyrate system. Foods and Chinese medicinal herbs from NPC high risk areas in China were screened by the same technique[50]. No EA inducing activity was found in 73 samples of 14 different foods tested including salted fish, pickle and oil. Ether extraction of 495 Chinese medicinal herbs from 106 families were studied. Fifteen herbs from 10 families were found to have inducing activity. They are Daphne genkwa, Stellera chamaejasme, wikstroemia chamaedaphne, wikstroemia indica, Edgeworthia Chrysantha, Caesalpinia sappan, Desmodium styracifolium, Premna fulva, Sparganium stoloniferum, Datura stramonium, Tinospora sp. , Aleuritopteris argentea, Clematis intricata, Knoxia valerianoides, and Angelica pubescens. Water extracts of herbs also have EA – inducing activity, but not so strong as that of ether extracts. Water extract of herbs plus n – butyrale markedly enhanced the induction of EA antigen in Raji cells. For example, the EA cell positive rate induced by n – butyrate plus water extract of Wistroemia Indica was 25%. This herb is produced in Guangdong, Guangxi, Fujian provinces and other places. It is the main component in making tables and ointment of " detoxication and antiphlogosis" in Guangdong province[51]. Tung oil is a strone inducer and people living in NPC high – risk areas where many tung oil trees are planted, come in close contact with tung oil trees. The relationship between the activation of EBV by these herbs and the development of NPC needs further study.

The Role of EB Virus in the Development of NPC

It was shown that NPC patients had various antibodies to EB Virus. Furthermore, EB Virus markers (DNA and EBNA) and virus production have been demonstrated in the epithelial tumors cells and the tumor cells passed in nude mice. These data indicate that there is a close relationship between EB Virus and NPC. Our studies further confirmed this association.

(1) The infection rate of EB Virus in NPC high and low risk areas showed no great differ-

ence. However the GMT of the complement fixing antibody to EB Virus in healthy population aged 20 and over in high risk areas was significantly higher than those in low risk areas. These data indicate that EB Virus was more active in persons living in high risk areas[7].

(2) The positive rate of VCA/IgA antibody in the sera from the general population aged 30 and over in high risk area increased with the increase in age. Propably this is due to some internal or external factors stimulating the activation of EB Virus[11,12].

(3) The IgA/VCA antibody level of persons with atypical hyperplasia or atypical metaplasia and patients with NPC is significantly higher than those with simple hyperplasia or simple metaplasia or normal mucosa. After 8 – 37 months follow – up study the detection rate of NPC from the atypical hyperplasia and atypical metaplasia group is 10 times that from simple hyperplasia, simple metaplasia and normal mucosa group. These data indicate the close relationship between the presence of IgA/VCA antibody and atypical hyperplasia, atypical metaplasia and NPC[52].

(4) The detection rate of NPC from IgA/EA antibody positive persons is very high (30%)[12].

(5) The detection rate of NPC from IgA/VCA antibody positive individuals is 38 times that of the general population in the same age group. The IgA/VCA antibody can be detected 8 – 30 months prior to the clinical and histological diagnosis of NPC in stage I[27].

(6) EBNA was not only found in carcinoma cells but also in normal epithelial cells and hyperplastic cells of the nasopharynx[38]. This seems to rule out the theory that EB Virus is the passenger of NPC. Some hyperplastic cells carrying EBNA in the nasopharynx might be the precancerous cells.

(7) EB Virus DNA was not only found in carcinoma cells but also in normal nasopharyngeal mucosa[42,43].

These data suggest that the EB Virus plays an important causative role in the development of NPC, but EB Virus is not the unique factor. Environmental factor, genetic factor and its synergistic effect with EB Virus might also play an important role. It needs further study.

Control and Prevention of NPC

The present data indicate that EB Virus may play an important role in the development of NPC. The detection of EBV IgA antibody and virus markers is helpful to the diagnosis of NPC, especially in their early stages. Therefore, it is possible to control or prevent NPC through early diagnosis and early treatment or through intervention against EB Virus activation, even before the etiology of NPC has been clarified.

(1) *Early detection and early treatment*: Radiotherapy is very efficient in the treatment of NPC, particularly in their early stages. Pan et al[53]. reported that ohemean 5 – yrs survival rate of NPC after radiotherapy was 47.7%. It was 61.3%, 55.0%45.8% and 23.6% for stage I, II, III, and IV respectively. Chang et al. reported 5 – yrs survival rate as high as 93% for stage I patients[54]. As mentioned above, by a combination of all methods including serology, anticomplement immunoenzymatic test, clinical, cytological and histological examination for a mass survey in Wuzhou city, all NPC patients were detected in their early stage, 70% of whom were in stage I[12]. Therefore, serological mass survey, periodic follow – up study on IgA/VCA or IgA/EA antibody positive and EBV marker

positive individuals, and education of the public on the early signs of NPC would lead to more NPC patients in their early stage. If all NPC patients are treated with radiotherapy in their early stage, the mortality rate can certainly be reduced, even before the etiology has been clarified.

(2) *Treatment and prevention with interferon*: Interferon has been shown to have antiviral and antitumoral activities. NPC is a viral associated tumor and is an ideal model to test both the antiviral and antitumoral effects of interferon. Treuner et al[55] reported that a NPC patient in its advanced stage with cranial invasion showed complete regression of tumor after treatment with fibroblast interferon, but no success was obtained in other patients. It is suggested that interferon would be more effective in the treatment of NPC patients in their early stages or in the prevention of NPC development in VCA/IgA antibody positive individual. An early stage (I) NPC patients was treated by our research group with human leukocyte interferon. 3×10^6 per day for 5 weeks, then 3 times a week with the same dose for another 5 weeks Tumor showed regresssion 3 – 4 weeks after treatment, then gradually progressed again, and a change was made to treat it with radiotherapy[56]. Another clinical trial was carried out by our research group and Dr. de The [57] with interferon from Dr. Cantell. Fifteen NPC patients at stage II were selected and randomized for interferon treatment with the same schedule as mentioned above followed by radiotherapy. Another 15 NPC patient in the same stage were treated with cobalt 60 alone as control. Among the 10 patients who completed their IF treatment, 3 patients showed noticeable regression of their primary NP tumor for some weeks. 3 patients exhibited no change in the primary tumor mass and 3 patients showed progression. In one further case, the tumor showed some regression for 5 weeks and then progressed. Histological examination showed that after interferon therapy some degenerative process of the tumor cells took place in 4 patients and noo rooio in 2 patients, these data indicate that leukocyte interferon exhibits some anti – tumor effect on NPC. Some tumor progressed 5 – 10 weeks after interferon treatment. It seems that the dose of interferon might not be enough. At the same time advanced or recurrent NPC cases were treated with interferon with the same dose and schedule in the Cancer Institute of Zhongshan Medical College. No effect on tumor was found, but some severe side effect occurred and most treatments were stopped 4 – 5 weeks after therapy.

In order to study the antiviral activity of interferon, experiments were carried out by using B95 – 8 cells and Raji cells carrying EBV genome[58]. An unexpected result was obtained since the interferon showed enhancement of spontaneous VCA – EA induction in B95 – 8 cells and EA induction in Raji cells (Fig. 14). Only when treated simultaneously with croton oil and n – butyrate did interferon enhance EA induction in Raji cells. The enhancing activity of interferon could only be partially inhibited by retinoids, but the induction of EA in Raji cells by croton oil and n – butyrate could be inhibited completely by retinoid. Thus it is suggested that the mechanism for the enhancement of EA induction by interferon is different from that of EA induction in Raji cells by croton oil and n – butyrate. We do not know whether the enhancing activity for EA and VCA anduction is favorable or not for the treatment of NPC patients with interferon. There was no change in EB virus specific IgG and IgA antibody level in the sera of NPC patients in the early stage before and after treatment with interferon for 10 weeks, and the tumor showed either temporary or progressive regression, necrosis or degene ration. Thus besides the direct anti – tumor activity of interferon, it is possible that the enhancing activityof interferon might cause the activation of EB Virus and lead to the induction of EA and VCA which would cause the death of cells and might be of benefit to the NPC patients. If interferon is effective in the treatment of early NPC, the next step is to try to apply interferon to the possible prevention of NPC in IgA/VCA antibody positive individuals.

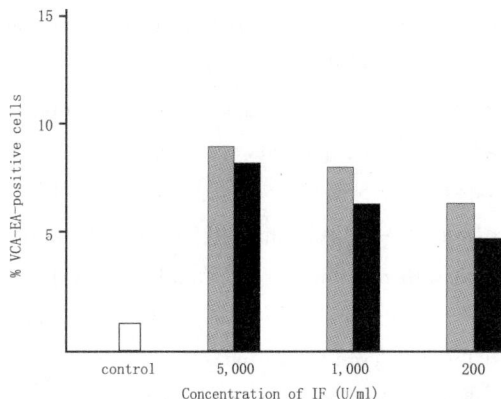

Stippled areas = IF, preparation A; darkened areas = IF
preparation B; open area = control

**Fig. 14   Effect of interferon (IF) on VCA
and EA induction in B95 – 8 cells**

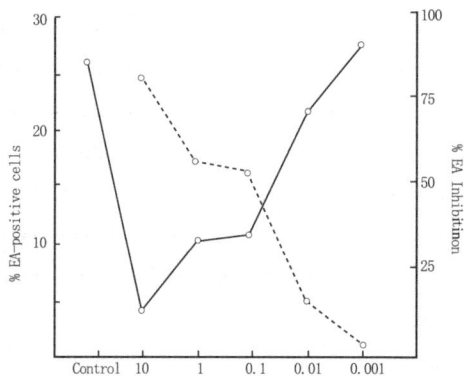

7901 (0.001 – 01 μmol/L) on EA induction in Raji cells. Raji cells were treated with croton oil (250 ng/ml) and n – butyrate (4mmol/L) for 48h ·—· EA – positive cells, ·····= % EA inhibition

Fig. 15   Inhibitory effect of retinoid

(*3*) *Prevention with retinoids*: Yamamoto et al. [59] first reported that retinoic acid interfers with EB Virus EA induction in Raji cells treated with TPA. Ito et al. [60] also reported that increase in EBV EA and VCA in Raji and P3HR – 1 cells treated with croton oil and n – butyrate was markedly inhibited by retinoic acid and retinoids. Retinoids made in China (7901, 7902) and those made by Hoffmann – LaRoche Company (Ro10 – 9359, Roll – 1430) were compared by Zeng et al[61]. The results further confirmed that the induction of EA in Raji cells treated with croton oil and n – butyrate was strikingly inhibited by these retinoids (Fig. 15). The toxictity of 7901 is very low, so a clinical trial was carried out in IgA/VCA antibody positive indivduals with it. No toxictity was found after giving orally 50 mg of retinoids pre day for 6 months[62]. There was no change in EB Virus IgA antibody level before and after administration of retinoids. The IgA/VCA antibody declined slowly after complete regression of tumor by radiotherapy, thus follow – up study on the IgA antibody level should be carride out.

Conclusion

Detection of EBV IgA antibody in the sera and EBV markers in nasopharyngeal carcinoma with viro – immunological tests in combination with clinical, cytological and histological examination is valuable for the early detection of NPC. Some atypical hyperplasia or atypical metaplasia in the nasopharynx might be precancerous lesions. Serologyhas become a routine test widely used for the detection of NPC in China and ti is possible that the mortality rate of NPC may be reduced through early detection and early treatment. The present evidences suggest that EB Virus plays an important role in the development of NPC, but it is not the unique factor. The environmental and genetic factors may also play an important role. In order to clarify the etiology of NPC, the role of EB Virus, evironmental and genetic factors in the development of NPC should be simultaneously studied. There are many persons having EBV IgA antibody and EBV markers. These markers can be used for the intervention against EB Virus by applying different measures such as interferon, retinoids, antiviraldrugs, vaccine, etc. If the intervention against EB Virus succeeds, it would lead to the prevention of NPC and may provide further evidences of the relationship between EB Virus and NPC.

〔原载《Etiology and pathogenesis of Nasopharyngeal Carcinoma》1987，18 – 47〕

References

1　De the, G.. Epidemiology of Epstein – barr and associated diseases in man. In: The Herpes viruses. Roizman, A. B. (ed), Vol. I, Plenum Press, New York, 1982

2　Old, L J, Boyse E A, Ostten H F, de Harven E, Geering G Williamson B, Clifford P. Precipitating antibody in human serum to an antigen present in cultured Burkitt's lymphoma cells. Proc Natl Acad Sci USA, 1966, 56: 1699 – 1704

3　Henle G, Henle W. Epstein – Barr Virus – specific lgA serum antibodies as an outstanding feature of nasopharynaeal carcinoma. Int J Cancer, 1976, 17: 1 – 7

4　Ho J H C, Ng M H, Kwan H C, Chau J C W. Epstein – Barr Virus – specific IgA and IgG serum antibodies in nasopharyngeal carcinoma. Br J Cancer, 1976, 34: 655 – 660

5　Desgranges, C, de Thé G. : IgA and nasopharyngeal carcinoma. In: Oncogenesis and herpesrviruses. de Thé, G., Henle, W. Rapp, F. (eds). Ⅲ, IARC Scientific publication, 1978, 24: 883 – 891

6　Wolf H, Zur Hausen H, Becker V. EB Viral genome in epithelial nasopharyngeal carcinoma cells. Nature (London) Biol, 1973, 244 – 245

7　Tumor Control Team of Zhongshan County, Department of Microbiology of Zhongshan Medical College, Laboratory of Tumor Viruses of Cancer Institute, Institute of Epidemilolgy of Chinese Academy of Medical Sciences: A study on the complement fixing antibody to EB virus in groups of normal individuals in Guangdong province and Beijing. Chinese J E N T, 1978ml: 23 – 25

8　Liu Y X, Zeng Y, Dong W P, Gao, G. R.: Detection of EB virus specific IgA antibody from patients with nasopharyngeal carcinoma by immunoenzymatic method, Chinese J Oncol, 1979, 1: 8 – 11

9　Zeng Y, Shi D F, Zhong J M, Wei J N, Fung C, Mo Y K. Detection of Antibodies to EB virus in sera from normal individuals in Beijing, Chengchun, Nanning and Zangwu county. Unpublished data.

10　De Thé G, Ho J H C, Muir C. Nasopharyngeal carcinoma. In: viral infections of human. Evans, A. S. (ed), 1976, 539 New York: Plenum

11　Zeng Y, Liu Y X, Zhen S W, Wei J N, Zhu J S, Zai H J. Application of immunoenzymatic method and immunoautoradiographic method for the mass survey of nasopharyngeal carcinome. : Chinese J Oncol, 1979, 1: 2 – 7

12　Zeng Y, Zhang L. G, Li H Y, Jan M G, Zhang Q, Wu Y C, Wang Y S, Su G R. Serological mass survey for early detection of NPC in Wuzhow city. Int. J. Cancer, 1982, 29: 139 – 141

13　Department of Microbiology and Cancer Hospital of Zhongshan Medical College. , Deaprtment of Virology of Cancer Institute, Department of Tumor Viruses of Institute of Virology. Cancer Institute of Zhongshan County. : Investigation of

Epstein – Barr Virus complement – fixing antibody levels in sera of patients with nasopharyngeal carcinoma and nasopharynageal mucosal hyperplasia. Chinese Med J, 1980, 93: 359 – 364

14 Laboratory of Tumor Viruses of Cancer Institute. , Laboratory of Tumor Viruses of Institute of Epidemiology. , Department of Radiotherapy of Cancer Institute. , Department of Otolaryngology of Beijing Worker – Peasant – Soldier Hospital. : Detection of EB virusspecific serum IgG and IgA antibodies from patients with naspharyngeal carciroma. Acta Microbiol Sin, 1978, 18: 253 – 258

15 Li S J Zhou, Y B, Hu X T, Zeng Y. Detection of IgG/EA antibody in sera from patients with nasopharyngeal carcinoma, Chinese J ENT, 1980, 71 – 74

16 Desgranges C, de Thé G, Ho J H C, Ellouz R. Neutralizing EB virus specific IgA in throat washings nasopharyngeal carinoma patients. Int J cancer, 1977, 19: 627 – 633

17 Zhong J M, Zeng Y, Liu Y X, Wei J N, Pi G H, Zhu J S, Mao Y K, Zhan J Y: Detection of EBV IgA/EA antibody to in saliva from patients with nasopharyngeal carcinoma and normal subjects. J. Epidmiol, 1980, 1: 225 – 226

18 Liu Z R, Shen, M. , Zeng Y, Dai H J, Du R S, Application of Immunoradioautograph for detection of IgA/VCA antibody in saliva from patients with NPC, Chinese J Med Exam, 1979, 2: 197 – 198

19 Wei J N, Zhang S, Tung S Z, Huang Z L: Detection of EB virus IgA/VCA antibody in sera from patient with nasopharyngeal carcinoma, Guangxi Yi Xue, 1980, 6: 5 – 6

20 Jian S W, Li Z W, Luo W L, Li M C, Pan W Z, Zhang X H. The application of Immunoenzymatic method for the diagnosis of Nnasopharyngeal carcinoma. Cancer Research Reports, Cancer Institute of Zhongshan Medical College, 1981, 2: 23 – 25

21 Li W J, LI C C, Liang Y R, Chen A M, Zhang F, Ho P Y. Study on clinico – serological Research I. Analysis of IgA/VCA antibody from 1006 NPC cases, Cancer (Chinese), 1982, 1: 43 – 48

22 Zeng Y, Zhang M, Liu Z R, Zheng Y H, Du R S, Li X H, Gan B W, Hu M C, Zhen M, He S A, Mum G P: Detection of IgA antibody to EB virus VCA from patients with nasopharyngeal carcinoma by immunofluorescence test. Chinese J Oncol, 1979, 1: 81 – 85

23 Liu Z R, Shan M, Han Z S, Dat H J, Hu Y L, Cao G R, Dong W P. Immunoradioautography and ist application in the detection of anti – EBV IgA antibody in NPC patients. Kexne Tongbao, 1979, 24: 715 – 720

24 Zeng Y, Gong C H, Jan M G, Fan C, Zhong LK. Detection of IgA/EA antibody to EB virus by immunoautoradiography. Int J Cancer, 1983, 31: 599 – 601

25 Zeng Y, Liu Y X, Wei J N, Zhu J S, Wang P Z, Zhong J M, Li R C, Pan W J, Li E J, Tan B F. Serological mass survey of nasopharyngeal carcinoma. Acta Acad Med Sin. 1979, 1: 123 – 126

26 Zeng Y, Liu Y X, Liu Z R, Zhen S W, Wei J N, Zhu J S, Zei H S. Application of an immunoenzymatic method and immunoautoradiographic method for a mass survey of nasopharyngeal carcinona. Intervirology, 1980, 13: 162 – 168

27 Zeng, Y, Wang P C, Zhong J M, Dan H, Li E J, Wei J N, Zhu Z S, Pan W J, Liu Y X, Tan P F. Follow – up studies on EBV IgA/VCA antibody positive persons. Intervirology, 1984, 20, 190 – 194

28 Ho J H C, Kwan H C, Ng M H, de Thé G. Serum IgA antibodies to EBV capsid antigens preceeding symptoms of nasopharyngeal carcinona. Lancet, 1978, (1): 436 – 437

29 Lanier A P, Henle W, Benderm T R, Henle G, Talbot M L. Epstein – Barr Virus – specific antibody titers in seven Alashan natives before and after diagnoses of nasopharyngeal carcinona. Int. J. Cancer, 1980, 26: 133 – 137

30 Ho J H C. An epidemiologic and clinical study of nasopharyngeal carcinoma, Int J Radiol Oncol

Biol, Phys, 1978, 4: 181 – 198

31　Zhu Z S, Pan W J, Zhong J M, Li L Y, Zeng Y, Tan M G, Fan G. Study on IgA/VCA antibody in sera from boat people in Zangwu county. Tumor prevention, treatment and study Cancer Research on Prevention and Treatment, 1983, 10: 189 – 190

32　de Thé G. , Ablashi D V, Liabecef A, Mourali N. : Nasopharyngeal carcinoma (NPC) VI. Presence of an EBV nuclear antigen in fresh tumor biopsies: Prelimirary result Biomedicine, 1973, 19: 349 –352

33　Huang D P, Ho J H C, Henle W, Henle G. Detection of Epstein – Barr Virus associated nuclear antigen in nasopharyngeal carcinoma cells from fresh biopsies. Int J Cancer, 1974, 14: 580 – 588

34　Klein G, Giovanella B C, Lindahl T, Fialkow P J, Singh S, Stehlin J. Direct evidence for the presence of Epstein – Barr Virus DNA and nuclear antigen in malignant epithelial cells from patients with anagpastic Carcinoma of nasopharynx, Proc. Natl. Acad. Sci, USA, 1976, 71· 4737 – 4741

35　Zeng Y, Pi G H, Zho W P. Establishement of an anticomplement immunoenzymatic test for detection of EB Virus nuclear antigen. Acta, Acad, Med, Sin, 1980, 2: 132

36　Pi G H, Zeng Y, Zhao W P, Zhang Q. Development of an anticomplement immunoenzyme test for detection of EB Virus nuclear antigen (EBNA) and antibody to EBNA. J. Immunol. Methods 44, 73 – 78 (1981).

37　Zhang jiang Medical College. Diagnosis of nasopharyngeal carcinoma by cytological examination of exfoliated cells taken by negative pressure suction. Chinese Med. J, 1976, 1: 45 –47

38　Zeng Y, Pi G H, Zhang Q, Shen S J, Zhao M L, Ma J L, Dong H J. Appplication of anticomplement immunoenzymatic method for the detection of EBNA in carcinoma cells and normal epithelial cells from nasopharynx, llth Int. Symp. Nasopharyngeal Carcinoma, Dusseldorf, West Germa-

ny. Grundmann et al （eds）: Cancer campaign, vol. 5, Nasopharyngeal carcinoma. Gustav Fischer Verlag. stuttgart. New York, 1981: 237 – 245

39　Zeng Y, Zhen S J, Dan H, Ma T L, Zhang Q, Zhu JS, Zheng T R, Tan B F. Detection of early stage NPC from IgA/VCA antibody positive individuals by anticomplement immunoenzymatic method. Acta Acad Med Sin, 1982, 4: 254 – 255.

40　Zhen S J, Zhen Z L, Ma T L, Zhan Q, Zeng Y. Further study on the detection of EBNA from the nasopharyngeal mucosa of NPC patients. Acta Zhangjiang Medical College, 1983, 1: 34 – 37

41　Chen J K, Jan S W, Wu B, Ru T C, Li B J, Chang F, Ro H L, Li S R, Li M C, Xiao CT. Study on EBNA antibody in sera and localization of EBNA in cancerous tissue from patients with nasopharyngeal carcinona, Cancer, 1982, 1: 40 –44

42　de Thé G, Desgranges C, Zeng Y, Wang P C, Bornkamm G W, Zhu T S, Shang M. Search for pre – cancerous Lesions and EBV markers in the nasopharynx of IgA positive individuals. Cancer Campaign, Vol. 5, nasopharyngeal carcinoma, Grundmann et al （eds）. Gustav Fischer, Stuttgart, New York, 1981, 111 – 117

43　Desgranges C, Bornkamm Zeng Y, Wang P C, Zhu J S, Shang M, de Thé G. Detection of Epstein – Barr Viral DNA internal repeats in the nasopharyngeal mucosa of Chinese with IgA/EBV – specific antibodies. Int, J. cancer, 1982, 29: 87 – 91

44　Huang D P, Ho J H C, Saw D, Teoh T B. Carcinoma of the nasal and paranasal region in rats fed cantonese salted marine fish, In: Nasopharyngeal carcinoma. Etiology and control. de Thé, G. , and Ito, Y. , （eds）. International Agency for Research on cancer, 1978: 315 – 318

45　De Thé G, Personal communication

46　Tao C C, Pan W J, Zhong L G, Zeng Y. Study on living and fool habit of people in Cangwu county and Wuzhou city, Unpublished data

47　Ito Y, Kawanishi M, Harayama T, Takalayashi

S. Combined effect of the extracts from croton Tiglium, Euphorbia Lathyfis or Euphorbia Tirucalli and n – Butyrate of Epstein – Barr Virus Expression in Human Lymphoblastoid P3Hr – 1 and Raji Cells, Cancer letter, 1981, 12: 175 – 180

48 Hirayama T, Ito Y. A New View of the etiology of nasopharyngeal carcinoma, Preventive medicine, 1981, 10: 614 – 622

49 Ito Y, Yanase S, Fujita J, Harayama T, Takashima M, Imanaka H. A shortterm in vitro assay for promotor substances using human lymphoblastoid cells latently infected with Epstein – Barr Virus. Cancer Letter, 1981, 13: 29 – 37

50 Zeng Y, Zhong J M, Mo Y K, Xiao X Q. Induction of Epstein – Barr Virus Early antigen in Raji cells by Chinese herb medicine, Intervirology, 1983, 19: 201 – 204

51 Jiang – Su New Medical College (Eds), In: Traditional Chinese Medicement Grand Dictionary Shanghai Scientific Technology press, 1977

52 Li E J, Tan B F, Zeng Y, Wang P Z, Zhong J M, Deng H, Zhu Z S, Wei J N, Pan W J. Some changes observed on nasopharyngeal mucosa of persons showing EB IgA/VCA antibody positive. Chinese M J in press

53 Pan G Y, Huang G Y, Mo J D, Zeng C S, Zhong J L. Radiotherapy of nasopharyngeal carcinoma, Chinese Med. J, 1974, 54: 687 – 691

54 Chang C P, Liu T F, Chang Y W, Cao S L. Radiation therapy of nasopharyngeal carcinoma, Acta Radiologica Oncology, 1980, 19: 433 – 438

55 Treuner J, Neithammer D, Dannecker G, Hagmann R, Neef V, Hofschneider P H: Sucessful treatment of nasopharyngeal carcinoma with interferon. Lancet, 1980, 1: 817 – 819

56 Wang P C, Dong H, Wu S H, Wang L Z, Zeng Y, Hu Y T. Preliminary treatment of nasopharyngeal carcinoma with human interferon, Interferon Sci. Menoranda A 1072/2 April, 1981

57 Wang P Z, Zeng Y, Dan II, Na E R, Li E J, Gresser I, Cantell K, de – The G. Progress report of alpha interferon therapeutic trial of early cases of nasopharyngeal carcinona in South China, Interferon Scientific Menoranda I – A1229 November, 1982

58 Zeng Y, Zhong J M, G de – The, Wu S H, Hou Y T, Miao X W. Enhancement of spontaneous VCA – EA induction in B95 – 8 cells and EA induction in Raji cells treated with human leukocyte interferon. Intervirology, 1982, 18: 33 – 37

59 Yamamoto N, Birter K, Zur Hansens H. Retinoie acid inhubition of Epstein – Barr incudtion. Nature (London) 553 – 555, 1979

60 Ito Y, Kishishita M, Morigalci T, Yanasa S, Hirayama T. Induction and intervention of EB virus antigens in human lymphoblastoid cell lines: a simulation model for study of cause and prevention of nasopharyngeal carcinoma and Burkitt lymphoma, Dusseldorf NPC Symp, 1980

61 Zeng Y, Zhow H M, Xu S P. Inhibitory effect of Retinoids on Epstein – Barr virus induction in Raji cells. Intervirlolgy, 1982, 16: 29 – 32

62 Zeng Y, Zhong L G. Treatment of EB virus IgA/VCA antibody positive persens with Retinoids 7901, unpublished data

63 Tang Z H, Wu Y T, Yang J, Rong B P, Xy Z G, Zhao C N, Meng Q C. Determination of serum UCA/IgA antibody by immunoenzymatic method and diagnosis of Nasopharyngeal carcinoma. J. Shanghai Medicine, 1981, 4: 720 – 724

44. Detection of EBV Specific IgA Antibodies to EA, MA and EBNA – 1 Recombinant Proteins in NPC Patients and Controls

ZENG Y[1], WOLF H[2], TAKADA K[3], ARRAND JR[4], de – The G[5]

1. Inst Virology, Chinese Acad Prev Med, Beijing, China; 2. Max von Pettenkofer Inst, Munchen FRG; 3. School Med, Nikon University Tokyo, Japan; 4. Paterson Inst for Cancer Res, Manchester, UK; 5. CNRS, Fac Med Alexis Carrel, Lyon, France

Henle and Henle first demonstrated that Epstein – Barr virus (EBV) IgA/VCA – EA was specific for nasopharyngeal carcinoma (NPC)[1], soon confirmed by other authors[2,3]. Serological screenings and follow – up studies were then implemented in NPC high risk areas (Zangwu county and Wuzhou city in the Guangxi Autonomous Region of the People's Republic of China). The NPC early stage detection rate was therefore increased from 18.1% to 81% and NPC could be predicted by serological screening and follow – up studies[4-7], using IgA antibodies to VCA and to EA.

In the course of following – up IgA positive asymptomatic individuals, IgA/VCA antibodies could be detected 8 years before NPC onset.

The different recombinant proteins of EBV, as produced in E. coil or in mammalian cells, recently became available for serological test. This preliminary paper reports the presence of antibodies in the IgA calss to different components of EA, MA and EBNA – 1 in sera from NPC patients, as compared to normal individuals.

Detection of IgA Antibodies to EA (P138 and P54)

Recombinant proteins P138 and P54 were made in E. coli and antibodies to such proteins were tested by western blot. After removing IgG antibody in sera with SPA, and as seen in Tab. 1 the proportion of NPC sera with IgA/EA (P138 + P54) antibodies increased from 87.6% to 95.0%.

As seen in Tab. 2, 44.5% of NPC patients at early stage had IgA antibodies to both P138 and P54, while only 33.3% had antibodies to P138 alone and 22.2% to P54 alone. These findings suggest that EA is a complex of EB virus antigens and that test with one peptide is not enough for the diagnosis of NPC.

Detection of IgA Antibody to MA

IgA/MA antibody in sera from NPC patients was tested by indirect immunofluorescence test using B95 – 8 cells (LMA) and Raji cells (EMA) treated with TPA and n – butyric as target cells. As seen in Tab. 3, 60% of NPC patients had IgA/LMA antibody and only 15% had IgA/EMA antibody, but no IgA antibody to LMA or EMA was found in normal individuals. It further shows that the IgA/MA antibody is mainly related to LMA[8-10].

P04 cells transfected with EB/MA gene was further used for detection of IgA/MA antibody in various types of sera, after removing the IgA/MA antibody with SPA. The proportion of IgA/MA antibody positive NPC sera was 86%. Three of 57 IgA/VCA antibody – positive but asymptomatic normal individuals exhibited IgA/EA and IgA/MA antibodies and two of them were found to have NPC in early stage (Tab. 4).

Tab. 1 Detection of IgA/EA (P138 + P54) antibodies by western blot in NPC sera and controls

sample	stage	No. sera	unabsorbed		absorbed	
			No. ⊕	% ⊕	N⊕	% ⊕
NPC	II	29	25	86.2	28	96.6
patients	III	52	44	84.6	48	92.3
	IV	32	30	93.7	31	96.9
sub – total		113	99	87.6	107	95.0
patients with tumor oter than NPC		20	2	10	–	–
normal indiv		42	0	0	–	–

Tab. 2 Comparison of IgA antibodies to P138 and P54 in sera from NPC patients at early stage (western blot)

No. sera	IgA antibody positive			Total
	P138	P54	P138 + P54	
117	39	26	52	117
positive rate (%)	33.3	22.2	44.5	100

Tab. 3 Detection of antibody to EBV/MA

target cells	sera	No. samples	IgA/MA		IgA/MA	
			No. ⊕	% ⊕	No. ⊕	% ⊕
B95 – 8 (LMA)	NPC	20	20	100	12	60
	normal indiv	20	18	90	0	0
Raji (EMA)	NPC	20	15	75	3	15
	normal indiv	20	8	40	0	0

Tab. 4 Comparison of IgA antibodies to MA, VCA and EA in Chinese and Algerian/Tunisian NPC cases and in controls

	No. samples	antibody positive		
		Iga/VCA	IgA/EA	IgA/MA
Algerian/ Tunisian NPC cases	33	33(100%) *1:413	30(91%) *1:75.4	28(85%) 1:25.5
Chinese NPC cases	118	118(100%)		103(87.2%)
Chinese IgA/VCA + normal ind.	57	57(100%)	3(5.2%)	3(5.2)%
Chinese IgA/VCA – normal ind.	24	0	0	0

* : GMT

Detection of IgA Antibody to EBNA – 1

As seen in Tab. 5, IgA antibodies to EBNA – 1 were tested by indirect immunofluorescence test, using K4 cells transfected with EBV DNA Bam H1 K fragment as target cells. After removing IgG antibody with SPA, the positive rate of IgA/EBNA – 1 antibody increased from 48% to 78% in NPC patients with GMT (1:21) while only 5.3% of normal individuals were positive with low GMT 1:5.2.

The positive rate of IgG antibodies to EBNA – 1 in sera from NPC patients and normal individuals was 100% and 92% respectively, with GMT of IgG/EBNA – 1 antibody for NPC being 4 times higher than that from normal individuals (Tab. 6).

Tab. 5 Detection of IgA/EBNA – 1 antibody in sera treated/untreated with SPA

Group	No. samples	Unabsorbed sera		Absorbed sera	
		No. ⊕	% ⊕	No. ⊕	% ⊕
NPC	50	24	48	39	78
patients		GMT	1:13	GMT	1:21
normal	38	1	2.6	2	5.3
individuals		GMT	1:5.1	GMT	1:5.2

Tab. 6 Detection of IgG/EBNA – 1 antibody in sera from NPC patients by IF

Group	No. samples	IgG/EBNA – 1		IgA/EBNA – 1	
		No. ⊕	% ⊕	No. ⊕	% ⊕
NPC	5	50	100	39	78
		GMT 1:89.4		GMT 1:21.0	
Normal	38	35	92	2	5.3
Indiv.		GMT 1:18.3		GMT 1:5.2	

These results suggest that IgA/EBNA – 1 antibody is specific for NPC and can be used for NPC diagnosis.

In conclusion, these results indicated that as IgA/VCA antibody, IgA/EA, MA and ENA – 1 antibodies are specific and valuable for NPC diagnosis. Yong et al[11] and Fahraeus et al[12] reported that EBV/EBNA – 1 and MLP but not EBNA – 2 were expressed in NPC tumors. It is likely that the IgA antibodies to different EBV antigens reflect the presence of EBV replication in tumors. Whether IgA antibody to LMP and further EBV antigens are present in the sera of NPC patients remain to be studied.

〔In 《Epstein – Barr Virus and Human Disease》 1988: 309 – 313〕

References

1　Henle G, Henle W. Int J Cancer, 1976, 17: 1

2　Desgranges C, de The G. in Oncogenesis and Herpesviruses Ⅲ. G. de The, Y. Ito F. Rapp, Eds. (IARC Sciencific Publication No 25, Lyon, France), 1978, 883

3　Zeng Y, Shang M, et al, Acta Microbiol Sinica, 1978, 253: 18

4　Zeng Y, Liu C, et al. Intervirology, 1980, 13: 162

5　Zeng Y, Zhang J M, et al. Intervirology, 1983, 20: 190

6　Zeng Y, Zhang L G, et al. Int J Cancer, 1982, 29: 139

7　Zeng Y, Zhang L G, et al. Int J Cancer, 1986, 36: 545

8　Zhu X X, Zeng Y, et al. Int J Cancer, 1986, 37: 689

9　Du B, Zeng Y, et al. Chinese J Virology, 1987, 3: 92

10　Zeng Y, Du B, et al. Chinese J Virology, 1987, 3: 396

11　Young L S, Dawson C W, et al. J Gen Virol, 1988, 69: 1051

12　Fahraeus R, Hu L F, et al. Int J Cancer, 1988, 42: 239

45. Epstein – Barr Virus Activation in Raji Cells by Extracts of Preserved Food from High Risk Areas for Nasopharyngeal Carcinoma

Shao Y M[1,2,3], Poirier S[1,2], Ohshima H[2], Malaveille C[2], Zeng Y[3], de The G[1,4], Bartsch H[2,4]

1. CNRS Laboratory of Epideniology and Immunology of Tumours, Faculty of Medicine A. Carrel, 69373 Lyon Cedex 08; 2. International Agency for Research on Cancer, Unit of Environmental Carcinogens and Host Factors, 150 cours Albert Thomas, 69372 Lyon Cedex 08. France; 3. Institute of Virology, Chinese Academy of Preventive Medicine, Beijing 100052, China

Summary

Epstein – Barr virus (EBV) activation of latent infection and traditional life styles, especially food habits, have been strongly associated with an increased risk of nasopharyngeal carcinoma (NPC) in humans. On the basis of anthropological studies in Tunisia, southern China and Greenland, extracts of representative preserved food items consumed frequently by the high – risk populations for NPC were assayed for the presence of EBV activators in Raji cells. A strong EBV activation activity was observed in aqueous extracts of some Cantonese salted dried fish from China, harissa (a spice mixture) and to a lesser extent qaddid (dry mutton preserved in olive oil) from Tunisia. These new data may support epidemiological evidence for the importance of Cantonese salted and dried fish and other food items in the etiology of NPC.

Introduction

Epstein – Barr virus (EBV*) DNA is regularly present in undifferentiated carcinoma of the nasopharynx (NPC) in humans[1]. IgA antibodies against EBV viral capsid antigens and early antigens (EA), reflecting EBV activation in vivo, have been reported in patients prior to tumour development[2-6]. In view of the restricted geographical distribution of NPC and the ubiquity of EBV, however, other cofactors must be considered in the etiology of NPC. Traditional life styles, especially consumption of Cantonese – style salted fish, has been found to be associated with the development of NPC with an increased relative risk of 37. 2 when exposure occurs in early age[7]. Further, an anthropological study on diet in southern China. Tunisia and Greenland suggested that preserved foods could be a risk factor for this tumour[8]. In a previous study, we analysed preserved foods frequently consumed in these three high – risk areas for NPC for their contents of volatile N – nitrosamines (VNA). We observed relatively high levels in some preserved food samples daily consumed[9]. such as salted and dried fish, fermented vegetables and qaddid (dried mutton preserved in olive oil). The aim of the present study was to examine extracts of the same food samples for the presence of substances that could activate EBV in EBV latently infected Raji cells.

Materials and Methods

On the basis of consumption frequency and method of preservation used, 28 representative food specimens were selected from among those analysed previously for VNA[9]. Aqueous, hexane and ethyl acetate extracts of these specimens were prepared as follows: 10 g of each food sample were homogenized with 20 – 50 ml of deionized water in a Polytron homogenizer for 3 min. After incubation in a shaking water – bath for 30 min at 4℃, the homogenate was centrifuged at 15 000 g for 20 min, and the pellet was re – extracted with the same amount of water. The combined supernatants were filtered through a 0. 45 μmol/L Millipore filter and lyophilized. The lyophilate was then dissolved in 6 ml of water. The residual material after aqueous extraction was further extracted in a Soxhlet apparatus, first with 100 ml n – hexane, then with 100 ml ethyl acetate (25 passages each). The solvent was evaporated using a rotary evaporator, and the residue was dissolved in 1. 6 ml dimethyl sulphoxide. All these aqueous and dimethyl sulphoxide extracts were stored at – 80℃ prior to the analysis.

The method used for assaying EBV activation was that described by Ito et al[10]. with a minor modification, i. e. treatment of foetal calf serum by pH shock[11]. In brief, human lymphoblastoid Raji cells, bearing latent EBV genomes, were cultured in RPMI 1640 medium supplemented with 10% fetal calf serum in a humidified atmosphere of 5% CO_2 at 37℃. The test for indnotion of EBV (EA) was performed by incubating Raji cells at a density of 5×10^5 cells/ml for 48 h at 37℃ in the presence of the food extracts, supplemented with 4 mmol/L sodium n – butyrate and 15% foetal calf serum. After incubation for 48 h, aliquots of sedimented cells were transfered onto glass slides, air dried and fixed with acetone for 20 min at 4℃. Cells that expressed EBV – EA were detected by an immunoenzymic test[12] instead of the usual immunofluorescence method (as in [10]). This assay was performed with IgG antibodies against EA (as contained in 2 – fold diluted, pooled NPC sera) and peroxidase – labelled protein A. In each assay, 500 cells were scored, and the percentage of EBV – EA – positive cells was recorded. The mean background value in the presence of 4 mmol/L sodium n – butyrate was 3. 05% ±0. 51 SD; (n = 11). Positive controls treated with 12 – O – tetradecanoylphorbol – 13 – acetate (TPA) (20 ng/ml) and 4 mmol/L sodium n – butyrate gave a mean of 43. 5% ±8. 96 SD; (n = 11). The mean values listed in Tab. 1. were calculated from two series of duplicate experiments. Under these assay conditions, the samples were considered to be positive when the percentage of EA – positive cells after treatment of 4 mmol/L sodium n – butyrate and test sample was >10% (i. e. three times background value in the presence of 4 mmol/L sodium n – butyrate only) showing a dose – dependent increase (Tab. 1).

Country/food item	EBV – EA – positive cells (%) Concentration of food extract (mg wet weight equivalent of food/ml medium)			
	0. 005 – 0. 13	0. 66	3. 32	16. 6
Southern China				
Cantonese salted, dried fish and shellfish:				
Japanese mackerel *	< 5	8. 4	10. 2	17. 8
Squid sample 1 *	< 4. 0	5. 0	13. 6	8. 6
Squid sample 2 *	< 3. 2	3. 4	8. 8	16
Eight other Cantonese salted, dried fish and shellfish	< 4. 0	< 5. 1	< 6. 0	< 5. 2
Fermented shrimp/fish paste	< 4. 1	3. 0	3. 2	4. 8
Fermented soya bean paste	< 2. 6	3. 0	3. 1	5. 1
Cabbage fermented in brine	< 3. 5	3. 3	4. 2	4. 1
Tunisia				
Harissa (spice mixture):				
Sample 1 *	< 3. 9	7. 7	13. 9	22. 6
Sample 2 *	< 4. 9	10. 0	14. 9	19. 9
Sample 3 *	< 4. 7	10. 9	12. 6	3. 2
Sample 4 (preserved in can)	< 3. 1	2. 6	3. 0	3. 3
Qaddid (dried mutton preserved in oil)	< 2. 6	4. 2	7. 8	9. 6
Touklia (stewing base)	< 5. 2	7. 2	6. 1	1. 8
Turnips fermented in brine	< 4. 1	5. 0	5. 5	6. 4
Salted anchovies	< 3. 0	2. 4	2. 8	2. 2
Louben (sap from the mastic tree)	< 3. 1	3. 2	4. 0	4. 1
Greenland				
Mikialak (dried Atlantic cod)	< 2. 4	3. 0	3. 2	2. 8
Uuvag (dried polar cod)	< 2. 5	2. 2	2. 4	6. 1
Amassat (dried capelin)	< 2. 4	2. 0	2. 1	4. 8
Panerteq (dried fjord seal meat)	< 2. 8	4. 2	6. 1	7. 2
Berries preserved in seal oil	< 4. 4	4. 3	4. 2	4. 4

Results and Discussion

The results on aqueous extracts are shown in Tab. 1. Of the preserved food samples from southern China, three out of 11 Cantonese salted dried fish and shellfish were found to contain substance (s) that activate EBV latency in Raji cells. Three out of four samples of harissa (a spice mixture) and, to a lesser extent, one sample of qaddid from Tunisia showed EBV activation activity. The activity found in these food extracts was comparable to those reported for extracts of Chinese vegetables, plants and medicinal herbs, some of which are known to contain phorbol ester – type compounds[13-15]. For most of the positive samples, a dose – response relationship between the amount of food extract tested and EBV activation in Raji cells was observed. Aqueous extracts of other selected preserved food samples from southern China. Tunisia and Greenland gave negative results; however, because of the wide variation in EBV activation by the same type of food (Tab. 1), further samples of each food are now being assayed. The variation may be due to different modes of preparation or even to individual differences in the mode of preparation. All n – hexane and ethyl acetate extracts up to 62. 5 mg wet weight equivalent of food/ml medium produced 1. 2% to 6. 1% EBV – EA – positive cells and thus were considered negative (data not shown).

Preliminary experiments were carried out to isolate and characterize the EBV – activating agents present in aqueous extracts of harissa. The substance (s) appears to be relatively stable, since no apparent loss of activity was observed when the harissa sample was heated at 100℃ for 10 min or incubated in 0. 1 mol/L NaOH solution at 37℃ for 30 min. However, about 50% of the activity was lost after incubation in 0. 1 mol/L HC1 solution at 37℃ for 30 min. Preliminary partition experiments suggest that at least two different compounds are present in the harissa extract, including one extractable by ethyl acetate under neutral conditions and another soluble onyl in water.

In conclusion, the results reported here represent, to our knowledge, the first demonstration that Cantonese salted fish, karissa and qaddid – preserved foods consumed very frequently in high – risk areas for NPC – contain substance (s) capable of strongly activating EBV in Raji cells. Recent case – control studies conducted in Hong Kong show that consumption of Cantonesestyle salted fish is significantly associated with an increased risk of NPC[7], Similarly, an anthropological study conducted in the same three high – risk areas[8] also suggested that NPC could be associated with eating habits, especially consumption of preserved foods, including harissa, qaddid and salt – dried fish. Further studies should be underaken to isolate and identify the active substances in order to better understand the role of these foods in the etiology of NPC. A case – control study is now underway in the three high – risk areas for NPC, to study the correlation between NPC and consumption of food items containing VNA, EBV – activating substance (s) and other genotoxic substance (s). Results of experiments to detect mutagens present in the same food extracts used in this study will be reported separately elsewhere.

Acknowledgements

The authors would like to thank G. Lenoir for his valuble comments on this manuscript. We also thank N. Roche for technical assistance, A. Hubert for collecting the food specimens, E. Heseltine and M. B. D´Arcy for editorial and secretarial help, respectively. This study was supported by the CNRS (GGS17), the Association pour Recherche sur le Cancer (ARC) (Contract 6071) and the Association for Virus Cancer Prevention.

[In 《Carcinogenesis》 1988, 9: 1455 – 1457]

References

1 De The G, Ho J H C, Muir CS. Nasopharyngeal carcinoma. In Evans, A S. (ed). Viral Infections of Humans Epidemiology and Control. John Wiley New York, 1982: 621 −622

2 Henle G, Henle W. Epstein − Barr virus − specific IgA serum antibodies as an outstanding feature of nasopharyngeal carcinoma. Int J Cancer, 1976, 17: 1 −7

3 Desgranges C De The G. (1978). IgA and nasopharyngeal carcinoma. In De The, G. Ito Y. and Rapp, F (eds), Oncogenesis and Herpesviruses

Ⅲ. International Agency for Research on Cancer, Lyon, 1978, 24: 883 −891

4 Wara W M, Wara D W, Phillips T L, Ammahh, A. Elevated Ig A in carcinoma of the nasopharynx. Cancer, 1975, 35: 1313 −1315

5 De The G, Zeng Y. Population screening for EBV markers toward improvement of nasopharyngeal carcinoma control. In Epstein, M. A. and Achong B. G (eds), The Epstein Barr Virus. Recent Advances. Willam, 1986

46. Wikstroemia Indica Promotes Development of Naso −pharyngeal Carcinoma in Rats Initiated by Dinitrosopiperazine

Tang W P[1], Huang P G[1], Zhao M L[1], Liao S L[1], Zeng Y[2]

1. Zhangjiang Medical College, Guangdong, China

2. Institute of Virology, Chinese Academy of Preventive Medicine, Beijing, China

Summary

Nasopharyngeal carcinoma was induced in an initiation/promotion model in rats by s. c. injection of dinitrosopiperazine in the nasopharyngeal cavity. This was followed by repeated 10 − cal administration of an extract of roots of the Chinese medicinal herb WI (botanical family: Thymelaeaceae). Three groups of rats were used: group −1 received DNP followed by repeated WI; group −2 received DNP once; group −3 received WI repeatedly. At 180 −205 days after DNP + WI administration 26% of the rats in that group exhibited NPC (two were carcinomas in situ and four were early infiltrating carcinomas). In the other two groups no carcinomas were found. In the group which received DNP followed by WI, other pathological changes, such as hyperplasia of nasopharyngeal epithelium, squamous metaplasia, and papillary hyperplasia, were also more frequent than that in the other two groups.

Introduction

Using a combination of viral and environmental factors, in what may be considered an initiation/promotion protocol in Raji cells vitro (Hecker 1979; zur Hausen et al, 1979), many extracts from plants and herbs of Euphorbiaceae, Thymelaeaceae, and other plant families have been tested for promoting activity (Ito 1981; Zeng et al. 1983, 1984; Zhong et al. 1986; Zi et al. 1985). It was found that many of the extracts induced an early antigen of EBV in the Raji cells. Also, in lymphocytes transformation by EBV was enhanced by such extracts (Hu and Zeng 1985). Furthermore, some of the extracts enhanced the development of Rous sarcoma, papilloma, and cervical carcinoma in vivo (Hu et al. 1986; Zeng 1987; Sun et al. 1987.) Whether or not the same extracts would have an effect on inducing NPC in an initiation/promotion model in vivo requires investigation. It is well – documented that the plant WI of the family of Thymelaeaceae is found in high – risk areas of NPC such as Guangdong, Guangxi, and other provinces of China; it is used as a Chinese medicinal herb. On the other hand it has been shown in various laboratories that in rats and mice NPC may be induced by certain chemical carcinogens (Druckrey et al. 1964, 1967; Wang 1965; Pan and Yao 1978; Yao et al. 1981; Tang and Juang 1978; Huang et al. 1977, 1978). For example, by a combination of nickel sulfate and DNP, Ou induced NPC in rats (Ou 1982). Here we report that NPC may be induced in rats by combined use of DNP and WI in an initiation/promotion protocol.

Materials and Methods

1. Initiator: DNP was synthesized in our laboratory, m. p. 157 – 158℃ (Fan and Wen 1985). It was mixed with a small amount of Tween 80 and distilled water to give a solution containing 9 mg DNP/ml suit able for injection.

2. Promoter: An ether extract of roots of WI was prepared by repeated digestion of the plant material at room temperature with ether for 1 week. The dry residue of the extract was dissolved in ethanol to make up a solution containing 10 mg WI/ml.

3. Animals: A total of 72 hybrid rats of unspecified strain, two months old, male: female were used 1: 1. They were divided into 3 groups and the initiation/promotion protocol was as rollows:

(1) DNP followed by WI: 32 rats were each given 1 ml DNP solution s. c. After 10 days 0. 1 ml of WI solution was dropped into the nasopharyngeal cavity twice a week, for 7 weeks.

(2) DNP once: 20 rats were given 1 ml DNP s. c. only.

(3) WI alone: 20 rats were given 0. 1 ml WI solution twice a week by dropping it into the nasopharyngeal cavity.

After 7 weeks, the administration of WI was stopped, and all rats were observed continously for 180 – 205 days, then they were sacrified. The entire nasopharygeal cavity was removed and fixed in Bouin's solution, using Jenkin solution to eliminate calcium. It was embedded in paraffin and sectioned following routine staining with H&E for investigation under a light microscope.

Results

Both of the control groups 2 and had no tumors with the exception of one case of an atypical dysplasia and one of papilloma of the nasopharyngeal epithelium in thi WI group. The main pathological change was squamous metaplasia and papillary hyperplasia (Tab. 1).

Tab. 1　Survey of pathological changes of nasopharyngeal epithelium in rats

Survivors/tumors	Group		
	DNP followed by WI	DNP alone	WI alone
Survivors of 180 – 205 days	23/32	14/20	15/20
Squamous metaplasia	19	10	10
Papillary hyperplasia	18	8	8
Papilloma	0	0	1
Dysplasia	4	0	1
Carcinoma in situ	2	0	0
Early infiltrating carcinoma	4	0	
Rate of carcinoma (%)	26. 1	0	0

In the group which received DNP followed by WI six rats developed NPC, a tumor incidence of 26%. The other pathological changes, such as hyperplasia of the nasopharyngeal epithelium, squamous metaplasia, and papillary hyperplasia, were more frequent than in the control groups. The tumor cells in the six cases grew towards the cavity as papilliform, gyrusform, or fungiform and even obstructed the cavity. Two of the NPC in the group were carcinomas in situ (Tab. 1), and four early infiltrating carcinomas. Among the four cases of early infiltrating carcinomas, three were poorly differentiated squamous cell carcinomas and showed no keratinization, another was cylindrical, and a few of them displayed an adenoid structure filled with mucin staining light blue. The cells of all four cases penetrated the basal membrane and infiltrated the masenchyme. Of the two carcinomas in situ, one was a carcinomatous change of cylindroepithelial cells, and the other was a carcinomatous change of squamous epithelial cells. Three cases of poorly differentiated squamous cell carcinomas were papilliform, gyrusform, and fungiform. The poorly differentiated cylindrocellular carcinoma was papilliform.

Discussion

Poirier et al. (1987) found nitrosamines in common foods in high – risk areas for NPC in Tunisia, South China, and Greenland. Huang et al. (1977, 1978) reported that extracts from salted fish containing nitrosamines could induce NPC in rats. Experimentally it was found that DNP is a solitary carcinogen with relative organotropy for the nasopharyngeal epithelium (Le et al. 1982). As known for other nitroso compounds large dose or many small doses, repeat dely administered for a long time, may elicit NPC. For example, Pan and Yao (1978) induced NPC in rats by giving 1. 5 mg DNP s. c. twice a week, with a minimum cumulative dose of 99 mg. Within 6 months after stopping the administration of DNP 64% of the rats had developed squamous carcinoma of the nasopharynx.

The plant WI is frequently used as a Chinese medicinal herb for treating infection. It is the main ingredient of the patent medicine "Jiedu – Yiaoyan Pian" made in Guangdong. Zeng et al. found (Zeng et al. 1983, 1984; Zhong et al. 1986; Ni et al. 1985) that the ether extract of the roots activated EBV in Raji cells and enhanced the EBV – induced transformation of lymphocytes.

With the background of the now classical initiation/promotion protocol with DMBA/diterpene ester in the mouse skin model (Hecker 1984; Hecker et al. 1984) and in view of the findings on systemic administration of the initiatior DMBA (Pyerin and Hecker 1980) our experiments show:

（1）the rats in the DNP group were given a subcarcinogenic dose by injection of 9 mg DNP s. c. (about $\frac{1}{10}$ of the minimum cumulative dose for inducing cancer）, as confirmed independantly by Lu （1983）, （2）the WI group, which was given 14 doses of WI also received subcarcinogenic exposure, and （3）giving the same dose of DNP once, followed by the same dose of WI 14 times may truly be considered an initiation/promotion protocol: the experiments showed that in nasopharyngeal epithelium of ratd DNP may play the role of a tumor initiator and WI the role of a tumor promoter. Since the experiments lasted only 180 – 205 days, the observation of metaplasia, hyperplasia, and one papilloma after administration of WI alone was remarkable. This effect appeared to indicate that WI may act as a complete carcinogen when applied over longer periods. It will be important to carry out further experimental studies on the effects of nitrosamines and EBV in conjunction with environmental tumor promoters detected in high – risk areas of NPC.

〔In《J Cancer Res Chin Oncol》1988, 114: 429 –431〕

References

1　Druckrey H, Ivankovic S, Mennel H, Preussmann R Selektive Erzeugung von Carcinogen der Nasnhohle bei Ratten durch N – Di – Nitrosopiperazin, Nitrosopiperidin, Nitrosomorohol in Methyl – allyl – , Dimethylund Methyl – vinyl – Nitrosamin. Z Krebsforsch, 1964, 66: 138 – 150

2　Druckrey H, Preussmann R, Ivankovic S, Schmahl D Organotrope carcinogen Wirkungen bei 65 verschiedenen Nitroso – Verbindungen an BD – Ratten. Z Krebsforsch, 1967, 69: 103

3　Fan JY, Wen HJ Chemical synthesis of several carcinogenic agents. Bull Hunan Med Coll, 1985, 10（4）: 317 –319, 3332

4　Hecker E. Diterpene ester type modulators of carcinogenesisnew findings in the mechanism of chemical carcinogenesis and in the etiology of human tumors. In: EC Miller et al. （eds）Naturally occurring carcinogens – mutagens and modulators of carcinogenesis. Japan Sci Soc Press, Tokyo/UNIV Park Press, Baltimore, 1979, 263 – 266

5　Hecker E. In: Kang et al. （translated）Scientific base of cancer. Academic Press, New York, 1984, 385 – 394

6　Hecker E. Cocarcinogens of the diterpene ester type as principal risk factors of cancer in Curacao and possibly in South China: identiication of second order risk factors of cancer in multifactorial carcinogenesis. In: Wagner G, Thang YH （eds）Cancer of the liver, esophagus, and nasopharynx. Berlin Heidelberg New York, Springer, 1987, 101 – 113

7　Hecker E, Adolf W, Hergenhahn M, Schmidt R, Sorg B Irritant diterpene ester promoters of mouse skin: contributions to etiologies of environmental cancer and to biochemical mechanisms of carcinogenesis. In: Fujiki H et al. （eds）Cellular interactions by environmental tumor promoters. Japan Sci, Soc Press, Tokyo, VNU Science Press, Utrecht, 1984, 3 – 36

8　Hu YL, Zeng YThe extracts from some Chinese herbs enhanced the transformation of human lymphocytes by EBV. Chin J Oncol, 1985, 8: 417 –418

9　Hu YL, Zeng Y, Ito Y Croton oil, Wikstroemia chamaedaphne and Wikstroemia indica enhanced rabbit papilloma induced by papilloma virus. Chin J Virol, 1986, 2: 81 – 82

10　Huang DP, Ho J HC, Gough TA, Webb KS Volatile nitrosamines in some traditional Chinese food products. J Food Saf, 1977, 1: 1 – 6

11　Huang DP, Ho JHC, Saw D, Theoh TB Carcinoma of the nasal and paranasal regions in rats fed Cantonese salted marine fish. In: de The G, Ito G, Davis W (eds). Nasopharyngeal carcinoma: etiology and control. IARC Scientific Publications No. 20, Lyon, 1978, 315–328

12 Ito Y Induction and intervention of Epstein–Barr virus expression in human lymphoblastoid cell line. Cancer campaign, vol 5, New York, Nasopharyngeal carcinoma, Fischer, Stuttgart, 1981

13 Le JY, Pan SC, Yao KT The mechanism of organ specific carcinogenicity of N, N–Dinitrosopiperazine in rats. Bull Hum Med Coll, 1982, 7 (2): 129–135

14 Lu YF The role of nickel sulfate in the induction of nasopharyngeal carcinoma in rats. Cancer, 1983, 2: 100–102

15 Ni HY, Zeng Y, Zhong JM Distribution of plants and herbs containing EBV inducer. J Ninbo Univ, 1985, 1: 86–88

16 Ou BX Trace elements and nasopharyngeal carcinoma. Cancer, 1982, 1: 86–89

17 Pan SC, Yao KT Induction of nasopharyngeal carcinoma in rats by nitroso compounds. KEXUE TONGBAO, 1978, 12: 756–760

18 Poirier S, Ohshima H, Bourgade MC, de The G, Bartsch H Volatile nitrosamine levels in common foods from Tunisia, South Cina and Greenland high risk areas for nasopharyngeal carcinoma. Int J Cancer, 1987, 39: 293–296

19 Pyerin WG, Hecker E Tumor initation in mouse skin by 7, 12–dimethyl–benz (a) anthracene: irrelevance of systemic activation. Cancer Lett, 1980, 8: 317–321

20 Shen SZ Preliminary report on interstisural reaction of nasopharyngeal carcinoma in situ in rats by methylcholanthrene, J Zhangjiang Medical College, 1978: 10–15

21 Sun Y, Chen MH, Zeng Y Tumor promoting effect

of the extracts of Wikstroemia chamaedaphne and HHPA on cervical cancer induced by Herpes simplex virus type 2 in mice. Chin J Virol, 1987, 3: 131–133

22 Tang WP, Huang SW Preliminary report on interstitial reaction of nasopharyngeal carcinoma in situ in rats by methylcholanthrene. J Zhangiang Medical College, 1978, 10–15

23 Wang HW Study on inducing nasopharyngeal carcinoma in mice. J Exp Biol, 1965, 10: 190–199

24 Yao KT, Pan SC, Huang JI, Wen DS Further investigation of experimental induction of nasopharyngeal carcinoma in rats by dinitrosopiperazine. Bull Hunan Med Coll, 1981, 6 (1): 1–6

25 Zeng Y Prospective studies on nasopharyngeal carcinoma and Epstein–Barr virus inducers. In: Wagner G, Zhang YH (eds) Cancer of the liver, esophagus, and nasopharynx. Berlin Heidelberg New York, Springer, 1987, 164–169

26 Zeng Y, Zhong JM, Mo YK, Miao XC Epstein–Barr virus early antigen induction in Raji cell by Chinese medicinal herbs. Intervirology, 1983, 19: 201–204

27 Zeng Y, Miao XC, Jiao BO, Ito Y Epstein–Barr virus activation in Raji cells, with ether extracts of soil from different areas in China. Cancer, Lett, 1984, 23: 53–59

28 Zhong JM, Mo YK, Ni HY, Huang CC, Zhin CZ, Tam ZT, Zeng Y Studies on EBV inducer from plants, herbs and foods. J Guangxi Med, 1986, 8: 145–146

29 Zur Hausen H, Bornkamm GW, Schmidt R, Hecker E Tumor initiators and promoters in the induction of Epstein–Barrvirus. Proc Natl Acad Sci USA, 1979, 76: 782–785

47. Aetiological Studies on Nasopharyngeal Carcinoma in China

ZENG Y[1], ZHANG J M[2]

1. Institute of Virology, Chinese Academy of Preventive Medicine, Beijing, China

2. Nasopharyngeal Carcinoma Research Unit of Zangwu County, China

Early Diagnosis of Nasopharyngeal Carcinoma

Henle and Henle first demonstrated that Epstein – Barr (EB) virus IgA antibodies to the viral capsid antigen (VCA) and early antigen (EA) are specific for nasopharyngeal carcinoma (NPC)[1]. This was confirmed by other authors, soon after[2,3]. The regular presence of such antibodies in patients with NPC and their absence in those with other tumours indicated that such antibodies are useful markers for the diagnosis of NPC.

Aserological mass survey of NPC was carried out in the Zangwu county of Guangxi Autonomous Region by the immunoenzymatic test and IgA/VCA antibody positive persons were followed up yearly for eight years (Tab. 1)[4-7]. The early detection rate for NPC was increased from 18.6% to 81%, while in Zhangshan county it was increased to 100%[8].

Tab. 1 Comparison of the NPC detection rate in zangwu county

Stage	I	II	III	IV	Total	Early detection rate (%)
NPC in outpatient clinics (%)	0	11 (18.6)	24 (40.7)	24 (40.7)	59 (100)	18.6
Screening (%)	5 (33.3)	5 (33.3)	3 (20.0)	2 (13.4)	15 (100)	66.6
B year follow – up (%)	5 (24.0)	12 (57.0)	4 (19.0)	0	21 (100)	81.0

NPC patients could be detected just by serological screening, and within 8 years of initial detection of IgA antibody, especially in persons in whom IgA antibody titres increased (Tab. 2).

IgA antibody titres were also followed up. The percentage of IgA/VCA antibody – positive persons losing antibody or showing fluctuating, stable, increasing or declining levels was 31.6, 7.2, 40.7, 7.3 and 13.2, respectively (Tab. 2).

Tab. 2 Relationship between IgA/VCA antibody change and NPC detection after 8 year follow up studies

	Loss	Fluctuating	Stable	Increase	Decline	Total
Positive persons (%)	295 (31.6)	67 (7.2)	378 (40.7)	68 (7.3)	123 (13.2)	931 (100)
NPC Detection (%)	0	0	6 (1.6)	15 (22.1)	0	21

It is of interest that most NPC cases were detected in the groups showing increasing or stable IgA antibody titres. Precancerous cells or early cancer should be present a few years before the cancer is detected. If intervention for the prevention of NPC is to be contemplated, it should be undertaken in individuals with increasing or persisting levels of IgA antibody. NPC could be detected within a period of 8 years after the first detection of IgA antibody. These findings may reflect important individual variations in the process of NPC development. The titres of IgA antibody to VCA and EA are proportional to the detection rate for NPC (Tab. 3).

Tab. 3   Relationship between IgA antibody and NPC detection rate

Item	Antibody Titer				
	10 – 20	40 – 80	160 – 320	640 – 1280	Total
IgA/VCA +	850	520	178	59	1607
NPC	8	12	10	11	41
NPC detection rate (%)	0.9	2.3	5.6	18.6	2.6
IgA/EA +	16	18	7		41
NPC	6	14	7		27
NPC detection rate (%)	37.5	77.8	100		65.9

All these findings indicate that EB virus plays an important role in the development of NPC.

A sensitive and specific western blot assay was used for the detection of IgA/EA antibody. Recombinant EA antigen was obtained from plasmids which expressed the EA proteins p138 and p54 separately. After treatment of sera with staphylococcal protein A (SPA), the positive rate for IgA/EA antibody detection increased from 87.6% to 95.1 % (Tab. 4).

Tab. 4   Detection of IgA/EA antibody by western blot

Sample	Clinical stage	No. of Sera	Unabsorbed		Absorbed	
			Positive number	Positive rate (%)	Positive number	Positive rate (%)
NPC patients	II	29	25	86.2	28	96.6
	III	52	44	84.6	48	92.3
	IV	32	30	93.6	31	96.9
	Total	113	99	87.6	115	96.9
Patients with tumors other than NP		20	2	10		
Normal individuals		42	0	0		

44.5% of NPC patients had IgA antibodies to both p138 and p53, while 33.3% had antibody to P138 alone and 22.2%, to p54 alone (Tab. 5). Patients with NPC hda a much higher percentage

Tab. 5 Comparison of IgA antibody to p138 and p54 in sera from NPC patients by western blot

No. of sera	Positive for IgA P138	P54	antibody p138 + p54	Total
117	39	26	52	117
Positive rate（%）	33. 3	22. 2	44. 5	100

of positives for IgA/EA（P138 + P54）antibody than did patients with other tumours or normal individuals. These findings auggest that EA consists of a complex of antigens and tests for only one protein in the complex is insufficient for the diagnosis of NPC.

These data indicate that the detection of IgA antibodies to EB virus VCA and EA is very useful for the early diagnosis of NPC and that this virus plays an important role in the development of nasopharyngeal carcinoma.

EB Virus Inducers

Ito et al reported that the extracts of plants from some members of the Euphorbiacae family in Japan efficiently induce EB virus EA in Raji cells[9-11]. Ether extracts from 1693 plants, including Chinese medicinal herbs, were studied for EB virus inducer activity and more than fifty were found to have inducing activity[12]. Many plants such as Aleurites fordii, Wikstroemia indica（WI）, Sapium sebiferum and Euphorbia antiquorum grow in high risk areas for NPC. EB virus inducers were also found in the soil and in vegetables growing under such plants in southern China.[13]. Extracts from plants and herbs containing EB virus inducers can enhance lymphocyte transformation by EB virus in vitro（Tab. 6）[14], as well as tumour growth by the Rous sarcoma virus, rabbit papillomavirus, DMBA, methylcholanthrene and dinitrosopiperazine（DNP）（Tab. 7）[15-18].

Tab. 6 Enhancing effect of extracts from some plants on the transformation of human lymphocytes by eb virus

Plant extract	Concentration(μg/ml)			
	0	0. 2	2	20
Stellaria chamaejasme	68	131(1. 9)	187(2. 8)	254(3. 7) *
Wikstroemia chamaedaphne	68	158(2. 3)	210(3. 1)	190(2. 9) *
Daphne genkwa	68	160(2. 4)	210(3. 1)	229(3. 4) *
Wikstroemia indica	68	111(1. 6)	200(2. 9)	280(4. 1) *
Sparganium stoloniferum	63	137(2. 2)	161(2. 6)	284(4. 5) *
Bulbophyllum inconspicuum	55	40(0. 7)	40(0. 7)	42(0. 7)
Abrus fruticulosus	55	44(0. 8)	42(0. 7)	48(0. 8)

Notes: Figures indicate number of colonies and（ ）ratio of colonies in experimental group to colonies in control group. Tumour Promother Control TPA（5 μg/ml）: 186 colonies *（$P < 0.05 \sim 0.01$）

Tab. 7 Survey of pathological changes in nasopharyngeal epithelium of rats

Survivors/tumors	Group		
	DNP * followed by WI * *	DNP alone	WI alone
Survivors for 180 – 205 days	23/32	14/20	15/20
Squamous metaplasia	19	10	10
Papillary hyperplasia	18	8	8
Papilloma	0	0	1
Dysplasia	4	0	1
Carinoma in situ	2	0	0
Early infiltrating carcinoma	4	0	0
Rate of carcinome（%）	26. 1	0	0

Notes: * Dinitrosopiperazine; * * Wikstroemia indica

We recently found that salted fish from southern China and harisa and qaddid from Tunisia contained EB virus inducers. These preserved foods are very frequently consumed in high risk areas for NPC (Tab. 8). Mutagens were also detected in the same food extracts[19], this being the first time that EB virus inducers and mutagens were found to be present in the same foods. Huang eg al[20] reported that salted fish contained appreciable quantities of nitrosodimethylamines which can induce tumours in rats. Case control studies conducted in Hong Kong showed that consumption of Cantonese – style salted fish was significantly associated with an increased risk for NPC[21]. Our data demonstrated that EB virus inducers can enhance lymphocyte transformation by the virus as well as promote the development of NPC in rats by DNP[15][16][17]. Together they may have a synergistic effect on the development of human naspharyngeal carcinoma.

Tab. 8 EBV – EA activators in aqueous extracts of
preserved food samples from southern China, Tunisia and Greenland

Country/food item	EBV – EA – positive cells(%)			
	Concentration of food extract(mg wet weight equivalent of food/ml medium)			
	0. 005 – 0. 13	0. 66	3. 32	16. 6
Southern China				
Cantonese salted, dried fish and shellfish:				
Japanese mackerel *	<5. 0	8. 4	10. 2	17. 8
Squid sample 1 *	<4. 0	5. 0	13. 6	8. 6
Squid sample 2 *	<3. 2	3. 4	8. 8	16. 0
Eight other Cantonese salted, dried fish and shllfish	<4. 0	<5. 1	<6. 0	<5. 2
Fermented shrip/fish paste	<4. 1	3. 0	3. 2	4. 8
Fermented soya bean paste	<2. 6	3. 0	3. 1	5. 1
Cabbage rermented in brine	<3. 5	3. 3	4. 2	4. 1
Tunisia				
Harissa(spice mixture):				
Sample 1 *	<3. 9	7. 7	13. 9	22. 6
Sample 2 *	<4. 9	10. 0	14. 9	19. 9
Sample 3 *	<4. 7	10. 9	12. 6	3. 2
Sample 4(preserved in can)	<3. 1	2. 6	3. 0	3. 3
Qaddid(dried mutton preserved in oil)	<2. 6	4. 2	7. 8	9. 6
Touklia(stewing base)	<5. 2	7. 2	6. 1	1. 8
Turnips fermented in brine	<4. 1	5. 0	5. 5	6. 4
Salted anchovies	<3. 0	2. 4	2. 8	2. 2
Louben(sap from the mastic tree)	<3. 1	3. 2	4. 0	4. 1
Greenland				
Mikialak(dried Atlantic cod)	<2. 4	3. 0	3. 2	2. 8
Uuvag(dried polar cod)	<2. 5	2. 2	2. 4	6. 1
Amassat(dried caplin)	<2. 4	2. 0	2. 1	4. 8
Panertep(dried fjord seal meat)	<2. 8	4. 2	6. 1	7. 2
Berries preserved in seal oil	<4. 4	4. 3	4. 2	4. 4

Note: * Samples considered to be positive

[In 《World Scientific》 1989, 92 – 102]

References

1　Henle G, Henle W. Epstein – Barr virus – specific IgA serum antibodies as an outstanding feature of nasopharyngeal carcinoma. Int J Cancer, 1976, 17: 1 – 7

2　Desgranges C, de – The G. IgA and nasopharyngeal carcinoma. In: Oncogenesis and Herpesviruses Ⅲ. Eds G de – The, Y Ito F. Rapp Lyon: IARC, (IARC Sciencific Pulication No. 25), 1978, 883 – 891

3　Zeng Y, Shang M. Detection of IgA and IgG antibodies to EBV in sera from nasopharyngeal carrinoma. Acta Microbiologic Sinica, 1978, 18: 253 – 258

4　Zeng Y, Liu Y C, Chen S, Wei J, Zhu J, Zai H. Application of immunoenzymatic method and immunoautoradiographic method for the mass survey of nasopharyngeal carcinoma. Intervirology, 1980, 13: 162 – 168

5　Zeng Y, Zhang J M, Li L Y, et al. Follow – up studies on Epstein – Barr virus IgA／VCA antibody – positive persons in Wangwu County, China. Intervirology, 1983, 20: 190 – 194

6　Zeng Y, Zhang L G, Li H Y, et al. Serological mass survey for early detection of nasopharyngeal carcinoma in Wuzhou city, China. Int J Cancer, 1982, 29: 139 – 141

7　Zeng Y, Zhang L G, Wu Y C, et al. Prospective studies on nasopharyngeal carcinoma in Epstein – Barr virus IgA／VCA antibody – positive persons in Wuzhou city, China. Int J Cancer, 1985, 36: 545 – 547

8　Liang J S. Serological screening of NPC in Zhangshen County. Symposium on early detection of NPC, Shaogun, Guangdong. April, 1988

9　Ito Y, Kawanishi M, Harayama T, Takabayashi S. Combined effect of the extracts from Croton tiglium, Euphorbia lathyris or Euphorbia tirucalli and n – butyrate on Epstein – Barr virus expression in human lymphoblestoid P3HR – 1 and Raji cells. Cancer Letter, 1981, 12: 175 – 180

10　Ito Y, Yanase S, Fujita J, Harayama T, Takashima M, and Imanaka H. A short – term in vitro assay for promoter substances using human lymphoblastoid cells latently infected with Epstein – Barr virus. Cancer Letters, 1981, 13: 29 – 37

11　Ito Y, Yanase S, Tokuda H, Kishishita M, Ohigashi H, Hirota M, Koshimizu K. Epstein – Barr virus activation by tung oil, extracts of Aleurites fordii and its diterpene ster 12 – 0 – hexadecanoyl – 16 – hydroxyphorbol – 13 – acetate. Cancer Letters, 1982, 18: 87 – 95

12　Zeng Y, Zhong T M, Mo Y K, et al. Epstein – Barr virus Early Antigen induction in Raji cells by Chinese medicinal herbs. Intervirology, 1983, 19: 201 – 204

13　Zeng Y, Miao X C, Jiao B. Epstein – Barr virus activation in Raji cells with ether extracts of soil from different areas in China. Cancer Letters, 1984, 23: 53 – 59

14　Hu Y L, Zeng Y. The extract from some Chinese herbs enhanced the transformation of human lymphocytes by EB virus. Chinese J Oncol, 1985, 1: 417 – 419

15　Hu Y L, Zeng Y. Wikstroemia chamaedaphne and Wisktromia indica enhanced the tumor growth by Rous sarcoma virus. (Unpublished data)

16　Hu Y L, Zeng Y. Wikstroemia chamaedaphne and Wikstroemia indica enhanced the tumor grwthe by methylcholanthrene in mice (Unpublished data)

17　Sun Y, Chen M H, Zeng Y, et al. Tumor – promoting effect of diterpene ester HHPA and extract of Wikstroemia chamaedaphne on cervical cancer in mice. Chinese Journal of Virology, 1987, 3: 131 – 133

18　Tang W P, Huang P G, Zhao M L, Liao S L, Zeng Y. Wikstroemia indica promotes development of nasopharyngeal carcinoma in rats initiated by dinitrosopiperazine. J of Cancer Research and Clinical Oncology , 1988, 114: 429 – 431

19　Shao Y M, Poirier S, Ohshima H, Malaveille C, Zeng Y, de The G, and Bartsch H. Epstein – Barr virus activation in Raji cells by extract of preserved food from NPC high – risk areas. Carcinogenesis, 1988, 9: 1455 – 1457

20　Huang D P, Ho J H C, Saw D, Tech T B. In: Naspharyngeal carcinoma. Etiology and control. Eds. de The, G and Ito, Y. International Agency for Research on Cancer, 1978, 315 – 318

21　Yu U C, Ho J H C, Lai S H, Henderson B E. Intake of Cantonese style salted fish as a cause of nasopharyngeal carcinoma: report of a case – control study in Hong Kong. Cancer Res, 1986, 46: 956 – 961

48.　Volatile Nitrosamine Levels and Genotoxicity of Food Samples from High – Risk Areas for Nasopharyngeal Carcinoma before and after Nitrosation

PLIRIER S[1,2], BOUVIER G[1,2], MALAVEILLE C[2], OHSHIMA H[2], SHAO Y M[2,3], HUBERT A[1], ZENG Y[3], de The G, BARTSCH H[2,4]

1. CNRS Laboratory of Epidemiology and Immunovirology of Tumors, Faculty of Medicine A. Carrel, 66373 Lyon Cedex 08; 2. International Agency for Research on Cancer, Unit of Environmental Carcinogens and Host Factors, 150 Cours Albert Thomas, 69372 Lyon Cedex 08, France; 3. Present address: Institute of Virology, Chinese Academy of Preventive Medicine, Beijing 100052, People's Republic of China; 4. To whom erprint requests should be sent

Summary

Traditional life – style, especially food habits, infection by Epstein – Barr virus (EBV) and genetic factors, have been associated with an increased risk of nasopharyngeal carcinoma (NPC). N – Nitroso compounds and other carcinogens either present in food or formed endogenously, as well as food constituents that acctivate EBV, have been suspected as etiological factors in NPC pathogenesis. For their characterization preserved food items, frequently consumed in NPC endemic areas in Tunisia, South China and Greenland, were sampled and screened for the presence of mutagens and volatile nitrosamines before and after nitrosation. Aqueous extracts as well as 2 organic extracts of the samples were assayed for genotoxicity in 2 *Salmonella typhimurium* strains and the SOS chromotest. The same extracts had previously been analyzed for volatile nitrosamines and for EBV – acctivating substances in Raji cells. In our study, 13 out of 16 food samples showed a weak, directly – acting genotoxicity in the SOS chromotest in at least one of the extracts, but only one sample from Greenland was found to be weakly mutagenic *in Salmonella* TA 98. Chemical nitrosation for 9 out of 15 samples of aqueous food extracts increased the genotoxic effect in the SOS chromotest. Levels of volatile nitrosamines were also elevated for 12 out of 15 samples; highest levels of N – nitrosodimetyhlamine were found in hard salted and dried fish from China (1200 μg/kg) and highest N – nitrosopyrrolidine levels in a Tunisian spice (3840 μg/kg). In non – nitrosated aqueous food extracts, the level of volatile nirosamines and genotoxic activities were not correlated with the EBV – inducing activity of the same samples. After chemical nitrosation, EBV – inducing activity was decreased or showed no change and was not correlated with increases in either the genotoxicity or the nitrosamine levels. Our results suggest that EBV – activating compounds belong to a different class of substances. However, there was an association between the changes in genotoxicity and nitrosamine levels due to nitrosation.

Introduction

Nasopharyngeal carcinoma (NPC) exhibits wide variations in incidence throughout the world, and is most common among Chinese in south – east Asia, Maghrebian Arabs in North Africa and Eskimos in the Arctic (de Thé et al. , 1982). Epstein – Barr virus (EBV) has been associated with NPC, because the EBV genome and Epstein – Barr nuclear antigen are regularly present in all epithelial tumor cells from NPC patients, taken from different geographical areas (Wolf et al. , 1973; Klein et al. , 1974; Desgranges et al. , 1975). In view of the restricted geographical distribution of NPC and the ubiquity of EBV prevalence, other genetic and environmental factors must be considered in NPC etiology (de Thé, 1982; de Thé and Zeng, 1986). Ho (1972) first postulated that traditional life – style, for example, consumption of Cantonese – style salted fish that may contain chemical carcinogens, is related to the development of NPC. Since then, several case – control studies conducted among southern Chinese (Armstrong et al. , 1983; Geser et al. , 1978; Henderson et al. , 1978; Yu et al. , 1985, 1986, 1988, 1989) have indicated that there is an association between consumption of Cantonese – style salted fish and the risk of developing NPC. In addition, Chinese who migrate from South China to low – risk or intermediate – risk areas for NPC, such as the USA, Canada and Singapore, continue to show a high rate of NPC (Armstrong et al. , 1974; Muir and Shanmugaratnam, 1967; Buell, 1974; Gallagher and Elwood, 1979). However, their descendants, who became adapted to a new life – style, are at lower risk of NPC (Zippin et al. , 1962; King and Haenszel, 1973; Buell, 1974). Several studies (Fong and Chan, 1973, 1977; Huang et al. , 1977, 1978, 1981; Tannenbaum et al. , 1985; Poirier et al. , 1987) have shown that some samples of Cantonese – style salted fish contained relatively high levels of volatile nitrosamines, all of which induce tumors in the nasal cavities of experimental animals (Druckrey et al. , 1967; Magee et al. , 1976).

A recent anthtopological study conducted in 3 high – risk areas for NPC (Tunisia, South China and Greenland) revealed that the widely different populations all commonly consume preserved food (Poirier et al. , 1987; Hubert and Robert-Lamblin, 1988). In order to study whether chemical carcinogens are present in these preserved food items, representative samples were collected in Tunisia, South China and Greenland. N – Nitrosodimethylamine (NDMA) was previously found at concentrations up to 113 μg/kg in several preserved fish samples from China and Greenland. ND-MA, N – nitrosopiperdine (NPIP) and N – nitrosopyrrolidine (NPYR) were also detected at lower concentrations in Tunisian stewing base (touklia), dried mutton preserved in olive oil (qaddid) and in several vegetables preserved in brine collected in Tunisia and South China (Poirier et al., 1978).

Since then, an epidemiological study conducted in Guangxi, South China, has shown that consumption of a number of preserved food before the age of 2, such as salted fish, salted mustard green and chung choi, a kind of salted root, is strongly associated with an increased risk of NPC (Yu et al., 1988). However, Japanese dried fish and vegetables in brine are also reported to contain relatively high concentrations of volatile nitrosamines (Kawabata et al., 1979), but the incidence of NPC in Japan is very low. In order to search for additional risk factors for NPC, we have further analyzed the food extracts from high – risk areas for the presence of substances that activate EBV *in vitro* (Shao et al., 1988) and for mutagens (this study). We found that aqueous extracts of Cantonese salted and dried fish, *harissa* (a Tunisian spice mixture), and *qaddid* (dried mutton preserved in oil) can induce EBV early antigen in latently infected Raji cells (Shao et al., 1988). IgA antibodies against EBV viral capsid antigen and early antigens reflecting an *in vivo* EBV activation are regularly present in NPC patients (Henle and Henle, 1976; Desgranges and de Thé, 1978) and appear in subjects before tumor development (Zeng et al., 1979, 1980; de Thé and Zeng, 1986).

In this study, we have screened extracts of the same food samples for the presence of mutagens. Since N – nitroso compounds could be formed endogenously from ingested foods and nitrated or nitrite, these aqueous food extracts were also examined, after acid – catalysed nitrosation *in vitro*, for their genotoxicity and levels of volatile nitrosamines. As an enhancing effect of chemical carcinogens such as N – methyl – N' – nitro – N – nitrosoguanidine on EBV replication and EBV – induced transformation has been reported (Henderson, 1988), we also examined whether the levels of genotoxicity and volatile nitrosamines are associated with the level of EBV – inducing activity in aqueous food extracts, before and after nitrosation.

Material and Methods

1. Food items

The most common types of preserved foods were collected from families or at local markets in various towns in Tunisia, in Macao and Guangxi (Wuzhou) in the PR China and on the west coast of Greenland. On the basis of frequency of consumption and method of preservation used, 16 representative food specimens were selected from among those analyzed previously for volatile nitrosamines (Poirier et al., 1987). Twelve out of 16 samples were consumed more than 3 times a week and the rest once or twice a week. Sample 16 was available in limited amounts and was investigated only for genotoxicity before nitrosation (Tab. 1).

Tab. 1 Genotoxicity of aqueous, n－hexang and ethyl acetate food extracts as determined in sos chromotest

Sample number	Country and origin of food sample	Induction factor[1]												Overall evaluation of the genotoxicity[3]
		Aqueous extract				n－Hexane extract				Ethyl acetate extract				
		1.8^2	5.4	1.8	54	6.7^2	20.7	67.3	202	6.7^2	20.2	67.3	202	
	China													
1	Hard salted and dried fish (grouper)	0.9	1.0	1.1	1.2	1.2	1.3	1.5	1.7	1.2	1.3	1.2	1.4	+
2	Soft salted and dried fish (Japanese mackerel)	1.0	1.2	1.3	1.0	1.6	1.6	1.9	2.2	1.3	1.4	1.5	1.6	+
3	Fermented shrimp paste	1.0	1.1	1.5	Tox^4	0.8	1.0	1.1	1.2	1.3	1.4	1.9	1.8	+
4	Fermented soya bean paste	1.2	1.0	1.0	0.8	0.9	1.0	1.0	1.2	1.0	1.2	1.6	Tox	+
5	Cabbage termented in brine	0.9	1.1	0.9	1.0	1.2	1.1	1.1	1.0	1.0	1.1	1.0	0.9	+
	Tunisia													
6	Spice mixture(harissa)	1.1	1.0	0.9	1.1	0.9	0.9	0.8	1.2	1.3	1.4	1.5	1.7	+
7	Dried mutton preserved in oil (qaddid)	1.0	1.0	0.9	0.9	1.0	1.2	1.6	1.7	1.3	1.3	1.4	1.4	+
8	Stewing base(toukia)	0.9	0.9	1.0	1.0	1.3	1.5	1.5	1.7	1.2	1.2	1.4	1.7	+
9	Turnips fermented in brine	1.0	0.9	0.9	0.9	0.9	1.0	0.9	0.9	0.9	0.8	0.7	0.9	－
10	Salted anchovies	1.0	1.0	1.1	1.5	1.4	1.7	.2	2.5	1.1	1.2	1.3	2.1	+
11	Sap from the mastic tree (louben)	1.0	1.0	1.0	1.0	1.1	1.6	2.0	2.0	1.1	1.3	1.7	2.5	+
	Greenland													
12	Dried atlantic cod(mikialak)	1.1	1.1	1.1	1.2	1.1	1.1	1.4	1.6	1.1	1.3	1.3	1.3	+
13	Dried polar cod(uuvaq)	1.0	1.0	1.1	1.6	0.9	0.9	1.0	1.3	1.0	1.0	1.0	1.0	+
14	Dried capelin(amassat)	1.1	1.2	1.2	1.3	1.0	1.1	1.4	1.6	0.8	1.1	1.4	1.5	+
15	Dried fiord seal meat (panerrteq)	1.1	1.1	1.4	2.0	1.0	1.2	1.5	Tox	1.1	1.4	1.6	2.1	+
16	Berries preserved in seal oil	1.1	1.1	1.1	1.2	0.9	1.1	1.1	0.9	0.9	0.9	1.0	1.1	－

Notes: [1] Mean values from 2 or 3 series of duplicate experiments. —[2] Concentrations mg wet weight equivalent of food/ml assay medium. —[3] Assigned positive when at least one of the extracts was genotoxic. —[4] Tox: less than 30% viable bacteria.

2. Preparation of the extracts

Aqueous, n－hexane and ethyl acetate extracts of these specimens were prepared as follows: 10 g of each food sample, uncooked, were homogenized in 20 － 50 ml of de－ionized water in a Polytron homogenizer for 3 min. After shaking for 30 min at 4℃, the homogenate was centrifuged at 15 000 g for 20 min, and the pellet was re－extracted with the same volume of water. The combined supernatants were filtered through a 0.45 μm Millipore filter and lyophilized. The lyophilisate was then dissolved in 6 ml of water. The residual material, after aqueous extraction, was further extracted in a Soxhlet apparatus, first with 100 ml n－hexane, and then with 100 ml ethly acetate (approx. 25 passages for each solvent). The solvents were evaporated separately using a rotary evaporator, and each of the residues was dissolved in 1.6 ml dimethylsulphoxide (DMSO). The aqueous and DMSO solutions were stored at －80 ℃ for less than 3 months, prior to analysis.

3. Nitrosation of the aqueous extracts

A 3－ml aliquot of aqueous extract, equivalent to a 5－g food sample, was adjusted to pH 1.5 with HCl and centrifuged at 2 000 g for 20 min at 4℃. The volume of the supernatant was made up to 4 ml with HCl/KCl buffer 0.2mol/L pH 1.5. To 3 ml of this solution were added 12 μl of

6. 5 mol/L sodium nitrite solution [or 3 μl in the case of fermented shrimp paste (*harma ha*) fermented soya bean paste, *touklia* (Tunisian stewing base) and berries in seal oil] and the pH was readjusted to 1. 5 (the final concentrations of nitrite were 26 mmol/L and 6. 5 mmol/L, respectively). The nitrite concentration was chosen in such a way that the induction factor increased linearly with time during a 20 – min nitrosation reaction. For genotoxicity assays, 1 – or 1. 5 – ml aliquots were removed at time 0 and after 20 min of incubation in a shaking water bath at 37℃ in the dark; residual nitrite was destroyed by adding ammonium sulfamate at a final concentration of 94 mmol/L of 23. 5 mmol/L, then after 3 min, the pH was adjusted to 7. Control assays without sodium nitrite were also carried out with 1 – ml aliquots of the aqueous food extract and incubated for 20 min at 37℃. The solutions to be tested in the SOS chromotest were immediately added to a buffered medium (pH 7. 4) containing bacteria. The stability of the genotoxic agents at pH 7. 0 was tested by incubating an aliquot of this solution at 37℃ in the dark for 30 min before genotoxicity testing.

4. Metabolic activation system

Adult male BD – VI rats received a single injection of Aroclor 1254 (500 mg/kg body weight) 5 days before being killed. A 9000 – g liver supernatant (S9) was prepared as described by Malaveille et al. (1982) and stored at – 80℃.

5. Salmonella mutagenicity assay

S. tyHIMURIUM TA 98 and TA 100 strains were provided by Dr. B. N. Ames (Berkeley, CA). Cultues (5 ml) were grown from frozen stocks (– 70℃) for 10 hr in Oxoid medium 2 at 37℃ with shaking. The presence of the R – factor was checked by seeding bacteria on agar that contained an ampicillin disk (10μg). The cultures were checked for mutabillity using methly methanesulfonate (MMS), 2 – nitrofluorene (NF) and benzo [a] pyrene (BaP) in the presence of an activation system. In the plate incorporation assay (Maron and Ames, 1983), 0. 2 μl MMS per plate induced 435 to 534 revertants (TA 100), 2. 5 μg NF per plate induced 302 – 436 revertants (TA 98) and 5 μmol/L BaP in the presence of 50 μl S9 per plate induced 660 – 799 revertants (TA 98).

The assay procedure described by Malaveille et al. (1982) was used with some modifications. Each assay consisted of 100 μl of 3. 3 – fold concentrated bacterial culture medium (3 to 6 × 10^8 cells), 100 μl of KCl mix or S9 mix (see below) and 10 μl of DMSO solution of a food extract. Incubation was carried out for 90 min at 37℃ with shaking. After addition of 2 ml of histidine – poor soft agar, the mixture was plated onto minimal glucose agar (Maron and Ames, 1983). One milliliter of S9 mix contained 100 μl 80 mmol/L $MgCl_2$, 100 μl 8 mmol/L $NADP^+$, 100 μl 50 mmol/L glucose 6 – phosphate, 400 μl 0. 25 mol/L. Sorensen phosphate buffer (SPB), pH 7. 4 10 μl or 50μl S9; the final volume was made up with 0. 15mol/L KCl – 5mmol/L SPB pH 7. 4; 1 ml of the KCl mix (without S9) consisited of 100 μl 80 mmol/L $MgCl_2$, 200 μl 0. 9% saline in 5mmol/L SPB pH 7. 4, 400 μl 0. 25 mol/L SPB, pH 7. 4 and 300 μl 0. 15 mol/L KCl – 5 mmol/L Sorensen phosphate buffer pH 7. 4.

6. SOS chromotest

Escherichia coli PQ 37 was provided by Dr. M. Hofnung, Institute Pasteur, Paris, France.

Cultures (5 ml) were grown overnight from frozen stocks (− 70℃) in L medium at 37℃ with shaking (Quillardet and Hofnung, 1985). Concurrent positive controls accompanied each experiment: 80 nmol MMS per assay increased the induction factor 2. 3 − to 4. 4 − told. In the presence of a metabolic activation system (5 μl S9 per ml S9 mix), 1 nmol of BaP increased the induction factor 3. 4 − to 6. 6 − fold.

The genotoxicity of the aqueous (before nitrosation) -hexane and ethyl acetate extracts of the food samples was measured using the SOS chromotest according to the method of Quillardet and Hofnung (1985) with minor modifications. Aqueous extacts were tested in the presence of 0%, 5% and 20% in S9 mix and organic extracts with 0%, 1% and 5% S9. Twelve milliliters of the S9 mix consisted of 200 μl salt solution (1. 65 mol/L KCl and 0. 4 mol/L $MgCl_2$), 50μl 1mol/L glucose 6 − phosphate, 400 μl 45 mmol/L $NADP^+$, 2500 μl 0. 4 mol/L Tris − HCl buffer pH7. 4, 5850 μl L.− medium, and contained 1%, 5% or 20% of S9; the final volume was adjusted with 0. 15 mol/L KCl − 5 mmol/L SPB, pH 7. 4. Buffer for β − galactosidase assay (B buffer), buffer for alkaline phosphatase assay (P buffer), o − nitrophenyl − β − D − galactopyranoside solution and p − nitrophenyl phosphate disodium solutions were prepared according to Quillardet and Hofnung (1985).

7. Assay procedure

Log − phase culture *of E. coli* PQ 37 (2 × 10^8 bacteria/ml), prepared from an overnight culture, was diluted 10 − fold with either L − emdium or S9 mix for assays with metabolic activation. Aliquots (300 μl) were distributed into a series of disposable plastic tubes containing 10 μl aqueous or DMSO solutions of the test sample. In order to avoid artefacts due to the presence of β − galactosidase or alkaline phosphatase in a queous food extracts, aliquots were also tested in the absence of bacteria. The absorbance obtained without bacteria (L medium only) was subracted from that obtained with bacteria. After 2 hr incubation at 37℃ with shaking, 150 − μl aliquots were removed from each tube and placed into a new series of tubes. Then, 1. 35 ml of B buffer or P buffer were added in the series of tubes for β − galactosidase and alkaline phosphatase assays, respectively. The tubes were shaken vigorously for 5 sec, then 250 − μl aliquots per tube were distributed into a 96 sterile wells on a microtiter tray. Fifty microliters per well of ONPG or PNPP solutions were added for β − galactosidase and alkaline phosphatase assays, respectively. The optical absorbance at 405 nm of each well was measured just after addition of enzyme substrate using an ELISA reader (Titertek R Multiscan MC, Flow, Irvine, Scotland). The microtiter trays were then placed in a ventilated oven at 37℃. After 5 min of temperature equilibration, microtiter trays for β − galactosidase assays and those of alkaline phosphatase assays were further incubated for 2 hr and up to 45 min, respectively. The absorbance at 405 nm of each well was read and enzyme activities were determined from the difference in absorbance and used to calculate the induction factor (Quillardet and Honfnung, 1985). This double reading of the absorbance avoids interference of the test material with the determination of enzyme activities.

For nitrosated aqueous extracts, the SOS chromotest was carried out as follows: 50 μl of the aqueous extracts were incubated with 50 μl log − phase culture of *E. coli* PQ 37 (2 × 10^8 bacteria/ml) and 50 μl 0. 4 mol/L Tris − HCl buffer pH 7. 4. After 1 hr of incubation in a shaking water bath in the dark at 37℃, 0.8 ml of L – medium were added and incubation was continued for a further 2 hr under the same conditions. Enzyme activities were measured using 100 – μl aliquots as described above. Preliminary experiments had shown that the test material contained no β – galactosidase or alkaline phosphatase and therefore no control assays without bacteria were performed.

The induction factor was calculated when ≥ 30% viable bacteria were present after incubation. When the percentage of bacterial survivors was lower, toxicity is mentioned in "Results" (Tab. Ⅰ and Ⅱ). According to the criteria of Quillardet and Hofnung (1985), only samples with induction factors above 1.5, showing a dose – response relationship, were considered as genotoxic. The SOS chromotest for each sample was carried out in duplicate.

8. Analysis of volatile nitrosamines

in vitro nitrosation of aqueous food samples was carried out as above and the reaction mixture was made up to 5 ml with distilled water. N – Nitrosomethylamine (100 ng) was added as an internal standard and the solution was extracted 3 times with 10 ml dichloromethane in the presence of 1.0 g NaCl. The combined dichloromethane extracts, to which 0.2 ml n – hexane was added, were dried over anhydrous $NaSO_4$ and concentrated to 0.5 ml in a Kuderna Danish evaporator. A 10 – μl aliquot was used for analysis of volatile nitrosamines by gas chromatography combined with a thermal enetgy analyzer (model 502 Thermoelectron, Waltham, MA) as described by Poirier et al. (1987).

Results and Discussion

The mutagenic activity of n – hexane and ethyl actante extracts from each of the 16 preserved foods collected from NPC endemic areas was measured in *S. tyhimurium* strains TA 98 and TA 100 in the absence and the presence of rat liver metabolic activation system. Concentrations tested ranged from 75 to 300 mg wet weight equivalent of food per ml of assay; higher concentrations could not be tested due to the limit of sllubility in DMSO. Only ethyl acetate extract of berries preserved in seal oil from Greenland (at 300 mg wel weight equivalent of food/ml assay medium in the presence of 5% S9 in the S9 mix) doubled the number of revertants above the spontaneous mutation level from TA 98 strain. Aqueous food extracts were not tested in *S. tyhimurium* strains because the presence of histidine leads to artefacts in the *Salmonella* assay. The results of SOS chromotest for the aqueous, n – hexane and ethyl acetate extracts of the 16 selected food samples are presented in Tab. Ⅰ. The induction factor, an index of the induction of the SOS DNA repair functions in *E. coli*, was greater than 1.5 in 13 out of 16 food samples, the highest value observed being 2.5 (samples 10 and 11). Although most of the tested samples were cytotoxic, no association was observed between induction factor of the samples and their toxicity. Addition of a rat liver metabolic activation system (up to 9.5% of S9) decreased the induction factor (data not shown). Thus, the results in Tab. Ⅰ revealed the presence of directly – acting genotoxic substances in many of the food items which, in general, were present mostly in the organic extracts.

In contrast with data, Fong et al. (1979) have reported that DMSO extracts of 2 samples of dried shrimps and 4 samples of different species of salted fish were mutagenic in *S. typhimurium* TA

100 and TA 98 in the presence of a metabolic activation system. This discrepancy may be due to the difference in extraction methods and/or the variability in the level of mutagens from one sample of a particular item to another. In view of the limited number of food items analyzed in our study, more fish samples and other Chinese preserved foods, reported by Yu et al. (1988) to be associated with an increased risk of NPC, should now be screened for the presence of genotoxic substances.

The endogenous formation of N – nitroso compounds in humans following ingestion of nitrate and amino precursors (proline) has been demonstrated by Ohshima and Bartsch (1981) and Bartsch et al. (1989). In order to determine the presence of nitrosamine precursors in the preserved food extract, the genotoxicity of the aqueous extracts was measured after acidcatalyxed nitrosation *in vitro* (Tab. 2). Nine out of 15 food samples were found to contain precursors that upon nitrosation yielded directly – acting genotoxic substances. Thus, nitrosation led to formation of genotoxic compounds not only in fish and meat, known to contain high concentrations of secondary and tertiary amnines, but also in vegetables fermented in brine as well as in spice mixtures from Tunisia. These preserved foods thus contain precursors that after nitrosation yield directlyacting, genotoxic (probable N – nitroso) compounds. We found that the genotoxic activity of the compounds formed after nitrosation was reduced after incubation for 30 min at 37℃, but was still detectable at 40% – 90% of its initial value (Tab. 2). It is therefore conceivable that these DNA – damaging compounds could be formed intragastrically form dietary precursors and could not only react in the digestive tract but, because of their relative stability, also reach distal targets such as the nasopharynx via the blood stream. Tab. 3 lists the levels of volatile nitrosamines detected before and after *in vitro* nitrosation of aqueous extracts of food samples. Before nitrite treatment only a trace amount of NDMA was detected in most samples, except for some fish samples, such as grouper from China, anchovies from Tunisia and *uuvaq* (dried polar cod) from Greenland and one sample of dried seal meat which were found to contain NDMA at concentration up to 388 μg/kg; NPYR was present in 7 out of 15 samples at concentrations up to 196 μg/kg. After nitrosation, volatile nitrosaminew were detected in all samples, in quantities up to 3856 μg/kg (sample 6). Seven out of 15 samples had an increased NDMA content and the highest level (1191 μg/kg) was observed for hard salted and dried fish (sample 1) from China. Similarly, NPYR content was increased in 12 samples with the highest level (3856 μg/kg) being in a spice mixture (sample 6) from Tunisia. Thus for 8 samples, levels of total volatile nitrosamines increased 2 – (sample 5) to 33 – fold (sample 9). In addition, an unidentified thermal energy analyzer – responsive peak was observed at a retention time after a peak of nitrosomorpholine (sample 1, 3, 4, and 14, Tab. 3) showing the presence of a precursor of an unknown nitroso compound. These results suggest that high concentrations of nitrosatable precursors for nitrosamines are present in preserved foods, especially in Japanese mackerel, cabbage fermented in brine, *uuvaq* and *amassat* (dried capelin from Greenland), and that carcinogenic volatile nitrosamines and other genotoxic substances could be formed endogenously in the human stomach after ingestion of such preserved foods.

Tab. 2　Cenotoxicity of aqueous food extracts due to nitrosation

Sample number	Country and origin of food sample	Induction factor[1]				Residual genotoxicity after 30 min at 37℃　(%)
		49[2]	98	196	392	
	China					
1	Hard salted and dried fish　(grouper)	1. 1	1. 4	1. 7	3. 8	71
2	Soft salted and dried fish　(Japanese mackerel)	0. 9	1. 3	1. 7	2. 3	4. 4
3	Fermented shrimp paste[3]	1. 3	1. 2	1. 5	Tox[4]	—[5]
4	Fermented soya bean paste[3]	1. 0	0. 7	1. 3	Tox	—[5]
5	Cabbage fermented in brine	1. 6	1. 7	3. 2	Tox	78
	Tunisia					
6	Spice mixture　(harissa)	0. 9	1. 4	3. 9	Tox	72
7	Dried mutton preserved in oil　(qaddid)	1. 3	1. 3	1. 7	2. 0	90
8	Stewing base　(touklia)[3]	0. 9	1. 0	1. 1	1. 8	83
9	Turnips fermented in brine	1. 1	1. 2	1. 6	3. 5	54
10	Salted anchovies	0. 9	0. 6	1. 1	Tox	—[5]
11	Sap from the mastic tree　(louben)	0. 9	0. 7	0. 9	1. 5	—[5]
	Greenland					
12	Dried atlantic cod　(mikialak)	1. 1	1. 0	1. 2	1. 1	—[5]
13	Dried polar cod　(uuvaq)	0. 9	1. 0	1. 3	1. 7	82
14	Dried capelin　(amassat)	1. 3	1. 2	1. 4	Tox	—[5]
15	Dried fiord seal meat　(panerteq)	1. 1	1. 2	1. 2	1. 7	82

Notes: [1] Values listed were calculated from the difference between induction factors measured at time 0 and 20 min plus1. Mean values from 1 – 3 series of duplicate experiments. —[2] Food extract in mg weight equivalent of food per ml assay medium. —[3] Nitrosation performed with 6. 5 mmol/L nitrite instead of 26 mmol/L nitrite. —[4] Tox: less than 30% surviving bacteria. —[5] Non – genotoxic compounds

Tab. 3　Levels of volatile nitrosamines in food samples before　(–)
and after　(+)　*in vitro* nitrosation of the aqueous food extracts

Sample number	Country and origin of food sample	Volatile nitrosamines　(VNA)　(μg/kg)							
		NDMA		NPIP		NPYR		Sum of VNA	
		–	+	–	+	–	+	–	+
	China								
1	Hard salted and dried fish　(grouper)[1]	388	1191	81	Traces	30	98	523	1289
2	Soft salted and dried fish　(Japanese mackere)	ND	377	ND	ND	ND	20	ND	397
3	Fermented shrimp paste[1,2]	Traces	Traces	ND	ND	97	48	105	48
4	Fermented soya bean paste[1,2]	ND	ND	ND	ND	ND	205	ND	205
5	Cabbage fermented in brine	ND	32	ND	66	62	95	85	193
	Tunisia								
6	Spice mixture　(harissa)	Traces	Traces	ND	ND	196	3856	196	3856
7	Dried mutton preserved in oil　(qaddid)	13	Traces	ND	ND	ND	326	13	326
8	Stewing base　(touklia)[2]	5	Traces	ND	ND	ND	35	5	35
9	Turnips fermented in brine	ND	ND	ND	ND	31	2024	62	2024
10	Salted anchovies	299	43	ND	ND	ND	ND	299	47
11	Sap from be mastic tree　(louben)	12	Traces	ND	ND	ND	10	12	10
	Greenland								
12	Dried atlantic cod　(mikiatak)	Traces	764	ND	ND	ND	423	ND	1187
13	Dried polar cod　(uuvaq)	17	324	ND	ND	ND	86	17. 4	410
14	Dried capelin　(amassat)	ND	277	ND	ND	24	316	24	593
15	Dried fiord seal meat	28	52	50	96	103	ND	190	118

Notes: [1] Contains unknown thermal energy analyzer – responsive peak at a retention time of 13. 3min. —[2] Nitrosation performed with 6. 5 mmol/L nitrite instead of 26 μmol/L nitrite. ND, not detected.

Since strong inducers of EBV early antigens in Baji cells have been found in aqueous extracts of soft Cantonese salted and dried fish, dried squid from China, *harissa* from Tunisia and to a lesser extent in *qaddid* from Tunisia (Shao et al. , 1988), we have examined whether such activities are associated with genotoxicity and volatile nitrosammine levels before and after chemical nitrosation (Tab. 4). EBV – inducing activity did not parallel either the genotoxicity or the levels of volatile nitrosamines in non – nitrosated samples analyzed in our study. As after chemical nitrosation of the aqueous food extracts, EBV – inducing activity was unchanged or decreased, also no association was observed between the changes in genotoxicity, EBV – inducing activities and levels of volatile nitrosamines. However, there appeared to be a moderate association between the increases in genotoxicity and levels of volatile nitrosamines after chemical nitrosation in 7 out of 10 samples (5 samples which were non – genotoxic before and after nitrosation were excluded from comparison).

Tab. 4 Comparison between genotoxicity, levels of volatile nitrosamines and Epstein – Barr virus early – antigen – inducing activity of food samples (aqueous extracts) before and after chemical nitrosation. data are compiled from this study (tables 1, 3) and from shao et al. (1988)

Sample number	Country and origin of food sample	Genotoxicity		Total volatile nitrosamines		EBV early – antigen [1] inducing activity	
		Before[2]	After[3]	Before[2]	After[3]	Before[2]	After[3]
	China						
1	Hard salted and dried fish (grouper)	–	↑	+	↑	–	=
2	Soft salted and dried fish (Japanese mackerel)	–	↑	–	↑	+	↓
3	Fermented shrimp paste	–	=	+	↓	–	=
4	Fermented soya – bean paste	–	=	–	↑	–	=
5	Cabbage fermented in brine	–	↑	+	↑	–	=
	Tunisis						
6	Spice mixture (*harissa*)	–	↑	+	↑	+	↓
7	Dried mutton preserved in oil (*qaddid*)	–	↑	+ / –	↑	+	↓
8	Stewing base (*touklia*)	–	↑	+ / –	↑	–	=
9	Turnips fermented in brine	–	↑	+	↑	–	=
10	Salted anchovies	–	=	+	↓	–	=
11	Sap from the mastic tree (*louben*)	–	=	+ / –	=	–	=
	Greenland						
12	Dried atlantic cod (*mikialak*)	–	=	–	↑	–	=
13	Dried polar cod (*uuvaq*)	+	=	+ / –	↑	–	=
14	Dried capelin (*amassat*)	+	↓	+	↑	–	=
15	Dried fiord seal meat (*panerteq*)	+	=	+	↓	–	=

Notes:[1] At variance with " Material and Methods" section, the nitrosation was performed in the presence of 50 mmol/L $NaNO_2$, pH 3.37℃ for 1 hr. —[2] Activities and levels of volatile nitrosamines before nitrosation. —[3] ↑, = or ↓: increased, unchanged or decreased biological activities or levels of volatile nitrosamines after nitrosation. —[4] Levels of volatile nitrosamines: —not detected; + / – ≤20 μg/kg; + >20μg/kg.

In conclusion, our studies have identified certain preserved food items that contain either high levels of precursors, that upon nitrosation yield volatile nitrosamines and mutagens, or substances with EBV – inducing activities. The relevance of these food items in NPC etiology should now be explored by case control studies in the high risk areas.

Acknowledgements

The work reported in this article was undertaken during the tenure of fellowships rewarded to Ms. S. Poirier by the Ligue Nationale Francaise Contre le Cancer and the Fondation pour la Recherche Médicale, and to Mr. G. Bouvier by the Ligue National Francaise Conter le Cancer. Partial support was provided by the Association pour la Recherche sur le Cancer. contract 6071. The authors are grateful to Dr. J. Cheney for editorial help and to Mrs. E. Bayle for secretarial assistance.

〔In 《Int J Cancer》 1989, 44: 1088 – 1094〕

References

1 Armstrong R W, Armstrong M J, Yu M C and Henderson B E, Salted fish and inhalants as risk factors for nasopharyngeal carcinoma in Malaysian Chinese. Cancer Res. , 1983, 43: 2967 – 2970

2 Armstrong R W, Kannan K M, and Dharmalingam S K, Incidence of nasopharyngeal carcinoma in Malaysia, with special reference to the state of Selangor. Brit J CANCER, 1974, 30: 86 – 94

3 Bartsch H, Ohshima H, Pignatelli B, Calmels S, Human exposure to endogenous N – nitroso compounds: quantitative estimates in subjects at high risk for cancer of the oral cavity, oesophagus, stomach and urinary bladder. Cancer Sury, 1989, 8 (In press)

4 Buell P, The effect of migration on the risk of nasopharyngeal cancer among Chinese. Cancer Res, 1974, 34: 1189 – 1191

5 Desgranges C, de The G. IgA and nasopharyngeal carcinoma. In: G. de Thé, Y. Ito and F. Rapp (eds.), Oncogenesis and herpes – viruses, Ⅲ, (IARC Scientific Publications 24), IARC, Lyon, 1978: 883 – 891

6 Desgranges C, Wolf H, de Thé G, Shanmugaratnam K, EllouzR, Cammoun M, Klein G, zur Hausen H. Nasopharyngeal carcinoma. X. Presence of Epstein – Barr genomes in epithelial cells of tumours from high and medium risk areas. Int J Cancer, 1975, 16: 7 – 15

7 De Thé G. Epidemiology of the Epstein – Barr virus and associated diseases. In: B. Roizman (ed), The herpes viruses, Vol. 1A, Plenum, New Yorkm, 1982, 25 – 103

8 De Thé G, Ho J H C, Muir C S. Nasopharyngeal carcinoma. In: A. S. Evans (ed.), Viral infections fo humans. Epidemiology and control, Wiley, New York, 1982, 621 – 652

9 De Thé G, Zeng Y, Population screening for EBV markers: toward improvement of masopharyngeal carcinoma control. In: M. A. Epstein and B. G. Achong (eds.), The Epstein – Barr virus. Recent advances, Heninemann, London, 1986: 236 – 249

10 Druckrey H, Preussmann R, Ivankovic S, ScmÄhl D. Organotrope carcinogene Wirkungen bei 65 verschi edenen N – nitroso – Verbindungen an BD – Ratt'en. Z Krebsforsch, 1967, 69: 103 – 201

11 Fong Y Y, Chan W C. Dimethylnitosamine in Chinese marine salt fish. Food Cosmet Toxicol, 1973, 11: 841 – 845

12 N – nitosopyttolidine in some Chinese food products. Food Cosmet Toxicol, 1977, 15: 143 – 145

13 Fong L Y Y, Ho J H C, Huang D P. Preserved foods as possible cancer hazards: WA rats fed salted fish have mutagenic urine. Int J Cancer, 1979, 23: 542 – 546

14 Gallagher R P, Elwood J M. Cancer motrality among Chinese, Japanese and Indians in British Columbia, 1964 – 73. Nat Cancer Inst Monogr, 1979, 53: 89 – 94

15 Geser A, Charnay N, Day N E, Ho J H C, de Thé G. Environmental factors in the etiology of nasopharyngeal carcinoma: report on a case – control study in the etiology of nasopharyngeal carcinoma: report on a case – control study in Hong Kong. In: G. de Thé, Y. Ito and W. Davis (eds.), Nasopharyngeal carcinoma: etiology

and control. IARC Scientific Publication 20, IARC, Lyon, 1978, 213 – 229

16　Henderson B E, Louie, E. Discussion of risk factors for naspharyngeal carcinoma. In: G. de Thé, Y. Ito and W. Davis (eds.), Nasopharygeal carcinoma: etiology and control. IARC Scientific Publications. IARC, Lyon. 1978, 20: 251 – 260

17　Henderson B E. Physicochemical – viral synergism during Epstein – Barr virus infection: A review. J nat Cancer Inst, 1988, 80: 476 – 483

18　Henle E, Henle, W. Epstein – Barr virus – specific IgA serum antibodies as an outstanding feature of nasopharyngeal carcinoma. Int J Cancer, 1976, 17: 1 – 7

19　Ho J H C. Nasopharyngeal carcinoma (NPC). In: G. Klein, S. Weinhouse and A. Haddow (eds.), Advanc Cancer Res, 1972, 15: 57 – 92

20　Huang D P, Ho J H C, Gough T A. Analysis for volatile nitrosamines in salt – preserved foodstuffs traditionally consumed by southern Chinese. In: G. de Thé, Y. Ito and W. Davis (eds.), Naspharyngeal carcinoma: etiology and control. IARC Scientific Pulications, IARC, Lyon, 1978, 20: 309 – 314

21　Huang D P, Ho J H C, Gough T A, Webb K S. Volatile nitrosamines in some traditional Chinese food products J Food Safety, 1977, 1: 1 – 6

22　Huang D P, Ho J H C, Webb K S, Wood B J, Gough T A. Volatile nitrosamines in salt – preserved fish before and after cooking. Food Cosmet. Toxical, 1981, 19: 167 – 171

23　Hubety H, Robert – Lamblin J. Apport de l'anthropelogie aux enquémiologiques: le cas du cancer du rhino – pharynx. Bult. Mem. Soc. Anthropol. Paris, 5, séris XIV, 1988, 35 – 46

24　Kawabata T, Ohshima H, Uibu J, Nakamura M, Matsui M, Hamano M. Occurence, formation and precursors of N nitroso compounds in Japanese diet. In: E. C. Miller, J. A. Miller, I. Hirone, T. Sugimura and S. Takayama

(edu.), Naturally occurring carcinogensmutagens and modulators of carcinogensis, pp. 195 – 209, Jniversity Park Press, Baltimore, 1979, 195 – 209

25　King H, Haenszel W. Cancer mortality among foreign – and native – born Chinese in the United States, J chron Dis, 1973, 26: 623 – 646

26　Klein G, Giovanella B C, Lindahl T, Fialkow P J, Singh S, Stehlin J S. Direct evidence for the presence of Epstein – Bary vinis DNA and nuclear antigen in maligant epithelial cells from patients with poorly differentiated carcinoma of the nasopharynx. Proc nat Acud Sol. (Wash.), 1974, 71: 4737 – 4741

27　Magee P N, Montesano R, Preussmann, R. N-nitroso compounds and related carcinogens. In: C. E. Searle (eds.), Chemical carcinogens (ACS Monogr, 173), pp. 491 – 625, American Chemical Society, Washington, DC, 1976: 491 – 625

28　Malaveille C, Brun G, Kolar G, Bartsoh, H., Mutagenic and alkylating activities of 3 – methyl – 1 – phenyltriazenes and their possible role as carcinogenic metabolites of the parent dimethyl compounds, Cancer Res, 1982, 42: 1446 – 1453

29　Maron D M, Ames B N, Revised methods for the Salmonelia mutagenicity test. Mutat Res, 1983, 113: 173 – 215

30　Muir C S, Shanmugaratnam K. The incidence of nasopharyngeal cancer in Singapore. In: C. S. Muir and K. Shanmugaratnam (eds.), Cancer of the nasopharynx. UICC Monogr. Ser., 1, Munksgaard, Copenhagen, 1967, 1: 47 – 53

31　Ohshima H, Bartsch H. Quantitative estimation of endogenous nitrosation in humans by monitoring N – nitrosoproline cxreted in the urine. Cancer Res, 1982, 41: 3658 – 3662

32　Poirier S, Ohshima H, de Thé G, Hubert A, Bourgade M C, Bartsch H, Volatile nitrosamine levels in common foods from Tunisia, South China and Greenland, hIgh – risk areas for nasopharyngeal carcinoma (NPC). Int J Cancer, 1987, 39: 293 – 296

33 Quillardet P, Hofnung M. The SOS chromotest, a colorimetric bacterial assay for genotoxins: procedures. Mutation Res, 1985, 147: 65 – 78

34 Shao Y M, Poirier S, Ohshima H, Malaveille C, Zeng Y, de Thé G, Bartsch H. Epstein – Barr virus activation in Raji cell by extracts of preserved food from high risk areas for nasopharyngeal carcinoma. Carinogenesis, 1988, 9: 1455 – 1457

35 Tannenbaum S R, Bishop W, Yu M C, Henderxon, B. E. Attempts to isolate N – nitroso compounds from Chinese – style salted fish. In: P. Greenwald (ed.), Fourth Symposium on Epidemiology and Cancer Registries in the Pacific Basin. Nat. Cancer Inst Monogr, 1985, 69: 209 – 211

36 Wolf H, zujr Hausen H., Becker V. EB – viral genomes in epithelial naspharyngeal carcinome cells. Nature (New Biol.), 1973, 244: 245 – 247

37 Yu M C, Ho J H C, Henderson B E, Armstrong R W. Epidemiology of nasopharyngeal carcinoma in Malaysia and Hong Kong. Fourth Symposium on Epidemiology and Cancer Registries in the Pacific Basin. Nat. Cancer Inst Monogr, 1985, 69: 203 – 207

38 Yu M C, Ho J H C, Lai S H, Henderson B E.

Cantonesestyle salted fish as a cause of nasopharyngeal carcinoma: report of a casecontrol study in Hong Kong. Cancer Res, 1986, 46: 956 – 961

39 Yu M C, Huang T B, Henderson B E. Diet, nassopharyngeal carcinoma: a case – control study in Guangzhou, China Int J Caccer, 1989, 43: 1077 – 1082

40 Yu M C, Mo C C, Chong W – Xi, Yeh F S, Henderson B E. Preserved foods and naspharyngeal cariinoma: a case – control study in Guanxi, China Cancer Res, 1988, 48: 1954 – 1959

41 Zeng Y, Liu Z R, Zhen S W, Wei J N, Zhu J S, Zei H S. Application of an immunoenzymatic method and immunoautoradiographic method for a mass survey of nasopharyngeal carcinoma. Intervirology, 1980, 13: 162 – 168

42 Zeng Y, Liu Y X, Wei J N, Zhu J S, Cai S L, Wang P Z, Zhong J M, Li R C, Pan W J, Li E J, Tan B F. Serologeical mass survey of nasopharyngeal carcinoma, Acta Acad med Sin, 1979, 1: 123 – 126

43 Zippin C, Tekawa I S, Bragg K U, Watson D A, Linden G. Studies on heredity and environment in cancer of the nasopharynx. J nat Cancer Inst, 1962, 29: 483 – 490

49. Linkage of a Nasopharyngeal Carcinoma Susceptibility Locus to the HLA Region

LU Sheng – jing[1], DAY Nicholas E[2], DEGOS Laurent[3], LEPAGE Virginia[4],
WANG Pei – Chung[1], CHAN So – Ha[6], IMONS Maicolm[7], MCKNIGHT Barbara[6],
EASTON Easton[9], Zeng YI[9], de – The G[8]

1. People's Regional Hospital, Nanning, Guangxi Autonomous Region, People's Republic of China

2. Medical Research Council Biostatistics Unit, 5 Shaftesbury Road, Cambridge CB2 2BW, UK

3. INSERM U 93, Hayem Research Center, Saint Louis Hospital, Paris, France

4. Microbiology Department, National University, Singapore

5. Immunogene Typing Laboratories, Immunoresearch, Melbourne, Australia

6. Department of Biostatistios, University of Washington, Seattle, Washington 98196, USA

7. Section of Epidemiology, Institute for Cancer Research, Sutton, Surrey SM2 5NG, UK

8. CNRS Laboratory of Epidemiology and Immunovirology of Tumors, Faculty of Medicine Alexis Carrel, 69372 Lyon Cedes 8, France; 9. Institute of Virology, Chinese Academy of Preventive Medicine, Beijing, People's Republic of China

The frequency of nasopharyngeal carcinoma is nearly 100 – fold higher in southern Chinese than in most European populations[1]. Earlier studies have suggested that an inceased risk of nasopharyngeal carcinoma is associated with specific haplotypes in the HLA region: relative risks slightly over twofold were found for haplotypes A2, Bw46 and the antigen B17 (refs 2 – 4). We now report a linkage study based on affected sib pairs which suggests that a gene closely linked to the HLA locus confers a greatly increased risk of nasopharyngeal carcinoma. The maximum Likelihood estimate is of a relative risk of approximately 21. The relationship between this suspected disease susceptibility gene (or genes) and known viral and environmental aetiological factors remains to be elucidated.

Our approach was to identify sibships that had more than one individual affected with nasopharyngeal carcinoma (NPC), and for which sufficient individuals could be typed to permit unambiguous assignment of haplotypes[5]. The study began in Singapore and Hong Kong in 1976 and extanded in 1983 to Nanning in the Guangxi Autonomous region of the People's Republic of China, where more sib pairs were accessible (see Tab. 1 legend). In each area, the members of each sibship, their parents and children were visited and blood samples taken. Thirty – four sibships with more than one case of NPC were identified, 31 with two cases and the remainder with three cases (Tab. 1). Of the sib – pair families, four had to be excluded, one because the pair were twins of unknown zygosity (but HLA identical) and three because an excess of haplotypes was seen in the family and parentage was ambiguous.

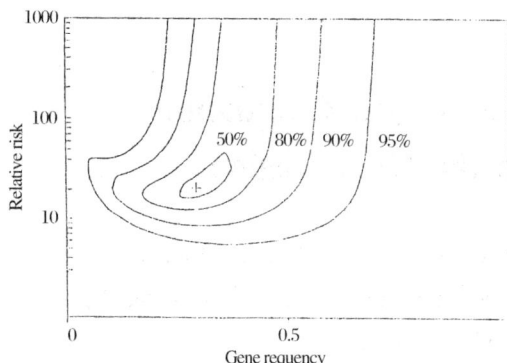

Fig. 1　For the model fitting, inference was based on the log – Likelihood difference conditional on the disease phenotypes, or lod score. The use of this conditional Likelihood allows inferences to be made free from ascertainment bial[6,8]. Likelihood calculations were performed using the program LINKAGE[7] for single – gene dominant and recessive models, varying the gene frequency relative risk associated with the disease susceptibility gene, and the recombination fraction θ between the susceptibility gene and the HLA region. In particular, the lower 95% confidence limit for the relative risk associated with the susceptibility gene, r_o, is given by:

$$2[\max_{r,p,\theta}I(r,p,\theta) - \max_{p,\theta}I(r_o,f,\theta)] = 3.84$$

where $I(r, f, \theta)$ denotes the log – Likelihood function and 3.84 is the upper 95% point for a x^2 distribution on 1 degree of freedom. The contours shown define confidence regions for r and f, at the 95%, 90%, 80% and 50% level with $\theta = 0$. The absolute maximum of the Likelihood occurs when $\theta = 0$, with values of $f = 0.29$, $r = 0.29$, and is marked by a cross (+).

Tab. 1　Sib – pair analysis

| | | 2 Expected under | | | | 1 Expected under | | | | 0 Expected under | | |
| | | | No | | | | No | | | | No | | |
	Observed	linkage	Dominant	recessive	Observed	linkage	Dominant	recessive	Observed	linkage	Dominant	recessive
Sib pairs[1]	16	6.75	10.83	13.96	8	13.5	13.18	10.40	3	6.75	2.99	2.64
Triplestz	2	2.25	3.30	4.01	4	4.50	4.39	3.91	3	2.25	1.31	1.08
Total	18	9.00	14.13	17.97	12	18.00	17.57	14.31	6	9.00	4.30	3.72

Number of shared haplotypes

Notes: Number of sib pairs sharing two, one or zero HLA haplotypes identical by descent, together with the expected number under the null hypothesis of no linkage, and under the best fitting dominant and recessive models. For the sib trios, all possible sib pairs are considered.

　　Goodness of fit[11] of the hypothesis of no linkage, X^2 (2d.f.) for sib pairs only = 17.0 $P < 0.001$; for all pairs (both sib paris and triples) = 12.0 $P < 0.005$. Goodness – of – fit tesls of the recessive and dominant models are each given by a 1 d.f. Likelihood ratio test against a three – parameter model, with relative risk r_2 for the DS heterozygote, and r_2 for the DS homozygote. x_1^2 for recessive, model = 0.32; x_1^2 for dominant model = 2.95. ($P = 0.08$).

　　[1]. 18 from China, 3 from Hong Kong, 5 from Singapore and 1 from Malaysia.

　　[2]. from China and 1 from Hong Kong.

　　HLA typing on the samples from Hong Kong and Singapore was performed in Singapore and restricted to locus A and B antigens, using a panel of over 200 antisera of Chinese, Malay, Janpanese, Filippino and Caucasian origin. HLA typing in Nanning used a panel of 143 allelic typing antisera (from Saint Louis Hospital, Paris) corresponding to HLA A, B, C and in most cases DR loci. Of the 38 NPC cases tested, the most common DR antigens were DR2 (18/38) and DR4 (16/38). No other DR antigen was present on more than seven individual cases.

The haplotype similarity of affected sibs and triplets is shown in Tab. 1, together with the expected numbers if no linkage were present. The total data show a clear divergence between observed and expected ($P < 0.005$). Among the sib pairs, comparing observed with expected gives a χ^2 value of 17.0 (2d. f.). This highly significant divergence demonstrates the existence of NPC disease susceptibility gene (s) linked to the HLA region.

For linkage analysis, single – gene models were explored[6-8], and as models assuming dominant inheritance fitted poorly (Tab. 1), the subsequent analyses were based on recessive models. Figure 1 shows the Likelihood of the observed results (lod score) for varying values of the relative risk (r) and the gene frequency (f) of the putative disease susceptibility gene when the recombination fraction (θ) is zero. The maximum lod score under the recessive model is + 2.39, and under the dominant model + 1.68, with P values of 0.004, and 0.024 respectively.

The maximum Likelihood estimate occurs when $\theta = 0$, $r = 20.9$ and $f = 0.29$. Infinite values of r are only slightly less likely for a range of values of f; the minimum value of r consistent with the data is 5.1 at the 95% level, and 9.8 at the 90% level. With values of θ greater than zero, all the Likelihood contours are further removed from the axis given by $r = 1$.

These results show the existence of an NPC disease susceptibility gene (s) conferring an increased risk for NPC soime tenfold larger than that associated with Bw46 or B17, with the upper 90% comfidence bound being infinite. The results thus provide strong support for the original hypotyesis[5] that much of the high risk for NPC among the southern Chinese is due to HLA – associated genes. The lifetime risk of NPC among Cantonese males in Singapore[1] is 1.6%, indicating that the frequency of a recessive disease susceptibility gene should be at least 0.13.

The mechanism by which an HAL – linked disease susceptibility gene might act is unknown. Two proposed environmental aetiological agents for NPC are the Epstein – Barr virus (EBV) and some constituents of preserved foods, including salted fish[8-10]. Insight into the mechanism of action of NPC disease susceptibility gene (s) should help to clarify the respective role of EBV, environmental cofactors and genetic susceptibility in the development of NPC.

[In 《Nature》 1990, 346: 470 –471]

References

1 Shanmugaratnam K. , Lee H P, Day N E. Cancer incidence in Singapore, 1968 – 1977, IARC Scientific Publication no. 47 (IARC, Lyon, 1983)

2 Simons M J et al. Int J Cancer, 1974, 18: 122 – 134

3 Simons M J et al. Lancet, 1975, 1: 142 – 143

4 Chan S H, Day N E, Kunaratnam N, Chia K B, Simons M J. Cancerm, 1982, 32: 171 – 176

5 Day N E, Simons M J. Tissue Antigens, 1976, 8: 106 – 119

6 Rish N. Am J hum, Genet, 1984, 38: 363 – 386

7 Lathrop G M, Laloual J M, Julies C, Ott. J Proc natn Acad Scl U S A, 1984, 81: 3443 – 3446

8 Ewens W J, Clarke C P. Am J hum Genet, 1984, 83: 853 – 372

9 Shanmugaratnam K. in Cancer Epidemiology and Prevention, 536 – 553 (eds Schottenfeid, D. & Fraumenl. J. F.) 536 –553 (Sanders, London, 1982)

10 Yu M C, Ho J H C, Lai S H, Henderson B E. Cancer Res, 1986, 46: 956 – 961

11 Stsckweider W C, Elston R C. Genet. Epid, 1985, 2: 85 – 97

50. Diagnosis of Nasopharyngeal Carcinoma by Means of Recombinant Epstein – Barr Virus Proteins

LITTLER Edward[1,3], BAYLIS Sally A[1], ZENG Yi[2], CONWAY Margaret J[1],
MACKETT Michael[3], ARRAND John R[1]

1. Cancer Resesrch Campaign Laboratories, Paterson Institute for Cancer
Research, Christie Hospital and Holt Radium Institute, Manchester, UK
2. Institute of Virology, Chinese Academy of Medical Sciences, Beijing, People's Republic of China
3. Correspondence to Dr E. Litter, Department of Moleculer Sciences, Wellcome
Research Laboratories, Langley Court, Beckenham, Kent BR3 3BS, UK.

Summary

The immune response of patients with nasopharyngeal carcinoma to Epstein – Barr virus (EBV) antigens is diagnostic of the tumour. Existing tests use EBV antigens produced in EBV – infected lymphoblastoid cells, but the virus replicates poorly in these cells. Serum samples from 18 patients diagnosed as having nasopharyngeal carcinoma were screened by western blot analysis, enzyme liked immunosorbent assay (ELISA), and immunofluorescence tests for antibodies to the EBV – coded alkaline deoxyribonuclease (DNase), thymidine kinase, and membrane antigen (gp340/220) produced in recombinant baculovirus of bovine papillomavirus systems. Each protein was a useful diagnostic marker for nasopharyngeal carcinoma, although in the gp340/220 ELISAs there was substantial overlap for both IgG and IgA antibodies between serum samples from nasopharyngeal carcinoma patients and those from healthy donors seropositiver for EBV. The EBV thymidine kinase was the most sensitive predictor of nasopharyngeal carcinoma; all such samples showed both IgG and IgA antibody responses to this protein and all gave clearly distinct titres from those of the EBV – seropositive donors in the IgA test. Each of the recombinant systems described is suitable for use in large – scale screening programmes for the early diagnosis of nasopharyngeal carcinoma.

Introduction

Nasopharyngcal carcinoma, a tumour of epithelial origin, is the commonest from of nasopharyngeal cancer.[1,2] Most nasopharyngeal carcinomas are radiosensitive, with survival rates of 33% after 3 years and 25% after 5 years for all cases (71% and 59%, respectively, for those confined to the nasopharynx).[2] In most countries cancer of the nasopharynx is rare, with rates of less than 1 per 100 000 per year. However, in several Chinese populations (Southern provinces of China, Hong Kong, Taiwan, Singapore, and Malaysia) the incidence is 15 – 30 per 100 000 per year,[3,4] In these Chinese populations it is the commonest tumour among men and the second most common among women. The incidence of nasopharyngel carcinoma is also high in Inuit populations, and there is evidence of high mortality associated with the disorder among black American men aged 60 – 80. [5,6]

There is much evidence to support an association between infection with Epstein – Barr virus (EBV) and nasopharyngeal carcinoma. EBV infection occurs in the vast majority of all human populations. Infection in childhood occurs without clinical symptoms, but if infection does not occur until adolescence the symptoms of infectious mononucleosis develop in about 50% of cases. It is possible to detect EBV DNA, RNA, and protein in the majority of biopsy samples from nasopharyngeal carcinomas, [7–10] but the immune response fo nasopharyngeal carcinoma patients to EBV antigens is unusual. Seroepidemiological surveys have shown that the titre of antibodies against the EBV capsid antigen (VCA) is about ten times higher in patients with nasopharyngeal carcinoma than in those with other tumours of the head and neck region. [11] These patients also have high concentration of antibodies to the EBV early antigen (EA), particularly to the diffuse component. EA antibody titres in nasopharyngeal carcinoma patients increase with the progression of the disease from stage I to stage IV. [12] The high VCA and EA antibody titres are reflected in the presence in both serum and saliva of IgA antibody to the same antigens. [13,14] The presence of IgA antibody to EBV – coded antigens is a diagnostic characteristic of serum from patients with naopharyngeal carcinoma.

EBV EA consists of several proteins including some virus – coded enzymes. The EBV – coded enzymes alkaline deoxyribonucease (DNase), DNA polymerase, and thymidine kinase are specifically neutralised serum from nasopharyngeal carcinoma patients. [15–17] The best characterised immune response is to the EBV – coded alkaline DNase, which is a good diagnostic and prognostic marker for nasopharyngeal carcinoma. [18,19]

An improtant obstacle to the use of EBV – coded proteins for diagnosis of nasopharyngeal carcinoma is that there is no fully permissive cell culture system for the replication of EBV. We have reported the expression of several EBV – coded proteins in recombinant systems and preliminary evidence of their value for the diagnosis of nasopharyngeal carcinoma. [20–22] We describe here the use of the EBV – coded alkaline DNase, gp340/220, and thymidine kinase in western blot, immunofluorescent tests, and enzyme – linked immunosorbent assays (ELISAs) to detect the presence of reactive IgG and IgA antibody in serum from patients with naspharyngeal carcinoma.

Methods

The construction of recombinant baculovirus expressing the EBV – coded alkaline DNase and thymidine kinase and bovinepapillomavirus – transformed cell lines expressing the EBV – coded membrane antigen have been described previously. [23,24] Cells producing the EBV recombinant proteins were resuspended in sodium dodecyl sulphate disruption buffer for polyactylamide gel electrophoresis; 0. 1% 'NP40' and 0. 1% sodium deoxycholate for ELISAs; or phosphate – buffered saline form immunofluotescence tests. For immunofluorescence cells were allowed to adhere to glass slides and then fixed in acetone for 10 min.

Conditions for polyactylamide gel electrophoresis and western blotting have been described previously. [21] Each strip contained about 10^5 cells infected with recombinant baculovirus. Serum samples were used at 1/50 dilution.

ELISA plates were sensitised with cells infected with recombinant baculovirus, resuspended as above and diluted in 0. 01 mol/L ammonium bicarbonate buffer at a concentration of 10^5 cells/ml, or with purified gp340/220 at 50 – 80 mg/ml. Serum samples diluted 1/10 were added and the reaction was detected by means of phosphatase – conjugated sheep antibody to human IgG or IgA.

The antisera used consisted of samples from 30 normal, healthy EBV seropositive (VCA +) subjects from the Manchester ares; 7 clinically normal EBV seronegative (VCA –) subjects (L. Young, University of Birmingham); and 17 patients with nasopharyngeal carcinoma from Wuzhou, Guangxi, China. The clinical stage of the tumour was define by Chinese criterria. [25] Serum samples were also examined for the presence of IgA antibody to EBV VCA and EA[11].

Results

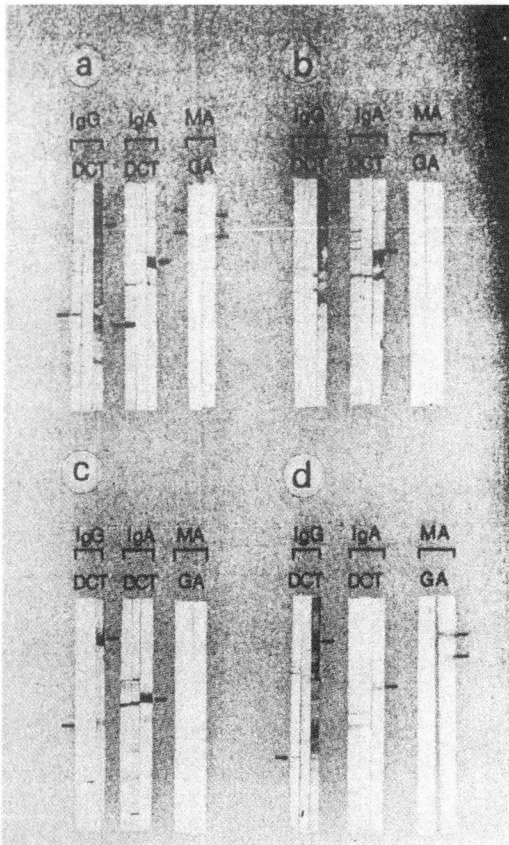

D = baculovirus – infected cells expressing EBV alkaline DNase; T = same cells expressing thymidine kinase; C = extract from control wild – type baculovirus – infected cells; MA = purified gp340/220 (G = IgG, A = IgA). Location of unique reactive bands shown ty bars. Panel a = patient 5; b = patient 12; c = patient 18; and d = patient 10. Since the strips used for western blots were obtained from different experiments, there is some variation in the abslute mobliity

Fig. 1 Western blot analysis of immunological reactivity with EBV proteins four samples from patients with nasopharyngeal carcinoma.

The serum samples from nasopharyngeal carcinoma patients showed several patterns of reactions with the recombinant EBV proteins (Fig 1, table 1). 3 serum samples (5, 6, and 14; table 1) had IgA and IgG antibody to all three proteins, although the concentration of IgG against gp340/220 was low (Fig 1A). 2 samples (12 and 15) had only IgG and IgA antibody against EBV thymidine kinase (fig 1B); sample 12 had a low antibody titre to the EBV VCA and sample 15 had no detectable IgA antibody to EBV EA. The remaining samples showed variable patterns of response to the three EBV proteins. Fig 1C shows the result for sample 18; it had IgG antibody to both DNase and thymidine kinase but IgA antibody only against thymidine kinase; this serum also had a low VCA titre. The remaining serum samples gave a variable reaction with both the EBV – coded alkaline DNase and gp340/220. For example, fig 1D shows the result for sample 10; it had IgG against DNase and thymidine kinase, but not gp340/20, and IgA against thymidine kinase and gp340/220, but not DNase.

Thus, on western bot analysis all serum samples from patients with nasopharyngeal carcinoma had IgG and IgA antibody against EBV thymidine kinase, including 2 (samples 3 and 15) which

Tab. 1 Immunoreactivity of nasopharyngeal carcinoma patients' serum samples to EBV proteins

| Patient | Extent of reactivity on western blot | | | | | | ELISA titre[1] | | | IF with TK | VCA IgA[2] | EA IgA | Clinical stage |
| | IgG | | | IgA | | | | | | | | | |
	DNase	TK	MA	DNase	TK	MA	IgA TK	IgG MA	IgA MA				
1	–	+ +	ND	–	+ +	ND	ND	ND	ND	ND	80	+	II
2	–	+ +	+ +	+ +	+ +	+ +	1.077	2.125	1.547	+	320	+	II
3	+ +	+ +	+	–	+ +	+	1.646	2.042	0.239	+	10	–	III
4	–	+ +	–	+	+ +	–	2.923	1.206	0.782	+	320	+	III
5	+ +	+ +	+	+ +	+ +	+ +	ND	1.885	1.508	+	320	+	III
6	+ +	+ +	+	+	+ +	+ +	0.981	1.890	1.115	+	80	+	III
7	–	+ +	–	–	+ +	+ +	2.098	1.209	0.443	W	80	+	I
8	+ +	+	+	–	+ +	–	1.727	1.808	0.361	+	20	+	I
9	+ +	+ +	–	+ +	+ +	+	1.160	1.526	0.514	+	80	+	III
10	+ +	+ +	–	–	+	+ +	1.085	2.045	0.673	+	80	+	III
11	+ +	+ +	+	–	+ +	+ +	2.007	2.168	0.533	+ +	640	+	III
12	–	+ +	–	–	+ +	–	0.992	0.251	0.169	+	10	+	III
13	–	+	+	+ +	+ +	+ +	2.780	1.300	1.997	+	160	+	II
14	+ +	+ +	+	+	+ +	+	2.065	1.192	0.662	+	320	+	III
15	–	+	–	–	+	–	ND	ND	ND	W	80	–	II
16	+ +	+ +	ND	–	+ +	ND	ND	ND	ND	ND	320	+	III
17	+ +	+ +	+ +	–	+ +	+	2.083	1.899	0.681	+	320	ND	III
18	+ +	+ +	–	–	+ +	–	1.978	2.109	0.828	+	320	ND	III

Notes: IF = immunofluorescence; TK = thymidine kinase; MA = gp340/220; ND = not done; W = week.

1. Absorbance at 450 nm

2. Reciprocal titre of IgA reaction with EBV VCA

had no detectable IgA anti – EA and low titres of IgA anti – VCA. 9 samples had IgG and 11 IgA antibodies against gp340/200. 11 samples had IgG and 7 IgA to the EBV DNase . These seemed to be no correlation with the histological findings on the tumour, but the reactivity of the serum samples to the three EBV proteins seemed to correlate broadly with the EBV serology based on VCA and EA. No substantial readtions were seen in serum samples from VCA + or VCA – donors with any of the EBV – coded proteins, except for IgG anti – gp340/220.

Western blot analysis is not suitable for screening purposes. To provide a more quantitative and rapid immunological assay for antibody against EBV thymidine kinase and gp340/ 220 we developed ELISAs. In the ELISA for thymidine binase we could not determine an IgG titre, since it used a crude lysate of baculovirus – infected cells and the baculovirus and host cell proteins are themselves targets for the human immune response. However, human serum does not seem to have large amounts of IgA antibody to these proteins. Fig 2 shows titres of IgA antibody to thymidine kinase in the three groups of subjects. The mean titre was higher in the nasopharyngeal carcinoma patients (1. 757 [SD 0. 639]) than in the VCA + (0. 510 [0. 136]) and VCA − (0. 366 [0. 045]) subjects. The lowest value for any nasopharyngeal carcinoma serum was 0. 981 (Tab. 1) and the highest for any VCA + serum was 0. 715.

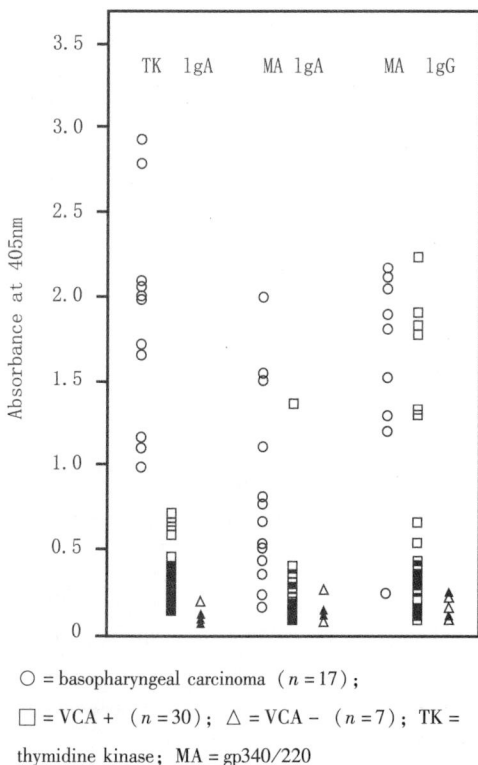

○ = basopharyngeal carcinoma (n = 17);

□ = VCA + (n = 30); △ = VCA − (n = 7); TK =

thymidine kinase; MA = gp340/220

**Fig. 2   Reactivity of serum samples
with EBV − coded proteins in ELISAs.**

Since purified gp340/220 was available, we could study the presence of both IgG and IgA antibody. The titres for IgG and IgA antibody against gp340/220 corresponded well to the presence of such antibody as determined by western blot analysis. There was also a good correlation with IgA antibody titres to the EBV VCA and EA. The titre of IgG anti − gp340/220 was not a good predictor of nasopharyngeal carcinoma (Fig. 2) − 6 of 30 VCA + samples had titres within the range of most of the nasopharyngeal carcinoma samples. The titre of IgA anti − gp340/220 was, however, a good indicator of nasopharyngeal carcinoma (mean [SD] titre 0. 806 [0. 523] vs 0. 162 [0. 080] for VCA + and 0. 141 [0. 050] for VCA −). Only 2 of the nasopharyngeal carcinoma samples had IgA anti − gp340/220 titres below the cutoff of 0. 4 and would be falsely considered negative. In addition 1 VCA + sample consistently had a very high titre.

One of the traditional techniques in southeast Asia for the diagnosis of nasopharyngeal carcinoma uses EBV − infected cell lines as a target to detect antibody (IgA) to EBV VCA and EA by indirect immunofluorescence tests. Although we believe that our ELISA system is a more quantitative, reliable, and efficient diagnostic test, we studied whether our recombinant systems would be useful as reagents in immunofluorescence tests. SF9 cells infected with a recombinant baculovirus expressing the EBV TK were used as targets for immunofluorescent analysis (Fig. 3). Although a good reaction was seen between serum from nasopharyngeal carcinoma patients and the target cells, only a low grade weak reaction was seen in the VCA + serum. All nasopharyngeal carcinoma patients' serum samples tested gave an immune response to the EBV thymidine kinase by immunofluorescence; 2 gave a weak response, 1 of which had no detetable IgA anti − EA.

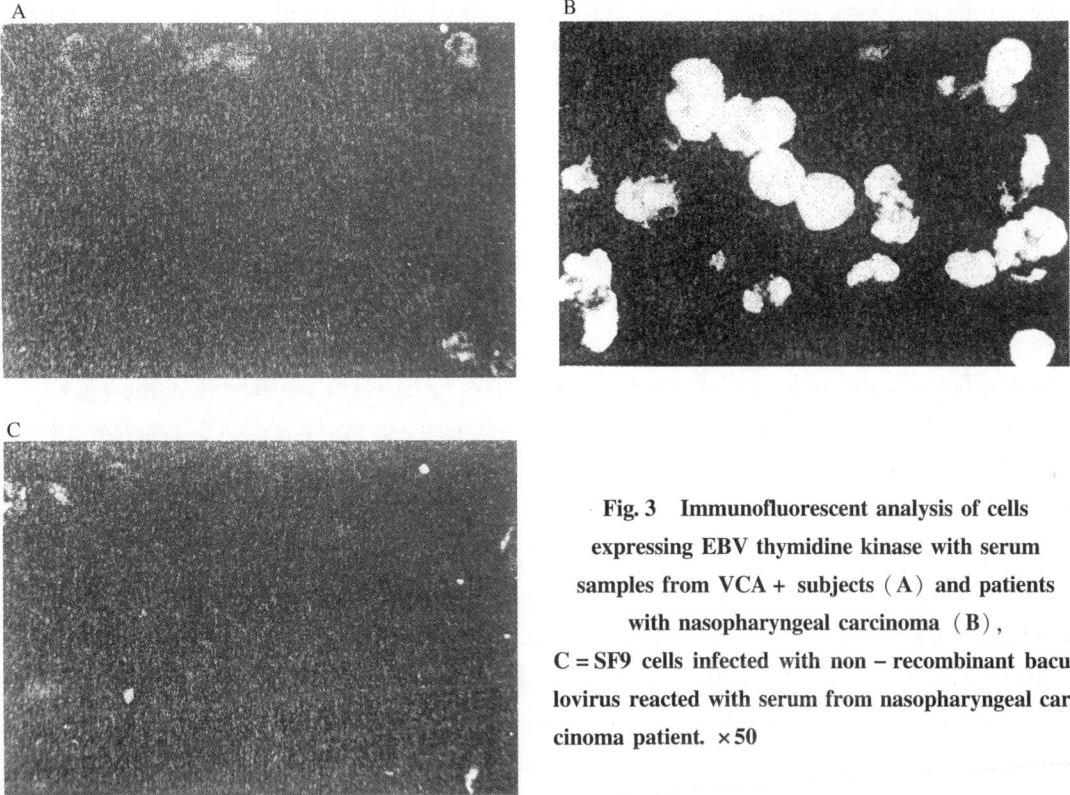

Fig. 3　Immunofluorescent analysis of cells expressing EBV thymidine kinase with serum samples from VCA + subjects (A) and patients with nasopharyngeal carcinoma (B), C = SF9 cells infected with non − recombinant baculovirus reacted with serum from nasopharyngeal carcinoma patient. ×50

Discussion

One requirement for improved treatment of nasopharyngeal carcinoma is a method of screening large, at − risk populations for the presence of tumour. The method should be able to detect tumours at an early stage of development, have low rates of false − positive and false negative results, and be rapid and economically viable. The association of EBV with nasopharyngeal carcinoma results in an unusual immunoloical response to a range of EBV proteins. Patients have high IgA tires to the EBV VCA and EA and can neutralise the activity of some EBV enzymes. [13,16 − 18] Already these responses are being used for the diagnosis and prognosis of nasopharyngeal carcinoma in the People's Republic of China and (including Taiwan)[15,18,19]. However, EBV replicates poorly in the lymphoblastoid cells used as the source of the diagnostic antigen, the cells are expensive to maintain, and the amount of EBV antigen present is variable. A consequence of the use of EBV − infected lymphoblastoid cells is that tests such as indirect immunofluorescence and enzyme neutralisation have to be used; these are difficult to apply in mass screening and give variable results owing to their subjective interpretation.

The three EBV − coded proteins used in our recombinant expression systems represent different classes of EBV protein: the EBV thymidine kinase and DNase are components of the restricted EA complex and gp340/220 is a late of VCA protein (unpublished). Each of the systems can produce milligram amounts of its respective protein in culture volumes of only one litre. Western blot analysis with the three proteins showed that each has a useful diagnostic capability, and the ELISA assays with thymidine kinase and gp340/220 as targets showed that a rapid diagnostic test is possible. Because some countries have already invested substantial resources in immunofluorescent and enzyme neutralisation tests, we have shown here and previously that the EBV thymidine kinase expressed in baculovirus and the gp340/220 expressed in bovine – papillomavirus transformed cells[20] may be used as sensitive replacements for immunfluorescent tests; the EBV alkaline DNase expressed in recombinant bculovirus can be used as a source of enzyme for neutralisation tests with the added bonus of ease of production and stability of the enzyme (unpublished).

Although all three proteins are useful as diagnostic targets, the EBV – coded thymidine kinase may be a particularly sensitive antigen. All of the serum samples from patients with nasopharyngeal carcinoma examined here (and a further 20 from other sources) show IgG and IgA antibodies to thymidine kinase; samples from VCA + and VCA – donors and patients with infectious mononucleosis do not show any such reactivity. We have previously shown that about 50% of patients with Burkitt's lymphoma have IgG antibody to the EBV thymidine kinase but do not have any IgA antibody[22]. A single sample from a patient with chronic infectious mononucleosis had IgG and IgA antibody against EBV thymidine kinase but at lower titres than in nasopharyngeal carcinoma (unpublished). We have previously suggested that the sensitivity of the EBV thymidine kinase is due to an unusual, hydrophilic streak of 290 aminoacids at the amino terminus[22]. The role of this region in the biochemistry of the EBV thymidine kinase is not known, but it does seem to be unique to the thymidine kinase of EBV and herpesvirus saimiri. The region may be necessary for replication in a lymphoid environment, or it may be simply that there is high evoltionary conservation in this protein between EBV and herpesvirus saimiri.

Only a proportion of the serum samples from patients with nasopharyngeal carcinoma showed a good reaction with EBV DNase in western blots; we believe this result is due to the lower expression of the EBV protein in the recombinant baculovirus (probably caused by the toxicity of the protein). This low level expression also precludes the use of crude lysates in ELISA tests, as is done for thymidine kinase. It will be necessary to produce purified EBV DNase from the recombinant baculovirus – infected cells before such tests can be developed. EBV – coded gp340/220 seems to be a good diagnostic reagent for nasopharyngeal carcinoma and the expression systems and purification procedures should be suitable for the establishment of diagnostic tests[25].

In its 1982 report on the biology of nasopharyngeal carcinoma the International Union against Cancer recommended that "screening programmes and reliable test systems for the early detection of individuals at risk by immunological and biochemical surveys should be extended as a high priority"[3]. We believe that the systems described here may from the basis of such screening programmes.

Acknowledgements

This work was supported by the Cancer Research Campaign (UK). S. B. was supported by the Science and Engineering Research Council and the Wellcome Research Laboratories.

[In 《The Lancet》 1991, 337 (874): 685 – 688]

References

1　Shanmugaratnam K. The pathology of nasopharyngeal carcinoma. In: Biggs PM, de The G, Payne LN, eds. Oncogenesis and herpes viruses. Lyon: IARC Scientific Publications no 2, 1972, 239

2　Shanmugaratnam K, Chan SH, de The G, et al. Histopathology of nasopharyngeal carcinoma: correlations with epidemiology, survival rates and other biological characteristics. Cancer, 1979, 44: 1029 – 1044

3　Simons MJ, Shanmugaratnam K, eds. The biology of nasopharyngeal carcinoma. Geneva: UICC Technical Report, 1982, 71

4　Waterhouse J, Muir CS, Shanmugaratnam K, Powell J, Caner incidence in five continents, vol Ⅳ. Lyon: IARC Scientific Publication no 42, 1982

5　Neilsen NH, Mikkelsen F, Hansen JPM. Nasopharyngeal cancer in Greenland: the incidence in an arctic Eskimo population. Acta Pathol Microbiol Scand 1977, 85A: 850 – 858

6　Cancer mortality in the US (1950 – 1977). Bethesda: National Cancer Institute, Monograph 59

7　Zur Hausen H, Schulte – Holthauzen H, Klein G, et al. EBV DNA in bioposies of Burkitt tumours and anaplastic carcinomas of the nasopharynx. Nature, 1970, 228: 1956 – 1958

8　Tugwood JD, Lau W – H, O S – K, et al. Epstein – Barr virus – specific transcription in normal and malignant nasopharyngeal bniopsies and in lymphocytes from healthy donors and infectious mononucleosis patients, Gen Virol, 1987, 68: 1081 – 1091

9　Young LS, Dawson CW, Clark D, et al. Epstein – Barr virus gene expression in nasopharyngeal carcinoma. Gen Virol, 1988, 69: 1051 – 1065

10　Raab – Traub N, Hood R, Yang C – S, Henry B, Pagano JS. Epstein – Barr virus transcription in nasopharyngeal carinoma. Virol, 1983, 48: 580 – 590

11　Henle W, Henle G, Ho HC, et al. Antibodies to Epstein – Barr virus in nasopharyngeal carcinoma, other head and neck neoplasms and control groups. Natl Inst, 1970, 44: 225 – 231

12　Henle W, Ho H – C, Henle G, Kwan HC. Antibodies to Epstein – Barr virus – related antigens in nasopharyngeal carcinoma. Comparison of active cases with long – term survivors. Natl Cancer Inst, 1973, 51: 361 – 369

13　Zeng Y, Gong CH, Jan MG, Fun Z, Zhang LG, Li HY. Detection of Epstein – Barr virus IgA/EA antibody for diagnosis of nasopharyngeal carcinoma by immunoautoradiography. Int Cancer, 1983, 31: 599 – 601

14　Zeng Y, Zhong JM, Li LY, et al. Follow – up studies of Epstein – Barr virus IgA/VCA antibody positive persons in Zangwu Couny, China. Intervirology, 1983, 20: 190 – 194

15　Cheng Y – C, Chen J – Y, Glaser R, Henle W. Frequency and levels of antibodies to Epstein – Barr virus – specific DN ase are elevated in patients with nasopharyngeal carcinoma. Proc Natl Acad Sci USA, 1980, 77: 6162 – 6165

16　Liu M – Y, Cou W – M, Nutter L, Hsu M – M, Chen J – Y. Antibody against Epstein – Barr virus DNA polymerase activity in sera of patients with nasopharyngeal carcinoma. Med Virol, 1989, 28: 101 – 105

17　Turenne – Tessier M, Ooka T, Calander A, de The G, Daille J. Relationship between nasopharyngeal carrinoma and high antibody titres to Epstein – Barr virus – specific thymidine kinase. Int Cancer, 1989, 43: 45 – 48

18　Chen J – Y, Chen C – J, Liu M – Y, et al. Antibody to Epstein – Barr virus – specific DNase as a marker for field survey of patients with nasopharyngeal carcinoma in Taiwanl. Med Virol, 1989, 27: 269 – 273

19　Tan RS, Cheng YC, Naegele RF, Henle W, Glaser R, Champion J. Antibody responses to Epstein – Barr virus – specific DNase in relation to the prognosis of juvenile patients with nasopharyngeal carcinoma. Int Cancer, 1982, 30: 561 – 565

20 Zeng Y, Du B, Miao X, Mackett M, Arrand JR. Detection of IgA/MA antibody in sera using PO₄ cells. Chin Virol, 1987, 3: 396 – 397

21 Baylis SA, Purifoy DJM, Littler E. The characterization of the EBV alkaline deoxyribonuclease cloned and expressed in E coli. Nucl Acids Res, 1989, 17: 7609 – 7622

22 Littler E, Newman W, Arrand JR. Immunological response of nasopharyngeal carcinoma patients to the Epstein – Barr Virus – coded thymidine kinase expressed in Escherichia coli. Int Cancer 1990, 45: 1028 – 1032

23 Baylis S, Purifoy DJ, Littler E. High level ex-

pression of the Epstein – Barr Virus alkaline deoxyribonuclease using a recombinant baculovirus: application to the diagnosis of nasopharyngeal carcinoma. Virology, 1991, 181: 390 – 394

24 Conway M, Morgan A, Mackett M. Expression of Epstein – Barr virus membrane antigen gp340/220 in mouse fibroblasts using a bovine papillomavirus vector. Gen Virol, 1989, 70: 729 – 734

25 Li S – L. Classfication of nasopharyngeal carcinoma by 5th National Nasopharyngeal Carcinoma Symposium, 1979. In: Tumours of the head and neck. Tianjin: Tianjin Science and Technology Publishing House, 1982, 251

51. Salivary and Serum IgA Antibodies to the Epstein – Barr Virus Glycoprotein gp340: Incidence and Potential for Virus Neutralization

YAO Q Y[1], ROWE M[1], MORGAN A J[2], SAM C K[3],
PRASAD U[3], DANG H[4], ZENG Y[5], RICKINSON A B[1,6]

1. Department of Cancer Studies, University of Birmingham; 2. Department of Pathology, University of Bristol, UK; 3. Institute for Advanced Studies, University of Malaya, Kualay Lumpur, Malaysia; 4. Cancer Institute, Wuzhou; 5. Institute of Virology, Beijing, China; 6. To whom correspondence and reprint requests should be sent, at the Department of Cancer Studies, University of Birmingham, Birmingham, B15 2TJ, UK.

Summary

Human antibody responses to the Epstein – Barr virus (EBV) glycoprotein gp340 have been measured using purified preparations of the native molecule as the substrate in ELISAs. This glycoprotein is the dominant component of the EBV envelope and a major target for the virus – neutralizing antibody response. Healthy virus carriers (both Caucasian and Chinese) regularly show detectable anti – gp340 IgG in serum and, unexpectedly, 21% – 30% of these individuals are also serum anti gp340 IgA positive. Chinese patients with the EBVgenome – positive malignancy nasopharyngeal carcinoma (NPC) show elevated serum IgA antibodies to gp340 but, given the background of responses among healthy virus carriers, anti – gp340 IgA titres are a poorer diagnostic indicator of NPC than serum IgA antibodies detectable by immunofluorescence against the multicomponent EBV early antigen (EA). Salivary IgA antibody responses to gp340 are potentially important as a means of neutralizing orally – transmitted virus. We detected salivary IgA (but not IgG) to gp340 in a minority (12% – 19%) of healthy virus carriers and in a higher proportion (49%) of NPC patients. Even saliva samples chosen for their relatively high anti – gp340 IgA titres showed only weak neutralizing activity against transforming EBV preparations whether or from B98. 5 cell culture supernatant or from the throat washing of an infectious mononucleosis patient. We conclude that in healthy virus carriers, salivary IgA responses to gp340 are unlikely to provide effective local immunity against re – infection with a second EBV strain.

Introduction

Epstein – Barr virus (EBV) infection is widepread in human populations and its incidence is usually monitored by immunofluorescence assays for serum antibodies to 1 of 3 multicomponent viral antigens. These are (i) the nuclear antigen EBNA which is expressed in latently – infected growth-transformed B lymphoblastoid cell lines and which is now known to be composed of 6 different gene products, EBNAs 1, 2, 3a, 3b, 3c and – LP, all of which are antigenically distinct (Kieff and Liebowitz, 1990); (ii) early antigen (EA) which is a composite of multiple viral proteins of the early lytic cycle, whose locations can be nuclear and/or cytoplasmic (Pearson et al. , 1983, 1987), and (iii) virus capsid antigen (VCA) which is a composite of several viral structural proteins and glacoproteins present within infected cells late in the lytic cycle (Kishishita et al. , 1984; Vroman et al. , 1985). Despite the complex nature of the antigens being detected, these immunofluorescence assays have played an important role in the identification and monitoring of EBV – related disease states (Henle and Henle, 1982), perhaps the best example being the virusgenome – positive malignancy nasopharygeal carcinoma (NPC) in which the prevalence and titre of serum IgA antibodies to EA and VCA have significant diagnostic and prognostic value (Henle and Henle, 1976; Zeng, 1985).

Methods for measuring antibody responses to a 4th multicomponent EBV antigen, the membrane antigen (MA) which is expressed on the surface of lytically – infected cells late in the cycle, have also been available for many years (Klein et al. , 1976). However, technical difficulties, not least the requirement for viable cells as the antigen – positive substrate, have greatly restricted anti – MA testing in large serological assays. Nevertheless, the MA complex, now known to be a composite of at least 3 virus – coded glycoproteins gp340/220 (Beisel et al. , 1985), gp85 (Heineman et al. , 1988) and gp78/55 (Mackett et al. , 1990), is of considerable scientific interest because it includes the dominant target antigens for virus – neutralizing antibody responses (Thorley – Lawson et al. , 1982). There is clearly a need to re – analyse the whole question of antibody responses to the individual virus envelope glycoproteins using either purified or recombinant antigen as the substrate and assay methods which are more quantitative and more sensitive than the original MA test.

An important step forward came with the development, in 2 independent laboratories (Luka *et al.*, 1984; Randile and Epstein, 1984; Uen, *et al.*, 1988), of ELISAs for antibodies to the EBV glycoprotein gp340. This glycoprotein is the most abundant component of the viral envelope, mediates virion binding to the EBV receptor on B cells, and is a major target for virus – neutralizing antibodies (Dolyniuk *et al.*, 1976; Nemerow *et al.*, 1987; Thorley – Lawson and Poodry, 1982). Despite the availability of these assays, there is still very, little published information on serological responses to gp340, particularly in EBV – related disease states, and no information on the potentially important question of salivary antibody responses to the glycoprotein. In the present work we have used fast protein liquid chromatography (FPLC) – purified gp340 (David and Morgan, 1988) as the substrate in ELISA assays (i) of serum IgA (and IgG) antibodies to the antigen in control conors and in NPC patients, to determine whether elevated IgA anti – gp340 reactivity is a better diagnostic indicator of the tumour – bearer state than IgA anti – EA or anti – VCA, and (ii) of salivary IgA antibodies to gp340, to determine the frequency of local antibody responses which have the potential to neutralize orallytransmitted virions and to identify suitable saliva samples for direct testing of virus – neutralizing activity.

Material and Methods

1. Serum and saliva samples

Serum and saliva samples were taken from the same groups of donors. These groups were (i) 20 healthy abult Caucasian donors from Birmingham, UK, known to be seronegative for anti – VCA IgG antibodies, (ii) 50 healthy adult Caucasian donors from Birmingham, known to be seropositive for anti – VCA IgG antibodies, (iii) 122 healthy adult Chinese donors from Wuzhou, China or from Kuala Lumpur, Malaysia, all of whom proved to be anti – VCA IgG seropositive, (iv) 87 NPC patients at primary presentation, again from Wuzhou , and (v) 20 NPC patients in remission from Kuala Lumpur. Saliva samples (from unstimulated saliva) were taken into salivettes (Sarstedt, Numbercht, Germany) and clarified by centrifugation, then (for Wuzhou and Kuala Lumpur donors) air – freighted on ice to Birmingham together with the serum samples. The salivary antibody assays were conducted within 3 days of receipt of the air – freighted material; preliminary work had shown that titres of anti – gp340 IgA in saliva samples were not altered by storage for up to a month at 4℃. Serum samples were also kept at 4℃ and titrated within 1 week of receipt.

2. ELISA for anti – gp340 antibodies

The method was adapted from an earlier protocol (Randle and Epstein, 1984) but now using as a substrate a purified preparation of the naturally glycosylated gp340 isolated by FPLC from the membranes of virus – producing B95. 8 cells (David and Morgan, 1988). Briefly, purified gp340 in bicarbonate buffer pH 9. 6 was coated on to ELISA plates by incubation overnight at 4℃. For the anti – gp340 IgG assay, each well on the ELISA plate received 50μl of buffer containing 1. 5 units of gp340, whereas for the IgA assay each well received 6. 5 units of gp340 (Northe *et al.*, 1982, for definition of the arbitrary units). Following blocking with PBS containing 1% BSA and 5% normal rabbit serum for 2 hr at room temperature, triplicate wells were incubated with one of a series of doubling dilutions of serum or saliva in blocking solution for 1 hr at 4℃. The plates were washed 3 times with PBS containing 0.05% Tween 20 and then specifically bound antibody was detected by incubation with peroxidase – conjugated rabbit antihuman IgG or IgA (Dakopatts, Glostrup, Denmark; 1:500 dilution) for 30 min at 4℃. Conventional development with orthophenylene diamine (OPD) was followed by measurement of absorbance at 492 nm in an ELISA reader. Serum samples were assayed at serial doubling dilutions starting at 1:20, and saliva samples at serial doubing dilutions starting at 1:2.

3. Immunofluorescence assays for anti – EA and anti – VCA antibodies

All sera were also assayed by serial doubling dilution starting at 1: 10 for IgG and IgA antibodies to EA and VCA by conventional indirect immunofluorescence, using acetonefixed slide preparations of P3HR1 virus – infected Raji cells and B95.8 cells respectively as the antigen – positive indicator cells (Henle and Henle, 1976).

4. Neutralization assays

Ten – fold dilutions of virus preparations, either B95.8 cell culture supernatant or throat washings from an infectious mononucleosis (IM) patient, were pre – incubated with an equal volume of the test saliva sample (or with medium as a control) for 1 hr at 37℃, then added to cord – blood indicator lymphocytes for a further 1 – hr incubation. The cells were then cultured in microtest plate wells as 6 – 8 replicates per test sample and observed over the next 6 weeks for the incidence of EBV – induced outgrowth of transformed cells.

Results

1. Serum IgG antibodies to gp340

In the anti – gp340 IgG ELISA, all 20 negative control sera (i, e., sera already shown to be anti – VCA negative by conventional immunofluorescence assay) gave absorbance readings below 0.5 units at the lowest dilution tested. Representative readings form 10 such control sera are shown in Figure 1a. Parallel assays regularly showed significant anti – gp340 IgG antibodies in sera from both Caucasian and Chinese seropositive (i, e., anti – VCA IgG – positive) donors, albeit sometimes at relatively low titres, representative results being shown in Figure 1b. Chinese NPC patients' sera assayed in the same way were also anti – gp340 IgG – positive, the tendency towards elevated tires being apparent from representative results (Fig. 1c). A summary of the incidence and mean titre of anti – gp340 IgG antibodies in sera from the various groups of donros is presented in Tab. I alongside corresponding data for anti – EA and anti – VCA IgG antibodies detected by conventional immunofluorescence testing of the same sera. All the antiVCA – positive sera from healthy donors showed anti – gp340 IgG detectable by ELISA, though not all were antiEA – positive. The elevation of anti – gp340 IgG titres in NPC patients versus healthy controls was not as marked as that observed when anti – VCA and particularly anti – EA IgG titres were compared between the same groups of sera.

A

B

C

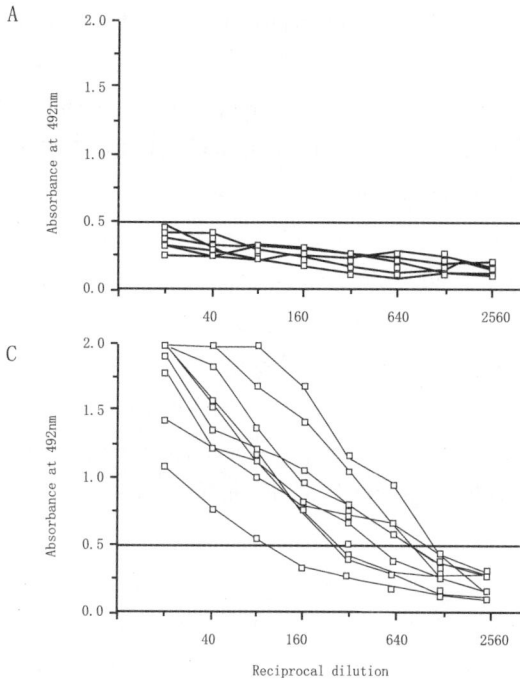

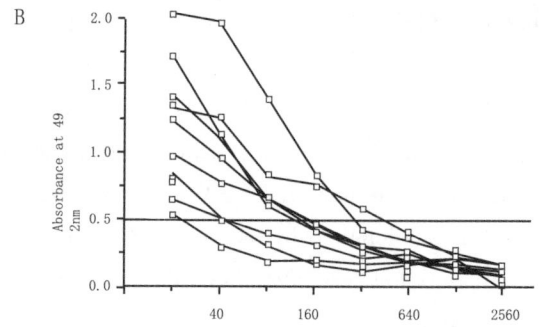

Fig. 1 – Serum anti – gp340 IgG ELISA readings for doubling dilutions of 10 representative sera from: (a) healthy Caucasian anti VCA IgG seronegative donors, (b) healthy Caucasian anti – VCA IgG seropositive donors, and (c) Chinese NPC patients at primary presentation. On this basis an absorbance value of 0.5 at 492 nm was set as background and ELISA titres were defined as the highest doubling dilution of serum which gave a reading above backround.

Tab. 1   Summary of serum antibody titres to gp340. EA and VCA[+]

Antibody	Assay[2]	Caucasian seropositive			Caucasian seropositive			Caucasian seropositive		
		Incidence	%	Mean titre[3]	Incidence	%	Mean titre[3]	Incidence	%	Mean titre[3]
Serum IgG	Anti – gp340	50/50	100	1 : 283	122/122	100	1 : 179	87/87	100	1 : 745
	Anti – EA	34/50	68	1 : 50	79/122	65	1 : 54	87/87	100	1 : 1221
	Anti – VCA	50/50	100	1 : 428	122/122	100	1 : 375	87/87	100	1 : 5023
Serum IgA	Anti – gp340	15/50	30	1 : 66	25/122	21	1 : 42	82/87	94	1 : 446
	Anti – gpEA	0/50	0	—	1/122	0.8	1 : 10	85/87	98	1 : 246
	Anti – VCA	5/50	10	1 : 20	5/122	4	1 : 10	84/87	97	1 : 608

Notes: [1] Sera from 20 Caucasian EAV – negative donors were included as controls in the same assay and gave uniformly negative results (data not shown). – [2] Anti – gp340 titres determined by ELISA assay: anti – EA and anti – VCA titres determined by conventional immunofluorescence using Ig class – specific second – step conjugated antibodies. – [3] Mean titre of positive samples.

2. Serum IgA antibodies to gp340

Representative results obtained on testing the same groups of sera in the anti – gp340 IgA ELISA are presented in Fig. 2 for seronegative controls(a), seropositive donors(b) and NPC patients (c). Clearly, only a minority of EBV – seropositive donors have detectable anti – gp340 IgA antibodies in serum, and these are at much lower titres than anti – gp340 IgA antibodies in the serum of NPC patients. Overall results for the various groups of donors are summarized in Tab. 1, again with corresponding data for IgA antibodies to EA and VCA detected in the same sera by immunofluorescence tests. Interestingly, anti – gp340 IgA was more prevalent among Caucasian and Chinese seropositive donors (21% – 30%) than were anti – EA or anti – VCA IgA antibodies. All 3 types of IgA reactivity showed a much higher prevalence among NPC patients, but the increase was most marked for anti – EA IgA since this was almost absent from healthy seropositive donor sera.

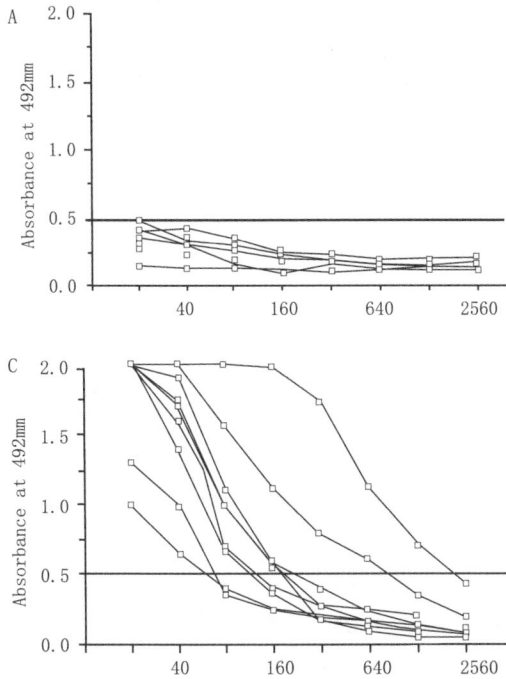

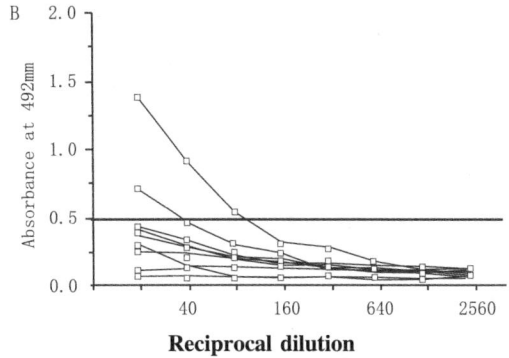

Fig. 2 Serum anti – gp340 IgA EKUSA reading for doubing dilutions of 10 representative sera from: (*a*) healthy Caucasian anti VCA IgG – seronegative donors, (*b*) healthy Caucasian anti – VCA IgG – seropositive donors, and (*c*) Chinese NPC patients at primary presentation. Titres were defined as in Fig. 1

3. Salivary IgA antibodies to gp340

Saliva samples were collected, without salivary stimulation, from the same groups of donors and tested by ELISA for salivary IgA antibodies to gp340. Again, representative results from 10 seronegative control donors, 10 seropostive donors and 10 NPC patients are shown in Fig. 3 to illustrate the specificity and sensitivity of the assay. Significant anti – gp340 IgA reactivity could be detected in saliva from a minority of serpositive donors and from a higher proportion of NPC patients. The overall results are summarized in Tab. 2.

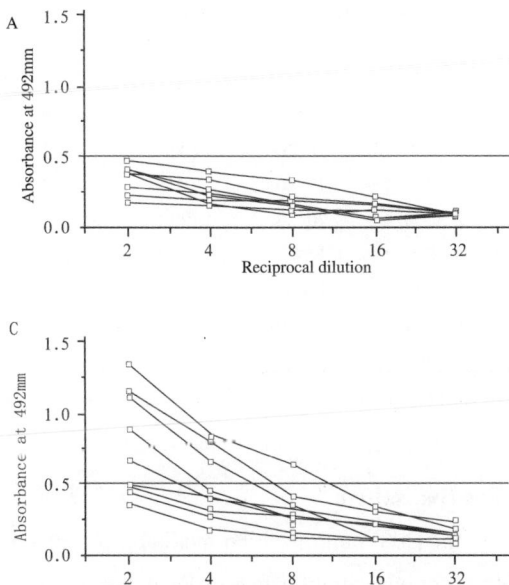

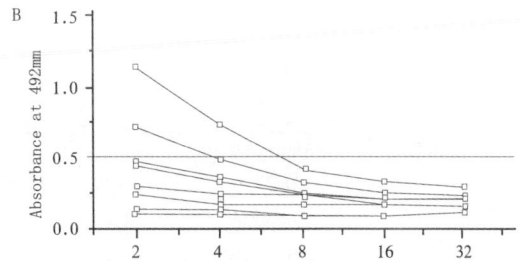

Fig. 3 Salivary anti-gp340 IgA ELISA reading for doubling dilutions of 10 representative saliva samples from: (*a*) healthy Caucasian anti-VCA IgG-seronegative donors, (*b*) healthy Caucasian anti-VCA IgG-seropositive donors, and (*c*) Chinese NPC patients at primary presentation. Titres were defined as in Fig. 1

Tab. 2　Summary of salivary antibody titres to gp340[1]

Antibody	Assay	Caucasian seropositive			Chinese seropositive			Chinese NPC patients		
		Incidence	%	Mean titre[2]	Incidence	%	Mean titre	Incidence	%	Mean titre
Salivary IgG	Anti-gp340	0/50	0	–	0/122	0	–	0/87	0	–
Salivary IgA	Anti-gp340	6/50	12	1:3.7	23/122	19	1:3.0	43/87	49	1:7.6

Notes: [1] Saliva from 20 Caucasian EBV-negative donors were included as controls in the same assays and gave uniformly negative results (data not shown) [2] Mean titre of positive samples

In parallel tests, no saliva sample ever gave significant anti – gp340 reactivity in the IgG ELISA (Tab. 2). These IgG assays were conducted as an additional control to check whether any saliva samples were continated with serum antibodies coming from serous secretions.

4. Anti – gp340 antibody status of newly diagnosed versus remission NPC patients

Earlier studies have emphasized the correlation between anti – EA and anti – VCA IgA serum antibody status and tumour burden in NPC patients (Henle and Henle, 1976; Zeng, 1985). In the present study we have extended the analysis to anti – gp340 antibody responses in serum and saliva samples taken at one referral centre (Kuala Lumpur) from 20 newly – diagbised patients and from 20 patients whose NPC was in remission following treatment. The complete analysis of relevant IgG and IgA antibodies in these patients is summarized in Tab. 3. While there was no significant change in the incidence of serum IgA antibodies to gp340, EA and VCA, there was some reduction in all 3 antibody titres following treatment. Changes in the corresponding serum IgA antibodies were more marked, with remission patients showing a reduced incidence of detectable responses and a lower mean titre for IgA antibody – positive sera in all 3 assays; salivary IgA responses to gp340 also showed a reduced incidence in remission patients.

Tab. 3　Antibody status in newly diagnosed vs. remission NPC patients

Antibody	Assay[1]	Newly diagnosed NPC patients			Remission NPC patients		
		Incidence	%	Mean titre[2]	Incidence	%	Mean titre
Sernm IgG	Anti-gp340	20/20	100	1:794	20/20	100	1:425
	Anti-EA	20/20	100	1:1137	19/20	95	1:570
	Anti-VCA	20/20	100	1:4000	20/20	100	1:2268
Serum IgA	Anti-gp340	20/20	100	1:428	13/20	65	1:85
	Anti-EA	19/20	95	1:269	10/20	50	1:142
	Anti-VCA	19/20	95	1:1012	15/20	75	1:356
Salivary IgA	Anti-gp340	16/20	80	1:8	9/20	45	1:8

Notes: 1 Anti-gp340 titres determined by ELISA assay. anti-EA and anti-VCA titres by conventional immunofluorescence assay [2] Mean titre of positive samples

5. Virus neutralization by anti – gp340 IgA – positive saliva

The final series of experiments was designed to look for EBV neutralization mediated by salivary antibodies to the virus, and thus to assess the possible functional significance of local IgA antibody responses. For this work, saliva specimens from 3 NPC patients, each with an anti – gp340 IgA titre of at least 1: 20 by ELISA assay, were pooled, as were 3 control saliva samples again from NPC patients but lacking detectable anti – gp340 IgA activity. Ten – fold dilutions of B95. 8 culture supernatants and of throat washings from an IM patient were used as sources of transfroming virus; these preparations were pre – exposed to medium alone or to the pooled saliva samples before being added to cord-blood indicator cells. Tab. 4 shows the results of the subsequent transformation assay from

which it is clear that the pooled anti-gp340 IgA-positive saliva samples mediated a significant, but incomplete, neutralization of EBV-transforming activity. This same partial nevtralization, both of B95. 8 virus and of IM throat washing virus, was observed at a second testing (data not shown).

Tab. 4 Nevtralzation of EBV by an anti-gp340 IgA positive saliva preparation

Virus dilutions pre-exposed to:	Incidence of transformation of cord cells exposed to:					
	B95. 8 virus dilutions			IM throat washing dilutions		
	10^{-3}	10^{-4}	10^{-5}	Neat	10^{-1}	10^{2}
Medium alone[1]	6/6	6/6	5/6	6/6	6/6	6/6
Anti-gp340 IgA negative salive[1]	6/6	6/6	4/6	6/6	6/6	6/6
Anti-gp340 IgA positive saliva[2]	6/6	5/6	0/6	6/6	6/6	1/6

Notes: [1] Pool of 3 NPC patient saliva samples lacking detectable anti-gp340 IgA [2] Pool of 3 NPC patient saliva samples with anti-gp340 IgA titre of >1: 20

Discussion

The ELISA upon which the present work is based uses as a source of antigen a fully glycosylated gp340 preparation of proven immunogenicity which has been FPLC purified from the membranes of lytically-infected cells (David and Morgan, 1988; Morgan et al. , 1989). There is a distinct advantage in using the native glycoprotein rather than the product of the gp340 coding sequence expressed from a recombinant vector in bacterial, yeast or even mammalian cells (Beisel et al. , 1985; Schultz et al. , 1987; Whang et al. , 1987; Emini et al. , 1988). It is likely that many anti-gp340 antibody reactivities in human serum are directed against conformational epitopes on the molecule and recombinant proteins may not faithfully reproduce all of these epitopes, particularly since the overall tertiary structure of native gp340 must be influenced by its pattern of glycosylation within infected cells (Whang et al. , 1987; Emini et al. , 1988). In this context, we were encouraged to find from our initial set of assays that all 172 healthy seropositive donors tested (i. e. , all individuals with detectable anti-VCA IgG, the usual criterion for prior EBV infection) showed detectable serum IgG responses to gp340, whereas all 20 seronegative control donors were also anti-· gp340 IgG-negative by ELISA (Fig. 1, Tab. 1). This bears witness to both the specificity and the sensitivity of the assay and strongly suggests that anti-gp340 IgG antibodies are maintained for life in the serum of virus-carrying individuals in a manner similar to anti-VCA IgA antibodies; by contrast, only a subpopulation of such individuals are detectably anti-EA IgG-positive.

By increasing approximately 4-fold the amount of purified gp340 substrate used in the ELISA, we were able to adapt the basic 2-step ELISA protocol used for analysing IgG responses to the measurement of gp340-specific IgA antibodies in both serum and saliva; other workers have reported the development of a 3-step ELISA to achieve the necessary sensitivity for anti-gp340 IgA antibody detection (Uen et al., 1988). On beginning the serological screening for IgA antibody responses, we were surprised to find that a significant minority of healthy seropositive donors (30% Caucasian donors, 21% Chinese donors) were anti-gp340 IgA-positive by ELISA (Fig. 2, Tab. 1). These numbers were significantly higher than the incidence of anti-VCA IgA positivity (< 10%) and of anti-EA IgA positivity (1%) in the same group of sera. The anti-gp340 IgA-positive status of these healthy donor sera was reproducible on repeated testing and was unrelated to the titre of co-resident anti-gp340 IgG (data not shown). Indeed, in additional experiments (data not shown), we were able to completely remove anti-gp340 IgG reactivity from selected sera by adsorption on protein A Sepharose without significantly affecting the anti-gp340 IgA titre measurable by ELISA. We are therefore confident that the serum IgA results are not artefactual.

One consequence of the relatively high incidence of serum anti-gp340 IgA antibodies among healthy virus carriers is to reduce the usefulness of this response as a diagnostic marker for NPC. Thus, although the incidence and mean titre of serum anti-gp340 IgA responses are significantly raised in NPC patients, the increase is less marked than noted for IgA antibodies to VCA and in particular to EA. We conclude therefore that serum IgA responses to EA (Henle and Henle, 1976; Zeng, 1985) remain the best single marker distinguishing NPC patients from healthy controls (Tab. 1). Furthermore, our some what limited data on newly-diagnosed versus remission NPC patients (Tab. 3) again suggest that serum anti-EA IgA levels are the most sensitive to a change in tumour load. Accordingly, the search for the best possible serological marker of early NPC diagnosis should perhaps now focus on the individual viral proteins which together constitute the EA complex recognized in conventional immunofluorescence testing (Pearson et al., 1983, 1987).

Using the same anti-gp340 IgA ELISA protocol as used in the serological assays, we were able to detect salivary IgA to gp340 in only 12% – 19% healthy seropositive donors, and then almost always at titres which were much lower than positive serum IgA titres (Fig. 3, Tab. 2). This is perhaps to be expected since total IgA levels in saliva are an order of magnitude lower than IgA levels in serum (Delacroix et al., 1982). It is nevertheless formally possible that EBV particles produced endogenously from sites of chronic virus replication in the oropharynx (Yao et al., 1985) could be engaging gp340-specific IgA and reducing the levels of free antibody detectable by ELISA. We are confident that the positive results which were obtained do reflect the existence of a local immune response and are not due to minor contamination of saliva samples by antibodies from serous secretions. Thus there was no correlation between the salivary IgA and serum IgA results for individual donors in the anti-gp340 ELISA (data not shown); furthermore there was never any detectable anti-gp340 IgG in saliva samples (Tab. 2) despite the fact that these same donors consistently displayed high titres of such antibodies in serum. Salivary IgA responses to gp340 were more frequent in NPC patients but generally the increases both in incidence and in mean titre were less marked than for serum IgA responses (compare Tab. 1 and Tab. 2). We provisionally conclude that in NPC patients EBV lytic antigen expression, eithe focally within the tumour itself of in adjacent normal epithelium, induces a strong systemic IgA response but a less marked secretory IgA response. A different picture appears to be emerging with respect to the EBV latent protein EBNA 1, one of the few viral gene products consistently expressed in NPC cells (Young *et al.* , 1988; Fahraeus *et al.* , 1988); a recent study measuring that subset of anti-EBNA 1 antibodies recognizing the immunodominant gly-ala repeat domain of the molecule found striking elevations of both serum *and* salivary IgA titres in NPC patients (Foong *et al.* , 1990).

One of our main objectives in this work was to assess the potentially protective function of EBV-specific IgA responses. In preliminary experiments using protein A adsorption to completely deplete the IgG antibodies from selected human sera, we found that sera with detectable anti-gp340 IgA antibodies retained weak EBV-neutralizing activity whereas anti-gp340 IgA-negative sera did not (data not shown). A more important question concerned the neutralizing capacity of salivary IgA antibodies since these might provide a means of neutralizing virus transmitted by the natural oral route. It is clear from the data in Tab. 4 that even pooled NPC saliva with a relatively high titre of anti-gp340 IgA antibodies possessed only weak neutralizing activity, being active for instance against a 1: 100 dilution but not against a 1: 10 dilution of an EBV-containing throat washing from an IM patient.

The overall conclusion from this part of the work is that healthy virus carriers either have no detectable salivary IgA to gp340 or have such weak reactivity as to offer little practical protection against orally-transmitted virus. This is interesting when set against our recent finding that healthy virus carriers appear to carry just one isolate of EBV in the circulating B-cell pool (Yao *et al.* , submitted for publication); at the B-cell level, therefore, such individuals seem to bc innune to reinfection with a second virus strain. It becomes very important to know whether such immunity to re-infection also extends to the virus' other target tissue, oropharyngeal epithelium (Sixbey *et al.* , 1984, 1989). If this is indeed the case, then such immunity is unlikely to be conferred by salivary antibody responses and must have some other, as yet unidentificd, basis.

Acknowledgements

This work was supported by the Cancr Research campaign, London, UK. M. R. is a Wcllcome Senior Research Fellow in Basic Biomedical Sciences. We thank Ms. D. Williams for excellent secretarial support.

[In 《Int J Cancer》 1991, 48: 45 – 50]

References

1 Beisel C, Tanner J, Matsuo T, Thokley-lawson D, Kezdy F, Kieff E, Two major outer envelope glycoproteins of Epstein-Ban virus are encoded by the same gene. J Virol, 1985, 54: 665 – 674

2 David E M, Morgan A J. Efficient purification of Epstein-Barrvirus membrane antigen gp340 by fast protein liquid chromatography. J immunol Meth, 1988, 108: 231 – 236

3 Delacroix D L, Dlve C, Rambaud J C, Vaerman J P. IgA subclasses in various secretions and is serum. Immunology, 1982, 47: 282 – 285

4 Doi Yniuk M, Wolff E, Kieff E. Proteins of Epstein-Barr virus. II. Electrophoretic analysis of the polypeptides of the nucleocapsid and the glucosamine-and polysaccharide-containing components of enveloped virus. J Virol, 1976, 18: 289 – 297

5 Emini E A, Schlief W A, Armstrong M C, Silberklang N, Schultz L D, Lehman D, Maigetter R Z, Qual Tiere L. F, Pearson G R, Ellis R W. Antigenic analysis of the Epstein-Barrvirus major membrane antigen (gp350/220) expressed in veast and mammalian cells; implica-

tions for the development of a sub-unit vaccine. Virology, 1988, 166: 387 – 393

6　Fahraeus R, Fu H L, Ernberg I, Finke J, Rowe M, Klein G, Falk K, Nilsson E, Yadav M, Busson P, Tursz T, Kalln B. Expression of Epstein-Barr virus-encoded proteins in nasopharyngeal carcinoma. Int J Cancer, 1988, 42: 329 – 338

7　Foong Y T, Cheng H M, Sam C K, Dillner J, Hingerer W, Prasad U. Serum and salivary IgA antibodies against a definedepitope of the Epstein-Barr virus, nuclear antigen (EBNA) are elevated in nasopharyngeal carcioma. Int J Cancer, 1990, 45: 1061 – 1064

8　Heineman T, Gong M, Sample J, Kieff E. ldentification of the Epstein-Barr virus gp85 gene. J Virol, 1988, 62: 1101 – 1107

9　Henle G, Henle W. Epstein – Barr virus-specific IgA serum antibodies as an outstanding feature of nasopharyngeal carcinoma. Int J Cancer, 1976, 17l: 1 – 7

10　Henle W, Henle G. Immunology of Epstein-Barr virus. In: B. Roizman (ed), The Herpesviruses, Vol. 1, pp. 2009 – 252. Plenum Press, New York, 1982

11　Kieff E, Liebowitz D. Epstein-Barr virus and its replication. In: B. N. Fields, D. M. Knipe, R. M. Sharock, M. S. Hirsch, J. L. Melnick, T. P. Morath and B. Roizman (eds.), Virology, pp. 1889 – 1920, Raven Press, New York, 1990

12　Kishishita M, Luka J, Vroman B, Poduslo J E, 0Pearson G R. Production of monoclonal antibody to a late intracellular Epstein-Barr virus-induced antigen. Virology, 1984, 133: 363 – 375

13　Klein G, Cclfford P, Klein E, South R T, Minowada J, Kourilsky F, Burchenal J H. Membrane immunofluorescence reactions of Burkitt lymphoma cells from biopsy specimens and tissue cultures J nat Cancer Inst, 1967, 39: 1027 – 1044

14　Luka J, CHASE R C, Pearson G R. AAsensitive enzyme-linked immunosorbent assay (Elisa) against the major EBV-associated antigens. I. Correlation between Elisa and immunofluorescence titers using purified antigens. J immunol

Meth, 1984, 67: 145 – 156

15　Mackett M, Conway M J, Arrand J R, Haddad R S, Huttfletcher L M. Characterisation and expression of a glycoprotein encoded by the Epstein-Barr virus BamHI I fragment. J Virol, 1990, 64: 2545 – 2552

16　Morgan A J, Allison A C, Finerty S, Scullion F T, Byers N E, Epstein MA. Validation of a first generation Epstein-Barr virus vaccine preparation suitable for human use. J med Virol, 1989, 29: 74 – 78

17　Nemfrow G R, Mold C, Schwend V K, Toliefson V, COOPER N R. Identification of gp350 as the viral glycoprotein mediating attachment of Epstein-Barr virus (EBV) to the EBV/C3d receptor of B cells: sequence homology of gp350 and C3 complement fragment C3d. J Virol, 1987, 61: 1416 – 1420

18　North J R, Morgan A J, Thompson J L, Epstein M A. Quantification of an Epstein-Barr virus-Barr virus-associated membrane antigen (MA) component. J virol Meth, 1982, 5: 55 – 65

19　Pearson G R, Luka J, Petti L, Sample J, Birkenbach M, Braun D, Kieff E. ldentification of an Epstein-Barr virus early gene encoding a second component of the restricted early antigen complex. Virology, 1987, 160: 151 – 161

20　Pearson G R, Vroman B, Chase B Scullfy, T Hummel M, Kieff E. Identification of polypeptide components of the Epstein-Barr virus early antigen complex using monoclonal antibodies. J Virol, 1983, 47: 193 – 201

21　Randle B J, Epstein M A. A highly sensitive enzyme. linked immunosorbent assay to quantitate antibodies to Epstein-Barr virus membrane antigen gp340. J virol Meth, 1984, 9: 201 – 208

22　Schultz L D, Tanner J, Hofmann K J, Emini ea., Condra J H, Jones R E, Kieff E, Ellis R W. Expression and secretion in yeast of a 400 kD envelope glycoprotein derived from epstein-Barr virus. Gene, 1987, 54: 113 – 123

23　Sixbey J W, Nedrud J G, Raab-Traub N, Hanes R A, Pagano J S. Epstein-barr virus replication in oropharyngeal epithelial cells. New Engl. J. Med, 1984, 310, 1225 – 1230

24 Sixbey J W, Shirley P, Chesney P J, Buntin D M, Resnick L. Detection of a second widespread strain of Epstein-Barr virus. Lancet, 1989, 11, 761 – 765

25 Thorley-Lawson D A, Edson C M, Geilinger K. Epstein-Barr virus antigens – a challenge to modern biochemistry Advanc. Cancer Res, 1982, 36, 295 – 348

26 Thorley-Lawson D A, Poodry C A, Identification and isolation of the main component (gp350 – gp220) of Epstein-Barr virus responsible for generating neutralising antibodies in vivo. J Virol, 1982, 43, 730 – 736

27 Uen W-C, Luka J, Pearson G R. Development of and enzyme-linked immunosorbent assay (Elisa) for detecting IgA antibodies to the Epstein-Barr virus. Int J Cancer, 1988, 41: 479 – 482

28 Vroman B, Luka J, Rodriguez M, Pearson G R. Character-isation of a maior protein with a mo-lecular weight of 160, 000 associated with the viral capsid of Epstein-Barr virus. J Virol, 1985, 52, 107 – 113

29 Whang Y, Silberklang M, Morgan A, Munshi S, Lenny A B, Ellis, R. W. and Kieff, E. , Expression of the Epstein-Barr virusgp350/220 gene in rodent and primate cells. J Virol, 1987, 61: 1796-1807

30 Yao Q Y, Rickinson A B, Epstein, M A. A re-examination of the Epstein-Barr virus carrier state in healthy seropositive individuals. Int J Cancer, 1985, 35: 35 – 42

31 Young L S, Dawson CW, CLlarkD, Rupani H, Busson P, Tursz T. Johnson, A. and Rickinson, A. B. , Epstein-Barr virus gene expression in nasopharyngeal carcinoma. J gen Virol, 1988, 69: 1051 – 1065

32 ZENG Y. Seroepidemiological studies on nasopharyngeal carcinoma in China. Advanc. Cancer Res, 1985, 44: 121 – 138

52. Detection of Anti-Epstein-Barr-Virus Transactivator (Zebra) Antibodies in Sera From Patients with Nasopharyngeal Carcinoma

JOAB Irène[1,6], NICOLAS Jean-Claude[2], SCHWAAB Guy[1],
de – The G[1], CLAUSSE Bernard[4], PERRICAUDET Michel[1], ZENG Yi[5]

1. Institute Gustave Roussy, CNRS, URA1301, 39 rue Camille Desmoulins 94800 Villejuif; 2. Hpital Rothschild, Boulevard de Picpus 75012 Paris; 3. Institute Pasteur, 25 rue du Docteur Roux 75015 Paris; 4. Institute Gustave Rouss, CNRS URA 1156, 39 rue Camille Desmoulins 94800 Villejuif France;

5. Chinese Academy of Preventive Medicine, 100 Ying Xin Jie, Beijing , China.

Summary

The Epsteln-Barr virus (EBV) is a ubiqultous Herpes virus which causes infectious mononu-cleosis and is associated with such different neoplasms as Burkitt's lymphoma and nasopharyngeal carcinoma. EBV latently infects its target cells; nevertheless, evidence of viral relication in NPC tumours has been uncovered. Among the EBV transactivators, the ZEBRA protein plays a crucial role in switching the virus from a latent to a productive mode. ZEBRA protein was produced using a eukaryotic expression vector: the open reading frame containing the BZFLI cDNA has revlously been inserted down-stream from the adenovirus major late promoter leading to expression of a 38×10^3 nuclear protein. We performed sero-logical studies by employing ZEBRA protein expressed in human cells for immunofluorescence and Western-blot assays. We were able to detect IgG anti-ZEBRA antibodies (IgG/ZEBRA) in 87% of NPC patients. These antibodies were absent in control sera; IgG/ZEBRA antibodies can be proposed as a useful marker for diagnosis of NPC tumors.

Introduction

The Epstein-Barr virus (EBV) is a ubiquitous human Herpes virus. It causes infectious mononucleosis and is closely associated with 2 neoplasms: Burkitt's lymphoma (BL) and naso-pharyngeal carcinoma (NPC) (de-Thé, 1982). In NPC epithelial cells the viral genome has been detected both in tumor biopsies by hybridization techniques and in tumors transplanted in nude mice (which are free of infiltrating B lymphocytes). EBV can latently infect its target cells, but different observations favor the hypothesis that viral replicaton occurs in the tumor: analysis of the structure of viral DNA termini has demonstrated the presence of linear viral DNA, specifically indicating that the viral productive cycle may occur in these tumors (Raab-Traub and Flynn, 1986); in NPC tumor biopsies different RNAs from the replicative phase of the EBV life cycle have been detected. InNPC patients, elevated IgG and IgA antibodies directed against EA and VCA viral protein may suggest that EBV reactivation occurs. The molecular mechanism leading to EBV activation is beginning to be understood and transactivators encoded by the EBV genome have been identified and have been shown to be involved in the activation of the lytic cycle (Chevalier-Greco et al. , 1986; Countryman and Miller, 1985; Countryman et al. , 1987; Hardwick et al. , 1988; Lieberman et al. , 1986; Wong and Levine, 1986). ZEBRA (BamHI Z EBV Replication Activator), also called Z or EBl, is encoded by BZLFl (BamHI Z left frame l); it switches the EB virus from the latent to the productive cycle. In NPC tumors where the EB virus seems to be partially reactivated, it is thus likely that viral transactivators are being produced; we describe here the detection of anti-ZEBRA antibodies in the sera of most NPC patients. EBV-specific IgA antibodies are characteristic of NPC patients: IgA/VCA, IgA/EA, IgA anti-thymidine kinase antibodies (IgA/TK), have been detected in serological analysis and have been proposed as useful markers for the early diagnosis of NPC tumors. The specific use of the detection IgG/ZEBRA antibodies is discussed.

Subjects and Methods

1. Plasmid

The plasmid pMLP BZLF1 has been previously described (Ronney et al. , 1986).

2. Cell line

Line 293, a human embryonic kidney fibroblast cell line transformed by adenovirus 5 (Harrison et al. , 1977), was transfeted with recombinant DNA by the calcium phosphate precipitation method (Graham and Van der Erb, 1973). Westernblot experiments have been described elsewhere (Joab et al. , 1987).

3. Serological tests

The sera were tested for specific IgG anti-VCA and anti-EBNA antibodies to EBV by an indirect immunofluorescence assay. The sera were tested for specific IgG anti-EA antibodies by an indirect immunoperoxidase assay. Anti-ZEBRA antibodies were assayed by indirect immunofluorescence experiments performed as described (Joabet al. , 1991).

4. Sera

Sera from 41 NPC patients were studied. The controls used for serological test consisted of 50 healthy French and 48 healthy Chinese individuals.

Results

The humoral immunological response to ZEBRA transactivator was studied in different EBV-related disorders. Tab. 1 shows the comparison of different antibodies to EB virus in sera of NPC patients: 87% of these patients contain anti-ZEBRA antibodies in their sera. This proportion is similar to that observed in NPC patients for IgA/VCA antibodies which represent the marker used for early diagnosis of NPC tumors in high-risk areas (Zeng et al. , 1983b). The anti-ZEBRA antibodies found in sera of NPC patients are mostly of IgG type and only 13% of NPC patients exhibit IgA/ZEBRA antibodies. The frequency of NPC patients with IgG/ZEBRA is higher than that of patients with IgA/EA antibodies. The same serological tests were performed with sera from healthy control individuals. Ninety-eight control individuals (50 French and 48 Chinese) were tested for the presence of anti-ZEBRA antibodies in their sera: these groups may have differd in their ages of primary EBV infection. The results are shown in Tab. 2: only one individual was positive for anti-ZEBRA antibodies although these subjects were positive for other anti-EBV antibodies to VCA and EA. The presence of anti-ZEBRA antibody-positivity could be related to an activation of the EBV lytic cycle. We tested sera from IM patients for the presence of different antibodies to EB virus: the results (Tab. 3) show that the percentage of IgG/ZEBRA antibody-positive patients is high (79%), although the GMT is lower in sera from IM than from NPC patients. Replication of the EB virus and the presence of IgG/ZEBRA in the sera of patients could be linked.

Tab. 1 comparison of different antibodies to EB virus in sera from NPC patients		
	Number positive/ number tested (%)	GMT
IgG/ZEBRA	36/41 (87)	1/85
IgG/VCA	28/28 (100)	1/409
IgG/EA	26/28 (92. 8)	1/25
IgG/EBNA	25/28 (89)	1/44
IgA/VCA	23/28 (82)	1/27
IgA/EA	12/28 (42)	1/4

Tab. 2 comparison of different antibodies to EBV in sera of control individuals		
	Nnmber positive/ number tested (%)	GMT
IgG/ZEBRA	1/98 (1)	1/5
IgG/VCA	45/50 (90)	1/162
IgG/EA	0/11 (0)	>1/5
IgG/EBNA	40/50 (90)	1/63

In order to check the specificity of anti-ZEBRA antibodies for EBV-related diseases, we performed serological tests on samples from a variety of different diseases (Tab. 4): we tested more than 700 subjects and only found anti-ZEBRA antibodies in sera of patients with EBV-related disorders: 50% of BL patients were IgG/ZEBRA-positive (GMT of 1/10). The highest titers were found in NPC sera. It is seen (Tab. 4) that in HIV-positive subjects and in AIDS patients in whom active EBV infection may occur. anti-ZEBRA antibodies are present in more than 30% of the sera tested (Joab *et al.*, 1991). We were surprised to observe that 13% of the patients with a head and/or neck tumor displayed anti-ZEBRA antibodies in their sera. This may suggest that EBV replication has occurred in those patients; this phenomenon may be unrelated to the presence of the tumor.

Tab. 4 detection of anti-EB virus transactivator (ZEBRA) antibodies in sera from patients with different pathologies

Pathologies	Number positive/ number tested (%)
Hepatocarcinoma	0/50 (0)
Chronic hepatitis	0/30 (0)
Acute lymphoblastoid leukemia	0/50 (0)
Berger's disease	1/43 (2)
Graft rejection	2/20 (10)
Multiple sclerosis	0/15 (0)
Sjögren's syndrome	0/15 (0)
Head and neck tumors	11/81 (13)
NPC	36/41 (87)
IM	26/31 (83)
BL (EBV [+])	11/22 (50)
Asymptomatic HIV [+]	75/229 (32)
AIDS	12/49 (24)
French controls	1/50 (2)
Chinese controls	0/48 (0)

Tab. 3 comparison of defferent antibodies to EB virus in sera from IM patients

	Number positive/ number tested (%)	GMT
IgG/ZEBRA	15/19 (79)	1/31
IgG/VCA	18/19 (94)	1/96
IgG/EA	18/19 (94)	1/3
IgG/EBNA	7/19 (36)	1/2.7
IgA/VCA	10/19 (52)	1/4.2

Discussion

Among EBV-associated antibodies, IgA/VCA has been considered to be a prominent feature of NPC (Henle and Henle, 1976). Serological surveys and follow-up studies of NPC in China have been used to show that IgA/VCA and IgA/EA are valuable markers for early detection of the disease (Zeng *et al.*, 1983*a*, *b*, 1985). In the present study we found a significantly high prevalence of EBV anti-ZEBRA transactivator antibodies in patients with NPC. In NPC patients the increase in anti-ZEBRA antibody production may be due to an over-production of ZEBRA protein either in epithelial cells of the tumor or in the (circulating or infiltrating) lymphocytes. IgG/ZEBRA antibodies were found in 87% of NPC patients, while only 1 healthy individual out of 100 displayed antibodies to this antigen; this result shows that the presence of IgG/ZEBRA in sera can be used to help in the diagnosis of NPC tumors. IgG/ZEBR has also been found in 85% of infectious mononucleosis (IM) patients and in 32 of the HIV-seropositive patients (Joab *et al.*, 1991). In IM patients, anti-ZEBRA antibodies may appear following intensive primary multiplication of the EB virus, before establishment of the latent infection. Only 50% of BL patients were positive for IgG/ZEBRA antibodies. The presence of anti-ZEBRA antibodies in the sera of HIV seropositive patients could be one of the consequences of EBV reactivation occurring in those subjects. High frequency and high level of antibodies to EBV DNase (Cheng *et al.* , 1980), to EBV DNA polymerase (Tan *et al.* , 1986) and to EBV thymidine kinase (de Turenne-Tessier *et al.* , 1989) are found in NPC sera. Other methods for detecting anti-ZEBRA antibodies in sera are now being studied in our laboratory to enable us to scale up the number of tests which can be performed. The use of anti-ZEBRA antibody detection for early diagnosis and posttherapeutic surveillance is under current investigation.

Acknowledgements

The authors are grateful to Drs. G. Lenoir, J. Gozlan, J. Leverger, J. D. Sraer, P. Lebon, L. Edelman, J. L. Lefrere and C. André for providing different sera.

〔In 《Int J Cancer》 1991, 48: 641 −649〕

References

1　Gheng Y C, Chen J Y, Glaser R, Henle W. Frequency and levels of antibodies to Epstein Barr virus specific DNase are elevated in patients with nasopharyngeal carcinoma. Proc nat Acad Sci, (Wash.), 1980, 77: 6161 −6165

2　Chevallier-Greco A, Manet E, Chavrier P, Mosnier C, Daillie J, Sergeant A. Both Epstein-Barr virus (EBV)-encoded transacting factors, EB1 and EB2, are required to activate transcription from an EBV early promoter. EMBO J, 1986, 5: 3243 −3249

3　Countryman J, Jenson H, Seibl R, Wolf H, Miller G. Polymorphic proteins encoded within Bzkfi of defective and standard Epstein-Barr viruses disrupt latency. Virology, 1987, 61: 3672 −3679

4　Gountryman J, MILLER G. Activation of expression of latent Epstein-barr herpes virus after gene transfer with a small cloned subfragment of heterogeneous viral DNA. Proc nat Acad Sci, (Wash.), 1985, 82: 4085 −4089

5　De-the G. Epidemiology of Epstein-Barr virus and associated diseases, In: B. Roizman (ed.), The herpesviruses, 1A, pp. 25 − 103, Plenum, New York, 1983

6　Deturenne-Tessier M, OOKA T, CALENDER A, DE-THE G, Daillie J. Relationship between nasopharyngeal carcinoma and high antibody titers to Epstein-Barr virus specific thymidine kinase. Int J Cancer, 1989, 43: 45 −48

7　Graham P L, Van Der Erb, A J. Anew technique for the assay of infectivity of human adenovirus 5 DNA. Virology, 1972, 52: 456 −467

8　Hardwick J M, Lieberman P M, Hayward S D. A new Epstein-Barr virus transactivator, R, induces expression of a cytoplasmic early antigen. J Virol, 1988, 62: 2274 −2284

9　Harrison T, Graham F L, Williams J. Host range mutant of Adenovirus type 5 defective for growth in HeLa cells. Virlolgy, 1977, 77: 319 −329

10　Henle G, Henle W. Epstein-Barr virus specific IgA serum antibody as an outstanding feature of nasopharyngeal carcinoma. Int J Cancer, 1976, 17: 1 −7

11　Joab I, Rowe D T, Bodescot M, Nicolas J C, Farrell P J, Perricaudet M. Mapping of the gene coding for Epstein-barr virus determined nuclear antigen Ebna3 and its transient overexpression in a human cell line by using an adenovirus expression vector. J Virol, 1987, 61: 3340 −3344

12　Joab I, Triki H, De Saint Martin J, Perricaudet M, Nicolass J C. Detection of anti-Epstein-Bar virus transactivator (Zebra) antibodies in sera from HIV patients. J infect Dis, 1991, 63: 53 −56

13　Liberman P M, O'hare P, Hayward G S, Hayward S D. Promiscuous transactivation of gene expression by an Epstein-barr virus encoded early nuclear protein. J Virol, 1986, 60: 140 −148

14　Raab-traub N, Flynn K. The structure of the termini of the Epstein barr virus as a marker of clonal proliferation. Cell, 1986, 47: 883 – 889

15 Ronney C M, Rowe D T, Ragot T, Farrell P J. The spliced BZLFI gene of Epstein-Barr virus (EBV) transactivates an early EBV promoter and induces the virus productive cycle. J Virol, 1986, 63: 3109 – 3116

16 Tan R S, LiJ S, Grill S P, Nutter L M, Cheng Y C. Demonstration of Epstein-Barr virus specific DNA polymerase in chemi-cally induced Raji cells and its antibody in serum from patients with naso-pharyngeal carcinoma. Cancer Res, 1986, 46: 5024 – 5028

17 Wong K M, Levine A J. Identification and mapping of Epstein Barr virus early antigens and demonstration of a viral gene activator that functions in trans J Virol, 1986, 60: 149 – 156

18 Zeng Y, Gong C H. Jan M G, Fun Z, Zhang L G, LI H Y. Detection of Epstein-Barr virus IgA EA antibodies for diagnosis of nasopharyngeal carcinoma by immunoautoradiography. Int J Cancer, 1983, 31: 599 – 601

19 Zeng Y, Zhang L G, Li H Y, Jan M G, Zhang Q, WU Y C, Wang Y S, SU G R. Serological mass survey for early detection of nasopharyngeal carcinoma in Wuzhou City, China. Int J Cancer, 1985, 29, 139 – 141

20 Zeng Y, Zhong J M, Li H Y, Wang P S, Tang H, Ma Y R, Zhu J S, Pan W J, Liu Y X, Wei Z N, Chen J Y, Mo Y K, Li E J, Tan B F. Follow-up studies on Epstein-Barr virus IgA VCA antibody-positive persons in Zangwu county, China. Intervirology, 1983, 20: 190 – 194

53. A Research for the Relationship between Human Papillomavirus and Human Uterine Cervical Carcinoma

I. The identification of viral genome and subgenomic sequences in biopsies of Chinese patients

SI Jing – yi[1], LEE Kun[1], HAN Ri – cai[1], ZHANG Wei[1], TAN Bing – bing[1], SONG Guo – xing[1], LIU Shi – de[1], CHEN Lian – fong[1], ZHAO Wei – ming[1], JIA Li – ping[1], MAI Yong – yan[2], ZENG Yi[3], ZHOU Yi – nan[4], WANG Yue – zhu[4], LING Jian[4], SUN Yu[5], MENG Xiang – jin[5], YU Zhang – fong[6], PU Li – ming[6]

1. Department of Biophysics, Instiute of Basic Medical Sciences, Chinese Academy of Mdedical Sciences and the Peking Union Medical College, 5 Dong Dan San Tiao, Beijing 100005, China; 2. Department of Gynecology and Obstetrics, The First Teaching Hospital, Beijing Medical University, Beijing 100034, China; 3. Department of Tumor Viruses, Institute of Virology, Chinese Academy of Preventive Medicine, 100 Yin Xin Jie, Beijing 100052, China; 4. Department of pathology, Gynecology and Obstetrics, The First Teaching Hospital, Xinjiang Medical College, Urumchi 830054, Xinjiang Uighur Autonomous Region, China; 5. Institute of Virology, Hubei Medical College, Wuuhan 430071, Hubei Province, China; 6. Department of Epidemiology, Faculty of Public Health, Harbin Medical University, Harbin 150001, Heilongjiang Province, China

Summary

Biopsies from 318 cases with squamous cell carcinoma of the uterine cervix, 48 with cervical and vulvar condylomata, 14 with cervical intraepithelial neoplasia (CIN), 34 with chronic cervicitis and 24 with normal cervical epithelium were collected from different geographic regions with different cervical cancer mortalities. The NDA. DNA dot – blot and Southern blot hypridization results show that there is a close relationship between HPV-16 and the uterine cervical squamous cell carcinoma in China. One very interesting observation is that the finding of HPV-16 – homologous DNA differs significantly among five geographic regions, and corresponds with the mortalities from cervical cancer of these five regions. HPV – 11 was found mainly in benign lesions. The rate of detection of HPV-16 in Chinese women increased from 8.3% in normal cervical epthelium to 20% in chronic cervicitis, 28% in cervical condyloma. 50% in CIN and 60.4% in cervical cancer. It is suggested that HPV-16 infection may be an etiological factor in the development of human cervical carcinoma. From the results of Southern blot hybridization, it appeared that HPV-16 DNA had been integrated into the genome of the host cell in cervical cancer. Whereas the HPV-16 DNA sequence was only present as an episome in normal cervical epithelium and cervical benign lesions. The rate of occurence of E6 – E7 genes is the highest (88.9%) compared with that of other subgenomic fragments of HPV-16 in specimens of human cervical cancer in China. This implies that E6 and E7 may be the oncogenic genes of HPV-16 and play an important role in the carcinogenesis of human cervical epithelial cells. The amplification and rear-rangement of the c-*myc* protooncogene are closely associated with the occurrence of cervical cancer. The results presented here revealed that the activated c-*myc* oncogene may cooperate with HPV-16 in the carcinogenic processes.

[**Key words**]　　Uterine cervical carcinoma; Human papillomavirus; Moleculat epidemilolgy; Carcinogenesis; Viral oncogene

Introduction

Many authors have shown that the risk of cervical carcinoma in women is strongly associated with hygiene (Beral 1974). This suggests that a sexually transmitted infectious agent may be involved. For over 20 years, much attention has been focused on herpes simplex virus type2 (HSV – 2) (McDougall et al. 1984; Vonka et al. 1984; Frenkel et al. 1972; Aurelian et al 1981), but its role in cervical carcinoma has not been confirmed (Eglin et al. 1981). In recent years, there has been more support for the view that papillomavirus (HPV) infection might be an etiological factor in the development of human cervical cancer (Howley 1989; Lee et al. 1988; Orth et al. 1977; zur Hausen 1989a). HPV is a subgroup of the papovaviruses and is a species-specific small DNA virus containing about 7.9×10^3 bases (7.9 kb) of double-stranded DNA in its genome. To date, more than 60 genotypes of HPV have been isolated and characterized, of which HPV – 6, 11, 16, and 18 appear in a high percentage of cervical, vulvar and penial cancer tissues, whereas HPV – 6 and HPV – 11 are found mainly in benign lesions of the female genital tract (Pecoraro et al. 1989; zur Hausen 1989b).

The squamous cell carcinoma of the uterine cervix is one of the most common cancers in Chinese women (Lee et al. 1958). As there are few reports concerning the relationship between Chinese patients and HPV, the present study is designed to detect the viral DNA sequences in the proliferative lesions of the female genital tract and in cervical cancer, using HPV – 11, 16 and 18 probes, and nucleic acid hybridization methods in order to ascertain whether the development of cervical cancer is associated with HPV infection in China.

Materials and Methods

Biopsics from 318 cases with squamous cell carcinoma of the uterine cervix, 48 with cervical and vulvar condylomata, 14 with cervical intracpithelial ncoplasia (CIN), 34 with chronic cervicitis and 24 subjects with normal cervical cpithclium werc diagnosced by the Department of Pathology, and were collected from different geographic regions of China Xinjiang Uighur Autonomous Region (western China), Beijing (northern China), Heilongjiang Province (northeastern China), Hubei Province (central China) and Guizhou Province (southwestern China).

All specimens were prepared for DNA · DNA dot-blot hyridization assay by using HPV – 11. 16, 18 whole genomic DNA labelled with $[^{32}p]$ dNTP as probes to detect viral homologous sequences in samples. Among those casses with positive results on dot-blot hybridization (with HPV-16 DNA as probe), 30 were randomly chosen for Southern blot hybridization with HPV-16 whole genomic DNA and its subgenomic sequences, E6, E7, E1, E4; as well as E5, L1 and L2 as separate probes.

1. Nucleic acid hybridization DNA preparation (Davis et al. 186; Cai1987)

Tissues were homogenizcd in an ice bath and resuspended in TRIS/EDTA buffer. After adding proteinase K to a final concentration of 100 μg/ml and sodium dodecyl sulphate (SDS) to a final concentration of 0. 5% , the samples were incubated at 55℃ for 5h and phenol/chloroform was added for cxtracting DNA followed by ethanol for DNA precipitation. Finally, DNA was disolved in TRIS/EDTA buffer and stored at – 20℃.

2. DNA · DNA dot-blot hybridization (Denhardt 1966)

To prepare the blot DNA, 2μg DNA of each sample, 2μg salmon sperm DNA for the negative control and 1μg HPV plasmid DNA from each type for positive control were transfered to a nitroccllulose filter; they were then denatured with 0. 5mol/L NaOH and the filter was dried in air for hybridization. To prepare the probes, HPV – 11, 16, 18 rccombinant DNA in pBR-322 plasmids was radiolabelled with $[^{32}p]$ dTTP or dCTP by the nick-translation method [specific activity, (1 – 3) $\times 10^{8}$ cpm/μg DNA] (Rigby et al. 1977), For the hybridization, filtcrs were incubated in 0. 25% low-fat milk and 6 × standard salinc citrate (SSC; Sambrook et al, 1989) for 2h at 60℃. ^{32}P-labelled HPV DNA probes were then added to the solution and incubated for 16h at 68℃.

The filter was washed three times with a mixture of 2 × SSC, 0. 1% SDS and 0. 2% low-fat milk for 30min. Filters were then dried, placed in a plastic mount and exposed to X-ray film with an intensifying screen for 12 – 72h at – 70℃.

3. Soutern blot hybridization (Southern 1975)

DNAs extracted from samples were digested with *Pst l*, and separated on a 0. 8% agarose gel (10μg DNA/sample) for electrophorcsis for 12 – 24h at room temperature with low voltage (1 – 2 V/cm length of gel). After denaturation with 0. 15mol/L NaOH for 1h, the neutralized gcl was transferred onto a nitrocellulose filter by the modificd method of Southern (Guo at al. 1987).

HPV-16 total DNA as well as its subgenomic fragments, such as the 1. 3-kb fragment (containing the E6 – E7 open reading frame) and the 1.0-kb (E2, E4) and 2.6-kb (E5, L1 and 1.2 open reading frames) fragments recovered from the digestion of EcoRI, $Pst1$ and $Ava11$ after electrophoresis, were radiolabelled with $[^{32}P]$ -CTP (10μCi/μl) by the nick-translation method to a specific activity of (1 – 3) $\times 10^8$ cpm/μg DNA to become the probes.

Hybridization was carried out by a modification of the procedure of Denhardt (Denhardt 1966). The filter was put into a sealed plastic bag containing prehybridization solution (100μg/ml salmon sperm DNA, 0.25% low-fat milk, 6 × SSC) for prehybridization at 68℃ 6 – 18h. Then, hybridization was performed in the solution of prehybridization plus the ^{32}P-radiolabelled probes at 68℃ for 16 – 24h. The filter was then washed three times in 2 × SSC plus 0.1% SDS (15 min/ time) at room temperature, and three times in 0.1 × SSC solution plus 0.5% SDS at 62℃ for 15min.

The filter was exposed to X-ray film in a plastic mount with an intensifying screen for autoradiography at – 70℃.

Results and Discussion

1. Identiftcation of HPV DNA sequences from lesions of the epithelial tissues of the female genital tract

The results of dot-blot hybridization of 318 cases of cervical carcinoma collected from five geographic regions of China, using ^{32}P-labelled HPV – 11, 16 and 18 DNA as probes, is shown in Figs. 1 and 2, and summarised in Tab. 1.

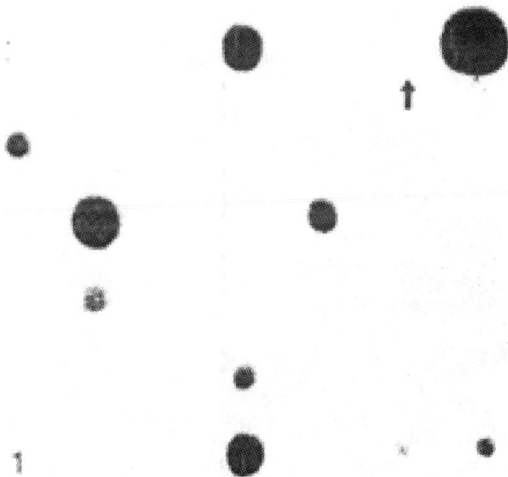

2

Fig. 1　Part of the results of dot-bolt hybridization of DNA from cervical cancer specimens using HPV-16 DNA as probe. HPV-16 DNA serves as the positive control (⇒), and salmon sperm DNA, as the negative control (➡)

Fig. 2　Part of the results of dot – blot hybridization of DNA of cervical cancer specimens from Guizhou Province showing less hybridization with HPV-16 DNA as probe. HPV-16 DNA serves as the positive control (⇒), and salmon sperm DNA, as the negative control (➡)

Tab. 1   The homologous sequences of human
papillomavirus (HPV) DNA in human uterine
cervical carcinoma tissue

Geographic regions	No. of patients	No. (percentage positive)		
		HPV – 11	HPV-16	HPV – 18
Xinjiang	99	4 (4%)	77 (77%)	7(7%)
Beijing	49	1 (2%)	32(65.3%)	3(6.1%)
Heilongjiang	50	0	28(56%)	2(4%)
Guizhou	21	0	10(47.6%)	0
Hubei	99	1 (1%)	45(45%)	2(2%)
Total	318	6 (1.9%)	192(60.4%)	14(4.4%)

Of the 318 cases analysed, 60.4% had detectable HPV-16 DNA. This result shows that there is a close relationship between HPV-16 and human cervical carcinoma in China. The high detection rate of HPV-16 – homologous DNA in cervical cancer in China coincides with some recent reports from different countries (Durst et al. 1983; Crum et al. 1984; Gissmann 1984; Coleman et al. 1986). Our findings also revealed that HPV-16 infection was more prevalent than that with PV – 11 and 18 in the Chinese cancer cases. However, the frequency of HPV-16 – homologous DNA was significantly different among five geographic regions. Biopsies from Xinjiang Uighur Autonomous Region showed the highest positive rate (77%). Further analysis revealed that there were significant variations in the rate between rural areas and Urumchi city in Xinjiang – a much higher percentage (88%) occurring in the former compared with only 66% in the latter. The positive rates were lowest in biopsies collected from Guizhou and Hubei. It is interesting that the occurrence of HPV-16 – homologous DNA corresponds with the mortality rate of cervical cancer in the five different regions, as illustrated in Fig. 3. Up to now, there has been no report on the above result. The reasons why different areas show different homlolgies with the viral genome, and why these correspond to the mortality rates of the areas are still unknown, and demand further study.

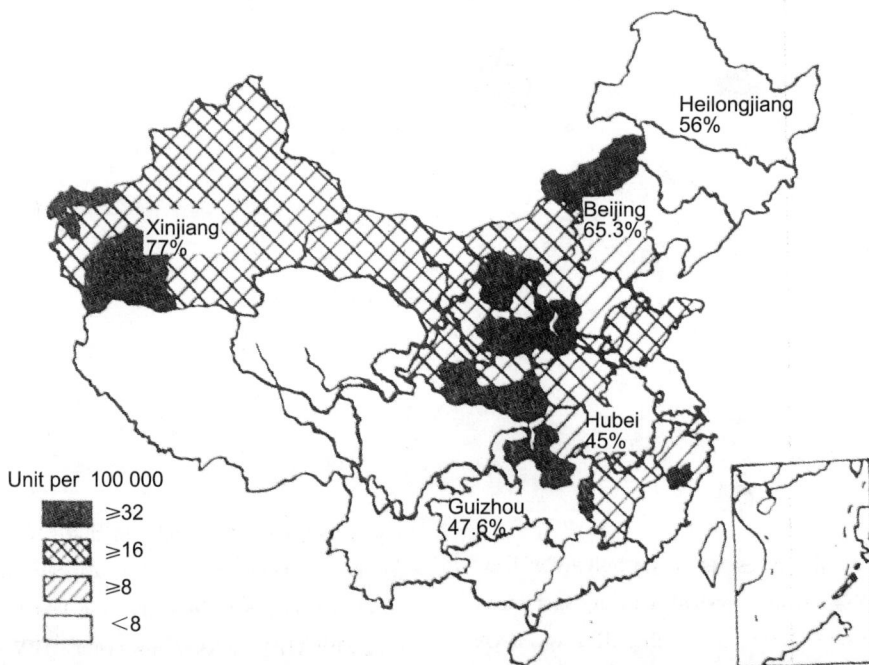

Fig. 3   The geographic distribution of mortality from uterine cervical carcinoma in
China. The percentages in this figure indicate the extent to which cervical
carcinoma tissue from different geographic regions is homologous to HPV – 16 DNA

2. HPV DNA sequences in other lesions of the female genital tract

The results with other lesions of the female genital tract, including CIN, Condyloma and chronic cervicitis, are summarised in Tab. 2.

The percentage of HPV-16 DNA detected in cancer cases was much higher than that in patients with other lesions of the female genital tract, whereas the rate of detection of HPV – 11 DNA in biopsies from other lesions was higher than that in carcinoma cases. These results showed that HPV – 11 existed mainly in benign lesions such as condyloma. The patients with CIN revealed as association with HPV-16 in fection the positive rate of HPV-16 – homologous DNA being 50% in the CIN cases.

The rate of detection of HPV-16 in Chinese women has increased from 8.3% in normal cervical epithelium, 20.0% in cervicitis, 28.0% in cervical condyloma and 50.0% in CIN to 60.4% in cervical cancer. The findings in our study support the widely accepted views that there is a close relationship between HPV infection, CIN and cervical carcinoma.

Sequences homologous to HPV-16 have been detected in 8.3% of normal cases. It is possible that normal women with no clinical evidence of cervical HPV infection could harbour HPV DNA in their cervical epithelium. The risk of cancer in these normal cases is not yet understood, and this can only be resolved by olngterm prospective studies.

The rate of HPV – 18 infection was decidedly

Tab. 2 The homologous sequences of HPV DNA in cervical intraepithelial neoplasia (CIN), condyloma, chronic cervicitis and cervical carcinoma compared with normal cervical tissue

Pathological diagnosis	No. of cases	No. (percentage positive)		
		HPV – 11	HPV-16	HPV – 18
Cervical carcinoma	318	6 (1.9%)	192(60.4%)	14(4.4%)
CIN	14	0	7 (50%)	0
Cervical condyloma	25	8 (32.0%)	7 (28.0%)	0
Vulvar condyloma	23	10 (43.5%)	2 (8.7%)	0
Chronic cervicitis	34	0	7 (20.6)	0
Normal cervical epithelium	24	0	2 (8.3%)	0

lower than that of HPV-16 in China, whereas a cervical cell line derived from a Chinese patient with cervical squamous epithelial cell carcinoma, CC801, was proved to be positive for HPV – 18 DNA by Southern blotting in our laboratory. This finding coincides with results for several other cervical cancer cell lines, including Hela cells (Han et al. 1989). Its meaning and significance deserve further study.

3. Detection of HPV-16 – DNA – homlolgous sequences from biopsies of cervical lesions by the Southern blot hybridization method

In order to analyse the physical state of HPV DNA in cervical benign lesions including condyloma, chronic cervicitis, CIN and cervical carcinoma as well as in normal cervical epithelium, Southern blot hybridization was employed.

A random selection of 32 cases among those that were positive on dot-blot hybridization (with HPV-16 DNA as probe) was used for Southern blot analysis, including 8 cases of condyloma, 2 of CIN, 20 of cervical cancer and 2 of normal epithelium. All the samples were digested by different restriction endonucleases and subjected to electrophoresis on a 0.8% agarose gel; Southern blot hybridization was then carried out with ^{32}P-labelled HPV-16 DNA as the probe.

From the results it appeared that the HPV-16 DNA sequence ould only be detected as an epi- some in normal cervical epithelium and cervical benign lesions such as condyloma. On the other hand, HPV-16 DNA was found to have been integrated into the genome of host cell of cervical canc- er. Several positive bands were shown in these cervical samples; that is , there was multisite inte- gration of HPV-16 DNA sequences into the host cell genome (Fig. 4).

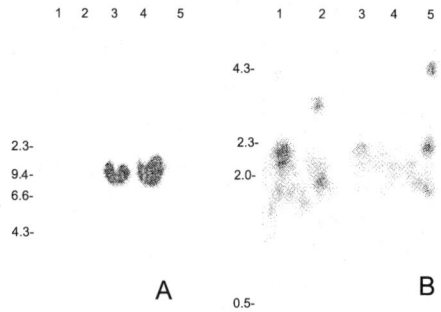

Fig. 4 a, b. Part of the results of Southern blot hybridization of the DNAs from cervical cancer spec- imens with HPV-16 DNA as probe. Cervical cancer DNAs were digested with (a) *Bam*HI and (b) with *Pst*I and subjected to electrophoresis on a 0.8% aga- rose gel. The specimen of *lane I* (a) serves as nega- tive control and is from a case of normal cervical epi- thelium showing no dot-blot hybridization with HPV- 16 DNA. The numbers at the *left* of each panel repre- sent the sizes (kb) of the fragments of λ phage DNA digested with *HindIII*

As for CIN, the situation differed between the benign lesions and cancerous tissue: in one case, the HPV-16 DNA sequence existed in the form of an episome, and in another case these was integration (Fig. 5).

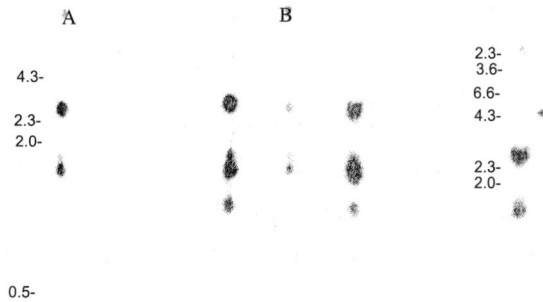

Fig. 5   A – C. Southern blot hybridization analysis of the DNA from different cer- vical lesions with HPV-16 DNA as probe. DNAs (10μg/sample) extracted from normal cervical epithelia (A), cevrical condyloma (B), and cervical intraepi- thelial neoplasia (CIN) (C) were digested with *Pst*I, separated on a 0.8% aga- rose gel and then subjected to Southern blot analysis. *Arrowhead* indicates the ad- ditional fragment in *Pst*I-digested DNA from CIN. In one case of CIN (C lane h), HPV-16 DNA sequences were integrated, and in another case (*Clane u*) they were present as an episome. The numbers at the *left* of each panel represent the si- zes (kb) of the fragments of λ phage DNA digested with *Hin*dIII

4. Determination of HPV-16 oncogenic genes

Inorder to determine which fragments of the HPV-16 genome are responsible for the carcinogenicity, three fragments of HPV-16 DNA, E6 – E7, E2 – E4 and E5 – L1 – L2, were collected (Fig. 6) and labelled with $[^{32}P]$ dTTP as probes after digestion of HPV-16 DNA by restriction endonucleases, PstI, EcoRI and AvaII.

A group of 29 specimens were chosen from those cases

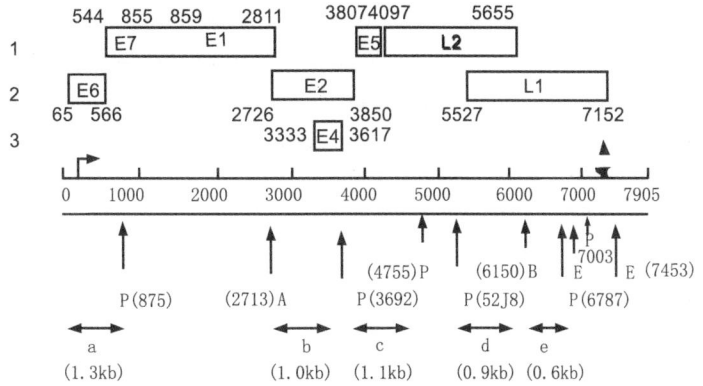

Fig. 6   Genomic organization of HPV-16 and subgenomic probes. Restriction sites of AvaII. BamHI, EcoRI are indicatcd by A. B. E and P. respeetively. Below the line, subgenomic probes are indicated by a, b, c, d and c

found to be positive by dot – blot hybridization using the whole HPV-16 genome as a probe. The highest rate of positivity was found with E6 – E7 as the probe (88.9%). This means that E6 – E7 open-reading frames were always retained completely in the cancer cell genome, and that other fragments may be defetive in the carcinogenic process (as shown in Tab. 3). Schwarz et al. (1985) and Smotkin and Wettstein (1986) indicated that the E6 and E7 early genes of papillomavirus were implicated in the induction and maintenance of ncoplasias. In vitro, E6 and E7 of HPV-16 were shown to induce anchorage independence in rodent fibroblasts (Kanda et al. 1988; Zhang et al. 1991). In our laboratory, NIH3T3 cells were transformed successfully by recombinant retrovirus containing E6 and E7 of HPV-16, and the specific mRNA of E6 and E7 was detected in transformed cell (Si et al. 1991).

Our studies reported here imply that the E6 and E7 genes of HPV-16 are required for establishment and maintenance of the transformed phenotype. Therefore, E6 and E7 may be the oncogenic genes of HPV-16, and play a key role in the transformation processes.

5. Analysis of protooncogenes in the genomes of cervical cancer specimens

The DNA of 15 cases of cervical cancer that hybridized with HPV-16 DNA, and DNA of 2 normal cases were digested by E. cor: or Hind III and run on an 0.8% agarose gel. After denaturing,

Tab. 3   The result of southern blot hybridization of HPV-16 subgenomic fragments from 27 cascs of cervical canccr specimens that were proved to be positive for HPV-16 whole genome

HPV-16 subgenomic probes	No. of patients	No. positive	Percentage positive
E6, E7	27	24	88.9
E2, E4	27	3	11.1
L1, L2, E5	27	15	55.6

the DNA was transferred from the gel to a nitrocellulose membrane, and finally hybridized with the radiolabelled probes c-mye, c-Ha-ras and c-fosby Southern blotting.

It was found that the protooncogenc c-myc was either amplified or rearranged, and that both alterations were present in 10 cervical carcinoma specimens, which contained integrated HPV-16 DNA in their genomes (Fig. 7). There were no similar alterations to be found in normal specimens. Only one sample from a cancer case appeared to have the c-Ha-*ras* protooncogene rearrangement and none appeared to have amplification of rearrangement of c-*fos* protooncogenc.

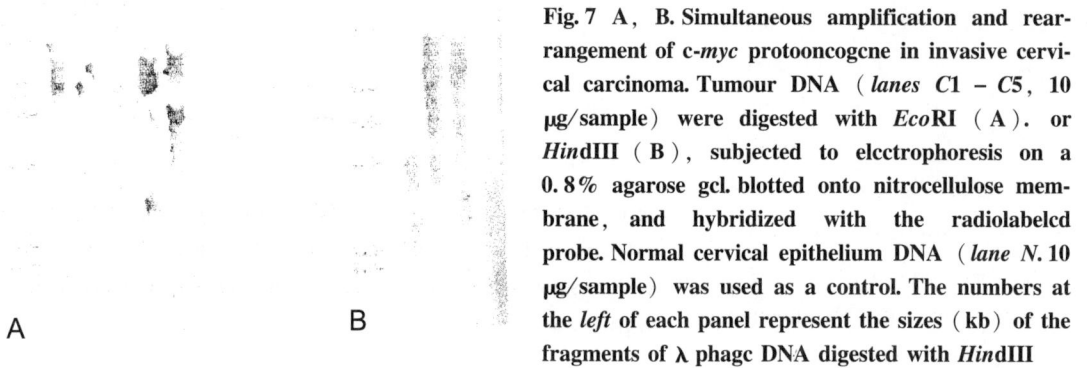

A   B

Fig. 7 A, B. Simultaneous amplification and rearrangement of c-*myc* protooncogcne in invasive cervical carcinoma. Tumour DNA (*lanes C*1 − *C*5, 10 μg/sample) were digested with *Eco*RI (A). or *Hin*dIII (B), subjected to elcctrophoresis on a 0.8% agarose gcl. blotted onto nitrocellulose membrane, and hybridized with the radiolabelcd probe. Normal cervical epithelium DNA (*lane N*. 10 μg/sample) was used as a control. The numbers at the *left* of each panel represent the sizes (kb) of the fragments of λ phagc DNA digested with *Hin*dIII

This finding suggests that amplification and rearrangement of the c-*myc* protooncogene arc closely associated with the occurrence of cervical cancer. The data presented here revealed that the activated c-*myc* oncogene could cooperate with HPV-16 in the carcinogenesis of human cervical cancer. Some other agents, such as HSV − 2, smoking and so on, may also be needed and cooperate in the oncogenic processes. The mechanism by which HPV, oncogenes and other agents exert their cooperative actions remains to be studied.

6. Conclusions

From the DNA · DNA dot-blot and Southern blot hybridization results, it shows that there is a close relationship between HPV-16 and uterinc cervical squamous cell carcinoma in China. One very interesting finding is that the occurrcnce of HPV-16 − homologous DNA differs significantly among the five geographic regions, and corresponds with the cervical cancer mortalitics of those regions.

HPV − 11 existed mainly in benign lcsions. The rate of detection of HPV-16 in Chinese women increased from 8.3% in normal cervical epithelium to 20.0% in chroruc cervicitis, 28.0% in cervical conodyloma, 50.0% in CIN and 60.4% in cervical cancer. It is suggested that HPV-16 infection may be an etiological factor in the development of human cervical carcinoma.

From the results of Southern blot hybridization, it appears that HPV-16 DNA was integrated into the genome of the host cell in cases of cervical cancer, whereas the HPV-16 DNA scquence was only present as an episome in normal cervical epithelium and cervical benign lesions.

The occurrence of E6 − E7 genes is highest (88.9%) compared with other subgenomic fragments of HPV-16 in the specimens of human cervical cancer in China. This implies that E6 and E7 genes may be the oncogenic genes of HPV-16 and play an important role in the carcinogenesis of the human cervical epithelial cell.

The amplification and rearrangement of c-*myc* protooncogene are closely associated with the occurrence of cervical cancer. The results presented here reveal that the activated c-*myc* oncogene may cooperate with HPV-16 in the carcinogenic processes.

Acknowledgements

This project was supported by the National Natural Science Foundation of China, the Science Foundation of the Ministry of Public Health and the Science Foundation of the National Seventh Five-Year Plan of China.

〔In 《J Cancer Res Clin Oncol》 1991, 117: 454 - 459〕

References

1 Aurelian L, Dessler II, Rosenshein NB, Barbour G. Viruses and gynecologic cancers: herpesvirus protein (ICP10/AG - 4), a cervical tumor antigen that fulfills the criteria for a marker of carcinogenecity. Cancer, 1981, 48: 455 - 471

2 Beral V. Cancer of the cervix: a sexual transmitted infection. Lancet, 1974, 11073

3 Cai L. In: Cai L (ed). The techniques in research for nuclcic acid. Science Press, Beijing, 1987

4 Coleman DV, Wickenden C, Malcolm ADB Association of human papillomavirus with squamous carcimoma of the uterine cervix. Ciba Founda Symp, 1986, 120: 175 186

5 Crum CP, lkenberg H, Richart RM, Gissmann L. Human papillomavirus type 16 and early cervical neoplasia. N Engl J Med, 1984, 310: 880 - 883

6 Davis LG, Dibner M, Battey JF. Guanidine isothiocyanate preparation of total RNA. In: Basic methods in molecular biology. Elscvier. Amsterdam, 1986, 130 - 135

7 Denhardt D. TA membrane-filter technique for the detection of complementary DNA. Biochem Biophys Res Commun, 1966, 23: 641 - 646

8 Durst M, Gissmann L, Ikenberg H, zur Hausen H. A papil-lomavirus DNA from a cervical carcinoma and its prevalence in cancer biopsy samples from different geographic regions. Proc Natl Acad Sci USA, 1983, 80: 3812 3815

9 Eglin RP, Sharp F, Maclean AB, Macnab JCM, Clements JB, Wilkie NM. Detection of RNA complementary to herpes simplex virus DNA in human cervical squamous cell neoplasms. Cancer Rcsm, 1981, 41: 3597 3603

10 Frenkel N, Roizman B, Cassai E, Nahmias A. A DNA fragcervical cancer tissue. Proc Natl Aead Sci USA, 1972, 69: 3784 - 3789

11 Gissmann L. Papillomaviruss and their association with cancer in animal and man. Cancer Surv, 1984, 3: 161 - 181

12 Guo Xiaojun, Wang Shenwu Low fat milk used for nucleic acid hybridization. Adv Biochm Biophys, 1987, 2: 7

13 Han R, Si J, Zhang W, Lee K. The detection of the human papillomaviruses type 18 and 11 homologous DNA sequences and their transcription in human cervical cancer and its cell line (CC - 801). Chin J Cancer, 1989, 8: 383 - 384

14 Hausen H zur. Papillomaviruses as carcinoma viruses. In: Klein G (ed) Advances in viral oncology, vol8. Raven, New York, 1989a, 1 - 26

15 Hausen H zur. Papillomaviruses in anogenital cancer as a model to understand the role of viruses in human cancer. Cancer Res, 1989b, 49: 4677 - 4681

16 Howley PM. The role of papillomaviruses in human cancer. In: Devita VT, Hellan S, Rosenberg SA (eds) Important advance in oncology. Lippincott, Philadelhia, 1989, 55 - 73

17 Kanda T, Furuno A, Yoshiike K. Human papillomavirus type 16 open reading frame E7 encodes an transforming gene for rat 3Y1 cells. J Virol, 1988, 62: 610 - 613

18 Lee K, Hu S, Chen M, Ding L, Liu T, Hu Z. A statistical analysis of 27149 biopsy cases of cancer in China. Chin J Pathol, 1958, 4: 258 - 260

19　Lee K, Si J, Han R, Wang S, Zhang W, Song G, Liu S, Chen L, Y, Sun Y, Meng X. An ultrastructural research for the relationship between human uterine cervical carcinoma and human papillomavirus: II. A combined gene molecular and ultrastructural study. In: Hashimoto H, Kuo KH, Lee K, Ogawa K (eds). Recent devclopment of electron microscopy. Nakanishi, Kyoto, 1988, 305 – 312

20　McDougall JK, Smith P, Tamimi Hk, Tolentino E, Galloway DA. Molecular biology of the relationship between Herpcs simplex virus-2 and cervical cancer. In: Giraldo G, Beth E (eds) The role of viruses in human cancer. Elscvier, Amsterdam, 1984, 59 – 71

21　Orth G, Brettburd F, Favre, Croissant O. Papillo-maviruses: possible role in human cancer. In: Hiatt HH, Watson JK, Winsten JA (eds) Origins of human cancer. Gold Spring Harbor Laboratories, New York, 1977, 1043 – 1068

22　Pecoraro G, Morgan D, Defendi V. Differential cffects of human papillomavirus type 6, 16 and 18 DNAs on immortalization and transformation of human cervical epithelial cells. Proc NatlAcad Sci USA, 1989, 86: 563 – 567

23　Rigby PWJ, Dieckmann M, Rhodes C, Berg P. Labelling deoxyribonucleic acid to high specific activity in vitro by nick translation with DNA polymcrase I. J Mol Biol, 1977, 113: 237 – 251

24　Sambrook J, Fritsch EF, Maniatis T. In: Sambrook J, Fritsch EF, Maniatis T (eds) Molecular cloning, a laboralory manual. 2nd edn, vol 3, B. 13, Preparation of reagents and buffers used in molecular cloning. Cold Spring Harbor Laboratory, New York, 1989

25　Schwarz E, Freese UK, Gissmann L, Mayer W, Roggenbuck B, Strcmalav A, Hauscn II zur. Structure and transcription of human papillomavirus sequences in cervical carcinoma cells. Nature, 1985, 314: 111 – 114

26　Si J, Lec K, Zhang W, Han R, Song G, Chen L, Zhao W, Jia L, Liu S, Mai Y, Zeng Y. A research for the relationship between human papillomavirus and human utcrine cervical carcinoma: II. Molecular genetic and ultrastructural study on the transforming activity of recombinant retrovirus containing human papillomavirus type 16 subgenomic scquences. J Cancer Res Clin Oncol, 1991, 117: 460 – 472

27　Smotkin D, Wettstein FO. Transcrption of human papillomavirus type 16 carly genes in a cervical cancer and a cancer-de-rived cell line and identification of the E7 protein. Proc Natl Acad Sci USA, 1986, 83: 4680 – 4684

28　Southern EM. Detection of specifie sequences among DNA fragments separated by gel clectrophoresis. J Mol Biol, 1975, 98: 503

29　Vonka V, Kanka J, Jelinck J, Subrt l, Zuchanek A, Havranbova A, Vachal M, Hirsch I, Domorazbova E, Zavadova H, Richterova V, Naprstkova J, Dvorakova V, Svoboda B Prospcctive study on the rclationship between cervical neoplasia and Herpes simplex type – 2 virus: I. Epidemiological characteristics. Int J Cancer, 1984, 33: 49 – 60

30　Zhang W, Si J, Lee K, Chen L, Zhao W, Han R. Cytological study on transforming activity of human papillomavirus type 16 subgenomic DNA in vitro. Chin J Med, 1991

54. A Research for the Relationship between Human Papillomavirus and Human Uterine Cervical Carcinoma

II. Molecular genetic and ultrastructural study on the transforming activity of recombinant retrovirus containing human papillomavirus type 16 subgenomic sequences

SI Jing – yi[1], LEE Kun[1], ZHANG Wei[1], HAN Ri – cai[1], SONG Guo – xing[1], CHEN Lian – fong[1], ZHAO Wei – ming[1], JIA Li – ping[1], LIU Shi – de[1], MAI Yong – yan[2], ZENG Yi[3]

1. Department of Biophysics, Institute of Basic Medical Sciences, Chinese Academy of Medical Sciences and the Peking Union Medical College, 5 Dong Dan Sam Tiao, Beijing 100005, China; 2. Department of Gynecology and Obstetircs, The First Taching Hospital, Beijing Medical University, Beijing 100034, China; 3. Department of Tumor Viruses, Institue of Virology, Chinese Academy of Preventive Medicine, 100 Yin Xin Jie, Beijing 100052, China

Summary

In order to elucidate the role of HPV-16 in the development of genital cancer, NIH3T3 cells were transfected by HPV-16 whole genome and its two early genes, E6 – E7. Besides ordinary calcium phosphate/DNA coprecipitation technique, a newly designed recombinant retrovirus containing the HPV-16 genome or subgenomes was used to infect cells for transfer of the target genes. The transforming activities have been demonstrated to be most efficient when a bioengineering technique of this kind is used. HBV – 16 DNA was proved to have transforming potential for NIH3T3 cells, and the DNA of HPV-16 was proved to undergo multisite integration into transformed cells and nude mice tumour cells. The E6 – E7 open reading frames are sufficient for transforming NIH3T3 cells independently in vitro, which implies that E6 – E7 open reading frames are transforming genes or even viral oncogenes of HPV-16. The RNA transcribed by the E6 – E7 of HPV-16 was expressed in transformed cells and in tumour cells of nude mice. The use of a recombinant retrovirus for gene transfer in this study is much more efficient than of calcium phosphate/DNA coprecipitation. The lack of a tissueculture system suitable for HPV replication in vitro makes HPV gene recombination into a specially engineered retrovirus for viral – mediated gene transfer of particular significance for the possible application of viral carcinogenesis, both in vitro and in vivo, for basic and clinical research.

[Key words] HPV; Transforming activity; Recombinant retrovirus; Viral oncogene

Introduction

In recent years, there has been increasing evidence that human papillomaviruses (HPV) are emerging as the major agents for the etiology of human cervical carcinoma through clinical, epidemiological, molecular biological and tumour virological studies (Stanbridge et al, 1981; Fukushima et al. 1985; Howley 1989; zur Hausen 1985; Gao et al, 1988). More than 60 HPV genotypes have been isolated and characterized, and the heterogeneity of HPV types is considered to reflect the adaptation of these viruses in specific differentiated tissues (zur Hausen 1989). HPV - 6, 11, 16, and 18 have been found in neoplastic lesions of the genital tract (Pecoraro et al. 1989).

In our laboratory, a molecular hybridization study showed that 60.4% of biopsies obtained from Chinese patients with cerviral cancer contained a sequence homologous to HPV-16 DNA; it is therefore the most prevalent viral type in Chinese women (Si et al. 1987). Our earlier work also showed that the early - region E6 and E7 genes of HPV-16 might play a key role in the carcinogenesis of cervical cancer (Si et al, 1990).

Clinical observations and recent experimental results have suggested that the development of cervical cancer, like other cancers, involves multistage processes and multiple etiological agents (Weinstein 1988). In order to elucidate the role of HPV-16 in the development of gential cancer , the present study was designed to transfect cultured NIH3T3 cells with HPV-16 whole genome and its two subhenomic genes, E6 and E7, separately. In addition to the ordinary calcium phosphate/DNA coprecipitation technique, the method of a newly designed recombinant retrovirus containing the HPV-16 genome or subgenomes was used to infect cells for transferring the target genes. The transforming activities have been demonstrated to be more efficient when this kind of bioengineering technique is used compared with the results of ordinary methods.

The transforming activity of HPV-16 was first observed in mouse NIH3T3 cells (Yasumoto et al, 1986; Noda et al. 1988; Bedell et al. 1987). Several laboratories have shown that much more recalcitrant cells, both of fibroblast and epithelial origin, are efficiently transformed by HPV DNA singly or cooperatively with other agents such as oncogene EJ - ras, etc. (Tsunokawa et al. 1986; Matlashewski et al. 1988). Up to now, no report has been found on the use of a recombinant retrovirus as a vector for HPV gene transfection.

Materials and Methods

1. Plasmid preparation

A plasmid pSV2 - neo/HPV16, containing the whole HPV-16 genome, cloned into the BanHI site of pSV2 - neo, and plasmids HZIP - 16 and HZIP - 16K, containing the total early genes (6.6kb) as well as the E6 - E7 (1kb) subgenomic sequence in plasmid pZIP - NeoSV (X) 1, were kindly provided by McCance (Matladhewski et al. 1987) (Fig. 1 and 2)

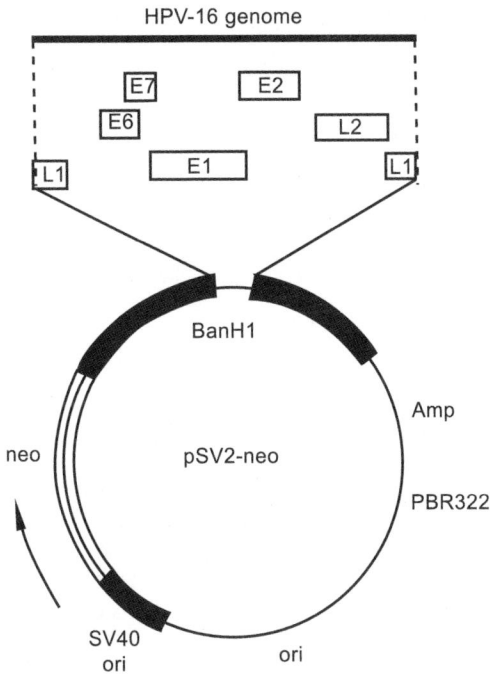

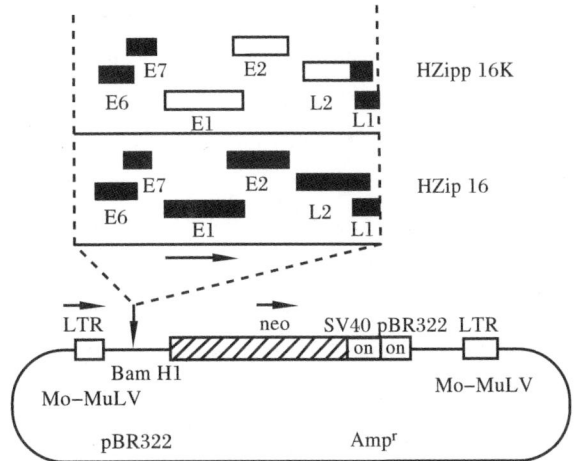

Fig. 2　Structure of recombinant plasmid HZip16, the HPV-16 6. 6 – kb fragment including nucleotides 0 – 6150 and 7454 – 7904. HZip16K is the HZip16 deletion mutant lacking the KpnI fragment from nucleotides 880 – 5377

Fig. 1　Structure of recombinant plasmid pSV 2 – neo/HPV16. The HPV-16 DNA was linearized at the unique BamHI site and cloned into the BamHI site of pSV2 – neo. Amp, ampicillin; SV40, simian virus 40; ori, origin

Recombinant plasmids pSV2 – neo/HPV16, HZip16, HZip16K and pZI P – NeoSV（X）l were transfected *into* E. coli by a modified method of Mandel (Mandel and Higa 1970).

2. Preparation of pSV2 – neo, pZIP – NeoSV（X）l vector plasmids

The pZIP – NeoSV（X）l vector consists of the Moloney murine leukemia virus（M – MuLV）transcriptional unit, derived from an integrated M – MuLV provirus, and PBR322 seqences necessary for the propagation of the vector DNA in *E. coil*. HZip16K was hydrolysed into two parts ［pZIP – NeoSV（X）l and E6 – E7］ by BamHI after electrophoresis. The pZIP – NeoSV（X）l part was collected and ligated into a circular plasmid by T4 ligase. pSV2 – neo was collected after pSV2 – neo/HPV16 had been digested by BamHT.

The amplification and purification of plasmids were performed as described by Mandel and Cai (Cai 1987).

NIH3T3 and ψ – 2 cells culture. NIH3T3 cells were maintained in Dulbecco modified eagle medium（DMEM）supplemented with 10% calf serum, 0. 2 mmol/L glutamine, 100 units/ml penicillin and 100 μg/ml streptomycin. Cells were grown in a humidified 5% CO_2 incubator at 37℃ and fed every 3 days.

ψ – 2 cells were cultured in the same conditions as NIH3T3 cells.

3. Transfections

pSV2 – neo/HPV16 plasmid transfection into NIH3T3 cells was performed by the calcium phosphate/DNA coprecipitation technique as described by Wigher (Wigher et al. 1978) and Davis (Davis et al. 1986). After transfection for 72h, NIH3T3 cells were digested by trypsin and cultured in DMEM with G418 (400 μg/ml) until anti-G418 clone formation.

HZIP-16, HZIP – 16K and pZIP-NeoSV (X) I were transfected into ψ2 cell by calcium phosphate/DNA coprecipitation. The ψ2 cell line contains integrated copies of the M-MuLV proviurs genome, which provides all the *trans* functions necessary for the encapsidation of a recombinant genome, whereas it is unable to encapsidate its own RNA. ψ2 cells were digested by trypsin 72 h after infection and cultured in DMEM medium containing G418 (400 μg/ml), until the formation of an anti-G418 cell clone that could produce defective retrovirus (pZIP-ψ2, HZIP16-ψ2 and HZIP16K-ψ2cells). After culture for 18h, the culture fluid of cloned ψ2 cells (PZIP-ψ2, PZIP16-ψ2, HZIP16K-ψ2 cells) containing a large number of virions (viral fluid) was harvested, filtered through a 0.45 – μm microporefilter, then added into NIH3T3 cell culture medium at 37℃ in a 5% CO_2 incubator. After 2 h, some fresh medium was added to the NIH3T3 cell culture flask. The transformed NIH3T3 cells (HZIP16-3T3 and HZIP – 16K – 3T3 cells) could be formed by infection with recombinant retrovirus after culture in G418 – containing (400 μg/ml) DMEM medium for selection of transformed cells.

The experimental design is illustrated in the schematic diagram shown in Fig. 3.

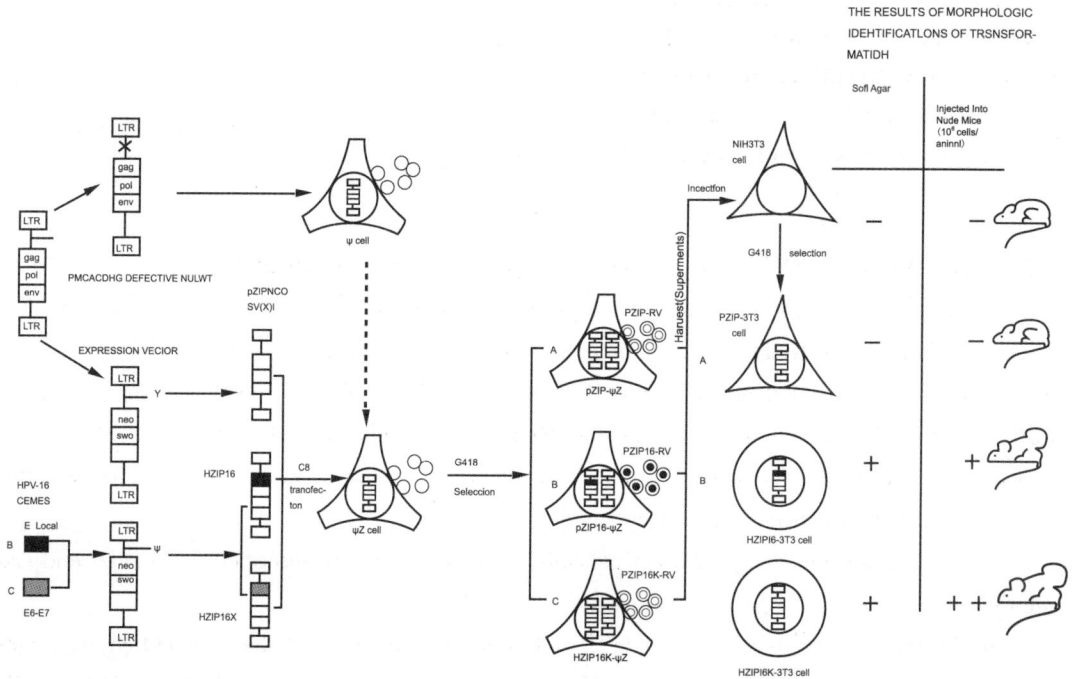

Fig. 3　A schematic diagram of the mechanism and results of the transformation of NIH3T3 cell by the recombinant retroviruses containing HPV-16 subgenomes

4. Transforming activity

Soft agar assay. Samples of 5×10^5 transformed cells were suspended in serum-free DMEM con-

taining 0. 3% agar (Difco Lab) , and the suspending mixture was plated onto the 0. 5% agar medium layer in culture dishes. After 15 min at room temperature, the dishes were removed into a CO_2
incubator at 37℃.

Tumorigenicity assay: HZIP16-3T3, HZIP16K – 3T3, pZIP – 3T3 and NIH3T3 cells were injected separately into nude mice subcutaneously at 1×10^6 cells/animal. pZIP – 3T3 and NIH3T3
cells were used for negative controlls. Five mice were injected with each kind of cell.

5. DNA · DNA hybridization

DNA extracted from the cells mentioned above was digested by restriction endonuclease
*Bam*HI, then run on an 0. 8% agarose gel in buffer containing 0. 04 mol/L Tris-acetic acid ,
0. 002 mol/L EDTA. In addition to the above buffer, 0. 2 μg/ml ethidium bromide was added. After
UV transilluminator photograping, the gel was placed in a denaturing solution (0. 25 mol/L NaOH
and 1. 5 mol/L NaCI) with shaking for 30 min (repeated twice) and, after neutralization, was
subjected to Southern blotting. The DNA was transferred to a nitrocellulose filter from the gel by the
method of Southern (Southern 1975).

The HPV-16 DNA was radiolabelled with $[^{32}P]$ dCTP by nicktranslation (Rigby et al. 1977)
[specific activity (0. 5 – 2) $\times 10^8$ cpm/μg"] as a probe. Hybridization was performed by a modified method of Cai (Cai 1990) under stringent conditions (42℃ for 16 – 24 h). The filter was
washed twice in 2 × SSC (150 mmol/L NaCI plus 15 mmol/L sodium citrate, pH 7. 0) containing
0. 5% sodium dodecyl sulphate (SDS) for 30 min at room temperature and twice in 0. 1 × SSC containing 0. 1% SDS at 65℃ for 20 min, and then exposed to X-ray film in a plastic mount with an
intensifying screen for 7 – 12 days at – 70℃.

6. DNA · RNA hybridization

Cell hybridization in situ (Paeratakul et al. 1988). NIH3T3 cells transformed by HZIP – 16
and HZIP – 16K (HZIP16-3T3 and HZIP16K – 3T3) were plated onto nitrocellulose filters in gra-
ded concentrations of 2×10^5, 1×10^5 and 5×10^4, then fixed with 1% glutaraldeyhde at 4℃ for
1h, washed with buffer (100 mmol/L TRIS, pH 8. 0; 50 mmol/L EDTA) three times and diges-
ted with 20 μg proteinase K/ml buffer at 37℃ for 30 min. Instead of proteinase K, 1 mg RNase/ml
buffer was used for the control filter. The filter was suspended in 3 × SSC solution (containing 35%
formamide, 3 × SSC, 1 × Denhard's solution, 0. 5% SDS and 100 μg salmon sperm DNA/ml)
and subjected to prehybridization at 65℃ for 12 h and then to hybridization (prehybridization solution plus ^{32}p – labelled HPV – 16 DNA probes and 10 mmol/L EDTA/L at 65℃ for 12h). The filter was washed in 2 × SSC, 0. 5% SDS at room temperature for 15 min, 1 × SSC, 0. 5% SDS at
55℃ for 15 min and 0. 1 × SSC, 0. 5% SDS until there was no evident background, and finally exposed to X-ray film for 7 – 15 days at – 70℃.

DNA · RNA Northern blot hybridization. Total cellular RNA was extracted with guanidine hydrochloride and by the hot phenol method . Samples of 20 μg RNA were electrophoresed throughagarose
gel containing 2. 2 mol/L formaldehyde. The gel was washed twice in 10 × SSC for 20 min. The procedures of gel transfer, hybridization and autoradiography were the same as those for Southern blot
hybridization described above.

7. Preparation of specimens for light and electron microscopy

Light microscopy. Tumour tissues obtained from nude mice were prepared for light microscopy by routine procedures.

Electron microscopy. The cells (pZIP-ψ2, HZIP16K – ψ2, the transformed HZIP16-3T3 and HZIP16K – 3T3 as well as the control NIH3T3 and pZIP – 3T3 cells) and tumour tissues from nude mice were fixed in 3.8% glutaraldehyde in phosphate-buffered saline, postfixed in 1% osmium tetroxide for 2 h, dehydrated in graded ethanol (50%, 70%, 90% and 100%) and embedded in Epon812. The ultrathin sections were stained with uranyl acetate and lead citrate and observed under a JEM-2000EX electron microscope.

Results

1. Estimation of the transforming activity of pSV2 – neo/HPV16

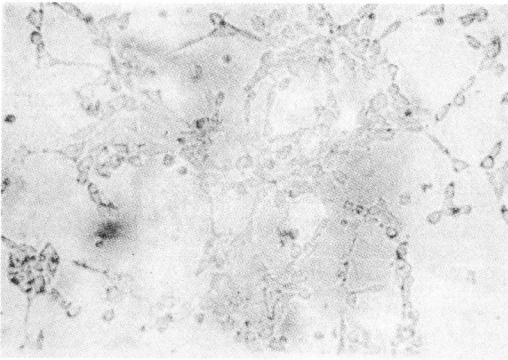

Fig. 4   Morphologically transformed cells induced by pSV2Nco/HPV16 transfectional DNA. ×845

Transformed foci were detected morphologically 4 – 6 weeks after the transfection of pSV2 – neo/HPV16 plasmid by the calcium phosphate/DNA coprecipitation method. The transfectants were selected with G418. Some cells survived and a number of colonies were found to be G418resistant. The transformed cells appeared more reflective and were larger and round-shaped with rapid growth. The mitotic figures could easily be seen throughout the culture. The transformed cells also showed loss of contact inhibition and a decrease in serum dependence (Fig. 4). Five monoclones were selected randomly from the transformed cell colonies and expended to become cell lines, namely, Z102, C202, Z303, Y16, and Q16.

Total cellular DNAs from five monoclones were extracted and subjected to Southern blot analysis by using ^{32}P-labelled HPV-16 DNA (7.9 kb) as probe. When non-digested cellular DNA was analysed, DNA sequences homologous to HPV-16 were found in all of the transformed cell lines (Fig. 5). No HPV-16 – homologous sequence could be deteced in nontransfected NIH3T3 cells. The number of copies of HPV-16 DNA per cell line was different among the five cell lines. Viral genes appeared with obvious amplification in both Q16 and Y16. All the detectable HPV-16 DNA sequences comigrated with high-molecular-mass cellular DNA. The migration of the HPV-16 – related sequences was not characteristic of pSV2 – neo/HPV16 form 1 (close supercoiled circular), form 2 (open circular) or form 3 (linear), and might represent the integration of HPV-16 DNA into host-cell chromosomal DNA.

To analyse further whether or not HPV-16 DNA had been integrated into the host-cell genome, DNA of five transformed cell linecs was digested by *Bam*HI. Because the HPV-16 DNA catenated into the *Bam*HI site of the pSV2 – neo vector, after the *Bam*HI digestion the unit length of the HPV-16 DNA (7.9 kb) was excised from the pSV2-neo sequence. The results of Southern blot hybridization revealed that in addition there was positive reaction at the 7.9-kb site; additional hybridized bands were also observed (Fig. 5). The fragments larger than 7.9 kb may represent the integration of HPV-16 sequences into host chromosomal DNA, the integrated HPV-16 segments being linked with the host sequence. The fragments less than 7.9 kb, on the other hand, may indicate that there was a partial gene junction between the host-cell and virus DNA, or rearrangements and deletions within HPV-16 dNA sequences.

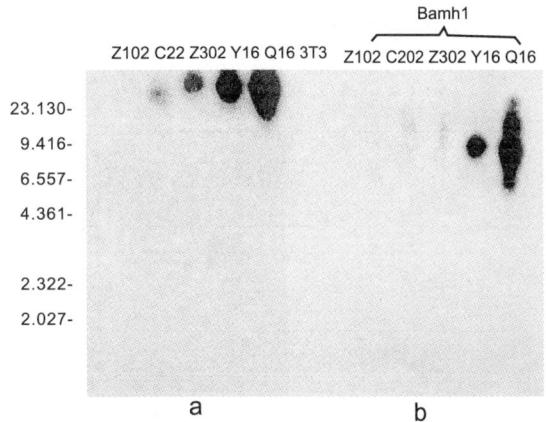

Fig. 5　A, B. Southern blot analysis of HPV-16 DNA sequences in transformed cells (Z102, C202, Z302, Y16 and Q16) that had been transfected with pSV2Neo/HPV16. A Nondigested DNAs; B DNA digested by *Bam*HI.

The presence of multiple hy bridized bands meant that there was multisite integration of HPV-16 DNA into the host chromosome and that the integrated sites were different from each other.

2. Estimation of the transforming activity ofthe whole HPV-16 early gene and E6 – E7

① Growth properties of transformed cells

Plasmids of HZIP16 and HIP16K (containing the whole early gene and E6 – E7 respectively) were transfected into NIH3T3 cells by infection of recombinant retrovirus (HZIP16-RV, HZIP16K-RV). The recombinant viruses are termed viral vectors, and are derived from retroviruses (Fig. 3).

Retroviruses are small viruses with a genome comprising RNA, and they consist of a peptide and a capsule surrounding the RNA genome. When a cell is infected by a retrovirus, the RNA genome is imjected into the cell and copied into DNA by an enzyme, reverse transcriptase. This DNA passes into the nucleus of the infected cell and becomes integrated into the host chromosomal DNA. This process is highly efficient and can yield stable integration of the viral gene into nearly all cells in culture.

Plasmid pZIPNeoSV (X) I was transfected into ψ-2 cells to form a kind of recombinant retrovirus containing no HPV-16 genome, and this virus was used to transfect NIH3T3 cells as a negative control by the same method as described above. One week after the infection of HZIP16-RV and HZIP16K-RV as well as the control recombinant retrovirus (pZIP-RV), there were three kinds of cells surviving the selection with G418, and these were termed HZIP16-3T3, HZIP16K-3T3 and pZIP-3T3 respectively (Fig. 3). Among them, HZIP16K-3T3 and HZIP16-3T3 showed evident signs of transformation. Under the light microscope, The HZIP16-3T3 and HZIP16K-3T3 cells became large and round-shaped, being strongly reflective and, in an unpolarized manner, grew into criss-cross multilayers. Mitotic figures could easily be seen throughout the culture. In addition, HZIP16K-3T3 were much larger and grew faster than HZIP16-3T3 cells, whereas the pZIP-3T3 cells retained the same morphology as the normal NIH3T3 cells, having a spindle shape, monolayer growth and being less reflective (Figs. 6 –8).

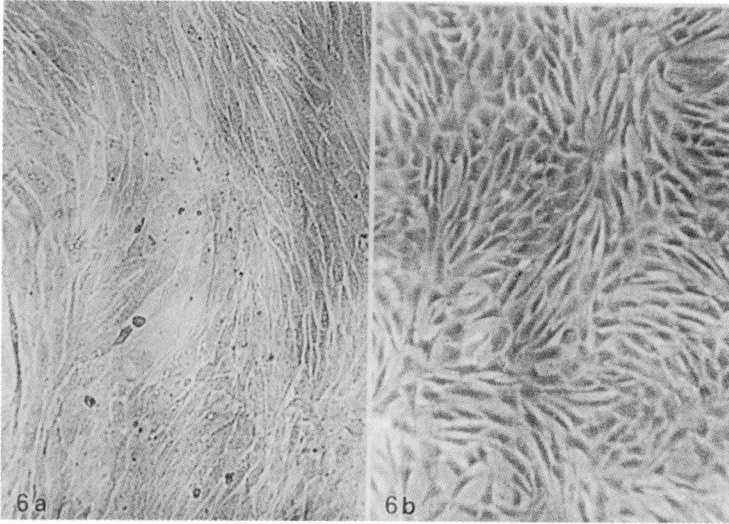

Fig. 6 Light micrographs of control groups: a. morphology of NIH 3T3 cells; b. morphology of pZIP 3T3 cells; the morphology and pattern of arrangement are similar to those of NIH3T3 cells. ×590

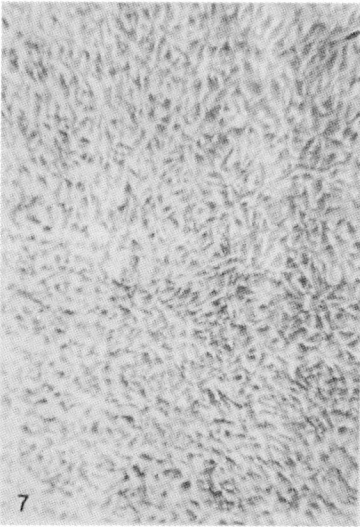

Fig. 7 Micrograph of the morphology of HZIP16-3T3 cells showing that the NIH3T3 cell was transformed by recombinant tetrovirus (HZIP16 – RV) containing HPV-16 whole early genes, and prominant alterations both in cell morphology and the pattern of arrangement. ×590

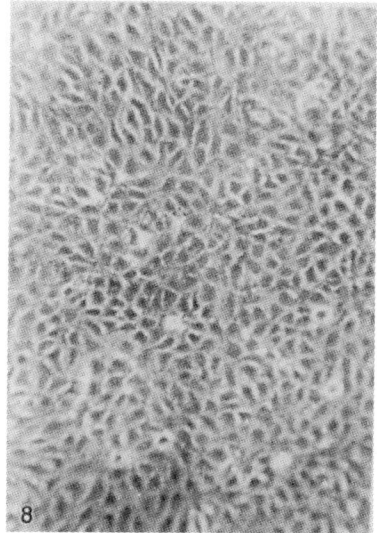

Fig. 8 The morphology of HZIP16K-3T3 cells showing that the NIH3T3 cell was evidently transformed by the recombinant retrovirus (HZIP16K-RV) containing HPV-16 E6 – E7 early genes, and remarkable alterations both in cell morphology and the pattern of arrangement. ×590

② Analysis of anchorage dependence

Samples of 5×10^5 cells of each kin (HZIP16-3T3, HZIP16K-3T3, pZIP-3T3 and NIH3T3 cells) were suspended in 0. 3% soft agar containing serum-free DMEM. After 2 weeks incubation, Many colonies of various sizes were seen in the HZIP16-3T3 and HZIP16-3T3 cell groups. This meant that the cells transformed by HPV-16 subgenomic fragments showed the characteristics of anchorage-independent growth. By contrast, pZIP-3T3 and NIH3T3 cells grew slowly in soft agar and

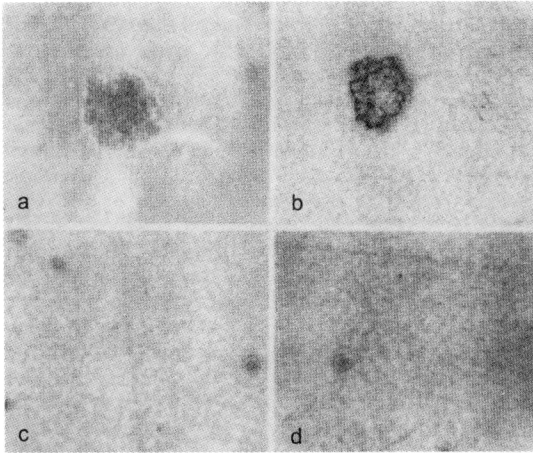

Fig. 9 a – d. Micrographs of the test for anchorage dependence. Cells were suspended in 0.3% soft agar, and incubated for 2 weeks. colonies were formed in a (HZIP16-3T3 cells) and b (HZIP16K-3T3). No colony was found in c (pZIP-3T3) and d (NIH3T3)

no colonies were discovered (Fig. 9). These results suggest that the cells receiving HPV-16 DNA are stimulated to grow in soft agar and this stimulation must be due to the direct effect of an HPV-16 genomic function. It is possible that HZIP16-3T3 and HZIP16K-3T3 show a malignant phenotype because of HPV-16 subgenomes integrating into NIH3T3 cells.

③ Tumorigenicity of the transformed cells

HZIP-3T3, HZIP16K-3T3 and controlled NIH3T3, pZIP-3T3 cells were injected into athymic nude mice (1 × 10^6 cells/animal), five mice per group, in order to assay for tumorigenicity. In the HZIP16-3T3 and HZIP16K-3T3 groups, tumours were formed in all mice within 15 days after inoculation, reaching more than 15mm in diameter, adn all the tumours transplanted with HZIP16K-3T3 were considerably larger than those with HZIP16-3T3. However, no tumour was observed in any of the ten animals of the control groups (inoculated with pZIP-3T3 and NIH3T3 cels) as long as 50 days after injection (Fig. 10). Although both HzIP16-3T3 and HZIP16K-3T3 appeared highly tumorigenic within nude mice, the latent trmorigenic ability of HPV-16 E6 – E7 early genes was stronger than that of HPV-16 total early genes.

④ The morphology of transformed cells in vitro and tumour tissues formed in nude mice

Electron microscopic observation of recombinant-retrovivus-producing cells. HZIP16-ψ2 and HZIP16K-ψ2 cells, harbouring and producing recombinant retroviruses with HPV-16 whole early genes and its E6 – E7 subgenomes, revealed oval or spindle shapes. Some nuclei were remarkably swollen with large amounts of euchromatin. Interchromatin granules and perichromatin granules, often seen in other vivus-infected cells (Lee et al. 1987), were easily observed in this kind or nucleus. There were some dense particles (45 – 50 nm in

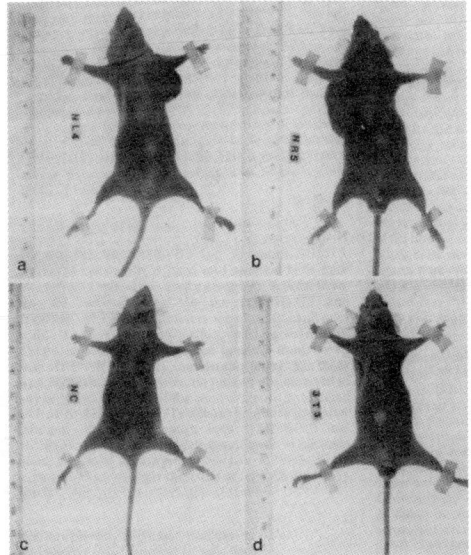

Fig. 10 Tumours formed in nude mice on the 14th day after injection of 1 × 10^6 cells of (a) HZIP16-3T3 and (b) HZIP16K-3T3. No tumour had formed in nude mice by the 49th day after injection of 1 × 10^6 cells of (c) pzIP-3T3 and (d) NIH3T3

diameter) dispersed singly or in clusters beneath the thickened endoplasmic reticulum or plasma membranes from which the mature retroviruses were budding, with the processes of encapsidation of the dense viral core. These dense particles were, therefore, proposed to be the precursors of the RNA core of the recombinant retrovirus. Fig. 11 shows that the viruses were budding from the plasma membrane and Fig. 12 shows some virus particles in the cavity of the endoplasmic reticulum.

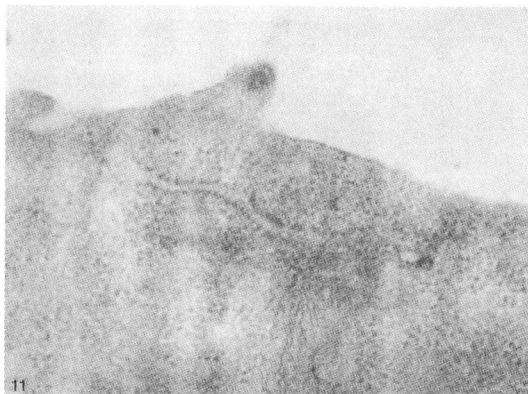

Fig. 11　Recombinant retrovirus budding from the thickened plasma membrane of virus-producing cell (HZIP16-ψ2). Dense precursor core can be seen beneath the membrane. ×80 000

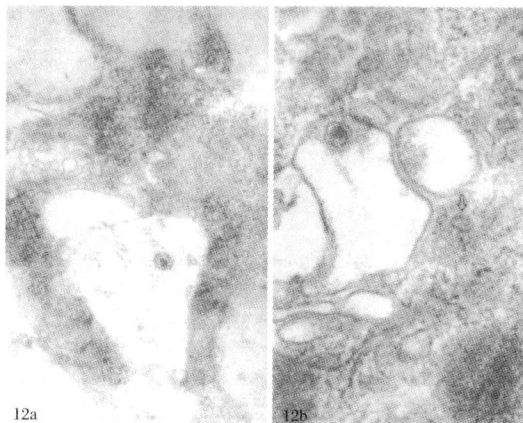

Fig. 12　Electron micrographs of mature recombinant retrovirus in dilated endoplasmic reticulum of virus-producing cell: a HZIP16-ψ2, × 30 000; b HZIP16K-ψ2, × 40 000. Precursor viral core (⇒) can be seen in the cytoplasm of HZIP16K-ψ2 cell

The morphology and morphogenesis of the recombinant retrovirus is very similar to that of other retroviruses described in previous reports (Fraenkel-Conrat et al. 1988).

The morphlogy of tumour tissues induced in nude mice. The whole tumours were round, oval or nodular in shape with a clear margin. The cross-section of the tumour appeared pink in shape with a clear margin. The cross-section of the tumour appeared pink in colour and homogeneous. Under a light microscope, a large number of spindle cells with different sizes were compacted in a disorderly manner. The nuclei of tumour cells appeared to be of different size and shape; some were very large with a lot of mitotic figure. The pathology of the pictures gave rise to the diagnosis of a typical fibrosarcoma (Fig. 13).

Under the electron microscope, the cell nuclei were very large with several large nucleoli and abundant euchromatin. Interchromatin gran-

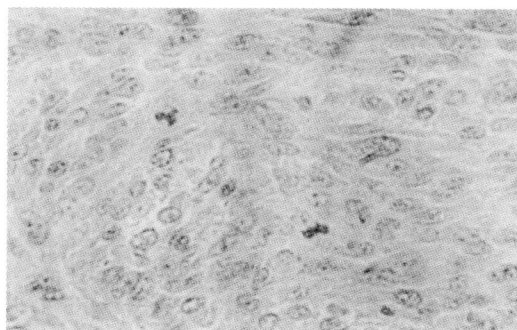

Fig. 13　Micrograph of the morphology of tumour tissue induced in nude mice (H&E stain). A large number of spindle cells of different size were compacted. The nuclei of tumour cells appeared to be of different sizes and shapes with a lot of mitotic figure. ×600

ules, perichromatin granules and nuclear bodies were frequently seen in some nuclei. As well as a few variousley sized, vacuous mito-chondria and endoplasmic reticulum, a great number of free ribosomes were scattered through out the cytoplasm of the tumour cells. Some collagen fibers were seen in the intercellular space of the tumour tissue. The mitotic figures were easily found (Fig. 14 and 15). All the evidence given by light and electron microscopy demonstrated that the tumour cells were of a malignant character with poor differentiation.

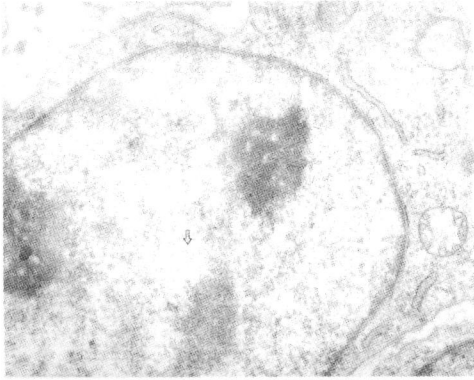

Fig. 14　Electron micrograph of tumour tissue in nude mice. A noticeably large nucleus with abundant euchromatin, two large nucleoli, interchromatin granules and a nuclear body (⇒) were seen. A great number of free ribosomes, scattered throughout the cytoplasm, and swollen mitochondria were observed

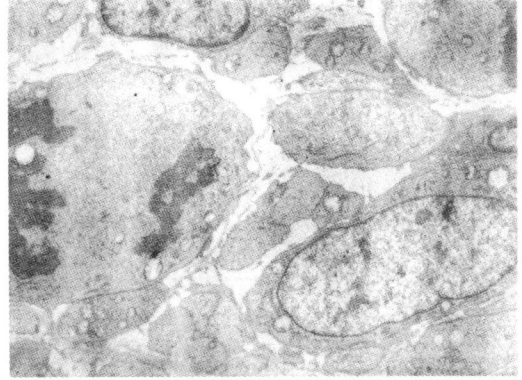

Fig. 15　Electron micrograph of tumour tissue in nude mice. Malignant cells with large nuclei and two cells in mitotic division stage were seen. Collagen fibers present in the intercellular space between tumour cells showed the fibrous cell origin of this malignant tumour

The ultrastructure of transformed cells. HZIP16-3T3 and HZIP16K-3T3 cells showed characteristics of low differentiation, like the tumour cells described above, such as an increased nuclear/cytoplasm ratio, abundant euchromatin in the nuclei and a large number of free ribosomes in the cytoplasm, etc.

No retrovirus particles could be observed in any of the malignant tumour cells and transformed cells although there were related homologous sequences of HPV-16 DNA or its E6 – E7 subgenomes in the DNA of these cells, as shown in our DNA · DNA hybridization. This again shows the fact, described in our previous paper (Lee et al. 1988), that once the carcinogenic viral genome or subgenomes are integrated into the cellular genome and have caused the cell to transform into a malignant state, the virus itself as a whole is no longer detectable.

3. The detection of HPV-16 subgenomes in transformed cells

The nondigested DNA obtained from HZIP16-3T3, HZIP16K-3T3 and tumour cells of nude mice were subjected to Southern blot hybridization with ^{32}P-labelled E6 – E7 subgenomes of HPV-16 as the probe, showing that the fragment in the position of high-molecular-weight DNA (23 kb) presented a clear hybridization signal. This means that HPV-16 DNA has been integrated into the host cell genome. The DNAs of these transformed cells, digested by *Bam*HI, gave rise to several hybridized bands in addition to the 23-kb fragment (Fig. 16). This again means that multisite integrations of the HPV-16 sequence into transformed cellular DNA have tave taken place. Considering these data, it is unlikely that the HPV-16 sequences are present in the episomal form in the transformants. It is suggested that the integration of the HPV-16 DNA is needed for the maintenance of the malignant phenotype of transformed cells.

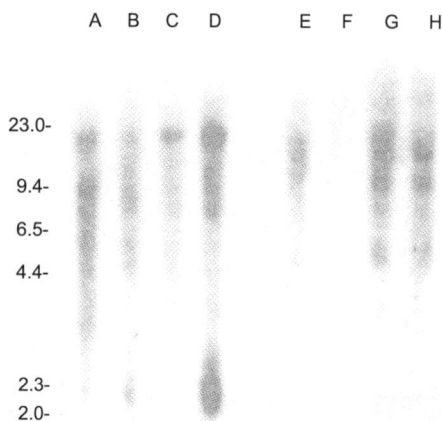

Fig. 16　Southern blot analysis of HPV-16 DNA sequence in transformants and their tumours. Each DNA was digested with *Bam*HI. *Lanes A and B* were from tumour produced by HZIP16K-3T3; *C* was from transformed cells, HZIP16K-3T3, and *D* was from HZIP16K-ψ2 cells. *Lanes E* abd *F* were from tumour produced by HZIP16-3T3. *G* was from transformed cells of HZIP16-3T3 and *H* was from HZIP16-ψ2 cells. All of them showed positive hybridization with HPV-16 E6 – E7 DNA as the probe and high-molecular-weight DNA (23 kb) gave clearly positive signals

4. The detection of transcriptional activity of HPV-16 subgenomes in transformed cells

DNA · RNA hybridization in situ was used for detecting the transcriptional activities of HPV-16 subgenomes in HZIP16-3T3 and HZIP16K-3T3 cells. This method is more sensitive and fast than that of Northern blot hybridization. All three concentrations of HZIP16-3T3 and HZIP16K-3T3 cells (2×10^5, 1×10^5 and 5×10^4) revealed positive hybridization to the ^{32}P-labelled HPV-16 E6 – E7 DNA probes, whereas the hybridization signals became much weaker after RNase digestion of transformed cells (Fig. 17). It is suggested that this result indeed represents DNA · RNA hybridization, and demonstrates that HPV-16 E6 – E7 early genes are expressed in transformed cells.

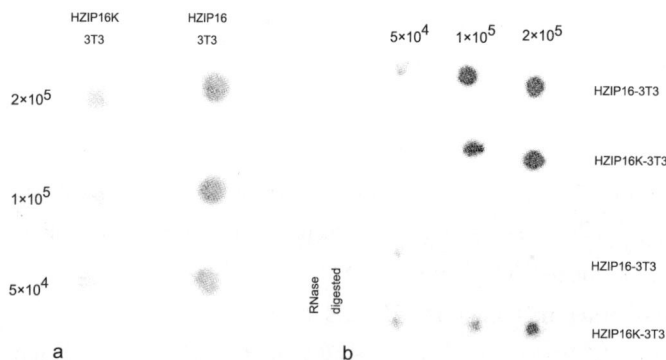

Fig. 17　a. The result of DNA · RNA hybridization in situ for detecting the transcriptional activity of HPV-16 subgenomes in HZIP16-3T3 and HZIP16K-3T3 cells revealed positive hybridization with HPV-16 E6 – E7 DNA as probe. b A comparison of RNA · DNA hybridization in situ between cells (HZIP16-3T3 and HZIP16K-3T3) digested and not digested by RNase. The hybridization signals become much weaker after RNase digestion

5. The comparison of gene-transfer effects between the methods using recombinant retrovirus and calcium phosphate/DNA coprecipitation

Hundreds of cell colonies appeared with NIH3T3 cells infected by recombinant retrovirus after

G418-selected culture for 1 week (Fig. 18). However, no clone formed in transformed cells when the calcium phosphate/DNA coprecipitation method was used, after the same time of selective culture mentioned above. There were only a few cell colonies formed after selective culturing for 3 weeks. The formation of cell colonies after the same recombinant plasmid had been used to transfer genes into NIH3T3 cells by different methods

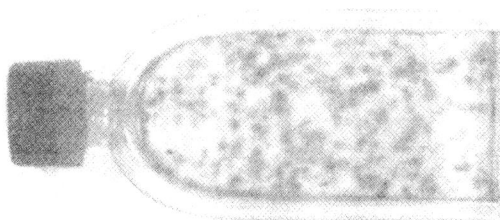

Fig. 18   Hundreds of cell colonies appeared with NIH3T3 cells in fected by recombinant retrovirus after G418-selective culture for 1 week

was quite different in time and number. From the results of our experiment, the retrovirus vector produced by the genetic engineering method is a more efficient method for gene transfer than others at present.

Discussion

The results presented here have indicated that the plasmids containing the HPV-16 genome were efficient for the transformation of NIH3T3 cells, and that the transformed cells have clearly undergone changes that alter morphology, growth characteristics and tumorigenicity in vivo. The transformation is indeed due to the effect of HPV-16 DNA, as the recombinant retrovirus pZIP-RV, containing no HPV-16 genes, is unable to transform HIH3T3.

The data from this study showed that the E6 – E7 open reading frames of HPV-16 were sufficient for the transformation of NIH3T3 cells. The efficiency of transformation by the E6 – E7 construct was higher than that observed with the entire early genes of HPV-16. The reason for this is still to be studied.

In our other paper (Si et al. 1990), the HPV-16 DNA was digested by *Pst* I to produce several pieces and these were labelled with ^{32}P to create probes to detect viral DNA sequences in biopsies of cervical cancer, which showed a positive reaction with a HPV-16 total DNA probe by dot-blot hybridization. The results showed that 89% of samples gave a positive reaction with the E6 – E7 subgenome probe. The results of our transforming experiment coincide with the earlier report. We conclude, therefore, that a major transforming function of HPV-16 is localized in the E6 – E7 genes, and that the E6 – E7 open reading frames encodes a transforming gene, which plays a pivotal role in the careinogenic processes.

The transformation experiments in this study have shown that, in nude mice, not only transformed cells, induced by either HPV-16 whole genome or its subgenomic fragments (total early genes and E6 – E7 genes), but also tumour cells all contain integrated HPV-16 DNA. It has been proved in our laboratory that HPV-16 DNA is ususally integrated into cellular DNA in biopsy tissues of cervical cancer, but that in cervical benign lesions such as chronic cervicitis and condyloma it is in an episomal form. The evidence from our study suggests that the integration of HPV-16 genes into cellular DNA may play an important role in the processes of malignant conversion of cervical epithelial cells.

Some reports of transforming experiments with HPV-16 DNA in vitro were describe the use of routine methods involving calcium phosphate/DNA coprecipitation, electrophoration, protoplast fusion and microinjection etc. Transfection by these methods is rather inefficient. The DNA will be stably integrated into the chromosomes of only 1 in 100 000 cells by calcium phosphate/DNA coprecipitation, and the other techniques of gene transfer have generally similar results (Ledley 1987).

In our method foreing genes are packaged into a specially engineered virus, and introduced into cells by infection with the recombinant viruses. This process is called viral-mediated gene transfer, and the recombinant viruses are termed viral vectors. The recombinant retrovirus vector for HPV-16 gene transfer was employed in this study to investigate the transforming effect of HPV-16. This method or retroviral-mediated gene transfer has proved to be highly efficient and can yield stable integration of genes into virtually all cells in culture, as shown by our experiments, up to now, no report of a recombinant retrovirus being used for the study of HPV-16 DNA transformation in vitro has been found.

A typical retroviral sequence includes the sequences necessary for encoding the *gag*, pol and *env* polypeptides, the long terminal repeats (LTRs) necessary for the initiation of viral transcription and polyadenylation of viral transcripts and for integration, and a descrete sequence (ψ site) near the LTR, required for packaging the RNA genome in the viral capsule, called the package signal sequence (Fig. 3).

By means of recombinant DNA techniques, two kinds of viral mutant have been obtained. One is capable of directing synthesis of retroviral core protains and assembly of a retroviral capsule, which, however, lacks the packaging sequence (ψ). The mutant can then be introduced into cultured cells to produce a packaging-defective cell line (ψ2) that secretes cmpty virus capsules. The second mutant is an expression vector called pZIPNeoSV (X) I containing a recombinant foreign gene in place of the viral genes but retaining the packaging sequence (ψ). If HPV-16 genes are recombined into this expression vector and introduced into cells that also contain the packaging-defective mutant (ψ2 cells), defective retrovirus-producing cells (HZIP16-ψ2, HZIP16K-ψ2) are formed. In these cells, the expression vector containing recombinant HPV-16 genes is packaged into empty virus capsules, producing defective retroviruses (HZIP16-RV and HZIP16K-RV), which retain the ability to infect cells efficiently and integrate the foreign genes into the host chromosome. These virus particles can be used to infect NIH3T3 cells and transform them to detect the transformation potential of HPV-16 genes. Since the defective retroviruses do not contain genes for any viral protains, a cell infected with this kind of defective retrovirus is incapable of inducing viral proliferation or viral disease.

Because of the lack of a tissue-culture system suitable for HPV replication in vitro, a method that uses recombination of HPV genes into a specially engineered virus is of significance for studying the biological behavior of HPV. Furthermore, a number of retroviruses carrying the HPV gene are obtained by this method, and these viruses will be available for universal studies both in vitro and vivo. This gene-transfer method may make it possible to infect live animals with viral vectors leading to transformation of cells in vivo. As a model for introducing recombinant genes into live animals, allowing experiments to develop from those from in vitro to ones vivo and from cell model to animal model, it is certainly of particular significance for basic medical research and clinical application.

Acknowledgements

This project was supported by the National Natural Science foundation of China, the Science foundation of the Ministry of Public Health and the Science Foundation of the National Seventh Five-Year Plan of China.

〔In 《J Cancer Res Clin》 1991, 117: 460 –472〕

References

1 Bedell MA, Jones KH, Laimins LA. The E6 – E7 region of human papillomavirus type 18 is sufficient for transformation of NIH3T3 and Rat-1 cells. J Virol, 1987, 61: 3635 – 3640

2 Cai L. In: Cai L (ed). The techniques in rescarch for nucleic acid. Science Press, Beijing, 1987, 32

3 Cai L. The application of detection for HBV DNA in the liver cell. In: Cai L (ed). The application of genomic engineering techniques in basic and clinic medicine. People's Health Press, Beijing, 1990, 191 – 192

4 Davis LG, Dibner MP, Battey JF. Calcium phosphate transfection of nonadherent and adherent cells with purified plasmid. In Basic methods in molecular biology. Elsevier, 1986, 286 – 289

5 Fraenkel-Conrat H, Kimball PC, Levy JA. Retroviridae. In: Fraenkel-Conrat H, Kimball PC, Levy JA (eds) Virology, 2nd edn. Prentice Hall, Englewood Cliffs, New Jersey, 1988, 108 – 125

6 Fukushima M, Okagaki T, Twiggs LB, Clark BA, Zachow KR, Ostrow RS, Faras AJ. Histological types of carcinoma of the uterine cervix and the detectability of human papilloma virus DNA. Cancer Res, 1985, 45: 3252 – 3255

7 Gao H, Si J, Lee K, Han R, Wang S, Gu S, Zeng Y, Zhou Y, Wang Y. Detection of human papillomavirus DNA in cervical carcinoma cells in Chinese patients. Acta Acad Med Sin, 1988, 10: 276 – 278

8 Hausen H zur. Genital papillomavirus infection. In: Rigby PWJ, Wilkie NM * eds (Viruses and cancer. Cambridge University Press, 1985, 83 – 90

9 Hausen H zur. Papillomaviruses in anogenital cancer as a model to undersand the role of viruses in human cancer. Cancer Res, 1989, 49: 4677 – 4681

10 Howley PM. The role of papillomaviruses in human cancer. In: Devita VT, Hellan S, Rosenberg SA (eds) Important advance in oncology. Lippincott, Philadelphia, 1989, 55 – 73

11 Ledley FD. Somatic gene therapy for human diseaseLBack-ground and prospects. Part 1. J Pediatr, 1987, 110: 1 – 8

12 Lee K, Bao J, Wang J, Zhao W, Liu S, Si J, Wang Y, Zhang W, Jiang J. Ultrastructural study of the morphogenesis of herpes simplex virus type 2 in organ cultured human uterine cervix and the interactions between virus and host cell. J Electron Microsc Tech, 1987, 7: 73 – 84

13 Lee K, Si J, Han R, Wang S, Zhang W, Song G, Liu S, Chen L, Zhao W, Jia L, Sheng Q, Mai Y, Zeng Y, Gu S, Sun Y, Meng X, Zhou Y, Wang Y. An ultrastructural research for the relationship between human uterine cervical carcinoma and human papillomavirus: I. An electron microscopic observation on human genital condyloma and the morphogenesis of human papillomavirus in host cells: II. A combined gene molecular and ultrastructural study. In: Hashimoto H, Kuo KH, Lee K, Ogawa K (eds) Recent development of electron microscopy 1987. Nakanishi, Kyoto, Japan, 1988, 293 – 312

14 Mandel M, Higa A. Calcium dependent bacterio-phage DNA infection. J MOl Biol, 1970, 53: 154

15 Matlashewski G, Schneider J, Banks L, Jones N, Murray A Crawford L. Human papillomavirus type 16 DNA cooperates with activated ras in transforming primary cells. EMBO J, 1987, 6: 1741 – 1746

16 Matlashewski G, Osborn K, Banks L, Stanley M, Crawford L. Transformation of primary human fibroblast cells with human papillomavirus type 16 DNA and EJ-*ras*. Int J Cancer, 1988, 42: 232 – 238

17 McCance DJ, Kopan R, Fuchs E, Laimisus L. Human papil-lomavirus type 16 alters human epithelial cell differentiation in vitro. Proc Natl Acad Sci USA, 1988, 85: 7169 – 7173

18 Noda T, Yajima H, Ito Y. Progression of the phenotype of transformed cells after growth stimu-lation of cells by a human papillomavirus type 16 gene function. J Viol, 1988, 62: 313 – 324

19 Paeratakul U, DeStasio PR, Taylor MW. A fast and senitive method for detecting specific viral RNA in mammalian cells. J Virol, 1988, 62: 1132 – 1135

20 Pecoraro G, Morgan D, Defendi V. Differential effects of human papillomavirus type 6, 16 and 18 DNAs on immortalization and transformation of human cervical epithelial cells. Proc Natl Acad Sci USA, 1989, 86: 563 – 567

21 Rigby PD, Rhodes MD, Berg P. Labelling deox-yribonucleic acid to high specific activity in vitro by nick translation with DNA polymerase I. J Mol Biol, 1977, 113: 237 – 351

22 Si J, Lee K, Han R, Wang S, Zhang W, Song G, Liu S, Chen L, Zhao W, Sheng Q, Jia L, Mai Y, Gu S, Zeng Y. Gene molecular and ul-trastructural studies on relationship between hu-man squamous epithelial carcinoma of uterine cer-vix and human papillomavirus. Acta Acad Med Sin, 1987, 9: 264 – 270

23 Si J, Luo W, Lee K, Han R, Tan B, Ling J, Zhang W, Zhao W, Liu S, Chen L, Jia L. Detection of human papill omavirus subgenom-ic fragments in the biopsies of Chinese cervical cancer cases. Acta Academic Medicine sinicae, 1990, 12: 136

24 Si J, Lee K, Han R, Zhang W, Tan B, Song G, Liu S, Chen L. The identification of viral ge-nome and subgenomic sequences in biopsies of Chinese patients. Acta Acad Med Sin (in press), 1991

25 Southern E. Detection of specific sequence among DNA fragment separated by gel electrophoresis. J Mol Biol, 1975, 98: 503

26 Standbridge CM, Mather J, Curry A, Butler EB. Demonstration of papillomavirus particles in cervical and vaginal scrape material: a report of 10 cases. J Clin Pathol, 1981, 34: 524 – 531

27 Tsunokawa Y, Takebe N, Kasamatsu T, Terada M, Sugimura T. Transforming actvity of human papillomavirus type 16 DNA sequence in a cervi-cal cancer. Proc Natl Acad Sci USA, 1986, 83: 2200 – 2203

28 Weinstein IB. The origins of human cancer: mo-lecular mechanisms of carcinogenesis and their im-plications for cancer prevention and treatment. Twenty-seventh G H A Clowes Memorial Award Lecture. Cancer Res, 1988, 48: 4135 – 4143

29 Wigher M, Pellicer S, Silverstein S, Axel R. Biochemical transfer of single-copy eukaryotic genes using total cellular DNA as donor. Cell, 1978, 14: 725 – 731

30 Yasumoto S, Burkhardt AL, Doniger J, Dipaolo JA. Human papillomavirus type 16DNA induced malignant transformation of NIH3T3 cells. J Vir-ol, 1986, 57: 572 – 577

55.   HIV Infection and AIDS in China

ZENG Yi

Institute of Virology, Chinese Academy of Preventive Medicine,

100 Ying Xin Jie, Xuan Wu Qu, Beijing , China

Summary

To date, a total of 305 280 sera from Chinese and foreigners were tested for HIV-1 antibody by ELISA, gelatin particle agglutination test, immunofluorescence test or immunoenzymatic test and the positive sera were confirmed by Western Blot. Altogether 379 sera from Chinese were positive. Four Chinese hemophiliacs had HIV-1 antibody in 1985. It was the first time to demonstrate that HIV-1 transmitted into China in 1983. One homosexual man and four subjects returned form Africa were asso HIV-1 antibody positive. Sera from 365 drug addicts and 2 spouse of them from south west border of China gave positive reaction. Sixty-eight foreigners and one Chinese hemophiliac from Hong Kong were also seropositive. Two Chinese HIV carriers developed AIDS and died in 1990. AIDS patients from U. S. A. were diagnosed clinically and serologically in China, one of them was overseas Chinese. A HIV-1 virus was isolated from an American patient.

Introduction

AIDS first appeared in 1981 among homosexual men in U. S. A. It was believed that HIV would certainly spread into China after carrying out the open policy since 1978. We started to screen the HIV antibody in 1984. The first AIDS case as a tourist from USA appeared in China in 1985 and 4 hemophiliacs were found to have HIV antibody in the same year (Zeng et al, 1986; Fang et al, 1989). Since then the public concern about the status of HIV infection and AIDS in mainland China has gradually increased. Here we reported the results of a large scale survey for the persons of HIV-1 infection and AIDS in general population.

Materials and Methods

1. Sera: Serum samples were obtained from different groups of Chinese population and some foreigners by my laboratory, provincial Health and antiepidemic station and other health institutions (Tab. 1). Sera were stored at-20℃ – 70℃ until testing.

2. HIV antibody assay: Sera were screened for HIV 1 antibody using ELISA, immunoenzymatic tests, immunofluorescence test or gelatin particle agglutination test. When positive results were obtained, testing was repeated using the same or other techniques. When the tests were repeatedly positive, the sera were confirmed by Western Blot.

3. Isolation of HIV-1 virus: Blood sample was collected form an American AIDS patient who was sick in China as a tourist. The serum was inoculated into MT-4 cells and cultivated at 37℃, and the medium was changed twice a week. The cells were tested for HIV antigen by immunofluorescence test, then confirmed by Western Blot. The control HIV virus (LAV) was kindly sent by Dr. Montagnier.

Results

1. HIV infection and AIDS: Up to the end of November 1990, a total of 305 280 sera from Chinese and foreigners wer screened by ELISA, immunoenzymatic test or gelatin particle test. The positive sera was confirmed by Western Blot. The results were shown in Tab. 1. Altogether, 378 sera from Chinese were positive. Four of 19 sera from Chinese hemophiliacs treated with factor Ⅷ produced in the United States were found to have HIV-1 antibody in 1985. One died of cerebral hemorrhage with AIDS related complex (ARC) in 1987 (Fang et al, 1989), the other three are still alive. It was the first time to demonstrate that HIV-1 had transmitted into China via imported factor Ⅷ from USA in 1983. One homosexual man and four subjects returned from Africa were also HIV-1 antibody positive. Sera from 365 drug addicts and 2 of their spouses from southwest border of China near Burma gave positive reaction. The drugaddicts were very young, 90% of them were 15 to 39 years of age (Tab. 2). The youngest one was 15 years old. Eighty-three percent of drugaddicts are minorities including Dai and Jing Po and 80% are farmers.

The incidence of HIV infection in the intravenous drug abusers was very high (60%). Many drug abusers claimed that they only used drug by smoking, but there are still 15% seropositive (Tab. 3). No HIV-1 antibody was found among drug abusers living far from the south west border. Sixty-seven foreigners and one Chinese hemophiliac from Hong Kong were also seropositive. No HIV-1 antibody was found in other groups including prostitutes, blood donors and etc.

Tab. 1 Serological Screening of HIV-1 Antibody in China

Group	Samples tested	Positive number
Chinese	273 187	378
AIDS	2	2
Drug abusers	2 567	365
Spouse of drug abusers	50	2
Hemophiliacs	250	4
Homosexual men	96	1
Persons returned from abroad	53 261	4
Prostitutes, STD patients	45 136	0
Persons injected with blood product	4 593	0
Persons served in hotel	15 740	0
Blood donors	83 109	0
Seamen	8 155	0
Prisoner	1 846	0
Patients other than AIDS	3 509	0
Others	54 873	0
Persons from outside of China	32 093	68
AIDS		3
Foreigner		64
Hemophiliacs from Hong Kong	1	
Total	305 280	446

Tab. 2 Age distribution of HIV antibody positive persons

Age (Years)	Positive number	Positive rate (%)
<15	nil	nil
15	29	11.9
20	131	53.7
30	60	24.6
40	21	8.6
50	3	1.2
Total	244	100

Tab. 3 HIV antibody positive rate in drug abusers

Drug abusers	Sample tested	Positive number	Positive rate
I. V	164	105	64
Smoking	265	40	15.1
Unknown	64	1	1.6

Altogether, 5 AIDS patients were found in China, 2 of them Chinese who developed AIDS and died in 1990, one was the drug abuser, the other was infected abroad 5 years ago. Three AIDS patients from USA were diagnosed clinically/serologically in China, one of them was overseas Chinese who developed AIDA and returned to China. More than 50 sera from African students and Chinese returned from Africa were also treated for HIV-2 antibody. All of them were negative.

2. Isolation of HIV-1 virus: An American AIDS patient as a tourist in China was diagnosed clinically/serologically. The serum from this patient was inoculated into the MT-4 cells. A few cells and 96% of cells showed positive HIV-1 antigen by immunofluorescence test at 11 and 14 days after cultivation, respectively. There were typical bands of HIV-1 from the antigens of this new isolate and it was designated as HIV-1 CA-1 virus.

Discussion

The incidence of HIV infection and AIDS continues to increase worldwide, it also increased rapidly. In Thailand, HIV was transimitted into China in 1983. Although it is not a severe problem in whole China now, but there are many HIV infected patients in the southwest border of China. The seropositive rate (64%) was very high among the intravenous drug abusers due to the frequent sharing of syringes or needles. Some of them claimed that they only had drug by smoking, but they became seropositive, it means that they had drug intravenously. No HIV-1 antibody was found in drug abusers living far from the southwest border. People living at the border between China and Burma can go and come freely, so the HIV transmitting into China was from Burma. It is the time to limit or stop the spread of HIV from the border to other areas of China.

Although there is no HIV antibody in sera from prostitutes or STD patients, it is still a high risk group and will be the problem in the future, because the number of patient with STD have been markedly increased in the recent years. HIV-1 virus, AC-1, was isolated and confirmed by Western Blot, this was the first isolate of HIV in China and can be used for preparation of diagnostic reagents.

The research of isolating HIV-1 virus from HIV antibody positive drug abusers are being carried out.

[In 《AIDS Research》 1992, 6: 1-5]

References

1 Zeng Y, Fan J, Zhang Q, Wang PC, Tang S, Zhou C, Zheng XW, Liu DP. Detection of antibody to LAV/HTLV-Ⅲ in sera from hemophiliacs in China. AIDS Research 2 (Supl). 1986, 147-150.

2 Fang D, Xu YH, Dai D, Han YJ, Wang PC, Lang YM, Lang Y, Zeng Y. Clinical analysis of four Chinese hemophiliacs with HIV infection. Chinese Med J, 1989, 102: 819-824

56. Suppression of Human Nasopharyngeal Carcinoma Cell Growth in Nude Mice by the Wild-Type p53 Gene

CHEN Wei – ping[1], LEE Yang[1], WANG Hui[1],

YU Geng-geng[1], JIAO Wei[2], ZHOU Wei – ya[2], ZENG Yi[1]

1. Department of tumor Virus and HIV, Institute of Virology, Chinese Academy of Preventive Medicine, Beijing 100052, People's Republic of China;

2. Guangxi Regional Hospital, Nanning, Guangxi, People's Republic of China

Summary

Wild-type and mutant human p53 genes were transfected into the nasopharyngeal carcinoma (NPC) cell line CNE-3. Tumorigenicity in nude mice showed that the tumor resulting from the cells transfected with the wild-type p53 gene grew more slowly and was smaller that from control CNE-3 cells. In contrast, the tumor from the cells transfected with the mutant p53 gene grew faster than that produced by cells transfected with the wild-type p53 gene and that produced by control CNE-3 cells. The results demonstrate that the wild-type p53 gene could inhibit the NPC cell growth in nude mice and the mutant p53 gene could enhance the NPC cell growth in nude mice. The p53 gene may also play an important role in the pathogenesis of NPC.

[**Key words**]　Nasopharyngeal carcinoma; Suppressor gene; Nude mice; Tumorigenicity

Introduction

The p53 gene is a nuclear protein. Because it is often overexpressed in transfected cells and can transform the primary rodent cells with oncogene *ras*, p53 has been suspected to be an oncoprotein for many years (Rogel et al. 1985). Recent studies showed that the wild-type p53 gene did not have a transforming function and, on the contrary, could inhibit the tumor cell growth. If the wild-type p53 gene mutated it would have transforming potential (Hinds et al. 1989; Finlay et al. 1989). Mutations of the p53 gene occur in many types of human tumors (Nigro et al. 1989). Baker et al. (1990) transfected the colorectal carcinoma cell lines with wild-type and mutant p53 gene. The results demonstrated that the wild-type p53 gene could inhibit cell growth in nude mice, but until now there has been no report on the p53 gene and its function in nasopharyngeal carcinoma (NPC) cells. In this work NPC cell line CNE-3 was transfected with wild-type and mutant p53 genes, and the tumorigenicity in nude mice was compared.

Materials and Methods

Plasmids pC53-SN3 (wild-type) and pC53-SCX3 (mutant) were obtained from Professor Bert

Vogelstein, John Hopkins University, USA; PAB 1801 mAb against p53 protein was obtained from Professor Lionel Crawford, Cambridge, UK; G418 and polybrene are the products of Sigma.

1. Cell line and DNA transfection: CNE-3 is the cell line established from a patient with metastatic nasopharyngeal carcinoma in the liver (W. Jiao, W. Y. Zhou, P. Z. wang, and Y. Zeng, unpublished data). The cells were cultivated in RPMI-1640 medium, 10% fetal bovine serum, 2 mmol/L l-glutamine, 100U/ml penicillin, 100μg/ml streptomycin, at 37℃ in 5% CO_2. Plasmids pC53-SN3 and pC53-SCX3 were transfected into the cells by polybrene-induced DNA-mediated transfer previously described by Rhim et al. (1989), and then selected with G418 (500μg/ml) for 12 days until the control CNE-3 cell died completely. Cultivation of the cells was then continued with 100 μg/ml G418. Cell colonies formed were counted.

2. Colony formation in soft agar: A 5-ml sample of cell suspension (2×10^5 cells/ml) in 0.35% agar was overlayed on a 60-mm dish containing a 0.6% agar base. The dish was incubated at 37℃ in 5% CO_2. Viable colonies were scored after 21 days.

3. Tumorigenicity in nude mice: Balb/c nude mice, 4 weeks old, were inoculated subcutaneously with 1×10^7 freshly trypsinized cells in order to determine tumorigenicity. Tumors were observed for 5 weeks. The speed and size of tumor growth were recorded.

Results

1. DNA transfection and screening

Colony formation after transfection with wild-type and mutant p53 genes and selection of geneticin were compared (Tab. 1). The cells transfected with pC53-SN3 formed twofold fewer colonies than those transfected with pC53-SCX3. The number of colonies in soft agar were similar for the two plasmids and no significant difference was found. But the CNE-3-p53-wt colonies appeared 1 week later than those of CNE-3-p53-mt and the CNE-3 control.

Cell size and speed of growth in vitro were also compared and no difference was found between the cell samples. In order to identify p53 protein expression in both transfected and non-transfected cells, an indirect immunofluorescent assay (IFA) with PAB 1801 mAb against p53 protein was carried out. Fluorescent intensity showed that both wild-type-and mutant-p53-transfected cells could express detectable p53 protein. In contrast, p53 protein could not be detected in control CNE-3 cells with IFA (data not shown).

Tab. 1 Colony formation of CNE-3 cells

Cells	Colonies
G418 screening	
CNE-3	0
CNE-3-p53-wt	55
CNE-3-p53-mt	108
Growth in soft agar	
CNE-3	70
CNE-3-p53-wt	65
CNE-3-p53-mt	87

2. Tumorigenicity in nude mice

Samples of 1×10^7 cells were inoculated subcutaneously into the back of nude mice. Tumorigenicity was determined and compared. Tumor from the control cells and tumor from the cells transfected with the mutant type p53 gene appeared 1 week later, but the tumor from cells transfected with the wild-type p53 gene appeared 3 weeks later. Tumor growth speed showed significant difference among them (Fig. 1).

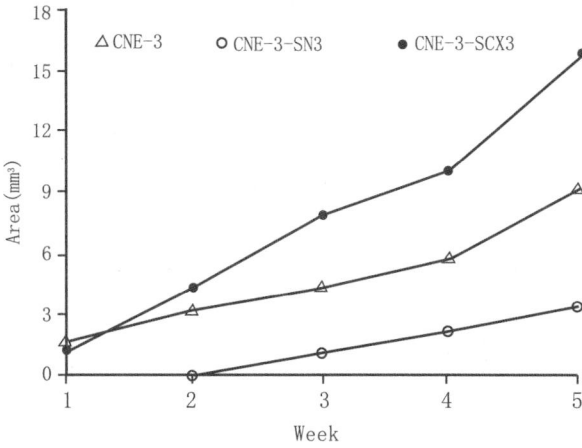

CNE-3, control cells; *CNE*-3 – SCX3, CNE – 3 cells transfected with mutant p53 gene. CNE-3 – SN3, CNE-3 cells transfected with wild-type p53 gene. Initially the tumor resulting from the control cells grew a little faster than that of the cells transfected with mutant p53 gene, but 2 weeks later, the control cells tumor was growing significantly more slowly than that resulting from cells lransfected with the mutant p53 gene ($P < 0.001$). In contrast, tumor from the cells transfected with the wild-type p53 gene grew more slowly than that of the control cells and that of the cells transfected with mutant p53 gene ($p < 0.001$). The volume and weight of tumor at the 5th week are shown in Tab. 2 and Fig. 2. An approximately threefold difference was found among them

Fig. 1 Comparison of tumor growth speeds in nude mice.

Discussion

In this work the nasopharyngeal carcinoma cell line CNE-3 was transfected both with mutant and wild-type p53 genes. The results of tumorigenicity assays in nude mice demonstrated that the wild-type p53 gene could inhibit cell growth in nude mice. This is the first report that the p53 gene can act as a suppressor gene in NPC cells.

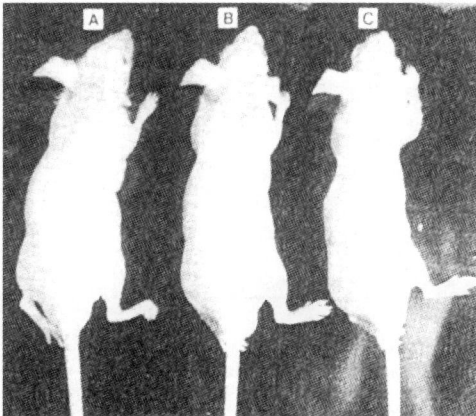

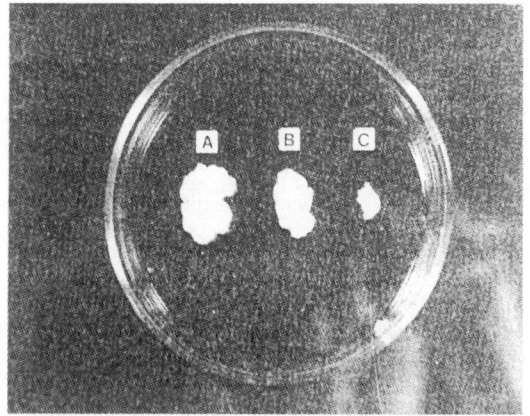

Fig. 2 Tumors in nude mice. *A*, **CNE-3 cells transfected with mutant p53 gene;** *B* **CNE-3 control cells;** *C*, **CNE-3 cells transfected with wild-type p53 gene**

Baker et al. (1990) transfected colorectal carcinoma cell lines with both mutant and wild-type p53 genes. The results suggest that the wild type p53 gene can specially suppress cell growth in nude mice and may act as a suppressor of growth of colorectal carcinoma cells; however, until now there has been no report of the p53 gene in nasopharyngeal carcinoma cells.

Mutations of the p53 gene occur commonly in many kinds of tumor such as haptocellular carcinoma (Hsu et al. 1991; Bressas et al. 1991), and colorectal carcinoma (Nigro et al. 1989). Whether or not hte p53 gene is mutated in NPC cells, deserves further study.

In cells transformed by simian virus (SV) 40, adenovirus and human papilloma virus, p53 protein could form a complex with SV40 large T antigen, EIB 55×10^3 protein, and HPV E6 protein respectively (Lane and Crawford 1979; Linzer and Levine 1979; Sarnow et al. 1982; Werness et al. 1990). It may be a mechanism for the functional inactivation of the p53 gene.

Cang Wu County and Wu Zhou city of Guang Xi Autonomous Region in south China are high-risk areas for NPC. Seroepidemiological studies have shown that NPC development is related to Epstein-Barr virus (EBV) infection (Zeng 1985). The EBNA and LMP genes were found in NPC tissues. But the relation between p53 and EBNA or LMP is not yet clear. Further studies on this subject will be very helpful in the study of the EBV transforming function and in research into the molecular mechanism of NPC pathogenesis.

Tab. 2　Comparison of tumors in nude mice at the 5th week

Parameter	Origin of tumors		
	CIN-3-p53-mt	CNE-3	CNE-3-p53-wt
Volume of tumors (mm³)	15.7	9.33	3.50
Weight of tumors (g)	1.27	0.42	0.12

Acknowledgements

We thank Prof. B. Vogelstein at John Hopkins University, USA, for providing plasmids PC53-SV3 and PC53-SCX3, and Prof. L. Crawford Cambridge, UK, for monoclonal antibody against p53 protein.

[In 《J Cancer Res Clin Oncol》 1992, 119: 46-48]

References

1　Baker SJ, Markowitz S, Fearon ER, Willson JKV, Vogestein B. Suppression of human colorectal carcinoma cell growth by wild type p53. Science, 1990, 249: 912-919

2　Bressac B, Kev M, Wands J, Ozturk M. Selective G to T mutations of p53 gene in hepatocellular carcinoma from southern Africa. Nature, 1991, 350: 429-431

3　Finlay CA, Hinds PM, Levine AJ. The p53 proto-oncogene can act as a suppressor of transformation. Cell, 1989, 57: 1083-1093

4　Hinds P, Finlay C, Levine AJ. Mutation is required to activate p53 gene for cooperations with the ras oncogene and transformation. J Virol, 1989, 63: 739-736

5　Hsu IC, Metcalf RA, Sun T, Welsh JA, Wang NJ, Harris CC. Mutational hotspot in the p53 gene in human hepatocellular carcinomas. Nature, 1991, 350: 427-428

6　Lane DP, Crawford LV. T antigen is found to host protein in SV40-transformed cells. Nature, 1979, 278: 261-263

7　Linzer DIH, Levine AJ. Characterzation of a 54 kdalton cellular SV40 tumour antigen present in SV40-tranformed cells and uninspected embryonal carcinoma cells. Cell, 1979, 17: 43 – 52

8　Nigro JM, Baker SJ, Preisinger AC, Jessuo JM et al. Mutations in the p53 gene occur in diverse human tumors types. Nature, 1989, 342: 705 – 708

9　Rhim JS, Park JB, Jay G. Neoplastic transformation of human keratinocytes by polybrene-induced DNA mediated transfer of an activated oncogene. Oncogene, 1989, 4: 1403 – 1409

10　Rogel A, Popliker M, Webb CG, Oren M. p53 cellular tumour antigen: analysis of mRNA levels in normal abult tissue, embryos and tumours. Mol Cell Biol, 1985, 5: 2851 – 2855

11　Sarnow P, Ho YS, Williams J, Levine AJ. Adenovirus Elb-58 kdtumour antigen and SV40 large tumour antigen are physically associated with the same 54 kd cellular protein in transformed cells. Cell, 1982, 28: 387 – 394

12　Werness BA, Levine AJ, Howley PM. Association of human papillomavirus 16 and 18 E6 proteins with p53. Science, 1990, 284: 76 – 79

13　Zeng Y. Seroepidemiological studies on nasopharyngeal carcinoma in China. Adv Cancer Res, 1985, 44: 121 – 138

57.　Antioncogenes in Nasopharyngeal Carcinoma Tissues

CHEN Wei – ping, LEE Yang, WANG Hui, YU Geng-geng, ZENG Yi

Department of Tumor Virus and HIV, Institute of Virology, Chinese Academy of Preventive Medicine, 100 Ying Xin Jie, Xuan Wu Qu, Beijing 100052, People's Republic of China

Retinoblastoma is a childhood cancer that occurs in both hereditary and nonhereditary forms. Rb gene was found in the research of retinoblastoma and was mapped to region q14 of human chromosome 13 (Sparkes RS et al, 1983). The deletion of Rb gene have been observed in somatic cells of hereditary retinoblastoma patients, retinoblastoma cells (Sparkes et al, 1983; Friend SH et al, 1986; Lee WH et al, 1987), and a variety of human tumors including osteosarcoma, synovial saracoma, and other soft-tissue sarcoma, small cell lung carcinoma, breast carcinoma, and bladder carcinoma (Weiberg RA, 1989). But until now there has been no report on the changes of Rb gene in nasopharygeal carcinoma. In this work two nasohparyngeal carcinoma culture cells in vitro, CNE-1 and CNE – 2, two nude mice NPC tumor, CNT – 1 and CNT – 5, and six biopsies from Wuzhou city and Cangwu county in Guangxi autonomous region, a high incidence area of NPC, Rb gene was studied with Southern-blot.

　　DNAS were extracted from the above NPC tissues, then were digested with Hind III and subjected to southern analysis. DNAS were sequentially probed with two fragments of Rb cDNA clone: a 0.9 kb fragment representing the 5'portion of the gene, and a 3.8 kb fragment representing the 3'portion of the gene.

　　The results demonstrated that Rb gene changed in NPC cells. The abnormalities contain completely or partially deleted of Rb gene. and rearrangement of the gene. It is the first report in the world that Rb gene changed in NPC cells. Rb gene may play an important role in the pathogenesis of NPC.

The p53 gene is a nuclear protein. Because it is often overexpressed in transfected cells and can transform the primary rodent cells with oncogene ras, p53 has been suspected to be an oncoprotein for many years (Rogel et al. 1985). Recent studies showed that the wild-type p53 gene did not have a transforming function and, on the contrary, could inhibit the tumor cell growth. If the wild-type p53 gene mutated it would have transforming potential (Hinds et al. 1989; Finlay et al. 1989). Mutations of the p53 gene occur in many types of human tumors (Nigro et al. 1989). Baker et al. (1990) transfected the colorectal carcinoma cell lines with wild-type and mutant p53 gens. The results demonstrated that the wild-type p53 gene could inhibit cell growth in nude mice, but until now there has been no report on the p53 gene and its function in nasopharyngeal carcinoma (NPC) cells. In this work NPC cell line CNE-3 was transfected with wild-type and mutant p53 genes, and the tumorigenicity in nude mice was compared.

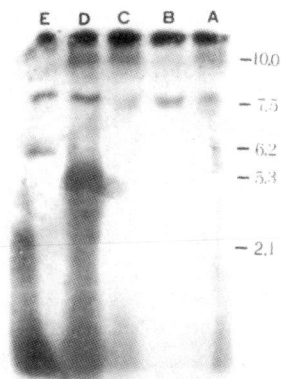

Fig. 1   Normal tissue (A), CNT-5 (B), CNT-1 (C), CNE-2 (D), CNE-1 (E) DNA were probed with Rb cDNA 3. 8kb fragment. CNE-2 has the normal DNA pattern. 10. 0kb fragment was deleted in CNE-1. 7. 5kb fragment in CNT-1 and 10. 0kb fragment in CNT-5 were partially deleted

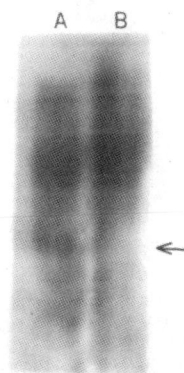

Fig. 2   CNE-1 and CNE-2 DNA were probed with RB cDNA 0. 9kb fragment. 6. 0 − 7. 0kb fragment was deleted in CNE-1 cell (arrowed)

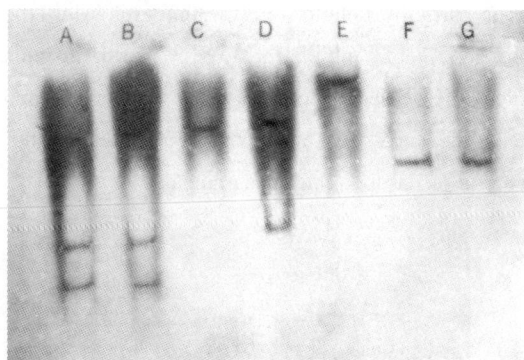

Fig. 3   Six NPC biopsies DNA (B – G) were probed with Rb cDNA 0. 9kb fragment. Tissue NO. 1 has normal DNA pattern. DNA bands of tissue No. 5 and 6 were different from others and Rb gene rearrangement was found in this two biopsies DNA of No. 4 may not completely digested and a large Mt. DNA bands was found. Low Mt. DNA bands were deleted in tissue No. 2 and 3

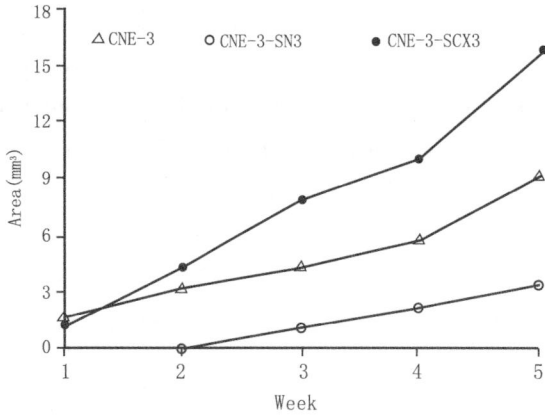

Fig. 4 Comparison of tumor growth speeds in nude mice

CNE-3, control cells; CNE-3-SCX3, CNE-3 cells transfected with mutant p53 mutant p53 gene; CNE-3-SN3, CNE-3 cells transfected with wild-type p53 gene. Initially the tumor resulting from the control cells grew a litter faster than that of the cells transfected with mutant p53 gene, but 2 weeks later, the control cells tumor was growing significantly more slowly than that resulting from cells transfected with the mutant p53 gene ($P < 0.001$). In contrast, tumor from the cells transfected with the wild-type p53 gene grew more slowly than that of the control cells and that of the cells transfected with mutant p53 gene ($P < 0.001$). The volume and weight of tumor at the 5th week are shown in Tab. 1 and Fig. 5. An approximately threefold difference was found among them.

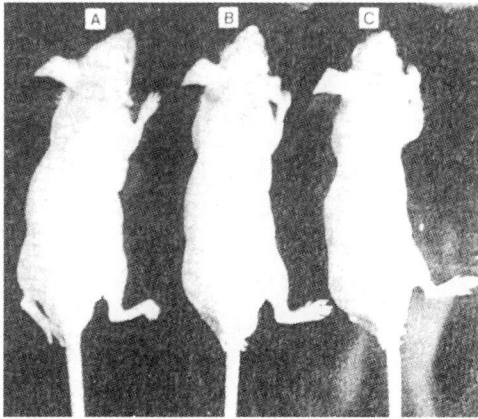

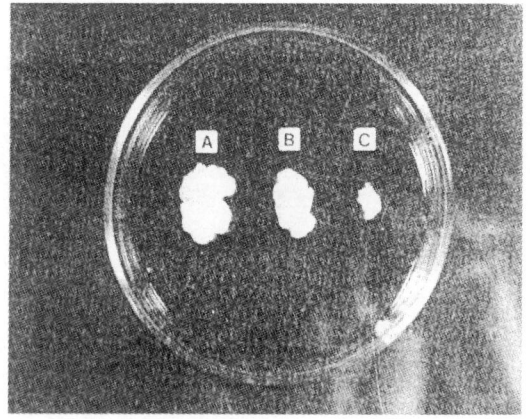

Fig. 5 Tumors in nude mice. A, CNE-3 cells transfected with mutant p53 gene; CNE-3 control cells; C, CNE-3 cells transfected with wild-type p53 gene

Tab. 1 Comparison of tumors in nude mice at 5th week

Parameter	Origin of tumors		
	CNE-3-p53-mt	CNE-3	CNE-3-p53-wt
Volume of Tumors (mm)^3	15.7	9.33	3.50
Weight of Tumors (g)	1.27	0.42	0.12

In conclusion, the results demonstrated that Rb gene has been changed in nasopharyngeal carcinoma tissues. The abnormalities contain completely or partiallly deletion of Rb gene. And wild-type p53 gene could inhibit the CNE-3 growth in nude mice. The tumor suppressor gene, Rb gene and p53 gene, may play an impotant role in the pathogenesis of nasopharyngeal carcinoma.

Acknowledgements

We thank Prof. Bert volelstein, The Johns Hookins University, USA, for providing plasmid pC53-SCX3 and pC53-SN3 and Prof. Lionel Crawford, Imperial Cancer Research Fund Laboratories, UK, for providing monoclon antibody PAB 1801 against p53 protein (Banks et al. 1986).

[In 《The Epstain-Barr Virus and associated Diseases》 1993, 255: 533-537]

References

1　Banks LM, Matlashewski G, Crawford L. Isolation of human p53-specific monoclonal antibodies and their use in the study of human expression. Eur J Biochem, 1986, 159: 529–534

2　Baker SJ, Markowitz S, Fearon ER, Willson JKV, Vogelstein B. Supression of human colorectal carcinoma cell growth by wild-type p53. Science, 1990, 249: 912–919

3　Finlay CA, Hinds PM, Levine AJ. The p53 proto-oncogene can act as a suppressor of transformation. Cell, 1989, 57: 1083–1093

4　Frinds SH, et al. A human DNA segment with properties of the gene that predisposes to retinblastma and osteosarcoma. Nature, 1986, 323: 623

5　Hinds P, and Crawford LV. Mutation is required to activate the p53 gene for cooperations with the ras oncogene and transformatron. J Virol, 1989, 63: 739–746

6　Lane DP, and Crawforld LV. T antigen is found to host protein in SV40 – transformed cells. Nature, 1979, 278: 261–263

7　Lee WH, et al. Human retinoblastoma suscepti-bility gene: cloning indentification and sequences. Science, 1987, 235: 1394

8　Nigro JM, Baker SJ, Preisinger AC, Jessup JM et al. Mutations in the p53 gene occur in diverse human tuomor types. Nature, 1989, 342: 705–708

9　Rogel A, Popliker M, Webb CG, and Oren M. p53 cellular tumour antigen: anlaysis of mRNA levels normal adult tissue, embryos and tumours. Mol cell Biol, 1985, 5: 2851–2855

10　Sparkes RS, et al. Gene for hereditory retinoblastoma assigned to human chromosome 13 by linkage to esterse D. Science, 1983, 219: 917

11　Weiberg RA. Oncogene, antioncogene, and the molecular bases of multistep carcinogenesis. Cancer Res, 1989, 49: 3713

58. A 10-Year Prospective Study on Nasopharyngeal Carcinoma in Wuzhou City and Zangwu County, Guangxi, China

ZENG Yi[1], DENG Hong[2], ZHONG Jian – ming[2], HUANG Nai – qin[2]
LI Ping – jun[3], HUANG Yu – ying[2], LI Yue[2], WANG Pei – zhong[4], de – The Guy[5]

1. Institute of Virology, Chinese Academy of Preventive Medicine, People's Republic of China;

2. Wuzhou Cancer Institute, Wuzhou City, Guangxi, People's Republic of China;

3. Zangwu Cancer Institute, Zangwu County, Guangxi, People's Republic of China;

4. Guangxi Autonomous Reginal Hospital, Nanning, Guangxi, People's Republic of China;

5. Institute Pasteur, 25, rue du Docteur Roux, 75015 Paris, France

Summary

Population ser-epidemiological surveys in Wuzhou City and Zangwu Cunty on South China established the value of the lgA/VCA and lgA/EA immunoenzymatic tests for an early detection of NPC at early clinical stages (1 and Ⅱ) in 87% of the cases, where radiotherapy ensures high survival rate (60% at 10 years, all clinical stages included). Furthermore such tests permitted to characterize the high risk group (lgA/VCA positive individuals) who are at more than 200 fold higher risk for later development of the tumor, than lgA negative persons. These data strongly support an etiological role of EBV in NPC pathogenesis.

Introduction

Serologic screening was implemented using an immunoenzymatic test in Zangwu County and Wuzhou city in 1978 and 1980 respectively, permitting an increase in detecting early clinical stages 1 and Ⅱ nasopharyngeal carcinoma (NPC) from 18. 6 – 31. 5% to 61% – 88. 8% respectively (Zeng et al 1980, 1982). Then a 4-year follow-up observation of EBV lgA/VCA antibody-positive persons was conducted in both Zangwu county and Wuzhou city (Zeng et al , 1983, 1988) leading to the detection of thirty-two new early cases of NPC increasing the early diagnosis rate to 91. 5% . These results indicate that serologic examination of the anti-EBV/lgA/VCA antibody does greatly increase the early diagnosis rate of NPC. However, there were some remaining problems to be clarififed, since the EBV lgA/VCA antibody persists for long time. ln such conditions, can the EBV lgA/VCA antibody test be used as an index for predicting the occurrence of NPC ? To answer this question, continuous clinical and histologic follow-up observations on the EBV lgA/VCA antibody positive persons were carried out. Results of a 10 year follow-up observation are reported herewith.

1. Ten year follow-up studies of NPC in Wuzhou City

As shown in Tab. 1 and Tab. 2 1136 out of 20 726 orginally screened in 1980 (aged over 40 years) were found positive for the an anti-EBV/lgA/VCA antibody immunoenzymatic test. Furthermore eighteen cases of NPC were originally found clinically and histologically confirmed, of which 16 (88. 9%) were at an early stage (Ⅰ and Ⅱ). Additional 29 cases were detected during a 10-year regular follow-up observation (1980 – 1990) among the lgA positive individuals regularly follows up, of whom 25 (86. 2%) were detected at an early stage of the tumor. Altogether, 47 patients were thus detected by serologic screening and follow-up observation, of which 41 (87. 2%) were at the early stage. ln contrast, among 3 374 cases of NPC which were detected in the outpatient clinics during the 1980 – 1989 period, without serological screening, only 873 (25. 8%) were at an early stage.

As seen in Tab. 2, there were 6 persons diagnosed as NPC and originally found to have EBV lgA/VCA antibody at serologic screening, but who lacked yearly follow-up. They were at an advanced stage when examined in hospital 3 – 6 years after original serologic screening. Among them, five were at stage Ⅲ and one at stage Ⅳ. Three cases showed a 2 to 5 fold rise in the antibody titer. 2 cases showed no significant rise , and 1 case who had a titer of 1 : 10 at the time of serologic screening showed no detectable antibody when the diagnosis was established.

Tab. 1　NPC early diagnostic rate from
serological screening and follow-up studies in Wuzhou City（1980 – 1990）

		Clinical Stage				Total	Early diagnostic rate
		I	II	III	IV		
Original screening,	No.	10	6	2	0	18	
（1980）	（%）	（55.5）	（33.3）	（11.2）		（100）	（88.8）
10 year Follow-up,	No.	5	20	4	0	29*	
（1980 – 1990）	（%）	（17.2）	（69.0）	（13.8）		（100）	（86.2）
Screening, and	No.	15	26	6	0	47	
Follow-up	（%）	（31.9）	（55.3）	（12.8）		（100）	（87.2）
Patients out clinic							
without screening,	No.	27	846	2043	458	3374	
（1980 – 1989）	（%）	（0.8）	（25）	（60）	（13.6）	（100）	（25.8）

Notes: ＊Among those, 23（79.3%）were detected within five years after serologic screening and 6（21.7%）between the 6th and 10th year（one case）. Furthermore 14（48.3%）showed a 4 fold increase of anti-Ebv/lgA/VCA antibody titer at time of diagnosis as compared to initial screening. The remaining cases showed no change or a fluctuation of only 2-fold in the antibody titer

Tab. 2　NPC patients from EBV lgA/VCA antibody
positive and negative persons without follow-up

Serum No.	Sex	llgA/VCA titer at screening	lgA/VCA titer at diagnosis	Clinical stage	Time interval between screening & diagnosis	
427	M	40	40	III	3	year
8797	M	10	< 50	III	3	
5839	M	80	40	III	3	
1338	F	10	160	IV	3	
9453	M	80	320	III	4	
462	F	10	320	III	6	
1280	M	< 5	80	II	4	
8483	F	< 5	160	IV	5	
11921	M	< 5	20	III	6	
13522	M	< 5	< 5	III	7	

Beside, as seen in Tab. 3, four cases of NPC were diagnosed among the original EBV lgA/VCA antibody-negative individuals group, who were not followed up. When the diagnosis was established, three cases had an antibody titer of 1 : 20 – 1 : 160 and one case had no detectabe antibody. They were diagnosed as NPC between 4 and 7 years after original serologic screening.

To sum up, and as seen in Tab. 3, 57 cases of NPC were detected among 20 726 persons during a 10 years period, giving an NPC detection rate of 275/100 000, with an average annual incidence of 27.5/100 000. Among them, 53 cases（93%）were detected among the EBV lgA/VCA antibody-positive individuals and only 4 in the lgA/VCA negative group, giving a NPC detection rate of 20/100 000 and a yearly incidence of 2. 0/100 000 in such a group. The difference in NPC risk between lgA/VCA positive and negative was 233 fold.

2. Survival rate of NPC after radiotherapy

A post-radiotherapy follow-up observation was conducted on NPC patients in Wuzhou City. Results are presented in Tab. 4. The 5 years and the 10 years survival rate of patients detected by serologic screening were 68. 7% and 59. 8% respectively, whereas, those patients undeted by serologic screening exhibited significantly lower survival rates (50. 9% and 32. 0% respectively at 5 and 10 years).

Tab. 3　Detection rate and incidence of NPC in lgA/VCA positive and negative persons

	Total	lgA/VCA positive	lgA/VCA negative
Persons	20726	1136	19590
NPC cases, No.	57	53 (93%)	4 (7%)
10 y detection rate (per 100 000)	275	4665. 5	20. 4
Yearly incidence (per 100 000)	27. 5	466. 5	2. 0

Tab. 4　Survival rate of NPC after radiotherapy (all stages)

Year	Survival Rate %		
	Original screened individuals	Non screened	Total
1	96. 49	88. 09	89. 68
2	87. 30	73. 79	76. 43
3	75. 54	59. 19	62. 45
4	71. 28	54. 85	58. 16
5	68. 69	50. 99	54. 62
6	68. 69	48. 02	52. 36
7	59. 83	44. 87	48. 00
8	59. 83	44. 87	48. 00
9	59. 83	44. 87	48. 00
10	59. 83	32. 05	39. 27

3. Ten-year prospective studies in Zangwu County

In the rural Zangwu county, serological screening was implemented in 1979 and 931 individuals with lgA/VCA antibody were also followed up for 10 years, As seen in Tab. 5, the early detection rate of NPC markedly increased from 18. 6% to 66. 7% due to the serological screening and kept high (80. 9%) during the 10 years follow-up.

Tab. 5　Serological screening and follow-up studies on NPC Zangwu County (1979 – 1989)

		Clinical Stages				Total	Early diagnostic rate
		I	II	III	IV		
Before screening	No.	0	11	24	24	59	
	%	(0)	(18. 6)	(40. 7)	(40. 7)	(100)	(18. 6)
At time of	No.	5	5	3	2	15	
screening (1979)	%	(33. 3)	(33. 3)	(20)	(13. 3)	(100)	(66. 7)
During follow-up	No.	5	12	4	0	21	
(1979 – 1989)	%	(23. 8)	(57. 1)	(19. 1)	(0)	(100)	(80. 9)

As seen in Tab. 6, NPC could be detected 7 years after initial serologycal calscreening especially in individuals where lgA/VCA antibody titer increased, with development of IgA/EA antibodies up to 3 years to dignosis.

Tab. 6　EBV lgA/VCA and lgA/EA antibody changes and detection of NPC during follow-up in Zangwu

Name	Sex	Age	lgA Ab. titers	0	1	2	3	4	5	6	7	NPC stage at diagnosis
				(year of follow up)								
Pan	M	60	VAC	20	20	20	20	20	40	80		I
			EA									
Liu	F	40	VAC	20	20	20	20	10	160	160		I
			EA									
Li	F	52	VAC	20	20	20		20	40	40	80	I
			EA									
Pan	F	50	VAC	20	20	10	10	10	10	10	10	II
			EA									
Wu	M	48	VAC	20	20	20	20	20	80	160	160	II
			EA					10	10	20	40	

Tab. 7　Relationship between increasing EBV/VCA antibody titers and development of NPC in Zangwu County

No. of Persons follow-up (10 years)	Negative Seroconversion	Fluctuation	No Change	4X increase	4X Decline
931	304	67	367	66	127
(100%)	(32.7%)	(7.3%)	(39.4%)	(7.1%)	(13.6%)
No. of NPC	0	0	6	15	0
NPC detection rate			(1.6%)	(22.7%)	

Tab. 8　Positive seroconversion rate (from lgA/VCA antibody negative to antibody positive) 10 year after initial screening

lgA/VCA Negative persons at original screening	lgA/VCA seroconversionant titers				Total
	1:5	1:10	1:20	1:40	
2300	47	40	4	3	94
Positive rate (%)	(2.0)	(1.74)	(0.17)	(0.13)	(4.1)

As seen in Tab. 7, when mean variations in lgA/VCA antibody titers were considered in the original lgA/VCA positive group of 931 individuals, the percentage of IgA/VCA antibody-positive persons either losing or showing fluctuating or having constantly low or stable titer or those increasing or declining their titer were 32.7%, 7.3%, 39.4%, 7.1% and 13.6% respectively. lt is of interest that only two sub-groups were at high risk for developing NPC: (a) individuals who increased their lgA/VCA antibody titer by fourfold or more and who exhibited a 22.7% chance of developing NPC; and (b) individuals who had stable lgA/VCA antibody titers with a chance of 1.6% of developing NPC.

As seen in Tab. 8, reexamination of 2300 EBV lgA/VCA antibody negative individuals at the 10th year after initial serologic screening showed that only 4.1% became positive.

Conclusion

EBV lgA/VCA and lgA/EA antibodies were found to be very valuable markers for early diagnosis of NPC. More than 90% of NPC could be detected in the lgA antibody-positive individuals, especially when the lgA/VCA antibody titer increased and lgA/EA antibody developed. The clinical stage of NPC at diagnosis can be reversed and the 5 years survival rate was found significantly better in these early detected cases. According to both lgA/VCA and lgA/EA antibodies markers the risk of developing NPC can be predicted 5 – 10 years before clinical diagnosis of NPC . These data strongly support EBV virus playing an etiological role in the development of NPC.

[In 《The Epstein-Barr virus and associated Diseases》 1993, 225: 735 – 741]

References

1 Zeng Y, Liu Yuxi, Chun ren, Chen Sanwen, Wei Jineng, Zhu, Jisong, Zai Huijong. Application of immunoenzymatic method and immuno-autoradiographic method for the mass survey of nasopharymgeal carcinoma. lntervirology, 1980, 13, 162 – 168

2 Zeng Y, Zhang lG, Li H Y, Tan M G, Zhang Q, Wu Y C, Wang Y S, Su G R. Serological mass survey for early detection of nasopharyngeal carcinoma in Wuzhou City, China lnt J Cancer, 1982, 29, 139 – 141

3 Zeng Y, Zhang J M Li, L Y Yang P Z, Tang H, Ma T R Zh, J S, Pan W J, Liu Y X, Wei, Z N, Chen J Y, Mo Y S Li, E J, Tan B F. Follow – up studies on Epstein – Barr lgA/VCA antibody positive persons in Zangwu county, China lntervirology, 1983, 20, 190 – 194

4 Zeng Y, Zhang L G, Wu Y C, Huang N Q, Li J Y, Wang Y B, Jian M K, Meng N N. Prospective strdies on nasopharyngeal carcinoma in Epstein – Barr virus lgA/VCA antibody positive persons, Wuzhou city , China lnt J Cancer, 1986, 36, 545 – 547

59. Urinary Excretion of Nitrosamino Acids and Nitrate by Inhabitants of High and Low Risk Areas for Nasopharyngeal Carcinoma in Southern China

ZENG Yi[1], OHSHIMA Hiroshi[2], BOUVIER Guy[2], ROY Pascal[2], ZHONG Jian – ming[3] LI Binjun[2], BROUET lsabelle[2], de The Guy[4], BARTSCH Helmut[2]

1. lnstitute of Virology, Chinese Academy of Preventive Medicine, Beijing, People's Republic of China; 2. lnternational Agency for Research on Cancer, 69372 Lyon Cedex 08, France;

3. Zangwu Cancer lnstitute, Zangwu, People's Republic of China; Unite d'Epidemologie des Oncogenes Institute Pasteur, 75724 Parts, France

Summary

The hypothesis that endogenous synthesis of nitrosamines from dietary precursors is a risk factor for nasopharyngeal carcinoma (NPC) in China was tested by applying the nitrosoprolint (NPRO) test to subjects living in high- and low-risk districts for NPC in Zangwu County, Guangxi region, in southern China. Samples of 12-h urine were collected from 77 subjects: (a) before any treatment;

(b) after ingestion of proline; and (c) after ingestion of proline together with vitamin C. NPRO, other nitrosamino acids, and nitrate were measured as indices of exposure to preformed and endogenouely formed nitrosamines of their precursors. The NPRO level after proline intake was significantly increased in subjects from the low-risk area ($P = 0.012$) and markedly reduced after ingestion of ascorbic acid ($P = 0.007$), but such an effect was not seen in subjects from the low-risk area. Levels of N-nitrosothiazolidine-4-carboxylic acid and the sum of nitrosamino acids in subjects in the high-risk area were significantly reduced by ascorbic acid ($P < 0.01$) but were not reduced in subjects from the low-risk area. The urinary nitrate level was about twice as high in subjects from the high-risk area. In subjects from high- and low-risk area combined, NPRO levels in any of the three dose groups were highly correlated with nitrate levels ($P = 0.0001$). These results demonstrate a higher potential for endogenous nitrosation in subjects living in the high-risk area of NPC and suggest the occurrence of nitrosation inhibitors in the diet consumed in the low-risk area. Thus, in addition to infection by Epstein-Barr virus and genetic predisposing factors, dietary habits that may entail higher nitrosamine exposure appear to play a role in NPC etiology.

Introduction

NPC exhibits wide variations in incidence throughout the world; it is most common in China and southeast Asia, and among Magrebian Arabs in north Africa and Eskimos in the Arctic[1]. Risk factors for NPC that have been identified include genetic predisposition (HLA haplotypes)[2,3], infection by EBV, and environmental factors, especially food consumption habits[4-7]. Earlier studies have shown that some samples of Cantonese-style salted fish contain relatively high levels of volatile nitrosamines[8,9], some of which induce tumors in the nasal cavities of experimental animals[10,11]. Recent studies in high-risk area for NPC in Tunisia, southern China, and Greenland revealed that these widely different populations all commonly consume preserved foods[12]. A study in GuangXi in southern China has shown that consumption before the age of 2 years of a number of presered foods, such as Cantonese-style salted fish, is strongly associated with an increased risk for NPC[6]. However, similar Japanese dried fish and vegetables were also reported to contain relatively high concentrations of volatile nitrosamines, but the incidence of NPC in Japan is very low.

We have searched , therefore, for additional hitherto unkown environmental risk factor for NPC and analyzed food extracts from high-risk areas for the presence of substances that activate EBV. Catonese-style soft salted dried fish and *harissa*, a Tunisian spice mixture, were found to contain agents with EBV-inducing activity[13]. Furthermore, our previous studies revealed that certain preserved food items contained high levels of precursors that , upon nitrosation *in vitro*, yielded volatile nitrosamines and direat-acting mutagens[14], suggesting that *in vitro* nitrosation could occur after ingestion of precursors and nitrosating agents in the diet . In the present study, we applied the NPRO test to inhabitants of high-and low-risk area for NPC in southern China to compare their endogenous nitrosation potential. The excretion of urinary mitrosamino acids and of nitrate was used as an index of individual exposure to nitrso compounds or their precursors, ingested in food or formed endogenously[15].

Materials and Methods

1. Subjects and Sample Collection: Samples of 12-h urine were collected in the spring of 1990 from 77 healthy subjects living in two villages in Zangwu county in the Guangxi region of southern China with contrasting incidence rates for NPC. The high-risk district had an incidence of NPC of 28/100 000/year, while for the low-risk district the corresponding figure was 2.9/100 000/year[16]. These figures are taken from the local cancer registry, which was established in 1976. Characterisics of the study subjects who were selected at random are shown in Tab. 1. Three urine specimens were collected (for 12 -h overnight starting 1 h after the evening meal) on three consecutive days from each subject according to the following protocols: (a) urine samples (groups H1, L1) were collected before dosing in order to determine the background levels of nitrosamino acids and nitrate; (b) proline specimens (groups H2, L2) were collected after subjects had ingested 300 mg L-proline 1 h after the evening meal; (c) proline plus vitamin C specimens (groups H3, L3) were collected after subjects had ingested 300 mg L – proline together with 300 mg ascorbic acid 1 h after the evening meal. Analyses of nitrosamino acids and nitrate were performed as reported by Lu *et al.* [17]. Study subjects were asked to complete a questionnaire to obtain information on demography, food items, beverages consumed, and number of cigarettes smoked during the urine collection period.

2. Statistical Analyses: Because the distribution of the variables was skewed, nonparametric tests were used. Descriptive analysis included mecian and its standard error (SE in Tab. 1-3) as proposed in the BMDP software[18]. To compare the distribution of the concentration of nitrosamino acids and of nitrate excreted by inhabitants from high-and low-risk ares, a Wilcoxon rank-sum statistical test was used. In each area the differences between samples 1, 2, and 3 were compared using the Wilcoxon signed-rank test for comparison of matched samples. Correlation between continuous variables was calculated using natural logarithmic transformation (to normalize the variables) or Box-Cox transformation[19] when the log transformed variables were still not normal. All the tests were two-sided, and the level of significance chosen was 5%.

Tab. 1   **Characteristics of study subjects**

	n	Median age (SE) (years)	No. of Smokers	No. of Nonsmokers
High-risk district				
Male (M)	18	40.5 (1.2)	12	6
Females (F)	19	39.0 (1.4)	0	19
Low-risk district				
Males	20	48.0 (2.9)	15	5
Females	20	44.0 (3.5)	1	19
Comparison between low- and high-risk areas				
High-risk (M + F)	37	40.0 (0.9)	12	25
Low-risk (M + F)	40	47.0 (2.9)	16	24
P value of comparison		0.004	Nonsignificant	

Tab. 2   Median (SE of median) for volume of 12-h urine and for amounts of N-nitrosamino acids, and creatinine detected in urine

Urine sample from	Excretion of N-nitrosamino acids (μ-g/12h)			Urine volume (ml/12h)	Creatinine (mmol/12)	Nitrate (mmol/12h)
	NPRO	NTCA	Sum[a]			
High-risk district[b]						
H1	3. 90 (0. 87)	21. 30 (4. 30)	33. 30 (5. 25)	350 (40. 41)	3. 18 (0. 40)	0. 87 (0. 14)
H2	7. 60 (2. 86)	39. 30 (10. 71)	68. 50 (13. 42)	300 (28. 87)	4. 14 (0. 35)	2. 01 (0. 31)
H3	3. 40 (0. 84)	18. 70 (3. 78)	28. 40 (4. 50)	350 (28. 87)	3. 65 (0. 40)	1. 13 (0. 24)
P (H1 vs. H2)[c]	0. 012	0. 073	0. 067	0. 530	0. 006	0. 004
P (H2 vs. H3)	0. 007	0. 004	0. 007	0. 812	0. 105	0. 030
Low-risk district[a]						
L1	3. 30 (0. 80)	23. 26 (4. 11)	31. 30 (7. 79)	300 (28. 87)	2. 94 (0. 37)	0. 62 (0. 10)
L2	3. 17 (1. 24)	13. 78 (2. 26)	20. 10 (4. 59)	300 (37. 53)	2. 82 (0. 30)	0. 39 (0. 08)
L3	2. 44 (0. 80)	13. 33 (3. 00)	18. 70 (4. 62)	250 (28. 87)	2. 68 (0. 33)	0. 45 (0. 07)
P (H1 vs. H2)	0. 485	0. 008	0. 035	0. 458	0. 553	0. 526
P (H2 vs. H3)	0. 936	0. 687	0. 545	0. 662	0. 936	0. 643
Comparison between areas						
P (H1 vs. L1)	0. 930	0. 839	0. 819	0. 751	0. 693	0. 027
P (H2 vs. L2)	0. 090	<0. 001	0. 002	0. 437	0. 018	<0. 001
P (H3 vs. L3)	0. 495	0. 737	0. 340	0. 078	0. 210	<0. 001

Notes: [a] Sum also includes N-nitrososarcosine and N-nitroso-2-methyl-thiazolidine 4-carboxylic acid (values not shown)

[b] Samples from undosed subjects (1). proline-dosed subjects (2). and proline-and vatamin C-dosed subjects (3)

[c] P values of comparison are also listed, with significant values underlined

Tab. 3   Medians (SE of mdeian) for volume of 12-h urine and amounts of N-nitrososomino acids and nitrate detected in urine (expressed per mmol creatinine levels)

Urine sample from	n	Excretion of N-nitrosamino scids (μg/mmol creatinine)			Nitrate (mmol/mmol creatinine)
		NPRO	NTCA	Sum[a]	
High-risk district[b]					
H1	37	1. 40 (0. 30)	7. 79 (1. 72)	11. 00 (2. 17)	0. 41 (0. 07)
H2	37	1. 83 (0. 47)	9. 85 (1. 65)	15. 30 (3. 35)	0. 46 (0. 13)
H3	37	0. 94 (0. 21)	4. 51 (0. 62)	7. 20 (0. 69)	0. 41 (0. 09)
P (H1 vs. H2)[c]		0. 225	0. 531	0. 656	0. 307
P (H2 vs. H3)		0. 018	0. 018	0. 032	0. 422
Low-risk district[b]					
L1	39	1. 09 (0. 16)	8. 44 (1. 54)	10. 60 (1. 36)	0. 16 (0. 02)
L2	40	1. 25 (0. 45)	4. 23 (1. 69)	8. 25 (2. 28)	0. 15 (0. 04)
L3	40	1. 21 (0. 18)	5. 78 (0. 87)	8. 15 (1. 96)	0. 21 (0. 03)
P (L1 vs. L2)		0. 413	0. 046	0. 120	0. 597
P (L2 vs. L3)		0. 872	0. 619	0. 773	0. 350
Comparison between areas					
P (H1 vs. L1)		0. 835	0. 759	0. 666	0. 002
P (H2 vs. L2)		0. 335	0. 019	0. 040	<0. 001
P (H3 vs. L3)		0. 771	0. 318	0. 714	0. 005

Notes: [a] See Footnote a in Tab. 2.

[b] Urine form (1) undosed. (2) proline-dosed, and (3) proline plus vitamin C-dosed subjects.

[c] P values of comparison are also listed, with significant values underlined

Results

The nitrosamino acids analyzed were NPRO, NTCA, *N*-nitrososarcosine, and *N*-nitroso-2-methylthiazolidin-4-carboxylic acid. The latter two are not individually listed but are inclued in the sum levels of all four nitrosamino acids. Tab. 2 summarizes urine volume , levels of nitrosaamino acids (μg/12 h/person), nitrates (mmol/12 h/person), and creatinine (mmol/12 h/person) that were detected in the sixsets of urine samples. The volumes of the 12-h urine samples showed no difference between the two areas. Median urinary levels of NPRO, NTCA, and the sum of nitrosamino acids in the samples from undosed subjects did not differ between subjects from the high-and low-risk areas. Intake of proline (300 mg) increased urinary NPRO excretion by the subjects in the high-risk area from 3. 9 to 7. 6 μg/day ($P = 0.012$) but did not significantly change NPRO levels among subjects in the low – risk area. Intake of ascorbic acid together with proline (300 mg) by the high-risk subjects significantly decreased the urinary level of NPRO, NTCA , and the sum of nitrosamino acids ($P < 0.01$). This inhibiting effect was not significant in subjects from the low-risk area. The marked reduction of urinary excretion of nitrosamino acids after intake of ascorbic acid indicates that endogenous formation of these compounds was inhibited by ascorbic acid; this decrease provides an indication of an individual's nitrosation potential *in vivo*, which was found to be higher in subjects living in the high-risk area.

Since , for logisitic reasons, no 24-h urines could be collected , we determined the creatinine concentration in the urine and expressed the levels of nitrosamino acids per mmol of creatinine (Tab. 3). However, as shown before, endogenously formed NPRO is almost totally excreted within 12 h[20]. Creatinine concentrations in the urine did not show any consistent difference between subjects from the two areas (Tab. 2). Comparisons of urinary nitrosamino acid levels, expressed per mmol of creatinine levels, led to essentially the same conclusions as for the uncorred values in Tab. 2, although some comparisons were not statistically different. For example, NPRO levels after proline intake did not increase when expressed per creatinine (H1: H2 in Tab. 2 *versus* Tab. 3). A correlation between creatinine and NPRO excretion has been observed earlier and could be attributable to an increased meat or protein intake[21]. However, the major findings presented in Tab. 2 are confirmed in Tab. 3: after proline intake, NTCA levels and the sum of nitrosamino acids were significantly higher in subjects from the high-risk area (by roughly a factor of 2). After ascorbic acid intake, the urinary levels of NPRO, NTCA, and the sum of nitrosamino acids were significantly reduced in subjects living in the high-risk area, while no such reduction was seen in the subjects in the low-risk area.

The effect of smoking on urinary excretion of nitrosamino acids was examined. In the group 1 samples (undosed subjects), the amounts of either NPRO ($P < 0.01$), NTCA ($P < 0.01$), and the sum of nitrosamino acids ($P < 0.001$) were higher in smoker than in nonsmokers. In group 2 (proline specimens) and group 3 (proline plus vitamin C specimens), there were no statistical differences between smokers and nonsmokers. The proportions of smokers in high-and low-risk ares were similar (Tab. 1).

The urinary levels of nitrate per 12 h or those corrected for creatinine are listed in Tab. 2 and 3. The nitrate levels excreted in subjects from the high-risk area in all three dose groups were significantly higher than those from the low-risk area. This approximately 2-fold difference remacined significant when nitrate levels were expressed per mmol of creatinine (Tab. 3). After Box-Cox transformation of data on urinary volumes from undosed subjects and those receiving proline, the urinary volume was positively correlated with log NPRO ($P < 0.01$); this positive correlation, however, became nonsignificant when the proline values were corrected for creatinine. The Pearson correlation coefficients (each at $P = 0.0001$) between urinary levels of NPRO and of nitrate (both log transformed) were 0.47 in undosed specimens (combined groups H1 and L1; 0.67 in proline specimens [combined groups H2 and L2 (Fig. 1)], and 0.42 in proline plus vitamin C specimens (combined groups H3 and L3). Similarly, in undosed specimens (groups H1 and L1 combined), log [sum of nitrosamino acids] *versus* log nitrate were correlated ($r = 0.33$ and $P < 0.01$; Fig. 2). The linear regression lines (shown in Fig. 1 and 2) had slopes of 0.75 and 0.38, respectively. Recorded food items consumed by subjects during the day of urine collection included meat, salted and fresh fish, vegetables, and fruits. Each type of nutrient was poorly represented, and the relationship between the levels of nitrosamino acids and the amount of any type of food item could not be adequately analyzed.

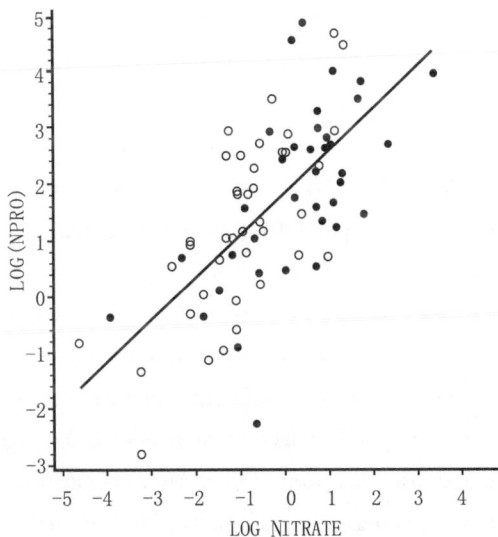

Fig. 1 A plot of log (NPRO) versus log (nitrate) concentration in the urine in subjects dosets with proline from high-risk (●) and low-risk (○) districts for NPC in southern China ($r = 0.67$, $P < 0.001$). The corresponding linear regression line is shown

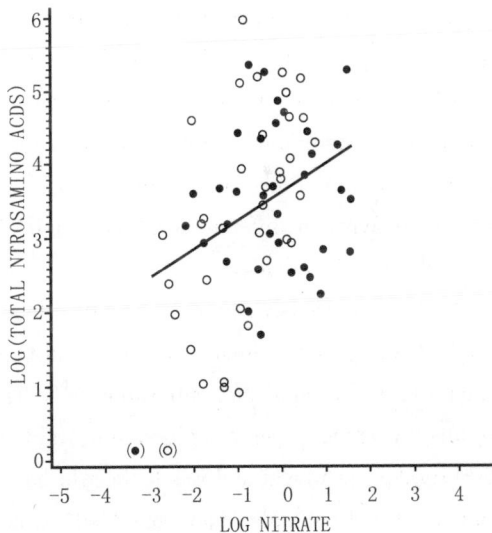

Fig. 2 A plot of log (sum of nitrosamino acids) versus log (nitratel concentration in the urine of undosed subjects from high-risk(●)and low-risk(○) districts for NPC in southern China ($r = 0.33, P < 0.001$). The corresponding linear regression line is shown. Two outliving points (in parentheses) with very low levels of mtrosamine acids were excluded for this analysis.

Discussion

The NPRO test has previously been applied to subjects living in high-and low-risk ares for stomach cancer in Japan, Poland, and Costa Rica and for esophageal cancer in northern China[22]. In general, the nitrate exposure and potential for endogenous nitrosation, measured by the increased levels of urinary NPRO after intake of proline, were much higher in the high-risk populations. Furthermore, low-risk subjects may ingest sufficient amounts of protective agents to suppress endogenous nitrosation. Results from this study provide for the first time an indication that intragastrically formed N-nitroso compounds (or other nitrite-derived mutagens/carcino-gens) may be risk factors for NPC in southern China. The roughly 2-fold higher nitrosation potential that we have observed in the high-risk subjects appears attributable to their higher nitrate intake and lack of nitrosatin inhibitors that are probably present in the diet of subjects living in the low-risk area. Our study was not designed to enable us to pinpoint a particular food item that could inhibit nitrosation. Vegetables and fruits contain not only nitrate but other constituents as well, such as vitamin C and phenolic compounds which generally inhibit N-nitrosation. The subjects from the high-risk area, after intake of vitamin C , had a drastically lowered level of nitrosamino acids, and no such decrease was observed in subjects from the low-risk area. These results strongly suggest that nitrosation of dietary precursor compounds ingested by high-risk subjects may generate mutagens or carcinogens that play a role in the etiology of NPC. This hypothesis fits with the ealier suggestion by Ho[23] , who proposed that risk factors for NPC involve an interaction between genetically determined susceptibility, early infection by the Epstein-Barr virus, and exposure to chemical carcinogens through the consumption of traditional preserved food, especially salted fish in southern China, from the weaning period onward. Since then, several case-control studies (4-7) have been conducted in southern Chinese living in different locations, and all of the results support Ho's hypothesis, that diet play a role in NPC etiology.

We have previously screened preserved food items that are frequently consumed in areas endemic for NPC in southern China, Tunisia, and Greenland for the presence of mutagens and volatile nitrosamines, before and after nitrosation[9,14]. The levels of preformed volatile nitrosamines (N-nitrosodimethylamine, N-nitrosopiperidine, and N-nitrosopyrrolidine) ranged from nondetectable up to 500 μg/kg wet weight and was highest in hard salted and dried fish from China. After chemical nitrosation, 9 of 15 samples (aqueous food extracts) showed increased mutagenicity, and the levels of volatile nitrosamines were elevated in 12 of 15 samples; the highest level of N-nitrosodimethylamine was found in nitrosated extracts of hsrd salted and dried fish from China (1200 μg/kg wet weight) and the highest N-nitrosopyrrolidine level in a nitrosated *harissa* sample (about 3800 μg/kg wet weight). These two food item are among those that recent studies in high-risk areas have implicated as NPC risk factors. Some volatile nitrosamines induce tumors in the nasal cavity of experimental animals[10,11] and are therefore candidates for involvement in NPC etiology.

Our observation that nitrosation potential is increased in subjects living in a high-risk area for NPC implies that volatile nitrosamines could be formed *in vivo* at high exposure levels in subjects who ingest dietary nitrosamine precursors from early childhood onward. In fact, Yu et al[5]. have shown in a case-control study on young NPC patients that consumption of Cantonese-style salted fish during childhood, especially between 1 and 10 years of age, is associated with NPC; the relative risk for weekly as compared to rare consumption, at the age of 10, was 37.7. It was estimated that over 90% of NPC cases in Hong Kong Chinese under the age of 37 could be linked to consumption of salted fish during childhood. The nature and levels of nitrosamines and nitrite-derived mutagens formed after nitrosation of high-risk food items such as salted fish have not been characterized, apart from the three volatile nitrosamines mentioned above. Two animal bioassays have demonstrated the carcino-genic effects of Cantonese-style salted fish: Huang and Ho[24] reported that 4 of 10 female Wistar albino rats which consumed steamed Cantonese-style salted fish developed carcinoma (both adenocarcinoma and squamous cell carcinoma) of the nasal and paranasal regions, and Yu et al.[25] found nasal cavity tumors (squamous cell carcinoma, undifferentiated and spindle-cell carcinomas) in 3 of 74 Wistar-Kyoto rate fed powdered diet containing salted fish.

In order to identify additional risk factors for NPC, we have further analyzed food iterms consumed by inhabitants from areas of high NPC risk for the presence of substants that activate EBV in vitro[13]. We found that aqueous extracts of Cantonese-style salted and dried fish, and harissa, a Tunisian spice mixture, can induce EBV early antigen in latently infected Raji cells[13,26]. This activity may be important, since it is known that lgA antibodies against EBV viral and capsid antigen and early antigens reflect in vivo EBV activation; these are regularly present in NPC patients and appear in subjects before tumor development[27]. However, since EBV-inducing activity did not parallel either mutagenicity of the level of volatile nitrosamines in the food extracts investigated[13], we are now undertaking the isolation and characterization of EBV-inducing substances in food item that have been associated with NPC risk.

In conclusion, our results support the hypothesis that diet plays a role in NPC etiology, as indicated by case-control studies[28]; the latter consistently demonstrated that consumption of Chinese-style salted fish is strongly related to risk for nasopharyngeal cancer. Nitrite-derived carcinogens (fromed from dietary precursors) and EBV-inducing substance are also implicated as risk factors.

Acknowledgments

We thank Dr. J. Cheney for editorial help and Mrs. Wrisez for typing the manuscript.

〔In 《Cancer Epidemiology Biomarkers and prevention》 1993, 2: 195 – 200〕

References

1　de Thé G. Epidemiology of the Epstein-Barr virus and associated diseases. In: B. Roizman (ed). The Herpes Viruses, New York: Plenum Press, 1982, 1 A. 25 – 103.

2　Simons M J, Chan S H, Wee G B. Shanmuga-ratnam K, Goh E H, Ho J H C, Chau J C W, Darmalingham S, Prasad U, Betual H, Day N

E, de Thé. Nasopharyngeal carcinoma and histo-compat-ibility antigens. In: G. de Thé and Y. lto (eds.) Nasopharyngeal Carcinoma: Etiology and Control, IARC Scientific Publication. Lyon: International Agency for Research on Cancer, 1978, 20: 271 – 282

3　Lu S J, Day N E, Degos L, Lepage V, Wang P

C, Chan S H, Simons M, McKnight B, Esaton D, Zeng Y, de ThéG. Linkage of a nasopharyngeal carcinoma susceptibility locus to the HLA region. Nature (Lond.), 1990, 346: 470 – 471

4　Geser A, Charnay N, Day N E, Ho H C, de Thé, G. Environmental factors in the etiology of nasopharyngeal carcinoma: Linkage of a case-control study in Hong Kong. ln: G. de The and Y. lto (eds.), Nasopharyngeal Carcinoma: Etiology and Control, IARC Scientific Publication. Lyon. lnternational Agency for Research on Cancer, 1978, 20: 213 – 229

5　Yu C, Ho J H C, Lai S H, Henderson B E. Cantonesestyle salted fish as a cause of nasopharyngeal carcinoma: report of a case-control study in Hong Kong. Cancer Res, 1986, 46: 956 – 961

6　Yu M C, Mo C C, Chong W X, Yeh F S, Henderson B E. Preserved foods and nasopharyngeal carcinoma: a case-control study in Guangxi, China. Cancer Res, 1988, 48: 1954 – 1959

7　Yu M C, Huang T B, Henderson B E. Diet and nasopharyngeal carcinoma: a case-control study in Guangzhou, China Int J Caner, 1989, 43: 121 – 138

8　Fong Y Y, Walsh W C. Carcinogenic nitrosamines in cantones salted-dried fish. Lancet, 1971, 2: 1032

9　Poirier S, Ohshina H, de ThéG, Hubert A, Bourgade M C, Bartsch H. Volatile nitrosamine levels in common foods from Tunisia, South China and Greenland, high risk areas for nasopharyngeal carcinoma (NPC). Int J Cancer, 1987, 39: 293 – 296

10　Druckrey H, Preussmann R, lvankovic S, Schmahl D. Organotropic carcinogenicity of 65 different N-nitroso compounds in BD rats. Z. Krebsforsch, 1967, 69: 103 – 210

11　Magee P N, Montesano R, Preussmann R. N-Nitroso compounds and related carcinogens. ln: C. E. Searle (ed.), Chemical Carcinogens, ACS Symposium Series. Washington, DC: American Chemical Society, 1976, 173: 491 – 625

12　Jeannel D, Hubert A, De-Vathaire F, Ellouz E, Camoun M, Ben Salem M, Sancho-Garnier H, de Thé, G. Diet, living conditions and nasopharyngeal carcinoma in Tunisia-a case-control study. Int J Cancer, 1990, 46: 421 – 425

13　Shao Y M, Poirier S, Ohshima H, Malaveille C, Zeng Y, de ThéG, Bartsh H. Epstein-Barr virus activation in Raji cells by extracts of preserved food from high risk areas for nasopharyngeal carcinoma. Carcinoma (Lond), 1988, 9: 1455 – 1457

14　Poirier S, Bouvier G, Malaveille C, Ohshima H, Shao Y M, Hubert A, Zeng Y, de Thé G, Bartsh H. Volatile nitrosamine levels and genotoxicity of food samples from high-risk areas for nasopharyngeal carcinoma before and after nitrosation. Int J Cancer, 1986, 44: 1088 – 1094

15　Ohshima H, Bartsh H. Quantitative estimation of endogenous nitrosation in humans by monitoring N-nitrosoproline excreted in the urine. Cancer Res, 1981, 41: 3658 – 3662

16　Pan W, Wu C, Li L, Liao J. Analysis of the epideminologicaltrend of nasopharyngeal carcinoma: Cangwu county, Guangxi (in Chinese). Cancer (Phila), 1988, 7: 345 – 348

17　Lu S-H, Ohshima H, Fu H-M, Tian Y, Li F M, Blettner M, Wahrendorf J, Bartsh H. Urinary excretion of N-nitrosamino acids and nitrate by inhabitants of high-and low-risk areas for esophageal cancer in northern China: endogenous formation of nitrosoproline and its inhibition by citamin C. Canner Res, 1986, 46: 1485 – 1491

18　BMDP Statistical Software. B MDP, 1990 Release. Los Angeles: BMDP Statistical Software, lnc, 1990

19　Box G E P, Cox D R. An analysis of transformations. J. R. Stat. Soc. , Ser. B (Methodol.), 1964, 26: 211 – 243

20　Stich H F, Ohshima H, Pignatelli B, Michelon J, Bartscn H. lnhibitory effect of betel nut extracts on endogenous nitrosation in humans. J Nat Cancer lnst, 1983, 70: 1047 – 1050

21　Mirvish S S, Grandjean A C, Moller H, Fike S, Maynard T, Jones L, Rosinsky S, Rosinsky

S, Nie G. N-Nitrosoproline excretion by rural Nebraskans drinking water of varied nitrate content. Cancer Epidemiol. , Biomarkers & Prev, 1992, 1: 455 – 461

22 Bartsch H, Ohshima H, Pignatelli B, Calmels S. Human exposure to endogenous N-nitroso compounds: quantitative estimates in subjects at high risk for cancer of the oral cavity, oesophagus, stomach and urinary bladder. Cancer Surv, 1989, 8: 335 – 362

23 Ho J H C Genetic, environmental factors in nasopharyngeal carcinoma. In: W. Nakahara, K. Nishioka, T. Hirayama, and Y. lto (eds.), Recent Advances in Human Tummunology. Tokyo: University of Tokyo Press, 1971: 275 – 295

24 Huang D P, Ho J H. Carcinoma of the nasal and paranasal regions in rats fed Cantonese salted marine fish. In G. de Théand Y. lto (eds.), Nasopharyngeal Carcinoma: Etiology and Control, IARC Scientific Publication no. 20, Lyon: lnternational Agency for Research on Cancer, 1978, 315 – 318

25 Yu M C, Nichols P W, Zou X N, Estes J, Henderson B E. lnduction of malkgnant nasal cavitv tumors in Wistar rats fed Chinese salted fish. Br J Cancer, 1989, 60: 198 – 201

26 Bouvier G, Poitier G, Poirier S, Shao Y M, Malaveille C, Ohshima H, Polack A, Bornkam G W, Zeng Y, de ThéG, Bartsch H. Epstein-Barr virus activators, mutages and volatile nitrosamines in preserved food samples from high-risk areas for nasopharyngeal carcinoma. ln: l. K. O'Neill, J. Chen, and H. Bartsch (eds.), Relevance to Human Cancer of N-Nitroso Compounds, Tobacco Smoke and Mycotoxins, lARC Scientific Publication. Lyon. lnternational Agency for Research on Cancer, 1991, 105: 204 – 209

27 de Vathaire F, Sancho-Garnter H, de Thé H, Pieddeloup C, Schwaab G, Ho J H C, Ellouz R, Micheau C, Cammoun Y, Cachin Y, de Thé G. Prognostic value of EBV markers in the clinical management of nasopharyngeal carcinoma (NPC): a multicenter follow-up study. lnt J Cancer, 1988, 42: 176 – 181

28 lnternational Agency for Research on Cancer. Some naturally occurring substances, some food items and constituents, heterocyclic aromatic amines and mycotoxins. lARC Monogr. Eval. Carcing. Risks Hum. , 56: in press, 1993

60. Environmental and Dietary Risk Factors for Nasopharyngeal Carcinoma: A Case-Control Study in Zangwu County, Guangxi, China

ZHENG Y M[1], TUPPIN P[2], HUBERT A[2]
JEANNEL D[2], PAN Y J[3], ZENG Y[4], de The G[2]

1. Cancer Institute of Wuzhou 543002, Guangxi Autonomous Region, People's Republic of China; 2. Unit on Epidemiology of Oncoviruses, Pasteur Institute, 28 rue du Dr Roux, 75724 Paris Cedex 15 , France: 3. Nasopharyngeal Carcinoma Institrte, Zangwu, Guangxi Autonomous Region, People's Repubic of China; 4. Institute of Virology, Chinese Academy of Preventive Medicine, Beijing 100052, People's Republic of China.

Summary

A case-control study was conducted on 88 incident cases of histologically confirmed undifferentiated nasopharyngeal carcinoma (NPC) in Zangwu County, China, and 176 age-sex-and neighbourhood-matched controls. The design of this study was defined after an anthropological survey on living habits in regions of high NPC incidence and the evidence of carcinogenic substances in some commonly consumed preserved foods. Subjects were interviewed regarding living conditions and diet in the year preceding the diagnosis of NPC and, with the help of their families, during childhood and weaning. After adjustment for a living conditions score to eliminate a confounding effect, an increased risk associated with consumption of salted fish during weaning and childhood was confirmed, especially for salted fish in rice porridge. The consumption of leafy vegetables was associated with a reduced risk for NPC, and consumption of melon seeds between 2 and 10 years of age with an increased risk. After multivariate analysis and adjustment according to the living conditions score, the consumption of salted fish in rice porridge before age 2 ($OR = 3.8$, $P = 0.005$), exposure to domestic woodfire ($OR = 5.4$, $P = 0.01$) and consumption of herbal tea ($OR = 4.2$, $P = 0.02$) were found to be independently related to the risk of NPC. The excess risk associated with the use of domestic wood fire increased if there were no windows in the house and with poor ventilation and cooking outside the house in a shack. As well as confirming the importance of the consumption of salted fish in childhood, this study has been the first to provide unequivocal evidence for two other factors implicated in increasing the risk of NPC in China, the adult consumption of traditional medicines (herbal tea) and exposure to domestic wood fumes.

Introduction

Nasopharyngeal carcioma (NPC) is common among Chinese (especially the Cantonese), with an age-standardised annual incidence rate of $30/10^5$ for males and $13/10^5$ for females (Muir et al., 1987); among the Maghrebian Arabs in North Africa (Parkin, 1986) ($3.4/10^5$ for males and $1.1/10^5$ for females in Algeria); and among the Eskimos in the Arctic (Lanier et al., 1976) ($10/10^5$ for males and $4/10^5$ for females). Elsewhere the incidence is low with an age-standardised annual incidence of less than $1/10^5$ reported in Europe and North America (Waterhouse et al., 1982).

The undifferentiated type of nasopharyngeal carcinoma (UCNT) seems to be associated with three aetiological factors: firstly, the Epstein-Barr virus (EBV), which is regularly present in the carcinomatous cells (Andersson-Anvret et al., 1978; see review by de Thé, 1982); secondly, a disease susceptibility gene, close to, but different from, an HLA gene, evidence for which was obtained in Chinese families with multiple NPC cases among sibs (Lu et al., 1990); thirdly, environmental factors associated with traditional preserved food (Ho, 1971; Geser et al., 1978). Several case-control studies conducted among southern Chinese have indicated an association between the consumptin of salted fish, especially during weaning, and the risk of developing NPC (Armstrong et al., 1983; Yu et al., 1986, 1989; Ning et al., 1990). More recent studies in China and Tunisia have suggested that the consumption in early youth of salted and preserved foods other than salted fish is also associated with an increased risk of NPC (Yu et al., 1988, 1989; Jeannel et al., 1990). In addition, NPC has been found to be associated with low socioeconomic level and a traditional lifestyle, and some potential risk factors associated with a traditional lifestyle, including the use of domestic wood fires, have been proposed (Armstrong *et al.*, 1978; Geser *et al.*, 1978; Jeannel *et al.*, 1990).

The present epidemiological study was part of a multidisciplinary NPC project (Hubert *et al.*, 1993). The first step was to conduct an anthropological study in the three high-risk groups for NPC (Cantonese Chinese, Mahrebian Arabs and Eskimos) with the aim of identifying common or comparable factors which could be linked to this tumour. This approach provided detailed background data on food habits and lifestyle, and after a comparative analysis the conclusion was that traditional preserved food preparations could represent the common factors (Hubert *et al.*, 1993). Food samples were then collected in South China, Macao, Tunisia and Greenland, and laboratory analysis revealed the presence of volatile nitrosamines and reactivants of EBV (Poirier *et al.*, 1987; Shao *et al.*, 1988). Case-control studies were carried out in Tunisia (Jeannel *et al.*, 1990), in Macao (Hubert *et al.*, 1993), and in Wuzhou City and Zangwu County (China), presented here. The aim was to investigate simultaneously a broad range of socioeconomic and environmental factors as well as dietary histary, with details of consumption frequencies and types of traditional food preparation which may increase the risk of NPC.

Population and Methods

1. Area: The population of Wuzhou City is 170 000 and that of Zangwu County, a rural area, is 550 000, predominantly Cantonese Chinese belonging to the Han ethnic group. This area belongs to the Guangxi Autonomous Region, which had the second highest mortality rates of NPC among all the Chinese provinces ($8.5/10^5$ age-standardised male mortality rates). These areas were selected because of the facilities offered by the cancer register in Wuzhou and clinical units specialising in the treatment of NPC in wuzhou and the county of Zangwu.

2. Subjects: This study included all incident cases of undifferentiated NPC diagnosed and histologically confirmed from the starting date of 1 January 1986, until 90 cases were accumulated. Wuzhou cases were recruited at the Wuzhou Cancer Institute, providing the patients were residents in Wuzhou at the time of diagnosis. Zangwu cases were identified from the Nasopharyngeal Carcinoma Institute of Zangwu, which specialises in NPC detection. No other institution in Wuzhou or Zangwu could diagnose or treat NPC. Eighty-eight patients (29 in Wuzhou and 59 in Zangwu) were included in the case-control study. In each case, the area of residence was ascertained and two controls who agreed to participate were selected by the interviewers in the immediate neighbourhood. Matching criteria were sex, age (plus or minus 4 years) and place of residence. These controls were interviewed within the same week as the patients and in the same conditions.

3. Data collection: Interviews were conducted at home using the local dialect in the presence of the family members and particularly parents, as far as when ever possible. In China many family members often live under the same roof as an extended family, so it was relative easy to collect data on childhood diet and weaning from subjects' mothers whenever possible, or from the female relative who took care of the subject during infancy. For 60% of cases and 60% of controls one or both parents were household members. For the other 40%, anther relative who cared for the subject during youth was present. Only one case and four controls had to older relatives at home and data comcerning their youth were noted as missing. The six interviewers were physicians at the Wuzhou Cancer Institute and Nasopharyngeal Carcinoma Institute of Zangwu, and they participated in several clinical and epidemiological studies carried out by those institutes. These interviewers had been trained by our team's nutritional anthropologist (A. H.) especially for this study.

4. *Lifestyle questionnaire*: This questionnaire was prepared by the anthropologist in our team (A. H.) and was submitted to preliminary field testing; it requested information on past and present socioeconomic conditions, housing and diets. Data on lifestyle, including educational levels, marital status, place of birth, residential history, personal or family income, housing, types of fuel used, kitchen and toilet equipment and sleeping conditions, were checked for two periods: childhood and the year preceding the diagnosis of NPC. This second period was chosen to investigate adult habits immediately before the onset of disease and deterioration of health (for the NPC cases) while limiting recally bias. Data on diet covered fourperiods: weaning, childhood and adolescence, and the year prior to diagnosis of NPC reflecting adult diet, with the same periods for matched controls. For all food categories, except some spices and condiments, subjects were asked to choose between six frequency categories (1 –2 times a day, 3 –4 times a week, 1 –2 times a week, 1 –2 times a month, 1 –2 times a

year or never). Food groups covered all dietary intake including drinks as well as methods of preparation and preservation and evolution of consumption over the past 20 years.

5. *Statistical analysis*: We used matched pairs and conditional logisitic regression to obtain for each study variable odds ratios (*ORs*) (estimates of the relative risk) and their *P*-value and 95% confidence intervals. In order to adjust for socioeconomic variables, we estimated aliving condition score using variables from the lifestyle questionnaire indicating poor socioeconomic conditions found to be linked with NPC. The selection of such variables was monitored using a conditional logistic regression procedure. The score was established by weighting each selected variable by coefficients obtained in this way. For each food item, the

Tab. 1　Odds ratio (*OR*) for sociodemographic factors used to create a sociodemogrphic score

	Cases (88)	Controls (176)	Matched odds ratio (*OR*)			
			Crude analysis		Logisic model	
			OR	*p*	*OR*[a]	*p*
In childhood						
Type of houseing						
Apartment or single-storey house	28	66	1		1	
Rural dwelling	60	110	2. 5	0. 09	2. 3	0. 1
In year before diagnosis						
House windows						
Yes	80	169	1		1	
No	8	7	2. 8	0. 1	3. 1	0. 04
Monthly income (yuans per month)						
>200	10	43			1	
101 –200	39	79	3. 2	0. 02	3. 5	0. 01
<101	39	54	5. 5	0. 001	6. 5	<0. 001
The sociodemographic score[b]						
0	5	29	1			
1 –5	21	35	3. 5			
>5 –8	24	69	2. 0			
>8	38	43	5. 1	trend test *P* =0. 006		

Notes: [a] *OR* adjusted for the other factors. [b] This score was established by the sum of the three variables above weighting by coefficients obtained with a conditional logistic model

OR was adjusted on this score. In each conditional logistic regression we inclued all variables associated with NPC with a *P*-value less than or equal to 0. 2.

Results

The 88 NPC cases were poorly differentiated or undifferentiated carcinomas. Four were stage Ⅰ according to Ho's classification (Ho. 1971), 27 were stage Ⅱ, 44 stage Ⅲ and two stage Ⅳ; one subclassifiable case was unknown. Sixty - four (73%) of the patients were males, with a mean age of 41. 6 years (95% *Cl* 31. 9 - 51. 3) for cases, and 41. 5 years (95% *Cl* 31. 9 - 51. 3) for controls. The age distribution among cases was: 15. 9% less than or equal to 30 years old, 32. 9% between 31 and 40 years, 34. 2% between 41 and 50 years, and 17% more than 50 years.

Two cases (2%) and ten controls (6%) were born outside the Guangxi Region, but all the cases and controls belonged to the Han ethnic community. There was no significant difference between cases and controls in their marital status or their level of education. Tab. 1 presents the sociodemographic variables linked with NPC and used to establish the living conditions score. The risk of NPC was higher for a monthly income between 101 and 200 yuan per month (*OR* = 3. 2, *P* = 0. 02) *P* = 0. 02) and greater still for an income less than 101 yuan per month (*OR* = 5. 5, *P* = 0. 001) as compared with income of more than 200 yuan per month (trend test. *P* = 0. 001). We included two further variables in the score: type of housing in childhood and lack of house windows during the preceding year, both variables having a weak association with NPC. There are clearly more NPC cases than controls with high score reflecting a low economic level (*P* = 0. 006).

As shown in Tab. 2, concering the consumption of salted fish during the three studied periods of life, significant associations with salted fish in rice porridge were ob-

Tab. 2 Odds ratio (*OR*) for nasopharyngeal carcinoma in relation to consumption of salted fish

	Cases (88)	Controls (176)	Matched odds ratio (*OR*) Crude analysis *OR*	*p*	Adjustment for the score *OR*[a]	*p*
During weaning						
Salted fish						
No	65	148	1		1	
Yes	22	23	2	0. 03	2. 4	0. 01
Before the age of 2						
Salted fish (steamed of fried)						
Rarely	52	115	1		1	
Monthly and weekly	35	57	1. 3	0. 3	1. 4	0. 3
Salt fish soup						
Rarely	76	150	1		1	
Monthly and weekly	11	12	1. 9	0. 2	1. 8	0. 2
Salt fish in rice porridge						
Rarely	71	158	1		1	
Monthly and weekly	16	14	2. 5	0. 02	3. 5	0. 006
Between the ages of 2 *and* 10						
Salted fish (steamed or fried)						
Rarely	33	77	1		1	
Monthly and weekly	55	99	1. 3	0. 3	1. 4	0. 2
Salt fish soup						
Rarely	73	159	1		1	
Monthly and weekly	15	17	2. 0	0. 8	2. 3	0. 05
Salt fish in rice porridge						
Rarely	66	156	1		1	
Monthly and weekly	22	20	2. 6	0. 006	3. 2	0. 003

[a] *OR* adjusted for the sociodemographic score

served with monthly and weekly consumption duing the three periods, when adjusted for the living conditions score: weaning ($OR = 2.4$, $P = 0.01$), before the age of 2 ($OR = 3.5$, $P = 3.5$, $p = 0.006$) and between the ages of 2 and 10 ($OR = 3.2$, $P = 0.003$). Consumption of salted fish during the year preceding NPC was very low for both cases (2.3%) and controls (0.6%) and decreased significantly for both over the pasr 20 years.

Among the studies food items, consumption of the following foods and condiments during the preceding year was shown to be significantly associated with a reduced crude risk for NPC (Tab. 3): leafy vegetables, beef, monosodium glutamate (MSG). But after adjustment for the living conditions score, only the consumption of leafy vegetables remained associated with a reduced risk for NPC. Consumption of salted, dried or tinned foods such as meat, eggs or vegetables in brine was not found to be significantly linked with risk for NPC except for consumption of melon seeds during childhood before and alcoholic drinks were not found to be significantly associated with NPC risk. Furthermore, the use of wood as domestic fuel during the preceding year was shown to be associated with increased risk for NPC ($OR = 3.7$, $P = 0.02$) (Tab. 3). After adjustment for the living conditions score, the risk associated with use of woodfire increased ($OR = 6.4$, $P = 0.003$). Herbal tea drinking was associated with an increased risk for NPC before and after adjustment for

score. In a multivariate matched logistic regression analysis taking into account the living conditions score, three variables remained significantly associated with NPC: use of wood as fuel, consumption of salted fish in rice porridge before the age of 2 and herbal tea drinking the year preceding diagnosis (Tab. 4). When considering separately urban and rural areas (Wuzhou City and Zangwu County), the independent risk factors in Zangwu County were consumption of salted fish in rice porridge before the age of 2 and herbal tea drinking in the year preceding diagnosis, whereas in Wuzhou City consumption of salted fish during weaning andmelon seeds between the age of 2 and 10 emerged as risk factors.

Tab. 3 Odds ratio (OR) for nasopharyngeal carcinoma by diet and environmental factors

	Cases (88)	Controls (176)	Crude analysis		Adjustment for the score	
			OR	p	OR^a	p
In year before diagnosis						
Wood fuel						
No	8	31	1		1	
Yes	80	145	3.7	0.02	6.4	0.003
In childhood (2 – 10 year)						
Melon seeds						
No	74	163	1		1	
Yes	14	11	2.6	0.02	2.8	0.02
In year before diagnosis						
Beef						
Rarely	55	97	1		1	
Monthly	21	40	0.7	0.3	0.8	0.6
Weekly	12	38	0.3	0.03	0.6	0.3
Leafy vegetables						
Monthly	10	4	1		1	
Weekly	40	88	0.2	0.007	0.2	0.014
Daily	38	84	0.2	0.006	0.1	0.008
Monosodium glutamate (MSG)						
NO	42	65	1		1	
Yes	46	111	0.6	0.05	0.7	0.3
Herbal tea						
No	55	130	1		1	
Yes	33	46	3.7	0.007	4.5	0.006

^{a}OR adjusted for the sociodemographic score

In Tab. 5, the risk for NPC of the use of wood fire was studied in conjunction with environmental factors which may modify the level of fumes. Absence of windows, poor ventilation and cooking outside the house in a shack increased the excess of risk for NPC associated with domestic woodfires with statistically significant trends.

Nosebleeds and buzzing in the ears, considered to represent early symptoms of NPC, were most frequent among cases (Tab. 6). Consumption of herbal tea in the year before diagnosis was more frequent for cases than for controls, but this association was not related to stage of NPC and it persistent when subjects with nosebleeds and buzzing (possible early symptoms of NPC) were excluded (Tab. 6). This indicates that the consumption of herbal tea was not a response to the onset of NPC. Moreover, consumption of herbal tea was not significantly associated with nosebleeds ($P = 0.3$) and buzzing ($P = 0.5$). Consumption of herbal mixtures during weaning and childhood was associated with a relative risk of 1.8 after adjustment for the living conditions score, but the excess risk was not significant ($P = 0.07$).

Tab. 4　Odds ratio (OR) for nasopharyngeal carcinoma, 95% confidence intervals (CI), in a multiple conditional logistic regression model

	OR^a	95% CI	P – value
Before the age of 2			
Salted fish in rice porridge	3.8	(1.5 – 9.8)	0.005
(monthly and weekly)			
In year before diagnosis			
Use of wood fuel	5.4	(1.5 – 19.8)	0.01
Consumption of herbal tea	4.2	(1.3 – 13.0)	0.02
Sociodemographic score	1.4	(1.2 – 1.7)	<0.001

Note: $^a OR$ adjusted for the other factors and the score

Tab. 5　Odds ratio (OR) for nasopharyngeal carcinoma by use of wood fire associated with factors which may modify the level of fumes

Factor	n	No woodfire %	n	Woodfire%	n	Woodfire%	Trend test P
Windows in house				*present*		*absent*	
Cases	8	(21)	73	(34)	7	(54)	
Controls	31	(79)	139	(66)	6	(46)	0.008
Total (264)	39		212		13		
Crude OR (P)	1		3.6	(0.03)	7.8	(0.009)	
Ventilation				*good*		*poor*	
Cases	8	(21)	24	(32)	56	(34)	
Controls	31	(79)	50	(68)	95	(66)	0.01
Total (264)	39		74		151		
Crude OR (P)	1		3.1	(0.07)	4.7	(0.01)	
Kitchen				*inside house*		*outside in a shack*	
Cases	8	(21)	66	(34)	14	(45)	
Controls	31	(79)	128	(66)	17	(55)	0.01
Total (264)	39		194		31		
Crude OR (P)	1		3.4	(0.04)	5.9	(0.01)	
Window in the kitchen				*present*		*absent*	
Cases	6	(19)	59	(33)	7	(41)	
Controls	26	(81)	118	(67)	10	(59)	0.07
Total (264)	32		177		17		
Crude OR (P)	1		3.4	(0.008)	5.4	(0.06)	

Tab. 6　Relationship between consumption of herbal tea and non – specific symptoms of NPC and stage

	Cases		Controls		P	OR	P
	n	%	n	%			
For all cases and controls							
Herbal tea							
Yes	33	(37)	46	(26)		4.1[a]	0.01
No	55	(63)	130	(74)	<0.01		
Bleeding from the nose							
Yes	10	(11)	4	(2)		2.7[a]	0.2
No	78	(89)	172	(98)	<0.01		
Buzzing in the ears							
Yes	15	(17)	7	(4)		2.4[a]	0.1
No	73	(83)	169	(96)	<0.001		
For cases without nasebleeds and buzzing and their controls only							
Herbal tea							
Yes	22	(37)	35	(29)		3.6[a]	0.06
No	38	(63)	85	(71)	<0.01		
For cases in stage Ⅰ *and* Ⅱ *their controls only*							
Herbal tea							
Yes	14	(45)	16	(26)		6.01[b]	0.03
No	17	(65)	46	(74)			
For cases in stage Ⅰ *and* Ⅱ *their controls only*							
Herbal tea							
Yes	19	(33)	30	(26)		3.9[b]	0.07
No	38	(67)	84	(74)			

Note:[a] OR adjusted for the two other factors. [b] OR only adjusted for the score.

Discussion

This study confirmed the role of consumption of salted fish as a risk facter for NPC and identified a specific risk associated with salted fish in rice porridge during weaning. Moreover, it established associations between the consumption of herbal tea, the use of domestic wood fire and an increased risk of NPC, and these were still significant after adjusting for a living conditions score and so were independent of socioeconomic status. These three risk factors were independently linked to an increased risk of NPC after selection by a stepwise logistic regression.

This study may be affected by some bias inherent in case-control studies, especially as data on diet taken almost 30 years ago were collected. To minimise recall bias, only close relatives who took care of the subject during youth were interviewed together with the subject. Recall bias due to cognition of disease status and risk factors should be minimal because no specific preventive campaign about risk factors had been conducted prior to the study in the area. This is also supported by the fact that among all the types of salted fish preparation investigated only salted fish in rice porridge emerged as a risk factor. The data on the frequency intake inevitably include a certain percentage of misclassification, particularly with older subjects recalling the past. If these errors can be assumed to be random and similar for cases and controls, they lead to an observed odds ratio closer to the unity than the true relative risk.

In previous studies, indicators of lower socioeconomic status and poor housing conditions were found to be positively associated with NPC in South-East Asia and Tunisia (Greser *et al.* , 1978; Armstrong *et al.* , 1983; Jeannel *et al.* , 1990). Few studies have analysed the risks associated with traditional dietary risk factors while adjusting for socioeconmic factors (Geer *et al.* , 1978; Jeannel *et al.* , 1990).

Our results highlighted in Zangwu Region the importance of salted fish in rice porridge eaten during weaning and childhood as a risk factor for NPC, and the relative risk was even higher when adjusted for living conditions score. The salted fish is usually steamed prior to mixing with rice porridge, Similarly, a study performed in a low-risk region for NPC, Tianjin (China), showed that the consumption of steamed salted fish at the age of 10 years carried a higher relative risk than consumption of fried, grilled or boiled salted fish at the same age (Ning *et al.* , 1990). Methods of cooking (duration, temperature, associated food) could have an effect on the amount and/or the activity of carcinogenic substances present in salted fish. Interestingly, the excess risk acquired during infancy persisted, although exposure to this risk factor dramatically decreased, with only 2% of cases and 0. 6% of controls continuing to eat salted fish, whatever the type of preparation, as compared with respectively 43% and 33% in childhood. The apparent evolution of diet in the Zangwu Region, characterised by a large decrease in consumption of preserved food, fish as well as vegetables or meat, and its replacement by fresh products has not yet affected the risk in the generations concerned by this change, as shown by the persistence of a high rate of NPC incidence in Zangwu Region, but an effect may be observed in future decades. Regular consumption of leafy vegetables was associated with a reduction of the risk for NPC, as reported in other studies for other vegetables (Yu *et al.* , 1989; Ning *et al.* , 1990).

An increased risk was found for melon seed consumption between the ages of 2 and 10 (especially in Wuzhou City, where this snack is sold on the streets). It would be interesting to investigate the presence of carcinogenic substances in this dried salted snack, and it should be mentioned that several reports on animal models for NPC have suggested aflatoxins as potential co-carcinogens (Levine *et al.* , 1990).

Two factors, not well established before, were independently linked to the risk of NPC: using wood fuel during adulthood and drinking herbal tea. A positive association with occupational exposure to smoke and fumes or working in poorly ventilated places has also been reported (Lin *et al.* , 1973; Henderson *et al.* , 1976; Jeannel *et al.* , 1990; Yu *et al.* , 1990). In two previous case-control studies, patients with NPC used firewood for cooking more frequently than controls, but this association was not studied together with diet and not adjusted for socioeconomic level (Djojapranata & Soesilowati, 1967; Shanmugaratnam *et al.* , 1978). In the present study , the risk associated with the use of wood fuel during adulthood was higher after adjustment for the living conditions score, and increased with lack of windows, poor ventilation or cooking outside in a shack. The lack of association during childhood is probably because everyone used firewood 40 years ago , as shown by our data. Thus, our observation might suggest the role of certain fumes in NPC develpploment; this has been under discussion since a high incidence of NPC has been observed in Hong Kong boat people who cooked in the open air, and so were supposed to be little exposed to fumes (Ho,

1967). Thus, it would be interesting to investigate which type of wood and dried plants are for woodfire in the Guangxi Region. Besides, the amounts of 3, 4-benzpyrene (3, 4-BP) in smoke samples collected from a high-risk area were higher than those from a low-risk area (Kai-Tai *et al.*, 1987).

Medicinal herbal preparations, such as herbal tea, are frequently used to prevent or to treat many diseases in China, and they have been postulated to be a risk factor for NPC (Zeng *et al*, 1983). One could object that a more frequent use of herbal tea among cases could have been a consequence of the appearance of symptoms of NPC. This is not supported by the present data: there was no significant difference between cases and controls with respect to the evolution of consumption 1 year before diagnosis compared with 20 years before. Furthermore, there was no association between the use of herbal tes and the presence of nasal discharge, nosebleed or buzzing in the ears during the year preceding diagnosis of NPC. Having no details on the types of preparation used by cases and controls, this point cannot be further evaluted. It is interesting to note that several species of plants of the Euphorbiance family used in common Chinese herbal mixtures contain diterpene, an EBV reactivator and tumor promoter (Hirayama & lto, 1981; Zeng *et al.*, 1983). But in Guangzhou, two members of the Euphorbiacese family (P. *emblica*and C. *crassifolius*) used as herbal ingredients were not found to present a risk of NPC (Yu *et al.*, 1990). In Tunisia the use of traditional remedies in youth such as poultices of castor plant leaves (*Ricinus communis*L, Euphorbiaceae) was found to be associated with an increased risk of NPC (Jeannel *et al.*, 1990). A study conducted by Hildesheim *et al.* (1992) in the Philippines suggested that if herbal medicines interact with EBV in the development of NPC, it is rather through a direct proliferative effect on EBV-transformed cells than through reactivation of EBV infection.

Furthermore, if we compare our results in Zangwu Region in China with that of a case-control study on 29 Chinese incident cases in Macau conducted by our laboratory using the same design and questionnaire, similar risk factors emerged: salted fish consumption before age of 2 years ($OR = 15.5$, $P = 0.02$) ($OR = 15.5$, $P = 0.02$) and fireplace in kitchen during childhood ($OR = 5.9$, $P < 0.01$). Besides, it was interesting to note a protective effect associated with the use of powdered milk before the age of 2 ($OR = 0.04$, $P < 0.01$) (Hubert *et al.*, 1993).

In summary, the data presented here confirm the role of environmental and dietary factors in NPC carcinogenesis. In addition to the known increased risk associated with early consumption of salted fish, this study pointed out that the type of salted fish preparation might be of importance. Moreover, the fact that herbal tea and use of woodfire appear to be risk factors opens a new area of research concerning the role of carcinogens and EBV reactivants of plant origin. When one considers the environmental risk factors associated with NPC in different parts of the world, one is left with the view that, besides EBV and genetic factors ethnic differences in lifestyle, particularly food preparation or consumption, remain the best aetiological hypothesis for explaining the geographical distribution of the disease.

We are grateful to the medical teams of the Cancer Registy of Wuzhou and the Cancer Unit of Zangwu Hospital for their active collaboration in the field study and Dr H. Sancho-Garnier and F. de

Vathaire, lnstitu Gustave Roussy, Villejuif, France, for their methodological help in statistical analysis. We wish to thank Dr R. Bomford for his helpful review of the manuscript. This study was supported by the Chinese National Sciences and Technology Committee, the CNRS (SDl 5660, URA 1157), ARC contract 6071 and Virus Cancer Prevention.

[In 《Br J Cancer》 1994, 69: 508 – 514]

References

1 Andersson-anvrest M, Forsby N, Klein G, Henle W. The association between undifferentiated nasopharyngeal carcinoma and Epstein-Barr virus shown by correlated nucleic acid hybridization and histo-pathological studies. ln Nasopharyngeal Carcinvma: Etiology and Control. de Thé G, lto Y. (eds). IARC: Lyon. 1978, 20: 347 – 357

2 Armstrong R W, Kannan Kuty M, Armstrong M J. Self-specific environments associated with nasopharyngeal carcinoma in Selangor, Malaysia Soc Sci Med, 1978, 12D, 149 – 156

3 Armstrong R W, Armstrong M J, Yu M C, Henderon B E. Salted fish and inhalants as risk factors, for nasopharyngeal carcinoma in Malaysian Chinese. Cancer Res, 1983, 43, 2967 – 2970

4 De The G. Epidemiology of Epstein-Barr virus and associated diseases . ln The Herpes Viruses, Vol. 1A. B. Roizman (ed.). Plenum Press: New York, 1982, 25 – 103

5 Djqjapranata M, Soesilowati S. Nasopharyngeal cancer in East Java (lndonesia). ln Cancer of the Nasopharynx, Vol. 1, UICC Mongraph Series, Muir, C. S. & Shanmugaratnam, K. S. (eds). Munksgaard: Copenhagen, 1967, 43 – 46

6 Geser A, Charnay N, Day N E, HO H C, De The G. Environmental factors in the etiology of nasopharyngeal carcinoma: report of a case control study in Hong Kong. ln Nasopharyngeal Carcinoma: Etiology and Control, Vol. 20, de Thé, G. &lto, Y. (eds). IARC. Lyon. 1978, 213 – 229

7 Henderson B E, Louie E, JING J S, Buell P, Garoner M B. Risk factors associated with naso

pharyngeal carcinoma. N Engl J Med, 1976, 295: 1101 – 1106

8 Hildesheim A, West S, De Veyra E, De Guzman M, Jurado A, Jones C, Imai J, Hinuma, Y. Herbal medicine use, Epstein-Barr virus, and risk of nasopharyngeal carcinoma. Cancer Res, 1992, 52: 3048 – 3051.

9 Hirayama T, Ito Y. A new view of the etiology of nasopharyngeal carcinoma. Prev. Med, 1981, 10: 614 – 622.

10 Ho H C. Nasopharyngeal carcinoma in Hong-Kong. ln Cancer of the Nasopharynx, Vol. 1, UICC Monograph Series, Muir, C. S. & Shanmugaratnam, K. S. (eds). Munksgaard: Copenhagen, 1967, 58 – 63

11 Ho, H. C. Genetic and environmental factors in nasopharyngeal carcinoma. ln Recent Advance in Human Tumor Virology and lmmunology Nakahara, W., Nishioka, K., Hirayama. T. & lto, T. (eds) Proceedings of the First lnternational Cancer Symposium of the Princess Takamatsu Cancer Research Fund, University of Tokyo, 1971, 275 – 295

12 Ho, H. C. Stage classification of nasopharyngeal carcinoma: a review. ln Nasopharyngeal carcinoma: etiology and control, Vol. 20, de Thé, G. & lto, Y. (eds). IARC: Lyon, 1978, 99 – 113

13 Hubert A, Jeannel D, Tuppin P, De The G. Anthropology and epidemiology: a pluridisplinary approach of environmental factors of nasopharyngeal carcinoma. ln: The Associated Epstein-Barr Virus and Diseases, Vol. 225, Colloque Inserm, Tursz, T., Pagano, J. S., Ablashi, G., de Thé, G. Lenoir, G. & Pearson, G. R. (eds). John Libbey Furotexi Ltd, 1993, 777 – 790

14 Jeannel D, Hubert A, De Vathaire F, Ellouz

R, Camoun M, Ben Salen M, Sancho-garnier, H, De The, G. Diet, living conditions and nasopharyngeal carcinoma in Tunisia: a case-control study. lnt J Cancer, 1990, 46: 421 – 425

15 Lanier A P, Bender T R, Blot W J F, Hurlbert W B. Cancer incidence in Alaska natives. lnt J Cancer, 1976, 18: 409 – 412

16 Kai-Tai Y, Peng-nian W, Ji-Wen J.. The role of promotion in the carcinogensis of nasopharyngeal carcinoma. ln Cancer of the Liver, Esophaus and Nasopharynx, Wagner, G. & Zhang, Y. H. (eds). Spring: Berlin, 1987, 187 – 193

17 Levine P H, Huang A T, Weiland L, Hildesheim A, Brizel D, Boyce cole T, Fisher S R, Panells T J, Jian J. Nasopharyngeal carcinoma: 1990. ln Epstein-Barr Virus and Human Disease, Ablashi, D., Huang, A., Pagano, J., Pearson, G. & Yang, C. (eds). Humana Press: Clifton, NJ, 1990, 313 – 330

18 Lin T M, Chen K P, Lin C C, Hsu M M, Chiang T C, Jung P F, Hirayama T. Retrospective study on nasopharyngeal carcinoma. J Natl Cancer lnst, 1973, 51: 1403 – 1408

19 Lu S J, Day N E, Degos L, Lepage V, Wang P C, Chan S H, Simons M, Mcknight B, Easton D, Zeng Y, De The G. Linkage of a nasopharyngeal carcinoma susceptibility locus to the HLA region Nature, 1990, 346: 470 – 471

20 Muir C, Waterhouse J, Mack T, Powell J, Whellen S. (eds). Cancerlncidence in Five Continents, Vol. 88. IARC: Lyon, 1987

21 Ning J P, Yu M C, Wang Q S, Henderson B E. Conaumption of salted fish and other risk factors for nasopharyngeal carcinoma (NPC) in Tianjin, a low-risk region for NPC in the People's Republic of China. J. Natl. Cancer lnst, 1990, 82: 291 – 296

22 Parkin D M. (ed.) Cancer Occurrence in Developing Countries. Vol. 75, IARC: Lyon, 1986

23 Poirier S, Oshima H, De The G, Hubert A, Bourgade M C, Bartsch H. Volatile nitrosamine levels in common foods from Tunisia, South Chi-

na and Greenland, high risk areas for nasopharyngeal carcinoma. lnt. J. Cancer, 1987, 39, 293

24 Shanmugaratnam K, Tyk C Y, goh E H, Chia K B. Etiological factors in nasopharyngeal carcinoma: a hospital based, retrospective, case-control, questionnaire study. lnNasopharyngeal Carcinoma: Etiology and Control, Vol. 20, de Thé, G. & lto, Y. (eds). IARC. Lyon, 1978, 199 – 212

25 Shao Y M, Poirier S, Oshima H, Malavielle C, Zeng Y, De The G., Bartsch, H. Epstein-Barr Virus activation on Raji cells by extracts of preserved foods from high risk areas for nasopharyngeal carcinoma. Carcinogenesis, 1988, 9: 1455 – 1457

26 Waterhouse J A H, Muri C, Shanmugaratnam K, 3 others. Cancer lncidence in Five Continents, Vol. 42. IARC. Lyon, 1982

27 Yu M C, Ho J H C, Lai S H, Henderson B E. Cantonese-style salted fish as a cause of nasopharyngeal carcinoma: report of a case-control study in Hong Kong. Cancer Res, 1986, 46, 956 – 961

28 Yu M C, Mi C C, Chong W X, Yeh F S, Henderson B E. Preserved foods and nasopharyngeal carcinoma: a case-control study in Guangxi China. Cancer Res, 1988, 48: 1954 – 1949

29 Yu M C, Huang T B, Henderson B E. Diet and nasopharyngeal carcinoma: a case-control study in Guangzhou China. lnt. J Cancer, 1989, 43: 1077 – 1082

30 Yu M C, Garabrant D H, Huang T B, Henderson B E. Occupational and other non-dietary risk factors for nasopharyngeal carcinoma: a case-control study in Guangzhou China. lnt. J Cancer, 1990, 45: 1033 – 1039

31 Zeng Y, Zhong J M, MIAO X C. Epstein-Barr Virus early antigen induction in Raji cells by Chinese medicinal herbs. lntervirology, 1983, 19: 201 – 204

61. Establishment of a Human Malignant T Lymphoma Cell Line Carrying a Retrovirus-like Particle with RT Activity

LAN Xiang – ying[1], ZENG Yi[1], ZHANG Dong[1], HONG Ming – li[2], WHANG De – xin[2]

ZHANG Yong – li[1], FENG Zi – jing[2], TANG Mei – hua[3], FENG Bao – zhang[3]

1. Institute of Virology, Chinese Academy of Preventive Medicine, Beijing; 2. Friendship Hospital, Beijing; 3. Institute of Hematology, Chinese Academy of Medical Sciences, Tianjin

Summary

We have established an IL – 2 independent malignant lymphoma line (CM – 1) from peripheral T lymphocytes donated by a female patient with nervous system disease, the biological characteristics of CM – 1 cells was studied in this paper, Another T lymphocytes, such as peripheral T lymphocytes donated by a male patient with multiple sclerosis, could be transformed into a malignant lymphoma line by using filtered supernatant of the CM – 1 cultured medium, thus the CM – 2 cell line was established. The CM – 1and CM – 2 cells were transplanted by subcutaneous inoculation into nude mice, and could cause the occurrence of typical malignant lymphoma. The observation of electron micrographs suggested the existence of virions in the CM – 1 and CM – 2 cells, and these virions were similar to retrovirus in the ultra-structure characteristice. It was found that this virus possesses reverse transcriptase activity. Results obtained from serological assay, molecular hybridization and PCR excluded the existence of other human viruses, which were commonly used in our laboralory. The unknown virus possesses strong transformation activity, and probably is a new retrovirus. Meanwhile, the work on the clone and sequence analysis of this virus are being carried out.

Introduction

Profound studies on human retroviruses led to the discovery that HTLV – I is the cause of adult T cell leukemias-lyphomas (Hinuma et al. , 1981) and also related to some nervous system diseases (Gessain et al. , 1985; Osame et al. , 1987; Bartholomew et al. , 1986). We had established the CL – 8 cell line from the blood cells of a Chinese patient with leukemia and demonstrated that is was the CL – 8 cells harbour HTLV – I by immunological and molecular biological methods (Lan et al. , in press). We had also found the antibody against HTLV – I from five patients with nervous system diseases (Lan et al. , 1993). So we attempted to determine if there are other retroviruses which are related to nervous system diseases and leukemias in China. Through ten years' hard working, we have established a human malignant T lymphoma cell line carrying a retrovirus from peripheral blood lymphocytes donated by a patient with nervous system disease (Lan et al. ,

1992; Hong et al. , 1992). This unknown retrovirus is able to transform normal human T lympho-cyte. Results are reported as follows.

Materials and Methods

1. Establishment of the CM – 1 cell line

The CM – 1 cell line was derived from peripheral blood lymphocytes of a 36 – yearold woman. She was admitted to the Beijing Friendship Hospital on 2nd February 1990 on account of limb paralysis for more than one month and coma for 4 days. On the next day, her body temperature went up to $39 - 40\,℃$ and she was complicated with pneumonia. After two weeks, the body tempera-ture returned to normal, but she was still in a state of semicoma. At that time, 5 ml venous blood was collected and the lymphocytes were separated by Ficoll-Conray gradient centrifugation, and cul-tured with RPMI – 1640 medium and supplemented with 20% fetal calf serum (FCS) containing 1% penieillin, 1% streptomycin and 1% glutamine. The culture was incubated at $37\,℃$ in 5% CO_2 atmosphere. No growth factor was used. The cell line thus established was designated as CM – 1 (Chinese malignant T lymphoma cel line – 1).

2. Biological characteristics of the CM – 1 cell

The growth and morphological characteristics of the CM – 1 cells were directly observed under an inverted optic microscope.

The determination of CM – 1 cell growth curve: The suspension of new passaged CM – 1 cells with 5×10^4 cell/ml concentration was distributed into 42 small culture bottles with 3 ml per bot-tle. The culture was incubated at $37\,℃$ in 5% CO_2 atmosphere. Three bottles were taken out every 24 hours and the supernatant was aspirated out. The cells growing on the wall of bottles dropped after treatment with trypsin solution. The number of cells was counted. The procedure was carried out for 14 days without change of the medium. The growth curve was plotted.

3. Studies on the CM – 1 cell surface marker

The CM – 1 cell surface marker was studied by indirect immunofluorescence technique with 5 systems and 28 kinds of anti-leukocyte monoclonal antibodies.

(1) Target cell: the CM – 1 cells and the CM – 1 clones.

(2) Monoclonal antibodies against the leucocyte:

a. Monoclonal antibodies against T lymphocyte CD1, DC2, CD3, CD4, CD5, CD6, CD7, CD8, CD27.

b. Monoclonal antibodies against the activated lymphocyte, CD25, CD71, HLA-DR.

c. Monoclonal antibodies against myelocyte CD11 (B), CD13, CD15, CD33, CD15 (H198).

d. Monoclonal antibodies against B lymphocyte CD9, CD19, CD20, CD21, CD22, CD10 (CALLA), Smig.

e. Monoclonal antibodies against other leukocyte CD38, CD45, CD45R, HLA – 1.

4. Establishment of the CM – 2 cell line

The suspension of new passaged CM – 1 cells was frozen and thawed three times, and then was centrifugalized at 3000r/min for 10 min. The supernatant was passed through a 0. 45 μm filter. The peripheral blood from a 67 – year – old woman with multiple sclerosis was treated by the same process as mentioned above. T lymphocyte was cultured at 37℃ in 5% CO_2 atmosphere with RPMI – 1640 medium supplemented with 20% FCS (1% penicillin G, 1% streptomycin and 1% glutamine), the filtered supernatant of 1/10 volume, 5 μg PHA and 20 U/ml IL – 2 was added into the medium. The cell line thus established was designated as CM – 2 (Chinese malignant T lymphocyte cell line – 2).

5. Studies on tumorigenesis in nude mice

The suspensions of new passaged CM – 1 and CM – 2 cells were separately collected and centrifugalizad at 2000r/min for 10 min. The pellet was resuspended in a small volume. The 2×10^7 cells of CM – 1 and of CM – 2 cells were separately transplanted into nude mice by subcutaneous inoculation, the mice were reared and preserved.

The tumors were taken out of the mice and cut into the size of $1 \times 1 \times 1mm$, and were transplanted into new nude mice again, thus the tumor was subcultured as such.

6. Electron microscopy

The subcultured CM – 1 and CM – 2 cells and the cells of CM – 1 and CM – 2 that were separately induced by 5 μg/ml PHA and 25 ng/ml TPA for 48 – 72 hours , were separately collected and centrifugalized at 1000r/min for 10 min. The pellets were fixed with 4% glutaraldehyde and 1% OsO_4, and were embedded in cpoxyresin (Epon 618). Ultrathin sections were nade by using an ultramicrotome (LKB NOVE), then were stained with uranyl acetate and lead citrate. Electron micrographs were tahen with a JEM 1200EX electron microscope.

7. Viral identification and assay

A. Serological Assay

a. CM – 1 cells acted as target cells, and were tested for antibodies against some known viruses by indirect immunofluorescence technique with standard antisera against these viruses.

b. 0. 1ml cells (CM – 2 cells, MT – 2 cells carrying HTLV – Ⅰ, MT – 4 cells infected by HIV – 1 and B9 – 58 cells carrying EBV) were separately placed on the microscope slide, and fixed by cold acetone at – 20℃ for 15 min, and dried in air. 0. 1 ml of diluted serum or cerbrospinal fluid (CSF) of CM – 1 donor was added on the slide, incubated at 37℃ for 45 min, washed three times in the washing solution (10 mmol/L pH7. 4 PBS, 0. 01% Triton X – 100). 0. 05 ml fluorescein – labelled mouse anti-human IgG diluted in 0. 01% Even's blue was added, incubated at 37℃ for 30 min, washed three times, observed under the Olympus fluorescence microscope.

c. The serum and cerebrospinal fluid were tested for HIV – 1 antibody by Western blot method with the standard kit (Bio – Rad).

B. Test for the Viral Genes in the CM – 1 Cells

The suspension of subcultured CM – 1 cells and the whole blood cells of the CM – 1 cell donor in convalescent phase were collected and centrifugalized. The cell DNA was extracted by phenol-chloroform.

a. Probe Hybridization

The cell DNA was digested and cleaved by Hing Ⅲ and tested for genes of some known viruses with Southern-blot method. The probes include W-fragment, LMP gene and EBNA – 1 gene of EBV DNA, and labelled by ^{32}P – dCTP with nick-translation method.

b. Test for HIV – 1 and HTLV – Ⅰ Genes with PCR

The concentration of CM – 1 cell DNA was 0. 1 μg/ml, the primer for HIV – 1 was gag-pol, and the primer for HTLV – Ⅰ was gag-env.

After an initial 5 – min denaturation ai 49℃, 30 cycles ao 95℃ for 1 min, 50℃ for 1 min, 72℃ for 1 min were performed (Perking, Automated Thermal Cycler), a final 7 – min extension at 72℃ followed. The PCR products were analyzed by electrophoresis on a 4% agarose gel and the bands were visualised by ethidium bromide staining.

8. Purification of the virus and test for RT activity

A. Purification of the virus: The subcultured CM – 2 cells activated yb PHA and TPA for 48 – 72 hours were collected and centrifugalized at 3000r/min and 4℃ for 20 min. The supernatant was placed in a dialyzer and was concentrated by PEG (22 000) at 4℃, then was centrifugalized at 10 000 r/min for 30 min to remove cell debris. The pellets of cells were suspended with a small a-mount of TNE buffer, the suspension was frozen and thawed three times, and then was centrifugal-ized at 10 000 r/min for 30 min. This supernatant was combined with the concentrated supernatant and passed through a 0. 45 μm fiter. The filtered solution was centrifugalized ai 35 000 r/min and 4℃ for 2 hours, in TFT70. 38 rotor of the Kontron T – 2080 ultracentrifuge. The pellets were sus-pended in TNE buffer to about 1/200 of the volume at beginning. 0. 4 ml of the concentrated prepata-tions were layered on the top of a 20% – 60% sucrose gradients and centrifugalized at 50 000 r/mol and 4℃ for 16 hours in the Kontron TST 60. 4 rotor. The gradients were collected dropwise (0. 5 ml per fraction) and the density of fractions were determined.

B. Test for reverse transcriptase activity: 50 μl of each fraction was added into round bottom wells of 12 – well plate. 50 μl of disruption buffer was added into each well with the sample. The mixtures were incubated at 37℃ for 15 min. The components of disruption buffer are 100 mmol/L Tris-HCl (pH 8. 0), 300 mmol/L KCl, 10 mmol/L DTT and 0. 1% Triton X – 100. Twenty-five μl of reaction mixture was added into each well. The mixtures were incubated at 37℃ for 20 hours. The components of the reaction mixture are 50 mmol/L Tris-HCl (pH 8. 0), 150 mmol/L KCl, 12 mmol/L $MgCl_2$, 5 mmol/L DTT, 0. 05% Triton X – 100, 50 μg/ml Poly (rA): oligo (dT)$_{15}$ and 10 μCi/ml ^{3}H – TTP. The reaction was stopped by adding 0. 1 ml of ice-cold 10% TCA and the reactants were placed at 4℃ for 20 min. The products of the reaction were collected with glass-fibre filters. and washed by 2 ml of ice-cold 10% TCA twice and 2 ml of ice-cold 95% ethanol once. The incorporated radioactivtity of the filters in the solution of liquid scintillation was ceasured with a liquid scintillation counter (Beckman LS – 5000 TA).

Result

1. Establishment and biological characterization of the CM – 1 cell line

On the second day, the cells grew very actively. On the third day, they could be subcultured. Since then, they were subcultured twice a week, and up to now they have been subcultured *in vitro* for more than two years and their growth is still vigorous. CM – 1 cells were preserved in liquid N_2, the rate of resuscitation was more than 90%.

CM – 1 cells can adhere and grow on the wall of wares and dissociate continuously into the suspension. The adhered cells are polymeric and aggregate to form clumps.

After the CM – 1 cell line was established, the blood of the patient in convalescent phase was collected. T lymphocytes were separated and cultured under the same condition as that in establishing the CM – 1 cell line, but it was failed.

The growth of CM – 1 cells in culture follows the patten depicted in Fig. 1. A lag following seeding is followed by a period of exponential growh (log phase, from the 2nd day to the 7th day). The amount of cells

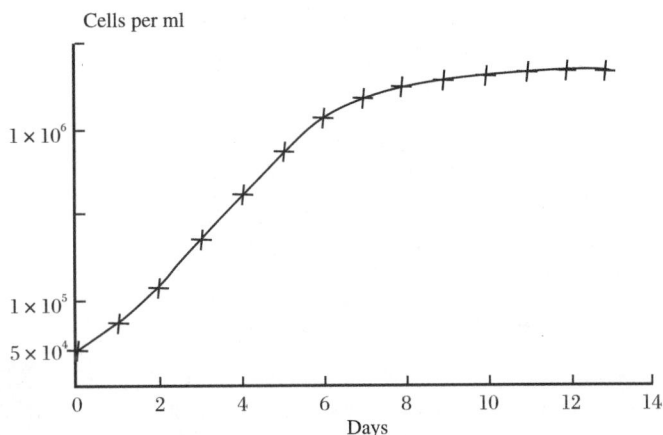

Fig. 1   The CM – 1 cells growth curve

reached 2.5×10^6 cells per ml of medium which is 50 times more than at the start. The plateau phase began from the 11th day. The CM – 1 cells could survive on the 14th day. They could be subcultured by changing new medium.

2. Surface marker of CM – 1 cells

The results of assay by indirect immunofluorescence technique with 28 kinds of anti-leucocyte monoclonal antibodies showed that characteristic marker of T lymphocyte was present on the membrane of all CM – 1 cells, and 20% – 30% of which also showed myclocyte marker.

All results are showed in Tab. 1.

3. Establishment of CM – 2 cell line

After three weeks, clones of the cells were visible to the naked eyes (Fig. 2). The cells could be subcultured and a continuous cell line was established designated as CM – 2. The characteristics and morphology were the same as those of the CM – 1 cells. The control lymphocytes without the filtered supernatant of the cultured CM – 1 cells were all broken and dead three weeks later.

4. Tumorigenic result

After the CM – 1 cells were transplanted into nude mice by subcutaneous inoculation, the tumors grew in the size of $5 \times 7 \times 3$ mm about 10 weeks later. The histopathological section showed that the tumor was a typical malignant T lymphoma.

The cultured cells of CM – 2 were transplanted into nude mice by subcutaneous inoculation. The tumors appeared in 75% of the mice 4 – 5 weeks later, and grew in size to $5 \times 6 \times 4$ mm about 10 weeks later (Fig. 3). The histopathological section showed that the tumor caused by CM – 2 cells was also a typical malignant lymphocytoma.

Tab. 1 Analysis of the surface markers of
CM − 1 cells and clonal cell lines by
antileukocyte monoclonal antibodies（McAbs）

McAbs	Kinds of cells				
	CM − 1	Colony5	Colony9	Colony A − 30	Suspended Colony 3
CD1	0	NT	NT	NT	NT
CD2	59	NT	NT	NT	NT
CD3	25	90	< 5	46	90
CD4	> 90	> 90	NT	90	> 90
CD5	> 90	> 90	> 90	> 90	> 90
CD6	0	NT	NT	NT	NT
CD7	> 90	> 90	> 90	> 90	> 90
CD8	0	NT	NT	NT	NT
CD27	95	NT	NT	NT	NT
CD25	0	NT	NT	NT	NT
CD71	95	> 90	NT	> 90	> 90
HLA − DR	0	0	0	0	0
CD11（B）	0	NT	NT	NT	NT
CD13	< 5	NT	NT	NT	NT
CD14	0	NT	NT	NT	NT
CD33	0	< 10	< 5	< 12	0
CD15（H198）	34	4	16	17	10
CD10(CALLA)	> 95	> 90	> 90	> 90	> 90
CD9	> 95	NT	NT	NT	NT
CD19	0	0	0	0	< 5
CD20	0	NT	NT	NT	NT
CD21	0	NT	NT	NT	NT
CD22	0	NT	NT	NT	NT
Smig	0	NT	NT	NT	NT
CD38	100	NT	NT	NT	NT
CD45	100	NT	NT	NT	NT
CD45R	< 10	NT	NT	NT	NT
HLA − 1	> 95	NT	NT	NT	NT

NT: no test

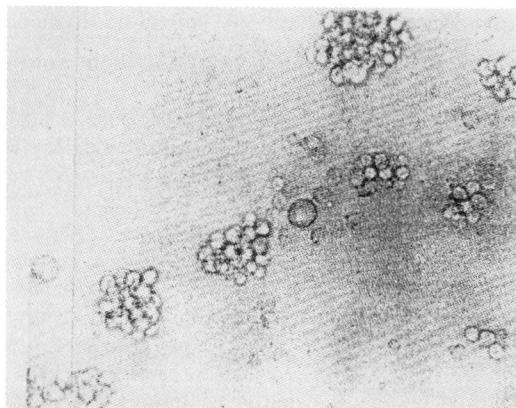

Fig. 2 The picture of clones of the CM − 2 cells

Fig. 3 The picture of the nude
mouse with T malignant lymphoma
caused by the CM − 2 cells

The tumors were subcutaneously transplanted into the new nude mice again, the tumors grew in all of the mice 7 − 8 weeks later. The growth period of the tumor was shortened. Up to now a tumor of nude mouse has been subcultured for six generations（more than one year）.

5. Electron micrograph

The clectron micrograph of the ultrathin section of the CM − 1 cells showed the following facts.

The CM-1 cell was poorly differentiated, and was polynuclear. The endoplasma reticulum was found thickened. Some globular virus particles of 80 − 120 nm in diameter were found in the cytoplasm, with big and loose nucleoid substance in the center. The diameter was similar to that of human retrovirus. The electron micrograph（Fig. 4）shows that a mature virus particle was releasing by means of budding.

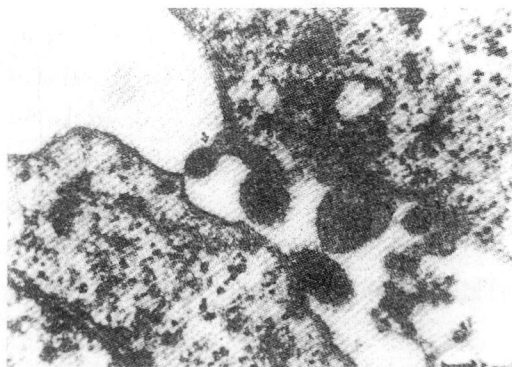

Fig. 4 The micrograph of ultrathin section of the
CM − 1 cell. Arrow shows a virus particle budding
from cytomembrane. （Bai: 100 nm）

The size of CM − 2 cells was 9 − 15 μm in diameter, and the diameter of a few cells was more than 20 μm. The nucleus of CM − 2 cells appeared polymorphic. The main part of endoplasma was incarnation. Abundant nucleoprotein was found in the cytoplasm. Growth annulate lemmas were found in some of the CM − 2 cells, and this phenomenon usually appears in malignant cells of tumor.

The vesicles of rough-surfaced endoplasmic reticulum (RER) enlarged, and most of them enclosed viral particles (Fig. 5 − A). This is a characteristic feature of the CM − 2 cells. Not only globular viral particles, but also some coalescence of two or more viral particles were found (Fig. 5 − B). This virus released by means of budding out from cytomembrane or budding from RER membrane into the cistern of RER. There were some clumps of electronic dense granular substance, which usually located next to RER, in the cytoplasm. Viral particles budded from these clumps into the cisterns of RER (Fig. 6). These clumps possibly were viral matrixes created when viruses grew in cytoplasm.

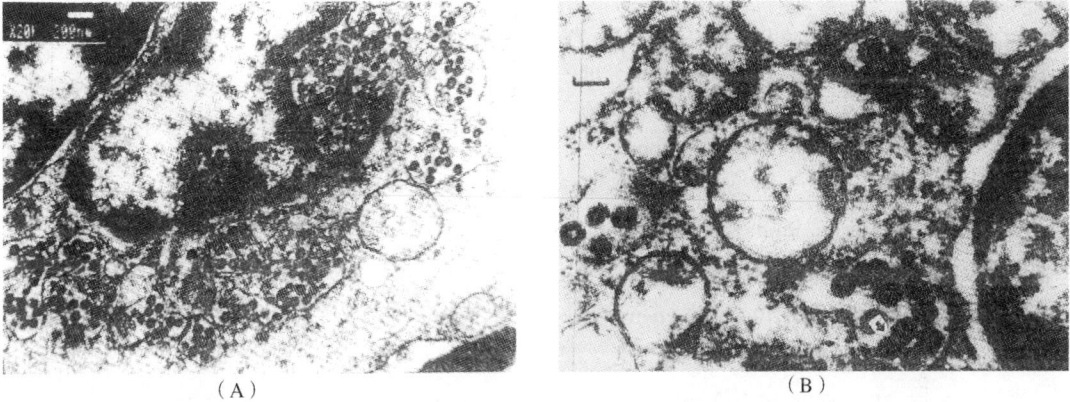

(A) | (B)

(A) The vesicles of rough-surfaced endoplasmic reticulum enclose many virus particles (Bar: 200 nm) 1;

(B) Globular virus particles and some coalescences of viruses are found (Bar: 100 nm)

Fig. 5　The micrograph of ultrathin section of the CM − 2 cell

The quantity of viruses was more in the induced cells than in the non-induced cells. The quantity of virus in the CM − 2 was obviously more than in CM − 1.

The particles looked like type C virus. The globular particle had two layers of membranous structure (like capsule and nucleocapsid) enclosing a electronic dense nucleus near the center (Fig. 7), there were also many viruses without dense nucleus. But in the cistern a lot of particles without dense nucleus were found, and only in a few particles the dotted dense nucleus was found. The size of the viral particle is 68. 3 − 94. 3 nm in diameter, with an average about 81. 2 nm.

6. Viral identification and assay

A. Serological assays

(a) The immunological reactions between the serum of cerebrospinal fluid of CM − 1 donor and the antigens of the above-mentioned viruses tested by Western Blot or immunofluorescence assay were all negative. This showed that the patient was not infected by the above-mentioned viruses.

(b) The assay for the virus antigens in the CM − 1 cclls by immunofluorescence assay showed that the virus in the CM – 1 cells did not possess the antigenicity of HTLV – Ⅰ, HIV – 1 and EBV, which were often used in our laboratory.

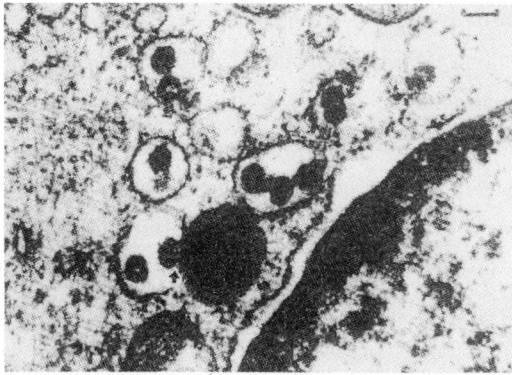

Fig. 6   The micrograph of ultrathin section of the CM –2 cell. Arrow shows that a virus particle is budding from the membrane of rough-surfaced endoplasmic reticulum into the cisterns (Bar: 100 nm)

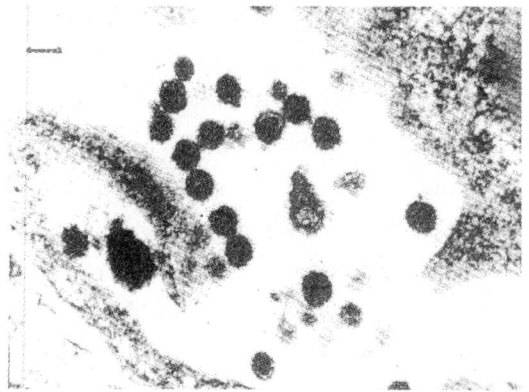

Fig. 7   The micrograph of ultrathin section of the CM – 2 cell. showing ultra-structure characteristics of the virus (Bar: 100 nm)

(c) There was no immunological reaction between the CM – 2 cells and the serum of its donor. The CM – 2 cells acted as target cells, but the CM – 2 cells could react with the serum of the CM – 1 donor. It seems that the antigen in CM – 2 cells (transformed cells) came from the filtered supernatant of the cultured CM – 1 cells.

B. Tests for viral gene in CM – 1 cell

(a) Molecular hybridization test: The results obtained from molecular hybridization test showed that the W-fragment, LMP and EBNA – 1 gene were all negative, this excludes the existence of EBV genes in the CM – 1 cells.

(b) The results of PCR confirmed that the viral gene and antiviral gene have no relationship with the certain human retrovirus, such as HIV – 1 and HTLV – Ⅰ.

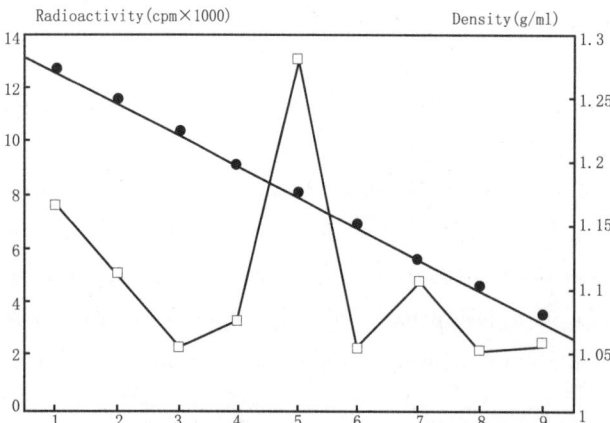

Fig. 8   RT-activity distribution according to density after eentrifugation in sucrose density gradient

7. Reverse transcriptase (RT) activity.

All fractions obtained by ultracentrifugation in sucrose density gradient were assayed for RT activity with poly (A): oligo (dT)$_{15}$ as template-primers. The results showed that RT activity appeared in the tube 5, its density was 1. 16 g/ml (Fig. 8).

Discussion

No breakthrough had been made in the long – term culture of T lymphocyte *in vitro*, untill the time between the late 1970s and the early 1980s. Since then, two methods for the long-term culture of T lymphocyte *in vitro* became available. First, the IL – 2 dependent T lymphocyte lines can be established by adding the growth factors such as IL – 2 into the culture medium (Steven and James, 1981); second, to immortalize the T lymphocytes by certain retroviruses infection (Miyoshi et al., 1981). The malignant T lymphoma cell line that we had established from the peripheral blood lymphocytes donated by a female patient with limb paralysis and coma, proliferated rapidly and actively. The number of cells could increase by 50 times in 10 days. Further more, they were able to grow spontaneously *in vitro* without addition of growth factors such as IL – 2. The above features showed that there were many growth factors produced by cells themselves. This is really a rare phonomenon.

The CM – 1 and CM – 2 cells were transplanted into nude mice by subcutaneous inoculation, and coule developed tumors that were typical malignant T lymphoma proved by the histopathological section. The electron micrographs suggested the existence of 80 nm diametered viral particles, which had two layers of membranous structure enclosing a electron dense nucleus near the center and released by budding out from cytomembrane, all these are similar to the ultra-structural characteristics of retrovirus. This virus has reverse transcriptase activities and is capable of transforming normal T lymphocytes *in vitro*. The results of serological assay, probe hybridization and PCR showed that the viruses in the CM – 1 and CM – 2 cells had no relation with the human viruses often used in our laboratory, these excluded the possibility of contamination from other human viruses. The conclusion drawn from these data is that a possible new kind of retrovirus existed in the CM – 1 and CM – 2 cells.

The observation of electron microscope showed that the CM – 2 cells carried more virions than the CM – 1 cell, and more antigen to the serum of the CM – 1 cell donor was found in the serological detection. The CM – 2 cells are more suitable to be used as the antigen to the virus in the immouno – fluorescence assay.

At present, the patient who donated the CM – 1 cells is still alive, no signs of concerned tumor, but degradation of intelligance and motion difficulties can be found. Whether the virus has any relation with diseases of the CM – 1 cells donor is not clear yet, more work is needed on this question. We are trying to screen more sera from patients with varions diseases by immounofuorescence assay, in order to make research on which disease relates to this virus. Meanwhile, the work on further purification, clone and sequence analysis are being carried out. The classification and name of this virus still need more intensive work.

[In 《Biomedical and enviromental Sciences》1994, 7: 1 – 12]

References

1 Bartholomew C, Cleghorn F, Charles W, Ratan P, Roberts L Maharaji K, Jankey N, Daisley H, Hanchard B, Blattner W. HTLV – I and tropical spastic paraparesis. Lancet, 1986, 2: 99 – 100

2 Gessain A, Barin F, Vernant J C, Gout O,

Maurs J C, Gout O, Maurs L, Calender A, de The G. Antibodies to HTLV – I in patients with tropical spastic paraparesis. Lancet, 1985, 2: 407 – 410

3　Hinuma Y, Nagata K Hanaoka M, Matsumoto T, Kinoshita K, Shirakakawa S, Miyoshi L. Adult T-cell leukemia: antigen in an ATL cell line and detection of antibodies to the antigen in human sera. Proc. Acad. Sci. USA, 1981, 78, 6476 – 6480

4　Hong ML, Lan XY, Zhang DF, Feng ZJ, Wang DX, Zeng Y. Morphology of type C retrovirus-like particles from a patient with cerebritis. Journal of Electron Microscopy Society, 1992, 11: 375 – 376. (in Chinese)

5　Lan XY, Zeng Y, WANG DX, Feng ZJ, Tang MH, Ji Y, Yu G, Li K. Establishment of a human malignant T lymphocytic cell line releasing retrovirus-like particles. Chinese Journal of Virology, 1992, 8: 187 – 190. (in Chinese)

6　Lan XY, Zeng Y, Wang DX, Chen Z, He SQ, Guo SS, Ouyang MQ, Du B. HTLV – I and the neuromyelopthics in Chinese, Chinese Journal of Virology, 1993, 9: 382 – 385. (in Chinese)

7　Lan XY, Zeng Y (In press). Establishment of a cell line carrying HTLV – I from the blood donated by a patient with lcukemia. Chinese Journal of Virology. (in Chinese)

8　Miyoshi I, Kubonishi I, Yoshimoto S, Akagi T, Ohtsuke Y, Shiraishi Y, Nagata K, Hinuma Y. Typc C virus particles in a cord T-cell line derived by co-cultivating normal human cord T-cell line derived by co-cultivating normal human cord lcukoeytes and human leukaemic T cells. Nature, 1981, 294, 770 – 771

9　Osame M, Matsumoto M, Usuku K, Izumo S, Ijichi N, Amitani H, Tara M, Igata A. Chronic progressive myclopathy with elevated antibodies to HTLV – I and ATL cells. Ann. Neurol, 1987, 21: 117

10　Steven G, James W. Lnterleukin – 2 dependent culture of cytolytic T cell times. Immunolog-ical Rev, 1981, 54: 81 – 109.

62.　Screening of Epstein-Barr Virus Early Antigen Expression Inducers from Chinese Medicinal Herbs and Plants

ZENG Yi[1], ZHONG Jian-ming[2], YE Shu-qing[1], NI Zhi-yu[1]
MIAO Xue-qian[1], MO Yong-kun[2], LI Zi-lin[3]

1. Institute of Virology, Chinese Academy of Preventive Medicine, Beijing; 2. Nasopharyngeal Control and Treatment Institute of Cangwu, Guangxi, Guangxi Herbs Botany Garden, Naning; 3. Institute of Chinese Materia Medica, China Academy of Traditional Chinese Medicine, Beijing

Summary

Ether extracts of 1693 Chinese medicinal herbs and plants from 268 families were studied for the induction of Epstein-Barr viral (EBV) early antigen (EA) expression in the Raji cell line, Fifty-two from 18 families were found to have inducing activity Twenty-five and seven of them were from Euphorbiaceae and Thymelaeaceae, respectively. Some of them, such as Croton tiglium, Euphorbia kansui, Daphne genkwa. Wikstroemia chamaedaphen, Wikstroemia indica, *Prunus mandshuricd* Koehne and *Achyramhes bidemdid* are commonly used drugs. The significance of these herbs in the activation of EBV *in vivo* and their relation to the development of nasopharyngeal carcinoma were discussed.

Introduction

EB virus is closely related to the development of nasopharyngeal carcinoma, while environmental carcinogens and tumor promotor may also play a role in the development of nasopharyngeal carcinoma. The known tumor promotor TPA extuaeted from croton oil can be an EB viral inducer for B lymphocyte transformation, and an enhancer of cancer growth caused by viral or chemical carcinogens. We used the method of induction for EB viral early antigen (EA) expression in Raji cell to detect the EB viral EA expression inducer in 495 species of Chinese medicinal herbs. Some medicinal herbs were found acting as EB viral EA indueers (Zeng et al. , 1984; Zeng et al. , 1985), and as a indueer of lymphocyte transformation by EB virus and rat embryonic cell transformation by adenovirus (Hu and Zeng, 1985), as well sa a tumor promotor (Hu et al. , 1986; Sun et al. , 1987a; Sun et al. , 1987b). It was also found that Chinese patent medicines such as croton cream formulation sold from drug stores in Beijing also contained tumor promotor (Zeng et al. , 1986). Application of Chinese medicinal herbs to treat various diseases are widespread, hence it is very necessary to continue screening tests for EB viral EA inducers. Among pharmacetitical plants and other plants, 1693 plant species have been screened, of them 52 species have the function in inducing EB viral EA expression.

Materials and Methods

1. Cell line

Raji cell carrying EB virus genome in RPMI 1640 culture medium containing 10% calf serum.

2. Plant origin

Plants were collected from the Guangxi Herbs Botany Garden; Cangwu County, Wuzhou City and Nanning City of Guangxi; and the Beijing Chinese Drug Store. All the plants had passed the evaluation by the Guangxi Herbs Botany Garden.

3. Method of extraction

The Ito method was taken. Plant extracts, after the plants were immersed in 100 ml egher for 72 h, were filtrated with filter paper and evaporated with ether. The extracts were dissolved in dehydrated alcohol to reach a concentration of 10 mg/ml. Then the alcoholic solution was kept in refrigerator at 4℃ for further dilution when in use.

4. Experimental steps

(1) Different concentrations of sample extracts were added into Raji cell culture containing 10^5 cells per ml reaching the final concentrations of 12. 5 μg cell/ml.

(2) The Raji cell culture contained 4 mmol $\cdot$ L^{-1} sodium butyrate. Another sample extract was added into a sodium butyrate-free culture as the control.

（3）Croton oil positive controls were also ser up.

（4）Incubation at 37℃ for 48 h.

（5）Cell smear preparations for the detection of EB viral EA using immunoenzymatic test.

（6）If the percentage of EA positive cells in the experimental group was 3 times higher than that in the sodium butyrate group, the result of the experiment was considered as positive for EA activation.

Results

Tab. 1 lists 52 species of EA induction positive plants, of which most belong to the family Euphorbiaceae. Strong positives (induced Raji Cells contain EA positives above 20%) include 34 species (Euphorbiaceae 21, Thymelaeaceae 7, Verbenaceae 2, Rubiaceae 1, Solanaceae 1, Amaranthaceae 1, Iridaceae 1). Moderate positives (EA positives between 10% and 19%) include 6 species (Euphorbiaceae 3, Sinopteridaceae 1, Spargoniaceae 1, Iridaceae 1). Weak positives (EA positives below 10%) include 11 species. Among 52 species some are in common use (such as *Croton tiglium*, *Euphorbia kansui*, *Daphne genkwa*, *Wikstroemia chamaedaphne*, *Wikstroemia indica*, *Belamcanda chinensis*, *Prunus mandshuria* Koehne and *Achyranthes bidentata*, etc.).

Tab. 1　List of Positive Herbs and Plants

No.	Family	Herbs and plants	EA positive cell rate	
			12.5 g/ml	2.5 g/ml
1	Euphorbiaceae	*Aleurites moluecana*	6.4	2.8
2		*Codiaeum veriegatum*	50	26
3		*Codiaeum veriegatum forma taeniosum*	37	26
4		*Codiaeum veriegatum CV*	19.2	26.4
5		*Croton calcarsus*	16.2	14.2
6		*Croton lachnocarpus*	9.6	22.4
7		*Croton tiglium*	42.8	26.0
8		*Euphorbia antiqnorum CV "cristata"*	28	36
9		*Euphorbia tumdata*	17.2	9.8
10		*Euphorbia helioseopia*	36.4	34.8
11		*Euphorbia kansui*	18.0	38.0
12		*Euphorbia lathvris*	20.0	10.0
13		*Euphorbia marginata*	7.2	20.8
14		*Euphorbia milli*	25.8	20.4
15		*Euphorbia thymifolia*	40	32
16		*Excoecaria cochinchinensis*	36.8	52.8
17		*Excoecaria venenata*	13	42.4
18		*Jatropha multifida*	21.6	35.8
19		*Pedilanthus tithvmaloides*	11.6	22.4
20		*Sapium discolor*	35	30.2
21		*Sapium sebiferum*	5.4	24.8
22		*Sapium rotundifolium*	21.4	1.5
23		*Vernicia fordii*	17.0	15.8
24		*Vernicia montana*	49.6	43.2
25		*Euphorbia antiquorum*	20.8	7.4

No.	Family	Herbs and plants	EA positive cell rate	
			12. 5 g/ml	2. 5 g/ml
26		*Daphne genkwa*	46	45
	Thvmclaeaceae			
27		*Edgeworthia ehrysantha*	43	28
28		*Stellera chamaejasme*	36	46
29		*Wikstroemia chamaedaphne*	32	53
30		*Wikstroemia indicd*	42	50
31		*Aquilaria sinensis*	25. 6	22
32		*Wikstroemia nutans*	28. 6	17. 2
33	Leguminasae	*Caesalpinia sappan*	8	4
34		*Desmodium stvrqcifolium*	6	4
35	Rubiaceae	*Knoxia valerianoides*	3. 2	0. 8
36		*Galium aparina var tenrum*	36	10
37	Verbenaceae	*Prenna fulva*	52	42
38		*Duranta repens*	24. 6	36
39	Iridaceae	*Belamcanda chinensis*	0. 8	36. 4
40		*Iris tectorum*	16	6
41	Sinopteridaceae	*Aleuritopteris argentea*	6	17
42	Ranunculaceae	*Clematis intricata*	3	2
43	Menispermaceae	*Tinospora sagittata*	7	3
44	Solanaceae	*Datura atramonium*	0	25
45	Spargoniaceae	*Spargonium stoloniferum*	16	2
46	Balsaminaceae	*Impatiens balsamina*	8. 6	7
47	Compositae	*Ixeris debilis*	6. 8	9
48	Caprifoloaceae	*Viburnum sempervirens*	8. 6	8
49	Actinidiaceae	*Actinidia latifolia*	3. 6	0
50	Piperaceae	*Piper hainaense*	7. 2	4
51	Rosaceae	*Prunus mandshurica* Koehne	18. 4	14
52	Amaranthaceac	*Achyranthes bidentata*	23. 9	10

Notes: Control: Sodium butyrate + croton oil EA cell positive rate 42. 8% – 45. 7% ; sodium butyrate EA positive cell rate 0. 4% – 1% ; croton oil EA positive cell rate 0. 8% – 1. 6%

Discussion

One-thousand six-hundred and ninety-three species of Chinese medicinal herbs and plants were screened for EB viral EA inducer with Raji cell line, and 52 species of 18 families were found having the function of inducing expression of EB viral EA. Among them some are in common use, such as *Croton tiglium*, *Euphorbia kansui*, *Daphne genkwa*, *Wikseroemia chamaedaphne*, *Wikstroemia indica*, *Belamcanda chinensis*, *Prunus mandshurica* Koehne and *Achyranthes bidentata*. EB viral EA inducers and even tumor promotors (Zeng et al. , 1986) were also found in 9 Chinese patent medicines. It was proved that some of them, such as W. indica used in Guangdong, could stimulate lymphocyte transformation induced by EB virus, and D. genkwa which was widely used in inducing labor could enhance rat nasopharyngeal carcinoma induced by chemical carcinogens (Tang et al. , 1988a; Tang et al. , 1988b; Zhong et al. , 1987). Among 50 more species screened as EA activation positive plants, 39 – 40 are distributed in nasopharyngeal carcinoma highly prevalent areas in Guangdong and Guangxi provinces. In the areas abundant in these plants, EB viral inducers or tumor promotors may exist in soil, vegetables and bees, and this may further explain the linkage of environmental factors with human life. The nasopharyngeal carcinoma is high in incidence along the banks of the Xijiang Valley, and the incidence in boat dwellers is also 2 – 3 times higher than in land rewidents (Zhu et al. , 1983). Whether EB viral inducer and tumor promotors exist in the Xijiang River is a problem to be solved, because the river water is turbid and muddy. Imai et al. (1988) reported that Burkit Lymphoma and nasopharyngeal carcinoma occurred in Kenya, Uganda and Eastern African countries, where many EB viral inducers and tumor promotors existed, even in pond water. Hence, the high incidence of nasopharyngeal carcinoma in the Xijiang Valley, especially in boat dwellers may be related to drinking of water containing EB viral inducers and tumor promotors. This conjecture should be further confirmed and studied. Tomei and Glaser (1988) reported that TPA and EB virus could stimulate SV40 virus inducing human epithelium multiplication. Thus, the *in vitro* synergistic action of EB viral inducer and tumor promotors in connection with EB virus occurred in the epithelial cells should be studied. It was also found in our laboratory that some plants or foods contained both EB viral inducers and tumor promoters, and also mutagens (Shao et al. , 1988). Faggioni et al. (1983) reported that N-methy1-N-nitroguanidine could activate lymphocyte transformation induced by EB virus. Hence, the possibility of the occurrence of synergism between EB viral inducers, tumor promotors carcinogens and EB virus, causing the transformation of epithelium of nasopharynx into cacer cells, is worthwhile for further exploration. Serological surveys and 10 – year follow-up observations in Wuzhou City and Cangwu County were carried out. Early diagnosis rate of nasopharyngeal carcinoma can be greatly raised by using EB viral IgA/VCA antibody and EA antibody markers; and the occurrence of nasopharyngeal cancer can be predicted 5-10-year in advance. It indicated that EB virus played an important role in the development of nasopharyngeal carcinoma (Deng et al. , 1992). Our study (Lu et al. , 1990) on genetic factors of nasopharyngeal carcinoma was suggesting the existence of nasopharyngeal carcinoma susceptibility gene linked with HLA. From the above, we put forwar the idea on the etiology of nasopharyngeal carcinoma, i. e. gene factor is the basis of the disease occurrence, EB virus plays an important role in the development of the disease; EB viral inducers, tumor promotors and carcinogens in environments act as synergists in the development of nasopharyngeal carcinoma. Further researches are undergoing.

〔In 《Biomedical and Environmental Sciences》 1994, 7: 50 – 55〕

References

1 Deng H, Zeng Y, Huang N, Huang Y, Li Y, Su LM, Zhong HX, Lian YX, Wang PZ, G. de The. Prospective studies of nasopharyngeal carcinoma for 10 years in Wuzhou City, Guangxi Chinese J Viology, 1992, 8: 32. (in Chinese)

2 Faggioni A, Ablashi D V, Armstrong G, Dahlberg J, Sundar S K, Rice J M, Donovan P J. Enhancing effect of N-Metgtl-N-Nitrosoguanidine (MNNG) on Epstein-Barr Virus Replication and comparison of continuous and discontinuous TRA treatment of ESV nonproducer and producer cells for antigen induction and/or stimulaiton. In Nasopharyngeal Carcinoma; Current Concepts (W. Prasad, D. V. Ablashi, P. H. Levinc, and G. R. Pearson, Eds.), University of Malaya, 1983, 333

3　Hu YL, Zeng Y. Enhanced transformation of human lymphocytes by Chinese herbs. Chinese Journal Oncology, 1985, 7: 417. (in Chises)

4　Hu YL, Zeng Y, lto Y. Croton oil, Wikstroemia chamaedaphne and Wikstroemia india cnhanced the rabbit papilloma induced by papilloma induced by papilloma virus. Chinese J Virology, 1986, 2: 81. (in Chinese)

5　Imai S, Kinoshita T, Koizumi, Aya T, Matsunra A, Sugiura M, Btay Kyi Mizuno F, Osato T, Yamaka T, Chiba S, Ohigashi H, Koshimizu K, Miyazaki T, Agishi Y. An environmental plant factor enhancing EBV-specific event in East Africa: Redution of killer T-cell function and its protection by hot spring water. In Epstein Barr and Human Disease (D. V. Ablashi, A. Faggioni, G. R. F. Krueer, J. S. Pagano, and G. R. Pearson Eds.), Hamana Press, 1988, 48

6　Lu S J, Day N E, Degos L, Lepage V, Wang P C, Chan S B, Simons M, McKnight B, Easton D, Zeng Y. Linkage of a nasopharyngeal carcinoma susceptibility locus to the HLA region. Nature, 1990, 346: 470 – 471.

7　Shao Y M, Poiricr S, Ohshima H, Malaveile C, Zeng Y, G de The, Bartsch, H. Epstein-Barr virus activation in Raji cells by extracts of preserved food from high risk areas for nasopharyngeal carcinoma. Carcinogenesis, 1988, 9: 1445 – 1447

8　Sun Y, Chen MH, Xiao H, Liu HY, Chen X, Zeng Y. Ito, Y. Tumor promoting effect of diterpene ester HHPA and extract of Wikstroemia chamaedaphne of HSV – 2 induced carcinoma in mice. Chinese J Virology, 1987, 3: 131. (in Chinese)

9　Sun Y, Chen Mh, Zang YX, Xıao H, Liu IIY, Chen X. Promoting effects of the Chinese medical herb. Wikstroemia chamaedaphne and the tung of extracts on carcinoma of uterine cervix induced by HSV – 2 or methylcholanthrene (MCA) in mice. Chiese J Oncology, 1987b, 9, 345. (in Chiese)

10　Tang WP, Huang PG, Zhao ML, Cai QZ, Liao SL, Zeng Y. The role of Wikstroemia chamaedaphne Meise in promoting experimental nasopharyngeal carcinoma (NPC) in rats. Cancer, 1988a, 7: 171 – 172. (in Chiese)

11　Tang WP, Huang PG, Zhao ML, Liao SL, Zeng Y. Wikstroemia indica promotes development of nasopharyngeal carcinoma in rats initiated by dinitrosopiperazine. J. Cancer Res, Clin Oncol, 1988b, 114: 429 –431

12　Tomei L D, Glaser R. Enhanced SV40 immortalization of primary human epidermal cells following phorbol Ester Dependent EBV Transformation. In Epstein-Barr and Human Disease (D. V. Ablashi, G Armstrong, J. Dahlberg, S. K. Sundar, J. M. Rice, and P. J. Donovan, Eds.). Hamana Press, 1988, 495

13　Zeng Y, Miao XQ, Jiao W, Hu YL. Epstein-Barr Virus early antigen expression induction and enhancement of lymphocyte transformation by Yuauhuadine II and Wikstroemia chamaedaphne. Chinese J Virology, 1985, 1: 229. (in Chinese)

14　Zeng Y, Wan Y, Ye SQ, Miao XQ, Zhong JM. Induction of Epstein-Barr Virus early antigen in Raji cell by some Chinese patent medicine. Chinese J Virology, 1986, 2: 306. (in Chinese)

15　Zeng Y, Zong JM, Mo YK, Miao XQ. Epstein-Barr Virus early antigen expression induction in Raji cell by Chinese medicinal herbs. Acta Academiae Medicinae Sinicae, 1984, 6: 84. (in Chinese)

16　Zhong JM, Cheng JY, Mo YK, Tang CT, Zeng Y. Study on the Epstein-Barr Virus Inducers in the ether extracts of soil and vegetables in Zangwu County. Cancer, 1987, 6: 35 – 37. (in Chinese)

17　Zhu JS, Pan WJ, Zhong JM, Li LY, Zeng Y, Jiang MK, Fang Z. A serological survey on nasopharyngeal cancer in boat dwellers of Cangwu County. Cancer Research on Prevention and Treatment, 1983, 10: 189 – 190. (in Chinese)

63. Serological Survey of Nasopharyngeal Carcinoma in 21 Cities of South China

DENG Hong[1], ZENG Yi[2], LEI Yi – ming[3], ZHAO Zheng – bao[3],
WANG Pei – zhong[4], LI Bing – jun[5], PI Zhi – ming[1], TAN Bi – fang[1],
ZHENG Yu – ming[1], PAN Wen – jun[5], ZHONG Zheng – yi[6], Wu Jue – yan[7]

1. Wuzhou Institute for Cancer Research. Wuzhou City 543002, Guangxi; 2. The Academy of Preventive Medical Sciences of China, Beijing; 3. Guangxi Public Health Department, Nanning, Guangxi; 4. Guangxi People's Hospital. Nanning, Guangxi; 5. Cangwu Institute for Prevention and Treatment of Nasopharyngeal Carcionma, Cangwu, Guangxy; 6. Wangning Public Health Bureau, Wanning, Hainan; 7. Fengkai Public Health Bureau, Fengkai, Guangdong

Summary

This pater reports the results of serological survey of 318 912 persons for nasopharyngeal carcinoma (NPC) in 21 cities and counties of south China. There were 8 441 persons with positive VCA-IgA antibody (sinle item positive) of EB virus (EBV), with a rate of 2.65%. In these VCA-IgA positive persons, 287 persons also had positive EA-IgA (double items positive) of EBV. The overall positive rate was 0.09%. 100 cases of NPC were found and 87 of them (87.0%) were in early stage. NPC found in the group with single item positive accounts for 1.19%, but the rate in the group with double items positive was 19.16% (55 cases). In NPC patients with double items positive, 49 cases were in early stage (89.1%). In 100 cases of NPC found, 45 cases appeared with negative EA-IgA, only with positive VCA-IgA, which indicated that for diagnosis of NPC, sensitivity of EA-IgA was lower than that of VCA-IgA, but its specificity was higher. Therefore, both can increase the detecting rate and early diagnosis rate of NPC. The age of people checked varied with different antibody positive rate and NPC detecting rate. The three items showed a positive correlation. The results are compatible sith those of the prospective study for NPC in Wuzhou City, Guangxi, China. The method for NPC serological diagnosis can be extended and applied to raise the NPC detecting rate and early diagnosis rate at secondary prevention, And, it is further proved theat there is a close relationship between NPC and EBV.

Introduction

We used the method for nasopharyngeal carcinoma (NPC) serological diagnosis to detect VCA-IgA antibody of EBV in the prospective study in Wuzhou City, Guangxi, China. NPC early diagnosis rate has increased from 20% to 49.4% at outpatient service and 94.74% in the mass survey. The 5 – year survival rate of NPC in population of the urban district has risen to 54.62% and the 10 – year survival rate 39.27%. In other words, we have attained the aim of secondary prevention for NPC. [1,2] In order to understand whether the results of the prospective study can be applied and promoted in other areas , we made an NPC serological mass survey of 318 912 persons in 21 cities and counties of Guangxi Province, Fengkai County of Guangdong Province, and Wanning County of Hainan Province, from January 1991 to July 1993. The data are reported below.

Material and Methods

1. Areas of the mass survey: There were altogether 5 cities: Guigang , Qinzhou, Yulin, Liuzhou and Beihai; and 16 counties: He, Rong, Teng, Lingshan, Cenxi, Hepu, Lipu, Luchuan, Bobai, Luzhai, Guiping, Mengshan, Zhongshan, and Fuchuan in Guangxi Province, Wanning in Hainan Province and Fengkai in Guangdong Province.

2. Subjects: Cadres and workers of offices, farms, forestry centres, factories, mines and other enterprises and institutions from above cities and counties.

3. Population: Altogether 318 912 persons were surveyed, including 194 118 men and 124 794 women. Proportion of male to female was 1.6 : 1. Age groups are shown in Tab. 1.

Tab. 1　The relationship among age, VCA – IgA and NPC

Item	Age (years)										Total
	<30	30 –	35 –	40 –	45 –	50 –	55 –	60 –	65 –	>70	
Population	89433	46089	46874	46013	30488	28271	18356	8314	3068	1992	318912
VCA-IgA positive	1800	916	1163	1223	964	938	776	400	159	102	8441
VCA-IgA positive rate (%)	2.01	1.99	2.48	2.66	3.16	3.32	4.23	4.81	5.18	5.11	
NPC cases	6	2	13	29	21	14	9	4	2	0	
NPC detecting rate (%)	6.7	4.3	27.73	63.0	68.9	49.5	48.03	48.1	65.2	0	

4. Sera: One to two drops of blood were collected with a plastic tube of 1.5 mm in diameter and sent to the laboratories of Wuzhou Institute of Cancer Research, Guangxi People's Hospital, and Cangwu Insitute of Prevention and Treatment of NPC. Sera were obtained there.

5. Immunoenzymatic method[3,4]: B95 – 8 cells were used for the detection of VCA-IgA antibody, and Raji cells were used for the detection of EA-IgA antibody. Sera diluted at 1 : 5 and 1 : 10 were added to cells in separate wells of slides. The slides were incubated at 37℃ for 30 min at a humid atmosphere and washed 3 times with PBS. Horseradish peroxidase labelled antihuman IgA antibody in appropriate dilution was added to the slides. The slides were incubated for 30 min, washed 3 times with PBS and flooded with diaminobenzene solution and H_2O_2 for 10 min. Positive and negative controls were included in each experiment. A serum dilution of 1 : 10 for VCA-IgA or 1 : 5 for EA-

IgA showing a characteristic brown colour of this test was considered positive and tested with further 6 dilutions. Tests were completed in 5 – 10 days after the collection of blood specimens.

6. Clinical examination: Persons being serologically positive received clinical examination at nasopharynx within 5 – 15 days after the collection of blood specimens. Biopsies were taken from suspected NPC cases.

7. Histological examination: Biopsies were sent to Wuzhou Cancer Research Insitute and Guangxi People's Hospital for histological examination.

Results and Discussion

Among the 318 912 subjects, 8441 had positive VCA – IgA antibody (single item positive) of EBV, the positive rate being 2. 65%, and geomatric mean titre (GMT) of the antibody was 1 : 33. 84. There were 287 persons being both VCA-IgA and EA-IgA (double items positive) of EBV positive. The positive rate of EA-IgA was 0. 09% and GMT of EA-IgA was 1 : 10. 52. A total of 100 cases of NPC were found in the group with single item positive (not including the patients diagnosed of treated before the survey). NPC detecting rate in the positive VCA-IgA cases was 1. 18%, In persons with double items positive, 55 cases of NPC were found (19. 16%), which was 16. 2 times as many patients as in the single item positive group. In 100 NPC cases in the group with single item positive, there were 87 cases in early stage (according to the TNM staging standard of Chinese Hunan Conference in 1979, stages I-II as early stage; T: primary tumor at nasopharynx; N: transferred to neck lymphaden; M: transferred to lymphaden in other parts; stage I: $T < 0.5$ cm, $N = 0$, $M = 0$; stage II: $T > 0.5$ cm, but not over cavum nasopharyngeum, $N = 0$ or $N < 3 \times 3$cm, $M = 0$; one or two of following is stage III or stage IV: T over cavum, $N > 3 \times 3$ cm, $M = 1$, basion destroyed, cranial nerve damaged), and early diagnosis rate was 87. 0%. On the other hand, in 55 NPC cases of the group with double items positive, there were 49 cases in early stage (89. 1%), In the above 100 NPC cases, 45 cases had negative EA-IgA, only with positive VCA-IgA. It shows that in diagnosis of NPC, sensitivity of EA-IgA is lower than that of VCA-IgA, but its specificity is higher. Therefore, both can increase the detecting rate and early diagnosis rate of NPC (Tabs. 2, 3).

Tab. 2 Antibody titre distribution and NPC cases in 8441 persons

Item	Titre							Toal	Positive rate (%)
	1 : 5	1 : 10	1 : 20	1 : 40	1 : 80	1 : 160	1 : 320		
VCA-IgA positive	0	6837	1161	327	89	25	2	8441	2. 65
NPC cases	0	25	19	24	24	8	0	100	
NPC detecting rate (%)	0	0. 37	1. 64	7. 34	26. 97	32. 0	0		
EA-IgA positive	110	131	36	10	0	0		287	0. 09
NPC cases	15	25	12	3	0	0		55	
NPC detecting rate (%)	13. 64	19. 68	33. 33	30. 0	0	0			

Tab. 3 Positive antibodies and NPC clinical stages and histological classification

| Item | Clinical stages | | | | | Detecting rate (%) | Early diagnosis rate (%) | Poorly diff ca. | Histological classification | | |
	I	II	III	IV	Total				Rate (%)	Vesicular nuc. ca.	Rate (%)
VCA-IgA cases	34	53	10	3	100	1.19	87.0	75	75.0	25	25.0
Rate (%)	34.0	53.0	10.0	3.0	100						
EA-IgA cases	14	35	4	2	55	19.16	89.1	42	76.4	13	23.6
Rate (%)	25.5	63.6	7.3	3.6	100						

In the group with VCA-IgA titre of 1 : 10 there were 25 NPC cases (0.37%). When VCA-IgA titre was 1 : 20, 1 : 40, 1 : 80 and 1 : 160, NPC finding rate was 1.64%, 7.34%, 26.97% and 30.0% respectively. This shows that there is a positive correlation between VCA-IgA titre and NPC detecting rate. When EA-IgA titre was 1 : 5, 1 : 10 and 1 : 20, NPC detecting rate was 13.64%, 19.08%, and 33.33% respectively. Both standards showed positive relationship. But when EA-IgA was 1 : 40, NPC detecting rate dropped to 30.0%, and when EA-IgA $\geqslant$ 1 : 80 the rate was zero. The result may be related with the lower sensitivity of EA-IgA in the NPC diagnosis. In 100 NPC cases with single item positive, 75 (75%) were poorly differentiated carcinoma and 25 (25%) were vesicular nucleus cell carcinoma. No highly differentiated carcinoma was found. In 55 NPC cases with positive EA-IgA, 42 (76.4%) were pooly differentiated carcinoma and 13 (23.6%) were vesicular nucleus cell carcinoma. Proportion of NPC histological classification was similar in the two groups (Tabs. 1, 2).

In the 318 912 persons, there were 89 433 aged < 30 year (28%). 14.5%, 14.7%, and 14.4% aged 30 – 34, 35 – 39, and 40 – 44 years respectively. The rate dropped gradually after 45 years of age. Age, VCA-IgA positive rate and NPC detecting rate showed a positive correlation. In the group aged < 30 years, VCA-IgA positive rate was 2.01%; in the group aged 30 – 65 years the rate increased with age; in the groups aged 65 – 69 years and over 70 years, the rate was 5.18% and 5.11% respectively. In other words, there is a positive relationship between age and VCA-IgA positive rate. NPC detecting rate in the group aged below 30 years was 6.7/10^5 and in the group aged 65 – 69 years was 65.2/10^5. Between the age 30 and 64 years, the rate increased with age. There were two peaks of NPC detecting rate between the age 30 and 69 years. The first occurred in the group aged 45 – 49 years (68.9/10^5) and the second in the group aged 65 – 69 years (65.2/10^5). In groups with 50 – 54, 55 – 59, and 60 – 64 years of age NPC detecting rate was 49.5/10^5, 49.03/10^5, and 48.1/10^5 respectively (Tab. 3).

The results of this survey coincide with one of the prospective studies on NPC in Wuzhou City Guangxi, as well as with the study in high incidence areas of NPC. This indicates that NPC serological diagnosis for mass survey can find NPC or early stage NPC for secondary prevention. [5,6] The method can be extended and applied to mass survey of NPC in other areas and people. And, it is further proved that there is a close relationship between NPC and EBV.

[In 《Chinese Medical Journal》 1995, 108 (4): 300 – 303]

References

1 Zeng Y, Liu YX, Wei JN, et al. Serological mass survey of nasopharyngeal carcinoma. Acta Academiae Medicinae Sinicae Sinicae 1979, 1 (2): 123

2 Deng H, Zeng Y, Wang NQ, et al. Follow-up studies on serological screening of NPC for 10 years in Wuzhou City in Guangxi Province. Proceedings of The 10th Asia Pacific Cancer Conference, Beijing, 1991, 86

3 Zeng Y, Liu YX, Liu CR, et al. Application of immunoenzymic method and immunoautoradiographic method for the mass survey of nasophayngeal carcinoma. Chin J Oncol, 1979, 1 (2): 81

4 Zeng Y, Zhang LG, Li JY, et al. Serological mass survey for early detection of nasopharyngeal carcinoma in Wuzhou city, China. Int J Cancer, 1982, 29: 139

5 Zeng Y, Zhang LG, Wu YC, et al. Follow-up studies on NPC in Epstein-Barr virus IgA/VCA antibody positive persons in Wuzhou city, China, Chin J Virol, 1985, 1 (1): 7

6 Deng H, Zeng Y, Wang NQ, et al. Prospective study of NPC scene in Wuzhou city, Guangxi, Chinese Medical Abstracts. Internal Medicine Supplement, 1993, 318

64. Detection of Epstein-Barr Virus DNA in Well and Poorly Differentiated Nasopharyngeal Carcinoma Cell Lines

ZHI Ping – teng[1], OOKA Tadamasa[2], HUANG Doll P[3], ZENG Yi[1]

1. Institute of Virology, Chinese Academy of Preventive Medicine, 100 Yin Xin Jie, Beijing, 100050, China; 2. Laboratoire de Virologie Moleculaire, IVMC, UMR30, Faculte de Medecine Alexis Carrel, 69372, Lyou, France; 3. Department of Anatomical and Cellular Pathology, Faculty of Medicine, Chinese University of Hong-Kong, Hong-Kong

Summary

Undifferentiated and poorly differentiated nasopharyngeal carcinoma (NPC) were known to be tightly associated with Epstein-Barr Virus (EBV). Its association with well differentiated NPC was also reported. In the present study, the presence of EBV was investigated by nucleic acid hybridization, Polymerase Chain Reaction (PCR), Immunoblot and *in situ* hybridization in two well differentiated NPC cell lines (CNE – 1 and HK – 1) and two other poorly differentiated NPC cell line (CNE – 2 and CNE – 3), Contrary to previous report indicating the absence of EBV in these cell lines, EBA DNA and proteins were present in all cell lines. The detection of EBV became more easily when the investigation was carried out on the nude mice tumor induced by transplantation of each NPC epithelial cell line. The EBV latent membrane protein (LMP1) was found by *in situ* hybridization to be intergrated partly in the chromosomal DNA of these cell lines. The observations indicate that EBV could persist for a long time in the carcinoma cells established directly from well and poorly differentiated tumor biopsies and from transplantable NPC tumor in nude mice.

[**Key words**]　Epstein-Barr Virus; Nasopharyngeal Carcinoma (NPC)

Introduction

Nasopharyngeal carcinoma (NPC) is one of the most prevalent malignant tumors in Southern China and South Asia. The presence of EBV genomes not only in poorly and undifferentiated, but also well differentiated NPC carcinoma biopsies and the serological evidence of EBV in NPC patients indicate that the virus plays an important role in the development of NPC[1,2,4-6]. There carcinoma cell lines were established directly from NPC biopsies in our laboratory, CNE – 2 from poorly differentiated NPC, CNE – 1 and HK – 1 from well differentiated NPC, and another one CNE – 3 cell line was established from transplantable poorly differentiated NPC tumor in nude mice. Among them, the presence of EBV genome in CNE – 1, CNE – 2 and HK – 1 was not evident, since nucleic acid hybridization and anticomplement immunoenzymatic test done in these cell lines were revealed negative[7,9-11]. CNE – 1 and CNE – 2 were generally used as EBV negative epithelials cells so far[3]. A possible explanation was the loss of EBV during a long culture *in vitro* or the absence of tight of association of EBV with well differentiated NPC.

We asked whether these cell lines are really negative to EBV. Four methods such as PCR, immunoblot, *in situ* hybridation and Southern blot were used to search for the presence of EBV. The data reported here showed that some of established cell lines contained both EBNA – 1 and LMP – 1 genes and expressed their proteins. Moreover EBV DNA or protein became easily detectable in the tumors induced by these cell lines in nude mice. Not only well differentiated, but also poorly differentiated cell lines contained EBV DNA. These data confirmed earlier report an association of EBV in both well and poorly differentiated NPC biopsies[6]. The NPC cell lines established since 15 – 18 years and considered as EBV negative cells contained EBV DNA sequence and expressed two EBNA – 1 and LMP – 1 latent proteins.

Materials and Methods

1. Cell Lines and Culture

Poorly differentiated human NPC cell line: CNE – 2[7] and two well differentiated NPC cell lines, CNE – 1 and HK – 1[9,11] were established directly from NPC biopsies in our laboratory. Another poorly differentiated NPC cell line (CNE – 3) was established from transplantable poorly differentiated NP tumor in nude mice[8]. They were maintained in RPMI – 1640 medium supplemented with 20% foetal calf serum (FCS) at 37℃ in 5% CO_2. The RHEK – 1 cell line, an immortalized human epithelial cell line kindly provided by Dr. Rhim (NIH, Washington), was grown in Dulbecco's Modified Eagle's Medium with 10% FCS and antibodies.

2. Transplantation of NPC Cell Lines in Nude Mice

The cells growing exponentially were harvested and washed three times with fresh RPMI – 1640 medium. The cell suspension (2×10^6) was transplanted nuder the dorsal skin of the nude mice. The tumors with masses of 2 – 3 cm^3 were obtained in about 30 days and they were frozen imediately at – 160℃.

3. Southern Blot Hybridization

For Southern blot, the DNAs were extracted by standard methods; briefly, tumor tissues from nude mice were weighed, washed with phosphate saline buffered (PBS) and ground in a little basin with liquid nitrogen. 10 ml of TEN buffer (15 mmol/L Tris-HCl, pH 8.0, 15 mmol/L EDTA, 15 mmol/L NaCl) was added per gram of tissue or 10^8 cultured cells. 50 μg/ml of proteinase K (Boehringer Mannheim) and 1/20 (v/v) of 20% sodium dodecyl sulfate (SDS) were then added in samples and incubated for 4-5 hrs at 55℃ followed by phenol-chloroform (V/V) extraction. 10 μg of the DNA precipitated by ethanol were suspended in TE buffer (10 mmol/L Tris-HCl, pH 7.5, 0.1 mmol/L EDAT), digested with Xho – 1 restriction enzyme (Sino-American Biotech. Co.) for 7 hrs at 37℃, electrophoresed through 0.8% agarose and transerred onto nitrocellulose membranes (Bio-Rad). Menbranes were baked for 2 hrs at 80℃. The hybridization was dons in 50 mmol/L sodium phosphate (pH 7.2), 7% SDS, 1% BSA and 100 μg/ml denatured salmon sperm DNA, with about 10^6 cpm/ml of probe at 65℃ overnight. Membranes were then washed in $1 \times$ SSC containing 0.1% SDS, and exposed to X-ray film at $-70℃$. Radiolabelled DNA probes were prepared with ^{32}P-dCTP (Amersham, England) by nick translation (specific activity = $> 5 \times 10^7$ cpm/μg DNA), using 3.1×10^3 EBNA – 1 and 2.9×10^3 full length LMP – 1 sequences prepared respectively from pBR322 – W plasmid (a gift from Dr. H. Wolf, Germany) and pUC-ly plasmid (a gift from Dr. E. Kieff, Harvard Medical School, Boston) after BamH1 digestion.

4. Immunoblotting

Pelleted cells were suspended in protein extraction buffer (1% SDS, 1% β-mercaptoethanol, 1 mmol/L PMSF, 20 mmol/L Tris-HCl, pH 7.0) and sonicated twice for 3 min at 5 Hz. Guanidine-HCl was then added to 5 mmol/L and samples were centrifuged for 10 min at 5000 g. The supernantants were electrophorsed on a 12.5% polyacrylamide gel containing SDS and the proteins were transferred onto nitrocellulose filter. Then the filter was incubated overnight at 4℃ with monoclonal anti-EBNA-1 and anti-LMP-1 antibodies (with 1 : 50 000 dilution) as well as with sera from NPC patients, After washing, the filters were incubated for 1 hr at room temperature with peroxidase labelled anti-rabbit antibodies. A monoclonal antibody S12 directed against LMP – 1 was from Dr. E. Kieff (Harvard Medical School, Boston). A monoclonal antibody against EBNA – 1 was obtained in our laboratory.

5. Polymerase Chain Reaction (PCR) Assay

PCR was carried out using PCR kit (Hua Mei Biotech Company) on Perkin Elmer thermal cycler (Perkin-Elmer). The reaction mixture was in total 25 μl containing 1 μg of DNA, 0.2 mmol/L dNTP, 3 mmol/L $MgCl_2$, 1U Taq polymerase (Amplitaq, Perkin-Elmer) and 2.5 μmol of each primers:

The sequence for BamH1-W fragment: 5′ primer: W1, 5′CCA GAG GTA AGA GGA CTT 3′; 3′ primer: W2. 5′GAC CGG TGC CTT CTT AGG 3′, respectively at position 1399 and 1520 on Bam H1-W giving 121 bp of amplified fragment. Amplification condition was denaturation at 94℃ for 35 sec, annealing at 50℃ for 40 sec, extension at 72℃ for 45 sec and 30 cycles.

The sequence for BARF1 open reading frame: BA1: 5′ – CCAGAGCAATGGCCAGGTTC – 3′, BA4: 5′ – CAAGGTGAAATAGGCAAGTGCG – 3′, respectively at positions 165 496 and

166 192on BARF1 sequence, giving 697 bp of amplified fragment. After 10 min incubation at 95 ℃, 1 unit of Taq polymerase (Perkin Elmer) was added. The samples were subjected to 35 cycles of : 1 min at 95 ℃, 1 min at 55 ℃ and 2 min at 72 ℃; after the last cycle, they were held at 72 ℃ for 7 min. Specificity of PCR product was tested by dotblot or Southern blot hybridization. For dot blot, 5 μl of the above PCR products was deposited onto the membrane of nitrocellulose and dried. 200 μl of denaturing buffer (0.5 mol/L NaOH, 1 mol/L NaCl) were then added to each well and ater 15 min, 200 μl of neutralizing buffer (0.5 mol/L TrisHCl, pH 7.4, 2 mol/L NaCl) was added. The membrane was baked for one hour at 80 ℃ and the hybridization was carried out in the abovementioned condition with BamH1 – W sequence as probe.

For Southern blot, 10 μl of the above PCR products was electrophoresed on 2% standard agarose gel and transferred onto Hybond nylon filter. The hybridization was carried out using BARF1 sequence as probe as previously described (see Southern blot hybridization).

6. Nonisotopic In Situ Hybridization

The cells recovered on the slide were pretreated for 15 min with proteinase K (0.9 U/ml) at room temperature. The slides were denatured for 5 min in70% formamide, 2 × SSC and then dehydrated by ethanol series. LMP – 1 probe labelled with Bio – 11 – dUTP by random priming DNA labelling kit (Beijing Medical Universith, Hematology Institute) were denatured for 10 min at 70 ℃ in the following hybridization mixture; 50% formamide, 2 × SSC and 10% dextran sulfate. Hybridization was carried out at 42 ℃ overnight. The slides were washed by 2 × SSC containing 50% formamide for 30 min at 42 ℃, then in 2 × SSC with avidin fluorescein for 30 min, After washing in PBS for 30 min, the slides were incubated with biotin labelled antiavidin antibody for 30 min. Cell nuclei were stained with PI and observed in an inverted fluorescence microscope.

Results

1. Detection of EBV DNA in Cells Lines by *In Situ* Hybridization

The detection of EBV DNA was carried out by *in situ* hybridization using LMP – 1 probe. The positive signal was obtained as pale yellow points in cell nuclei under an inverted fluorescence microscope (Fig. 1). A hybridization ratio of 95% was obtained with B95 – 8 cells. CNE – 3 as a poorly differentiated NPC, and CNE – 1 and HK – 1 as a well differentiated NPC had hybridiaztion ratio of 20% and 5% – 10% respectively.

2. Detection of EBV Genes Products by Immunoblot

The detection of both EBNA – 1 and LMP – 1 proteins was carried out by immunoblot on CNE – 2 cell line, TPA – SB – treated CNE – 1 cell line or the tumors biopsies induced by CNE – 1 and CNE – 2 cell lines, named as CNE – 1 and CNE – T2 respectively using a monoclonal antibody against EBNA – 1 or sera from NPC patients (containing a high titer of anti – EBNA – 1). A band of approximately 60 × 10^3 was detected in CNE – 2, CNE – T1, CNE – T2 and in NPC – T5 (extracted from NPC tumor in nude mice, used as positive control) by a monoclonal anti – EBNA – 1 antibody (Figs. 2A; 1 – 5). The same size of protein was identified when the immunoblots were tested by polyclonal human sera from NPC patients (Fig. 2B; 6 – 9). EBNA – 1 expression in CNE – 1 cells was only found by the same antibody when the cells were activated with TPA – SB (Fig. 3; 1).

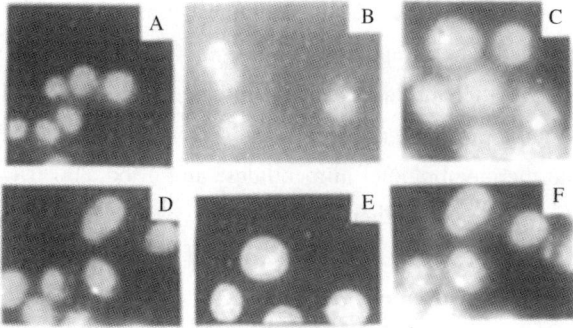

A: EBV negative CEM cell. B: EBV positive positive B95 − 8 cell. C: CNE − 2 cell. D: CNE − 1 cell. E: CNE − 3 cell. F: HK − 1 cell

Fig. 1　Detection of EBV DNA in cell lines by *in situ* hybridization. Nonisotopic *in situ* hybridization was carried out using LMP1 probe labelled with Bio − 11 − dUTP. The cells were stained with PI, then observed by microscopy at 10×40

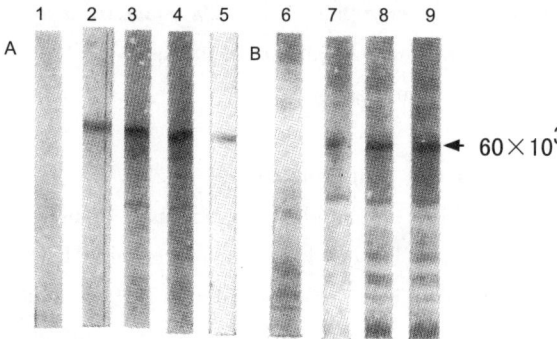

60×10^3

A. monoclonal anti − EBNA1 antibody
1. Colon cancer (as negative control)　　2. CNE − 2
3. CNE − T2　4. CNE − T1　5. NPC − T5 (as positive control)
B. Sera from NPC patients
6. CNE − 1　　7. CNE − 2　　8. CNE − T1　　9. CNE − T2

Fig. 2　Detection of EBNA1 protein by immunoblot. The cellular extracts from cell lines and tumors were electrophoresed on 12% SDS − polyacrylamide. The proteins were transferred onto nitrocellulose filter and the filters were incubated with monoclonal adti − EBNA antibody or sera from NPC patients

A monoclonal S12 antibody directed against LMP − 1 protein was then tested on these samples. Two bands of $63 \times 10^3 − 66 \times 10^3$ were found in all CNE − 2, CNE − T1, CNE − T2, TPA − SB − treated CNE − 1 and NPC − T5 (Fig. 4, A to D), while CNE − 1 and CNE − 3 as well as the biopsy from colon cancer (Fig. 4. E) were negative to this antibody.

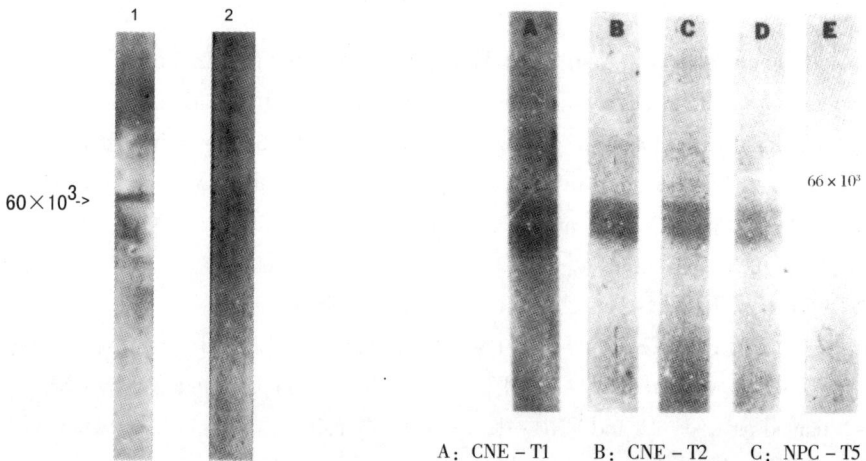

60×10^3->

1. CNE − 1 treated with TPA plus sodium butyrate
2. RHEK − 1 cell (as EBV negative control)

Fig. 3　Detection of EBNA1 protein NPC cell line treated with TPA − SB. The immunoblot was incubated with sera from NPC patients

66×10^3

A: CNE − T1　　B: CNE − T2　　C: NPC − T5
D: CNE − 2　　E: Colon cancer (as negative control)

Fig. 4　Detection of LMP1 antigen by immunoblot. The cellular extracts from cell lines and tumors were electrophoresed on 12% SDS − polyacrylamide. The proteins were transferred onto nitrocellulose filter and the filters were incubated with S12 monoclonal antibody directed against LMP1 protein

3. Detection of EBV DNA by southern blot hybridization

The investigation of LMP – 1 nucleotide sequence appeared negative when NPC cell lines were examined on Southern blot hybridization. We therefore searched for the LMP – 1 sequence in Xho – 1 digested DNA from tumor biopsies CNE – T1, CNE – T2 and CNE – T3 (Fig. 5). A typical LMP – 1 band around 6. 1 kb became detectable in all tumors as well as in B95 – 8 (used as positive control), but negative in CNE – 2 cell line (lane 4) and in a biopsy from colon cancer (lane 5).

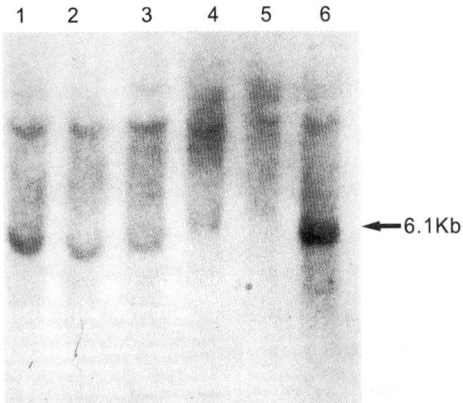

Fig. 5　Detection of LMP1 sequence in CNE – 2, CNE – T1, CNE – T2 and CNE – T3 by Southern blot hybridization. DNAs extracted from three tumors, CNE – T1 (1), CNE – T2 (2) and CNE – T3 (3), NPC cell line CNE – 2 (4), Colon cancer (5) and B95 – 8 cell line (6) were digested by Xho – 1 restriction enzyme, electrophoresed, transfered onto nitrocellulose filter and hybridized with radiolabelled LMP1 sequence used probe

4. Detection of EBV DNA by PCR

In order to further confirm the presence of EBV DNA in NPC cells, a sensitive PCR assay was carriy out on the tumor DNAs extracted from CNE – T1, CNE – T2, and CNE – T3. A viral sequence, Bam H1 – W fragment was subjected to investigation by PCR. Amplified products were examined on 6% polyacrylamide gel electrophoresis (PAGE). The 121 bp (Fig. 6. B and D – F) sequence amplified by the primers directed to Bam H1 – W fragment was identified in B95 – 8, CNE – T1, CNE – T2 and CNE – T3 DNA, whereas no such sequence was present in EBV negative RHEK – 1 epithelial cell DNA (Fig. 6, lane C). In order to verify whether the amplified products from CNE – T1, CNE – T2, CNE – T3 and B95 – 8 DNA is specific, the dot – blot assay was carried out with the amplofied sequences. The probes of Bam H1 – W hybridized specifically with the amplified products (data not shown). The BARF1 (one of EBV oncogenes[11], was also examined by PCR on NPC cell lines (Fig. 7). A major 679 bp BARF1 sequence was present in all CNE cell lines as well as the positive controls (BamHIA and P3HR1 DNA) (lanes 1 to 3), while DNAs from Raji (deleted by BARF1) and RHEK – 1 cells (an EBV negative epithelial cell line) failed to show any positive signal (lane 4 and 5). Among the NPC cell lines, CNE3 showed a high hybridization signal (lane 3), whereas a slight signal was obtained with CNE2 DNA (lane 2).

Discussion

The data presented here showed and four NPC cell lines established from well and poorly differentiated NPC contained EBV DNA. This was demonstrated by different methods and summarized in Tab. 1. Particular attention should be taken concerning the identification of viral DNA in two cell lines established from well differentiated NPC, because these cell lines were cultured over 15 – 18 years in the laboratories and considered most of cases as EBV negative cell lines. This result could reinforce the previous observations not only on a tight relationship between EBV and undifferentiated NPC, but also its association with well differentiated NPC[1,6]. The diverse methods used here for the detection of EBV DNA or protein did not give a constant positive signal in *in vitro* – cultured cell lines. It however is worth to notice that LMP – 1 gene was already cloned from CNE – 1 and CNE – 3 cell lines[10] and moreover EBV EBERs molecules have been detected in more than 90% of cells of all cell lines by *in situ* hybridization[14]. In taking together the observations, EBV DNA is present in these cell lines.

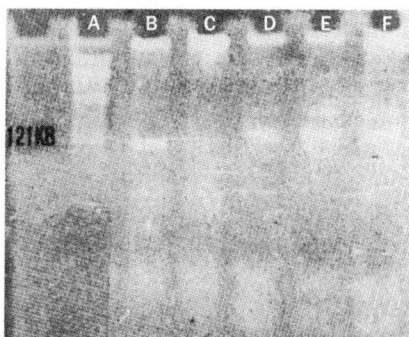

A: Marker; B: B95 – 8 DNA; C: EBV negative RHEK – 1 DNA K – 1 DNA; D: CNET1; E: CNE – T2; F: CNE – T3

Fig. 6　Detection of EBV BamH1 – W fragment by PCR. PCR reaction was carried out as described in Materials and Methods. A 121 bp of amplified fragment correspond to the position between 1399 and 1520 on BamH1 – W fragment

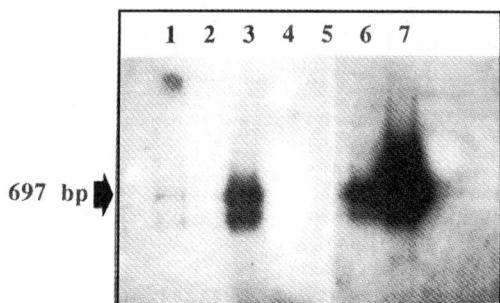

1: CNE – 1 cell line; 2: CNE – 2 cell line; 3: CNE – 3 cell line; 4: Raji DNA (BARF1 negative control); 5: RHEK – 1 cell line (EBV negative control); 6: Namalwa cell line (EBV positive control); 7: B95 – 8 (EBV positive control)

Fig. 7　Detection of BARF1 sequence by PCR in NPC cell lines. A 697 bp of amplified fragment correspond to the position between 165 496 and 166 192 on EBV genome

Tab. 1　Detection of EBV DNA and proteins in NPC cell lines and in nude mouse tumors – derived from NPC cell lies

	CNE – 1	CNE – 1 – TS*	CNE – 2	CNE – 3	CNE – T1	CNE – T2	CNE – T3	HK – 1	B95 – 8
In situ hybrid.									
LMP1 (%)	5 – 10	ND	5 – 10	20	ND	ND	ND	5 – 10	95
Immunoblot									
EBNA – 1	–	+	+	–	+	+	ND	ND	+
LMP – 1	–	+	+	–	+	+	ND	ND	+
Southern blot									
LMP – 1	–	–	–	–	+	+	+	ND	+
PCR									
BamH – W	+	ND	ND	ND	+	+	+	+	+
BARF1	+	ND	+	+	ND	ND	ND	ND	+

Note: * NPC cell line was treated with TPA and sodium butyrate for three days

Interestingly the 60×10^3 LMP – 1 protein became detectable with CNE – 1 cells were treated with TPA – SB. These activators may act on the transcriptional regulation of the LMP – 1 gene to trigger expression, so that a productive EB virus cycle is activated *in vitro*.

In comparison with the results from NPC cell lines, relatively constant EBV positive signal was obtained in nude mice tumors. This may come from EBV DNA replication which could occur more easily in tumor in nude mice than in cell lines *in vitro*[15,16]. Whereas our and previous[6] reports showed a tight association of EBV with well differentiated NPC, there is no EBV IgA/VCA antibody in sera from many NPC patients with well differentiated carcinoma. This may be due to the restriction of the EBV DNA amplification and expression by cellular differentiation. It is suggested that our results stress the need to investigate another well differentiated head and neck cancer and other cancers in relation with EBV.

NPC cell lines with EB virus could consititute as a model for studying the molecular event *in vitro* and *in vivo* between EBV and tumor epithelial cells at different states of differentiation.

Acknowledgments

We thank Prof. Sir A. M. Epstein and Dr. G. de The for revising this manuscript. This work was supported in part by research grants from the European Communities contract N° CI1CT930010 for T. O, Z. Yand the Chinese National Scientific Committee for Z. P. T, D. P. H, Z. Y.

[In 《Virus Genes》 1996, 13 (1): 53 –60]

References

1 Wolf H, Zur Hausen H, Becker Y, Nature, New Biol, 1973, 244: 245 –247

2 Desgranges C, de The G in de The G, Henle W, Rapp F. (eds). Oncogenesis and Herpesvirus Ⅲ. IARC Scientific Publications N° 25, Lyon, 1978, 883 –891

3 Lerman M I, Sakai A, Kai – Tai Y, Colburn N H. Carcinogenesis, 1987, 8: 121 –127

4 Desgranges C, Bornkamm G, Zeng Y, Wang P C, Zhu J S, Shang M, de The G. Int J Cancer, 1982, 29: 87 –91

5 Zeng Y, Liu Y X, Liu C R, Chen S W, Wei J N, Zai H J. Intervirology, 1980, 13: 162 –168

6 Raab – Traub N, Flynn K, Pearson G, Huang A, Levine P, Lanier A, Pagano J. Int J Cancer, 1987, 39: 25 –29

7 Gu S Y, Tan B G, Zeng Y, Zhou W P, Li K. Nasopharyngeal Carcinoma Current Concept (Prasad U. et al. eds) University of Malaya, 1978: 273 –276

8 Jiao W, Zhou W Y, Teng Z P, Wang P C, Zeng Y, Chinese J Virology in press

9 Huang D P, Ho J H C, Poon Y F, Chew E C, Saw D, Lui M, Li C L, Mak L S, Lai S H, Lan W H. Int J Cancer, 1980, 26: 127 –132

10 Su L, Teng Z P, Zeng Y. Chinese J. Virology, 1994, 11: 114 –118

11 Laboratory of Tumor Viruses of Cancer Institute, Laboratory of Tumor Viruses of Institute of Epidemiology Chinese Academy of Medical Sciences, Sci Sin. 21, 127 –134, 1978.

12 Wei M. X. and Ooka T., The EMBO. J, 1989, 8: 2897 –2903

13 Yuan F, Zeng Y. Journal of Nanjin Medical College, 1986, 9: 3 –7

14 Liu C S, Teng Z P, Zeng Y. Chinese J. Virology in press

15 Trumper P A, Epstein M A, Giovanelia B C, Finerty S, Int J Cancer, 1977, 20: 655 –662

16 Trumper P A, Epstein M A, Giovanelia B C. Int J Cancer, 1976, 17: 578 –587

65. Epstein-Barr Virus in Synergy with Tumor-Promoter-Induced Malignant Transformation of Immortalized Human Epithelial Cells

LI Bao-min[1], JI Zhi-wu[1], LIU Zhen-Sheng[1,2], ZENG Yi[1]

1. Institute of Virology, Chinese Academy of Preventive Medicine, 100 Ying Xin Jie, Xuan Wu Qu, Beijing 100052, P. R. China;

2. Xijing Hospital, the Fourth Military Medical University, Xian Shanxi 710032, P. R. China

Summary

It is difficult to study how Epstein – Barr virus (EBV) causes transformation of human epithelial cells. The major difficulty is that cultured human epithelial cells do not express EBV receptor (complement receptor 2, CR2), hence EBV cannot infect such epithelial cells directly. In order to investigate the role of EBV in the transformation of human epithelial cells, pSG – CR2 – Hyg carrier was transfected into immortalized human epithelial cells (293 cells) to express EBV receptor. EBV could infect these CR2 – positive cells directly, and expressed EBV antigens. EBV – infected epithelial cells grew in piles with multiple cellular layers and lost contact inhibition in vitro. In soft – agar culture containing 12 – O – tetradecanoylphorbol 13 – acetate (TPA), EBV – infected 293 cells formed more and larger colonies. When EBV – infected 293 cells were transplanted subcutaneously into nude mice, and treated with TPA, poorly differentiated carcinoma was induced. Thses results suggest that EBV could induce the malignant transformation of immortalzed human epithelial cells in synergy with TPA.

[**Key words**]   Epstein – Barr virus; EBV receptor; Tumor promoter

Introduction

It is known that Epstein – Barr virus (EBV) is closely associated with the development of naso-pharyngeal carcinoma (NPC) (Wolf et al. 1973; Trumper et al. 1977) but it is not clear how EBV gets into human epithelial cells (Shapiro and volsky 1983; Volsky et al. 1980) and how it in-duces malignant transformation of epithelial cells, since there is no or minimal expression of EBV receptor after *in vitro* passage of epithlial cells (Graessann et al. 1980; Sixbey et al. 1987), thus it has not been possible, up to now, for epithlial cell lines to be infected by EBV *in vitro*. In order to overcome this barrier, it was reported heat EBV van infect human epithelial cells transfected with a complement receptor 2 (CR2) carrier, express EBV early antigen, repliate in these cells and pro-duce virions (Li et al. 1992). Ito et al. showed that the synthesis of EBV antigens in EBV – carrying cells was significantly increased when these cells were treated with a combination of n – butyrate and 12 – O – tretradecanoylphorbol 13 – acetate (TPA) (Ito et al. 1983). Zeng et al. found that many kinds of plants, including Chinese medicinal herbs, contain TPA – like substances which, in combination with n – butyrate, induced the synthesis of EBV antigens, enhanced the transformation of B lymphocytes by EBV and promoted the development of NPC in rats by dinitrosopiperazine (Zeng et al. 1983, 1984; Tang et al. 1988). Thus, EBV inducers and tumor promoters may play important roles in the development of human carcinoma. In spite of these finding, there has been no *in vitro* evidence that EBV can directly induce neoplastic rtansformation of human epithelial cells. In order to investigate further the transformation of human epithelial cells by EBV, CR2 – positive human immortalized epithelial cells (293 cells) were established and infected directly by EBV, and poorly differentiated carcinoma in nude mice could be induced by EBV in synergy with tumor promoter.

Materials and Methods

1. Animals

Male or female nude mice (Balb/c) aged 4 – 5 weeks and weighing 20 – 25 g were used for transplantation. The breeding and housing of the experiments wre carried on under aseptic conditions.

2. Cell lines and culture

Immortalized 293 cells were derived from human embryonic kidney cells. The 293 cells were maintained in Dulbecco's modified Eagle (DME) medium supplemented with 10% fetal bovine serum (FBS), 100 μg/ml streptomycin and 100 units/ml penicillin. B95 – 8 cells were from EBV – transformed marmoset B cells, which were cultured in RPMI – 1640 medium supplemented with 10% FBS, 100 μg/ml streptomycin and 100 units/ml penicilin in an incubator under a humidified atmosphere of 5% CO_2.

3. Antibodies and plasmids

HB5 is an IgG2a mose monoclonal antibody that specifically recognizes human CR2. Fluorescein – conjugated goat anti – (mouse IgG) and fluorescein – conjugated goat anti – human (human IgG) wew purchased from Sigma Chemical Co. NPC patient sera were obtained from the Caner Institute, Wuzhou City, Guangxi Zhuang Autonomous Region, China. pSG – 5, a eukaryotic expressing vector containing simian virus 40 (SV40) early promoter, and pBluscriptKs (—) – CR2 containing the full – length human RC2 cDNA were provided by professor Takata (Yamaguchi University, Yamaguchi, Japan).

4. Construction and transfection of plasmid

A 3219 – bp restriction fragment from plasmid pBS – CR2 was cloned into the EcoRI site of the pSG – 5 vector to generate pSG – CR2, and a 1600 – bp Hyg fragment, consisting of SV40 early promoter and the hygromycin phosphotransferase (Hyg) geme, was inserted into pSG – CR2 to create plasmid pSG – CR2 – Hyg. A 15 – μg sample of pSG – CR2 – Hyg DNA was introduced into 293 cells by calcium phosphate precipitation (Kingston 1987). After a 6 – h incubation with DNA, the cells were shocked with 20% dimethylsulfoxide for 2 min. Transfected cells were maintained in DME medium containing 400 μg/ml hygromycin B for 48 h after transfection. After 2 weeks of culture, drug – resistant colonies were picked and expended in DME medium containing hygromycin B.

5. Measurement of CR2 expression

Drug – resistant colonies were screened for expression of CR2 by indirect immunofluorescence staining by incubation of the cells at 37℃ for 30 min with HB5 at dilution of 1/50, followed by fluorescein – isothiocyanate (FITC) – conjugated goat anti – (mouse IgG) at a dilution of 1/10. EBV – positive Burkitt lymphoma cell line Raji cells expressing CR2 were used as a positive control. By counting 500 cells, the percentage of positive cells was calculated.

6. EBV preparation and infection

B95 – 8 cells were cultured in medium containing 20 ng/ml TPA and 4 mmol/L butyrate, at 37℃ for 48 h. The supernatant was collected and cellular debris was eliminated by low – speed centrifugation, and the virus suspension was concentrated by centrifugation at 45 000 g for 120 min. The virus pellet was then resuspended in RPMI – 1640 medium. Filtering through a 0. 45 – μm filter produced 150 – fold concentrated B95 – 8 EBV. CR2 – 293 cell monolayer cultures at 60% – 70% confluence were washed twice in phosphate – buffered saline, (PBS) incubated with EBV concentrate for 4 h at 37℃, then fed again with fresh medium. Cultures were re – fed with fresh medium every 2 – 3 days.

7. Detection of EBV antigens

The CR2 – 293 cells were harvested 4, 8, 12, 16, 24, 28 days after addition of EBV and assayed for the expression of EBV antigen as follows. The cells were scraped and smeared onto the slides, then fixed in cold acetone, incubated with a 1 : 50 dilution of NPC sera at 37℃ for 30 min, washed with PBS, and incubated with FITC – conjugated goat anti – (human IgG) at a dilution of 1/50 at 37℃ for 30 min. After washing, the cells were examined by fluorescence microscopy. Alternatively, some of hte cells were cultured in medium with 5 ng/ml TPA, and the expression of EBV antiens was examined. The percentages of positive cells were calculated by counting 500 cells.

8. Amplification of EBV DNA

EBV – infected CR2 – 293 cells were cultured for 15 or 30 days then separated by addition of trypsin and collected by centrifugation. These cells were digested with sodium dodecyl sulfate/proteinase K, and the cellular DNA was extracted with phenol/chloroform/isopentanol and precipitated with ethanol. The following primers were used for amplification: primer sequences of the W fragment of EBV DNA: 5'CCAGAGGTAAGAGGACTT3'; 5'GACCGGTGCCTTCTTAGG3'. The polymerase chain reaction (PCR) regime was 25 cycles (30 s at 94℃, 30 s at 55℃, and 1 min at 72℃) followed by 10 min at 72℃. PCR products were analyzed on 2% agarose gels stained with ethidium bromide.

9. Colony formation in soft agar

Agarose was dissolved in disitilled water and autoclaved at 120℃ for 30 min. The base layer was 0. 66% agarose in 1. 0 ml DME medium containing 5 ng/ml TPA, and the top layer was 0. 33% agarose in 1. 0 ml DME medium containing 5 ng/ml TPA and 1×10^4 293 cells. Culture dishes were incubated at 37℃ in 5% CO_2.

10. Tumor formation

Suspensions of 2×10^6 cells infected with EBV in 0. 4 ml growth medium containing 20 ng TPA were subcutaneously injected into Bald/c nude mice; thereafere 50 ng TPA was injected weekly into the transplanted side. Nude mice were inspected every other day for the appearance and progressive growth of tumors, and sacrificed after 6 – 7 weeks, Tumor tissues were dissected, fixed in 10% (v/v) buffered formaldehyde, and embedded in paraffin for histological examination. For negative controls, 293 cells and CR2 – 293 cells were injected subcutaneously.

11. Detection of EBV DNA in tumor tissues by *in situ* hybridization

DNA complementary to EBV-encoded earl RNA(EBER)was labeled by random hexanucleotide priming using dogoxigenin-11-dUTP to produce DNA probes. Tumor specimens were fixed with formaldehyde, embedded in paraffin and sectioned; the sections were then dewaxed by toluene and hydrated with 100%, 70%, 50% alcohol. The DNA of the sections was denatured. The specimen slides were hybridized with a digoxigenin-labeled EBER DNA probe under high stringency as described by Wu et al. (1990).

Results

1. Expression of CR2 in 293 cells

293 cells were transfected with pSG – CR2 – Hyg and selected for 2 weeks in medium contaning hygromycin B. All of six resistant cell colonies were analyzed for the expression of CR2 by indirect immunofluorescent staining with HB5 monoclonal antibody. The CR2 transfectants showed membrane – associated staining exactly as did the control Raji cells, and the highest positive rate was 32% (Figs. 1a, b). The levels of CR2 expression varied between different transfected colonies. Therefore, drug – resistant colonies expressing relatively large amounts of CR2 were picked and expanded in medium containing hygromycin B. The control of 293 cells was negative in this assay (Fig. 1c).

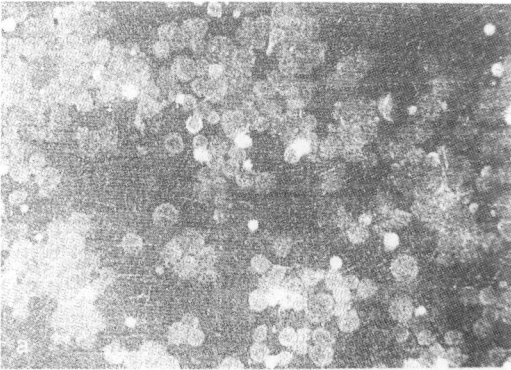

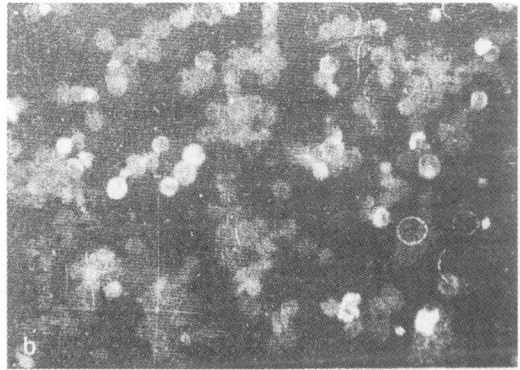

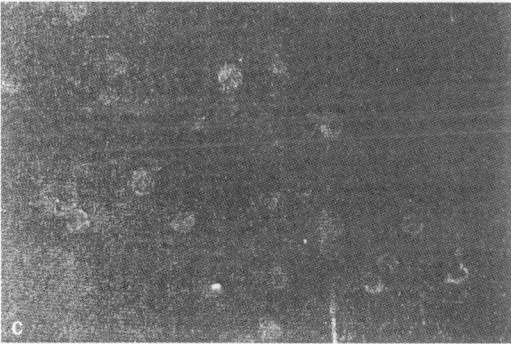

a: 293 cells were transfected with pSG – CR2 – Hyg vector; drug – resistant cells were screened with hygromycin B then smeared, fixed with cold acetone, reacted with 1 : 50 HB5 (CR2) monoclonal antibody, and stained with fluorescein – labeled sheep antibody to mouse IgG. Positive reactions were observed; b: Raji cells showing expression of CR2; c: Untransfected 293 cells did not show CR2 expression

Fig. 1   a – c Expression of CR2 – vector – transfected 293 cells

2. EBV infection of CR2 – 293 cells

CR2 – 293 cells were exposed to 150 – fold concentrates of B95 – 8 EBV and were incubated for 4, 8, 12, 16, 20, 24 or 28 days. Some EBV – infected CR2 – 293 cells expressed the EBV antigens, detectable by indirect immunofluorescence staining with NPC patient sera (Fig. 2a). By 8 days after infection, 36% of EBV – antigen – positive cells had peaked and 2% of EBV – antigen – positive cells were still detectable at day 28 (Fig. 3). No staining with NPC patient serum as seen in 293 cells preincubated with EBV or in CR2 – 293 cells in the absence of EBV infection (Fig. 2b). To examine further whether EBV DNA existed in the cells 2 and 4 weeks after infection, the cells were collected after trypsin treatment and total cellular DNA was extracted: this was then analyzed by PCR for a 120 – bp sequence from BanHI W fragments of the EBV genome. The results of product amplification revealed that the EBV sequence was present only in EBV – infected CR2 – 293 cells and B95 – 8 cells. PCR analysis did not reveal the presence of the EBV sequence in CR2 – 293 cells or in 293 cells exposed to EBV (Fig. 4).

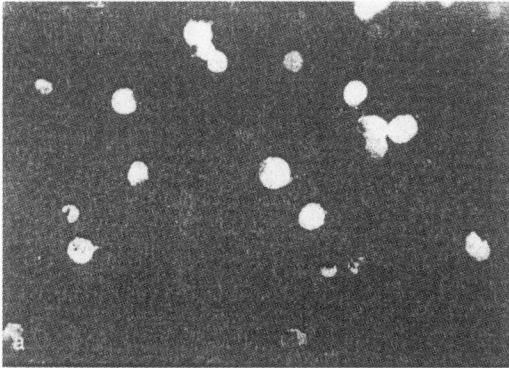

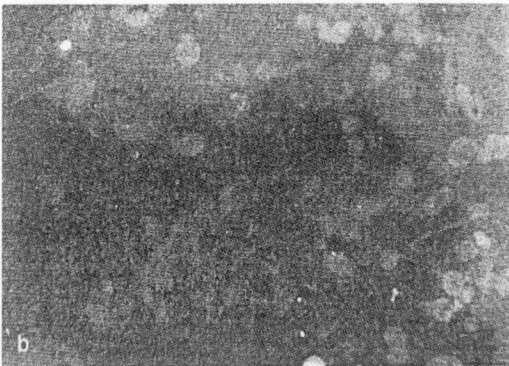

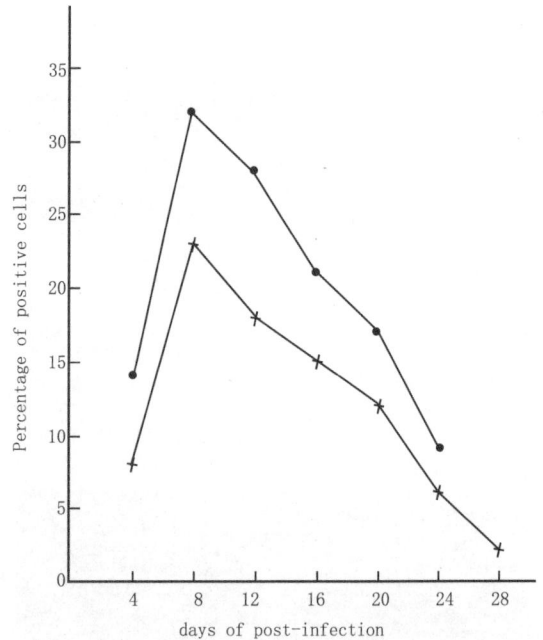

a: CR2 – 293 cells were infected with B95 – 8 EBV and continuously cultured for 8 days. Cultured cells were collected by trypsinization, the cells were washed in phosphate – buffered saline and spread on slides as cell smears. Slides were fixed and stained for EBV antigens by indirect immunofluorescence using sera of patients with nasopharyngeal carcinoma; b: EBV – infected 293 cells

Fig. 2 a, b Epstein – Barr virus (EBV) infection of CR2 – positive 293 cells

+ EBV – infected CR2 – 293 cells; ■ EBV – infected CR2 – 293 cells activated by 5 ng/ml 12 – O – tetradecanoylphorbol 13 – acetate (TPA)

Fig. 3 Percentage of cells expressing EBV antigens

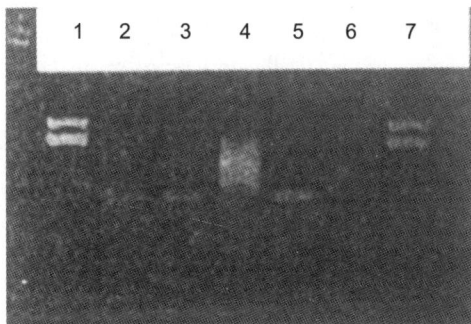

On days 15 and 30 after infection, the cells were treated with trypsin and washed; cell DNA was obtained by sodium dodecyl sulfate lysis/ proteinase K digestion and phenol extraction. Primers recognized EBV W fragment. The polymerase chain reaction was performed as described in Materials and methods Lanes: 1, and 7: pBR322 DNA/BstnI marker; 2: CR2 – 293 cells infected with EBV for 30 days 3: CR2 – 293 cells infected with EBV for 15 days; 4: CR2 – 293 cells; 5: B95 – 8 cells; 6: 293 cells infected with EBV

Fig. 4　Amplification of EBV W fragment from EBV – infected CR2 – 293 cells

3. Growth properties of EBV – infected CR2 – 293 cells

The CR2 – 293 cell culture was infected with concentrates of B95 – 8 EBV. Two days after the EBV infeciton, the cells were plated in soft agar with 5 ng TPA. These cells formed more and larger colonies in soft agar after 10 days. The CR2 – 293 cells that received either EBV exposure alone or TPA treatment alone only formed a few small colonies. EBV – infected 293 cells cultured in medium with 0. 5 ng TPA grew in tight, disorganized bundles that became partly multilayered and formed foci before they reached confluence. EBV – infected CR2 – 293 cells without TPA treatment and CR2 293 cells treated with TPA showed no detectable change compared to the 293 cells.

4. Tumor formation of EBV – infected CR2 – 293 cells

CR2 – 293 cells or EBV – infected CR2 – 293 cells were inoculated subcutaneously into Balb/c nude mice. 293 cells alone or 293 cells treated with TPA and exposed to EBV did not form tumor in five and six nude mice in each group, and no tumor was observed in any of four nude mice transplanted with EBV – infected CR2 – 293 cells without TPA treatment. However, in there of the six nude mice subcutaneously transplanted with EBV – infected CR2 – 296 cells and subcutaneously injected with TPA in the inoculated side at weekly intervals, marked tumor nodules were observed after 2 – 3 weeks, which grew progressively. The size of the tumor nass was 1. 4 cm × 1. 7 cm × 0. 4 cm, 0. 6 cm × 0. 4 cm × 0. 3 cm, 1. 3 cm × 0. 9 cm × 1. 2 cm in the three mice. Tumor – bearing nude mice were sacrificed after 6 – 7 weeks. Histopathological examination of hematoxylin/eosin – stained paraffin sections from tumor tissues showed poorly differentiated carcinoma (Fig. 5). Tumor samples were fixed, denatured and hybridized with EBER DNA probes. In situ hybridizaiton demonstrated that EBV early RNA were present in tumor tissues (Fig. 6). This observation confirmed that the tumor developed as a consequence of malignant transformation of the EBV – infected CR2 – 293 cells following treatment of TPA.

Discussion

The association of EBV with NPC has been known for many years, but the role of EBV in the development of this tumor in poorly understood, partly because of the absence of a demonstrable EBV – receptor molecule on the epithelial cell lines available and the inability of EBV to infect human epithelial cells in vitro. In the present study, human epithelial cells (292 cells) were transfected with a CR2 – expression vector. Immunofluoresecence assays on drug – resistant colonies confirmed that 32% of CR2 – 293 cells expressed CR2 on their surfaces. Following exposure to EBV, many cells expressed the EBV antigens, and still retained EBV DNA for 4 weeks after EBV infection, proving that CR – positive 293 cells were susceptible to EBV. Treatment of EBV – infected CR2 – 293 cells with TPA induced a striking morphological transformation. CR2 – 293 cells that were infected with EBV in combination with TPA treatment showed a marked increase in colony – formation ability in soft agar. However, the cells that received either TPA treatment alone or EBV exposure alone did not have a significantly increased ability to form colonies. These data indicate that transformation of immortalized human epithelial cells by increased soft – agar colony formation and morphological alteration is dependent on infectious virus and treatment with TPA.

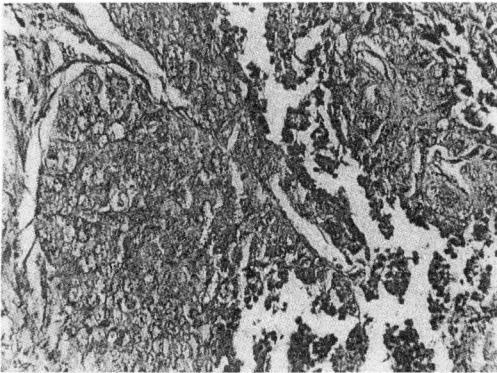

Fig. 5   Micrographs of paraffin sections from tumors formed in nude mice. EBV-infected CR2-293 cells were cultured for 2 days. Samples containing 2×10^6 cells and 50 ng TPA were transplanted subcutaneously into nude mice, and 50 ng TPA was injected into the side of cell transplantation each week thereafter. Marked tumor nodules formed in the nude mice, which were sacrificed after 6 – 7 weeks for histological examination of subcutaneous tumors

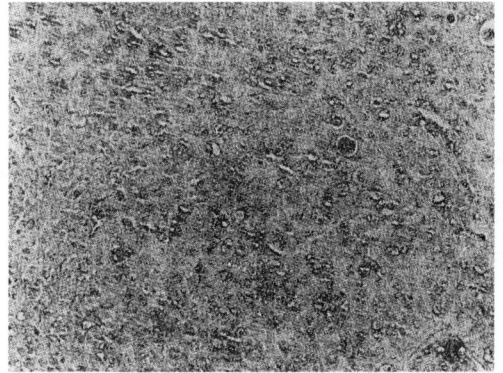

Fig. 6   Detecion of EBV – encoded early RNA (EBER) in tumor tissues by in situ hybridization. Tumor specimens were fixed with formaldehyde, embedded in paraffin and sectioned. The probes were digoxigenin – labeled EBER DNA fragments

Our results further indicated that tumors formed in three out of six nude mice inoculated with EBV – infected CR2 293 cells subcutaneously and treated with TPA, in situ hybridization revealed that the expression of EBER occurred in the tumor tissues. However, EBV – infected 293 cells or 293 cells treated with TPA could not induce carcinoma. Further work has also shown that carcinoma can not be induced by EBV infecting fresh fetal nasopharyngeal tissues in the presence of TPA, except under the synergetic effect of TPA and n – butyrate (Liu et al. 1996). Other studies have revealed that dual exposure of human lymphocytes to EBV and purified 4 – deoxyphorbol ester induced chromosome rearrangements and development of lymphoma in nude mice (Aya 1991). Griffin and Karran also found that the growth properties of monkey kidney epithelial cells transfected with BamH1 D – A fragments of EBV DNA were dramatically altered, and cell cultures with viral DNA were maintained for more than 1 years. However, the cells were incapable of producing tumors in nude mice (Griffin and Karran 1984). Clearly EBV alone is not sufficient to induce tumor and malignat tranformation of epithelial cells.

It has been known for sometime that NPC is prevalent in Southern China. Some studies have presented strong epidemiological evidence that phorbol esters are related of NPC in southern regions of China, where many herbs and plants contain diterpene esters (Ito et al. 1981; Mizuno et al. 1983; Zeng et al. 1984; Hirayama and Ito 1981). Our results demonstrate that tumor promoters could play a synergetic role with EBV in the development of NPC.

In conclusion, this study showed that EBV can infect 293 cells transfected with CR2 carrier and express EBV antigens, EBV DNA exists in EBV-infected 293 cells for more than 1 month. We have reported evidence that EBV was capable of increasing the clonability and causing morphological transformation of human immortalized epithelial cells, and that EBV could also induce human immortalized epithelial cells to from poorly differentiated carcinoma; however, all these changes depend on the presence of TPA.

Acknowledgments

This work was supported by grants from the Commission of the European Communities and from the National Scientific Committee of China

[In 《J Cancer Res Clin Oncol》 1997, 123: 441 –446]

References

1 Aya T. Chromosome translocation and c – MYC activation by Epstein – Barr virus and Euphorbis tirucalli in lymphocytes. Lancet, 1991, 337: 1190

2 Graessmann A. Wolf H, Bomkamm GW. Expression of Epsitein – Barr virus genes in different cell types after microinjection of viral DNA. Proc Natl Acad Sci USA, 1980, 77: 433 – 436

3 Griffin BE, Karran L. Immortalization of monkey epithelial cells by specific fragments of Epstein – Barr virus DNA. Nature, 1984, 309: 78 – 82

4 Hirayama T, Ito Y. A new view of the etiology of nasopharyngeal carcinoma. Rev Med, 1981, 10: 614 – 622

5 Ito Y, Yanase S, Fuji J, Harayama T, Takashima M, Imanaka H. A short – term in vitro assay for promoter substances using human lymphoblastoid cells latently infected with Epstein – Barr virus. Cancer Lett, 1981, 13: 29 – 37

6 Ito Y, Yanase S, Tokuda H. Epstein – Barr virus activated by tug oil, extracts of Aleurites fordii and its diterpene exter 12 – O – hexadecanoly –

16 – acetate. Cancer Lett, 1983, 18: 87 – 95

7 Kingston RE. Transfection using calcium phosphate – DNA precipitate formed in Hepes. In Ausubel FM, Brent R, Kingston RE, Moore DD, Seidman JG, Smith JA, Struhi K (eds) Current protocols in molecular biology. Wiley, New York, 1987, 9. 1. 1 – 9. 1. 3

8 Li QX, Young LS, Niedobitek G, Dawson CW, Birkenbach M, Wang F, Rickinson AB. Epstein – Bar virus infection and replication in human epithelial cell system. Nature, 1992, 356: 347 – 350

9 Liu ZS, Li BM, Liu YF, Zeng Y. Studies on human nasopharyngeal malignant lymphoma and-undifferentiated carcinoma induced the synergetic effect of EB virus and tumor promoter. Chin J Virol, 1996, 12: 1 – 8

10 Mizuno F, Koizumi S, Osato T, Kokwaro JO, Ito Y. Chinese and African Euphorbiaceae plant extracts: markedly enhancing effect on Epstein – Barr virus – induced transformation. Cancer Lett, 1983, 19: 199 – 205

11 Shapiro IM, Volsky DJ. Infection of normal human epithelial cells by Epstein – Barr virus. Science, 1983, 219: 1225 – 1228

12 Sixbey JW, Davis DS, Young IS, Hutt – Fletcher I, Tedder TF, Rickinson AB. Human epithelial cell expression of an Epstein – Barr virus receptor, J Gen Virol, 1987, 68: 805 – 811

13 Tang WP, Huang PG, Zhao ML, Liao SL, Zeng Y. Wikstroemia indica promoter development of nasopharyngeal carcinoma in rats initiated by dinitrosopiperazine. J Cancer Res Clin Oncol, 1988, 114: 429 – 431

14 Trumper PA, Epstein MA, Giovanella BC, Finerty S. Isolation of infections EB virus from the epithelial tumor cells of nasopharyngeal carcinoma. Int J Cancer, 1977, 20: 655 – 662

15 Volsky DJ, Shapiro IM, Klein G. Transfer of Epstein – Barr virus receptors to receptor negative cells permits virus penetration and antigen experssion. Proc Natl Acad Sci USA, 1980, 77: 5453 – 5457

16 Wolf H, Zur Hausen H, Becker V. EB viral genomes in epithelial nasopharyngeal carcinoma cells. Nat New Biol, 1973, 244: 245 – 247

17 Wu TC, Mann RB, Charache P. Detection of EBV gene expression in Reed – Stemberg cells of Hodgkin's disease. Int J Cancer, 1990, 46: 801 – 804

18 Zeng Y, Zhong JM, Mo YK, Miao XC. Epstein – Barr virus early antigen induction in Raji cells by Chinese medicinal herbs. Intervirology, 1983, 19: 201 – 204

19 Zeng Y, Miao XC, Jaio B, Li HY, Ni HY, Ito Y. EpsteinBarr virus activation in Raji cells with ether extracts of soil from different areas in China. Cancer Lett, 1984, 23: 53 – 59

66. Epstein-Barr Virus Strain Variation in Nasopharyngeal Carcinoma from the Endemic and Non – Endemic Regions of China

SUNG Nancy S[1], EDWARDS Rachel H[1], SEILLIER-MOISEIWITSCH Francoise[2],
PERKINS Ashley G[1], ZENG Yi[3], RAAB-TRAUB Nancy[1]

1. Lineberger Comprehensive Cancer Center, University of North Carolina, Chapel Hill, NC, USA; 2. Department of Bisostatistics, School of Public Health, University of North Carolina. Chapel Hill, NC, USA; 3. Instivute of Virology, Chinese Academy of Preventive Medicine, Beijing, People's Republic of China

Summary

Nasopharyngeal carcinoma (NPC) occurs with a striking geographic incidence and is endemic in parts of southern China, where it is the major cause of cancer death. Epstein – Barr virus (EBV) is detected in all cells of the majority of NPC cases regardless of geographic origin. A small subset of EBV genes is expressed in NPC, including the latent membrane protein (LMP – 1). LMP – 1 is essential for transformation of B lymphocytes and is considered to be the EBV oncogene. This analysis of the DNA sequence variation within the LMP – 1 gene reveals a consensus sequence for a strain, denoted China1, which predominates in East Asis where NPC is endemic. The China1 strain is characterized by nucleotide changes at 13 loci in the amino terminal portion of the LMP − 1 gene when compard with the B95 − 8 prototype, including a point mutation resulting in the loss of an Xho1 restriction site. This strain was present in 9 of 15 NPC biopsy specimens from the endemic region and in 7 of 13 from northern China, where NPC is non − endemic. A second strain, China2, was detected in 4 of 15 endemic isolates and in 2 of 13 nonendemic isolates; this strain was characterized by a cluster of 5 nucleotide changes in the amino terminal portion of LMP − 1 in addition to those seen in China1. It was also marked by distinct changes in the carboxy terminal region of LMP − 1 including the retention of amino acids 343 − 352. All China1 isolates were EBV type 1, whereas the China2 isolates did not correlate with EBV type. Phylogenetic relationships between these 2 strains were determined, as were signature amino acid alterations that discriminate between them.

Introduction

Nasopharyngeal carcinoma (NPC) occurs with a remarkable geographic pattern of incidence and is considered to be endemic in parts of southern China, where it is the major cause of cancer death. Mediterranean African and the Alaskan Inuit people also have elevated incidences of NPC (de The, 1982). An etiologically complex disease, NPC has been linked to several environmental, dietary, viral and genetic co − factors. Nitrosamines and other tumor − promoting agents have been identified as co − factors present in food products in the high − incidence areas (Armstrong et al., 1983). Perhaps the most complling contributing factor is the Epstein − Barr virus (EBV), which is detected in all cells of undifferentiated NPC biopsy specimens regardless of geographic origin (reviewed in Raab − Traub, 1996). Although EBV is ubiquitous in the human population, it is possible that one contributing factor to the endemic incidence of NPC could be the presence of a particular viral variant predominating in the endemic region.

All EBV isolates can be classified as type 1 or type2 (or types A and B) based on sequence divergence in the EBNA −2, 3A, 3B and 3C genes (Adldinger et al., 1985; Sample et al., 1990). Type 1 EBV is more prevalent in Chinese NPC (Abdel − Hamid et al., 1992; Zimber et al., 1986), although none of the type − discriminating EBNA genes are expressed in NPC. Sorting independently of these 2 EBV types are several strains defined by DNA sequene polymorphisms in the LMP − 1 gene (Miller et al., 1994). The LMP − 1 protein, detected in at least 65% of NPC tumors (Young et al., 1988), is able to transform rodent fibroblasts in viro and is essential for EBV − mediated transformation of B lymphocytes (reviewed in Kieff, 1996). In addition to the induction of several B − cell activation antigens and adhesion molecules, LMP − 1 protects infected cells from apoptosis by upregulaiton of the bel −2 and A −20 genes (reviewed Kieff, 1996). It also interacts with and engages the TRAF signal transduction pathway, resulting in the induction of EGFR expression and activation of NK − kB (Hammarskjold and Simurda, 1992; Miller et al., 1995; Mosialos et al., 1995).

The predominant EBV starin in China, designated here as China, is characterized by a cluster of 13 nucleotide changes with respect to the B95 − 8 prototype strain in the amino terminal region of LMP − 1 (Miller et al., 1994). These include a point mutation resulting in the loss of an Xho restriction site. The China 1 strain is also distinguished by changes in the carboxy terminal region of

LMP – 1, most notable the deletion of amino acids 343 – 352 (Abdel – Hamid et al. , 1992; Chen et al. , 1992; Hu et al. , 1991; Miller et al. , 1994). A related strain, previously found in Alaskan isolater, shares 14 of the 15 amino terminal changes with China 1, including the Xhol polymorphism, but at the carboxy terminus retains amino acids 343 – 352 and harbors 15 additional nucleotide changes not found in China 1 (Miller et al. , 1994).

Our present study was undertaken to determine the prevalence of EBV strains, based on sequence variation in the LMP – 1 gene, in NPC tumors from patients from northern China, where the incidence of this malignancy is much lower than in southern China and where EBV typing has not been done. For comparison, tumors from patients from the NPC endemic area of southern China were also examined, and the relatedness of the various EBV strains found in China was determined by phylogenetic analysis.

Material and Methods

1. Patient tissue specimens

NPC tissue biopsies were obtained from patients at the Cancer Hospital in Beijing (specimens 108, 423, 509, 525, 614 and 615), the Bai Qiu En Medical Univerity Hospital in Changchun, Jilin Province (specimens 14 – 27 and 121 – 126) and the Guangxi Regional Hospital in Nanning, Guangxi Autonomous Region (specimens GX1 – 6, 127 – 129, 132 – 133, 136 – 145) in the People's Republic of China. Patients were interviewed to verify whether their ancestral home was within the endemic or non – endemicregion.

Tissues were Dounce – homogenized on ice in buffer containing 15 mmol/L NaCl, 15 mmol/L Tris – HCl (pH 8. 0) and 1 mmol/L EDTA, subjected to 4 rounds of freeze/thaw and digested with Proteinase K/SDS for 4 hr at 56℃. Following extracton in phenol/chloroform and ethanol precipitation, DNA was resuspended in 10 mmol/L Tris – HCl (pH 8. 0) /1 mmol/L EDTA.

2. DNA sequencing

The DNA sequence corresponding to the amino terminal portion of LMP – 1 was determined by first amplifying 0. 3 μg of tumor DNA with the polymerase chain reaction (PCR) using the primers LMPEXC and LMP9747 (Tab. 1), followed by asymm etric amplification and dideoxy sequen-

Tab. 1 Primers used in PCR and Sequencing

Primer	EBV coordinates	Sequence
LMPEXC	168813 – 168832	5'CAACCAATAGAGTCCACCAG 3'
LMP9747	169747 – 169767	5'TTCTGTTGCACTTGGC 3'
LMP9550	169570 – 169550	5'GCCCTACATAAGCCTCTCAC 3'
LMP9233	169233 – 169250	5'TCCAGTGGACAGAGAAG 3'
LMPFUB	168736 – 168756	5'ACAACGACACAGTGATGAACACCACC 3'
LMPFUB2	168328 – 168348	5'GAAGAGGTTGAAAACAAAGGA 3'
LMPFUE	168163 – 168183	5'GTCATAGTAGCTTAGCTGAC 3'
LMP8808	168808 – 168830	5'GTGGACTCTATTGGTTGATCTC 3'
EBNA3C – 5'	99938 – 99960	5'AGAAGGGGAGCGTGTGGTTGTGT 3'
EBNA3C – 3'	100092 – 100071	5'GGCTCGTTTTTGACGTCGGC 3'
EBNA3C – type 1 – Pr		5'GAAGATTCATCGTCAGTG 3'
EBNA3C – type 2 Pr		5'CCGTGATTTCTACCGGGAGT 3'

cing using the primers LMP9550 and LMP9233, as previously described (Miller et al. , 1994). For sequencing of the carboxy terminus, 10 of the isolates (127, 133, 142, 145, 16, 26, GX1, 132, 614, 615 and 126) were subjected to PCR amplification using primers LMPFUE and LMP8808 (Tab. 1). PCR products were subcloned into the pGem3Z vector. One clone from each isolate was selected for dideoxy sequencing using the LMPFUB primer, with the exception of isolate 145, for which 3 positive clones were sequenced.

3. LMP – 1 deletion analysis

Genomic DNA (100 – 200 ng) from each isolate was subjected to PCR amplification for 30 cycles, at an annealing temperature of 55℃ and an extension time of 1 min, using the primers FUE and FUB2 (Tab. 1) which flank the region of the deletion. In isolates that retain or delete amino acids 343 – 352, the PCR yielded 186 bp and 156 bp products, respectively. Products were resolved on 10% acrylamide/TBE gels and transferred to Hybond N$^+$ membrane (Amersham, Arlingto Heights, IL) for 15 min using a gel dryer with no heat. Following denaturation in 0. 4mol/L NaOH for 10 min, membranes were baked for 2hr at 80℃, pre – hybridized for 15 min at 65℃ in Rapid – Hyb solution (Amersham), hybridized for 2 hr to a riboprobe containing a 1. 9 kb XhoI fragment from the LMP – 1 gene and visualized by autoradiography.

4. EBNA typing

Genomic DNA from NPC isolates was amplified by PCR using the primers EBNA3C – 5′ and EBNA3C – 3′ (Tab. 1) (Sample et al. , 1990). For each reaction, 100 – 200 ng of DNA was amplified for 35 cycles with an annealing temperature of 55℃ and an extension time of 1 min. Products were resolved on agarose gels, blotted to Hybond N$^+$ membrane, and hybridized at 40℃ as described above to each of 2 eng – labeled oligonucleotide probes, EBNA3C – type 1 – Pr and EBNA3C – type2 Pr (Tab. 1)   (Sample et al. , 1990). These probes distinguish EBV types 1 and 2 based on size heterogeneity of the PCR products, yielding a 246 bp product in type 1 EBV and a 153 bp product in type 2 EBV.

5. Phylogenetic and statistical analyse

Using 12 Chinese isolates, a consensus sequence for the N – terminal region of LMP – 1 was generated using the PRETTY program from the Wisconsin Package (Genetics Computer Group, Madison, WI). The 12 sa-

Tab. 2 Consensus definiton of EBV China strain based on Lmp – 1 N – Terminal region[1]

EBV coordinate	LMP-1 codon	Change	Frequency (%)
169467	3	CAC/His > CGC/Arg	19/26 (73. 1)
169437	13	CGA/Arg > CCA/Pro	17/26 (65. 4)
169425	17	CGA/Arg > CTA/Leu	27/27 (100)
169402	25	CTA/Leu > ATA/Ile	27/27 (100)
169379	32	C > G	26/27 (96. 3)
169361	38	T > C	26/27 (96. 3)
169352	41	C > T	20/27 (74)
169339	46	GAC/Asp > AAC/Asn	24/27 (88. 9)
169322	51	C > G	22/26 (84. 6)
169286	63	A > T	22/24 (91. 6)
169280	65	A > C	19/22 (86. 4)
169276	67	T > C	17/20 (85)
169274	67	G > C	17/20 (85)

Notes: [1] China l strain consensus was compiled fron 5 previously published isolates (Chen et al. , 1992; Hu et al. , 1991; Miller et al. , 1994) and from 22 isolates reported in Figure la, b. A nucleotide change, relative to the B95 – 8 strain, at a particular locus was defined as characteristic of of the China l strain if it appeared in at least 65% of the isolates. Not included in the consensus are 5 Chinese isolates that were identical to B95 – 8

mples included the previously published Cao, 1510, C11 and C13 isolates (Chen et al. , 1992; Hu et al. , 1991; Miller et al. , 1994) as well as 8 samples first reported here (GX1, 132, 126, 127, 133, 142, 145, 16). Loci that diverge from the B95 – 8 prototype strain are reported in Tab. 2.

To determine the phylogenetic relationships within various EBV isolates from China, the LMP – 1 sequence, including codons 1 – 70 and 192 – 386, was compiled from the 12 samples described above as well as from the previously pubished AL, pOT and B95 – 8 isolates (Miller et al. , 1994). Sequences were aligned using the Wisconsin Package Version 9. 0 sequence editor (Genetics Computer Group), and all constant positions were eliminated. Distance matrices were calculated with the PHYLIPI DNADIST program (Felsenstein, 1995), using both the Jukes and Cantor (1969) and Kimura (1983) two – parameter methods with a transition/transversion fixed at 2. Phylogentic trees were constucted from these distance matrices by both the neighbor – joining (Saitou and Nei, 1987) and UPGMA methods, using the NEIGHBOR and DRAWTREE programs from the PHYLIP package.

Classification and Regression Tree (CART) methodology (Wadsworth international Group, Belmont, CA) was used to construct regression trees to determine predictive amino acid positions for a particular strain, using sequence from the 15 isolates described above, A total of 34 amino acid positions were analyzed, which included only those loci at which an amino acid change had occurred in more than one isolate.

Results

1. Analysis of the LMP – 1 amino terminus: a consensus definition of the predominant Chinese strain, China l

NPC biopsy specimens were collected from the Guangxi Autonomous Region of southern China, where NPC is endemic, as well as from the cities of Beijing and Changchun, regions of northern China where NPC incidence is much lower, Strain prevalence was first determined by DNA sequencing of the amino terminal portion of the LMP – 1 gene. In this region, spaning from EBV genone coordinates 169474 to 169280, variations from the B95 – 8 prototype strain were noted at 28 loci, resulting in changes at 16 amino acid residues (Fig. 1).

Whereas 5 of 28 isolates (17. 8%), including 1 from the endemic region and 4 from the non – endemic region, were identical to the B95 – 8 prototype strain, 22 of 28 isolates examined (78. 5%) harbored a cluster of 13 nucleotide alterations resembling the predominat Chinese strain previously reported (Miller et al. , 1994). A consensus definition of this strain, designated Chinal, was then derived from these data from 4 previously reported Chinese isolates (Chen et al. , 1992; Hu et al. , 1991; Miller et al. , 1994). The DNA sequence variation defining this strain is reported in Tab. 2 and is also indicated by grey shading in Fig. 1c. This cluster of changes always includes the loss of an Xhol restriction enzyme site at nucleotide position 169425 and an amino acid substitution at codon 25. Other changes – including amino acid substitutions at codons 3, 13, 17 and 46 as well as silent changes within codons 32, 38, 41, 51, 63, 65 and 67 – are detected in a majority but not all of the isolates. Two of the isolates, 129 from the endemic region and 125 from the non – endemic region, may have undergone recombination within this portion of the LMP – 1 gene, as both are identical to the B95 – 8 genotype following codons 25 and 38, respectively.

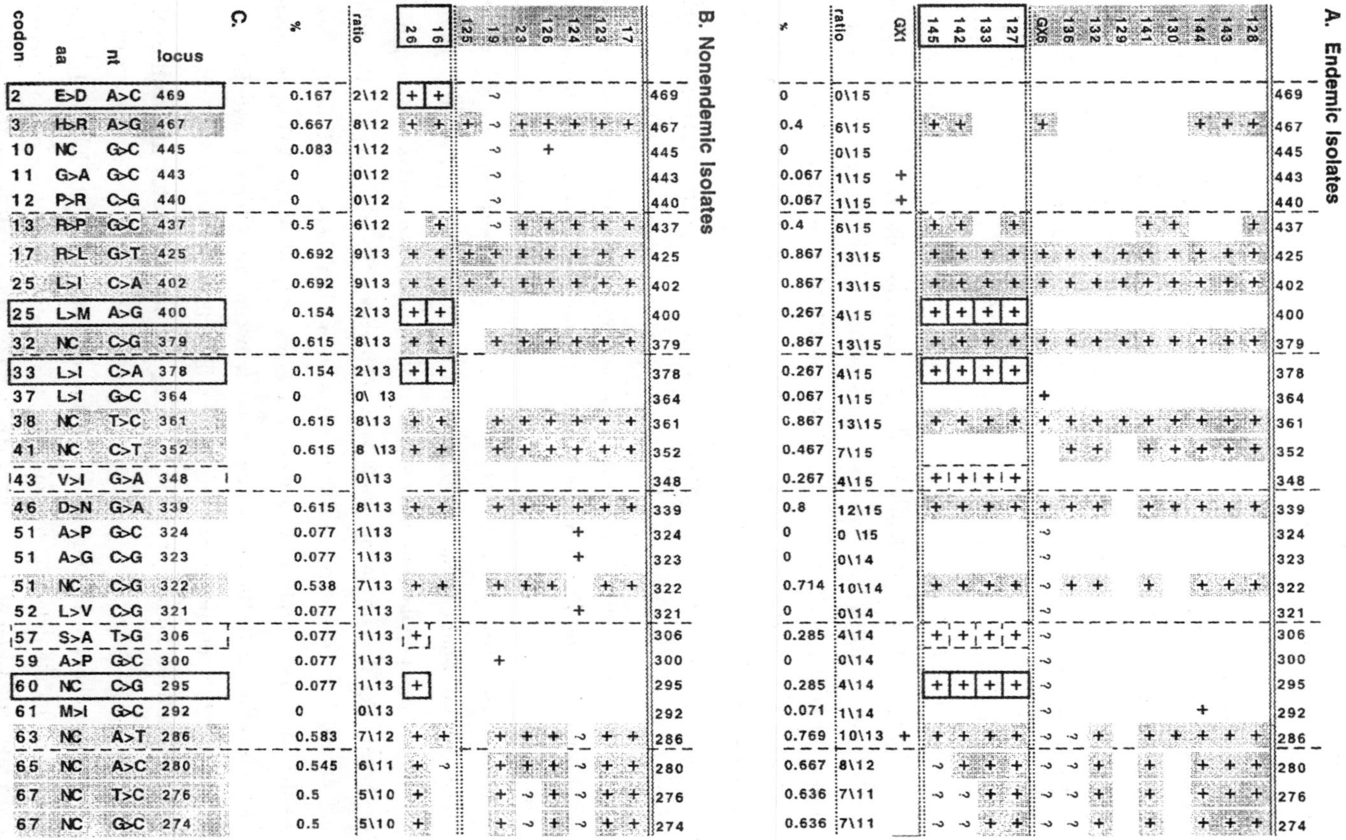

A. Endemic Isolates

B. Nonendemic Isolates

C.

Fig.1 – Sequence variation in the amino terminal portion of LMP – 1. Numbers actrss top tow correspond to EBV genome coordinates 169469 to 16274; numbers in left – hand colume refer to individual isolates. A plus sign indicates variaton from the B95 – 8 prototype strains at this locus; a question mark indicates sequence beyond this point is unknown. Cluster of base changes associated with the China1 strein is indicated by grey shading; base changes associated with the China2 strain are boxed; base changes associated with the Alaskan strain are marked with a dashed box. "Ratio" includes results from 1 endemic and 4 non – endemic isolates that were identical to the B95 – 8 prototype strain. (a) LMP – 1 sequence variation in isolates from the NPC endemic region of southern China. (b) Sequence variation in isolates from the NPC non – endemic region of northern China. (c) Consensus profile of nucleotide(nt) and amino acid(aa) changes associated with the China1, China2 and Alaskan strains

2. Identification of a distinct strain of EBV, China 2

DNA sequence analysis of the LMP – 1 amino terminus also revealed a distinct cluster of unique nucleotide changes, in addition to those found in the China 1 strain, in 6 of 28 isolates (21.4%). These included 4 from the endemic region and 2 from the non – endemic region (Fig. 1a, b, boxed). Of these 5 nucleotide changes, 3 have to been previously reported in NPC from any geographic location, including a change from CTA/Leu to ATG/Met at codon 25, CTC/Leu to ATC/Ile at amino acid 33 and a silent change (CTC→CTG) within codon 60. Interestingly, the 2 other changes, GTT/Val to ATT/Ile at codon 43 and a change from TCC/Ser to GCC/Ala at codon 57, have been previously observed only in EBV – associated tumor isolates from Alaska (dashed boxes in Fig. 1; Miller et al., 1994). All of the changes are conservative, with the exception of the substitution of the non – polar alanine residue within the transmembrane domain at position 57. This new strain is designated China 2.

Four of 6 NPC biopsy specimens harboring the China 2 strain originated in the endemic area of southern China, whereas the remaining 2 were from the non – endemic area in northeastern Chian. Surprisingly, the 2 China 2 isolates from northern China both lacked the Alaskan – associated change at amino acid 43. They were further distinguished by an additional change from GAA/Glu to GAC/Asp in codon 2. Taken together, isolates bearing either the China 1 or China 2 strains and having lost the Xho 1 restriction site account for 22 of 28 isolates (78.6%), and are distributed fairly evenly, being found in 13 of 15 isolates (86.7%) from the endemic area and in 9 of 13 isolates (69.2%) from the non – endemic area. One other sample from the endemic region (GX1) lacked the China 1 or China 2 changes but had 2 nucleotide changes, compared with B95 – 8, that were not detected in any of the other isolates. Both rexulted in non – conservative amino acid changes, from Gly to Ala at position 11; and Pro to Arg at position 12, and did not include loss of the Xho1 restriction site (Fig. 1a).

Although previous studies reported the presence of the Xhol polymorphism in 88 of 89 (98.9%) of Chinese NPCs (AbdelHamid et al., 1992; Chen et al., 1992; Hu et al., 1991; Miller et al., 1994), in the patient population analyzed here, the prevalence of this polymorphism is approx, 22% lower, weakening its correlation with NPC incidence, In this survey, the wild type or B95 – 8 strain of EBV is more prevalent in Asia than has been previously reported. Additionally whereas the Xhol polymorphism was foud in the majority of Asian NPCs, analysis of this polymorphism does not allow a distinction to be made between the China 1 and China 2 strains, and its use may obscure the genetic diversity existing within the LMP – 1 gene.

3. Prevalence of the deletion of amino acids 343 – 352: China 2 isolates are uniformly undeleted

In addition to the nucleotide changes present in the amino terminal portion of LMP – 1, Chinese NPCs are also characterized by the delection of amino acids 343 – 352 within the carboxy terminal cytoplasmic domain of LMP – 1 (Chen et al., 1992; Hu et al., 1991; Miller et al., 1994). This deletion was also detected in a significant portion of EBV – positive Hodgkin's tumors from Europe (Knecht et al., 1993), with the suggestion that it is associated with a more aggressively transforming phenotype (Hu et al., 1993; Knecht et al., 1993; Li et al., 1996). The Alaskan strain is undeleted, as is the B95 – 8 prototype strain (Miller et al. , 1994). To further define the relationshop between the China 2 and other strains, the prevalence of the deletion was determined in 42 NPC isolates from China.

The deletion was detected in 16 of 21 (76. 2%) NPC biopsy specimens from the endemic area, in 14 of 21 (66. 1%) from the non – endemic area (Tab. 3), but in none of the 6 China. 2 isolates analyzed. In isolate 145, PCR revealed the deleted form; however, analysis of individual clones revealed 2 distinct forms present in this 1 isolate (Fig. 2). Subsequent sequencing revealed that the undeleted China 2 form would not be readily amplified from genomic DNA due to base pair changes in the primer region and that the co – present deleted form was actually a distinct strain (discussed in the following section). The China 2 strain is therefore further distinguished from the China 1 strain by the uniform retention of amino acids 343 – 352. This deletion can also sort inde-

pendently from the Xho1 polymor-phism in the amino terminus of LMP – 1, found in both China 1 and China 2 isolates. Additional-ly, as evidened by sample 145, it is clear that both deleted and un-deltetd forms of the LMP – 1 gene can co – exist in the same tumor.

Tab. 3 Prevalence of the Deletion of LMP – 1 amino Acids 343 – 352 in Chinese NPC

	Endemic isolates	Non-endemic isolates	Total
aa343 – 352 deleted	16/21 (76. 2%)	14/21 (66. 1%)	30/42 (71. 4%)
aa343 – 352 undeleted	5/21 (23. 8%)	7/21 (33. 3%)	12/42 (28. 5%)

A

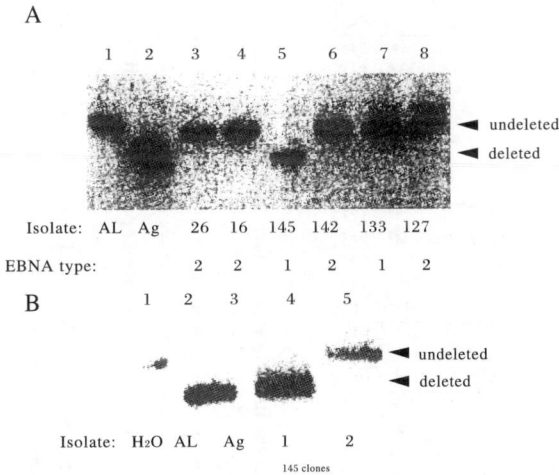

(A) Detection of deletion of LMP – 1 amino acids 343 – 352 by PCR of genomic DNA extracted from tumor biopsy material. EBNA type for each isolate is also shown. (B) Detection of deletion of LMP – 1 amino acids 343 – 352 in 2 separate clones of China 2 isolate 145.

Fig. 2 Deletion status of China 2 isolates. PCR products were resolved on polyacrylamide gels, blotted to nylon membrane and probed with an oligonucleotide specific for sequence flanking the deleted region. AL, Alaskan liver biopsy speci-men containing undeleted LMP – 1 sequence, and Ag, from he AG876 cell line carrying dele-ted LMP – 1 sequence, were included as con-trols. Other isolates are as shown

4. The China 2 strain can also be distinguished by nucleotide changes at the carboxy ter-minus of the LMP – 1 gene

To determine whether the China 2 isolates identified in Fig. 1 also have characteristic changes within the LMP – 1 carboxy terminus, the DNA sequence of 5 of these isolates was determined and analyzed from nucleotides 168755 to 168163. Distinct, consistent nucleotide changes were detected at 9 nucleotide loci, resulting in 7 alterations at the amino acid level (Fig. 3). Analysis of 2 addi-tional isolates from the non – endemic region (614 and 615), which were undeleted by the PCR as-

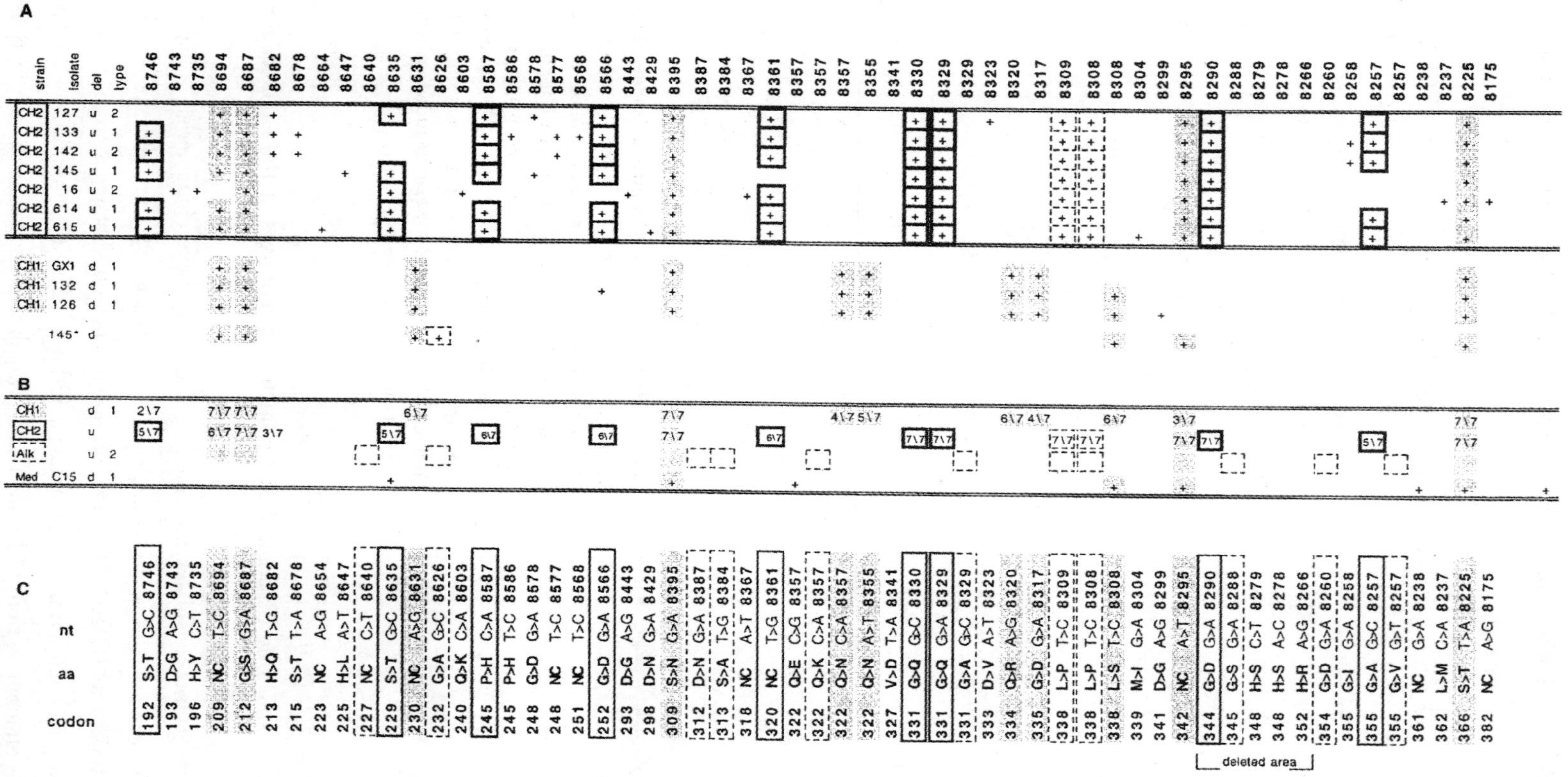

Fig. 3 – LMP –1 sequence variation in the carboxy terminus. Numbers across top row correspond to EBV genome coordinates 168746 to 168175. Variation from the B95 –8 prototype strain at the locus shown is indicated by a plus sign. Deletion of amino acids 343 –352 is indicated by "d "in "del" column; "u "denotes undeleted. EBNA typing results are shown in "type" column. Base changes associated with the China 1 strain are indicated by grey shading; variations associated with China2 strain are boxed; Alaskan changes are in dashed boxes. (a) Strain variation in the 7 China2 isolates, also 3 China1 isolates, and 1 co –infectiong strain from isolate145 (145*). (b) Consensus profile of nuclotide changes characteristic of the China1, China2 and Alaskan strina as well as a Mediterranean isolate, China1 consensus was compiled from 7 isolates, 4 of which were published elsewhere (Chen et al ., 1992; Hu et al ., 1991; Miller et al ., 1994). A change is noted if it appeared in at least 4 of the 7. Alaskan consensus includes changes noted in 2 isolates from Alaska (Miller et al ., 1994). (c) Nucleotide and amino acid coordinates of variations characteristic of China1, China2 and Alaskan strains indicated as described above

say, revealed the same pattern of changes and are therefore also classified as China 2. These changes, within codons 192, 229, 245, 252, 331, 344 and 355, were all found in at least 5 of the 7 China 2 isolates analyzed. Another change within codon 338 (TTG/Leu→CCG/Pro), fond in all 7 samples, is identical to that detected in Alaskan isolates. The China 1 isolates also are altered at this locus, but the substituted amino acid is serine rather than proline.

Of the 9 changes in China 2, 1 (within codon 320; CCT→CCG) was silent, and 3 (codon 192, AGT/Ser to ACT/Thr; codon 229, AGT/Ser to ACT/Thr; and codon 331 GGA/Gly to CAA/Gln) were conservative. The change at aa 192 was also detected in 2 previously reported sequences (Chen et al., 1992; Hu et al., 1991), and the change at aa 229 has been detected in only 1 other isolate, C15 from the Mediterranean region (Miller et al., 1994). Another conservative change at residue 213, found in only 3 of 7 isolates, lies within the TNF – receptor – associated factor (TRAF) – interacting domain of LMP – 1 but is outside the core TRAF binding motif, PX-QXT (Mosialos et al., 1995). Other non – conservative changes included CCT/Pro to CAT/His at codon 245, GGC/Gly to GAC/Asp at codon 252, GGC/Gly to GAC/Asp at codon 344 and GGT/ Gly to GCT/Ala at codon 355. Isolate 16, however, harbored changes at only 3 of he 7 codons.

For the purpose of comparison, a consensus profile of changs characteristic of the China 1 strain is shown in Fig. 3b. This consensus was derived from a total of 7 isolates; 4 previously reported (Chen et al., 1992; Hu et al., 1991; Miller et al., 1994) and 3 isolates (GX1, 126 and 132) sequenced in this study (Fig. 3a). Also included for comparison are sequence changes derived from 2 previously analyzed Alaskan isolates and 1 Mediterranean NPC xenograft, C15 (Miller et al., 1994). As was the case in the amino terminal portion of LMP – 1, China 2 and China 1 share common nucleotide variation from B95 – 8 at several loci. Analysis of the carboxy terminal region, however, reveals a clearer divergence between these strains, as there ate consistent changes in China 1 that are not found in China 2 (codons 322, 334, 335 and 338). China 2 shares the substitution of proline for leucine at codon 338 with the Alaskan isolates and the retention of codons 343 – 352, but none of the other changes in the carboxy terminus are in common between the China 2 and Alaskan strains (Fig. 3b). The nucleotide and corresponding amino acid changes associated with the China 2, China 1 and Alaskan strains are shown in Fig. 3c.

Isolate GX1 had 2 unique changes but none of the China 1 – associated changes in its amino terminus, yet had 8 of the 9 China 1 – associated nucleotide changes in the carboxy terminus (Fig. 3a), suggesting that interstrain recombination had occurred within the transmembrane domain of LMP – 1. Two other isolates, 125 and 129, may also be chimeras of the predominant China 1 strain and the prototype B95 – 8 strain, as they both appear to flip from one strain to the other following codons 25 and 38, respectively.

The sequence of hte co – infecting deleted strain in isolate 145 was also determined and is shown in Figure 3a (145*). It did not contain any of the China 2 – associated changes but did contain 1 Alaskan – associated change (in amino acid residue 232), indicating that it is a unique strain distinct from the China 2 found in the same tumor biopsy specimen. It also provides evidence of recombination within LMP – 1, as it resembles China 1 from codons 192 to 230 and from codons 338 to 382, but in the intervening sequence is identical to B958. These results suggest that recombination within the LMP – 1 gene may occur, but does so infrequently in NPC and, somewhat unexpectedly, does not necessarily occur across the internal repeat region (amino acids 250 – 298) within the carboxy terminus.

5. Phylogenetic analysis of viral nucleotide sequence differences

To determine the phylogenetic relationships among various EBV isolates from China, the LMP – 1 DNA sequences from both the amino and carboxy terminal regions of 15 isolates were aligned and constant positions removed. Using he one – parameter model (Jukes and Cantor, 1969), in which the rate of substitution between all nucleotides is assumed to be equal, a distance matrix was constructed. A phylogenetic tree was then drawn using the neighborjoinjing method (Saitou and Nei, 1987). Shown in Fig. 4, the China 2 isolates form a clear cluster and were more closely related to the Alaskan isolates than to the other China 1 isolates, which was not unexpected because both groups retain amino acids 343 – 352 and share several other nucleotide changes. The China 1 isolates also form a cluster; however, it is not as well defined as that formed by the China 2 isolates. Using the Kimura (1983) two – parameter method, which assumes that transition – type changes are more likely to occur than transversion – gype changes, a tree with an identical branching pattern resulted (data not shown). A very similar result was obtained when the UPGMA method was used to construct the trees and when using sequence data from either the amino or carboxy terminal portion of LMP – 1 (data not shown). Therefore, the China 2 cluster is robust to different methods of both calculating distances and constructing trees.

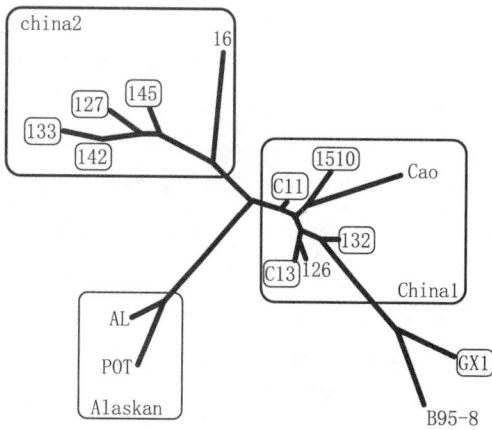

Fig. 4 Phylogenetic analysis of China 1, China 2 and Alaskan strains. Isolates from the endemic region of China are circled

None of the phylogenetic reconstructions indicated strain sorting based on geographic location within China, although the 2 Alaskan isolates included in the analysis were quite distinct. Whereas most Chinese isolates are clearly different from the B95 – 8 prototype, the EBV strains isolated from the NPC endemic area of southern China are not markedly different from those found in the non – endemic area of northern China.

6. Classification of strains based on amino acid signature patterns

Because the China 1, China 2 and Alaskan strains share much of their divergence from B95 – 8, particularly in the amino terminus, the CART methodology was applied to determine the amino acid positions that could categorize them. In 15 isolates for which sequence from both the amino and carboxy terminal portions of LMP – 1 had been determined, 34 amino acid loci at which changes were found in more than 1 isolate were analyzed. Each of the 4 loci in Tab. 4 showed perfect discrimination between the China 2 strain and the others. These include 2 positions in the amino terminus and 2 in the carboxy terminus. Also, a combination of positions 245 and 229 rerfectly classify the China 2 strain. At position 245, 4 of 5 China 2 isolates had substituted histidine for proline, and the other China 2 isolate remained unchanged at this position. The substitution of threonine for serine at position 229 then classifies it as China 2. Positions capable of perfectly classifying the China 1 strain included the substitution of isoleucine at position 25, serine at position 338, and also the deletion of amino acids 343 – 352. The Alaskan strain could be discriminated perfectly at 9 different loci (Tab. 4), mostly clustered in the carboxy terminus. Further definition of the Alaskan strainu awaits sequence information from more than the 2 isolates included here.

Tab. 4 Aminoacid positions within LMP – 1 that discriminate among the China 2 and alaskan strains

Strain	Locus	Change
China 1	25	Leu to Ile
	338	Leu to Ser
	343 – 352	deleted
China 2	25	Leu to Met
	33	Leu to Ile
	331	Gly to Gln
	344	Gly to Asp
Alaskan	22	Ser to Pro
	63	Ile to Val
	232	Gly to Ala
	312	Asp to Asn
	313	Ser to Ala
	331	Gly to Ala
	345	Gly to Ser
	354	Gly to Ser
	355	Gly to Val

7. LMP – 1 sequence variation in the China 2 strain sorts independently of EBV EBNA type

All EBV isolates can be classified as type 1 or type 2 (or types A and B) based on sequence divergence in the EBNA – 2, 3A, 3B and 3C genes (Adldinger et al., 1985; Sample et al., 1990). China 1 isolates uniformly contained type 1 EBV, and both Alaskan isolates were type 2 (Abdel – Hamid et al., 1992); however, the deletion within the LMP – 1 gene has been shown to sort independently of EBNA type (Miller et al., 1994). Of the 8 China 2 isolates analyzed, 4 carried type 1 and 4 carried type 2 (Fig. 3a). Furthermore, the EBNA types were equally distributed among the isolates from the endemic and non – endemic regions despite the closer geographic proximity of the non – endemic region (northeastern China) to Alaska, where only type 2 was previously described. This finding further substantiates the lack of correlation between EBNA type and LMP – 1 genotype originally suggested by Miller et al. (1994).

Dicussion

This study represents the first comparison of EBV strains in isolates from the NPC – endemic region of southern China with those from the NPC non – endemic region of northern China, and reports a new strain, China 2, based on DNA sequence variation in the LMP – 1 gene.

Based on DNA sequence analysis of the amino – terminal 68 codons of LMP – 1, the isolates could be divided into 3 strains, summarized in Tab. 5. Five of the 28 (17.8%) were indistinguishable from the B95 – 8 prototype strain. Sixteen of the 28 (57.1%) had the predominant China 1 strain, characterized by a cluster of 13 nucleotide changes in the amino terminal portion of LMP – 1, including loss of an Xho1 restriction enzyme site. Six of the 28 (21.4%), although sharing some common nucleotide changes with China 1 including the XhoI polymorphism, carry a previously unrecognized viral strain here designated China 2, The divergence between China 1 and China 2 was more apparent in the carboxy terminal portion of LMP – 1, most notably in that amino acid residues 343 – 352 were retained in all the China 2 isolates.

Characterized by 3 distinct amino acid changes in the amino terminus and 7 in the carboxy terminal portion of LMP – 1, China 2 was found in 26.6% of the primary tumor isolates from the endemic area of China and in 15.4% from the non – endemic region. China 2 appears to be closely related to the " Minor strain 2", recently reported in 5 of 74 (6.8%) primary NPCs from Hong Kong, also carrying changes within LMP codons 331, 338, 344 and 355 (Cheung et al., 1996) A similar cluster of changes has also been reported in other EBV – associated diseases in Asia, including 1 infectious mononucleosis (IM) and 1 chronic EBV case in Japan (Itakura et al., 1996), as well as 4 cases of chronic tonsillitis, 1 lung carcinoma and 1 sino – nasal carcinoma from Hong Kong (Leung et al., 1997). It is likely that these isolates are in fact China 2; however, it is not known if they also carry the distinct China 2 residues in the amino terminus of LMP – 1. Because previous determinations of EBV strain prevalence in NPC were based primarily on the Xhol polymorphism that does not distinguish among the China 1, China 2 and Alaskan strains, it is likely that LMP – 1 sequence is more divergent within Asian NPC and other EBV – associated diseases than has been reported previously. There was no marked difference in the distribution of EBV strains in the endemic and non – endemic regions (Tab. 5), with the exception of the B95 – 8 – like isolates, which were more prevalent in northern China, Several previous studies have suggested that the strains detected in EBV-associated malignancies may reflect

Tab. 5 Distribution of EBV strains Based on Variation in LMP – 1 N – Terminus

Strain	Endemic region	Non – endemic region	Total
China 1	9/15 (60.0%)	7/13 (53.8%)	16/28 (57.1%)
China 2	4/15 (26.6%)	2/13 (15.4%)	6/28 (21.4%)
B95 – 8	1/15 (6.7%)	4/13 (30.8%)	5/28 (17.9%)
Other	1/15 (6.7%)	0/13	1/28 (3.6%)

the prevalence of strains found in the general population (Cheng et al., 1996; Khanim et al., 1996). The detection of China 1, China 2 and a surprising number of B95-8-like isolates in primary NPC from the non-endemic region suggests that all strains of EBV can contribute to the etiology of NPC.

Functional analysis of the LMP – 1 protein has focused on the carboxy terminal region. Notably, amino acids 204 – 208 contain a PXQXT motif, through which LMP – 1 binds to the TNF receptorrassociated factors (TRAF) (Mosialos et al., 1995). It is through activation of this TRAF signaling pathway that LMP – 1 induces expression of the EGFR (Miller et al., 1997). This TRAF – binding motif, as well as a similar PXQXS motif at amino acids 379 – 383, are conserved in all of the isolates examined here (Fig. 3) and are also conserved within the LMP – 1 homologue in simian EBV (Franken et al., 1996), underscoring the importance of TRAF signaling for LMP – 1 funciton. On the other hand, an additional perfect PXQXT motif at amino acids 320 – 324 is disrupted in the China 1 and Alaskan strains but is conserved in China 2.

It is likely that some of the changes in cellualr gene expression induced by LMP – 1 are a result of the activation of NF – kB through at least 2 independent activating regions on the LMP – 1 C – terminus (Huen et al. , 1995). The proximal region, extending from amino acids 187 to 231, is fairly conserved among the isolates examined here. The only non – conservative change found was CAT/His to CAG/Glu at codon 213 in 3 of the China 2 isolates. In the more distal NF – kB activating region, extending from amino acids 352 to 386, a conservative change from TCT/Ser to ACT/Thr at codon 366 was found in all of the isolates, as well as in the Alaskan and Mediterranean isolates previously reported (Miller et al. , 1994). A non – conservative change, from GGT/Gly to GCT/Ala at codon 355, was found in 5 of 7 China 2 isolates.

That EBV genomes can undergo interstrain recombination during productive replication is well established (Walling and Raab – Traub, 1994), and some areas of the genome that contain repetitive sequence ate highly susceptible to homologous recombination. In addition, inter – typic recombinants of EBV have been isolated (Burrows et al. , 1996). Within the carboxy terminus of LMP – 1 is an 11 amino – acid repeat region, extending from codons 250 to 308, which is present in varying numbers in different EBV isolates and is thought to be a site at which homologous recombination might occur between different strains (Miller et al. , 1994). Four of the NPC isolates reported here (isolates 125, 129, 145* and GX1) appear to have undergone interstrain recombination within the LMP – 1 gene, as does the previously reported C11 isolate (Miller et al. , 1994). C11 carries all of the China 1 – associated changes in the amino terminus of LMP and up to codon 309 of the carboxy terminus, is identical to the B95 – 8 strain between codons 309 and 338, then flips back to China 1 from codons 338 to 386, as does isoate 145*. Interestingly, in only one of these isolates (C11) has the apparent recombination occurred at the repeat region within the carboxy terminal portion of LMP – 1.

It has been proposed that the region around and including amino acids 343 – 352 may constitute a deletional hot spot and that the 30 bp deletion occurs as a result of mispairing of the short repeats flanking the deleted region during replicaiton (Sandvej et al. , 1994). This hypothesis would also explain why the 7 China 2 isolates examined remain undeleted, as they carry a change from G to A at nucleotide position 168290, and 5 of the 7 carry an additional change from G to C at position 168257. Both of these changes disrupt the 9 – bp repeat unit and would presumably prevent any mispairing from occurring (Fig. 5). In the deleted China 1 isolates, not only are these point mutations not present, but also in 6 of the 7 isolates analyzed, a change from A to T at position 168295 extends the perfect repeat unit from 9 to 10 base pairs (Fig. 5). This hypothesis fails to explain, however, why the B95 – 8 strain has remained undeleted despite having 2 perfect repeat units. The identification of homologous elements flanking the site of deletion suggests that a deletion variant coule theoretically be produced from an undeleted form. In our study, a single sample (isolate 145) contained both deleted and undeleted form, and the sequence analysis revealed the presence of two distinct strains. This suggests that a rare NPC may harbor 2 strains of EBV. Although the clonality of this tumor was not determined, the detection of 2 strains may indicate that this is either a biclonal tumor or that the progenitor cell was simultaneously infected by 2 different strains. Alternatively, one

EBV strain may be present in the tumor cells and an additional strain in the infiltrating lymphocytes.

```
B95-8      aGGCGGCGGT   catagtcatgattccggcca  tGGCGGCGGT   undeleted
China 1    TGGCGGCGGT  (catagtcatgattccggcca  TGGCGGCGGT)  deleted
China 2    TGGCGACGGT   catagtcatgattccggcca  TGGCGGCGCT   undeleted
       168295                                      168256
```

Fig. 5 DNA sequence flanking the 30 bp deletion in the C-terminus of *LMP* – 1 DNA sequence from EBV genome coordinates 168 295 to 168 295 is shown. Perfect repeat units (Sandvej *et al.* , 1994) are in uppercase; deleted portion of China 1 strain is in parentheses. Point mutations that either enhance or disrupt the repeat structure are underlined and bold

In conclusion, we report a new EBV strain, China 2, which is related to both China 1, the predominant strain in Asia, and the Alaskan strain; we furthermore identified specific amino acid positions that distinguish between them. The distribution of the China 1 and China 2 strains is not markedly different in the NPC non – endemic region of northern China compared with the NPC endemic region in southern China, with the exception that the B95 – 8 prototype strain that was found more often in NPC from the non – endemic region. This finding suggests that other risk factors for NPC, such as exposure to environmental carcinogens and genetic susceptibility, may be more likely than EBV strain variation to account for the high incidence of NPC in southern China.

Acknowledgements

We gratefully acknowledge the help of Dr. Cai Wei Ming of the Beijing Cancer Hopital, Dr. Du Bo of the Bai Qiu En Medical University Hospital in Changchun, Jilin Province, and the Guangxi Regional Hospital in Nanning, Guangxi Autonomous Region, People's Republic of China, for supplying patient tissues. We also thank Ms. Xiaomei Zhang, Ms. Ling Zhou and Mr. Zhiwu Ji of the Instiute of Virology in Beijing for helpful advice and discussions; Mr. B. Lesser for help running the Wisconsin Package Version 9. 0; Ms. M. DeLuca for assistance with the PHYLIP package; Mr. R. Budrevich and Mr. M. Jensen assistance with the CART package; and Dr. W. Miller for critical review of the manuscript. N. S. was supported by a postdoctoral fellowship from the WHO International Agency for Research on Cancer and by a National Research Service Award (5F32 – CA68751 – 02) from the National Cancer Institute, United States Public Health Service. These studies were also supported by NIH grants CA32979 and CA67384 to N. R. T.

〔In 《Int J Cancer》 1998, 76: 207 – 215〕

References

1 Abdel – Hamid M, Chen J J, CONSTANTNE N, MASSOUD M, RAAB – TRAUB N. EBV strain variation: geographical distribution and relation to disease state. Virology, 1992, 190: 168 – 175

2 Adldinger H K, Delius H, Freese U K, Clarke J, Bornkamm G W. A putative transforming gene of Jijoye virus differs from that of Epstein – Barr virus prototypes. Virology, 1985, 141: 221 – 234

3 Armstrong R W, Armstrong M J, Yu M C, Henderson B E. Salted fish and inhalants as risk factors for nasopharyngeal carcinoma in Malaysian Chinese. Cancer Res, 1983, 43, 2967 – 2970

4 Burrows J M, Khanna R, Sculley T B, Alpers M P, Moss D J, BURROWS S R. Ientification of a naturally occurring recomvbinant EpsteinBarr virus isolate from New Guinea that encodes both type 1 and type 2 nuclear antigen sequences. J. Virol, 1996, 70: 4829 – 4833

5 Chen M L, Tsal C N, Liang C L, Shu C H,

Huang C R, Sulitzeanu D, Liu S T, Chang Y S. Cloning and characterization of the latent membrane protein (LMP) of a specific Epstein – Barr virus variant derived fron the nasopharyngeal carcinoma in the Taiwanses population. Oncogene, 1992, 7: 2131 – 2410

6   Cheung S T, Lo K W, Leung S F, Chan W Y, Choi P H, Johnson P J, Lee J C, Huang D P. Prevalence of LMP1 deletion variant of Epstein – Barr virus in nasopharyngeal carcinoma and gastric tumors in Hong Kong [letter]. Int J Cancer, 1996, 66: 711 – 712

7   De The G. Epidemionlogy fo Epstein – Barr virus and associated diseases in man. In: B. Roizman (ed.), Herpesvirus, 25 – 87, Plenum Press, New York, 1982, 25 – 87

8   Felsenstein J. PHYLIP (Phylogeny Inference Package), University of Washington, Seattle, 1995

9   Franken M, Devergne O, Rosenzweig M, Annis B, Kieff E, Wang F. Comparative analysis identifies conserved tumor necrosis factor receptor – associated factor 3 binging sites in the human and simian Epstein – Barr virus oncogene LMP1, J Virol, 1996, 70: 7819 – 7826

10   Hammarskjold M L, Simurda M C. Epstein – Barr virus latent membrane protein transactivates the human immunodeficiency virus type 1 long terminal repeat through induction of NFkB activity. J Virol, 1992, 66: 6496 – 6501

11   Hu LF, Chen F, Zheng X, Ernberg I, Cao S L, Christensson B, Klein G, Winberg G. Clonability and tumorigenicity of human epithelial cells expressing the EBV encoded membranc protein LMP1. Oncogene, 1993, 8: 1575 – 1583

12   Hu L F, Zabarovsky E R, Chen F, Cao S L, Ernberg I, Klein G, Winberg g. Isolation and sequencing of the Epstein – Barr virus BNLF – 1 gene (LMP1) from a Chnese nasopharnageal carcinoma. J gen Virol, 1991, 72: 2399 – 2409

13   Huen D S, Henderson S, Croon – Carter S, Rowe M. The Epstein – Barr virus latent protein 1 (LMP – 1) mediates activation of NF – kB and cell – surface phenotype via two effector regions in its carboxy terminal cytoplasmic domain. Oncogene, 1995, 10: 549 – 560

14   Itakura O, Yamada S, Narita M, Kikuta H. High prevalence of a 30-base pair deletion and single-base mutations within the carboxy terminal end of the LMP-1 oncogene of Epstein-Barr virus in the Japanese population. Oncogene, 1996, 13: 1549 – 1553

15   Jukes T H, Cantor C R. In: H. N. Munro (ed.), Mammalian protein metabolism Ⅲ, Academic Press, New York, 1969, 21

16   Khanim F, Yao Q Y, Niedobitek G, Sihota S, Rickinson A B, Young L S. Analysis of Epstein-Barr virus gene polymorphisms in normal donors and in virus – associated tumors from different geographic locatlions. Bolld, 1996, 88: 3491 – 3501

17   Kieff E. Epstein – Barr virus and its replication. In: B. Fields, D. Knipe and P. Howley (eds.), Fields virology. Lippincott-Raven, Philadelphia, 1996, 2: 2343 – 2396

18   Kinura M. The neutral theory of molecular evolution, Cambridge University Press, New York, 1983

19   Knecht H, Bachmann E, Brousset P, Sandvej K, Nadal D, Bachmann F, Odermatt B F, Delsol G, Pallesen G. Deletions within the LMP1 oncogene of Epstein – Barr virus are clustered in Hodgkin's disease and identical to those observed in nasopharyngeal carcinoma. Blood, 1993, 82: 2937 – 42

20   Leung S Y, Yuen S T, Chung L P, Chan A S, Wong M P. Prevalence of mutations and 30 – bp deletion in the C-terminal region of Epstein-Barr virus latent membrane protein – 1 oncogene in reactive lymphoid tissue and non-nasopharyngeal EBV-associated carcinomas in Hong Kong Chinese. Int J Cancer, 1997, 72: 225 – 230

21   Li S N, Chang Y S, Liu S T. Effect of a 10 – amino acid deletion on the oncogenic activity of latent membrane protein 1 of Epstein – Barr

virus. Oncogene, 1996, 12: 2129 – 2135

22　Miller W E, Earr H S, Raab – Traub N. The Epstein – Barr virus latent membrane protein 1 induces expression of the epidermal growth factor receptor. J Viroil, 1995, 69: 4390 – 4398

23　Miller W E, Eowards R H, Walling D M, Raab – Traub N. Sequence variation in the Epstein – Barr virus latent membrane protein 1 [erratum in J gen Virol. , 1995 May; 76: 1305]. J gen Virol, 1994, 75: 2729 – 2740

24　Miller W E, Mosialos G, Kieff E, Raab – Traub N. Epstein – Barr virus LMP1 induction of the epidermal growth factor receptor is mediated through a TRAF signaling pathway distinct from NF – kappa B activation. J Virol, 1997, 71: 586 – 594

25　Mosialos G, Birkenbach M, Yalamanchili R, Van Arsdale T, Ware C, Kieff E. The Eptein – Barr virus transforming protein LMP1 engages signaling proteins for the tumor necrosis factor receptor family. Cell, 1995, 80: 389 – 399

26　Raab – Traub N. Pathogenesis of Epstein – Barr virus and its associated malignancies, Semin Virol, 1996, 7, 315 – 323

27　Saitou N, Nei M. The neighbor – joining method: a new method for constructing phylogenetic trees, Mol biol Evol, 1987, 4: 406 – 425

28　Sample J, Young L, Martin B, Chatman T, Kieff E, Rickinson A, Kieff E. Epstein – Barr virus types 1 and 2 differ in their EBVA – 3A, EBNA – 3B, and EBNA – 3C genes. J Virol, 1990, 64: 4084 – 4092

29　Sandvej K, Peh S C, Andresen B S, Pallesen G. Identification of potential hot spots in the carboxy – terminal part of the Epstein – Barr virus (EBV) BNLF – 1 gene in both malignant and benign EBV – associated diseases: high frequency of a 30 – bp deletion in Malaysian and Danish peripheral T – cell lymphomas, Blood, 1994, 84: 4053 – 4060

30　Walling D M, Raab – Traub N. Epstein – Barr virus intrastrain recombination in oral hairy leukoplakia. J Virol, 1994, 68: 7909 – 7917

31　Young L S, Dawson C W, Clark D, Rupani H, Busson P, Tursz T, Johnson A, Rickinseon A B. Epstein – Barr virus gene expression in nasopharyngeal carcinoma. J gen Virol, 1988, 69: 1051 – 1065

32　Zimber U, Adldinger H K, Lenoir G M, Vuillauem M, Knebel – Doeberitz M B, Laux G, Desgranges C, Wrrtman P, Freese U K, Schneider U, Bornkamm G. Geographical prevalence of two types of Epstein – Barr virus. Virology, 1986, 154: 56 – 66

67.   Amino-acid Change in the Epstein-Barr-Virus Zebra Protein in Undifferentiated Nasopharyngeal Carcinomas from Europe and North Africa

GRUNEWALD Virginie[1] , BONNET Mathilde[1] , BOUTIN Sylvie[1] , YIP Timothy[2] ,

LOUZIR Hechmi[3] , LEVRERO Massimo[4] , SEIGNEURIN Jean Marie[5] ,

RAPHAEL Martion[6] , TOUITOU Robert[1] , MARTEL-PENOIR Dominique[1] , COCHET Chantal[1] ,

DURANDY Anne[7] , ANDRE Patrice[8] , LAU W[2] , ZENG Yi[9] , JOAB Irene[1]

1. CNRS URA 1301, Villejuif, France; 2. Queen Elizabeth Hospital, Kowloon, Hong – Kong;

3. Institute Pasteur, Tunis, Tunisia; 4. Universita di Roma" La Sapienza", Rome, Italy;

5. CHRU de Grenoble, France; 6. Hopital Avicenne, Bobigny, France; 7. INSERM U429,

Hopital des Enfants Malades, Paris, France; 8. CHR Pontchaillou, Rennes, France;

9. Institute of Virology, 100 Ying Xin Jie, Beijing, China

Summary

Different Epstein – Barr – virus (EBV) variants were found to be associated with nasopharyngeal carcinoma (NPC). The type – C variant lacks the BamHI site between the BamHI WI and I regions and the type – f variant has an extra BamHI site in the BamHI F fragment. The BNLF1 gene (which encodes the LMP1 protein) from a nude – mouse – passaged CAO strain and from NPC biopsies from Taiwanese patients also exhibits variations resulting in structural and functiona differences in the protein. The BZLF1 gene encodes the ZEBRA protein which triggers the EBV lytic cycle. A difference has been observed in 8 amino acids in the ZEBRA sequence in B95 – 8 (Z95) and P3HRI (ZP3) cell lines. EBV found in NPC biopsies and peripheral – blood cells from Asians was predominantly of the ZP3 type (75%), while 81% of samples from different EBV – associated diseases and peripheral – blood cells from North Africa of Europe were of the Z95 type. We found that an alanine 206 had been replaced by a serine in the Z95 sequence in 72% of the NPC biopsies from European and North African patients. The Zser 206 variant is found in a significantly lower percentage ($P < 0.001$) of other EBV – positive tissues from individuals in the same region (10% – 33%). In contrast, a 30 – bp deletion is observed near the 3' end of the LMP1 gene in the majority of EBV (86%) from NPC and peripheral – blood cells from Asians, whereas a significantly lower percentage ($P < 0.001$) of NPC biopsies from European and North African patients (56%) have this deletion, as do lymphocytes from control individuals from the same region (36% and 55% respectively).

Introduction

Epstein – Barr virus (EBV) causes infectious mononucleosis (IM) and is highly associated with various tumors such as endemic Burkitt's lymphoma, lymphomas in immunocompromised patients, 30% −40% of Hodgkin's disease, rare T − cell lymphoma, and most nasopharyngeal carcinomas (NPC) (Rickinson and Kieff, 1996). Natural isolates of the Epstein − Barr virus can be classified into 2 major strains, type A and type B, based on differences in coding − region sequences in EBNA2, EBNA3A, B and C antigens and in EBER small non − polyadenylated RNA sequences (Arrand et al., 1989; Rowe et al., 1989). In addition, variations have also been reported in the EBV EBNA1 sequence in NPC biopsies and xenografts (Snudden et al., 1995; Wrightham et al., 1995). Other variants have been found to be associated with NPC: the type − C variant, which lacks the BamHI site between the BamHI W1 and I regions, and the type − f variant, which has an extra BamHI site in the BamHI F fragment (Lung et al., 1991). The BNLF1 gene, from a passaged nude − mouse CAO strain (Hu et al., 1991) and from NPC biopsies from Taiwanese patients (Chen et al., 1992) also showed variations leading to structural and functional differences in the LMP1 protein (Hu et al., 1993). Such variations have also been observed in other EBV − related tumors (Knecht et al., 1993) and in different cell lines (Miller et al., 1994). A key gene in the virus has undergone changes which may be altering its functions.

The BZLF1 gene encodes the ZEBRA (Z Epstein − Barr − virus Replication Activator) protein, which triggers EBV into the lytic cycle. The basic leucine Zipper (bZip) domain in the ZEBRA protein may be interfering with cellular functions via proteinprotein interactions with p53 (Zhang et al., 1994), the p65 sub − unit of NF − kB (Gutsch et al., 1994), retinoc acid receptors (Sista et al., 1993) or with the G_0/G_1 arrest in the ell cycle (Cayrol and Flemington, 1996). The BZLF1 sequence has been described in Taiwanese NPC (Chang et al., 1992) and in different cell lines (Packham et al., 1993). Differences have been observed in B95 − 8 (Z95) and P3HR1 or Akata (ZP3) cell lines which are independent of the types A and B virus (Packham et al., 1993). The P3HR1 and Akata BZLF1 sequences are in fact quite similar, although they harbour the B and A viruses respectively. The BZLF1 gene in P3HR1 exhibits differences from that found in B95 − 8 giving rise to changes in 8 amino acids. Two or three copies of a 29 − bp tandem repeat sequence are found in the first intron in B95 − 8 and in P3HR1 respectively.

This study has analyzed the BZLF1 sequence in control peripheral − blood − cell (PBC) pellets from healthy blood donors (HBD) (from France, Italy, North Africa and Asia) and in different tissue specimens of EBV − associated diseases. These include NPC biopsies (from Europe, North Africa, Asia and Oceania), tissue samples from immunocompromised patients, oral hairy leucoplakia (OHL) and lymphocytes from patients with infectious mononucleosis (IM). We report an amino − acid change from an alanine 206 to a serine in the dimerization dimain of the ZEBRA protein. This difference was far more frequently observed in NPC tissues from European and North African patients than in tissues from other EBV − related diseases or in healthy individuals from the same regions.

Material and Methods

1. Tissue samples

NPC biopsies from Europe (18 from Italy, 6 from France), 7 from North Africa, 38 from

Asia and 2 from Oceania were collected in the Institut Gustave Roussy, France, in the Institute of Virology, Beijing and in the Queen Elizabeth Hospital, Hong – Kong. Tissues from immunocompromised patients [biopsy samples and posttransplant – lymphomas (PTL) – derived cell lines] were obtained at the Hopital des Enfants Malades, Hopital Pitie Salpetriere (Paris) and Hopital Bicetre (Kremlin Bicetre). Non – Hodgkin's lymphoma biopsy specimens from HIV – positive patients were collected by the French study group for HIV tumors. Tissues from OHL were obtained from the Ludwig Institute for Cancer Research (London): Cell lines and PBC pellets derived from IM were collected in the Centre Hospitalier Regional et Universitaire de Rennes, the Faculte de Medecine de Grenoble and the Laboratoire do 1'Epidor (Virrljuif) and from HBD (at the Queen Elizabeth Hospital, Hong – Kong, the Institut Pasteur in Tunis, the blood bank, Hopital Saint Louis, Paris, and the Universita di Roma, Italy).

2. Preparation of blood lymphocytes

Blood lymphocytes were separated by centrifugation through a Lymphoprep solution (Nycomed Pharma, Oslo, Norway) (5.7g/ml Ficoll 400, 9g/ml diatrioate sodium, density 1.077g/ml) and lymphocyte pellets were stored at $-20\,^\circ\!C$.

3. Extraction of DNA

Pulverized frozen biopsy specimens, thick paraffin sections of pellets from 10^7 B lymphocytes were treated for 3 hr at $55\,^\circ\!C$ with 10 vol of a solution containing 200 μg/ml of proteinase K (Appligene, Illkirch, France) dissolved in 50 mmol/L Tris – Cl, pH 8.5, EDTA 1 mmol/L. Tween – 20 0.5% (Sigma, Saint Quentin Fallavier, France). The enzyme was then heat – inactivated at $95\,^\circ\!C$ for 10 min. Samples were then stored at $-20\,^\circ\!C$.

4. Amplification of DNA

The DNA samples (2 μl) were subjected to PCR, using Taq DNA polymerase (Promega Charbonnieres, France) and specific oligonucleotides, in a 480 DNA Thermal cycle (Perking Elmer, Roissy, France). The standardized cycle procedure for all reactions was denaturation for 5 min at $95\,^\circ\!C$ followed by the addition of 2.5U of the enzyme, denaturation at $95\,^\circ\!C$ for 30 sec, annealing at $58\,^\circ\!C$ for 1 min, and a 2 – min extension at $72\,^\circ\!C$ for 30 cycles. Cycling was followed by a 7 – min extension at $72\,^\circ\!C$. When necessary, 2 μl of the PCR product were taken for a second round of PCR, using internal primers. The PCR products were analyzed by electrophoresis through a 1% agarose gel.

5. Sequening analysis of PCR products

PCR products were purified using the Qiaex gel extraction kit (Qiagen, Courtaboeuf, France), in conditions specified by the manufacturer. PCR amplified fragments were sequenced with the Prism Ready Reaction Dyedeoxy Terminator Cycle Sequencing Kit (Applied Biosystems, Roissy, France) on a Model 373 A automatic sequencer.

6. Restriction – fragment – length polymorphism analysis (RFLP)

The PCR products were digested by restriction enzyme Bsm1 (GIBCO – BRL, Cergy – Pontoise, France) at $65\,^\circ\!C$ for 2 hr and the length of the fragments sized on a 2% agarose gel (Bioprobe Systems, Montreuil – sous – Bois, France).

Results

1. Most samples from Europe and North Africa harbour a type – Z95 EBV, while samples from Asia are predominantly of type – ZP3

The BZLF1 gene was investigated by sequencing or by RFLP of 98 PBC pellets from HBO from Europd, North Africa and Asia and tissues representing different EBV – associated diseases: 68 NPC biopsy specimens, lymphoma tissues from 12 post – transplanted patients, 27 non-Hodgkin's lymphomas from HIV-positive patients, 6 OHL, 8 cell lins and 3 PBC pellets derived from patients with IM. The results (Tab. 1) show that 140 of the 162 samples from Europe and North Africa (86%) harbour a type – Z95 EBV, while 46 of the 66 samples (72%) from Asia are predominantly of the type ZP3.

2. Analysis of the BZLF1 open-reading-frame sequence fron NPC specimens

We completely sequenced the BZLF1 gene in EBV from 17 NPC biopsy specimens, and compared the sequences with those described by Packham et al.

Tab. 1 Distribution of Z95 and ZP3 viruses among different pathologies determined by sequencing or by RFLP

Pathologies	ZP3	Z95	Total
NPC (Europe + North Africa)	2	29	31
NPC (Asia)	26	9	35
OHL	0	6	6
Post – transplant lymphomas	3	9	12
Lymphomas of HIV patients			
Burkitt's lymphomas	4	5	9
B-cell lymphomas	1	5	6
Primary brain lymphomas	4	8	12
IM	1	10	11
PBC from HBD (France)	0	11	11
PBC from HBD (Notrh Africa)	6	23	29
PBC from HBD (Italy)	9	26	35
PBC from HBD (Asia)	17	6	23
PBC from diabetics (Asia)	5	3	8

(1993). One NPC biopsy sample (from Italy) exhibited a ZP3-type gene with no amino-acid change. The BZLF1 open-reading-frame sequence was strictly identical to the Z Akata sequence in 2 samples from Asia and 1 from Oceania. The remaining 13 EBV genomes were all identified as type Z95.

The variations in the Z95 sequence resulted in changes in only 3 aa, 2 conservative and 1 non – conservative (Tab. 2). Out of 13 specimens, 7 had a variation in the first exon. In 3 cases, there was a change from a C to a G at position 102 994, and, from a T to a G at position 102 975 in the other 4. These changes led to conservative amino – acid changes, with glutamic acid 54 being replaced by aspartic acid and leucine 61 by isoleucine respectively. In the type – Z95 gene, a single non – conservative amino – acid change that occurs in the third exon was produced by the replacement of a C by an A at position 102332 in 10/13 cases. Thi change led

Tab. 2 Amino-acid changes in the zebra protein of Z95-type EBV from NPC

NPC patients	First exon		Third exon
	54 glu to asp C-to-G 102 994	61 leu to ileu T-to-G 102 975	206 ala to ser C-to-A 102332
BP8		+	
BP14			+
RUG			+
VIS		+	+
STA	+		+
BP27		+	+
BP20		+	+
BP40			+
BP46			+
CAN	+		+
FOI	+		+
BP24		No change	
LU		No change	

to the replacement of the amino − acid alanine 206 by a serine （Zser206）.

3. Variation in the sequence of the Z95 introns

In addition to the mutations observed in the open reading frame, sequence differences were present within both introns in the BZLF1 gene. We have already described a variation in the splice donor site of the second intron in the BZLF1 gene in an NPC tumor （Cochet et al. , 1993）. Different changes were found in the first intron of the Z95 sequence: an A was substituted for a T at position 102609 （also seen in the ZP3 sequence） in 7 NPC, and a G for a T at position 102621 in 4 NPC. Deletion of a 29 − bp repeat element （extending from 102607 to 102636） was observed in 3 NPC cases after sequencing and in a further case after RFLP analysis （described later）. Furthermore, RLFP allowed us to detect 2 NPC cases with four 29 − bp units wthin the first intron. These changes apparently had no effect on the splicing phenomenon, since fully processed BZLF1 transcripts were found in the corresponding NPC tissues described （Cochet et al. , 1993）.

All the viruses exhibiting the variation Zser206 also had a separate variation in the second intron at position 102 407 （a G replaced by a T on the coding strand）. In addition to the 31 samples （NPC, lymphomas from immunocompromised patients, OHL, IM and cells from HBD） in which the BZLF1 gene had already been sequenced, the amplified fragments （102 187 − 103 226） of 8 additional samples were also sequenced to verify that the Zser206 variation was always associated with the variation at 102 407. This G − to − T change at 102 407 results in the loss of a BsmI restriction site. Consequently, RFLP of a 1040 − bp fragment amplified with the ZES2 − ZEAS primers （Tab. 3） exhibited a restriction pattern for Z95 DNA （359, 319, 221, 117, 24 bp） different from that for Zser206 DNA （259, 338, 319, 24 bp） （Fig. 1）. The extra 29 − bp sequence present in the ZP3 BZLF1 sequences was easily detected by the same analysis, since the 359 − bp fragment was replaced by a fragment of 388 bp （Fig. 1）. Following BsmI restriction analysis, RFLR was performed on 51 additional NPC samples （Tab. 4）. The Zser206 variant was found in 72% of Z95 viruses in NPC samples from Europe and Notrh Africa.

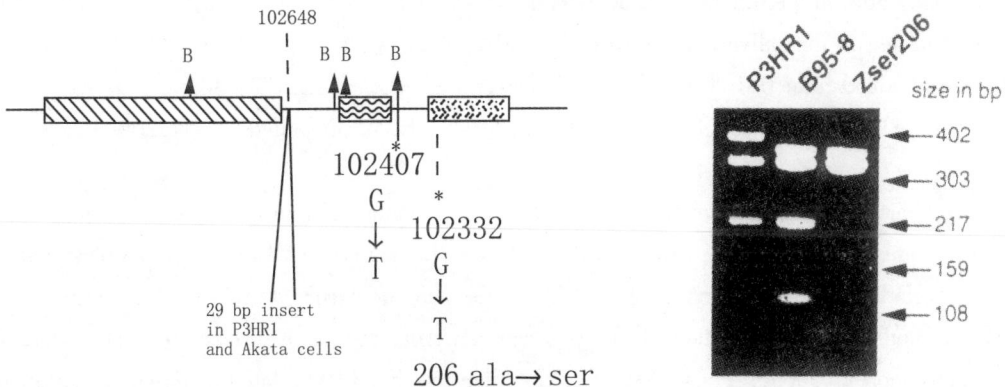

Fig. 1　Restriction-fragment-length polymorphism analysis of the BZLF1 gene. RFLP of a 1040 − bp fragment amplified with ZES2-ZEAS primers, performed with BsmI. The changes indicated are on the coding strand

Tab. 3　primers used in PCR amplification[1]

Target	Sequence（5′to3′）	Genomic coordinated
ZAS2	AGTATGCCAGGAGTAGAACA	102041 – 102060
ZEAS	AGGCGTGGTTTCAATAACGG	102187 – 102207
ZES	GCCACCTTTGCTATCTTTGC	103180 – 103161
ZES1	TGACATCACAGAGGAGGCTG	103261 – 103242
ZES2	AGGGGAGATGTTAGACAGGT	103226 – 103207
LMP3AS	CTGCCAATTCTCGCATGTCC	167967 – 167986
LMP1CS	GGTCATAGTCATGATTCCGG	168269 – 168288
LMP2CS	CTAGCGACTCTGCTGGAAAT	168373 – 168392
LMP1AS	AATGGAGGGAGAGTCAGTCA	168099 – 168118

[1] PCR amplification was performed with LMP2CS – LMP1AS or LMP1CS – LMP3AS, and with ZES1 – ZAS2 or ZES2 – ZEAS followed by amplification with ZES2 – ZAS2 or ZES – ZEAS primers. Sequencing was carried out with the same and internal primers.

Tab. 4　Observation of zser 206 in the EBV zebra protein from Z95-type EBV isolated from different tissues

Tissues	Number/total（%）
NPC	
NPC（Europe + Notrh Africa）	21/29（72）
NPC（Asia + Oceania）	2/9（22）
OHL	2/6（33）
Post-transplant B-cell lymphomas	3/9（33）
LCL	3/6
Biopsies	0/3
Lymphomas of HIV patients	3/18（17）
Burkitt's lymphomas	2/5
B – cell lymphomas	1/5
Brain lymphomas	0/8
IM	1/10（10）
PBC from HBD（France）	2/11（18）
PBC from HBD（Italy）	4/26（15）
PBC from HBD（North Africa）	3/23（13）
PBC from HBD（Asia）	0/6（0）
PBC from diabeties（Asia）	0/3（0）

4. Analysis of the BZLF1 sequence in different EBV – associated diseases

We also examined the BZLF1 sequence in other EBV – associated diseases, by sequencing or by RFLP. These included lymphomas from immunocompromised patients, 6 biopsy specimens from PTL, 6 cell lines established from PTL and 27 lymphomas from HIV patients（6 B – cell lymphomas, 9 Burkitt's lymphomas and 12 primary brain lymphomas）. Most of the specimens were of the Z95 type（Tab. 1）. Tab. 5 shows that, among the Z95 viruses, 33% of PTL and 17% of lymphomas from HIV patients exhibited the Zser206 change. The Zser206 variant was also found in 33% of OHL samples. EBV was examined in IM samples（8 LCL and 3 PBC pellets from patients with IM）. In 3 cases, paired cell lines（CRE 8 – 13, GRE23 – 25 and GRE 79 – 80）were obtained from spontaneously growing peripheral – blood lymphocytes and cordblood lymphocytes immortalized with the virus present in the saliva. The other LCLs（MA, HA, GRE 71, GRE 82, GRE 88 and GRE 90）were obtained from PBL. One of the IM cases, from whom GRE 79 – 80 were derived, had a chronicc form of the disease. In this partioclar case, EBV presented a Zser206 BZLF1 gene（Tab. 4）.

Specimens derived from one patient（GRE 8 – 13）contained multiple changes, and the same variations were observed in both cell lines derived from this patient, All the other samples contained very few variations（Tab. 5）. Among the Z95 viruses in the HBD, 0 to 18% presented the Zser206 variation（Tab. 4）. Statistical analysis was performed using the Fisher test. The Zser206 is significantly more frequent（$P < 0.001$）in NPC than in other EBV-related diseases or in PBC from HBO of the same region.

5. Analysis of the type of virus in the PBC of 14 NPC patients

We also analyzed the type of virus in the PBC of 14 NPC patients. In 13/14 of the cases, the same type of EBV was identified in the NPC tumor and in the PBC from the same patient, suggesting that the Zser206 – variant EBV is not restricted to the oropharynx region. The remaining patient had a type – Z95 EBV in PBC and a Zser206 in the tumor.

6. Analysis of the 30 – bp deletion in the 3′ end of the LMP1 gene

Apolymorphism in the 3′ end of the LMP1 gene has been reported in different EBV – associated tumors and cell lines (Miller et al., 1994). The variation includes a 30 – bp deletion covering an NcoI, site, In order to investigate possible segregation of both ZEBRA and LMP1 mutation in EBV isolated from NPC, we

Tab. 5 Sequence variations in the introns of EBV BZLF 1gene from different pathologies

Genomic coordinate		Changes	Pathologies	Patients
102637	1 st intron	G to A	IM	GRE 8 – 13
102621	1 st intron	G to T	NPC	VIS,RUG,BP20,BP40
			PTL	41, 48, 35
			IM	GRE 79 – 80
102612	1 st intron	T to C	IM	GRE 82
102594	1 st itron	G to A	IM	GRE 8 – 13
del 102636 to 102607	1 st intron		NPC	CAN, FOI, STA
102407	2nd intron	G to T	NPC	CAN, FOI, RUG, VIS, STAR, BP14, BP20, BP40, BP46
			PTL	25, 41, 48
			IM	GRE 79 – 80

amplified a region (168099 – 168392) encompassing the 30 – bp deletion. Amplification of a type – B95 – 8 virus will produce a fragment of 293 bp which, when digested by NcoI, will yield 2 bands of 169 and 124 bp. The results are shown in Figure 2. Restriction fragments from viruses from 3 HBD, 1 IM, 2 OHL, 3 PTL. 3 lymphomas from HIV patients and 6 NPC exhibited the same fragment lengths as the B95 – 8 strain. Viruses from 14 HBD, 5 IM, 1 OHL, 5 PTL, 9 lymphomas from HIV patients and 9 NPC exhibited an amplified fragment of 293 bp, which does not change after digestion by NcoI, Sequence analysis showed that the loss of the NcoI site was due to the replacement of a T by a C at position 168266. A deletion of 30 bp leading to the deletion of amino acids 346 to 355 in LMP1, described by Miller et al. (1994), was observed in 27 HBD, 6 diabetics, 3 cases of IM, 3 lymphomas from HIV patients, 2 PTL and 46 NPC (Fig. 2). The results show that the 30 – bp deletion in the LMP1 gene can also be observed in non – tumor specimens. In addition, a 69 – bp deletion was observed, leading to the loss of amino acids 334 to 356 in 1 HBD, 1 case of IM, 1 OHL and 3 NPC specimens. In 4 cases (2 OHL and 2 PTL), NeoI – restriction analysis of the LMP amplified fragment produced a pattern compatible with the presence of 2 different EBV strains (Fig. 3, lanes 11, 14, 17, 23). DNA from these 4 cases was amplified using 2 sets of primers, one of which lies within the 30 – bp and the 69 – bp deletions. Figure 3 shows that, in addition to a B95 – 8 pattern, samples LP5, LP6 and OHL7 also harbor a 30 – bp deletion, whereas OHL 5 harbors a 69 – bp deletion. Extensive studies of the polymorphism of the LMP1 gene, described by Miller et al. (1994), was not undertaken.

(a)

ATG   Ncol

LMP2CS   ►▼◄   LMPIAS   pA

a   b   c

▌ Δ in CAO strain

(b)   IM   PTL   PTL   NPC   NPC

B 1 2 3 4 5 6 7 8 9   10 11 12   13 14 15 16 17 18 19 20 21 22 23 24   25 26 27 28

◄ 293
◄ 263
◄ 169
◄ 124

293 ►
263 ►

◄ 224

263 ►

Tissues	(number of cases)	L95-8	L95-8, Ncol-	D 346-355	D 334-356
NPC (Europe and North Africa)	(30)	1	9	17	3
NPC (Asia)	(34)	5	0	29	0
OHL	(6)	2	1	1	2
Post-transplant B-cell lymphoma	(12)	3	5	4	0
Lymphomas of HIV patients	(15)	3	9	3	0
IM	(10)	1	5	3	1
PBC from HBD (France)	(11)	2	4	4	1
PBC from HBD (North Africa)	(20)	1	8	11	0
PBC from HBD (Asia)	(14)	0	2	12	0
PBC from Diabetics (Asiaa)	(6)	0	0	6	0

(c)

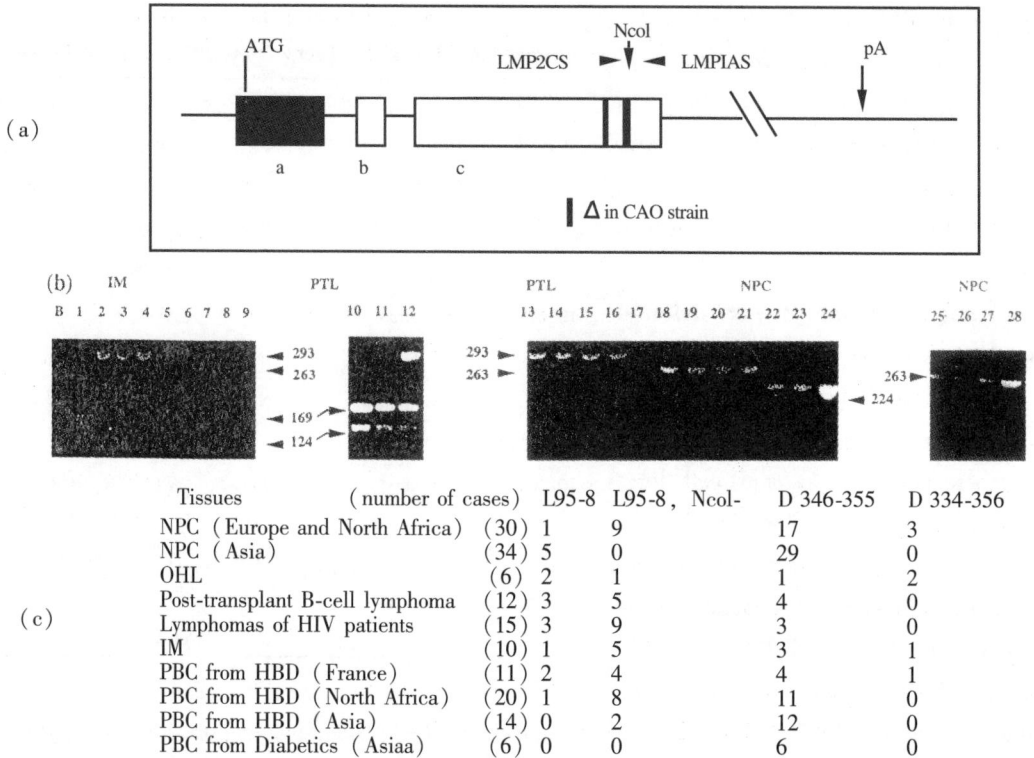

(a) Schematic representation of the BNLF1 gene indicating the position of primers and deletions. (b) Genomic DNA was PCR – amplified with LMP1 AS and LMP2CS primers, digested with NcoI restriction enzyme, separated on 2% agarose gels and visualized through ethidium – bromide staining. Lane B, B95 – 8; lanes 1 – 9, IM (respectively GRE 79, GRE 71, GRE 88, GRE 90, HA, MA, GRE 8, GER 23, GRE 82); lanes 10 – 14, PTL (respectively 41, 48, 35, 60, 22); lanes 15 – 28, NPC (respectively Vis, BP38, BP46, BP27, BP14, BP20, STA, RUG, BP40, BP24, CAN, YS, LC, VX). (c) Variations in the 3′end of the LMP1 open reading frame isolated from different pathologies

Fig. 2 Detection of deletions in the carboxy terminal region of the LMP protein in EBV isolated from different diseases.

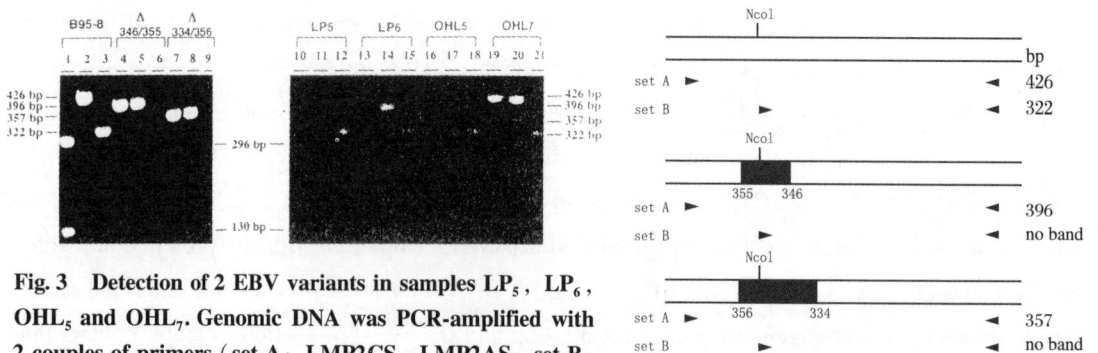

B95-8   Δ 346/355   Δ 334/356   LP5   LP6   OHL5   OHL7

1 2 3 4 5 6 7 8 9   10 11 12 13 14 15 16 17 18 19 20 21

426 bp —
396 bp —
357 bp —
322 bp —

— 296 bp

— 130 bp

— 426 bp
— 396 bp
— 357 bp
— 322 bp

Ncol

set A ►   ◄ 426
set B   ►   ◄ 322

Ncol

355 346

set A ►   ◄ 396
set B   ►   ◄ no band

Ncol

356 334

set A ►   ◄ 357
set B   ►   ◄ no band

Fig. 3 Detection of 2 EBV variants in samples LP₅, LP₆, OHL₅ and OHL₇. Genomic DNA was PCR-amplified with 2 couples of primers (set A; LMP2CS – LMP2AS; set B; LMP1CS – LMP3AS), one of which (LMP1AS) was lying within Δaa 334 – 356 and Δaa346 – 355. Lanes 1 – 3, B95 – 8; lanes 4 – 6, Δaa346 – 355; lanes 7 – 9, Δaa334 – 345; lanes 10 – 12, LP₅; lanes 13 – 15, LP₆; lanes16 – 18, OHL₅; lanes 19 – 21, OHL7, Lanes 1, 2, 4, 5, 7, 8, 10, 11, 13, 14, 16, 17, 19, 20, set A; lanes3, 6, 9, 12, 15, 18, 21, set B. Lanes 1, 4, 7, 10, 13, 16, 19 PCR product digested by NcoI

Discussion

Analysis of the 3'end of the LMP1 open reading frame showed that the 30 – bp deletion frequentloy observed in NPC (86% in Asia and 56% in Europe and North Africa) can also be found in non – tumor samples. This confirms the observations of Sandvej et al. (1994) on tonsils from patients with IM and of Khanim et al. (1996) in blood donors and virus – associated tumors from different geographic locations. No difference was observed in the frequency of the EBV carrying the 30 – bp – deleted LMP1 gene in tumor vs. non – tumor cells or in epithelial vs. lymphocyte cells within the same geographical region. These variations near the 3' end of the LMP1 gene do not co – segregate with any particular mutation of the BZLF1 gene.

Amplification of the LMP1 sequences revealed the presence of 2 viral strains in 4 immunocompromised patients. The co – infection with nultiple EBV strains has been reported in OHL (Walling et al. , 1992).

The 31 BZLF1 genes that were throughly examined exhibited limited variation which resulted in only a small number of changes in the amino – acid sequence. The variations seen in the first exon led to conservative changes in EBV from NPC and IM. and nonconservative changes in 3 EBV strains from IM. No changes were observed in the second exon and all variations in the third exon led to the change from an alanine to a serine at codon 206. EBV with this variation also had a change within the second intron at position 102407. The fact that the 2 mutations were present simultaneously suggests that this virus is a variant. This variant might also exhibit other mutations conferring specific biological properties. EBV Zser206 variant was more frequently observed in NPC from Europe and North Africa (72%) than in the other EBV – related diseases studied (10% – 33%) or in control PBC from HBD (13% – 18%) in the same region.

The tropism of the Zser206 EBV for epithelial cells does not seem to be involved, since in 92% of the cases, the same virus was recovered from NPC biopsy specimens and from PBC.

The ZP3 type harbors an additional copy of a 29 – bp repeat sequence within the first intron. Chen et al. (1996) described a difference between epithelial – and blood – tissue distribution of Z95 and ZP3 genotypes. In the specimens we examined the frequency with which type – ZP3 viruses were identified in both NPC biopsy samples and PBC from HBD from Hong Kong and China was identical (74%). Of the type – ZP3 from NPC that we analyzed, 85%, harbored type – A virus.

Amplification on the BZLF1 and the LMP1 sequences showed different strains of EBV – infected individuals from Europe and North Africa vs. Asia, since most of the samples from Europe and North Africa (93%) had an EBV of the Z95 type, while most of the specimens from Asia (73%) exhibited a ZP3 type of EBV. Moreover. 86% of the samples from Asia analyzed had the 30 – bp deletion in the LMP1 gene, whereas this deletion was less frequent (55%) in Europe and North Africa. If EBV plays a role in the mechanism leading to NPC, then the different types of viruses may be involved in cellular transformation.

It should be noted that ZP3 contains a serine at position 205 instead of an alanine in Z95 strains. Amino acids 205 and 206, which are within the dimerization domain, do not seem to be involved in dimerizaton activity (Flemington and Speck. 1990). The coiled – coil model described by Flemington and Speck (1990) suggests that they are located symetrically on the rear side of the α helix and might be contributing to interactons between ZEBRA and different proteins. The introduction of a polar residue might alter these properties. The isolate in the OHL sample used by Lau et al. (1992) also had a ser205 in a Z95 sequence. Although these serines are not fund in any identified consensus phosphorylation sequence, the presence of serine residues at either 205 of 206 may not be negligeable. A change from an ala to ser may be important, as illustrated by the observation of Francis et al. (1997). who reported that the change from a ser to an ala in the basic domain of ZEBRA was able to separate its function of transcriptional activation and disruption of latency. The specific properties of Zser206 are currently under investigation.

Acknowledgements

We thank Dr. Laplanche and Dr. Benhamou for their help in the statistical analysis and Dr. P. Farrell for providing DNA from OHL biopsy specimens. We are grateful to Dr. A. Alberga and Mrs. L. SAINT Ange for careful reading of the manuscript. V. G. was supported by a grant from "La Ligue Nationale Contre le Cancer".

[In 《Int J Cancer》 1998, 75: 497 – 503]

References

1　Arrand J R, Young L S, Tugwood J D. Two families of sequences in the small RNA-encoding region of Epstein-Barr virus (EBV) correlate with EBV typesa A and B. J Virol, 1989, 631: 983 – 986

2　Cayrol C, Flemingtog E. G_0/G_1 growth arrest mediated by a region encompassing the basic leucine zipper (bZIP) domain of the Epstein-Barrvirus transactivator Zat J biol Chem, 1996, 271: 31799 – 31802

3　Chang Y, Chen M I, Wu R, Chang P, Shu C, Liu S. Sequence analysis of the BZLF1 gene of Epstein-Barr virus derved from a new variant isolated in Taiwan, Nucleic Acids Res, 1992, 20: 139

4　Chen H L, Lung M L, Chan K H, Griffin B E, NG M H. Tissue distribution of Epstein-Barr-virus genotypes. J Virol, 1996, 70: 7301 – 7305

5　Chen M L, Tsai C N, Liang C L, Shu C H, Hung C R, Sulttzeanu D, Liu S T, Chang Y S. Cloning and characterization of the latent membrane protein (LMP) of a specific Epstein-Barr-virus variant derived from nasopharyngeal carcinoma in the Taiwanese population. Oncogene,

1992, 7: 2131 – 2140

6　Cochet C, Martel-renoir D, Grunewald V, Bosq J, Cochet G, Schwaab G, Bernaudin J F, Joab I. Expression of the Epstein-Barrvirus imediate-early gene, BZLF1, in nasopharyngeal-carcinoma tumor cells, Virology, 1993, 197: 358 – 365

7　Flemington E, Speck S H. Evidence for coiled-coil dimer formation by an Epstein-Barr-virus transactivator that lacks a beptad repeat of leucine residuesl. Proc, nat. Acad. Sci. (Wash), 1990, 87: 9459 – 9463

8　Francis A L, Gradoville L, Millre G. Alteration of a single serine in the basic domain of the Epstein-Barr-virus Zebra protein separates its functions of transcriptional activation and disruption of latency. J Virol, 1997, 71: 3054 – 3061

9　Gutsch D E, Holleyguthrie E A, Zhang Q, Stein B, Blanar M A, Baldwin AS, Kenney S C. The B-Zip transactivator of Epstein-Barr virus. Bzkf1, functionally and physically interacts with the p65 sub-unit of NF-kappa b. Mol, cell Biol, 1994, 14: 1939 – 1948

10　Hu L F, Chen F, Zheng X, Ernberg I, Ggo S

L, Christensson B, Klein G, Winberg G. Clongability and tumorigenicity of human epithelial cells expressing the EBV-encoded membrane protein LMP1, Oncogene, 1993, 8: 1575 – 1583

11 Hu L F, Zabarovsky E R, Chen F , Cao S L, Ernberg I, Klen G, Winberg G. Isolation and sequencing of the Epstein-Barr-virus BZLF1 gene (LMP1) from a Chinese nasopharyngeal carcinoma. J gen Virol, 1991, 72: 2399 – 2409

12 Khanim F, Yao Q Y, Niedobitek G, Sihota S, Rickinson A B, YOUNG L S. Analysis of Epstein-Barr-virus-gene polymorphisme in normal donors and in virus-associated tumors from different geographic locations, Blood, 1996, 88: 3491 – 3501

13 Knecht H, Bachmann E, Brousset P, Sandvej K, Nadal D, Bachmann F, Odermatt B F, Delsol G, Pallesen G. Deletions within the LMP1 oncogene of Epstein-Barr virus are clustered in Hodgkin's disease and identical to those observed in nasopharyngeal carcinoma. Blood, 1993, 82: 2937 – 2942

14 Lau R, Packham G, Farell P J, Differenial splicing of Epstein-Barrvirus immediate-early RNA. J Virol, 1992, 66: 6233 – 6236

15 Lung M L, Lam W P, Sham J, Choy D, Yong-Sheng Z, Guo H Y, Ng M H. Detection and prevalence of the" f" variant of Epstein-Barr virus in Southern China. Virology, 1991, 185: 67 – 71

16 Miller W E, Edwards R H, Walling D M, Raabtraub N. Sequence variation in the Epstein-Barr-virus latent membrane protein I. J gen Virol, 1994, 75: 2729 – 2740

17 Packham G, Brimmell M, Cook D, Sinclair A J, Farrell P J. Strain variation in Epstein-Barr-virus immediate-early genes. Virology, 1993, 192: 541 – 550

18 Rickinson A B, Kieff E. Epstein-Barr virus, Int B. N. Fieles, Knipe. D. M. and A. L. Howley Er (eds.). Virology, Lippincott-Raven Press,

Philadeephia, 1996, 2394 – 2446

19 Rowe M, Young L S, Cadwallader K, Pettl L, Kief E, Rickinson A B. Distinction between Epstein-Barr-virus-type-a (EBNA-2A) and type-b (EBNA-2B) isolates extends to the EBNA-3 family of nuclear proteins. J Virol, 1989, 63: 1031 – 1039

20 Sandvej K, Peh S C, Andrese B S, Pallesen G. Identification of potential hot spots in the carboxy-terminal prar of the Epstein-Barr-virus (EBV) BNLF-1 gene in both malignant and benign EBV-associated diseases; high frequency of a 30-bp deletion in Malaysian and Danish peripheral T-cell lymphomas. Blood, 1994, 84: 4053 – 4060

21 Sista N D, Pagano J S, Liao W, Kenney S. Retinoic acid is a negative regulator of the Epstein-Barr-virus protein (BZLF1) that mediates disruption of latent infecton. Proc, nat. Acad, Sci. (Wash.), 1993, 90: 3894 – 3898

22 Snudden D K, Smith P R, Lai D, Ng M H, Griffn B E. Alterations in the struture of the EBV nuclear antigen, EBNA1, in epithelial-cell tumors. Oncogene, 1995, 10: 1545 – 1552

23 Walling D M, Edmiston S N, Sixbey J W, Abdel-Hamid M, Resnick L, Raab-Traub N. Co-infection with multiple strains of the Epstein-Barr virus in human-immunodeficiency-virus-asociated hairy leukoplakia. Proc, nat, Acad, Sci. (Wash.), 1992, 89: 6560 – 6564

24 Wrightham M N, Stewart J P, Janjua N J, Pepper S D, Sample C, Rooney C M, Arrand J R. Antigenic and sequence variation in the C-terminal unique domain of the Epstein-Barr-virus nuclear antigen EBNA-1. Virology, 1995, 208: 521 – 530

25 Zhang Q, Gutsch D, Kenney S. Functional and physical interaction between p53 and BZLF1-implications for Epsein-Barr-virus latency. Mol. cell. Biol, 1994, 14: 1929 – 1938

68. Synergistic Effect of Epstein-Barr Virus and Tumor Promoters on Induction of Lymphoma and Carcinoma in Nude Mice

LIU Zhen – sheng, LIU Yan – fang, ZENG Yi

Summary

Balb/c nude mice were subcutaneously transplanted with fetal nasopharyngeal mucosa infected with B95 – 8 Epstein – Barr virus (EBV), n – Butyrate and/or 12 – O – tetradecanoylphorbol 13 – acetate (TPA) were injected subcutanously on the third day and once a week thereafter. About 10 days later, tumor masses gradually grew in these mice. Histopathological examination was carried out 15 weeks later. Three cases of lymphomas (two T cell lymphomas and one B cell lymphoma) were observed in the group receiving EBV and TPA, and one T cell lymphoma and three cases of undifferentiated carcinoma were found in the group receiving EBV, TPA and n – butyrate, but to case was found in the control groups that were transplanted with fetal nasopharyngeal tissue infected with EBV, or TPA and n – butyrate alone. Polymerase chain reaction amplification and in situ hybridization revealed that lymphoma and carcinoma cells contained the EBV LMP1 and EBERs genes. LMP1 protein was also found in the carcinoma. The T and B cell lymphomas and the nasopharyngeal carcinoma in nude mice were derived from human nasopharyngeal mucosa; this was proved by using human specific monoclonal antibodies to CD3 for T cells, to CD20 for B cells, and to epithelial membrane antigen for epithelial cells. Nucleotide sequence analysis indicated that the homologies of EBV LMP1 genes in the induced malignant lymphomas and undifferentiated carcinomas to the B95 – 8 cell gene were around 96% and 99% respectively. The results showed that EB virus can infect nasopharyngeal mucosa of the human fetus and consequently induce malignant transformation by the synergistic effect of the tumor promoters, and that EBV DNA can persist in the lymphomas and carcinomas.

[**Key words**] EBV Tumor promoters; Lymphoma Carcinoma; Nude mice

Introduction

Epstein – Barr virus (EBV) is known to be the agent causing human infectious mononucleosis and to be closely related to nasopharyngeal carcinoma (Wof et al. 1973; Zeng 1985; Zeng et al. 1979, 1980; Pi et al. 1981). Transfection of the EBV latent membrane protein 1 (LMP1) gene into immortalized cells can induce tumors in nude mice (Wang et al. 1985; Fahnaeus et al. 1990; Christopher et al. 1990; Hu et al. 1993). Nevertheless, there is no direct evidence yet showing that EBV is the etiological cause of nasopharyngeal carcinoma. However, Ito et al. (1981) reported that the synthesis of EBV antigens in EBV – carrying cells was significantly increased when these cells were trested with a combination of 12 – O – tetradecanoylphorbol 13 – acetate (TPA) and n – butyrate. Zeng et al. also reported that some Chinese medoconal herbs contanining TPA – like substances in combination with n – butyrate could induce the synthesis of EBV antigens enhance the transformation of EBV – infected lyphocytes and promote the development of nasopharyngeal carcinoma in rats by dinitrosopiperazing (Zeng et al. 1994; HU and Zeng 1985. 1986; Tang et al. 1988). Huang et al. found that cantonese salted fish contains nitrosamine and could induce carcinoma in the nasal and paranasal regions of rats (Huang et al. 1978). Shao et al. (1988) reported that salted fish also contains EBV inducers. Aya et al. (1991) showed that dual exposure of human lymphocytes to EBV and purified 4 – deoxyphorbol ester could induce chromosome rearrangement and development of lymphoma. Ji et al. (1990) demonstrated that the metabolic products of anaerobic vacteria isolated from the nasopharyngeal cavity contain butyric acid which could markedly induce the expression of EBV antigens in Raji and P3HR – 1 cells in synergy with TPA. Shao et al. (1995, 1997) reported that there was an EBV receptor gene (CR2) in human fetal nasopharyngeal mucosa and nasopharyngeal carcinoma cells. Our previous work also showed that EBV nuclear antigen was present in the normal and hyperplastic epithelial cells of human nasopharyngeal mucosa (Zeng 1981). Therefore, attempts were made to induce lymphoma and carcinoma in nude mice by simultaneously administering EBV – infected human fetal nasopharyngeal mucosa and tumor promoters. When EBV – infected human fetal nasopharyngeal mucosa was transplanted subcutaneously into nude mice treated with TPA and n – butyrate, lymphomas and carcinomas were successfully induced. The experiments provided evidence for the etiological role of EBV in the development of nasopharyngeal carcinoma in man.

Materials and Methods

1. Animals

A group of 26 Balb/c nude mice (4 – 6 weeks old), fed with food and water sterilized by autoclave, were obtained from the Animal Center, Chinese Academy of Medical Sciences.

2. Cell, plasmids, and reagents

B95 – 8 cells were obtained from the Institute of Virology, Chinese Academy of Preventive Medicine. pRV2 – EBERs and pUC – LMP1 were provided by Dr. Irene Joab (Institute Gustave Roussy, Paris). A digoxigenin labelling and detection kit (Boehrnger Mannheim GmbH Mannheim Germany), Apal nuclease, the four nucloside triphosphates and Taq polymerase (Chinese – American Bioengi – neering Company) were purchased.

3. Preparation of EBV

B95 –8 cells were cultivated in RPMI – 1640 medium supplemented with 2 mmol/L glutamine, 100 μg/ml streptomycin, 100 IU/ml penicillin and 15% fetal bovine serum. The cells were treated with 20 ng/ml TPA and 4 mmol/L n – butyrate for 48 h. More than 90% of the cells were alive at a density of 10^6 cells/ml. After centrifugation at 1000 r/min for 20 min, the cell pellet was removed. The supernatant was collected and centrifuged at 20 000 r/min for 2 h. The pellet was immediately resuspended in fresh medium. All procedures were carried out at 4℃. The B95 – 8 EBV solution was concentrated 150 times through a 0. 45 – μm – pore – size filter (virus titer 10^1 log/ml) and stored in liquid nitrogen.

4. Tumor formation

Under aseptic condition, nasopharyngeal mucosa from a human fetus (4 – 5 months old) was separated and cut into pieces of 0.5 – 1.0mm^3; 1.5 ml concentrated B95 – 8 EB viral suspension was then added to the tissue pieces and incubated at 37℃ for 2 h. After virus adsorption, the viral suspension was removed by centrifugation at 1500 r/min for 10 min. The tissue pieces were subcutaneously transplanted into nude mice and 200 μg n – butyrate and /or 50 ng TPA were injected subcutaneously into each mouse on the 3rd day after transplantation and once a week thereafter. Nude mice were also transplanted with either nasopharyngeal mucosa alone, or nasopharyngeal mucosa infected with EBV, or nasopharyngeal mucosa plus TPA and n – butyrate as controls. All animals were observed for 15 weeks.

5. Extraction and amplification of tissue DNA

A piece of tumor tissue was frozen and ground, tissue DNA was obtained by sodium dodecyl sulfate lysis/proteinase K digestion and phenol extraction. The following primers were used for amplification, Primer sequences for exon 1, 2 and intron 1, 2 of the LMP1 gene were primer 1: 5′ – GCCAGAGCATCTCCAATAA – 3′, and primer 2: 5′ – GGTCGTGTTCCATCCTCAG – 3′. The LMP1 geme was amplified in a 50 – μl polymerase chain reaction (PCR) mixture for 30 cycles at 94℃ for 1 min, 55℃ for 1 min and 72℃ for 1 min, and was followed by an estension at 72℃ for 10 min. A 10 – μl sample from each PCR product was analyzed by electrophoresis through 2% agarose and stained with ethidium bromide.

6. Immunohistochemical studies

Tumor tissue sections were incubated with a 1 : 50 diution of CD3 momoclonal antibody, a 1 : 100 dilution of CD20 (Dako), or monoclonal antibodies to epithelial membrane antigen (EMA), cytokeratin AEI/EA3 and EBV LMP1 (Zymed) at 37℃ for 30 min separately. The sections were washed with phosphate – buffered saline and incubated with horseradish – peroxidase – conjugated goat anti – (mouse IgG) at 37℃ for 30 min. After that, the substrate solution was applied for 5 min to yield a brown reaction product in tumor samples.

7. *In situ* hybridization

A 600 – bp Accl restriction fragment was obtained from pRV – 2 EBERs containing the EBV EBERI /and EBER2 genes and a 1800 – bp LMP1 fragment was from pUC – LMP1. DNA fragments were labelled by random hexanucleotide priming with digoxigenin to create EBERs and LMP1 probes. Sections 5 μm thick were cut from the blocks of formalin – fixed and paraffin – embedded tissue specimens, dewaxed in xylene, dehydrated in serially graded ethanol washes (100%, 95% and 75%), digested with proteinase K for 30 min at 37℃ and hybridized with the probes. After that, the hybridization procedure was completed by the use of sheep antidioxin antibody and the nitroldue tetrazolium 5 – Bromo – 4 – Chloro – 3 – Indolyl – phosphate system.

8. Cloning and analyzing PCR products

The amplified products were ligated to the pGME – T vector and transformed into JM109 bacteria. The transformants were plated on 5 – bromo – 4 – chloro – 3 – indolyl β – D – galactoside selective defined medium. The single white colony was picked out, and the insertion of PCR products into plasmids was verified with Apal restriction digestion. The positive clone was cultivated under shaking, and the plasmid DNA was extracted and purified by the polyethyleneglycol PEG 8000 precifptation method. PCR products were sequenced by a DNA sequencer 373a – 18 system.

Results

1. Tumor formation in nude mice

EBV – infected fetal nasopharyngeal mucosa were transplanted subcutaneously onto the backs of nude mice. On the third day after transplantation and subsequently once a week, 200 μg n – butyrate and 50 ng TPA were injected subcutaneously into each mouse. After 10 days, transplanted tissues began to enlarge and, within a period of about 2 months, they reached 2.1 ± 0.39 cm^3, at which stage they were movable nodules. After 7 – 15 weeks, tumor tissues were removed from nude mice and examined histopathologically. Tumor from 1 mouse was diagnosed at T cell lymphoma and tumors from the other 3 mice were undifferentiated carcinomas (Figs. 1, 2). After EBV – infected fetal nasopharyngeal mucosa had been transplanted into nude mice and they had been treated with TPA only, 3 of the 6 mice developed malignant lymphomas, with 2 T cell lymphomas and 1 B cell lymphoma. The nude mice that received either EBV – infected fetal nasopharyngeal mucosa or normal fetal nasopharyngeal mucosa alone did not develop tumor within 15 weeks. Also, no tumors were observed in any nude mice transplanted with tissues from fetal nasopharyngeal mucosa treated with TPA and n – butyrate within 13 – 15 weeks. Human nasopharyngeal mucosa formed cysts in 2 of the 6 nude mice treated with EBV alone. The mucosa transplanted into other control groups was degraded (Tab. 1).

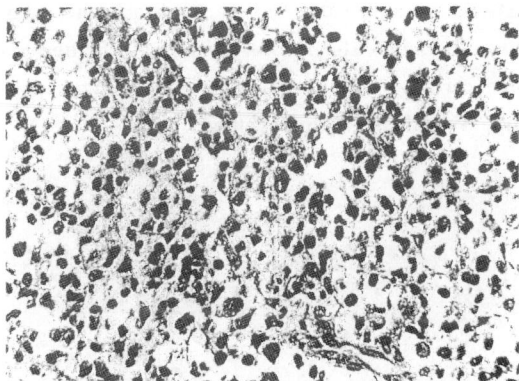

Fig. 1 Micrographs of paraffin section of lymphoma in nude mice formed by transplantation of Epstein-Barr virus (EBV) -infected fetal nasopharyngeal mucosa and treated with 12-O-tetradecanoylphorbol 13-ace-tate (TPA). Malignant lymphoma stained by hematoxylin eosin (H&E); ×200

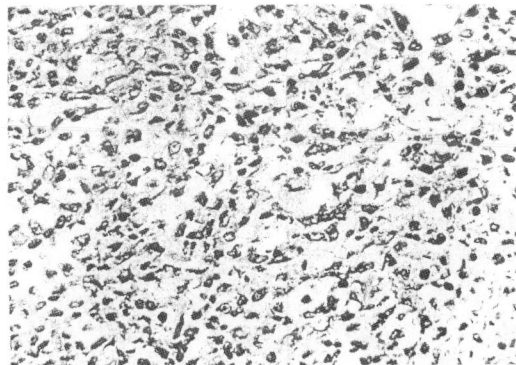

Fig. 2 Micrographs of paraffin section of carcinoma in nude mice formed by transplantation of EBV-infected fetal nasopharyngeal mucosa and treated with TPA and n-butyrate. Undifferentiated carcinoma with irregular nuclei and rich cytoplasm; mitosis can be seen. H&E; ×200

Determination of the human origin of undifferentiated carcinomas and lymphomas

Immunohistochemical staining of carcinoma tissues from nude mice with human specific monoclonal antibody to EMA on human epithelial cells showed a positive reaction in undifferentiated carcinoma (Fig. 3) but was negative in the control rat exophageal cancer cells. Both the undifferentiated carcinoma cell and the rat esophageal cancer cells showed a positive reaction to nonspecific monoclonal antibody cytokeratin AE1/AE3 to epithelial cells. Immunohistochemical staining of lymphona cells with human specific CD3 and CD20 monoclonal antibodies further demonstrated that three malignant lymphomas were originally from human T cells and one was from human B cells (Fig. 4a, b). However, there was no reaction of undifferentiated carcinoma cells with CD3 or CD20 monoclonal antibodies.

Tab. 1 Tumor formation from fetal nasopharyngeal mucosa infected with Epstein-Barr virus (EBV). NPT nasopharyngeal tissues. TAP 12-O-tetradecanoylphorbol 13/-acetate

Group	Numbers of nude mice	Time after transplantation (week)	Cyst formation	Mallignant T (B) lymphoma	Undifferentiated - carcinoma
NPT	4	15	0	0	0
NPT + EBV	6	15	2	0	0
NPT + TPA + n-butyrate	4	13 – 15	0	0	0
NPT + EBV + TPA	6	8 – 15	0	2 (1)	0
NPT + EBV + TPA + n – butyrate	6	7 – 15	0	1	3

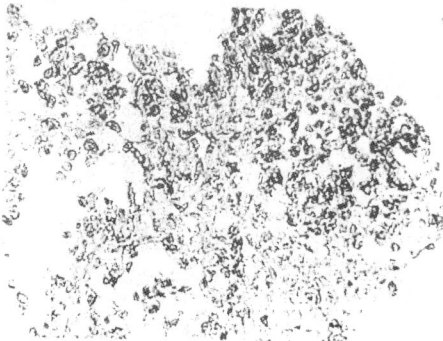

Fig. 3 Immunohistochemical staining of undifferentiated carcinoma from human fetal nasopharyngeal mucosa and esophageal carcinoma of rat. Routine peroxidase/antiperoxidase immunohistochemical staining. Immunohistochemical staining of undifferentiated carcinoma Epithelial membrane antigen positive. Peroxidase/antiperoxidase; ×200

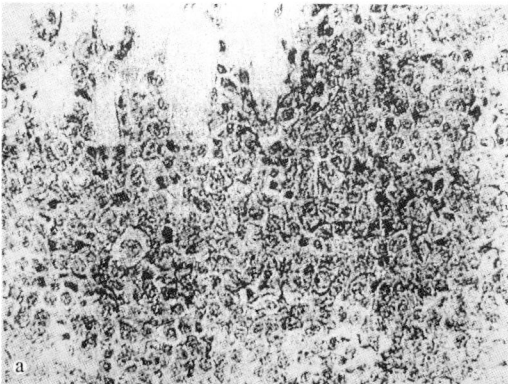

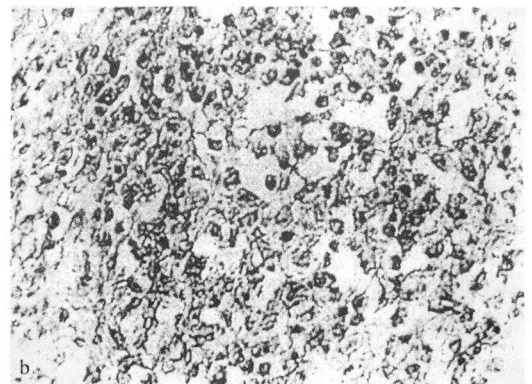

Fig. 4 a, b Immunohistochemical staining of malignant lymphoma. Routine peroxidase/antiperoxidase immunohistochemaical staining. a: Malignant T lymphoma. immunohistochemical stainin CD3 positive. Peroxidase/antiperoxidase, ×200. b: Malignant B lymphoma, immunohistochemical staining CD20 positive. Peroxidase/antiperoxidase, ×200

2. Detection of EBV LMP1 protein in undifferentiated carcinoma cells

Undifferentiated carcinoma cells, produced in nude mice from human fetal nasopharyngeal mucosa, were stained with monoclonal antibody to LMP1. A positive reaction was observed in these cells (Fig. 5).

3. Determination of EBV genes by *in situ* hybridization

Digoxigenn-labelled LMP1 and EBERs gene probes were used for *in situ* hybridization. The results showed hte presence of the EBERs gene in the cellular nuclei of lymphoma tissues and of the LMP1 gene

Fig. 5 Immunohistochemical staining of undifferentiated carcinoma. EBV LMP1 positive. Peroxidase/antiperoxidase; ×200

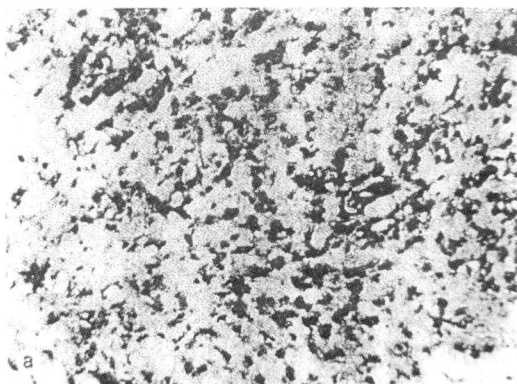

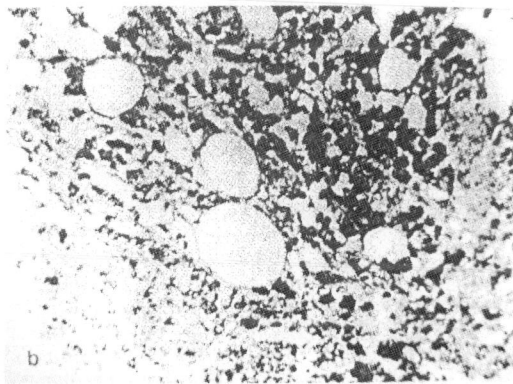

in undifferentiated carcinoma (Fig. 6a, b). There were no specific hybridization signals in the cellular nuclei of the controlled breast carcinoma. These indicated that EBV genes did exist in malignant lymphoma and undifferentiated carcinoma originating from EBV-infected human fetal nasopharyngeal mucosa.

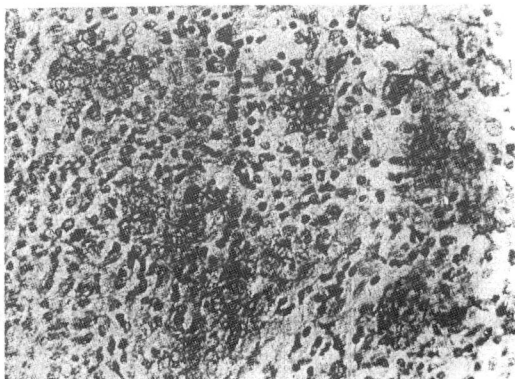

Fig. 6 a, b *In situ* hybridization of EBV genes in tumor tissues. a: *In situ* hybridization of EBERs gene in malignant lymphoma. Purpleblue minute granules, as positive signal. located in the cell nucleus. b: *In situ* hybridization of LMP1 gene in undifferentiated carcinoma. Purple-blue minute granules as positive signal, mainly located in the cell nucleus and cytoplasm

4. PCR amplification

Amplification of EBV DNA from lymphoma and undifferentiated carcinoma was performed with LMP1 primers, Electrophoresis of the PCR products in 1.0% agarose gels showed a 553-bp DNA fragment from lymphona and carcinoma (Fig. 7a. b), and then the PCR products were ligated to the pEGM-T vector, and Apal digestion showed that the PCR DNA fragment LMP-1 had been inserted into the vector.

5. DNA sequence analysis of the LMP-1 gene of B95-8 cells and tumor tissues

Analyses of EBV LMP1 exon 1, 2 and intron 1, 2, and DNA of B95-8 cells and tumor tissues showed the homologies of LMP1 sequences from malignant lymphoma and undifferentiated carcinoma to B95-8 EBV LMP1 to be about 96% and 99% respectively (Fig. 8). The results revealed that EBV in tumor tissues was concordant with B95-8 EBV.

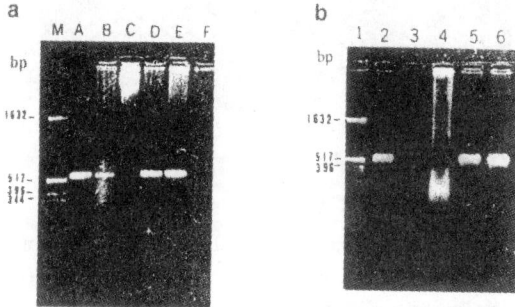

Fig. 7 a, b Result of the polymerase chain reaction (PCR) detection of tumor DNA, a: Result of the PCR detection of malignant lymphoma DNA. M: PBR322/hinf 1 DNA marker; A: B95-8 cell DNA; B, D, E: malignat lymphoma DNA; C: malignant lymphoma DNA (not amplified); F: 293 cell DNA. b: Result of the PCR detection of undifferentated carcinoma DNA. 1: PBR322/Hinf 1 DNA marker; 2: B95-8 cell DNA; 3, 5, 6: undifferentiated carcinoma DNA; 4: 293 cell DNA

1   CTG AGG ATG GAA CAC GAC CTT GAG AGG GGC CCA CCG GGC CCG CGA CGG CC
2
3 primer2(169480-169462)
CCT CGA GGA CCC CCC CTC TCC TCT TCC CTA GGC CTT GCT CTC CTT CTC CTC   CTC

TTG GCG CTA CTG TTT TGG CTG TAC ATC GTT ATG AGT GAC TGG ACT GGA GGA   GCC

CTC CTT GTC CTC TAT TCC TTT GCT CTC ATG CTT ATA ATT ATA ATT TTG ATC   ATC

TTT ATC TTC AGA AGA GAC CTT CTC TGT CCA CTT GGA GCC CTT TGT ATA CTC   CTA
CTG ATG AGT   AAG TAT TAC ACC CTT TGC CCC ACA CCC CCT TTC CCT TAC TCT   TCC

TTC TCT AAC GCA CTT TCT CCT CTT TCC CCA GTC ACC CTC CTG CTC   ATC GCT   CTC

TGG AAT TTG CAC GGA CAG GCA TTG TTC CTT GGA ATT GTG CTG TTC ATC TTC   GGG
TGC TTA CTT GGT AAG ATC TAA CAT TCC CTA GGA ATT ATT TAC CAC ACC CCC   ACT

TTT CCA ACC CTA ACA CTC TTT TTT CAA CGC AGT CTT AGG TAT CTG GAT CTA   CTT

ATT GGA GAT GCT CTG GC
primer 1   (168945-168927)

Fig. 8 Comparison of the nucleotide sequences of EBV-LMP 1 gene (168927-168945) from (1) B95-8, (2) malignant lymphoma and (3) undifferentiated carcinoma

Discussion

Our study revealed that EBV infection of nasopharyngeal mucosa alone can not induce lymphoma or undifferentiated carcinoma, but that these can be induced under the synrgetic effect of inducers and tumor promoters (TPA and/or n-butyrate). An interesting finding is that lymphomas are induced by the synergistic effect of TPA; T cell lymphoma is more active than B cell lymphoma and undifferentiated carcinoma induction needs another inducer, such as n – butyrate. In another study, we induced poorly differentiated carcinoma in nude mice by EBV-infected immortalized 293 cells (carrying CR2 receptors) together with TPA, showing that the process involved in EBV induction of lymphoma and carcinoma may be somewhat different (Li et al. 1997).

The origin of lymphoma and carcinoma induced in transplanted nude mice was identified by immunohistochemical staining with human specific monoclonal antibodies. Lymphomas of T and B cells reacted positively with monoclonal antibodies CD3 and CD20 respectively nd carcinoma reacted positively with EMA. The results clearly demonstrated that the tumors in transplanted nude mice are originally from the EBV-infeted human fetal nasopharyngeal mucosa.

Zeng (1981) found EBV nuclear antigen in normal and hyperplastic epithelial cells of nasopharyngeal mucosa. Shao et al. (1995, 1997) had detected the gene for the EBV receptor CR2 in nasopharyngeal mucosa. These data indicate that there are CR2 receptors located on the epighelial cells of human nasopharyngeal mucosa. The inability of EBN to infect *in vitro* culticated monolayer epithelial cells may be due to there beijing no of little expression of CR2, but more expression *in vivo*. EBV spreads widely in the human population and, in China, more than 95% of EBV infection occurs in childhood. In areas of southern China where there is a high risk of nasopharyngeal cancer there are a lot of Chinese herbs, plants and foods containing EBV inducers and tumor promoters (Zeng et al. 1983, 1994; Hu and Zeng 1985, 1986; Tang et al. 1988). Besides, the anaerobic bacteria producing butyric acid occur frequently in the human nasopharynx (Ji et al. 1990). These factors, together with EBV nay play an important role in the development of nasopharyngeal carcinoma. A previous study by us showed that a 21 times higher risk will occur when there is a HLA linkage to a nasopharyngeal carcinoma susceptibility gene (Lu et al. 1990). Our studies lead us to suggest that, for nasopharyngeal carcinoma development, EBV, genetic factors and host immunity play an etiological role, and tumor promoters and/or chemical carcinogens may have a synergistic effect.

In brief, our study describe the synergetic effect of EBV and tumor promoters on the induction of lymphomas and cad carcinomas in transplanted nude mice and also provides an important animal model for studing the etiology and mechanism of nasopharyngeal carcinoma development. The data further suggest that EBV plays an etiological role in the development of nasopharyngeal carcinoma.

Acknowledgements

We thank Professor Sir Athony Epstein and Professor G. Dethe for comments and revisisons on the manuscripts. This work was suppotred by grants from the Commission of the European Communittee and National Scientific Committee of China.

[In 《J Cancer Res Oncol》 1998, 124: 541 – 548]

References

1 Aya T, Kingoshita T, lmai S, Koizumi S, Mizuno F, Osato T, Satoh C, Oikawa T, Kuzumaki N, Ohigashi H, Koshimizu K. Chromosome translocation and c-myc activation by EpsteinBarr virus and Euphorbia tirucalli in B lymphocytes. The Lancet, 1991, 337: 1190

2 Christopher WD, Alan BR, Lawrence SY. Epstein-Barr virus latent membrane protein inhibits human epithelial cell differ entiation. Nature, 1990, 344: 777 – 780

3 Fahraeus R, Rymo L, Rhim JS, Klein G. Morphological transformation of human keratinocytes expressing the LMP gene of Epstein – Barr virus, Nature, 1990, 345: 447 – 449

4 Hu YL, Zeng Y. Enhanced transformation of human lymphocytes by Chinese herbs. Chin J Oncol,

1985, 7: 471 – 419

5 Hu YL, Zeng Y. The enhancing effects of sodium butyrate on the transformation of human lymphocytes by EBV. Chin J Cancer, 1986, 5: 243 – 246

6 Hu LF, Chen F, Zheng X, Ernberg I, Cao SL, Christensson B, Klein G, Winberg G. Clonability and tumorigenicyty of human epithelial cells expresing the EBV encoded membrane protein LMP1. Oncogene, 1993, 8: 1575 – 1583

7 Huang DP, Ho JH, Saw D, Teo TB. Carcinoma of the nasal and paranasal region in rats fed Cantonese salted marinefish. Scientific publication 20. ARC, Lyon, 1978, 315

8 Ito Y, Kawanishi M, Harayama T, Takabayashi S. Combined effect of the extracts from Croton tig-

lium, Euphorbia lathyris or Euphorbia tirucalli and n – butyrate on Epstein – Barr virus expression in human lymphoblastoid P3HR1 and Raji cells. Cancer Lettm, 1981, 12: 175 – 180

9　Ji ZW, Zeng Y, Wang PZ, Tan HZ. Induction of antigens in Raji cells and P3HR – 1 cells by anaerobiec bacterium isolated from nasopharynx of patients with nasopharyngeal carcinoma and other diseases. Chin J Cancer, 1990, 1: 1 – 3

10　Li BM, Ji ZW, Liu ZS, Zeng Y. Epstein – Barr virus in synergy with tumor – promoter – induced malignant transformation of immortalized human epithelial cells, J Cancer Res Clin Oncol, 1997, 123: 441 – 446

11　Lu SL, Day NE, Degos L, Lepage Y, Hung PC, Chan SHM. Mcknight B, Easton D, Zeng Y, The G de. The genetic basis for nasopharyngeal carcinoma linkage to HLA region. Nature, 1990, 346: 479 – 481

12　Pi GH, Zeng Y, Zhao WP, Zhao Q. Development of an anticomplement immunoenzyme test for detection of EB virus nuclear antigen (EBNA) and antibody to EBNA. J Immunol Methods, 1981, 44: 73 – 78

13　Shao YM, Poiries S, Oshima H, Malaveille C, Zeng Y, The G de, Bartsch H. Epstein – Barr virus activation in Raji cells by extract of preserved food from high risk areas for nasopharyngeal carcinoma. Carcinogenesis, 1988, 9: 1455 – 1457

14　Shao XY, Chen ZC, Yao KT. DNA sequencing of the Epstein – Barr virus binding site fo EBVR/CR2 gene in nasopharyngeal carcinoma, Chin J Viro, 1995, 11: 15 – 20

15　Shao XY, He ZM, Chen ZC, Yao KT. Expresion of an Epstein – Barr virus receptor and Epstein – Barr virus dependent transrormation of human nasopharyngeal epithelial cells. Int J Cancer, 1997, 71: 750 – 755

16　Tang WP, Huang PG, Zhao ML, Liao SL, Zeng Y. Wikstroemia indica promoters development of nasopharyngeal carcinoma in rats initiated by dinitrosopiperazine. J Cancer Res Clin Oncol, 1988, 114: 429 – 431

17　Wang D, Laebowitz D, Kieff E. An EBV membrane protein expressed in immortalized lymphocytes transforms established rodent cells. Cell, 1985, 43: 831 – 840

18　Wolf H, Zur Hansen H, Becker V. EB viral genomes in epithelial nasopharyngeal carcinoma cells. Nat New Biol, 1973, 244: 245 – 247

19　Zeng Y, Shen JJ, Pi GH, Ma JL, Zhang Q, Zhao ML. Application of anticomplement immunoenzymatic method for the detection of EBNA in carcinoma cells and normal epithelial cells from the nasopharynx, 1 lth International Symposium on Nasopharyngeal Carcinoma, Dusseldorf, West Germany, Grundman et al (eds). Fischer, Stuttgart, New York Cancer Campaign, 1981, 5: 237 – 245

20　Zeng Y. Seroepidemiological studies on nasopharyngeal carcinoma in China. Adv Cancer Res, 1985, 44: 121 – 139

21　Zeng Y, Liu YX, Wei JN, Zhu JS, Cai SL, Wang PZ, Zhong JM, Li RC, Pan WJ, Li EJ, Tan BF. Serological mass survey of nasopharyngeal carcinoma (in Chinese). Acta Acad Med Sin, 1979, 1: 123 – 126

22　Zeng Y, LIU YX, Liu CR, Chen SW, Wei JN, Zhu JS, Zai HJ. Application of an immunoenzymatic method and an immunoautoradiographic method for a mass survey of nasopharyngeal carcinoma. Intervirology, 1980, 13: 162 – 168

23　Zeng Y, Zhong JM, Miao XO. Epstein – Barr virus early antigen induction in Raji cells by Chinese medicinal herbs, Intervirology, 1983, 19: 201 – 204

24　Zeng Y, Zhong JM, Ye SQ, Ni ZY, Miao XQ, Mo YK, Li ZL. Screening of Epstein – Barr virus early antigen expression inducers from Chinese medicinal herbs and plants. Biomed Environ Sci, 1994, 7: 50 – 55

69.　Control of AIDS Epidemic in China

ZENG Yi[1] , WU Zun – you[2]

1. Institute of Virology, Chinese Academy of Preventive Medicine;

2. National Center for AIDS Prevention and Control, Chinese Academy of Preventive Medicine

Summary

It is predicted that AIDS will be one of the major barriers impeding social and economic development in the 21st century. The HIV/AIDS epidemic has gone through the entry and expansion phases and entered into a rapid increase phase in China. This paper discusses the challenges and strategies for controlling the HIV/AIDS epidemic in China.

[Key words] AIDS, Control and Prevention

Introduction

Since the Acquired Immunodeficiency Syndrome (AIDS) was first reported in 1981,[1-3] the virus has spread rapidly to most parts of the world. The United Nation Program on HIV/AIDS (UN-AIDS) and the World Health Organization (WHO) estimate that 49.9 million people have been infected with the HIV virus.[4] The spread of HIV/AIDS has had an impact on populations affected, significantly altering family and social structures, and incurring substantial economic costs for indi – viduals, private companies and government.

AIDS is still a deadly but preventable disease. It is unlikely that an effective vaccine will be produced to prevent people becoming infected with HIV, or a cure discovered for treating people diagnosed with AIDS in the near future. It is predicted that AIDS will be one of major barriers for social and economic development in the 21st century.

The HIV/AIDS epidemic has gone through the entry and expansion phases and entered into a rapid increasing phase in China.[5] It is predicted that China may become a country with the largest number of HIV infected people unless correct intervention strategies and measures are taken promptly. At present, the HIV/AIDS in China is at the crossroads between a controlled and uncontrolled epidemic.

At the 131st Xiangshan Scientific Meeting on Strategies for Control of AIDS in China, 6 – 8 December 1999, organized by Zeng Yi, Chen Keyi and Qing Boyi, and supported by the Ministry of Science and Technology and the Chinese Academy of Science. Scholars from various fields, like social science, law, health, and medicine, discussed epidemic trends of HIV/AIDS, strategies and measures to control its transmission through unsafe use of blood, and its products, injecting drug use, and high risk sexual behaviors. Social and economic impacts of HIV/AIDS were discussed. Recent developments made in the prevention of vertical transmission, reconstruction of immune systems and vaccine were also presented at the meeting.

Experiences and Lessons

The HIV/AIDS epidemic follows a common model, developing rapidly within high risk groups and spreading to the general population. High risk groups usually refer to homosexual men, injecting drug users (IDUs), and prostitutes. People in high risk groups are often not in the social mainstream and governments therefore usually ignore the HIV/AIDS epidemic among these groups, particularly in the early stages of the epidemic. It is often mistakenly believed that the HIV/AIDS epidemic will be limited within these high risk groups, and will not affect the general public. However, the HIV/AIDS epidemic will spread to the general population, if governments do not take action to control the epidemic among high risk groups in its early stages, or if their commitment and actions are not sufficient, appropriate, or specific enough. Unfortunately, it is difficult to control the epidemic once it has been reported among the general public.

HIV mainly infects people aged between 20 and 49 years, a group that are in the prime of their reproductive years and the most economically productive compared to other age groups. Therefore, the consequences of the epidemic seriously affect families as well as society at large. Africa has suffered due to inefficient government responses to the HIV/AIDS epidemic in its early stages. For instance, in Africa, the cumulative number of people infected with HIV and who have died of AIDS was more than 30 million. HIV infection prevalence rate has increased to 20% – 50% among general adults. Life expectancy has also been reduced dramatically. In Zimbabwe, average life expectancy has been reduced from 64.9 years before the epidemic to 39.2 years in 1998. At same time, Zimbabwe's working population was reduced from 76.49% before the epidemic to 61.62% in 1998. This also affected agriculture production, which decreased from 30% to 60%. A considerable number of families and villages have been destroyed completely by AIDS. The epidemic overall has hampered the social and economic development in most Sub – Sahara African countries.

In Asia, the epidemic has spread very rapidly. Thailand has reported 80 000 AIDS eases cumu – latively, becoming the fourth leading country in terms of reported AIDS cases in the world. In northern Thailand, some families have disappeared and villages dramatically reduced in size. Negative population growth rates have been observed in some districts of northern Thailand between 1990 and 1997. India, the second largest country in the world in terms of population, has become the leading country in terms of the number of people infected with HIV. It is estimated that more than four million people have become infected with HIV in India. In Cambodia, HIV/AIDS has already spread between high risk groups and the general public. The HIV infection rate in some regions in Burma was reported to be surprisingly high. The HIV epidemic remains widespread and uncontrolled in Russia.

Experience from countries with low HIV/STD rates shows that early implement of effective inter – vention strategies and control measures can prevent, or at least control, the numbers of people becoming infected with HIV. In Australia, nationwide AIDS campaigns and other specific intervention strategies and measures launched in the early stage of the epidemic have been successful in controlling the numbers of people infected with HIV. In effect, the HIV epidemic in Australia did not become wide – spread among the general populace, and HIV rates among high risk groups, such as IDUs and sex workers, remains very low. At present, the HIV infection rate in Australia is about 5% among IDUs and below 0.1% among sex workers. Among the homosexual population, the incidence of HIV infection has been reduced from 10% in 1984 to less than 1% in 1990, and presently remains stable at this low level. [6-7]

Even if the epidemic reached the general public, the epidemic could still be controlled if governments had the political will and took sufficient and appropriate action. In Thailand, the Thai Prime Minister responded by chairing the national AIDS committee and encouraging his ministries and all government sectors to participate in AIDS prevention campaigns, thus leading the national response to the AIDS epidemic. In addition, nationwide anti – AIDS campaigns and condom promotions were launched, and a 100% condom policy implemented in brothels around the country. Consequently, annual sexually transmitted disease (STD) cases since 1989 have been drastically reduced from 200 000 per year before 1989 to less than 10 000 in 1996. At the same time, the number of annual HIV infections has dropped dramatically. [8]

Nevertheless, for HIV/AIDS prevention programs to be effective, some essential components need to be implemented. Intravenous drug users need to have access to clean needles and syringes and methadone maintenance programs, and sex workers to condoms and the ability to use these condoms effectively. Sexual health care services also need to be provided.

Evidence from countries around the world demonstrated that corporate or capital punishment for drug use and prostitution is ineffective, forcing drug users and sex workers underground. This makes it difficult for health education prevention programs to access these populations, promoting the rapid spread of HIV among these hidden groups and from them to others. Some countries have taken a realistic approach to controlling the HIV epidemic, and drug use and sex workers. Law enforcement departments have focused on reducing drug supply and demand. At same time, needle exchange and methadone maintenance programs have been implemented among IDUs. To control prostitution and sexual transmission of HIV, young women are provided with more job opportunities and encouraged to leave the sex trade. For existing sex workers, health education, condom promotion and sexual health care service programs have been provided. These strategies have effectively controlled HIV/AIDS and STDs without increasing drug use and prostitution in countries that have adopted these measures.

Experience over the past two decades has demonstrated that AIDS epidemics can be controlled if governments take AIDS seriously and develops pragmatic strategies to control it. Otherwise, AIDS epidemics will place a great burden on the affected population producing enormous social change and economic losses and instability.

Epiemic of HIV/AIDS in China

Asia has become a new center for the AIDS epidemic. Will China becomes the country with the largest number of HIV cases in the 21st century?

HIV was introduced into China in 1982 while the ftrst AIDS case was reported in 1985. Since 1994, the annual reported number of new HIV infections has increased rapidly, In 1998, HIV has been reported in 31 provinces, autonomous regions and municipalities in China. It was estimated that the actual number of HIV infection was about 400 000 at the end of 1998. So far, there is no indication that the spread of HIV in China will be controlled or slowed down in the near future.

1. Injecting drug use

Injecting drug use is the major mode of HIV transmission in China. It accounts for about 70% of total reported cumulative HIV infections. The number of provinces that have reported HIV infec – tions among IDUs increased from one province in 1994 to 21 provinces in 1999. Outbreaks of HIV infections have been reported among IDUs in Yunnan, Xinjiang, Guangxi, Sichuan, and Guang- dong provinces and autonomous regions, [9] HIV infection rates among IDUs has reached 20% – 30% in some areas in these provinces, and about 60% – 80% in Ruili, Yunnan Province and Yin- ing, Xinjiang Automonous Region. Unfortunately, the HIV epidemic continues to spread among and from IDUs in China.

2. Blood and blood products

Transmission of HIV through blood and blood products and plasma donation is a severe problem in China. A study, conducted by the Chinese Academy of Preventive Medicine. has shown that HIV infection rate among former plasma donors in one village ranged from 16.1% to 63%. It is difficult to estimate the actual number of HIV infections caused by blood transfusions but recent years have seen dozens of lawsuits related to HIV infections caused by transfusions. It was reported that after a family member was initially infected by HIV from blood transfusions; other members were infected by secondary sexual transmission and tertiary vertical transmission.

Since 1995, the Ministry of Health has requested that all blood donations must be tested for HIV, however, the majority of county and sub – county hospitals do not have the capacity to per- form HIV screening tests. The proportion of HIV infections caused by blood supply and plasma dona- tions is considerably higher in China compared to other developing countries.

3. Sexual contact

In recent years, the number of HIV infections caused by sexual contact has increased. A cohort study showed that HIV prevalence among spouses of HIV infected persons increased from 3.1% in 1990 to 12.3% in 1997. The rate of increase for annual reported STDs was about 30%. In 1999, 860 000 STD cases were reported although the actual number of new STD cases was estimated at about four to eight million people. In addition, the current increases in the number of documented arrests among prostitutes and their clients suggests that there maybe a greater potential for a larger AIDS epidemic in China, in that, HIV sexual transmission will gradually increase and Will become the major mode of HIV transmission among the general population.

4. Mother – to – child transmission

In recent years, the number of women infected with HIV has increased in China. In Xinjiang, 40 pregnant women have been identified as HIV positive. In 1999, in a Guangzhou clinic, five women who had undergone pregnancy terminations have tested HIV positive over a 3 – month peri- od. It is predicted that mother – to – child transmission will increase with more women becoming HIV infected in China.

HIV/AIDS has spread from China's borders to inland areas, and from metropolitan to rural areas, The majority of HIV infected persons and people around them, however, are ignorant of their infection status. It is critical to take action now, otherwise this window of opportunity to control the AIDS epidemic in China will be lost.

Law enforcement practices to combat drags and prostitution over the last two decades have warned us that epidemics stemming from drug use and prostitution cannot be eradicated in the near future, although efforts can be made to reduce the number of high risk drug use and sexual behaviors. Governments at all levels should face these realities and be pragmatic in their approach to taking appropriate and effective strategies and measures to control the STD/HIV among and from high risk groups. Voluntary blood donation is still not prevalent in China although the "Law of Blood Donation" was issued in 1997. Unsafe use of blood products and illegal collection of plasma is still problems in some areas.

The consequences of the HIV/AIDS epidemic are apparent in areas where the HIV epidemic started early. In one village with a population of 200 in Yunnan Province, more than 20 young adults have died of AIDS. There are some villages where four to ten young adults have died of AIDS. The epidemic has serious social implications. In one study conducted by the Chinese Academy of Preventive Medicine, 16.8% of HIV infected people had children under 5 years and predicts that these children may grow up to be one or two parent orphans.

The economic impact of H1V/AIDS in China can be severe. In 1999, the Chinese Academy of Preventive Medicine investigated the medical costs of HIV infected persons in several hospitals in Beijing. It was found that the annual cost in the outpatient department was 6971 yuan (about US $840) on average, annual costs for hospitalization costs was 47 577 yuan (about US $5730) on average, and medical costs for a person diagnosed with HIV about 300 000 yuan (about US $36 000) on average. It was estimated that the average cost for loss of productivity due to HIV infection would be about 470 000 yuan (about US $56 000) per person, given an average working age of up to 60 years, annual production of 15 000 yuan (about US $1800) per working person, and average age at death of 28.4 years for an AIDS infected person as observed in Yunnan Province. It was estimated that the total health care costs for China would be about 462 to 770 billion yuan (about US $56 to 93 billion), given the estimates of 600 000 to 1 000 000 cumulative infections in China until the end of the year 2000.

Challenges

Although the Chinese government has paid a great deal of attention to AIDS control, there are some barriers that need to be removed urgently. The main barrier is insufficient awareness by some government officials of future trends, impacts, and inputs to control the HIV/AIDS epidemic. For instance, some local government leaders have inappropriate attitudes towards AIDS prevention education campaign and this has restricted local AIDS campaign programs and promoted further HIV transmission. Some local government leaders have also hampered scientists' efforts to collect epidemiologieal data on HIV prevalence in their areas. They are concerned that these data will affect economic development in their local area, and their political status. Evidence however indicates that AIDS campaigns do not affect economic development and tourism. United States remains the strongest economic development though many AIDS education and intervention programs have been carried out. In Thailand and Yunnan province, tourism is one of the major industries in these areas but neither respectively has been affected by the national AIDS campaign in Thailand or AIDS prevention education programs in Yunnan. The naivete among local leaders has resulted in insufficient responses to the epidemic in terms of implementing AIDS prevention education campaigns, policies to support effective AIDS control measures, implementation of effective specific intervention programs, and researches on prevention and control strategies, medicine, vaccines, and basic science research. If these problems cannot be surmounted then the goals and objectives set in the "China Long/Medium Term Plan for AIDS Control 1998 – 2010"[10] is impossible to achieve and hence, the epidemic will remain uncontrolled. It is predicted that there will be about 10 million HIV cases in China by the year 2010 if the government does not take appropriate and sufficient action to the AIDS epidemic.

Because of low awareness of AIDS among the general public, discrimination against people with HIV/AIDS and their families is still a big problem in China. One man, for example, was kicked out of school because he was infected with HIV as are many similarly HIV infected persons who are kicked out of their villages and ostracized by their families. Some families have attempted to murder infected family members by forcing them to swallow large amounts of sleeping pills.

Discrimination of HIV infected people is also popular among health workers in China. Once patient is diagnosed with HIV in a hospital, he/she will be quickly discharged with many excuses provided by health workers.

Worldwide practices have demonstrated that HIV/AIDS epidemics can be controlled by imple – menting effective intervention measures. These are:

- prevention education;
- safe blood supply;
- condom promotion among high risk groups and sexually active people;
- needle exchange and/or methadone maintenance programs for injecting drug users;
- management of STDs;
- treatment of pregnant women to prevent mother – to – child transmissions.

Studies have shown that health education by itself has limited impacts on changing risk behaviors. Health education must be combined with programs distributing condoms, and needles and syringes. As long as these measures can be implemented, the HIV/AIDS epidemic is controllable. However, these strategies have not been implemented in China because of current legal regulations and policies. IDUS and prostitution are two key issues in the HIV/AIDS epidemic and the behaviors and practices of both these groups are prohibited. Therefore, needle exchange/methadone maintenance programs for IDUs and condom promotion among prostitutes and their clients cannot be legally implemented at this point in time. To implement these strategies among IDUs and sex workers represents the biggest challenge for controlling AIDS in China today.

Suggestions

The HIV/AIDS epidemic will affect the social and economic development, even the national sta－bility, if we do not respond to the epidemic appropriately. Control of AIDS is a complicated issue. The making of intervention strategies and measures must be based on scientific studies, and experience from around the world, Intervention activities moreover must be sufficient enough to address specific risk behaviors promoting HIV transmission.

In order to effectively control the AIDS epidemic in China, we suggest the following：

1. The State Council takes a leading role in AIDS control.

The State Council creates and establishes a National AIDS Committee with the Premier or Vice Premier as chair；

An office be established under the direction of the National AIDS Committee.

2. Prevention should be the first priority of the National AIDS Committee；

The National AIDS Committee launches a National intensive AIDS campaign.

3. Policies need to be developed for adopting intervention strategies and measures that have been proved effective in controlling the AIDS epidemic worldwide, to encourage and protect health workers and social worker carrying out needle exchange and methadone maintenance programs for IDUs, and to promote condom use among prostitutes and their clients.

4. The safety of blood and blood products needs to be improved by enhancing management, clarifying responsibilities, and protecting the human rights of patients.

5. More researches on AIDS control and prevention needs to be completed, including epidemiological, behavioral science, social science, dissemination science, pharmaceutics, and vaccine studies.

6. Funding for AIDS control and prevention efforts needs to be increased to achieve the goals and objectives set in the "China Long/Medium Term Plan for AIDS Control 1998 – 2010. "

［In 《Bulletin of the Chinese Academy of Sciences》 2000, 14（2）: 106 –110］

70. Study of Immortalization and Malignant Transformation of Human Embryonic Esophageal Epithelial Cells Induced by HPV18 E6E7

SHEN Zhong – ying[1], CEN Shan[2] SHEN Jian, CAI Wei – jia[1], XU Jin – jie[1],
TENG Zhi – ping[2], HU Zhi[1], ZENG Yi[2]
1. Department of Tumor Pathology, Medical College of Shantou University；
2. Institute of Virology, Chinese Academy of Preventive Medicine

Summary

In order to study the effect of viruses and tumor promoters on the tumorigenicity of the esophagus, human embryonic esophageal epithelial cells were infected with human papilloma virus HPV18 E6E7 – AAV in synergy with 12 – O – tetradecanoylphorbol 13 – acetate (TPA) to observe their malignant transformation. The cultured esophageal epithelial cells incubated with HPV18 E6E7 – AAV were divided into two groups: the SHEEC1 group was exposed to TPA (5 ng/ml) for 4 weeks at the 5th passage of the cells; the SHEE group served as the control and was cultured in the same medium without TPA. The morphological phenotype, the DNA content during the cell cycle and the chromosomes were analyzed. The tumorigenicity was assessed by colony formation after cultivation in soft agar and transplanting the cells into nude mice. HPV18 E6E7 DNA was assayed by fluorescent *in situ* hybridization (FISH) and the polymerase chain reaction (PCR). The SHEE group, at its 20th passage, grew as a monolayer with the cells showing anchorage dependence and contact inhibition. The chromosome analysis showed diploidy, and soft – agar cultivation and injection into nude mice showed the cells to be non – tumorigenic. They were therefore immortalized cells. In contrast, the SHEEC1 group (TPA group) showed increased DNA synthesis and a proliferative index that was higher (45%) than that of the SHEE group (34%). The number of large colonies of dense multilayer cells (positively transformed foci) in soft agar was high in SHEEC1 group (4.0%) but low in the SHEE group (0.1%). Tumors resulting from transplantation were observed in all six nude mice injected subcutaneously with cells of the SHEEC1 group but no tumor developed in mice receiving cells of the SHEE group. In both groups of cells, HPV18 E6E7 DNA was positively detected by FISH and PCR. The malignant transformation of human embryonic epithelial cells was induced *in vitro* by HPV18 E6E7 in synergy with TPA. This is a good evidence for the close relationship between HPV and the etiology and pathogenicity of esophageal carcinoma. It is also a reliable model for studying the cellular and molecular mechanisms of carcinogenesis of esophageal carcinoma.

〔**Key words**〕 Human embryonic esophageal epithelium; HPV18 E6E7 genes; Immortalization; TPA; Malignant transformation

Introduction

Induction of malignant transformation in cells cultured *in vitro* is an important way to study carcinogenesis. It is used to study not only the etiology of carcinogenesis, but also the tumor – promoting factors, and it is more feasible than an animal model. We have previously successfully induced squamous cell carcinoma by subcutaneously transplanting human embryonic esophageal cells into nude mice in synergy with benzopyrene (Shen et al. 1997). Lu et al. (1989) also induced squamous cell carcinoma from human embryonic esophageal epithelium cultivated *in vitro* with N – methyl-N – benzylnitrosamine. These results demonstrated that strong chemical carcinogens may induce carcinogenesis in cultured cells. Liu et al. (1996) induced human nasopharyngeal carcinoma in nude mice by infecting fetal nasopharyngeal mucosa with Epstein Barr virus in combination with tumor promoters 12 – O – tetradecanoylphorbol 13 – acetate (TPA) and n – butyrate. On the basis of the theory that there are many factors and many stages in tumorgenesis, we used human papilloma virus HPV18 E6E7 – AAV to infect human embryonic esophageal mucosa to establish an immortalized epithelial cell line SHEE (Shen et al. 1999). This experiment was then repeated with epithelial cells exposed to the tumor promoter (TPA) to induce malignant transformation, thus establishing a carcinogenic model of human embryonic esophageal epithelial cells. This work has both theoretical and practical significance in the study of viral etiology and mechanisms involved in the carcinogenesis of esophageal carcinoma.

Materials and Methods

Construction and identification of HPVI8

1. E6E7 PAAV3 vector

E6E7 genes were amplified by the polymerase chain reaction (PCR) from the template PGEM/HPV18 (provided by Prof. Zeng Yi) in which the E6E7 genes were ligated to the PGEM – T vector. The E6E7 genes were cleaved from the vector and inserted into the PAAV3 vector. The involvement of the E6E7 genes in the PAAV3 vector was demonstrated by Southern blot hybridization. PAAV – E6E7 and PAd8 were transfected into HEK 293 cells to obtain recombinant virus containing E6E7.

2. Cultivation of human embryonic esophageal mucosa

One esophagus, obtained from a 4 – month – old embryo, which was proved to be normal, was cut into small pieces and cultivated in 199 medium (Gibco) with 10% calf serum and antibiotics (100 U/ml penicillin, 100 U/ml streptomycin).

3. HPV18 E6E7 – AAV infection and TPA treatment

Pieces of human embryonic esophageal tissue were cultivated in serum – free medium for 2 h, incubated with HPV18 E6E7 – AAV for 2 h after removal of the medium, and cultured again in 199 medium with calf serum. The growing human embryonic esophageal epithelium infected with HPV18 E6E7AAV was divided into two groups: cells of the SHEEC1 group were exposed to TPA at their 5th and 13th passages, for 2 weeks in each case; TPA was added to the culture medium at the dosage of 5 ng/ml. The SHEE group served as the control without TPA. The assays for the two groups were the same. In the another group, the original cultured epithelium was exposed to TPA only, but it could not passaged continuously.

4. Morphological observation

The cultured epithelial cells were observed under a phase – contrast microscope, under a light microscope with Giemsa staining, and in the electron microscope (Hitachi 300).

5. Cell cycle and chromosome analyses

Cells of the 20th passage were digested, washed twice with phosphate – buffered saline PBS, fixed by 70% alcohol, prepared as a single – cell suspension and stored at 4℃. The cells were stained with propidium iodide (Sigma) and analyzed by flow cytometry (FACSort, B – D Co.). The cells in the proliferative phase, the percentage of cells that were more than tetraploid and the proliferation index ($S + G_2M/G_0G_1 + S + G_2M$) were calculated. The mitotic index of the cells in the two groups at the 20th passage was calculated by counting the number of cells undergoing mitosis in their exponential phase. The cultured cells that showed more mitosis were chosen for colchicine treatment (10 μg/ml) and cultured for 2 – 3 h. The cells were collected for routine Giemsa staining for chromosome analysis.

6. Cell colony formation in soft agar

The exponential – phase cells of the two groups were trypsinized and stained with trypan blue to count the number of living cells. The living single – cell suspension [10^3 cells/ml in 0. 35% agar (Agarose, V312A, Promega)] was overlayed on 0. 7% agar in petri dishes. Five dishes for each group were incubated in 5% CO_2 in a 37 ℃ incubator for 40 days and the cell colonies were then counted.

7. Tumorigenesis in nude mice

Six – week – old BALB/C nude mice (supplied by the Experimental Animal Center of Zhongshan Medical University) were bred in isolated conditions. Six nude mice were subcutaneously injected with SHEEC1 cells in the 20th passage (1×10^6 cells/mouse). Another six nude mice were injected with SHEE cells in the same manner. They were observed every 3 days for 2 months and then killed for histopathological examination.

8. Fluorescent *in situ* hybridization (FISH)

Cells grown on cover slides were fixed with 4% paraformaldehyde, pretreated, digested with proteinase K and hybridized with an HPV18 E6E7 probe overnight at 42℃, The hybridized cells were treated with formamide and ftuorescently labeled (fluoresceinisothiocyanate linked to Avidin – D). The cell nucleus was stained with propidium iodide (10 μg/ml) and observed under a fluorescence microscope.

9. HPV18 E6E7 detection by PCR

The PCR primers for HPV18 E6E7 were designed according to oligo software, synthesized by the Shanghai Bioengineering Company.

Upstream primer: 5' – GAC ACT AGT ACT ATG GCG CGC TTT GAG – 3'

Downstream primer: 5' – AGT ACT AGT TTA CAA CCC GTG CCC TCC – 3'

The SpeI sita is shown in bold type.

Template DNAs were extracted from SHEE cells, SHEECI cells and the tumors developing following transplantation in nude mice. The PCR kit was purchased from the Sai – Bai – Sheng Biocompany and the samples were amplified by PCR (GTC – 2; Applied Res Co. USA). In 40 automated cycles, denaturing, annealing and extension time and temperature were as follows: 60 s at 94℃; 30 s at 50℃ and 120 s at 72℃. The PCR products were analyzed by agarose gel electrophoresis.

Results

1. Morphological observation

Under a light microscope, the cells in the SHEE group were uniform in size and shape (Fig. 1A), and grew as an even monolayer showing anchorage dependence and attachment inhibition. Cells in the SHEEC1 group crowded together and were of different sizes; many more cells with giant nuclei cells and several nucleoli (Fig. 1B) were seen. Under the electron microscope, cells with an ovoid nucleus and a small nucleolus were seen in the SHEE group and there were tonofilaments in the cytoplasm (Fig. 2A). In the SHEE1 group, however, the nucleus was full of folds and hollows with an enlarged nucleolus and a lack of tonofilaments (Fig. 2B). This show that the cells were over – proliferating and poorly differentiated.

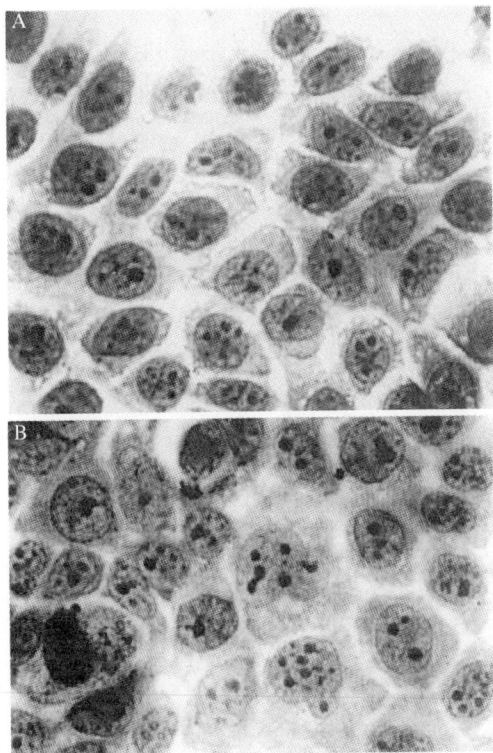

A. The immortalized cells of the SHEE group were grown in monolayers and had a uniform cell nucleus and small nucleolus (Giemsa; ×400). B. The malignantly transformed cells of the SHEEC1 group had enlarged nuclei of different sizes with several nucleoli (Giemsa; ×400)

Fig. 1 A, B Morphology of cultured cells

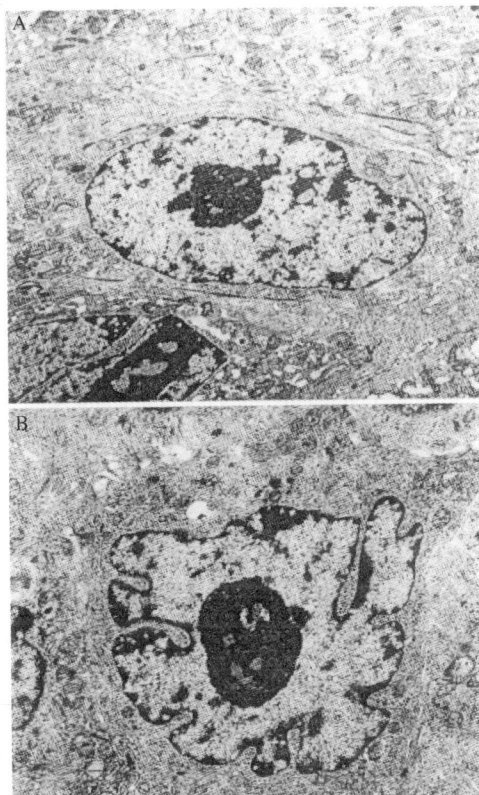

A. The immortalized cells of the SHEE group had an ovoid nucleus and tonofilaments in the cytoplasm (arrow) (electron microscope, EM; ×7000). B. The malignantly transformed cells of the SHEEC1 group had irregularly shaped nuclei with folds and hollows on the nuclear membrane; the nucleoli were enlarged and there were few tonofilaments. EM; ×7000

Fig. 2 A, B Micrographs of cultured cells

2. Chromosome analysis

The number of chromosomes in the SHEE group ranged from 45 to 54 per nucleus, nuclei containing more than 46 chromosomes accounting for 23.08%. The modal number of chromosomes was still diploid (Fig. 3A). In the SHEEC1 group, the number of chromosomes increased to 96 per nucleus (Fig. 3B), the proportion of nuclei containing more than 46 chromosomes rising to 55.56% (Tab. 1); the modal chromosome content hyperdiploid.

3. Kinetics of cell proliferation assayed by flow cytometry

The DNA histograms (Fig. 4A, B) show the proliferation index of the SHEEC1 group to be greater (45%) than that in the SHEE group (34%). A more than tetraploid cell content was found to be more frequent in the SHEEC1 group than in the SHEE group (5.70%, 1.53%).

4. Cell colony formation in soft agar

Cells were cultivated in soft agar and were observed once every 10 days. Small colonies (fewer than 10 cells) were found in the SHEE group and large colonies (more than 20 cells), which grew rapidly to form multilayer colonies, swelling in the center and protrusions at the margin (Fig. 5), were found in SHEEC1 group.

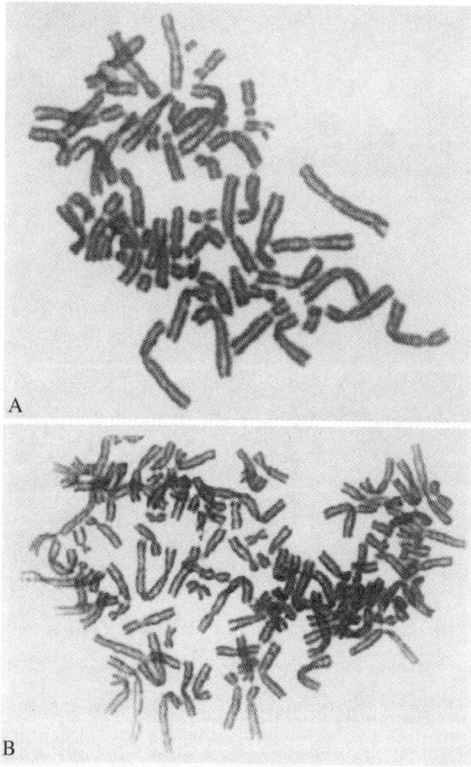

A. The chromosome number of cells in the SHEE group was in the diploid range. B. In the SHEEC1 group, the chromosome number was more than tetraploid. Giemsa; ×1000

Fig. 3   A, B Chromosome analysis

Tab. 1   Chromosome analysis of the SHEE and SHEEC1 groups

Cell group	Number of cells						Percentagegroup of cells with >46 chromosomes (%)
	Dividing	Having a chromosome number					
		<46	46	−52	−72	−96	
SHEE	26	6	14	6	0	0	23. 08
SHEEC1	27	4	8	7	5	3	55. 56

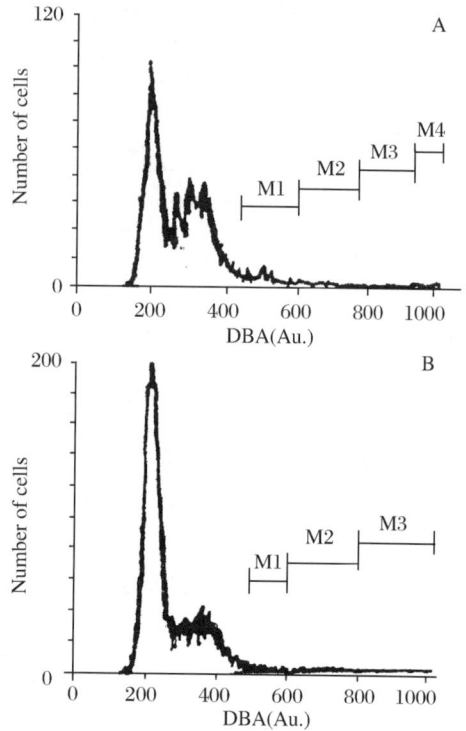

A. DNA histogram of the SHEEC1 group: M1 – M4 DNA > 4n. B. DNA histogram of the SHEE group: M1, M2, M3, DNA > 4n

Fig. 4   A, B Flow – cytometric assay

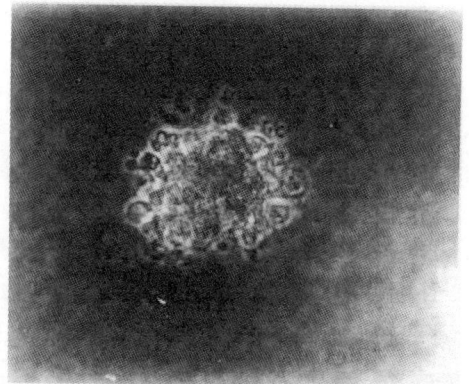

Fig. 5   A large colony formation in soft agar. Phase contrast; ×100

5. Tumor development in nude mice

When SHEEC 1 cells were injected into the axilla of nude mice, they grew rapidly to form tumor in 20 days (Fig. 6A). In histological examination on the 30th day, the cells showed a large nucleus, less cytoplasm, a large nucleolus, and infiltration and destruction of muscular fibers (Fig. 6B). The tumor could be passaged continuously in nude mice and a cell line was established that grew faster than the primary SHEEC1 cells. But when the cells of the SHEE group were injected into the nude mice, they gradually reduced and disappeared in 20 days.

6. HPV18 E6E7 detection by FISH

Scattered hybridized spots were seen under a fluorescent microscope in cells of both the SHEEC1 (Fig. 7A) and SHEE groups (Fig. 7B), proving the existence of HPV18 E6E7 genes in the cell nucleus.

7. Detection of HPV18 E6E7 by PCR

Fig. 8 showed the bands of PCR DNA products of the SHEE group, the SHEEC1 group and of tumors growing following transplantation into nude mice, analyzed by agarose gel electrophoresis. A specific band fragment of 875 bp was found as a marker in each of them, proving the existence of an HPV18 E6E7 gene fragment in the cells of all three groups.

Discussion

To assess the cell transformation, this experiment was based on four approaches: (a) the morphological changes of the cell; (b) analysis of the chromosome ploidy; (c) cell colony formation in soft agar; (d) tumorigenesis in nude mice. In the course of 20 passages, the SHEE group grew as a monolayer and still retained the characteristics of anchorage dependence and contact inhibition. In soft – agar cultivation and when transplanted into nude mice, SHEE proved to be nontumorigenic; they were therefore immortalized cells (Hopfer et al. 1996). Under the transmission electron microscope, they showed tonofilaments in the cytoplasm indicating a certain degree of differentiation of the epithelial cells.

HPV16 E6E7 can induce cell immortalization in human oral epithelial cells (Oda et al. 1996), mammary epithelial cells (Wazer et al. 1995), bronchial epithelial cells (Viallet et al. 1994) and in epithelial cells of human pancreatic ducts (Furukawa et al. 1996). The induction of immortalization of epithelial cells by HPV18 E6E7 is considered to be caused by the action of products expressed by the HPV E6E7 genes on the antioncogenes, since the E6 protein may degrade the p53 – encoded protein (Demers et al. 1994) and the E7 protein may act on retinoblasloma protein (Boyer et al. 1996). The degradation and inactivation of the products of the tumor – suppressor genes promote entry into the cell cycle, thus leading to cell proliferation.

TPA is a strong tumor – promoting compound. The dosage used may vary from 0. 1 ng/ml (Bessi et al. 1995) to 300 ng/ml (Sakai et al. 1995) and the time of induction may be from several hours to several weeks. In any event, TPA exerts a synergetic effect in the malignant transformation of cells. In this work, we have induced the malignant transformation of the SHEEC1 group of cells, using a low dosage (5 ng/ml) and long – term cultivation. As reported previously, the synergetic effect of TPA is due to its action upon the cytokine signal system, specifically acting on protein kinase C through a diglyceride to promote protein synthesis and cell proliferation (Wolf 1985).

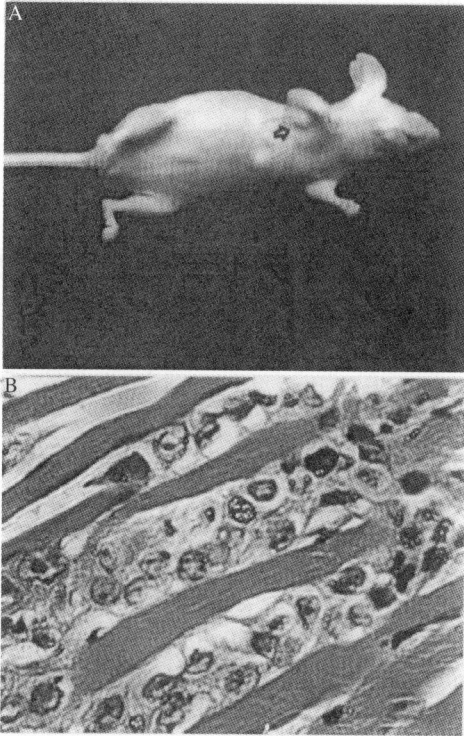

Fig. 6 A. Two tumor masses developing 20 days after 10^6 SHEEC1 cells were injected into the axilla of a nude mouse (arrow). B. Infiltration of tumor cells and the damaged muscular stratium seen microscopically. Hematoxylin/eosin; ×400

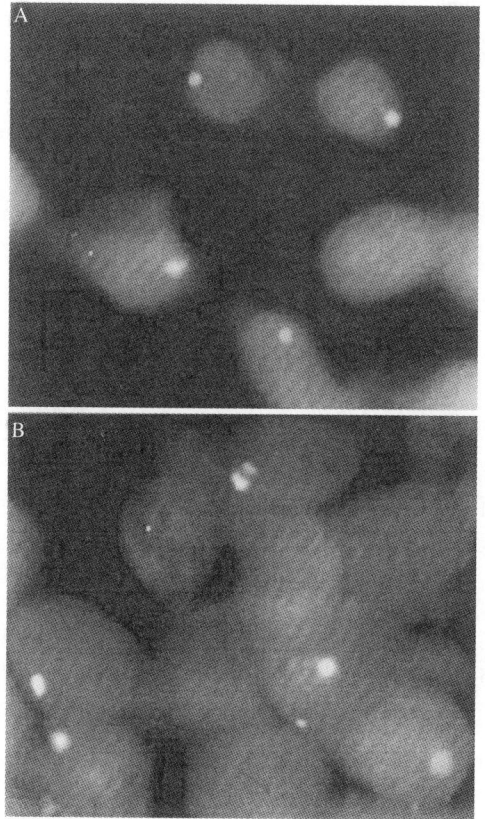

A. Cells of the SHEE group. B. Cells of the SHEEC1 group. FISH; ×1000

Fig. 7 A, B Detection of HPV18 E6E7 by fluorescent *in situ* hybridization (FISH) in two groups of cells, shown as light hybridized spots within the cell nucleus

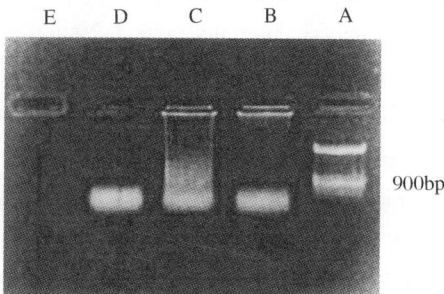

900bp

A. PBR322/BstNI, B. SHEE group, C. SHEEC1 group, D. tumors developing after transplantation into a nude mouse, E. negative control

Fig. 8 Agarose gel electrophoresis of polymerase chain reaction products

Carcinogenesis is a prolonged event with many etiologies and many stages, and our experiment was designed on the basis of two etiologies and two stages (IARC/NCI 1985). The HPV18 E6E7 infection, as the initiating factor, was the first stage and adding the tumor – promoting factor TPA was the second stage. Giving alow dosage of TPA for 4 weeks induced the malignant transformation in 10 weeks. The delayed transformation, which involved a quantitative change and qualitative change of cellular characteristics and a small number of transformed cells becoming a large number of transformed cells, needs a definite period of time. So, given that people are usually exposed to frequent small doses of carcinogens and tumor promoters, carcinogenesis has to be experienced for a long time. Our *in vitro* experimental design, using a low dose and a long time for it to take effect, therefore corresponds to the actual conditions as far as possible.

The human embryonic epithelial cells used in this experiment were free from various external factors, were infected with HPV18 E6E7 to induce immortalization and were infected with HPV18 E6E7 in synergy with TPA to induce malignant transformation. This provides direct evidence for the close relationship between HPV and the etiology and pathogenesis of esophageal carcinoma. It is also a reliable model for studying the viral etiology of esophageal carcinoma, environmental carcinogens, the molecular biological changes of cell carcinogenesis, and the biology of cell proliferation, differentiation and reversion. This work has both theoretical and practical value.

〔In 《J Cancer Res Clin Oncol》 2000, 126: 589 – 594〕

References

1　Bessi H, Rast C, Rether B, Nguyen Ba G, Vasseur P. Syn – ergistic effects of chlordane and TPA in multistage morpho – logical transformation of SHE cells. Carcinogenesis, 1995, 16: 237 – 244

2　Boyer SN, Wazer DE, Band V. E7 protein of human pap – illoma virus – 16 induces degradation of retinoblastoma protein through the ubiquitin – proteasome pathway. Cancer Res, 1996, 56: 4620 – 4624

3　Demers GW, Halbert CL, Galloway DA. Elevated wild – type p53 protein levels in human epithelial cell lines immortalized by the human papilloma virus type 16 E7 gene. Virology, 1994, 198: 169 – 174

4　Furukawa T, Duguid WP, Rosenberg L, Viallet J, Galloway DA, Tsao MS, Long – term culture and immortalization of epithelial cells from normal adult human pancreatic ducts transfected by E6E7 gene of human papilloma virus 16. Am J pathol, 1996, 148: 1763 – 1770

5　Hopfer U, Jacobberger JW, Gruenert DC, Ecker RE, Jat PS, Whitsett JA. Immortalization of epithelial cells. Am J Physiol, 1996, 270: CI – 11

6　IARC/NCI/EPA working group Cellular and molecular mechanisms of cell transformation and standardization of transformation assays of established cell lines for the prediction of carcinogenic chemical. Overview and recommended proto – cols. Cancer Res, 1985, 45: 2395 – 2399

7　Liu ZS, Li BM, Liu TF, Zeng Y. Studies on human nasopharyngeal malignant lymphoma and undifferentiated carcinoma induced the synergetic effect of EB virus and promotors (in Chinese). Chinese J Virol, 1996, 12: 1 – 8

8　Lu SY, Cui XX, Xie JG. Esophageal carcinoma in human fetus induced by N – methyl – N – benzyl – nitrosamine (NMBZN). Chin J Oncol, 1989, 11: 401 – 403

9　Oda D, Bigler L, Lee P, Blanton R, HPV immortalization of human oral epithelial cells: a model for carcinogenesis. Exp Cell Res, 1996, 226: 164 – 169

10　Sakai A, Miyata N, Takahashi A. Initiating activity of quinones in the two – stage transformation of BALB/3T3 cells. Carcinogenesis, 1995, 16: 477 – 484

11　Shen ZY, Cai W J, Shen J, Xu J J, Cen S, Ten ZP, Hu Z, Zeng Y. Immortalization of human fetal esophageal epithelial cells induced by E6 and E7 genes of human papilloma virus 18 (in Chi-

nese). Chin Exp Clin Virol, 1999, 13: 121 –
123

12 Shen ZY, Xu J J, Fang D, Shen J. A study on
human fetal esophagus heterotransplantation and
induced carcinoma in nude mice (in Chinese).
In: Li CH (ed) Current advances in tumor biology. MMS, Beijing, 1997, 185

13 Viallet J, Liu C, Emond J, Tsao M. Characterization of human bronchial epithelial cells immortalized by the E6 and E7 genes of human papilloma virus type 16. Exp Cell Res, 1994, 212:

3641

14 Wazer DE, Liu XL, Chu Q, Gao Q, Band V.
Immortalization of distinct human mammary epithelial cell types by human papilloma virus 16 E6
or E7. Proc Natl Acad Sci USA, 1995, 92:
3687 3691

15 Wolf M. A model for intracellular translocation of
protein kinase C involving synergism between calcium and phorbol ester. Nature, 1985, 315:
546 – 549

71.　The Genetic Events of HPV – Immortalized Esophageal Epithelium Cells

SHEN Zhong – ying[1], XU Li – yan[1], CHEN Xiao – hong[2], CAI Wei – jia[1],

SHEN Jian[1], CHEN Jiong – yu[3], HUANG Tian – hua[2], ZENG Yi[4]

Departments of 1. Pathology and 2. Biology, 3. Central Laboratory of Tumor Hospital, Medical College
of Shantou University; 4. Institute of Virology, Chinese Academy of Preventive Medicine

Summary

We studied cytogenesis, telomere and telomerase, and c – myc, ras, bcl – 2, and p53 genes
of cells in the progressive process of immortal epithelial cells from embryonic esophagus induced by
human papillomavirus (HPV). The SHEE cell line, established by us, consist ot immortalized epithelial cells from the embryonic esophagus induced by genes E6E7 of HPV type 18. It was in initial
malignant transformation when cultivated over 60 passages without co – carcinogens. Cells of the
10th, 31st, and 60th passages were represented in the progressive process within the immortal period. In these three stages of the cell line, the modal number of chromosome and karyotypes were analyzed. The telomere length was assayed by Southern blot methods, and the telomerase activity was
analyzed by hTR and hTERT assay. C – myc, p53, bcl – 2, ras genes were assayed by the multi –
PCR method. The morphology of the 10th passage cells exhibited good differentiation, the 60th passage cells were relatively poorly differentiated, and the 31st passage cells differentiated in two distinct ways. The growth characteristics of the 31st and 60th passage cells were weakened at contact –
inhibition and anchorage – dependent growth. Karyotypes of three cell passages belonged to hyperdiploid and hypotriploid with abnormal'chromosomes +1, +3, +7, +9, +17, +18; del (1)
(p32); der (4), t (4;?) (q31;?); der (5), t (5;?) (q31;?); der (13), t (13; 13)
(pll; qll) and others. Bimodal distribution of chromosomes with more aberrant chromosomes appeared in the 31st and 60th passage cells. Telomere length sharply shortened from normal fetal
esophagus to the 10th and 31st passage step by step, but was stable from the 31st to the 60th passage and the telomerase activities measured were expressed at late two passages. p53 mutant was positive in three passages, c – myc was positive in the 31st and the 60th passage K – ras only in the
last. The results reveal that changes of chromosomes, telomere length, telomerase activity and certain gene expressions are important events of HPV – immortalized esophageal epithelium cells. All of
these changes occurred in dynamic progressive process. This cell line may be useful for the elucida-

tion of the genetic mechanism of cellular immortalization.

[**Key words**] Esophagus; Human papillomavirus; Immortalization; Cytogenesis; Molecular genesis

Introduction

Recent evidence indicates that several cytogenetic abnormalities are implicated in the development and progression of esophageal carcinoma[1]. But the molecular and genetic alterations involved in the progression of immortal cells remain poorly understood. The immortal epithelial SHEE cell line induced by genes E6E7 of human papillomavirus (HPV) type 18 in our laboratory[2], were proliferative squamous epithelial cells in property[3]. At early passages, there had been malignant transformation by promoter TPA[4]. In continual cultivation, the cells at the 31st passage begun to differentiate in two directions, nest cells in good differentiation, and nest cells of poor differentiation. Part of the cells in the 61st passage (SHEE61A) were at initial malignant transformation[5].

The range of genetic events include chromosomal aberration, telomeric shortening and telomerase activity, and expression of certain genes. In general, the immortal or transformed cells caused by carcinogens are accompanied by chromosomal abnormality and mutation of genes[6]. The cytogenetic alterations manifested themselves as abnormal structure and number of chromosomes, as well as changes in the telomere and telomerase[7,8]. The genetic changes, first in oncogenes and anti−oncogenes[9], involved ras, myc, bcl−2 and p53.

To determine the genetic changes in immortalized SHEE cell line from the 10th to the 60th passages we studied the cytogenetic changes including the modal number, karyotype of chromosomes, the telomere and telomerase, and the molecular changes in c−myc, ras, bcl−2, p53 genes.

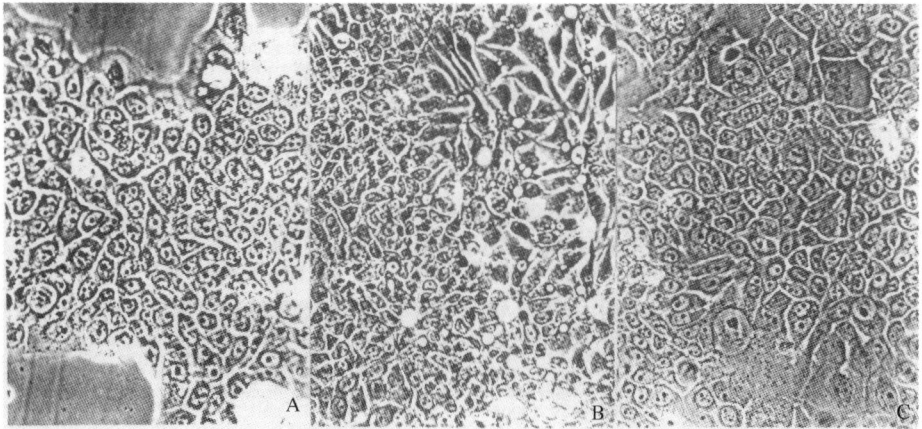

A. the cells of SHEE10 were of even monolayer distribution. B. the cells of SHEE31 were of biphasic differentiation. C. the cells of SHEE60 grew closely together, varying in shape and size (phase contrast microscope, ×400)

Fig. 1 Morphology of living cells in three passages of SHEE

Materials and Methoods

1. Cell culture: The SHEE cell line, a kind of immortal epithelial cell from the embryonic esophagus induced by genes E6E7 of human papilloma virus (HPV) type 18, was established by us[1]. It was routinely cultivated in culture medium 199 (Gibco) with 10% calf serum, 100 units of penicillin and streptomycin, in an incubator with a humidified atmosphere of 5℃ CO_2 and 95% air. Selected different generations at the 10th passage (SHEE10), the 31st passage (SHEE31) and the 60th passage (SHEE60) were inoculated in culture flasks.

2. Living cell examination: The cell shape, anchorage – dependent and contact – inhibited growth were examined by phase contrast microscopy.

3. Cytogetnetic analysis: Metaphase spreads of SHEE10, SttEE31 and SHEE60 cells were obtained using standard cytogenetic methods. Briefly, the cultured cells of each passage were preserved at 3 – 4℃ for 3 – 4 h to make cells at synchronous stage, then cultivated at 37℃ for 3 – 4 h, added in 0. 05 μg/ml colchicine for 1 h. Harvesting was by standard method and stained with Giemsa. One hundred metaphases were scored for each line.

4. G – band of chromosome: The prepared chromosome specimens, after 2 – 3 days at room temperature, were treated with 0. 125% trypsin under 37℃ for 1 rain, rinsed, Giemsa stained, and 20 nicely dispersed mitotic nuclei were chosen for analysis of G – band karyotypes.

5. Telomere and telomerase subunits: The genomic DNA from normal fetal esophagus, SHEE10, SHEE31 and SHEE60 were extracted. Telomeric restriction fagments (TRF) were measured by Southern blot[11]. Briefly, 20 μg of genomic DNA was digested with HinfI (MBI) and run on 0. 7% agarose gel with marker DNA/HindIII. After electrophoresis the gel was blotted to nylon membrane (Hybond N$^+$ nylon membrane, Amersham Life Science) and hybridized to the digoxin – labelled probes (CCCTAA)$_3$ at 50℃ in 5 × SSC, 0. 1% SLS, 0. 02% SDS for 12 – 16 h and washed twice at room temperature in 2 × SSC, 0. 1% SDS for 5 min, once at 50℃ in 1 × SSC, 0. 1% SDS for 10 min, twice at 50℃ in 1 × SSC, 0. 1% SDS for 10 min. twice at 50℃ in 0. 1 × SSC, 0. 1% SDS for 5 min, stained with NBT/BCIP and the middle points were measured to get the mean telomere length[12].

The telomerase activities were performed by expression of telomerase RNA (hTR) and human telomerase reverse transcriptase (hTERT). Cellular RNA extracts were assayed with reverse transcription – PCR (RT – PCR) assays in contrast with house – keeping gene GAPDtt[13]. For RT – PCR, total RNA was prepared from cells by using Gstract^{TM} RNA Isolation Kit (Maxim Biotech, Inc.). Briefly, eDNA was synthesized from 10 μg of total RNA using Ready – to – use First Strand eDNA Synthesis Kit (Maxim Biotech, Inc.). The PCR reaction was performed using 2μl aliquots of the reverse – transcribed eDNA, 2. 5 μl 10 × PCR buffer, 1. 5 μl 25 mmol/L $MgCl_2$, 2. 5 μl 1. 0 mmol/L dNTP, 2. 0 μl specific primers, 14 μl water, 0. 5 μl taq (2 U/μl). The primer pairs of hTR, hTERT and GAPDH were: hTR, 5' – TCTAACCCTAACTGAGAAGGGCGT AG – 3', 5' – GTTTGCTCTAGAATGAACGGTGGAAG – 3' for 22 cycles of PCR (94℃, 45 sec; 55℃, 45 sec; 72℃, 90 sec). hTERT, 5' – CGGAAGAGTGTCTGGAGCAA – 3', 5' – GGATGAAGCGGAGTCTG–

GA – 3' for 31 cycles (94℃ for 45 sec, 55℃ for 45sec. 72℃ for 90 sec). GAPDH, 5' – GAAG-GTGAAGGT CGGAGTC – 3', 5' – GAAGATGGTGATGGGATTTC – 3' for 33 cycles of PCR (94℃, 60 sec; 55℃, 60 sec; 72℃, 60 sec). PCR products of each sample were subjected to electrophoresis in a 1.5% agarose gel containing 0.5 μg/ml ethidium bromide (EB, Sigma).

6. *Multiplex PCR (mPCR) detection*: C – myc, bcl – 2, and p53 were detected by eDNA mPCR and ras was detected by DNA mPCR[10]. Total RNA of three stages of the SHEE cell lines were extracted by extract track kit II (Roche Co., USA), which had been reverse transcribed into eDNA. C – myc, bcl – 2, p53 and house – keeping gene GAPDH were analyzed with detected mPCR kit (APO – MO50G. Maxim Biotech Inc.). The ras mPCR kit (RAS – MO50, Maxim Co.) contains ras positive contrast, H – ras, K – ras primer and DNA marker. Each product of PCR was electrophoresed with 1.5% agarose gel containing EB (0.5 μg/ml), epi – illuminating fluorescence detected EB fluorescent band, with the molecular weight of 921 bp, GAPDH; 371 bp, c – myc; 233 bp, bcl – 2; 204 bp, p53; in the procedure of APO – MO50G kit, 133 bp, K – ras gene exon – 2; 103 bp, H – ras gene exon – 1 in the procedure of RAS – MO50 kit.

Tab. 1　Number of chromosomes in SHEE line series

No. of passage	No. of cell	No. of chromosomes				Modal number
		≤46 (%)	47 – 57 (%)	58 – 68 (%)	≥69 (%)	
SHEE10	91	17 (18.7)	45 (49.5)	22 (27.7)	4 (4.3)	58 – 62
SHEE31	100	12 (12.0)	38 (38.0)	46 (46.0)	4 (4.0)	55 – 57, 61 – 63
SHEE60	102	10 (9.6)	26 (25.5)	44 (43.1)	22 (21.6)	58 – 60, 63 – 65

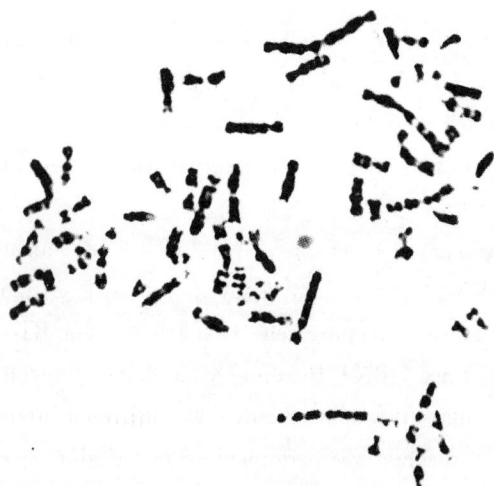

Fig. 2　Hyperdiploid of SHEE31
(Giemsa, ×1000)

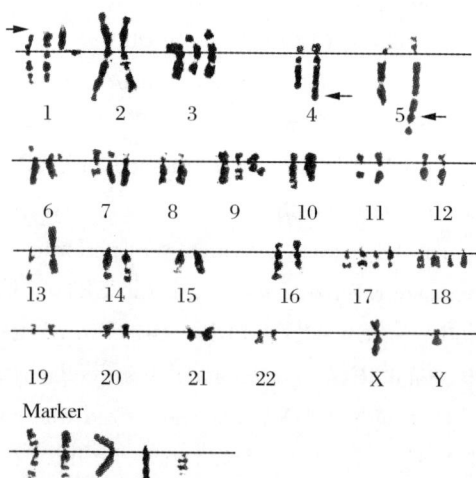

Fig. 3　Karyotype of SHEEIO with many trisomy or tetrasomy.

58, xy. +1, +3. +7, +9, + +17, + +18

Results

1. Morphology of cells: Early immortal SHEE10 cells grew evenly on the dish. Cells appeared typical of squamous epithelium with multi – angular outline and oval nucleus (Fig. 1A). SHEE31 cells were biphasic in morphology, partially well differentiated and partially undifferentiated basal cells (Fig. 1B). The cells of SHEE60 were different in shape and size and grew densely packed (Fig. 1C). The poorly differentiated cells in SHEE60 were weakened during contact – inhibition growth.

2. Cytogenetic abnormality: The number of chromosomes in SHEE10, SHEE31, and SHEE60 (Tab. 1) ranged between 32 and 196, these chromosomes were mainly hyperdiploids and hypotriploids (Fig. 2), most hypertriploid cells were found at SHEE60. The modal number of chromosomes at SHEE10 was 58 – 62 with abnormal chromosomes +1, +3, +7, +9, +17, + +18 (Fig. 3). SHEE31 and SttEE60 had a bimodal distribution, 55 – 57, 61 – 63 at SHEE31 with the abnormal chromosomes, +1, +3, +7, +9, +11, +12, +14, +17, +18. and 58 – 60, 63 – 65 at SHEE60 with the abnormal chromosomes, +1, +3, + +7, +9, +11, +12, +13, −14, +17, +18, +22. The modal number from the 10th to the 60th passage increased slowly.

3. Structural aberration of chromosomes was identified as follows: The composite karyotypes at SHEE10 were del (1) (q 12). del (1) (p32); der (4), t (4;?) (q31;?); der (5), t (5;?) (q31;?) der (13). t (13; 21), (pll; qll). At SHEE31 were del (1) (ql2) del (1), (p32); der (4), t (4;?) (q31;?); der (13), t (13; 21) (pl 1; ql 1) At SHEE61A were dei (1) (p32); der (4), t (4;?) (q31;?); der (5). t (5;?) (q31;?); der (13), t (13: 13) (pl 1; ql 1); der (13), t (13; 14) (pll: qll). In these three cell lines, karyotypic aberrations were mainly partial deletion of chromosomes lq12 and p32. and translocation of 4q, 13p. According to these results, the instability and increased imbalance of chromosomes, and the appearance of preneoplastic aneuploidy[14], were probably the genetic background for cell immortalization.

4. p53, bcl – 2, c – myc, ras and GAPDH: The highest p53expression was in SHEE60, while the lowest or negative in SHEE10. In SHEE31, p53 and ras were positive at low level and c – myc and bcl – 2 were very weak (Fig. 4). It was probable that the p53 was a mutant.

5. Telomere length and telomerase activity:

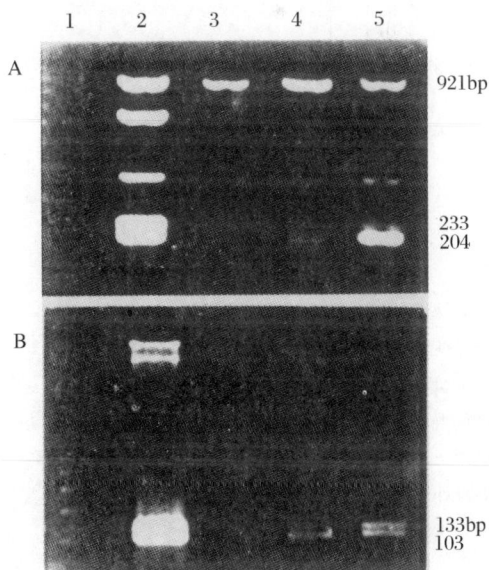

A, p53, bcl – 2, c – myc and GAPD // : B, H – ras and K – ras; lanes: 1, marker; 2, positive contrast: 3, SHEE10: 4. SHEE31: 5, SHEE60

Fig. 4　Elctrophoretograph of PCR products

The TRF length of the three cell lines of SHEE and normal fetal esophageal tissue were measured by Southern blot analysis. The mean length of normal fetal esophageal epithelial telomere was 30.0 kb with band concentration: SHEE10 telomere bands shortened to 17.3 kb, SttEE31 shortened again to 3.5 kb, and the SHEE60 bands were 3.8 kb (Fig. 5). Telomere length sharply decreased in the cells of SHEE10 and SHEE31, and it was stable after the 31st passage. By hTR and hTERT determination (Fig. 6), telomerase activation was absence in fetal esophageal tissue but appeared in cells of 31st and 60th passages, as it was strongly positive in SHEE60.

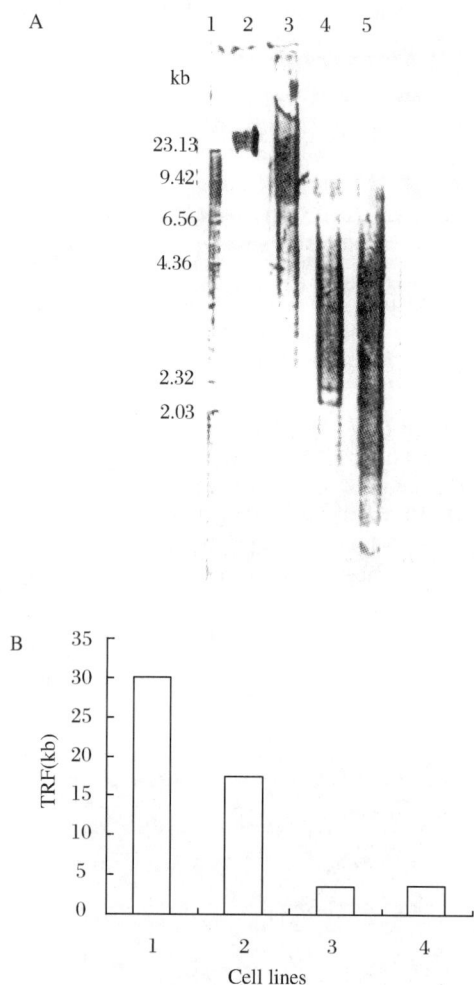

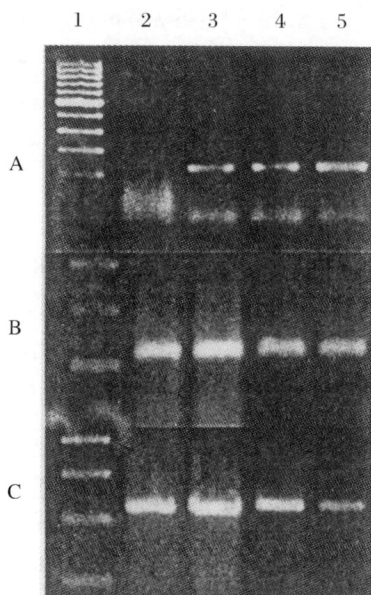

Lanes: 1, marker; 2, fetal esophagus: 3, SHEE10; 4, SHEE31; 5, SHEE60. Column: 1, fetal esophagus: 2, SHEE10; 3, SHEE31: 4, SHEE60 (bar graph was plotted from the data in Fig. 5A)

Fig. 5   A, the length of TRF in SHEE series. B, alterations of TRF in SHEE series

A. hTR (A); B. hTERT; C. GAPDH. 1, marker; 2, normal fetal esophageal tissue; 3, SHEE10: 4, SHEE3 I; 5, SHEE60

Fig. 6   Gel electrophoretogram of hTR and hTERT. The activity of telomerase in SHEE series

Discussion

This study in the genetic events of immortalized esophageal epithelial cells shows that aberration of chromosomes, alteration of telomere length, telomerase activity and the expression of certain genes in three passages were found to varying extents. There were more hyperdiploid and hypotriploid in the chromosomes of three SHEE immortal cell lines, the separate modal number of chromosomes first appeared in SHEE31 and continued to SHEE60. The number of chromosomal sets and the percentage of hyperploids varied from the SHEE10 to the SHEE60, and the SHEE60 had more hypertriploid cells than the others lines, thus, showing that chromosomes of SHEE series cell lines were unstable. Chromosome instability, often referred to as karyotypic instability, is one of the major characteristics of the SHEE cells. Chromosome instability in most cases reflects the occurrence of defective mitosis, including unequal distribution of chromosomes to daughter cells, which leads to generation of aneuploid cells[15]. Aneuploidy destabilized the karyotype and thus initiated an autocatalytic karyotype evolution generating preneoplastic and eventually neoplastic karyotypes. An aneuploid being a discrete chromosome mutation will tend to lead immortal cells to progressively change toward malignant transformation. The changes of cytogenetics will control the proliferation and differentiation of cells[16], which will easily undergo malignant transformation by promoters[17].

In our data, the telomere of the normal fetal esophageal cell 30 kb in length, shortened to 17 kb in SHEE10, then shortened again to 3.5 kb in SHEE31, and maintained this level thereafter. The telomerase first appeared weak in the 10th passage. It could not prevent continual shortening of telomere. This dissociation between telomere dynamics and telomerase activity supports the existence of additional controls on TRF in SHEE cells [18]. Therefore, in practice, the cells of SHEE had difficulty surviving in cultivation before the 20th passage. In 1965, Hayflick reported that the culture life of human diploid fibroblasts was limited to 50 – 100 generations, the same as epithelial cells[19]. In 1985, Greider discovered the activity of teromerase[20]. In 1994, Kim also found a specific association between the telomerase activity with immortal cells and cancer cells[21]. The telomerase is perhaps a prerequisite to immortal cells. Weitzman and Hahn believed T – antigen of SV40 and ras gene induce transformation of normal epithelium cells which require expression of telomerase[22], so activity of the telomerase is the early event of immortalization[23].

Our results indicated that immortalization of SHEE might require both activation of telomerase and other genetic alterations that promote cell hyperplasia and abrogate normal differentiation. Some genes, such as c – myc, ras, p53 and bcl – 2, which we analyzed are related to immortalization and malignant transformation of cells. In our data, c – myc, ras, bcl – 2 and p53 (a probable mutation type) were shifted – up in SHEE60, the morphology and tumorigenicity in some of the cells showed the phenotype of malignant transformation, so the cells SHEE60 were in the premalignant stage. In SHEE10 cells which were in an early stage of immortalization, c – myc and ras gene were negative and, p53, bcl – 2 were at a low level. In general, c – myc and p53 (mutant) promote proliferation of cells[24,25] and bcl – 2 can prevent cell apoptosis[26]. Ras is a transforming growth signal causing rapid aneuploid and malignant transformation of cells[27,28]. It has been reported that when the cell transformation induced by SV40 occurred, ras was a required gene.

Our data demonstrated that multiple stages in the immortalization process in SHEE cells are associated with different genetic and phenotypic characteristics, just as the immortal prostate cell line or skin keratinocytes reported elsewhere[30,31].

The SHEE cell line was induced by ttPVI8 E6E7 and viral DNA was demonstrated in the cells[17]. Infection of HPV can cause confusing karyotypes[32], such as chromosomal breakage, abnormalities in structure and number of chromosomes[33]. It also hints that immortal esophageal cells induced by HPVI8 E6E7 may cause chromosomal changes, and liabilities in genetic characteristics[33]. The virus genome inserts and integrates with the chromosomes of host[34], and causes the activation and expression of oncogenes[35]. HPV E6E7 protein, in conjugation with anti – oncoprotein p53[36] and pRb[37], thus loses control of cellular growth, and promotes the phenotypic production of cellular transformation[38,39]. Exposure of the early passage conditionally immortal cell line to the viral oncogenes, HPV or SV40T, results in conversion of these telomerase negative to expression of high levels of telomerase activity[40]. Previous reports also have shown that activation of telomerase can be achieved by the E_6 and E_7 proteins of HPV[41,42].

In summary, the frequency of these numerical chromosomal alteration have led us to propose that a crucial event in the process of immortalization may be the induction of genetic instability and preneoplastic aneuploidy at an early stage. In the progressive developmental stage, the immortalized epithelial cells may require both activation of telomerase and amplification of some genes such as c – myc, ras, bcl – 2 and p53 (mutant) which abrogate normal differentiation and promote dysplasia in some cells.

Acknowledgements

Contract grant sponsor, National Natural Science Foundation of China (39830380).

〔In 《Int J Mol Med》 2001, 8: 537 – 542〕

References

1 Barrett MT, Sanchez CA, Galipeau P, Neshat K, Emond M and Reid BJ. Allelic loss of 9p21 and mutation of the CDKN2/pI6 gene develop as early lesions during neoplastic progression in Barrett's esophagus. Oncogene, 1996, 13: 1867 – 1873

2 Shen ZY, Cai W J, Shen J, Xu J J, Cen S, Ten ZP, ttu Z and Zeng Y. Immortalization of human fetal esophageal epithelial cells induced by E6 and E7 genes of human papilloma virus 18. Chin J Exp Clin Virol, 1999, 13: 121 – 123

3 Shen ZY, Shen J, Cai W J, Cen S and Zeng Y. Biological characteristics of human fetal esophageal epithelial cell line immortalized by the E6 and E7 gene of HPV type 18. Chin J Exp Clin Virol, 1999, 13: 209 – 212

4 Shen ZY, Cai W J, Shen J, Xu J J, Cen S. Ten ZP, Hu Z and Zeng Y. Human papilloma virus 18E6E7 in synergy with TPA induced malignant transformation of human embryonic esophageal epithelial cells. Chin J Virol, 1999, 15: 1 – 6

5 Shen ZY, Chen XH, Shen J, Cai W J, Chen JY, Huang TH and Zeng Y. Malignant transformation of immortalized human embryonic esophageal epithelial cells induced by human papillomavirus. Chin J Virol, 2000, 16: 97 – 101

6 Wang X, Xiao F and Wang M. Establishment of two human esophageal carcinoma cell lines and their cytogenetic analysis. Chung Hua Chung Liu Tsa Chih, 1998, 20: 5 – 8

7 Tsao SW, Zhang DK, Cheng RYS and Wan JS. Telomerase activation in human cancer. Chin Med J, 1998, 111: 745 – 750

8 Zhang DK, Ngan HY, Cheng RY, Cheung AN, Liu SS and Tsao SW. Clinical significance of telomerase activation and telomeric restriction fragment (TRF) in cervical cancer. Eur J Cancer, 1999, 35: 154 – 160

9 Balmain A, An – See K, Brown K, Bryson S,

Clarke M, Crombic R and Fee F. Oncogenes and tumour suppressor genes in multistage carcinogenesis. In: Scientific Report 1993. Cancer Research Campaign Beatson Laboratories, Bell and Ltd, Glasgow, 1993, pp37 – 41

10 Uchida K. Recombination and amplification of multiple portions of genomic DNA by a modified polymerase chain reaction. Anal Biochem, 1992, 202: 159 – 161

11 Allsopp RC, Vaziri H, Patterson C, Goldstein S, Younglai EV, Futcher AB, Greider CW and Harley CB. Telomere length predicts replicative capacity of human fibroblasts. Proc Natl Acad Sci USA, 1992, 89: 10114 – 10118

12 Ohyashiki JH, Ohyashiki K, Sano T and Toyama K. Non – radioisotopic and semi – quantitative procedure for terminal repeat amplification protocol. Jpn J Cancer, 1996, 8: 329 – 331

13 Nakamura TM, Morin GB, Chapman KB, Weinrich SL, Andrews WH, Lingner J, Harley CB and Cech TR. Telomerase catalytic subunit homologs from fission years and human. Science, 1997, 277: 955 – 959

14 Li R, Sonik A, Stindl R, Rasnick D and Duesberg P. Aneuploidy vs. gene mutation hypothesis of cancer: recent study claims mutation but is found to support aneuploidy. Proc Natl Acad Sci USA, 2000, 97: 3236 – 3241

15 Tarapore P and Fukasawa K. p53 mutation and mitotic infidelity. Cancer Invest, 2000, 18: 148 – 155

16 Hahn WC, Counter CM, Lundberg AS, Beijershergen RL, Brooks MW and Weinberg RA. Creation of human tumour cells with defined genetic elements. Nature, 1999, 400: 464 – 468

17 Shen ZY, Cen S, Shen J, Cai WJ, Xu J J, Teng ZP, Hu Z and Zeng Y. Study of immortalization and malignant transformation of human embryonic esophageal epithelial cells induced by HP-VI8E6E7. J Cancer Res Clin Oncol, 2000, 126: 589 – 594

18 Kiyozuka Y, Asai A, Yamamoto D, Senzaki H, Yoshioka S, Takahashi H, Hioki K and Tsubura

A. Establishment of novel human esophageal cancer cell line in relation to telomere dynamics and telomerase activity. Dig Dis Sci, 2000, 45: 870 – 879

19 Hayflick L. The limited in vitro lifetime of human diploid cell strains. Exp Cell Res, 1965, 37: 614

20 Greider CW and Blackburn EH. Identification of a specific telomere terminal transferase activity in tetrahymena extracts. Cell, 1985, 43: 405 – 413

21 Kim NW, Piatyszek MA, Prowse KR, Harley CB, West MD, Ho PL, Coviello GW, Wright WE, Weinrich SL and Shay JW. Specific association of human telomerase activity with immortal cells and cancer. Science, 1994, 266: 2011 – 2015

22 Weitzman JB and Yaniv M. Rebuilding the road to cancer. Nature, 1999, 400: 401 – 402

23 Hiyama T, Yokozaki H, Kitadai Y, Haruma K, Yasui W, Kajiyama G and Tahara E. Overexpression of human telomerase RNA is an early event in oesophageal carcinogenesis. Virchows Arch, 1999, 434: 483 – 487

24 Tan LJ and Shen ZY. A study on quantitative analysis of c – myc in situ hybridization (DNA – DNA) in epithelial cancerization of human esophagus. Cancer, 1997, 16: 109 – 110

25 De – Miglio MR, Simile MM, Muroni MR Pusceddt, S, Calvisi D, Carru A, Seddaiu MA, Daio L. Deiana L, Pascale RM and Feo F. Correlation of c – myc overexpression and amplification with progression of preneoplastic liver lesions to malignancy in the poorly susceptible Wistar rat strain. Mol Carcinog, 1999, 25: 21 – 29

26 Jost M, Class R, Kari C, Jensen PJ and Rodeck U. A central role of BcI – X (L) in the regulation of keratinocyte survival by auto – crine EGFR ligands. J Invest Dermatol, 1999, 112: 443 – 449

27 Mok SC, Bell DA, Knapp RC, Fishbaugh PM, Welch WR, Muto MG, Berkowitz RS and Tsao SW. Mutation of K – ras protoncogene in human ovarian epithelial tumors of borderline malignan-

cy. Cancer Res, 1993, 53: 1489 – 1492

28　Mutter GL, Wada H, Faquin VC and Enomoto T. K – ras mutations appear in the premalignant phase of both microsatellite stable and unstable endometrial carcinogenesis. Mol Pathol, 1999, 52: 257 – 262

29　Chin L, Tam A, Pomerantz J, Wong M, Holash J, Baldeesy N, Shen Q, O'Hagan R, Pantginis J, Zhou H, Horner JW 11, Cordon – Cardo C, Yanwpoulos GD and Depinho RA: Essential role for oncogenic Ras in turnout maintenance. Nature, 1999, 400: 468 – 472

30　Fusenig NE and Boukamp P. Multiple stages and genetic alterations in immortalization, malignant transformation, and tumor progression of human skin keratinocytes. Mol Carcinog, 1998, 23: 144 – 158

31　Hukku B, Mally M, Cher ML, Peehl DM, Kung H and Rhim JS. Stepwise genetic changes associated with progression of non – tumorigenic HPV – 18 immortalized human prostate cancer – derived cell lin to a malignant phenotype. Cancer Genet Cytogenet, 2000, 120: 117 – 126

32　Steenbergen RD, Hermsen MA, Walboomers JM, Meijer GA, Baak JP, Meijer CJ and Snijders PJ. Non – random ailelic losses at 3p 11p and 13p during HPV – mediated immortalization and concomitant loss of terminal differentiation of human keratino – cytes. Int J Cancer, 1998, 76: 412 – 417

33　Duensing S, Lee LY, Duensing A, Basile J, Piboonniyom S, Gonzalez S, Crum CP and Munger K. The human papillomavims type 16E_6 and E_7 oncoproteins cooperate to induce mitotic defects and genomic instability by uncoupling centrosome duplication from the cell division cycle. Proc Natl Acad Sci USA, 2000, 29: 10002 – 10007

34　Weijerman PC, van Drunen E, Konig JJ, Teubel W, Romijn JC, Schroder FH and Hagemeijer A. Specific cytogenetic aberrations in two novel human prostatic cell lines immortalized by human papillomavirus type 18 DNA. Cancer Genet Cytogenet, 1997, 99: 108 – 115

35　Mullokandov MR, Kholodilov NG, Atkin NB, Burk RD, Johnson AB and Klinger HP: Genomic alterations in cervical carcinoma losses of chromosome heterozygosity and human papilloma virus tumor status. Cancer Res, 1996, 56: 197 – 205

36　Choo KB, Chen CM, Han CP, Cheng WT and Au LC. Molecular analysis of cellular loci disrupted by papillomavims 16 integration in cervical cancer: frequent viral integration in topologically destabilized and transcriptionlly active chromosomal regions. J Med Virol, 1996, 49: 15 – 22

37　Demers GW, Halbert CL and Galloway DA. Elevated wild – type p53 protein levels in human epithelial cell lines immortalized by the human papillomavirus type 16 E7 gene. Virology, 1994, 198: 169 – 174

38　Boyer SN, Wazer DE and Band V. E7 protein of human papilloma virus – 16 induces degradation of retinoblastoma protein through the ubiquitin – proteasome pathway. Cancer Res, 1996, 56: 4620 – 4624

39　Garbe J, Wong M, Wigington D, Yaswen P and Stampfer MR. Viral oncogenes accelerate conversion to immortality of cultured conditionally immortal human mammary epithelial cells. Oncogene, 1999, 18: 2169 – 2180

40　Song S, Liem A, Miller JA and Lambert PF. Human papilloma – virus types 16 E6 and E7 contribute differently to carcinogenesis. Virology, 2000, 26: 141 – 150

41　Nowak JA: Telomerase, cervical cancer, and human papilloma – virus. Clin Lab Med, 2000, 20: 369 – 382

42　Klingelhutz A J, Foster SA and McDougall JK. Telomerase activation by the E6 gene product of human papillomavirus type 16. Nature, 1996, 380: 79 – 82

72. A Comparative Study of Telomerase Activity and Malignant Phenotype in Multistage Carcinogenesis of Esophageal Epithelial Cells Induced by Human Papillomavirus

SHEN Zhong – ying[1], XU Li – yan[1], LI Chun[1], CAI Wei – jia[1],
SHEN Jian[1], CHEN Jiong – yu[2], ZENG Yi[3]

1. Department of Pathology; 2. Central Laboratory of Tumor Hospital, Medical College of Shantou
University, Shantou; 3. Institute of Virology, Chinese Academy of Preventive Medicine

Summary

To examine certain characteristics of multistep carcinogenesis, we studied telomerase activity and malignant phenotypes in the immortal, premalignant and malignant stages of esophageal epithelial cells induced by HPV. An immortalized human fetal esophageal epithelial cell line (SHEE) was induced by E6E7 genes of human papillomavirus (HPV) type 18. Cells in the 10th passage, (SHEE10), 31st passage (SHEE31), 61st passage (SHEE61) and SHEE61A which were selected and expanded from anchorage – independent growth colonies of SHEE61, were examined as follows: cell morphology by electron – microscopy; the cell cycle by flow cytometry, telomerase activity by TRAP assay, tumorigenic detection including anchorage – independent growth by soft agar culture and tumor formation by inoculating cells into SCID and nude mice, and detection ot HPV 18 E6E7 oncoprotein by Western blot. The morphology ot the SHEE10 cells exhibited good differentiation, the SHEE6C and SHEE61A cells were relatively poorly differentiated, and the SHEE31 cells were differentiated in two distinct ways. The telomerase was activated in SHEE31, SHEE61 and SHEE61A, but not in SHEE10 cells. SHEE61 and SHEE61A cells were weakened in contact – inhibition and increased in anchorage – independent growth. Inoculated into SCID and nude mice, the cells of the earlier two passages could not develop tumors; the SHEE61 developed one tumor in four SCID mice, but not in nude mice, and the SHEE61A cells developed tumors in both strains of immunodeficient mice. HPV 18 E6E7 DNA detection by Western blotting was positive in all cell passages. In the process of carcinogenesis by HPV, the cells of SHEE31 are in an immortalized state with telomerase activity. The fact that SHEE61 cells remained immortalized and also demonstrated anchorage – independent growth, reveals premalignant character; the cells of SHEE61A exhibited malignant transformation with tumor formation in mice. The results revealed that the telomerase activity, anchorage – independent growth and tumor formation in nude mice are the indicators for immortalization, premalignancy and malignancy, respectively.

[Key words] Esophagus; Human papillomavirus; Carcinogenesis telomerase; Anchorage – independent growth

Introduction

The fundamental aspects of cancer biology consist of the proliferation and differentiation of cells, cell senescence[1] and cell apoptosis, instability of chromosome[2], gene aberration[3], telomere length[4], telomerase activity[5], anchorage – independent growth (AIG)[6] and tumor formation[7].

Recent studies have demonstrated a higher frequency of telomerase activity in immortal cell lines[8,9] and cell malignancies[10,11]. It is clear that immortalization and the activation of a telomere maintenance mechanism are closely associated; every immortalized human cell line examined to date has been found to have telomere maintenance activity. In addition this higher frequency of telomerase expression has been discovered in HPV – positive esophageal squamous cell carcinoma[12,13].

Malignant transformation of cells is normally detected by AIG *in vitro* and tumor formation *in vivo*[14]. AIG usually correlates with transformation of cells as well as tumorigenesis[15], and the transformed phenotype can be associated with AIG[16]. Therefore, AIG is a pivotal method to detect the ability of tumor cells to survive and metastasize *in vivo* and allows transformed cells to form colonies in a semisolid medium *in vitro*[16].

By HPV18 E6E7 inducement, we established an immortalized human embryonic esophageal epithelial cell line (SHEE)[17]. At early passages, SHEE10 kept the characteristics of esophageal squamous epithelial cells[18]. In continual cultivation, the cells at the 31st passage (SHEE31) began to differentiate in two directions, nest cells of good differentiation and nest cells of poor differentiation[19]. When this cell line propagated over 60 times (SHEE61), the passage intervals shortened, cells grew overlapped, and cell morphology was atypical. It was suspected to have malignant transformation[20,21].

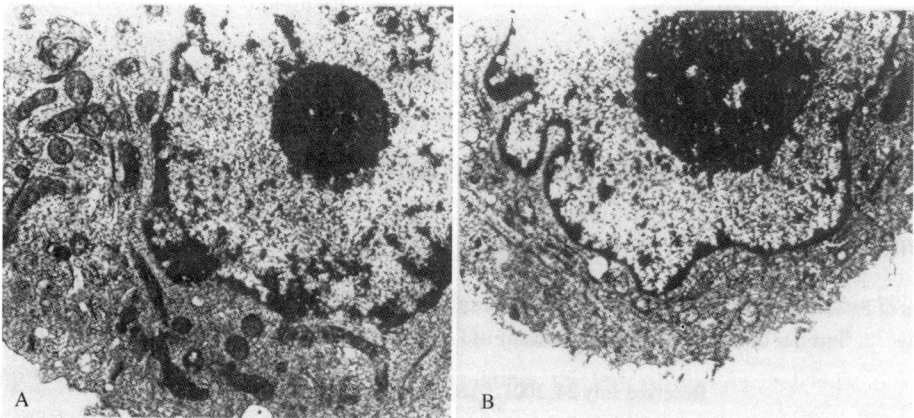

A, SHEE10 revealed differentiated epithelial cells (EM, ×10K); B, SHEE61A cells displayed proliferative pattern (EM, ×10K)

Fig. 1   Electron – micrography of SHEE

According to our previous studies on SHEE, we found that studies of human cell carcinogenesis *in vitro* were hampered by the perplexing problem of identifying immortal, premalignant and malignant cells. In this study we describe the SHEE in terms of cell morphological phenotype, telomerase activity, anchorage – independent growth and tumor formation, to ascertain whether the characteristics of SHEE cells in various stages were in immortalization or in malignant transformation, and to confirm a multistep procession of tumorigenicity in esophageal epithelium induced by human papillomavirus (HPV).

Materials and Methods

1. Cell culture: The immortalized esophageal epithelial cell line SHEE induced by HPV 18 E6E7 AAV was established by us[17], and cultured in medium 199 (Gibco) supplemented with 10% calf serum and 100 units/ml each of penicillin and streptomycin. Before taking various passages of SHEE10, SHEE31, and SHEE61 cells for examination, they were cultured in flasks and in a 24 – well plate within cover – slides.

2. Selective clonal expansion from soft agar culture[7]: The cells of SHEE10, SHEE31 and SHEE61 were cultured in soft agar dishes with 0.7% agarose (Promega) at the bottom and 0.35% agarose on top. After 2 weeks the rapidly growing cell clone of SHEE61 was collected and recultured in flasks. The selected cells grew and were named SHEE61A.

3. Cell colony formation in soft agar: The exponential – phase cells of the four groups, SHEE10, SHEE31, SHEE61 and SHEE61 A, were trypsinized and stained with trypan blue to count the number of living cells. The living single – cell suspension (10^3 cells/ml) was cultured in 0.35% agar (agarose, V312A, Promega) over 0.7% agar in petri dishes. Two dishes for each group were incubated in 5% CO_2, in a 37℃ incubator for 4 weeks and the cell colonies (AIG) were counted. AIG in four groups of SHEE cells using the Automated Region of Interest (ROI) Finder of Kodak ID image analysis system (IAS) was calculated. In brief, images were captured of colonies cultured on the soft – agar dish using Kodak DC290 digital camera, and colonies in the ROIs evaluated using Edge Detection software. Using the minimum and maximum size field to define objects between 20 – 200 pixels (– 5 to 50 cells), two values were acquired by automated ROI measurements. The first selected value identified ROI and the second value reported the total number of AIG foci in varying sizes of ROIs.

4. Electron – microscopic examination: For electron – microscopic examination, SHEE10, SHEE31, SHEE61 and SHEE61A cells were collected, respectively, centrifuged and the cell pellet fixed with 2.5% glutaraldehyde. The tissue of the transplanted tumor in nude mice was cut into 1 mm^3 pieces, and fixed in the same way. Both samples were routinely prepared and observed under electron – microscope (Hitachi 300).

5. Flow cytometry analysis: Cells of SHEE10, SHEE31, SHEE61 and SHEE61A were collected respectively and washed twice with PBS, fixed with 70% alcohol and filtered through a nylon net of 360 meshes to make a single cell suspension (10^6 cells/ ml). The cells were stained by propidium iodide (PI) for 15 min and analyzed by flow cytometry (FACSort, B – D Company). In a DNA histogram, the percentage of cells in each phase of the cell cycle, and the cells containing hyper-ploidy DNA were calculated. The proliferative index was calculated by the formula of (S + G_2M/ G_0G_1 + S + G_2 M).

6. *Telomerase activity assay*: Telomerase activity was measured using TRAP – eze Telomerase Detection Kit (Oncor Inc.). Frozen samples were homogenized in 10 – 50 μl of ice – cold lXC-HAPS lysis buffer (10 mmol/L Tris – HCl, pH 7.5, 1 mmol/L $MgCl_2$, 1 mmol/L EGTA, 0.1 mmol/L benzamidine, 5 mmol/L β – mercaptoethanol, 0.5% CHAPS, 10% glycorol). After 30 min of incubation on ice, the lysate was centrifuged at 12 000 g for 20 rain at 4℃.

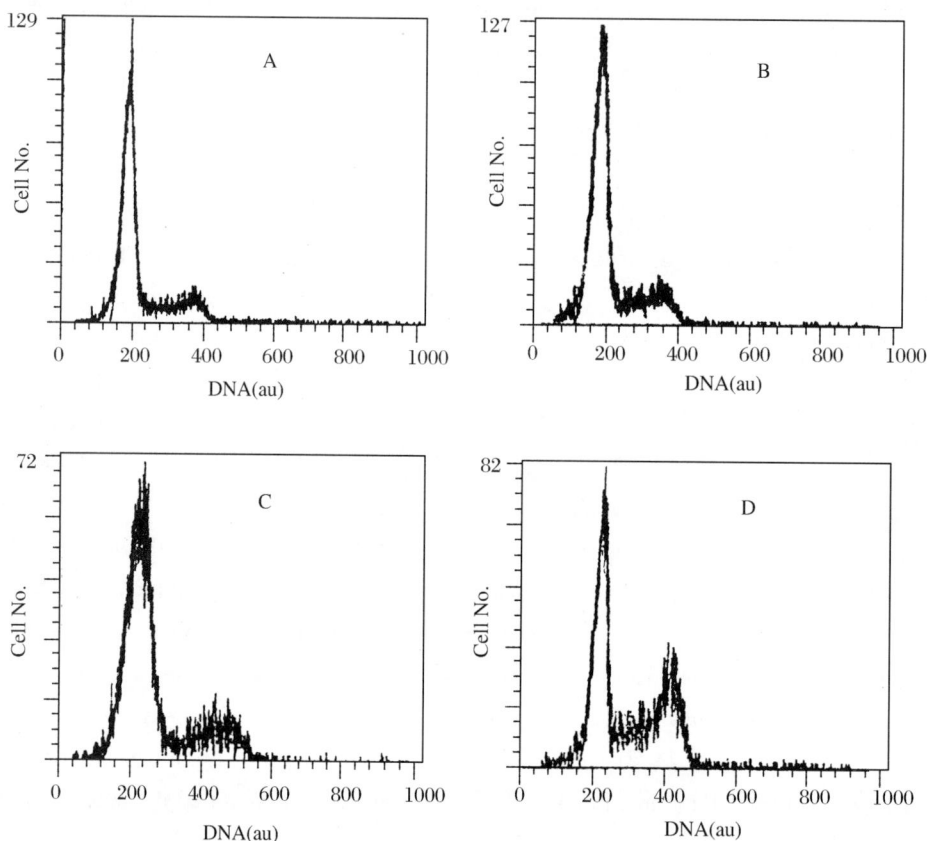

A, SHEE10; B, SHEE31; C, SHEE61 and D. SHEE61A

Fig. 2 DNA histogram of SHEE

The telomeric repeat amplification protocol (TRAP) reaction was performed using 1.0 μl lysate or 1/10 diluted lysate, 2.5 μl 10 × TRAP buffer (200 mmol/L Tris – HCl, pH 8.3, 15 mmol/L $MgCl_2$, 630 mmol/L KCl, 0.5% Tween – 20, 10 mmol/L EGTA, 0.1% BSA), 0.5 μl 2.5 mmol/L dNTP, 0.5 μl TS primer, 0.5 μl TRAP primer mix, 19.5 μl water, and 0.5 μl taq (2 U/μl). After incubation at 30℃ for 30 min, the reaction mix was immediately transferred to 94℃ and PCR performed at 94℃ for 30 sec, 55℃ for 30 sec, 72℃ for 30 sec for 35 cycles. PCR products were separated in a non – denaturing 12% PAGE in 1 × TBE at 5 V/cm. The gel was stained using $AgNO_3$ and was photographed.

7. *Tumorigenicity*: SHEE10, SHEE31, SHEE61 and SHEE61A cells (10^6 cells/0.2 ml) were inoculated separately into the right axilla of four nude mice (BALB/c, supplied by Experimental Animal Center of Zhongshan Medical University) and four severely combined immunodeficiency (SCID) mice for each group, and the tumor generation was observed for two months.

8. *HPV18 E6E7 assays by Western blot*: The protein expression of HPVI8 E6E7 was detected by the Western blot method. After digestion in flask with 0.25% trypsin and 0.2% EDTA, the cells were collected and washed three times with ice – cold PBS, then lysed in buffer [50 mmol/L Tris – HCl, pH 8.0, 150 mmol/L NaCl, 100 μg/ml phenyl – methyl – sulfonyl fluoride (PMSF), 1% Triton X – 100] for 30 min on ice. After removal of cell debris by centrifugation (12 000 g, 5 min), the protein concentration of lysates was measured by the Bradford method. Fifty μg of proteins of different passages were boiled for 5 min in sample buffer and were separated by 10% SDS – PAGE, then transferred onto nitrocellulose membrane (BioRad). Non – specific reactivity was blocked by incubation overnight at 4 in buffer (10 mmol/L Tris – HCl, pH 7.5, 150 mmol/L NaCl, 2% Tween – 20. 4% bovine serum albumin). The membrane was then incubated with mouse anti – HPVl8 E6E7 antibody, (SC – 460, HPV – CIP5, Santa Cruz Biotechnology Inc.), followed by reaction with antimouse IgG – HRP antibody. Reactive protein was detected by ECL chemilumines – cence system (Amersham).

Results

1. *Cell morphology*: Under the electron microscope, SHEE10 and SHEE31 cells had round or ovoid nuclei and a moderate nucleolus and there were tonofilaments in the cytoplasm (Fig. IA) and showed good differentiation. In the SHEE61 and SHEE61A groups, however, the nucleus was full of folds and hollows with an enlarged nucleolus and a lack of tonofilaments (Fig. 1B). This shows that the cells were proliferated and poorly differentiated.

2. *Proliferative cycle of cells*: In the DNA histogram, the proliferative index of SHEE10, SHEE31, SHEE61 and SHEE61A cells were 25.5%, 33.3%, 43.0% and 52.5%, respectively (Fig. 2). Hyperploidy cells were seen in SHEE10 cells, 2.76%; SHEE31, 4.2%; SHEE61 cells 6.1 and SHEE61A, 8.30%.

3. *Telomerase activity*: In the early passages, telomerase activity could not be detected in SHEE10 cells and it appeared in the cells of 31st passage and was maintained continuously over 60 passages. Comparing the variation in expression of telomerase in three passages of cells, the cells of SHEE31, SHEE61 and SHEE61A, had been detected at the same level. The positive control of human esophageal squamous cell carcinoma expressed the highest telomerase activity, and the negative control of normal mucosa of esophagus did not express this activity (Fig. 3).

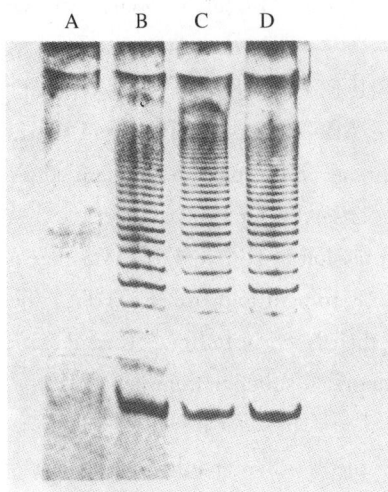

A, SHEE10; B, SHEE31; C, SHEE61; D, SHEE61A

Fig. 3　Telomerase ladder of SHEE

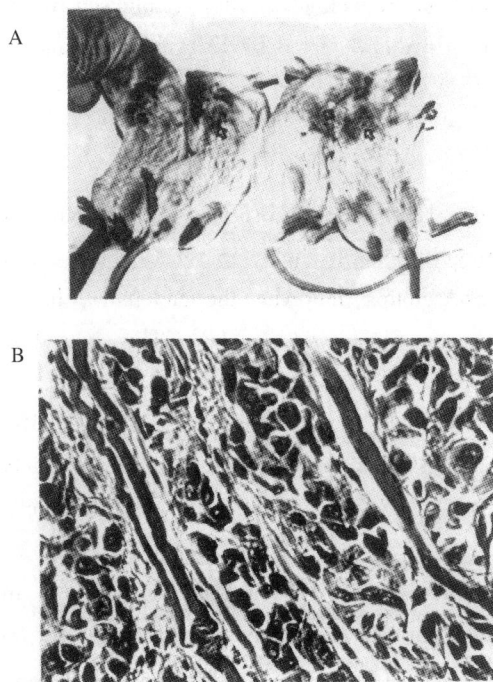

A large colony is shown in SHEE64A (phase contrast microscope, ×200)

Fig. 4　SHEE61 cells were cultured on soft agar

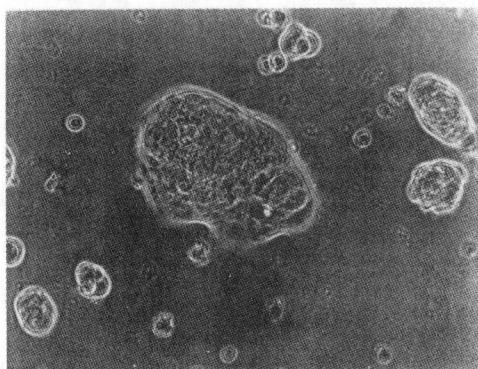

Tab. 1　Quantitative calculation of AIG foci in soft agar culture (%)

Cells	<5	6 – 10	11 – 20	21 – 50
SHEE 10	3	0	0	0
SHEE31	10	20	0	0
SHEE61	20	15	10	5
SHEE61A	15	20	25	30

Fig. 5　Heterotransplanting SHEE61A cells into SCID mice. tumors in right axilla (arrow) of four SCID mice were found (A), and the histopathology of tumor was shown (B) (H&E. ×400)

4. Anchorage – independent growth (AIG): Cells of four passages were cultivated in soft agar and were observed weekly. In SHEE61 and SHEE61A cells grew rapidly in soft agar cultured with anchorage – independent foci formation. Small colonies (fewer than 10 cells) were found in the SHEE10 and SHEE31 group, and large colonies (more than 20 cells) were found in SHEE61 and SHEE61A. The colonies in SHEE61A grew rapidly with protrusions in the center (Fig. 4). The quantitative AIG was calculated in Tab. 1.

5. Tumor development in nude and SCID *mice*: In SCID mice inoculated with SHEE10, SHEE31, SHEE61 and SHEE61A cells, SHEE10 and SHEE31 cells did not develop tumors. SHEE61 cells developed only one tumor as a hetero – transplant in four SCID mice with a long latency period, while the cells, which were injected into the nude mice, produced no tumor. When SHEE61A cells were injected into the axillae of two strains of mice, they grew rapidly to form a tumor in 20 days (Fig. 5A). In histological examination on the 30th day, the cells showed a large nucleus, less cytoplasm, a large nucleolus, and infiltration and destruction of muscular fibers (Fig. 5B). The tumor could be propagated continuously in nude mice and a cell line was established that grew faster than the primary SHEE61 cells.

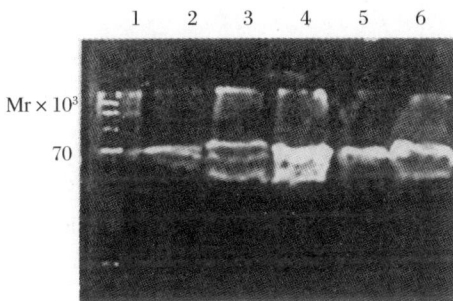

1, marker; 2, positive control; 3, SHEE10; 4, SHEE31; 5, SHEE61; 6, SHEE61A

Fig. 6　The expression of HPV18 E6E7 was detected by Western blot

Tab. 2　The telomerase activity and malignant phenotypes in four stages of SHEE

Cells	SHEE10	SHEE31	SHEE61	SHEE61A
Telomerase activity	−	+ +	+ +	+ +
Anchorage independent growth	−	−	±	+ +
Tumorigenicity	−	−	+	+ +

± , weak; + , mild; + + , moderate.

6. Expression of HPV18 E6E7: Figure 6 shows the expression of HPV18 E6E7 in SHEE series. Three passages of immortal cells and the SHEE60A cells were analyzed by Western blot. A specific band fragment of 70×10^3 was found as a marker in each of them, proving the existence of HPV18 E6E7 gene expression in the cells of various stages in SHEE series.

7. Comparison of telomerase activity and malignant phenotype in SHEE *series* (Tab. 2): In the multistages of carcinogenesis, the cells of SHEE10 were negative in telomerase activity, AIG and tumor formation. The cells of SHEE31 were positive for telomerase activity without AIG and tumor formation. The cells of SHEE61 were positive for telomerase activity and AIG, but weak in tumorigenesis. In the cells of SHEE61A, all the three criteria were positive. The four groups of SHEE cells all showed they were in the proliferative state and all expressed the HPV 18 E6E7 protein.

Discussion

Carcinogenesis is a multistep process[22] – immortalization, premalignancy and malignancy[23]. The pathway from immortalization to full malignancy may be long involving the accumulation of genetic (including molecular genetic and cytogenetic) alterations and changes of biological behavior. Recently Weitzman and Yaniv[24] and Hahn et al[25] claimed that three mutant genes converted normal human cells into tumorigenic cells. Some genes are important in carcinogenesis. Moreover, Li et al[26] have suggested aneuploidy vs. gene mutation hypothesis of cancer. Aneuploidy destabilizes the karyotype and causes tumor formation.

In this study, an attempt was made to identify the morphologically distinct type of immortalization and transformed cells in the SHEE series. The changes in the manner of growth were as follows: growth patterns of the SHEE10 cells were uniform in size and shape, and grew as an even monolayer showing anchorage – dependence and attachment inhibition; cells of the SHEE31 were differentiated in two directions, partially well differentiated and partially undifferentiated; the SHEE61 cells grown on slides differed in shape and size. In some areas, the cells grew overlapped and their anchorage growth and contact inhibition weakened; in SHEE61A most of the cells resembled the cells of SHEE61 in morphology and growth pattern.

A few cells in the immortalized cell population may have undergone malignant transformation in the SHEE61 cell line. Using the selective clone expansion method, it is possible to search for cells, which had been subject to malignant transformation from the immortalized cell line. After repeated proliferation and amplification, the number of selected cells were increased and exhibited a novel characteristics (SHEE61A) including malignant phenotype. In SHEE61, before it became tumorigenic, there was accumulation of all the above – mentioned changes, such as telomerase activity, AIG and other growth patterns.

There may be two roles for telomerase in the immortal and malignant lesions[1]. The first one will be physiological and can be present in benign lesions regardless of their HPV status[10]. The second one is associated with cancer development and is activated during the late stage of multistep carcino – genesis. Furthermore, studies of telomerase activity in premalignant lesions assist in the definition of associating telomerase with multistep carcinogenesis[8,9]. If detectable, telomerase assay can be applied for early detection of cancer, it will be a biomarker for premalignancy[27].

The role that high – risk HPV plays in epithelial carcinogenesis[28] will be ascribed to the effects of protein expression[29,30] that interact with and change cellular phenotypes[31,32]. One of the most crucial changes is activation of telo – merase by HPV[10] leading to immortalization[33,34]. Although HPV oncogenes can induce immortalization of normal cells[35], it can be difficult to induce cell malignancy without promoter. Oda et al reported the establishment of an immortalized oral epithelial cell line induced by HPV16, which had been cultured for four years and 350 passages[36]. When it was inoculated into nude mice, it did not develop cancer, despite progressive chromosome changes[37]. They questioned whether malignant transformation can be induced simply by HPV. In our studies, we found that the selective cell clone (SHEE61A) produced by the optimal seeking method from the immortalized esophageal epithelium SHEE61 induced by HPV, can further be malignantly transformed. From immortalization to full malignancy, the process may be a long period of progressive states, and requires more chromosomal aberrancy with some oncogene disturbances.

By the evidence of changes in cell morphology, telomerase, AIG, and tumorigenicity, we have proven that the SHEE10 was in a preimmortal stage without telomerase activity, AIG and tumorigenesis in mice: the SHEE31 was in an immortal stage and telomerase positive but without AIG and tumorigenesis. The SHEE61 were premalignant with telomerase, AIG positive and weak tumorigenicity. Differing from SHEE10, SHEE31 and SHEE61 cells, the phenotype of SHEE61A cells was characterized by rapid growth, different cell size, crowded multiplication, enlarged nuclei and nucleoli, especially the AIG *in vitro* and tumorigenesis *in vivo*. The results demonstrate that the SHEE61A cells were in fully malignant transformation. The immortalized cells of esophageal epithelium induced by HPV may be in slowly progressive malignant transformation.

In summary, based on our studies, telomerase, anchorage – independent growth and tumorigenicity of nude mice may be used for detection of immortal, premalignant and malignant characteristics respectively. The SHEE cell line is a model of multistage carcinogenesis, it will continue to serve for studying the sequential and stepwise evolution of the cancer process by HPV. Anchorage – independent growth is an indicator of premalignant or malignant changes. The SHEE61 may also provide a good *in vitro* model system to investigate the cellular and molecular events involved in premalignant stages of the esophageal epithelium induced by HPV.

Acknowledgements

This work was supported by Dr Richard Potter, PhD, Professor of Biology, California University. Contract grant sponsor: National Natural Science Foundation of China (39830380).

〔In 《Int J Mol Med》 2001, 8: 633 –639〕

References

1　Redel RR. The role of senescence and immortalization in carcinogenesis. Carcinogenesis, 2000, 21: 477 –484

2　Heselmeyer K, Hellstrom AC, Blegen H, Schrock E, Silfversward C, Shah K, Auer G and Ried T. Primary carcinoma of the fallopian tube: comparative genomic hybridization reveals high genetic instability and a specific, recurring pattern of chromosomal aberrations. Int J Gynecol Pathol, 1998, 17: 245 –254

3　Rhim JS. Molecular and genetic mechanism of prostate cancer. Radiat Res, 2001, 155: 128 –132

4　Yang XY, Kimura M, Jeanclos E and Aviv A. Cellular proliferation and telomerase activity in CHRF – 288 – I 1 cells. Life Sci, 2000, 66: 1545 –1555

5　Kim S, Kaminker P and Campisi J. TIN_2 a new regulator of telomere length in human cells. Nat Genet, 1999, 23: 405 –412

6　Cifone MA and Fidler I J. Correlation of patterns of anchorage – independent growth with in vivo behavior of cells from a murine fibrosarcoma. Proc Natl Acad Sci USA, 1980, 77: 1039 – 1043

7　Harris CC. Human tissues and cells in carcinogenesis research. Cancer Res, 1987, 47: 1 – 10

8　Mutirangura A, Sriuranpong V, Termrunggraunglert W, Tresukosol D, Lertsaguansinchai P, Voravud N and Niruthisard S: Telomerase, activity and human papillomavirus in malignant, premalignant and benign cervical lesions. Br J Cancer, 1998, 78: 933 –939

9　Avilion AA, Piatyszek MA, Gupta J, Shay JW, Bacchetti S and Greider CW. Human telomerase RNA and telomerase activity in immortal cell lines and tumor tissues. Cancer Res, 1996, 56: 645 – 650

10　Snijders P J, van – Duin M, Walboomers JM, Steenbergen RD, Risse EK, Helmerhorst TJ, Verheijen RH and Meijer CJ. Telomerase activity exclusively in cervical carcinomas and a subset of cervical intraepithelial neoplasia grade III lesions:

strong association with elevated messenger RNA levels of its catalytic subunit and high – risk human papillomavirus DNA. Cancer Res, 1998, 58: 3812 – 3818

11　Kim NW, Piatyszek MA, Prowse KR, Harley CB, West MD, Ho PL, Coviello GM, Wright WE, Weinrich SL and Shay JW. Specific association of human telomerase activity with immortal cells and cancer. Science, 1994, 266: 2011 – 2015

12　Hiyama T, Yokozaki H, Kitadai Y, Haruma K, Yasui W, Kajiyama G and Tahara E. Overexpression of human telomerase RNA is an early event in oesophageal carcinogenesis. Virchows Arch, 1999, 434: 483 – 487

13　Kiyozuka Y, Asai A, Yamamoto D, Senzaki H, Yoshioka S, Takahashi H, Hioki K and Tsubura A. Establishment of novel human esophageal cancer cell line in relation to telomere dynamics and telomerase activity. Dig Dis Sci, 2000, 45: 870 – 879

14　Miller AC, Blakely WF, Livengood D, Whittaker T, Xu J, Ejnik JW, Hamilton MM, Parlette E, John TS, Gerstenberg HM and Hsu H. Transformation of human osteoblast cells to the tumorigenic phenotype by depleted uranium – uranyl chloride. Environ Health Perspect, 1998, 106: 465 – 471

15　Tokumitsu Y, Nakano S, Ueno H and Niho Y. Suppression of malignant growth potentials of v – Src – transformed human gall – bladder epithelial cells bv adenovirus – mediated dominant negative H – Ras. J Cell Physiol, 2000, 183: 221 – 227

16　Moore SM, Rintoul RC, Walker TR, Chilvers ER, Haslett C and Sethi T. The presence of a constitutively active phos – phoinositicle 3 – kinase in small cell lung cancer cells mediates anchorage – independent proliferation via a protein kinase B and p70s6k – depende. Cancer Res, 1998, 58: 5239 – 5247

17　Shen ZY. Cen S, Cai WJ, Xu J J, Ten ZP, Shen J, Hu Z and Zeng Y. Immortalization of human fetal esophageal epithelial cells induced by E6 and E7 genes of human papilloma virus I8. Chin Exp

Clin Virol, 1999, 13: 121 – 123

18　Shen ZY, Shen J, Cai W J, Cen S and Zeng Y. Biological characteristics of human fetal esophageal epithelial cell line immortalized by the E6 and E7 gene of HPV type 18. Chin Exp Clin Virol, 1999, 13: 5 – 8

19　Shen ZY. Xu LY, Chen MH, Cai WJ, Shen J, Chen JY, Hon CQ and Zeng Y. Biphasic differentiation of immortalized eso – phageal epitheliums induced by HPV18E6E7. Chin J Virol, 2001, 17: 210 – 214

20　Shen ZY, Cen S, Shen J, Cai W J, Xu J J, Teng ZP, Hu Z and Zeng Y. Study of immortaliztion and malignant transformation of human embryonic esophageal epithelial cells induced by HPVISE6E7. J Cancer Res Clin Oncol, 2000, 126: 589 – 594

21　Shen ZY. Chen XH, Shen J, Cai WJ, Chen JY, Huang TFH and Zeng Y. Malignant transformation of immortalized human embryonic esophageal epithelial cells induced by human papiilomavirus. Chin J Virol, 2001, 16: 97 – 101

22　Hennings H, Glick AB, Greenhalgy DA, Morgan DL, Strickland JE, Tennenbaum T and Yuspa SH. Critical aspects of initiation, promotion, and progression in multistage epidermal carcinogenesis. Proc Soc Exp Biol Med, 1993, 202: 1 – 8

23　Hukku B, Mally M, Cher ML, Peehl DM, Kung H and Rhim JS. Stepwise genetic changes associated with progression of non – tumorigenic HPV – 18 immortalized human prostate cancer – derived cell line to a malignant phenotype. Cancer Genet Cytogenet, 2000, 120: 117 – 126

24　Weitzman JB and Yaniv M. Rebuilding the road to cancer. Nature, 1999, 400: 401 – 402

25　Hahn WC, Counter CM, Lundberg AS, Beijersbergen RL, Brooks MW and Weinberg RA. Creation of human tumour cells with defined genetic elements. Nature, 1999, 400: 464 – 468

26　Li R, Sonik A, Stindl R, Rasnick D and Duesberg P. Aneuploidy vs. gene mutation hypothesis of cancer: recent study claims mutation but is found to support aneuploidy. Proc Natl Acad Sci

USA, 2000, 97: 3236 – 3241

27 Wisman GB, Hollema H, de – Jong S, ter – Schegget J, Tjong – A – Hung SP, Ruiters MH. Krans M, de – Vries EG and van – der – Zee AG. Telomerase activity as a biomarker for (pre) neoplastic cervical disease in scrapings and frozen sections from patients with abnormal cervical smear. J Clin Oncol, 1998, 16: 2238 – 2245

28 Alani RM and Munger K. Human papillomaviruses and associated malignancies. J Clin Oncol, 1998, 16: 330 – 337

29 Duensing S. Lee LY, Duensing A, Basile J, Piboonniyom S, Gonzalez S. Crum CP and Munger K. The human papillomavirus type 16 E6 and E_7, oncoproteins cooperate to induce mitotic defects and genomic instability by uncoupling centrosome duplication from tile cell division cycle. Proc Natl Acad Sci USA, 2000, 97: 10002 – 10007

30 Filatov L, Golubovskaya V, Hurt JC, Byrd LL, Phillips JM and Kaufmann WK. Chromosomal instability is correlated with telomere erosion and inactivation of G2 checkpoint function in human fibroblasts expressing human papillomavirus type 16 E6 oncoprotein. Oncogene, 1998, 16: 1825 – 1838

31 Tsao SW, Mok SC, Fey EG, Fletcher JA, Wan TS, Chew EC, Muto MG, Knapp RC and Berkowitz RS. Characterization of human ovarian surface epithelial cells immortalized by human papilloma viral oncogenes (HPV – E6E70RFs).

Exp Cell Res, 1995, 218: 499 – 507

32 Wan TS, Chart LC, Ngan HY and Tsao SW. t (High) frequency of telomeric associations in human ovarian surface epithelial cells transformed by human papilloma viral oncogenes. Cancer Genet Cytogenet, 1997, 95: 166 – 172

33 Klingelhutz AJ, Foster SA and McDougall JK. Telomerase activation by the E6 gene product of human papillomavirus type 16. Nature, 1996, 380: 79 – 82

34. Steenbergen RD, Walboomers JM, Meijer C J, van – der – Raaij – Helmer EM, Parker JN, Chow LT, Broker TR and Snijders PJ. Transition of human papillomavirus type 16 and 18 transfected human foreskin keratinocytes towards immortality activation of telomerase and allele losses at 3p, 10p, 1 lq and/or 18q. Oncogene, 1996, 13: 1249 – 1257

35 Cottage A, Dowen S, Roberts I, Pett M, Coleman N and Stanley M. Early genetic events in HPV immortalized keratino – cytes. Genes Chromosomes Cancer, 1996, 30: 72 – 79

36 Oda D, Bigler L, Lee P and Blanton R. HPV immortalization of human oral epithelial cells: a model for carcinogenesis. Exp Cell Res, 1996, 226: 164 – 169

37 Oda D, Bigler L, Mao EJ and Disteche CM. Chromosomal abnormalities in HPV – 16 – immortalized oral epithelial cells. Carcinogenesis, 1996, 17: 2003 – 2008

73.　Mother to Child Transmission of HIV – Antiretroviral Therapy and Therapeutic Vaccine: A Scientific and Community Challenge

Joint Working Group Report of AIDS/Infectious Diseases Pmp and Mother and Child Pmp

BIBERFELD G. (Solna, Sweden), BIBERFELD P. (Stockholm, Sweden), BUONAGURO F. (Naples. Italy), CHARPAK N. (Bogota, Colombia), THÉ de G. (Paris, France), REA M. Ferreira (Sao Paulo, Brazil), GRAY G. (Soweto, South Africa), HURAUX Ch (Paris, France), LINDBERG A. (Marcy l'Etoile, France), SAMUEL N. M. (Guindy – Chennai, India), SCARLATTI G. (Milan, Italy), TLOU S. (Gabarone, Botswana), Ph. Van de Perre (Montpellier, France), ZENG Yi (Beijing, China), ZETTERSTRÖM R. (Stockholm, Sweden)

Since our last Erice workshop in August 2000, (Acta Paediatrica 89: 1385 – 6. 2000) significant progress has been achieved in both HIV antiretroviral therapy (ART) and in HIV vaccine research and development, with the emerging concept of therapeutic vaccine considered as a possible complement to ART. We reviewed these two areas in the context of mother to child transmission, and appreciated the increasing political awareness concerning the HIV epidemic now considered as a planetary emergency, which led to the engagement of UN and of G8 to provide funds to implement access to ART in low income and most severely affected countries.

In spite of these efforts, UNAIDS estimates that in 2000 around 36 million people were living with HIV infection, 95% of them in the developing world, that at least 5. 3 million being infected during the year 2000 alone, and that more than 22 million AIDS patients died since 1985.

Mother to child transmission of HIV (MTCT) represents a particularly dramatic aspect of this HIV epidemic with an estimated 600 000 of newborns infected yearly, 90% of them in Africa. Mother and child health, being a key factor for any sustainable development, avoidance of MTCT HIV must be a priority and become achievable in many countries.

Part one

1. Antiretroviral therapy

(1) *Access to care*

①In many parts of the world, HIV is already the leading cause of adult and child mortality. The HIV pandemic compromises the gains made in recent decades in terms of quality of life and life expectancy[1].

②Due to civil society pressure, national initiatives (generic drugs) and high level negotiation with pharmaceutical companies, the cost of ARV for developing countries has been reduced considerably (about one tenth of the initial sale price). A recent decision of the United Nations to prioritise allocation of sufficient resources for HIV, tuberculosis and malaria control should hopefully make a comprehensive package of prevention, care and research including ARV more readily available in the near future.

③However, in resource – limited countries, human, financial and logistic resources needed for programmes of HIV care may compete with other sectors of the health care systems.

④There is a wide discrepancy in access to care including ARV according to different settings in the developing world. The range of people having access to adequate ARV therapy is from a few hundred in many countries to several tens of thousands such as in Botswana, South Africa, Thailand, India and Brazil among others[2-4].

⑤Important prerequisites for access to care implementation involve the whole health system and structure (accessibility, social acceptability, VCT (Voluntary counselling and testing) structures and competence, education of health professionals...), as well as economic and political commitment.

⑥A new comprehensive and socially acceptable concept of taking care of households instead of individuals is emerging in some countries such as South Africa[2], Botswana[3] and India[4]. This concept could decrease the economic fragility of affected households and mitigate the impact of HIV on vulnerable children and orphans.

⑦Innovative and appropriate technologies are developed or are already available, such as plasma and salivary rapid test for HIV diagnosis in adults as well as CD4 counts by alternatives to flow cytometry[5] and modified p24 antigen measurement technologies for monitoring of therapeutic efficacy[6]. These techniques could render monitoring and diagnosis more accessible and affordable in resource – limited settings.

⑧New drugs that may be active against HIV, including new ARV families (integrase inhibitors, inhibitors of viral assembly, cytokine as adjuvants, etc.) and drugs from the traditional pharmacopoeia (China, India, some African countries) are currently under evaluation[5]. These compounds may improve the efficacy of current ARV regimens.

(2) *Perinatal and postnatal mother – to – ehild transmission of* HIV

①Antenatal care and VCT are entry points fbr prevention and care.

②Disclosure of the test results to the husband/partner and significant others varies considerably from one area to another (50% –80% in Soweto, South Africa, 17% in Dares Salaam, Tanzania, 15% in Namakal, India, less than 10% in Abidjan, Cote d'Ivoire and Bobo – Dioulasso, Burkina Faso) and is a frequent limiting factor for maternal interventions (ARV, feeding practices, ...)[2-4,8].

③Efficacy of short regimen of perinatal prophylactic ARV in reducing MTCT lessens over time if breastfeeding is prolonged and may even be lost[8-9].

④Maternal CD4 count is a strong predictor of the efficacy of perinatal prophylactic ARV. Indeed, in a combined analysis of two clinical trials evaluating short zidovudine regimens in the perinatal period, no efficacy was demonstrable at any time during the follow up in women who had less than 500 CD4 cells per μl at delivery[8,9]. However, in women with more than 500 CD4 cells per μl at delivery, breastfeeding had only a minimal, non – significant, impact on transmission.

⑤Interruption of maternal ARV administration around the time of lactation may increase short – term breast milk viral load and, putatively, infant transmission[8].

⑥Breastfeeding by HIV – infected women has been reported to be associated with an excess maternal mortality in a clinical trial performed in Nairobi, Kenya[10].

(3) *Plea For Action*

①The success of HIV prevention and care programmes depends on good access to and the performance of the primary health care system. The strengthening of the necessary infrastructure and human resources to deliver HIV prevention and care is of the utmost priority.

②Access to care/ARV should not be restricted to ARV therapy alone but should be considered as a continuum of medical and psychosocial support. Voluntary counselling testing and care should be regarded as components of a comprehensive package of prevention and care.

③ARV and drugs for prophylaxis and treatment of opportunistic infections should be made available, affordable and sustainable and distributed in an equitable way.

④HIV prevention and care programmes should include also availability of reliable and inexpensive tests to diagnose and monitor the treatment of HIV infection and associated conditions, as well as appropriate training for health care workers in management. Urgent recommendations are needed for criteria for initiation of therapy, scheduling, switching, interrupting and monitoring regimens. In order to ensure success of such programmes, joint decision making involving the whole therapeutic team and the household/family are mandatory. Health care workers should be provided with training on occupational hazards, appropriate equipment and management of all accidental exposures.

＊ Pilot country programme of this nature will help to establish policy and allocate adequate resources.

①In order to improve MTCT intervention programme coverage, the disclosure by women of their HIV test result to husband/partner and significant others should be encouraged by learning from successful experiences (such as in South Africa and India), with respect to local socio – cultural mores. Disclosure of HIV status may further encourage husband/partner to get tested and improve overall efficacy of prevention programmes. The social consequences of disclosure should be carefully elucidated in all settings prior to implementation.

②In programmes of prevention of MTCT, CD4 counts, as well as other surrogate markers still to be validated, may become a critical criterion for adapting prophylaxis, maternal treatment and appropriate infant feeding options. More research is required for ensuring transmission risk reduction in mothers who benefited from a perinatal prophylactic ARV but have no acceptable alternative to breast feeding, including ARV regimens covering the lactation period.

③By no means should an HIV – infected pregnant woman eligible for ARV therapy be deprived of adequate ARV therapy for herself, where available.

④Access to appropriate family planning services must be guaranteed.

⑤A possible association between an excess maternal mortality in HIV – infected mothers and breastfeeding should be urgently scrutinised in existing data sets (retrospectively) and in new research projects.

⑥Considerable efforts are still needed to optimise safety of all potential feeding practices by appropriate education of both health professionals and mothers and identify adequate standardised indicators to assess infant feeding practices (formula feeding, animal milk, exclusive breastfeeding, early cessation of breastfeeding, etc.).

⑦More research is urgently needed on differential transmission risks associated with breastfeeding practices, pathophysiology of breast milk transmission and viral/host relationships related to MTCT. Social sciences should also contribute considerably to our understanding of HIV transmission by breastfeeding.

Part two

Since 1985, attempts to develop an efficacious HIV vaccine have been as numerous as have the failures. The barrier has been the lack of sufficient scientific knowledge. A major problem in the development of an HIV vaccine is the high variability of HIV. There are two types of HIV: HIV – 1 and HIV – 2. The HIV – 1 is further divided into 3 groups (M, N and O) and each group is subdivided into several subtypes. In addition there are sub – subtypes and inter – subtype as well as intergroup recombinants. The distribution of the various HIV – 1 subtypes differs in different parts of the world. Furthermore there are virus variants which differ in their phenotype according to the chemokine receptor usage, the major of which being CCR5 and CXCR4.

Can we make a vaccine? It is still not certain when it will be possible to develop an effective and safe preventive HIV vaccine for use in humans since no efficacy trial has yet been completed. However, knowledge gained over the last decades in studies of the natural infection of humans, in particular long – term survivors, combined with pre – clinical non – human primate vaccine studies have led us to believe that an efficacious vaccine can be developed. Furthermore, approximately 70 phase I/II clinical trials done in humans using different vaccine constructs and modalities of immunization have shown that the tested vaccine candidates were safe and able to induce specific immune responses of varying intensity. An ideal sterilizing prophylactic vaccine should be safe and induce: i) cross – neutralizing antibodies against primary HIV – 1 isolates from divergent HIV subtypes; ii) strong and broad $CD4^+$ T – cell responses; iii) poly – epitopic, cross – clade reactive CD8 + CTL responses; iv) mucosal immune defenses; and v) long – term protection.

Prophylactic vaccine approaches There are several types of possible HIV vaccine candidates including live attenuated virus, whole inactivated virus, recombinant produced subunits, synthetic peptides, live recombinant vaccines and viral DNA. Live attenuated SIV vaccines have been the most efficient in eliciting protective immunity in the SIV/macaque model. However, these vaccines were found to induce disease in macaques and this approach is not applicable in humans for safety reasons. Envelope subunits of HIV – 1 elicit antibodies which neutralize laboratory strains but fail to efficiently neutralize field isolates, and usually do not induce cytotoxic T lymphocytes (CTL) responses.

Recent experiences indicates the advantages of using mixed modality immunization, i. e. immunizing with several HIV – 1 antigens (env, *gag*, pol, nef, tat, etc.) either as genetic information in

DNA (plasmids) or in live expressing vectors (pox, adeno, salmonella, BCG, etc.), or as virus – like particles, recombinant proteins, peptides, peptide – conjugates. Typically an individual will be primed with one construct (DNA or vector) and boosted with another construct (vector or proteins). Several regimens are currently in phase I/II studies.

Therapeutic vaccine approaches　Before 1996 it was not realistic to have an effective therapeutic vaccine. This is because the HIV – 1 infection seriously reduces the number and impairs the function of CD4 + cells, which are central to the immune system. HAART (highly active anti – retroviral therapy), which besides limiting the virus replication and improving the quality of life gives a partial, if not full, restoration of the immune system, allows the development of a therapeutic vaccine controlling HIV infection. An initial phase I study, using a pox vector coding for env and *gag* protein plus recombinant gp160 protein, delayed viral rebound for four months in 2 out of 4 patients who stopped HAART treatment after termination of vaccine alone. These encouraging results initiated a series of phase II trials with different mix – modality regimens. Data from these trials will be available in the second half of 2002.

Can we make a mother – child vaccine? Current vaccine trials focus on the adult population but do not yet address the children. Even if an efficacious prophylactic vaccine was available there may be an insufficient time to elicit a protective HIV – 1 immune response in the newborn against the perinatal HIV infection. Therefore the pregnant woman is the obvious target to vaccinate.

However, pregnant women are usually excluded from vaccination. Indeed it is only the tetanus toxoid that is given to pregnant women in the third trimester in regions where neonatal tetanus is a serious threat. The obvious need of HIV vaccination in pregnant women raises new challenging ethical questions, in particular whether the benefit to the child and mother is greater than potential risks. Therefore we need to evaluate the use of replicating versus non – replicating vectors as well as the use of DNA in experimental animal systems, before starting a clinical trial. Furthermore it will be necessary to design immunization protocols for the pregnant mother, preferably during the third trimester, followed by a prophylactic vaccination for the newborn. Although the primary goal is to prevent HIV – 1 transmission to the fetus/child, it is imperative that the mother will continue to be a vaccine recipient.

When can mother – child vaccine be available? As soon as risk factors are evaluated and regulatory approval is obtained phase I trials can be started. However, needed pre – clinical animal studies and the difficult ethical, safety and liability concerns with respect to the pregnant woman are likely to slow development. Therefore it is unrealistic to expect that a comprehensive vaccine program will be available in the coming decade.

〔In 《A Scientific and Community Challenge》 2001, 412 – 418〕

References

1   Adetunji J. Trends in under − 5 mortality rates and the HIV/AIDS epidemic. Bull World Health Organ, 2000, 78: 1200 − 6

2   Gray G. AIDS pediatric epidemics in South Africa. World Federation of Scientists. Planetary Emergencies Conference. Erice, Italy, August, 2001

3   Tlou S. Mother − to − child transmission of HIV in Botswana. World Federation of Scientists. Planetary Emergencies Conference. Erice, Italy, August, 2001

4   Samuel N M. AIDS in India. World Federation of Scientists. Planetary Emergencies Conference. Erice, Italy, August, 2001

5   Lyamuya E F, Kagoma C, Mbena E C, Urassa W K, Pallangyo K, Mhalu F S, Biberfeld G. Evaluation of the FACSCount, TRAx CD4 and Dynabeads methods for CD4 lymphocyte determination. J Immunol Meth, 1996, 195: 103 − 112

6   Bush C E, Donovan R M, Manzor O, Baxa D, Moore E, Cohen F, Saravolatz L D. Comparison of HIV type 1 RNA plasma viremia, p24antigenemia, and unintegrated DNA as viral load markers in pediatric patients. AIDS Res Hum Retroviruses, 1996, 12: 11 − 5

7   Zeng Yi. AIDS in China. World Federation of Scientists. Planetary Emergencies Conference. Erice, Italy, August, 2001

8   Van de Perre, P. Mother − to − child transmission of HIV with special emphasis on breastfeeding transmission. World Federation of Scientists. Planetary Emergencies Conference. Erice, Italy, August, 2001

9   Wiktor S Z, Leroy V, Ekpini E R, et al. 24 − month efficacy of short − coursematernal zidovudine for the prevention of mother − to − child HIV − 1 transmission in a breast feeding population. A pooled analysis of two randomized clinical trials in West Africa. XⅢ International AIDS Conference. Durban, South Africa. July 2000

10   Nduati R, Richardson BA, John G, et al. Effect of breastfeeding on mortality among HIV − 1 infected women: a randomised trial. Lancet, 2001, 357: 1651 − 1655

74. Detection of Human Papillomavirus in Esophageal Carcinoma

SHEN Zhong − ying[1], HU Sheng − ping[1], LU Li − chun[1], TANG Chun − zhi[1],

KUANG Zhong − sheng[1], ZHONG Shu − ping[1], ZENG Yi[2]

1. Shantou University Medical College, Shantou, Guangdong;

2. The Virus Research Institute Chinese Academy of Preventive Medicine

Summary

The aim of the study was to assess the prevalence of human papillomavirus (HPV) in the esophagus in the coastal region of Eastern Guangdong, Southern China, an area with a high incidence of esophageal carcinoma. Fresh surgical resection esophageal specimens were obtained from 176 esophageal carcinoma patients admitted to the Tumor Hospital of Shantou University Medical

College. The samples were subjected to polymerase chain reaction (PCR) to detect HPV infection using consensus and type – specific primers for HPV type 6, 11, 16, and 18. The incidence rate was 65.5%, 69.1%, and 60% in tissues of cancerous, paracancerous and normal mucosa, respectively. Further analysis of the distribution of HPV types in the three sections of tissues showed that the high – risk HPV types 16 and 18 were found mainly in the cancer cells (43.2%), whereas the low – risk HPV types 6 and 11 were seen mainly in the normal mucosa (52.3%). The total infection rate of the high – risk HPV types 16 and HPV 18 was the highest in cancerous tissues (54.5%), followed by paracancerous tissues (19.5%), and the lowest in normal mucosa (11.7%). There was high incidence of HPV infection in the esophageal epithelium in Eastern Guangdong, Southern China, where esophageal carcinoma is prevalent. HPV was seen in the normal, paracancerous and cancerous tissues, with the high – risk HPV type 16 and 18 more common in cancerous tissues. The results indicate that the high incidence of esophageal carcinoma in this area is associated with HPV infection.

[**Key words**]　Epidemiology; Esophageal squamous cell carcinoma; Human papillomavirus

Introduction

Esophageal squamous cell carcinoma is common in China as well as in some parts of the world, but is rare in the occidental countries, and the different distribution of esophageal carcinoma between high – and low – incidence areas can be as high as 300 – fold (Day, 1984). Although most high – incidence areas of esophageal carcinoma are seen inland (He et al. , 1997), the region of the Eastern Guangdong is, however, the only coastal area with high – incidence esophageal carcinoma with morbidity of 197. 82/10^5 world standardized population for males and 81. 32/10^5 world standardized population for females (Chen et al. , 1996).

The etiology of esophageal carcinoma remains unclear. Studies of esophageal carcinoma have suggested that genetic predisposition (Tada et al. , 2000), dietary (Ren and Han, 1991) or environmental factors (Ribeiro et al. , 1996), such as nitrosamine (Siddiqi et al. , 1991; Gurski et al. , 1999), tobacco smoking (Zambon et al. , 2000), alcohol consumption (Talamini et al. , 2000), spicy food (Sharma, 1999), malnutrition (Franceschi et al. , 2000), trace element deficiency (Newberne et al. , 1997), and fungal toxin (Liu et al. , 1992) could be important factors in the carcinogenesis of this tumor. Since the first report of human papillomavirus (HPV) in esophageal carcinoma in 1982, implicating a potential risk factor of HPV in the development of esophageal carcinoma (Syrjanen, 1982), the existence of HPV in esophageal carcmoma was confirmed further by methods of immunohistochemistry, serology, Southern hybridization, polymerase chain reaction (PCR), *in situ* hybridization (ISH) and others. Recent evidence has shown that esophageal infection with HPV, particularly high – risk types 16 and 18, increased esophageal carcinoma morbidity 13 – fold (Dillner et al. , 1995), indicating that HPV may have pathogenic significance in esophageal carcinoma (Bjorge et al. , 1997; Poljak et al. , 1998; Takahashi et al. , 1998). The detection rate of HPV in esophageal lesions is varied geographically (Sur and Cooper, 1998), however, ranging from 0% (Smits et al. , 1995; Kok et al. , 1997) to 60% (Chen et al. , 1994). The role

· 486 ·

of HPV in the pathogenesis of esophageal carcinoma remains to be determined.

The objective of the present study was to determine whether HPV infection in esophageal tissues was common in esophageal carcinoma patients residing in the coastal region of the Eastern Guangdong area. The study was also designed to explore the role of HPV infection, most notably HPV 16 and 18, in esophageal carcinogenesis.

Materials and Methods

1. Specimens

During the period from 1994 – 1997, 176 fresh specimens of resected esophagus were obtained from patients, who were treated for esophageal cancer at the Affiliated Tumor Hospital, Shantou University Medical School, China, while living in the high – incidence area for esophageal carcinoma in the coastal region of the Eastern Guangdong. All specimens had esophageal squamous cell carcinoma confirmed histologically. Every specimen was cut into three parts: cancerous, paracancerous, and normal tissues, and each cut used a new microtome blade to avoid contamination of the samples.

The specimens were subjected to the following tests: 1) 165 samples from 55 patients were tested by PCR using HPV consensus primers for general HPV infection; 2) 132 samples from 44 patients were tested by PCR using type – specific primers to determine the infection rates of high – risk types 16 and 18 and low – risk types 6 and 11, respectively; and 3) 231 samples from 77 patients were tested by PCR using type – specific primers for infection with HPV types 16 and 18, respectively.

2. Cell Lines

Cell lines Ec/CUHK1 and Ec/CUHK2 were human esophageal carcinoma cell lines free of HPV infection (gifts from Prof. Y. Chew of Hong Kong Chinese University). Cells were cultured in a humidified incubator at 37℃ with 5% CO_2 in air in M199 medium (Gibco BRL, Gaithersburg, MD) supplemented by 10% calf serum, 100 U/ml each of penicillin and streptomycin, pH 7.0. Cells were harvested at confluency by 0.25% trypsin. DNA extracted as described below was used as HPV negative control in the experiments.

3. DNA Extraction

Tissues were homogenized followed by proteinase K (200 μg/ml; Promega, Madison, WI) digestion in thepresence of 0.5% SDS at 37℃ overnight. Samples were then subjected twice to phenol – chloroform – isoamyl alcohol extraction. DNA was precipitated with absolute ethanol followed by washing twice with 70% ethanol. After air drying at room temperature, DNA was dissolved in TE buffer (10 mmol/L Tris HCl, 1 mmol/L EDTA, pH 7.8) and stored at 4℃ until used.

Recombinant plasmid DNA HPV 16 – pBR322 was obtained from Prof. S. Lu, the Cancer Research Institute, Chinese Academy of Medical Sciences and HPV 18 – pBR322 was a gift from Prof. K. Yao, the Cancer Research Institute, Hunan Medical University. Plasmid DNA extracted as above was used as positive control and the plasmid DNA extracted from cell lines Ec/CUHK1 and Ec/CUHK2 was used as negative control in PCR.

4. PCR Analysis

Five different sets of primers (Tab. 1) were used in this study. The consensus primers L1C1 and L1C2, synthesized by the Cancer Research Institute (Chinese Academy of Medical Science, Beijing), were designed as described [Yoshikawa et al. , 1991] to amplify HPV types 6, 11, 16, and 18 and targeted a segment of 144 bp in the highly conserved HPV L 1 gene (90% homologous among the HPV types). The primer sets 2 – 4, synthesized by the Department of Biology, Fudan University, Shanghai, were type – specific primers for detection of HPV type 6, 11, 16, and 18, respectively (Kiyabu et al. , 1989).

Tab. 1　Sequence of Primers Used

Set	Primers	Sequence[a]
1	L 1C 1	CGTAAACGTTTTCCCTATTTTTTT
	L 1C2	GTTATGTCTCATAAATCCCAT
2	HPV 6	GCACGTCTAAGATGTCTTGTTTAG
		AGACCAGTTGTGCAAGACATTTAA
3	HPV 11	AGACCAGTTGTGCAAGACATTTAA
		AAGGGAAAGTTGTCTCGCCACACA
4	HPV 16	ATGAACTAGGGTGACATTT
		GCTGTTAGGCACATATTT
5	HPV 18	GCTGGTTAGGCACATATTT
		ATGTATGCACAGCTTAGTC

Note: [a] All the sequences are shown from 5' to 3'.

Samples of 10 ng DNA were mixed with one set of primers flanking the DNA fragments to be amplified. Reactions were set in 25 μl × PCR buffer containing 10 mmol/L Tris HCl, pH 8. 4, 50 mmol/L KCl, 1. 5 mmol/L MgCl2, 4 mmol/L dNTP and 1. 5 U Taq DNA polymerase (Promega). DNA was subjected to 40 cycles of amplification in PCR (GeneAmp PCR System 2400, Perkin – Elmer, Foster City, CA) with denaturing at 94℃ for 30 sec, annealing at 50℃ for 30 sec, and elongating at 72℃ for 1 min. In each experiment, 10 ng plasmid DNA (HPV 16 – pBR322 or HPV 18 – pBR322), 20 ng DNA from Ec/CUHK1 or Ec/CUHK2, and a no – DNA reaction were included as positive, negative and blank controls, respectively. The amplified product was resolved by electrophoresis on a 2% agarose gel (Promega) containing 0. 5% μg/ml of ethidium bromide in 1 × TAE buffer. Samples were considered positive if bands of 263 bp (HPV 6), 144 bp (HPV 11), 130 bp (HPV 18), and 100 bp (HPV 16) were observed under the UV light.

5. Statistical Analysis

The difference was tested by χ^2 tests. The HPV infection rate was analyzed by χ^2 analysis. Statistical significance is assumed if P – value < 0. 05.

Results

1. Overall detection rate of HPV DNA: The overall detection rate in 165 samples from 55 patients by PCR using the consensus primers for all HPV types DNA was greater than 60% , with 60% for normal mucosa (33/55), 69. 1% for paracancerous mucosa (38/55), and 65. 5% (36/55) for cancerous tissues, respectively (Tab. 2). The results indicated that HPV infection was common in esophagus in the patients residing in the local region, and the infection was widely distributed all over the esophageal mucosa.

Tab. 2 Detection of HPV in 55 Specimens

Location	Number of positive specimens	Percent
Normal	33	60. 0
Paracancerous	38	69. 1
Cancerous	36	65. 5

2. Different detection rates of high and low risk HPV *types in tissues with different pathological states*: HPV is classified broadly in terms of its carcinogenesis into two groups: high risk (HPV16 and 18) and low risk (HPV 6 and 11). PCR amplification in 132 samples from 44 patients using type – specific primers was performed to determine the prevailing group. As shown in Tab. 3, the detection rates in both groups were similar. The distribution of the two groups differed in the tissues of various pathology. For the high risk group, the highest detection rate was seen in cancerous tissues (19/44, 43. 2%), decreasing in paracancerous mucosa (17/44, 38. 6c2), and the lowest in normal mucosa (6/44, 13. 6%). The difference in detection rates between the cancerous tissues and the normal mucosa was highly significant ($P < 0. 01$). In the low risk group, the order of the detection rates in tissues of different pathological states was contrasted to that of the high risk group: highest in normal mucosa (23/44, 53. 3%), intermediate in paracancerous mucosa (17/44, 38. 6%), and lowest in cancerous tissues (15/44, 34. 1%). There was no significant difference between the cancerous tissues and the normal mucosa ($P > 0. 05$). These results suggested that high – risk group HPV types 16 and 18 might have a closer association with the development of esophageal carcinoma. There were five patients with mixed HPV infection of high and low risk types (data not shown).

Tab. 3 Comparison of HPV High Risk and Low Risk Type Infection in 44 Specimens

HPV type	Number of positive specimens (%)		
	Normal	Paracancerous	Cancerous
HPV 6 and 11	23 (52. 3) *	17 (38. 6)	15 (34. 1) *
HPV 16 and 18	6 (13. 6) * *	17 (38. 6)	19 (43. 2) * *

Notes: * $\chi^2 = 0. 198$, P > 0. 05.

* * $\chi^2 = 7. 22$, P < 0. 008.

3. Detection rates of HPV *16 and 18*: To determine further whether infection of one type in the high risk group was prevalent over the other, PCR was applied in 231 samples from 77 patients using type – specific primers for HPV 16 and 18, respectively. The results showed (Tab. 5) that HPV 16 infection rate in all the tissues of various pathological states was higher than that of HPV 18. There were eight patients with mixed HPV infection of types 16 and 18 (data not shown).

Discussion

It is well known that HPV infection of squamous cells can lead to hyperplasia and papilloma [Sandvik et al. , 1996] . Since its first identification of HPV as a causative factor of human warts in 1907, it has been recognized that HPV is an important human carcinogen for various cancers of the skin, oral cavity, pharynx, larynx, lung, cervix, and anogenital system (Zur Hausen, 1987). The role of HPV in the etiology of esopheal carcinoma is now attracting attention. There is a close relationship between papilloma and squamous cell carcinoma of the esophagus (Sandvik et al. , 1996) considering that the multi and micropapillary lesion in the paracancerous mucosa displayed HPV

DNA (Shen et al. , 2000). We established recently an immortalized human fetal esophageal epithe-lial cell line by introducing the HPV 18 E6E7 genes into the cells (Shen et al. , 1999a) and a transformed human fetal esophageal epithelial cell line by infecting the cells with the HPV 18 E6E7 genes in synergy with phorbol acetate (TPA) treatment (Shen et al. , 1999b). These results strong-ly suggest an important carcinogenic role of HPV in the development of esophageal carcinoma. HPV detection rate in esophageal lesions varies geographically. An obvious phenomenon is that the HPV detection rate is absent (Morgan et al. , 1997; Saegusa et al. , 1997) or significantly lower in areas with moderate or low incidence of esophageal carcinoma (Lam et al. , 1997), 4.3% in Beijing and 4.4% in Cincinnati, Ohio (Suzuk et al. , 1996); in high incidence areas of the disease, however, the detection rate is much higher (43.1% in Linxian, China (Chang et al. , 1990) and 60% in Fuzhou, China (Zambon et al. , 2000). It appears that the role of HPV in esophageal carcinogene-sis might be more pronounced in areas of the world with a high prevalence of esophageal carcinoma (Chang et al. , 2000). Our study patients all came from a high incidence area for esophageal carci-noma in the Eastern Guangdong, Southern China. The overall HPV detection rate in these specimens was up to 60% or higher, with the high risk HPV types16 and 18 being predo-minant in cancerous tis-sues (43% –54%). Our result was consistent with the findings reported indicating that HPV infec-tion, particularly the high risk types, may be one of the major risk factors in esophageal carcinogenesis in this high frequency area of esophageal carcinoma (Chang et al. , 2000).

Tab. 4 PCR for HPV 16 and 18 in 77 Specimens

Location	Number of positive specimens (%)		
	HPV 16 (+)	HPV 18 (+)	Total
Cancerous	30 (39.0)	17 (22.1)	42[a] (54.5)
Paracancerous	14 (18.2)	3 (3.9)	15[a] (19.5)
Normal	8 (10.4)	2 (2.6)	9[a] (11.7)

Note: a Mixed infection, total eight cases.

As shown in Tab. 2 – 4, HPV infection is not limited to cancerous and precancerous tissues. Normal epithelial tissues are also HPV positive. High risk types are predomi-nant in cancerous tissues, whereas low risk types are more common in healthy tissues, indicating that high risk types may play a more important role in carcinogenesis. Mixed infections of high and low risk types are also seen (Chang et al. , 2000). Because low risk types can cause proliferation of epithelium, mixed infections of high and low risk types 16, 18, 6, and 11 may have synergic effects in the enhancement of cell proliferation and transformation (deVilliers et al. , 1999).

The Eastern Guangdong coastal region of Southern China is one of the 6 high incidence areas for esophageal carcinoma in China, and also of the few coastal regions with high frequency of esophageal carcinoma in the world. The morbidity and mortality of esophageal carcinoma is ranked as number one in the overall malignant tumors in this region. Our results suggest that the high incidence of esophageal carcinoma in this particular region may be associated with the high infection rate with HPV, and HPV may be one of the major risk factors in the development of this tumour. Multiple risk factors have been implicated in the development of esophageal carcinoma in this region, such as ni-trosamine (Shen et al. , 1987; Lin et al. , 1985) and trace element deficiency (Shen et al. , 1997).

Acknowledgements

This work was supported by Dr. R. Potter, PhD, Professor of Biology, California University. The authors would like to thank Prof. H. Lin of Sun Yi – shun Medical University; Prof. S. Lu of the Cancer Research Institute, Chinese Academy of Medical Sciences; Prof. K. Yao of Hunan Medical University; and Prof. Y. Qiu of Hong Kong Chinese University who kindly donated HPV plasmid DNA and cell lines.

[In 《J Med Viral》 2002, 68: 412 – 416]

References

1 Bjorge T, Hakulinen T, Engeland A, Jellum E, Koskela P, Lehtinen M, Luostarinen T, Paavonen J, Sapp M, Schiller J, Thoresen S, Wang Z, Youngman L, Dillner J. A prospective, seroepide miological study of the role of human papillomavirus in esophagea cancer in Norway. Cancer Res, 1997, 57: 3989 – 3992

2 Chang F, Syrjanen S, Shen Q, Ji HX, Syrjanen K. Human papillomavirus (HPV) DNA in esophageal precancer lesions and squamous cell carcinomas from China. Int J Cancer, 1990, 45: 21 – 25

3 Chang F, Syrjanen S, Shen Q, Cintorino M, Santopietro R, Tosi P Syrjanen K. Human papillomavirus involvement in esopha – geal carcinogenesis in the high incidence area of China: a study of 700 cases by screening and type – specific in situ hybridization. Scand J Gastroenterol, 2000, 35: 123 – 130

4 Chen B, Yin H, Dhurandhar N. Detection of human papilloma – virus DNA in esophageal squamous cell carcinomas by the polymerase chain reaction using general consensus primers Hum Pathol, 1994, 25: 920 – 923

5 Chen WS, Cai SS, Qiu JW. Lin K, Yan H, Zhang C. Epidemiologic features of esophageal cancer in Nanao county Guangdong Province from 1987 – 92 (in Chinese). Aizheng, 1996, 15: 274 – 276

6 Day NE. The geographic pathology of cancer of the esophagus. Br Med Bull, 1984, 40: 329 – 334

7 deVilliers EM, Lavergue D, Chang F, Syrjanen K, Tosi P, Cintorino M, Santopietro R, Syrjanen S. An interlaboratory study to determine the pres-ence of human papillomavirus DNA in esophageal carcinoma from China. Int J Cancer, 1999, 81: 225 – 228

8 Dillner J, Knekt P, Schiller JT, Hakulinen T. Prospective seroepidemiological evidence that human papillomavirus type 16 infection is a risk factor for oesophageal squamous cell carcinoma. Br Med J, 1995, 311: 1346

9 Franceschi S, Bidoli E, Negri E, Zambon P. Talamini R, Ruol A, Parpinel M, Levi F, Simonato L, La Vecchia C. Role of macronutrients, vitamins and minerals in the aetiology Of squamous – cell carcinoma of the oesophagus. Int J Cancer, 2000, 86: 626 – 631

10 Gurski RR, Schirmer CC, Kruel CR, Komlos F. Kruel CD, Edelweiss MI. Induction of esophageal carcinogenesis by diethylnitro – samine and assessment of the promoting effect of ethanol and N-nitrosonornicotine: experimental model in mice. Dis Esophagus, 1999, 12: 99 – 105

11 He D, Zhang DK, Lam KY, Ma L, Ngan HY, Liu SS, Tsao SW. Prevalence of HPV infection in esophageal squamous cell carcinoma in Chinese patients and its relationship to the p53 gene mutation. Int J Cancer, 1997, 72: 959 – 964

12 Kiyabu M, Shibata D, Arnheim N, Martin WJ, Fitzgibbons PL. Detection of human papillomavirus in formalin fixed, invasive squamous carcinomas using the polymerase chain reaction. Am J Surg Pathol, 1989, 13: 221 – 224

13 Kok TC, Nooter K, Tjong – A – Hung SP, Smits HL, Ter – Schegget JJ. No evidence of known types of human papillomavirus in squamous cell cancer of the oesophagus in a low – risk area. Eur

J Cancer, 1997, 33: 1865 – 1868

14 Lam KY, He D, Ma L, Zhang D, Ngan HY, Wan TS, Tsao SW. Presence of human papillomavirus in esophageal squamous cell carcinomas of Hong Kong Chinese and its relationship with p53 gene mutation. Hum Pathol, 1997, 28: 657 – 663

15 Lin K, Shen ZY, Cai SS. The preliminary detection of nitrosamines in the fish juice pickle vegetable and dry pickle radish in the high incident area of the esophageal carcinoma (in Chinese). Chung – Hua Zhongliu Zazhi, 1985, 7: 32 – 33

16 Liu GT, Qian YZ, Zhang P, Dong WH, Qi YM, Guo HT. Etiological role of Alternaria alternata in human esophageal cancer. Chin Med J (Engl) , 1992, 105: 394 – 400

17 Morgan RJ, Perry AC, Newcomb PV, Hardwick RH, Alderson D. Human papillomavirus and oesophageal squamous cell carcinoma in the UK. Eur J Surg Oncol, 1997, 23: 513 – 517

18 Newberne PM, Schrager TF, Broitman S. Esophageal carcinogenesis in the rat: zinc deficiency and alcohol effects on tumor induction. Pathobiology, 1997, 65: 39 – 45

19 Poljak M, Cerar A, Seme K. Human papillomavirus infection in esophageal carcinomas: a study of 121 lesions using multiple broad – spectrum po] ymerase chain reactions and literature review. Hum Pathol, 1998, 29: 266 – 271

20 Ren A, Han X. Dietary factors and esophageal cancer: a case – control study. Chung Hua Liu Hsing Ping Hsueh Tsa Chih, 1991, 12: 200 – 204

21 Ribeiro U Jr, Posner MC, Safatle Ribeiro AV, Reynolds JC. Risk factors for squamous cell carcinoma of the oesophagus. Br J Surg, 1996, 83: 1174 – 1185

22 Saegusa M, Hashimura M, Takano Y, Ohbu M, Okayasu I. Absence of human papillomavirus genomic sequences detected by the polymerase chain reaction in oesophageal and gastric carcinomas in Japan. Mol Pathol, 1997, 50: 101 – 104

23 Sandvik AK, Aase S, Kveberg KH, Dalen A, Folvik M, Naess O. Papillomatosis of the esophagus. J Clin Gastroenterol, 1996, 22: 35 – 37

24 Sharma D. Carcinoma of oesophagus – aetiological factors and epidemiology: an overview. J Indian Med Assoc, 1999, 97: 360 – 364

25 Shen ZY, Chen ZP, Lu SX. Investigation on nitrosamines in the diets of the inhabitants of high – risk area for esophageal cancer in the southern China and analysis of the correlation factors (in Chinese). J Hygiene Res, 1987, 26: 266 – 269

26 Shen WY, Shen ZY, Chen MH. A multivariable discriminant analysis on the trace elements content of hair in the districts of high, middle and low incidence of the esophageal cancer. World Elemental Med, 1997, 4: 5 – 8

27 Shen ZY, Cen S, Cai WJ, Teng ZP, Shen J, Hu Z, Zeng Y. Immortalization of human fetal esophageal epithelial cells induced by E6 and E7 genes of human papillomavirus 18. Chinese J Exp Clin Virol, 1999a, 13: 121 – 123

28 Shen ZY, Cai WJ, Shen J, Xu JJ, Cen S, Ten ZP, Hu Z, Zeng Y. Human papilloma virus 18E6E7 in synergy with TPA induced malignant transformation of human embryonic esophageal epithe – lial cells. Chinese J Virol, 1999b, 15: 1 – 6

29 Shen J, Shen ZY, Zheng RM, Li LC. Study on micropapilloma – tosis of the esophageal mucosa adjacent to the cancer. Shijie Huaren Xiaohua Zazhi, 2000, 8: 1289 – 1290

30 Siddiqi MA, Tricker AR, Kumar R, Fazili Z, Preussmann R. Dietary sources of N – nitrosamines in a high – risk area for oesophageal cancer: Kashmir, India. LARC Sci Publ, 1991, 105: 210 – 213

31 Smits HL, Tjong – A – Hung SP, ter – Schegget J, Nooter K, Kok T. Absence of human papillomavirus DNA from esophageal carcinoma as determined by multiple broad spectrum polymerase chain reactions. J Med Virol, 1995, 46: 213 – 215

32 Sur M, Cooper K. The role of the human papilloma virus in esophageal cancer. Pathology, 1998, 30: 348 - 354

33 Suzuk L, Noffsinger AE, Hui YZ, Fenoglio - Preiser CM. Detection of human papillomavirus in esophageal squamous cell carcinoma. Cancer, 1996, 78: 704 - 710

34 Syrjanen KJ. Histological changes identical to those of condylomatous lesions found in esophageal squamous cell carcinomas. Arch Geschwulstforsch, 1982, 52: 283 - 292

35 Tada K, Oka M, Hayashi H, Tangoku A, Oga A, Sasaki K. Cytogenetic analysis of esophageal squamous cell carcinoma cell lines by comparative genomic hybridization: relationship of cytogenetic aberration to in vitro cell growth. Cancer Genet Cytogenet, 2000, 117: 108 - 112

36 Takahashi A, Ogoshi S, Ono H, Ishikawa T, Toki T, Ohmori N, Iwasa M, Iwasa Y, Furihata M, Ohtsuki Y. High - risk human papillomavirus infection and overexpression of p53 protein in squa-

mous cell carcinoma of the esophagus from Japan. Dis Esophagus, 1998, 11: 162 - 167

37 Talamini G, Capelli P, Zamboni G, Mastromauro M, Pasetto M, Castagnini A, Angelini G, Bassi C, Scarpa A. Alcohol, smoking and papillomavirus infection as risk factors for esophageal squamous - cell papilloma and esophageal squamous - cell carcinoma in Italy. Int J Cancer, 2000, 86: 874 - 878

38 Yoshikawa H, Kawana T, Kitagawa K, Mizuno M, Yoshikura H, Iwamoto A. Detection and typing of multiple genital human papillomaviruses by DNA amplification with consensus primers. Jpn J Cancer Res, 1991, 82: 524 - 531

39 Zambon P, Talamini R, La Vecchia C, Dal - Maso L, Negri E, Tognazzo S, Simonato L. Smoking type of alcoholic beverage and squamous - cell oesophageal cancer in northern Italy. Int J Cancer, 2000, 86: 144 - 149

40 Zur Hausen H. Papillomaviruses in human cancer. Cancer, 1987, 59: 1692 - 1696

75. Telomere and Telomerase in the Initial Stage of Immortalization of Esophageal Epithelial Cell

SHEN Zhona - yina[1], XU Li - yan[1], LI En - min[1], CAI Wei - jia[1], CHEN Min - hua[1], SHEN jian[1], ZENG Yi[1], LI En - min[2], ZENG Yi[3]

1. Department of Tumor Pathology, Medical College of Shantou University;
2. Department of Biochemistry and Molecular Biology, Medical College of Shantou University;
3. Institute of Virology, Chinese Academy of Preventive Medicine

Summary

Objective　To search for the biomarker of cellular immortalization, the telomere length, telomerase activity and its subunits in cultured epithelial cells of human fetal esophagus in the process of immortalization.

Methods　The transgenic cell line of human fetal esophageal epithelium (SHEE) was established with E6E7 genes of human papillomavirus (HPV) type 18 in our laboratory. Morphological phenotype of cultured SHEE cells from the 6th to 30th passages, was examined by phase contrast microscopy, the telomere length was assayed by southern blot method, and the activity of telomerase was analyzed by telomeric repeat amplification protocol (TRAP). Expressions of subunits of telomerase, hTR and hTERT, were assessed by RT – PCR. DNA content in cell cycle was detected by flow cytometry. The cell apoptosis was examined by electron microscopy (EM) and TUNEL label.

Results　SHEE cells from the 6 th to 10 th passages showed cellular proliferation with a good differentiation. From the 12 th to the 15 th passages, many senescent and apoptotic cells appeared, and the telomere length sharply shortened from 23 kb to 17 kb without expression of hTERT and telomerase activity. At the 20 th passage, SHEE cells overcame the senescence and apoptosis and restored their proliferative activity with expression of telomerase and hTERT at low levels, but the telomere length shortened continuously to the lowest of 3 kb. After the 30 th passage cells proliferation was restored by increment of cells at S and G2M phase in the cell cycle and telomerase activity expressed at high levels and with maintenance of telomere length.

Conclusion　At the early stage of SHEE cells, telomeres are shortened without expression of telomerase and hTERT causing cellular senescence and cell death. From the 20 th to the 30 th passages, the activation of telomerase and maintenance of telomere length show a progressive process for immortalization of esophageal epithelial cells. The expression of telomerase may constitute a biomarker for detection of immortalization of cells.

Introduction

Telomerase activity was demonstrated in cancer of digestive tract[1-3], such as gastric[4-7], hepatic[8-10] colorectal[11-13] and esophageal cancers[14-19]. Inhibition of telomerase activity will be a new therapeutic for cancer[20-26]. Telomerase activity can be used as a diagnostic marker for cancer[27-30]. Normal mammalian cells grow in cultural medium with a limited number of passages before entering senescence and death[31], which are associated with shortening of telomere. Telomeres are specialized structures at chromosomal ends that are composed of TTAGGG DNA repeats[32]. Telomerase is a ribonuclear protein complex, which contains human telomerase reverse transcriptase (hTERT) as a catalytic domain, and human telomerase RNA component (hTR)[33]. Telemetries cap chromosomal ends perform the function of preventing abnormal chromosomal fusions and rearrangemen[34]. However, each time a cell divides, the most distal part of the chromosome is incompletely duplicated and the telomere becomes shorter. Critically short telomeres enable the formation of aberrant chromosomal structures resulting in growth arrest or senescence[35]. With expression of telomerase or hTR and hTERT, the length of telomere extends to maintain the life span of the cells. There are other roles of telomerase in immortal and malignant lesion, such as proliferative potential[36], delaying senescence[37,38], promoting cell cycle, cell immortalization and carcinogenesis[39].

Recently, many papers have indicated that human papillomavirus (HPV) are the important etiological factor in esophageal carcinoma[40-42]. Induced by HPV 18 E6E7 genes, we establised an immortalized cell line (SHEE) from the esophageal epithelium which underwent[43-45] malignant transformation[47,48]. Changes of telomere length and telomerase activity in the cell line are not clear at this early stage, nor is which criteria to use to detect immortalization of cells and what the relationship between telomerase and cell phenotype in SHEE cells is. The goal of this study is to explore when telomerase activity appears in the immortalized progressive process arid study the relationship between telomerase and cellular phenotype.

Materials and Methods

1. Cell culture and EM examination

The SHEE cell line was a kind of immortal embryonic esophageal epithelium induced by $E_6 E_7$ genes of human papillomavirus (HPV) type 18 in our laboratory[43]. The continual growth cells from the 6 th to 30 th passages were routinely cultivated in flasks and the 24 – well plates (Coming Co.) with culture medium 199 (Gibco), 100 ml $\cdot$ L^{-1} bovine serum, 100 U penicillin and streptomycin in a humidified atmosphere of 50ml $\cdot$ L^{-1} CO_2 and 950 ml'LE air. The cell shape and size, anchorage – dependent growth and contact – inhibited growth were examined by phase contrast microscopy. For electron microscopic assessment, cells were spun to fonn a pellet and fixed with 25 g $\cdot$ L^{-1} glutaraldchwle. They were dehydrated in graded ethanol and embedded in Araldite. Ultra – thin sections were cut with glass knifes and mounted on copper grids. They were contrasted for 15 min with uranyl acetate and for 3 min with lead citrate. The sections were examined by electron microscope (Hitachi, H – 300).

2. Cell proliferative cycle and apoptosis

Cells cultured in the flask were digested, washed twice with PBS, fixed by 70 % alcohol, prepared as single – cell suspension and stored at 4 ℃. Cells were stained with propidium iodide (Sigma) and analyzed with flow cytometry (FACSort, B – D Co.). The percentage of cells in various stages of the cell cycle, the apoptotic cell rate (AI) and proliferation index (PI = S + G_2M/ C_0G_1, + S + G_2M) were calculated. These cells on the glass coverslips within the 24 – well plate were incubated with 10 mg $\cdot$ L^{-1} proteinase K for 15 min at room temperature. After the quenching of endogenous peroxidase, labeled nuclei with TUNEL (*In – Situ* Death Detection kit, Boehringer Mannhein Co.) were detected according to the instructions of the manufacturer. The brownish nucleus was considered positive apoptotic nucleus.

3. Telomere length analysis[49]

The genomic DNA of $10^6 - 10^8$ cells was extracted. The telomeric restriction fragment (TRF) was measured by southern blot. Briefly, 20 μg of genomie DNA was digested with Hinf I and run on 7 g $\cdot$ L^{-1} agarose gel with marker DNA/Hind Ⅲ. After electrophoresis the gel was blotted to nylon membrane (Hybond[TM] N^+, Amersham, Life Science) and hybridized to the Dig – labeled probes (CCCTAA)$_3$ at 50 ℃ in 5 × SSC, 1 g $\cdot$ L^{-1} Sod. n – Lauroylsarcosine (SLS), 0. 2g $\cdot$ L^{-1} SDS for 12 ~ 16h and washed twice at room temperature in 2 × SSC, 1 g $\cdot$ L^{-1} SDS for 5 min, once at 50 ℃ in 1 × SSC, 1 g $\cdot$ L^{-1} SDS for 10 min, twice at 50℃ in 1 × SSC, 1 g $\cdot$ L^{-1} SDS for 10 min, twice at 50 ℃ in 0. 1 × SSC, 1 g $\cdot$ L^{-1} SDS for 5 min, stained with NBT/BCIP and the median points were measured to obtain the mean telomere length[50].

4. Telomerase activity assay

Telomerase activity was measured using the telomeric repeat amplification protocol (TRAP).

Frozen samples were homogenized in $10 - 50 \mu l$ of ice – cold lyses buffer ($10 \ mmol \cdot L^{-1}$ Tris – HCl, pH7.5, $1 \ mmol \cdot L^{-1}$ EGTA, 0.1 $mmol \cdot L^{-1}$ Benzamidine, $5 \ mmol \cdot L^{-1} \beta$ – mercaptotha-nol, $5 \ g \cdot L^{-1}$ CHAPS, $100 \ ml \cdot L^{-1}$ glycerol). After 30 min of incubation on ice, the lysate was centrifuged at 12 000 g for 20 min at 4 ℃. TRAP – eze Telomerase Detection Kit (Oncor Inc.) reac-tion was performed using 1 μl lysate or 1/10 diluted lysate, 2.5 μl 10 × TRAP buffer (200 $mmol \cdot L^{-1}$ Tris – HCl, pH8.3, 15 $mmol \cdot L^{-1}$ $MgCl_2$, 630 $mmol \cdot L^{-1}$ KCl, 0.5 % Tween 20, 10 $mmol \cdot L^{-1}$ EGTA, $1 \ g \cdot L^{-1}$ BSA), 0.5 μl 2.5 $mmol \cdot L^{-1}$ dNTP, 0.5 μl Ts primer, 0.5 μl TRAP prim-er mix, 19.5 μl water, 0.5 μl taq ($2 \times 10^6 U \cdot L^{-1}$). After incubation at 30 ℃ for 30 min, the reaction mix was immediately transferred to 94 ℃ and performed PCR (GeneAmp PeR System 2400, PE, USA) at 94 ℃ for 30 s, 55 ℃ for 30 s, for 30 cycles. PCR products were separated in a non – denaturing $120 \ g \cdot L^{-1}$ PAGE in 1 × TBE at $5 \ V \cdot cm^{-1}$. The gel was stained using $AgNO_3$ and was photographed.

5. Subunits of telomerase analysis

The activities of telomerase were performed by hTERT and hTR analysis. Analysis of expression of hTR and hTERT was determined by reverse transcription – PCR (RT – PCR) amplification in contrast with house – keeping gene GAPDH. Total RNA was isolated from the cell using Gstract™ RNA Isolation Kit (Maxim Biotech, Inc.). cDNA was synthesized from 10 μg of total RNA using Ready – to – use First Strand cDNA Synthesis Kit (Maxim Biotech. Inc.). PCR reaction was per-formed using 2 μl aliquots of the reverse – transcribed cDNA 2.5 μl, 10 × PCR buffer, 1.5 μl, 25 $mmol \cdot L^{-1}$ $MgCl_2$. 2.5 μl, 1.0 $mmol \cdot L^{-1}$ dNTP, 2.0 μl specific primers, 14 μl water, 0.5 μl taq ($2 \times 10^6 U \cdot L^{-1}$). hTERT mRNA was amplified using the primer pair: 5' – CGGAAGAGT-GTCTG – GAGCAA – 3' and 5'GGATGAAGCGGAGTCTGGA – 3' for 31 cycles (94 ℃ for 45 s, 55 ℃ for 45 s, 72 ℃ for 90 s). hTR was amplified using the primer pair: 5' – TCTAACCCTAACT-GAGAAGGGCGTAG – 3' and 5' GTTTGCTCTAGAATGAACGGTGGAAG – 3' GAPDH was amplified using the primer pair: 5' – GAAGGTGAAGGTCGGAGTC – 3' and 5' – GAAGATGGTGATGG-GATITC – 3' for 33 cycles (94 ℃ for 60 s, 55 ℃ for 60 s, 72 ℃ for 60 s). PCR products of each sample were subjected to electrophoresis in a $15 \ g \cdot L^{-1}$ agarose gel containing $0.5 \ mg \cdot L^{-1}$ ethidium bromide.

Results

In the initiated passages, the cells in the 6th to the 10th passages were uniform in size and shape (Fig. 1A), and grew as an even monolayer with characteristics of anchorage – dependent and attach-ment – inhibited growth. Cells continuously cultured in the 12th to the 16th passages exhibited mor-phologic changes in which cells were enlarged arid flattened and exhibited differentiation and senes-cence. When many cells had shrunk, were round and floated freely, the majority underwent apoptosis and cell death with a few cells surviving (Fig. 1B). Overcoming senescence and apoptosis, the cells of the 20th passage restored their proliferation capacity (Fig. 1C). After 30 passage the cells prolifera-ted again and exhibited diphasic differentiation, a portion of cells displayed the undifferentiated basal epithelium and the other portion displayed differentiated squamous epithelitm (Fig. 1D).

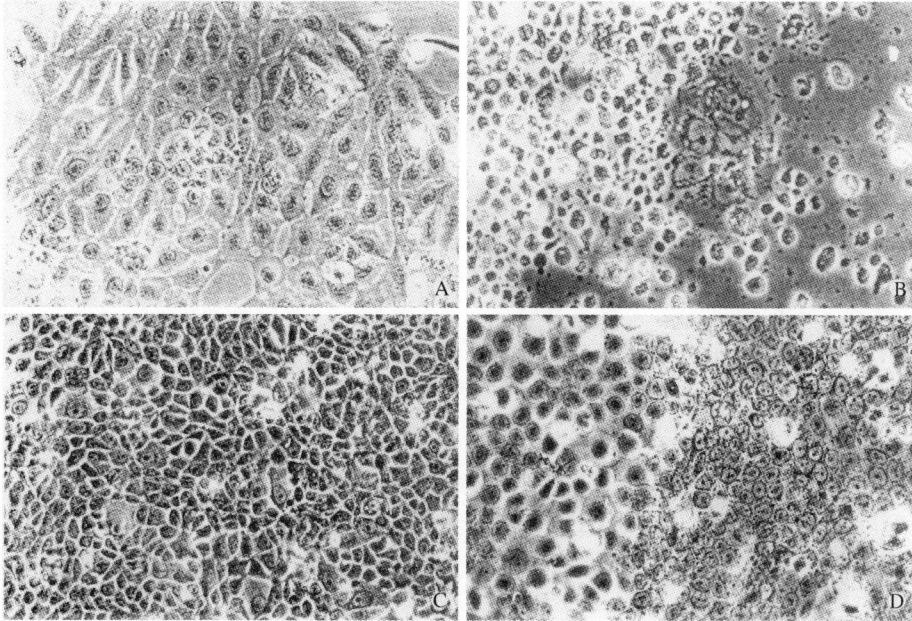

A: Cell in 6 – 10th passages showed differential phenotype (phase – contrast microscopy, Ph ×400) ;

B: Cells of 16th passage displayed apoptosis with a few of cells survived (Ph ×400);

C: Cells of 20th passage displayed hyperplasia (Ph ×200) ;

D: Cells of 30th passage displayed proliferative activity with diphasic differentiation (Ph ×400)

Fig. 1　Morphologic changes in living cells of SHEE

1. Proliferative index (PI) and apoptosis

The PI of cell in 10, 16, 20, 30 the passage were 25.5 %, 17.3 %, 43.3 %, and 43.0 % respectively (Fig. 2, A, B, C, D). Apoptotic cells index (AI) was in 10th passage, 7.3 %; 16th, 57.5 %; 20th, 5.7 % and 30th, 7.5 %. The cells in the 16th passage were at the stage of senescence and death, and the 10th, 20th and 30th were at their proliferated stages at various levels. Cells in $C_0 G_1$, phase were identified as the differentiated cells containing, some senescent cells.

The TUNEL assay was also used to characterize the biological features of cells apoptosis. Many TUNEL – positive nuclei were observed in cells of passage 16th (Fig. 3), and in a few cells in other passages. By EM examination, many apoptotic cells revealed rounded and shrunken nuclei with condensed chromatin stuck closely to the nuclear membrane (Fig. 4).

2. Telomere length

Following continuous growth of the cells, the telomere length of the cells in the 6th to 10th passages exhibited shortening from a mean size of 23kb in the normal esophageal mucosa to 17kb at passage 10. At 20th passage telomere length was shortened continually to the shortest 3kb, but maintained till the 30th passage (Fig. 5).

3. Telomerase activity, expression of hTR and hTERT

Expression of hTR and hTERT in SHEE cells was determined by RT – PCR. The hTERT expression was positive in the 20th and 30th passages, but negative in the 10th passage (Fig. 6A). Cells of the 10th, 20th and 30th passage showed positive expression of hTR (Fig. 6B). House – keeping protein GAPDH was used as a control (Fig. 6C). Comparing the variation of telomerase activity in SHEE from the 10th to the 30th passage, cells of the 10th passage were negative, while the 20th was weak, and the 30th was apparent. The positive control of hunman esophageal squamous cell carcinoma expressed the highest telomerase activity (Fig. 7).

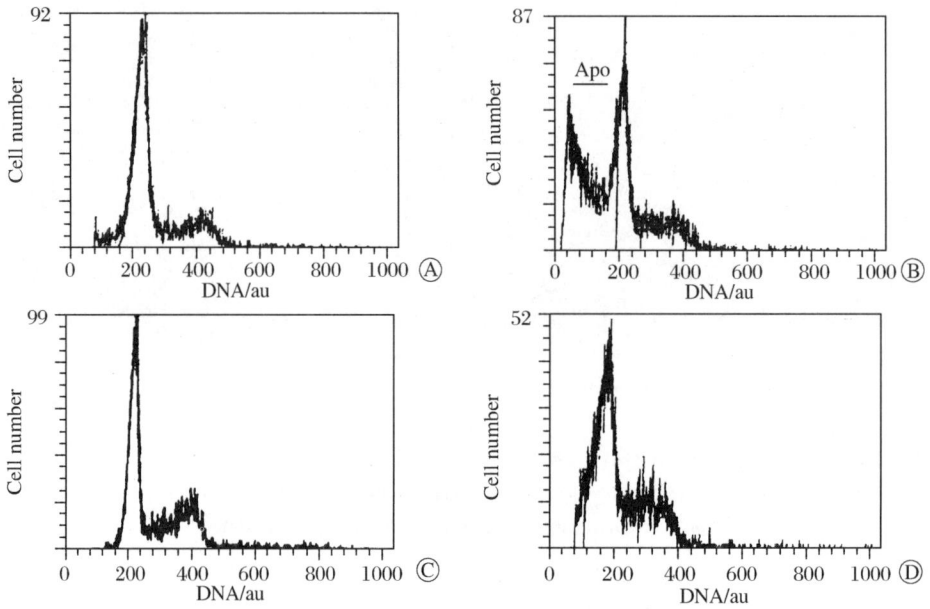

A: 10th passage; B: 16th passage; C: 20th passage; D: 30th passage.

Apo: Apoptotic peak; AU. Arbitrary unite

Fig. 2　DNA histogram of SIIEE cells

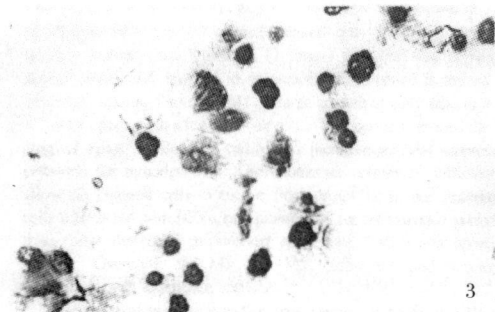

Fig. 3　TUNEL positive apoptotic nuclei in 16th passage of SHEE. (× 400)

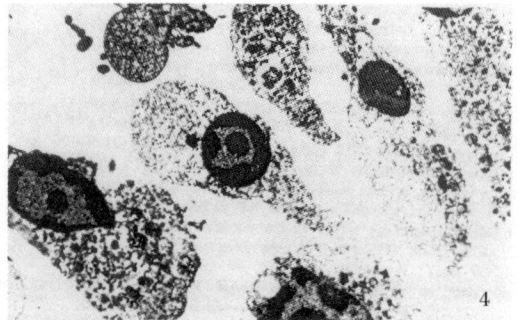

Fig. 4　Electron – photomicrograph of apoptotic cells in 16th passage Shrunken cells and rounded nuclei with marginated condensed chromatin. (EM × 3000)

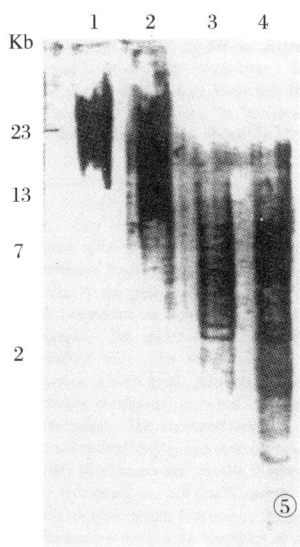

1: Normal esophagus; 2: Cells of 10th passage; 3: Cells of 20th passage; 4: Cells of 30th passage

Fig. 5 Telomere length of SHEE series using southern blot

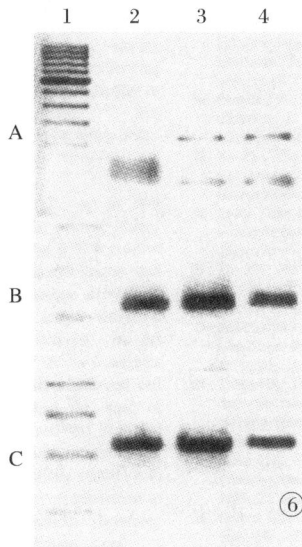

Lane 1: marker; Lane 2: 10th passage; Lane 3: 20th passage; Lane 4: 30th passage

Fig. 6 Gel electrophoretogram of hTERT (A), hTR (B) and GAPDH (C)

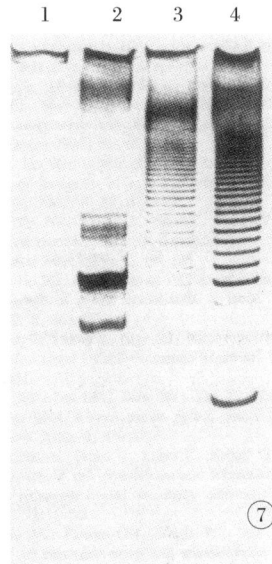

Lane 1: 10th passage, negative; Lane 2: 20th passage, weak; Lane 3: 30th passage, positive; Lane 4 – human esophageal squamous cell carcinoma, strong positive;

Fig. 7 Measurements of telomerase activity using TRAP assay

4. The relationship between telomerase activity and cellular proliferation

At an early stage in the immortalization process, SHEE cells could be divided into three stages. At the primary stage, telomerase activity and expression of hTERT was absent and the telomere length shortened. The cells of SHEE were proliferated and differentiated and then their were senescent and apoptotic cells in the culture. At the early immortal stage, cells exhibited telomerase activity and its subunits hTERT at low levels with telomere length being continually shortened and proliferation restored. At the immortal stage, telomerase and hTERT were expressed in high level and telomere length maintained, accompanied by cell proliferation. From 6th to 30th passages the cells expressed hTR.

Discussion

After transferring the HPV18E6E7 genes to the fetal esophageal epithelial cells, we established an immortal cell line designated SHEE. In order to focus on the process of immortalization, we monitored the dynamic changes of telomere length and telomerase activity of SHEE cells for extended periods of time (from the 6th passage to the 30th passage). At the primary stage the cells of passages 6 – 10 appeared as differentiated squamous epithelial cells with telomere length shortened without telomerase expression, after which the cells of passages 12 – 16 became senescent and underwent cell death. After overcoming the senescence and cell death, cells of the 20th passage expressed telomerase activity at low levels, where the length of shortened telomere was not maintained, but the proliferation of cells was restored. At the 30th passage, the cells exhibited a higher level of telomerase activity, hTR and hTERT, with maintained telomere length and continual proliferation of cells (immortalization). Therefore, one can suggest that telomerase activity and maintenance of telomere length might be necessary for immortalization of human esophageal epithelium *in vitro*. To investigate the immortalization process, we used the SHEE cell model to define steps in aging of cells. After transduction of E6E7 genes from HPV type 18 normal cultured cells proliferate until they reach a discrete point (passage 12 times) in which the population growth ceases and develops to senescence. This period is termed the MI stage of aging[51]. After M1, a large amount of cells dead to reach a " crisis" point with a few cells survival. This period is termed the M2 stage of aging[52]. Both M1 and M2 are therefore potential suppression pathways for tumorigenesis. The telomerase activity is sufficient to allow the cultured cells to escape from crists[53]. In our experiment cells transfected with HPV, cell proliferated for an extended period of time, after that cells encountered senescence (M1) and apoptosis (M2). Overcome the M1 and M2, cells exhibited accelerant hyperplasia with telomerase activity.

Recent evidence suggests that viral oncogenes might directly up – regulate telomerase activity[54], such as HCV core protein, EBV –, HPV – and SV40 T antigen – expressing cell clones[55-58]. In our data the increase of hTR in the primary cells of SHEE might be directly or indirectly affected by the HPV viral oncogenes. Previous reports also have shown that telomerase activity can be achieved by the E_6 and E_7 protein of HPV[59]. HPV 16 E6 oncoprotein was capable of inducing telomerase activity in monolayer cultures of proliferating keratinocyte[60]. It was more likely that HPV E6/E7 transcription or other additional alterations, as chromosome instability[61,62] was prerequisite for induction of telomerase activity in proliferating cells. This hypothesis fits very well with our data on HPV – mediated immortalization of cells *in vitro*.

In summary, immortal cell line of the SHEE, up to 30th passage, may be divided into three stages. At the primary stage, telomerase activity and hTERT of cells were absent but hTR is positive with telomere length shortening and the cells became senescent and apoptotic. At the early – immortalized stage, the telomerase activity and hTERT expressed at low level and telomere length shortened continuously, but underwent cell hyperplasia occurred. At the immortalized stage, the telomerase activity and its two subunits expressed at a high level, telomere length was maintained and cell proliferation continued, in which the cells reached the stage of immortalization. The shortened telomere length , and the activated telomerase activity display in a dynamic process. Our results prove that shortening of telomeres and absence of telomerase activity contribute to cellular senescence and cell death, and the activation of telomerase to maintain telomere length is necessary for immortalization. Therefore, the telomerase activity is the biomarker of immortalization of cell.

[In 《World J Gastroenterol》 2002, 8 (2): 357 – 362]

References

1 Yakoob J, Hu GL, Fan XG, Zhang Z. Telomere, telomerase and digestive cancer. Worm J Gastroenterol, 1999, 5: 334 – 337

2 He XX, Wang JL. Activity of telomerase and oncogenesis. HuarenXiaohua Zazhi, 1998, 6: 1100 – 1101

3   Yang SM, Fang DC, Luo YH, Lu R, Liu WW. Telomerase activity in gastroeintestind submucosal tumors and its clinical significance. Huaren Xiaohua Zazhi, 1998, 6: 765 – 767

4   Zhan WH, Ma JP, Peng JS, Gao JS, Cai SR, Wang JP, Zheng ZQ, Wang L. Telomerase activity in gastric cancer and its clinical implications World J Gastroenterol, 1999, 5: 316 – 319

5   He XX, Wang JL, Wu JL, Yuan SY, Ai L. Telomerase expression, Hp infection and gastric mucosal carcinogenesis. Shijie Huaren Xiaohua Zazhi, 2000, 8: 505 – 508

6   He XX, Wang JL, Wu JL, Yuan SY, Ai L. Telomere, cellular DNA content and gastric mucosal carcinogenesis. Shijie Huaren Xiaohua Zazhi, 2000, 8: 509 – 512

7   Yao XX, Yin L, Zhang SY, Bai WY, Li YM, Sun ZC. hTERT expression and cellular immunity in gastric cancer and precancerosis. Shijie Htuaren Xiaohua Zazhi, 2001, 9: 508 – 512

8   Meng ZQ, Yu EX, Song MZ. Inhibition of telomerase activity of human liver cancer cell SMMC 7721 by chemotherapeutic drags. Shijie Huaren Xiaohua Zazhi, 1999, 7: 252 – 254

9   Fu JM, Yu XF, Shao YF. Telomerase and primary liver cancer. Shijie Huaren Xiaohua Zazhi, 2000, 8: 461 – 463

10 Qt, B, Li BJ, Lu ZW, Pan HL. Clinical significance of telomerase activity detected in fine – needle aspiration speciments to liver cancer diagnosis. Shijie Huaren Xiaohua Zazhi, 2001, 9: 538 – 541

11 Qiu SL, Huang JQ, Wang YF, Peng ZH. Analysis of telomerase activity in colorectat cancer, precancerous lesions and cancer washings. Shijie Huaren Xiaohua Zazhi, 1998, 6: 992 – 993

12 Sobti RC, Kochar J, Singh K, Bhasin D, Capalash N. Telomerase activation and incidence of HPV in human gastrointestinal tumors in North Indian population. Mol Cell Biochem, 2001, 217: 51 – 56

13 Jia L, Li YY. Telomerase activity of exfoliated cancer cells in colonic luminal washings. Huaren Xiaohua Zazhi, 1998, 6: 955 – 957

14 Koyanagi K, Ozawa S, Ando N, Takeuchi H, Ueda M. Clinical significance 1o telomerase activity in the cancerous epithelial region of oesophageal squaous carcinoma. Br , I Surg, 1999, 86: 674 – 679

15 Hiyama T, Yokozaki H, Kitadai Y, Haruma K, Yasui W, Takara F. Overexpression of human telomerase RNA is an early oesophageal carcinogenesis. Virchows Arch, 1999, 434: 483487

16 Kiyozuka Y, Asai A, Yamamoto D, Senzaki H, Yoshioka S, Hioki K, Tsubura A. Establishment of novel human esophageal cancer cell relation to telomere dynamics and telomerase activity. Dig Dis Sci, 2000, 45: 870 – 879

17 Koyanagi K, Ozawa S, Ando N, Mukai M, Kitagawa Y, Ueda MM. Telomerase activity as an indicator of malignant in iodine – nonreactive lesions of the esophagus. Cancer, 2000, 88: 1524 – 1529

18 Xu LY, Shen ZY, Li EM, Cai W J, Shen J, Li C, Chen JY, Zeng Y. Telomere length and telomerase activity in immortalized and malignantly transformed human embryonic esophageal epithelial cell lines by E6 and E7 genes of HPV 18 type. Aibian Qibian Tubian, 2001, 13: 137 – 140

19 Morales CP, Lee EL, Shay JW. In situ hybridization for the detection of telomerase in the progression from Barrett's esophagus to esophageal adenocarcinoma. Cancer, 1998, 83: 652 – 659

20 Xia ZS, Zhu ZH, He SG. Effects of ATRA and 5 Fu on growth and telomerase activity of xenografts of gastnc cancer in nude mice. Shijie Huaren Xiaohua Zazhi, 2000, 8: 674 – 677

21 Li ZS, Zhu ZG, Yin HR, Chen SS, Lin YZ. Diversity of telomerase activity in human and murine tumor cells transfected with cytokine genes. Shijie Huaren Xiaohua Zazhi, 1999, 7: 194 – 196

22 Zhu ZH, Xia ZS, He SG. The effects of ATRA and 5 Fu on telomerase activity and cell growth of gastric cancer cells in vitro. Shijie Huaren Xiaohua Zazhi, 2000, 8: 669 – 673

23 Sarvesvaran J, Going J, Milroy R, Kaye SB, Nicol Keith W. Is small cell lung cancer the perfect target for anti – telomerase treatment? Carcinogenesis, 1999, 20: 1649 – 1651

24 Zhang FX, Zhang XY, Fan DM, Deng ZY, Yan Y, Wu HP, Fan JJ. Antisense telomerase RNA induced human gastric cancer cell apoptosis. World . I Gastroenterol, 2000, 6: 430 – 432

25 Bednarek A, Shilkaitis A, Green A, Lubet R, Kelloff G, Christov K, AIdaz CM. Suppression of cell proliferation and telomerase activity in 4 – (hydroxyphenyl) retinamide – treated mammary tumors. Carcinogenesis, 1999, 20: 879 – 883

26 Herbert BS, Wright AC, Passons CM, Wright WE, Ali IU, Kopelovicl Shay JW. Effects of chemopreventive and antitelomerase agents on the spontaneous immortalization of breast epithelial cells. J Natl Cancer Inst, 2001, 93: 39 – 45

27 Yeh TS, Cheng A J, Chen TC, Jan YY, Hwang TL, Jeng LB, Chen MF, Wang TC. Telomerase activity is a useful marker to distinguish malignant pancreatic cystic tumors from benign neoplasms and pseudocysts. J Surg Res, 1999, 87: 171 – 177

28 Shroyer KR, Thompson LC, Enomoto T, Eskens JL, Shroyer AL, McGregor JA. Telomerase expression in normal epithelium, reactive atypia, squamous dysplasia, and squamous cell carcinoma of the uterine cervix. Am. l Clin Pathol, 1998, 109: 153 – 162

29 Wisman GBA, Hollema H, de – Jong S, ter – Schegget J, Tjong – A – Hung SP, Ruiters MHJ, Krans M, de – Vries EGE, van – der – Zee AGJ. Telomerase activity as a biomarker for (Pre) neoplastic cervical disease in scrapings and frozen. sections from patients with abnormal cervical smear. J Clin Oncol, 1998, 16: 2238 – 2245

30 Rudolph P, Schubert C, Tanmm S, Heidom K, Hauschild A, Michalska i, Majewski S, Krupp G, Jablonska S, Parwaresch R. Tclomerase activity in melanocytic lesions: A potential marker of tumor biology. Am , I Path, 2000, 156: 1425 – 1432

31 Jones CJ, Kipling D, Morris M, ltepburn P, Skinner J, Bot, nacer A, Wyllie FS, Ivan M,

Bartek J, Wynford – Thomas D, Bond JA. Evidence for a telomere – independent " clock " limiting RAS oncogene – driven proliferation of human thyroid epithelial cells. Mol Cell Biol, 2000, 20: 5690 – 5699

32 Griffith JD Comeau L. Rosenfield S, Stansel RM, Bianchi A, Moss II, de Lange T. Mammalian telomers end in a large duplex loop. Cell, 1999, 97 : 503 – 514

33 Nakano K, Watney E, McDougall JK. Telomerase activity and expression of telomerase RNA component and telomerase catalytic subunit gene incervical cancer. Am J Pathol, 1998, 153: 857 – 864

34 Schwartz JL, Jordan R, Liber H, Murnane JP, Evans HH. TP53 – dependent chromosome instability is associated with transient reductions intelomere length in immortal telomerase – positive cell lines. GenesCITromosomes Cancer, 2001, 30: 236 – 244

35 Chin L, Artandi SE, Shen Q, Tam A, Lee SL, Gottlieb GJ, Greider CW, DePinho RA. p53 deficiency rescues the adverse effects of telomere loss and cooperates with telomere dysfunction to accelerate carcinogenesis. Cell, 1999, 97: 527 – 538

36 Yang . XY, Kimura M, Jeanclos E, Aviv A. Cellular proliferation and telomerase activity in CHRF – 288 – 11 cells. Dfe Sci, 2000, 66: 1545 – 1555

37 Nickoloff BJ, Chaturvedi V, Bacon P, Qin JZ, Denning MF, Diaz MO. Id – 1 delays senescence but does not immortalize keratinocytes. J Biol Chem, 2000, 275: 27501 – 27504

38 MacKenzie KL, Franco S, May C, Sadelain M, Moore MA. Mass cultured human fibroblasts overexpressing hTERT encounter a growth crisis following and extended period of proliferation. Exp Cell Res, 2000, 259: 336 – 350

39 Sugihara M, Ohshima K, Nakamura H, Suzumiya J, Nakayama Y, Kanda M, Haraoka S, Kikuchi M. Decreased expression of telomerase – associated RNAs in the proliferation of stem cells in com-

parison with continuous expression in malignant tumors, lnt J Oncol, 1999, 15: 1075 – 1080

40　Wang DX, Li W. Advances in esophageal neoplasms etiology. Shijie Huaren Xiaohua Zazhi, 2000, 8: 1029 – 1031

41　Ma QF, Jiang H, Feng YQ, Wang XP, Zhou YA, Liu K, Jia ZL. Detection of human papillomavims DNA in squamous cell carcinoma of the esophagus. Shijie Huaren Xiaohua Zazhi, 2000, 8: 1218 – 1224

42　Zou SY, Liu XM, Tang XP, Wang P. Immunohistochemical and electron microscopic observation on positive HPV 16 – E6 protein in esophageal cancer. Shijie Huaren Xiaohua Zazhi, 1998, 6: 47 – 48

43　Shen ZY, Cen S, Cai W J, Xu JJ, Ten ZP, Shen J, Hu Z, Zeng Y. Immortalization fetal esophageal epithelial cells induced by E6 and E7 genes of human papilloma vires to. 7onghua Shiyan He Linchuang Bingduxue Zazhi, 1999, 13: 121 – 123

44　Shen ZY, Shen J, Cai WJ, Cen S, Zeng Y. Biological characteristics of human fetal esophageal epithelial cell line inmmortalized by the E6 and E7 gene of HPV type 18. Zhonghua Shiyan He Linchuang Bingduxue Zazhi, 1999, 13: 209 – 212

45　Shen ZY, Xu LY, Chen MH, Cai WJ, Shen J, Chen JY, Hon CQ, Zeng Y. Biphasic differentiation of immortalized esophageal epitheliums induced by HPW18E6E7. Bingdu Xuebao, 2001, 17: 210 – 214

46　Shen ZY, Xu LY, Chen XH, Cai WJ, Shen J, Chert JY, Huang TH, Zeng Y. The genetic events of HPV – immortalized esophageal epitheliumcells, lnt J Mol Med, 2001, 8: 537 – 542

47　Shen ZY, Cen S, Shen J, Cai WJ, Xu JJ, Teng ZP, Hu Z, Zeng Y. Study of immortaliztion and malignant transformation of human embryonic esophageal epithelial cells induced by HPVI8E6E7. J Cancer Res ClinOncel, 2000, 126: 589 – 594

48　Shen ZY, Chen XH, Shen J, Cai WH, Chen JY, Huang TFH, Zeng Y. Malignant transformation of immortalized human embryonic esophageal epithe-

lial cells induced by human papillomavirus. Bingdu Xuebao, 2000, 16: 97 – 101

49　Xu LY, Li EM, Shen ZY, Cai WJ, Shen J. A nonradio – labelled method assays to measure the telomere length of human chromosome. Aibian Qibian Tubian, 2001, 13 : 1 – 4

50　Hou M, Xu D, Bjorkhohn M, Gruber A. Real – time quantitative telomeric repeat amplification protocol assay for the detection of telomerase activity. Clin Chem, 2001, 47: 519 – 524

51　Ouellette MM, Liao M, flerbert BS, Johnson M, Holt SE, Liss HS, Shay JW, Wright WE. Subsenescent telomere lengths in fibroblasts immortalized by limiting amounts of telomerase. J Biol Chem, 2000, 275: 10072 – 10076

52　Lustig AJ. Crisis intervention: The role of telomerase. Proc Natl Acad Sci USA, 1999, 96: 3339 – 3341

53　Halvorsen TL, Leibowitz G, Levine F. Telomerase activity is sufficient to allow transformed cells to escape from crisis. Mol Cell Biol, 1999, 19: 1864 – 1870

54　Zhu J, Wang H, Bishop JM, Blackburn EH. Telomerase extends the lifespan of virus – transformed human cells without net telomere lengthening. Proc Natl Acad Sci USA, 1999, 96: 3723 – 3728

55　Nowak JA. Telomerase, cervical cancer, and human papillomavims. Clin Lab Med, 2000, 20 – 369 – 382

56　Ray RB, Meyer K, Rayt R. Hepatitis C virus core protein promotes immortalization of primary human hepatocytes. Virology, 2000, 271: 197 – 204

57　Guo W, Kang MK, Kim HJ. Park NH. Immortalization of human oral keratinocytes is associated with elevation of telomerase activity and shortening of telomere length. Oncol Rep, 1998, 5: 779 – 804

58　Mutirangura A, Sriuranpong V, Termrunggraunglert W, Tresukosol D, Lertsaguansinchai P, Voravud N, Niruthisard S. Telomerase activity and human papillomavirus in malignant, premalignant and benign cervical lesions. Br J Cancer,

1998, 78 – 933 – 939

59　Nair P, Jayaprakash PG, Nair MK, Pillai MR. Telomerase, p53 and human papillomavirus infection in the uterine cervix. Acta Oncol, 2000, 39: 65 – 70

60　Snijders PJ, van – Duin M, Walboomers JM, Steenbergen RD, Risse, EK, Helmerhorst TJ, Verheijen, RH, Meijer PJ. Telomerase activity exclusively in cervical carcinomas and a subset of cervical intraepithelial neoplasia grade Ⅲ lesions: strong association with elevated messenger RNA levels of its catalytic subunit and high – risk human papillomavirus DNA. Cancer Res, 1998,

58: 3812 – 3818

61　Golubovskaya VM, Filatov LV, Behe CI, Presnell SC, Hooth MJ, Smith GJ, Kaufmann WK. Telomere shortening telomerase expression, and chromosome instability in rat hepatic epithelial stem – like cells. Mol Carcinog, 1999, 24: 209 – 217

62　Filatov L, Golubovskaya V, Hurt JC, Byrd LL, Phillips JM, Kaufmann WK. Chromosomal instability is correlated with telomere erosion and inactivation of G2 checkpoint function in human fibroblasts expressing human papillomavirus type 16 E6 oncoprotein. Oncogene, 1998, 16: 1825 – 1838

76.　Immortal Phenotype of the Esophageal Epithelial Cells in the Process of Immortalization

SHEN Zhong – ying[1], XU Li – yan[1], LI En – min[2], SHEN Jian[1], ZHENG Rui – ming[1], CAI Wei – jia[1], ZENG Yi[3]

1. Departments of Pathology; 2. Biochemistry, Medical College of Shantou University; 3. Institute of Virology, Chinese Academy of Preventive Medicine

Summary

To search for potential biomarkers used to monitor the process of immortalization, we investigated the relative level of telomerase activity and other immortal phenotypes in the SHEE esophageal epithelial cell line. This human fetal esophageal epithelial cell line, induced by human papilloma virus (HPV) 18 E6E7, was continually propagated over 100 passages. Fourteenth passage cells (SHEE14) were cultured in a flask with a serum – free medium and continually cultured to the 30th passage (SHEE30), Cells of SHEE14, SHEE20 and SHEE30 were examined according to cell morphology, cell cycle, apoptosis, contact – inhibition growth, anchorage – dependency, dose – dependency to epithelial growth factors (EGF), telomerase activity and tumorigenicity. The SHEE14 cells exhibited good differentiation with contact – inhibition and anchorage – dependent growth. The SHEE20 cells exhibited increase of senescent and apoptotic cells, and difficulty in propagation. The SHEE30 cells exhibited a higher proliferative index and some undifferentiated cells, with weakened contact – inhibition and anchorage – dependent growth. The telomerase was activated in cells of SHEE30, but not in SHEE14 and SHEE20 cells. The different response to dose – dependency to

EGF was not statistically different in SHEE14 and SHEE30. Three groups of cells displayed lack of tumor formation in nude mice. Compared with SHEE14 and SHEE20, SHEE30 cells were of immortalized status with immortal phenotype, which consisted of telomerase activity, increase of cell proliferation, weakened contact – inhibition and anchorage – dependent growth, dose dependency to EGF and lack of tumor formation. From passage 14th to 30th passage, SHEE cells went through cellular senescence, apoptosis and immortalization. With a view toward diagnostic and biological aspects, telomerase activity is a crucial step and a cardinal requirement for immortalization. The telomerase activity and other immortal phenotypes are potential markers for monitoring the process of immortalization.

[**Key words**]　Esophageal epithelial cell; Immortalization; Senescence; Apoptosis; Immortal phenotype

Introduction

The human fetal esophageal epithelial cell line SHEE was previously established by E6E7 genes of HPV18 in our laboratory[1] and propagated over 100 passages. In early passages, SHEE retained the characteristics of esophageal epithelial cells such as monolayer growth, contact inhibition, presence of tonofilament and keratoprotein, those of which were manifested in the squamous epithelium origin, and presence of HPV18 E6E7 genes by Southern blot hybridization and FISH methods[2]. When this cell line had been passaged over 85 times, cells grew overlapped, and cell morphology was atypical with tumor formation after cells were transplanted into SCID mice[3]. It was determined that some cells displaying malignant transformation at the 61st passage[4] were in premalignant state[5]. The genetic event and telomere length were studied at the initial passage of SHEE[6,7]. The phorbol ester (TPA) and co – factors can promote these cells toward malignant transformation[8,9]. We found that studies of human cell carcinogenesis *in vitro* were hampered by the perplexing problem of identifying immortal, premalignant and malignant cells. We carried out further studies on the process of immortalization of SHEE to find out the immortal phenotype, which was used to monitor immortalization of cells.

In recent years, some scientific effort has gone to explore the immortal phenotype for defining immortalization, and many biological criteria were studied. Detection of genetic events[10], immortalizing genes[11], chromosome aberration[12], loss of heterozygosity of chromosomes[13] and dose – dependency to growth factor[14] were investigated as immortal phenotypes. This research has tremendously increased our knowledge of biological behaviors of immortal cells. Until now, common characteristic criteria for different cells to detect immortal status in fidelity and in practice could not be found. Nevertheless, it is very useful to explore the process of immortalization development and to determine the variability between immortalized cells and malignant transformed cells.

In this report, we studied telomerase activity and other immortal phenotypes, consisting of morphological changes, DNA profiling in cell cycle, anchorage – dependent growth, inhibition – dependent growth, dose – dependency to epithelial growth factors (EGF) and tumor formation at the early stage of SHEE.

Materials and Methods

1. Cell line and cell culture: The SHEE cell line was established from embryonic esophageal epithelium induced by genes 18 E6E7 of human papillomavirus (HPV) type 18 in our laboratory with a long – term *in vitro* passage (over 100 passages). The cells in 10 – passage intervals were stored in a liquid nitrogen container. The cells of the 10th passage were reviewed and were continually cultivated toward the 30th passage in serum – free cultural medium without containing calf serum, with 100 units of penicillin and streptomycin in a humidified atmosphere of 5% CO_2 and 95% air. The serum – free medium consisted of the basal medium (MCDB151, Sigma) and trace elements, additionally added transferring, 10 μg/ml; hydrocortisone, 0.15 μg/ml; epidermal growth factor (EGF), 5 ng/ml; insulin (Sigma Chemical Co.), 5 ng/ml; and extracts of bovine hypophysis (Gibco BRL) 40 μg/ml. The cells of 14th (SHEE14), 20th (SHEE20) and 30th passage (SHEE30) were examined as follow. Cell shape and size, anchorage – dependent growth and contact – inhibited growth were examined by phase – contrast microscopy.

2. Analysis of cell proliferative cycle: Cells in the flasks were trypsinized and collected, washed twice with PBS, fixed by 70% alcohol, prepared as single – cell suspension and stored at 4℃. Cells were stained with propidium iodide (Sigma) and analyzed with flow cytometry (FACSort, B – D Co.). The percentage of cells in various stages of the cell cycle, the apoptotic cell rate (AI) and proliferation index ($PI = S + G_2M/G_0G_1 + S + G_2M$) were calculated. Chicken erythroid nuclei were spiked into standardized DNA content measure – ments. In these analyses CVs were always less than 5.

3. Telomerase activity assay: Telomerase activity was measured using TRAP – eze Telomerase Detection Kit (Oncor Inc.). Frozen samples were homogenized in 10 – 50 μl of ice – cold lysis buffer (10 mmol/L Tris – HCl, pH 7.5, 1 mmol/L EGTA, 0.1 mmol/L benzamidine, 5 mmol/L β – mercaptoethanol, 0.5% CHAPS, 10% glycerol). After 30 min of incubation on ice, the lysate was centrifuged at 12 000 g for 20 min at 4℃. The telomeric repeat amplification protocol (TRAP) reaction was performed using 1 μl lysate or 1/10 diluted lysate, 2.5 μl 10 × TRAP buffer (200 mmol/L Tris – HCl, pH 8.3, 15 mmol/L $MgCl_2$, 630 mmol/L KCl, 0.5% Tween 20, 10 mmol/L EGTA), 0.5 μl 2.5 mmol/L dNTP, 0.5 μl Ts primer, 0.5 μl TRAP primer mix, 19.5 μl water, 0.5 μl Taq (2 U/μl). After incubation at 30℃ for 30 min, the reaction mix was immediately transferred to 94℃ and PCR was performed (GeneAmp PCR system 2400, Biosystems, USA) at 94℃ for 30 sec, 55℃ for 30 sec, 72℃ for 30 sec for 35 cycles. PCR products were separated in a non – denaturing 12% PAGE in IX TBE at 5 V/cm. The gel was stained using $AgNO_3$.

4. Cell colony formation in soft agar cultivation: The exponential growth cells were trypsinized, and the percentage of living cells was counted by trypan blue stain. The living single cell suspension (10^3 cells/ml) was cultured in 0.35% agar (Agarose, V312A, Promega) overlaid on 0.7% agar in a 6 – well plate (Coming), incubated in 5% CO_2, 37℃ for 20 days and the cell colonies were then counted. Colony efficiency (CE) = number of colonies counted/number of cells plated.

5. Contact inhibition of growth: In three passages of SHEE, cells were cultured in the original flasks without propagation to other flasks, to observe whether or not the cells stopped replicating when they come into contact with other cells.

6. Dose dependency to EGF: EGF in various concentration, 5 ng/ml, 3 ng/ml and 1 ng/ml, was added to the serum – free culture medium. About 100 cells of SHEE14 and SHEE30 were cultured respectively in a 6 – well culture plate (Coming Co.) with three concentrations of cultural medium for 3 days at the exponential growth period. Colony formation was measured every day by image analysis system (Q520, Cambridge). Colony formation efficiency was calculated (same as above).

7. Tumorigenesis in nude mice: Six – week – old BALB/C nude mice (supplied by the Experimental Animal Center of Zhongshan Medical University) were bred in isolated conditions. Four nude mice were subcutaneously injected with SHEE14 or SHEE30 in 1×10^6 cells/mouse, respectively. They were observed every 10 days for 2 months and then sacrificed, and the tumors histopathologically examined.

8. Statistical methods: The results of colony efficiency in cells of different passages were assessed by the χ^2 test. The statistical analyses were performed using the SPSS 10.0 for Windows software (SPSS, Inc., Chicago, IL).

Results

1. Morphology of cells: SHEE14 cell grew evenly in flasks and appeared typical of squamous epithelium with multi – angular outline and oval nucleus with contact – inhibition growth and anchorage – dependency (Fig. 1A). Cells from the 17th to the 22nd passage exhibited growth retardation. SHEE20 cells displayed large, flattened terminal – mature senescent cells and abundant apoptotic cells which had small rounded cells with coagulated chromatin in nucleus and were always detached from the culture flask (Fig. 1B). SHEE30 cells grew exuberantly and displayed biphasic morphology, partially well differentiated and partially undifferentiated basal cells (Fig. 1C) with weakened contact – inhibition growth.

2. Contact – inhibition growth: In SHEE14 and SHEE20, cells grew in a single layer. When they came into contact with each other, proliferation of cells arrested, and a significant number of cells remained quiescent and some cells died, a phenomenon called contact – inhibition growth. In SHEE30, cells underwent crowded growth and connecting, some cells died and detached, retained a network outlook in the culture flask, and some cells revealed proliferation and were weak in contact – inhibition growth and some cells revealed differentiation.

3. Cell cycle analysis: To characterize the altered growth kinetic in early passage cultures, DNA analyses in cell cycle was performed in cells of initial passage (SHEE14), cells of senescence and apoptotic stage (SHEE20) and in immortalized cells (SHEE30). Analysis of SHEE14 cells demonstrated about 12.6% in S phase and 36% in G_0G_1 phase (Fig. 2A). SHEE20 cells accumulated in G_1G_0 and in sub – G_1G_0 phase (total 83.7%), which showed an apoptotic peak and cells of synthesized DNAs phase were 8.6% (Fig. 2B). Cells analyzed during this periods exhibited senescence and apoptosis in morphology. The DNA synthesized cells in the SHEE30 were increased to 31.6%, indicating over – growth of the proliferalive cells (Fig. 2C).

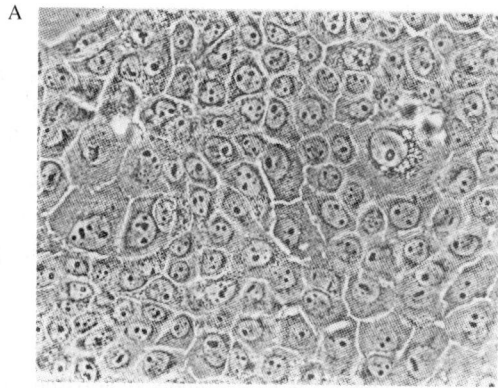

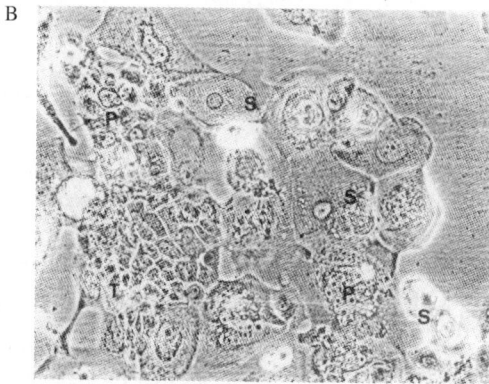

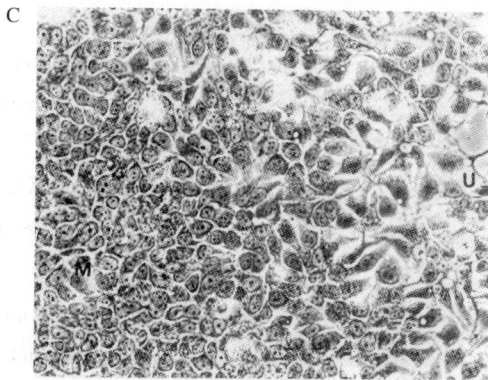

(A), Cell in 14th passages showed differentiative pheno-
typc [phase - contrast microscopy (Ph, × 320)].
(B), Cells of 20th passage displayed cell proliferation
(P), cell senescence (S) and apoptosis (T) (Ph, ×
200). (C), Cells of 30th passage displayed diphasic differentiation, partially well differentiated (u), partially
undifferentiated (m) (Ph, ×200)

**Fig. 1 Morphological changes in living coils of
SHEE**

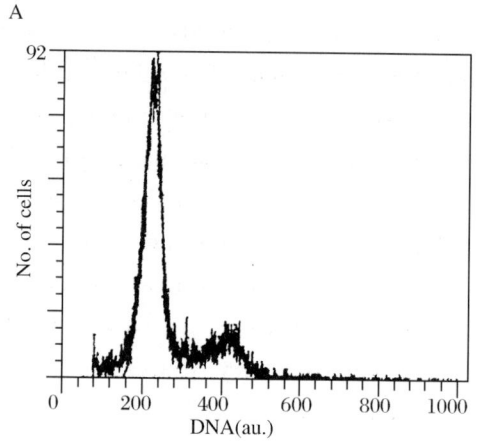

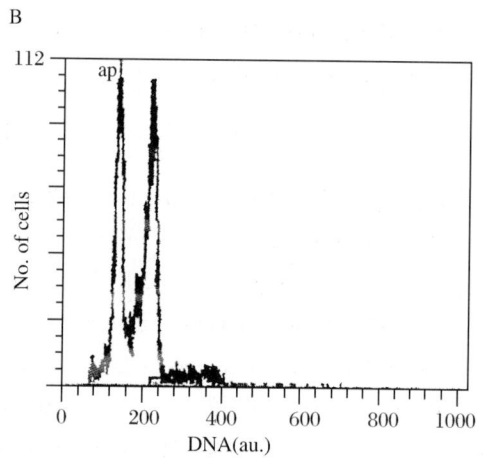

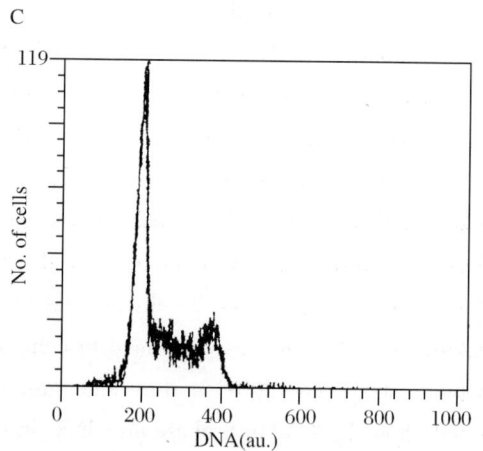

(A), SHEE14; (B), SHEE20; (C), SHEE30. ap, apoptotic peak; au, arbitrary unit

**Fig. 2 DNA histogram of three passages of
SHEE cells**

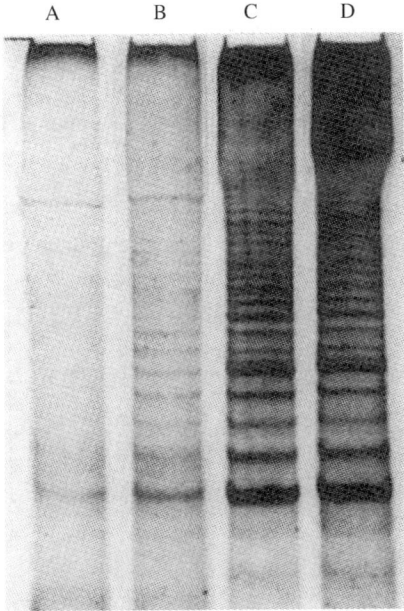

Lane A; 14th passage, negative; lane B: 24th passage, weak; lane C: 30th passage, positive; lane D: human esophageal squamous cell carcinoma, strong positive

Fig. 3   Measurements of telomerase activity using TRAP assay

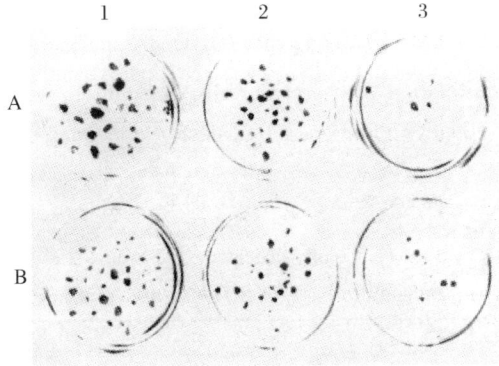

(A), SHEE30 cell colonies on the 5th day were confluent; (B), SHEE14 cells; 1, 5 ng/ml; 2, 3 ng/ml; 3, 1 ng/ml

Fig. 4   Colony formation of SHEE14 and SHEE30 on the 5th day in culture plate with various doses of EGF. Decreasing colony formation was found in lower concentration of EGF

4. Telomerase activity: Telomerase activation was absent in cells of fetal esophageal tissue in SHEE14. The activity of telomerase first appeared in SHEE24, but it was weakly positive, and was positive in SHEE30 (Fig. 3). Our results showed that telomerase activity appeared in low levels after senescent and apoptotic stages.

5. Dose – dependency to EGF: Cells of SHEE14 and SHEE30 were cultured in the serum – free medium with additional different concentration of EGF. The number of colony formation both in SHEE14 and SHEE30 decreased in low concentrations of EGF (Fig. 4). The colony efficiency (colony number/number of seeded cells) at the 5th day is listed in Tab. 1. The response to variant doses of EGF in both passages was not significantly different ($P > 0.1$).

6. Cell colony formation in soft agar: Cells of three passages were cultivated in soft agar for 30 days. The small colonies (cells < 20) and a few of the large colonies (cells > 20) were found in SHEE30. The SHEE14 and SHEE20 could not be found in the large colony formation in the soft agar (Tab. 2). The data demonstrated that anchorage – dependent growth was decreased in some cells of SHEE30.

7. Tumor formation: When cells of SHEE14, SHEE20 and SHEE30 were transplanted into the axilla of 4 nude mice respectively, none of the three cell passages formed tumors in mice within 40 days. Thus, these cells were non – malignant.

8. Comparison of the biological behavior in the three groups of SHEE: The general characters of the three passages of SHEE cells are listed in Tab. 3. Without tumor formation in nude mice, the cells of these three groups from SHEE were non – malignant. SHEE30 are immortalized cells and

SHEE14 and SHEE20 pre – immortalized cells. We suggest the following criteria of immortalization. Apart from activation of telomerase activity, the immortal phenotype of SHEE showed increase of cell proliferation with a higher DNA synthesis, weakened contact – inhibition growth and anchorage – dependency, maintenance of dose – dependency to EGF, but not tumor formation in immunodeficient mice.

Tab. 1 Colony efficiency (%) of dose dependency to EGF in SHEE14 and SHEE30ᵃ

EGF ng/ml	5	3	1
SHEE14	45	30	10
SHEE30	42	23	9

ᵃX^2 test (R × C) ; $p > 0.1$

Tab. 2 Colony efficiency (%) of soft agar in three passages of cells

Cells	Small colony	Large colony
SHEE14	0.5 ~ 1.0	0
SHEE20	0.1 ~ 0.5	0
SHEE30	5.0	0.5

Tab. 3 Biological behavior in preimmortal and immortal SHEE cells

	SHEE14	SHEE20	SHEE30
Telomerase activity	–	±	+ +
DNA synthesis	+	±	+ +
Undifferentiating	–	–	±
Apoptosis	+	+ +	+
Contact – inhibition growth	+ +	+ +	+
Anchorage – dependency	+ +	+ +	+
Dose – dependency to EGF	+ +	+ +	+ +
Tumor formation	–	–	–

–, negative; ±, weakly positive; +, moderately positive; + +, strongly positive.

Discussion

The process of immortalization of SHEE went through stages of cellular proliferation, cell senescence, cell apoptosis, restoring proliferation followed by cell immortalization. At the early stage after transduction of HPV18 E6E7, cells of SHEE do not express telomerase activity. Cells in the 17th – 22nd passages tend to become senescent and undergo crisis with decrease of DNA synthesis and increase of cell apoptosis. With apparently strong telomerase activity, cells overcame senescence and crisis, and entered a proliferative stage.

Senescence and crisis are the two checkpoints to safeguard against excessive proliferation[15,16], which is the hallmark of immortalization and cancer cells[17]. The first checkpoint of this process, known as senescence, referred to as mortality stage 1 (M1)[18], has implications for cell aging and suppresses tumorigenicity[19]. The second checkpoint, known as crisis, referred to as mortality stage 2 (M2)[20,21], causes the cells to enter apoptosis and die. Both M1 and M2 are potential suppression pathways for tumorigenesis. With the start of telomerase activity, both checkpoints M1 and M2 were overcome and cells became immortalized[15,22]. These observations suggest that telomerase and/ or telomere maintenance are crucial to overcome both the M1 and M2 checkpoints[21,23].

Telomerase activity correlated with cell cycle regulator[24,25] to promote cell proliferation[26,27].

We studied the correlation between telomerase activity and the proliferation status of the cells, the high levels of telomerase activity in SHEE30 may be linked to the proliferative stage of the cells, which have a high level of synthesized DNA and high proliferative indices (PI > 30%). Alternatively, cells of SHEE14 and SHEE20 have a negative telomerase activity or a weak positive with a low DNA − labeling index (PI < 30%). It suggested that an examination of proliferation indices was an important criterion for immortalization of esophageal epithelial cells, and telomerase activity is a regulator of cell proliferation, but not a biomarker of malignant transformation[28,29]. Telomerase could also affect genetic and epigenetic changes in human epithelial cell[30] and have an effect on cell proliferation and differentiation[31] accompanying decrease of anchorage − dependent growth.

Telomerase expression has been discovered in HPV − positive cervical squamous cell carcinoma[32−34]. Previous reports have also shown that the E6 and E7 protein of HPV[35] can achieve telomerase activity. HPV16 E6 onco − protein was capable of inducing telomerase activity in mono − layer cultures of proliferating keratinocyte [32]. In additional alterations it was more likely that the HPV E6/E7 synergized with chromosome instability[36,37] were a prerequisite for induction of telomerase activity in proliferating cells.

Previous studies have demonstrated a higher frequency of telomerase activity in immortalized cell lines[38] and cell malignancies[39,40]. It is clear that immortalization and the activation of a telomere maintenance mechanism are closely associated, and telomerase activity has been found to have telomere maintenance activity. Therefore, telomerase activity is a useful marker for cell proliferation (29), but not malignant transformation[41]. We revealed that cells of SHEE30 were in immortalization by displaying activation of the telomerase accompanied by elevation of synthesized DNA indices, weak anchorage − dependent growth, dose − dependency to EGF, undifferentiation in some cells and lack of tumor formation. These expressions of biological characteristics were designated as the immortal phenotype. Although these criteria are useful for the immortal markers, they depend on a number of factors: i) the markers should be biologically relevant to the growth of cells and display strong proliferation; ii) the marker must be an accurate and reproducible assay method; iii) the markers should be a biological predictor of immortalization; iv) it must be a simple and feasible procedure.

In summary, the SHEE cells in passage 20 underwent senescence and crisis, which were at the pre − immortalization phase. In the 30th passage, telomerase became activated, cells overcame senescence and crisis, restored proliferation and displayed immortalized state. The immortal phenotype consisted of activity of telomerase, the cellular proliferation, weakened contact − inhibition and anchorage − independent growth, dose dependency to EGF and lack of tumor formation. Therefore, activation of telomerase activity, and its immortal phenotype are crucial criteria to detect the acquisition of immortalization.

Acknowledgements

Contract grant sponsor: National Natural Science Foundation of China (39830380).

[In 《Int J Mol Med》 2002, 10: 641 −646]

References

1 Shen ZY, Cai W J, Shen J, Xu J J, Cen S, Ten ZP, Hu Z and Zeng Y. Immortalization of human fetal esophageal epithelial cells induced by E6 and E7 genes of human papilloma virus 18. Chin J Exp Clin Virol, 1999, 13: 121 – 123

2 Shen ZY, Shen J, Cai WJ, Cen S and Zeng Y. Biological characteristics of human fetal esophageal epithelial cell line immortalized by the E6 and E7 gene of HPV type 18. Chin J Exp Clin Virol, 1999, 13: 209 – 212

3 Shen ZY, Shen J, Cai WJ, Chen JY and Zeng Y. Malignant transformation of the immortalized esophageal epithelial cells. Chin J Oncol, 2002, 24: 107 – 109

4 Shen ZY, Xu LY, Chen MH, Cai W J, Shen J, Chen JY, Hon CQ and Zeng Y. Biphasic differentiation of immortalized esophageal epitheliums induced by HPV18E6E7. Chin J Virol, 2001, 17: 210 – 214

5 Shen ZY, Chen XH, Shen J, Cai W J, Chen JY, Huang TH and Zeng Y. Malignant transformation of immortalized human embryonic esophageal epithelial cells induced by human papillomavirus. Chin J Virol, 2000, 16: 97 – 101

6 Shen ZY, Xu LY, Li EM, Cai W J, Chen MH, Shen J and Zeng Y. Expression of telomere, telomerase in the initial stage of immortalization of esophageal epithelial cell. World J Entero – gastrol, 2002, 8: 357 – 362

7 Shen ZY, Xu LY, Chen XH, Cai W J, Shen J, Chen JY, Huang TH and Zeng Y. The genetic events of HPV – immortalized esophageal epithelium cells. Int J Mol Med, 2001, 8: 537 – 542

8 Shen ZY, Cai WJ, Shen J, Xu J J, Cen S, Ten ZP, Hu Z and Zeng Y. Human papilloma virus 18E6E7 in synergy with TPA induced malignant transformation of human embryonic eso – phageal epithelial cells. Chin J Virol, 1999, 15: 1 – 6

9 Shen ZY, Cen S, Shen J, Cai W J, Xu J J, Teng ZP, Hu Z and Zeng Y. Study of immortalization and malignant transformation of human embryonic esophageal epithelial cells induced by HP-VI8E6E7. J Cancer Res Clin Oncol, 2000, 126: 589 – 594

10 Kaul SC, Wadhwa R, Sugihara T, Obuchi K, Komatsu Y and Mitsui Y. Identification of genetic events involved in early steps of immortalization of mouse fibroblasts. Biochim Biophys Acta, 1994, 1201: 389 – 396

11 Pereira – Smith OM, Stein GH, Robetorye S and Meyer – Demarest S. Immortal phenotype of the Hela variant D98 is recessive in hybrids formed with normal human fibroblasts. J Cell Physiol, 1990, 143: 222 – 225

12 Bertram MJ, Berube NG, Hang – Swanson X, Ran Q, Leung JK, Bryce S, Spurgers K, Bick RJ, Baldini A, Ning Y, Clark LJ, Parkinson EK, Barrett JC, Smith JR and Pereira – Smith OM. Identification of a gene that reverses the immortal phenotype of a subset of cells and is a member of a novel family of transcription factor – like genes. Mol Cell Biol, 199919: 1479 – 1485

13 Loughran O, Edington KG, Berry I J, Clark LJ and Parkinson EK. Loss of heterozygosity of chromosome 9 p21 is associated with the immortal phenotype of neoplastic human head and neck keratinocytes. Cancer Res, 1994, 54: 5045 – 5049

14 Franco S, MacKenzie KL, Dias S, Alvarez S, Rafii S and Moore MA. Clonal variation in phenotype and life span of human embryonic fibroblasts (MRC – 5) transduced with the catalytic component of telomerase (hTERT). Exp Cell Res, 2001, 268: 14 – 25

15 Lustig AJ. Crisis intervention. the role of telomerase. Proc Natl Acad Sci USA, 1999, 96: 3339 – 3341

16 Reddel RR. The role of senescence and immortalization in carcinogenesis. Carcinogenesis, 2000, 21: 477, 484

17 Wisman GB, Hollema H, de Jong S, ter – Schegget J, Tjong – A – Hung SP, Ruiters MH, Krans M, de Vries EG and van der Zee AG. Telomerase

activity as a biomarker for (pre) neoplastic cervical disease in scrapings and frozen sections from patients with abnormal cervical smear. J Clin Oncol, 1998, 16: 2238 – 2245

18 Ouellette MM, Liao M, Herbert BS, Johnson M, Holt SE, Liss HS, Shay JW and Wright WE. Subsenescent telomere lengths in fibroblasts immortalized by limiting amounts of telomerase. J Biol Chem, 2000, 275: 10072 – 10076

19 Smith JR and Pereira – Smith OM. Replicative senescence. implications for in vivo aging and tumor suppression. Science, 1996, 273: 63 – 67

20 Wright WE, Brasiskyte D, Piatyszek MA and Shay JW. Experimental elongation of telomeres extends the lifespan of immortal normal cell hybrids. EMBO J, 1996, 15: 1734 – 1741

21 MacKenzie KL, Franco S, May C, Sadelain M and Moore MA. Mass cultured human fibroblasts overexpressing hTERT encounter a growth crisis following an extended period of proliferation. Exp Cell Res, 2000, 259: 336 – 350

22 Litaker JR, Pan J, Cheung Y, Zhang DK, Liu Y, Wong SC, Wan TS and Tsao SW. Expression profile of senescence – associated beta – galactosidase and activation of telomerase in human ovarian surface epithelial cells undergoing immortalization. Int J Oncol, 1998, 13: 951 – 956

23 Halvorsen TL, Leibowitz G and Levine F. Telomerase activity is sufficient to allow transformed cells to escape from crisis. Mol Cell Biol, 1999, 19: 1864 – 1870

24 Hsieh HF, Ham HJ, Chiu SC, Liu YC, Lui WY and Ho LI. Telomerase activity correlates with cell cycle regulators in human hepatocellular carcinoma. Liver, 2000, 20: 143 – 151

25 Engelhardt M, Kumar R, Albanell J, Pettengell R, Hart W and Moore MA. Telomerase regulation, cell cycle, and telomere stability in primitive hematopoietic cells. Blood, 1997, 90: 182 – 193

26 Yang XY, Kimura M, Jeanclos E and Aviv A. Cellular proliferation and telomerase activity in CHRF – 288 – 11 cells. Life Sci, 2000, 66:
1545 – 1555

27 Holt SE, Wright WE and Shay JW. Regulation of telomerase activity in immortal cell lines. Mol Cell Biol, 1996, 16: 2932 – 2939

28 Mokbel K, Parris CN, Ghilchik M, Williams G and Newbold RF. The association between telomerase, histopathological parameters, and KI – 67 expression in breast cancer. Am J Surg, 1999, 178: 69 – 72

29 Belair CD, Yeager TR, Lopez PM and Reznikoff CA. Telo – merase activity: a biomarker of cell proliferation, not malignant transformation. Proc Natl Acad Sci USA, 1997, 94: 13677 – 13682

30 Farwell DG, Shera KA, Koop JI, Bonnet GA, Matthews CP, Reuther GW, Coltrera MD, McDougall JK and Klingelhutz AJ. Genetic and epigenetic changes in human epithelial cells immortalized by telomerase. Am J Pathol, 2000, 156: 1537 – 1547

31 Shroyer KR, Thompson LC, Enomoto T, Eskens JL, Shroyer AL and McGregor JA. Telomerase expression in normal epithelium, reactive atypia, squamous dysplasia, and squamous cell carcinoma of the uterine cervix. Am J Clin Pathol, 1998, 109: 153 – 162

32 Nowak JA. Telomerase, cervical cancer, and human papilloma – virus. Clin Lab Med, 2000, 20: 369 – 382

33 Nair P, Jayaprakash PG, Nair MK and Pillai MR. Telomerase, p53 and human papillomavirus infection in the uterine cervix. Acta Oncol, 2000, 39: 65 – 70

34 Riethdorf S, Riethdorf L, Schulz G, Ikenberg H, Janicke F, Loning T and Park TW. Relationship between telomerase activation and HPV 16/18 oncogene expression in squamous intraepithelial lesions and squamous cell carcinomas of the uterine cervix. Int J Gynecol Pathol, 2001, 20: 177 – 185

35 Klingelhutz AJ, Foster SA and McDougall JK. Telomerase activation by the E6 gene product of human papillomavirus type 16. Nature, 1996, 380: 79 – 82

36 Golubovskaya VM, Filatov LV, Behe CI, Presnell SC, Hooth MJ, Smith GJ and Kaufmann WK. Telomere shortening, telomerase expression, and chromosome instability in rat hepatic epithelial stem – like cells. Mol Carcinog, 1999, 24: 209 – 217

37 Filatov L, Golubovskaya V, Hurt JC, Byrd LL, Phillips JM and Kaufmann WK. Chromosomal instability is correlated with telomere erosion and inactivation of G2 checkpoint function in human fibroblasts expressing human papillomavirus type 16 E6 oncoprotein. Oncogene, 1998, 16: 1825 – 1838

38 Kim NW, Piatyszek MA, Prowse KR, Harley CB, West MD, Ho PL, Coviello GM, Wright WE, Weinrich SL and Shay JW. Specific association of human telomerase activity with immortal cells and cancer. Science, 1994, 266: 2011 – 2015

39 Hiyama T, Yokozaki H, Kitadai Y, Haruma K, Yasui W, Kajiyama G and Tahara E. Overexpression of human telomerase RNA is an early event in oesophageal carcinogenesis. Virchows Arch, 1999, 434: 483 – 487

40 Kiyozuka Y, Asai A, Yamamoto D, Senzaki H, Yoshioka S, Takahashi H, Hioki K and Tsubura A: Establishment of novel human esophageal cancer cell line in relation to telomere dynamics and telomerase activity. Dig Dis Sci, 2000, 45: 870 – 879

41 Yeh TS, Cheng AJ, Chen TC, Jan YY, Hwang TL, Jeng LB, Chen MF and Wang TC. Telomerase activity is a useful marker to distinguish malignant pancreatic cystic tumors from benign neoplasms and pseudocysts. J Surg Res, 1999, 87: 171 – 177

77. Progressive Transformation of Immortalized Esophageal Epithelial Cells

SHEN Zhong – ying[1], XU Li – yan[1], CHEN Min – hua[1], SHEN Jian[1], CAI Wei – jia[1], ZENG Yi[2], SHEN Zhong – ying, XU Li – yan, CHEN Min – hua, SHEN Jian, CAI Wei – jia
1. Department of Tumor Pathology; Medical College of Shantou University;
2. Institute of Virology, Chinese Academy of Preventive Medicine

Summary

Objective: To investigate the progressive transformation of immortal cells of human fetal esophageal epithelium induced by human papillomavirus, and to examine biological criteria of sequential passage of cells, including cellular phenotype, proliferative rate, telomerase, chromosome and tumorigenicity.

Methods: The SHEE cell series consisted of immortalized embryonic esophageal epithelium which was in malignant transformation when cultivated over sixty passages without co – carcinogens. Cells of the 10th, 31st, 60th and 85th passages were present in progressive development after being transfected with HPV. Cells were cultivated in a culture flask and 24 – hole cultural plates. Progressive changes of morphology, cell growth, contact – inhibition, and anchorage – dependent growth characteristics were examined by phase contrast microscopy. The cell proliferation rate was assayed by flow cytometry. The modal number of chromosomes was analyzed. HPV18E6E7 was detected by Western blot methods and activities of telomerase were analyzed by TRAP. Tumorigenicity of cells was detected with soft agar plates cultivated and with tumor formation in SCID mice.

Results: In morphological examination the 10th passage cells were in good differentiation, the 60th and 85th passages cells were in relatively poor differentiation, and the 31st passage cells had two distinct differentiations. The characteristics of the 85th and 60th passage cells were weakened at contact – inhibition and anchorage – dependent growth. Karyotypes of four stages of cells belonged to hyperdiploid or hypotriploid, and bimodal distribution of chromosomes appeared in the 31st and 60th passage cells. All of these characteristics combined with a increasing trend. The activities of telomerase were expressed in the latter three passages. Four fourths of SCID mice in the 85th passage cells and one fourth of SCID mice in the 60th passage cells developed tumors, but the cells in the 10th and 31st passage displayed no tumor formation.

Conclusion: In continual cultivation of fetal esophageal epithelial cells with transduction of HPV18E6E7, cells from the 10th to the 85th passage were changed gradually from preimmortal, immortal, precancerous to malignantly transformed stages. All of these changes were in a dynamic progressive process. The establishment of a continuous line of esophageal epithelium may provide a *in vitro* model of carcinogenesis induced by HPV.

Introduction

The cell line SHEE was derived from immortalized embryonic esophageal epithelium induced by gene E6E7 of HPV 18 in our laboratory[1,2] being cultivated and propagated over 100 passages. The 31 st generation (SHEE31) had begun to express partial cell differentiation into two directions with some nests of cells with good differentiation and some with poor differentiation[3]. The 61st generation cells (SHEE61) were premalignant cells[4], and displayed a fully malignant transformation with a strong invasive potency at the 85th passage (SHEE85)[5]. We believe that this established cell line (SHEE), continually affected by expression of HPV, would change its biological characters such as cell proliferation, differentiation, chromosome and telomerase, and that this might be controlled by cytogenesis (chromosomes) and molecular genetics (genes).

In general, the immortalized or transformed cells caused by carcinogens are always accompanied by chromosome abnormality and mutation of gene[6]. The chromosome's changes are manifested in structure and the number of chromosomes[7,8]. All of these changes appear in the procedure from quantitative to qualitative changes. The length of telomere in living cells was continually shortened after cellular mitosis[9]. Because the somatic cells have no or lower levels of telomerase activity, telomere would be shortened, so it limits the division and lifespan of cells[10,11]. Immortal or malignant cells manifest telomerase activities, which can maintain the telomere length[12,13], so they will be immortal. With exposure of the early passage of immortal cell line to viral oncogenes, HPV or SV40T, conversion of these telomerase from negative expression to high levels of telomerase activity resulted[4]. Telomerase would be present in benign lesions and activated during the late stage of carcinogenesis[15].

Changes occurred in SHEE cells from the 10th to the 85th passage, with emphasis on their phenotypes, cytogenetic changes, telomerase activity and tumorigenicity, were studied in this paper. Phenotype of cells included the morphological changes of proliferation and contact – inhibition growth, and the modal number of chromosomes and the tumorigenicity, especially soft – agar culture and tumor formation in severely combined immuno – deficient (SCID) mice.

Materials and Methods

1. Cell culture

The SHEE cells were routinely cultivated in culture medium 199 (GIBCO) with 10 % bovine serum, 100 units of penicillin and streptomycin in a humidified atmosphere of 5 % CO_2. Selected generations at the 10th passage (SHEE 10), 31st passage (SHEE31 60th passage (SHEE60) and 85th passage (SHEE85) were inoculated in culture flask and 24 – hole culture plate with glass slide inside.

2. Living cell examination

The cell shape, anchorage dependent and contact – inhibited growth were examined by phase – contrast microscopy.

3. Cell cycle analysis

Cultured cells of each passage were collected from suspended and digested cells, fixed by 70 % alcohol, then filtered through nylon mesh, to generate single – cell suspension (1×10^6/ml). The cells were stained with propidium iodide (Sigma) and were analyzed using flow cytometery (FCM) (FACSorte Becton – Dickinsn). Data of DNA of cells were collected and analyzed with Lysis II software, then a histogram was drawn and the cell percentages of each proliferatot ion stage in the cell cycle and the number of cells more than 4n of DNA were calculated. The proliferation index formula: $S + G_2M/ G_0G_1 + S + G_2M$ and the cell amount at $preG_0G_1$ stage, the apoptotic cells, were calculated.

4. Cytogenetic analysis

Metaphase spreads were obtained using standard cytogenetic methods. Briefly, the culturing cells of each passage were preserved at $3 - 4$ ℃ for $3 - 4$ h to make cells on synchronous stage, then cultivated at 37 ℃ for $3 - 4$ h, and added in 0.05 ng/ml colchicine for 1 h. Harvesting was by standard method and stained with Giemsa, $50 - 100$ metaphases were scored for each line.

5. Telomerase activity assay[16]

Telomerase activity was measured using the telomeric repeat amplification protocol (TRAP). Frozen samples were homogenized in $10 - 50$ μl of ice – cold lyses buffer (10 mmol · L^{-1} Tris – HCl, pH7.5, 1 mmol · L^{-1} EGTA, 0.1 mmol · L^{-1} Benzamidine, 5 mmol · L^{-1} βmercaptothanol, 5 g · L^{-1} CHAPS, 100 ml · L^{-1} glycerol). After 30 min of incubation on ice, the lysate was centrifuged at 12 000 g for 20 min at 4℃. TRAP – eze Telomerase Detection Kit (Oncor Inc.) reaction was performed using 1 μl lysate or 1/10 diluted lysate, 2.5μl 10 × TRAP buffer (200 mmol · L^{-1} Tris – HCl, pH8.3, 15 mmol · L^{-1} $MgCl_2$, 630 mmol · L^{-1} KCl, 0.5 % Tween 20, 10 mmol ·

L^{-1} EGTA, 1 g · L^{-1} BSA), 0.5 μl 2.5 mmol. L^{-1} dNTP, 0.5 μl Ts primer, 0.5 μl TRAP primer mix, 19.5 μl water, 0.5 μl taq (2 μ · L^{-1}). After incubation at 30 ℃ for 30 min, the reaction mixture was immediately transferred to 94 ℃ and performed PCR (GenAmp PCR System 2400, PE, USA) at 94 ℃ for 30 s, 55 ℃ for 30 s, for 35 cycles. PCR products were separated in a non – denaturing 125 g · L^{-1} PAGE in 1 × TBE at 5V · cm^{-1}. The gel was stained using $AgNO_3$ and was photographed.

6. Soft agar assays

Four passages of cells (1 × 10^4) were cultivated in each hole of the 6 – hole plastic plate (Coming Co.) which was covered with two layers agarose (Agarose, V312A, Promega), the bottom, 1% and the upper. 0.5 %. The cells were incubated in 5 % CO_2 at 37 ℃ incubator for 40 d and the cell colony formations were scored every ten days. Each soft – agar cloning experiment was carried out at least in duplicate.

7. Tumorigenicity assays

In vivo tumor graft experiments were performed on the severely combined immunodeficient (SCID) mice (C. B – 17/IcrJ – scid, Animal Lab of Chinese Academy of Medical Sciences). Cells of each passage were injected into the subaxillary skin of four mice with 1 × 10^6 cells for every one. Mice were observed weekly for two months and the tumor tissues were examined histopathologically.

8. HPV18E6E7 assays

The protein expression of HPV 18E6E7 was detected by Western blot method. The cells were washed three times with ice – cold PBS, then were lysed in buffer [50 mmol/L Tris – HCl, pH8.0, 150 mmol/L NaCl, 100 μg/ml phenyl – methyl – sulfonyl fluoride (PMSF), 1% TritonX – 100] for 30 min at ice. After removal of cell debris by centrifugation (12 000 g, 5 min), the protein concentration of lysates was measured by Bradford method. 50 μg proteins of different passage boiled for 5 min in sample buffer were separated by 10 % SDS – PAGE, transferred onto nitrocellulose membrane (Bio – Rad). Nonspecific reactivity was blocked by incubation overnight at 4 ℃ in buffer (10 mmol/L Tris – HCl, pH7.5, 150 mmol/L NaCl, 2 % Tween – 20, 4 % bovine serum albumin). The membrane was then incubated with antibody of mouse anti HPV $18E_6$, (SC – 264, Zhong Shan Biotech Co.), followed by reaction with anti – mouse IgG antibody. Reactive protein was detected by ECL chemiluminescence system (Amersham).

Results

1. Proliferation and differentiation of SHEES

The cells of SHEE10 grew evenly on the flask. Cells appeared to have the characteristics of squamous epithelium (Fig. 1, A) with multiangular outline and oval nucleus. SHEE31 were attached to the dish with partial differentiating into squamous epithelium and partial undifferentiated basal cells (Fig. 1, B). The cells of SHEE60 and SHEE85 were differently shaped and sized and cells were crowded and overlapped (Fig. 1c, D).

2. FCM analyzed cell cycle

In the DNA histogram (Fig. 2), the distribution of DNA content of SHEE31 was similar to that of SHEE10, the proliferative indexes of SHEE 10 and SHEE 31 were at the same level (32.0%, 35.2%), but different from SHEE60 (47.5%) and SHEE85 (54.3%). Of all DNA > 4n cells there were SHEE10, 2.5%; SHEE31, 4.7%; SHEE60, 6.1% and SHEE85,

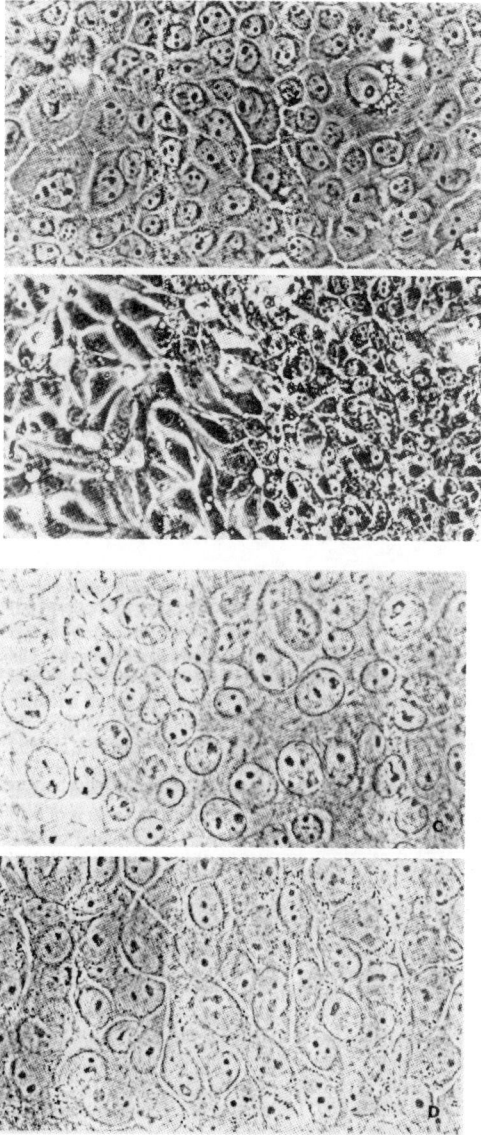

A, SHEE10, good differentiation (×400); B, SHEE31, differentiated to two directions, well differentiated (left), and poorly differentiated (right) (×200); C, SHEE60, poor differentiation (×400); D, SHEE85 shows different shape and size with larger nucleolus (×400)

Fig. 1 Morphology of SHEE cell (photographs of phase contrast microscope)

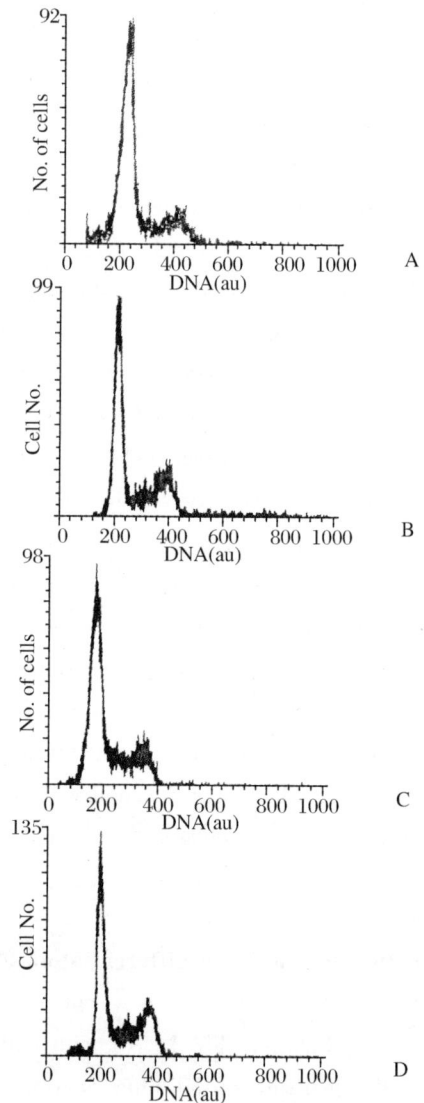

A, SHEE10; B, SHEE31; C, SHEE60; D, SHEE85; au, arbitrary unit

Fig. 2 DNA histograms of SHEE

7.2 %. This showed that heteroploid and hyperploid tumor cells increased with progressive culture of SHEE.

3. The modal number of chromosome

The number of chromosomes in SHEE10, SHEE31, SHEE60 and SHEE85 (Tab. 1) ranged between 32 and 196, and these chromosomes were mainly hyperdiploids and hypotriploids. Most cells of hypertriploids were found at SHEE60. Modal number of chromosomes at SHEE10 was 58 – 62, at SHEE31 and SHEE60 were bimodal distribution, 55 – 57, 61 – 63, and 58 – 60, 63 – 65 respectively, at SHEE85 was 59 – 65. The modal number from the 10th passage to the 85th passage increased slowly.

Tab. 1　Number of Chromosome in SHEE Series

Number of passage	Number of cell	≤46 (%)	47 – 57 (%)	58 – 68 (%)	≥69 (%)	modal number
SHEE10	91	17 (18.7)	45 (49.5)	22 (27.7)	4 (4.3)	58 – 62
SHEE31	100	12 (12.0)	38 (38.0)	46 (46.0)	4 (4.0)	55 – 57, 61 – 63
SHEE60	52	5 (9.6)	13 (25.0)	22 (42.3)	12 (23.1)	58 – 60, 63 – 65
SHEE85	85	4 (4.7%)	16 (18.8%)	54 (63.5)	11 (12.9)	59 ~ 65

4. Telomerase activity

Telomerase activation was absent in normal esophageal epithelium and SHEE10. The activity of telomerase appeared in the 31st passage, and it was strongly positive in SHEE60 and SHEE85.

5. Expression of HPV18E6E7

The expression of HPV18E6E7 was examined by Western blot method. The figure (Fig. 4) showed the expression of protein of HPV18E6E7 at cells of four stages of SHEE cell lines.

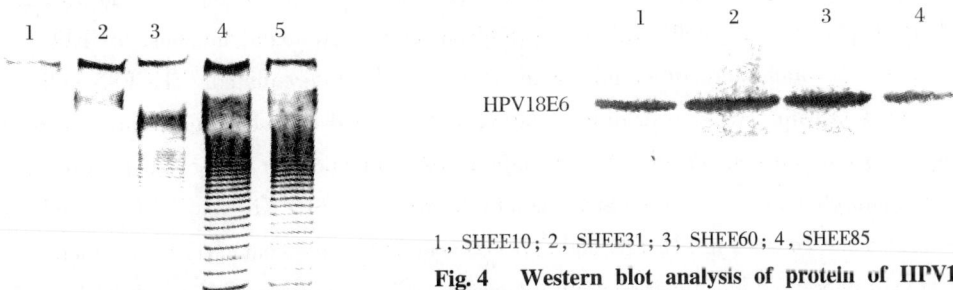

series 1, normal esophageal epithelium;
2, SHEE10; 3, SHEE31; 4, SHEE60;
5, SHEE85

Fig. 3　Activity of telomerase SHEE

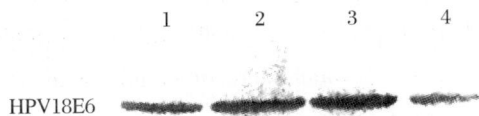

1, SHEE10; 2, SHEE31; 3, SHEE60; 4, SHEE85

Fig. 4　Western blot analysis of protein of HPV18E6 in SHEE

6. Tumorigenicity

SHEE 10 could not grow on soft agar, but SHEE31, SHEE60 and SHEE85 could. SHEE31 formed small colonies, (less than 20 cells in a colony) compared to SHEE60 and SHEE85 in which large colonies (more than 50 cells in a colony) were formed. SCID mice inoculated with cells of SHEE10 and SHEE31 did not form tumors, but one quarter of SCID mice with inoculation of SHEE60 cells formed tumors with a latency period of over 2 months (data not shown). It was determined that a percentage of cells of SHEE60 manifested malignancy. All SCID mice inoculated with SHEE85 cells manifested tumor formation, which infiltrated into muscular layer histopathologically (Fig. 5).

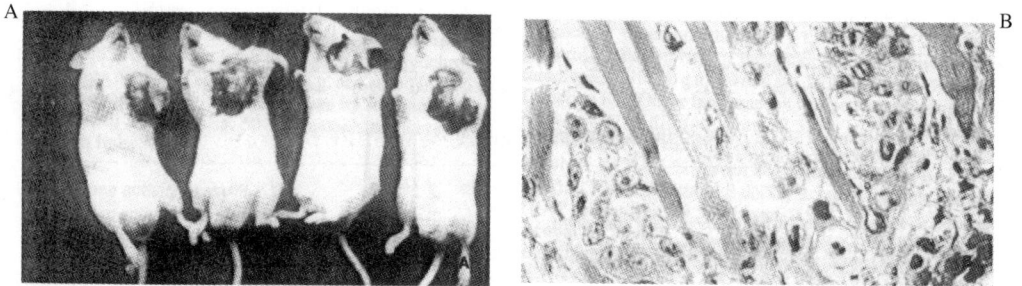

Fig. 5 Heterotransplanting SHEE85 cells into SCID mice Tumors in right axiua (arrow) of SCID mice are found (A). Invasion of tumor cells is found in muscular layer (B). HE, × 400

Discussion

Cellular proliferation, differentiation and apoptosis are fundamental life activities, and are also the growth markers of immortal cells. According to the DNA content and proliferation index, the cells of SHEE10, SHEE31, SHEE60 and SHEE85 all show proliferative characteristics. The proliferation index and cell numbers of polyploid (DNA > 4n) of each passage were compared as a result of SHEE10 < SHEE31 < SHEE60 < SHEE85. SHEE31 cells showed differentiation in two directions: one displayed relatively large and multiangular cells with abundant cytoplasm and oval nucleus; the others displayed small cells with less cytoplasm and small round nucleus. SHEE60 cells overlapped to grow, which were differently shaped and poorly differentiated. SHEE85 cells were crowed in cultivation with cells of poor differentiation and received less contact – inhibition. Detecting anchorage growth and contact inhibition specificity by cultivation on the soft agar is of help to judge its malignant character[17,18]. The small colony formation of SHEE10 and SHEE31 cultivated in soft agar, and large colonies in SHEE85 and SHEE60 showed that the characteristics of anchorage – dependent growth decreased but tumorigenicity increased. A few tumors are formed in SCID mice incubated with SHEE60. So we judged that they were not at a fully malignant stage but at a premalignant stage. SHEE85 cells, which were transplanted into 4 SCID mice and developed tumors in all mice, expressed malignant transformation.

There were more hyperdiploid and hypotriploid in the chromosomes of four stages of SHEE cell lines. The separate modal number of chromosomes first appeared in SHEE31 and continued to SHEE60. The number of chromosomal sets and the percentage of hyperploid in SHEE31 varied between SHEE10 and SHEE60, and SHEE85 has more hypotriploid cells than the others. All above showed that chromosomes of SHEE series cell lines were unstable and more susceptible to malignant transformation by promoters[19, 20]. The changes of cytogenetics will control the proliferation and differentiation of cells[21].

In 1965, Hayrick reported that the culture life of human diploid fibroblast was limited to 50 – 100 generations, the same to epithelial cells. In 1985, Greider discovered the activity of telomerase. In 1994, Kim also identified a specific association between the telomerase activity and immortal cells or cancer cells. Telomerase activity was demonstrated in cancers of the digestive tractt[22 – 24], such as gastric[25 – 29], hepatic[30 – 32] colorectal[33 – 35] and esophageal cancer[36 – 38]. The telomerase activity is possibly both a prerequisite and a diagnostic criterion for immortal cells[39]. Weitzman and Hahn believed T – antigen of SV40 and ras gene induce transformation of normal epithelium cells which require expression of telomerase activity[40,41], so over – expression of telomerase related to proliferation is the early event of cancer[42]. In our data the telomere was 30kb in length in the normal fetal esophageal cell, shortened to 17kb in SHEE 10, then further shortened to 3.5kb in SHEE 31, and then maintained at this level continually[43]. The telomerase, first appeared in the 20th passage, could not prevent shortening of telomere, because it was in a low – or noncatalytic function, therefore cells had difficulty to survive in cultivation before the 20th passage. The cells grew stably alter the 31st passage. Our results indicated that immortalization of SHEE might require activation of telomerase.

Infection of HPV can cause karyotype confusion, such as breaking of chromosome, abnormal structure and number of chromosomes[44]. It also suggests that immortal esophageal cells induced by HPV18 E6E7 may affect the changes of chromosomes, and cause instability of genetic characters[45]. Viruses can cause loss of contact inhibition, decrease of adhesion between cells, confusion of cellular skeletal structure, and loss requiring to growth factr[46]. The virus genome inserts and integrates with the chromosome of host, causing the activation and expression of oncogenest[47]. HPV E6E7 protein is conjugated with anti – oncoprotein p53 and pRb[48], thus causing loss of control of cellular growth, and encouraging the phenotyping production of cellular transformation. Previous reports have shown that activation of telomerase can be achieved by the E_6 and E_7 proteins of HPV[49]. It has been reported that E_6 could promote malignant change, while E_7 may cause benign neoplasm[50]. It has been indicated elsewhere that virogene HPV16 E6 alone can cause cellular malignant change[51]. In this experiment, we found that immortal cells contained HPV18 E6E7 and that, therefore, the SHEE series can gradually lead to the malignant transformation. It is also postulated that HPV is likely be a major risk factor for esophageal cancer[52].

In summary, of all these cells, some underwent aging, apoptosis and died, whereas others proceeded to malignant transformation. To search the direction of development of cells, we evaluated these changes by referring to the cytogenetic index. In these four series of immortal cells, the chromosomes presented characteristics of hyperdiploid and hypotriploid along with separation of modal number. The positive activity of telomerase can help determine the cells that are progressing from preimmortal toward immortal stages. The SHEE60 passage showed initial partial malignant change which could be regarded as premalignancy. The cells of SHEE85 were in fully malignant transformation with tumor formation and invasive potential. In conclusion, it is possible to demonstrate multiple stages in the transformation process that are associated with different genetic and phenotypic characteristics.

[In 《World J Gastroenterol》 2002, 8 (6): 976 – 981]

References

1　Shen ZY, Cen S, Cai WJ, Ten ZP, Shen J, Hu Z, Zeng Y. Immor – talization of human fetal esophageal epithelial cells induced by E6 and E7 genes of human papilloma virus. Zhonghua Shiyan He Linchuang Bingduxue Zazhi, 1999, 13: 121 – 123

2　Shen ZY, Shen J, Cai WJ, Cen S, Zeng Y. Biological characteristics of human fetal esophageal epithelial cell line immortalized by the E6 and E7 gene of HPV type 18. Zhonghua Shiyan He Linchuang Bingduxue Zazhi, 1999, 13: 209 – 212

3　Shen ZY, Xu LY, Chen MH, Cai WJ, Chen JY, Hon CQ, Shen J, Zeng Y. Biphasic differentiation of immortalized esophageal epitheliums induced by HPV 18E6E7. Bingdu Zuebao, 2001, 17: 210 – 214

4　Shen ZY, Chen XH, Shen J, Cai WH, Chen JY, Huang TH, Zeng Y. Malignant transformation of immortalized human embryonic esophageal epithelial cells induced by human papillomavirus. Bingdu Xuebao, 2000, 16: 97 – 101

5　Shen ZY, Shen J, Cai WJ, Chen JY, Zeng Y. Malignant transformation of the immortalized esophageal epithelial cells. Zhonghua Zhongliuxue Zazhi, 2002, 24: 107 – 109

6　Evan G, Littlewood T. A mattor of life and cell death. Science, 1998, 281: 1317 – 1322

7　Shen ZY, Xu LY, Chen XH, Cai WJ, Shen J, Chen JY, Huang TH, Zeng Y. The genetic events of HPV – immortalized esophageal epithelium cells. Int J Mol Med, 2001, 8: 537 – 542

8　Fusenig NE, Boukamp P. Multiple stages and genetic alterations in immortalization, malignant transformation and tumor progression of human skin keratinocytes. Mol Carcinog, 1998, 23: 144 – 158

9　Shen ZY, Xu LY, Li EM, Cai WJ, Chen MH, Shen J, Zeng Y. Telomere and telomerase in the initial stage of immortalization of esophageal epithelial cell. World J Gastroenterol, 2002, 8: 357 – 362

10　Tsao SW, Zhang DK, Cheng RY, Wan TYS. Telomerase activation in human cancer. Chin Med J, 1998, 111: 745 – 750

11　Jones CJ, Kipling D, Morris M, Hepbum P, Skinner J, Bounacer A, Wyllie FS, Ivan M, Bartek J, Wynford – Thomas D, Bond JA. Evidence for a telomere – independent " clock" limiting RAS oncogene – driven proliferation of human thyroid epithelial cells. Mol Cell Biol, 2000, 20: 5690 – 5699

12　Zhang DK, Ngan HY, Cheng RY, Cheang AN, Liu SS, Tsao SW. Clinical significance of telomerase activation and telomeric restriction fragment (TRF) in cervical cancer. Eur J Cancer, 1999, 35: 154 – 160

13　Hsieh HF, Ham HJ, Chiu SC, Liu YC, Lui WY, Ho LI. Telomerase activity correlates with cell cycle regulators in human hepatocel – lular carcinoma. Liver, 2000, 20: 143 – 151

14　Mutirangura A, Sriuranpong V, Termrung graunglert W, Tresukosol D, Lertsaguansinchai P, Voravul N, Niruthisard S. Telomerase activity and human papillomavirus in malignant, premalrgnant and benign cervical lesions. Br J Cancer, 1998, 78: 933 – 939

15　Nowak JA. Telomerase, cervical cancer, and human papillomavirus. Clin Lab Med, 2000, 20: 369 – 382

16　Hou M, Xu D, Bjorkholm M, Gruber A. Real –

time quantitative telomeric repeat amplification protocol assay for the detection of telomerase activity. Clin Chem, 2001, 47: 519 – 524

17 Sakaguchi M, Miyazaki M, Inoue Y, Tsuji T, Kouchi H, Tanaka T, Yamada H, Namba M. Relationship between contact inhibition and intranuclear S100C of normal human fibroblasts. J Cell Biol, 2000, 149: 1193 – 1206

18 Calaf G, Russo J, Tait L, Estrad S, Alvarado ME. Morphological phenotypes in neoplastic progression of human breast epithelial cells. J Submicrosc Cytol Pathol, 2000, 32: 83 – 96

19 Shen ZY, Cai WJ, Shen J, Xu JJ, Cen S, Ten ZP, Hu Z, Zeng Y. Human papilloma virus 18E6E7 in synergy with TPA induced malignant transformation of human embryonic esophageal epi – thelial cells. Bingdu Xuebao, 1999, 15: 1 – 6

20 Shen ZY, Shen J, Cai WJ, Wu XY, Zheng RM, Zeng Y. The promtor effects of malignant transformation of sodium butyrate on the immortalized esophageal epithelium induced by human papillomavirus. Zhonghua Binglixue Zazhi, 2002, 31: 39 – 41

21 Weitzman JB, Yaniv M. Rebuilding the road to cancer. Nature 1999, 400, 401 – 402

22 Yakoob J, Hu GL, Fan XG, Zhang Z. Telomere, telomerase and digestive cancer. World J Gastroenterol, 1999, 5: 334 – 337

23 He XX, Wang JL. Activity of telomerase and oncogenesis. Huaren Xiaohua Zazhi, 1998, 6: 1100 – 1101

24 Yang SM, Fang DC, Luo YH, Lu R, Liu WW. Telomerase activity in gastroeintesind submucosal tumors and its clinical significance. Huaren Xiaohua Zazhi, 1998, 6: 765 – 767

25 Zhan WH, Ma JP, Peng JS, Gao JS, Cai SR, Wang JP, Zheng ZQ, Wang L. Telomerase activity in gastric cancer and its clinical implications. World J Gastroenterol, 1999, 5: 316 – 319

26 He XX, Wang JL, Wu JL, Yuan SY, Ai L. Telomerase expression, Hp infection and gastric mucosal carcinogenesis. Shijie Huaren Xiaohua

Zazhi, 2000, 8: 505 – 508

27 He XX, Wang JL, Wu JL, Yuan SY, Ai L. Telomere, cellular DNA content and gastric mucosal carcinogenesis. Shijie Huaren Xiaohua Zazhi, 2000, 8: 509 – 512

28 Yao XX, Yin L, Zhang SY, Bai WY, Li YM, Sun ZC. hTERT expression and cellular immunity in gastric cancer and precancerosis. Shijie Huaren Xiaohua Zazhi, 2001, 9: 508 – 512

29 Xia ZS, Zhu ZH, He SG. Effects of ATRA and 5 Fu on growth and telomerase activity of xenografts of gastric cancer in nude mice. Shijie Huaren Xiaohua Zazhi, 2000, 8: 674 – 677

30 Meng ZQ, Yu EX, Song MZ. Inhibition of telomerase activity of human liver cancer cell SMMC 7721 by chemotherapeutic drugs. Shijie Huaren Xiaohua Zazhi, 1999, 7: 252 – 254

31 Fu JM, Yu XF, Shao YF. Telomerase and primary liver cancer. Shijie Huaren Xiaohua Zazhi, 2000, 8: 461 – 463

32 Qu B, Li BJ, Lu ZW, Pan HL. Clinical significance of telomerase activity detected in fineneedle aspiration speciments to liver cancer diagnosis. Shijie Huaren Xiaohua Zazhi, 2001, 9: 538 – 541

33 Qiu SL, Huang JQ, Wang YF, Peng ZH. Analysis of telomerase activity in colorectal cancer, precancerous lesions and cancer washings. Shijie Huaren Xiaohua Zazhi, 1998, 6: 992 – 993

34 Sobti RC, Kochar J, Singh K, Bhasin D, Capalash N. Telomerase activation and incidence of HPV in human gastrointestinal tumors in North Indian population. Mol Cell Biochem, 2001, 217: 51 – 56

35 Jia L, Li YY. Telomerase activity of exfoliated cancer cells in colonic luminal washings. Huaren Xiaohua Zazhi, 1998, 6: 955 – 957

36 Koyanagi K, Ozawa S, Ando N, Takeuchi H, Ueda M. Kitajima M. Clinical significance of telomerase activity in the non – cancerous epithelial region of oesophageal squamous cell carcinoma. Br J Surg, 1999, 86: 674 – 679

37 Hiyama T, Yokozaki H, Kitadai Y, Haruma K,

Yasui W, Kajiyama G, Takara E. Overexpression of human telomerase RNA is an early oesophageal carcinogenesis. Virchows Arch, 1999, 434: 483 – 487

38　Kiyozuka Y, Asai A, Yamamoto D, Senzaki H, Yoshioka S, Takahashi H, Hioki K, Tsubura A. Establishment of novel human esophageal cancer cell relation to telomere dynamics and telomerase activity. Dig Dis Sci, 2000, 45: 870 – 879

39　Koyanagi K, Ozawa S, Ando N, Mukai M, Kitagawa Y, Ueda M, Kilajima M. Telomerase activity as an indicator of malignant in iodinenonreactive lesions of the esophagus. Cancer, 2000, 88: 1524 – 1529

40　Hahn WC, Counter CM, Lundberg AS, Beijersbergen RL, Brooks MW, Weinberg RA. Creation of human tumour cells with defined genetic elements. Nature, 1999, 400: 464 – 468

41　Xu LY, Shen ZY, Li EM, Cai WJ, Shen J, Li C, Hong CQ, Chen JY, Zeng Y. Telomere length and telomerase activity in immortalized and malignantly transformed human embryonic esophageal epithelial cell lines by E6 and E7 genes of HPV 18 type. Aibian Qibian Tubian, 2001, 13: 137 – 140

42　Morales CP, Lee EL, Shay JW. In situ hybridization for the detection of telomerase RNA in the progression from Barrett's esophagus to esophageal adenocarcinoma. Cancer, 1998, 83: 652 – 659

43　Shen ZY, Xu LY, Li C, Cai WJ, Shen J, Chen JY, Zeng Y. A comparative study of telomerase activity and malignant phenotype in multistage carcinogenesis of esophageal epithelial cells induced by human papillomavirus. Int J Mol Med, 2001, 8: 633 – 639

44　Steenbergen RD, Hermsen MA, Walboomers JM, Meijer GA, Baak JP, Meijer CJ. Non – random allelic losses at 3p llp and 13p during HPV – mediated immortalization and concomitant loss of terminal differentiation of human keratinocytes. Int J

cancer, 1998, 76: 412417

45　Duensing S, Lee LY, Duensing A, Basile J, Piboonniyom S, Gonzalez S, Crum CP, Munger K. The human papillomavirus type 16E_6 and E_7 oncoproteins cooperate to induce mitotic defects and genomic instability by uncoupling centrosome dupli – cation from the cell division cycle. Proc Natl Acad Sci USA, 2000, 29: 10002 – 10007

46　Garbe J, Wong M, Wigington D, Yaswen P, Stampfer MR. Viral oncogenes accelerate conversion to immortality of cultured conditionally immortal human mammary epithelium cells. Oncogen, 1999, 18: 2169 – 2180

47　Wang P, Peng ZL, Wang H, Liu SL. Study on the carcinogenic mechanism of human papilomavirurs type 16 E7 protein in cervical carcinoma. Zhonghua Shi Yah He Linchuang Bing Du Xue Zazhi, 2000, 14: 117 – 120

48　Zur Hausen H. Papillomaviruses in human cancers. Proc Assoc Am Physicians , 1999, 111: 581 – 587

49　Song S, Liem A, Miller JA, Lambert PF. Human papillomavirus types 16 E6 and E7 contribute differently to carcinogenesis. Virology, 2000, 26: 141 – 150

50　Song S, Pitot He, Lambert PF. The human papillomavirus type 16 E6 gene alone is sufficient to induce carcinomas in transgenic animals. J Virol 1999, 73, 5887 – 5893

51　Shen ZY, Cen S, Shen J, Cai WJ, Xu JJ, Teng ZP, Hu Z, Zeng Y. Study immortalization and malignant transformation of human embryonic esophageal epithelial cells induced by HPV18 E6E7. JCancer Res Clin Oncol, 2000, 126: 589 – 594

52　Shen ZY, Hu SP, Shen J, Lu LC, Tang CZ, Kuang ZS, Zhong SP, Zeng Y. Detection of human papillomavirus in esophageal carcinoma. J Med Virol, 2002, 68: 412416

78. Prospects of Vaccination as a Means of Preventing Mother – to – Child Transmission of HIV – 1

BIBERFELD G[1], BUONAGURO F[2], LINDBERG A[3], de – The G[4], ZENG Yi[5], ZETTERSTRÖM R[6]

1. Microbiology and Tumor Biology Center, Karolinska Institute and Swedish Institute for Infectious Disease Control, Stockholm, Sweden; 2. Viral Oncology and AIDS Reference Center, Istituto Nazionale Tumori " Fond. G. Pascale ", Naples, Italy; 3. Aventis Pasteur, Lyon, France and Division of Clinical Bacteriology, Huddinge Hospital, Karolinska Institute, Stockholm, Sweden; 4. Institute Pasteur, Paris, France; 5. Institute of Virology, Beijing, China; 6. Acta Pœdiatrica, Karolinska Hospital, Stockholm, Sweden

[Key words]　　HIV – 1; Mother – to – infant; Prevention; Transmission; Vaccination

This report addresses the issue of HIV – 1 vaccination as a possible complement to antiretroviral therapy in preventing mother – to – child transmission (MTCT) of HIV – 1. The topic was discussed at the 2001 Erice workshop on the prevention of MTCT of HIV – 1, but was not included in the report published in the November 2001 issue of this journal[1].

Attempts to develop an effective HIV vaccine, ongoing since 1985, have so far been unsuccessful, a major problem being the high variability of HIV. In addition to the two main types (HIV – 1 and HIV – 2), there are three groups of HIV – 1 (M, N and O), each in turn further subdivided into several subtypes, and sub/subtypes, inter – subtypes and inter – group recombinants are emer – ging in the highly endemic areas of Africa and Asia[2-4]. The different global distribution of the various HIV – 1 subtypes and the need to adapt vaccines to the subtypes in different continents are factors exacerbating development of vaccine.

Possible Vaccine Candidates

An ideal sterilizing prophylactic vaccine must be safe and induce: (i) cross – neutralizing antibodies against primary HIV – 1 isolates from divergent HIV subtypes; (ii) a strong and broad CD4 + T – cell response; (iii) poly – epitopic, cross – clade reactive CD8 + CTL responses; (iv) mucosal immune responses; and (v) long – term protection.

There are various types of HIV vaccine candidates, including live attenuated virus, whole inactivated virus, recombinant produced subunits, synthetic peptides, live recombinant vaccines and viral DNA. Live attenuated vaccines have been shown to be the most efficient in eliciting protective immunity in the SIV/macaque model[5], but live HIV vaccines cannot be used in humans for safety reasons. HIV – 1 envelope subunits, which have so far been the most frequently tested vaccine candidates, elicit neutralizing antibodies to laboratory HIV – 1 strains but not to field isolates, and usually fail to induce cytotoxic T lymphocyte (CTL) responses[6, 7].

Recent data indicate the advantage of mixed modality immunization, i. e. immunization with several HIV – 1 antigens (env, *gag*, pol, nef, tat) as genetic information in DNA (plasmids), in live expressing vectors (poxvirus, adenovirus, salmonella, BCG), or as virus – like particles, recombinant proteins, peptides or peptide – conjugates. Subjects may be primed with one construct (DNA or vector) and boosted with another (vector or proteins). Several mixed modality regimens are currently in phase I / II human studies[6].

It cannot yet be estimated when an effective and safe prophylactic HIV – 1 vaccine for use in humans will be available. However, the demonstration of HIV – specific cellular immune responses in a proportion of highly exposed non – infected individuals and the observation that strong cellular immune responses are correlated with long – term survival in HIV – 1 infected individuals, as well as results of HIV/SIV vaccine studies in non – human primates, indicate that an efficacious HIV vaccine can be developed[5-8]. Furthermore, approximately 70 phase I / II clinical trials, including more than 30 different vaccine candidates, have been conducted in non HIV – infected adult volunteers[6, 7]. The tested vaccines have been shown to be safe and to induce immune responses of varying intensity. The first phase III efficacy vaccine trials, which use HIV – 1 envelope subunits, are ongoing in the USA and Thailand. There are also plans to launch efficacy trials involving priming with canarypox virus expressing HIV – 1 env (subtype B and E), *gag*, protease, pol and nef and boosting with recombinant gp 120[9].

Vaccine Trials in Newborn Infants

A few phase I/II HIV – 1 vaccine trials have been conducted in neonates in the USA. Recombinant HIV – 1 envelope vaccines have been shown to be safe and immunogenic, as measured by a lymphocyte proliferation assay, in neonates and infants born to HIV – 1 infected women[11, 12].

A canarypox viral vector – based HIV – 1 vaccine expressing gp120, *gag* and protease (ALVAC vcp 205) has also been shown to be safe and to induce HIV – 1 – specific cellular immune responses in about 50% of vaccinated neonates born to HIV – 1 infected women (cf. 12).

The important issue of an HIV vaccine to neonates as a means of preventing MTCT of HIV through breastfeeding has recently been reviewed[12]. In view of the size of the problem in developing countries, where breastfeeding remains the only alternative because of the lack of clean water to provide formula feeding, the urgency to test available HIV vaccine candidates is high. A critical question is whether an HIV vaccine given to newborns will induce strong immune responses rapidly enough to protect against HIV – 1 transmission during the first 2 – 3 mo of breastfeeding.

The first phase I HIV – 1 vaccine trial in infants in an African setting where breastfeeding is the norm is planned to be conducted at the Mulago Hospital in Kampala, Uganda[12]. Mother – infant pairs will receive short – course nevirapine treatment and the infants will be vaccinated with an ALVAC HIV – 1 vaccine expressing gp120, *gag*, protease, pol and nef (vcp 1452). However, if ongoing studies aimed at preventing breast milk transmission of HIV – 1 by prophylactic antiretroviral therapy are found to be successful, it may become difficult for ethical reasons to conduct phase III efficacy HIV vaccine trials in infants during the breastfeeding period.

Passive Immunization of Newborn Infants

Passive immunization using HIV immunoglobulin is another approach towards preventing MTCT of HIV – 1 peripartum and postpartum during breastfeeding. This will be tried in Kampala in combination with nevirapine treatment.

Therapeutic vaccination of HIV – infected pregnant women

Therapeutic vaccination might also be considered as a way of reducing MTCT of HIV – 1, although pregnant women are usually excluded from vaccinations for safety reasons. Before highly active anti – retroviral therapy (HAART) became available in 1996, the development of an effective therapeutic vaccine was considered unrealistic, since infection with HIV – 1 seriously reduces the number and impairs the functions of CD4 + T cells. However, therapeutic vaccination of patients on HAART, which limits virus replication and thereby partly restores the immune system, may be a way of further controlling HIV infection. In an initial phase 1 study, vaccination using a pox vector coding for env and *gag* protein followed by recombinant gp 160 protein delayed the viral rebound for 4 mo in 2 out of 4 patients who had stopped HAART treatment. A series of therapeutic phase II vaccine trials using different HIV – 1 vaccine regimens is ongoing.

The possibility of therapeutic vaccination of HIV – 1 – infected pregnant women raises ethical concern about the risk/benefit to the " infant – mother" pair, and is an issue which has to be carefully evaluated. Results from animal experiments must be evaluated before human clinical trials are contemplated.

[In 《Acta Paediatr》 2002, 91: 241 – 243]

References

1 Biberfeld G, Biberfeld P, Buonaguro F, Charpak N, Blaudin de Thé G, Rea MF, et al. Mother – to – child transmission of HIV – 1. Acta Paediatr, 2001, 90: 1337 – 1340

2 Carr JK, Foley B, Leitner T, Salminen M, Korber B, McCutchan FE. Reference sequences representing the principal genetic diversity of HIV – I in the pandemic. In: Human retrovirus and AIDS. Los Alamos: Los Alamos National Laboratory, 1998

3 Hoelscher M, Kim B, Maboko L, Mhalu F, von Sonnenburg F, Birx DL, et al. High proportion of unrelated HIV – 1 intersubtype recombinants in the Mbeya region of southwest Tanzania. AIDS, 2001, 15: 1461 – 1470

4 Peeters M, Sharp PM. Genetic diversity of HIV – 1: the moving target. AIDS, 2000, 14 Suppl 3: 129 – 140

5 Almond NM, Heeney JL. AIDS vaccine development in primate models. AIDS, 1998, 12 Suppl A: 133 – 140

6 Mulligan M J, Weber J. Human trials of HIV – 1 vaccines. AIDS 1999, 13 Suppl A: 105 – 112

7 Esparza J, Bhamarapravati N. Accelerating the development and future availability of HIV – 1 vaccines: why, where, and how? Lancet, 2000, 355: 2061 – 2066

8 Rowland – Jones S, Tan R, McMichael, A. Role of cellular immunity in protection against HIV infection. Adv Immunol, 1997, 65: 277 – 346

9 Cohen J. AIDS research: debate begins over new vaccine trials. Science, 2001, 293: 1973

10 Borkowsky W, Wara D, Fenton T, McNamara J, Kang M, Mofenson L, et al. Lymphoproliferative responses to recombinant HIV – 1 envelope antigens in neonates and infants receiving gpl20 vaccines. J Infect Dis, 2000, 181: 890 – 896

11 Cunningham CK, Wara DW, Kang M, Fenton T, Hawkins E, McNamara J, et al. Safety of 2 recombinant human immunode – ficiency virus type 1 (HIV – 1) envelope vaccines in neonates born to HIV – 1 – infected women. Clin Infect Dis, 2001, 32: 801 – 807

12 Bass E. Aids vaccines and HIV transmission by breastfeeding. In: Newsletter of the International AIDS Vaccine Initiative (IAVI), 2001, 5: 314

79. Construction and Characterization of a Chimeric Virus (BIV/HIV – 1) Carrying the Bovine Immunodeficiency Virus *gag* – pol Gene

CHEN Guo-min[1], WANG Shu-hui[2], Xiong Kun[2], WANG Jin-zhong[2], YE Tao[1],
DONG Wen-ping[1], WANG Qi[1], CHEN Qi-min[2], GENG Yun-qi[2], WOOD Charles[3], ZENG Yi[1]

1. Department of Tumor Virus and HIV, Institute of Virology, CAPM;
2. College of Life Sciences, Nankai University; 3. Nebraska Center
for Virology School of Biological Sciences, University of Nebraska

Summary

HIV – 1$_{HXB2}$ 5'LTR region, most of BIV$_{R29}$ *gag* – pol segment and HIV – 1$_{HXB2}$ pol IN – 3'LTR region were respectively amplified. A chimeric clone, designated as pHBIV$_{3753}$, was constructed by cloning three fragments sequentially into pUC18. MT4 cells were transfected with pHBIV$_{3753}$. The replication and expressions of the chimeric virus (HBIV$_{3753}$) were monitored by RT activity and IFA. The results firstly demonstrated that it is possible to generate a new type of the BIV/HIV – 1 chimeric virus containing BIV *gag* – pol gene.

The development of safe, effective HIV vaccines is considered to be one of the most important ways to control the incidence of HIV infection[1]. Among the various types of vaccines developed, attenuated vaccines have the advantage that they mimic natural virus infection in the host. Although an HIV vaccine for humans has not yet been produced, SIV or SHIV vaccines attenuated by gene modification have protected monkeys from homologous and heterologous challenge viruses[2]. However, their safety for human use is questionable because they can cause fatal immune system disease[3].

Bovine immunodeficiency virus (BIV), a non – primate lentivirus, more closely resembles human and non – human primate immunodeficiency viruses in structure, immunology and genetics, and is unable to infect homans or human cells[4]. There is, therefore, the possibility of constructing a new chimeric virus to be used as an attenuated vaccine against HIV. In this work, we sought to construct a series of chimeric viruses of BIV and HIV − 1 to generate a chimeric virus that can be used as a candidate for attenuated vaccine. We successfully generated a himeric virus/H (BIV) that contains the *gag* − pol gene from BIV.

The complementary DNA sequence data of BIV_{R29} and HIV − 1_{HXB2} are from GenBank. The following gene segments were obtained by polymerase chain reaction (PCR), using p_{HXB2} or pBIV plasmids as templates. The first segment was the entire HIV − 1 5' long − term repeat (LTR) region (1 − 19 nt and 788 − 776 nt). The second segment contained most of the BIV_{R29} *gag* − pol segment (701 − 727 nt and 4444 − 4421 nt). The third was the HIV − 1_{HXB2} pol IN − 3'LTR region (4377 − 4411nt and 9719 − 9696 nt). Two ends of three pairs of primers included different enzyme sites, respectively. PCR products were then digested by respective enzymes and ligated sequentially into pre-digested pUC18 vector (Fig. 1a). The recombinant chimeric clone was designated as $pHBIV_{3753}$. It was verified by $_{Xbal}$, BamHI, HhldlII, BglI, and PstI enzyme digestion.

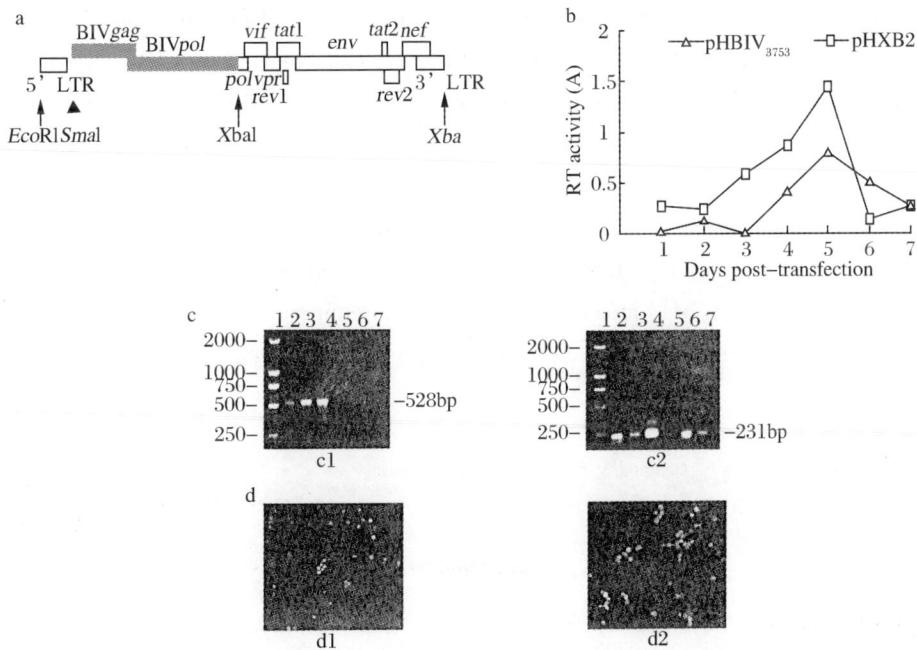

(a) Genome organization of the chimeric clone. The arrowheads indicate the positions of enzyme sites used to make the recombinant. White boxes represent long − term repeats (LTR) and open reading frames from p_{HXB2}. Black boxes represent *gag* − pol, not including the IN region of pol, from primate bovine immunodeficiency virus (pBIV). (b) Kinetics of $HBIV_{3753}$ replication in MT4 cells. Viral replication was monitored by reverse transcriptase (RT) activity of culture supernatants. (c) Reverse transcriptase − polymerase chain reaction (RT − PCR) analysis of BIV *gag* and HIV − 1 tat reverse transcripts of $HBIV_{3753}$. Lanes 2 − 4: RT − PCR amplification of BIV *gag* fragments; lanes 5 − 7: PCR amplification of BIV *gag* fragments used as control; using the cellular RNA as templates at 2, 4, 7 days post − transfection, respectively (c1). Lanes 2 − 4: RT − PCR amplification of the full − length HIV − 1 tat fragments, the cellular RNA from cells transfected with p_{HXB2} used as controls; lanes 5 − 7: RT − PCR amplification of $HBIV_{3753}$; collected 5, 6, 7 days post − transfection, respectively (c2). (d) BIV or HIV − 1 antigen expressions in MT4 cells transfected with the chimeric clone. The smears were prepared at 48 h post − transfection and analysed by indirect immunofluorescence assay with either mouse anti − BIV p26 monoclonal antibody (d1) or human anti − HIV − 1 antibody (d2)

Fig. 1   Genetic structure and biological activity of $HBIV_{3753}$

Equivalent quantities of $pHBIV_{3753}$ or p_{HXB2} plasmid were introduced into MT4 cells by electro – operation. The culture supernatant from transfected cells was collected from 1 to 7 days. To assess virus replication kinetics, the virion – associated reverse transcriptase (RT) activity was determined by a BrdUTP incorporation assay (Cavidi Tech AB, Sweden). The level of RT activity peaked at 5 days post – transfection (Fig. 1b). BIV replicative capacity was lower than that of HIV – 1. Although the promoters and regulatory protein of HIV – 1 were used in $HBIV_{3753}$, the *gag* – pol gene came from BIV. Therefore, $HBIV_{3753}$ replicative ability was lower than that of HIV_{HXB2}.

At 2, 4, and 7 days post – transfection, the cells were harvested and the total RNA of transfected cells was extracted. BIV *gag* gene transcript of $HBIV_{3753}$ was amplified by RT – PCR (Fig. 1cl). For 5 – 7 days post – transfection, the HIV – 1 tat gene transcript of $HBlV_{3753}$ was obtained by RT – PCR. (Fig. 1c2). Our results showed that despite the HIV – 1 , *gag* – pol region being replaced by BIV *gag* – pol, $HBIV_{3753}$ could form full – length and splicing transcripts.

The antigenicity of $HBIV_{3753}$ was analysed by indirect immunofluorescent assay (IFA). Forty – eight hours post – transfection, the cells exposed to mouse anti – BIV p26 monoclonal antibody showed a fluorescence signal only in the cytoplasm. The cells also reacted with HIV – I – positive serum, and a fluorescence signal appeared both on the cell membrane and in the cytoplasm. These data indicate correct $HBIV_{3753}$ gene expression and also show correct antigenicity for two kinds of viral proteins (Fig. 1d).

This is the first demonstration to show the use of a BIV gene to replace the structural gene of HIV – 1, and thereby create a new type of $HBIV_{3753}$, which contains biological activity. The final goal of authors is to obtain a safe, effective, and attenuated chimeric virus. In our laboratory, a series of BIV/HIV – 1 chimeric clones are being constructed and analysed.

Acknowledgements

The authors are indebted to Drs S. Cen and X. F. Yu for critical reading.

[In 《AIDS》 2002, 16 (1): 123 – 125]

References

1 Clements JE, Montelaro RC, Zink MC, et al. Cross – protective immune responses induced in rhesus macaques by immunization with attenuated macrophage – tropic simian immunodeficiency virus. J Virol, 1995, 69: 2737 – 2744

2 Hayami M, Igarashi T, Kuwata T, et al. Gene – mutated HIV – 1/SIV chimeric viruses as AIDS live attenuated vaccines for potential human use. Leukemia, 1999, 13 (Suppl. 1): S43 – S45

3 Baba TW, Liska V, Khimani AH, et al. Live attenuated, multiply deleted simian immunodeficiency virus causes AIDS in infant and adult macaques. Nat Med, 1999, 5: 194 – 203

4 Gonda MA, Luther DG, Fong S, et al. Bovine immunodeficiency virus: molecular biology and virus – host interactions. Virus Res, 1994, 32: 1555 – 1581

80. Morphological and Functional Changes of Mitochondria in Apoptotic Esophageal Carcinoma Cells Induced by Arsenic Trioxide

SHEN Zhong – ying[1], SHEN Jian[1], LI Qiao – shan[1], CHEN Cai – yun[2], CHEN Jiong – yu[3], ZENG Yi[4]

1. Department of Pathology, Medical College of Shantou University;

2. Central Lab. Medical College of Shantou University;

3. Central Lab. of Tumor Hospital, Medical College of Shantou University;

4. Institute of Virology, Chinese Academy of Preventive Medicine

Summary

Objective　To demonstrate that mitochondrial morphological and functional changes are an important intermediate link in the course of apoptosis in esophageal carcinoma cells induced by As_2O_3.

Methods　The esophageal carcinoma cell line SHEEC1, established in our laboratory, was cultured in 199 growth medium, supplemented with $100ml \cdot L^{-1}$ calf serum and $3 \mu mol \cdot L^{-1} As_2O_3$ (the same below). After 2, 4, 6, 12, 24 h of drug adding, the SHEEC1 cells were collected for light – and electron – microscopic examination. The mitochondria were labeled by Rhodamine fluorescence probe and the fluorescence intensity of the mitochondria was measured by flow cytometer and cytofluorimetric analysis. Further, the mitochondrial transmembrane potential (MTP, $\Delta\Psi m$) change was also calculated.

Results　The mitochondrial morphological change after adding As_2O_3 could be divided into three stages. In the early – stage (2 – 6 h) after adding As_2O_3, an adaptive proliferation of mitochondria appeared; in the mid – stage (6 – 12 h) a degenerative change was observed; and in the late – stage (12 – 24 h) the mitochondria swelled with outer membrane broken down and then cells death with apoptotic changes of nucleus. The functional change of the mitochondria indicated by fluorescent intensity, which reflected the MTP status of mitochondria, was in accordance with morphological change of the mitochondria. The fluorescent intensity increased at early – stage, declined in mid – stage and decreased to the lowest in the late – stage. 24 h after As_2O_3 adding, the cell nucleus showed typical apoptotic changes.

Conclusion　Under the inducement of As_2O_3, the early apoptotic changes of SHEEC1 cells were the apparentmorphological and functional changes of mitochondria, afterwards the nucleus changes followed. It is considered that changes of mitochondria are an important intermediate link in the course of apoptosis of esophageal carcinoma cells induced by As_2O_3.

Introduction

Esophagus cancer is common in China[1-11]. The treatment is still a focus of research[12-17].

Induction of cell apoptosis is a novel therapeutic strategies for cancer[18-25]. In our previous work, we used As_2O_3 to induce apoptosis of esophageal carcinoma cells[26]. The pathomorphological changes evinced that cells became smaller, the cells shrank, the nuclei rounded up, chromatin agglutinated and marginated, nuclear membrane broke down and then followed by the degenerative changes of the cells. All these changes indicated typical morphological changes of apoptosist[27]. The necrotic changes were also found with a large dosage of As_2O_3[28]. We discovered that in the early-stage of cell apoptosis, prior to the obvious change of cell nuclei, the mitochondria showed proliferation. The detailed morphological changes of mitochondria of esophageal carcinoma cells induced by As_2O_3 were firstly described in our paper[29]. We also found that nitri coxide (NO) was released from the cultured esophageal carcinoma cell line after administration of As_2O_3 with increasing amounts at the early apoptotic stage[30]. Furthermore, down regulated expression of bcl-2 and over expression of bax were always found in apoptotic cells induced by As_2O_3[31].

Some authors hold that apoptosis is a programmed cell death (PCD); the death signal originates from the inside of cells; the change chiefly involves the cell nucleus with no apparent changes seen in the cytoplasm and cell organelle[32-33]; making it different from cell necrosis[34]. In our studies, the morphological changes of apoptotic cells induced by As_2O_3 were different from the programmed cell death in which the latter showed the nuclear changes at first and then cytoplasm, and the former were vice versa[35]. In recent years, it has been explained that apoptosisis related to certain factors, such as Bcl-2/Bax[36-39], Ca^{2+}[40] and cytochrome c[41-42], which are all located on mitochondria[43]. When they are released from mitochondria, they can inhibit or promote cell apoptosis. Therefore, mitochondria are thought to be the apoptosis regulation center[44]. Mitochondria are also an important organell. They are concerned with cell breathing, oxygen metabolism, enzyme activity and energy supply. All of those functions relate to the permeability of the mitochondria and mitochondrial transmembrane potential (MTP, $\Delta\Psi m$). When MTP decreases, the mitochondria generate morphological and functional changes[45-47]

Rhodamine 123 (Rho123), a kind of fluorescent dye, is traditionally used as a mitochondria probe[48]. Rho 123 can quickly gather on living cell mitochondria. The fluorescence intensity of Rho123 represents MTP which reflects the cell in a quiescent or active condition, and in a proliferative or differentiative manner[49]. Flow cytometer and fluorescent microphotometry are the satisfactory instruments to measure Rho123 fluorescent intensity. The purpose of this paper is to study the mitochondrial morphological and functional changes during the cell apoptosis of esophageal carcinoma cell line induced by As_2O_3, thus demonstrating that mitochondrial changes play an important role in the course of cell apoptosis.

Materials and Methods

1. Cell line and As_2O_3 adding

The esophageal carcinoma cell line SHEEC1 is the human embryonic esophageal epithelial cells malignantly transformed by HPV 18 E6 E7 in synergy with TPA[50]. It is cultured in 199 growth medium, supplemented with 100 ml · L^{-1} calfserum andantibiotics. In experiments, SHEEC1 cells were cultured separately in culture flasks and on 24 – well culture plates (Corning Co.) with the cover slide inside the well, in every well 10^4 SHEEC1 cells were inoculated. As_2O_3 (Sigma, St. Louis, Mo; Lot A 1010) was prepared in concentration of $3\,\mu mol \cdot L^{-1}$ with 199 growth medium. The experimental group and the control group without As_2O_3 administered were examined at definite times. The experiments were repeated once.

2. Examination under light – and electron – microscope

At 2, 4, 6, 12, 24 h after As_2O_3 adding, one culture flask of SHEEC1 cultured cells was taken for examination. The floating cells in the flasks were collected by centrifugation (CytospinⅢ, Shandon Co.), Giemsa stained and examined by light – microscope. Cells attached to flask were digested with $2.5g \cdot L^{-1}$ trypsin, centrifuged, the cell pallet was fixed with $25g \cdot L^{-1}$ glutaraldehyde, and were routinely prepared for electronmicroscopic examination.

3. Rhodamine fluorescent probe labeling and cytofluorimetric analysis (CFA) [51,52]

SHEEC1 cells were placed on the slide after reacting with As_2O_3 at various times, stained by Rhodamine 123 (Rho 123, MW381, Molecular Probe Inc. Eugene) at the concentration of $10mg \cdot L^{-1}$, and the cells were incubated in 37 ℃, $50ml \cdot L^{-1}$ CO_2 incubator for 15 min. It was examined by fluorescent microscopy and cytofluorimetry. Using the Nikon fluorescent microscope (Fluophot, Nikon) with Low – cost cooled digital CCD camera system and software STARI (Photometrics LTD. USA), the fluorescent image of mitochondria of SHEEC1 cells labeled by Rho123 were displayed on the screen of monitor, the fluorescent intensity of cells was measured by scanning method, and the average amount of cellular fluorescence was calculated by software.

4. Flow cytometer (FCM) examination [53]

Following As_2O_3 treatment, SHEEC1 cell cultured in flasks were harvested with trypsinization, washed once with PBS, resuspended in PBS, and incubated with Rho123 ($10mg \cdot L^{-1}$) at 37 ℃ for 15 min, stained cells were wash twice with PBS, dispersed, filtered through a 360 mesh nylon net to make single cell suspension, 10^9 cell $\cdot L^{-1}$ were detected by flow cytometer (FACSort, B – D Co. USA) using exciting light 488nm and emission light 515nm to detect Rho123 fluorescent intensity. The histogram managed by the computer was drawn according to the fluorescent intensity value of one cell. Partial of SHEEC1 cells were fixed with $700ml \cdot L^{-1}$ alcohol, stained with propidium iodide (Sigma) and analyzed with flow cytometer. The cell cycle and apoptotic cell rate were calculated.

5. Calculation of mitochondrial transmembrane potential (MTP. ΔΨm) [46]

Examining 10^4 cells by FCM, the average fluorescent intensity of the cells labeled by Rho 123 before and after As_2O_3 adding were drawn as histograms for comparing. By cytofluorimetric analysis the average fluorescent intensity value ($\bar{x} \pm s$) was calculated from one cell.

Results

1. Cell apoptosis

Twenty – four hours after As_2O_3 acting on SHEEC1 cells, the apoptotic peak (28% of the cells) before G_1G_0 in DNA histogram of FCM examination appeared (Fig. 1). Collecting the floating cells by cytospin and Giesma staining, the cell nuclei showed typical cell apoptotic changes with chromatin agglutinated and marginated (Fig. 2).

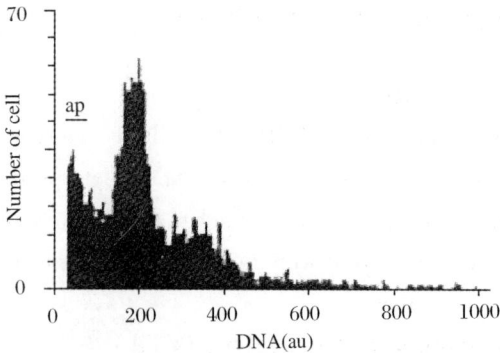

Fig. 1 DNA histogram of SHEEC1 cells 24h after Fig. 2 Apoptotic changes 24h after As₂O₃

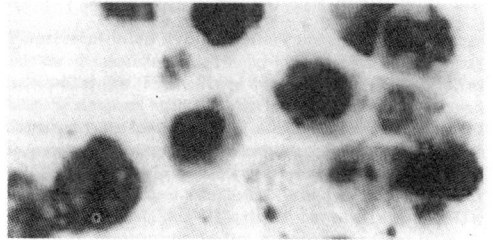

Fig. 2 As₂O₃ adding, ap, apoptotic peak.
adding, HE ×400

2. Morphological changes of mitochondria under transmission electron – microscope

Before adding As_2O_3 the mitochondria were located around the nucleus in one or two arrays (Fig. 3A). There were fixed intervals between mitochondria, in which other organelles were present. When adding As_2O_3 2 – 4 h, the mitochondria increased, which showed either concentration in certain areas or in one pole of the cytoplasm or distributed in inner, middle or outer layer of the cytoplasm (Fig. 3B). Mitochondria were oval in shape and different in size. The newly proliferated mitochondria were smaller with dense matrix. Some mitochondria were condensed with indistinct ridges and some mitochondria were crowded closely together. After 6 h, the high electron dense and irregular shaped substances precipitated in the mitochondrial matrix, even filled up the whole mitochondria (Fig. 3C). The autophagosomes resulting from wrapping of condensed mitochondria by the lysosomes were frequently seen. After 12 h, the mitochondria swelled, its outer membrane broke down, left a single layer of membrane, which were seen like a balloon or a vacuole. After 24h, the cell nucleus shrank and chromatin agglutinated locating near the nuclear membrane with mitochondria swelling, or becoming vacuole – like or broken down (Fig. 3D).

3. Functional changes of mitochondria in cell apoptosis: the dynamic changes of MTP ($\Delta\Psi m$)

(1) *Mitochondrial fluorescence intensity detected by FCM*: After As_2O_3 was added to SHEEC1 cells, the changes of mitochondria fluorescence intensity from different reacting times were seen in histogram (Fig. 4 A, B, C, D). A slight increase of mitochondrial fluorescence intensity was observed at 2h after added As_2O_3. With treatment of As_2O_3 for 4 – 6h, fluorescent intensity of mitochondria was decreased sharply. After 12 – 24h fluorescent intensity was the lowest.

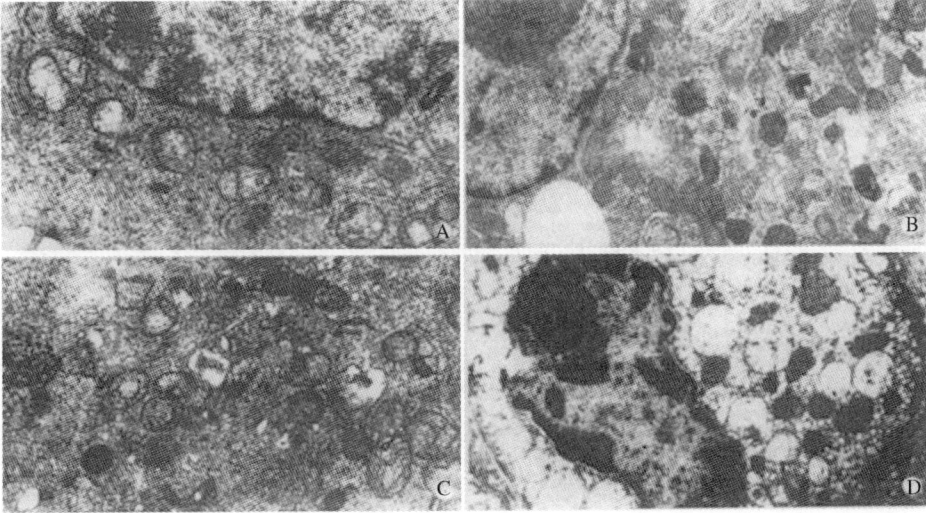

Mitochondria in 1 – 2 arrays located around the cell nucleus, not adding As_2O_3; Increment of mitochondria 2 – 4 h afterAs$_2$O$_3$ adding; Dense substances deposition in mitochondria 4 – 6h after As_2O_3 adding; Apoptotic cell showed cell nucleus shrank, chromatin agglutinated, mitochondria increased and swelled as balloon – like 24h after As_2O_3 adding

Fig. 3   Apoptotic cells (EM × 15 000)

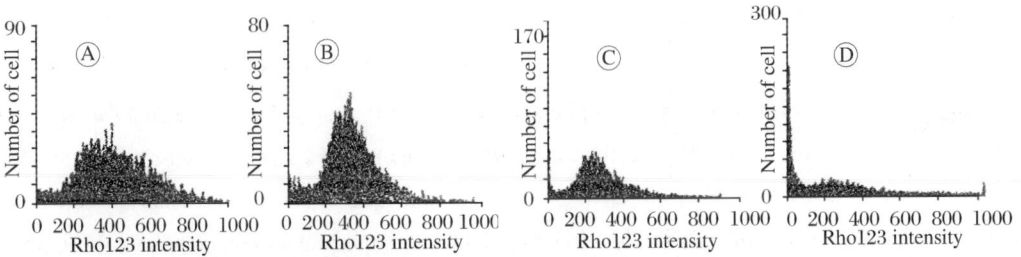

A: Control; B: 2 – 4 h; C: 4 – 6 h; D: 12 – 24 h

Fig. 4   The histogram of mitochondrial fluorescent intensity by FCM after As_2O_3 adding

(2) *Fluorescent intensity by cytofluorimetric analysis*: Under fluorescent microscope, the number of mitochondria of cells was increased at first (Fig. 5) and then decreased. The fluorescent intensity increased in 2h after As_2O_3 added, declined in 4 – 6h and decreased to the lowest in the 12 – 24h (Tab. 1). An increment offluorescence intensity in partial early – stage apoptotic cells after 2h of As_2O_3 adding and the intensity decreased hereafter. Following fluorescence associated with the uptake of dye Rho123 allows to evaluate $\Delta\Psi m$ modifications, the results showed the dynamic MTP changes in the apoptotic process induced by As_2O_3.

The Rho 123 fluorescence intensity of the labeled mitochondria differed from different reacting times after adding As_2O_3. At first fluorescent intensity increased and then the rapidly declining value of fluorescence intensity was in accordance with both results of FCM and CFA. It taking cell morphology into account, the fluorescence intensity changes may reflect the consequence of As_2O_3 stimulation to mitochondria for different times. 2 h after As_2O_3 was added, the mitochondria proliferated and the fluorescent intensity increased, soon after the intensity swiftly declined and went to the lowest at 24 h, which indicated that morphological and functional changes of mitochondria induced by As_2O_3 represented the process cell apoptosis.

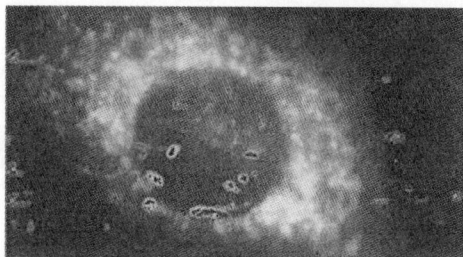

Fig. 5 Increment of mitochondria with Rho123 labeled in cytoplasm of SHEEC1 after 2 – 4h As_2O_3 adding. ×1000

Tab. 1 Average fluorescence intensity value of SHEEC1 after As_2O_3 adding (arbitrary unit ×10^{-4}/cell)

T (after As_2O_3) h	Fluorescence intensity ($\bar{x} \pm s$)
Control	180. 3 ± 75. 7
2	206. 4 ± 93. 2
4	170. 2 ± 80. 3
6	168. 2 ± 72. 2
12	114. 4 ± 70. 3
24	90. 7 ± 85. 6

Discussion

Reports about As_2O_3 inducement of apoptosis of cancer cells have been seen frequently in hemopoietic stem cells and leukemia cells[54-60], but rarely in epithelial tumor cells[61-64]. We have tried to explore the possibility of curing esophageal carcinoma by using As_2O_3 treatment *in vitro*. The experimental results have shown that As_2O_3 can induce cancer cell apoptosis, large doses of As_2O_3 can even induce cell necrosis. Our previous works indicated that at the early – stage of cell apoptosis, morphological changes of themitochondria might be an important phenomenon in the course of esophageal carcinoma cell apoptosis induced by As_2O_3[29,31]. Our results showed that morphological and functional changes of mitochondria of SHEEC1 cells were induced by As_2O_3. It could divide into three stages. Two to four h after As_2O_3 administration, the mitochondria proliferated with a lot of new small mitochondria, distributing from the inner layer to the outer layer of cytoplasm. This was the early reaction of mitochondria of SHEEC1 cells to the effect of As_2O_3. 6 h after As_2O_3 inducement, many ridges on mitochondria were seen. The dense substances began to precipitate in the matrix of mitochondria and the condensed or damaged mitochondria were engulfed by lysosomes to form autophagosomesas seen in lymphocytes[65] Twelve hours after As_2O_3 inducement, the mitochondria were swelling, or vacuolation with mitochondria ridges decreased or disappeared. Twenty – four h after As_2O_3 inducement, apoptotic cells appeared with coagulating chromatin in nucleus and shrinking in the whole cell. The mitochondria swelled like a balloon. During the whole course of cell apoptosis, changes of mitochondria preceded the changes in nuclei.

The fluorescent intensity value detected by CFA and FCM reflects the function of mitochondria[66]. The change of Rho 123 fluorescent intensity under As_2O_3 treatment may be divided into 3 time phases: 2 – 4h after As_2O_3 inducement, mitochondria increased fluorescent intensity, but began to decline after 4 – 6 h and decreased to the lowest after 12 – 24 h. These functional changes of mitochondria were in accordance with mitochondrial morphological changes.

The functional changes of mitochondria may be accompanied with decreasing the formation of ATP, reducing the activity ofdehydrogenase[67], thus influencing cell respiration, cell metabolism, energy supply and even the cell death. If the mitochondrial release cytochrome Cor apoptotic inducement factors (AIF), they may activate the caspases enzyme system, which further act upon cell nucleus and cell keratinoprotein to induce irreversible apoptotic changes[68]. If the mitochondrial changes resulted in lowering of $\Delta\Psi m$, increase of oxygen free radical and blocking up the formation of ATP, the cells will be finally undergo necrosis, because they lose the ability of electron bond transmission. Therefore, the mitochondrial changes may induce cell apoptosis and also cell necrosis[69]. When the inducement factor is strong or highly concentrated it induces cell necrosis. If less in amount and strength, it may give times to activate the caspases enzyme system[70], the cell apoptosis will develop. Mitochondrial fluorescent probe Rho123 is a very useful tool, which may specifically conjugate with mitochondria to indicate cells living state or metabolic statet[25]. Detecting Rho123 fluorescence intensity of mitochondria may reveal mitochondrial quantity and function under different kinds of stimuli. The Rho123 fluorescence intensity is stronger in proliferative cells than in quiescent cells, and the intensity decreases in damaged mitochondria caused by harmful stimulit[48]. The amount of Rho123 conjugated with mitochondria differs in different types of cells and in different cell functionalstatus[66]. The mitochondrial changes of SHEEC1 cells induced by As_2O_3 occurred 2 – 4 h after drug adding. Under the same cultured conditions, mitochondria were supposed to be the firstly targeting site in the course of cell apoptosis. Therefore, under As_2O_3 inducement, the morphological and functional in mitochondria of SHEEC1 cells, which happened prior to cell nuclear DNA change, may be regarded as the important link in cell apoptosis.

[In 《World J Gastroenterol》 2002, 8 (1): 31 – 35]

References

1　He LJ, Wu M. The distribution of esophageal and cardiac carcinoma and precancerous of 2238. World J Gastroenterol, 1998, 4 (Suppl 2): 100

2　Yu GQ, Zhou Q, DING Ivan, Gao SS, Zheng ZY, Zou JX, Li YX, Wang LD. Changes of p53 protein blood level in esophageal cancer patients and normal subjects from a high incidence area in Henan, China. World J Gastroenterol, 1998, 4: 365 – 366

3　Gao SS, Zhou Q, Li YX, Bai YM, Zheng ZY. Zou JX, Liu G, Fan ZM, Qi YJ, Zhao X, Wang LD. Comparative studies on epitheliallesions at gastric cardia and pylori c antrum in subjects from a high incidence area for esophageal cancer in Henan, China. World. Gastroenterol, 1998, 4: 332 – 333

4　Wang LD, Zhou Q, Wei JP, Yang WC, Zhao X, Wang LX, Zou JX, Gao SS, Li YX, Yang CS. Apoptosis and its relationship with cell proliferation, p53, Waflp21, bcl – 2 and c – myc in esophageal carcinogenesis studied with a high risk population in northern China. World J Gastroenterol, 1998, 4: 287 – 293

5 Qiao GB, Han CL, Jiang RC, Sun CS, Wang Y, Wang YJ. Overexpression of P53 and its risk factors in esophageal cancer in urban areas of Xi'an. World J Gastroenterol, 1998, 4: 57–60

6 Jiao LH, Wang LD, Xing EP, Yang GY, Yang CS. Frequent inactivation of p16 and p15 expression in human esophageal squamous cellcarcinoma detected by RT PCR. World J Gastroenterol, 1998, 4 (Suppl 2): 105

7 Bai YM, Wang LD, Seril DN, Liao J, Yang GY, Yang CS. Expression of hMSH2 in human esophageal cancer from patients in a high incidence area in Henan, China. World J Gastroenterol, 1998, 4 (Suppl 2): 107

8 Qi YJ, Wang LD, Nie Y, Cai C, Yang GY, Xing EP, Yang CS. Alteration of p19 mRNA expression in esophageal cancer tissue from patients at high incidence area in northern China. World J Gastroentero, 1998, 4 (Suppl 2): 108

9 Zhang X, Geng M, Wang YJ, Cao YC. Expression of epidermal growth factor receptor and proliferating cell nuclear antigen in esophageal carcinoma and precancerous lesions. Huaren Xiaohua Zazhi, 1998, 6: 229–230

10 Wang D, Su CQ, Wang Y, Ye YK. Deletion of p16 gene at a highfrequency in esophageal carcinoma. Huaren Xiaohua Zazhi, 1998, 6: 1052–1053

11 Zou JX, Wang LD, SHI Stephanie T, Yang GY, Xue ZH, Gao SS, Li YX, YANG Chung S. p53 gene mutations in multifocal esophageal precancerous and cancerous lesions in patients with esophageal cancer in high risk northern China. ShifieHuaren Xiaohua Zazhi, 1999, 7: 280–284

12 Deng LY, Zhang YH, Xu P, Yang SM, Yuan. Expression ofinterleukin 1β converting enzyme in 5–FU induced apoptosis in esophageal carcinoma cells. World J Gastroenterol, 1999, 5: 50–52

13 Xiao ZF, Zhang Z, Wang Z, Zhang HZ, Wang M, Shi ML, Yin WB. Value of CT Scan on radiotherapy of esophageal cancinoma. HuarenXiaohua Zazhi, 1998, 6 (Suppl 7): 181–184

14 Fu JH, Rong TH, Huang ZF, Yang MT, Wu YL. Comparative assessment of three prosthesis types of palliative intubation for late stage esophageal carcinoma. Huaren Xiaohua Zazhi, 1998, 6: 984–986

15 Chen KN, Xu GW. Diagnosis and treatment of esophageal cancer. Shijie Huaren Xiaohua Zazhi, 2000, 8: 196–202

16 Wu XY, Zhang XF, Yin FS, Lu HS, Guan GX. Clinical study on surgical treatment of esophageal carcinoma in patients after subtotal gastrectomy. World J Gastroenterol, 1998, 4 (Suppl 2): 68–69

17 Gao ZD, Xu XY, Mao AW, Zhou XF, Jiang H. Combination of arterial infusion chemotherapy and radio therapy in the treatment of 36 cases of middle and late stageesophageal cancer. World J Gastroenterol, 1998, 4 (Suppl 2): 72

18 Guo WJ, Yu EX, Zheng SG, Shen ZZ, Luo JM, Wu GH, Xia SA. Study on the apoptosis and cell cycle arrest in human liver cancer SMMC7721 cells induced by Jianpili qi herbs. Shifie Huaren Xiaohua Zazhi, 2000, 8: 52–55

19 Tu SP, Jiang SH, Qiao MM, Cheng SD, Wang LF, Wu YL, Yuan YZ, Wu YX. Effect of trichosanthin on cytotoxicity and induction of apoptosisof multiple drugs resistence cells in gastric cancer. Shijie HuarenXiaohua Zazhi, 2000, 8: 150–152

20 Liang WJ, Huang ZY, Ding YQ, Zhang WD. Lovo cell line apoptosisinduced by cyclo heximide combined with TNF α. Shijie huarenXiaohua Zazhi, 1999, 7: 326–328

21 Lu XP, Li BJ, Chen SL, Lu B, Jiang NY. Effect of chemotherapy ortargeting chemotherapy on apoptosis of colorectal carcinoma. ShijieHuaren Xaiohua Zazhi, 1999, 7: 332–334

22 Shen YF, Zhuang H, Shen JW, Chen SB. Cell apoptosis and neoplasms. Shijie Huaren Xiaohua Zazhi, 1999, 7: 267–268

23 Sun ZX, Ma QW, Zhao TD, Wei YL, Wang GS, Li JS. Apoptosisinduced by norcantharidin in human tumor cells. World J Gastroenterol, 2000,

6: 263 – 265

24　Zhu HZ, Ruan YB, Wu ZB, Zhang CM. Kupffer cell and apoptosis inexperimental HCC. World J Gastroenterol, 2000, 6: 405 – 407

25　Evan G and Littlewood T. A matter of life and cell Death. Science, 1998, 281: 1317 – 1320

26　Shen ZY, Tan LJ, Cai WJ, Shen J, Chen C, Tang XM, Zheng MH. Arsenic trioxide induces apoptosis of oesophageal carcinoma in vitro. Intern J Mol Med, 1999, 4: 33 – 37

27　Shen ZY, Tan LJ, Cai WJ, Shen J, Chen CY, Tang XM. Morphologicstudy on apoptosis of esophageal carcinoma cell line induced byarsenic trioxide. Shijie Huaren Xiaohua Zhahi, 1998, 6: (Suppl 7) 226 – 229

28　Shen J, Wu MH, Cai WJ, Shen ZY. The effects of arsenite trioxide invarious concentration on the esophageal carcinoma cell line. ZhongguoZhongliu Shengwu Zhiliao Zazhi, 2001, 8: 106 – 109

29　Shen ZY, Shen J, Cai WJ, Hong C, Zheng MH. The alteration ofmitochondria is an early event of arsenic trioxide induced apoptosisin esophageal carcinoma cells. Intern J Mol Med, 2000, 5: 155 – 158

30　Shen ZY, Shen WY, Chen MH, Hong CG, Shen J. Alterations of nitricoxide in apoptosis of esophageal carcinoma cells induced by arsenite. Shijie Huaren Xiaohua Zhahi, 2000, 8: 1101 – 1104

31　Shen ZY, Shen J, Chen MH, Li QS, Hong CQ. Morphological changes of mitochondria in apoptosis of esophageal carcinoma cells inducedby As_2O_3. Zhonghua Binglixue Zazhi, 2000, 29: 200 – 203

32　Deng LY, Zhang YH, Zhang HX, Ma CL, Chen ZG. Observation of morphological changes and cytoplasmic movement in apoptosisprocess. World J Gastroenterol, 1998, 4: 66 – 67

33　Floryk D, Ucker DS. Molecular mapping of the physiological celldeath process. Mitochondrial events may be disordered. Ann N YAcad Sci, 2000, 926: 142 – 148

34　Shen ZY, Chen CY, Shen J, Cai WJ. Ultrastructural study of apoptosisand necrosis in the esopha-

geal carcinoma cell line induced by arsenictrioxide. Zhongguo Yixue Wulixue Zazhi, 1999, 16: 91 – 94

35　Shen ZY, Chen MH, Li QS, Shen J. An ultrastructural study on theprogrammed cell death of human amniotic epithelium. Dianzi XianweiXuebao, 2000, 19: 259 – 260

36　Zhang CS, Wang WL, Peng WD, Hu PZ, Chai YB, Ma FC. Promotion of apoptosis of SMMC7721 cells by bcl – 2 ribozyme. Shijie HuarenXiaohua Zazhi, 2000, 8: 417 – 441

37　Yuan RW, Ding Q, Jiang HY, Qin XF, Zou SQ, Xia SS. Bcl – 2, p53protein expression and apoptosis in pancreatic cancer. Shijie HuarenXiaohua Zazhi, 1999, 7: 851 – 854

38　Wang LD, Zhou Q, Wei JP, Wang WC, Zhao X, Wang LX, Zou X, Gao SS, Li YX, Yang CS. Apoptosis and its relationship with cellproliferation, p53, waflp21, bcl – 2 and c – myc in esophageal carcinogenesis studied with a high risk population in northern China. WorldJ Gastroenterol, 1998, 4: 287 – 293

39　Pastorino JG, Chen ST, Tafani M, Snyder JW and Farber JL. Theoverexpression of Bax produces cell death upon induction of the mito – chondrial permeability transition. J Biol Chem, 1998, 273: 7770 – 7775

40　Fang M, Zhang HQ, Xue SB, Li N, Wang L. Intracellular calcium distribution in apoptosis of HL – 60 cells induced by harringtonine: intranuclearaccumulation and regionalization. Cancer Letters, 1998, 127: 113 – 121

41　Mootha VK, Wei MC, Buttle KF, Scorrano L, Panoutsakopoulou V, Mannella CA, Korsmeyer SJ. A reversible component of mitochondrial respiratory dysfunction in apoptosis can be rescued by exogenous cytochrome c. EMBO J, 2001, 20: 661 – 671

42　Zimmermann KC, Waterhouse NJ, Goldstein JC, Schuler M, GreenDR. Aspirin induces apoptosis through release of cytochrome c frommitochondria. Neoplasia, 2000, 2: 505 – 513

43　Li H, Kolluri SK, Gu J, Dawson MI, Cao XH,

Hobbs PD, Lin BZ, ChenGQ, Lu JS, Lin F, Xie ZH, Fontana JA, Reed JC, Zhang XK. Cyto – chrome c release and apoptosis induced by mito- chondrial targeting of nuclear orphan receptor TR3. Science, 2000, 289: 1159 – 1164

44 Brenner C and Kroemer G. Apoptosis Mitochon- dria – the death signalintergrators. Science, 2000, 289: 1150 – 1151

45 Heerdt BG, Houston MA, Anthony GM, Augenli- cht LH. Mitochondrial membrane potential ($\Delta\Psi$mt) in the cordination of p53 – inde – pen- dent proliferationand apoptosis pathways in human colonic carcinoma cells. Cancer Res, 1998, 58: 2869 – 2875

46 Wakabayashi T, Karbowski M. Structural changes of mitochondriarelated to apoptosis. Biol Signals Recept, 2001, 10: 26 – 56

47 Hail N J, Lotan R. Mitochondrial permeability transition is a centralcoordinating event in N – (4 – hydroxyphenyl) retinamide – inducedapopto- sis. Cancer Epidemiol Biomarkers Prey, 2000, 9: 1293 – 1301

48 Shapiro HM. Membrane potential estimation by flow cytometer. Methods, 2000, 21: 271 – 279

49 Buckman JF, Reynolds IJ. Spontaneous changes in mitochondrial membrane potential in cultured neurons. J Neurosci, 2001, 21: 5054 – 5065

50 Shen ZY, Cen S, Shen J, Cai WJ, Xu JJ, Ten ZP, Hu Z, Zeng Y. Studyof immortaliztion and malignant transformation of human embryonic esoph- ageal epithelial cells induced by HPV18E6ET. J Cancer ResClin Oncol, 2000, 126: 589 – 594

51 Canete M, Juarranz A, Lopez – Nieva P, Alonso – Torcal C, VillanuevaA, Stockert JC. Fixation and permanent mounting of fluorescent probesafter vi- tal labeling of cultured cells. Acta Histochem, 2001, 103: 117 – 126

52 Follstad BD, Wang DI, Stephanopoulos G. Mito- chondrial membrane potential differentiates cells resistant to apoptosis in hybridomacultures. Eur J Biochem, 2000, 267: 6534 – 6540

53 Bedner E, Li X, Gorczyca W, Melamed MR, Darzynkiewicz Z. Analysis of apoptosis by laser scanning cytometry. Cytometry, 1999, 35: 181 – 195

54 Li YM, Broome JD. Arsenic targets tubulins to induce apoptosis inmyeloid leukemia cells. Canc- er Res, 1999, 59: 776 – 780

55 Bazarbachi A, EI – Sabban ME, Nasr R, Quignon F, Awaraji C, KersualJ, Dianoux L, Zermati Y, Haidar JH, Hermine O, de – The H. Arsenictrioxde and interferon – alpha synergize to induce cell cycle arrest andapoptosis in human T – cell lymphotropic virus type I – transformedcells. Blood, 1999, 93: 278 – 283

56 Jing Y, Dai J, Chalmers – Redman RM, Tatton WG, Waxman S. Arsenictrioxide selectively in- duces acute promyelocytic leukemia cell apopto- sisvia a hydrogen peroxide – dependent pathway. Blood, 1999, 94: 2102 – 2111

57 Lallemand – Breitenbach V, Guillemin MC, Janin A, Daniel MT, DegosL, Kogan SC, Bishop JM, de – The H. Retinoic acid and arsenic synergizeto eradicate leukemic cells in a mouse model of acute promyelocyticleukemia. J Exp Med, 1999, 189: 1043 – 1052

58 Rousselot P, Labaume S, Marolleau JP, Larghero J, Noguera MH, Brouet JC, Fermand JP. Arse- nic trioxide and melarsoprol induceapoptosis in plasma cell lines and in plasma cells from myelo- mapatients. Cancer Res, 1999, 59: 1041 – 1048

59 Huang XJ, Wiernik PH, Klein RS, Gallagher RE. Arsenic trioxideinduces apoptosis of myeloid leukemia cells by activation of caspases. Med On- col, 1999, 16: 58 – 64

60 Zhu XH, Shen YL, Jing YK, Cai X, Jia PM, Huang Y, Tang W, ShiGY, Sun YP, Dai J Wang ZY, Chen SJ, Zhang TD, Waxman S, ChenZ, Chen GQ. Apoptosis and growth inhibition in ma- lignant lympho – cytes after treatment with arsenic trioxide at clinically achievable concentrations. J Nat Cancer Inst, 1999, 91: 772 – 778

61 Chen HY, Liu WH, Qin SK. Induction of arsenic trioxide on apoptosis ofhepatocarcinoma cell lines. Shijie Huaren Xiaohua Zazhi, 2000, 8: 532 –

535

62 Gu QL, Li NL, Zhu ZG, Yin HR, Lin YZ. A study on arsenic trioxide inducing in vitro apoptosis of gastric cancer cell lines. World J Gastroenteral, 2000, 6: 435 – 437

63 Tu SP, Jiang SH, Tan JH, Jiang XH, Qiao MM, Zhang YP, Wu YL, Wu YX. Proliferation inhibition and apoptosis induction by arsenic trioxide on gastric cancer cell SGC 7901. Shijie Huaren Xiaohua Zazhi, 1999, 7: 18 – 21

64 Tan L, Chen X, Shen ZY. Study on the proliferative inhibition of human esophageal cancer cells with treatment DMSO and As$_2$O$_3$. Shanghai Di – er Yike Daxue Xuebao, 1999, 19: 5 – 8

65 Huo X, Piao YJ, Huang XX, Quao DF. Ultrastructural observation of mitochondria in apoptotic lymphocytes induced with cycloheximide. Dianzi Xianwei Xuebao, 1998, 17: 702 – 705

66 Hu Y, Moraes CT, Savaraj N, Priebe W, Lampidis TJ. Rho (0) tumor cells: a model for studying whether mitochondria are targets for rhodamine 123, doxorubicin, and other drugs. Biochem Pharmacol, 2000, 60: 1897 – 1905

67 Green DR, Reed JC, Mitochondria and apoptosis. Science, 1998, 281: 1309 – 1312

68 Sugrue MM, Tatton WG. Mitochondrial membrane potential in aging cells. Biol Signals Recept, 2001, 10: 176 – 188

69 Lee HC, Yin PH, Lu CY, Chi CW, Wei YH. Increase of mitochondria and mitochondrial DNA in response to oxidative stress in human cells. Biochem J, 2000, 348: 425 – 432

70 Seol JG, Park WH, Kim ES, Jung CW, Hyun JM, Lee YY, Kim BK. Potential role of caspase – 3 and – 9 in arsenic trioxide – mediated apoptosis in PCI – 1 head and neck cancer cells. Int J Oncol, 2001, 18: 249 – 255

81. Nitric Oxide and Calcium Ions in Apoptotic Esophageal Carcinoma Cells Induced by Arsenite

SHEN Zhong – ying[1], SHEN Wen – ying[1], CHEN Ming – hua[1], SHEN Jian[1], CAI Wei – jie[1], ZENG Yi[3], SHEN Wen – ying[2]

1. Department of Pathology, Department of Chemistry; 2. Medical College of Shantou University; 3. Institute of Virology, Chinese Academy of Preventive Medicine

Summary

Objective To quantitatively analyze the nitric okide (No) and Ca^{2+} in apoptosis of esophageal carcinoma cells induced by arsenic trioxide (As$_2$O$_3$).

Methods The cell line SHEEC1, A malignant esophageal epithelial cell induced by HPV in synergy with TPA in our laboratory, was cultured in a serum – free medium and treated with As$_2$O$_3$. Before and after administration of As$_2$O$_3$, NO production in cultured medium was detected quantitatively using the Griess Colorimetric method. Intracellular Ca^{2+} was labeled using the fluorescent dye Fluo3 – AM and detected under confocal laser seanning microscope (CLSM). Which was able to acquire data in real – time enabling Ca^{2+} dynamics of individual cells *in vitro*. The apoptotic cells were examined under electron microscopy.

Results Intracellular concentration of Ca^{2+} increased from 1. 00 units to 1. 09 – 1. 38 units of fluorescent intensity at As_2O_3 treatment and NO products subseuqently released from As_2O_3 – treated cells increased from 0. 98 – 1 . 00 × 10^{-2} mol · L^{-1} up to 1. 48 – 1. 52 × 10^{-2} mol · L^{-1} and maintained in a high level continuously. Finally apoptosis of cells occurred, chromatin being agglutinated, cells shrunk, nuclei became round and mitochondria swelled.

Conglusion Ca^{2+} and NO increased with cell damage and apoptosls in cells treated by As_2O_3. The Ca^{2+} is an initial messenger to the apoptotic pathway. To investigate Ca^{2+} and NO will be a new direction for stuclying the apoptotic signaling messenger of the esophageal carcinoma cells induced by As_2O_3.

Introduction

Arsenic trioxide (As_2O_3) has been proved to be a genotoxic and a carcinogenic agent[1-6]. Previous studies also showed that As_2O_3 induced cellular apoptosis in leukemia[7-15], in cancer cells of head and neck[16] and other cancer cells[17-22], Sb. As_2O_3 has antitumoral effect. We found that As_2O_3 induced apptosis in esophageal squamous carcinoma cells[23]. The pathomorphological changes induced by As_2O_3 revealed that cells became smaller and shrank, nucleus rounded up, chromatin agglutinated and marginated, the nuclear membrane broke down followed by degenerative changes and cell mortality. All these changes indicated typical morphological changes of apoptosis[24,25]. Mitochondria, an important celluar apparatus, is related to cell breathing, oxygen metabolism, enzymeactivity and energy supply. Our data demonstrated that the primary target of As_2O_3 inducing apoptosis of esophageal carcinoma cells might be the mitochondria[26]. It is likely that As_2O_3 is a mitochondriotoxic agent[27, 28]. At the early stage of cellular apoptosis induced by As_2O_3, the mitochondria generated morphological and functional changes[29.30].

NO exerts a wide range of its biological properties via its interaction with mitochondria and NO mediated mitochondria damage[31]. In our previous data, an increase level of nitrite, a stable product of NO, was detected in the culture medium of esophageal carcinoma cells in arsenite – treated apoptosis[32]. Calciumions (Ca^{2+}) act as a universal second messenger in a variety of cells. Numerousfunctions of all types of ceils are regulated by Ca^{2+} to a greater or lesser degree. Because of the importance of Ca^{2+} in biology, numerous methods of analyzing cellular Ca^{2+} activity have been established. Confocal laser scanning microscopy (CLSM) allows the precise spatial and temporal analysis of intracellular Ca^{2+} activity at the subcellular level. This optical technique has enabled scientists to document the dynamic changes of intracellular Ca^{2+} *in vitro*[33].

Arsenic may generate reactive oxygen species to exert its toxicity, which is implicated in DNA damage, signal transduction and apoptosis. What we are interested in is to see if NO and Ca^{2+} are involved in arsenic – induced apoptosis and to observe the changes of its target organeUe – mitochondria. This study is to investigate the original messengers that initiate apoptosis and to detect quantitatively Ca^{2+} and NO in the apoptotic process of esophageal carcinoma cell line induced by As_2O_3.

Materials and Methods

1. Cell line generation and cell culture

The esophageal carcinoma cell line (SHEFC1) was a malignantly transformed cell line of human embryonic esophageal epithelium induced by HPV18 E6E7 in synergy with TPA (12 – 0 – tetradecanoyl – phorbol – 13 – acetate)[34]. Calls were cultured in 50 ml flasks and 24 – well plate (Coming) With serum – free meditum. The culture medium contained the basal medium (MCDB151) with trace elements (M – 6645 Sigma) and added transferrin, hydrocorticosone, epidermal growth factor (EGF), insulin (Sigma Chemical Co.) and extracts of bovine hypophysis (Gibco, BRL), but without calf serum, nitrite and nitrate, while containing streptomycin and penicilline (50 mg $\cdot$ L^{-1} for each).

2. The administration of arsenic

Arsenic trioxide (As$_2$O$_3$) obtained from Signa Chemieal Co, (st. Louis MO, Lot A 1010), at concentration of 0, 1. 3 and 5 μmol $\cdot$ L^{-1} was added into the culture flask and 24 well platese, respectively. for 0, 2, 4, 8, 12 and 24 h. The experiment was repeated once.

3. Transmission electron – microcopy (EM) examination

At the endpoints of As$_2$O$_3$ (24h), cells of each group were digested with 0. 25% trypsin, centrifuged, fixed with 2. 5% glutaraldehyde, and routinely prepared for electron microscopic examination. The samples were observed under transmission electronmicroscope (Hitachi 300).

4. Cell cycle and apoptotic rate analyzed by flow cytometry (FCM)

Cells of repeated experiment were harvested to measure the ratio of apoptotic cells to survived cells. The cells were washed twice with PBS, dispersed, and filtered through 360 mesh nylon net to make a single cell suspension. It was fixed with 700 ml $\cdot$ L^{-1} precooled alcohol in ice. Before analysis, cells were suspended in PBS and stained with propidium iodide. Cells of $1 \times 10^9 \cdot$ L^{-1} were detected by flow cytometry (FACsort, B – D Co. , USA). The DNA histogram was drawn according to the fluorescent intensity value of 10^4 cells.

5. Procedure of NO detection[35]

The nitrite/nitrate colorimetfic method, using the kit purchased from Boehfinger Mannheirn Co, was used to detect NO in culture medium. The culture medium of 0. 2 ml from flask was regularly deactivated at 80℃ for 5 min and deproteinated by centrifugation in 12 000 r $\cdot$ min^{-1} for 30 min, and the supematant was determined. The procedure for NO determination was as follows: Sample solution of 100 μl, 50 μl of nicotinamide adenine dinucleotide phosphate (NADPH) and 4 μl ofthe enzyme nitrate reductase (NR), were placed into a 3 ml test robe mixed, incubated for 30 min at room temperature, and added 50 μl color reagent I & Ⅱ, respectively, mixed, and allowed to stand in thedark at room temperature for 10 to 15 min. The NO content of the samples and the blank was estimated with Shimadzu UV/120 spectrophotometry by λ450 nm and was calculated by calibration curve of standard addition method. The standards were prepared from known amounts of stock NO$_3$ and NO$_2$ and run in parallel with test samples in each assay.

6. Determination of intracellular calcium level using CLSM[33, 36]

The cells were cultured on the coverslips within the glass bottom of a small cultured dish (No. 0, uncoated, and irradiaed, MatTek Co. , USA). At the exponential growth period, the cells were stained with 10 μmol $\cdot$ L^{-1} flue – 3/AM (Molecular Probe) for 30 min at 37℃, and washed with (135 mol $\cdot$ L^{-1}NaCl, 10 mol $\cdot$ L^{-1}HEPES, 0.4 mol $\cdot$ L^{-1} MgCl$_2$, 1 mol $\cdot$ L^{-1} CaCl$_2$, 1 g $\cdot$ L^{-1} D – glucose, 1 g $\cdot$ L^{-1} bovine serum albumin, pH 7.3 at least 3 times. Then the cells were placed in the culture medium 199 to maintain them in living state. Before and after administration of As$_2$O$_3$ the fluorescence intensity was determined by CLSM in dynamic changes for up to 900 s. Using scan – time series menu, time series was used to scan some definite cells repeatedly to monitor the dynamic changes in fluorescent intensity of intracellular Ca^{2+} content over time. The parameters of the CLSM (Ultima 312, Meridian Instalments Inc. , USA) were as follows: the excited light 488 nm, the emission light 530 nm and pinhole 10 – 40 nm. The fluorescent intensity of pixel was collected and managed with the software of the instrument.

Results

1. Cell apoptosis

Ultrastructural morphological changes of mitochondda in As$_2$O$_3$ treated cells were described in the previous reports[25,26]. Cells treated with As$_2$O$_3$ at different concentrations for 24h displayed an apoptotic appearance. Under electron microscope, condensed and marginated chromatins in most of the nuclei appeared accompanying swelling mitochondria (Fig. 1). By flow cytometry, time course study on As$_2$O$_3$ induced apoptosis revealed that apoptotic peak can be observed as early as 12h, after the incubation of arsenic trioxide in 3 μmol $\cdot$ L^{-1}. The apoptotic cells accounted for 5.0 % of total cell population at 12h and 28.3% at 24h (Fig. 2).

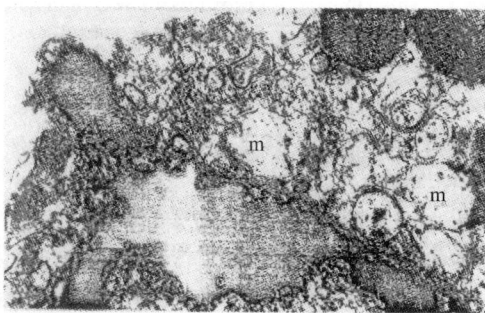

m, mitochondria, n, nucleus. EM × 7000

Fig. 1   Ultrastructure of SHEECl cell treated with 3 μmol $\cdot$ L^{-1} As$_2$O$_3$. Apoptotic appearance displayed with swelling of mitochondria and nuclear chromatin coagulating and mergination

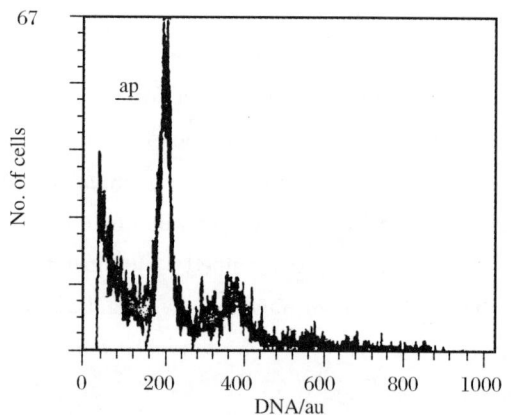

Fig. 2   DNA histogram of SHEEC1 cell in 24 h after adding 3 μmol $\cdot$ L^{-1} of As$_2$O$_3$. ap, apoptotic peak

2. NO determination

When As_2O_3 acted on the SHEFC1 for 2 – 24h, in 0, 1, 3, and 5 $\mu mol \cdot L^{-1}$ As_2O_3, NO in cultured medium was increased at the time points. The amount of NO released from SHEEC1 was increased from the basal condition ($0.98 - 1.00 \times 10^{-2} \mu mol \cdot L^{-1}$) up to the high level ($1.48 - 1.52 \times 10^{-2} \mu mol \cdot L^{-1}$) (8h) and maintained for 24h (Fig. 3). The concentration of NO in different groups varied, high concentration of NO in 5 $\mu mol \cdot L^{-1}$ of As_2O_3 and low concentration of NO in 1 $\mu mol \cdot L^{-1}$ of As_2O_3.

3. Dynamic change calcium of intracellular calcium

To show the time course of changes in Ca^{2+} in individual cells, the changes in fluorescence intensity (arbitrary unit, au) at different representative cells were measured. Upon the initiation of

Fig. 3　NO determination of SHEECl treated with different concentrations of As_2O_3. NO increased markedly in 5 $\mu mol \cdot L^{-1}$ of As_2O_3 group (A), intermediately in 3 $\mu mol \cdot L^{-1}$ of As_2O_3 group (B) and lowly in 1 $\mu mol \cdot L^{-1}$ of As_2O_3 group (C). The control group, 0 $\mu mol \cdot L^{-1}$ of As_2O_3, was remained on the basal lines (D).

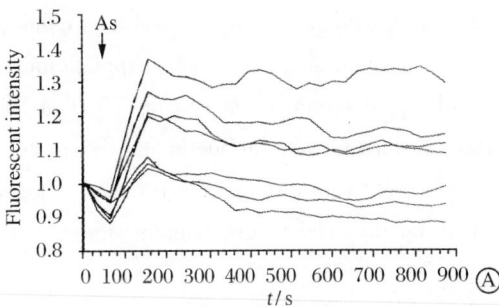

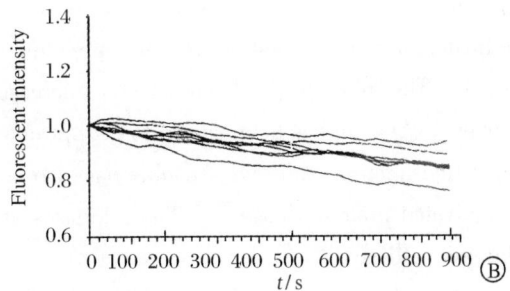

stimulation by As_2O_3, all the cells responded with a rapid rise in $[Ca^{2+}]$ from 1.00 au to 1.09 – 1.38 au of fluorescent intensity. The peak levels of Ca^{2+} in all cells were consistently reached at about 900s after stimulation (Fig. 4A). In the control group, without being treated with As_2O_3, the fluorescent intensity of cells were remained on the basal line (Fig. 4B).

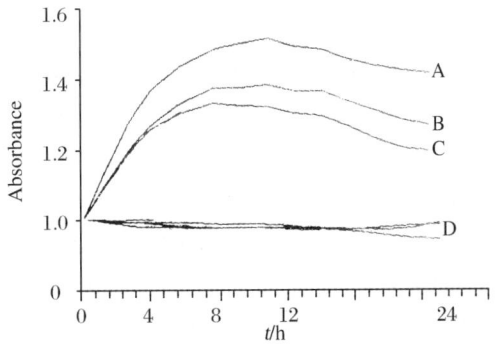

Fig. 4　Dynamic changes of intracellular calcium in 7 cells of SHEEC1 treated with As_2O_3. A, SHEEC1 cells treated with As_2O_3 in 3 $\mu mol \cdot L^{-1}$. B, Control group without adding As_2O_3

Discussion

In general, the process of cell apoptosis involved three phases: the initiation phase, the effector phase and the degradation phase[37]. The initiation (or signal transduction) phase is the stage in which specific or non – specific pro – apoptotic signal transduction pathways are activated. The effecter (or central control) phase mainly occurs in the mitochondria[38] where mitochondria membranes are unstable as a result of the action of the permeability alternation. Some genes such as p53 and bcl – 2, activate in this phase[39–41]. The degradation (or morphological and biochemical changes) phase manifest the postmitochondrial features of apoptosis, in which soluble intermembrane proteins released from mitochondria played an active role in the activation of proteolytic destruction. In our previous reports, we investigated the early changes of the apoptotic cells induced by As_2O_3 and defined the phase in which As_2O_3 was involved[26, 27]. Our results demonstrated that As_2O_3 acted directly on mitochondria for the early stage of apoptosis. The alteration of mitochondria in arsenic trioxide treated tumor cells could be observed as early as 2 h after the treatment[27,30]. In this study we investigated signal messengers of apoptosis, by first selecting both messengers of NO and Ca^{2+} in the apoptotic pathway.

Experiments on the effects of various moduators (dose and time lag) of arsenic in the level of Ca^{2+} and NO were carried out. Nitric oxide (NO) is a free radical generated in cells by nitric oxide synthases (NOS)[42]. It is a gaseous inter – and intra – cellular messenger that plays as a signaling molecular in many physiological and pathological processes and it is also a cytotoxic agent involved in many diseases, which has been elaborated extensively during the last decade. Various intra – or extra – cellular facors act on mitochondria to produce NO. NO binds to cytochrome oxidase[43], blocks respiratory chain and induces apoptosis[44, 45].

Cells themselves control intracellular Ca^{2+} concentration ($[Ca^{2+}]$) strictly with several Ca^{2+} regulatory mechanisms, such as Ca^{2+} channels, Ca^{2+} pumps. and Ca^{2+} exchangers. The role of calcium is as the important intracellular signal element in regulating cell death[46]. As revealed by previous reports, it seems that calcium changes in apoptosis vary with stimuli and cell lines[47]. This data suggested that an early, gradual and sustained increase in intracellular Ca^{2+} is necessay for the appearance of apoptotic characteristics. In the examination of CLSM with Fuo – 3. AM as a calcium indicator, we found that a rise in intracellular calcium was elicited at once after application of As_2O_3. The mechanism of how arsenic increase esintracellular calcium levels was not clear at this moment. Arsenic has been shown to disrupt mitochondria and may elevate intracellular calcium via a sigal transduction pathway. Arsenite has also been reported to activate protein kinase C and mitogen – activated protein kinase[48]. These kinases are known to be involved in the calcium signal transduction pathway.

According to the previous reports, the relationship between NO. Ca^{2+} and mitochondria in apoptosis is as follows: various extracellular factors can induce the increase of intracellular Ca^{2+} levels ($[Ca^{2+}]$ i), modulating cellular signaling and gene expression, and the increased ($[Ca^{2+}]$ i) effect on NO production through the iNOS pathway[49,50]; mitochondria are a source of NO[51], the production of which may affect energy metabolism, O_2 consumption and O_2 free radical formation[52] mitochondrial Ca^{2+} uptake in combination with NO production triggers the collapse of mitochondrial membrane potential, affecting rnitochondrial respiration and culminating in delayed cell death[53].

In conclusion, our data proved that increased calcium ions and nitric oxide triggered by As_2O_3

may play an important role in arsenite – induced apoptosis in esophageal carcinoma cells. The demonstration of the involvement of Ca^{2+} and NO in arsenite – induced apoptosis suggests a new direction for studying the apoptotic pathway.

［In《World J Gastroenterol》2002, 8（1）：40 – 43］

References

1 Matsui M, Nishigo C, Tpyokuni S, Takada J, Akaboshi M, Ishikawa M, Imamura S, Miyachi Y. The role of oxidative DNA damage in humanarsenic carcinogenesis: detection of 8 – hydroxy – 2' – deoxyguanosine inarsenic – related Bowen's disease. J Invest Dermatol, 1999, 113: 26 – 31

2 Goering PL, Aposhian HV, Mass MJ, Cebrian M, Beck BD, WaalkesMP. The enigam of arsenic carcinogenesis role of metabolism. Toxicol Sci, 1999, 49: 5 – 14

3 Schaumloffel N, Gebel T, Heterogenicity of the DNA damage provoked byantimony and arsenic. Mutagenesis, 1998, 13: 281 – 286

4 Ho IC, Yih LH, Kao CY, Lee TC. Tm – protoporphyrin potentiates arsenite – induced DNA stromatid breaks and kinetochore – negative micronuelei in hunman fibroblasts. Mutat Res, 2000, 452: 41 – 50

5 Gebel T. Suppression of arsenic – induced chromosome mutagenicity by antimony. Mutat Res, 1998, 412: 213 – 218

6 Gebel T, Birkenkamp P, Luthin S, Dunkelberg H. Arsenic（Ⅲ）, but not antimony（Ⅲ）, induced DNA – protein crosslinks. Anticancer Res, 1998, 18; 4253, 4257

7 Zhu XH, Shen YL, JingO, Cai X, Jia PM, HuangY, Tang W, Shi GY, Sun YP, Dai J, Wang ZY, Chen SJ, Zhang TD, Waxman S, Chen Z, Chen GO. Apoptosis and growth inhibition in malignant lymphocytes after treatment with arsenic trioxide at clinically achievable concentrations. J Nail Cancer Inst, 1999, 91: 772 – 778

8 Look AT. Arsenic and – apoptosis in the treatment of acute promyelocytic leukemia. J Natl Cancer Inst, 1998, 90: 86 – 88

9 Shao W, Fanelli M. Ferrara F. F, Riccioni R, Rosenauer A, Davison K, Lamph WW, Waxman S, Pelicci PG, Lo – Coco F, Avvisati G, Testa U, Peschle C. Gambacorti – Passerini C, Nervi C, Miller WH. Arsenic trioxide as an irtducer of apoptosis and loss of PML/RAR alpha protein in acute promyelocytic leukemia ceils. J Natl Cancer Inst, 1998, 90: 124 – 133

10 Soignet SL, Maslak P. Wang ZG, Jhanwar S, Calleja E, Dardashti LJ, Corso D, DeBlasio A, Gabrilove J, Scheinberg DA, Pandolfi PP, Warrell RP. Complete remission after treatment of acute promyelocytic leukemia with arsenic trioxide. N Engl J Mesd, 1998, 339: 1341 – 1348

11 Tamm I, Paternostro G, Zapata JM. Treatment of acute promyelocytic leukemia with arsenic trioxide. N Engl J Med, 1999, 340: 1043 – 1045

12 Jing Y, Dai J, Chalmers – Redman RM, Tatton WG, Waxman S. Arsenic trioxide selectively induces acute promyelocytic leukemia cell apoptosis via a hydrogen peroxide – dependent Pathway. Blood, 1999, 94: 2102 – 2111

13 Huang XJ, Wiernik PH, Klein RS, Gallagher RE. Arsenic trioxide induces apoptosis of myeloid leukemia cells by activation of caspases. Med Oncol, 1999, 16: 58 – 64

14 Lallemand – Breitenbach V, Guillemin MC, Janin A, Daniel MT, Degos L, Kogan SC, Bishop JM, de The H. Retinoic acid and arsenic synergize to eradicate leukemic ceils in a mouse model of acute promyelocytic leukemia. J Exp Med, 1999, 189: 1043 – 1052

15 Li YM, Broome JD. Arsenic targets tubulins to induce apoptosis in myeloid leukemia cells. Cancer Res, 1999, 59: 776 – 780

16 Seol JG, Park WH. Kim ES, Jang CW, Hyun

JM, Lee YY, Kim BK. Potential role of caspase – 3 and – 9 in arsenic trioxide – mediated apoptosis in PCI – 1 head arid neck cancer ceils, Int J Oncol, 2001, 18: 249 – 255

17　Chen HY, Liu WH, Qin SK. Induction of arsenic trioxide on apoptosis of hepatocarcinoma cell lines. Shijie Huaren Xiaohua Zazhi, 2000, 8: 532 – 535

18　Gu QL, Li NL, Zhu ZG, Yin HR, Lin YZ. A study on arsenic trioxide inducing in vitro apoptosis of gastric cancer cell lines. World J Gastroenrerol, 2000, 6: 435 – 437

19　Tu SP, Jiang SH, Tan JH, Jiang XH, Qiao MM, Zhang YP, Wu L, Wu YX. Proliferation inhibition and apoptosis induction by arsenic trioxide on gastric cancer cell SG – C 7901. Shijie Huaren Xiaohua Zazhi, 1999, 7: 18 – 21

20　Xu HY, Yang YL. Gao YY, Wu QL, Gao GQ. Effect of arsenic trioxide on human hepatoma cell line BEL – 7402 cultured in vitro. World J Gastroen terol, 2000, 6: 681 – 687

21　Wang W, Qin SK. ChenBA, Chen HY. Experimental study on antitumor effect of arsenic trioxide in combination with cisplatin or doxorubicin on hepatocellular carcinoma. World J Gastroenterol, 2001, 7: 702 – 705

22　Xu HY, Gao YY, Wu QL, Gao GQ, Yang YL, Chen SX, Liu TF. Proliferation inhibition arid apoptosis induction by arsenic trioxide on human hepatoma cell line in vitro. Shijie Huaren Xiaohua Zazhi, 2000, 8: 1233 – 1237

23　Shen ZY, Tan LJ, Cai WJ, Shen J, Chen C, Tang XM, Zheng MH. Arsenic trioxide induces apoptosis of oesophageal carcinoma in vitro. Int J Mol Med, 1999, 4: 33 – 37

24　Shen ZY, Tan LJ, Cai WJ, Shen J, Chen CY, Tang XM. Morphologic study on apoptosis of esophageal carcinoma cell line induced by arsenic trioxide, Shijie Huaren Xiaohua Zhahi, 1998, 6: 226 – 229

25　Shen ZY, Chen CY, Shen J, Cai WJ. Ultrastretural study of apoptosis and necrosis in the esophageal cearcinoma cell line induced by arsenictrioxide. Zhongguo Yixue Wulixue Zazhi, 1999, 16: 91 – 94

26　Shen ZY, Shen J, Chen MH, Li QS, Hong CQ. Morphological changes of mitochondria in apoptosis of esophageal carcinoma cells induced by As_2O_3. Zhonghua Binlixue Zazhi, 2000, 29: 200 – 203

27　Shen ZY, Shen J, Cai WJ, Hong CQ, Zheng MH. The alteration of mitochondria is an eatly event of arsenic trioxide induced apoptosis in esophageal carcinoma cells. Int J Mol Med, 2000, 5: 155 – 158

28　Kroemer G, de The H. Arsenic trioxide, a novel mitochondriotoxic anticancer agent? J Natl Cancer Inst, 1999, 91: 743 – 774

29　Shen ZY, Chen MH, Li QS, Shen J, An ultrastructural study study on the programmed cell death of human amniotie epithelium. Dianzi Xianwei Xuebao, 2000, 19: 259 – 260

30　Shen ZY, Shen J, Li QS, Chert CY, Chen JY, Zeng Y. Morphological and functional Changes of mitochondria in apoptotic esophageal carcinoma cells induced by arsenic trioxide. World J Gastroenterol, 2002, 8: 31 – 35

31　Rachmilewitz D. Role of nitric oxide in gastrointestinal tract. World J Gastroenterol, 1998, 4 (Suppl 2), 28 – 29

32　Shen ZY, Shen WY, Chen MH, Hong CQ, Shen J. Alterations of nitric oxide in apoptosis of esophageal carcinoma cells induced by arsenite. Shijie Huaren Xiaohua Zhahi, 2000, 8: 1101 – 1104

33　Takahashi A, Camacho P, Lechleiter JD, Herman B. Measurement of intracellular calcium. Physiol Rev, 1999, 79: 1089 – 1125

34　Shen ZY, Cen S, Shen J, Cai WJ, Xu JJ, Teng ZP, Hu Z. Zeng Y. Study of immortalization and malignant transformation of human embryonic esophageal epithelial cells induced by HPV18 E6E7. J Cancer Res Clin Oncol, 2000, 126: 589 – 594

35　Shen WY. Chen MH, Shen ZY, Zhang LM, Microspectrophotometric determination of nitric oxide. J Shantou Univ Med College, 2000, 13: 10 – 11

36 Satoh Y, Nishimura T, Kimura K, Mori S, Saino T. Application of real – time confocal microscopy for observation of living cells in tissue specimens. Hum Cell, 1998, 11: 191 – 198

37 Kroemer G, Dallaporta B, Resch Rigon M. The mitochondrial death/life regulator in apoptosis arid necrosis. Annu Rev Physiol, 1998, 60: 619 – 642

38 Brenner C. Kroemer G. Apoptosis Mitochondria – the death signal intergrators. Science, 2000, 289: 1150 – 1151

39 Zhang CS. Wang WL, Peng WD, Hu PZ, Chai YB, Ma FC. Promotion of apoptosis of SMMC7721 cells by bcl – 2 ribozyme. Shijie Huaren Xiaohua Zazhi, 2000, 8: 417 – 419

40 Yuan RW. Ding Q, Jiang HY, Qin XF, Zou SQ, Xia SS. Bcl – 2, p53 protein expression and apoptosis in pancreatic cancer. Shijie Huaren Xiaohua Zazhi, 1999, 7: 851 – 854

41 Wang LD. Zhou Q, Wei JP, Wang WC, Zhao X, Wang LX, Zou X, Gao SS, Li YX, Yang CS. Apoptosis and its relationship with cell proliferation, p53, waflp21, bcl – 2 and c – myc in esophageal carcinogenesis studied with a high risk population in northern China. World J Gastroenterol, 1998, 4: 287 – 293

42 Kuai XL. Ge ZJ, Meng XY, Ni RZ. Expression of nitric oxide synthase in human gastric carcinoma. Shijie Huaren Xianohua Zazhi, 2000, 8: 22 – 24

43 Li H, Kolluri SK, Gu J, Dawson MI, Cao X, Hobbs PD. Lin B, Chen G. Lu J, Lin F, Xie Z, Fontana JA, Reed JC, Zhang X. Cytochrome c release and apoptosis induced by mitochondrial targeting of nuclear orphan receptor TR3. Science, 2000, 2S9: 1159 – 1164

44 Brown GC Nitric oxide and mitochondrial respiration. Biochim Biophys acta, 1999, 1411: 351 – 369

45 Brown GC. Regulation of mitochondrial respira-

tion by nitric oxide inhibition of cytochrome c oxidase. Biochim Biophys Acra, 2001, 1504: 46 – 57

46 Duchen MR. Mitochondria and calcium: from cell signaling to cell death. J Physiol, 2000, 529 (Pt1) : 57 – 68

47 Fang M. Zhang H, Xue S. Role of calcium in apoptosis of HL – 60 cells induced by harringtonine. Sci China, 1998, 41: 600 – 607

48 Jun CD. Oh CD, Kwak HJ, Pae HO, Yoo JC, Choi BM. Chun JS, Park RK. Chung HT. Overexpression of protein kinase C isoforms protects RAW 264. 7 macrophages from nitric oxide – induced apoptosis: involvement of c – Jun N – terminal kinase stress – activated protein kinase, p38 kinase, and CPP – 32 protease pathways. J lrnmunol, 1999, 162: 3395 – 3401

49 Korhonen R, Kankaanranta H, Lahti A, Lah E M, Knowles RG, Moilanen E. Bi – directional effects of the elevanon of intracellular calcium on the expression of inducible nitric oxide synthase in J774 macrophages exposed to low and to high concentration of endotoxin. Biochem J, 2001, 354: 351 – 358

50 Gurr JR. Liu F, Lynn S, Jan KY. Calcium – dependent nitric oxide production is involved in arsenite – induced micronulei. Mutat Res, 1998, 416: 137 – 148

51 Giulivi C. Poderoso JJ, Boveris A. Production of nitric oxide by mitochondria. J Biol Chem, 1998, 273: 11038 – 11043

52 Nishikawa M, Takeda K, Sato EF, Kuroki T, Inoue M. Nitric oxide regulates energy metabolism and Bct – 2 expression in intestinal epithelial cells. Am J Physiol, 1998, 274 (5pt 1): G797 – 801

53 Umansky V, Ushmorov A, Ratter F, Chlichlia K, Bucur M, Lichtenauer A, Rocha M. Nitric oxide – mediated apoptosis in human breast cancer cells requires changes in mitochondrial functions and is independent of CD95 (APO – 1/Fas). Int J Oncol, 2000, 16: 109 – 117

82. Mitochondria, Calcium and Nitric Oxide in the Apoptotic Pathway of Esophageal Carcinoma Cells Induced by As$_2$O$_3$

SHEN Zhong – ying[1], SHEN Wen – ying[2], CHEN Ming – hua[1],

SHEN Jian[1], CAL Wei – jie[1], ZENG Yi[3]

Departments of 1. pathology and 2. Chemistry, Medical College of Shantou University;

3. Institute of Virology, Chinese Academy of Preventive Medicine

Summary

In order to explore the early apoptotic signal messengers and to search the apoptotic pathway, the morpho logical and functional changes of mitochondria were examined, and nitric oxide (NO) and calcium ions (Ca^{2+}) were measured in the course of apoptosis in esophageal carcinoma cells induced by As$_2$O$_3$. The esophageal carcinoma cell line SHEEC1, established by HPV in synergy with TPA in our laboratory, were cultured with serum – free medium in a culture flask, 24 – well plate and small culture chambers, and added with As$_2$O$_3$ at 1, 3, 5 μmol/L. After 0, 2, 4, 8, 12, 24 h of drug adding the NO were measured from extracellular cultured medium and the SHEEC1 cells were collected from flasks for electron microscopic examination. Fluorescent intensity (FI) of rhodamine 123 (Rho123) labeled cells was detected using laser confocal scanning microscope (LCSM) for evaluation of mitochondrial membrane potential. Intracellular Ca^{2+} of cells in small culture chambers labeled with Fluo – 3 dye were measured using LCSM over time. After adding As$_2$O$_3$, SHEEC1 cells revealed characteristic morphological and functional changes of mitochondria such as hyperplasia, swelling and disruption, accompanying decrease of trans – membrane potential (FI of Rho123 decreased). The Ca^{2+} level increased at once after adding As$_2$O$_3$ and the NO concentration increased step by step till 24 h, then apoptotic morphology of cells occurred. The results suggest that by inducement of As$_2$O$_3$ increasing Ca^{2+} and NO, the apoptotic signal messengers, will initiate the mitochondria – dependent apoptotic pathway.

〔**Key words**〕 Arsenite; Calcium; Nitric oxide; Apoptotic pathway; Esophageal carcinoma

Introduction

Recent studies have suggested that intracellular calcium (Ca^{2+}) and nitric oxide (NO) are induced by arsenite[1]. Calcium ions (Ca^{2+}) and nitric – oxide (NO) are key signaling molecules that are implicated in the regulation of numerous cellular processed, i. e. physiological and pathological phenomena that occur in cells are accompanied by ionic changes such as NO and Ca^{2+}[2]. Mitochondrial Ca^{2+} uptake in combination with NO production triggers the cell death[3].

Mitochondria is an important cell apparatus. It is concerned with cell breathing, oxygen metabolism, enzyme activity and energy supply[4]. All of those functions relate to the permeability of the mitochondria and mitochondrial transmembrane potential (MTP). When MTP ($\Delta\Psi m$,) decreases, the mitochondria generate morphological and functional changes[5].

The facts that apoptotic pathways vary among different cells and in different stimuli, indicates that there is cell type – specific control of the pathway. There are many pathways in cell apoptosis, such as p53 – dependent apoptotic pathway[6], glucose – dependent apoptotic pathway[7] and EGF – mediated apoptotic pathway[8]. On the downstream, caspase cascade has been shown to be involved in the apoptotic pathway. For example, in the case of caspase – dependent apoptotic pathways[9], caspases appear to be required for certain aspects of apoptosis and are activated to cleave some substrates, the targeted proteins, including poly (ADP – ribose) polymerase (PARP) (DNA cleavage), actin and cytoskeleton of cells[10]. On the upstream, many factors which initiate apoptosis of cells act on cells through signal transduction pathways, such as the TGF – betal signaling pathway of apoptosis[11], the superoxide radical apoptotic signaling pathway[12] and the PKC apoptotic signaling pathway[13].

In our previous work, we used As_2O_3 to induce apoptosis of esophageal carcinoma cells[14]. The pathomorphological changes evidenced that cells became smaller, the cells shrank, the nuclei rounded up, chromatin agglutinated and marginated, nuclear membrane broke down followed by degenerative cell changes. All these changes indicated typical morphological changes of apoptosis[15]. The necrotic changes were also found with a large dosage of As_2O_3[16]. We observed that in the early – stage of cell apoptosis, prior to the obvious change of cell nuclei, the mitochondria showed proliferation. The detailed morphological changes of mitochondria of esophageal carcinoma cells induced by As_2O_3 have been described by us[17]. Furthermore, down regulated expression of bcl – 2 and overexpression of Bax was always found in apoptotic cells induced with As_2O_3[18]. Our data indicate that the severe damage of the cytoskeleton was accompanied with apoptotic process induced by As_2O_3 and that the disrupture of tonofilaments and cleavage of actin could play an important role in the morphological changes of apoptosis[19].

To further define the mechanisms of initiated signals acting on the apoptotic pathway, we investigated NO, Ca^{2+} and their target organelle mitochondria. The main objectives of the present study were to examine the morphological and functional changes of mitochondria and to evaluate the apoptotic signaling pathway of NO and Ca^{2+} and their relation with mitochondria – dependent apoptotic pathway.

Materials and Methods

1. Cell culture and treatment: The esophageal carcinoma cell line (SHEEC1) is a malignantly transformed cell line of human embryonic esophageal epithelium induced by HPV18 E6E7 in synergy with TPA[20]. Cells were cultured in 50 ml flasks with serum – free medium. The culture medium consisted of the basal medium MCDB151 with trace elements (M – 6645, Sigma) and added transferrin, hydrocorticosone, epidermal growth factor (EGF), insulin (purchased from Sigma Chemical Co.) and extracts of bovine hypophysis (Gibco BRL) etc., without containing calf serum, nitrite and nitrate, but containing streptomycin and penicillin (50 μg/ml for each). Arsenic trioxide (As_2O_3) obtained from Sigma Chemical Co. (St. Louis, MO, Lot A 1010). Zero, 1, 3, 5 μmol/L in concentration was added into the culture flasks and 24 – well plate (coverslips inside), respectively for 0, 2, 4, 8, 12 and 24 h. The experiment was repeated once.

2. Morphological examination of mitochondria: Following the addition of As_2O_3, SHEEC1 cells of each group at various intervals were digested with 0.25% trypsin, centrifuged, fixed with 2.5% glutaraldehyde, and routinely prepared for electron microscopic examination. The samples were observed under transmission electron – microscope (Hitachi 300).

3. Mitochondrial transmembrane potential (MTP $\Delta\Psi m$)[21]: SHEEC1 cells cultured on the slide, after reacting with 3 μmol/L As_2O_3 at various times, stained by rhodamine 123 (Rho123, MW381, Molecular Probe Inc. Eugene) at the concentration of 10 μg/ml, and the cells were incubated at 37℃, in 5% CO_2 incubator for 15 min, then examined by laser confocal scanning microscopy. The average fluorescent intensity of the 10^2 cells was measured before and after As_2O_3 adding.

By examination of LCSM the average fluorescent intensity value (average $\bar{x} \pm s$) of 10^2 cells was calculated.

4. Procedure of NO detection[22]: The nitrite/nitrate colorimetric method, using the kit from Boehringer Mannheim Co, was used for detection of NO in culture medium. The culture medium (0.2 ml) at time intervals was deactivated at 80℃ for 5 min and deproteinated by centrifugation at 12 000 r/min for 30 min, and the supernatants were determined. The procedure for nitric oxide determination was as follows: in a 3 ml test tube, 100 μl of sample solution, 50 μl of nicotinamide adenine dinucleotide phosphate (NADPH) and 4 μl enzyme nitrate reductase (NR), was mixed, incubated for 30 min at room temperature, and 50 μl color reagent Ⅰ & Ⅱ was added, respectively, mixed, allowed to stand in the dark at room temperature for 10 to 15 rain. The nitric oxide content of the samples and the blank was estimated with Shimadzu UV/120 spectrophotometry by λ 450 nm and was calculated by calibration curve of standard addition method. The standards were prepared from known amounts of stock NO_3 and NO_2 and run in parallel with test samples in each assay.

5. Intracellular calcium level detected by LCSM[23,24]: The cells were cultured on the coverslips within the glass bottom small cultured dish (NO. 0, Uncoated, irradiation; MatTek Co. USA) for vital staining of cytosolic – free calcium in dynamic changes. At the exponent growth period, the cells were stained with 10 μmol/L fluo – 3/AM (Molecular Probe) for 30 rain at 37℃, and washed with a standard medium (135 mmol/L NaCl, 10 mmol/L HEPES, 0.4 mmol/L $MgCl_2$, 1 mmol/L $CaCl_2$, 0.1% D – glucose, 0.1% bovine serum albumin, pH 7.3) at least 3 times. Then the cells were placed in the standard medium to remain viable. Before and after administration of As_2O_3 the fluorescent intensity were determined by CLSM in dynamic changes for up to 900 sec. Scan – Time Series Menu was used to scan some cells repeatedly to monitor the dynamic changes in fluorescent intensity of intracellular Ca^{2+} content over time. The parameters of the CLSM (Ultima 312, Meridian Instruments Inc., USA) were as follows: the excited light 488 nm, the emission light 530 nm and pinhole 10 – 40 nm. The fluorescent intensity of pixel was collected and managed with the software of the instrument.

Results

1. Morphological changes of mitochondria under transmission electron microscope: Before adding As_2O_3 the mitochondria of SHEECl cells were located around the nucleus in one or two arrays. Other organelles were present between mitochondria. When adding As_2O_3 2 – 4h later, the number of mitochondria increased, which were either concentrated in certain area or in one pole of the cytoplasm. Mitochondria were crowded closely together showing hypertrophy with abundant cristae (Fig. 1A). After 6 h, the high electrically dense and irregularly shaped substances precipitated in the mito – chondrial matrix, even filling the whole mitochondria. After 12h, the mitochondria swelled like a balloon or a vacuole (Fig. 1B), and then its outer membrane broke down, leaving a single layer of membrane and the cristae disappeared (Fig. 1C). The cell nucleus shrank, with agglutinated chromatin locating near the nuclear membrane, and showed typical apoptotic nuclear changes.

2. Uptake of Rho123 in living cells: Uptake of Rho123 dye in cytoplasm of SHEECl cells increased after 2 h of As_2O_3 addition (Fig. 2A), but decreased continually from 8 to 24h after addition of As_2O_3 (Fig. 2B and C).

3. The dynamic changes of mitochondrial transmembrane potential (MTP $\Delta\Psi m$): The changes of mitochondrial fluorescence labeled with Rho123 in SHEEC1 were measured by LCSM. The average $\bar{x} \pm s$ of the mitochondrial fluorescent intensity in 0, 1, 3, 5 $\mu mol/L$ As_2O_3 groups are seen in Fig. 3. The results showed an increment of fluorescent intensity in early – stage, 2h of As_2O_3 adding, and the FI decreased thereafter, indicating MTP change of the mitochondria membrane.

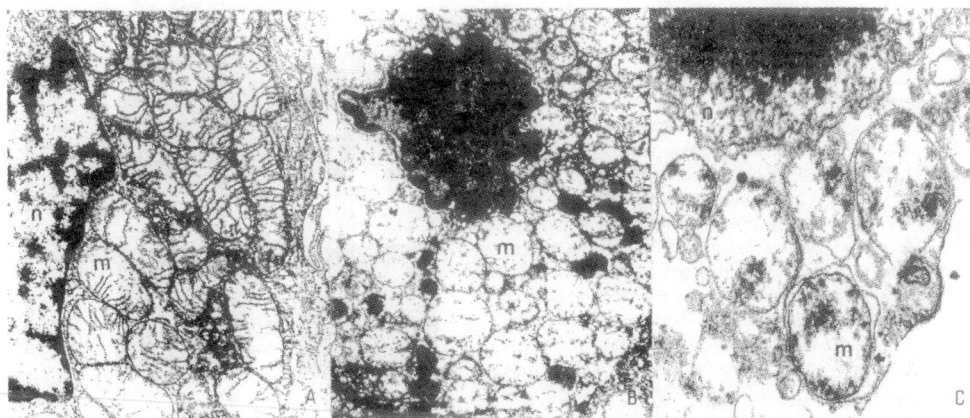

A, at 2 – 4 h after treatment hypertrophic mitochondria were aggregated with abundant cristae (m) (EM, ×15 k); B, at 12 h of As_2O_3 treatment, almost all mitochondria (m) were swollen like balloon with a pyknotic nucleus (n) (EM, ×10 k); C, at 24 h of As_2O_3 treatment the outer layer of membrane in several mitochondria broke (m) and chromatin in nucleus was coagulated (n) (EM, ×30 k). m, mitochondrial; n, nucleus; arrow head, broken outer membrane of mitochondria

Fig. 1 Electron micrograph of SHEEC1 cells treated with As_2O_3 at 3 $\mu mol/L$

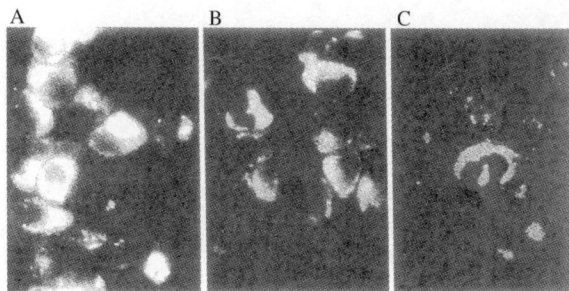

A, 2 h after As_2O_3; B, 8 h after As_2O_3; C, 12 h after As_2O_3 (LCSM, ×500)

Fig. 2　Uptake of Rho123 in SHEEC1 cells

4. NO examination: Without adding As_2O_3, the culture medium of SHEEC1 contained low concentration of NO ($0.38 \times 10^{-2} \pm 0.12 \times 10^{-2}$ μmol/L). After adding As_2O_3 (1, 3 and 5 mol/L), NO contents of culture medium was increased accompanying treatment time (Fig. 4). NO increment began at 2 h, $1.28 \times 10^{-2} \pm 3.7 \times 10^{-2}$ μmol/L (5 μmol/L group), $1.30 \times 10^{-2} + 3.2 \times 10^{-2}$ μmol/L (3 μmol/L group), and $0.8 \times 10^{-2} + 0.17 \times 10^{-2}$ μmol/L (1 μmol/L group) and was sustained to the end point in 3 μmol/L and 5 μmol/L groups, but in 1 μmol/L group, NO decreased. In statistical analysis by Wilcoxon test, the differences between data of means ± s calculated in As – treated and in non – As – treated groups, were statistically significant ($p < 0.05$). There was no difference between 3 μmol/L and 5 μmol/L groups.

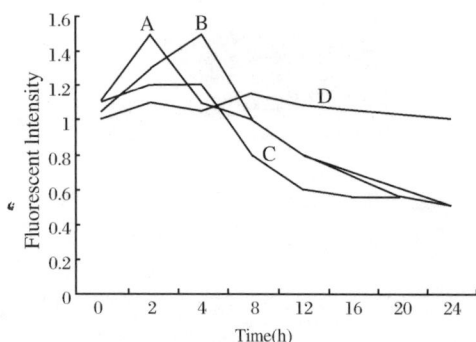

A, 5 μmol/L As_2O_3; B, 3μmol/L As_2O_3 C, 1 μmol/L As_2O_3; D, 0 μmol/L As_2O_3

Fig. 3　Following the As_2O_3 addition, the average fluorescence intensity of Rho123 in SHEECl from 2 to 24 h

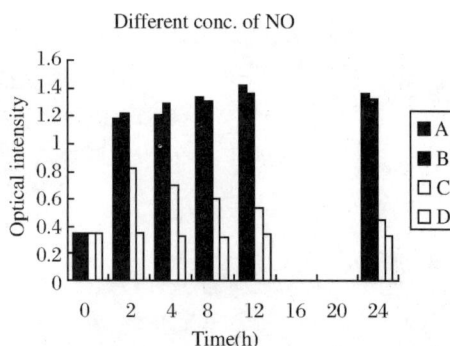

Fig. 4　Optical intensity of NO. NO determination of SHEEC1 treated with different concentration of As_2O_3. NO markedly increase in 5 μmol/L of As_2O_3 group (A), 3 μmol/L group (B), and low increase in 1 μmol/L of As_2O_3 group (C). The control group, 0 μmol/L of As_2O_3, remained on the basal lines (D)

5. The calcium change over time: To show the time course of changes in Ca^{2+} in individual cells (Fig. 5A), the changes influorescent intensity at representative area of 7 cells were studied as displayed in Fig. 5B, with control group in Fig. 5C. Upon the initiation of stimulation by As_2O_3, all the cells responded with a rapid rise in concentration of calcium (Ca^{2+}), which was sustained for 900 sec after stimulation.

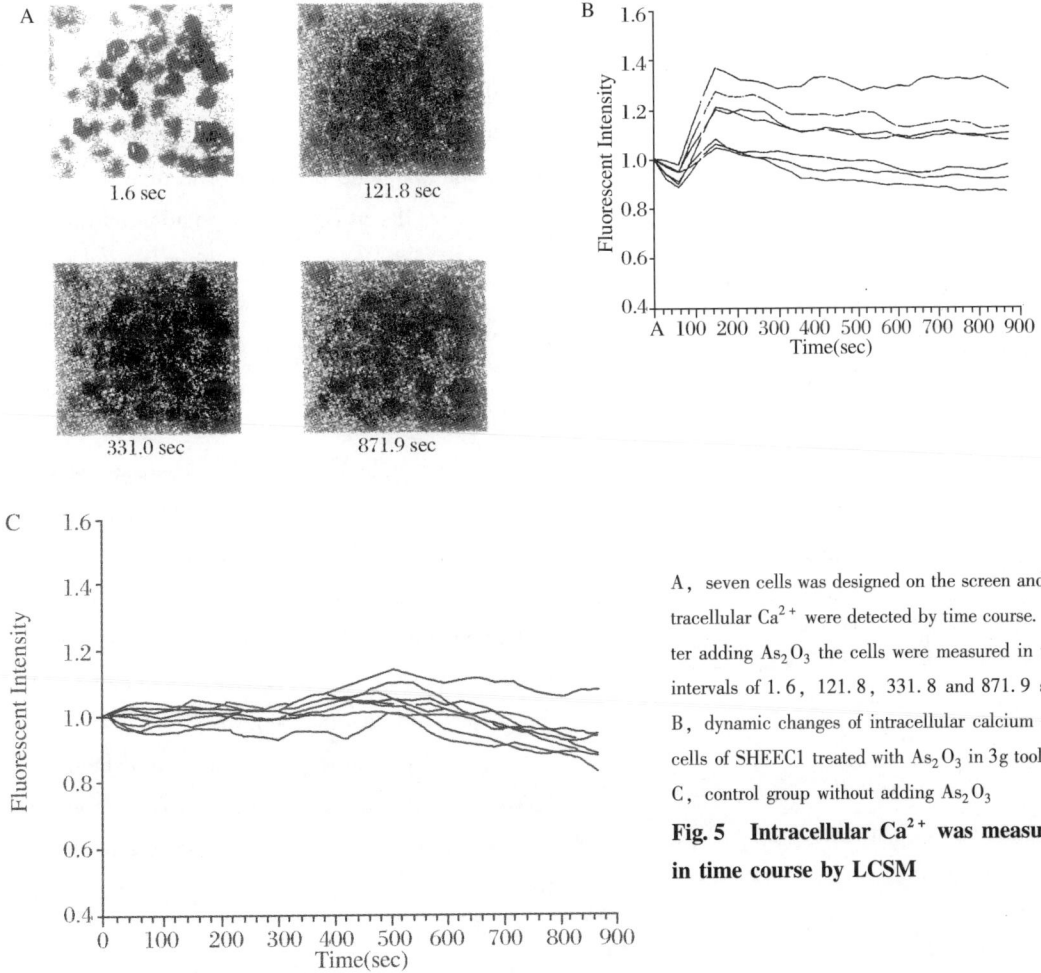

A, seven cells was designed on the screen and intracellular Ca^{2+} were detected by time course. After adding As_2O_3 the cells were measured in time intervals of 1.6, 121.8, 331.8 and 871.9 sec. B, dynamic changes of intracellular calcium in 7 cells of SHEEC1 treated with As_2O_3 in 3g tool/L, C, control group without adding As_2O_3

Fig. 5 Intracellular Ca^{2+} was measured in time course by LCSM

Discussion

As a result of these experiments, we found that intracellular calcium level increased at once after adding As_2O_3 and then extracellular nitric oxide increased. We have shown morphological and functional changes of mitochondria of SHEEC1 induced by As_2O_3: 2 – 4 h after As_2O_3 administration the mitochondria proliferated and hypertrophied accompanying elevation of Rho123 uptake. In 8 – 12 h after As_2O_3 administration, the mitochondria underwent damage with swelling and balloon – like appearance, and Rho123 uptake decreased. Finally, the outer layer of membrane of mito – chondria broke down and cristae disappeared accompanying Rho123 uptake at the lowest. The functional changes of mitochondria were in accordance with mitochondrial changes.

Arsenic induced up – regulation of Ca^{2+} and NO caused damage of mitochondria and apoptosis in SHEEC1 cells. While a pathway for Ca^{2+} accumulation into cytosol has long been established, the second messenger Ca^{2+} can enter from the outside through channel influx into cells and it can also be released from internal stores[25]. Ca^{2+} can also enter the nucleus through the signal transduction pathway. It is important that elevated levels of Ca^{2+} in nucleus, especially if maintained for long periods, can be cytotoxic for cells[26]. It seems that the signaling system and cytotoxicity of Ca^{2+} effects the apoptotic pathway. The increase in Ca^{2+} of nuclei results in DNA fragmentation and intracellular Ca^{2+} plays an essential role in the induction of apoptosis[27]. It is well known that the typical morphological and biochemical characteristics of apoptosis, i. e. chromatin condensation and cleavage of host chromatin into oligonucleosome – length fragments, by endonuclease via caspase activated DNase (CAD)[29] take place in the nucleus. NO triggered such a time – dependent caspase cascade in these cells[30]. Moreover, NO caused a release of mitochondrial protein cytochrome c into the cytosol, NO and Ca^{2+} converge on mitochondria to trigger major pathways in apoptosis[31].

As a key event in apoptosis, mitochondria release caspase activators (such as cytochrome c), changes in electron transport, loss of mitochondrial transmembrane potential[32], and produces a high level of reactive oxygen species[33]. bcl – 2 in the mitochondria will regulate some caspase activators, cytochrome c and/or apoptosis – inducing factor (AIF)[34] released from mitochondria. Although for a long time the mitochondrial changes were considered as hallmarks of apoptosis, mitochondria appear today as the central executioner of programmed cell death[35].

Taking into account our data and cited reports, we suggest that in the schema of arsenic – derived

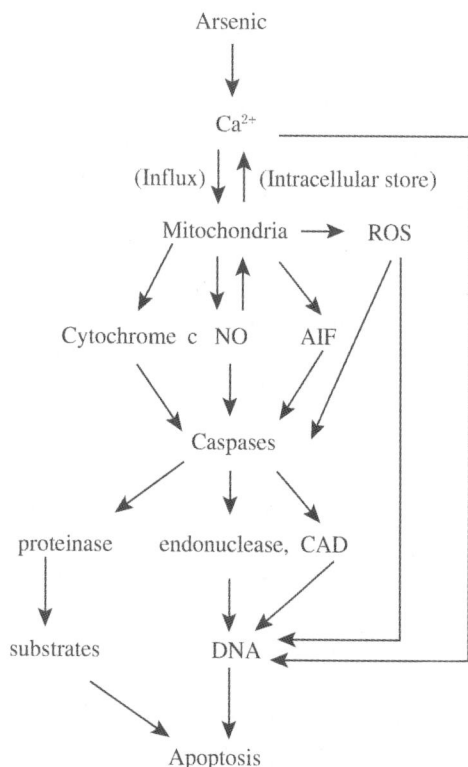

At the first stage (the signal and transcluction), stimuli and intracytoplasmic messengers increased the concentration of Ca^{2+} and NO. At the second stage (functional activation and damage stage of mitochondria), cytochrome c, caspases, apoptosis inducing factor (AIF), and reactive oxygen species (ROS) were released. At the third stage (morphologic and biochemical changes), degradation of DNA, actin and other constituent proteins occurred, through activation of proteinase, endonuclease and caspase activated DNase (CAD) which lead to structural alternation and cell death

Fig. 6　Schematic representation of As_2O induced apoptotic pathway

apoptotic pathways, calcium can stimulate a constitutive apoptotic pathway in SHEEC1 cells, which could cooperate with NO and other apoptotic pathways via the mitochondria (Fig. 6).

The process of cell apoptosis can be divided into three stages[35,36]. The first stage is multiple signaling pathways leading from death – triggering extracellular agents to intracellular mitochondria through membrane traffic. The second stage is a central control and execution stage. In this stage, releasing of cytochrome c, apoptosis inducing factor (AIF) and reactive oxygen species (ROS) occur, which is controlled by mitochondria[37]. The execution stage thus requires at least four components, caspase, bcl – 2s[38], mito – chondria and endonucleases[39,40]. These factors together contribute to the cells entering death pathway. The third stage, cells undergo morphological and biochemical changes, accompanying degradation of DNA, actin and other constituent proteins, which lead to cell death. NO, Ca^{2+}, proteases, and nucleases are considered mediators of apoptosis in the mito – chondrial apoptotic pathway of As_2O mediated apoptosis of SHEEC1 cells.

In summary, we have shown that SHEEC1 cells undergo apoptosis induced by As_2O_3. *In vitro* these cells revealed characteristic morphological and functional changes of mitochondria and elevation of Ca^{2+} and NO. They primarily acts as signal transducers, but can also confer cytotoxicity if calcium and NO products exist in large quantities. Cultured SHEEC1 cells respond to NO and Ca^{2+} as the modulators of the cell death pathway. Thus, crosstalk between NO and Ca^{2+} signaling pathways through the mitochondria is probably an important mechanism for apoptosis of SHEEC1. We designed a scheme that Ca^{2+} and NO initiated mitochondria – dependent apoptotic pathway, which may be exploited in the future for cancer therapy.

Acknowledgements

Contract grant sponsor: National Natural Science Foundation of China (39830380).

[In 《Int J Mol Med》 2002, 9 (4): 385 – 390]

References

1 Gurr JR, Liu F, Lynn S and Jan KY. Calcium – dependent nitric oxide production is involved in arsenite – induced micronuclei. Mutat Res, 1998, 416: 137 – 148

2 Rachmilewitz D. Role of nitric oxide in gastrointestinal tract. World J Gastroenterol, 1998, 4 (Suppl. 2): 28 – 29

3 Korhonen R, Kankaanranta H, Lahti A, Lahde M, Knowles RG, Moilanen E. Bi – directional effects of the elevation of intracellular calcium on the expression of inducible nitric oxide synthase in J774 macrophages exposed to low and to highconcentration of endotoxin. Biochem J, 2001, 354: 351 – 358

4 Green DR and Reed JC. Mitochondria and apopto-sis. Science, 1998, 281: 1309 – 1312

5 Petit PX, Lecoeur H, Zorn E, Dauguest C, Mignotte B and Gougeon ML. Alterations in mitochondrial structure and function are early events of dexamethasone – induced thymocyte apoptosis. J Cell Biol, 1995, 130: 157 – 167

6 Vaisman A, Varchenko M, Said I and Chancy SG. Cell cycle changes associated with formation of Pt – DNA adducts in human ovarian carcinoma cells with different cisplatin sensitivity. Cytometry, 1997, 27: 54 – 64

7 Shim H, Chun YS, Lewis BC and Dang CV. A unique glucose – dependent apoptotic pathway induced by c – Myc. Proc Natl Acad Sci USA, 1998, 95: 1511 – 1516

8 Lei W, Mayotte JE and Levitt ML. EGF – dependent and independent programmed cell death pathways in NCI – H596 non – small cell lung cancer ceils. Biochem Biophys Res Commun, 1998, 245: 939 –945

9 Siro S, Minoru T, Kazuo U and Masaya I. Requirement of caspase – 3 (– like) protease – mediated hydrogen peroxide production for apoptosis induced by various anticancer drugs. J Biol Chem, 1998, 273: 26900 –26907

10 Van – Engeland M, Kuijpers HJ, Ramaekers FC, Reuteling sperger CP and Schutte B. Plasma membrane alterations and cytoskeletal changes in apoptosis. Exp Cell Res, 1997, 235: 421 –430

11 Ohta S, Yanagihara K and Nagata K. Mechanism of apoptotic cell death of human gastric carcinoma cells mediated by transforming growth factor beta. Biochem J, 1997, 324: 777 –782

12 Ray RB, Meyer K, Steele R, Shrivastava A, Aggarwal BB and Ray R. Inhibition of tumor necrosis factor (TNF – alpha) – mediated apoptosis by hepatitis C virus core protein. J Biol Chem, 1998, 273: 2256 –2259

13 Ferraris C, Cooklis M, Polakowska RR and Haake AR. Induction of apoptosis through the PKC pathway in cultured dermal papilla fibroblasts. Exp Cell Res, 1997, 234: 37 –46

14 Shen ZY, Shen J, Chen MH, Li QS and Hong CQ. Morphological changes of mitochondria in apoptosis of esophageal carcinoma cells induced by As203. Chin J Pathol (in Chinese), 2000, 29: 200 –203

15 Shen ZY, Cen S, Shen J, Cai WJ, Xu JJ, Teng ZP, Hu Z and Zeng Y. Study of immortalization and malignant transformation of human embryonic esophageal epithelial cells induced by HPVI8E6ET. J Cancer Res Clin Oncol, 2000, 126: 589 –594

16 Shen ZY, Tan LJ, Cai W J, Shen J, Chen C, Tang XM and Zheng MH. Arsenic trioxide induces apoptosis of oesophageal carcinoma in vitro. Int J Mol Med, 1999, 4: 33 –37

17 Shen ZY, Shen J, Cai WJ, Hong CQ and Zheng MH. The alteration of mitochondria is an early event of arsenic trioxide induced apoptosis in esophageal carcinoma cells. Int J Mol Med, 2000, 5: 155 –158

18 Shen ZY, Tan LJ, Cai WJ, Shen J, Chert CY and Tang XM. Morphologic study on apoptosis of esophageal carcinoma cell line induced by arsenic trioxide. Shijie Huaren Xiaohua Zhazhi (in Chinese), 1998, 6: 226 –229

19 Shen ZY, Shen J, Cai WJ, Chen CY and Chen JY. Changes in cytoskeleton of apoptosis of esophageal carcinoma cells induced by As_2O_3. Chin J Pathol, 2001, 30: 369 –370

20 Shen ZY, Shen WY, Chen MH, Hong CQ and Shen J. Alteration of nitric oxide in apoptosis of esophageal carcinoma cells induced by arsenite. Shigie Huaren Xiaohua Zazhi, 2000, 8: 1101 – 1104

21 Heerdt G, Houston MA, Anthony GM and Augenlicht LH. Mitocondrial membrane potential ($\Delta\Psi$mt) in the coordination of p53 – independent proliferation and apoptosis pathway in humancolonic carcinoma cells. Cancer Res, 1998, 58: 2869 –2875

22 Shen WY, Chen MH, Shen ZY and Zhang L1M. Microspectro – photometric determination of nitric oxide. J Shantou Univ MedCollege, 2000, 13: 10 –11

23 Takahashi A, Camacho P, Lechleiter JD and Herman B. Measurement of intracellular calcium. Physiol Rev, 1999, 79: 1089 –1125

24 Satoh Y, Nishimura T, Kimura K, Mori S and Saino T. Application of realtime confocal microscopy for observation of living cells in tissue specimens. Hum Cell 1, 1998, 1: 191 –198

25 Fang M, Zhang H and Xue S. Role of calcium in apoptosis of HL –60 cells induced by harringtonine. Sci China, 1998, 41: 600 –607

26 Boullerne AI, Nedelkoska L and Benjamins JA. Role of calcium in nitric oxide – induced cytotoxicity. J Neurosci Res, 2001, 63: 124 –135

27 McConkey PDJ, Hartzell S, Duddy H and Hakanson S. Orrenius, 2, 3, 7, 8 – tetrachlorod-

ibenzo – p – dioxin kills immature thymocytes by Ca2 + mediated endonuclease activation. Science, 1989, 242: 256 – 259

28 Fang M, Zhang HQ, Xue SB, Li N and Wang LM. Intracellular calcium distribution in apoptosis of HL – 60 cells induced by harringtonine: intranuclear accumulation and regionalization. Cancer Lett, 1998, 127: 113 – 121

29 Lechardeur D, Drzymala L, Sharma M, Zylka D, Kinach R, Pacia J, Hicks C, Usmani N, Rommens JM and Lukacs CL. Determinants of the nuclear localization of the heterodimeric DNA fragmentation factor (ICAD/CAD). J Cell Biol, 2000, 150: 321 – 334

30 Seoi JG, Park WH, Kim ES, Jang CW, Hyun JM, Lee YY and Kim BK. Potential role of caspase – 3 and – 9 in arsenic trioxide – mediated apoptosis in PCI – 1 head and neck cancer cells. Int J Oncol, 2001, 18: 249 – 255

31 Richter C. Nitric oxide and its congeners in mitochondria: implications for apoptosis. Environ Health Perspect, 1998, 106 (Suppl. 5): 1125 – 1130

32 Ormerod MG, Sun XM, Snowden RT, Davies R, Fearnhead Hand Cohen GM. Increased membrane permeability of apoptotic thymocytes: a flow cytometric study. Cytometry, 1993, 14: 595 – 602

33 Oridate N, Suzuki S, Higuchi M, Mitchell MF, Hong WK and Lotan R. Involvement of reactive oxygen species in N – (4 – hydroxyphenyl) retinamide – induced apoptosis in cervical carcinoma ceils. J Natl Cancer Inst, 1997, 89: 119 1 – 1198

34 Daugas E, Nochy D, Ravagnan L, Loeffler M, Susin SA, Zamzami N and Kroemer G. Apoptosis – inducing factor (AIF): a ubiquitous mitochondrial oxidoreductase involved in apoptosis. FEBS Lett, 2000, 476: 118 – 123

35 Golstein P. Controlling cell death. Science, 1997, 275: 1081 – 1082

36 Kroemer G, Dallaporta B and Resch – Rigon M. The mitochondrial death/life regulator in apoptosis and necrosis. Annu Rev Physio, 1998, 160: 619 – 642

37 Brenner C and Kroemer G. Apoptosis mitochondria-the death signal intergrators. Science, 2000, 289: 1150 – 1151

38 Chhieng DC, Ross JS and Ambros RA. bcl – 2 expression and the development of endometrial carcinoma. Mod Pathol, 1996, 9: 402 – 406

39 Nagata S. Apoptotic DNA fragmentation. Exp Cell Res, 2000, 256: 12 – 18

40 Arends M J, Morris RG and Wyllie AH. Apoptosis. the role of the endonuclease. Am J Pathol, 1990, 135: 593 – 607

83. The Inhibition of Growth and Angiogenesis in Heterotransplanted Esophageal Carcinoma via Intratumoral Injection of Arsenic Trioxide

SHEN Zhong – ying[1], SHEN Jian[1], CHEN Ming – hua[1], WU Xian – ying[1],
WU Min – hua[1], ZENG Yi[2]

1. Department of Pathology, Medical College of Shantou University
2. Institute of Virology, Chinese Academy of Preventive Medicine

Summary

To investigate the antitumor action of arsenic trioxide (As_2O_3) by intratumoral injection into solid tumors, tumor growth inhibition (TGI) and angiogenesis of hetero – transplanted esophageal carcinoma in mice was carried out. The cultured human esophageal carcinoma cells were inoculated into both laterals of the abdominal wall of severe combined immunodeficient (SCID) mice. When both lateral tumors had grown to about 10 mm × 8 mm × 5 mm, the right tumors were treated with an intratumoral injection of As_2O_3 in dosage of 1, 5 and 10 μg per day, respectively, for 10 days sequentially. Left tumors were treated with PBS (phosphate buffer solution) as control. The weight of transplanted tumor masses were measured and counted for TGI. The tissue of tumor, liver, kidney, heart, lung and brain was examined histopathologically and tumor tissues were examined by light – or electron – microscope. Ki – 67 and CD34 were assessed by immunohistochemistry and positive nuclei of Ki – 67 and microvessel density (MVD) labeled by CD34 were measured. The results revealed that on the 20th day after the first injection, As_2O_3 – treated tumors were suppressed markedly as compared with the Contrarily situated tumor, accompanied by a marked apoptosis and necrosis in tumor cells. The tissue of liver, kidney, heart, lung and brain was unaffected by As_2O_3. MVD in tumor tissue was decreased in the right side tumor with the significant difference in the 5 μg and 10 μg group ($P < 0.01$). TGI was 5.80 ($P > 0.05$), 58.66 ($P < 0.01$) and 73.97% ($P < 0.01$) in the 1, 5 and 10 μg groups respectively, but 2.21% ($P > 0.05$) in the control group. Conclusively, a repeated administration of As_2O_3 (5 and 10 μg × 10) induced an increase of tumor growth inhibition and decrease of angiogenesis in the solid tumor in tumor progressive periods. These results suggest that intra – tumoral injection of As_2O_3 may be investigated as a modality to treat some solid tumors.

[**Key words**]   Arsenic trioxide; Esophageal carcinoma; Intratumoral injection; Aangiogenesis; SCID mice

Introduction

Arsenic trioxide (As_2O_3) induces clinical remission of patients with acute promyelocytic leukemia[1,2]. As a novel anticancer agent a few *in vivo* experimental investigations of its efficacy on solid cancers have been done[3]. The mechanisms of cell death and growth – inhibition induced by As_2O_3 have not yet been clarified, especially in solid cancers. In our previous studies, the esophageal carcinoma cell line was examined as a cellular model for As_2O_3 treatment[4], and As_2O_3 – induced cell death and inhibition of cell growth were evaluated[5].

The method of intratumoral injection has been investigated in experimental animal models to increase the localization of As_2O_3 in tumors and to reduce their up – take in normal tissues, allowing higher and more frequent doses of As_2O_3 to be used for treatment of tumors[6]. Using an animal model permits the assessment of a wide range of drugs and methods of administration before their use in clinical trials. Heterotopic models of tumors, therefore, seem very useful for the development of effective therapies and are expected to become the new standard method. But the real significance of drug efficacies *in vivo* models will remain to be validated through the clinical studies.

Angiogenesis is an important factor in the progression and enlargement of solid neoplasms and is in close relation to invasion and metastases[7,8]. In a number of specific tumor types, including neoplasms of the lung[9], colon[10], oral[11], and prostate carcinoma[12], the development of a growing tumor requires abundant blood supply and the vascularity may be an important prognostic indicator[13,14] and also a biomarker in cancer therapy[15]. The mechanism whereby neoplasms induce these changes in their supporting stroma is not well understood. The work of Folkman *et al* has demonstrated the critical role of angiogenesis in neoplasia[16,17]. These ideas have been applied to a variety of tumor types. The solid neoplasms can be divided into two distinct phenotypic stages in a stepwise model of tumor progression: the initiative (prevascular) period and the progressive (vascular) period. The former phase would correlate with the establishment of tumors and the latter with rapid tumor growth and potential for metastasis. The current investigations focus on the roles of As_2O_3 to therapeutically manipulate or abolish the angiogenic properties of tumors and to limit their potential for growth and spread.

This study was designed to further define the relationship of microvessel density (MVD) with As_2O_3 – treatment in the progressive period, and further investigate the growth inhibition and angiogenesis in esophageal cancer cell after As_2O_3 treatment.

Materials and Methods

1. Animals: Severely combined immunodeficient mice (SCID), 6 weeks of age and weighing 20 – 25 g, from the Center of Experimental Animal Center, Zhongshan Medical University, were used in this study. All experiments were carried out according to the guidelines of the Laboratory Protocol of Animal Handling, Medical College of Shantou University.

2. Tumor cell and models: The esophageal carcinoma cell line (SHEEC) is a malignantly transformed cell line of human embryonic esophageal epithelium induced by HPV18E6E7 in synergy with TPA in our laboratory. SHEEC cells (10^6 in 0. 2 ml) were inoculated into the subcutaneous tissues of both abdominal walls of SCID mice and solid tumors formed.

3. Drug and treatment: Arsenic trioxide (As_2O_3) in ampoule (0. 1% As_2O_3, 10 ml) was supplied by the Pharmacy of Chinese Traditional Medicine of the First Hospital affiliated to Harbin Medical University (Harbin, P. R. China). Tumors were selected for As_2O_3 treatment when they reached volumes of about 1. 0 cm $\times$ 0. 8 cm $\times$ 0. 5 cm. After diluting the As_2O_3 with buffered saline, the concentration of 1 μg, 5 μg and 10 μg/0. 2 ml of As_2O_3 solution was injected into tumor mass per day in the right tumor region for 10 days sequentially and the left side was injected with PBS as the control tumor. Ten animals in each group were injected with As_2O_3 and ten without treatment as a control group, Four groups of mice were sacrificed at 20th day after the first injection As_2O_3. Tumor weight was measured after the mice were sacrificed.

4. Histopathological and ultrastructural examination: The samples of the experiment including tumor tissues and hearts, lungs, spleens, livers, kidneys and brains were fixed in 10% fortualin and prepared section routinely for pathological examination. The sections of tumor with surrounding connected tissue were also used for immunohistochemical analysis by light – microscope. The tissues of the transplanted tumors in SCID mice were cut into 1 mm^3 pieces, and fixed in 2. 5% glutaraldehyde. The samples were routinely prepared and observed under electron – microscope (Hitachi 300).

5. Inhibition and regression of tumor: Tumor growth inhibition (TGI), which reflects the retardation in tumor growth of the arsenic – treated tumor relative to the contrary side tumor, is widely used to assess the efficacy of therapeutic agents against xenografts. Using tumor weight as a measure of size for illustration of TGI expressed as a percentage is calculated as follows: % TGI = 100 (W_L – W_R) $/W_L$, where W_L is the mean tumor weight of the left side tumor (mg) (PBS – treated group) and W_R is the mean tumor weight of the right side tumor (mg) (arsenic – treated group).

6. Immunohistochemistry for microvessels: For identification of microvascular structures, 4μm thick sections were cut and stained with mouse monoclonal anti – CD34 (MAB – 0034, Maixim Biotech Co.). The antibody of CD34 reacts with vascular endothelial cells in normal tissues and in benign and malignant proliferations. The sections were counter – stained with Mayer's haematoxylin for 2 min.

7. Microvessel count: Microvessel density (MVD) was determinedly light microscopy in area of tumor tissues section. The micro – vessels were carefully counted on 20 fields ($\times$ 100). The mean and SE was expressed as the number of microvessels identified within the tumor area.

8. Growth fraction by Ki – 67 *staining:* For this growth fraction determination, tissues sections were immunostained with the Ki – 67 monoclonal antibody (Dakopatts Ltd. , High Wycombe, England). Negative controls included sections stained with omitting the primary antibody but following the same procedure. The number of cells with nuclear staining was determined by image analysis system (Cambridge, Q520) with the master software in 10 fields ($\times$ 200). An average of 1000 nuclei per section was counted.

9. Statistical analysis: All statistical analysis was performed by the Statistical software system (Stat – Soft PSSA 10. 0). Univariate analysis by the Mann – Whitney U test for different groups and the x^2 test for discrete variables were used to assess differences between the lateral tumors. The difference was considered to be statistically significant when $P < 0. 05$.

Results

1. Regression of established tumors upon intratumoral injection of arsenic: All mice of four groups were sacrificed on the 20th day after the first As_2O_3 treatment. There was a significant reduction of tumor weight in the 5 μg ($t = 7.16$, $P < 0.01$) and 10 μg ($t = 12.66$, $P < 0.01$) groups of As_2O_3-treated tumors compared with the tumors of the contrary sides (Tab. 1). Suppression of tumor weight was statistically significant. TGI was low in As_2O_3 in the 1 μg group and high in the 5 μg ($u = 3.03$, $P < 0.01$) and 10 μg groups ($u = 3.55$, $P < 0.01$). The difference of TGI in various dosage of arsenic compared with control group was of statistical significance.

Tab. 1   Tumor growth inhibition (TGI) of tumors treated with arsenic

Dosage (μg)	W_L (mg)		W_R (mg)		TGI (%)	p – value (R/L)
	n	mg	n	mg		
1	10	1223 ± 244	10	1152 ± 193	5.80	> 0.05
5	10	1270 ± 312	10	525 ± 105	58.66[a]	< 0.01
10	10	1178 ± 207	9	307 ± +67	73.94[a]	< 0.01
0 (control)	10	1307 ± 397	10	1278 ± 401	2.21	> 0.05

Notes: a Significant difference ($P < 0.01$) from the control. Each value indicates the $\bar{x} \pm s$ mean ± SE (n, number of mice). Significant difference p – value (R/L) of tumor with arsenic administration compared with contrary tumor; TGI (%) = ($W_L - W_R$) /W_L

Repeating the experiment with the same number of SCID mice and the same treatment schedules again demonstrated significant reduction in SHEEC tumor growth as in the original experiment. The treated tumors had marked growth inhibition compared to that of the opposite side tumors (Fig. 1).

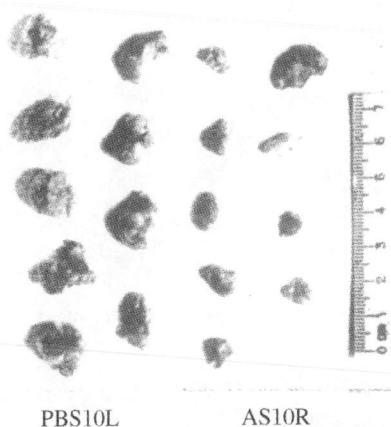

PBS10L   AS10R

Fig. 1   The solid tumor mass of transplanted SHEEC in SCID mice were treated with or without As_2O_3. In 10 μg As_2O_3 group the weight of tumor masses in right lateral (AS 10R) were less than that in tumor masses in left lateral (PBS 10L).

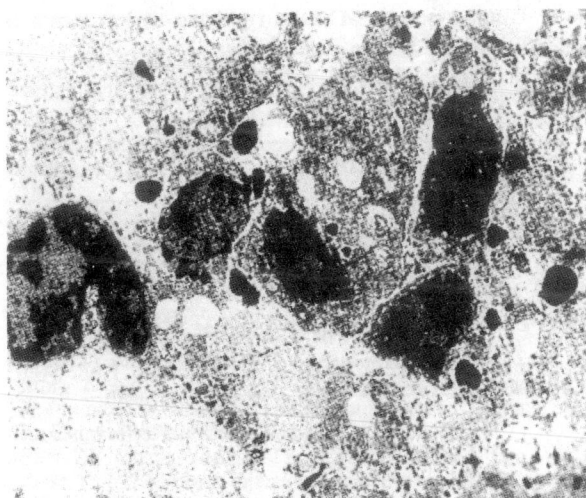

Fig. 2   Ultrastructure of SHEEC cells treated with 10 μg of arsenic. Mass of necrotic cells appeared with a few apoptotic nuclei (EM, ×4000).

2. Histological and ultrastructural examination: In 5 μg and 10 μg groups As_2O_3 induced a pronounced central necrosis in the tumor. More than 70% of the tumor tissue underwent central necrosis with scattered apoptotic area, and retained peripheral zone of tumor tissue unaffected with a well-demarcated interface between the viable and necrotic tissues. Tumor regrowth appeared to be initiated from the peripheral zone. Untreated control tumors consisted of undifferentiated round cells with evidence of focal necrosis, The tumor tissues ot three distinct dosage groups and the control group were observed by electron microscopy. In the 1 μg group cells were intact without the chromatin agglutinant. In the 5 and 10 μg groups chromatin of nuclei coagulated into coarse mass attaching near the nuclear membrane or flowed out of the nucleus, and cell organelles were swollen with secondary degeneration, which revealed the pathological changes of apoptosis and necrosis (Fig. 2).

3. Quantitative analysis of microvessels: A single microvessel was defined as any brown immunostained endothelial cell distinguished from adjacent tumor cells and other connective tissue elements. These studies demonstrated the decreased capillary density of right side tumor with treatment of As_2O_3. In the right sided tumors capillaries were merely restricted to the stroma immediately adjacent to the tumor cell nests and the bulk of the stroma between carcinoma cell nests contained a few capillaries (Fig. 3A). In the left sided tumors without treatment with As_2O_3 the number of capillaries was abundant (Fig. 3B).

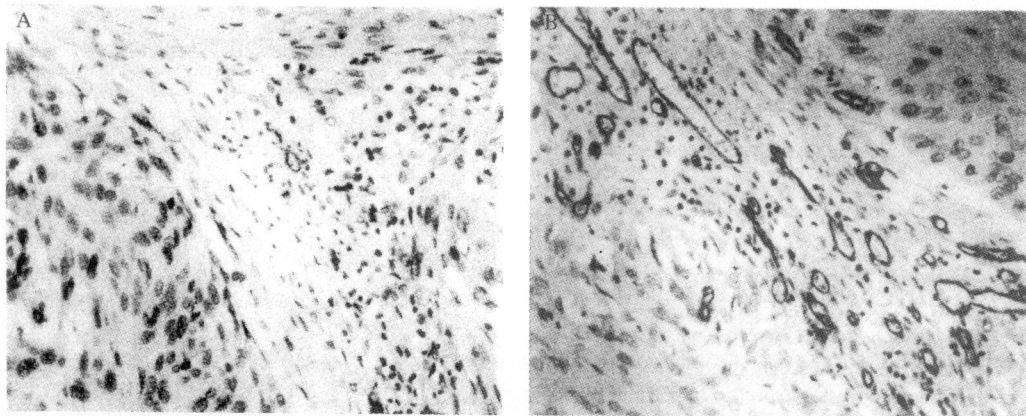

A, In the tumor with As_2O_3 administration, there were a few microvessels, in dense stroma; B, In the tumor tissue without arsenic administration there were abundant microvessels in the stroma (CD34 immunostain, ×100)

**Fig. 3   Capillaries stained with antibodies of CD34 are shown localized
at the stroma interface between carcinoma cell nests**

The microvessel density (MVD) was higher in the areas of the left tumor than in the right tumor in each group of the mice, the difference was statistically significant at dosage of 5 μg and 10 μg (Tab. 2).

4. Proliferative activity of tumor cells: Proliferative activity of the tumor cells was determined by image analysis system to estimate the expression of the Ki − 67 in the areas of the section. The mean of two percentages of Ki − 67 − labeled nuclei in tumors of both sides was calculated and used in the statistical analysis. The positive rate of Ki – 67 (%) in left side tumors was higher than that of right side tumors (Tab. 3).

Tab. 2   Comparison of MVD in right versus left tumor on 20 fields in various dosage of arsenic

Dosage	Right		Left		t – value	p – value
(μg)	n	MVD	n	MVD		
1	10	194. 7 ± 80. 7	10	230. 0 ± 107. 6	0. 83	> 0. 05
5	10	126. 4 ± 55. 3	10	201. 5 ± 50. 6	3. 17	< 0. 01
10	9	87. 5 ± 27. 7	10	182. 3 ± 58. 8	4. 61	< 0. 01
0 (control)	10	182. 0 ± 23. 5	10	220. 0 ± 67. 6	1. 68	> 0. 05

Tab. 3   Comparison of Ki – 67 labeled nuclei in tumors of both sides at various dosage of As_2O_3

Dosage	Right		Left		t – value	p – value
(μg)	n	%	n	%		
1	10	25. 7 ± 11. 4	10	30. 7 ± 10. 5	1. 02	> 0. 05
5	10	10. 5 ± 3. 5	10	25. 3 ± 11. 5	3. 89	< 0. 01
10	9	7. 2 ± 1. 7	10	18. 5 ± 8. 8	3. 97	< 0. 01
0 (control)	10	22 ± 10. 3	10	25. 6 ± 12. 7	0. 70	> 0. 05

5. Correlation of capillary density with growth fraction of Ki – 67: To determine whether angiogenesis is associated with cell proliferation status, we compared the relationship between MVD and rate of Ki – 67 at various dosages of arsenic. It was noted that decreasing Ki – 67 rate and MVD accompanied increasing concentration of arsenic and Ki – 67 rate were parallel to the MVD in right (treated) tumors (Fig. 4).

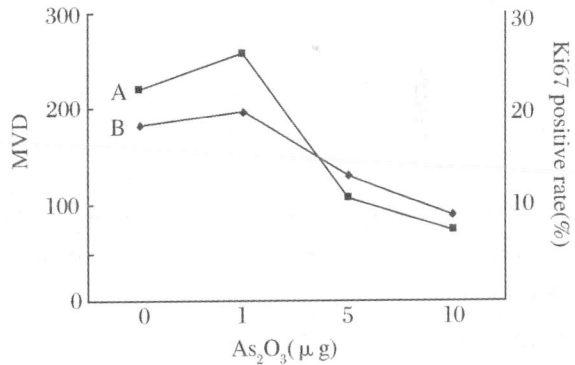

Fig. 4   Correlation of MVD and Ki – 67 positive rate was parallel to the MVD. A, Ki – 67; B, MVD

Discussion

This report suggests the approach of intratumoral injection in experimental animal models to increase the localization of arsenic trioxide level in tumors, to reduce their uptake in normal tissues, and thus to improve the antitumoral effects. Experimental animal models permit the assessment of a wide range of dose concentration and methods of administration before their use in clinical trials. Although many articles on the effects of As_2O_3 on tumors *in vitro* have been published (4, 18, 19), the reports, emphasized that the relationship between increasing As_2O_3 uptakes in tumors and improving therapeutic results *in vivo*, were rare. To achieve this improvement in therapeutic efficacy of As_2O_3, it is necessary to increase concentration in the localization in tumor while minimizing uptake in normal tissues at the same time, thereby achieving a maximum therapeutic effect. However, As_2O_3 have different biodistributions and pharmacokinetics in animal models compared to humans, animal models are useful for developing and evaluating new experimental approaches to increase the localization of As_2O_3 level in tumors.

Angiogenesis, the formation of new blood vessels from existing vasculature, is essential to the progression of tumor and allowing tumors to invade surrounding tissue and metastasize. The concept of treatment of solid tumors by inhibiting tumor angiogenesis was first report almost 30 years ago[20], and more than two decades ago angiogenesis that proceeded neoplastic transformation was studied[21]. In recent studies, some investigators showed interest in the proangiogenic molecule such as basic fibroblast growth factor[22] and vascular endothelial growth factor (VEGF) and the approaches of growing vascular endothelial cells in culture[23], *in vivo* assays of angiogenesis and antiangiogenic drug development[24]. In this study we measured the MVD in tumor tissues by labeled CD34, which marked endothelial cells of microvessels. The results revealed that increasing concentration of As_2O_3 decreased MVD and Ki – 67 labeled cells, and up regulated TGI.

The histological analysis of our study demonstrated that neovascularization is formed adjacent to the tumor cell mass. An angiogenic capability developed as a consequence of some localized change in the population of SHEEC cells. The neovascularization was initiated during early tumor development from surrounding tissue. The resulting *in vivo* studies confirm that onset of angiogenic activity occurs in surrounding connective tissue at an early stage of tumor mass formation. This feature revealed that induction of angiogenesis is a local event, performed by tumor cells and environmental stroma cells. After tumor was established, the decrease of the MVD in tumor tissue was related to treatment of As_2O_3. A high degree of tumor angiogenesis takes place *in vivo* without treatment of As_2O_3, and decrease of MVD appeared in As_2O_3 – treated tumors. Angiogenic activity first appears in the early stages after inoculation of SHEEC before the onset of tumor formation. These findings suggest that induction of angiogenesis is an important step in carcinogenesis.

Arsenic is a cytotoxic drug, which effects tumor cells and host cells. We used As_2O_3 to induce apoptosis *in vitro* of esophageal carcinoma cells[25]. The necrotic changes were also found to have a large dosage of As_2O_3[26]. Prior to the obvious change of cell nuclei, the mitochondria were swollen balloon – like and disrupted[27] and the functional changes of mitochondria helped promote production of NO and Ca^{2+}, which are related to the mitochondria – dependent apoptotic pathway[28]. Our data indicate that the severe damage of the cytoskeleton was accompanied with apoptotic process induced by As_2O_3[29].

In this study we have confirmed that As_2O_3 might have an antivascular effect on solid tumors. The mechanisms of vascular shut down are explained by the known biological effects of As_2O_3. One possible mechanism is that As_2O_3 can induce selective endothelial cell injury, resulting in vascular shut down in tumor tissue. Our recent studies indicate that arsenic may generate reactive oxygen species to exert its toxicity (unpublished data). The reactive oxygen species induced by arsenic are known to induce poly – ADP – ribosylation, which is implicated in DNA repair, signal transduction, and apoptosis[30]. Our studies have suggested that intracellular increase of nitric oxide synthase (NOS) and nitric oxide (NO) are induced by arsenite[28], and the induction of nitric oxide may be important to the etiology of arsenic – induced vascular disorders in humans. NOS has been demonstrated to have a direct correlation with MVD in human cancer[24].

In summary, we have demonstrated that at higher dosage (>5 μg) As_2O_3 via intratumoral injection causes apoptosis and necrosis in transplanted esophagealcancer and induced inhibition of tumoral growth and angiogenesis. In the progressive stage tumor development was dependent upon microvessel density. Therefore, suppression of angiogenesis is an important strategy for tumor therapy.

Acknowledgements

This study was supported by a grant sponsor from the National Natural Science Foundation of China (39830380), Research and Development Foundation of Shantou University (L00012) and by the Chinese National Human Genome Center, Beijing, P. R. China.

[In 《Oncol Rep》 2003, 10: 1869 – 1874]

References

1 Cai X, Shen YL, Zhu Q, Jia PM, Yu Y, Zhou L, Huang Y, Zhang JW, Xiong SM, Chen S J, Wang ZY, Chen Z and Chen GQ. Arsenic trioxide – induced apoptosis and differentiation are associated respectively with mitochondrial transmembrane potential collapse and retinoic acid signaling pathways in acute promyelocytic leukemia. Leukemia, 2000, 14: 262 – 270

2 Shen ZX, Chert GQ, Ni JH, Li XS, Xiong SM, Qiu QY, Zhu J, Tang W, Sun GL, Yang KQ, Chen Y, Zhou L, Fang ZW, Wang YT, Ma J, Zhang P, Zhang TD, Chen SJ, Chen Z and Wang ZY. Use of arsenic trioxide (As_2O_3) in the treatment of acute promyelocytic leukemia (APL): II. Clinical efficacy and pharmacokinetics in relapsed patients. Blood, 1997, 89: 3354 – 3360

3 Lew YS, Brown SL, Griffin RJ, Song CW and Kim JH. Arsenic trioxide causes selective necrosis in solid murine tumors by vascular shutdown. Cancer Res, 1999, 59: 6033 – 6037

4 Shen ZY, Tan LJ, Cai WJ, Shen J, Chen C, Tang XM and Zheng MH. Arsenic trioxide induces apoptosis of oesophageal carcinoma *in vitro*. Int J Mol Med, 1999, 4: 33 – 37

5 Shen ZY, Tan LJ, Cai WJ, Shen J, Chen CY and Tang XM. Morphologic study on apoptosis of esophageal carcinoma cell line induced by arsenic trioxide. Shijie Huaren Xiaohua Zhazhi (in Chinese), 1998, 6: 226 – 229

6 Shen ZY, Zhang Y, Chen JY, Chen MH, Shen J, Luo WH and Zeng Y. Intratumoral injection of arsenic to enhance antitumor efficacy in human esophageal carcinoma cells xenograft. Int J Cancer (In press)

7 Dekel Y, Koren R, Kugel V, Livne PM and Gal R. Significance of angiogenesis and microvascular invasion in renal cell carcinoma. Pathol Oncol Res, 2002, 8: 129 – 132

8 Edel MJ, Harvey JM and Papadimitriou JM. Comparison of vascularity and angiogenesis in primary invasive mammary carcinomas and in their respective axillary lymph node metastases. Clin Exp Metastasis, 2000, 18: 695 – 702

9 Tsoli E, Zacharatos P, Dasiou – Plakida D, Peros J, Evangelou K, Zavras AI, Yannoukakos D, Konstantopoulou I, Asimacopoulos PJ, Kittas C and Gorgoulis VC. Growth index is independent of microvessel density in non – small cell lung carcinomas. Hum Pathol, 2002, 33: 812 – 818

10 Tabara H, Kohno H, Dhar DK, Kotoh T, Yoshimura H, Masunaga R, Tachibana M, Kubota H and Nagasue N. Concurrent expression of angiogenic growth factors and neo – vasculariza-

tion during tumorigenesis in colorectal carcinoma patients. Acta Oncol, 2001, 40: 622 - 628

11 Artese L, Rubini C, Ferrero G, Fioroni M, Santinelli A and Piattelli A. Microvessel density (MVD) and vascular endothelial growth factor expression (VEGF) in human oral squamous cell carcinoma. Anticancer Res, 2001, 21: 689 - 695

12 Matsushima H, Goto T, Hosaka Y, Kitamura T and Kawabe K. Correlation between proliferation, apoptosis, and angiogenesis in prostate carcinoma and their relation to androgen ablation. Cancer, 1999, 85: 1822 - 1827

13 Lenczewski A, Terlikowski SJ, Sulkowska M, Famulski W, Sulkowski S and Kulikowski M. Prognostic significance of CD34 expression in early cervical squamous cell carcinoma. Folia Histochem Cytobiol, 2002, 40: 205 - 206

14 Aim MJ, Jang SJ, Park YW, Choi JH, Oh HS, Lee CB, Paik HK and Park CK. Clinical prognostic values of vascular endo - thelial growth factor, microvessel density, and p53 expression in esophageal carcinomas. J Korean Med Sci, 2002, 17: 201 - 207

15 Sharma RA, Harris AL, Dalgleish AG, Steward WP and O'Byrne KJ. Angiogenesis as a biomarker and target in cancer chemoprevention. Lancet Oncol, 2001, 2: 726 - 732

16 Folkman J, Watson K, Ingber D and Hanahan D. Induction of angiogenesis during the transition from hyperplasia to neoplasia. Nature, 1989, 339: 58 - 61

17 Folkman J and Klagsbrun M. Angiogenic factors. Science, 1987, 235: 442 - 447

18 Gu QL, Li NL, Zhu ZG, Yin HR and Lin YZ. A study on arsenic trioxide inducing *in vitro* apoptosis of gastric cancer cell lines. World J Gastroenterol, 2000, 6: 435 - 437

19 Xu HY, Yang YL, Gao YY, Wu QL and Gao GQ. Effect of arsenic trioxide on human hepatoma cell line BEL - 7402 cultured *in vitro*. World J Gastroenterol, 2000, 6: 681 - 687

20 Kerbel RS. Tumor angiogenesis. past, present and the near future. Carcinogenesis, 2000, 21: 505 - 515

21 O'Byrne KJ, Dalgleish AG, Browning MJ, Steward WP and Harris AL. The relationship between angiogenesis and the immune response in carcinogenesis and the progression of malignant disease. Eur J Cancer, 2000, 36: 151 - 169

22 Compagni A, Wilgenbus P, Impagnatiello MA, Cotten M and Christofori G. Fibroblast growth factors are required for efficient tumor angiogenesis. Cancer Res, 2000, 60: 7163 - 7169

23 Marrogi AJ, Travis WD, Welsh JA, Khan MA, Rahim H, Tazalaar Pairolero P, Trastek V, Jett J, Caporaso NE, Liotta LA and Harris CC. Nitric oxide synthase, cyclooxygenase 2, and vascular endothelial growth factor in the angiogenesis of non - small cell lung carcinoma. Clin Cancer Res, 2000, 6: 4739 - 4744

24 Lu J and Jiang C. Antiangiogenic activity of selenium in cancer chemoprevention. metabolite - specific effects. Nutr Cancer, 2001, 40: 64 - 73

25 Shen ZY, Shen J, Chen MH, Li QS and Hong CQ. Morphological changes of mitochondria in apoptosis of esophageal carcinoma cells induced by As_2O_3. Chin J Pathol (in Chinese), 2000, 29: 200 - 203

26 Shen ZY, Chen CY, Shen J and Cai WJ. Ultrastructural study of apoptosis and necrosis in the esophageal carcinoma cell line induced by arsenic trioxide. Chin J Med Physics, 1999, 16: 91 - 94

27 Shen ZY, Shen J, Cai WJ, Hong CQ and Zheng MH. The alteration of mitochondria is an early event of arsenic trioxide induced apoptosis in esophageal carcinoma cells. Int J Mol Med, 2000, 5: 155 - 158

28 Shen ZY, Shen WY, Chen MH, Shen J, Cai WJ and Zeng Y. Mitochondria, calcium and nitric oxide in the apoptotic pathway of esophageal carcinoma cells induced by As_2O_3. Int J Mol Med, 2002, 9: 385 - 390

29 Shen ZY, Shen J, Cai WJ, Chen CY and Chen JY. Changes in cytoskeleton of apoptosis of

esophageal carcinoma cells induced by As$_2$O$_3$. Chin J Pathol, 2001, 30: 369 – 370

30 Lynn S, Shiung JN, Gurr JR and Jan KY. Arse-

nite stimulated poly (ADP – ribosylation) by generation of nitric oxide. Free Radic Biol Med, 1998, 24: 442 – 449

84. Ezrin, Actin and Cytoskeleton in Apoptosis of Esophageal Epithelial Cells Induced by Arsenic Trioxide

SHENG Zhong – ying[1], XU Li – yan[1], LI En – min[2], LI Jin – tao[1], CHEN Ming – hua[1], SHEN Jian[1], ZENG Yi[3]

Department of 1. pathology and 2. Biochemistry, Medical College of Shantou University,

3. Institute of Virology, Chinese Academy of Preventive Medicine

Summary

Ezrin is a key protein in membrane – cytoskeleton interaction. Expression of ezrin in actin – rich cell surfaces may play a role in the modulation of cell shape and adhesion. The aim of this study was to detect ezrin, actin and cytoskeleton and to explore their relationship in the apoptosis of esophageal epithelial cells (SHEE) induced by arsenic trioxide (As$_2$O$_3$). The SHEE is an immortalized human fetal esophageal epithelial cell line, and the cells were treated by administering 5, 10 and 20 μmol of As$_2$O$_3$. The proliferation and apoptosis of SHEE cells were examined by flow cytometry with propidium iodide staining. Ezrin expression was detected by immunocytochemistry and Western blotting. Actin filament was stained by FITC – labeled phalloidin and detected quantitatively by fluorescent microscopy. Cell morphology and microfilaments were examined by electron microscopy. The results revealed that As$_2$O$_3$ induced an inhibition of proliferation and the promotion of apoptosis in SHEE cells. The ezrin, actin and cytoskeleton were decreased after As$_2$O$_3$ treatment and the cellular morphology of apoptosis developed. Our results suggested that the morphological changes of arsenic – induced apoptosis of human esophageal epithelial cells were initiated by ezrin and actin – cytoskeletal aberrance.

[Key words] Esophageal epithelium; Aarsenic; Eezrin; Actin; Cytoskeleton

Introduction

Arsenic trioxide (As$_2$O$_3$) inhibits growth and induces apoptosis in many malignant cells[1-7]. As$_2$O$_3$ is used as an effective antitumor agent, not only for anti – leukemia[8-10], but also used in solid tumors[11-14]. In our previous studies, As$_2$O$_3$ altered the cell morphology of esophageal cancer cells, reduced cell viability and increased the number of apoptotic cells[15,16]. The pathomorphological changes induced by As$_2$O$_3$ revealed that cells became smaller and shrank, mitochondria swelled, nuclear chromatin agglutinated and marginated, the nuclear membrane broke down followed by degenerative changes and cell mortality[17,18]. All these changes indicated typical morphological changes of apoptosis. In recent literature, studies on the morphological changes of apoptotic cells were focused on nuclear alternation, in which condensed and marginated chromatins in shrunk and rounded nuclei were caused by DNA fragmentation[19,20]. Little is known about the mechanism causing alteration of the shape and size of apoptotic cells, which become smaller, condensed and rounded.

Actin – containing microfilaments play an essential role in determining cell shape, and in cell locomotion and contractility[21]. Microtubules and actin filaments are involved in mitosis, cell signaling, and mobility[22]. Filamentous actin and actin – associated proteins, such as ezrin and CD44, can assemble into higher order structures, such as filopodia, microspikes, lamellipodia, and stress fibers[23]. In various cell types, microfilaments interact with other proteins to provide machinery for their proper morphology.

Molecular biological studies have demonstrated the function of the various components of the cytoskeleton system and enabled the characterization of protein – protein interactions involved in their assembly[24]. It has become clear that the ezrin family, a major component of adhesion protein, is able to transfer signals from the extraceUular matrix to the interior of the cell, to critically modulate the organization of the cyto – skeleton, and to engage to proliferation, apoptosis, and cell differentiation[25].

Nevertheless, our knowledge of the mechanisms of ezrin regulating the functional state of the cytoskeletion and, hence, the dynamics of cell alteration, a process of crucial importance in development of apoptosis, remains limited. In this study we investigated arsenic inhibiting cell proliferation and inducing cell apoptosis, and the mechanisms of cell shape and size alteration to involve ezrin, actin filament and cytoskeleton.

Materials and Methods

1. Cell culture and As_2O_3 administration: The SHEE cell line was established from embryonic esophageal epithelium induced by genes E6E7 of human papillomavirus (HPV) type 18 in our laboratory with a long – term *in vitro* passage (over 100 passages). The cells in 10 – passage intervals were stored in liquid nitrogen. The SHEE10 cells, an early passage of immortalized cell, were cultured in a water – jacked incubator, maintained at 37℃ with a humidified atmosphere of 5% CO_2 and 95% air. Cells were grown in medium 199 supplemented with 10% fetal calf serum, 50 μg/ml penicillin and 50 μg/ml streptomycin in 50 ml culture flasks. Arsenic trioxide (As_2O_3) in ampoule (0.1% As_2O_3, 10 ml) was supplied by the Pharmacy of Chinese Traditional Medicine of First Hospital affilated to Harbin Medical University (Harbin, P. R. China). Various concentrations of As_2O_3, 5, 10 and 20 μmol, were added into the culture for a period up to 24 h. To measure the effect of As_2O_3 on cell growth, cultured cells were examined at different time points and doses.

2. Examination by electron microscope: For electron microscopic assessment, cells were collected by trypsinization and immediately fixed with 2.5% glutaraldehyde. The cells were dehydrated in graded ethanol and embedded in Araldite. Ultra – thin sections were cut with glass knives and mounted on copper grids. They were contrasted for 15 min with uranyl acetate and for 3 min with lead citrate. The sections were examined using a transmitting electron microscope (H – 300, Hitachi).

3. Flow cytometry analysis: To measure the ratio of apoptotic cells and normal cells in As_2O_3 treated and untreated cells, SHEE cells were washed with PBS after treatment of As_2O_3 and then fixed with 70% precooled alcohol on ice. Cells were suspended in PBS contained 100 μl RNase (1 μg/ml) and then stained with propidium iodide at a dose of 30 μg/ml for 15 min. DNA content of cells was measured by flow cytometry (FACSort, B – D Co.). In a DNA histogram, the percentage of cells in each phase of the cell cycle, and the cells containing hypoploidy DNA (apoptotic cells) were calculated. The proliferative index (PI) was calculated by the formula of $(S + G_2M/G_0G_1 + S + G_2M)$.

4. Immunocytochemical analysis of ezrin: The SHEE cells cultured on the cover slips were fixed in 10% formalin. Immunoperoxidase staining by the avidin – biotin – peroxidase complex method was performed with a Maxim ABC kit (Maxim Co.). The samples were immersed for 30 min in methanol containing 0. 3% H_2O_2, and washed in PBS (pH 7. 4). After autoclaving, the slips were incubated for 10 min with normal swine serum to block non – specific binding of the antibody. The slips were exposed to mouse anti – human ezrin antibody (MAB – 0351, p81, Maxim – Bio) at a dilution of 1 : 200, at 4 ℃ overnight. The samples were then incubated with biotinylated secondary antibody for 30 min at room temperature, and with the Maxim ABC reagent for 30 min. For visualization slips were immersed in 0. 05% diaminobenzidine tetrahydrochloride solution containing 0. 01% H_2O_2 with a light counterstain.

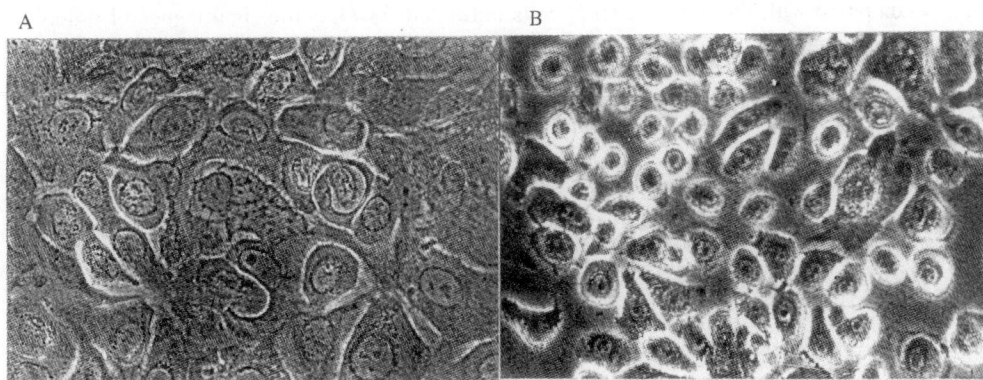

A, before As_2O_3 was added, the cells showed a single layer of squamous appearance with gaps between cells; B, after As_2O_3 added, the cells became shrunken, round shaped and dissociation cellular attachment occurred (phase – contrast, ×320)

Fig. 1   Phase – contrast micrograph of living SHEE cells

5. Ezrin assay by Western blot: The protein expression of ezrin was detected by the Western blot method. Cells were collected and were washed three times with precooled PBS, then were lysed in buffer [50 mmol/L Tris – HCl, pH 8. 0, 150 mmol/L NaCl, 100 μg/ml phenyl – methyl – sulfonyl fluoride (PMSF), 1% Triton X – 100] for 30 min on ice. After removal of cell debris by centrifugation (12 000 r/min, 5 min), the protein of different samples was boiled for 5 min in sample buffer and was separated by 10% SDS – PAGE, transferred onto nitrocellulose membrane (Bio – Rad). Non – specific reactivity was blocked in buffer (10 mmol/L Tris – HCl, pH 7.5, 150 mmol/L NaCl, 2% Tween 20, 4% bovine serum albumin) by incubation overnight at 4℃. The membrane was then incubated with monoclonal antibody of mouse anti – human ezrin (p81, MAB – 0351, Maxim – Bio) followed by reaction with anti – mouse IgG – IRP antibody. Reactive protein was detected by ECL chemi – luminescence system (Amersham).

6. *Fluorescent analysis of F – actin*[26]: SHEE cells, grown on glass cover slips, were washed twice with prewarmed PBS, pH 7.4. The sample was fixed in 4% formaldehyde solution in PBS for 10 min at room temperature, washed two or more times with PBS, placed on each cover slip in a glass petri dish and extracted with a solution of acetone for 3 to 5 min. It was then washed two or more times with PBS. Fixed cells were pre – incubated with PBS containing 1% BSA for 20 – 30 min. FITC labeled phalloidin (F –432, Molecular Probe Co.) was diluted, 5 μl methanol stock solution into 200 μl PBS for each cover slip to be stained for 20 min at room temperature, washed with PBS twice, sample was examined under fluorescent microscopy (Optiphot, Nikon). Fluorescent intensity of phalloitin in cells was performed by camera head (start I) and controller software (v. 1.5, Photometrics Ltd). The average fluorescent intensity of cells was measured with or without adding As_2O_3 and the average fluorescent intensity value (average $\bar{x} \pm s$) of 10^3 cells was calculated.

7. *Statistical analysis*: Semiquantitative measurement of ezrin in cells in various concentrations of As_2O_3 was scanned and data quantified using 1D Image Analysis software (Version 3.5, Kodak Co.). As compared with the data of SHEE cells untreated As_2O_3, the significance of difference was evaluated by t – test at $P < 0.05$.

Results

1. *Morphology of living cells*: The cell morphology showed no significant changes at 2h after As_2O_3 treatment. SHEE cells appeared polygonal with micro – projection and cell – to – cell attachment (Fig. 1A). Within 12 –24h of 10 μmol of As_2O_3 treatment, the cells had shrank significantly and displayed a ball – like appearance with loss of microvilli, membrane ruffles and cell – to – cell contact (Fig. 1B). In parallel studies without As_2O_3 treatment, no significant changes of cell morphology could be observed up to 24 h. Therefore, a high concentration of As_2O_3 (over 10 μmol) induced significant morphologic changes in SHEE.

2. *Ultrastructure of SHEE cells before and after As_2O_3 administration*: SHEE cells displayed irregular to round shaped nuclei with random distribution of heterochromatin, and a large nucleolus. Cellular tonofilaments were distributed evenly and surrounded the nucleus (Fig. 2A). After 10 or 20 μmol of As_2O_3 was added for 4 –8h, tonofilaments were pressed and resulted in dislocation and rearrangement between the hyperplastic and swollen mitochondria (Fig. 2B). When SHEE cells were treated with 10 or 20 μmol/L of As_2O_3 for 12 –24h, balloon – like mitochondria were observed, visualized tonofilaments disappeared and a condensed shrunken nucleus was evident (Fig. 2C).

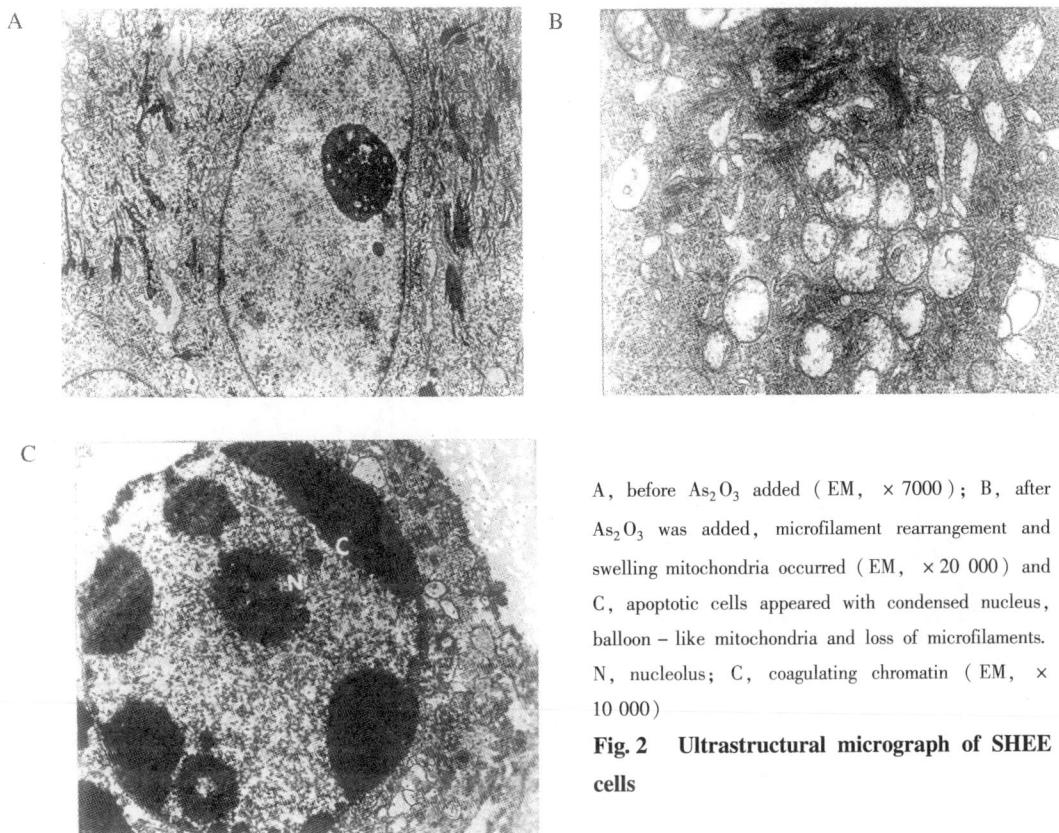

A, before As_2O_3 added (EM, ×7000); B, after As_2O_3 was added, microfilament rearrangement and swelling mitochondria occurred (EM, ×20 000) and C, apoptotic cells appeared with condensed nucleus, balloon – like mitochondria and loss of microfilaments. N, nucleolus; C, coagulating chromatin (EM, × 10 000)

Fig. 2   Ultrastructural micrograph of SHEE cells

3. Proliferative index and apoptotic rate: Flow cytometric analysis of propidium iodide labeled SHEE cells showed the typical G_0/G_1 peak and G_2/M peak at double fluorescence peaks before As_2O_3 treatment (Fig. 3A). However, the sub – G_1 peak, representative of apoptotic cells, emerged and the G_2/M peak decreased at 12 h after 10 μmol As_2O_3 administration (Fig. 3B). It was shown that the growth rate was inhibited. In dose – response experiments, it was evident that concentration over 10 μmol of As_2O_3 for 12 – 24 h would induce apoptosis in SHEE, while having an effect only on cell necrosis in SHEE cells during treatment with a large dosage, 20 μmol of As_2O_3. Apoptotic rate varied at 5, 10 and 20 μmol As_2O_3 for 24 h. The basal rate of apoptosis (RA) in the absence of As_2O_3 was 2.5% ±0.5% and the proliferation index (PI) was 35.3% of the total SHEE cell population; at 5.0 μmol As_2O_3, the RA was 5.2% ±1.7%; PI, 30.9% ± 10.3%. However, at 10 μm As_2O_3 the RA was increased to 20.5% ±7.6%, and PI decreased to 18.2% ±7.2%, and at 20 μmol As_2O_3 RA was significantly increased, 52.5% ±13.5%, and PI significantly decreased, 10.5% ±3.2%.

4. Localization of F – actin: The subcellular localization of actin of cultured cells was studied by FITC – labeled phalloidin in examination under a fluorescent microscope. The effect of As_2O_3 on intracellular actin varied in different concentration of As_2O_3. In SHEE cells, a fibrillar punctate staining pattern was observed in the cytoplasm (Fig. 4A). After As_2O_3 administration, the FITC –

labeled actin decreased accompanied by changes in cellular morphology, in which the polygonal shape containing filopodial projections was transformed into a ball – like shape (Fig. 4B). The fluorescent intensity (FI) of F – actin was different in treated and untreated SHEE cells.

When As_2O_3 induced apoptosis of SHEE cells, the actin fluorescent intensity were decreased in SHEE cells according to the concentrations of As_2O_3 added (Fig. 5).

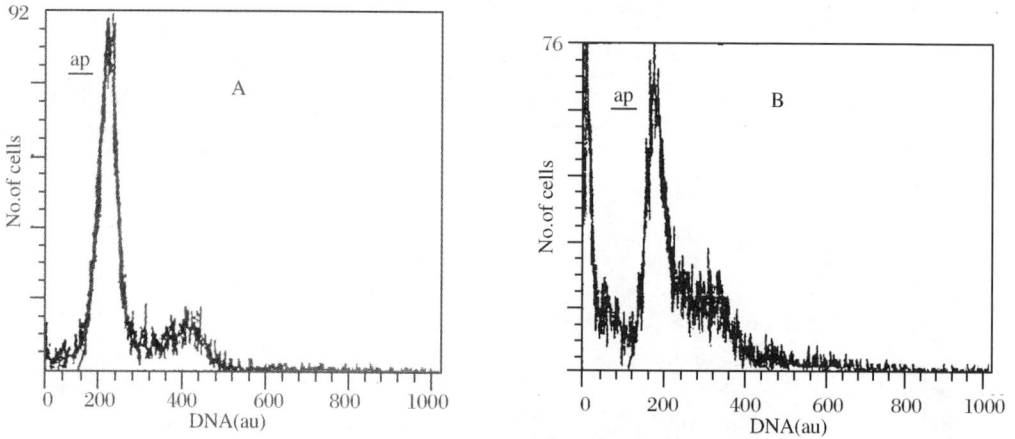

A, cells before adding As_2O_3 ; B, cells in 24 h after adding 10 μmol/L of As_2O_3. ap, apoptotic peak; au, arbitrary unit

Fig. 3   DNA histogram of SHEE

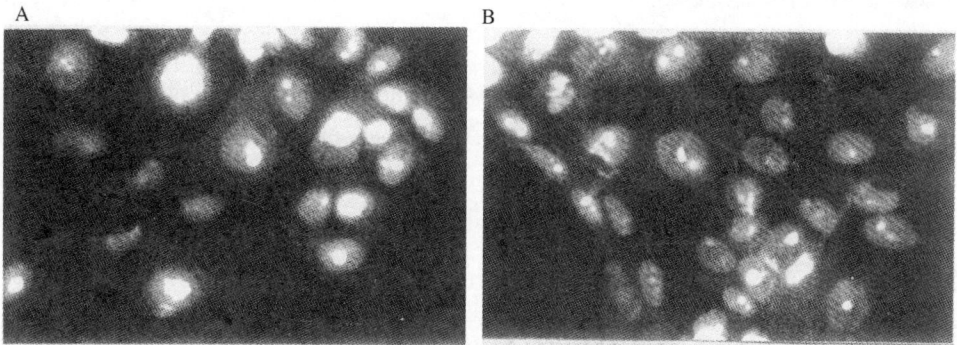

A, before As_2O_3 was added, F – actin can be seen as fibrillar punctate structures. B, after adding 10 μmol of As_2O_3, there were a few fibrils in the cytoplasm of the shrunken cells (fluorescent microscope, ×400)

Fig. 4   Localization of actin in SHEE cells (FITC – phalloidin labeled)

5. *Ezrin expression by immunocytochemistry*: On single – layered culture, ezrin protein expressed in a honey comb – like appearance preferentially located in the plasma membranes from non – treated cells (Fig. 6A). Whereas from As_2O_3 – treated cells immunostaining of ezrin was often weakened (Fig. 6B). Decrease of ezrin proteins in cytoplasmic membrane probably was responsible for the As_2O_3 treatment in early phase of apoptosis.

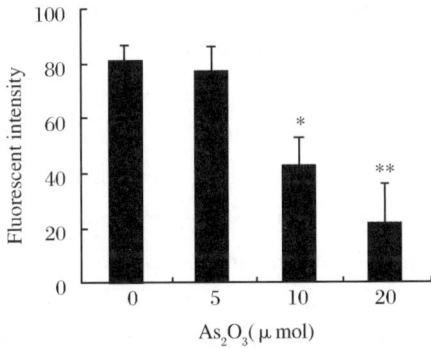

Data are expressed as the percent of the total cellular actin measured in similar condition. The columns represent the rheans ± S ($n = 3$). * $P < 0.05$, * * $P < 0.01$, versus control

Fig. 5   Effect of As$_2$O$_3$ on Factin in various dosage and time

A   B

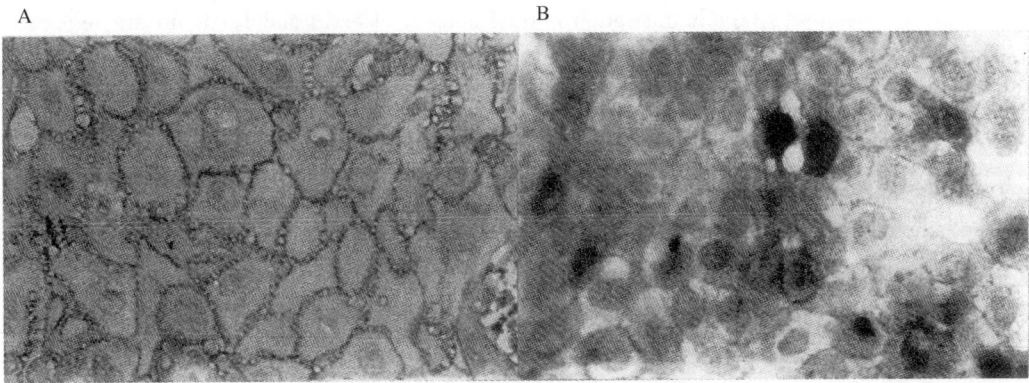

A, before As$_2$O$_3$ was added, ezrin protein expressed strongly in the plasma membrane; B, after As$_2$O$_3$ – treatment ezrin protein expressed weakly on plasmic membrane (Immunostain, ×400)

Fig. 6   Expression of ezrin in SHEE cells before and after 10 μmol of As$_2$O$_3$ administration for 6h

Ezrin expression by Western blot. During the As$_2$O$_3$ – induced apoptosis we studied the expression of ezrin by Western blot. SHEE cells contained an abundance of ezrin, whereas ezrin proteins of SHEE cells showed an immunoreactive migrating 8. 1 × 10^3. The ezrin was detected semi – quantitatively (Fig. 7), after 10 and 20 μmol of As$_2$O$_3$ administration the lysates showed decreased expression in the SHEE cells. In the 5 μmol of As$_2$O$_3$ group, no differences were detected in the ezrin expression level.

Discussion

Previously, we revealed that As$_2$O$_3$ induced

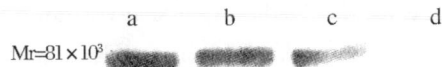

The ezrin protein bands of SHEE cells located in 81 × 10^3. Lane a, ezrin protein from SHEE cells before As$_2$O$_3$ treatment, the ezrin band is clearly evident; lane b, ezrin protein from SHEE cells after 5 μmol As$_2$O$_3$ treatment; lane c, ezrin protein from SHEE after 10 μmol As$_2$O$_3$ treatment; lane d, ezrin protein from SHEE after 20 μmol As$_2$O$_3$ treatment. Densitometric scanning analysis shows the intensity of the ezrin bands

Fig. 7   Western blot analysis of pattern of ezrin protein electrophoresis

inhibition of proliferation of cells[27] and As$_2$O$_3$ can induce apoptosis in the SHEE cell line, in a dose – dependent manner[28]. We also showed that SHEE cells are susceptible, to As$_2$O$_3$ – mediated apoptosis and exhibit a pyknotic morphology, cell rounding with shrinking of nucleus accompanying by fragmentation of nuclear DNA[29]. In this study, apoptosis was induced by administration of 10 and 20 μmol of As$_2$O$_3$, characterized by marked morphological alterations of the SHEE cells, such as pyknotic nucleus with chromatin condensation and fragmentation.

Ezrin, a protein of 81×10^3 is a cytoplasmic protein enriched in microvilli and cell surface structures. Ezrin has a membrane – cytoskeleton linker role[30]. Its NH$_2$ – terminal domain is associated with the plasma membrane and the COOH – terminal domain with the cytoskeleton[31]. Ezrin is expressed in epithelial cells and mesenchymal cells and is also expressed by certain tumors[32-35].

In this study, ezrin may be an important mechanism for changes of SHEE cell shape after As$_2$O$_3$ is added. Before As$_2$O$_3$ administration normal content of ezrin and F – actin are required for the maintenance of normal cellular morphology. When the co – localization between ezrin and F – actin is disrupted by As$_2$O$_3$, SHEE cells undergo a dramatic morphological change. The dissociation of connection between ezrin and actin or ezrin translocation caused major changes in cell shape[23], which will result in substantial cytoskeletal rearrangement. Actin cytoskeleton and its components, the microfilaments, intermediate filaments, microtubules and other cytoskeleton – related proteins, play an important role in the dynamics of apoptosis[36].

Actin – containing microfilaments control cell shape, adhesion, and contraction[21]. The actin – based cytoskeleton plays a central role in cell motility and cortical morphogenesis[36], and is reorganized during cell transformation[37]. It has been reported that cytoskeletal filaments are the targets of some anticancer drugs[37], As$_2$O$_3$ induces extensive cytoskeletal protein breakdown and significantly increases cell bleb, rounding – up, and overall size modification, modulating morphogenesis by tethering actin microfilamentsto the plasma membrane. We identified phalloidin in this study as an actin – associated protein. Actin is an abundant component of the cytoplasmic cytoskeleton of SHEE cells. Phalloidin expression was down regulated in SHEE cells coinciding with major cytoskeletal and morphological alterations. The relationship between actin and micro – filaments was further verified by phalloidin staining and electron microscopic observation. Our results indicate that the cytoskeleton also undergoes marked morphological changes during administration of As$_2$O$_3$. Among these changes, the rearrangement and disappearance of actin – cytoskeleton has been recognized as one of the common events of the apoptosis of SHEE cells. The morphological changes may actually be induced by cytoskeleton rearrangement and actin degradation after As$_2$O$_3$ treatment.

We hypothesized that cytoskeleton changes of apoptosis induced by arsenic could be mediated by two different stages. The first stage, an ezrin – mediated association of cyto – membrane with the actin cytoskeleton was blocked, including a dissociation of proteins from F – actin, a loss in ezrin and actin – binding, membrane – anchoring proteins, and the actin – filament rearrangement in the early stage of As$_2$O$_3$ – mediated apoptosis. The cytoskeletal rearrangement is likely to be a critical event in the pathway to apoptosis[38]. The second stage involves mitochondria to release cytochrome c, inducing activity of the apoptotic caspase cascade by the activation of caspases enzyme family members of cysteine proteases[39,40]. The caspases, which mediated cleavage of cytoskeletal actin[41], are the key effector proteins of apoptosis. At subsequent stages microfilament breakdown or disappearance occurred along with cell rounding. Two stages always strictly intermingled. But in some reports actin is not shown as a direct substrate for the ICE – like family of proteases[42].

In our study morphology of arsenic – induced apoptosis of human immortalized esophageal epithelial cells is regulated by a mechanism involving its association with ezrin and actin. Damage to the actin may play an important role in the morphology of apoptosis after As_2O_3 administration. We provide evidence of association of ezrin to arsenic – triggered shape alteration. In the early stages microfilaments were rearranged which probably dissociated from attachment of cytomembrane through ezrin degeneration, and the severe damage of the cytoskeleton resulting in apoptosis was a subsequent event.

Acknowledgements

This investigation was supported by an award by National Nutural Science Foundation of China (39830308); by Development Foundation of Shantou University (L00012), and by the Chinese National Human National Genome Center, Beijing.

〔In 《Int J Mol Med》 2003, 12: 341 – 347〕

References

1 Shen ZY, Tan LJ, Cai W J, Shen J, Chen C, Tang XM and Zheng MH. Arsenic trioxide induces apoptosis of esophageal carcinoma *in vitro*. Int J Mol Med, 1999, 4: 33 – 37

2 Li X, Ding X and Adrian TE. Arsenic trioxide inhibits proliferation and induces apoptosis in pancreatic cancer cells. Anticancer Res, 2002, 22: 2205 – 2213

3 Shi Y, Liu Y, Huo J and Gao G. Arsenic trioxide induced apoptosis and expression of p53 and bcl – 2 genes in human small cell lung cancer cells. Zhonghua Jie He He Hu Xi Za Zhi, 2002, 25: 665 – 666

4 Pu YS, Hour TC, Chen J, Huang CY, Guan JY and Lu SH. Cytotoxicity of arsenic trioxide to trasitional carcinoma cells. Urology, 2002, 60: 346 – 350

5 Zhu XH, Shen YL, Jinag YK, Cai X, Jia PM, Huang Y, Tang W, Shi GY, Sun YP, Dai J, Wang ZY, Chen S J, Zhang TD, Waxman S, Chen Z and Chen GQ. Apoptosis and growth inhibition in malignant lymphocytes after treatment with arsenic trioxide at clinically achievable concentrations. J Natl Cancer Inst, 1999, 91: 772 – 778

6 Chen HY, Liu WH and Qin SK. Induction of arsenic trioxide on apoptosis of hepatocarcinoma cell lines. Shijie Huaren Xiaohua Zazhi, 2000, 8: 532 – 535

7 Tu SP, Jiang SH, Tan JH, Jiang XH, Qiao MM, Zhang YP, Wu YL and Wu YX. Proliferation inhibition and apoptosis induction by arsenic trioxide on gastric cancer cell SGC 7901. Shijie Huaren Xiaohua Zazhi, 1999, 7: 18 – 21

8 Shim MJ, Kim HJ, Yang S J, Lee IS, Choi HI and Kim T. Arsenic trioxide induces apoptosis in chronic myelogenous leukemia K562 cells: possible involvement of p38 MAP kinase. J Biochem Mol Biol, 2002, 35: 377 – 383

9 Tamm I, Paternostro G and Zapata JM. Treatment of acute promyelocytic leukemia with arsenic trioxide. N Engl J Med, 1999, 340: 1043 – 1045

10 Jing Y, Dai J, Chalmers – Redman RM, Tatton WG and Waxman S. Arsenic trioxide selectively induces acute promyelocytic leukemia cell apoptosis via a hydrogen peroxide – dependent pathway. Blood, 1999, 94: 2102 – 2111

11 Seol JG, Park WH, Kim ES, Jung CW, Hyun JM, Lee YY and Kim BK. Potential role of caspase – 3 and – 9 in arsenic trioxide – mediated apoptosis in PCI – 1 head and neck cancer ceils. Int J Oncol, 2001, 18: 249 – 255

12 Gu QL, Li NL, Zhu ZG, Yin HR and Lin YZ. A study on arsenic trioxide inducing in vitro apoptosis of gastric cancer cell lines. World J

Gastroenterol, 2000, 6: 435 – 437

13　Xu HY, Yang YL, Gao YY, Wu QL and Gao GQ. Effect of arsenic trioxide on human hepatoma cell line BEL – 7402 cultured *in vitro*. World J Gastroenterol, 2000, 6: 681 – 687

14　Munshi NC, Tricot G, Desikan R, Badros A, Zangari M, Toor A, Morris C, Analssie E and Barlogie B. Clinical activity of arsenic trioxide for the treatment of multiple myeloma. Leukemia, 2002, 16: 1835 – 1837

15　Shen ZY, Tan LJ, Cai WJ, Shen J, Chen CY and Tang XM. Morphologic study on apoptosis of esophageal carcinoma cell line induced by arsenic trioxide. Shijie Huaren Xiaohua Zhahi (in Chinese), 1998, 6: 226 – 229

16　Shen ZY, Chert CY, Shen J and Cai WJ. Ultrastructural study of apoptosis and necrosis in the esophageal carcinoma cell line induced by arsenic trioxide. Chin J Med Physics (in Chinese), 1999, 16: 91 – 94

17　Shen ZY, Shen J, Chen MH, Li QS and Hong CQ. Morphological changes of mitochondria in apoptosis of esophageal carcinoma cells induced by As203. Chin J Pathol (in Chinese), 2000, 29: 200 – 203

18　Shen ZY, Shen J, Cai WJ, Hong CQ and Zheng MH. The alteration of mitochondria is an early event of arsenic trioxide induced apoptosis in esophageal carcinoma cells. Int J Mol Med, 2000, 5: 155 – 158

19　Nagata S. Apoptotic DNA fragmentation. Exp Cell Res, 2000, 256: 12 – 18

20　Arends MJ, Morris RG and Wyllie AH. Apoptosis. The role of the endonuclease. Am J Pathol, 1990, 136: 593 – 608

21　Mykkanen OM, Gronholm M, Ronty M, Lalowski M, Salmikangas P, Suila H and Carpen O. Characterization of human palladin, a microfilament – associated protein. Mol Biol Cell, 2001, 12: 3060 – 3073

22　Jordan MA and Wilson L. Microtubules and actin filaments: dynamic targets for cancer chemotherapy. Curt Opin Cell Biol, 1998, 10: 123 – 130

23　Lamb RF, Ozanne BW, Roy C, McGarry L, Stipp C, Mangeat P and Jay DG. Essential functions of ezrin in maintenance of cell shape and lamellipodial extension in normal and transformed fibroblasts. Curt Biol, 1997, 7: 682 – 688

24　Borradori L and Sonnenberg A. Structure and function of hemidesmosomes: more than simple adhesion complexes. J Invest Dermatol, 1999, 112: 411 – 418

25　Kondo T, Takeuchi K, Doi Y, Yonemura S, Nagata S and Tsukita S. ERM (Ezrin/Radixin/ Moesin) – based molecular mechanism of microvillar breakdown at an early stage of apoptosis. J Cell Biol, 1997, 139: 749 – 758

26　Endresen PC, Prytz PS and Aarbakke J. A new flow cytometric method for discrimination of apoptotic cells and detection of their cell cycle specificity through staining of F – actin and DNA. Cytometry, 1995, 20: 162 – 171

27　Shen ZY, Shen WY, Chen MH, Shen J, Cai WJ and Zeng Y. Nitric oxide and calcium ions in apoptotic esophageal carcinoma cells induced by arsenite. World J Gastroenterol, 2002, 8: 40 – 43

28　Shen ZY, Shen J, Li QS, Chen CY, Chen JY and Zeng Y. Morphological and functional changes of mitochondria in apoptotic esophageal carcinoma cells induced by arsenic trioxide. World J Gastroenterol, 2002, 8: 31 – 35

29　Shen ZY, Shen WY, Chen MH, Shen J, Cai WJ and Zeng Y. Mitochondria, calcium and nitric oxide in the apoptotic pathway of esophageal carcinoma cells induced by As_2O_3. Int J Mol Med, 2002, 9: 385 – 390

30　Andreoli C, Martin M, Borgne R, Reggio H and Mangeat P. Ezrin has properties to self – associate at the plasma membrane. J Cell Sci, 1994, 107: 2509 – 2521

31　Parlato S, Giammarioli AM, Logozzi M, Lozupone F, Matarrese P, Luciani F, Falchi M, Malorni W and Fais S. CD95 (APO – 1/Fas) linkage to the actin cytoskeleton through ezrin in human T lymphocytes: a novel regulatory mechanism of the CD95 apoptotic pathway. EMBO J,

2000, 19: 5123 – 5134

32   Vaheri A, Carpen O, Heiska L, Helander TS, Jaaskelainen J, Majander – Nordenswan P, Sainio M, Timonen T and Turunen O. The ezrin protein family: membrane – cytoskeleton interactions and disease associations. Curr Opin Cell Biol, 1997, 9: 659 – 666

33   Ohtani K, Sakamoto H, Rutherford T, Chen Z, Satoh K and Naftolin F. Ezrin, a membrane – cytoskeletal linking protein, is involved in the process of invasion of endometrial cancer ceils. Cancer Lett, 1999, 147: 31 – 38

34   Jiang WG and Hiscox S. Cytokine regulation of ezrin expression in the human colon cancer cell line HT29. Anticancer Res, 1996, 16: 861 – 865

35   Bohling T, Turunen O, Jaaskelainen J, Carpen O, Sainio M, Wahlstrom T, Vaheri A and Haltia M. Ezrin expression in stromal cells of capillary hemangioblastome. An immunohistochemical survey of brain tumors. Am J Pathol, 1996, 148: 367 – 373

36   Atencia R, Asumendi A and Garcia – Sanz M. Role of cytoskeleton in apoptosis. Vitarn Horm, 2000, 58: 267 – 297

37   Geiger KD, Stoldt P, Schlote W and Derouiche A. Ezrin immunoreactivity is associated with increasing malignancy of astrocytic tumors but is absent in oligodendrogliomas. Am J Pathol, 2000, 157: 1785 – 1793

38   DeMeester SL, Cobb JP, Hotchkiss RS, Osborne DF, Karl IE, Tinsley KW and Buchman TG. Stress – induced fractal rearrangement of the endothelial cell cytoskeleton causes apoptosis. Surgery, 1998, 124: 362 – 371

39   Janicke RU, Sprengart ML, Wati MR and Porter AG. Caspase – 3 is required for DNA fragmentation and morphological changes associated with apoptosis. J Biol Chem, 1998, 273: 9357 – 9360

40   Kojima H, Endo K, Moriyama H, Tanaka Y, Alnemfi ES and Slapak CA. Abrogation of mitochondrial cytochrome c release and caspase – 3 activation in acquired multidrug resistance. J Biol Chem, 1998, 273: 16647 – 16650

41   Mashima T, Naito M and Tsuruo T. Caspase – mediated cleavage of cytoskeleted actin plays a positive role in the process of morphological apoptosis. Oncogene, 1999, 18: 2423 – 2430

42   Brown SB, Bailey K and Saviii J. Actin is cleaved during constitutive apoptosis. Biochem J, 1997, 323: 233 – 237

85.   Reactive Oxygen Species and Antioxidants in Apoptosis of Esophageal Cancer Cells Induced by As₂ O₃

SHEN Zhong – ying[1], SHEN Wen – ying[2], CHEN Ming – hua[1], SHEN Jian[1], ZENG Yi[3]

1. Department of Pathology, Medical College of Shantou University

2. Institute of Virology, Chinese Academy of Preventive Medicine

Summary

To explore the relationship between the reactive oxygen species (ROS) and apoptosis in esophageal carcinoma cells (SHEE85) induced by arsenic trioxide (As₂O₃), we focused on changes of apoptosis, ROS, and antioxidants. Apoptosis of SHEE85 was confirmed by means of DNA fragmentation stained by Hoechst 33342, Sub – G$_1$ cells scored by flow cytometry and ultrastructure of cells by electron microscopy. To evaluate the level of ROS, the chemiluminescent method was used for measuring the production of superoxide anion ($O - '_2$). Lipid peroxide (malondialdehyde, MDA), superoxide dismutase (SOD) and glutathione peroxidase (GSH – Px) were measured respectively by the photometry method. In the cells treated with As$_2$O$_3$ at a concentration of 5.0 μmol/L for 2 – 24 h, the content of cellular $O - '_2$ and MDA was increased, but SOD and GSH – Px were significantly lower in the process of apoptosis in SHEE85. As$_2$O$_3$ at concentration of 0.5 μmol/L did not cause cell apoptosis but promoted cell proliferation. These results suggest that As$_2$O$_3$ at a high dosage (5 μmol/L) causes cell apoptosis and at a low dosage (0.5 μmol/L) causes cell proliferation. The essential mechanisms of cell apoptosis induced by As$_2$O$_3$ may be related to the increase of ROS and decrease of anti – oxidation. ROS and antioxidants participate in the apoptotic pathway of esophageal carcinoma cells.

[**Key words**]   Esophageal carcinoma cell; Arsenic trioxide; Reactive oxygen species; Antioxidant

Introduction

Arsenic agents, compounds of realgar – a drug of traditional Chinese medicine, have been used as anticancer herbs for over a thousand years[1]. Subsequently some of these pure chemical compounds, such as As$_2$O$_3$, have been used in medicine[2]. Because of their toxic and oncogenic effects, arsenic agents were abandoned for a long time. It was not until the 1970s, by way of long – term clinical trials, that arsenic agents were shown to be highly effective in the treatment of several types of leukemia[3]. It has been documented that arsenic can induce apoptosis in leukemic cells by activating apoptotic pathways[4,5]. Following clinical observations of their effect on leukemia, arsenic compounds have received renewed attention in recent years[6]. To date, arsenic trioxide is considered to be one of the more potent drugs for chemotherapy of cancer[7-9].

In our previous studies we have examined the action of As$_2$O$_3$ on an esophageal carcinoma cell line[10]. The results demonstrated that arsenic trioxide markedly inhibits growth and survival of the esophageal epithelial ceils (SHEE) by induction of apoptosis in cells[11,12]. Our data also demonstrated that the primary target of arsenic trioxide – induced apoptosis of SHEE cells was probably the mitochondria[13,14]. It is now possible that arsenic trioxide is a mito – chondriotoxic agent. The alteration of mitochondria induced by arsenic trioxide seems to occur before the condensation of chromatin in the early stage of apoptosis[15].

Arsenic trioxide (As$_2$O$_3$) causes serious damage to cells, especially to mitochondria leading to apoptosis, but its damage and responsible mechanisms of cells are not fully understood. Some researchers have suggested that the genotoxicity of arsenic may be mediated by the reactive oxygen species (ROS)[16-18], the product of oxidative stress[19]. ROS contains hydrogen peroxide, superoxide hydroxy radicals, lipid peroxide and so on. In normal and malignant cells ROS are generated through a respiratory chain of mitochondria but are scavenged by antioxidant defense systems[20]. When the antioxide system is weakened, oxidative stress is considered to produce varying damage to cells. Previous reports in some studies on effects of ROS on cell apoptosis showed contradicting out – comes such as: ROS may prevent caspases from functioning optimally in neutrophils cells[21]; ROS is involved in trans – criptional regulation of genes and a survival factor in cancer cells[22]; ROS is involved in prevention of apoptosis and maintenance of proliferation in lymphoma cells[23]; and ROS intermediates in spontaneous apoptosis of neutrophils[24]. Malignant cells can be sensitized to under- go growth inhibition and apoptosis by arsenic trioxide through modulation of the glutathione redox system[25]. The most critical mechanism in As_2O_3 – induced apoptosis and its interactions with the ROS are still not clear, especially in epithelial cancer. In order to explore the relationship between ROS and apoptosis in epithelial cancer cells we studied the apoptosis of esophageal malignant epithe- lial cells induced by As_2O_3 and the dynamic changes of ROS and antioxidant.

Materials and Methods

1. Cell and drug: The SHEE cell is a kind of immortal esophageal epithelial cell, induced by genes E6E7 of human papilloma virus (HPV) type 18 in our laboratory[26]. In continual cultivation, the cells at the 85th passage (SHEE85) displayed poor differentiation with malignant transforma- tion[27]. It was routinely cultivated in culture medium M199 (Gibco) with 10% calf serum, 100 units of penicillin and streptomycin, in an incubator with a humidified atmosphere of 5% CO_2 and 95% air. Cells were cultured in 50 ml culture flasks or on the cover slips in small culture chamber (Coming Co.). They were then treated with arsenic trioxide (As_2O_3) (Sigma Co., St. Louis MO, Lot A 1010), in different concentrations, 0.5 and 5.0 μmol/L, for different times of 2, 4, 8, 12, 24 h, and were subjected to further examination. The control group of SHEE85 was cultured in the same conditions without As_2O_3 treatment.

2. Transmission electron microscopy: For electron microscopic assessment, cells were digested and spun to form a pellet, and fixed with 2.5% glutaraldehyde. They were dehydrated in graded ethanol and embedded in Araldite. Ultra – thin sections were cut with glass knives and mounted on copper grids. They were contrasted for 15 min with uranyl acetate and for 3 min with lead citrate. The sections were examined with an electron microscope (Hitachi H – 300).

3. Flow cytometry examination: Cells were collected and washed with PBS, and fixed with 70% pre – cooled alcohol in ice, stained with propidium iodide at a dose of 80 μg/ml for 20 min. The DNA ploidy was determined by a flow cytometer (FACSort, B – D Co.) and apoptotic peak of Sub – G_1 cell and the proliferative index (PI) ($S + G_2M/G_1 + S + G_2M$) was determined.

4. DNA fragmentation: Cells cultured in the flask were labeled with Hoechst 33342 (H342, Sigma) in 1 μg/ml at 37℃ for 20 min. After washing with PBS, the living cells were observed with a fluorescent phase – contrast microscope using a mercury lamp with excitation and emission settings of 360 and 450 nm respectively.

5. Chemiluminescence (CL) measurements of super oxidates[28]: The hypo – xanthinoxidase (Ho – xo) reactive system for superoxide ions ($0 – '_2$); 5×10^5 cells/ml of SHEE85 cells were cultured in a small dish and treated with As_2O_3. Before measurement the cells were washed with PBS and kept in the dark at 4℃ until use. The kinetic curve of CL in the system was immediately recorded with a computerized high – sensitivity single – photon counter (SPC, type BPCL – 4, manufactured at the Institute of Biophysics, Academia Sinica, China). The voltage in the photomultiplier was kept at 875 V, and the spectral range of CL that can be recorded is around 340 – 800 nm.

6. Colorimetric method for MDA, SOD *and* GSH – Px *assay*: These kits from Nanjing Jiancheng Biotechnology Co. were used for detection of SOD, MDA and GSH – Px: the reaction of glutathione (GSH) – oxidized glutathione (GSSG) for glutathione peroxidase (GSH – Px); thio – barbital acid (TBA) reaction for lipid peroxidation malonaldehyde, (MDA); Fenton reaction for OH – ; and reaction of xanthinoxidase for superoxide dismutase contents. At interval the cells of SHEE were trypsinized from flask and counted, 10^5 cells were deactivated at 80℃ for 5 rain and deproteinated by centrifugation at 12 000 r/min for 30 min, and the supernatants were determined. Briefly, the procedure was as follows: in a 3 ml test tube, 100 μl of sample solution was mixed with the reagents of the kit, incubated for 30 min at room temperature, and 50 μl color reagentI&II was added respectively, mixed, allow to stand in the dark at room temperature for 10 to 15 min. The samples for SOD, MDA, GSH – Px and the blank were estimated with spectrophotometry (Shimadzu UV/120) by λ 530, 412 and 532 nm, respectively, and were calculated by calibration curve of standard addition method. The standards were prepared by kits and run in parallel with test samples in every assay.

Results

1. DNA *fragmentation*: Treatment of SHEE85 cells with 5. 0 μmol/L As_2O_3 induced morphological change of typical apoptosis. H342 staining – cells showed nuclear shrinkage with chromatin condensation and fragmentation, indicative that these cells were on the way to death showing apoptotic morphology (Fig. 1A). In the 0. 5 μmol/L and control group of SHEE85 cells, the neucleus was plump with even staining of H342 (Fig. 1B and C).

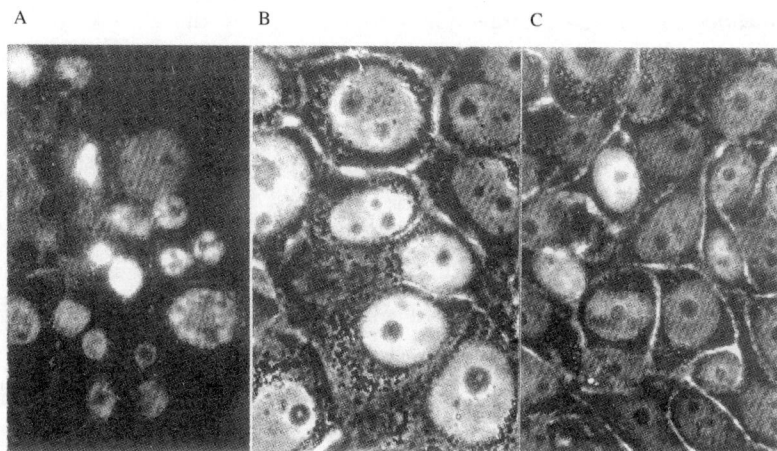

The living cells were stained with Hoechst 33342. A, cells treated with 5 μmol As_2O_3 showed shrunken nuclei with coagulation and degradation of chromatin (×400); B, cells treated with 0. 5 μmol As_2O_3 showed increased fluorescent intensity (×400); C, untreated SHEE85 (×400)

Fig. 1 Morphological changes in neucleus under fluorescent phase – contrast microscope

2. Cell cycle and apoptotic peak: Fig. 2 shows the DNA histogram of SHEE85. The proliferative index of SHEE85 cells treated with 5.0 μmol/L As_2O_3 was 15.7%; 0.5 μmol/L group, 49.2% and control group, 27.5% (Fig. 2). The SHEE85 cells treated with 0.5 μmol/L As_2O_3 were in proliferative manner. Following As_2O_3 treatment, the apoptotic cells can be counted by their lower DNA content (Sub – G_1 cells) with detection of flow cytometry. These data suggest that an apoptotic cell death occurred at higher concentration of arsenic treatment after 12 h incubation, but did not occur in a low concentration (0.5 μmol/L).

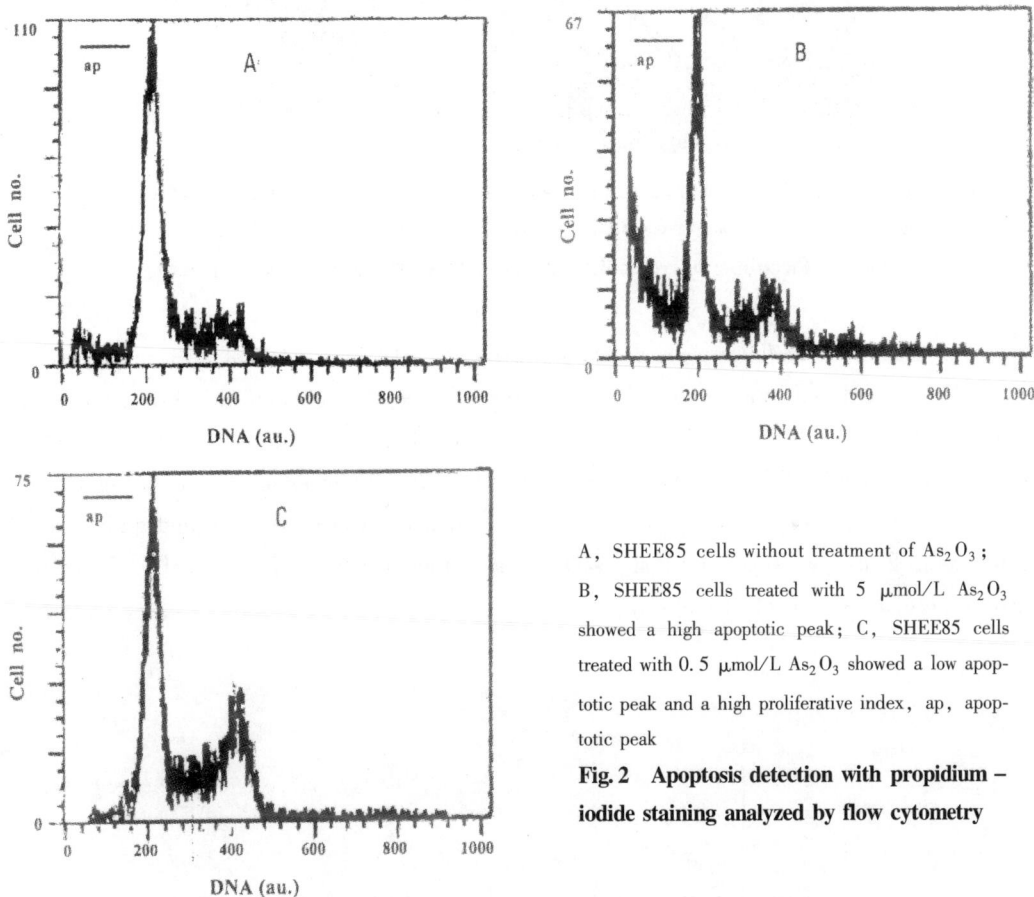

A, SHEE85 cells without treatment of As_2O_3; B, SHEE85 cells treated with 5 μmol/L As_2O_3 showed a high apoptotic peak; C, SHEE85 cells treated with 0.5 μmol/L As_2O_3 showed a low apoptotic peak and a high proliferative index, ap, apoptotic peak

Fig. 2 Apoptosis detection with propidium – iodide staining analyzed by flow cytometry

3. Ultrastructural changes: By electron microscopy at 0.5 μmol/L of As_2O_3 no obvious morphological changes were observed in mitochondria, but nucleus displayed hyperplasia (Fig. 3A). At 5.0 μmol/L As_2O_3 the size and the shape of mitochondria varied. Some of mitochondria were swollen and their cristae had disappeared, with the outer membranes of mitochondria disrupted. The nuclear chromatin condensed to block and attached to the nuclear membrane. At 24 h, cells began to shrink, and the nucleus became round with condensed chromatin around the nuclear membrane (Fig. 3B).

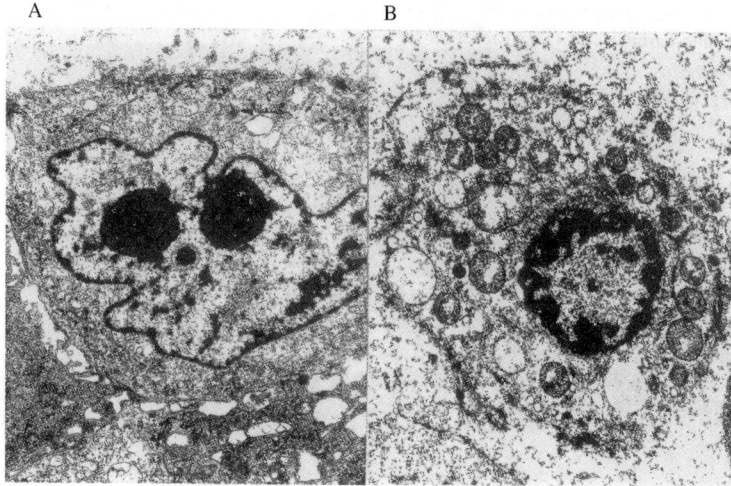

A, SHEE85 cells treated with 0.5 μmol/L showed a proliferative nucleus (×7000); B, SHEE85 cells treated with 5 μmol/L As$_2$O$_3$ for 24 h showed the patterns of apoptotic nucleus including swollen mitochondria (×10 000)

Fig. 3   Electronic microscopic features of SHEE85 cells treated with As$_2$O$_3$

4. The level of intracellular peroxide and antioxidant: At 5.0 μmol/L of As$_2$O$_3$, the optical intensity of O – '$_2$ and MDA was increased, with extended time of incubation with As$_2$O$_3$ and the intracellular antioxidants SOD and GSH – Px were decreased (Figs. 4 and 5). The data suggested that decreasing intracellular SOD and GSH – Px were not sufficient to suppress increasing O – '$_2$ and MDA. In the 0.5 μmol/L As$_2$O$_3$ group the dynamic changes of both peroxide and antioxidant were not apparent. Thus, we suggested that As$_2$O$_3$ induced the oxidative toxicity in SHEE85 cells via a hydrogen peroxide – dependent pathway.

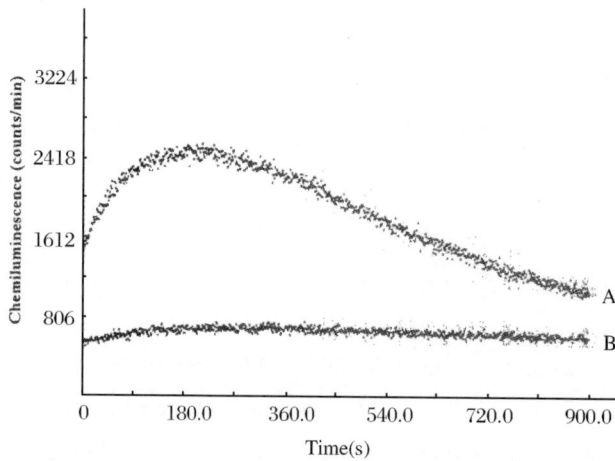

A, 5 μmol/L; B, 0.5 μmol/L

Fig. 4   Chemiluminescence responses of SHEE85 cells (5 ×10^5 cells/ml) following treatment of As$_2$O$_3$

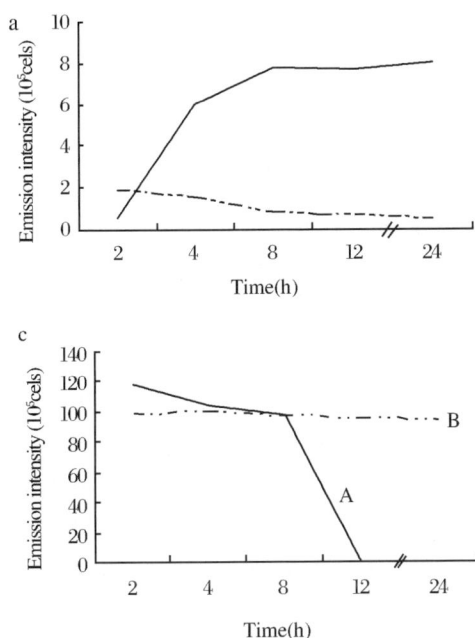

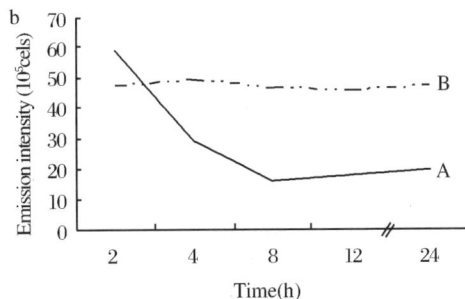

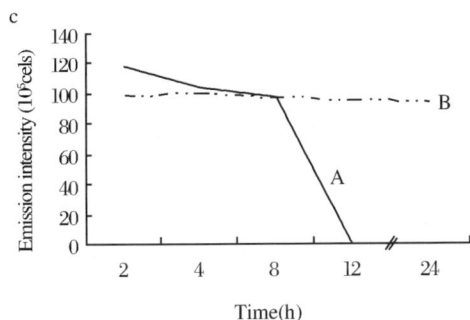

a, MDA; b, SOD; c, GSH – Px. A, 5.0 μmol/L;
B, 0.5 μmol/L

Fig. 5   Measurement of MDA, SOD and GSH – Px in various dosage of As₂O₃

Discussion

In our previous studies, arsenate trioxide can induce apoptosis in the malignant esophageal cells. This apoptosis is an active form of cell death that is involved in a number of biological processes, such as NO, calcium and DNA[29-32]. In this experiment As_2O_3 also significantly induced $0^-{}_2$ and MDA production and reduced SOD and GSH – Px. These results suggested that As_2O_3 induced apoptosis of cells by increasing amounts of superoxide anion and lipid peroxidation, and decreasing amounts of superoxide dismutase and glutathione peroxide. As_2O_3 may activate the mitochondrial apoptotic pathway by the redox disequilibrium and releasing cyto – chrome c into the cytosol further triggering the caspase cascade leading cells to apoptosis[33,34].

Our data show that SOD and GSH – Px form part of the adaptive response of SHEE85 to oxidative stress. At 12 h after administration of 5.0 μmol/L As_2O_3 the level of GSH – Px drops down to zero. Alterations in the activity of this enzyme may reflect reduced cellular defense against oxidative stress, as SOD and GSH – Px are also involved in the regulation of apoptosis. Also, when constitutive SOD and GSH – Px activity is inhibited, accumulation of products of $0 -'_2$ and lipid peroxidation occurs resulting in increased cellular apoptosis. SOD and GSH – Px provides protection to the cells from oxidative injury induced by different endogenous or exogenous ROS. When the production of ROS overrides the antioxidant capability of the target cell, oxidative stress results and oxidative damage of cells occurs. ROS involved in apoptosis in our cellular systems is mainly based on the excessive formation of ROS as well as the depletion of cellular antioxidants.

The effect of As_2O_3 on cells is dependent on the dosage. Our previous data indicated that small amounts of As_2O_3 could promote cell proliferation[35]. In this study, low level of intracellular ROS induced by low concentration of As_2O_3 (0.5 μmol/L) which are insufficient to induce cellular apoptosis, but probably play a physiological role as second messengers by regulating gene expression and cell proliferation[23]. Also, it has been reported that pretreatment of cells with a mild oxidative stress can result in their protection against apoptotic stimuli[36]. Evidence has been forwarded that in some cases the exposure of cells to low, non – toxic levels of the superoxide and hydrogen peroxide can exert a stimulatory effect on their proliferation, rather than promoting apoptosis or cell necrosis[37].

A recent hypothesis of the generation of superoxide is that the electron leak in the oxygen – side at the cytochrome c of the respiratory chain causes the induction of $O-'_2$ and H_2O_2[38]. The electron leaks of the respiratory chain make up a metabolic path of $O-'_2 \rightarrow H_2O_2 \rightarrow H_2O$ in mitochondria[39]. The major source of ROS in most cell types is probably the leakage of electrons from the mitochondrial electron transport, which results in the formation of superoxide anions. Superoxide anions can be converted to hydrogen peroxide by SOD. Hydrogen peroxide is then detoxified by glutathione peroxidase and catalyses[17]. The reactive pathways initiated by $O-'_2$ have been defined as the radical metabolism of mitochondria, because many different kinds of ROS can be generated in the process of the reactions of $O-'_2$. As_2O_3 causing decrease of mitochondrial efficiency to regulated $O-'_2$ generation may be demonstrated by $\Delta\Psi m$ disrupture[14]. These data further support our hypothesis that inducement of DNA damage by ROS is one of the apoptotic pathways in As_2O_3 treatment[40].

In summary, after exposure to As_2O_3, cellular content of $O-'_2$ and MDA is increased, and SOD and GSH – Px are decreased. As_2O_3 at various dosages causes different results: 5.0 μmol/L induce apoptosis, and 0.5 μmol/L probably promote cellular proliferation. These results also suggested that increased $O-'_2$ and MDA and decreased SOD and GSH – Px are the essential mechanisms of apoptosis induced by As_2O_3. It was concluded that there might be an intrinsic relationship between apoptosis and free radicals *in vitro*.

Acknowledgements

Contract grant sponsor: National Natural Science Foundation of China (39830380) and Research and development foundation of Shantou University (L00012).

[In 《Int J Mol Med》 2003, 11: 479 –484]

References

1 Mervis J. Ancient remedy perform new trick (news). Science,1996,273: 578

2 Sears DA. History of the treatment of chronic myelocytic leukemia. Am J Med Sci, 1988, 296: 85 – 86

3 Sun HD, Hu XC and Zhang TD. At – Lin treated 32 cases of acute promyelocytic leukemia. Chin J Integrate Chin West Med, 1992, 12: 170 – 171

4 Gianni MKM, Chelbi – Alix MK, Benoit G, Lanotte M, Chen Z and Dec – The H. Combined arsenic and retinoic acid treatment enhances differentiation and apoptosis in arsenic – resistant NB4 cells. Blood, 1998, 91: 4300 – 4310

5 Shen ZX, Chen G, Ni JH, Li XS, Xiong SM, Tang W, Yang KQ, Chin QY, Chen Y, Zhou L, Zeng XY, Fang ZW, Wang YT, Ma J, Zhang P, Zhang TD, Chen S J, Chen Z and Wang ZY. Use of arsenic trioxide (As_2O_3) in the treatment of acute promyelocytie leukemia (APL). II. Clinical efficacy and pharmacokinet-

ics. Blood, 1997, 89: 3354 – 3360

6 Chen Z, Wang ZY and Chen S J. Acute promyelocytic leukemia, cellular and molecular basis of differentiation and apoptosis. Pharmacology, 1997, 76: 141 – 149

7 Miller WH, Schipper HM, Lee JS, Singer J and Waxman S. Mechanisms of action of arsenic trioxide. Cancer Res, 2002, 64: 3893 – 3903

8 Ohnishi K, Yoshida H, Shigeno K, Nakamura S, Fujisawa S, Naito K, Shinjo K, Fujita Y, Matsui H, Sahara N, Takeshita A, Satoh H, Terada H and Ohno R. Arsenic trioxide therapy for relapsed or refractory Japanese patients with acute promyelo – cytic leukemia: need for careful electrocardiogram monitoring. Leukemia, 2002, 16: 617 – 622

9 Qian J, Qin S, He Z, Wang L, Chen Y, Shao Z and Liu X. Arsenic trioxide for the treatment of medium and advanced primary liver cancer. Zhonghua Gan Zang Bing Za Zhi, 2002, 10: 63

10 Shen ZY, Tan LJ, Cai WJ, Shen J, Chen C, Tang XM and Zheng MH. Arsenic trioxide induces apoptosis of esophageal carcinoma in vitro. Int J Mol Med, 1999, 4: 33 – 37

11 Shen ZY, Tan LJ, Cai W J, Shen J, Chen CY and Tang XM. Morphologic study on apoptosis of esophageal carcinoma cell line induced by arsenic trioxide. Shijie Huaren Xiaohua ZhaZhi (in Chinese), 1998, 6: 226 – 229

12 Shen ZY, Chen CY, Shen J and Cai WJ. Ultrastructural study of apoptosis and necrosis in the esophageal carcinoma cell line induced by arsenic trioxide. Chin J Med Physics (in Chinese), 1999, 16: 91 – 94

13 Shen ZY, Shen J, Chen MH, Li QS and Hong CQ. Morphological changes of mitochondria in apoptosis of esophageal carcinoma cells induced by As$_2$O$_3$ Chin J Pathol (in Chinese), 2000, 29: 200 – 203

14 Shen ZY, Shen J, Li QS, Chen CY, Chen JY and Zeng Y. Morphological and functional changes of mitochondria in apoptotic esophageal carcinoma cells induced by arsenic trioxide. World J Gastroenterol, 2002, 8: 31 – 35

15 Shen ZY, Shen J, Cai WJ, Hong CQ and Zheng MH. The alteration of mitochondria is an early event of arsenic trioxide induced apoptosis in esophageal carcinoma cells. Int J Mol Med, 2000, 5: 155 – 158

16 Applegate LA, Luscher P and Tyrrell RM. Induction of heme oxygenase. A general response to oxidant stress in cultured mammalian cells. Cancer Res, 1991, 51: 974 – 978

17 Bauer MK, Vogt M, Los M, Siegel J, Wesselborg S and Schulze – Osthoff K. Role of reactive oxygen intermediates in activation – induced CD95 (APO – l/Fas) ligand expression. J Biol Chem, 1998, 273: 8048 – 8055

18 Wang TS, Kuo CF, Jan KY and Huang H. Arsenite induces apoptosis in Chinese hamster ovary cells by generation of reactive oxygen species. J Cell Physiol, 1996, 169: 256 – 268

19 Maeda H, Hori S, Nishitoh H, Ichijo H, Ogawa O, Kakehi Y and Kakizuka A. Tumor growth inhibition by arsenic trioxide (As$_2$O$_3$) in the orthotopic metastasis model of androgen – independent prostate cancer. Cancer Res, 2001, 61: 5432 – 5440

20 Dai J, Weinberg RS, Waxman S and Jing Y. Malignant cells can be sensitized to undergo growth inhibition and apoptosis by arsenic trioxide through modulation of the glutathione redox system. Blood, 1999, 93: 268 – 277

21 Fadeel B, Ahlin A, Henter JI, Orrenius S and Hampton MB. Involvement of caspases in neutrophil apoptosis: regulation by reactive oxygen species. Blood, 1998, 92: 4808 – 4818

22 Ha HC, Thiagalingam A, Nelkin BD and Casero RA Jr. Reactive oxygen species are critical for the growth and differentiation of medullary thyroid carcinoma cells. Clin Cancer Res, 2000, 6: 3783 – 3787

23 Del Bello B, Paolicchi A, Comporti M, Pompella A and Maellaro E. Hydrogen peroxide produced during gamma – glutamyl transpeptidase activity is involved in prevention of apoptosis and

maintainance of proliferation in U937 cells. FASEB J, 1999, 13: 69 – 79

24 Kasahara Y, Iwai K, Yachie A, Ohta K, Konno A, Seki H, Miyawaki T and Taniguchi N. Involvement of reactive oxygen intermediates in spontaneous and CD95 (Fas/APO – 1) – mediated apoptosis of neutrophils. Blood, 1997, 89: 1748 – 1753

25 Jing Y, Dai J, Chalmers – Redman ME, Tatton WG and Waxman S. Arsenic trioxide selectively induces acute promyelocytic leukemia cell apoptosis via a hydrogen peroxide – dependent pathway. Blood, 1999, 94: 2102 – 2111

26 Shen ZY, Cen S, Cai W J, Xu J J, Ten ZP, Shen J, Hu Z and Zeng Y. Immortalization of human fetal esophageal epithelial cells induced by E6 and E7 genes of human papilloma virus 18. Chin Exp Clin Virol, 1999, 13: 121 – 123

27 Shen ZY, Shen J, Cai WJ, Chen JY and Zeng Y. Malignant transformation of immortalized esophageal epithelial cells. Zhonghua Zhongliu Zazhi, 2002, 24: 107 – 109

28 Nishinaka Y, Aramaki Y, Yoshida H, Masuya H, Sugawara T and Ichimori Y. A new sensitive chemiluminescence probe, L – 012, for measuring the production of superoxide anion by cells. Biochem Biophys Res Commun, 1993, 193: 554 – 559

29 Shen ZY, Shen WY, Chen MH, Hong CQ and Shen J. Alterations of nitric oxide in apoptosis of esophageal carcinoma cells induced by arsenite. World Chin J Digest, 2000, 8: 1101 – 1104

30 Shen ZY, Shen WY, Chen MH, Shen J, Cai WJ and Zeng Y. Quantitative analysis of nitric oxide and calcium ions in apoptosis of esophageal carcinoma cell induced by arsenite. World J Gastroenterol, 2002, 8: 40 – 43

31 Shen J, Wu MH and Shen ZY. Evaluating of the chromatin condensation in living cells during apoptotic process. Chin J Stereol Image Anal, 2000, 5: 114 – 118

32 Shen ZY, Shen J, Cai W J, Chen ZY and Chen JY. The change of the cytoskeleton in apoptosis of esophageal carcinoma cell induced by AszO3. Chin J Pathol, 2001, 30: 369 – 370

33 Zoratti M and Szabo I. The mitochondria permeability transition. Biochim Biophys Acta, 1995, 1241: 139 – 176

34 Skulachev VP. Cytochrome c in the apoptotic and antioxidant cascades. FEBS Lett, 1998, 423: 275 – 280

35 Shen J, Wu MH, Cai WJ and Shen ZY. The effects of arsenite trioxide in various concentrations on the esophageal carcinoma ceil line. Chin J Cancer Biother, 2001, 8: 106 – 109

36 Maellaro E, del Bello B and Comporti M. Protection by ascorbate against apoptosis of thymocytes: implications of ascorbate – induced non – lethal oxidative stress and poly (ADP – ribosyl) action. Exp Cell Res, 1996, 226: 105

37 Schulze – Osthoff K, Los M and Baeuerle PA. Redox signalling by transcription factors NF – kappa B and AP – 1 in lymphocytes. Biochem Pharmacol, 1995, 50: 735 – 741

38 Zhao YG and Xu JX. Mitochondria, reactive oxygen species and apoptosis. Prog Biochem Biophys, 2001, 28: 168 – 171

39 Xu JX, Li X and Zhang YX. Mitochondrial respiratory chain. a Self – defense system against oxygen toxicity. In: Proceedings of the International Symposium on Natural Antioxidants: Molecular Mechanisms and Health Effects. AOCS Press, Champaign, Illinois, 1996, 530 – 539

40 Shen ZY, Shen WY, Chen MH, Shen J, Cai WJ and Zeng Y. Mitochondria, calcium and nitric oxide in the apoptotic pathway of esophageal carcinoma cells induced by As2O3. Int J Mol Med, 2002, 9: 385 – 390

86. Upregulated Expression of Ezrin and Invasive Phenotype in Malignantly Transformed Esophageal Epithelial Cells

SHEN Zhong – ying[1], XU Li – yan[1], CHEN Ming – hua[1], LI En – min[2],
LI Jin – tao[1], WU Xian – ying[1], ZENG Yi[3]

1. Department of Pathology, Medical College of Shantou University;
2. Department of Biochemistry, Medical College of Shantou University;
3. Institute of Virology, Chinese Academy of Preventive Medicine

Summary

Objective: To investigate the correlation between ezrin expression and invasive phenotype formation in malignantly transformed esophageal epithelial cells.

Methods: The experimental cell line employed in the present study was originated form the progressive induction of a human embryonic esophageal epithelial cell line (SHEE) by the E6E7 genes of human papillomavirus (HPV) type 18. The cells at the 35th passage after induction called SHEE-IMM were in a state of immortalized phase and used as the control, while that of the 85th passage denominated as SHEEMT represented the status of cells that were malignantly transformed. The expression changes of ezrin and its mRNA in both cell passages were respectively analyzed by RT – PCR and Western blot. Invasive phenotype was assessed *in vivo* by inoculating these cells into the severe combined immunodeficient (SCID) mice via subcutaneous and intraperitoneal injection, and *in vitro* by inoculating them on the surface of the amnion membranes, which then was determined by light microscopy and scanning electron microscopy.

Results: Upregulated expression of ezrin protein and its mRNA was observed in SHEEMT compared with that in SHEEIMM cells. The SHEEMT cells inoculated in SCID mice were observed forming tumor masses in both visceral organs and soft tissues in a period of 40 days with a special propensity to invading mesentery and pancreas, but did not exhibit hepatic metastases. Pathologically, these tumor cells harboring larger nucleus, nucleolus and less cytoplasm could infiltrate and destroy adjacent tissues. In the *in vitro* study, the inoculated SHEEMT cells could grow in cluster on the amniotic epithelial surface and intrude into the amniotic stroma. In contrast, unrestricted growth and invasiveness were not found in SHEEIMM cells in both *in vivo* and *in vitro* experiment.

Conclusion: The upregulated ezrin expression is one of the important factors that are possibly associated with the invasive phenotype formation in malignantly transformed esophageal epithelial cells.

Introduction

It has been well known that cancer cells can invade and destroy surrounding tissues by their disseminative potency. This acquired malignant property is believed recently to be determined by the abnormal changes of expression patterns of certain genes such as c – met, urokinase type plasminogen activator receptor or ezrin and so on[1-9]. Therefore, the correlation between the invasive phenotype of tumor cells and the aberrant expression of these genes has become a focus of attention by the worldwide oncologists.

Induced by E6E7 genes of human papilloma virus (HPV) type 18, we have established both kinds of immortalized and malignantly transformed cells from a human embryonic esophageal epithelial cell line (SHEE)[10]. At early passages after induction, the cell did not display any abnormal growth ability in soft agar and nude mice[11-13]. At the 20th passage, the cell was noted to be immortalized with the appearance of telomerase[14,15]. At the 30th passage, the cell manifested as a phenotype of biphasic differentiation[16]. While cultivated over 60 passages, the cell exhibited atypical morphological changes and chromosomal alterations that were suspected to be a kind of premalignant transformation[17]. Over 85 passages, the cell was observed having harbored a property of malignant transformation shown by its unrestricted growth in the nude mice and invasion to the surrounding tissues[18,19]. Obviously, this process of malignant transformation of SHEE cells induced by E6E7 genes of HPV type 18 possesses a progressive and stepwise characteristic, which is no doubt a good cell model suitable for the study of correlations between abnormal gene expression and corresponding malignant phenotype formation in tumorigenesis.

Differential display analysis of RNA samples isolated from the SHEEIMM and SHEEMT cells has been performed in our previous study. There were 15 up – regulated and 6 down – regulated genes being identified by the cDNA microarray method (Data to be published), in which ezrin was one of the up – regulated genes that have been found existed in certain diseases and tumors. Being a membrane – cytoskeleton linker, ezrin protein is located in cytoplasm and is rich in microvilli and cell surface structures with the function involved in the formation of microvilli and intercellular junctions, as well as the cell motility and invasive behavior of malignant tumors[20-28]. Up to the present, however, there have been few reports concerned about how the ezrin expressed in the malignant transformation of esophageal epithelial cells and what was its relation to the invasiveness formation of tumor cells, which are of both theoretical and practical importance in the investigation of cancerogenesis. Thus, our present study was conducted to identify the correlation between the ezrin gene expression and invasive phenotype formation in malignantly transformed esophageal epithelial cells.

Materials and MethodsS

1. *Cell lines and cell culture*

The experimental cells came from our laboratory, and were established by the progressive induction of SHEE cell line with the E6E7 genes of HPV type 18. The 35th, 85th passages of induced cells were employed in the present study. The cells at the 35th passage called SHEEIMM were in a state of immortalized phase, while that of the 85th passage denominated as SHEEMT represented the status of cells that were malignantly transformed. Both passages of cells were continuously cultivated in flasks with medium 199 (GIBCO) supplemented by 10 % fetal bovine serum (FBS), 100 U · m^{-1} penicillin and 100 UU · m^{-1} streptomycin at 37 ℃ in a humidified atmosphere containing 5 % of CO_2.

2. *RT − PCR*

Total RNA was extracted from SHEEIMM and SHEEMT cells using the Trizol reagent (Invitrogen). The first − strand cDNA synthesis was performed with 1 μg total RNA and carried out at 42℃ for 1 h followed by at 95℃ for 5 min and at 0 − 5℃ for 5 min according to protocol of reverse transcription system (Promega). The synthesized cDNA was diluted to 100 μl with TE and stored at − 20℃ until use. In the following experiment, 5 μl of cDNA was amplified in a 25 μl PCR reaction volume with Advantage 2 PCR Kit (CLONTECH). Both of the ezrin primer 5'CGGGCGCTCTAAGGGTTCT3' (sense), 5' TGCCTTTGCAAAGCTTTTATTTCA3' (antisense) and GAPDH primer 5' GAAGGTGAAGGTCGGAGTC 3' (sense), 5' GAAGATGGTGATGGGATTTC 3' (antisense) were synthesized by Genecore Company (Shanghai). After prepared by routine procedures, PCR products were visualized by electrophoresis on 1% agarose gel stained with ethidium bromide and quantitated with Gelworks 1D Intermediate software (version 3.51, Kodak).

3. *Western blot*

Western blot was used to detect ezrin protein expressed by experimental cells. Confluent cells of 3 flasks were washed three times with ice − cold PBS and then lysed i'n buffer containing 50 mmol/L Tris − HCl (pH 8.0), 150 mmol/L NaCl, 100 μg · ml^{-1} phenyl − methyl − sulfonyl fluoride (PMSF) and 1% Triton X − 100 for 30 rain on ice. After removal of cell debris by centrifugation at 12 000 g for 5 min, 50 μl of supernatant was boiled for 5 min in the sample buffer and separated by 10 % SDS − PAGE, which was then transferred onto the nitrocellulose membrane (Pall Corporation). After non − specific reactivity was blocked by 5 % fat − free milk in TBST (10 mmol/L Tris − HCl, pH 7.5, 150 mmol/L NaCl, 0.05 % Tween 20) for 1 h at room temperature, the membrane was incubated in turn with monoclonal antibody of mouse against human ezrin p81 (Maixin − Bio) and anti − mouse IgG − HRP antibody. Reactive protein was finally detected by ECL chemiluminescence system (Santa Cruz).

4. *Oncogenesis and invasive potency of SHEE in vivo*

SCID mice (C. B − 17/lcrJ − scid nu/nu) were from Animal Laboratory Center, Chinese Academy of Medical Sciences. For the determination of their oncogenesis and invasive potency, SHEEIMM and SHEEMT cells were cultured in flasks with fresh medium to reconstitute their surface protein. After digested and washed twice with PBS, they were counted and resuspended in PBS solution (10×10^6 · m^{-1}) until use. Then SCID mice were anesthetized with 7 % chloralhydrate followed by intraperitoneal and subcutaneous inoculation with the both passages of cells. That is, 2 × 10^6 SHEEMT or SHEEIMM cells in 0.2 ml PBS were injected into the peritoneal cavity in 5 mice and into the right axilla of the same number of animals. Instead, the control mice were injected only with 0.2 ml of PBS. The experimental animals were checked daily and all were killed on after day 40 inoculation by an overdose of anesthetic. Tissues from tumor mass, mesentery, pancreas and gastrointestinal tract were sampled and prepared with the routine method to produce thin paraffin sections (5 μm) stained by hematoxylin and eosin for the assessment of tumor invasion.

5. *Invasive potency of SHEE in vitro*

Tumor cell invasion *in vitro* was assessed by using a fresh fetal amnion, which was cultured in 199 medium supplemented with 10 % FBS. In the experiment, a piece of amnion and 50 000 SHEEIMM or SHEEMT cells were in turn added to one well of a 24 – well plate and incubated together for 24 h and 72 h, and then were washed for several times with PBS. Then the samples, including the piece of amnion and cells adhering to it, were fixed with 2.5 % glutaraldehyde and post – fixed with 2 % osmium tetraoxide. After full dehydration with gradient concentrations of ethanol, the samples were adhered to an aluminium stub and sprayed plating with gold for 3 minutes (EIKD, IB – 3, Hitachi), which were further examined by Hitachi H300 electron microscope with the attachment of scanning apparatus.

Results

1. *Expressive alterations of ezrin and its mRNA*

The ezrin mRNA detected by electrophoresis on 1% agarose gel was shown as a band of 3032 bp segment, whose expression was observed to be significantly up – regulated in SHEEMT by RT – PCR assay compared with that expressed in SHEEIMM cells as displayed in Fig. 1. The same difference was also noted in the expression of ezrin protein between both of the cell passages (Fig. 2).

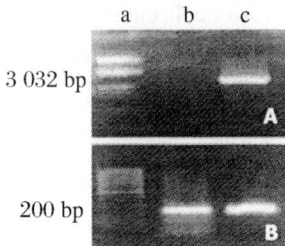

A. Ezrin; B. GAPDH; a. Marker; b. SHEEIMM; c. SHEEMT

Fig. 1　RT – PCR assay of ezrin mRNA. The expression of ezrin mRNA was significantly increased in SHEEMT compared with that in SHEEIMM cells

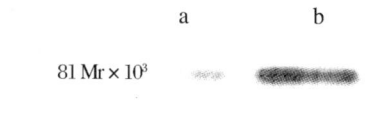

Lane a: SHEEIMM; lane b: SHEEMT

Fig. 2　Western blot analysis of ezrin protein expression. The ezrin protein was exhibited as a band of 81×10^3 segment, whose expression was up – regulated in SHEEMT compared with that in SHEEIMM cells

2. *Oncogenesis and invasiveness in SCID mice*

After inoculated into the SCID mice, SHEEMT cells were noted growing rapidly and forming tumor masses in 40 days. In peritoneal cavity, tumors were observed occurring on the mesentery, pancreas, urinary cyst, sub – diaphragm etc. with a special propensity for invading mesentery and pancreas, but did not exhibit any signs of hepatic metastases. Histological examination revealed that these tumor cells did not only grow on the organ surface, but also invade into adjacent tissues (Fig. 3). In the subcutaneous tissue of the right axilla, tumor mass was observed macroscopically forming on the thoracic wall and penetrating into the thoracic cavity. Pathologically, these tumor cells harboring larger nucleus, nucleolus and less cytoplasm could infiltrate and destroy adjacent muscular fibers (Fig. 4). Once transplanted, tumors could keep be passed to other SCID mice. In contrast, SHEEIMM cells were not found to form tumors in inoculated tissue by gross examination.

3. *Invasive potency in vitro*

After inoculated on the amnion, SHEEMT cells were shown growing in cluster on the epithelial surface with the formation of pseudopod that intruded into the gap between intercellular conjunctions (Fig. 5A). On the cutting surface, SHEEMT cells were observed invading into the amnion stroma (Fig. 5B). It was not found that the inoculated SHEEIMM cells could adhere to or colonize on the amniotic epithelium.

Discussioon

Invasive potency is one of the most important features of malignant tumors and is involved in a critical cascade of events such as extracellular matrix degradation, cell migration and colonization in the assaulted tissue. The invasive phenotype formation of malignant cells requires up – regulated expression of certain adhesive molecules, enzymes and related genes responsible for the interaction between cancer cells and extracellular matrix[29]. Several co – factors have been reported to be involved in the process of invasion and metastasis[30-32] besides ezrin that has been considered as an important molecule contributing to the malignantly transformation of cells. In addition to combining with adhesion molecules such as E – cadherin and catenin implicated in cell – cell and cell – matrix adhesion[33], ezrin plays a critical role in the determination of invasiveness of cancer cells[34].

In the present study, we investigated the expression of ezrin and its mRNA in a malignantly transformed esophageal epithelial cell line SHEEMT and further demonstrated its unrestricted growth and invasive potency in both *in vivo* and *in vitro* circumstances. The experimental results revealed that the expression ofezrin protein and its mRNA was significantly upregulated in SHEEMT compared with that in SHEEIMM cells. The SHEEMT cells inoculated to SCID mice could form tumor masses in both visceral organs and soft tissues in a period of 40 days. Pathologically, these tumor cells harboring larger nucleus, nucleolus and less cytoplasm could infiltrate and destroy adjacent tissues. In the *in vitro* study, the inoculated SHEEMT cells could grow in cluster on the amniotic epithelial surface and intrude into the amniotic stroma. In contrast, unrestricted growth and invasive property were not found in the SHEEIMM ceils that were used as the control in both *in vivo* and *in vitro* experiments. As to our knowledge, this is one of the few reports concerned about how the ezrin gene expresses and what is its relation to the invasiveness formation in malignantly transformed esophageal epithelial cells.

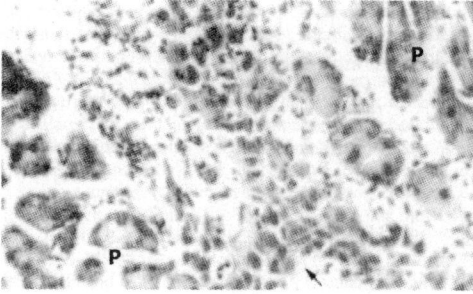

Fig. 3　The inoculated SHEEMT cells (arrow) invaded into the parenchyma of pancreas (P) (HE, ×200)

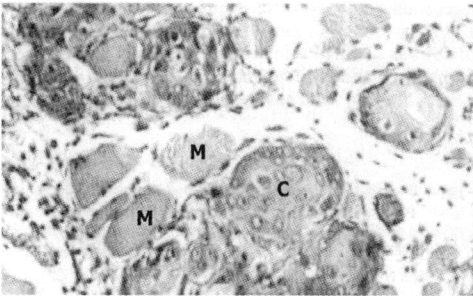

Fig. 4　Tumor formation in subcutaneous tissue inoculated with SHEEMT cells (C), the latter was also shown to invade and destroy nearby muscle fibers (M). (HE, ×200)

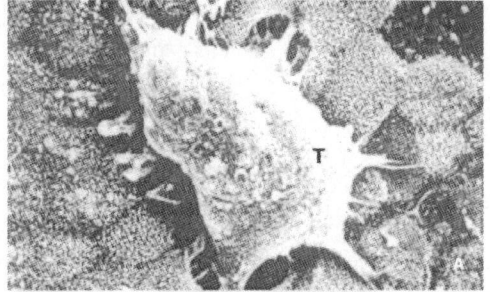

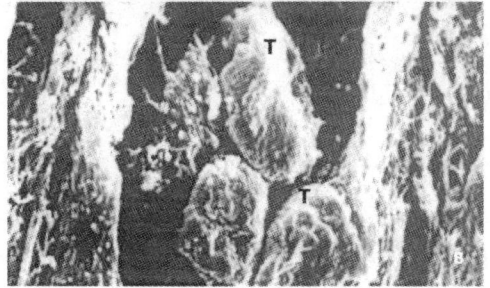

A. A cluster of SHEEMT cells (T) grew on the amniotic epithelial surface (bar, 5 μ); B. On the cutting surface, SHEEMT cells (T) were observed invading into the amnion stroma (bar, 50 μm)

Fig. 5　The invasiveness of inoculated SHEEMT cells on amniotic epithelium was demonstrated by scanning electron microscopy

New technology in molecular medicine allows global descriptions of complex expression patterns of genes responsible for the malignant properties of tumors[35-38] With the two representative passages of cells described above, we have defined the profile of genes involved in the invasiveness of tumor cells using a sensitive cDNA microarray in a previous study. We found that increased expression of ezrin protein in SHEEMT cells was in accordance with their acquisition of the invasive potency[39]. Based on these observations, we therefore speculated that ezrin might be an important candidate of genes in charge of the invasive behaviors of malignantly transformed esophageal epithelial cells, as reported in lymphoma and astrocytic tumors[40,41]. However, recent work in this field has also focused on the identification of ezrin – binding molecules, including CD44, CD43, intercellular adhesion molecule[42-44] and syndecan – 2[45]. Moreover, modulation of the ERM protein ezrin by Merlin and NF – 2 has been reported in other literatures[46-48]. In the meanwhile, we have demonstrated that following expression of MMP_2 and MMP_9, highly invasive potency was developed in SHEEMT cells (Data to be published). All these suggest the invasiveness of tumor cells is determined by multiple genes and co – factors with complicated cellular signal passways. Therefore, future works are necessitated to demonstrate more exactly the roles of ezrin and related molecules in the formation ofinvasive potency of cancer ceils by the gene knockout technique and other powerful tools.

[In 《World J Gastroenterol》 2003, 9 (6): 1182 – 1186]

References

1　Ueda M, Terai Y, Yamashita Y, Kumagai K, Ueki K, Yamaguchi H, Akise D, Hung YC, Ueki M. Correlation between vascular endothelial growth factor – C expression and invasion phenotype in cervical carcinomas. Int J Cancer, 2002, 98: 335 – 343

2　Nestl A, Von Stein OD, Zatloukal K, Thies WG, Herrlich P, Hofmann M, Sleeman JP. Gene expression patterns associated with the metastatic phenotype in rodent and human tumors. Cancer Res, 2001, 61: 1569 – 1577

3　Comoglio PM, Tamagnone L, Boccaccio C. Plasminogen – related growth factor and semaphorin receptors: a gene superfamily controlling invasive growth. Exp Cell Res, 1999, 253: 88 – 99

4　Janneau JL, Maldonado – Estrada J, Tachdjian G, Miran I, Motte N, Saulnier P, Sabourin JC, Cote JF, Simon B, Frydman R, Chaouat G, Bellet D. Transcriptional expression of genes involved in cell invasion and migration by normal and tumoral trophoblast cells. J Clin Endocrinol Metab, 2002, 87: 5336 – 5339

5　Guo Y, Pakneshan P, Gladu J, Slack A, Szyf M, Rabbani SA. Regulation of DNA methylation in human breast cancer. Effect on the urokinase – type plasminogen activator gene production and tumor invasion. J Biol Chem, 2002, 277: 41571 – 41579

6　Jiang Y, Xu W, Lu J, He F, Yang X. Invasiveness of hepatocellular carcinoma cell lines: contribution of hepatocyte growth factor, c – met, and transcription factor Ets – 1. Biochem Biophys Res Commun, 2001, 286: 1123 – 1130

7　Orian – Rousseau V, Chen L, Sleeman JP, Herrlich P, Ponta H. CD44 is required for two consecutive steps in HGF/c – Met signaling. Genes Dev, 2002, 16: 3074 – 3086

8　Loktionov A, Watson MA, Stebbings WS, Speakman CT, Bingham SA. Plasminogen activator inhibitor – 1 gene polymorphism and colorectal cancer risk and prognosis. Cancer Lett, 2003, 189: 189 – 196

9　Tokunou M, Niki T, Eguchi K, Iba S, Tsuda H, Yamada T, Matsuno Y, Kondo H, Saitoh Y, Imamura H, Hirohashi S. c – MET expression in myofibroblasts: role in autocrine activation and prognostic significance in lung adenocarcinoma. Ant J Pathol, 2001, 158: 1451 – 1463

10　Shen ZY, Xu LY, Chen MH, Shen J, Cai WJ, Zeng Y. Progressive transformation of immortalized esophageal epithelial cells. World J Gastroenterol, 2002, 8: 976 – 981

11　Shen ZY, Cen S, Cai WJ, Teng ZP, Shen J, Hu Z, Zeng Y. Immor – talization of human fetal esophageal epithelial cells induced by E6 and E7 genes of human papilloma virus 18. Zhonghua Shiyan He Linchuang Bingduxue Zazhi, 1999, 13: 121 – 123

12　Shen ZY, Xu LY, Chen Xh, Cai WJ, Shen J, Chen JY, Huang TH, Zeng Y. The genetic events of HPV – immortalized esophageal epithelium cells. Int J Mol Med, 2001, 8: 537 – 542

13　Shen ZY, Shen J, Cai WJ, Cen S, Zeng Y. Biological characteris – tics of human fetal esophageal epithelial cell line immortalized by the E6 and E7 gene of HPV type 18. Zhonghua Shiyan He Linchuang Bingduxue Zazhi, 1999, 13: 209 – 212

14　Shen ZY, Xu LY, Li C, Cai WJ, Shen J, Chen JY, Zeng Y. A comparative study of telomerase activity and malignant phenotype in multistage carcinogenesis of esophageal epithelial cells induced by human papillomavirus. Int J Mol Med, 2001, 8: 633 – 639

15　Shen ZY, Xu LY, Li EM, Cai WJ, Chen MH, Shen J, Zeng Y. Telomere and telomerase in the initial stage of immortalization of esophageal epithelial cell. World J Gastrointerol, 2002, 8: 357 – 362

16　Shen ZY, Xu LY, Chen MH, Cai WJ, Shen J, Chen JY, Hon CQ, Zeng Y. Biphasic differentiation of immortalized esophageal epitheliums induced by HPV18E6E7. Bingdu Xuebao, 2001, 17: 210 – 214

17 Shen ZY, Cen S, Shen J, Cai W, Xu J, Teng Z, Hu Z, Zeng Y. Study of immortalization and malignant transformation of human embryonic esophageal epithelial cells induced by HPV18E6E7, J Cancer Res Clin Oncol, 2000, 126: 589 – 594

18 Shen ZY, Cai WJ, Shen J, Xu JJ, Cen S, Teng ZP, Hu Z, Zeng Y. Human papilloma virus 18E6E7 in synergy with TPA induced malignant transformation of human embryonic esophageal epithelial cells. Bingdu Xuebao, 1999, 15: 1 – 6

19 Shen ZY, Shen J, Cai WJ, Chen JY, Zeng Y. Identification of malignant transformation in the immortalized esophageal epithelial cells. Zhonghua Zhongliu Zazhi, 2002, 24: 107 – 109

20 Mangeat P, Roy C, Martin M. ERM proteins in cell adhesion and membrane dynamics. Trends Cell Biol, 1999, 9: 187 – 192

21 Scherer SS, Xu T, Crino P, Arroyo EJ, Gutmann DH. Ezrin, radixin, and moesin are components of Schwann cell microvilli. J Neurosci Res, 2001, 65: 150 – 164

22 Ohtani K, Sakamoto H, Rutherford T, Chen Z, Satoh K, Naftolin F. Ezrin, a membranecytoskeletal linking protein, is involved in the process of invasion of endometrial cancer cells. Cancer Lett, 1999, 147: 31 – 38

23 Makitie T, Carpen O, Vaheri A, Kivela T. Ezrin as a prognostic indicator and its relationship to tumor characteristics in uveal malignant melanoma. Invest Ophthalmol Vis Sci, 2001, 42: 2442 – 2449

24 Ohtani K, Sakamoto H, Rutherford T, Chen Z, Kikuchi A, Yamamoto T, Satoh K, Naftolin F. Ezrin, a membrane – cytoskeletal linking protein, is highly expressed in atypical endometrial hy – perplasia and uterine endometrioid adenocarcinoma. Cancer Lett, 2002, 179: 79 – 86

25 Tokunou M, Niki T, Saitoh Y, Imamura H, Sakamoto M, Hirohashi S. Altered expression of the ERM proteins in lung adenocarcinoma. Lab Invest, 2000, 80: 1643 – 1650

26 Johnson MW, Miyata H, Vinters HV. Ezrin and moesin expres – sion within the developing human cerebrum and tuberous sclerosis – associated cortical tubers. Acta Neuropathol, 2002, 104: 188 – 196

27 Mykkanen OM, Gronholm M, Ronty M, Lalowski M, Salmikangas P, Suila H, Carpen O. Characterization of human palladin, a microfilament – associated protein. Mol Biol Cell, 2001, 12: 3060 – 3073

28 Gautreau A, Poullet P, Louvard D, Arpin M. Ezrin, a plasma membrane – microfilament linker, signals cell survival through the phosphatidylinositol 3 – kinase/Akt pathway. Proc Natl Acad Sci USA, 1999, 96: 7300 – 7305

29 Gonzalez RR, Devoto L, Campana A, Bischof P. Effects of leptin, interleukin – lalpha, interleukin – 6, and transforming growth factor – beta on markers of trophoblast invasive phenotype: integrins and metalloproteinases. Endocrine, 2001, 15: 157 – 164

30 Chert Z, Fadiel A, Feng Y, Ohtani K, Rutheford T, Naftolin F. Ovarian epithelial carcinoma tyrosine phosphorylation, cell proliferation, and ezrin translocation are stimulated by interleukin lalpha and epidermal growth factor. Cancer, 2001, 92: 3068 – 3075

31 Tran Quang C, Gautreau A, Arpin M, Treisman R. Ezrin function is required for ROCK – mediated fibroblast transformation by the Net and Dbl oncogenes. EMBO J, 2000, 19: 4565 – 4576

32 Stapleton G, Malliri A, Ozanne BW. Downregulated AP – 1 activity is associated with inhibition of Protein – Kinase – C – dependent CD44 and ezrin localization and upregulation of PKC theta in A431 cells. J Cell Sci, 2002, 115: 2713 – 2724

33 Si HX, Tsao SW, Lam KY, Srivastava G, Liu Y, Wong YC, Shen ZY, Cheung AL. E – cadherin expression is commonly downregulated by

CpG island hypermethylation in esophageal carci-noma cells. Cancer Lett, 2001, 173: 71 – 78

34　Hiscox S, Jiang WG. Ezrin regulates cell – cell and cell – matrix adhesion, a possible role with E – cadherin/beta – catenin. J Cell Sci, 1999, 112: 3081 – 3090

35　Schindelmann S, Windisch J, Grundmann R, Kreienberg R, Zeillinger R, Deissler H. Expression profiling of mammary carcinoma cell lines: correlation of in vitro invasiveness with expression of CD24. Tumour Biol, 2002, 23: 139 – 145

36　Zhan F, Cao L, Hu C, Li G. Differentially expressed cDNA sequences homologous with known genes in human nasopharyngeal carcinoma. Hunan Yike Daxue Xuebao, 1999, 24: 103 – 106

37　Khanna C, Khan J, Nguyen P, Prehn J, Caylor J, Yeung C, Trepel J, Meltzer P, Helman L. Metastasis – associated differences in gene expression in a murine model of osteosarcoma. Cancer Res, 2001, 61: 3750 – 3759

38　Wang KC, Cheng AL, Chuang SE, Hsu HC, Su IJ. Retinoic acid – induced apoptotic pathway in T – cell lymphoma: Identification of four groups of genes with differential biological functions. Exp Hematol, 2000, 28: 1441 – 1450

39　Shen J, Chen MH, Zheng RM, Chen JJ, Shen ZY. Detection of tumor cell invasion in vitro by scanning electron microscopy. Dianzi Xianwei Xuebao, 2000, 19: 313 – 314

40　Geiger KD, Stoldt P, Schlote W, Derouiche A. Ezrin immunore – activity is associated with increasing malignancy of astrocytic tumors but is absent in oligodendrogliomas. Am J Pathol, 2000, 157: 1785 – 1793

41　Akisawa N, Nishimori I, Iwamura T, Onishi S, Hollingsworth MA. High levels of ezrin expressed by human pancreatic adeno – carcinoma cell lines with high metastatic potential. Biochem Biophys Res Commun, 1999, 258: 395 – 400

42　Harrison GM, Davies G, Martin TA, Jiang WG, Mason MD. Distribution and expression of CD44 isoforms and Ezrin during prostate cancer – endothelium interaction, Int J Oncol, 2002, 21: 935 – 940

43　Legg JW, Lewis CA, Parsons M, Ng T, Isacke CM. A novel PKC – regulated mechanism controls CD44 ezrin association and directional cell motility. Nat Cell Biol, 2002, 4: 399 – 407

44　Guan XQ, Wang CJ, Li YY. Effects of Ezrin on differentiation and adhesion of hepatocellular carcinoma. Ai Zheng, 2002, 21: 281 – 284

45　Granes F, Urena JM, Rocamora N, Vilaro S. Ezrin links syndecan – 2 to the cytoskeleton. J Cell Sci, 2000, 113: 1267 – 1276

46　Gronholm M, Sainio M, Zhao F, Heiska L, Vaheri A, Carpen O. Homotypic and heterotypic interaction of the neurofibromatosis 2 tumor suppressor protein merlin and the ERM protein ezrin. J Cell Sci, 1999, 112: 895 – 904

47　Meng JJ, Lowrie DJ, Sun H, Dorsey E, Pelton PD, Bashour AM, Groden J, Ratner N, Ip W. Interaction between two isoforms of the NF2 tumor suppressor protein, merlin, and between merlin and ezrin, suggests modulation of ERM proteins by merlin. J Neurosci Res, 2000, 62: 491 – 502

48　Gutmann DH, Sherman L, Seftor L, Haipek C, Hoang Lu K, Hendrix M. Increased expression of the NF2 tumor suppressor gene product, merlin, impairs cell motility, adhesion and spreading. Hum Mol Genet, 1999, 8: 267 – 275

87. Cytogenetic and Molecular Genetic Changes in Malignant Transformation of Immortalized Esophageal Epithelial Cells

SHEN Zhong – ying[1], XU Li – yan[1], CHEN Min – hua[1], CAIT Wei – jia[1],
SHEN Jian[1], CHEN Jiong – yu[2], ZENG Yi[3]

1. Department of Pathology; 2. Central Laboratory of Tumor Hospital, Medical College of
Shantou University; 3. Institute of Virology, Chinese Academy of Preventive Medicine

Summary

The purpose of this study was to evaluate the extent to which the expression of $p53$, $c – myc$, $bcl – 2$, ras genes and chromosomes, along with activity of hTERT, impacts on the malignant transformation of immortalized esophageal epithelial cells. The SHEE cell line was established from an embryonic esophageal epithelial cell induced by transduction of E6E7 genes of human papillomavirus type 18 (HPV18E6E7). In cells of the 85th passage (SHEE85), the malignant transformation of SHEE was confirmed by morphology, cell proliferative index and tumor formation in SCID mice. $C – myc$, $p53$, $bcl – 2$ and ras genes were assayed by the multi – PCR method with house – keeping gene GAPDH as control. The modal number of chromosomes was analyzed and its expression of subunit of telomerase, hTERT, was assessed by RT – PCR. Expression of HPV18E6E7 was assayed by Western blotting. The results showed that cells of SHEE85 were atypical and exhibited proliferative status with a proliferation index of 45.70%. Tumors formed in SCID mice with invasion of adjacent tissue. The karyotype belonged to hypotriploid and displayed expression of hTERT. $C – myc$, $K – ras$, $bcl – 2$ and $p53$ (expression of phosphoprotein) were positive in SHEE85. Expression of HPV18E6E7 was positive. Taken together, SHEE85 cells were in fully malignant transformation and their molecular mechanism involved the expression of cellular genes, such as $p53$, $bcl – 2$, $c – myc$ and ras, and aberrance of chromosomes. It is probable that all of these changes were related with HPV18E6E7.

〔**Key words**〕 Esophagus; Human papillomavirus; Genetics; Telomerase; Malignant transformation

Introduction

In our survey the majority of esophageal carcinoma in the Shantou area was found to be associated with infection of human papillomavirus (HPV)[1]. By inducement of E6E7 genes of human papillomavirus type 18 (HPV18E6E7) we established immortalized human fetal esophageal epithelial cell line (SHEE)[2,3] and used them as a model to investigate the genesis and progression of immortalization[4,5]. In the early passages, SHEE had a limited ability to replicate, and would eventually reach senescence (a non – dividing state) *in vitro*, which was usually accompanied by extensive cell death (crisis)[6]. When the activation of telomerase appeared, some cells escaped senescence and passed through a crisis phase. The cell continued to proliferate indefinitely and entered immortalization[7]. With increased passage, however, only a limited number of the cells were tumorigenic, mimicking the human esophageal carcinoma on the conversion from premalignant to invasive carcinoma[8]. They might then acquire the characteristics of cells that were fully cancerous or transformed[9].

Chromosome analysis of cancer cell lines provides important information about possible genetic alterations underlying the development of cancer. Detailed analyses of the SHEE series, the 10th passage (SHEE10), the 31st passage (SHEE31) and the 60th passage (SHEE60), have been reported[4]. Bimodal distribution of chromosomes with more aberrant chromosomes appeared from the 31st to the 60th passage cells[7] and the aneuploid chromosome numbers varied according to cell passages[9].

Some oncogenes and anti – oncogenes are important in the malignant transformation of tumors. The tumor suppressor P53 inhibits tumor formation, in part by inducing apoptosis, which is inhibited by anti – apoptotic $bcl – 2$[10]. $C – myc$ gene expression is an important event in cell proliferation, which induced anchorage – independent cell growth and led to transformation[11]. The ras family of oncogenes has been extensively studied for its implication in several types of human malignancies[12-14]. Bcl – 2 and $p53$ protein expression are useful discriminating prognostic factors in the evaluation of carcinoma[15]. These genes are involved in the carcinogenesis of cells.

To explore cytogenetic and molecular genetic changes in the malignant transformation from immortalized esophageal epithelial cells, we studied here the ectopic expression of $c – myc$, $bcl – 2$, ras, $p53$; the cytogenetic changes with modal number of chromosomes; and the expression of telomerase catalytic subumit (hTERT), all of which may refoect specific molecular events in the malignant transformation of SHEE.

Materials and Methods

1. Cell culture: The immortalized esophageal epithelial cell line SHEE, which was induced by HPV18E6E7 AAV in our laboratory was propagated continually. Cells of the 85th passage (SHEE85) were cultured in 50 ml flasks in culture medium 199 (Gibco) with 10% calf serum, 100 units of penicillin and streptomycin, in an incubator with a humidified atmosphere of 5% CO_2 and 95% air.

2. Light – microscopic examination: SHEE85 cells grown in flask were examined under phase – contrast microscope for cell morphology in living status.

3. Preparation of single – cell suspension: To prepare SHEE85 cells for assays, cells in the exponential growth phase were harvested by exposure to 0.25% trypsin/0.02% EDTA solution (w/v). The flask was sharply tapped to dislodge the cells, and supplementary medium was added. The cell suspension was pipetted to produce a single – cell suspension. The cells were washed and resuspended in PBS to the desired cell concentration. Cell viability was determined by trypan blue exclusion, and only single – cell suspensions of >90% viability was used.

4. Cytogenetic analysis: Metaphase spreads of SHEE85 cells were obtained using standard cyto-genetic methods. Briefly, the cultured cells of each passage were preserved at 3 – 4℃ for 3 – 4 h to produce synchronous stage cells and then cultivated at 37℃ for 3 – 4 h, added in 0. 05 μg/ml col-chicines for 1 h. Harvesting was by standard method and stained with Giemsa. One hundred meta-phases were scored.

5. Flow cytometry analysis: Single cell suspension was washed twice with PBS, fixed with 70% al-cohol and filtered through nylon net of 360 meshes to make a single cell resuspension (10^6 cells/ml). The cell was stained by propidium iodide for 15 rain and analyzed by flow cytometry (FACSort, B – D Company). In a DNA histogram, the percentages of cells in every phase of the cell cycle and cells in pre – G_0G_1 stage (apoptotic cells) were calculated. The proliferative index was calculated by the formula of ($S + G_2M/G_0G_1 + S + G_2M$).

6. Tunlorigensis and invasive potency in SCID mice: SHEE85 cells (10^6 cells/0. 2 ml) were in-oculated into the left axillae of four severely combined immunodeficient (SCID) mice; tumor genera-tion was observed for four weeks. Cells of SHEE85 (10^6 cells/0. 2 ml) were inoculated into the in-traperitoneal cavity for 6 weeks, and the invaded organs were examined macro – scopically and mi-croscopically.

7. HPV18E6E7 assays: The protein expression of HPV18E6E7 was detected by Western blotting method. The cells were washed three times with ice – cold PBS, and then were lysed in buffer [50 mmol/L Tris – HCl, pH 8. 0, 150 mmol/L NaCL, 100 ug/ml phenyl – methyl – sulfonyl fltuoride (PMSF), 1% TritonX – 100] for 30 rain on ice. After removal of cell debris by certrifugation (12 000 g, 5 min), thc protein concentration of lysates was measured by Bradford method. Protein (50 μg) of SHEE85 boiled for 5 min in sampls buffer buffer was separated by 10% SDS – PAGE, transferred onto nitrocellulose membrane (Bio – Rad). Non – specific reactivity was blocked by in-cubation overnight at 4℃ in buffer (10 mmol/L Tris – HCl, pH 7. 5, 150 mmol/L NaCl, 2% Tween – 20, 40% bovine serum albumin). The membrane was then incubated with antibody of mouse anti – HPV18E6E7 (SC – 264, Zhong Shah Biotech Co.), followed by reaction with anti – mouse IgG antibody. Reactive protein was detected by the ECL chemiluminescence system (Amer-sham).

8. C – myc, bcl – 2, p53 and ras genes assay by multi – PCR (mPCR): Preparaiton of the to-tal RNA was performed by Maxim MPCR Kits (APO – MO5OG, and RAS – MO50 Maxim Biotech. Inc.) including the primers of c – myc, bcl – 2, p53; ras and a house – keeping gene GAPDH. The isolation of undegraded and intact RNA of SHEE85 was prerequisite without RNase contamina-tion of buffers and containers. PCR thermocycle profile: 96℃ 1 min and 57℃ 4 min for 2 cycles; 94℃ 1 min and 57℃ 2. 5 min for 30 cycles; 70℃ 10 min and 25℃ soak. The ras mPCR Kit con-tains ras positive contrast, K – ras, H – ras primer and DNA marker. The electrophoresis of PCR products was used in the visual inspection of an agarose gel containing 0. 5 μg/ml ethidium bromide (Sigma).

9. Subunits of telomerase analysis: The activated telomerase was assayed by human telomerase reverse transcriptase (hTERT). hTERT mRNA was amplified using the primer pair: 5' – CG-GAAGAGTGTCTGGAGCAA – 3' and 5' – GTTGCTCTAGAATGAACGTGGAAG – 3' for 31 cycles (94℃ for 45 sec, 55℃ for 45 sec, 72℃ for 90 sec). The house – keeping gene GAPDH was amplified using the primer pair: 5' – GAAGGTGAAGGTCGGAGTC – 3' and 5' – GAAGATGGT GATGG-GATTTC – 3' for 33 cycles (94℃ for 60 sec, 55℃ for 60 sec, 72℃ for 60 sec). PCR products of SHEE85 were subjected to electrophoresis in a 1.5% agarose gel containing 0.5 g/ml ethidium bromide.

Results

1. Morphological observation: Light microscopy showed that the cells in SHEE85 crowded together with different sizes. considerably more giant nucleus cells with multinucleoli were seen (Fig. 1). indieating that the cells of SHEE85 were in over proliferation and in poor differentiation.

2. Kinetics of cell proliferation: The DNA histogram of SHEE85 cells (Fig. 2). The DNA content of cells revealed the cell distribution in the cell cycle consisted of $G_0 G_1$, 54.1%; S, 21.3%, G2 M, 22.5%; pre – $G_0 G_1$ 2.1%. An increase in cell proliferation (proliferative index was 43.8%), a decrease in apoptosis and an increase in hyperploidy cells occurred.

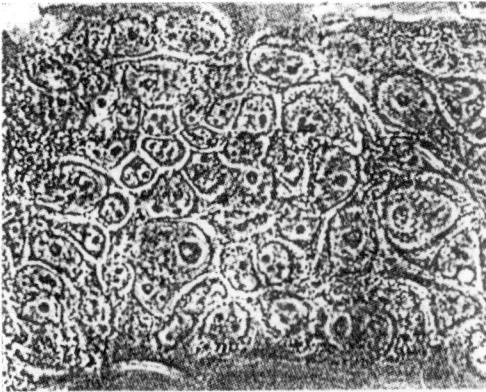

Fig. 1 **Morphology of cultured cells. The malignant transformed cells of SHEE85 had enlarged nucleus in different sizes (phase contrast, ×200)**

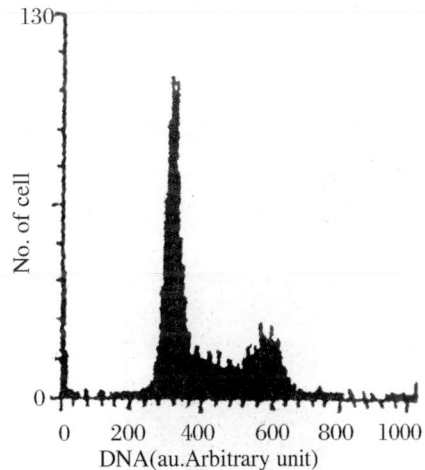

Fig. 2 **DNA histogram of SHEE85**

3. Tumor development in SCID mice: When SHEE85 cells were injected into the axilla of SCID mice, they grew rapidly to form a tumor in 40 days. In histological examination the cells showed large nucleus, fewer cytoplasm and large nucleolus with infiltration and destruction of muscular fibers (Fig. 3). The tumor could be passaged continuously in SCID mice.

4. Cytogenetic abnormality: The number of chromosomes in SHEE85 cells ranged between 45 – 108. These chromosomes were mainly hyperdiploid and hypotriploid (Fig. 4). In karyotype of SHEE85 chromosomes, the modal number of chromosomes was 59 – 65. The distribution of chromosomes was diploidy and hypotriploidy 4.9%; hyperdiploidy 18.8%; hypotriploidy 58.8%; hypertriploidy 17.7%. The chromosomes of SHEE cells were hyperdiploid in the early passages[4], but became hypotriploid in the 85th passage, showing the property of chromosome instability in SHEE cell lines.

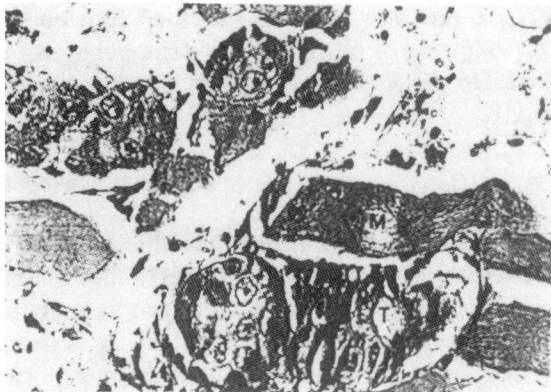

M, muscle fiber; T, tumor cells

Fig. 3 Tumor development in SCID mouse with invasion of muscular layer. Tumor cells invaded and damaged muscular stratium microscopically (H&E: ×400)

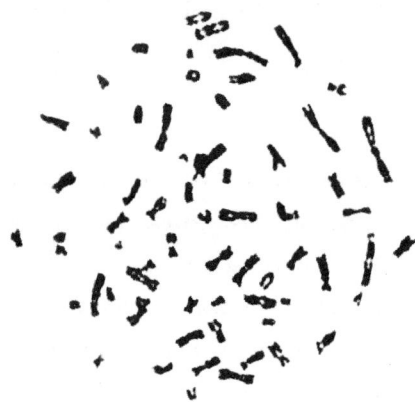

Fig. 4 Hypotriploid of SHEE85 (Giemsa, ×1000)

5. *Expression of HPV*18*E6E7*: Fig. 5 shows the expression of HPV18E6E7 in SHEE85. A specific band fragment of 70×10^3 was found as a marker, proving the existence of HPV18E6E7 gene expression in the cells of SHEE85 cells (Fig. 5 brief).

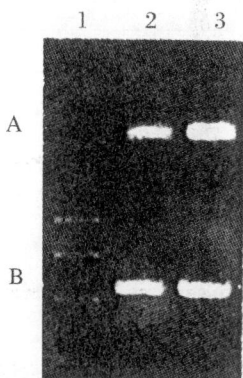

A, hTERT; B, GAPDH. 1, Maker; 2, SHEE85; 3, SHEE85

Fig. 6 Gel electrophoretogram of hTERT

6. *Detection of hTERT*: Expression of hTERT in SHEE85 cells was determined by RT – PCR. The hTERT expression was positive in the 85th passage. The house keeping protein GAPDH was used as a control (Fig. 6).

7. *mPCR assay of* p53 *c* – *myc*, *bcl* – 2 *and ras*: In passages 85 of SHEE the *p53*, *c* – *myc*, *bcl* – 2 *and ras* were seen (Fig. 7). The *p53* sequence analysis was align to phosphorylation, comparing with *p53* gene from GenBank (gbM22894) (not shown here).

Discussion

Our results determined that: i) a significant chromosomal aberrance with hTERT appearance exists in SHEE85, ii) up – regulation of *c* – *myc*, *bcl* – 2, *p53 and ras* oncogenes, iii) expression of HPV18E6E7. The mechanisms of malignant transformation of SHEE may require a synergistic effect generated by the coexistence of these factors[16,17]. In our studies an increasing body of evidence suggests that progression from immortalization of SHEE to malignant change is accompanied by morphological and genetic alterations. The malignant transformation is a gradual imbalance of normal tissue homeostasis involving many factors including chromosome aberrance and some spontaneous oncogenes[18].

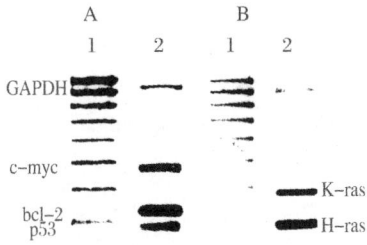

A. p53, bcl − 2, c − myc and GAPDH; B, H − ras and K − ras. Lanes: 1, marker; 2, SHEE85

Fig. 7　Electrophoretograph of PCR products

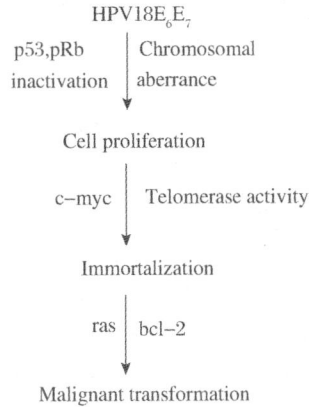

Fig. 8　Illustration of the transformed pathway of esophageal epithelial cells induced by HPV18E6E7

In vitro studies have revealed that high − risk HPV, types 16 and 18, induce immortalization of primary human keratinocytes by means of their E_6 and E_7 oncogene functions[19]. However, the process of immortalization requires host gene alterations induced by the expression of the viral oncoproteins[20]. The role of HPV in esophageal carcinogenesis could be due to effects of E6E7 protein expression, which interacts on the most crucial changes of aberrant chromosomes, and changes of cellular genetics and molecular genetics, both of which lead to cell proliferation[21]. Chromosome instability provides a predisposing background to malignancy, contributing to crucial genetic changes[22].

In our esophageal epithelial cell line SHEE the tumor suppressor gene p53 was expressed in phosphoprotein (unpublished data). Post − transcriptional modification of p53 by phosphorylation has been proposed to be an important mechanism of p53 stabilization and functional changes[23]. phosphorylation of p53 in particular activite specific DNA buinding function by stabilizing p53 tetromer formation[24], The stabilized p53 is in a wild − type conformation and the same number of phospho − forms is present, by which SHEE cells contributed to proliferation.

The enhanced bcl − 2 expression to prevent apoptosis seems to occur from the late stages and may play an important role in the carcinogenesis of SHEE[25]. The antiapoptosis gene bcl − 2 expression can promote cell growth and aggressiveness, bcl − 2 may protect epithelial cells from apoptosis[26]. Bcl − 2 expressed cells might continue to proliferate in over − crowded cultures[27]. Bcl − 2 might affect differentiation, inhibited cell from entering the process of terminal differentiation and conferred a survival advantage after the cells reached confluence[28]. Therefore, bcl − 2 can act as a key determinant in cell proliferation, differentiation and tumorigenesis and regulates the normal flow of cells through cycles of proliferation[29]. As such, it plays an important role in the maintenance of tissue homeostasis and function. The association of $bcl-2$ and $p53$ expression in SHEE supported the prediction that the coexistence of enhanced $bcl-2$ expression and altered $p53$ function in SHEE may create crucial synergistic effects in the carcinogenesis of SHEE[30].

C – myc gene is implicated in a large number of human solid tumors, leukemia and lymphomas as well as in a variety of animal neoplasm[31-33]. Chromosomal rearrangements and integration of oncogenic viruses frequently target myc locus, causing structural or functional alterations to the gene[34].

We illustrated $c-myc$ to be up – regulated in SHEE85, playing a central role in a cell – cell contact – mediated switch mechanism by which cell division vs. differentiation is determined[35]. $N-myc$ and $bcl-2$ cooperate to increase the expression, secretion, and activation of MMP – 2, which likely leads to a more tumorigenic phenotype due to increased MMP – 2 mediated invasion[36].

Our studies also show that cellular ras genes may be activated at different stages in immortalization and progression with different consequences for the cells. In the early passages it was probably an expression of wild – type ras, which had onto – suppressive properties[37,38]. In the later passages cell transformation $in\ vitro$ by ras mutant genes depended on the establishment of cells for indefinite growth, which is an independent step in tumorigenesis[39].

Further study is required on the construction and function of ras in our cell line.

The step – wise accumulation of genetic and cytogenetic alterations in SHEE development included viral integration – mediated chromosome rearrangements and genetic alterations. $bcl-2$ promoted cell proliferation in concert with $p53$, $c-myc$ and ras was involved in cell – cycle regulation. When hTERT appeared and cell immortalization occurred, in this way SHEE was responsive to malignant transformation by expression $H-ras$ or $K-ras$ oncogene[40,41]. From these studies in conjunction with our previous studies of the genetic evens of immortalization of SHEE[4], we postulated the pathway of carcinogenesis of esophageal epithelial cell line as shown in Fig. 8. After infection of HPV the integration and expression of E6E7 genes induced chromosomal aberrance and inactivation of $p53$ and pRb, which regulated epithelial cell growth and differentation. When telomerase activity appeared, the SHEE cells would be immortalized with up – regulation of $c-myc$. Combined with HPV18E6E7, activated ras and $bcl-2$ induced malignant transformation.

Acknowledgements

This investigation was supported by an award from the National Nutural Science Foundation of China (39830308), Research and Development Foundation of Shantou University (L00012) and Chinese National Human Genome Center, Beijing.

〔In 《Int J Mol Met》 2003, 12: 219 – 224〕

References

1 Shen ZY, Hu SP, Lu LC, Tang CZ, Kuang ZS, Zhong SP and Zeng Y. Detection of human papillomavirus in esophageal carcinoma. J Med Virol, 2002, 68: 412 – 416

2 Shen ZY, Cen S, Cai W J, Xu J J, Ten ZP,

Shen J, Hu Z and Zeng Y. Immortalization of human fetal esophageal epithelial cell induced by E_6 and E_7 genes of human papilloma virus 18. Chin Exp Clin Virol, 1999, 13: 121 – 123

3 Shen ZY, Shen J, Cai WJ, Cen S and Zeng

Y. Biological characteristics of human fetal esophageal epithelial cell line immortalized by the E_6 and E_7 gene of HPV type 18. Chin Exp Clin Virol, 1999, 13: 209 – 212

4 Shen ZY, Xu LY, Chen XH, Cai WJ, Shen J, Chen JY, Huang TH and Zeng Y. The genetic events of HPV – immortalized esophageal epithelium cells. Int J Mol Med, 2001, 8: 537 – 542

5 Shen ZY, Cen S, Shen J, Cai WJ, Xu J J, Teng ZP, Hu Z and Zeng Y. Study of immortalization and malignant transformation of human embryonic esophageal epithelial cells induced by HPV18E6E7. J Cancer Res Clin Oncol, 2000, 126. 589 – 594

6 Shen ZY, Xu LY, Li EM, Cai WJ, Chen MH, Shen J and Zeng Y. Telomere and telomerase in the initial stage of immortalization of esophageal epithelial cell. World J Gastroenterol, 2002, 8: 357 – 362

7 Shen ZY, Xu LY, Chen MH, Cai WJ, Shen J, Chen JY, Hong CQ and Zeng Y. Biphasic differentiation of immortalized esophageal epitheliums induced by HPV18E6E7. Chin J Virol, 2001, 17: 210 – 214

8 Shen ZY, Cai WJ, Shen J, Xu JJ, Cen S, Ten ZP, Hu Z and Zeng Y. Human papilloma virus 18E6E7 in synergy with TPA induced malignant transformation of human embryonic esophageal epithelial cells. Chin J Virol, 1999, 15: 1 – 6

9 Shen ZY, Xu LY, Chen MIi, Shen J, Cai WJ and Zeng Y. Progressive transformation of immortalized esophageal epithelial cells. World J Gastroenterol, 2002, 8: 976 – 981

10 Sulkowska M, Famulski W, Chyczcwski L and Sulkowski S. Evaluation of p53 and bcl – 2 oncoprotein expression in precancerous lesions of the oral cavity. Neoplasma, 2001, 48: 94 – 98

11 De Miglio MR, Simile MM, Muroni MR, Pusceddu S, Calvisi D, Carru A, Seddaiu MA, Daino L, Deiana L, Pascale RM and Feo F. Correlation of c – myc overexpression and amplification with progression of preneoplastic liver lesions to malignancy in the poorly susceptible

Wistar rat strain. Mol Carcinog, 1999, 25: 21 – 29

12 Mutter GL, Wada H, Faquin WC and Enomoto T. K – ras mutations appear in the premalignant phase of both microsatellite stable and unstable endometrial carcinogenesis. Mol Pathol, 1999, 52: 257 – 262

13 Jacobsen K, Groth A and Willumsen BM. Ras – inducible immortalized fibroblasts: focus formation without cell cycle deregulation. Oncogene, 2002, 21: 3058 – 3067

14 Mok SC, Bell DA, Knapp RC, Fishbaugh PM, Welch WR, Muto MG, Berkowitz RS and Tsao SW. Mutation of K – ras protooncogene in human ovarian epithelial tumors of borderline malignancy. Cancer Res, 1993, 53: 1489 – 1492

15 Staibano S, Lo Muzio L, Pannone G, Scalvenzi M, Salvatore G, Errico ME, Fanali S, De Rosa G and Piattelli A. Interaction between bcl – 2 and P53 in neoplastic progression of basal cell carcinoma of the head and neck. Anticancer Res, 2001, 21: 3757 – 3764

16 Weitzman JB and Yaniv M. Rebuilding the road to cancer. Nature, 1999, 400: 401 – 402

17 Hahn WC, Counter CM, Lundberg AS, Beijersbergen RL, Brooks MW and Weinberg RA. Creation of human turn out cells with defined genetic elements. Nature, 1999, 400: 464 – 468

18 Rhim JS. Molecular and genetic mechanisms of prostate cancer. Radiat Res, 2001, 155: 128 – 132

19 Tsao SW, Mok SC, Fey EG, Fletcher JA, Wan TS, Chew EC, Muto MG, Knapp RC and Berkowitz RS. Characterization of human ovarian surface epithelial cells immortalized by human papilloma viral oncogenes (HPV – E6E7ORFs). Exp Cell Res, 1995, 218: 409 – 507

20 Eichten A, Westfall M, Pietenpol JA and Munger K. Stabilization and functional impairment of the tumor suppressor p53 by the human papillomavirus type 16 E7 oncoprotein. Virology, 2002, 295: 74 – 85

21 Duensing S, Lee LY, Duensing A, Basile J,

Piboonniyom S, Gonzalez S, Crum CP and Munger K. The human papillomavirus type 16 E6 and E7 oncoproteins cooperate to induce mitotic defects and genomic instability by uncoupling centrosome duplication from the cell division cycle. Proc Natl Acad Sci USA, 2000, 97: 10002 – 10007

22 Heselmeyer K, Hellstrom AC, Blegen H, Schrock E, Silfversward C, Shah K, Auer G and Ried T. Primary carcinoma of the fallopian tube: comparative genomic hybridization reveals high genetic instability and a specific, recurring pattern of chromosomal aberrations. Int J Gynecol Pathol, 1998, 17: 245 – 254

23 Zhang J, Krishnamurthy PK and Johnson GV. Cdk5 phosphorylates p53 and regulates its activity. J Neurochem, 2002, 81: 307 – 313

24 Furihata M, Kurabayashl A, Matsumoto M, Sonobe H, Ohtsuki Y, Terao N, Kuwahara M and Shuin T. Frequent phosphorylation at serine 392 in overexpressed p53 protein due to missense mutation in carcinoma of the urinary tract. J Pathol, 2002, 197: 82 – 88

25 Fan XQ and Guo Y J. Apoptosis in oncology. Cell Res, 2001, 11: 1 – 7

26 Wang W, Johansson H, Bergholm U, Wilander E and Grimelius L. Apoptosis and expression of the proto – oncogenes bcl – 2 and p53 and the proliferation factor Ki – 67 in human medullary thyroid carcinoma. Endocr Pathol, 1996, 7: 37 – 45

27 Watanabe J, Kushihata F, Honda K, Mominoki K, Matsuda S and Kobayashi N. Bcl – xL overexpression in human hepatocellular carcinoma. Int J Oncol, 2002. 21: 515 – 519

28 Vaskivuo TE, Stenback F and Tapanainen JS. Apoptosis and apoptosis – related factors Bcl – 2. Bax. tumor necrosis factor – alpha. and NF – kappaB in human endometrial hyperplasia and carcinoma Cancer, 2002, 95: 1463 – 1471

29 Craig RW. MCLI provides a window on the role of the BCL2 family in cell proliferation, differentiation and tumorigenesis. Leukemia, 2002, 16: 444 – 454

30 Qian J and Shi Y. bcl – 2 and p53 protein expressions in the malignant transformation of ovarian endometriosis. Zhonghua Zhong Liu Za Zhi, 2001, 23: 403 – 405

31 Groco C, Alvino S. Buglioni S, Assisi D, Lapenta R, Grassi A. Stigliano V, Mottolese M and Casale V. Aetivation of c – MTC andc – NYB proto – oncogenes is associated with decreased apoptosis intumorcolon progression. Anticancer Rer, 2001, 21: 3185 – 3192

32 Devictor B, Bonnier P, Piana L, Andrac L, Lavaut MN, Allasia C and Charpin C. c – myc protein and Ki – 67 antigen immunodetection in patients with uterine cervix neoplasia: correlation of microcytophotometric analysis and histological data. Gynecol Oncol, 1993, 49: 284 – 290

33 Lossos IS, Alizadeh AA, Diehn M, Warnke R, Thorstenson Y. Oefner PJ, Brown PO, Botstein D and Levy R. Transformation of follicular lymphoma to diffuse large – cell lymphoma: alternative patterns with increased or decreased expression of c – myc and its regulated genes. Proc Natl Acad Sci USA, 2002, 99: 8886 – 8891

34 Popescu NC and Zimonjic DB. Chromosome – mediated alterations of the MYC gene in human cancer. J Cell Mol Meal, 2002, 6: 151 – 159

35 Demeterco C, Itkin – Ansari P, Tyrberg B, Ford LP, Jarvis RA and Levine F. c – Myc controls proliferation versus differentiation in human pancreatic endocrine cells. J Clin Endocrinol Metab, 2002, 87: 3475 – 3485

36 Noujaim D, van Golen CM, van Golen KL, Grauman A and Feldman EL. N – Myc and Bcl – 2 coexpression induces MMP – 2 secretion and activation in human neuroblastoma cells. Oncogene, 2002, 21: 4549 – 4557

37 Spandidos DA, Sourvinos G, Tsatsanis C and Zafiropoulos A. Normal ras genes: their onco – suppressor and pro – apoptotic functions (Review). Int J Oncol, 2002, 21: 237 – 241

38 Diaz R, Aim D, Lopcz – Barcons L, Malumbres M, Perez de Castro 1, Luc J, Ferrer – Miralles N, Mangues R, Tsong J, Garcia R, Perez –

Soler R and Pellicer A. The N – ras proto – onco-gene can suppress the malignant phenotype in the presence or absence of its oncogene. Cancer Res, 2002, 62: 4514 – 4518

39 Casanova ML, Latchet F, Casanova B, Murillas R. Fernandez – Acencero MJ, Villanueva C, Martinez – Palacio J, Ullrich A, Conti CJ and Jorcano JL. A critical role for ras – mediated, epidermal growth factor receptor – dependent an-giogenesis in mouse skin carcinogenesis. Cancer Res, 2002, 62: 3402 – 3407

40 Lundberg AS, Randell SH. Stewart SA, Elen-

baas B, Hartwell KA, Brooks MW, Fleming MD, Olsen JC, Miller SW, Weinberg RA and Hahn WC. Immortalization and transformations of primary human airway epithelial cells by gene transfer. Oncogene, 2002, 21: 4577 – 4586

41 MacKenzie KL, Franco S, Naiyer A J, May C, Sadelain M, Rafii S and Moore MA. Multiple stages of malignant transformation of human endo-thelial cells modelled by co – expression of telo – merase reverse transcriptase, SV40 T antigen and oncogenic N – ras. Oncogene, 2002, 21: 4200 – 4211

88.   E6/E7 Genes of Human Papilloma Virus, Type 18 Induced Immortalization of Human Fetal Esophageal Epithelium

SHEN Zhong – ying[1], CEN Shan[2], XU Li – yan[1], CAI Wei – jia[1],
CHEN Ming – hua[1], SHEN Han[1], ZENG Yi[2]

1. Department of Pathology. Medical College of Shantou University;

2. Institute of Virology, Chinese Academy of Preventive Medicine

Summary

To study the role played by human papilloma virus (HPV) in carcinogenesis, immortalized esophageal epithelial cells were induced by E_6 and E_7 genes of HPV type 18 and the biological be-havior was studied. Human fetal esophageal epithelial cells were transfected with recombined HPV18E6E7 AAV and were cultured and passaged in medium M199. In both the 10th passage (SHEE10) and the 31st passage (SHEE31), their proliferative rates by flow cytometry and their abilities to grow and form colonies in soft agar or to form tumors in SCID mice were examined. The HPV18 genes of E6E7 and its expression were determined using PCR methods. Cellular telomerase activity was detected by TRAP and chromosomes were analyzed by standard method. Immortalized cell lines of esophageal epithelium induced by the HPV18E6E7 were successfully established and cultured for >100 passages over 4 years. The result of PCR showed that the E6E7 gene of HPV18 was detectable in both cell clones. Both of them were unable to grow in soft – agarose medium and failed to produce tumors in SCID mice. Flow cytomerry demonstrated an average of 43% prolifera-tion index in SHEE31, but 28% in SHEE10. Telomerase activity was clearly identified in SHEE31 but not in SHEE10. Cytogenetic analysis demonstrated progression of chromosomal abnormalities with increasing trisome. Our data indicated that genes E6E7 of the HPV18 were capable of inducing immortalization in fetal esophageal epithelial cells. The immortal phenotype requires both activation of telomerase and genetic alterations that abrogate normal differentiation and promote celluar proliferation. This cell line can assist us to characterize the role played by HPV in carcinogenesis.

[**Key words**]　　Human papillomavirus; Esophageal epithelium; Immortalization

Introduction

Many viruses have been shown to have oncogenic potential, such as hepatitis B virus linked to cancer of the liver, Epstein – Barr virus to nasopharyngeal carcinoma, human papilloma virus (HPV) to carcinoma of the cervix and so on. Viral transformation refers to the changes in the function and morphology of a cell. These changes confer properties on the infected cell characteristic of neoplasia, which is always proceeded by immortalized and premalignant stage. Human papillomaviruses are common pathogens in the human population and are a major cause of human cancers[1]. In our local region 40% – 60% of esophageal carcinoma are HPV positive[2]. HPVs are able to transform human cells and are oncogenic in humans[3,4]. However, the detection rate of HPV in esophageal cancer is geographically varied[5-8], ranging from 40% to 60% in the high – incidence areas of esophageal cancer[9], which include the Far East and South Africa[10-12]. Our data indicated that HPV has an etiologic role inesophageal carcinogenesis at least in the high – incidence area of esophageal cancer[13].

In recent years humam adenoviruses have been attracting considerable attention because of their potential utility in gene transfer and gene therapy, in development of live viral vectored vaccines, and in protein expression in mammalian cells[14,15]. This is due to the virions of adenovirus being able to package other genome lengths[16], and because their defective vectors may be imserted with E6E7 gene of HPV. Therefore, we were able to make a recombined HPV18E6E7 with an associated adenovirus (HPV18E6E7AAV).

In previous studies, much scientific effort has gone into exploring the link between immortal phenotype nard immortalization and many biological criteria have been revealed[17]. Detection of genetic events, immortalizing gene[18] chromosome aberration[19], loss of heterozygosity of chromosome[20] and dose – dependency to growth factor[21] were investigated as immortal phenotype. To define the phenotype of immortalization it is considered extremely useful to determine how the process of immortalization developed and the variability between immortalized cells and malignantly transformed cells: in addition to determining the variability between immortalized calls and preimmortalized cells.

To explore the effect of HPV on the carcinogenesis of esophageal carcinoma, the recombinant adenovirus carrying the E6E7 genes of HPV type 18 was carried out and HPV18E6E7 AAV was transfected into human fetal esophageal epithelium to establish a new immortalized cell line. The biological characteristic of the cell line *in vitro* was observed to define the phenotype of immortalization.

Materials and Methods

1. Construction of recombinant plasmids: Recombinant plasmids were constructed by standard protocols[22]. pGEM/HPV18 plasmid was prepared by the alkaline lyses method from *E. coli* HMS174. The E6E7 gene of HPV18 was amplified with PCR. PCR product was lengthened in 0.9 kb and ligated with pGEM – T vector (Promega), generating pGEM – E6E7, pGEM – E6E7 was digested with Spet enzyme. The fragment was ligated with pAAV3 vector, generating pAAV – E6E7[15] (Fig. 1). The adenoviral vectors, which have been well characterized for transferring foreign genes into mammalian cells, are easy to manipulate, and can be grown to high titers PAAV – E6E7 was cotransfected with plasmid pAd8 into human 293 cell line, a hunnan kidney cell line (HK293), to produce the infectious virus (HPV18E6E7AAV). The virus was collected from the supernatant of the HK293 culture medium and the titers estimated of these recombinant viruses.

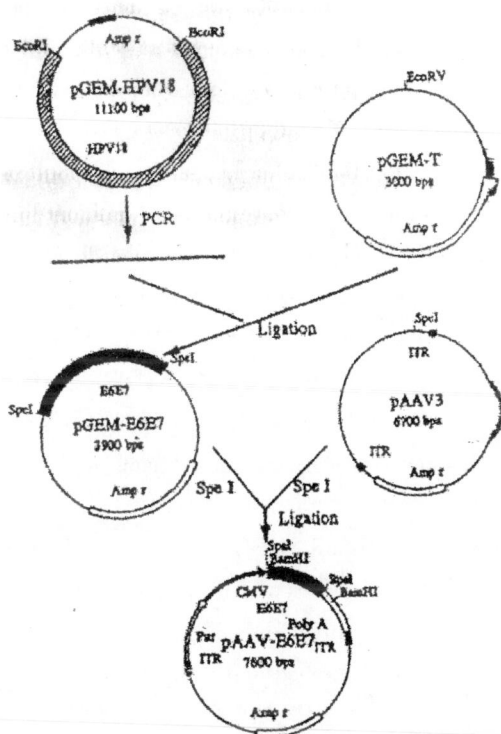

For insertion of E6E7 gene of HPV type 18, pGEM – HPV18 were amplified with PCR and ligated with pGEM. T, generating pGEM – E6E7 pGEM – E6E7 was digested with Spel and ligated with pAAV3 (also digested with Spel), generating pAAV – E6E7

Fig. 1   Coustruction of pAAV – E6E7

1, Marker; 2. PBR322/BstNI;
3. SHEE10; 4. SNEE31

Fig. 2   Aasay of HPV18E6E7 genes and its protein expression. Agarose gel electrophoresis of PCR products

2. Cell line: Esophageal tissue from an abortive five – month ferns was cut into small pieces, which were cultured in the flask using culture medium M199 with 10% bovine serum, 200 U/ml penecillin and 200 U/ml streptomycin. When small pieces of the esophageal tissue were adhered to the wall of the flask, the infectious virus of HPV18E6E7AAV was added and were incubated for 2 h, and then refreshed with normal culture medium. The epithelial cells grew from the margin of esophageal tissue and were continually cultured and propagated.

3. HPV18E6E7 detection by PCR: HPV18E6E7 primers were designed according to oligo software, synthesized by Shanghai Bioengineering Company. Upstream primer: 5' – GAC ACT AGT ACT ATG GCG CGC TTT GAG – 3'. Downstream primer: 5' – AGT ACT AGT TTA CAA CCC GTG CCC TCC – 3'.

Template DNA was extracted from SHEE10 and SHEE31 cells. PCR kit was purchased from Sai – Bai – Sheng Bio – company. The samples were amplified on PCR instrument (GTC – 2, Applied Res. Co., USA), and the PCR products were analyzeal by agarose gel electrophoresis.

4. Cell cycle analysis: SHEE10 and SHEE31 cells were collected respectively and washed twice with PBS, fixed with 70% alcohol and filtered through a nylon net of 360 meshes to make a single cell suspension (10^6 cells/ml). The cells were staind by propidium iodide for 15 min and anatvzed by flow cytometry (FACSort, B – D Company). In a DNA histogram the percentage of cells in every phase of the cell cycle, and the cells containing hyperploidy DNA were calculated. The proliferative index (PI) was calculated by the formula of ($S + G_2M/G_0G_1 + S + G_2M$).

5. Telomerase activity assay: Telomerase activity was measured using TRAP – eze Telomerase Detection Kit (Oncor Inc.). Frozen samples were homogenized in 10 – 50 μl of ice – cold 1x CHAPS lysis buffer (10 mmol/L Tris – HCl pH 7.5, 1 mmol/L $MgCl_2$, 1 mmol/L EGTA, 0.1 mmol/L benzamldine, 5 mmol/L β – mercaptoethanol, 0.5% CHAPS, 10% glycorol). After 30 min of incubation on ice, the lysate was centrifuged at 12 000 g for 20 min at 4℃. The telomeric repeat amplification protocol (TRAP) reaction was performed using 1.0 μl lysate or 1/10 diluted lysate, 2.5 μl 10 × TRAP buffer (200 mmol/L Tris – HCl, pH 8.3, 15 mmol/L $MgCl_2$, 630 mmol/L KCl, 0.5% Tween 20, 10 mmol/L EGTA, 0.1% BSA). 0.5μl 2.5 mmol/L dNTP, 0.5 μl TS primer, 0.5 μl TRAP primer mix. 19.5 μl water, 0.5 μl Taq (2U/μl). After incubation at 30℃ for 30 min the reaction mix immediately transferred to 94℃ and PCR performaed at 94℃ for 30 sec, at 55℃ for 30 sec, at 72℃ for 30 sec for 35 cycles. PCR products were separated in a non – denaturing 12% PAGE in 1x TBE at 5 volts/cm. The gel was stained using $AgNO_3$ and photographed.

6. Cytogenetic analysis: Metaphase spreads of SHEE10 and SHEE31 cells were obtained using standard cytogenetic methods. Briefly, the cultured cells of each passage were preserved at 3 – 4℃ for 3 – 4 h to make cells at: synchronous stage, and then cultivated at 37℃ for 3 – 4 h, added in 0.05μg/ml colchicines for 1 h. Harvesting was by standard method and stained with Giemsa. 50 – 100 metaphases were scored for each line.

7. Cell colony formation in soft agar cultivation: The exponentially grown cells of SHEE10 and SHEE31 were trypsinized, and the percentage of living cells was counted by trypan blue stain. The living single cell suspension (10^3 cells/ml) was cultured in 0.35% agar (Agarose, V312A, Promega) overlayed on 0.7% agar in a 6 – well plate (Coming), incubated in 5% CO_2, at 37℃ for 30 days and the cell colonies counted.

8. Tumorigenesis in nude mice: Six – week – old severe combined immutzodeficient mice (SCID C. B – 17/IcrJ – scid) (supplied by Experimental Animal Center of Zhongshan Medical University) were bred in isolated conditions. Four SCID mice were subcutaneoudy injected with both SHEE10 and SHEE31 cells in 5×10^6 cells/mouse respectively. They were, observed every 10 days for 2 months and then sacrificed to investigate the tumor by histopathological examination.

Results

1. Generation of the cell line: The new epithelial cells were cultivated from small pieces of esophageal tissues. The epithelial cells were propagated over 100 passages for a four – year period. The cell line SHEE was grown through preimmortalization, up to the 15th passage, cellular crisis in the 16th – 20th passages (data not shown) and then immortalization was established after 30 passages.

2. Cell morphology: The early immortal cells of SHEE10 grew evenly on the flasks, appeared like typical squamous epithelial cell with multiangular outline and oval nucleus and grew in monolayer with contact inhibition. SHEE31 cells were biphasic in morphology, partially well – differentiated squamous epithelial and partially undifferentiated basal cells.

3. Detection of HPV18E6E7 by PCR: Fig. 2 shows the bands of the PCR DNA products of both SHEE groups. Specific band fragment of 875 bp was revealed as a marker in SHEE10 and SHEE31 groups, proving the existence of HPV18E6E7 gene fragment in cells of both groups.

4. Cell prolifera tive cycle: Both SHEE10 and SHEE31 cells had similar DNA histograms (Fig. 3A and B). The proliferative indexes of SHEE10 and SHEE3t cells were 28.0% and 43.0%. respectively. Cells of polyploid and anenploid were frequently seen in SHEE31 cells (2.06%) but were less – in SHEE10 cells (0.56%).

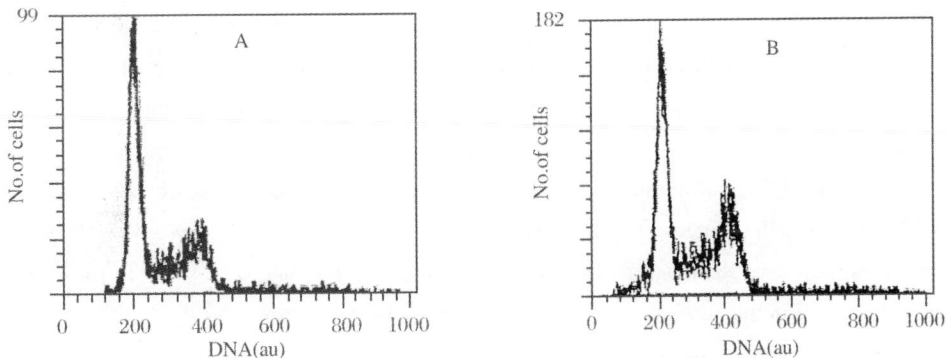

Fig. 3   DNA histogram of SHEE (A) , SHEE10; (B) . SHEE31

5. Cytogenetic abnormality: The number of chromosomes in SHEE10 and SHEE31 ranged between 32 and 196. These chromosomes were mainly hyperdiploids. Most modal numbers of chromosomes at SHEE10 was 59 – 62 with trisome of chromosomes +1, +3, +7, +9, +13, +17, +18. At SHEE31 the distribution was bimodal, 55 – 57. 6t – 63 with more trisomes of chromosomes, +1, +3, +7, +9, +11, +12, +14, +17, +18, +22, respectively. The modal number and trisome from the 10th to the 31st passage increased slowly. The data showed that genetic instability appeared.

6. Telomerase activity: In the early passages, telomerase activity could not be detected in SHEE10

(A) SHEE10; (B), SHEE31;
(C), Normal mucosa of esophagus

Fig. 4 Telomerase ladder of SHEE

cells and it appeared in the cells of 31st passage (Fig. 4).

7. Cell colony formation in soft agar: Cells of both passages were cultivated in soft agar for 30 days. The small colonies (cells <20) and a few of the large colonies (cells >20) were found in SHEE10 and SHEE31.

8. Tumorigenicity: In SCID mice inoculated with cells of SHEE10 and SHEE31 no tumors developed.

Discussion

In this study an immortalized human fotal esophageal epithelial cells line (SHEE) was established by E6E7 genes of HPV18 in our laboratory. The biological characteristics of the immortalized esophageal epithelial cells were revealed as follows: higher proliferation rate, chromosomal aberrance and instability and activity of telomerase but without tumorigenicity. We carried out further studies on the process of immortalization of SHEE to find out the immortal phenotype, which we used to monitor immortalization of cells. The proliferation of cells, the chromosome aberrance and telomerase activity were the phenotype of immortalization *in vitro*.

In HPV – induced immortalization of esophageal epithelium, both E6 and E7 oncoproteins appear to transform cells by different molecular mechanisms. It was probable that the viral genes were integrated into the bost genome, and expression of some viral genes (E_6 and E_7 of HPV) was found consistently. The E6 oncoproteins derived from oncogenic HPV inhibit the biological activities of the tumor suppressor gene p53[23]. The E7 oncoproteins bind to and inhibit the function of retinoblastoma gene product pRb[24], which are the common cellgrowth regulatory proteins. As an effect of both E6 and E7 on host cells immortalization of cells resulted.

Recent studies have demonstrated a highest frequency of telomerase activity in the immortalized cell line[25] and cell malignancies[26,27]. It is clear that immortalization and the activation of a telomerase are closely associated and telomerase activity has been found to have telomere maintenance activity. While telomerase activation is required for immortalization, telomerase activity is a useful marker for cell proliferation and immortalization, but not for malignant transformation[28,29]. Telomerase activity correlated with the cell cycle regnlator[30,31] to promote cell proliferation[32,33]. Telomerase could also effect genetic and epigenetic changes in human epithelial cells[34] and impact on cell proliferation and differenfiation[35,36]. We studied the correlation between telomerase activity and the proliferation status of the cells. High levels of telomerase activity in SHEE31 may be linked to the proliferative stage of the cells, which have a high level of synthesized DNA and high proliferative indices (PI, 43%). Alternatively, cells of SHEE10 have negative telomerase acavities with low DNA – labeling indices (PI, 28%).

The number of chromosomal sets and the percentage of hyperploid varied from SHEE10 to

SHEE31. The modal numbers were 58 – 62 in the SHEE10 and 55 – 57, 61 – 63 bimodal distribution in the SHEE31. The trisome were greater in SHEE31 than that in SHEE10. The phenomenon showed that chromosomes of SHEE series cell lines were unstable. Chromosome instability, in most cases reflects the occurrence of defective mitosis, including distribution of chromosomes to daughter cells, which leads to generation of aneuploid cells[37,38]. We revealed the chromosome aberrance to be an early event[39,40]. Aneuploidy destabilized the karyotype and thus initiated an autocatalytic karyotype evolution, generating preneoplastic and eventually neoplastic karyotypes.

Since the first report of human papillomavirus (HPV) in esophageal carcinoma in 1982, a potential risk factor of HPV in the development of esophageal carcinoma[41] has been an important research goal. By *in situ* hybridization assay (ISH) and polymerase chain reaction (PER), we identified HPV DNA in esophageal epithelium and carcinoma[42]. Therefore, the role of HPV in the carcinogenesis of esophageal carcinoma is important in etiology and pathogenesis. Telomerase expression has been discovered in HPV – positive squamotts cell carcinoma[43,44]. Previous reports have also shown that activity of telomerase can be achieved by the E6 and E7 protein of HPV[45,46]. HPV16 E6 oncoprotein was capable of inducing telomerase activity in monolayer cultures of proliferating keratiocyte[47]. Infection of HPV can cause karyotype confusion, such as fracture of chromosome, abnormal structure and number of chromosomes[18,48]. It also suggests that immortal esophageal cells reduced by HPV18E6E7 may effect the changes of chromosomes, and cause lability of genetic character[49]. The virus genome inserts and integrates with chromosomes of the host[50], which frequently occurred in common fragile sites[51]. causing the activation and expression of oncogenes. In other additional aterations it was more likely that HPVE6E7 also induced chromosome instability[52,53], which was also caused by telomere shortening[54]. Viruses can cause loss of contact inhibition, decrease of adhesion between cells, confusion of cellular skeletal structure, and less need for growth factors[55]. Therefore, all phenotypes of immortalization were induced by HPV.

In summary, transduction of human papillomavirus type 18 E6E7 to primary culture of human esophageal keratinocytes using a recombinant adenovirus vector were performed. Immortalization of the cells occurred at the 31st passage through cells senesced within 15 – 20 population doublings. It demonstrated that expression of E6 or E7 is sufficient to induce cellular immortalization. Cellular proliferation, telomerase activity and chromosome instability are the major characteristics of these immortalized esophageal cell lines.

Acknowledgements

Contract grant sponsor: National Natural Science Foundation of China (39930380); Research and Development Foundation of Shantou University (L00012), and supported by the Chinese National Human Genome Center, Beijing.

[In 《Oncol Rep》 2003, 10: 1431 – 1436]

References

1 Zur Hanson H. Papillomavirus infections – a major cause of human cancers. Biochim Biophys Acia, 1996, 1288: F55 – 78

2 Shen ZY, Hu SP. Lu LC. Tang CZ, Kuang ZS,

Zhong SP and Zeng Y. Detection of human papillomavirus in esophageal carcinoma. J Med Virol, 2002, 68: 412 – 416

3　McGlennen RC. Human papillomavirus oncogenesis. Clin Lab Med, 2000, 20: 383 – 406

4　Zhang L Liu T, Liu H and Gu C. HPV16 E6E7 fragments transform immortalized human bronchial epithelial cells into neoplastic cells. Zhonghua Bing Li Xue Za Zhi, 2000, 29: 350 – 353

5　Sur M and Cooper K. The role of the human papilloma vitus in esophageal carcer. Pathology, 1998, 30: 348 – 354

6　De Viltiers EM. Lavergne D, Chang F, Syrjanen K, Tosi P, Cinorino M, Santopietro R and Syrjanen S. An interlaboratory study to determine the presence of human papillomavirus DNA in esophageaI carcinoma from China. Int J Cancer, 1999, 81: 225 – 228

7　Takahashi A. Ogoshi S. Ono H, Ishikawa T, Toki T, Ohmon N, Iwasa M, Iwasa Y, Furihata M and Ohtsuki Y. High – risk human papillomavirus infection and overexpression of p53 protein in squamous call carcinoma of the esophagus from Japan. Dis Esophagus, 1998, 11: 162 – 167

8　Lambot MA. Pony MO and Noel JC. Human papillomavirus infection in esophageal squamous cell carcinoma in Belgium. Hum Pathol, 1998, 29: 1175 – 1176

9　Chang F. Syrjanen S, Shen Q, Cintorino M, Santopietro R, Tosi P and Syrjanen K. Human papillomavirus involvement in esophageal carcinogenesis in the high – incidenace area of China, A study of 700 cases by screening and type – specific in sint hybridization. Scand J Gastroenterol, 2000, 35: 123 – 130

10　Lavergne D and DeVilliers EM. Papillomavirus in esophageal papillomas and carcinomas. Int J Cancer, 1999, 80: 681 – 684

11　Cooper K, Taylor L and Govind S. Human papillomavirus DNA in oesophageal carcinomas in Sonth Africa. J Pathol, 1995, 175: 270 – 277

12　Poljak M Cerar A and Seme K. Human papilloma-

virus infection in esophageal carcinonmas a study of 121 lesions using multiple broad – spectrum polymerase chain reactions and literamre review. Hum Pathol, 1998, 29: 266 – 271

13　Lu LC, Shen ZY, You SJ, Shen J and Cai WJ. Detection of human papillomavrirus (EPV) on esophageal carcinoma by PCRand in – situ hybridization. Chin J Pathol, 1997, 26: 166 – 167

14　Zhou H and Beauder AL. A new vector system with inducible E2a cell line for production of higher titer and safer adenoviral vectors. Virology, 2000, 275: 348 – 357

15　Routes JM. Ryan S, Li H, Steinke J and Cook JL. Dissimilar immunogenicities of human papillomavirus E7 and adenovires E1A proteins influence primary tumor development. Virology, 2000, 277: 48 – 57

16　Bett AJ, Haddara W, Prevec L and Graham FL. An efficient and flexible system for consetruction of adenovirus vectors with insertions or deletions in early regions I and 3. Proc Natl Acad Sci USA, 1994, 91: 8502 – 8806

17　Schwab TS, Stewart T. Lehr J, Pienta KJ, Rhim JS and Macoska JA. Phenotypic characterization of immortalized normal and primary tuinor – derived human prostate epithelial cell cultures. Prostate, 2000, 44: 164 – 171

18　Macoska JA Beheshti B, Rhim JS, Hukkn B. Lehr J, Pienta KJ and Squire JA. Genetic characterization of immortalized human prostate epithelial cell cultures. Evidence for structural rearrangements of chromosome 8 and i (8q) chromosome formation in primary tumor – derived cells. Cancer Genet Cytogenet, 2000, 120: 50 – 57

19　Graham DA and Hemngton CS. HPV – 16 E2 gone disruption and sequence variation in CIN3 lesions and invasive squamous cell carcinomas of the cervix. relation to numerical chromosome abnormahties. Mot Pathol, 2000, 53: 201 – 206

20　Steenbergen RD. Hennsen MA, Walboomers JM Meijer GA Bank JP. Meijer CJ and Snijders PJ. Non – random allelic losses at 3p, 11p and

13p during HPV – mediated immortalization and concomitant loss of terminal differentiation of hmuan keratinocytes. Int. J Cancer, 1998, 76: 412 – 417

21 Akerman GS, Tolleson WH, Brown KL, Zyzak LL, Mourateva E, Engin TS, Basaraba A, Coker AL, Creek KE and Pirasi L. Human papillomavirus type 16 E5 and E7 cooperate to increase epidermal growth factor receptor (EGFR) mRNA levels, overcoming mechanisms by which excessive EGFR signaling shortens the lift span of normal human keratinocyes. Cancer Res, 2001, 61: 3837 – 3843

22 Sambrook J and Russell DW. Congtruction of genomic DNA libraries in cosmid vector. Molecular cloning: A Laboratory Manual Vol. 1. 3rd edition. Cold Spring Harbor Laboratory Press, New York, 2001, 411 – 423

23 He D, Zhang DK. Lam KY, Ma L, Ngan HY, Liu SS and Tsao SW. Prevalence of HPV infection in esophageal squanous cell carcinoma in Chinese patients and its relationship to the p53 gone mutation. Int J Cancer, 1997, 72: 959 – 964

24 Bover SN, Wazer DE and Bang V. E7 protein of human papilloma virus – 16 induces degradation of retinoblastoma protein through the ubiquitin – proteasome pathway. Cancer Res, 1996, 56: 4620 – 4624

25 Tsao SW Zhang DK. Cheng RYS and Wan TS. Telomerase activation in human cancers. Chin Med J, 1998, 111: 745 – 750

26 Hivama T, Yokozaki H, Kitadai Y, Haruma K, Yasui W, Kajiyama G and Tahara E. Overexpression of human telomerase RNA is an early event in oesophageal carcinogenesis. Virchows Arch, 1999, 434: 483 – 487

27 Kiyozuka Y, Asai A, Yamamoto D; Senzaki H, Yoshioka S, Takahashi H, Hioki K and Tsubura A. Establishment of novel human esophageal cancer cell line in relation to telomere dynamics and telomerase activity. Dig Dis Sci, 2000, 45: 870 – 879

28 Yeh TS, Cheng AJ, Chen TC, Jan YY. Hwang TL, Jeng LB, Chen MF and Wang TC. Telomerase activity is a useful marker to distinguish malignant pancreatic cystic tumors from benign neoplasms and pseudocysts. J Surg Keg, 1999, 87: 171 – 177

29 Belair CD. Yeager TR, Lopez PM and Reznikoff CA. Telomerase activity: a biomarker of cell proliferation, not maligant transformation. Proc Natl Acad Sci USA, 1997, 94: 1367 – 13682

30 Hsieh HF, Harn HJ, Chiu SC, Liu YC, Lui WY and Ho LI. Telomerase activity correlates with cell cycle regulators in human hepatocellular carcinoma. Liver, 2000, 20: 143 – 151

31 Engelhardt M. Kumar R, Albanell J. Pettengell R Han W and Moore MA. Telomerase regulation, cell cycle, and telomere stability in primitive hematopoietic cells. Blood, 1997, 90: 182 – 193

32 Yang XY. Kimura M, Jeanclos E and Aviv A. Cellular proliferation and telomerase activity in CHRF – 288 – 11 cells. Life Sci, 2000, 66: 1545 – 1555

33 Mokbel K, Parris CN, Ghilchik M, Williams G and Newbold RF. The association between telomerase hisopathological parameters, and KI – 67 expression in breast cancer. Am J Surg, 1999, 178: 69 – 72

34 Farwell DG. Shera KA, Koop JI, Bonnet GA, Matthews CP, Reuther GW, Coltrera MD. McDougall JK and Klingelhutz AJ. Geneic and epigenetic changes in human epithelial cells immortalized by telomerase. Am J Pathol, 2000, 156: 1537 – 1547

35 Shroyer KR Thompson LC, Enomoto T, Eskens JL, Shroyer AL and McGregor JA. Telomerase expression in normal epithelium, reactive atypia, squamous dysplasia, and squamous cell carcinoma of the uterine cervix. Am J Clin Pathol, 1998, 109: 153 – 162

36 Franco S, MacKenzie KL, Dias S, Alvarez S. Rafii S and Moore MA. Clonal variation in phenotype and life span of human embryonic fibroblasts (MKC – 5) transduced with the catalytic compo-

nent of telomerase (hTERT). Exp Cell Res,
2001, 268: 14 – 25

37 Rihet S. Bellaich P. Lorenzato M, Bouttens D,
Bernard D, Birembaut P and Clavel C. Human
papillomaviruses and DNA ploidy in anal condvlo-
mata acuminata. Histal Histopathol, 2000, 15:
79 – 84

38 Kashyap V and Das BC. DNA aneuploidy and infection of human papillomavirus type 16 in preneoplastic lesions of the uterine cervix: correlation with progression to malignancy. Cancer Lett,
1998, 123: 47 – 52

39 Shen ZY, Xu LY. Chen XH, Cai WJ. Shen J,
Chen JY. Huang TH and Zeng Y. The genetic
events of HPV – immortalized esophageal epithelium cells, Int J Mol Med, 2001, 8: 537 – 542

40 Cottage A. Dowen S Roberts J, Pett M. Coleman
N and Stanley M. Early genetic events in HPV im-
moralised keratinocytes. Genes Chromosomes
Cancer, 2001, 30: 72 – 79

41 Svrianen KJ. Histological changes, identical to
those of condylomatous lesions found in esophageal squamous cell carcinomas. Arch Geschwulstforsch, 1982, 52: 283 – 292

42 Lu LC, Shen ZY, You SJ, Shen J and Cai
WJ. Detection of human papillouavirus (HPV) on
esophageal carcinoma by PCR and in – stiu hybn-
dization (in Chinese). Ai Zheng, 1999, 18:
162 – 164

43 Veldman T, Horikawa I. Barrett JC and Schlegel
R. Transcriptional activation of the telomerase
hTERT gene by human papillomavirus type 16 E6
oncoprotein. J Virol, 2001, 75: 4467 – 1472

44 Oh ST Kvo S and Laimins LA. Telomerase activation by human papillomavirus type 16 E6 protein
induction of human telomerase reverse transcriptase expression throught Myc and GC – rich
Splbinding sites J Virol, 2001, 75: 5559 –
5566

45 Nowak JA. Telomerase, cervical cancer, and human papiltomavirus. Clin Lab Meal, 2000, 20:
369 – 382

46 Nair P, Jayaprakash PG, Nair MK and Pillai

MR. Telomerase, p53 and human papillomavirus
infection in the uterine cervix. Acta Oncol,
2000, 39: 65 – 70

47 Guo W Kang MK, Kim HJ and Park NH. Immortalization of human oral keratinocytes is associated
with elevation of telomerase activity and shortening
of telomere lengh. Oncol Rep, 1998, 5: 799 –
804

48 Hidalgo A, Schewe C, Petersea S. Salcedo M,
Gariglio P, Schluns K. Dietel M and Petersen I
Human papilloma virus status and chromosomal
imbalances in primary cervical carcinomas and
tumour cell lines. Eur J Cancer, 2000, 36: 542 –
548

49 Mian C. Bancher D, Kohlberger P, Kainz C,
Haitel A, Czerwenka K, Stani J, Breitenecker
G and Wiener H. Fluorescence in situ hybridization in cervical smears detection of numerical aberrations of chromosomes 7, 3, and X and relationship to HPV infection. Gynecol OncoI,
1999, 75: 41 – 46

50 Koopman LA Szuhai K, van Eendenburg JD,
Bezrookove V, Kenter GG, Schuuring E, Tanke
H and Fleuren GJ. Recunent integration of human
papillomaviruses 16, 45. and 67 near transloca-
tion breakpoints in new cervical cancer cell lines.
Cancer Res, 1999, 59: 5615 – 5624

51 Thorland EC, Myers SL, Persing DH, Sarkar
G, McGovem RM, Gostoat BS and Smith
DI. Human papillomavints type 16 integrations in
cervical tumors frequently occur in commnon fragile sites. Cancer Res, 2000, 60: 5916 – 5921

52 Duensing S, Lee LY, Duensing A, Basile J,
Piboonniyon S. Gonzalez S Crum CP and Manger
K. The human papillomavirus type 16 E6 and E7
oncoproteins cooperate to induce mitotic defects
and genomic instability by uncoupling centrosome
duplication from the cell division cycle. Proc Natl
Acad Sci USA, 2000, 97: 10002 – 10007

53 Filatov L, Golubovskaya V, Hurt JC, Byrd LL,
Phillips JM and Kaufmann WK. Chromosomal in-
stabilitty is correlated with telomere erosion and
inactivation of G2 checkpoint function in human

fibroblasts expressing human papillomavims type 16
E6 oncoprotein. Oncogene, 1999, 16: 1825 – 1838

54 Golubovskaya VM, Filatov LV, Behe CI, Presnell
SC, Hooth MJ, Smith GJ and Kaufmann
WK. Telomere shortening, telomerase expression,
and chromosome instability in rat hepatic epithelial

stem – like cells. Mol Carcinog, 1999, 24: 209 –
217

55 Song S, Liem A. Miller JA and Lambert PF.
Human papillomavirus tpes 16 E6 and E7 contribute differently to carcinogenesis. Virology, 2000,
267: 141 – 150

89.　The Multistage Process of Carcinogenesis in Human Esophageal Epithelial Cells Induced by Human Papillomavirus

SHEN Zhong – ying[1], XU Li – yan[1], LI En – min[2], CAI Wei – jia[1], SHEN Jlan[1],
CHEN Ming – hua[1], CEN Shan[4], TSAO Sai – wah[3], ZENG Yi[4]

Departments of 1 Pathology; 2. Biochemistry, Medical College of Shantou University,

3. Department of Anatomy. The University of HongKong;

4. Institute of Virology, Chinese Academy of Preventive Medicine

Summary

To investigate the multistage process of carcinogenesis, the progressive alteration of the morphology, telomerase, cytogenesis, oncogenes and tumorigenicity in the process of immortalization
and malignant transformation of the human fetal esophageal epithelial cell (SHEE) was studied. The
SHEE cells were immortalized by gene E6E7 of human papilloma virus (HPV) type 18 in our la-
boortory and continually cultivated over 100 passages, which had been malignantly transformed.
Calls at the 11th, 35th, 65th and 100th passage were examined according to the following criteria:
morphological changes of cell growth, contact – inhibition and anchorage – independent growth
(AIG); the cell proliferative and apoptotic index; the modal number of chromosomes; c – myc,
p53, bet – 2, ras; telomere length and activities of telomerase and tumorigenicity in nude mice or
severe combined immunodeficient (SCID) mice. The cells of the 11th passage were well differentiated and the cells of 100th passage were relatively poorly differentiated with polymorphism, while the
cells of 35th and 65th had two distinct differentiations. The proliferative indexes were 21.1%,
32.5%, 33.2%, and 40.9% and the apoptotic indexes were 3.3%, 2.7%, 3.5%, 2.7% in
the 11th, 35th, 65th and 100th passage respectively. Karyotypes of four cell passages belonged to
hyperdiploidy and hypotriploidy. C – myc, ras, p53 genes were low in the 10th and 35th, and
high in the 65th and 100th passage, but bcl – 2 was low in 4 passages. Telomere length sharply decreased from normal fetal esophagus cells until the 35th passage, but it was stably expressed in the
65th and 100th passage. The activities of telomerase were expressed in cells of the 35th, 65th and
100th passages. The efficiency of AIG varied in different passages of the SHEE cell and was absent

in the 11th passage, low efficiency in the 35th passage and 65th passage, and high efficiency in the 100th passage. Transplanted cells of the 65th and 100th passage into SCID mice resulted in tumor formation, but only the 100th passage calls could grow in nude mice. All of these characteristic changes were in dynamic progressive process. These data demonstrate that carcinogenesis of esophageal epithelial cells induced by HPV is the multistage process, which goes through the initial, immortal, premalignant and malignant transformation stages. The generation of esophageal carcinoma is caused by the accumulation of cellular, genetic and molecular changes.

[**Key words**] Esophagus; Carcinogenesis; Human papillomavirus; Immortalization; Malignant transformation

Introdution

Tumorigenesis in humans is a multistep process, as reported in lung[1,2] gastric cancer[3,4], oral[5,6], bladder[7] and skin[8] cancers. Carcinogenesis is a multistep phenomenon, beginning with precancerous conditions through stages of cell hyperplasia and dysplasia and ending in fully maliguant transformation. Cell lines can provide powerful model systems for the study of human tumorigenesis[9,10]. The transformation of a normal into a malignant cell is a multistep mechanism, which involves various alterations at the cellular, molecular and genetic level[11,12]. For neoplstic transformation of normal human cells, they must be first immortalized and then be converted into neoplastic cells. It is well known that the immortalization is a critical step for the neoplastic transformation of cells.

It is well accepted that cancer arises in a multistep fashion in which exposure to environmental carcinogens is a major etiological factor. Human papillomaviruses (HPV), more than 120 genotypes have been identified, cause certain common human cancers, most notably carcinoma of the cervix. HPVs are also associated with a broad spectrum of cutaneous and mucosal lesions. In 1982 Syrjanen firstly reported the relationship of HPV and esophageal cancer in the light of pathomorphological data and suggested that HPV might be a cause of esophageal cancer[13]. Extensive investigation has provided insights into the relationship between HPV and esophageal cancer in last two decades[14-16]. In our previous study the positive rate of HPV in samples of esophageal cancers was as high as 60% − 66% in areas with a high incident rate[17] and most researchers agree that HPV might play an important role in esophageal carcinogenesis[18,19].

BY HPV18E6E7 inducement, we have established an immortal esophageal epithelial cell line SHEE[20]. which keeps the characteristics of monolayer growth, contact inhibition, and the squamous epithelium origin[21,22]. The coatinually cultivated cells of SHEE developed into the malignant transformation[23,24]. The cell line offered a unique system for investigating the multistage process of carcinogenesis in human esophageal epithelial cells.

The mechanism underlying the process towards malignant conversion, usually covering a long latency period between primary infection of HPV and cancer emergence is presently not fully understood. The aim of this study was to investigate the cellular genetic and molecular biological characteristics in various passages of SHEE to explore the multistage process of carcinogenesis of esophageal epithelial cells induced by HPV18E6E7.

Materials and Methods

1. Cell culture: The SHEE cell line was a kind of immortal embryonic esophageal epithelium induced by genes E6E7 of human papilloma virus (HPV) type 18 in our laboratoty and has been propagated over 100 passages. From the liquid nitrogen storage, different generation of cells, the 11th passage (SHEE11), 35th passage (SHEE35), 65th passage (SHEE65) and 100th passage (SHEE100), were revived and inoculated to culture flask. The cell line was routinely cultivate in culture medium 199 (Giboo) with 10% bovine serum, 100 units, of penicillin and streptomycin in a humidified atmosphere of 5% CO_2 and 95% air. The cell shape and size, anchorage – dependent growth and contactinhibited growth were examined by phase – contrast microscopy.

2. Analysis of cell prolifferetiove index and apoprotic index: Cells of SHEE11, SHEE35, SHEE65 and SHEE100 in the flask were trysinized, washed twice with PBS, fixed by 70% alcohol, prepared as single – cell suspension and stored at 4℃. Cells was stained with propidium iodide (Sigma) and analyzed with flow cytometry (FCM FACSort, B – D Co.). In DNA histogram, the percentage of cells in various stages of the call cycle, the apoptotic index (AI) and prcoliferation index ($PIx = S + G_2M/G_0G_2 + S + G_2M$) were calculated.

3. Telomere length analysis: The DNA of samples containing fetal esophagus, SHEE11, SHEE35, SHEE65 and SHEE100 were extracted. Telomeric restriction fragment (TRF) was measured by Southern blot. Briefly, 20μg of genomic DNA was digested with Hinff and run on 0.7% agarose gel with marker DNA/HindⅢ. After electrophoresis the gel was blotted to nylon membrane (Hybond™ N⁺ nylon membrane, Amersham Life Science) and hybridized to the Dig – labeled probes $(CCCTAA)_3$ at 50℃ in 5 × SSC, 0.1% Sodium N-lauroylsarcosine (SLS), 0.02% SDS for 12 – 16 h and washed twice at room temperature in 2 × SSC, 0.1% SDS for 5 min, once at 50℃ in 1 × SSC, 0.1% SDS for 5 min twice, at 50℃ in 0.1 × SSC, 0.1% SDS for 5 min, stained with NBI/BCIP and the middle points were measured to obtain the mean tetomere length.

4. Telomerase activity assay: Telomerase activity was measured using TRAP – eze Telomerase Detection Kit (Oncor Inc.). Cells were added 10 – 50 μl of ice – cold lysis buffer (10 mmol/L TrisHCl pH 7.3, 1 mmol/L EGTA, 0.1 mmol/L benzamidine, 5 mmol/L β-mereaptoethanol, 0.5% CHAPS, 10% glycerol). After 30 min of incubation on ice the lysate was centrifuged at 12 000 g for 20 min at 4℃. The telomeric repeat amplification protocol (TRAP) reaction was performed using 1 μl lysate or 1/10 diluted lysate, 2.5 μl 10 × TRAP buffer (200 mmol/L Tris – HCl, pH 8.3, 15 mmol/L $MgCl_2$, 630 mmol/L KCl, 0.5% Tween 20, 10 mmol/L EGTA), 0.5 μl 2.5 mmol/L dNTP, 0.5 μl Ts primer, 0.5 μl TRAPprimer mix, 19.5 μl water, 0.5 μl Taq (2 U/μl). After incubation at 30℃ for 30 min the reaction mix was immediately transferred to 94℃ and performed PCR at 94℃ for 30 sec, 59℃ for 30 sec, for 30 cycles. PCR products were separated in a non – denaturing 12% PAGE in 1 × TBE at 5 V/cm. The gel was stained using AgNO₃.

5. Preparation of chromosome: The cells of SHEE11, SHEE35, SHEE65 and SHEE100 at their exponential phase of growth appeared as mitotic cells. These cultured cells were added with 0. 05 μg/ml colchicine (Fluka AG, CH – 9470 Buchs.) for 1 – 1. 5 h before harvest, and then digested with 0. 25% trypsin to obtain cell suspension. After centrifugating, harversted cells were added with 0. 075 mol/L KCl for a hypoosmotis treatment at 37℃ for 30 min. And then, 1 ml of fixed solution (methyl alcohol: glacial acetic acid = 3: 1) was added to fix cells for 30 min and repeated twice. The cells were then dropped on slide and Giemsa stained. The chromosome number of metaphase of cells in four groups was scored.

6. *C – myc, bcl – 2, p53 and ras genes assay by multi – PCR (mPCR)*: mPCR was performed by Maxim mPCP Kit (APO – MO5OG, Maxim Biotech Inc.) including the primers of c – myc, bcl – 2, p53, ras and a house – keeping gene GAPDH. The isolation of undegraded and intact RNA of SHEE series without RNase contamination of buffers and containers was a prerequisite. PCR thermocycle profile: 96℃ 1 min and 57℃ 4 min for 2 cycles; 94℃ 1 min and 57℃ 25 min for 30 cycles; 70℃ 10 min and 25℃ soak. The ras mPCR Kit (RAS – MO5OG, Maxim) contains ras positive contrast, K – ras, H – ras primer and DNA marker. The electrophoresis of PCR products was used in the visual inspection of an EB – stained agarose gel containing 0. 5 μg/ml ethidium bromide (Sigma).

7. *Cell cultured in soft agar cultivation*: The efficiency of anchorage – independent growth (AIG) was determined by two different approaches to assess the efficiency of colony formation. It is feasible in the case of two layers of soft agar conditions or in liquid medium over – agar, namely over – agar condition[25]. The exponential growth cells of SHEE series were trypsinized, stained with trypan blue to count the number of living cells. The living single cell suspensions (10^3 cells/ml) were cultured in 0. 35% agar (Agarose, V312A, Promega) overlaid on 0. 7% agar in a 6 – well plate (Coming), and in another plate of over – agar condition cells were cultured in liquid cultured medium over 0. 7% agar and incubated in 5% CO_2 at 37℃ for 20 days. The cell colonies with more than 20 cells were then counted.

8. *Tumorigenaais in nude mice and SCID mice*: Six – week – old BALB/C nude mice (supplied by the Experimental Animal Center of Zhongshan Medical University) and severe combined immunodeficient mice (SCID, C. B – 17/IcrJ – scid, supplied by Animal Center of Chinese Academy of Medical Science) were bred in isolated conditions respectively. Four groups of SCID mice and made mice were injected subcutaneously with SHEE series cells in 1×10^6 cells/mouse respectively. They were observed every 10 days for 2 months and then sacrificed for histopathological examination.

9. *Western blot analysis*: Cells cultured in flasks were washed three times with ice – cold PBS, and then were lysed in buffer (50 mmol/L Tris – HCl, pH 8. 0, 150 mmol/L NaCl, 100 μg/ml PMSF, 1% Triton X – 100) for 30 min on ice. After removal of cell debris by centrifugation (12 000 g, 5 min), the protein concentration of lysates was measured by Bradford method. Proteins (50 μg) of different groups were boiled for 5 min in sample buffer and were separated in 10% SDS – PAGE and transfered onto a nitrocellulose membrane (Bio – Rad). Nonspecific reactivity was blocked by incubation overnight at 4℃ in buffer (10 mmol/L Tris – HCl, pH 7. 5, 150 mmol/L NaCl, 2% Tween – 20, 4% bovine serum albumin). The membrane was then incubated with primary antibody (mouse anti HPV18E_6, MAB – 0308 Maixin Bio.). The secondary antibody

was used to detect bound primary antibody. Reactive protein was detected by ECL Western blot analysis Amersham Pharmacia Biotech).

Results

1. Cell morphology: In the initial passages, the cells in the SHEE11 group were uniform in size and shape, and grew as an even monolayer with characteristics of squamous epithelium (Fig. 1A). SHEE35 cells proliferated and exhibited diphasic differentiation, some cells displayed the undifferentiated basal epithelium and others displayed differentiated squamous epithelium (Fig. 1B). Cells in SHEE65 crowded together with undifferentiated, and mere mitotic nuclei (Fig. 1C). SHEE100 cells lost the pattern of maturation and cells displayed evidence of polymorphism with much more giant nucleus cells and multinucleoli (Fig. 1D).

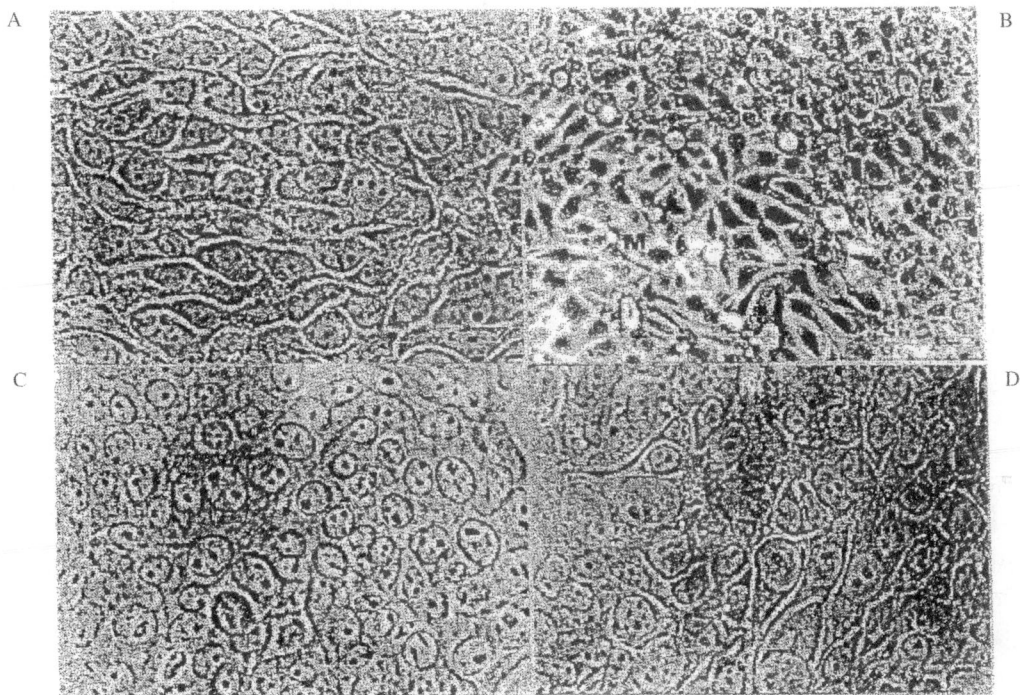

(A), Cells of SHEE11 displayed good differetiation and squamous epithelium in outlook (×400); (B), Cells of SHEE35 displayed two differentiated directions, partilly well differentiated (M) and partial poorly (N) (×200); (C), Cells of SHEE65 crowded together (×400); (D), Cells of SHEE100 displayed poor differentiation with polymorphism and meganuclei (×400)

Fig. 1 Morphology of living cells in the SHEE cell series was observed by phase contrast microscopy

2. Cell cycle analyzed by FCM: In the DNA histogram the distribution of DNA content of SHEE series cells is shown in Fig. 2. The proliferative indexes were 21.1%, 32.5%, 33.2% and 40.9% in SHEE11, SHEE35, SHEE65, and SHEE100 respectively. Accounting DNA <2 cells, the apoptotic indexes were 3.3%, 2.7%, 3.5%, 2.7% in SHEE11, SHEE35, SHEE65,

SHEE100 respectively.

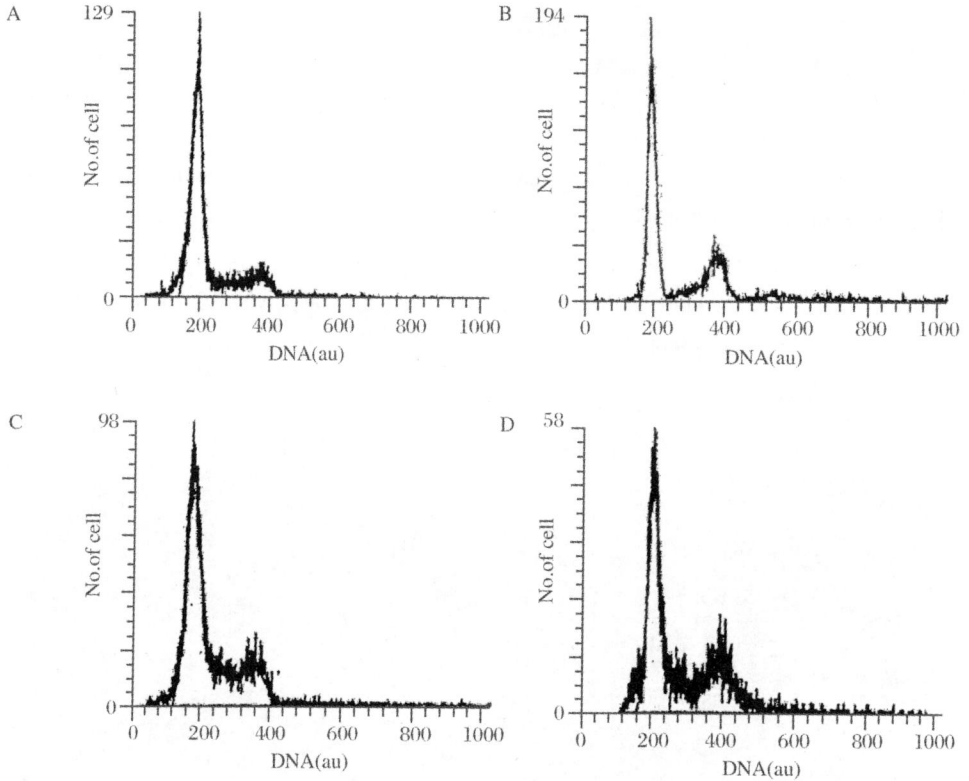

(A), SHEE11; (B), SHEE35; (C), SHEE65; (D), SHEE100; an, arbitraly unit

Fig. 2　DNA histograms of cell cyste

3. Cytogenesis abnormality: The number of chromosomes in SHEE11, SHEE35, SHEE65 and SHEE100 ranged between 30 and 178. These chromosomes were mainly hyperdiploids (Fig. 3A) and hypotriploids (Fig. 3B). More hypertriploid cells were found at SHEE65 and SHEE100. Modal number of chromosomes at SHEE11 was 51 – 54, at SHEE35 and SHEE65 were bimodal in distribution, 57 – 55, 62 – 64; and 59 – 61, 63 – 65 respectively, at SHEE100 was 60 – 67. The modal number from SHEE11 to SHEE100 increased slowly.

4. Telomere length and telomerase activity: Telomeric restriction fragment (TRF) length of SHEE series cells and fetal esophageal tissue was measured by Southern blot analysis. The mean telomere length of fetal esophageal tissue was 30. 0 kb; telomere length of SHEE11 shortened sharply to 17. 0 kb; SHEE35 to 3. 2 kb; SHEE65 and SHEE100, to 3. 5 and 3. 7 kb (Fig. 4A). Telomerase activation was absent in fetal esophageal tissue and SHEE11. Our results proved that the shortening telomere and absence of telomerase activity contributes to cellular senescence and cell death. The activity of telomerase first appeared in SHEE35, and appeared strongly positive in SHEE65 and SHEE100 (Fig. 4B).

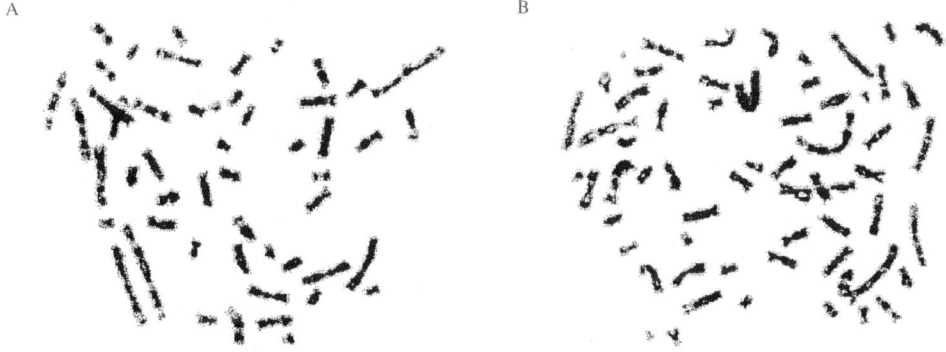

（A），Hyperdiploid of chromosome in SHEE11；（B），Hypotriptoid of chromosome in SHEE100

Fig. 3　Hyperploid karyotype of SHEE（Giemsa，× 1000）

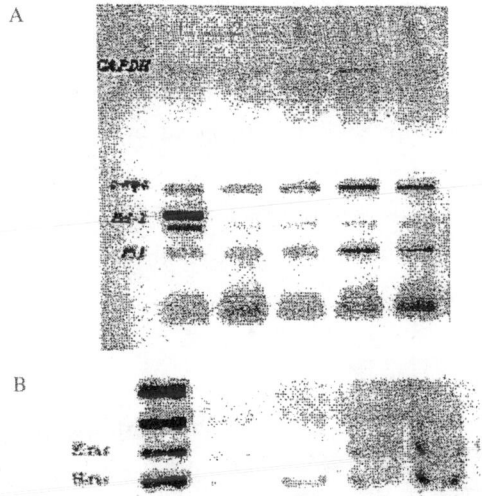

（A），p53，bel－2，c－myc GAPDH（B），H－ras and K－ras. Lanes：1，marker；2，SHEE11；3，SHEEE35；4，SHEE65；5，SHEE100

Fig. 5　The products of mPCR

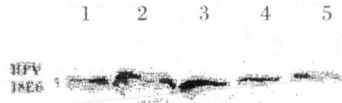

Lanes：1，marter；2，SHEE11；3，SHEE35；4，SHEE65；5，SHEE100

Fig. 6　Western blot analysis of protein of HPV18E

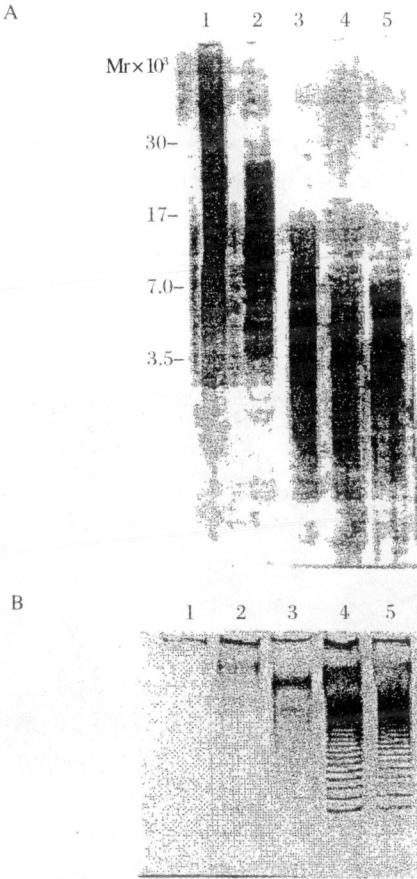

（A），Telemere length；（B），Telomerase activity. Lanes：1，normal esophagus；2，SHEE11；3，SHEE33；4，SHEE65；5，SHEE 100

Fig. 4　Telomere and telomerse of SHEE

5. p53, bd－2, c－myc, ras and GAPDH: The *p53* gene (mutant) appeared in all cells of four passages, but shifted－up in SHEE65 and SHEE100. The c－myc expressed in cells of SHEE11 and 35 in low level and expressed apparently in SHEE65 and SHEE100. Bel－2 expression in the 4 passages was low (Fig. 5A). The electrophoresis bands of ras mPCR products, including H－ras and K－ras were low in SHEE11 and SHEE35, intermediate in SHEE－65 and high in SHEE100 (Fig. 5B). The housekeeping gene GAPDH was expressed at a similar level in the four groups.

6. HPV18E6: The expression of HPV18E$_6$ was examined by Western blot analysis. The results showed the presence of HPV18E$_6$ in the four groups of SHEE cells (Fig. 6).

7. Colony formation on soft agar: The culture method of over－agar as an ancharage－independent growth (AIG) condition permitted more efficient recovery of cells compared with those in semisolid soft agar. The efficiency of AIG varied in different passages of SHEE cells and was absent in SHEE11, low in SHEE35 (0.05%－0.1%) and SHEE65 (1.0%－3.0%); and high in SHEE100 (8.0%－12.0%) (Fig. 7). The over－agar culture can be considered valid for the study of malignant transformation and to search for colonigenic cells.

8. Tumor formation: In order to observe the tumorigenicity, the ability of tumor formation and invasive potency were observed by means of cells in various passages transplanted into nude mice and SCID mice subcutaneously, Cells of SHEE11 and SHEE35 did not develop tumors in either group cells of SHEE65 developed a tumor in one of four SCID mice but not in nude mice. Cells of SHEE100 formed rumors in both trade and SCID mice (Fig. 8). In the tissue sections tumor cells invaded the adjacent tissues.

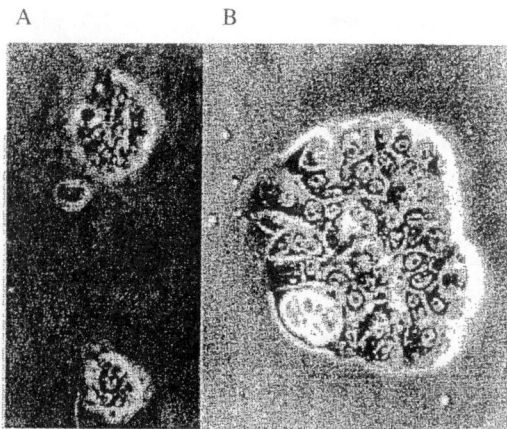

(A), A few cells grew on the soft agar in SHEE65; (B),
A large colony grew on over－agar in SHEE100

**Fig. 7　Colony formation on agar medium
(phase contrast microscopy, ×200)**

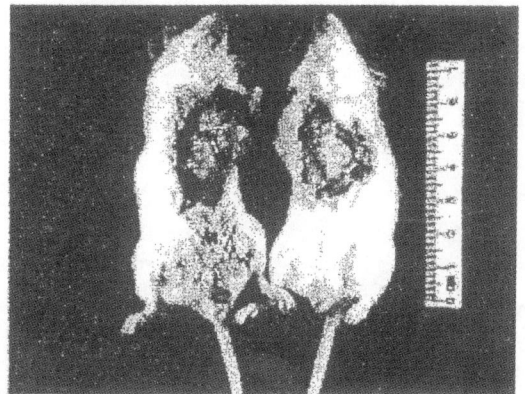

**Fig. 8　Cells of SHEE100 transplanted into SCID
mice and the tumor formed subcutaneously**

9. Comparison of biological criteria in four stages of SHEE: The biological criteria in various passage of SHEE are shown in Tab. 1, SHEE11 cells revealed proliferative morphological phenotype with good differentiation and shortening of telomere length without telomerase activity and the modal number of chromosome increased from 51 to 54. It was designated as the initial stage. In middle stage, SHEE35 revealed cell proliferation and two differentiated phenotypes accompanying the telomerase activity and the telomere length was sustained, but the phenotype was non – tumorigenic. The changes of cytogenesis were bi – modal chromosomes 57 – 58, 62 – 64. It was designated as the immortal stage. Cells of SHEE65 displayed bipotenial differentiation, bi – modal chromosome, 59 – 61, 63 – 65 and telomerase activation. Cultured on soft agar, cells of SHEE65 grew in different size of cooloies in which large, ones could develop into malignant tumorigenic clones. SHEE65 was judged in premalignant stage. In transformed stage, SHEE100 cells overlapped to grow with different shape and poor differentiation and less contact – inhibition with colony formation in soft agar. According to the invasive tumor formation in nude and SCID mice. SHEE100 cells revealed fully malignant tranformation.

Tab. 1 Comparison of biological criteria of variant passages of SHEE

Cell passages	SHEE 11	SHEE35	SHEE65	SHEE100
Proliferative index (%)	21. 1	32. 5	33. 2	40. 9
Apoptotic index (%)	3. 3	2. 7	3. 5	2. 7
Differentiation	Good	Biphasic	Biphasic	Poor
Contact inhibition	+	+	±	−
Telomere length ($\times 10^3$)	17	3. 2	3. 5	3. 7
Telomerase activits	−	+	+ +	+ +
Modal no. of chromosomes	51 – 54	57 – 58, 62 – 64	59 – 61, 63 – 6. 5	60 – 67
C – myc	+	+	+ +	+ +
p53	+	+	+ +	+ +
Ras	±	±	+	+
Bcl – 2	±	±	+	+
Colony growing in soft agar	−	−	+	+ +
Tumor formation in SCID mice	0/4	0/4	1/4	4/4
HPV18E6	+	+	+	+

Discussion

We first established an *in vitro* multistep esophageal carcinogenesis model by exposure of normal human fetal esophageal epithelium to HPV18E6E7 genes. The newly established esophageal epithelial cell line SHEE was cultured over 100 passages and went through a multistage process: the initial, immortal, premalignant and malignant stages. Upon introduction of the HPV genome, the cells bypassed the senescence checkpoint and commened proliferating, but without immortal life span during which telomere DNA continued to shorten, called the initiate stage (SHEE11). In SHEE35 clones surviving beyond the crisis, the immortalization was formed while telomerase activity appeared and telomere length stabilized. Large colonies on soft agar in SHEE65 and SHEE100 showed that the characteristics of anchorage – dependent growth decreased but tumorigenicity increased. A few tumors formed in SCID mice incubated with SHEE65 cells but not in nude mice. So we judged that they were at a premalignant stage. SHEE100 cells, which were transplanted into nude and SCID mice and developed tumors in all mice with invasive pathern, expressed malignant transformation.

In 1994. Kim et al. also found specific association among the telomerse activity with immortal cells and cancer cells[26]. The telomerase may be a prerequisite and the diagnostic criteria to immortal cells[27]. In our data the telomere length of the normal fetal esophageal cell was 30 kb in length shortened to 17 kb in SHEE11, and 3.5 kb in SHEE35, and then maintained this level continually to SHEE65 and SHEE100. The telemerase activity appeared im SHEE35, SHEE65 and SHEE100. The activation of telomerase to maintain telomere lengh was necessary for immortalization arid malignancy but was not sufficient for malignant transformation in the SHEE cell line. Our results indicated that immortalization and malignant transformation of SHEE cells might require activation of telomerase and other genes. which abrogated normal proliferation and differentiation of the cells.

Strong evidence in favor of the aberrant chromosome of cancer is early and essential events in tumor development. There was more aneuploidy (hyperdiploidy and hypotriploidy) in the chromosomes of the faur passages of SHEE. The separate modal number of chromosomes first appeared in-SHEE35 and continued to SHEE65, SHEE65 and SHEE100 had more hypotriploid cells than the others. The above showed that chromosomes of the SHEE series were unstable, which would mean easier malignant transformation by promoters[28]. It was possible that in the initial stage progessive chromosomal changes induced by papillomavirus transfection might result in the genetic instability and phenotypic variability as described elsewhere[29,30]. Aneuploidy renders the chromosome structure and segregation error – prone, because it unbalances mitosis proteins and the many teams of enzymes that synthesize and maintain chromosomes. The resulting karyotype instability sets off a Chain reaction of aueuploidizations, which generate ever more abnormal and eventually cancer – specific rearrangements of chromosomes[31]. Thus carcinogenesis is initiated by a random aneuploidy and the drivimg force of carcinogenesis is the inherent instability of the aneuploid karyotype. This dynamic process might result in the acquisition of full transformation with the parade of sufficient time.

The changes of cytogenetics will control a lot of gents, which affect proliferation, differentiation, apoptosis and other phenotype of cells. Some genes, such as c – mye, ras, $p53$ and bel – 2, are related to immortalization and malignant transformation of cells. In our data, c – myc, K – ras, H – ras and $p53$ (mutant) were shifted – up in SHEE65 and SHEE100. In SHEE11 and SHEE35 c – myc, ras, bcl – 2 and $p53$ were at a low level. In general, c – myc and $p53$ (mutant) promote proliferation of cells[32,33] and bcl – 2 can prevent apoptosis[34]. Ras is a transforming growth signal causing rapid anenploid and malignant transformation of cells[35,36]. It has been reported that the cell transformation induced by virus ras was a required gene[37].

The alterations of telomere, chromosome and some genes mentioned above were closely related with HPV infection. Infection of HPV can cause karyotype confusion[38]. such as break of chromosome, abnormal structure and number of chromosomes[39]. Out data also hints that the immortal esophageal cells induced by HPV18E6E7 may effect the changes of chromosomes. Abnormal chromosomes cause liability of genetic character[40,41]. The virus genome inserts and integrates with chromosomes of the host[42], and causes the activation and expression of oncogenes[43]. HPV E6E7 protein is conjugation with anti – oncoprotein $p53$ and pRb[44,45], thus losing control of cellular growth and urging the phenotyping production of cellular transfomation[46]. Previous reports have shown that activation of telomerase can be achieved by the E6 and E7 proteins of HPV[47]. Therefore, E6 could promote malignant change, while E7 may cause benign neoplasm[48]. The virogene HPV16E6 alone can cause cellular malignant change[49]. It has also been suggested that the viral oncoproteins E6 and E7 are essential components in malignant conversion, although, they are not sufficient for the development of the malignant phenotype[50]. In this experiment, we find that SHEE series cells containing HPV18E6E7 can gradually be the way to malignant stage. It is also postulated that HPV will be a major risk factor for esophageal cancer.

In summary, we have developed a malignant transformed cell model to facilitate the study of carcinogenesis in esophageal carcinoma *in vitro* through immortalization of human fetal esophageal epilthelium induced with HPV18E6E7. The muttistep process of malignant transformation consists of initial (transduction of E6E7 genes), preimmortalized (by pass senescence), immortalized (activity of telomerase). premalignant (anchorage – independent growth) and malignant transformation (tumor formation in immunodeficient mice). Our data reveal that the viral oncogene of HPV type 18 is one of important causes in etiology of esophageal cancer and it can initiate cell immortalization and malignant transformation *in vitro* through molecular and genetic alterations.

Acknowledgements

Contract grant sponsor: National Natural Science Foundation of China (39830380, 39800069, 30170428); Research and Devetopment Foundation of Shantou University (L00012), and supported by the Chinese National Human Genome Cemer, Beijing.

[In 《Oncol Rep》 2004, 11: 647 – 654]

References

1　Jeanmart M, Lantuejoul S, Fievet F, More D, Stum N, Brambilla C and Brambilla E. Value of immunohistochemical markers in preinvasive bronchial lesions in risk assessment of lung cancer. Cancer Res, 2003, 9: 2195 – 2203

2　Chyczewski L, Niklinski J, Chyczewska E, Niklinska W and Naumnik W. Morphotogical aspects of carcinogenesis in the lung. Lung Cancer 34 (Suppl 2), 2001, 517 – 525

3　Kang GH Lee S, Kim JS and Jung HY. Profile of aberrant CpG island methylation along multistep gastric carcinogenesis. Lab Invest, 2003, 83: 519 – 526

4　Yasui W, Yokozaid H, Fujimoto J, Naka K, Kuniyasu H and Tahara E. Genetic and epigenetic alterations in multistep carcinogenesis of the stomach. J Gastroenterol, 2000, 12: 111 – 115

5　Shintani S, Mihara M, Nakahara Y, Kiyota A, Ueyama Y, Matsumura T and Wong DT. Expression of cell cycle control proteins in normal epithelium, premnalignant and malignant lesions of oral cavity. Oral Oncel, 2002, 38: 235 – 243

6 Kang MK and Park NH. Conversion of normal to malignant pherotype. telomere shotening, telomerase activation, and genomic instability during immortalization of human oral keratinocytes. Crit Rev Oral Biol Med, 2001, 12: 38 – 54

7 Brandau S and Bohle A. Bladder cancer. I. Molecular and genenc basia of carcinogenesis. Eur Urol, 2001, 39: 491 – 497

8 Kub Q Y, Murao K, Matsumoto K and Arase. S Molecular carcinogenesis of squamous cell carcinomas of the skin J Med Invest, 2002, 49: 111 – 117

9 Schwab TS, Stewart T, Lehr J, Pienta KJ, Rhim JS and Macoska JA. Phenotypic characterization of immortalized normal and primaty tumor – derived human prostate epithelial cell cultures. Prostate, 2000, 44: 164 – 171

10 Rajah N. Pruden DL. Kaznari H, Cao Q, Anderson BE, Duncan JL and Schaeffer AJ. Characterization of an immortalized human vaginal epithelial cell line. J Urol, 2000, 163: 616 – 622

11 Ming SC. Cellular and molecular pathology of gastric carcinoma and precursor lesions: a critical review. Gastric Cancer, 1998, 1: 31 – 54

12 Fusenig NE and Boukamp P. Multiple stages and genetic alterations in immnortalization, malignant transformation and tumor progression of human skin keratinocytes. Mol Carcinog, 1998, 23: 144 – 158

13 Syrjanen KJ. Histological changes identical to those of coudylomatous lesions found in esophageal squamous cell carcinomas. Arch Geschwulstforsch, 1982, 52: 283 – 292

14 He D, Zhang DK. Lam KY, Ma L. Ngau HY, Liu SS and Tsao SW. Prevalence of HPV infection in esophageal squamous cell carcinoma in Chinese patients and its relationship to the p53 gene mutation. Int J Cancer, 1997, 72: 959 – 964

15 Sur M and Cooper K. The rote of the human papilloma vires in esophgeal cancer. Pathology, 1998, 30: 348 – 354

16 Poliak M Ceras A and Sene K. Human papilloma virus infection in esophageal carcinomas: a study of 121 fesions using multiple broad – spectrum polymerase chain reactions and literature review. Hum Pathol, 1998, 29: 266 – 271

17 Shear ZY, Hu SP, Shen J, Lu LC. Tang CZ, Kuang ZS and Zeng Y. Detection of human papillomavirus in esophageal carcinoma. J Med Virol, 2002, 68: 412 – 416

18 Chang F Svrianen S Shen Q Cintorino M, Santopietro R, Tosi P and Syrianen K. Human papillomavirus involvement in esophageal carcinogenesis in the high – incidence area of China. A study of 700 cases by sereerring and type – specific in situ hybridization. Scand J Gastroenterol, 2000, 35: 123 – 130

19 Lavergne D and De – Villiers EM. Papillomavirus in esophageal papillomas and carcinomas. Int J Cancer, 1999, 80: 651 – 684

20 Shen ZY, Cen S, Xu LY, Cai WJ Chen MH, Shen J and Zeng Y. E6/E7 genes of human papilloma virus type 18 induced immortization of human fetal esophageal epithelium. Oncol, 2003, 10: 1431 – 1436

21 Shen ZY, Cen S, Cai WJ, Xu JJ, Ten ZP, Shen J, Hu Z and Zeng Y. Immortalization of human fetal esophageal epithelial cells induced by E6 and E7 genes of human papilloma vires 18. Chin J Exp Clin Virol, 1999, 13: 121 – 123

22 Shen ZY, Shen J, Cai WJ, Cen S and Zeng Y. Biological characteristics of human fetatl esophageal epithelial cell line immortalized by the E6 and E7 gene of HPV type f8. Chin 1 Exp Clin Virol, 1999, 13: 209 – 212

23 Shen ZY, Xu LY, Chen MH, Shen J, Cai WJ and Zeng Y. Progressive transfomation of immortalized esophageal epithelial cells. World J Gastroenterol, 2002, 8: 976 – 981

24 Shen ZY, Shen J, Cai WJ, Chen JY and Zeng Y. Identification of malignant transformalion in the immortalized esophageal epithelial cells. Chin J Oncol, 2002, 24: 107 – 109

25 Donz Z, Cmaik JL. Wendel EJ and Colburn NH. Differential transformation efficiency but not

AP – 1 induction under anchoragedependent and independent conditions. Carcinogenesis, 1994, 15: 1001 – 1004

26　Kim NW, Piatyszek MA, Prowse KR. Harley CB, West MD Ho PL Coviello GM Wright WE, Weinrich SL and Shay JW. Specific association of human telomerase activity with immortal cells and cancer. Science, 1994, 266: 2011 – 2015

27　Farwell DG, Shera KA, Koop JL, Bonnet GA, Matthews CP, Reuther GW, Coltrera MD, Mc-Dougall JK and Klingelhutz AJ. Genetic and epigenetic changes in human epithelial cells immortalized by telomerase. Aan J Pathol, 2000, 156: 1537 – 1547

28　Shen ZY, Cen S, Shen J Cai WJ, Xu JJ, Teng ZP, Hu Z and Zeng Y. Study of immortalizaion and malignant tranformation of human embryonic esophageal epithelial cells induced by HPV18E6E7 J Cancer Res Clin Oncol, 2000, 126: 589 – 594

29　Shen ZY Xu LY, Chen MH, Cai WJ, Shen J, Ceng JY and Zeng Y. Cytogenetic and molecular genetic changes in malignant transformation of immortalized esophageal epithelial cells. Int J Mol Med, 2003, 12: 219 – 224

30　Shen ZY, Xu LY, Chen XH, Cai WJ Shen J, Chen JY, Huang TH and Zeng Y. The genetic events of HPV immnortalized esophageal epithelium cells. Int J Mol Med, 2001, 3: 537 – 542

31　Duesberg P and Li R. Multistep carcinogenesis. a chain reaction of aneuploidizations. Cell Cyele, 2003, 2: 202 – 210

32　De – Mighio MR, Simile MM, Muroni MR, Pussceddu S, Cahvisi D, Carru A, Seddaiu MA, Daino L, Deiana L, Pascale RM and Feo F. Correlation of c – myc oyerexpression and amplification with progression of preneoplastic liver lesions to malignancy in the poorly susceptible Wistar rat strain. Mol Carcinog, 1999, 25: 21 – 29

33　Tarapore P and Fukasawa K. p53 mutation and mitotic infidelity. Cancer Invest, 2000, 18: 148 – 155

34　Adams JM and Cory S. The Bcl – 2 protein family: arbiters of cell survival Science, 1998, 281: 1322 – 1326

35　Mutter GL, Wada H, Faquin WC and Enomoto T. K – ras mutations appear in the premalignant plase of both microsatellite stabie and unstable endometrial carcinogenesis. Mot Pathol, 1999, 52: 257 – 262

36　Chin L. Tam A, Pomerantz J, Wong M, Holash J, Bandeesy N, oncogenic Ras in tumor maintenance. Nature, 1999, 400: 468 – 472

37　Weitzman JB and Yaniv M. Rebuilding the road to cancer Nature, 1999, 400: 401 – 402

38　Duensing S, Lee LY, Duening A, Basile J, Piboonniyom S, Gonzalez S, Cmm CP and Munger K. The human papillomavirus type 16 E6 and E7 oncoproteins cooperate to induce mitotic defects and genomic instability by uncoupling centresome duplication from the call division cycle. Proc Nail Acad Sci USA, 2000, 97: 10002 – 10007

39　Villa LL. Human papillomaviruses and cervical cancer. Adv Cancer Res, 1997, 71: 321 – 341

40　Weijerman PC van Drunen E, Konig JJ Teubel W, Romijn JC, Schroder FH and Hagemeijer A. Specific cytogenetic aberrations in two novel human prosiatic cell lines immortalized by human papillomavirus type 18 DNA. Cancer Genet Cytogenet, 1997, 99: 108 – 115

41　Mullokandov MR, Kholodilov NG, Atkin NB, Burk RD, Johnson AB, Klinger HP. Genomic alterations in cervical carcinoma: losses of chromosome heterozygosity and human papilloma virus tumor status. Cancer Res, 1996, 56: 197 – 205

42　Pfeffer A, Schubbert R, Orend G, Hilger – Eversheim K and Doerfler W. Integrated viral genomes can be lost from adenovirus type 12 – induced hamster tumor cells in a clone – specific, multistep process with retention of the oncogenic phenotype. Virus Res, 1999, 59: 113 – 127

43　Choo KB, Chen CM, Han CP, Cheng WT and Au LC. Molecular analysis of cellular loci disrupted by papillomavirus 16 integration in cervical cancer: frequent viral integration in topologically

destabilized and transriptionally active chromosomal regions, J Med Virol, 1996, 49: 15 – 22

44　Demers GW, Halbert CL and Galloway DA. Elevated widtyge p53 protein levels in human epithelial cell lines immortalized by the human papill-
lomavirus type 16 E7 gene. Virology, 1994,
198: 169 – 174

45　Bover SN, Wazer DE and Band V. E7 protein of
human papilloma virus – 16 induces degradation
of retinblastoma protein through the ubiquitin –
proteasome pathway. Cancer Res, 1996, 56:
4620 – 4624

46　Itakura M, Mori S, Park NH and Bonavida
B. Both HPV and carcinogen contribute to the development of resistance to apoptosis during oral
carcinogenesis. Int J Oncol, 2000, 16: 591 –
597

47　Shen ZY, Xu LY, Li C, Cai WJ Shen J, Chen

JY and Zeng Y. A comparative study of telomerase
activity and malignant phenotype in multistage
carcinogenesis of esophageal epithelial cells induced by human papiIomavirus. Int J Mol Med,
2001, 8: 633 – 639

48　Klingelhutz AJ, Foster SA and McDougall
JK. Telomerase activation by the E6 gene product
of human papillomavius type 16. Nature, 1996,
380: 79 – 82

49　Song S, Pitot HC, and Lambert PF. The human
papillomavirus type 16 E6 genc alone is sufficient
to induce carcinomas in transgenic animals. J
Virol, 1999, 73: 5887 – 5893

50　Zur Hausen H. Immortalization of human cells and
their malignant conversion by high risk human
papillomavirus genotypes. SeminCancer Biol,
1999, 9: 405 – 411

90.　Intratumoral Injection of Arsenic to Enhance Antitumor Efficacy in Human Esophageal Carcinoma Cell Xenografts

SHEN Zhong-ying[1], ZHANG Yuan[2], CHEN Jiong-yu[3],
CHEN Ming-hua[1], SHEN Jian[1], LUO Wen-hong[2], ZENG Yi[4]

1. Department of Pathology; 2. Central Laboratory; 3. Tumor Hospital, Medical College
of Shantou University; 4. Institute of Virology, Chinese Academy of Preventive Medicine

Summary

To enhance the therapeutic efficacy of anticancer agents and to reduce systemic side effects, it
was decided to study the effect of arsenic trioxide directly on solid tumors to observe the anticancer
effect of arsenic on tumors and the distribution of arsenic in tumors and other organs. Esophageal
carcinoma cells were heterotransplanted in severe combined immunodeficient (SCID) mice in both
laterals of the abdominal wall. When both lateral tumors had grown to – 10mm × 8mm × 5mm, tumor –
bearing mice were used for 2 experiments. The right tumors were treated with intratumoral injection of
As_2O_3 in 1, 5 and 10 μg per day for 10 days sequent. The left tumors were treated with phosphate
buflfer solution as controls. To explore the distribution of As_2O_3 remaining in tumor and some organs,
a single intratumoral injection of As_2O_3 was studied with quantitative measurement of arsenic by

means of atomic absorption spectrometry. The results revealed that on the 17th day after the lst injection As_2O_3 – treated tumors were suppressed markedly compared to that of the contrarily lateral tumor accompanied by marked apoptosis and necrosis in tumor cells. The tumor growth inhibition (TGI) was 13.56%, 62.37% and 76.92% in 1, 5 and 10μg group, respectively. There were no pathological changes in heart, lung, spleen, liver, kidney or brain after arsenic administration. Distribution of As_2O_3 revealed that As_2O_3 remained at higher concentration in arsenic – treated tumor tissue than in other organs. Our data suggest that intratumoral delivery of As_2O_3 efficiently suppresses growth of transplanted esophageal carcinoma without systemic side effects. The protocol of As_2O_3 intratumoral injection will be its potential clinical utility for therapy of solid tumors.

〔**Key words**〕 Arsenic; Intratumoral injection; Esophageal carcinoma

Introduction

In previous experiments arsenic trioxide (As_2O_3) has been proved to be a genotoxic and a carcinogenic agent[1,2]. Recently arsenic trioxide has been an effective treatment for patients with acute promyelocytic leukemia[3,4], and other cell types of tumors, including malignant lymphocytes[5], myeloma[6], gastric cancer[7] and hepatoma cell line[8]. We have already noted effects of arsenic trioxide on an esophageal carcinoma cell line[9]. We also demonstrated that As_2O_3 mediates anticancer activity *in vitro* because of its ability to influence mitochondria[10,11], and to generate nitric oxide (NO) and reactive oxygen species (ROS)[12,13]. Accordingly As_2O_3 is known to exhibit antitumor effects on the basis of its highly cytotoxic nature, however, systemic distribution *in vivo* of these drugs causes undesirable side effects[14]. As demonstrated previously, after administration of As_2O_3 to intraperitoneal cavity of mice, it has a binding affinity for liver and kidneys, and it generates damage in both organs[15]. Arsenic used as an anticancer drug *in vivo* usually poses a problem that systematically delivered As_2O_3 has high toxicity and low effective antitumor action in local solid tumors.

Despite new developments in clinical oncology through establishment of new potent anticancer drugs, there are still significant problems in treating majority of solid tumors. In particular, alternative treatment modalities are required among conventional methods of therapy, however, intratumoral injection protocol may be a viable treatment for early local tumors. Approaches to intratumoral (i.t.) injection have been developed in animal models to increase the localization of drugs in tumors, to reduce their uptake in other tissues and to thus improve the tumor/normal tissue uptake ratios. This would permit higher and more frequent doses of anticancer drug use for tumor therapy[16].

It is unclear whether arsenic via i.t. injection directly impacts on the xenograft of esophageal cancer cells and how it effects *in vivo* redistribution and rebalance of arsenic concentration between tumor tissue and other organs. In the current study, we investigated As_2O_3 in SCID mice with i.t. injection inhibited esophageal tumor growth and *in vivo* distribution of arsenic in tumors and other organs.

Material and Methods

1. Animals: Severely combined immunodeficient mice (scm), 6 weeks of age, weighing 20 –

25 g, from the Center of Experimental Animal (Zhongshan Medical University), were used in this study. All experiments were carried out according to the guidelines of the Laboratory Protocol of Animal Handling, Shantou University of Medical College.

2. *Tumot modeIs*: The esophageal carcinoma cell line (SHEEC) is a malignantly transformed cell line of human embryonic esophageal epithelium induced by HPV18E6E7 in synergy with TPA in our laboratory[17]. SHEEC cells were inoculated subcutaneously into both abdominal walls of SCID mice and solid tumors formed. Tumors were selected for As_2O_3 treatment when they reached volumes of about 1. 0 cm ×0. 8 cm ×0. 5 cm.

3. *Drug and treatment*: Arsenic trioxide (As_2O_3) in ampoule (0. 1% As_2O_3, 10 ml) was supplied by the Pharmacy of Chinese Traditional Medicine of the First Hospital affiliated to Harbin Medical University (Harbin, P. R. China). In the 1st experiment, 30 mice were injected intratumorally in the right side with 1, 5 and $10\mu g$ of As_2O_3 in 0. 2 ml, respectively, once a day for 10 days, and tumors of the contrary side were injected with phosphate buffer solution (PBS) alone. Equal volumes of PBS were injected into both tumors in 10 mice as the control group under the same conditions. Mice were sacrificed 17 days after the lst injection and tumor masses were removed and weighed. The experiments were repeated at least once.

4. *Inhibition and regression of tumor*: Tumor growth inhibition (TGI), which reflects the retardation in tumor growth of arsenic – treated tumor relative to the contrary side tumor, is widely used to assess the efficacy of therapeutic agents against xenografts. Using tumor weight as a measure of size for illustration of TGI expressed as a percentage was calculated as follows:

$$\% TGI = 100 \ (W_L - W_R) \ /W_L$$

W_L is the mean tumor weight (mg) of the left side tumors (PBS – treated group) and W_R is mean tumor weight (mg) of the right side tumors (arsenic – treated group).

Tab. 1 Growth inhibition of tumors treated with arsenic

Dosage	W_L (mg)	W_R (mg)	TGI (%)
1 μg	1313 ±323	1135 ±243	13. 56
5 μg	1217 ±4 17	458 ±119[a]	62. 37[a]
10 μg	1131. 5 ±274	261. 5 ±58[a]	76. 92[a]
Control (NS)	1371 ±349	1336 ±412	2. 55

Notes: Each value indicates the mean ± SE (n = 10). TGI (%) = ($W_L - W_R$) /W_L. [a]Significant difference (P <0. 01) from the control.

5. *In* vivo *distribution of arsenic after f. t. injection*: In the second experiment, As_2O_3 (10 μg/mouse) was administered to another 21 tumor – bearing mice with i. t. injection of the right lateral tumors and 3 mice was administered with PBS as controls. At 0. 5, 1, 2, 4, 8, 16, 32 h following treatment, blood samples were drawn by ophthalmic vein under deep anesthesia and the mice were sacrificed. Reperfusion subjected into inferior vena cava with 10 ml of PBS containing heparin (4 units/ml) to remove blood components in blood vessels of tissues. Both tumor and normal tissues, including liver, kidney and brain from 3 mice in the 1st group, were collected and weighed. The arsenic concentration of these tissues was measured by use of an atomic absorption spectrometer.

6. *Arsenic measurement by atomic absorption spectrometry*: Samples of brain, liver, kidney,

blood and both sides of tumors from arsenic – treated mice were quick – frozen in acid – free vials and stored at –70℃. Samples weighed within the range of 0.2 – 0.6 g. Concentrated HNO_3 – H_2SO_4 – $HClO_4$ (1 : 1 : 2) (2 ml) was added to each sample, the vessel was capped with a stopper, and samples were heated in water – bath (90℃) for 2 h for digestion. Each sample was then added with 1 ml 30% H_2O_2 shaking, and deionized water was added to a final volume of 5 ml. Samples were analyzed by atomic absorption spectroscopy using an AA660 Graphite Furnace Atomic Spectrometry (GFAAS) (Shimadzu, Japan) set to 193.7 nm. The instrument was calibrated using a six – point 0.1 – 10 Pub standard curve prepared from an arsenic standard reference solution, and quality control standards were run during analysis to confirm the calibration.

7. Histopathological and ultrustructural examination: The samples of both experiments including tumor tissue and heart, lung, spleen, liver, kidney and brain were fixed in 10% formalin and the section were prepared routinely. Tissues of the transplanted tumors in SCID mice were cut into 1 mm^3 pieces, and fixed in 2.5% glutaraldehyde. Both samples were routinely prepared and observed under electron microscope (Hitachi 300) or light – microscope.

8. Statistical analysis: Student's t – test and U test was used to determine the significance between each experimental group. The difference was considered to be statistically significant when $P < 0.05$.

Results

1. Regression of established tumors upon i. t. injection of arsenic: In the 1st experiment all mice were sacrificed on the 17th day after the 1st As_2O_3 treatment. There was a significant reduction of tumor weight in groups treated with 5 ($t = 6.47$, $P < 0.01$) and 10 µg ($t = 8.17$, $P < 0.01$) of As_2O_3 compared with controls (Tab. 1). The suppression of the tumors was statistically significant. TGI was low in As_2O_3 in 1 µg group and high in 5 ($U = 2.58$, $P < 0.01$) and 10 µg group ($U = 2.86$, $P < 0.01$). The difference of TGI was of statistical significance.

Repeating the 1st experiment with the same number of SCID mice and the same treatment schedules againdemonstrated significant reduction in SHEEC tumor growth as in the original experiment. The treated tumors (right) had more marked growth inhibition than the opposite side (left) (Fig. 1).

2. Arsenic concentration in tumor tissues: The content of arsenic in tissues of right and left tumors was measured by atomic absorption spectrometry. During drug administration, the peak increase of concentration in arsenic – treated tumors was 4.23 µg/g, which was 3.36 times more than that of the left PBS – treated tumors (1.26 µg/g). But after drug withdrawal, at 32 hours, arsenic concentration decreased gradually to 0.07 (right) and to 0.04 µg/g (left) (Fig. 2). These results suggest that As_2O_3 is delivered to tumor tissue and retained for 4 h atconcentration >2 µg/g.

3. Arsenic distribution in other organs: The arsenic accumulation was observed in liver, kidney, brain and blood. As_2O_3 concentration showed a 2.8 – fold higher accumulation in arsenic – treated tumor compared with that of blood 1 h after injection. Arsenic contents in both liver and kidney increased, the peak concentration was 2.15 µg/g and 2.68 µg/g at 1h in tumor – bearing mice.

As Ⅱ L   As Ⅱ R

AS Ⅱ L, tumor masses on the left treated with PBS; AS Ⅱ R, tumor masses on the right treated with As_2O_3 (10 μg). The latter appeared smaller than that of the former

Fig. 1 The solid tumor mass of transplanted SHEEC in SCID mice were treated with As_2O_3

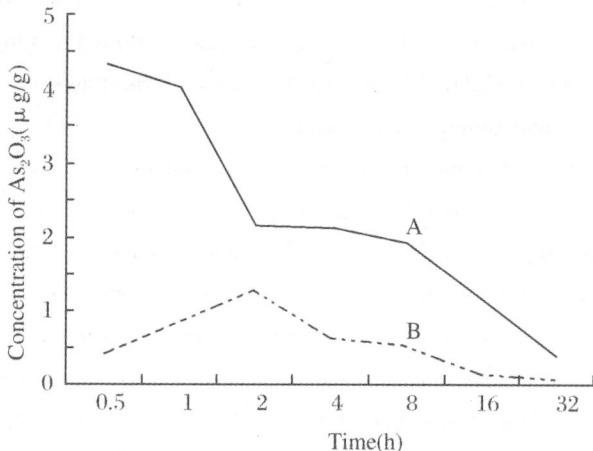

A, Arsenic – treated tumor; B, PBS – treated tumor

Fig. 2 Concentration curve of arsenic in both tumor tissues of the same mouse

Tab. 2 Distribution of arsenic（μg/g）in the organs.

Time（h）	Liver	Kidney	Brain	Tumor R	TumorL	Blood
0. 5	0. 47	1. 03	0. 30	4. 23	0. 41	0. 72
1	2. 15	2. 68	0. 68	4. 01	0. 86	1. 40
2	1. 45	2. 59	0. 98	2. 14	1. 26	0. 70
4	1. 01	0. 74	1. 56	2. 11	0. 49	0. 48
8	0. 66	0. 54	0. 68	1. 90	0. 35	0. 42
16	0. 52	0. 30	0. 23	1. 11	0. 11	0. 40
32	0. 21	0. 19	0. 05	0. 36	0. 04	0. 33
Control[a]	0. 03	0. 14	0	0	0	0. 12

Note: [a]Without administration of arsenic.

The normal value of liver and kidney was 0. 03 μg/g and 0. 14 μg/g, the arsenic content in brain was lower, the peak concentration was 1. 56 μg/g at 4 h, its peak concentration was delayed followed by blood, kidney and liver（Tab. 2）. These organs eliminated arsenic almost to the normal level within 32 h.

　　4. Histological and ultrastructural examination: In histological examination at the 17th day after 1st injection, all tumors in the 10 μg group presented mass necrosis and were partly replaced by fibrous tissue. The same histological features in the 5 μg group were also found but to a moderate degree. Only few focal necroses were observed at the center in the 1 μg and the control group. Samples of the 3 distinct groups with the control group were observed by electron microscopy. In the 1 μg group the nuclear and cytoplasmic membrane of cells was intact without the chromatin agglutinant. In the 5 and 10 μg group chromatin coagulated into coarse mass attaching near the nuclear membrane or flowed out of the nucleus, and cell organelles were swollen with secondary degeneration and necrosis （Fig. 3）.

Discussion

Our assumption that it might be possible to induce remarkable tumor regression via i. t. injection of arsenic trioxide was confirmed in this study. These cellular changes, apoptosis and necrosis, of arsenic – treated tumors were realized by achieving a characteristic morphology demonstrated by pathologic examination of light – and electron microscope. *In vivo*, TGI in the 5 and 10 μg group was significantly different to that of the control group. In the 10 μg group TGI reached up to

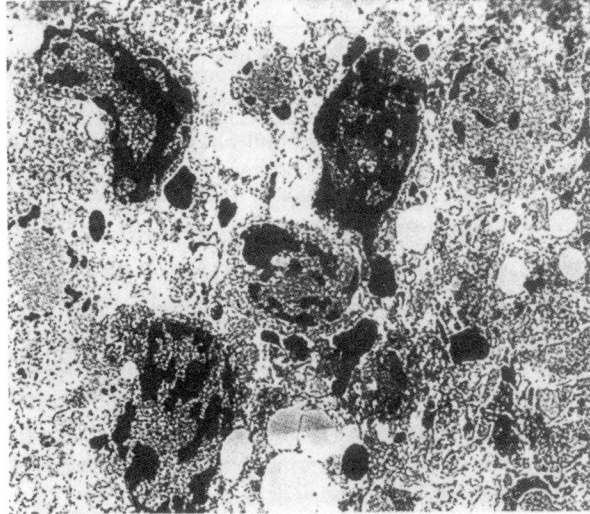

Fig. 3   Ultrastructure of SHEEC cells treated with 10 μg of arsenic. Masses of necrotic cells appeared (EM, ×4000)

76. 92% within 17 days. Using the i. t. injection technique we have provided evidence that high response rates could be achieved with direct effect on the cancer cells using arsenic without any significant side effects. In the 1 μg group, where tumor regression failed to occur, the kinetics of tumor growth was not significantly different from that of PBS control group.

We measured the *in vivo* distribution of arsenic after i. t. injection in tumor – bearing mice. Arsenic concentration increased immediately in the tumors of direct injection. The peak concentration of arsenic was 4. 23 μg/g and maintained at >4 μg/g for 1 h and at >1. 9 μg/g for 8 h, at this level arsenic would inhibit tumor growth. In the opposite lateral tumors, arsenic achieved peak concentration of 1. 26 μg/g, at 1 h and was then eliminated quickly. These results suggest that As_2O_3 in the concentration of 2 – 4 μg/g will be effective on tumor suppression but at the concentration <1. 26 μg/g and in short time was not effective as antitumor therapy.

The arsenic concentration in blood did not immediately reach the peak level at the time of arsenic administration. The time to attain peak concentration was 1h after drug administration. Released arsenic from tumor tissue can be redistributed into other organs and tissues via blood circulation. It was revealed that arsenic accumulated in liver and kidney was eliminated relatively rapidly, and no arsenic accumulation in liver or kidney was evident within 32h after single arsenic administration. However, multiple injection arsenic accumulation in some tissues has been reported[18]. Chronic toxicity of arsenic and the possible long – term effects of arsenic accumulation in the body will need further investigation.

In the course of our experiments no acute arsenic toxicity or death appeared. No severe heart, lung, liver, kidney or nervous system impairments were found. The amount of arsenic in most of the normal organs and tissues were far below the level of As_2O_3 in the arsenic – treated tumors. The arsenic elimination from tumors, blood and other organs was rapid. Thus, toxicity caused by arsenic is anticipated to be slight, and As_2O_3 is relatively safe and effective in treatment of solid tumors via i. t. injection.

Some studies have suggested that the effect of arsenic is also attributed to the extracellular matrix, which provides the support and nutrients to solid tumors. Arsenic trioxide causes selective necrosis in solid murine tumors by vascular shutdown[19]. In arsenic treatment, an increase of NO or ROS was found[20] which effects the permeability of the endothelium, defects in vascular growth[21] and caused oxidative DNA damage in vascular smooth muscle cells and endothelial cells[22,23].

Pretreatment of the cells with As_2O_3 also synergistically enhanced radiation – induced apoptosis[24]. If As_2O_3 can sensitize tumor cells to ionizing radiation in solid tumors, As_2O_3 in combination with ionizing radiation have a synergistic effect in the regression of the tumor mass.

In summary, the protocol of intratumoral injection provided a sufficient arsenic dose for the treatment of local tumors. Our data showed that As_2O_3 as intratumoral injection inhibited growth of xenografts of esophageal carcinoma cells in SCID mice and induced apoptosis and necrosis in solid tumors. Characteristics of this optimal regional treatment method are the lack of persistent systemic or local side effects, thereby increasing the localization of drug concentration and reducing uptake in normal tissues. Therefore, As_2O_3 is a relatively safe and effective remedy in the treatment of solid tumors via intratumoral injection. These findings should be useful in designing strategies for the use of arsenic in anticancer therapy.

Acknowledgements

This study was supported by National Natural Science Foundation of China (39830308), Development Foundation of Shantou University (L00012), and the Chinese National Human Genome Center, Beijing.

[In 《Oncology Reports》 2004, 11: 155 – 159]

References

1   Matsui M, Nishigori C, Toyokuni S, Takada J, Akaboshi M, Ishikawa M, Imamura S and Miyachi Y. The role of oxidative DNA damage in human arsenic carcinogenesis: detection of 8 – hydroxy – 2 – deoxyguanosine in arsenic – related Bowen's disease. J Invest Dermatol, 1999, 113: 26 – 31

2   Ho IC, Yih LH, Kao CY and Lee TC. Tin – protoporphyrin potentiates arsenite – induced DNA strand breaks, chromatid breaks and kinetochore-negative micronuclei in human fibroblasts. Mutat Res, 2000, 452: 41 – 50

3   Zhang TD, Chert GQ, Wang ZG, Wang ZY, Chen SJ and Chen Z. Arsenic trioxide, a therapeutic agent for APL. Oncogene, 2001, 20: 7146 – 7153

4   Look AT. Arsenic and apoptosis in the treatment of acute promyelocytic leukemia. J Natl Cancer Inst, 1998, 90: 86 – 88

5   Zhu XH, Shen YL, Jing YK, Cai X, Jia PM, Huang Y, Tang W, Shi GY, Sun YP, Dai J, Wang ZY, Chen S J, et al. Apoptosis and growth inhibition in malignant lymphocytes after treatment with arsenic trioxide at clinically achievable concentrations. J Natl Cancer Inst, 1999, 91: 772 – 778

6   Rousselot P, Labaume S, Marolleau JP, Larghero J, Noguera MH, Brouet JC and Fermand JP. Arsenic trioxide and melarsoprol induce apoptosis in plasma cell lines and in plasma cells from myeloma patients. Cancer Res, 1999, 59: 1041 – 1048

7　Gu QL, Li NL, Zhu ZG, Yin HR and Lin YZ. A study on arsenic trioxide inducing in vitro apoptosis of gastric cancer cell lines. World J Gastroenterol, 2000, 6: 435 – 437

8　Xu HY, Yang YL, Gao YY, Wu QL and Gao GQ. Effect of arsenic trioxide on human hepatoma cell line BEL – 7402 cultured in vitro. World J Gastroenterol, 2000, 6: 681 – 687

9　Shen ZY, Tan LJ, Cai WJ, Shen J, Chert C, Tang XM and Zheng MH. Arsenic trioxide induces apoptosis of esophageal carcinoma in vitro. Int J Mol Med, 1999, 4: 33 – 37

10　Shen ZY, Shen J, Cai WJ, Hong C and Zheng MH. The alteration of mitochondria is an early event of arsenic trioxide induced apoptosis in esophageal carcinoma cells. Int J Mol Med, 2000, 5: 155 – 158

11　Shen ZY, Shen J, Li QS, Chen CY, Chen JY and Zeng Y. Morphological and functional changes of mitochondria in apoptotic esophageal carcinoma cells induced by arsenic trioxide. World J Gastroenterol, 2002, 8: 31 – 35

12　Shen ZY, Shen WY, Chen MH, Shen J, Cai WJ and Zeng Y. Nitric oxide and calcium ions in apoptotic esophageal carcinoma cells induced by arsenite. World J Gastroenterol, 2002, 8: 40 – 43

13　Woo SH, Park IC, Park MJ, Lee HC, Lee SJ, Chun YJ, Lee SH, Hong SI and Rhee CH. Arsenic trioxide induces apoptosis through a reactive oxygen species – dependent pathway and loss of mitochondrial membrane potential in HeLa cells. Int J Oncol, 2002, 21: 57 – 63

14　Ohnishi K, Yoshida H, Shigeno K, Nakamura S, Fujisawa S, Naito K, Shinjo K, Fujita Y, Matsui H, Sahara N, Takeshita A, Satoh H, Terada H and Ohne R. Arsenic trioxide therapy for relapsed or refractory Japanese patients with acute promyelocytic leukemia: need for careful electrocardiogram monitoring. Leukemia, 2002, 16: 617 – 622

15　Shen J, Wu MH, Chen MH, Li QS and Shen ZY. The effects of arsenite trioxide on the transplanted hepatic carcinoma in mice. Chin J Cancer Biother, 2002, 9: 450 – 454

16　Gochi A, Orita K, Fuchimoto S, Tanaka N and Ogawa N. The prognostic advantage of preoperative intratumoral injection of OK – 432 for gastric cancer patients. Br J Cancer, 2001, 84: 443 – 451

17　Shen ZY, Cen S, Shen J, Cai WJ, Xu JJ, Teng ZP, Hu Z and Zeng Y. Study of immortalization and malignant transformation of human embryonic esophageal epithelial cells induced by HPV18E6E7. J Cancer Res Clin Oncol, 2000, 126: 589 – 594

18　Hood RD, Vedel – Macrander GC, Zaworotko M J, Tatum FM and Meeks RG. Distribution, metabolism, and fetal uptake of pentavalent arsenic in pregnant mice following oral or intraperitoneal administration. Teratology, 1987, 35: 19 – 25

19　Lew YS, Brown SL, Griffin RJ, Song CW and Kim JH. Arsenic trioxide causes selective necrosis in solid murine tumors by vascular shutdown. Cancer Res, 1999, 59: 6033 – 6037

20　Barchowsky A, Klei LR, Dudek EJ, Swartz HM and James PE. Stimulation of reactive oxygen, but not reactive nitrogen species, in vascular endothelial cells exposed to low levels of arsenite. Free Radic Biol Med, 1999, 27: 1405 – 1412

21　Ding I, Sun JZ, Fenton B, Liu WM, Kimsely P, Okunieff P and Min W. Intratumoral administration of endostatin plasmid inhibits vascular growth and perfusion in MCa – 4 murine mammary carcinomas. Cancer Res, 2001, 61: 526 – 531

22　Liu F and Jan KY. DNA damage in arsenite – and cadmium – treated bovine aortic endothelial cells. Free Radic Biol Med, 2000, 28: 55 – 63

23　Lynn S, Gurr JR, Lai HT and Jan KY. NADH oxidase activation is involved in arsenite – induced oxidative DNA damage in human vascular smooth muscle cells. Circ Res, 2000, 86: 514 – 519

24　Chun YJ, Park IC, Park MJ, Woo SH, Hong SI, Chung HY, Kim TH, Lee YS, Rhee CH and Lee SJ. Enhancement of radiation response in human cervical cancer cells in vitro and in vivo by arsenic trioxide (As_2O_3). FEBS Lett, 2002, 519: 195 – 200

91. Malignant Transformation of Human Embryonic Liver Cells Induced by Hepatitis B Virus and Aflatoxin B$_1$

GUO Xc[1], LAN Xy[1], ZHOU L[1], ZHANG Yl[1], ZENG Y[1], TENG Zp[2], SHEN Zy[3]

1. Institute for Virus Disease Control and Prevention; 2. Molecular Biology Lab,
Institute of Hematology; 3. Medical College of Shantou University

Summary

In order to investigate the effect of hepatitis B virus (HBV) and aflatoxin B$_1$ (AFB$_1$) on hepatocarcinogenesis, the human embryonic liver cells infected with HBV were transplanted to nude mice by subcutaneous route and the transplanted mice were divided into 4 groups for study, in which the group A of mice was injected with HBV – infected human embryonic liver cells and followed by injections of AFB$_1$ once a week (HBV AFB$_1$); the group B was treated with HBV as group A, but no AFB$_1$ was given (HBV$^+$); the group C was injected with normal human embryonic liver cells and AFB$_1$ was used as group (AFB$_1$$^+$) and the group D or control group was injected with normal embryonic liver cells without addition of AFB$_1$. The experimental results showed that the incidences of tumor formation in different groups were 27.3 % (6/22) in group A; 0% (0/13) in group B; 13.3% (2/15) in group C and 0% (0/14) in group D respectively. All the tumors formed were proved to be human hepatocellular carcinoma (HCC) by pathological examinations and the tumor tissues were anthrogenetic as demonstrated by EMA monoclonal antibody. The HBV – X and HBV – S genes could be detected in the tumor tissues by means of slot hybridization and PCR amplification, indicating that the HBV – DNA genes had integrated into DNA of host cells. Thus, we have successfully induced the human HCC through HBV infection and introduction of AFB$_1$ with a synergistic effect between HBV and AFB$_1$ in hepatocarcinogenesis.

[Key words]　　Human embryonic liver cells; HBV; Aflatoxin; Malignant; Transformation; Cell line

Introduction

Hepatocellular carcinoma (HCC) is a worldwide distributed disease accounting for 473 000 newly developed cases of HCC annually in the world and 235 000 cases in China[1]. Epidemiological and laboratory investigations have demonstrated that hepatitis B or C virus infection and afliatoxin exposure are the major and possibly synergistic risk factors for the development of HCC. Individuals with chronic hepatitis B virus (HBV) infection have a 200 – fold of greater risk to develop HCC than the age – matched uninfected controls[2]. Therefore, it is important to explore the role of HBV infection and the aflatoxin (AFB$_1$) exposure in hepatocarcinogenesis. Although more and more experiments support a direct effect of HBV on the hepatocarcinogenesis, but there is still no direct evidence in this regard. It has been shown that Epstein – Barr virus (EBV) can directly induced malignant transformation of the nasopharyngeal mucosa cells from human fetus through the synergistic effect of EBV and the tumor promoters[3]. The objective of the present investigation is to explore the synergistic effect HBV and AFB1 in hepatocarcinogenesis and to develop an experimental model for further studies.

Material and Methods

1. Mice

BALB/c nude mice (4 – 6 week old) were obtained from the Animal Center, Chinese Academy of Medical Science. Female and male were all used. They lived in SPF (specific pathogen free) clear room and was fed with food and water sterilized by autoclave.

2. Hepatitis B virus and reagents

Serum that contained $10^6 - 10^7$ copies of HBV DNA was collected from hepatitis B patients. The serum was supplied by Beijing 2nd Infectious Disease Hospital. Aflatoxin B_1 came from Institute of Nutrition and Sanitation, Chinese Academy of Preventive Medicine.

3. Tumor formation

The liver specimens were obtained from a human fetus (3 – 4 months old) under aseptic condition. Liver cells suspensions were prepared by mincing specimens through steel mesh, and then subjecting the cells to a concentration of 2×10^7/ml with RPMI 1640. The experiment was divided into 4 groups. Group A (HBV + AFB$_1$): The liver cells were infected with serum that contained HBV and incubated at 37℃ for 2 h. After virus adsorption, cells were centrifugation for 5 – 10 min by 1500 r/rain. And then about $0.5 \times 10^7 - 1 \times 10^7$ cells were xenografted subcutaneously to the back of nude mice. At the same time 400 ng AFB$_1$ was injected subcutaneously into the other side of back and once a week thereafter. Group B (HBV$^+$): The liver cells and nude mice were treated with HBV as group A, but no AFB$_1$ was used. Group C (AFB$_1$$^+$): The liver cells without HBV were incubated at 37℃ for 2 h too and then were transplanted subcutaneously into the back of nude mice. And AFB$_1$ was used as group A. Group D: Control group, the liver cells and nude mice were treated as group C and no AFB$_1$ was used. The whole experiment was finished by 3 times and 64 nude mice were used. All animals were observed for 3 – 4 months.

4. Histological procedures and HBV status

The tumor tissues obtained from group A and group C were fixed in 10% neutral formalin, embedded in paraffin, sectioned, and stained in hematoxylin and eosin.

To prepare tumor tissues DNA, the fresh tumor tissues obtained from group A were treated for 12 h at 55℃ with 500 volumes (μl) of a solution containing 200 μg/ml of proteinase K (Merck) dissolved in 10 mmol/L Tris – HCl (pH8.0), 2 mmol/L EDTA, 40 mmol/L NaCl and 0.5% SDS, then followed phenol extraction. The samples were stored at – 20℃ until used. Hepatitis B status was assessed by polymerase chain reaction (PCR) and by slot hybridization for the presence HBV S and X genes. Primers for HBV S gene were 5' – GGTATGTTGCCGTTTGTCCTCT – 3' and 5' – GGTATGTTGCCCCGTTTGTCCTCT – 3' and primers to amplify the X region were 5' – CCATGGCT-

GCTCGGGTGTG – 3' and 5' – GCTCTAGATGATFAGGCAGAGG – 3'. The length of the expected amplified product is 228 bp for S gene, 465 bp for X gene. The PCR conditions were 1 cycle at 94℃ for 3 min; 35 cycles at 94℃ for 45 s, 55℃ for 1 min, and 72℃ for 1 min; and 1 cycle at 72℃ for 10 min. Aliquots of each reaction were electrophoresis in 1.5% agarose gels which was stained with ethidium bromide.

Using HepG2 DNA that contains HBV DNA as template, HBV S and X genes were amplified by PCR. The products were purified with DNA retrieved kit and then labeled with Digoxin. Dig – labeled gene probes were hybridized with tumor tissues DNA blotted in nylon membrane by slot hybridization. The method was followed by directions of DNA test kit (BOEHRINGER MANNHEIM).

5. Establishment of cell lines

Three tumor tissues from group A were immediately placed in a modified Eagle medium respectively and eliminated necrosis tissues and contaminating fibroblasts. About $0.5cm^3$ of pieces were got by cutting specimens and separately transplanted to nude mice by subcutaneous again. When transplanted tumor developed about $1.5 cm^3$, the nude mice were put to death and the tumor tissues were treated as above. After 5 generations in nude mice the tumor tissues were taken out and washed twice by Hanks medium that contained double antibiotic, then were washed twice by RPMI 1640 medium. After that the tissues were sheared into tiny pieces and cultured in RPMI 1640 medium plus 15 % fetal bovine *in vitro*. Cultures were maintained in a humidified incubator at 37℃ in an atmosphereof 5 % CO_2. Cells were passaged when they grown at 90% confluence. Cells were passaged once a week at beginning and after 6 generation they were passaged once about 3 – 4 d. For cell line development, containating fibroblasts were initially removed by differential subculturing.

6. Morphology, immunochemistry, and growth kinetics of cell lines

For the morphological study of CBH – la, CBH – lb and CBH – 2, cultured cells were observed by phase – contrast microscopy. The ultrastructure of cell was observed by transmission and scanning electron microscope. Cells were cultured in cover slice and fixed, then were stained for human early membrane antigen by indirect immunofluorescence assay (Genezene, USA) and Ki67 was detected by immunochemistry. Ki67 positive nuclei were counted in 500 cells and the percentage of proliferative nucleus was statistic. Alpha fetoprotein was determined in the supematant of the culture, using complete medium as a blank, by radioimmunoassy (Abbott Laboratories, USA).

For cell growth studies, 5000 cells were seeded in different columns across 96 – well microtiter plates. Growth rate estimated by the measurement of absorbance following the MTT (Sigma Chemical Co.) assay in 3 cell lines[4]. All results represent the average of a minimum of 8 wells.

7. HBV status in cell lines

HBV S and X genes in 3 cell lines were detected by fluorescence *in situ* hybridization (FISH) and PCR. The cells were collected from 10 subculture cells. The method of PCR was described above. NBT/Bcip test kit and random biotin – labeled primer kit were purchased from Institute of Hematology, Beijing Medical University. Pd (N) 6 primer was purchased from Promega Company. Avidin D and anti – avidin D – were purchased from Vector Company. The detail method of FISH was used as described previously[5] .

8. Tumnorgenesis in nude mice

After the 60 passages, the cell lines of CBH – 1a, CBH – 1b and CBH – 2 were harvested separately. About $10^6/0.2$ ml cells were transplanted subcutaneously to the back of nude mouse. Four nude mice were used in each cell line and 12 nude mice were observed.

Results

1. Tumor formation in nude mice

There was an about $0.5 - 1.0$ cm^3 bulge in back of mouse after just injection of embryonic liver cells and it would be gradually disappeared during $3 - 5$ d. The tumors of group A occurred during $1 - 2$ months and increased gradually. They reached 1.5 cm × 1.5 cm × 1.0 cm, the substance was hard and the surface was uneven. The tumors of group C occurred during $1.5 - 2.0$ months and they grew slowly. The size of tumor was smaller than group A and there was necrosis in center of tumor. There were no tumors observed in the group B and D. The comparison of tumor formation in different groups was summarized in Tab. 1.

Tab. 1 Developing tumor induced by HBV and AFB$_1$ in different group

Groups (Times of experiment)	A (HBV + AFB$_1$)	B (HBV)	C (AFB$_1$)	D (Control)
1	2/6	0/4	1/5	0/4
2	3/9	0/5	1/6	0/6
3	1/7	0/4	0/4	0/4
Sum total (%)	6/22 (27.3)	0/13 (0)	2/15 (13.3)	0/14 (0)

The tumors of groups A and C were confirmed as hepatocellular carcinomas (HCC) by histopathological examination. The most cancer cells were square round and line up. The cell plasma was abundant and nuclear size was not slightly identical. There were some fibrocytes around the cancer nest (Fig. 1).

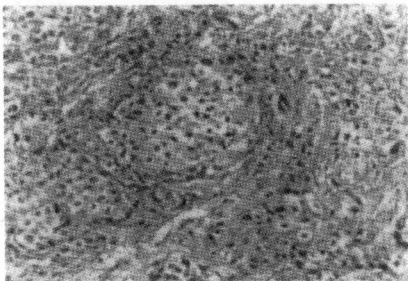

Fig. 1 Micrographs of paraffin section of hepatocellular carcinoma in nude mice H and E (×400)

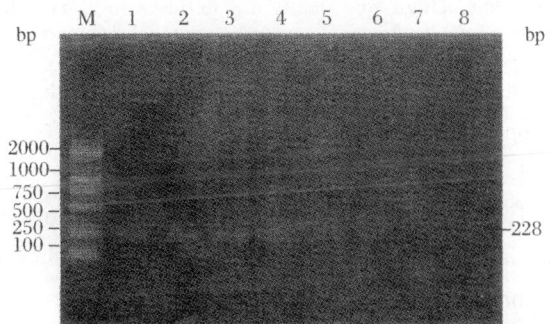

M: DNA marker; 1: Positive control; 2 – 7: Tumor tissue DNA; 8: Negative control

Fig. 2 S gene result of PCR in tumnor tissues

2. HBV status in tumor tissues

DNAs of the tumors from group A were amplified by 2 pairs of primers covering 2 different regions of the HBV genome: S and X genes. PCR results revealed that the strip in 228 bp was amplified in 5 tumors and positive control for S gene and the strip in 465 bp was in all samples for X gene, except for negative control (Figs. 2 and 3). For the further confirming the results above, Dig – labeled S and X genes probes were hybridized with tumor tissues DNA by slot hybridization. The findings were consistent with PCR results.

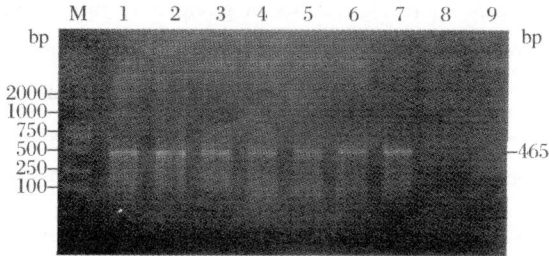

M: DNA marker; 1: Positive control; 2 – 7: Tumor tissue DNA;
8: Negative control; 9: Bland control

Fig. 3　X gene result of PCR in tumor tissues

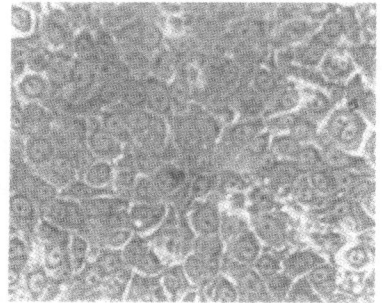

Fig. 4　Phase contrast microscope (×200)

3. CBH – 1a, CBH – 1b and CBH – 2 Cell lines

The tumor tissues from group A were cultured and passaged *in vitro* in a long time and 3 cell lines were established. Three cell lines were named CBH – 1a, CBH – 1b and CBH – 2. CBH – 1a and CBH – 1b were rooted in first time of experiment, and CBH – 2 was from second time of experiment. The cell lines were subcuhured nearly 100 generations. The cell shape was typical epithelioid cell. The limit of cell was clear and cells were tightly arranged. The cell size was equal, plasma was plentiful and binucleate cells were easy to see under the phase contrast microscope (Fig. 4). The Cell surface had microvilli, nuclear was round and located in the middle by transmission electron microscope. There were many organelles, such as mitochondrion, endoplasmic reticulum, ribosome and glycogen in plasma (Fig. 5). The shape of cell displayed oval or fusiform during growth movement state. There were many microvilli and pseudopodia in cell surface under scanning electron microscope. The cell was boll in cell division (Fig. 6).

4. Biological characteristic of cell lines

The cell from 3 cell lines was anthropogenetic by test EMA monoclonal antibody. The positive cells showed that the cell membrane was yellow fluorescence (Fig. 7). Counting the cell proliferative nucleus revealed that Ki67 positive nuclear was 38.2%. The positive nuclear was stained by tawny (Fig. 8). Alpha fetoprotein in supernatant of culture medium was 0.8 – 1.0 mmol/L and lower than normal level of serum. Growth curve of cell line indicated that cell growth rapidly and cell number went up straight line.

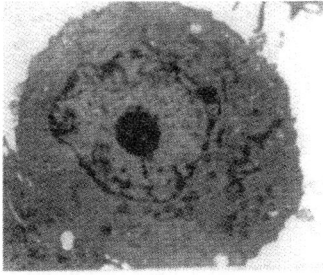

Fig. 5   TEM （ ×7000 ）

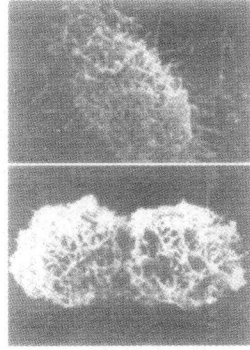

Fig. 6   SEM （ ×7000 ）

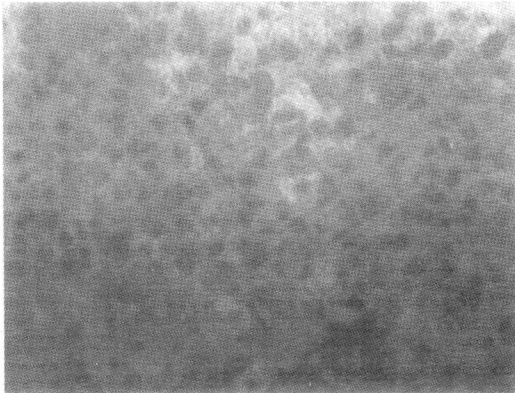

Fig. 7   Hmnan early membrane antigen is positive in tumor cell

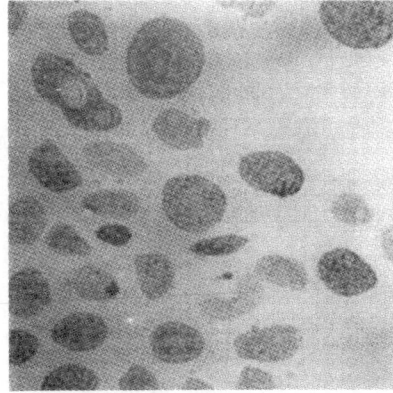

Fig. 8   Immunohistochemical staining of CBH – la
ceil, the positive nuclei of Ki67 are tawny （ ×200）

5. HBV status in cell lines

HBV S and X genes could be detected in nuclei of cell lines by FISH. The positive signal was green yellow bright spot on red – apricot nucleus （Figs. 9 and 10）. PCR amplification indicated that there was HBV X gene in cell lines （Fig. 11）.

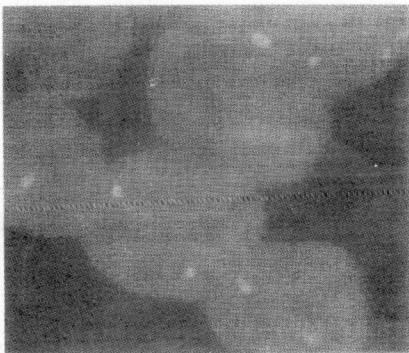

Fig. 9   The result of X gene by hybridization *in situ*
（fluorescence microscope ×400）

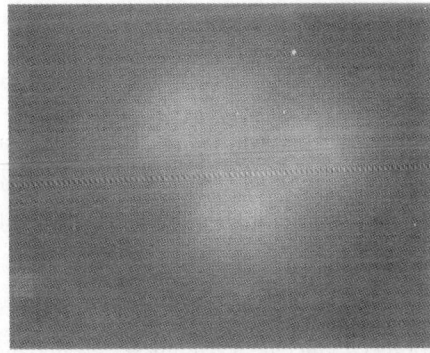

Fig. 10   The result of S gene by hybridization *in situ* （fluorescence microscope ×400）

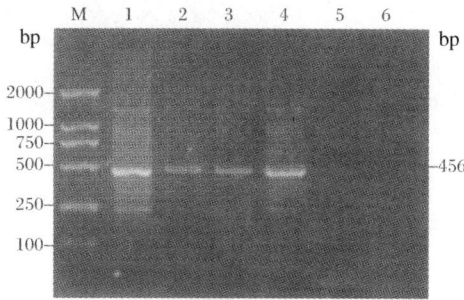

M: DNA marker; 1: Positive control; 2: CBH-1a; 3: CBH-1b;
4: CBH-2; 5: Negative control; 6: Bland control.

Fig. 11   The result of X gene by PCR amplification

6. Tumorgenesis of cell lines

The cells of CBH-1a, CBH-1b and CBH-2 were respectively inoculated to nude mice. The lesser tubercle under skin in nude mice occurred after 2-3 wk that cells of celllines were inoculated. Since then the tumors developed rapidly and reached 1.5 cm × 1.5 cm × 1.0 cm after 6 wk. The rate of tumorgenesis was 100% and all the tumors were hepatocellular carcinomas (HCC) by histopathological examination.

Discussion

Hepatitis B virus is an important etiological agent of hepatitis, cirrhosis and hepatocellular carcinoma (HCC) and it is suggested that HBV infection and the aflatoxin exposure are the major synergistic risk factors for the development of HCC[6-8]. In the present study, the development of malignant tumor formation in nude mice was observed in experiments with HBV-infected human embryonic liver cells plus the synergistic effect of aflatoxin B1 (AFB$_1$). It was found that the incidences of tumor formation were 27.3% in group of HBV-infected cells with addition of AFB$_1$; 13.3% in group with single AFB1 and 0% in group of single HBV or control group. These results indicate that there is synergism with HBV and AFB$_1$ and HBV infection can increase the incidence of tumorgenesis. Many factors are associated with tumorgenesis in which the tumor viruses play an important role. It is well known that tumor viruses cause transformation as a consequence of their ability to integrate their genetic information into DNA of host cells. Most often they also induce the chronic production of the oncoproteins that maintain the infected cells in a transformed state. Most oncoproteins encoded by oncogenes in viral genomes are intracellular products. The X gene of HBV seems to play an important role in the HBV-associated hepatocarcinogenesis. Recent studies have shown that the promoter activity of the human p53 gene is strongly repressed by HBV X protein (HBV X)[9]. In the present study, the HBV X gene was detected in the tumor tissues of group of nude mice and it was illustrate that the HBV-DNA had reached to the fetus liver cells. These results were consistent with those reported by Li et al[10], in which HCCs were induced by HBV infection and dietary AFB in tree shrews with the incidence of 67% in group of HBV+AFB$_1$, 30% in group of AFB$_1$ and 0% in group of HBV respectively.

On the basis of tumor formation induced by HBV and AFB$_1$, we established 3 HCC cell lines with a clear inducement. By analyzing the biological characteristics of these 3 cell lines, we found that the cells grew rapidly and were anchorage-independent, and the Ki67 positive cell was 38.2% by counting the nucleus of the proliferative cells. The incidence of tumor formation was

100% when the cell lines were inoculated into nude mice. The HBV X and S genes were found to be positive by fluorescence *in situ* hybridization, and the PCR amplification, revealed the X gene in early subcultures. These results suggest that the viral DNA had been integrated into DNA of host cells.

Qidong area of Jiangsu province is one of the highly prevalent regions of HCC in China the annual incidence rate HCC in this area is 910. 89/100 000 in population with chronic carriers of HBV, and this figure is significantly higher than that of 24. 24/100 000 in the control group. It had been reported that the aflatoxin exposure could increase the danger for developing HCC in those people who had infected with HBV[11]. The present experiment supports the previous survey on the cellular and molecular levels.

So far, it is still uncertain how HBV and AFB_1 interact within liver cells in the course of hepatocarcinogenesis. More studies show that X gene is an important oncogene in this process[12,13]. Recent studies suggest that persistent infection of HBV is essential for the induction of multi – stage genetic mutations in the chromosomes through immuno – mediated injuries of hepatocytes and the resulting hyperplasia[14]. The present study demonstrated that there was synergistic effect between HBV and AFB_1 in human hepatocarcinogenesis, and this would offer for an animal model and cell lines for further studies on the molecular mechanism during hepatocarcinogenesis. Insight for the hepatocarcinogenesis process should come from a multidisciplinary collaboration to explore important viral and host genes so that new approaches to diagnosis and treatment can be developed.

[In 《J Microbiol Immunol》 2004, 2 (3): 185 – 190]

References

1　Parkin DM. The global burden of cancer. Semin Cancer Biol, 1998, 8: 219 – 235

2　Ghebranious N, Sell S. Hepatitis B injury, male gender, aflatoxin, and p53 expression each contribute to hepatoeareinogenesis in transgenic mice. Hepatology, 1998, 27 (2): 383 – 391

3　Liu Z, Liu Y, Zeng Y. Synergistic effect of Epstein – Barr virus and tumor promoters on induction of lymphoma and carcinoma in nude mice. J Cancer Res Clin Oncol, 1998, 124 (10): 541 – 548

4　James C, William G. D, Adi FG, et al. Evaluation of a tetrazohum – based semiautomated colorimetric assay: assessment of chemosensitivity testing. Cancer Res, 1987, 47: 936 – 942

5　Teng ZP, Zeng Y. Detection of LMP gene of Epstein – Barr virus in nasopharyngeal carcinoma by in situ hybridization with biotin labeled probes. Chinese J Virology, 1994, 10 (2): 184 – 186

6　Guo XC, Wu YQ. Progress of prevention and control on viral hepatitis in China. Biomed and Environ Sci, 1999, 12: 227 – 323

7　Qian GS, Ross RK, Yu MC, et al. A follow – up study of urinary markers of aflatoxin exposure and liver cancer risk in Shanghai, People's Republic of China. Cancer Epidemiol Biomarkers Prev, 1994, 3: 3 – 10

8　Wang LY, Hatch M, Chen CJ, et al. Aflatoxin exposure and risk of hepatocellular carcinoma in Taiwan Int J Cancer, 1996, 76: 620 – 625

9　Lee SG, Rho HM. Transcription repression of the human p53 gene by hepatitis B viral X protein. Oncogene, 2000, 19 (3): 468 – 471

10　Li Y, Su JJ, Qin LL, et al. Synergistic effect of hepatitis B virus and aflatoxin 131 in hepatocarcinogenesis in tree shrews. Ann Acad Med Singapore, 1999, 28 (1) : 67 – 71

11 Lu SX, Zhang QN, Wang JB, et al. Hepatitis B virus, aflatoxin B and hepatocellular carcinoma. Zhongguo Zhongliu, 1999, 8: 305 – 306

12 Kim CM, Koike, Saito L, et al. HBx gene of hepatitis B virus induces liver cancer in transgenic mice. Nature, 1991, 351: 317 – 320

13 Yu DY, Moon HB, Son JK, et al. Incidence of bepatocellular carcinoma in transgenic mice ex-

pressing the hepatitis B virus X – protein. J Hepatol, 1999, 31 (1): 123 – 132

14 Ogden SK, Lee KC, Barton MC. Hepatitis B viral transactivator HBx alleviates p53 – mediated repression of alpha – fetoprotein gene expression. J Biol Chem, 2000, 275 (36): 27806 – 27814

92. Infectious Diseases in China

ZENG Yi[1], XU Hua[1], ZHANG Jia – xi[2]

1. Chinese Foundation for Prevention of STD and AIDS; 2. China Preventive Medicine Association

History of Infectious Disease

Before 1949, many epidemic diseases occurred in China due to corruption in the government, foreign invasions, and poor technologies. According to Chinese history, up to 305 years prior to the founding of the People's Republic of China (P. R. China), a plague epidemic affecting a total of 2. 6 million people caused 2. 4 million deaths. Smallpox epidemics occurred every year. Between 1933 and 1944, about 380 000 people contracted smallpox. In 1820, cholera was introduced into China; since then there have been about a hundred of serious cholera epidemics in China. Schistosomiasis once reached a two million square kilometer epidemic area, with 11 million people contracting the disease. There were hundreds of thousands of people who died or became disabled from measles, malaria, diphtheria, pertussis, meningococcal meningitis, and polio. Due to the lack of a basic infectious disease reporting system, statistics were very difficult to accurately obtain and monitor. Epidemics of sexually transmitted infections (STI) were rampant in both urban and rural areas; these were regarded as a social issue. In China's five thousand years of history, no governments have had the capa bility to control infectious epidemics and to establish a national epidemic preventive system.

In the early days of the founding of P. R. China, the Chinese government faced tough challenges. For example, in 1950, there were about 3400 human plague cases with 1200 deaths; 43 000 and 61 000 smallpox cases were reported respectively in 1951 and in 1952. In 1952, there were 1. 04 million measles cases with 41 000 deaths, and 2. 93 million malaria cases with 39 000 deaths reported (Ministry of Public Health, 1990; 2002). Pertussis, kala – azar, diphtheria, and relapsing fever were rampant. In 1951, a relapsing fever epidemic occurred, affecting 100 000 Chinese with relapsing fever and causing more than 10 000 deaths. STI epidemics were very severe; the incidence rate of syphilis in large cities was 3.8% – 4.5%, and in some minority areas the rate reached from 21.7% to 48%. Among the female sex workers (FSW) in Beijing, Shanghai, Wuhan, and Guishui, the incidence rate of STI was between 89.3% and 100%. To confront these challenges, the Chinese government carried out a "Put prevention first" principle, established an anti – epidemic system nationwide, and concentrated its efforts to prevent and control severe and life threatening infectious diseases. With a focus on infectious disease prevention and treatment, the Chinese government developed a national health campaign, and enforced occupational health, food hygiene, school hygiene, and radiation protection regulations. Through these strategies, the sanitation situation in urban and rural areas has greatly changed, and citizens' health status has been greatly improved.

By the end of the 1950s, China had basically eliminated smallpox, plague, kalaazar, typhus, relapsing fever, and STI. By the end of 2002, the morbidity rate of notifiable infectious diseases had decreased from 20 000/10^5 at the beginning of the founding of ER. China to 180/10^5 in 2002. The total incidence number for measles, pertussis, and diphtheria has decreased from 11.83 million to 56 252 (Ministry of Public Health, 1990; 2002). There have now been four consecutive years without a reported case of wild virus strain polio. The morbidity rates for schistosomiasis, filariasis, malaria, and leprosy have also decreased significantly. However, infectious diseases are key health issues within specific regions of China. In recent years, some infectious diseases have re-emerged, such as tuberculosis, malaria, and STI. Other new diseases like AIDS and SARS have emerged to threaten the lives and health of the Chinese, as well as China's economic development.

Health Care Institutions

Health care institutions in China originated in 1873 when the first health Quarantine institutions were constructed. In 1919, the first anti – epidemic department was started in the North of China.

In 1949, based on the experience and practices of the former USSR, a health antiepidemic station was set up in the former Zhong Chang Railway Administration in the Northeast of China. After the founding of the P. R. China, according to the principle of the Chinese government "To put prevention first," health anti – epidemic stations were set up in several provinces and cities of China. By the end of 1952, there were 147 such stations, 188 special preventive and treatment services, with 20 500 health anti – epidemic workers, including 19 750 health technical personnel, By the end of 1956, the health anti – epidemic stations covered most of China, except the areas of minority nationalities and remote areas. The health anti – epidemic stations were also set up within the railway system, and in large factories, mines, and enterprises. In addition, preventive and care sections were set up in the county hospitals and the health anti – epidemic groups were set up in the township health institutions. By 1965, there were 2499 health anti – epidemic stations at all levels and 822 preventive and treatment institutions, with 63 879 health technical personnel throughout China. This represents a 16 – fold increase compared with 1952 (Chinese Association of Preventive Medicine, 2003).

During the Cultural Revolution, the health anti – epidemic system and its services were seriously impacted. After the Cultural Revolution, the health anti – epidemic system and its services were resumed and further developed.

In 1983, in order to increase the capabilities of the health system, the National Preventive Medical Center was set up by the ministry of Health (it was changed to the Chinese Academy of Preventive Medicine in 1985, and changed to National Centers for Disease Prevention and Control in 2002). In 2002, there were 3580 Centers for Disease Prevention and Control, including 31 Centers at the provincial level, 383 at the prefecture level, 2519 at the county level, and 637 in other fields. There were a total of 204 444 personnel, including 156 838 health technical workers. China, in 2002, had as well 1839 special prevention and treatment institutions, and 571 health supervision services (Chinese Association of Preventive Medicine, 2003).

HIV/AIDS Prevention and Control

1. Epidemic situation

In 1982, Clotting Factor VIII with HIV – 1 was sent from the USA to China; during 1983 – 1985, four Hemophilia patients were infected with HIV – 1 through the injection of this Factor VIII from the USA. In 1985 the first AIDS patient came to China from the USA and died in China.

The cumulative reported number of people living with HIV/AIDS (PLWHA) had reached 40 560 by the end of 2002, with a total of 2639 AIDS cases and 1216 AIDS – related deaths in the 31 provinces, autonomous regions, and municipal cities in China except Taiwan, HongKong and Macao (Fig. 1). In 2002, 9824 new HIV infections were reported; reflecting a 19.5% increase over the same period in 2001. The number of infections has increased in 26 provinces, autonomous regions, and municipal cities in China. Among these, Hainan, Guizhou, Chongqing, Hubei, Hunan, and Anhui provinces reported that the number of PLWHA had an increase compared to the figure reported in the same period of 2001. In 2002, 1045 AIDS cases were reported, with 363 AIDS – related deaths, another increase from the previous year, Among reported cases, 4928 were infected through intravenous drug use, accounting for 50.2% of infections; infection through blood and blood products accounted for 10% of infections; heterosexual contact for 11%; maternalto – child transmission (MTCT) for 0.4%; and 28.3% of infections occurred through unknown transmission modes. Statistics also show that 78.4% of reported infections are among young people aged 20 to 39 (China HIV/AIDS Information Network, 2002).

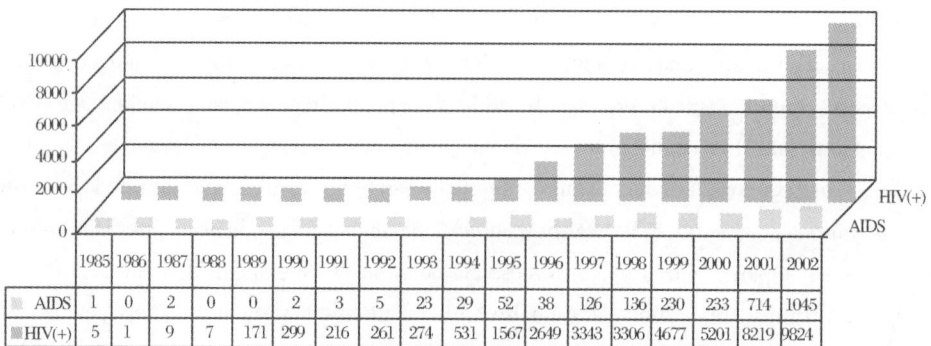

	1985	1986	1987	1988	1989	1990	1991	1992	1993	1994	1995	1996	1997	1998	1999	2000	2001	2002
AIDS	1	0	2	0	0	2	3	5	23	29	52	38	126	136	230	233	714	1045
HIV(+)	5	1	9	7	171	299	216	261	274	531	1567	2649	3343	3306	4677	5201	8219	9824

Fig. 1   Annual Reported Number of HIV/AIDS Cases in China (1985 – 2002)

Between 1985 and 2002, most HIV/AIDS cases were reported in the Yunnan province, followed by Xinjiang, Guangxi, Guangdong, Sichuan, and Henan provinces (Fig. 2). HIV/AIDS cases have been mainly distributed in the rural areas.

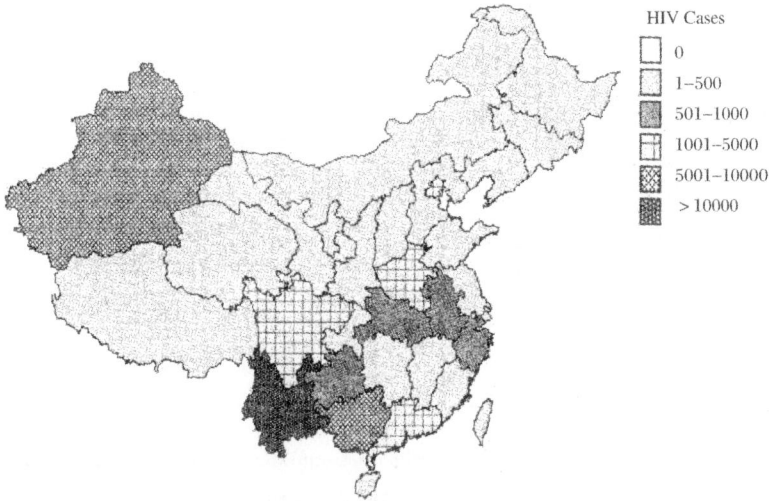

Fig. 2   Geographic Distribution of HIV/AIDS in China (1985 – 2002)

The proportion of male and female cases is 4 : 1 (Fig. 3).

HIV infection among different age groups is: 1% for age < 15; 5.9% for ages 15 – 20; 51.1% for ages 20 – 29; 28.9% for ages 30 – 39; 10% for age > 40; 3% unknown (Fig. 4).

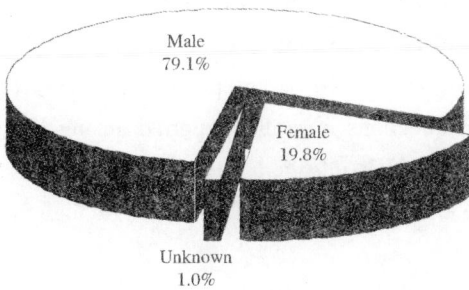

Fig. 3   Gender Distribution of HIV Infection in China (1985 – 2002)

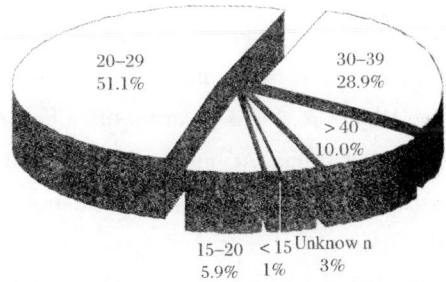

Fig. 4   Age Distribution of HIV Infection in China (1985 – 2002)

The major modes of transmission for HIV/AIDS in China since 1985 have been: intravenous drug use (IDU), accounting for 63.7% of infections; infection through the whole blood supply accounted for 9.3% and blood products accounted for 1.6%; heterosexual contact for 8.1%; MTCT for 0.2%, and unknown for 17.1% (see Figure 5).

This year, according to a joint HIV/AIDS epidemiological survey conducted by China, WHO, and UNAIDS, China has an estimated 840 000 persons living with HIV/AIDS, including 80 000

AIDS cases. Experts estimate there will be 10 million HIV/AIDS cases without effective control (see Fig. 6).

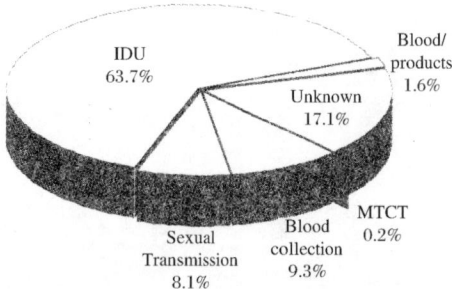

Fig. 5 Reported HIV Infections in China by Transmission Route (1985 – 2002)

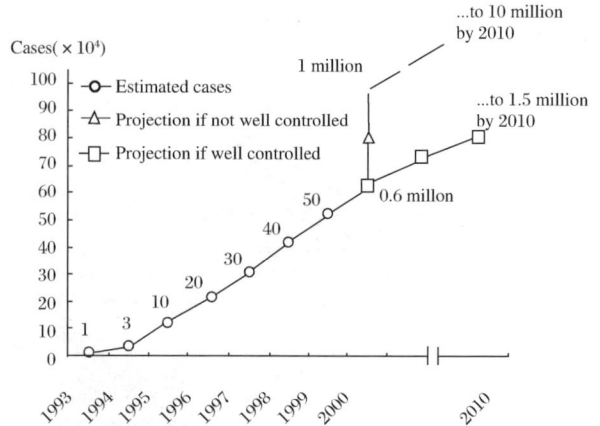

Fig. 6 Estimated HIV/AIDS Cases, 1993 – 2000, and Predicted Cases, 2010

2. Three phases of the HIV epidemic in China

(1) *The first phase*: The first phase, which began in 1985 and end in 1988 was marked by a small number of imported cases. The majority of infected persons during this time were foreigners or overseas Chinese, and the cases appeared only sporadically in coastal cities. Four hemophiliac patients infected with HIV through imported factor VIII were reported from Zhejiang.

(2) *The second phase*: The second phase, from 1989 to 1994, might be termed a limited epidemic. It began in October 1989 with the identification of HIV infection in 146 drug users in southwest Yunan (these were imported cases of HIV), and was followed by an epidemic in a few counties in Dehong prefecture. During this phase some HIV infections were also reported among IDU in other parts of China. At the same time, a small number of HIV infections were reported among laborers returning from abroad, STI patients, and FSW; the number of cases increased gradually during this time.

(3) *The third phase*: The third phase began in 1995 when HIV transmission among IDU spread beyond one prefecture in Yunnan province and into various regions such as Xinjiang, Guangxi, and Sichuan. Simultaneously, a large number of HIV infections were reported among blood/plasma donors in central China, mainly rural blood donors who moved frequently and were more likely to transmit HIV to different regions. Furthermore, cases of HIV infection among STI patients and FSW also increased.

3. Characteristics of the HIV epidemic in China

(1) *The epidemic is increasing dramatically*: Currently, there is a low prevalence of HIV nationally but a high prevalence in specific populations and certain regions. The main transmission routes are through intravenous drug use and, in the past, through the sale of blood and plasma. In

2003, according to a China CDC survey supported by WHO, UNAIDS, and US CDC, China has 840 000 people living with HIV/AIDS, among which 80 000 are AIDS cases. Although the adult prevalence rate is less than 0. 1%, the epidemic has spread to 31 provinces and the number of reported HIV/AIDS cases has increased significantly.

(2) *Increasing number of AIDS cases and AIDS – related deaths*: According to the annual case reports on HIV/AIDS, there has been a significant increase in the cumulative number of HIV/AIDS cases reported. Since 2001, China has approached a peak of AIDS cases and AIDS related deaths. In the 15 years from 1985 to 2000, the cumulative number of AIDS cases reported was 880, with 496 deaths, while the number of AIDS cases and related deaths between 2001 and 2002 were 1742 and 716 respectively. The reported number of AIDS cases in 2002 increased by 44% compared with 2001. In some villages in counties of Henan and Anhui provinces where the HIV epidemic is acute, concentrations of AIDS cases and deaths have occurred since 2001.

(3) *The HIV/AIDS epidemic is spreading from high – risk populations to the general population*: Although sharing injection equipment among IDU is the main transmission route, the proportion of sexually transmitted HIV infections also increased from 5. 5% in 1997 to 10. 9% at the end of 2002. Data from sentinel surveillance indicated that the HIV prevalence among FSW is increasing, and Mother to Child Transmission (MTCT) has continued to increase since the first reported case in 1995.

(4) *Risk factors exist for a generalized HIV epidemic*: High risk behaviors among IDU has increased, including sharing needles and sytinges. There still is a low rate of condom use in China. There is a persistent lack of HIV/AIDS related knowledge among Chinese, coupled with severe social discrimination against PLWHA. A high reproductive tract infection (RTI) prevalence rate among rural women is increasing the risk of HIV transmission in some areas. There remains a risk of HIV infection through blood transfusions and iatrogenic infections. Significantly, there is an imbalance in economic development, which has produced poverty in different areas and increased the risk of HIV infections. If China is not able to address these risk factors, the HIV/AIDS epidemic will not be controlled and there will be both serious individual and socio – economic repercussions. The AIDS epidemic will become a serious problem threatening national security and prosperity, social stability, and economic development.

4. High risk factors

(1) *Migrant population*: The current internal movement of temporary and permanent migration across China is without precedent. Estimates suggest that the total number of migrants, both temporary and permanent, may be as high as 120 million; that is, some 15% of the total labor force. This number, while extremely large, is actually increasing. A number of additional factors make the migrant labor force particularly vulnerable to HIV infection.

The mere fact that they are mobile makes it extremely difficult to reach this population, Migrants easily escape efforts at education and health promotion. AIDS knowledge within the migrant population is, therefore, very low.

Second, migrants in China are often young and live for extended periods of time away from the social pressure of their families and communities. They are frequently unmarried or are living apart from their spouses and children. Nevertheless, they are still sexually active and often have casual sex, often with FSW, without protection.

(2) *Sexually transmitted infections* (*STI*): STIs have risen sharply from 23 534 reported cases in 1985 to a level of 859 040 reported cases in 2000 (Fig. 7). Again, experts estimate that the actual number of STI cases in the country is 6 – 8 million. In most urban areas, the majority of STI patients come from within the non – resident population.

(3) *Drug use*: The most recent and severe increase in both drug trafficking and drug use in cities throughout China is closely connected with the migrant and non – resident population. There are some one million known drug users recorded by public security in Mainland China. The actual number is estimated to be between 3 – 5 million (Jianhua et al. , 2003).

STI Cases

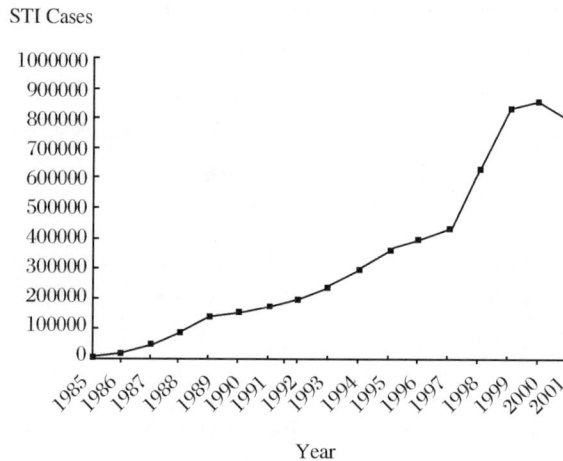

Year

Fig. 7   STI Cases in China by Year, 1985 – 2001

(4) *Commercial sex work*: Underground sex work is becoming a very serious issue. FSW working in most cities are often not permanent residents, but are migrants from distant regions. The number of FSW and clients arrested in 1985 was just 25 000, but in 1998 this number increased to 398 000 (Fig. 8). One estimate of the actual number of FSW in China is about six million (Jianhua et al. , 2003). Sexual transmission of of HIV is expected to gradually increase and will likely become the major route of HIV transmission among the general Population.

(5) *Men who have sex with men* (*MSM*): In China, homosexual activities are not illegal, but they are not acceptable in most of society. Most homosexual people are still under social pressure to hide their sexual orientation and to be married. Very few studies on homosexuality have been carded out in China. It is estimated that the total number of MSM in China is over 18 million. Because of high – risk behaviors and relatively little protection in this population, rates of HIV and STI infection among MSM are higher than in the general population. Based on one survey, the HIV infection rate in this population increased from 2. 5% in 1998 to 5. 5% in 2001 (Zeng, 2003).

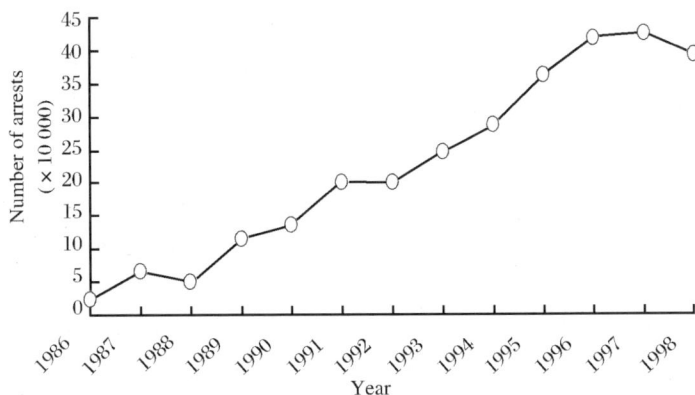

Fig. 8 Female Sex Workers (FSW) and Their Clients Arrested in China (1986 – 1998)

(6) *Hazards related to blood and blood products*: There are two main sources of donated blood in China: unpaid donations (either requested or voluntary donations), and paid donations. A small number of Chinese donate blood without being paid, but in general, people in China are reluctant to donate blood because of deep – rooted cultural beliefs. At present, there are about three million paid blood donors in China, many of them selling plasma during the 1990s. Most of these paid donors live in the poor provinces of central and southern China such as Henan, Hebei, Anhui, Shanxi and Guizhou. Some of them are migrants or are unemployed and rely on plasma sales for their primary income.

(7) *Inadequate infection control in health care settings*: In some rural and remote areas, hygiene is poor, and due to a lack of equipment and, in some cases, inadequate training, the sterilization of medical instruments is difficult. In addition, even in big cities disposable needles and syringes are often not destroyed, leaving the possibility that they will make their way into the larger community and be reused.

5. Challenges

With a population of 1. 3 billion in China, the effective control of HIV/AIDS is not only a benefit to China but also to the world. The Chinese government has issued The Chinese Mid – long Term Plan for HIV/AIDS Prevention and Control (1998 – 2010) and The Five – Year Action Plan for HIV/AIDS Prevention and Control (2001 – 2005). Many provinces and ministries have carried out strategic planning and a great deal of practical work has been done. Now a comprehensive preventive and control system has been developed. The whole society has been mobilized and many sectors from central to local levels all have become involved in the related activities. In cooperation with the Chinese government, NGOs in China (especially the Chinese Foundation for Prevention of STD and AIDS, and the Chinese Association for STD and AIDS Prevention and Treatment) have played a very important role in HIV/AIDS prevention and control. Many effective efforts have been made, such as health education, communitybased care, social mobilization, and high – risk population interventions. However, there are a lot of challenges and difficulties related to HIV/AIDS prevention and control in China. Support and assistance from various areas both within China and abroad are seriously needed.

6. Countermeasures

It is critical to take strong action against this epidemic now, otherwise, this window of opportunity to control the HIV/AIDS epidemic in China will be lost. HIV/AIDS education interventions are the most important strategy for controlling the HIV/AIDS epidemic. If countries fail to adopt education and intervention programs, about 46 million people, mostly in Sub – Saharan Africa, China, and India, would become infected (Zeng, 2003; Altman, 2002). A wider application of education and intervention programs would prevent 29 million people from becoming infected worldwide (Altman, 2002). This means that 70% of HIV infections in countries with rapidly growing epidemics (like China) would be prevented (Zeng, 2003). Thailand is a good example, where 5 – 6 million HIV infections have been prevented (Zeng, 2003). The cost of an aggressive prevention program would be USD $1000 for each infection prevented, much less than the cost of treating people once they have become infected (Zeng, 2003).

According to our experiences in the Shandong Weifang City, the cost of an HIV/AIDS education and intervention program would be RMB 2 yuans for each person in China, especially in counties and rural areas. Thus, the cost for carrying out an education and intervention program throughout China would be RMB 2.6 billion yuans for 3 – 5 years. Such a program may prevent 6 million people from becoming infected with HIV by 2010. Conquering HIV/AIDS requires a combination of prevention and treatment programs, yet aggressive efforts to prevent the spread of HIV/AIDS are far less expensive than treating those infected in an epidemic.

Recently, the Chinese government proclaimed that it would strengthen its efforts to fulfill the responsibility of containing and controlling the HIV/AIDS epidemic. The Chinese government has recognized HIV/AIDS prevention and treatment to be important tasks. The government will further define its goals in combating the disease, identify governmental responsibilities, and improve evaluation, supervision, and monitoring of activities. Accountability for the success or failure of prevention programs will be maintained and enforced.

Second, China is committed to providing free treatment and medicines to PLWHA who have economic difficulties. This health program will cover low – income PLWHA in urban areas, and all patients in rural areas. Moreover, central and local governments will invest more than 10 billion yuan to strengthen the medical assistance system for infectious diseases, and train professionals in comprehensive HIV/AIDS prevention and treatment modalities.

Thirdly, the Chinese government will improve its laws and regulations, and intensify interventions against dangerous behaviors that contribute to the spread of HIV/AIDS. Public awareness campaigns will be launched to educate the public and encourage citizens to participate in HIV/AIDS prevention and treatment efforts. Laws against illegal acts like drug trafficking, drug use, commercial sex work, and illegal blood collecting and distribution will be enforced, while drug – free communities and healthy sexual behaviors will be promoted.

The Chinese government will also work to protect the legitimate rights of PLWHA and oppose social discrimination. Throughout the 124 counties where the Chinese "care" project will be established, integrated measures including antiretroviral treatment, health care, and social programs will be adopted. The government will provide economic aid to PLWHA living in poverty, and their children will be able to attend school tuitionfree.

Finally, the government will be more active in international cooperation relating to HIV/AIDS. The HIV/AIDS situation in China has caused both concern and commitment from the international community. China welcomes the continued financial and technical support from the governments of other nations and international organizations. Meanwhile, the Chinese government is ready to be active in the global fight against HIV/AIDS and fulfill its responsibilities and obligations. The Chinese government has pledged USD $10 million to the Global Fund to Fight AIDS, Tuberculosis and Malaria. This money will support the HIV/AIDS prevention and treatment efforts of developing countries, fulfill the development goals of the UN Millennium Declaration, and contribute to the control of HIV/AIDS prevalence and the realization of global HIV/AIDS control objectives.

Sars Epidemic and Control

1. SARS epidemic in China and worldwide

Severe acute respiratory syndrome (SARS) originated in November 2002 in the Guangdong Province of China and has subsequently spread to more than 30 countries and regions, including Hong Kong, Taiwan, Vietnam, Canada, and Singapore, infecting approximately 8459 patients and resulting in more than 800 deaths. In China alone, 5327 cases and 349 SARS – related deaths have been reported.

SARS is an emerging infectious disease, and an effective treatment remains unavailable. The underlying causes that led to the SARS epidemic in China are complicated, yet two critical factors are China's large transitory population and the unpreparedness of some Chinese agencies to recognize and fight the epidemic.

Although the transmission mechanism of the agent or agents causing SARS has yet to be fully understood, it is believed that SARS is transmitted probably mainly by droplet secretions, fomites, or close person – to – person contact. The incubation period of SARS ranges from one day to 12 days, with a median of four days. The SARS epidemic in China was characterized with both hospital clustering and family clustering. The disease affected people in all age groups, with patients aged between 2 months and 92 years. Nevertheless, 71% of patients were between 20 and 49 years old. The male – to – female ratio was 1 : 1.6. There was no statistically significant difference between men and women in the mortality rate; however, the mortality rate was associated with age and co – morbidities. According to WHO, the fatality rates are as follows: <1% for below age 24, 6% for ages 25 – 44; 15% for ages 45 – 64; >50% for above age 65 (Zheng, 2003).

2. SARS epidemic control in China

The outbreak and rapid spread of SARS in China caused fear and concern in ordinary citizens and caught the immediate attention of both the Chinese government and the world. The responses taken were as follows:

① To protect people's health, the Chinese government set up a national command center to coordinate the efforts of disease treatment, control, and prevention. Thus, disease control became a national priority.

② A massive nationwide campaign to educate people to take precautions was launched and decisive measures were mandated by all levels of the government to control the epidemic throughout China.

③ Both medical experts and the media played extremely important roles in disseminating scientific knowledge to the public, and raising awareness of SARS prevention.

④ Soon after the national command center was set up, effective strategies based on the experience in fighting against SARS epidemic in Guangdong Province were formulated and adopted.

⑤ Four effective principles against the outbreak were summarized and adopted: early identification and diagnosis, early reporting, early isolation, and early treatment. To prevent further cases, patients were treated locally and controls against nosocomial infection were instituted. Thus, in a surprisingly short period of time, SARS was controlled and daily normal life was restored.

⑥ During the SARS epidemic, public health professionals played an extremely important role in education and prevention. Epidemiologists worked very closely with physicians treating patients, helped identify local sources of the epidemic, and conducted outbreak investigations. These epidemiologists visited patients' families and co – workers, and identified both suspected cases and super – spreaders. Numbers of identified cases were reported to the government daily, and the government then informed the public about the status of the outbreak. Under the leadership of the national command center, public health professionals helped government agencies strengthen the safety of public places by improving ventilation and reducing unnecessary activities. These professionals also educated people about how to minimize their risk of being infected, and how to enhance their immune system to fight SARS. Based on the information collected from hospitals and communities, public health professionals determined that both definite and suspected cases of SARS should be isolated for 12 to 14 days.

⑦ To control the epidemic, the Chinese government instituted scientific research on SARS etiology, treatment, and prevention, and encouraged international collaborations in these areas. On Mareh 12 2003, the WHO issued a global warning about the epidemic, and took unprecedented measures to effectively prevent the disease from spreading to unaffected countries and regions. The WHO dispatched a team of experts to China and set up a global disease control network in 10 countries including China. Chinese health professionals made important contributions toward identifying a novel coronavirus as the causal agent of SARS.

3. SARS and the public health system

The SARS epidemic has simultaneously evolved as both a significant challenge to humans' health, as well as a social and public health management problem. As an emergency crisis, the SARS epidemic has had at least the following 4 characteristics:

① Unpredictability: SARS infections were unpredictable and, at least initially, the mode of transmission was in question.

② Exigency: Swift attention to the infections, and quick treatment of cases was necessary to prevent further infections.

③ Social consequences: The SARS epidemic caused an imbalance in the social order, and damaged economic development, through rampant fear and anxiety. For a time, it was the most important issue facing China and overshadowed other priorities.

④ Worldwide implications: The SARS epidemic has spread from China and affected the entire world.

The SARS epidemic has created a significant challenge for the Chinese public health system. The Chinese government has responded by establishing and strengthening a prevention system to ensure an effective response to any future epidemic. Scientific studies on the origin of SARS, its pathogenesis and immunology, potential anti – SARS drugs and vaccines, and animal models are being conducted.

[In 《AIDS in Asia》 2004, 295 – 305]

References

1 Altman L A. Modest Anti – AIDS Efforts offer Huge Payoff, Studies Say, New York Times. 2002

2 China HIV/AIDS Information Network. Situation of HIV/AIDS Epidemic and its Prevention and treatment in China in 2002. Beijing, China

3 Chinese Association of Preventive Medicine. Review and Prospects for Chinese Health Antiepidemic System, 1993

4 Chinese Association of Preventive Medicine. Review and Prospects for Chinese Health Antiepidemic System, 2003

5 Jianhua Y, Qi X, Tao J, Houbao X. Simulation of Impact of Policy and Law Environment on HV/AIDS Spreading in China. United Nations Development Program, 2003

6 Ministry of Public Health. A Brief Introduction on China's Medical and Health Services. Beijing, China, 1990

7 Ministry of Public Health. National Situation of Morbidity and Mortality of Infectious Diseases in China. Beijing, China, 2002

8 Zeng Yi. Health Education and Intervention Are Major Strategies for AIDS Control Chinese Journal of Health Education, 2003, 11 (19): 846 – 848

9 Zheng L. SARS and Response to Emergency Public Health Events. Sciences Publishing House, Beijing, 2003

93. The Linkage Disequilibrium Maps of Three Human Chromosomes Across four Populations Reflect Their Demographic History and a Common Underlying Recombination Pattern

VEGA Francisco M. De La[1,12], ISAAC Hadar[1], COLLINS Andrew[2], SCAFE Charles R[1], HALLD6RSSON[1,9], SU Xiaoping[1,10], LIPPERT Ross A[1,11], WANG Yu[1], WEBSTER Marion Laig[1], KOEHLER Ryan T[1], ZIEGLE Janet S[1], WOGAN Lewis T[1], STEVENS Junko F[1], LEINEN Kyle M[1], OLSON Sheri J[1], GUEGLER Karl J[1], YOU Xiao-qing[1], XU Lily H[1], HEMKEN Heinz G[1], KALUSH Francis[3], ITAKURA Mitsuo[4], ZHENG Yi[5], Guy de The[6], O'BRIEN Stephen J[7], CLARK Andrew G[8], LSTRAIL Sorin[1], HUNKAPLLER Michael W[1], SPIER Eugene G[1], GILBERT Dennis A[1]

1. Applied Biosystems, Foster City, California 94404, USA; 2. Human Genetics Division, University of Southampton, Southampton, SO16 6YD, United Kingdom; 3. Celera Genomics, Rockville, Maryland 20850, USA; 4. Institute for Genome Research, The University of Tokushima, Tokushima 770 – 8503, Japan; 5. Institute of Virology, Chinese Academy of Preventive Medicine, Beijing 100052, China; 6. Department of Viral Oncology – Epidemiology, Institute Pasteur, Centre National de la Recherche Scientifique, 75015 Paris, France; 7. Laboratory of Genomic Diversity, National Cancer Institute, Frederick, Maryland 21702, USA; 8. Molecular Biology and Genetics, Cornell University, Ithaca, New York 14853, USA

Summary

The extent and patterns of linkage disequilibrium (LD) determine the feasibility of association studies to map genes that underlie complex traits. Here we present a comparison of the patterns of LD across four major human populations (African – American, Caucasian, Chinese, and Japanese) with a high – resolution single – nucleotide polymorphism (SNP) map covering almost the entire length of chromosomes 6, 21 and 22. We constructed metric LD maps formulated such that the units measure the extent of useful LD for association mapping. LD reaches almost twice as far in chromosome 6 as in chromosomes 21 or 22, in agreement with their differences in recombination rates. By all measures used, out – of – Africa populations showed over a third more LD than African – Americans, highlighting the role of the population's demography in shaping the patterns of LD. Despite those differences, the long – range contour of the LD maps is remarkably similar across the four populations, presumably reflecting common localization of recombination hot spots. Our results have practical implications for the rational design and selection of SNPs for disease association studies.

Introduction

Recently, there has been tremendous interest in empirically establishing the patterns of allelic association, also known as linkage disequilibrium (LD), among polymorphic variants of the human genome. When two alleles at adjacent loci co – occur in a chromosomal segment more often than expected if they were segregating independently in the population, the loci are in linkage disequilibrium (Weir 1996). The profile of LD depends on the age of the mutations, genetic drift, and the demographic history of a given population. It is also eroded by recombination (Jeffreys et al. 2001) and gene conversion (Ardlie et al. 2001). The extent of LD across genomic regions is a crucial parameter for defining the statistical power of association studies utilizing singlenucleotide polymorphisms (SNP) as surrogate genetic markers (Schork 2002), and for guiding the selection and spacing of such polymorphisms to create marker maps useful in candidate gene, candidate region, and eventually whole – genome association studies (De La Vega et al. 2002).

The surveys of LD performed to date with SNPs have been generally limited to small samples of the genome (Gabriel et al. 2002) or to single populations (Tsunoda et al. 2004) or chromosomes (Patil et al. 2001; Dawson et al. 2002; Phillips et al. 2003). Previous studies in a small number of loci have shown marked differences in the extent of LD observed between African and non – African populations (Kidd et al. 1998; Service et al. 2001). Service et al. (2001) performed a genome – wide survey of background LD with microsatellites in a population isolate showing LD extending on a cm range and suggesting that a population's demographic history contributes significantly to the observed patterns of LD. A large project to genotype hundreds of thousands of SNPs across the genome in four major populations is underway (The International HapMap Consortium 2003); at the time of this writing, full analysis is pending completion of the genotyping. Therefore, it remains to be examined in detail how LD differs among population groups and between distinct chromosomes.

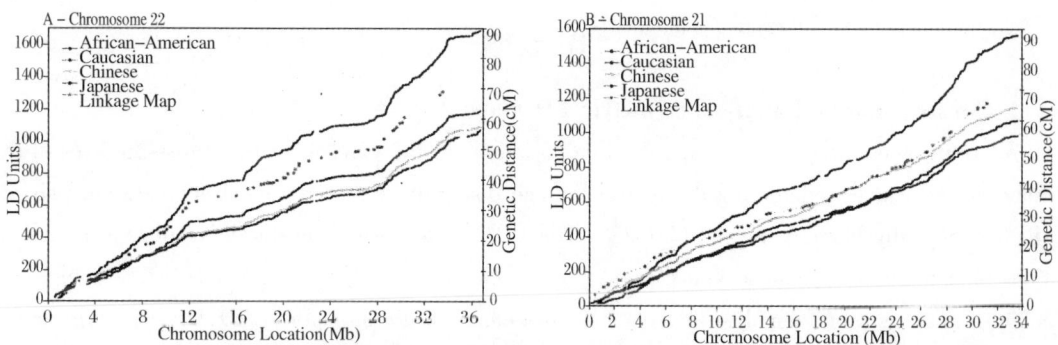

Fig. 1 (**Continued on next page**)

With the aim of developing a SNP map for candidate – gene and candidateregion association studies useful across multiple populations, we identified SNPs with a median spacing of less than 8. 4 kb covering almost the entire length (>211 Mb) of three human autosomes: chromosomes 6,

21 and 22 （see Supplemental Tab. 1）. We developed a set of 5' nuclease assays that are available （De La Vega et al. 2002） to genotype 24 940 SNPs selected from the Celera Human RefSNP database （Kerlavage et al. 2002） （v 3. 6） in 180 DNA samples from African – American, Caucasian （EuropeanAmerican）, Chinese, and Japanese unrelated individuals.

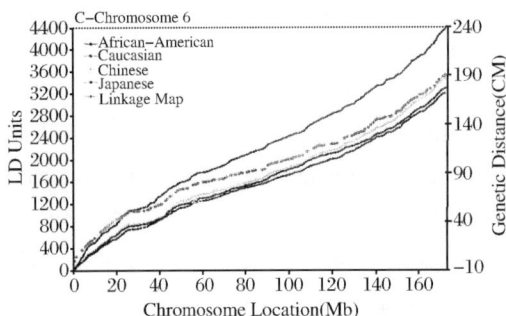

（A）Chromosome 22；（B）chromosome 21；（C）chromosome 6. SNP locations in LDUs （left vertical axis）and physical coordinates in Mb （horizontal axis）for African – American （blue）；Caucasians （red）；Chinese （turquoise）；Japanese （purple）. The location of the markers' part of the high – resolution linkage map of Kong et al. （2002）in the physical and the genetic maps is shown in green （cm scale, right vertical axis）

Fig. 1 Population – specific metric LD maps of the three chromosomes

Tab. 1 Statistics of blocks and steps in the LDU maps

Chromosome	Population	Mean block[a] sizes （kb）	Number of blocks	Mean step sizes （LDU）
6	African – American	23. 0	2110	2. 1
	Caucasian	28. 6	1965	1. 6
	Chinese	33. 8	2076	1. 7
	Japanese	35. 6	2045	1. 6
21	African – American	14. 1	747	2. 1
	Caucasian	17. 3	751	1. 4
	Chinese	20. 3	664	1. 7
	Japanese	21. 3	651	1. 5
22	African – American	5. 6	875	1. 9
	Caucasian	19. 6	827	1. 4
	Chinese	24. 4	516	2. 1
	Japanese	26. 9	488	2. 2

Note：[a]Block defined as continuous region with $\triangle LDU = 0$

Results

1. Construction and analysis of metric LD maps

A useful methodology to describe the fine variations in the patterns of LD across the axes of the chromosomes involves calculating the metric linkage disequilibrium units （LDUs）between pairs of SNPs described by Maniatis et al. （2002）. The LDU scale, which has additive distances and locations monotonic with physical and genetic maps （Zhang et al. 2002）, provides a coordinate system whose scale is proportional to the regional differences in the strength of LD, in a fashion analogous to the recombination maps constructed in cm used to guide linkage studies （Kong et al. 2002）. Fig. 1 shows the metric LD map of the three chromosomes plotted against physical location of the SNPs. A pattern of "plateaus," which correspond to regions of high LD, and "steps," which correspond to regions of increased recombination （Zhang et al. 2002）, is evident. Also plotted in the figure are the genetic locations, on the cm scale, and the physical locations of the markers used in construction of the high – resolution linkage map described by Kong et al. （2002）, revealing that the steps and plateaus of the LD map are mostly in concordance with the hot and cold spans of re-combination. On a large scale, multi – megabase segments with fewer than 100 LDUs can be ob-served even close to the telomeres in the small acrocentric chromosomes, but not in the large meso-centric chromosome 6, suggesting that elevated recombination rates may not be ubiquitous near all telomeres (Baird et al. 2000). As reported by Stenzel et al. (2004) in a study with larger sample size of European descent, a particularly large " cold span" is observed near the MHC region in chromosome 6p (31. 2 – 34. 6 Mb; Fig. 1C). It is also noteworthy that a cold span is found across the centromere of chromosome 6 (around 60 Mb), which is consistent with the notion that recombi-nation is suppressed in this region (Sun et al. 2004).

A remarkable property of the LDU maps for the four populations studied is that their overall contour is rather similar—most of the differences are found in the magnitude of the steps in regions of low LD/high recombination. The close correspondence between long – range LD patterns in these populations, as is evident from the many shared plateau and step regions, presumably reflects the common distribution of underlying recombination hot spots across populations (Kauppi et al. 2003). In finer scale, chromosomal segments with extensive LD and low haplotype diversity (i. e. , haplo-type "blocks"; Gabriel et al. 2002) can be identified as plateaus in the LD map where the increase in LDUs is very small or zero (Tapper et al. 2003). The latter block definition, although arbitrary, is more robust than those based on heuristics that typically yield ambiguous boundaries (Gabriel et al. 2002; compare with Schwartz et al. 2003). Tab. 1 describes the mean sizes and number of LD blocks (ΔLDU $= 0$) and the mean LDU length of the inter – block regions. The block sizes in the AfricanAmerican sample are consistently smaller. This suggests that because of the increased length of the population history, reflecting African origin, more recombination events have accumulated and blocks have been consequently split. Supporting this suggestion, we observe that the African – American population generally has more blocks than the other three populations. However, there are substantial differences across the three chromosomes and four populations with, for example, only 488 blocks in the Japanese population for chromosome 22, where there are 875 in African Ameri-cans. The differences between the same populations for chromosome 6 are, in contrast, relatively minor. The extent to which sample size differences and marker density influence the fine – scale structure of the maps is not entirely clear, but the trends observed are consistent with those expected given differences in length of population history and, consequently, historical recombination inten-sity. Increased length of population history and recombination intensity is likely to account for the o-verall increased step height in African Americans, although this is not true for chromosome 22 where African-American, Chinese, and Japanese samples show similar mean step heights. Cumulatively, blocks of relatively high LD (LDU $< 0. 3$) account for up to 54% of the chromosome span in the out – of – Africa populations, and up to 44% in African – Americans, with 15% – 18% of the chromosome segments being in recombination cold spans in all populations.

The extrapolation of the LDU length of the genome is interesting because LD maps are formula-ted such that one LDU corresponds to the "swept radius," which is the physical distance within which LD is likely to be useful for positional cloning (Morton et al. 2001). Therefore, the LDU length offers a lower limit of the number of SNPs required to cover the genome when spaced evenly on the LD map. Tab. 2 gives the LDU map lengths for the three chromosomes and four populations. The total LDU lengths of a chromosome in the out – of – Africa populations are always lower than in the African – American population and consistent with other metrics of LD. In all populations the total number of LDUs is proportionally lower in chromosome six than in chromosomes 21 and 22, where the LDUs per Mb increase over 80% with respect to the large chromosome. The LDU length of the genome has been extrapolated from each chromosome, and for all three together from the LDU/ cm ratio, using figures from Kong et al. (2002). The genome – wide estimate of LDUs calculated for each population is remarkably consistent across the chromosomes, with a mean of 88 957 for African – Americans and between 59 342 to 64 049 for the other three populations. These results suggest that the African – American whole – genome LD map is 30% longer, again reflecting the longer population history of African ancestry. Furthermore, these analyses suggest that a whole – genome scan would need to type SNPs in about 60 000 LD units in Caucasians and 85 000 in African – Americans. The total number of SNPs required will be some multiple of these figures because it will be important to perform scans using SNPs over a series of allele frequencies in order to capture ranges of haplotype diversity.

Tab. 2　Metric LD map lengths

Chromosome	Chromosomal map lengths (LDUs)				Mb	cM	Extrapolated genome – wide map lengths(LDUs)[a]			
	African – American	Caucasian	Chinese	Japanese			African – American	Caucasian	Chinese	Japanese
6	4362	3190	3464	3283	171. 7	189. 6	83 168	60 822	66 046	62 595
21	1560	1066	1141	979	33. 2	61. 9	91 105	62 255	66 636	57 174
22	1688	1183	1084	1062	35. 0	65. 9	92 597	64 894	59 464	58 257
						Mean:	88 957	62 657	64 049	59 342

Note: [a] Assuming the genome has 3615 cm (Kong et al. 2002)

2. Correlation of LD with sequence features

We investigated the correlation between LDUs and sequence descriptors looking for predictors of LD distance: GC content, density of SNPs discovered by random shotgun sequencing (Venter et al. 2001), repetitive elements (LINEs and SINEs), CpG islands, and the recombination rate estimated by the map of Kong et al. (2002). We calculated Pearson correlations across a variety of bin sizes ranging from 50 kb to 5 Mb. The results of this analysis show that at a large scale (bin sizes of 1 – 5 Mb) the recombination rate strongly correlates with LDUs (Pearson correlation $r \geqslant 0.8$ at 2 Mb bins, $P < 0.001$, see Tab. 3), whereas at a short scale the correlation decreases (see Supplemental Figure 2). The latter fact could be interpreted to mean that at short scale, gene/loci demographic history and drift dominate (Reich et al. 2002), however this could simply be due to the resolution limit of the linkage map of Kong et al. (2002) – about lcM. Shotgun SNP density, which here is used as a proxy for nucleotide diversity, is the second strongest correlating metric with

Pearson's coefficients ranging from 0. 55 – 0. 78. In addition, we observe a correlation of LDU with GC content (r =0. 16 – 0. 48). The previous two observations appear to be secondary to correlations between recombination and diversity (Lercher and Hurst 2002) and recombination and GC content (Montoya – Burgos et al. 2003), supporting the notion that recombination can be mutagenic (Strathern et al. 1995). SINEs, LINEs, and the density of CpG islands do not appear to be consistently correlated with LD. Since the presence of outliers can affect the Pearson correlation values, we also performed Spearman rank correlations (values in parenthesis on Tab. 3) without obtaining significant differences with the Pearson correlation results. Recombination perse explains 66% – 74% of the variance of LDUs, whereas including all six descriptors only increases the explained variance up to 87% (Tab. 4), a modest increase. Therefore, the smaller correlation values obtained for the descriptors other than recombination might be the result of finer – scale variation in the recombination rate than the resolution of the map of Kong et al. (2002) can measure. Most of the correlations obtained with LDUs can be replicated utilizing sliding windows of averaged | D' | or r^2 (Tab. 3), but they are always smaller (and, as expected, of inverse sign). The latter suggests that LDUs are more effective reporters of local variance of LD and historical recombination than simple averages of | D' | and r^2, and potentially a more suitable metric with which to position markers and develop standard SNP maps. Furthermore, because of its high correlation with recombination rate, the LD map could be potentially used to increase the resolution of the genetic map, predicting with good confidence genetic distances at intervals of less than 1 cm.

Tab. 3　Correlations of different metrics of LD with sequence descriptors of chromosomal segments

Chromosome	Recombination (cM/Mb)	Polymorphism density	GC content	LINES density	SINES density	CpG islands
Correlations[a] with LDU						
Chr. 6	0. 82(0. 81)	0. 48(0. 42)	0. 33(0. 4)	−0. 37(−0. 43)	0. 32(0. 41)	0. 23(0. 36)
Chr. 21	0. 88(0. 77)	0. 77(0. 69)	0. 50(0. 31)	−0. 26(−0. 16)	0. 001(0. 07)	0. 43((0. 31)
Chr. 22	0. 92(0. 9)	0. 69(0. 66)	0(0. 06)	0. 22(0. 12)	−0. 45(−0. 45)	−0. 38(−0. 38)
Correlations[a] with |D'|						
Chr. 6	−0. 68(−0. 73)	−0. 35(−0. 34)	−0. 24(−0. 33)	0. 33(0. 42)	−0. 30(−0. 37)	−0. 09(−0. 26)
Chr. 21	−0. 79(−0. 7)	−0. 57(−0. 49)	−0. 36(−0. 22)	0. 04(−0. 1)	0. 03(0. 001)	−0. 36(−0. 21)
Chr. 22	−0. 75(−0. 78)	−0. 63(0. 63)	0. 13(0. 04)	−0. 19(−0. 18)	0. 58(0. 43)	0. 34(0. 47)
Correlations[a] with r^2						
Chr. 6	−0. 62(−0. 72)	−0. 37(0. 43)	−0. 30(−0. 42)	0. 39(0. 45)	−0. 37(−0. 48)	−0. 14(−0. 28)
Chr. 21	−0. 81(−0. 79)	−0. 45(0. 38)	−0. 36(−0. 32)	0. 15(0. 1)	−0. 10(−0. 12)	−0. 29(−0. 27)
Chr. 22	−0. 79(−0. 79)	−0. 69(0. 67)	0. 1(0. 01)	−0. 22(−0. 2)	0. 6(0. 48)	0. 37(0. 44)

Notes: The correlations are reported for 2 – MB bins generated as described in Methods for the Caucasian samples. Similar results were obtained for the other populations studied. All correlations above 0. 2 are significant with p < 0. 001 derived by bootstrapping. Density of SNPs, LINES, SINES, and GpC islands is calculated as number of features per window.

[a]Pearson correlation (Spearman rank correlation).

3. Breakdown of LD with physical distance

Tab. 5 summarizes the half – length of the decay of LD: the distance over which LD decays to half of its maximum value to the asymptotic value (see Supplemental Fig. 1A, B, for a plot of the decay of | D' | and r^2 depending on the physical distance between markers). Chromosome 6 exhibits from 26% to 58% slower decay of LD with distance as compared with the smaller chromosomes. The metric | D' | commonly shows greater differentials than r^2. Another metric of the decay of LD is the chromosome – wide swept radius (Morton et al. 2001), which is equivalent to the average kb distance of one LDU. As shown in Tab. 5, this metric yields greater differences between the chromosomes in terms of decay of LD: The estimated swept radii are over 53% longer for chromosome 6 than for chromosomes 21 and 22. Thus, the overall extent of LD for the three chromosomes follows the comparative ranking: 6 > 21 ≈ 22, which is consistent with the known genetic lengths of these chromosomes. Since the number of recombination events per meiosis is rather similar across chromosomes, the large differences in their lengths are expected to result in lower overall recombination rates (Kong et al. 2002) and thus slower breakdown of LD in the larger chromosomes. Here we show that potentially useful LD extends up to 50 – 56 kb in chromosome 6 for the out – of – Africa populations (Caucasian, Chinese, and Japanese), whereas the African – Americans exhibit at least 25% smaller swept radius. In the smaller chromosomes this reduction is even larger, at least over 40%, where useful LD extends over 30 kb out – of – Africa, but only to 22 kb in African – Americans. When the decay of common pairwise metrics of LD is plotted versus the genetic distance of the markers expressed in LDUs, the curves show no significant differences between chromosomes or populations (see Supplemental Fig. 1C, D). The later is expected as the LD map already normalizes the LD pattern between populations and loci. The half – length of decay in this coordinate system provides a rough equivalence between the LDUs and the pairwise metrics (| D' | half – length ≈ 0.7 LDUs; r^2 half – length ≈ 0.3 LDUs) and supports the suggestion (Morton et al. 2001) that beyond one LDU, the intensity of LD would be too weak to provide sufficient power for association mapping under most genetic models.

Tab. 4 Fraction of the LDU variance explained by sequence descriptors

Chromosome	Recombination (cM/Mb)	All (six) descriptors	All (five) descriptors except recombination	Polymorphism density[a]	GC content	LINES density[a]	SINES density[a]	CpG islands[a]
6	0.673	0.725	0.431	0.232	0.110	0.137	0.102	0.054
21	0.766	0.890	0.744	0.590	0.249	0.068	0.000	0.186
22	0.840	0.917	0.548	0.472	0.000	0.047	0.206	0.141

Notes: The correlations are reported for 2 – MB bins generated as described in Methods for the Caucasian samples. Similar results were obtained for the other populations studied.

[a]Density calculated as number of features per window.

Discussion

Until recently, patterns of LD have been studied in small random samples of the genome (Daly et al. 2001; Gabriel et al. 2002), or in a few single chromosome studies (Dawson et al. 2001; Patti et al. 2001; Phillips et al. 2003), and mostly only for a single population (Tsunoda et al. 2004) or cosmopolitan sample aggregates (Patti et al. 2001). These investigations, although they provide an important contribution to our understanding of the underlying structure of LD in the human genome, have not allowed for a large – scale comparison of the differences in the strength and distribution of LD between major human populations. Here we have presented more – extensive studies of the patterns of LD for two previously studied chromosomes and, to our knowledge, the first high – resolution LD study of a large mesocentric chromosome (chromosome 6).

The partly cumulative effect of numerous population bottlenecks has had an important impact on the patterns of LD that we see today. Migration out of Africa approximately 100 000 years ago (Stringer and Andrews 1988) had, presumably, the greatest influence, but other bottlenecks of smaller magnitude have taken place over many generations. African – Americans, a population derived mainly from West Africans, show reduced LD decay distances and longer LDU maps. The samples of European ancestry, as well as the Asian populations show significantly shorter LDU maps and more extensive LD, presumably the result of acute population bottlenecks. Zhang et al. (2004a) defined the effective bottleneck time' (EBT) by analogy with effective population size. The LDU/ Morgan ratio estimates the effective number of generations over which LD has been declining consistent with the extent of LD we observe today. From the LD maps for the three chromosomes given here (Tab. 2), and assuming 25 years per generation, the EBT for the Caucasian population is 43 325 years (range for three chromosomes 42 050 – 44 879), the Chinese population has 44 300 years (range 41 123 – 46 075) and the Japanese population 41 039 years (range 39 550 – 43 300). It is noteworthy that these times are less than half of the time to the presumed out – of – Africa event (Stringer and Andrews 1988), consistent with the compound effect of subsequent bottlenecks in creating LD. To the extent that these three populations share a common history, the EBT ranges are all overlapping. By contrast the EBT for the African – American sample is 61 525 years (range 57 041 – 64 036), which reflects the very different demographic history. The African – American population is somewhat admixed (Collins – Schramm et al. 2002) and thus a native African population might show even longer LD maps and EBT. However, because of the ascertainment bias to SNPs of high heterozygosity across multiple populations in our study, it is unlikely that admixture effects are strongly influencing our results. It will be of interest to examine other populations with partial or complete African ancestry to determine the impact of recent admixture on the EBT.

Tapper et al. (2003) have constructed LD maps of chromosome 22 for the data of Dawson et al. (2002). LDU map lengths in UK – unrelated and CEPH samples are very similar, being in the range 818 LDUs to 841 LDUs over 62. 8 cm, implying a genome of approximately 48 000 LDUs in these populations. This is somewhat shorter than that seen in our current analysis; the discrepancy may be due to the lower resolution of the Dawson et al. (2002) map, where the SNP density may be insufficient to precisely estimate the length of the map, in particular in areas of high recombination. In contrast to haplotype block boundaries, additivity of LDU maps has been shown recently for a range of SNP densities (2 – 10 kb mean interval size on chromosome 20) (Ke et al. 2004).

However, additivity must be lost locally below critical SNP densities that depend to a large extent on recombination intensity. We were able to reproduce the two large recombination cold spans observed in the data of Dawson et al. (2001) (Tapper et al. 2003) (Fig. 1). Moreover, we were able to identify a third region with extensive LD close to the q – telomere of chromosome 22 in both populations, as well as other smaller recombination cold spans. Dawson et al. (2001) previously reported a correlation between the strength of LD and recombination. We observe a much stronger positive correlation between LDU and recombination rate, but we are unable to find a consistent correlation between LD and repetitive elements similar to those reported by these authors.

Tab. 5　Chromosome – wide SNP and LD summary statistics

Population	Statistic	Chr 6	Chr 21	Chr 22
African – American	No. of SNPs	9129	3418	3688
	SNP spacing on covered segments (kb)[a]			
	Mean	10. 1	7. 9	7. 3
	Median	6. 9	4. 5	4
	Minor allele frequency			
	Mean	0. 29	0. 3	0. 29
	Median	0. 28	0. 3	0. 29
	Decay of LD (kb)			
	∣ D' ∣　half length	28. 9	19	18. 2
	r^2 half length	16. 4	11. 8	9. 9
	Swept radius	41	22. 1	21. 8
Caucasian	No. of SNPs	9274	3551	3931
	SNP spacing on covered segments (kb)[a]			
	Mean	9. 9	7. 6	7
	Median	6. 7	4. 1	3. 7
	Minor allele frequency			
	Mean	0. 31	0. 31	0. 31
	Median	0. 31	0. 32	0. 32
	Decay of LD (kb)			
	∣ D' ∣　half length	46	29. 5	29. 3
	r^2 half length	26. 7	19. 2	19. 3
	Swept radius	56. 3	31. 2	30. 9
Chinese	No. of SNPs	10 916	3567	3496
	SNP spacing on covered segments (kb)[a]			
	Mean	11. 7	8. 2	7. 7
	Median	8. 3	5. 8	4. 7
	Minor allele frequency			
	Mean	0. 3	0. 3	0. 3
	Median	0. 3	0. 3	0. 3
	Decay of LD (kb)			
	∣ D' ∣　half length	41. 7	31. 2	31. 8
	r^2 half length	27. 5	20. 4	21. 8
	Swept radius	50. 2	31. 3	31. 6
Japanese	No. of SNPs	10 825	3536	3483
	SNP spacing on covered segments (kb)[a]			
	Mean	11. 8	8. 3	7. 8

Population	Statistic	Chr 6	Chr 21	Chr 22
	Median	8. 4	5. 9	4. 6
	Minor allele frequency			
	Mean	0. 3	0. 3	0. 31
	Median	0. 31	0. 31	0. 3
	Decay of LD (kb)			
	| D' | half length	44	32. 8	34
	r^2 half length	28. 9	21. 1	21. 1
	Swept radius	53. 1	34. 6	33

Note: [a] Covered segments are defined as intervals where inter – SNP distances are ≤50 Kb.

The gross – scale variation in local recombination rate is probably responsible for much of the consistency of the LDU maps across populations. Differences are mostly in the overall map length, as suggested by Maniatis et al. (2002), and suggest the possibility of developing a "standard" LD map that is efficient for association mapping in all populations if suitably scaled (Zhang et al. 2002). These scaling factors could be obtained from studies in more populations and would reflect different population history lengths. Nevertheless, Kauppi et al. (2003) have shown that in spite of a common distribution of recombination hot spots in the MHC class II region, haplotype composition in the cold spans is considerably divergent between populations. This is corroborated by the results of others (Crawford et al. 2004; Liu et al. 2004) and by an analysis of haplotype frequencies in LD blocks of the data sets of the present study (F. De La Vega, H. Isaac, C. Scafe, and E. Spier, unpubl.) showing a significant proportion of unshared haplotypes between African-Americans and out – of Africa populations, in spite of the similar shape of the LD maps. Therefore, our results do not necessarily imply that the similarity in the LD maps translates to an identical choice of optimal markers for association studies in different populations. Subsets of markers that attempt to capture the haplotype diversity of the genomic regions may need to be somewhat different between populations, even if the LD map can be extremely useful to guide their selection, for example, by determining sensible neighborhoods to select haplotype – tagging SNPs (Halld6rsson et al. 2004). However, the relevance of selecting markers that preserve haplotype diversity in terms of the power of an association study remains controversial (Zhai et al. 2004; Zhang et al. 2004b).

The role of LD maps in disease mapping is twofold. Firstly, LD maps indicate regions of LD breakdown within which higher SNP densities may be required for identification of some causal polymorphisms. Designing studies assuming a constant or average level of LD across the genome is clearly flawed (Schork 2002). Instead, marker selection, statistical power, and sample size estimations could now be based on the empirically determined LD map of the population of interest. Zhang et al. (2004b) have argued for a multi – stage design where an initial screen at relatively low SNP density is followed by a higher SNP density scan to fully elucidate relationship (s) with the disease phenotype. For the initial screen and for the addition of further SNPs, uniform spacing on the LD scale is optimal. Many studies have associated autoimmune disease phenotypes to the MHC region of chromosome 6p (Horton et al. 2004). The availability of a detailed LD map of this region, showing that

LD extent is not constant across the entire MHC span (Stenzel et al. 2004), should allow for better design and interpretation of association studies in this region. For example, the PSORS1 locus on 6p21 has been implicated with psoriasis by linkage (Leder et al. 1998). Our map shows extensive LD across the locus implying that the power to detect association should be high, which is consistent with the studies performed to date (Nair et al. 2000; Veal et al. 2002). Our results also suggest that, everything else being equal, studies in larger chromosomes would have more power, and studies in African – Americans will require about 30% more markers and/or larger sample size. Previous surveys of LD on population isolates have shown that "background" LD extends over large genetic distances (Service et al. 2001). Thus, younger population isolates should exhibit significantly shorter LD maps, making these populations an ideal target for testing the feasibility of whole – genome association studies. The second major role of LD maps is in multi – locus analysis where the LD map has the equivalent function for association mapping as the linkage map for multi-point analysis of major genes. Maniatis et al. (2005) demonstrated huge increases in power and a much reduced confidence interval when localizing a causal polymorphism on an underlying LDU map rather than a physical (kb) map. This dual role for the LDU map is, of course, applicable for association mapping studies in general and not just for the chromosomes we describe here.

In summary, our results illustrate the interplay between recombination and demography as the major forces shaping the patterns of LD in the human genome. While demography strongly impacts the overall extent of LD manifested in the LD map lengths, an underlying pattern of recombination appears to dominate at the chromosomal scale, defining the major features of the LD maps.

Methods

1. SNP ascertainment

All SNP data was obtained from the Celera Human RefSNP database (version 3.6), part of the Celera Discovery System (Kerlavage et al. 2002). This version of the database included about 2.4 million Celera SNPs, as well as 2.2 million publicly available SNPs uniquely mapped to the Celera Human Genome assembly, release 27. We developed a "triage" process for selecting SNPs that requires evidence of two independent discoveries of the minor allele (De La Vega et al. 2002). See Supplemental Methods for details.

Tab. 5 shows the statistics of the number of SNPs typed per chromosome and population, their minor allele frequency, and the SNPs density across "covered segments" —contiguous segments where inter – SNPs distances are $\leqslant 50$ kb (see Supplemental Tab. 1). The chromosomal regions not covered in our study include heterochromatin, highly repetitive regions, and duplicated regions, where it is difficult to develop genotyping assays. The mean (7 – 1.8 kb) and median (3.7 – 8.4 kb) SNP spacing indicate that we achieved a high – resolution coverage of the chromosomes. All SNPs genotyped successfully in our study are listed in Supplemental Tab. 2.

2. DNA samples

The African – American and Caucasian DNA samples, 45 each, were obtained from the Coriell Institute/National Institute of General Medical Sciences Human Variation Panels (http://

locus. umdnj. edu/ccr/). Supplemental Tab. 3 lists the Coriell IDs for these samples. The Chinese samples were obtained from 45 Hart Chinese patients enrolled in a cohort study of individuals at risk for nasopharyngeal carcinoma in Guangxi Province, China. An additional set of 45 Japanese DNA samples were obtained from unrelated healthy Japanese volunteers at the University of Tokushima, Japan. All the samples were collected with proper informed consent and IRB approval (see Supplemental Methods).

3. Assay development and genotyping

We used a high – throughput assay design pipeline fo develop 5' nuclease allelic discrimination assays used in the TaqMan Custom SNP Genotyping Assay design service (Applied Biosystems, Foster City, CA). After the design of primers and TaqMan probes, a computational quality – control step was performed to ensure uniqueness of the predicted amplicons in the genome assembly. This allowed us to eliminate potentially problematic SNP targets that may arise from repeated genomic regions, pseudo – SNPs, and other possible assembly artifacts (Heil et al. 2002).

Genotyping was performed with commercially available TaqMan Validated SNP Genotyping Assays from Applied Biosterns (http://myscience. appliedbiosystems. com), following the standard protocol suggested by the manufacturer (see the Supplemental Methods for more detail). The average genotyping error rate in our production lab was estimated at about 0.1% by routinely running duplicate control plates. The genotypes for the samples typed in the study are available for download within the SNPbrowser Software provided freely by Applied Biosystems (http:// www. allsnps. com/snpbrowser/), from which they can be easily exported. Genotypes and SNP records are also available from dbSNP (Sherry et al. 2001) under the submitter handle " ABI," and from the PharmGKB database (Hewett et al. 2002) under project code " LDCHRS. "

4. Construction of metric LD maps

The LDMAP program (http://cedar. genetics. soton. ac. uk/pub/PROGRAMS/LDMAP; Maniatis et al. 2002) was used to construct LD maps from phase – unknown diplotypes describing the variation in the extent of LD between adjacent SNPs expressed in LDUs. LDMAP estimates a Malécot ε parameter in each map interval. The length of the Ith interval is ε_i, d_i LDUs, where ε_i is the Malécot parameter, and d_i is the length of the interval on the physical map in kb. A chromosome has $\sum \varepsilon_i d_i$ LDUs. The model is formulated such that one LDU equals one swept radius so that equal spacing of SNPs on the LDU scale is required for coverage of a region.

5. Correlation with chromosomal features

The average values of LD were computed for equally sized basepairbins, uniformly spaced at 100 Kb intervals along each of the three chromosomes. This computation was repeated for the different measures of LD. The average values of each of six commonly known chromosomal descriptors were also computed for the same base – pair bins (recombination rate; GC content; SNP density; density of LINEs, SINEs, and CpG islands) from the Celera Human Genome assembly, release 27. The Pearson and Spearman's rank correlations between the LD measures and each descriptor were calculated (Mendenhall et al. 1986) and its significance was assessed calculating p – values obtained by bootstrapping. A linear minimum squared error model was also computed to predict each of the LD measures using each of the descriptors individually, or using all of the descriptors combined, or using all of the descriptors excluding the recombination rate. See the Supplemental Methods section for more details.

Acknowledgements

We are indebted to Leila Smith; Helen Belcastro; Annie Titus; Joanna Curlee; the production genotyping, LIMS, and IT teams of Services Development and Delivery; and the Global Oligo Operations teams of Applied Biosystems for their support in the generation of the data used in this paper. Thanks are also due to Mark Adams, Sam Broder, David Dailey, Penny Dong, Nelson Freimer, Derek Gordon, Kenneth Kidd, Kit Lau, Adam Lowe, Newton Morton, Michael Rhodes, Stefan Schreiber, Sue Service, John Sninsky, and Trevor Woodage for many helpful discussions and/or comments on the text and to Mignon Fogarty for assistance with the manuscript. The development of LDMAP was supported by the Medical Research Council, UK.

〔In 《Genome Research》 2005, 15: 454 – 462〕

References

1 Ardlie K, Liu – Cordero S N, Eberle M A, Daly M, Barrett J, Winchester E, Lander E S, and Kruglyak L. Lower – than – expected linkage disequilibrium between tightly linked markers in humans suggests a role for gene conversion. Am J Hum Genet, 2001, 69: 582 – 589

2 Baird D M, Coleman J, Rosser Z H, Royle N J. High levels of sequence polymorphism and linkage disequilibrium at the telomere of 12q: Implications for telomere biology and human evolution. Am J Hum Genet, 2000, 66: 235 – 250

3 Collins – Schramm H E, Phillips C M, Operario D J, Lee J S, Weber J L, Hanson R L, Knowler W C, Cooper R, Li H, Seldin M F. Ethnic – difference markers for use in mapping by admixture linkage disequilibrium. Ant. Hum. Genet, 2002, 70: 737 – 750

4 Crawford D C, Carlson C S, Rieder M J, Carrington D P, Yi Q, Smith J D, Eberle M A, Kruglyak L, Nickerson D A. Haplotype diversity across 100 candidate genes for inflammation, lipid metabolism, and blood pressure regulation in two populations, Am J Hum Genet, 2004, 74: 610 – 622

5 Daly M J, Rioux J D, Schaffner S F, Hudson T J, Lander E S. High – resolution haplotype structure in the human genome Nat Genet, 2001, 29: 229 – 232

6 Dawson E, Chen Y, Hunt S, Smink L J, Hunt A, Rice K, Livingston S, Bumpstead S, Bruskiewich R, Sham P, et al. A SNP resource for human chromosome 22: Extracting dense clusters of SNPs from the genomic sequence. Genome Res, 2001, 11: 170 – 178

7 Dawsoo E, Ahecasis G R, Bumpstead S, Chen Y, Hunt S, Beare D M, Pabial J, Dibling T, Tinsley E, Kirby S, et al. A first – generation linkage disequilibrium map of human chromosome 22. Nature, 2002, 418: 544 – 548

8 De La Vega F M, Dailey D, Ziegle J, Williams J, Madden D, Gilbert D A. New generation pharmacogenomic tools: A SNP linkage disequilibrium map, validated SNP assay resource, and high – throughput instrumentation system for large – scale genetic studies. Biotechniques Suppl, 2002, 48 – 50, 52 – 54

9 Gabriel S B, Schaffner S F, Nguyen H, Moore J M, Roy J, Blumenstiel B, Higgins J, DeFelice M, Lochner A, Faggart M, et al. The structure of haplotype blocks in the human genome. Science, 2002, 296: 2225 – 2229

10 Halld6rsson B V, Bafna V, Lippert R, Schwartz R, De La Vega F M, Clark A G, lstrail S. Optimal haplotype block – free selection of tagging SNPs for genome – wide association studies. Genome Res, 2004, 14: 1633 – 1640

11 Hell J, Gianowski S, Scott J, Winn – Deen E, McMullen I, Wu L, Gire C, Sprague A. An au-

tomated computer system to support ultra high throughput SNP genotyping. Pac. Symp. Biocomput, 2002, 7: 30 – 40

12 Hewett M, Oliver D E, Rubin D L, Easton K L, Stuart J M, Altman R B, Klein T E. PharmGKB: the Pharmacogenetics Knowledge Base. Nucleic Acids Res, 2002, 30: 163 – 165

13 Horton R, Wilming L, Rand V, Lovering R C, Bruford E A, Khodiyar V K, Lush M J, Povey S, Talbot Jr C C, Wright M W, et al. Gene map of the extended human MHC. Nat. Rev. Genet, 2004, 5: 889 – 899

14 The International HapMap Consortium. The International HapMap Proiect. Nature, 2003, 426: 789 – 796

15 Jeffreys A J, Kauppi L, Neumann R. Intensely punctate meiotic recombination in the class II region of the major histocompatibility complex. Nat. Genet, 2001, 29: 217 – 222

16 Kauppi L, Saiantila A, Jeffreys A J. Recombination hotspots rather than population history dominate linkage disequilibrium in the MHC class II region. Hum. Mol. Genet, 2003, 12: 33 – 40

17 Ke X, Hunt S, Tapper W, Lawrence R, Stavrides G, Ghori J, Whittaker P, Collins A, Morris A P, Bentley D, et al. The impact of SNP density on fine – scale patterns of linkage disequilibrium. Hum. Mol. Genet, 2004, 13: 577 – 588

18 Kerlavage A, Bonazzi V, di Tommaso M, Lawrence C, Li P, Mayberry F, Mural R, Nodell M, Yandell M, Zhang J, et al. The Celera Discovery System. Nucleic Acids Res, 2002, 30: 129 – 136

19 Kidd K K, Morar B, Castiglione C M, Zhao H, Pakstis A J, Speed W C, Bonne – Tamir B, Lu R B, Goldnmn D, Lee C, et al. A global survey of haplotype frequencies and linkage disequilibrium at the DRD2 locus. Hum. Genet, 1998, 103: 211 – 227

20 Kong A, Gudbjartsson D F, Sainz J, Jonsdottir G M, Gudjonsson S A, Richardsson B, Sigurdardottir S, Barnard J, Hallbeck B, Masson G, et al. A high – resolution recombination map of the human genome. Nat. Genet, 2002, 31: 241 – 247

21 Leder R O, Mansbridge J N, Hallmayer J, Hodge S E. Familial psoriasis and HLA – B: Unambiguous support for linkage in 97 published families. Hunt. Hered, 1998, 48: 198 – 211

22 Lercher M J, Hurst L D. Human SNP variability and mutation rate are higher in regions of high recombination. Trends Genet, 2002, 18: 337 – 340

23 Liu N, Sawyer S L, Mukherjee N, Pakstis A J, Kidd J R, Kidd K K, Brookes A J, Zhao H. Haplotype block structures show significant variation among populations. Genet. Epidemiol, 2004, 27: 385 – 400

24 Maniatis N, Collins A, Xu C F, McCarthy L C, Hewett D R, Tapper W, Ennis S, Ke X, Morton N E. The first linkage disequilibrium (LD) maps: Delineation of hot and cold blocks by diplotype analysis. Proc. Natl. Acad. Sci, 2002, 99: 2228 – 2233

25 Maniatis N, Morton N E, Gibson J, Xu C F, Hosking L K, Collins A. The optimal measure of linkage disequilibrium reduces error in association mapping of affection status. Hum. Mol. Genet, 2005, 14: 145 – 153

26 Mendenhall W, Scheaffer R L, Wackerly D D. Mathematical statistics with applications. Duxbury Press, Boston, MA, 1986

27 Montoya – Burgos J I, Boursot P, Galtier N. Recombination explains isochores in mammalian genomes. Trends Genet, 2003, 19: 128 – 130

28 Morton N E, Zhang W, Taillon – Miller P, Ennis S, Kwok P Y, Collins A. The optimal measure of allelic association. Proc. Natl. Acad. Sci, 2001, 98: 5217 – 5221

29 Nair R P, Stuart P Henseler T, Jenisch S, Chia N V, Westphal E, Schork N J, Kim J, Lim H W, Christophers E, et al. Localization of psoriasis – susceptibility locus PSORS1 to a 60 – kb interval telomeric to HLA – C. Am. J. Hum. Genet, 2000, 66: 1833 – 1844. [Erratum appears in Am. J. Hum. Genet, 2002, 70 (4): 1074].

30　Patil N, Bemo A J, Hinds D A, Barrett W A, Doshi J M, Hacker C R, Kautzer C R, Lee D H, Marjoribanks C, McDonough D P, et al. Blocks of limited haplotype diversity revealed by high – resolution scanning of human chromosome 21. Science, 2001, 294: 1719 – 1723

31　Phillips M S, Lawrence R, Sachidanandam R, Morris A P, Balding D J, Donaldson M A, Studebaker J F, Ankener W M, Alfisi S V, Kuo F S, et al. Chromosome – wide distribution of haplotype blocks and the role of recombination hot spots. Nat. Genet, 2003, 33: 382 – 387

32　Reich D E, Schaffner S F, Daly M J, McVean G, Mullikin J C, Higgins J M, Richter D J, Lander E S, Altshuler D. Human genome sequence variation and the influence of gene history, mutation and recombination. Nat. Genet, 2002, 32: 135 – 142.

33　Schork N J. Power calculations for genetic association studies using estimated probability distributions. Am J Hum Genet, 2002, 70: 1480 – 1489

34　Schwartz R, Halld6rsson B V, Bafna V, Clark A G, Istrail S. Robustness of inference of haplotype block structure. J Comput Biol, 2003, 10: 13 – 19

35　Service S K, Ophoff R A, Freimer N B. The genome – wide distribution of background linkage disequilibrium in a population isolate. Hum. Mol. Genet, 2001, 10: 545 – 551

36　Sherry S T, Ward M H, Khoiodov M, Baker J, Phan L, Smigielski E M, Sirotkin K. dbSNP: The NCBI database of genetic variation. Nucleic Acids Res, 2001, 29: 308 – 311

37　Stenzel A, Lu T, Koch W A, Hampe J, Guenther S M, De La vega F M, Krawczak M, Schreiber S. Patterns of linkage disequilibrium in the MHC region on human chromosome 6p. Hum Genet, 2004, 114: 377 – 385

38　Strathern J N, Sharer B K, McGill C B. DNA synthesis errors associated with double – strand – break repair. Genetics, 1995, 140: 965 – 972

39　Stringer C B, Andrews P. Genetic and fossil evidence for the origin of modern humans. Science, 1988, 239: 1263 – 1268

40　Sun F, Oiiver – Bonet M, Liehr T, Starke H, Ko E, Rademaker A, Navarro J, Benet J, Martin R H. Human male recombination maps for individual chromosomes. Am J Hum. Genet, 2004, 74: 521 – 531

41　Tapper W J, Maniatis N, Morton N E, Collins A. A metric linkage disequilibrium map of a human chromosome. Ann Hum Genet, 2003, 67: 487 – 494

42　Tsunoda T, Lathrop G M, Sekine A, Yamada R, Takahashi A, Ohnishi Y, Tanaka T, Nakamura Y. Variation of gene – based SNPs and linkage disequilibrium patterns in the human genome. Hum Mol Genet, 2004, 13: 1623 – 1632

43　Veal C D, Capon F, Allen M H, Heath E K, Evans J C, Jones A, Patel S, Burden D, Tillman D, Barker J N, et al. Family – based analysis using a dense single – nucleotide polymorphism – based map defines genetic variation at PSORS1, the maior psoriasis – susceptibility locus. Am. J. Hum. Genet, 2002, 71: 554 – 564

44　Venter J C, Adams M D, Myers E W, Li P W, Mural R J, Sutton G G, Smith H O, Yandell M, Evans C A, Holt R A, et al. The sequence of the human genome. Science, 2001, 291: 1304 – 1351

45　Weir B S. Genetic Data Analysis II. Sinauer Associates, Sunderland, MA, 1996

46　Zhai W, Todd M J, Nielsen R. Is haplotype block identification useful for association mapping studies? Genet Epidemiol, 2004, 27: 80 – 83

47　Zhang W, Collins A, Maniatis N, Tapper W, Morton N E. Properties of linkage disequilibrium (LD) maps. Proc Natl Acad Sci, 2002, 99: 17004 – 17007

48　Zhang W, Collins A, Gibson J, Tapper W J, Hunt S, Deloukas P, Bentley D R, Morton N E. Impact of population structure, effective bottleneck time, and allele frequency on linkage disequilibrium maps. Proc Natl Acad Sci, 2004, 101: 18075 – 18080

49　Zhang W, Collins A, Morton N E. Does haplo-

type diversity predict power for association mapping of disease susceptibility? Hum Genet,
2004, 115: 157 – 164.

50 http: //locus. umdnj. edu/ccr/; Coriell Institute, for details of the NIGMS Human Variation
Panels used in this study

51 https: //myscience. appliedbiosystems. com/; Applied Biosystems' myScience research environment,

for the TaqMan Validated SNP Genotyping Assays
used in this study

52 http: //cedar. genetics. soton. ac. uklpub/PROG-
RAMS/LDMAP; LDMAP Program

53 http: //www. allsnps. com/snpbrowser/; SNPbrowser Software, to obtain the genotypes used in
this study

94. Construction and Characterization of Chimeric BHIV (BIV/HIV – 1) Viruses Carrying the Bovine Immunodeficiency Virus *gag* Gene

ZHU Yi – xin[1], LIU Chang[1], LIU Xin – lei[2], QIAO Wen – tao[1], CHEN Qi – min[1],
ZENG Yi[3], GENG Yun – qi[1]

1. College of Life Sciences, Nankai University; 2. College of Life Sciences and Bioengineering,
Beijing University of Technology; 3. National Institute for Viral Disease Control and Prevention

Summary

Objective To explore the possibility of the replacement of the *gag* gene between human immunodeficiency virus and bovine immunodeficiency virus, to achieve chimeric virions, and thereby
gain a new kind of AIDS vaccine based on BHIV chimeric viruses.

Methods A series of chimeric BHIV proviral DNAs differing in the replacement regions in *gag*
gene were constructed, and then were transfected into 293T cells. The expression of chimeric viral
genes was detected at the RNA and protein level. The supernatant of 293T cell was ultra centrifuged
to detect the probable chimeric virion. Once the chimeric virion was detected, its biological activities
were also assayed by infecting HIV – sensitive MT4 cells.

Results Four chimeric BHIV proviral DNAs were constructed. Genes in chimeric viruses expressed correctly in transfected 293T cells. All four constructs assembled chimeric virions with different degrees of efficiency. These virions had complete structures common to retroviruses and packaged
genomic RNAs, but the cleavages of the precursor *gag* proteins were abnormal to some extent. Three
of these virions tested could attach and enter into MT4 cells, and one of them could complete the
course of reverse transcription. Yet none of them could replicate in MT4 cells.

Conclusion The replacement of partial *gag* gene of HIV with BIV *gag* gene is feasible. Genes
in chimeric BHIVs are accurately expressed, and virions are assembled. These chimeric BHIVs

（proviral DNA together with virus particles）have the potential to become a new kind of HIV/AIDS vaccine.

[**Key words**]   *gag* gene; Human immunodeficiency virus; Bovine immunodeficiency virus

Introduction

The HIV epidemic continues to expand at an alarming rate and is predicted to be the worst infectious disease ever to affect human beings, and the situation in Asia is also alarming[1]. Historical experience indicates that vaccines continue to be the most cost efficient and effective intervention available for preventing infectious diseases.

Up to now, there have been 80 candidate HIV/AIDS vaccines. Forty – seven of them have completed trials, but found to be either unsafe or ineffective, while 33 of them are in ongoing trials: one in phase III, three in phase II, two in phase I/II, and the rest in phase I. These vaccines include virus-like particles, peptides, recombinant protein subunits, recombinant bacterial vectors, recombinant viral vectors, and DNA vaccines[2]. However, the chances of getting an effective HIV/AIDS vaccine with current approaches are still very negligible. We still need new vaccine strategies against HIV/AIDS[3].

The chimeric HIV might just be such a new kind of HIV/AIDS vaccine, because it can mimic the natural infection of HIV[4-6], but out of concern about safety, current chimeric SHIVs are unfit for vaccines[5-7]. Actually, they are used as challenge viruses in AIDS/rhesus models to evaluate the efficacy of HIV/AIDS vaccines. HIV – 1 and SIV are so closely related phylogenetically that the chimeric virus of SIV/HIV is too dangerous to be a vaccine[5,6]. Then how about a chimeric virus between HIV and non – primates lentivirus, such as the bovine immunodeficiency virus?

Bovine immunodeficiency virus (BIV) is a lentivirus, which resembles HIV in its structural, genomic, antigenic, and biological properties[8]. Unlike HIV, BIV is a relatively mild lentivirus, and never causes severe acquired immunedeficiency syndrome in its host, the cow[9,10]. BIV seropositivity has been correlated with decreased milk production in dairy cattle, but has not been directly linked with clinical disease in naturally infected cattle[11]. Considering the similarity of genome to HIV and its low pathogenicity, BIV may be an appropriate candidate to combine with HIV to create a new kind of HIV chimeric virus – BIV.

Primary work showed the expression of BIV *gag* – pol gene in HIV backbone in human original MT4 cells[12]. This time we focus on the *gag* gene. We tried to explore the possibility of the replacement of the *gag* gene between human immunodeficiency virus and BIV. In doing so, a series of chimeric BHIV (BIV/HIV – 1) proviruses carrying the BIV *gag* gene have been constructed. In order to improve the expression of chimeric genes in human cells, the CMV promoter is introduced.

Materials and Methods

1. Materials

Infectious cDNA of BIV127 and HIV – 1 HXB$_c$2 were kept in our laboratory, and pcDNA3. 1 (–) vector was from Invitrogen.

2. Methods

(1) *Construction of chimeric BHIV proviral DNA* Modification of pcDNA3. 1 （−） vector: The pcDNA3. 1 （−） plasmid was double digested with EcoRI and BamHI, then the ends were blunted with T4 DNA polymerase and self－ligated by T4 DNA ligase, so that only one SacI site remained. Then the BssHII site was removed by cutting, blunting and religating.

(2) *Replacement of 5' HIV LTR U3 region by CMV promoter* The fragment 1 of 5' HIV LTR R U5 and partial *gag* was amplified by PCR, using sense primer from HIV－1 HXB$_c$2 positions 441nt－459nt (5'TCG AC－T TTT GCC TGT ACT GGG TCT3'), and anti－sense primer from HIV－1 HXB$_c$2 positions 2045－2026nt (5'TCC TTT CCA CAT TTC CAA CA3'). The fragment 2 of 3' HIV LTR and partial env was amplified by PCR, using sense primer from HIV－1 HXB$_c$2 positions 8824－8842nt (5'GTG ATT GGA TGG CCT ACT G3') and anti－sense primer from HIV－1 HXB$_c$2 positions 9719－9700nt (5'TGC TAG AGA TTT TCC ACA CT3').

Modified pcDNA3. 1 （−） vector was excised with SacI, blunted, excised with ApaI, then was ligated with the PCR product of fragment 1, which was already digested with ApaI. Thus, the 5' HIV LTR U3 region was replaced by the CMV promoter of the modified pcDNA3. 1 （−） vector. We got the first intermediate plasmid pCMV1.

Plasmid pCMV1 was double excised with XhoI and EcoRV, which was ligated with the PCR product of fragment 2, which was already digested with XhoI. Thus, the 3' HIV LTR was transferred into the vector. We got the second intermediate plasmid pCMV2.

Plasmid pCMV2 was double digested with ApaI and XhoI, and was ligated with the 7－kb HIV ApaI－XhoI fragment, then the complete genome of HIV was transferred to the modified pcDNA3. 1 （−） vector, along with the transfer of CMV promoter to the 5' HIV LTR U3 region. It was named pCHIV.

(3) *Construction of pCG1 and pCG2* Chimeric *gag* genes of pCG1 and pCG2 were derived from C2 and C3[13]. The chimeric *gag* genes from C2 and C3 were excised with BssHII/ApaI, and ligated into the vector of pCHIV, which was already digested with BssHII/ApaI. So we got the complete genome of chimeric HIV in pCG1 and pCG2.

(4) *Construction of pCG3* Two rounds of PCR produced the chimeric BHIV *gag* fragments of pCG3. The first round PCR used BIV127 as template, and the product of the first round together with pCG1 were used as templates in the second round PCR.

(5) *First round PCR sense primer* 5'TAA GGT TAG GGT GAC AC3', from BIV127 positions 741－757nt. Anti－sense primer: 5'CC－T ACA ATT CCT CTT CAA ATG3', from BIC127 positions 1963－1945nt.

(6) *Second round PCR sense primer* 5'TCG AC－T TTT GCC TGT ACT GGG TCT3', from HIV－1 H×B$_c$2 positions 441－459nt. Anti－sense primer was same as in the first round PCR.

(7) *The complete genome of pCG3 proviral DNA was constructed as follows* pCHIV was excised with ApaI, blunted, excised with BssHII, then ligated with the product of the second round PCR that was already digested with BssHII.

(8) *Construction of pCG5* The proviral genome DNA was constructed by blunt end liga-

tion. Plasmid pCHIV was double – digested with BssHll/ApaI. The ends were blunted with T4 DNA polymerase, and then were dephosphorylated by calf intestine alkaline phosphatase. The inserted fragment was a PCR product that was already phosphorylated by T4 polynucleotide kinase. After ligation, the clone of corrected insert orientation was selected by restriction enzyme mapping.

The sense primer of the PCR was 5'CAG AAG ACT CCG GAC AG3', from BIV127 667 – 683nt. Anti – sense primer was 5'CC – T ACA ATT CCT CTT CAA ATG3', from BIC127 1963 – 1945nt.

3. Transfection of 293T cells with BHlV proviral DNA

In a level 3 biosafety (P3) laboratory, four BHIV clones were individually transfected into 40% – 50% confluence 293T cells that were grown in 10% FBS DMEM in a six – well plate by using Polyfect (Qiagen). In order to reduce the side effect of remnant proviral DNA and Polyfect, cells were washed with PBS thrice and fresh culture medium was added 24 h after transfection. The cell RNA and protein were extracted 72 h post – transfection by Tri reagent (Sigma).

4. Electron microscopy (EM)

To perform electron microscopy (EM), supernatants were fixed with 2.5% glutaraldehyde, followed by treatment with 4% osmium tetroxide, and then routinely processed. Samples were stained with lead citrate and uranyl acetate and then visualized by using a JEOL JEM – 2000 FX transmission electron microscope.

5. Collection of chimeric BHIV virions

The culture supernatant was collected 72 h post – transfection and filtered with 0.45 – μm filter to remove cell fractions. The cell – free supernatant was ultra centrifuged at 30 000 g for 3 h. The pellet was stored at – 80 ℃.

6. Component assays of chimeric BHIV virion pellets

In Western blot, anti – BIV capsid (CA) polyclone antibody was used to detect the chimeric *gag* precursor and mature CA protein in these virions. Reverse transcriptase activities in these virions were assayed by Reverse Transcriptase Assay colorimetric kit (Roche). The genomic RNA in these virions were also extracted as follows: virion pellets were resuspended, DNase added to digest the remnant proviral DNA at 37℃ for 30 min, then the mixture was extracted with phenol/chloroform and RNA precipitated with ethanol.

7. Infection of MT4 with chimeric BHIV viral stock

The culture supematant was collected 72 h post – transfection and filtered with 0.45 μm filter to remove cell fractions. The cell – free supernatant was added into 5×10^6 in 5 ml 10% PBS RPMI 1640 medium. After incubation at 37 ℃ for 18 h, MT4 cells were washed thrice with PBS, then fresh medium was added and returned to incubator. After 8 h, cells were again washed thrice with PBS, and then their RNA, DNA and protein were extracted by Tri reagent (Sigma).

8. Primers in PCR or RT – PCR for detecting DNA or RNA

RT – PCR was performed to detect the early (tat gene) and late (*gag* gene) RNA transcripts in transfected 293T cells. RT – PCR was also performed to detect the genomic RNA (env gene) as a representative (Tab. 1).

Tab. 1 Primers in detecting PCR or RT – PCR

Primer name	Sequence (5' – 3')	Location	Product length (bp)
tatRT – PCR up	GACTCGGCTTGCTGAAG	HIV 695 – 711	250
tatRT – PCR low	GCTGTCTCCGCTTCTTC	HIV 5993 – 5977	
*gag*RT – PCR up	GGAGGCCAGAGCTGATAAG	BIV 1044 – 1062	560
*gag*RT – PCR low	GTCTGTGTACGGCTCCTTG	BIV 1608 – 1590	
envPCR up	AATGACGCTGACGGTACAGG	HIV 7826 – 7845	660
env PCR low	GTGCCAAGGATCCGTTCAC	HIV 8487 – 8469	

Results

1. Four chimeric BHIV proviral DNAs were constructed

Four chimeric BHIV proviral DNAs were constructed. Both restriction enzyme mapping and DNA sequencing confirmed the sequences of these proviruses. The genomic structure of each provirus is illustrated in Fig. 1.

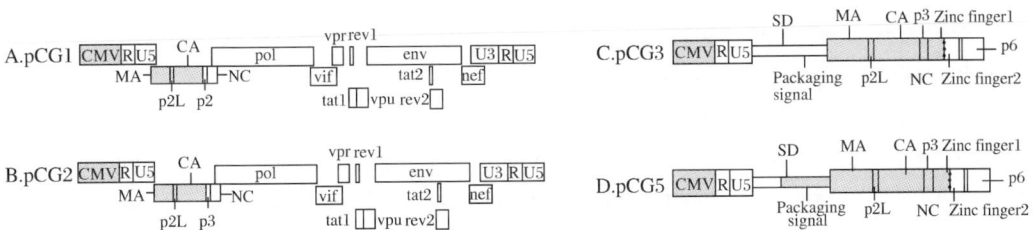

Fig. 1 The genomic structure of four chimeric BHIV DNAs. Yellow: CMV promoter; white: HIV – 1 gene; blue: BIV gene

2. Early and late genes were transcribed in 293T cells

After transfection, RNA from 293T cells was extracted. The detection of early transcript of tat gene and late transcript of *gag* gene was performed by RT – PCR. Both early and late gene transcripts of RNA were detected in all four chimeric viruses. The results of RT – PCR are shown in Fig. 2.

3. All four chimeric *gag* precursors were expressed and partially cleaved in 293T cells

Western blot analysis (Fig. 3) showed that in 293T cells all four chimeric *gag* precursors were expressed. There was no significant difference in the amount of *gag* expression. All fours *gag* precursors were partially cleaved. There were two cleavage profiles among these four *gag* precursors: pCG1 belonged to one cleavage profile; pCG2, pCG3, pCG5 belonged to another. The spectrum of the latter profile was depicted on the right side of Fig. 3.

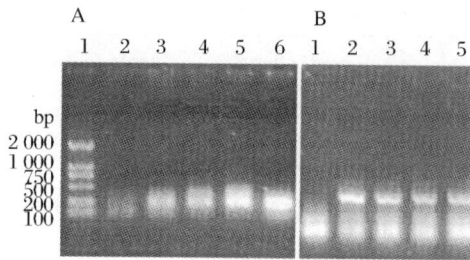

A: HIV - 1 tat: 1. DL2000, 2. Negative control 3. peG1, 4. pcG2, 5. pcG3, 6. pcG5; B: BIV *gag*: 1. Negative control, 2. pCG1, 3. pCG2, 4. pCG3, 5. pCG5

Fig. 2　RT - PCR results in 293T cells

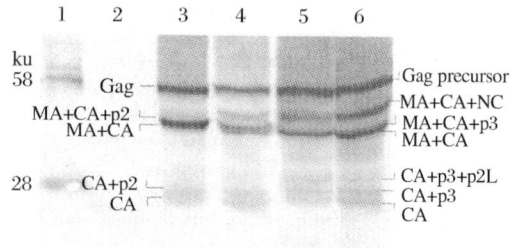

Lane 1: Marker; lane 2: 293T cell protein as negative control; lane 3: pCG1; lane 4: pCG2; lane 5: pCG3; lane 6: pCG5

Fig. 3　Western blot of 293T cells transfected by chimeric proviral DNAs. Western blot with rabbit anti - BIV CA serum

4. All four chimeric proviruses assembled virions in 293T cells

The cell - free supernatant was visualized by electron microscopy. Fig. 4 shows the results observed by EM. All four chimeric proviruses assembled virions in 293T cells. These virions have a diameter of about 100 nm, the typical scale of lentivirus. Unlike mature HIV virions, these particles do not have spindle cores, indicating that they are in the immature status.

5. Cleavage profiles of *gag* in chimeric virions were abnormal

The cell - free supernatant was ultra centrifuged. The pellet of viral particles was analyzed by Western blot analysis, as shown in Fig. 5. As it shows, very few *gag* precursors in virions were left, but the cleavage was still not complete. A large proportion of BIV CA existed in intermediate cleavage products, especially in CA + p3. A small part of BIV CA was still cut by HIV - 1 protease and the products of p15 and p10 of CA were generated.

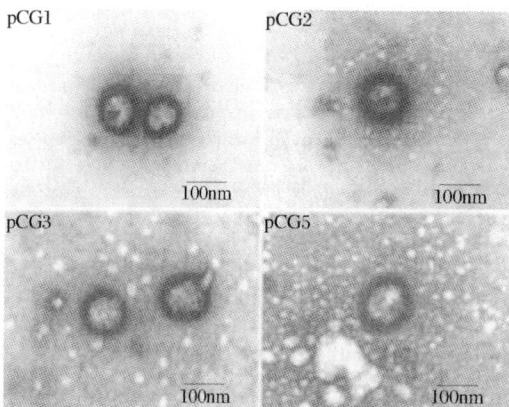

Fig. 4　Chimeric virus particles observed by EM

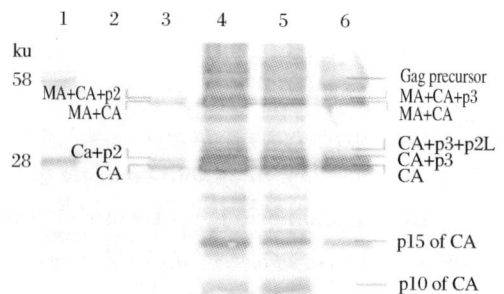

Lane 1: Marker; lane 2: 293T cell protein as negative control; lane 3: peG1; lane 4: pCG2; lane 5: peG3; lane 6: peG5

Fig. 5　Western blot of chimeric virions assembled from 293T cells after transfection. First antibody was anti - BIV CA

6. All four chimeric virions involved reverse transcriptase activities, while the RT activity of pCG1 virion was lower than the other three

The relative RT activities of four chimeric virions are shown in Figure 6. Each measure was taken with virions centrifuged from 1 ml supernatant of 293T cells. The transfection efficiency and cell confluence will also affect the release of chimeric virions, and the data of virion RT activities were semi – quantitative. We can conclude that the RT activity of pCG1 virion was lower than that of HIV positive control ($P < 0.01$ vs HIV), the other three do not have any significant difference from that of HIV control.

7. Genomic RNAs were packed in all four chimeric virions

Although the package signals and the package signal recognizers (mainly the nucleic capsid, NC, in *gag*) were different in these four chimeric virions, they all included genomic RNAs as the results of RT – PCR shown in Fig. 7.

8. Virions of pCG2, pCG3 and pCG5 can attach and enter MT4 cells

Virions of pCG2, pCG3 and pCG5 in supernatant were incubated with human lymphocyte primary MT4 cells for 24 h, then the remnant virions and culture medium were carefully eliminated. The protein of chimeric *gag* from virions can be detected in MT4 cells using the Western blot analysis shown in Fig. 8.

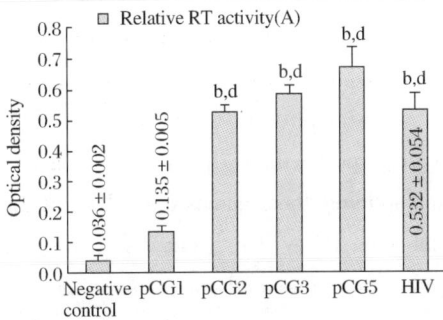

All samples are positive to negative control ($^{\circ}P < 0.01$ vs negative control). The RT activity of pCG1 virion was lower than the other three chimeric virions and HIV positive control (^{d}1P < 0.01 vs pCG1)

Fig. 6   Relative reverse transcriptase activities of chimeric virions in 1 ml supernatant of 293T cells ($\bar{x} \pm s$, n = 4)

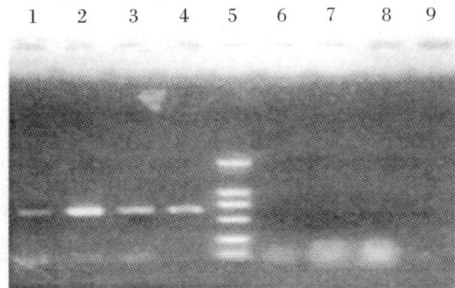

1. pCG1; 2. pCG2; 3. pCG3, 4. pCG5; 5. DL2000 marker; 6 – 9: Sample of pCG1, pCG2, pCG3, pCG5 RNAs without adding reverse transcriptase as negative control

Fig. 7   HIV – 1 env gene RT – PCR results of chimeric virions

9. Virons of pCG3 can complete reverse transcription in MT4 cells

After incubation of chimeric virions with MT4 cells, we carefully eliminated the remnant virions and traced proviral DNA in the culture medium, then extracted the genomic DNA of MT4 cells. Detection of genomic DNA of chimeric virions was performed by PCR. Only in pCG3 virion infected MT4 cells we can find positive result as illustrated in Fig. 9. That means virions of pCG3 can complete reverse transcription in MT4 cells.

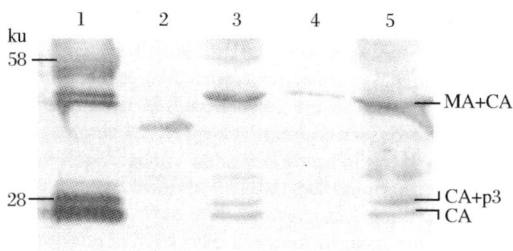

Lane 1: pCG5 virion as positive control; lane 2: MT4 cell protein as negative control; lane 3: peG2; lane 4: peG3; lane 5: pCG5

Fig. 8 Western blot of chimeric virions infecting MT4 cells. First antibody was anti – BIV CA

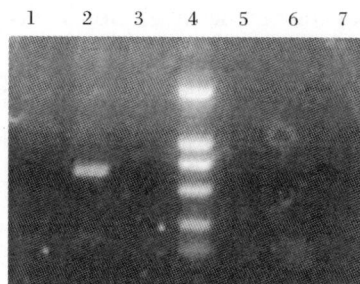

1. pCG2; 2. pCG3; 3. pCG5; 4. DL2000 marker; 5 – 7: Culture medium of pCG2, pCG3, pCG5 infecting MT4 (in last 8 h) as negative control

Fig. 9 HIV – 1 env gene PCR results of MT4 genomic DNA after infecting chimeric virions

Discussion

The constitutionally strong promoter of CMV can replace the HIV – 1 LTR U3 region and promote the RNA transcription of HIV – 1. Generally speaking, the transcription efficiency of the CMV promoter is higher than that of HIV – 1 LTR in human cells[14]. So, the replacement of promoter will enhance the expression of chimeric viral genes. Incidentally, the replacement took place only on the 5' LTR U3 region, not on the 3' LTR U3 region, nor any of LTR. That means the replacement only promotes the quantity of transcription, and does not change the quality of transcripts.

The expression of early genes in HIV relies on the correct post – transcriptional splicing of RNA in the nucleus of host cells[15]. In the genome of pCG4, the splicing donor comes from BIV, while the splicing receptor belongs to HIV – 1. From the result of tat gene RT – PCR (Fig. 2A), we can see that the heterogeneous donor and receptor matched each other and generated correct splicing.

The chimeric *gag* gene as a late gene can be transcribed and translated in 293T cells. That means the early gene can work and trigger the expression of late genes[16]. The expression of *gag* precursor in four chimeric BHIVs was of the same magnitude, and all four chimeric BHIVs assembled virions. However, the virions assembled in pCG1 were much less than the other three. Some research pointed out that the p2 plays an important role in HIV – 1 virion assembly[17 – 19]. The results here suggest that p3 – the counterpart of p2 – in BIV *gag* is also important in efficient assembly of BIV *gag*. Virions of chimeric *gag* (BIV *gag* as a major part) can still be assembled if BIV p3 is replaced by HIV – 1 p2, but the assembly efficiency will be very low. This is not congruent with the result of Guo et al.[13]. We deduce that this is because of the high transcription efficiency of CMV promoter counteracting the low assembly efficiency of pCG1 *gag*.

The primate lentivirus proteases have been intensively investigated[20], and it was found that the specificity is conserved among HIV – 1, HIV – 2, and SIV. For instance, synthetic HIV – 2 protease deaves the *gag* precursor of HIV – 1 with the same specificity as HIV – 1 protease[21,22], and the proteolytic processing of the HIV – 2 *gag* precursor is very similar to the processing of the

SIVMne gag precursor[23]. However, the substrate specificity of protease between HIV – 1 and the non – primate lentivirus BIV has not been studied previously. In these experiments, the chimeric gag precursors not only in cells but also in virions were cleaved by HIV – 1 protease. Among these cleaving sites in chimeric gags, some have come from HIV – 1 gag, some from BIV gag, and some are even mosaics from both HIV – 1 and BIV. Our results indicate that all these cleavage sites can be recognized and cleaved by HIV – 1 protease although in different efficiency (Fig. 3 and 5). The relatively low cleaving efficiency may lead to the immature morphology of virions (Fig. 4).

Previous study indicated that a small percentage of the CA protein undergoes secondary proteolysis to generate additional peptides during the natural cleavage of BIV gag precursor. A 10×10^3 C-terminal peptide of CA, has been immunologically identified; another putative 16 – ku N terminal peptide of CA has not been immunologically identified yet[24,25]. From Fig. 4 we can see that HIV – 1 protease also can recognize and cleave this secondary alternate site inside BIV CA. Besides the alternate CA cleavage product of p10, the other putative product of a 16 – ku peptide is also immunologically identified as p 1 5. This adds to the conservative character of protease substrate specificity among lentivirus.

There are four cleavage sites within the HIV – 1 gag precursor (six within chimeric gag, including the secondary alternate cleavage site inside CA), which on cleavage gives rise to the mature core proteins MA, CA, NA, and p6. (For simplicity and in the absence of specific data in our experiments, we exclude the HIV – 1 gag spacer peptide of pl between NC and p6 from the discussion.) The cleaving efficiency among the four sites is different. The site of p2 NC was cleaved first, NC – p6 second, MA – CA third, and CA – p2 last[26]. Sequential cleavage of gag precursor generates certain kinds of intermediates, and from the profile of intermediates, we can deduce the cleavage sequence. In the precursor of pCG1 gag, junctions of p2NC and NC – p6 have come from HIV – 1; they can surely be cleaved by HIV – 1 protease efficiently. So they are the first and second cleavage sites. The junction of CA – p2 is a chimeric junction; the segment to the left of scissible bond comes from BIV, to the right of the scissible bond comes from HIV – 1. It is the last site cleaved, and is responsible for the accumulation of CA + p2 intermediate (Fig. 5). In the precursor of pCG2 or pCG3 or pCG5, the p3 – NC junction no longer comes from HIV – 1. So, the cleavage efficiency is reduced, and a new intermediate of MA + CA + NC emerged (Fig. 3). The deduced cleavage sequence of chimeric gag precursor is illustrated in Fig. 10.

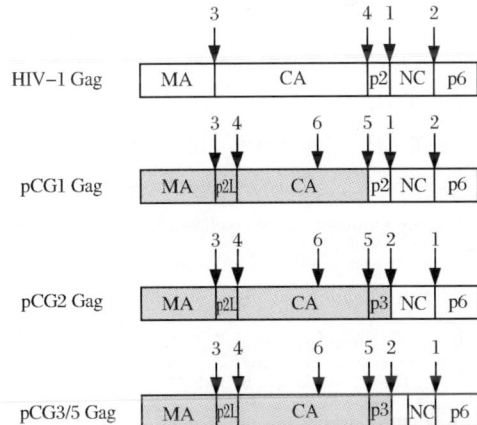

Fig. 10 Sketch map of deduced sequential cleavage of chimeric gag precursor by HIV – 1 protease (pl between NC and p6 is not concerned); numbers show the sequence of the cleavage. Blue: from BIV; white: from HIV

The structure of these chimeric virions is complete. They have genomic RNA, reverse transcriptase, core proteins, and envelope together with glycoprotein of *gp*120 and *gp*41 (this is deduced from the fact that they can attach and enter into MT4 cells). One of them even can complete reverse transcription in MT4 cells. But none of them can replicate in MT4 cells. Because of these characteristics, they have the potential of being a new kind of HIV AIDS vaccine.

They are relatively safe. They can express HIV antigen efficiently. They can assemble virions, and these virions can enter human lymphocytes. We anticipate that they can mimic the natural infection of HIV and stimulate the human immune system to elicit both cellular and humoral immune responses against HIVs.

In summary, the replacement of partial *gag* gene of HIV with BIV *gag* gene is feasible. Genes in chimeric BHIVs are accurately expressed, and virions are assembled. Chimeric BHIVs (proviral DNA together with virus particles) are expected to be a new kind of HIV/AIDS vaccine candidate.

Acknowledgements

We thank Liang Chen for providing the plasmid DNA of C2 and C3.

[In 《World J Gastroenterol》 2005, 11 (17): 2609 – 2615]

References

1 Roger D. HIV Surveillance, Prevention, Intervention, and treatment in Asia. The XV International HIV/AIDS Conference. New York: Guilford Press, 2004

2 IAVI database of AIDS vaccines in human trials, updated. Available from: URL: http://www.iavireport.org/trialsdb/default.asp, 14 December, 2004

3 erzofsky JA, Ahlers JD, Janik J, Morris J, Oh S, Terabe M, Belyakov IM. Progress on new vaccine strategies against chronic viral infections J Clin Invest, 2004, 114: 450 – 462

4 Ui M, Kuwata T0 Igarashi T, Ibuki K, Miyazaki Y, Kozyrev IL, Enose Y, Shimada T, Uesaka H, Yamamoto H, Miura T, Hayami M. Protection of macaques against a SHIV with a homologous HIV – 1 Env and a pathogenic SHIV – 89.6P with a heterologous Env by vaccination with multiple gene – deleted SHIVs. Virology, 1999, 265: 252 – 263

5 Silverstein PS, Mackay GA, Mukherjee S, Li Z, Piatak M Jr, Lifson JD, Narayan O, Kumar A. Pathogenic simian/human immunodeficiency virus SHIV (KU) inoculated into immunized macaques caused infection, but virus burdens progressively declined with time. J Virol, 2000, 74: 10489 – 10497

6 Kumar A, Lifson JD, Li Z, Jia F, Mukherjee S, Adany I, Liu Z, Piatak M, Sheffer D, McClure HM, Narayan O. Sequential immunization of macaques with two differentially attenuated vaccines induced long – term virus – specific immune responses and conferred protection against AIDS caused by heterologous simian human immunodeficiency Virus SHIV89.6 P. Virology, 2001, 279: 241 – 256

7 Whitney JB, Ruprecht RM. Live attenuated HIV vaccines: pitfalls and prospects. Curr Opin Infect Dis, 2004, 17: 17 – 26

8 Gonda MA. Bovine immunodeficiency virus. AIDS, 1992, 6: 759 – 776

9 Gonda MA, Luther DG, Fong SE, Tobin GJ. Bovine immunodeficiency virus: molecular biology and virus – host interactions. Virus Res, 1994, 32: 155 – 181

10 Carpenter S, Vaughn EM, Yang J, Baccam P,

Roth JA, Wannemuehler Y. Antigenic and genetic stability of bovine immunodeficiency virus during long – term persistence in cattle experimentally infected with the BIV (R29) isolate. J Gen Virol, 2000, 81: 1463 – 1472

11 McNab WB, Jacobs RM, Smith HE. A serological survey for bovine immunodeficiency – like virus in Ontario dairy cattle and association between test results, production records and management practices. Can J Vet Res, 1994, 58: 36 – 41

12 Chen G, Wang S, Xiong K, Wang J, Ye T, Dong W, Wang Q, Chen Q, Geng Y, Wood C, Zeng Y. Construction and characterization of a chimeric virus (BIV/HIV – 1) carrying the bovine immunodeficiency virus gag – pol gene. AIDS, 2002, 16: 123 – 125

13 Guo X, Hu J, Whitney JB, Russell RS, Liang C. Important role for the CA – NC spacer region in the assembly of bovine immunodeficiency virus gag protein. J Virol, 2004, 78: 551 – 560

14 Paya CV, Virelizier JL, Michelson S. Modulation of T – cell activation through protein kinase C – or A – dependent signalling pathways synergistically increases human immunodeficiency virus long terminal repeat induction by cytomegalovirus immediate – early proteins. J Virol, 1991, 65: 5477 – 5484

15 Feinberg MB, Jarrett RF, Aldovini A, Gallo RC, Wong – Staal F. HTLV – Ⅲ expression and production involve complex regulation at the levels of splicing and translation of viral RNA. Cell, 1986, 46: 807 – 817

16 Malim MH, Hauber J, Le SY, Maizel JV, Cullen BR. The HIV – 1 rev trans – activator acts through a structured target sequence to activate nuclear export of unspliced viral mRNA. Nature, 1989, 338: 254 – 257

17 Krausslich HG, Facke M, Heuser AM, Konvalinka J, Zentgraf H. The spacer peptide between human immunodeficiency virus capsid and nucleocapsid proteins is essential for ordered assembly and viral infectivity. J Virol, 1995, 69: 3407 – 3419

18 Accola MA, Hoglund S, Gottlinger HG. A putative alphahelical structure, which overlaps the capsid-p2 boundary in the human immunodeficiency virus type 1 gag precursor, is crucial for viral particle assembly. J Virol, 1998, 72: 2072 – 2078

19 Morikawa Y, Hockley D J, Nermut MV, Jones IM. Roles of matrix, p2, and N – terminal myristoylation in human immunodeficiency virus type 1 gag assembly. J Virol, 2000, 74: 16 – 23

20 Pettit SC, Henderson G J, Schiffer CA, Swanstrom R. Replacement of the P 1 amino acid of human immunodeficiency virus type 1 gag processing sites can inhibit or enhance the rate of cleavage by the viral protease. J Virol, 2002, 76: 10226 – 10233

21 Wu JC, Carr SF, Jarnagin K, Kirsher S, Barnett J, Chow J, Chan HW, Chen MS, Medzihradszky D, Yamashiro D. Synthetic HIV – 2 protease cleaves the gag precursor of HIV – 1 with the same specificity as HIV – 1 protease. Arch Biochem Biophys, 1990, 277: 306 – 311

22 Pichuantes S, Babe LM, Barr PJ, DeCamp DL, Craik CS. Recombinant HIV2 protease processes HIV1 Pr53gag and analogous junction peptides in vitro. J Biol Chem, 1990, 265: 13890 – 13898

23 Henderson LE, Benveniste RE, Sowder R, Copeland TD, Schultz AM, Oroszlan S. Molecular characterization of gag proteins from simian immunodeficiency virus (SIVMne). J Virol, 1988, 62: 2587 – 2595

24 Battles JK, Hu MY, Rasmussen L, Tobin G J, Gonda MA. Immunological characterization of the gag gene products of bovine immunodeficiency virus. J Virol, 1992, 66: 6868 – 6877

25 Tobin G J, Sowder RC 2nd, Fabris D, Hu MY, Battles JK, Fenselau C, Henderson LE, Gonda MA. Amino acid sequence analysis of the proteolytic cleavage products of the bovine immunodeficiency virus gag precursor polypeptide. J Virol, 1994, 68: 7620 – 7627

26 Tritch RJ, Cheng YE, Yin FH, Erickson – Vitanen S. Mutagenesis of protease cleavage sites in the human immunodeficiency virus type 1 gag polyprotein. J Virol, 1991, 65: 922 – 930

95.　Chinese NGOs in Action against HIV/AIDS

XU Hua[1], ZENG Yi [1,2], ANDERSON Allen F[3]

1. The US – China AIDS Alliance Foundation; 2. Institute of Virology, Chinese
Academy of Preventive Medicine; 3. School of Public and
Environmental Affairs, Indiana University, Bloomington, Indiana

Summary

Chinese nongovernmental organizations (NGOs) have played a significant role in the battle against AIDS in the People's Republic of China. This article provides a brief overview of the structure of these organizations, as well as an analysis of their principle accomplishments. Of great significance in this analysis is the fact that Chinese NGOs have effectively dealt with many sensitive health education areas that government authorities have felt reluctant to handle directly. As such, they have provided an indispensable component in the HIV/AIDS prevention and control calculus on the mainland.

[Key words] HIV/AIDS; NGO; Health education; Health intervention

Introduction

The AIDS epidemic continues its seemingly inexorable spread throughout the world. It is now very clear that the virus represents not only a medical problem, but also a challenging and multifaceted social problem. Because of this fact, it is imperative that nongovernmental organizations outside of, or tangential to, the medical arena become involved in prevention and control efforts. The Chinese government is supportive of the development of such organizations on the mainland, and has recognized the key role that they can play against the deadly virus. Vice Premier Wu Yi stated that "we should mobilize all the partners in the society to participate in the fight against HIV/AIDS. We need to improve our policies and strategies to build a better environment for all forces in the society to participate in the response, and to try our best to facilitate the involvement of all sectors[1]. " It was found early on in the epidemic in China that NGOs could deal with many sensitive health education interventions that the government felt reluctant to handle directly. As a result, nongovernmental organizations on the mainland that are focused on HIV issues are growing in both number and impact.

There are currently over fifty large NGOs involved in AIDS control. These large organizations can be divided into two groups. The first group is composed of mass organizations such as, inter alia, the All China Women's Federation, the All China Youth League, the Red Cross, the All China Federation of Trade, and the Chinese Working Committee for Caring for the Younger Generation. Each of these organizations has branches at different levels throughout China, and most of their work is done by professional social workers. The second group is composed of professional civil groups such as the Chinese Association of Medicine, the Chinese Association of Preventive Medicine, the China Family Planning Association, the Chinese Foundation for the Prevention of STD &AIDS (CFPSA), and the Chinese Association of STD/AIDS Prevention and Control (CASAPC), among others. The organizations in this group typically have branches at the provincial level, and most of their work is conducted by medical and public health personnel.

The number of small NGOs is even greater, and includes such organizations as the "Home for Loving Care" in Beijing, the "Mangrove Support Group", the "Love Care Family" in Wenxi county of Shanxi Province, the "Health Club" in Jilin City of Jilin Province, and the "Chengdu Gay Com munity Care Organization" in Sichuan Province. Most of their work is conducted by volunteers and persons living with HIV/AIDS.

Whatever their levels, most Chinese AIDS NGOs get their financial support from the government and international organizations; however, many also get support from social donations.

Specific Areas of Activity and Impact

Since the first case of AIDS was discovered on the mainland in 1985, Chinese NGOs have played a key role in prevention and control efforts. Associated activities fall into the following categories.

1. Health education

Without a vaccine or cure for HIV/AIDS, educational interventions are especially important in the prevention and control of the disease. Chinese NGOs have assisted government at various levels to develop policy and material for health education. Successes have been achieved in the broad dissemination of AIDS prevention knowledge, as well as antidiscrimination education. Specific activities have included:

(1) Various AIDS education booklets, pamphlets, and videos have been produced. For example, the educational video "AIDS Track" was produced by the CFPSA and disseminated nationwide. This video was awarded a Special Prize by the Chinese Ministry of Public Health. The video "Warning of the Century" was also produced by the CFPSA and aired on CCTV in 1999. During 2001, the CFPSA cooperated with the Beijing Television Station in airing a quiz show in which high school students competed in answering questions about HIV/AIDS.

(2) During 1994, five national newspapers ran an STD/AIDS knowledge competition in which more than one million people participated.

(3) In cooperation with local governments, the CFPSA held large – scale AIDS exhibitions in Guangzhou and Nanning during December, 1997, and May, 1998. With the support of the Ministry of Health, the CFPSA and the CASAPC collaborated in holding a large AIDS exhibition in Beijing during December, 1998. More than 200 000 people attended these exhibitions[2]. Thereafter, this exhibition format was expanded to many other provinces and cities. More than 500 000 people were exposed to prevention and control knowledge.

(4) AIDS information campaigns for railway workers was developed by the CASAPC and implemented along the Beijing – Hong Kong, Beijing – Ulaanbaatar, and Kunming Pingxiang railway lines.

(5) Peer education for young people was developed in Yunnan Province and expanded to other provinces.

(6) The CFPSA cooperated with the governments and health departments in the Weifang region of Shandong Province, especially the cities of Shouguang and Zhucheng, to develop a large – scale and multifaceted HIV/AIDS campaign for rural areas. This was accomplished using the following approach:

① A leadership group that included the governmental head of the county and heads of departments concerned with STD/HIV/AIDS prevention and control was established in every county in the region;

② Health professionals and social workers who were to be involved in survey and education efforts were trained;

③ Knowledge, attitude, and practice surveys were conducted prior to the design of the educational interventions;

④ In light of survey findings, multifaceted educational interventions were developed and implemented at the county and township levels;

⑤ Thereafter, intervention assessments were conducted. Preliminary results showed a coverage rate of the target population of 90 percent (about 100 000 persons), and the HIV/AIDS awareness rate rose from 35 percent to 60 percent. This inspired the Weifang region governmental leadership continue to build the region into a model area for HIV/AIDS prevention and control[3].

(7) China's principle mass organizations are members of the State Council AIDS Working Committee, and have actively joined in the response against HIV/AIDS. The All China Women's Federation and the All China Youth League, in cooperation with the Ministry of Health, conducted a "Face to Face" publicity campaign that targeted women and yong people in China Cares program sites. In 2003, the "Youth Red Ribbon" program supported 250 training workshops, knowledge contests, and plays. It also printed and distributed 1 100 000 copies of twenty – six different types of educational materials. The All China Federation of Trade Unions AIDS added education into the training of its members. For example, education on general sanitation and health care were supplemented with AIDS Education. By the end of 2003, more than 8 000 lecture courses had been conducted for over one million participants.

(8) Since the early 1990s, the Chinese Family Planning Association (CFPA) has paid special attention to AIDS prevention, and has taken an active part in the work of prevention and control. The CFPA has made great strides in IEC work, and has conducted a series of educational activities in diverse groups, ranging from Family Planning Association members, young people, and farmers to those in the floating (migrant) population, active military personnel, and service providers in entertainment establishments. Through these interventions, the CFPA has gained great experience in disseminating HIV/AIDS prevention knowledge.

2. Professional training and communication

Chinese NGOs organized various training courses and symposia on both sexually transmitted diseases and HIV/AIDS, and published journals to increase the professional knowledge of health system personnel. These activities included:

(1) The China International Symposium on AIDS in Beijing was hosted by the China Academy of Traditional Chinese Medicine and the CFPSA during December, 1995;

(2) The CASAPC produced educational materials for a ten province campaign, developed a training course on AIDS prevention and control for counselors on the mainland, and developed a training course on AIDS program management for Hong Kong;

(3) During 1996, the CASAPC developed a television film entitled "The Same World. The Same Hope" to educate viewers about HIV/AIDS, and organized a national level workshop on AIDS diagnosis and treatment;

(4) In 1997, the CASAPC jointed Hong Kong – mainland seminar on AIDS, and created an anti AIDS campaign along the Beijing – Hong Kong railway line;

(5) In 2001, the CASAPC assisted the Ministry of Health in organizing China's first large – scale AIDS conference, which attracted more than 1000 participants.

(6) From 2003 to 2005, four workshops on HIV/AIDS surveillance and diagnosis were organized for health professionals and workers in the systems of the Ministry of Health. Ministry of Justice and Ministry of Public Security in Jilin City, Baoding, Dalian, and Chengdu by the CFPSA and local Centers for Disease Control and Prevention. These workshops were supported by the University of Illinois at Chicago, Fogarty AITRP, the University of Nebraska, and the US – China AIDS Alliance Foundation;

(7) An International Medical Forum on Infectious Disease was held in Beijing by the CASAPC;

(8) The Working Committee of the Caring for the Younger Generation organization held several seminars and training classes at various governmental levels that included HIV/AIDS prevention and control knowledge. The trainees included cadres from various governmental units at the county, town, and village levels.

(9) The CASAPC held a national symposium during 2004 on the China – UNICEF Cooperation Project on AIDS Prevention and Control;

(10) The Chinese Journal for STD and AIDS Prevention and Control was started in 1995 by the CASAPC. The Journal on International Information on STD and AIDS went into publication in 1996. AIDS prevention and treatment information and advice on STDs and AIDS was posted on the internet, and the website "AIDS On Line" was established in November, 2001.

3. Social mobilization

Chinese NGOs have worked actively to involve all sectors of society in prevention and control efforts. These activities include:

(1) With the support of over thirty Chinese ministries and national departments, forty – seven national civil associations, forty – one media organizations, and representative from international or-

ganizations such as UNAIDS an the World Health Organization, the CFPSA launched the "121 Joint Action Plan" on March 28, 2003. The goal of the plan is to foster governmental and nongovernmental cooperation so that social resources can be utilized in an efficient and effective manner in fighting HIV/AIDS. The Joint Action Plan received broad social acceptance and media acclaim. National level scientists and academicians from the Chinese Academy of Sciences, the Chinese Academy of Engineering, and leaders from industry and commerce participated in activities that raised public awareness of both AIDS and the need for broad social involvement in battling the epidemic. On 1 December, 2003, a celebration was held in the Great Hall of the People in that recognized the achievements of, the Joint Action Plan. Both ZHOU Tienong, the Vice Chairman of the CCPCC, and WU Jieping, the Vice Chairman of the Standing Committee of the 9th NPC were in attendance. Individuals and organizations that had made donations to the plan were awarded certificates of merit. In January, 2004, the plan was recognized as being among the "Top 10 Public Relations Events in China During 2003. "

(2) From 2000 to 2005, the CFPSA engaged celebrities in literature and art as goodwill ambassadors. Among these individuals were film stars Jiang Wenli and Liu Xin, and singers Kris Phillips, Zheng Xulan, Zhang Huamin, and Cai Guoqing. These artists played a major role in publicizing AIDS knowledge, offering care and solace to orphans and persons living with HIV/AIDS, sending a message of nondiscrimination to the broader public, and conducting other AIDS – related activities.

4. Behavioral interventions and research targeting high risk populations

Over the years, Chinese NGOs have employed various mechanisms to carry out interventions among high risk populations; indeed, most needle exchange programs, interventions targeting drug users and men who have sex with men (MSM), and the promotion and distribution of condoms have been conducted by, or involved, Chinese NGOs.

(1) The Chinese Preventive Medicine Association (CPMA) has been involved in HIV/AIDS prevention and control since 1995. The CPMA implemented a condom promotion project, sponsored by the World Health Organization, among high risk populations. This was the first condom distribution project in China.

(2) The promotion and distribution of condoms at STD clinics in Shanghai and Henan Province was started in 1996.

(3) Behavioral investigations of commercial sex workers and long – distance truck drivers in Yunnan Province and Inner Mongolia were conducted in 1995 and 1997.

(4) An intervention targeting the drug user population in Liangshan Prefecture in Sichuan Province was conducted in 1998.

(5) The CASAPC conducted interventions among intravenous injection drug users, commercial sex workers, MSM, and migrants.

5. The AIDS network

Four joint meetings for NGOs on HIV/AIDS prevention and control were held in Beijing (1995), Zhuhai (2000), Harbin (2003), and Congqing (2005). At these meetings, NGO representatives from across China shared prevention and treatment information. At the first meeting, a Code of Action was passed by thirty – six organizations, including the CASAPC, the CFPSA, the Chinese Association of Medicine, the Chinese Association of Preventative Medicine, the China Family Planning Association, and the All China Women's Federation[4]. The code was reviewed and revised at subsequent meetings.

6. Care for people living with AIDS and AIDS orphans

Chinese NGOs have given great attention to the problems of social discrimination against people living with HIV/AIDS and AIDS orphans.

(1) The CFPSA and the Shanxi Provincial Health Department developed an educational intervention in Xia County during August – December, 2001. A preliminary assessment indicated a target population coverage rate of 90 percent, and a rise in the HIV/AIDS awareness rate from 5.7% to 34.6%. The awareness rate on HIV/AIDS transmission modes was raised from 41.2% to 81.1%. The awareness rate on HIV/AIDS non – transmission modes was raised from 12.32% to 45.07%. Thereafter, local NGOs in Xia County established a "Love Care Home" for people living with HIV/AIDS[5].

(2) The Working Committee of the "Caring for the Younger Generation" organization investigated the mental health status, general quality of life, and educational status of AIDS orphans in Henan, Shanxi, and Yunnan provinces. On the 16th World AIDS Day, the Working Committee and UNICEF held a joint seminar on "Caring for Orphans Affected by HIV/AIDS." Chinese state leaders, officials from UNICEF, and principals from relevant Chinese ministries attended the seminar and appealed to the public to be attentive to the living conditions of affected orphans.

(3) The CASAPC developed care and support programs for cities and counties in twelve provinces and the cities of Beijing and Shanghai.

(4) The Home of Loving Care has long been recognized for its care and treatment of people with HIV/AIDS, as well as its prevention and control efforts. Through the Home of Loving Care, those with AIDS have received medical care, disease consultation, guidance, and psychological support. It is a true bridge between people with HIV/AIDS and the broader society.

(5) More and more people living with HIV/AIDS are willing to publicize their status and participate in prevention and control activities. These individuals develop self – help groups and assist in antiretroviral treatment education.

(6) International NGOs also play an important role in antiretroviral and opportunistic infection treatment in China.

(7) The Positive Art Workshop (PAW) was established by persons living with HIV/AIDS in December, 2002. PAW activities include art events, interactive exhibitions in China and abroad, charity fundraisers, and AIDS awareness events. Of particular importance is their effort to document on film and still photography images that capture the hopes, fears, frustrations, and ideas for the future held by persons living with HIV/AIDS.

(8) Xia Shuqing, a woman from rural Jilin Province in northeast China, was infected with HIV in 2001. Initially despondent about her conditions, she received help from a local

NGO. Thereafter, she actively participated in AIDS – related activities, and organized a health club for others infected with HIV in her area. This club serves as a place for study, communication, mutual support, and self – help activities. In 2004, under the sponsorship of the CFPSA, Mrs. Xia attended the Chinese NGO satellite forum at the 15th International AIDS Conference in Bangkok and made a presentation to those in attendance. United Nations Secretary General Kofi Annan called her a hero in the fight against AIDS.

7. International exchange

Chinese NGOs have carried out many collaborative programs on AIDS prevention and control with international counterparts.

(1) Chinese NGOs actively participate in international AIDS conferences.

(2) Chinese NGOs participated in the AIDS meeting for Northeast Asian Countries in November, 2000. Those in attendance shared experiences in fighting the virus and strengthened their spirit of cooperation and common purpose.

(3) During the 15th International AIDS Conference in Bangkok, Chinese NGOs successfully organized various programs, including an NGO forum with the theme "Exchange, Cooperation, Friendship, and Facing the Challenge Together." More than 300 participants attended the forum, and heard key – note speeches by Dr. WANG Longde, the Vice – Minister of the Ministry of Health of the PRC, and Dr. Peter Piot, Executive Director of UNAIDS. The press reported that this was the first time that Chinese NGOs had made such a strong statement about HIV/AIDS.

(4) The 7th International Congress on AIDS in Asia and the Pacific was held in Kobe, Japan, on 1 – 5 July, 2005. During the Congress, a Chinese NGO symposium on HIV/AIDS prevention and control was held in the International Conference Center on the evening of 2 July 2005. This symposium was organized by the CFPSA, the CASAPC, and the Hong Kong AIDS Foundation. Financial support was obtained from numerous internal and external organizations and individual donators. Drs. Wang and Piot also attended this session and gave keynote addresses to more than 100 delegates from around the world. Members of vulnerable populations from the mainland, Hong Kong, and Taiwan also made presentations. A Chinese NGO exhibition was also held in the NGO exhibition area of the Congress. The news media once again noted the active role that Chinese NGOs are taking in the battle against HIV/AIDS.

(5) At the UN Special Assembly on AIDS on 2 June 2005, Vice – Minister Wang Long – de praised the contributions made by Chinese NGOs and others who are working hard against AIDS on the mainland. He encouraged the all sectors to participate in the battle. He stated that China is at a critical juncture in the battle against AIDS, and Chinese NGOs are key players in the effort. Indeed, he concluded that Chinese NGOs are "becoming an indispensable force in AIDS prevention and control[6]."

Challenges for the Future

In spite of the successes of Chinese NGOs in prevention and control efforts, several changes in approach need to be addressed:

（1）Chinese NGOs need a greater focus on capacity building.

（2）Chinese NGOs need increased financial investment.

（3）There is a need for more NGOs to participate in AIDS programs.

（4）There is a need for more cooperation and less competition between Chinese NGOs.

（5）There is a need for greater cooperation between government sectors and Chinese NGOs to avoid duplication of effort or gaps in coverage.

These changes will help to insure that the invaluable role that Chinese NGOs have played in the battle against AIDS will not only be maintained, but will actually grow in importance and impact.

[In 《Cell Research》 2005, 15 （11 -12）: 914 -918]

References

1　A Joint Assessment of HIV/AIDS Prevention, Treatment, and Care in China. Jointly prepared by the State Council AIDS Working Committee Office and the UN Theme Group on HIV/AIDS in China, 2004

2　Xu H, Anderson AF, Xu XH. AIDS knowledge exhibitions as an effective way to spread AIDS knowledge. Proceedings of the XIII International AIDS Conference, Durban, South Africa, 2000

3　Xu H, Zeng Y, Xu H, et al. "Developing Health Education on STD/AIDS Prevention Through Using Multi - Kind and MultiSector Methods is an Effective Way to Address the AIDS Epidemic," AIDS and Common Hygiene, China Medical Publishing House, 2005, 160 - 168

4　Wang GY. "Chinese NGOs will Jointly Fight HIV/AIDS," Chinese AIDS Prevention and Treatment, 1995, 1: 62 -63

5　Xu H. "Strengthening Interventions on HIV/ AIDS and Creating an Environment without Discrimination against PLWHA in Risk Areas" Poster, XIV International AIDS Conference, Barcelona, Spain. 2002

6　Wang LD. Presentation at the Chinese NGO Symposium at the CAAP, Kobe, Japan, 1 - 5 July 2005

96. Genetic Factors Leading to Chronic Epstein – Barr Virus Infection and Nasopharyngeal Carcinoma in Southeast China

GUO Xiu-chan[1,5], SCOTT Kevin[1], LIU Yan[2], DEAN Michael[2], DAVID Victor[2], NELSON George W[1], JOHNSON Randall C[1], DILKS Holli H[2], LAUTENBERGER James[2], KESSIN Bailey[1], MARTENSON Janice[2], GUAN Li[1], SUN Shan[2], DENG Hong[3], ZHENG Yu – ming[3], de The Guy[4], LIAO Jian[5], ZENG Yi[6], O'BRIEN Stephen J[2], WINKLER Cheryl A[1]

1. Laboratory of Genomic Diversity, SAIC Frederick, National Cancer Institute – Frederick, Frederick, USA; 2. Laboratory of Genomic Diversity, National Cancer Institute – Frederick, Frederick, USA; 3. Cancer Institute of Wuzhou, China; 4. Institute Pasteur, France; 5. Cangwu Institute for Nasopharyngeal Carcinoma Control and Prevention, China; 6. Institute for Viral Disease Control and Prevention, Chinese Center for Disease Control and Prevention, China

Summary

Nasopharyngeal carcinoma (NPC) is a complex disease caused by a combination of Epstein – Barr virus chronic infection, the environment and host genes in a multi – step process of carcinogenesis. The identity of genetic factors involved in the development of chronic Epstein – 8arr virus infection and NPC remains elusive, however. Here, we describe a two – phase, population – based, case – control study of Han Chinese from Guangxi province, where the NPC ~ incidence rate rises to a high of 25 – 50 per 100 000 individuals. Phase I. powered to detect single gene associations, enrolled 984 subjects to determine feasibility, to develop infrastructure and logistics and to determine error rates in sample handling. A microsatellite screen of Phase I study participants, genotyped for 319 alleles from 34 microsatellites spanning an 18 – megabase region of chromosome 4 (4p15. I – q12), previously implicated by a linkage analysis of familial NPC, found 14 alleles marginally associated with developing NPC or chronic immunoglobulin A production ($P = 0.001 – 0.03$). These associations lost significance after applying a correction for multiple tests. Although the present results await confirmation, the Phase II study population has tripled patient enrolment and has included environmental covariaes, offering the potential to validate this and other genomic regions that influence the onset of NPC.

[**Key words**] Nasopharyngeal carcinoma; Chromosome 4; Microsatellite; Association study; Epstein – Barr virus

Introduction

Nasopharyngeal carcinoma (NPC) is a disease with distinct racial and geographical distributions. In southern China, Taiwan, Vietnam and the Philippines, the incidence of NPC is 15 – 20 per 100 000 individuals per year, and in some local Chinese regions bordering the Xijiang River drainage in Guangdong and Guangxi provinces, the incidence is as high as 25 – 50 per 100 000 individuals. [1,2] An intermediate incidence is observed among the Arab populations of Northern Africa, [3] including Saudi Arabia; [4] in the Caribbean; and in the Eskimo populations of Alaska and Greenland. [5] Elsewhere, NPC is rare, with an incidence of less than 1 per 100 000. In the USA, NPC comprises only 0.2 per cent of all malignancies, with an incidence is 1 per 100 000. The male : female ratio t'or NPC is usually 2 or 3 : 1, with an incidence peak between 50 and 59 years of age. [6]

A link between NPC and Epstein – Barr virus (EBV) was reported in 1966[7]. Ten years later, the presence of immunoglobulin (Ig) A antibodies to EBV viral capsid antigens (EBV/IgA/VCA) was found to serve as a predictive marker for the development of NPC in Chinese populations[8]. More than 95 per cent of adults in all ethnic groups across the world are healthy carriers of EBV. In high NPC incidence regions, EBV infection of the nasopharyngeal epithelium induces IgA antibodies against VCA, suggesting that reactivation of EBV replication at the mucosal surface precedes the development of NPC. Consistent with this, approximately 2.5 per cent of the general population are EBV/IgA/VCA antibody positive. Of these, less than 3 per cent will develop NPC, while > 95 per cent of all NPC patients are EBV/IgA/VCA antibody positive[9-14]. In addition to EBV infection, case control studies have indicated a role for environmental factors, including food preservatives (carcinogenic nitrosamines), salt – preserved fish and phorbol esters in herbs and plants that are commonly consumed among ethnic populations with the highest NPC rates[15,16].

Evidence for genetic modulation of NPC risk has accumulated recently. Familial aggregation of NPC has been observed in China and in other countries[17-19]. Familial aggregation of NPC is uncommon in low – risk or non – Chinese populations. The proportion of NPC with affected first – degree family history is > 5 per cent in south China, 7.2 per cent in Hong Kong, 6.0 per cent in Yulin and 5.9 per cent in Guangzhou[20]. Descendants of south Chinese immigrants to western countries show progressively lower risk. but their NPC incidence remains higher than that of the indigenous population[21], suggesting both environmental and genetic components to disease susceptibility. Several studies have shown associations between HLA genes and NPC,[22-28] and the D6S1624 microsatdfite within the HLA class 1 region has been associated with NPC[29]. Studies comparing age of NPC onset report conflicting results for familial versus sporadic NPC. In a study comparing 200 probands with and without NPC – affected first – degree relatives from Singapore, the age of onset was 48 and 49 years, respectively[30]. In another Chinese study the average age of onset was 35.5 years in 32 Guangdong families with 4 – 5 relatives with NPC compared with 46.6 wears for sporadic cases[20], In a third study, however, the age of onset decreased from 44.5 years to 40.4 as the number of NPC – affected relatives increased from one to four[31]. There is, therefore, some suggestion that age of onset may be lower in families with one or more NPC – affected first – degree relatives.

A genome – wide linkage analysis of 20 NPC families from a high incidence region in Guangdong identified a susceptibility region on the short arm of chromosome 4[32]. Two chromosome 4p15. 1 – q12 markers. D4S405 and D4S3002, yielded high logarithm of the odds (LOD) scores (> 3. 5) by both parametric and multipoint non – parametric analysis in 70 per cent of the NPC families studied. A subsequent study of 18 families from Hunan province genotyped a panel of markers on the short arms of chromosomes 3, 9 and 4 that included D4S405 and D4S3002 and failed to detect an obvious susceptibility locus on 4p15. 1 – q12[33]. A region on chromosome 3p21. 31 – 21. 2 containing a tumour suppressor gene cluster, however, showed a modest association with NPC incidence[33].

Here, we describe the design of a new case – control study population recruited for the discovery of genetic factors that are involved in the development of chronic EBV infection and in the development of NPC. In a preliminary test to resolve the discrepancy between the two family – based studies, we performed a population – based case – control association analysis of 34 microsatellite markers within 4p15. 1 – q12 (Fig. 1) to determine if specific alleles within the region: 1) were associated with a propensiy to develop chronic EBV replication, as evidenced by IgA antibodies against EBV viral capsid antigen (EBV/I – gA/VCA); or 2) were associated with NPC susceptibility.

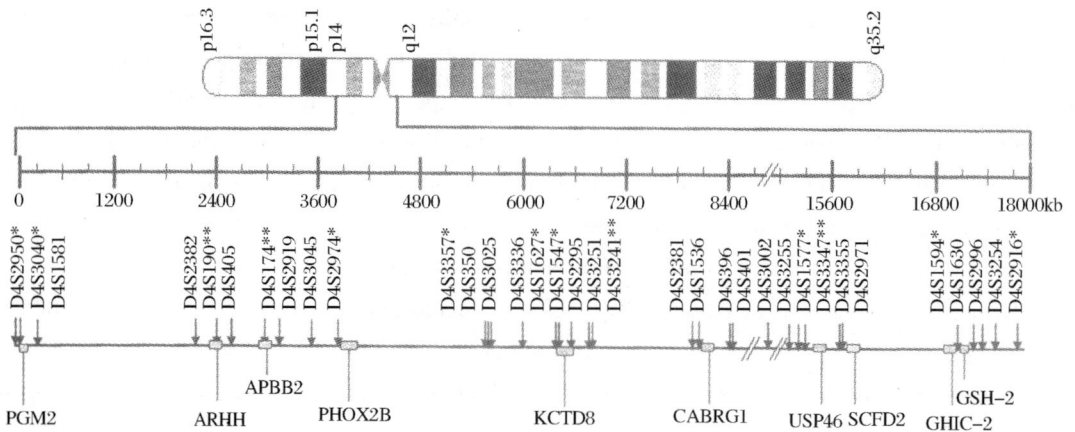

Fig. 1 **Map of markers for the nasopharyngeal carcinoma susceptibility locus on chromosome 4[28]. Thirty – four short tandem repeat markers distributed across an 18 megabase region were selected between D4S2950 to D4S2916, with intervals of 10 – 3 500 kilobases. The positional relationship among markers and selected genes are indicated below the map. Asterisks indicate levels of statistical significance: ∗ P < 0. 05 and ∗ ∗ P < 0. 01**

Materials and Methods

1. Study design

Enrohnent into the study occurred in two collection phases. The Phase I pilot was powered to detect single gent associations and to determine feasibility for meeting recruitment goals, accuracy of data collection and sample handling, and to develop the infrastructure for a large international collaboration. Cases and controls (n = 984) were recruited in 2000 from Wuzhou City and Cangwu County. bordering the Xijiang River in the Guangxi province of South East China. An effort was made to enrol triads consisting of a proband, an unaffected spouse and an adult child or parent. Family triads were enrolled for haplotype inference and for quality control assessment. Three clinically described disease categories were collected: 1) incident or prevalent NPC biopsy – confirmed (NPC$^+$) cases ($n = 350$) who were EBV/IgA/VCA antibody positive (IgA$^+$); 2) IgA$^+$ cases ($n = 288$) who were defined as EBV/IgA/VCA antibody positive and NPC free at the time of study enrolment (EBV/IgA/VCA titres were confirmed by serological testing at the time of study enrolment): 3) IgA$^-$ controls ($n = 346$). For each case, his or her spouse was tested for EBV/IgA/VCA antibodies, and the spouse and parent or adult child were invited to enrol. The IgA$^-$ group consisted of 346 spouses who were IgA$^-$ at the time of study enrolment (Tab. 1). A dominant model was selected for power calculations for two reasons: 1) if the true model is additive, there is little difference in power using either an additive model or a dominant model for power calculations (data not shown): 2) if, however, the true model is dominant, a dominant model is the most powerfial. Assuming a dominant genetic model and at least a 10 per cent allele frequency, this number of NPC, IgA$^+$, and IgA$^-$ cases and controls provided >90 per cent power to detect associations with an odds ratio (OR.) $\geqslant 3$, at the $p = 0.01$ level for a two – tailed test (Tab. 2).

Tab. 1 Characteristics of the Phase I study groups

Item	NPC positive	NPC negative	
		EBV/IgA/VCA positive	EBV/IgA/VCA negative
No. study participants	350	288	346
Male/female (% male)	233/117 (67%)	142/146 (49%)	129/217 (59%)
Age *	48 ± 10 (16 – 79)	44 ± 9 (20 – 77)	47 ± 10 (18 – 75)
IgA/VCA titre	1: 10 – 1: 640	1: 5 – 1: 80	<1: 5
IgA/EA titre	1: 10 – 1: 160 (57.4%)	1: 5 – 1: 20 (5%)	<1: 5

Notes: Abbreviations: EBV Epstein-Barr virus; IgA. ImmunolglobulinA; NPC Nasopharyngeal carcinoma; VCA riral capsid antigens * Age at NPC diagnosis and at study enrolment for IgA Serostatus

Phase II enrolment was initiated in 2004 and after the completion of Phase I collection. The Phase II design is a cross – sectional, case control study: family members were not recruited. A questionnaire capturing environmental factors, including occupational, dietary and tobacco exposures, was administered to each study participant at enrolment (Tab. 3). NPC cases were recruited from the Wuzhou Red Cross Hospital in collaboration with the Cancer Institute of Wuzhou, Wuzhou City and the Cangwu Institute for Nasopharyngeal Carcinoma Control and Prevention, Cangwu County. NPC cases, IgA$^+$ subjects and IgA$^-$ participants were recruited from cities and villages bordering the Xijiang River. Power was determined for single gene and gene environment interactions for participants in each group (Tab. 2). For single – gene asssociations at a 10 per cent allele frequency, power will range from 83 per cent to > 99 per cent and from 35 per cent to > 99 per cent to detect associations with an odds ratio (OR) of 1. 5 – 3. 0 at $P < 0.05$ and $P < 0.001$, respectively, for the dominant genetic model and a two – sided significance level. For gene – environment interactions, there is power to detect gene – environment effects for genotype and exposures with frequencies ≥ 0.1 for genotype and exposure. if the main exposure effect and genotype have an $OR \geq 1$ and an interaction effect of $OR \geq 3$ [34].

Tab. 2　Phase I and phase II sample power

Group	OR		Phase I		Phase II		All	
			Case/control	Power(%)	Case/control	Power(%)	Case/control	Power(%)
NPC + vs IgA +			350/288		1024/1009		1374/1297	
	0. 75	1. 5		17		64		79
	0. 50	2. 0		62		>99		>99
	0. 33	3. 0		98		>99		>99
IgA + vs IgA –			288/346		1009/1022		1297/1368	
	0. 75	1. 5		18		64		79
	0. 50	2. 0		64		99		>99
	0. 33	3. 0		99		>99		>99
NPC +/IgA + vs IgA –			638/346		2033/1022		2671/1368	
	0. 75	1. 5		25		77		90
	0. 50	2. 0		78		>99		>99
	0. 33	3. 0		>99		>99		>99

Note: Allele frequency = 10 per cent; Type I error = 1 per cent.

Power calculated For a 10% allele frequency, dominant genetic model and a p value of ≤ 0.01 for given case and control numbers using a two – railed test. For example, with the Phase I cases/controls available, we would discover a genetic association of strength $OR = 1.5$, only 14% of the time, but a stronger gene ($OR = 3.0$) would be detected 96% of the time with available patient numbersAbbreviations: IgA. immunoglobulin A; NPC, nasopharyngeal carcinoma

Tab. 3　Phase II study design and non – genetic covariates

Exposure	EBV/IgA/VCA status		
	NPC/IgA + (%)	IgA + (%)	IgA (%)
No. enrolled[1]	1024	1009	1022
Male: female (% male)	735/289 (72)	555/454 (55)	684/338 (67)
Consume dried meat[2]	292 (23.5)	172 (17.0)	283 (27.7)
Wood cooking fires[2]	995 (97.2)	960 (95.1)	914 (89.4)
Occupational exposure to solvents[2]	78 (07.6)	31 (03.1)	87 (08.5)
Smoking > 10 years	531 (51.9)	408 (40.4)	372 (36.4)

Notes: EBV, Epstein – Barr virus; IgA, immunoglobulin A; NPC. nasopharyngeal carcinoma; VCA, viral capsid antigens.

　　1 Greater than 99 per cent of IgA + and IgA – participants and 100 per cent of the NPC cases were born in Guangdong or Guangzhou provinces. 2 Participants reporting any level exposure.

Exclusion criteria for Phases I and II were ethnicity other than Han Chinese, birth or residency for more than six months outside of the NPC endemic region or failure to provide informed consent. Internal review board approval was obtained from all participating institutions and informed consent was obtained from each study participant or their guardian for subjects between 16 and 18 years of age.

2. Sample and data handling

A total of 10 – 20ml of blood was collected in acid citrate dextrose (ACD) vacutainers for serology testing, direct DNA extraction and for cryopreservation of peripheral blood mononuclear cells (PBMCs). Blood samples were separated into plasma and PBMCs. Serum was tested at the Wuzhou Centre for EBV/IgA/VCA antibodies and antibodies to EBV early antigen by immunoenzymatic assay. The PBMCs were EBV – transformed to establish lymphoblastoid cell lines (LCLs) as a renewable DNA source. In addition, 3 cc of whole blood were preserved in DNA Tris, ethylene diaminc tetraacetic acid and sodium dodecyl sulphate extraction buffer as a back – up DNA source. Questionnaires capturing demographic, laboratory and social history were administered at enrohnent. Two individuals entered responses to the questionnaire and laboratory results into a FileMaker Pro database independently as a method of capturing data entry errors.

3. Genomic DNA extraction

DNA was extracted from whole blood or lymphoblastoid cell lines using the QIAamp DNA blood maxi kit (Qiagen, Valencia, CA, USA, catalog #51194). More than 80 per cent of the genotypes were determined from DNA directly extracted from whole blood.

4. Microsatellite genotyping

Microsatellite loci (n = 34) containing 319 alleles were selected between D4S2950 and D4S2916 (18 megabases [Mb]) on chromosome 4 (Fig. 1). The markers consist of 22 dinucleotide repeats, two trinucleotide repeats and ten tetranucleotide repeats. The genetic and physical distances between marker pairs are as follows: mean = 0. 51 centimorgans (cM), 562 kilobases (kb); median = 0. 34 cM, 230 kb; and range = 0. 00 – 2. 79 cM, 11 – 4185 kb. The primer sequences were obtained from the University of Santa Cruz Genome Bioinformatics database[35]. All of the forward primers were 5' – tailed with the M13 sequence 5' – CACGACGTTGTAAAACGAC – 3'. The M13 – forward primers were used in combination with an M13 primer that had the same sequence but was labelled at its 5' end with a fluorescent reagent from Applied Biosystems (ABI) Foster City. CA, USA such as 6 – FAM, VIC or NED. The latter primer is the sole source of label and can be used with any M13 – forward primer to generate a labelled amplified allele[36]. The polymerase chain reaction (PCR) amplifications of individual microsatellite loci were performed in 10 μl volumes containing 10 mmol/L Tris – HCl (pH 8. 3), 50 mmol/L KCl, 2. 0 mmol/L $MgCl_2$, 0. 2 mmol/L of each dinucleotide triphosphate, 1 μmol/L labelled M13 primer and reverse primer, 0. 07 mmol/L M13 – tailed primer (15 : 1 molar ratio of labelled M13 primer versus a M13 – tailed forward primer), 25 ng genomic DNA, 0. 5 U TaqGold DNA polymerase (Applied Biosystems, Foster City, CA, USA). PCR amplification was performed in a PE Applied Biosystems Model 9700 using 384 high – throughput format plates. The PCR conditions were a modified touchdown PCR procedure: 95℃, 10 minutes; two cycles of 95℃, 15 seconds; annealing temperature, 30 seconds; 72℃, 45 seconds, at annealing temperatures of 60℃, 58℃, 56℃. 54℃, 52℃; 30 cycles at an annealing temperature of 50℃; 72℃, 30 minutes. Six PCR products were pooled together for multiplex loading according to the label colour and marker size. Samples were diluted appropriately pooled and then 3 μl of sample was mixed with 9 μl of formamide containing Liz 350 size standard (ABI). Samples were electrophoresed in a 22 cm capillary array using POP5 polymer and 3700 running buffer (ABI) on an ABI Model 3100 Automated DNA Sequencer using data collection software version 1. 0. 1 and Genescan Analysis software version 3. 7. Genotyping was performed using Genotyper Version 2. 5 and allele sizes were binned using Allelogram (Carl Manaster, available at http: //s92417348. onlinehome. us/software/allelogram/index. html). For quality control between plates, DNAs from 22 per cent of subjects were duplicated across plates. Mendelian errors were tested within the triad famlies using the PedChek program.

5. Genetic association analyses

Allele frequencies were computed and compared between cases and controls using Pearson's x^2 test or Fisher's exact test. ORs, 95 per cent confidence intervals (CIs) and p values were computed for dominant and recessive genetic models adjusted for age and sex. Logistic regression adjusted for age and sex was used to compute ORs using SAS PROC LOGISTIC software (SAS Institute, Cary, NC, USA). ORs were computed for a domnant model, comparing the combined homozygous and heterozygous genotypes against all other genotypes. When the allele frequency of the minor allele was ≥ 5 per cent, ORs were calculated for the recessive model, comparing the homozygous genotype against all other genotypes. Conformance to Hardy – Weinberg equilibrium expectations was calculated for all loci. Tests for D' as a measure of linkage disequilibrium (LI) were conducted for allele pairs using SAS Genetics software (SAS Institute, Cary. NC. USA).

Results

The Phase I pilot study enrolled participants from the Cancer Institute in Wuzhou City and the Cangwu Institute for Nasopharyngeal Carcinoma Control and Prevention. Cangwu County in Guangxi province in the autumn of 2000. For NPC cases, 71. 3 per cent of spouses and 81 per cent of adult children were enrolled. For cases with EBV/IgA/VCA titres consistent with chronic EBV infection, 72. 4 per cent of spouses and 67. 4 per cent of adult children or parents were enrolled. Complete triad sets were available for 366 NPC probands. As predicted for this higly endemic NPC region. 71. 8 per cent of the NPC cases were male, PBMCs cryopreserved on – site were transported to the Laboratory of Genomic Diversity – National Cancer Insitute (LGD – NCI) for EBV immortalisation: 83 per cent of 633 transformation attempts resulted in LCls.

Sample and genotyping errors were estimated by including 10 per cent duplicate sampling with one sample derived from DNA isolated directly from peripheral blood and the second from DNA isolated from LCLs. Less than 0. 5 per cent mismatches within duplicate samples were observed, all of which were resolved using family trios, indicating that tubes collected from a single individual were appropriately labelled (data not shown) and that error was not introduced during cell line development or sample handling. A second test for Mendelian errors using PcdChek was performed using the chromosome 4 microsatellite data (described below). Two unresolved Mendelian errors were observed within the 366 family triads. Near – complete genotyping and complete clinical data were available for 350 NPC cases, 288 IgA seropositives and 346 IgA seronegatives (Tab. 1).

Phase II enrolment occurred between November 2004 and July 2005 in Guangxi province. Subjects were enrolled if at least one parent was from the Guangxi or Guangdong provinces. NPC cases were identified as seroincident or seroprevalent cases presenting at Red Cross hospitals and IgA$^+$ and IgA$^-$ controls were identified from field stations in cities and villages bordering the Xijiang River drainage. Table 3 presents summary data of environmental exposures for the Phase II NPC$^+$, IgA$^+$ and IgA$^-$ groups and the numbers of participants enrolled.

We have addressed the questions of whether a locus within the chromosome 4p 15. 1 – q12 region leads to the development of NPC or the development of EBV/IgA/VCA in response to EBV replication using the Phase I cases and controls. Micro – satellite loci ($n = 34$) were distributed over an 18 Mb region on chromosome 4p15. 1 – q12, with intervals of10 – 3500 kb and an average distance of 530 kb. Four Phase I genetic association comparisons were made: 1) NPC cases versus EBV/IgA/VCA seropositive controls (Tab. 4); 2) EBV/IgA/VCA seropositive cases without NPC versus EBV/IgA/VCA seronegative controls (Tab. 5); 3) NPC cases plus EBV/IgA/VCA seropositive cases versus EBV/IgA/VCA seronegative controls (Tab. 6); and 4) NPC cases versus EBV/IgA/VCA seronegative controls (data not shown). No distortions in Hardy – Weinberg equilibrium were observed. Alleles with at least one significant result ($P < 0.05$) for either the dominant or recessive genetic models are reported in Tab. 4 – 6. The results arc presented without correction for multiple comparisons because the interrogated 4p 15. 1 – q12 regi on was previously implicated as a susceptibility locus in a family – based study and we were specifically testing the prior hypothesis that markers within the region would also be associated with NPC in a population – based study[32]. It should be noted that associations with $P > 0.0015$ would not remain significant after correction for multiple comparisons considering the 34 independent loci.

Tab. 4 Significant allele frequencies between NPC versus IgA$^+$ groups

cM *	Locus	Individuals#			Allen freq (%)			Dominant			Recessive		
		Allgle	NPC	IgA +	NPC		IgA +	OR	p value	95% CI	OR	p value	95% CI
37. 64	D4S2950	141	334	270	12. 9	<	19. 6	0. 70	0. 06	0. 48 – 0. 10	0. 30	0. 20	0. 11 – 0. 84
37. 74	D4S3040	213	324	261	17. 4	>	14. 0	1. 52	0. 03	1. 04 – 2. 22	1. 47	0. 42	0. 57 – 3. 75
37. 74	D4S3040	215	324	261	15. 6	>	12. 5	1. 21	0. 34	0. 82 – 1. 79	5. 36	0. 03	1. 17 – 24. 53
41. 57	D4S2974	135	347	281	60. 7	<	64. 9	1. 22	0. 41	0. 76 – 1. 97	0. 67	0. 02	0. 48 – 0. 93
41. 57	D4S2974	137	347	281	16. 9	>	11. 4	1. 51	0. 03	1. 03 – 2. 20	3. 43	0. 07	0. 92 – 12. 85
43. 27	D4S3357	271	340	271	29. 7	<	33. 4	0. 71	0. 04	0. 51 – 0. 99	1. 01	0. 98	0. 60 – 1. 71
43. 33	D4S350	256	346	275	64. 6	>	58. 7	1. 61	0. 05	1. 01 – 2. 56	1. 24	0. 21	0. 89 – 1. 7
44. 11	D4S1547	251	349	276	56. 4	>	52. 7	0. 91	0. 64	0. 62 – 1. 35	1. 53	0. 02	1. 07 – 2. 19
44. 28	D4S2295	222	345	278	76. 1	<	81. 3	0. 52	0. 12	0. 23 – 1. 19	0. 71	0. 05	0. 50 – 1. 0
45. 71	D4S2381	277	347	277	61. 8	<	65. 9	0. 59	0. 03	0. 36 – 0. 95	0. 92	0. 60	0. 66 – 1. 28
53. 64	D4S2971	165	344	275	45. 1	>	43. 3	1. 43	0. 05	1. 00 – 2. 05	0. 76	0. 20	0. 50 – 1. 1
55. 73	D4S2916	204	336	265	13. 8	>	10. 6	1. 55	0. 03	1. 03 – 2. 33	0. 57	0. 39	0. 16 – 2. 07

Notes: Abbreviations: CI, confidence interval; cM, centimorgan; IgA, immunoglobulin A; NPC, nasopharyngeal carcinoma; OR, odds ratio. * See Fig. 1.

Tab. 5　Group 2: Markers and alleles showing significant allele frequencies among IgA$^+$ cases without NPC and IgA$^-$ subjects

cM	Locus	Individuals#			Allen freq (%)			Dominant			Recessive		
		Allgle	IgA +	IgA –	IgA +		IgA –	OR	p value	95% CI	OR	p value	95% CI
37. 64	D4S2950	137	230	319	13. 3	<	18. 5	0. 56	0. 004	0. 38 – 0. 83	0. 75	0. 56	0. 28 – 2. 00
40. 14	D4S190	170	236	332	23. 9	>	18. 7	1. 40	0. 06	0. 98 – 1. 99	2. 38	0. 04	1. 03 – 5. 49
40. 75	D4S174	202	237	331	23. 4	<	32. 6	0. 59	0. 003	0. 42 – 0. 84	0. 33	0. 002	0. 16 – 0. 67
43. 27	D4S3357	271	232	326	34. 3	>	27. 8	1. 51	0. 02	1. 06 – 2. 13	1. 46	0. 20	0. 82 – 2. 61
44. 09	D4S1627	218	239	335	33. 3	<	38. 1	0. 81	0. 24	0. 58 – 1. 15	0. 57	0. 039	0. 33 – 0. 97
44. 48	D4S3241	136	233	325	13. 7	>	8. 9	1. 91	0. 004	1. 24 – 2. 94	2. 11	0. 28	0. 54 – 8. 27
45. 71	D4S2381	301	238	330	18. 9	<	23. 0	0. 70	0. 05	0. 49 – 1. 0	0. 90	0. 80	0. 42
45. 84	D4S1536	284	233	327	33. 7	<	40. 7	0. 79	0. 19	0. 56 – 1. 13	0. 48	0. 01	0. 27 – 0. 84
52. 95	D4S1577	143	238	332	34. 2	<	41. 6	0. 70	0. 05	0. 49 – 1. 0	0. 52	0. 01	0. 31 – 0. 86
53. 03	D4S3347	213	237	332	34. 8	>	29. 5	1. 65	0. 004	1. 17 – 2. 33	0. 89	0. 67	0. 50 – 1. 56
53. 03	D4S3347	217	237	332	45. 4	<	50. 0	0. 87	0. 48	0. 59 – 1. 28	0. 66	0. 05	0. 43 – 1. 0
54. 86	D4S1594	266	236	330	66. 1	>	64. 2	1. 76	0. 04	1. 02 – 3. 03	1. 01	0. 95	0. 71 – 1. 43

Notes: CI, confidence interval; cM, centimorgan; IgA, immunoglobulin A; OR, odds ratio.

Tab. 6　Group 3: Markers and alleles showing significant allele frequencies among NPC cases plus IgA$^+$ cases and IgA$^-$ subjects

cM	Locus	Individuals#			Allen freq (%)			Dominant			Recessive		
		Allgle	NPC	IgA –	NPC + IgA +		IgA –	OR	p value	95% CI	OR	p value	95% CI
37. 64	D4S2950	137	604	319	15. 0	<	18. 5	0. 69	0. 01	0. 51 – 0. 93	1. 06	0. 88	0. 50 – 2. 26
40. 14	D4S190	170	627	332	24. 1	>	18. 7	1. 50	0. 005	1. 13 – 1. 99	1. 99	0. 07	0. 96 – 4. 14
40. 75	D4S174	202	616	331	27. 4	<	32. 6	0. 79	0. 10	0. 60 – 1. 05	0. 46	0. 001	0. 28 – 0. 74
40. 75	D4S174	204	616	331	11. 5	>	8. 5	1. 45	0. 04	1. 02 – 2. 08	3. 17	0. 15	0. 67 – 15. 0
43. 27	D4S3357	275	611	326	21. 4	<	24. 8	0. 72	0. 02	0. 55 – 0. 96	0. 95	0. 87	0. 51 – 1. 75
44. 09	D4S1627	218	625	335	31. 8	<	38. 1	0. 76	0. 05	0. 58 – 1. 0	0. 57	0. 007	0. 38 – 0. 86
44. 48	D4S3241	136	611	325	13. 4	>	8. 9	1. 63	0. 007	1. 14 – 2. 31	1. 85	0. 30	0. 58 – 5. 89
45. 84	D4S1536	284	615	327	34. 5	<	40. 7	0. 78	0. 09	0. 59 – 1. 04	0. 55	0. 004	0. 37 – 0. 83
46. 25	D4S401	213	628	333	10. 8	<	11. 1	1. 09	0. 63	0. 77 – 1. 53	0. 33	0. 05	0. 11 – 1. 00
52. 72	D4S3255	189	614	328	8. 5	>	6. 1	1. 50	0. 05	1. 0 – 2. 25	3. 18	0. 30	0. 36 – 28. 0
52. 95	D4S1577	143	624	332	35. 8	<	41. 6	0. 72	0. 02	0. 54 – 0. 95	0. 67	0. 03	0. 46 – 0. 96
53. 03	D4S3347	213	623	332	34. 3	>	29. 5	1. 58	0. 001	1. 20 – 2. 08	1. 02	0. 93	0. 66 – 1. 58
53. 03	D4S3347	217	623	332	45. 2	<	50. 0	0. 90	0. 48	0. 66 – 1. 22	0. 62	0. 004	0. 45 – 0. 86
54. 86	D4S1594	266	620	330	66. 5	>	64. 2	1. 57	0. 03	1. 04 – 2. 38	1. 03	0. 86	0. 78 – 1. 35

Notes: CI, confidence interval; cM, cendmorgan; IgA, immunoglobulin A; OR, odds ratio.

1. Linkage disequilibrium among the 34 loci

The spacing of markers varied from 10 – 3500 kb, with denser coverage flanking the microsatellite markers with the highest LOD scores from the family study (Fig. 1). We calculated two-point D' as a measure of LD between all alleles at neighbouring short tandem repeat loci; however, a D' value of 1 (complete LD) was observed for only 60 two – point allele combinations. Using HapMap single nucleotide polymorphism (SNP) data (http://www.hapmap.org), we examined whether the microsatellites were included in reasonably strong LD blocks. The r^2 between any given marker pairs were set at a 0.8 cut – off threshold to determine the LD blocks. Only 11 of the 34 microsatdlite markers occurred within an LD block: D4S396 and D4S401 occurred within the same 17 kb block. Of the two NPC – linked markers, [32] D4S405 was not witlfin a block and D4S3002 occurred within an 8 kb block. The mean size of the blocks was 17.9 kb (range 8 – 50 kb).

2. Genetic association with NPC

Tab. 4 presents the locus name, location, allele length, allele frequencies, *ORs*, *p* values and 95 per cent Cls for 350 NPC cases and 288 EBV/IgA/VCA seropositive controls (93.0 – 99.7 per cent of NPC cases and 91.0 – 97.6 per cent of IgA $^+$ subjects were genotyped successfully). The genotype frequencies among NPC cases were significantly higher than those among control subjects for five alleles (*OR.* 1.51 – 5.36; *P* = 0.01 – 0.03): for the recessive model, D4S3040 – 215 and D4Sl547 – 251; and for the dominant model, D4S3040 – 213, D4S2974 – 137 and D4S2916 – 204. The genotype frequencies among NPC cases was statistically lower than among control subjects for four alleles (*OR* 0.3 – 0.71: *P* = 0.02 – 0.045): for the recessive model, D4S2950 – 141 and D4S2974 – 135; and for the dominant model, D4S3357 – 271 and D4S2381 – 277.

3. Genetic association with persistent IgA $^+$ status

To test the hypothesis that genetic factors may influence EBV/IgA/VCA formation in response to EBV infection, we compared genotype frequencies between 288 lgA $^+$ cases and 346IgA $^-$ controls (Tab. 1). Tab. 5 provides the allele frequencies. *p* values, *ORs* and 95 per cent *CIs* in cases and controls for significant results. Eleven alleles were significantly associated with lgA $^+$ persistence: five risk alleles (*OR* 1.51 – 2.38; *P* = 0.004 – 0.040) and six protective alleles (*OR.* 0.33 – 0.70; *P* = 0.002 – 0.050).

Because all NPC cases in our study were IgA $^+$, we then pooled NPC and IgA $^+$ cases together to increase power, with the hypothesis being that the alleles associated with IgA $^+$ serostatus would be shared among NPC $^+$ IgA $^+$ and NPC $^-$ IgA $^+$ individuals. Significant associations are presented in Table 6: four alleles were associated with risk for lgA + (*OR* 1.5 – 1.63; *P* = 0.001 – 0.030) and seven were protective (*OR* 0.46 – 0.76; *P* = 0.001 – 0.050). Based on the two comparisons (Tab. 5 and 6), ten alleles associated with IgA were shared in both comparisons. Five were highly significant (*P* < 0.01) associations with IgA $^+$ serostatus. Alleles D4S190 – 170 (*P* = 0.005; *OR* 1.5, 95% *CI* 1.13 – 2.0). 134S3241 – 136 (*P* = 0.004: *OR* 1.91, 95% *CI* 1.2 – 3.0) and D4S3347 – 213 (*P* = 0.001: *OR* 1.58, 95% *CI* 1.2 – 2.1) significantly increased the risk of developing EBV/IgA/VCA. Alleles D4S174 – 202 (*P* = 0.001: *OR* 0.46, 95% *CI* 0.3 – 0.7) and

D4S2950 – 137 (P = 0.0036; OR 0.56, 95 per cent CI 0.38 – 0.83) significantly decreased the risk of EBV/IgA/VCA. Within a single locus (D4S3357), one allele increased susceptibility (D4S3357 – 271) and the other allele was protective (D4S3357 – 275).

Discussion

We have described the design and recruitment efforts for a genetic association study to investigate the role of host genetic factors in the development of chronic EBV infection leading to NPC in subjects born and living in a region with one of the world's highest incidence rates of NPC. This study was conducted in two phases. Phase I was a pilot study to explore the feasibility of conducting a cross – sectional study in China (Tab. 1). The pilot provided strong support for expanding the study in several important ways: export perntits for genetic material were obtained, sample handling was excellent – with few detectable errors – and recruitment goals were attainable. Upon the successful completion of Phase 1, we increased the catchment area for IgA$^+$ cases to cities and villages along the Xijiang River and tributaries, expanded the study to include more subjects and added a detailed questionnaire to capture environmental exposures that may interact with host genes in the development of NPC (Tab. 2). Complementing previous studies, we also attempted to determine if Phase II of the study was powered for the detection of both gene – gene and gene – enviromnent interactions.

To revisit the recent linkage analysis in NPC families implicating a susceptibility locus linked to chromosome 4p15. 1 – q12, we selected 34 microsatellite loci spanning the 18 Mb region at intervals of 1. 0 – 3500 kb. Unlike in previous studies, we first also attempted to determine if the chromosome 4 region was associated with EBV/IgA/VCA antibody formation and, secondly, if the chromosome 4 region was associated with NPC incidence in the setting of EBV replication as indicated by EBV/IgA/VCA. We identified several loci that showed significant associations with either EBV/IgA/VCA or NPC status. The associations tended to be marginally significant for NPC (Tab. 4), with somewhat stronger associations observed for EBV/IgA/VCA (IgA$^+$) (Tab. 5 and 6).

Few NPC families have been identified outside of NPC endemic areas. More than 90 per cent of all NPC cases do not show fanfilial aggregation or family history, imply either environmental causes or geographical family clustering. Two family – based NPC linkage studies implicated different chromosomes as harbouring an NPC susceptibility locus[32,33], Although the studies differed in strategy, both used multiple families with two or more NPC cases from two separate high NPC incident provinces in China and included similar numbers of families and affected cases. Although it is possible that enviromnental exposures may differ between the two provinces, it is unlikely that different environmental factors account for the lack of concordance between the studies. More likely, multiple genes predispose to chronic EBV replication and the development of NPC, each of which may contribute only a small part of the total genetic influence. Family – based linkage Studies are ideal for identifying single genes with large effects, but are relatively insensitive for localising genetic factors with small erects. By contrast, case – control association studies are ideal for identifying genetic factors with small or moderate effects once a candidate gene or region has been identified[38].

We cannot exclude the possibility that there may be causal alleles in the chromosome 4 region that may be associated with chronic EBV replication or a predisposition to develop NPC. Marker associations within this region (Tab. 4 – 6) may be tracking a susceptible locus through LD. Because included alleles predominantly occurred at very low frequencies and haplotypc inferences were unreliable, we could not reliably assess associations with either EBV persistence or NPC (data not shown). A denser placement of polymorphic markers is required to survey the genetic variation content of the region more thoroughly.

Although this study did not find associations with robust p values for NPC, making conclusions tentative, a number of loci did show moderate to strong risk, suggesting that this region warrants further attention, particularly for chronic EBV replication. For one of the microsatellite loci D4s3347 (Tabs. 5 and 6), two alleles were associated with EBV/IgA/VCA, suggesting that these , alleles may be tracking a potential causative allele (Fig. 1). Of potential interest is the , association of two microsatellites with IgA incidence: D4S3347. which shows three significant associations with $P < 0.01$ and one with $P < 0.05$ for two alleles (213 and 217), and the tightly linked (<20 kb) D4S1577 locus, which also shows four significant associations ($P < 0.05$) (Tab. 5 and 6. Fig. 1). Microsatellite D4S190 occurs within the oncogene ARHH. D4S190 was associated with risk for EBV/IgA/VCA seropositive status but not with NPC. ARHH, a member of the ras hornolog gene family, encodes a small GTP – bindmg protein belonging to the RAS superfamily and is transcribed by only haemopoietic cells. A RHH non – coding variants that may afflict expression are observed in 46 per cent of diffuse large – cell lymphomas[39]. It is possible that one or more variant aleles of ARHH in LD with associated D4S190 – 170 may modify EBV replication.

Given the similar geographical distribution of familial and non – familial NPC, it is likely that both forms share similar aetiological risk factors, particularly environmental and viral factors; however, it is likely that the genetic factors underpirming familial, early – onset and non – familial NPC susceptibility may also overlap. It is also possible that different genes contribute to familial NPC cases, analogous to the situation in breast cancer, where BRCA 1 and BRCA2 account for only a small proportion of non – familial breast cancer cases[40,41]. The best approach to identifying NPC susceptibility factors may be the organisation of well – designed and highly powered case – control studies for whole – genome and targeted candidate gene association investigations, as we describe here.

[In 《Human Genomics》 2006, 2 (6): 365 – 375]

Acknowledgements

We gratefully acknowledge Beth Binns – Roemer and Maidar Jamba for excellent technical assistance and Dr Michael Smith for valuable discussions. This project has been funded, in whole or in part by federal funds from the National Cancer Instittnte, National Institutes of Health, under contract NOI – CO – 12400. The content of this paper does not necessarily reflect the views or policies of the Department of Health and Human Services. nor does mention of trade names, commercial products or organisations imply endorsement by the US Government. The publisher or recipient acknowledges the right of the US Government to retain a non – exclusive, royalty – free licence in and to any copyright covering the paper.

References

1 de The, G. Epideufiology of Epstein Barr Virus and associated diseases in man. in Roizman, B. (ed.), The Herpesviruses, Springer, New York, NY, 1982, 25 – 103

2 deThe G. Viruses and human cancers: ChaLlenges for preventive strategies, Environ. Health Perspcct, 1985, 103 (Suppl. 8): 269 – 273

3 Jeannel D, Hubert A, de Vathaire E, et al. Diet, living conditions and nasopharyngeal carcinoma in Tunisia-A case-control study. Int J Cancer, 1990, 46: 421 – 425

4 Laramore G E, Clubb B, Quick C, et al. Nasopharyngeal carcinoma in Sandi Arabia: A retrospective study of 166 cases treated with curative intent. Int J Radiat Oncol Biol Phys, 1988, 15: 1119 – 1127

5 Johansen L V, Mestre M, Overgaard J. Carcinoma of the nasopharynx: Analysis of treatment results in 167 consecutively admitted patients, Head Neck. 1992, 14: 200 – 207

6 Lee A W, Foo W, Mang O, et al. Changing epidemiology of nasopharyngea] carcinoma in Hong Koug over a 20 – year period (1980 – 1999): An encouraging reduction in both incidence and mortality. Int J Canter, 2003, 103: 680 – 685

7 Old LJ, Boyse E A, Oettgen HE, et al. Precipitating antibody in human serum to an antigen present in cultured Burkitt's lymphoma cells. Proc Natl Acad Sci USA, 1966, 56: 1699 – 1704

8 Henle G, Henle, W. Epstein – Barr virus – specific IgA serum antibodies as an outstanding feature of nasopharyngeal carcinoma. Int J Cancer, 1976, 17: 1 – 7

9 Deng H, Zeng Y, Lei Y, et al. Serological survey of naso – pharyngeal carcinoma in 21 cities of south China. Chin Med J (Engl.), 1995, 108: 300 – 303

10 Sham J S, Wei W l, Zong Y S, et al. Detection of subclinical nasopharyngeal carcinoma by fibre-optic endoscopy and multiple biopsy. Lancet, 1990, 335: 371 – 374

11 Zeng Y, Zhaug L G, Li H Y, et al. Serological mass survey for early detection of nasopharyngeal carcinoma in Wuzhon City, China. Int J Cancer, 1982, 29: 139 – 141

12 Zeng Y, Zhong J M, Li L Y, et al. Follow – up studies on Epstem – Barr virus IgA/VCA antibody – positive persons in Zangwu County. China Intervirology, 1983, 20: 190 – 194

13 Zeng Y, Zhang L G, Wu Y C, et al. Prospectivc studies on nasopharyngeal carcinoma in Epstein – Barr virus IgA/VCA antibody – positive persons in Wuzhou City. China Int J Cancer, 1985, 36: 545 – 547

14 Zong Y S, Sham J S, Ng M H, et al. Immunoglobulin A against viral capsid antigen of Epstein – Barr virus and iudirect mirror exanfination of the nasopharynx in the detection of asymptomatic nasopharyngeal carcinoma, Cancer, 1992, 69: 3 – 7

15 Jalbout M, Bel Hadj Jrad B, Bouaouina N, et al. Autoantibodies to tubulin are specifically associated with the young age onset of the nasopharyngeal carcinoma. Int J Cancer, 2002, 101: 146 – 150

16 Yu M C, Yuan J M. Epidenfiology of nasopbaryngeal carcinoma. Semin Cancer Biol, 2002, 12: 421 – 429

17 Drown T M, Heath C W, Lang R M, et al. Nasopharyugeal cancer in Bermuda. Cancer, 1976, 37: 1464 – 1468

18 Coffin C M, Rich S S, Dehner L P. Familial aggregation of nasopharyngeal carcinoma and other malignancies. A clinicopatbologic description. Cancer, 1991, 68, 1323 – 1328

19 Yu M C, Garabrant D H, Huang T B, et al. Occupational and other non – dietary risk factors for nasopharyngeal carcinoma in Guangzhou. China, Int J Cancer, 1990, 45: 1033 – 1039

20 Jia W H, Feng B J, Xu Z L, et al. Familial risk and clustering of nasopharyngeal carcinoma Guangdong, China. Cancer, 2004, 101: 363 – 369

21 Buell P. The effect of migration on the risk of , nasopharyngeal cancer among Chinese. Cancer Res, 1974, 34: 1189 – 1191

22 Hidesheim A, Apple R J, Chen C J, et al. Association of HLA class I and II alleles and extended haplotypes with nasopharyngcal carcinoma in Taiwan. J Natl Cancer Inst, 2002, 94: 1780 – 1789

23 Li P K, Poou A S, Tsao S Y, et al. No associatiun between HI. A – DQ and – DR genotypes with nasopharyngeal carcinoma in southern Chinese. Cancer Genct Cytogenet, 1995, 81: 42 – 15

24 Lu C C, Chen J C, Jiu Y T. Genetic susceptibility to nasopharyngeal carcinoma within the HLA – A locus in Taiwanese. Int J Cancer, 2003, 103: 745 – 751

25 Mokni – Baizig N, Ayed K, Ayed F B, et al. Association between HLA – A/ – B antigens and – DRBI alleles and nasopharyngeal carcinoma in Tunisia. Oncology, 2001, 61, 55 – 58

26 Pimtanothai N, Charoenwongse P, Mutirangura A, Hurley C K. Distribution1 of HLA – B alleles in nasopharyngeal carcinoma patients and normal controls in Thailand. Tissue Antigens, 2002, 59, 223 – 225

27 Thomas J A, lliescu V, Crawford D H, et al. Expression of HLA – DIL antigens in nasopharyngeal carcinoma: An immunohistological aualysis of the tumour cells and infiltrating lymphocytes. Int J Cancer, 1984, 33: 813 – 819

28 Wu S B, Hwang S J, Chang A S, et al. Human leukocyte antigen (HLA) frequency among patients with nasopharyngeal carcinoma in Taiwan. Anticacer Res, 1989, 9, 1649 – 1653

29 Ooi E E, Ren E C, Chan S H. Association between microsatellites within the human MHC and nasopharyngeal carcinoma, Int J Cancer, 1997, 74: 229 – 232

30 Loh K S, Goh B C, Lu J, et al. Familial naasopharyngeal carcinoma in a cohort of 200 patients, Arch. Otolaryngol. Head Neck Surgery, 2006, 132: 82 – 85

31 Zeng YX, Jia W H, Familial nasopharyngeal carcinoma. Serum Cancer Biol, 2002, 12, 443 – 450

32 Feng B J, Huang W, Shugart Y Y, et al. Genome – wide scan for familial nasopharygeal carcinoma reveals evidence of linkage to chromosome 4, Nat Cenet, 2002, 31: 395 – 399

33 Xiong W, Zeng Z Y, Xia J H, et al. A susceptibility locus at claromosome 3p21 linked to familial nasopharyngeal carcinoma. Cancer Res, 2004, 64: 1972 – 1974

34 Sannders C L, Barrett J H. Flexible matching in case – control studies of gene – environment interactions, Am J Epidemiol, 2004, 159: 17 – 22

35 hrtp: //genome. ucsc. edu/cgi – bin/hgGateway

36 Boutitn – Ganache I, Raposo M, Paymond M, Deschepper C E. Ml3 tailed printers improve the readability and usability, of microsatellite analyses performed with two different allele – sizing methods, Biotcchniques, 2001, 31, 24 – 26: 28

37 O'Connell J R, Weeks D E. PedCheck. A program for identification of genotype incompatibilities in linkage analysis, Am J Hum Genet, 1998, 63: 259 – 266

38 Risch N, Merikangas K. The fnture of genetic studies of complex human diseases, Science, 1996, 273: 1516 – 1517

39 Preudhomme C, Roumier C, Hildebrand M P, et al. Nourandom 4p13 rearrangements of the RhoH/TTF gene encodling a GTP – binding protein, in non – Hodgkin's lymphoma and multiple myeloma. Oncogene, 2000, 19: 2023 – 2032

40 Malone K E, Daling J R, Neal C, et al. Frequency of BRCA1/BRCA2 mutations iu a population – basetl sample of young breast carcinoma cases. Cancer, 2000, 88: 1393 – 1402

41 Peto J, Collins N, Barfoot R, et al. Prcevalence of BRCAI and BR CA2 gene mutations in patients with early – onset breast cancer. J Natl Cancer Inst, 1999, 91: 943 – 949

97. Induction of Cytotoxic T Lymphocyte Respones *in vivo* after Immunotherapy with Dendritic Cells in Patients with Nasopharyngeal Carcinoma

ZUO Jing-min[1], ZHOU Ling[1], CHEN Zhi-jian[2], LI De-rui[2], WANG Qi[1],
CHEN Jiong-yu[2], WANG Zhan[1], YE Shu-qing[1], ZENG Yi[1]

1. State Key Laboratory for Infections Disease Prevention and Control, National
Institute for Viral Disease Control and Prevention, Chinese Center for Disease
Control and Prevention; 2. Tumor Hospital of Shantou University

Summary

The aim of the present study was to determine the efficacy of immunotherapy with dendritic cells to elicit EBV – specific CTL – immunity in advanced cases of EBV – positive patients with nasopharyngeal carcinoma (NPC) and to determine the safety and toxicity of this preparation. Nine cases of histologically confirmed patients with NPC undergoing treatment with radiological therapy were enrolled in this study. Dendritic cells, generated *in vitro* from blood monocytes of patients were cultured and matured with cytokines and then infected with recombinant adenovirus vaccine containing EBV – latent membrane protein – 2 (Ad – LMP2). On 9 days' cultivation of cells, the matured DCs were harvested, irradiated with ^{60}Co and then injected intradermally to patients with NPC. The injections were performed 3 times totally. After immunization, the CTL responses were assayed by means of cytotoxicity and epitope – specific IFN – γ production. The results of this trial showed that all patients could tolerate this kind of treatment without any side effect, during which marked increase of LMP2 – specific CTL – responses could be demonstrated in 5 patients of this group. And the level of IgA/VCA antibody decreased in 8 of 9 patients, thus accounting for a better prognosis for these patients. All patients will be followed up for another one year. At least, the present work shows that intradennal vaccination with autologous DCs infected with recombinant Ad – LMP2 adenovirus is a safe procedure in NPC patients, in which this procedure can enhance the LMP2 – specific CTL responses in patients. These data are encouraging to develop more effective vaccine strategies for the treatment of nasopharyngeal carcinoma.

[**Key words**] Nasopharyngeal carcinoma (NPC) ; Dendritic cells; Immunotherapy ; CTL – responses.

Introduction

Nasopharyngeal carcinoma (NPC) is a highly prevalent cancer in Southern China, with a yearly incidence rate between 10 and 50 per 100 000[1,2]. Currently, the treatment for NPC is radically external radiotherapy, which can cure about 80% cases during early stage. However, only 10% – 40% cases during advanced stage can survive more than 5 years[3,4]. Once metastases developed, 85% of patients can not survive for more than 1 year[5]. There is therefore a need to develop additional forms of treatment for NPC.

There is considerable evidence that EBV plays an important role in the progression of NPC[6-8]. Epstein – Sarr vires (EBV) is a ubiquitous gamma – herpesvirus that can establish both latent and lytic infection. The primary EBV infection occurs mostly during childhood without obvious symptoms, but the EBV could persist in the body through the whole life. Once the EBV is animated by some inducement, it will become a pathogen of many diseases, including malignant diseases. EBV is associated with many human malignant diseases including lymphoproliferative disorder associated with immunocompromise, Burkitt's lymphoma (BE) [9], Hodgkin' s disease (HD) [10], undifferentiated nasopharyngeal carcinoma (NPC), and various T – cell lymphomas[11]. The definite function of EBV in the NPC etiology is not well understood, however the existence of EBV in the tumor tissue provides a potential target for gene therapy.

The EBV proteins expressed in tumor cell are very limited. There are only several antigens of EBV, such as EBNA1, LMP1, and LMP2 which can be detected in NPC and HD[12-15]. Of these three antigens, EBNA1 contains a Gly – Ala repeat sequence, which will interrupt the presentation of this antigen to T cells through MHC class I restricted pathway[16]. LMP2 is the most frequently recognized protein by CTL. Also many MHC class I restricted epitopes of LMP2 have been identified, which are conserved during different population[17-20]. So it becomes the target of immunotherapy for NPC.

MHC class I restricted CTL plays an important role in controlling the status of EBV infection. EBV specific CTL can be present at a high level in the blood during primary infection and last for the whole life accompanying with the virus. If the level of CTL responses is reduced, for example in the transplanted patients or HIV infected individuals, some EBV – driven lymphoma proliferation will occur. But these diseases can regress when the cellular immune response is recovered after the relaxation of immunosuppression or infusion of autologous EBV – specific CTL expanded *in vitro*[21]. Given these observations, it is likely to target the virus – specific immune response to the EBV positive human tumors.

Adenoviruses can transfer genes into a broad spectrum of cell types. Additionally, high titers of viruses and high level of transgene expression can generally be obtained. In order to develop a NPC therapeutic vaccine, we construct the AdEasy – LMP2 containing the full – length cDNA of EBV – LMP2[22]. DCs are specialized antigen – presenting cells that can prime T cell responses, playing important roles in the anti – infection and anti – tumor processes. In 1990s, along with the establishment of method of isolation and cultivation of DC, and with the development of correlative immunology, molecular biology and tumor pathology, various means of antigen delivery to modify DCs have been used to activate T lymphocytes *in vitro* and *in vivo*[23]. Now the clinical trials of DC vaccines on malignant melanoma, renal cell carcinoma, prostate cancer and colorectal cancer are carried on extensively[24-26], and the results from these studies are encouraging. The aim of this study is to transfer recombinant adenovirus containing LMP2 gene into autologous DC and this preparation is to be used to immunize NPC patients. The safety, toxicity, and efficacy in eliciting EBV specific CTL immunity are then to be evaluated.

Material and Methods

1. Patients

Nine patients with ages younger than 70 years were enrolled in this study. All patients were histologically proven NPC and had a high serum IgA/VCA. The radiotherapy had been completed 6 months prior to trial. Patients were required to have adequate hepatic and renal function, and have a life expectancy of more than 6 months.

2. DC preparation

Peripheral blood mononuclear cells (PBMCs) were isolated by Ficoll/Paque density gradient centrifugation. Isolated PBMCs were plated (2×10^7 cells/5ml per well) into 6 – well plates (Costa, Cambridge, MA) in serum – free RPMI 1640 medium. After 2 h of incubation at 37 ℃, non-adherent cells were removed and the adherent cells were cultured in RPMI 1640 containing 10% fetal calf serum (FCS), GM – CSF (50 ng/mL) and IL – 4 (25 ng/mL). On day 3 and day 6, half of the medium was replaced with fresh medium containing GM – CSF, IL – 4. Cell differentiation was monitored by light microscope. On day 7, the DCs were collected by centrifuge in 100 μl serum – free RPMI 1640, and infected by Ad – LMP2 in MOI of 1000 for 2 h at 37℃. Then the DCs were cultured for another 2d in RPMI 1640 containing 10% FCS, GM – CSF (50 ng/mL), IL – 4 (25 ng/mL) and TNF – α (20 ng/mL).

On day 9, DCs were washed 3 times and resuspended in sterile saline (total volume = 0.2 mL). The expression of LMP2 on infected DCs was detected by indirect immunofluorescence staining. The purity of DCs was analyzed by immunofluorescence staining for CD80, CD83 and CD86. After the DCs were irradiated by ^{60} Co, the DCs were delivered to patients intradermally. Two further preparations of autologous antigen – pulsed DCs were produced in the same manner. The two further injections are on day 14 and day 28. The average number of DCs is about 2×10^6 each time. The immunological responses were monitored with patients' blood sample on day 56 after the first injection.

3. Intracellular stain of IFN – γ

The 5×10^5 PBMC were stimulated with mixed LMP2 specific peptides (100 μg/mL, LL-WTLVV – LL, LTAGFLIFL, LLSAWILTA, SSCSSCPLSKI, TYGPVFMCL, IEDPPFNS) for 6 h. The Brefeldin A (1 μg/mL Sigma) was added during the last 3 h. Positive controls were performed by stimulating the cells with PMA (50 ng/mL, Sigma) and ionomycin (500 ng/mL, Sigma). Cells were stained with a FTTC – conjugated anti – CD8. After washing, cells were fixed with 4 % formaldehyde/PBS for 15 min at room temperature, permeabilized in 0.5 % saponin (Sigma), 2% bovine serum albumin in PBS, and stained with a PE – conjugated anti – IFN – γ (BD Pharmingen). Subsequently, cells were washed with PBS and detected by FACS.

4. Cytotoxicity assays

The suspension cells were collected after the PB – MCs were incubated in 37℃ for 2 h. Part of these suspension cells were pulsed with mixed peptides（100μg/ml, LLWTLVVLL, LTAGFLIFL, LLSAWILTA, SSCSSCPLSKI, TYGPVFMCL, IEDPPFNS）for 1 h to act as stimulator cells. Then the stimulate cells were incubated with autologous PBMCs in RPMI 1640 containing 10% FCS and IL –2（20 IU/ml）at the rate of 1 : 10 for 5 d. The target cells were autologous PHA blast coated with the mixes peptides by exposing to 100μg/ml peptide for 2 h, then washed and incubated in 96 – well Vbottom plates with effector cells at effector target ratio（E : T）of 10 : 1, 20 : 1, 50 : 1. The percentage specific lysis（means of triplicate wells）was detected through LDH method（Promega）.

5. Detection of IgA/VCA antibody

The B95. 8 cell were stimulated by TPA for 48 h, and fixed on slices. Sera of patients were diluted serially, and incubated on the slice at 37℃ for 45 min. The slices were washed by PBS for three times, and incubated with HRP conjugated anti – IgA antibody at 37℃ for 45 min. Then the slices were washed, substrate was added and the formation of color was monitored.

Results

1. Expression of LMP2 in DC

The DCs were infected with Ad – LMP2 at the MOI of 1000 for 2 h, and then cultured at 37℃ for another two days. After the DCs were collected by centrifuge and resuspended with saline, the DCs were incubated with anti – LMP2 rabbit serum（kindly provided by Professor Middeldorp JM, Vrije Universiteit Medical Centre, Netherlands）and FITC conjugated anti – rabbit – IgG（BD Pharmingen）. After the DCs were washed, they were detected by FACS. From the result we can see, there are about 60% DCs which expressed LMP2 when they were infected with Ad – LMP2 at the MOI of 1000（Fig. 1）.

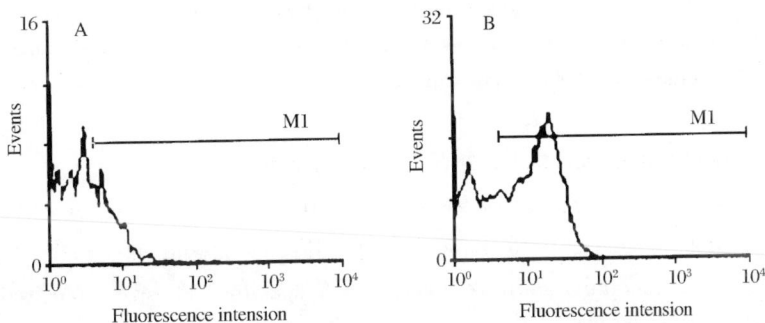

A: DCs were infected with Ad5 wild type virus, then stained with LMP2 antibody and tested through FACS. The results show that the DCs infected with wild type virus have no expression of LMP2. B: DCs were infected with Ad – LMP2, then stained with LMP2 antibody and examined by FACS. The results show that there are about 60% DCs expressing LMF2 when infected with Ad – LMP2 at the MOI of 1000.

Fig. 1　Expression of LMP2 in DC infected with Ad – LMP2

2. Adverse effects

All patients tolerated the immunization well without obvious side effects. They all completed the whole therapy. No obvious local swelling, local rigor, fever and fatigue were observed in any patients.

3. Cytotoxicity assay

To detect the LMP2 specific cytotoxicity in patients received DC immunotherapy, the patients were bled to detect the lytic activity against target cells in week 0 and week 8 through LDH method. As shown in Figs. 2 and 3, 5 immunized patients showed significant enhanced epitope – specific killing activity in week 8 compared with the result in week 0. On the other hand, no patient in the control patients showed enhanced epitope – specific killing activity. In the instance of E : T being 50 : 1, the lysis rate of patient 1, 2, 4, 7 and 8 increased from 28.90% to 54.77%, 30.00% to 51.01%, 42.60% to 67.45%, 22.00% to 32.10%, 10.90% to 35.60%, respectively.

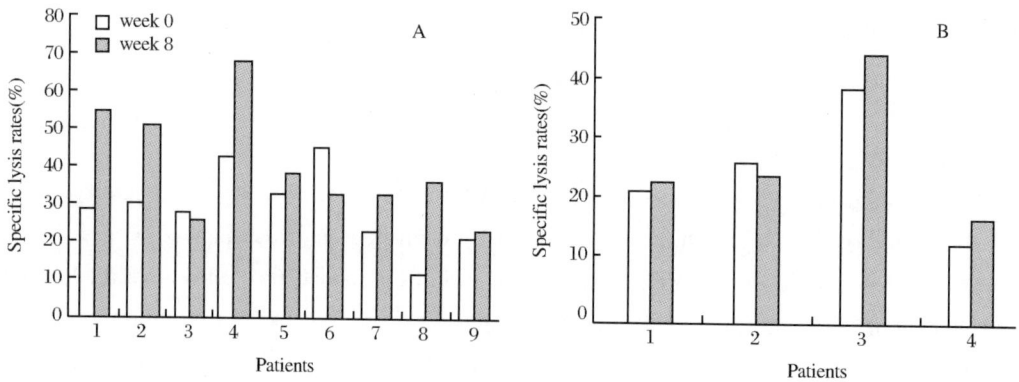

A: The PBMC of immunized patients were stimulated with mixed peptides, and autologous PHA blast coated with the mixed peptides as targets, then the specific lysis was detected at effector target ratio of 10 : 1, 20 : 1, 50 : 1. They were bled in week 0 and week 8 after immunization separately. The data being expressed as percentage specific lysis are at an effector target ratio of 50 : 1. B: The PBMC of control patients were tested in cytotoxicity assay as Fig. 2A. They were also bled in week 0 and week 8 separately. The data being expressed as percentage specific lysis is at an effector target ratio of 50 : 1

Fig. 2   Comparation of the lysis rate of CTL between pre – and post – therapy.

4. Flow cytometry of intracellular IFN$^+$/CD8$^+$ T cells

To detect the LMP2 specific cytotoxicity level in patients, the peptide – stimulated PBMC was immunostained for IFN – γ and CD8 to characterize the IFN – γ producing T cells. In the 4 patients who received IFN – γ intracellular staining assay, LMP2 specific CTL level increased markedly in 2 patients. The percentages of CD8$^+$ cells which produced IFN – γ in response to LMP2 specific pep – tides in whole lymphocyte of patient 7 and 8 increased from 0.09 % and 0.04% to 0.46 % and 0.38% (Figs. 4 and 5).

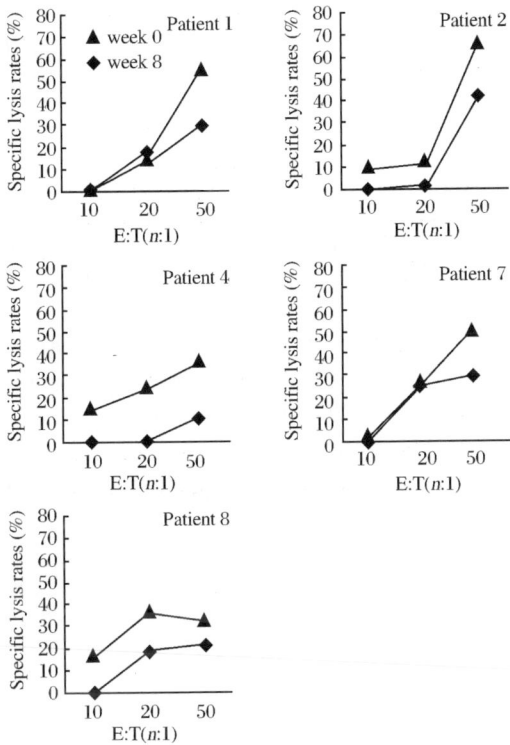

The PBMC of the 5 patients were tested in cytotoxicity assay as Fig. 2A. The data being expressed as percentage specific lysis are at an effector target ratio of 10 : 1, 20 : 1 and 50 : 1. The specific lysis between week 0 and week 8 was compared

Fig. 3 Detection of fluctuation of CTL level by LDH method

A: The PBMC of immunized patients were stimulated by mixed LMP2 specific peptides and then stained with an FTTC – conjugated anti – CD8 and PE – conjugated anti – IFN – γ, and detected by FACS. They were bled in week 0 and week 8 after immtmization separately. B: The PBMC of control patients were tested in intracellular staining of IFN – γ as Fig. 4A. They were bled in week 0 and week 8 separately

Fig. 4 Comparation of the IFN$^+$ CD8$^+$ cell ratio between pre – and post – therapy

5. Detection of IgA/VCA antibody

The IgA/VCA antibody level in patients was detected in week 0 and week 8, the result showed that the antibody level of IgA/VCA decreased to below 10 : 1 in 7 of 9 patients, which accounts for a good prognosis of these patients. The antibody level of other two patients did not have much changes (Fig. 6).

Disscusion

DCs are widely distributed in the human body as the most potent antigen presenting ceils (APCs), in which the costimulating molecules, adherent molecules and MHC – molecules are highly expressed on their cell surface, and they can initiate the primary T lymphocyte – mediated responses, such as immune responses against microbes, antitumor immune responses and transplant rejection processes[27]. The studies on activation of T lymphocytes *in vitro* or *in vivo* through tumor

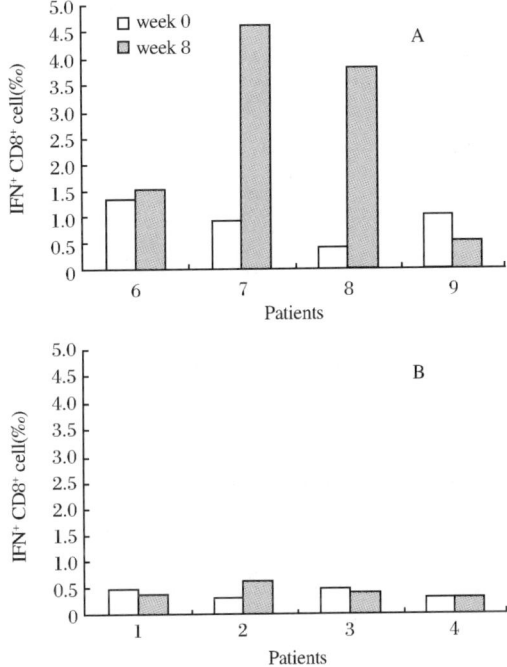

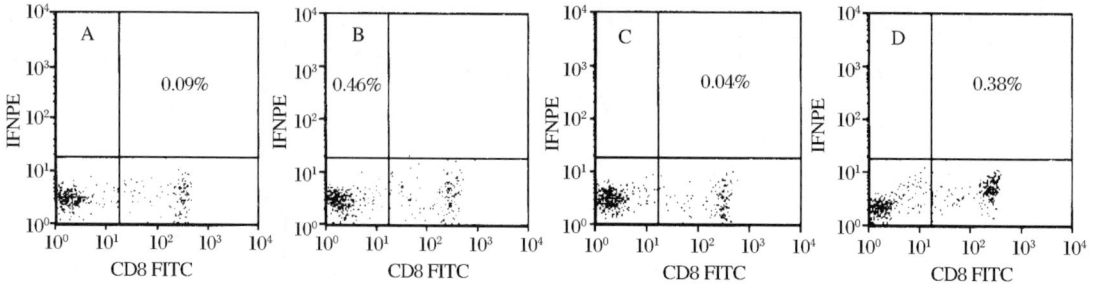

The PBMC of patient 7 and patient 8 were stimulated with mixed LMP2 specific peptides and then stained with a F1TC – conjugated anti – CD8 and PE – conjugated anti – IFN – γ, and detected by FACS. They were bled in week 0 and week 8 after immunization separately. A: IFN+ CD8+ cell ratio of patient 7 prior to vaccination. B: IFN+ CD8+ cell ratio of patient 7 after 3 vaccinations, C: IFN+ CD8+ cell ratio of patient 8 prior to vaccination. D: IFN+ CD8+ cell ratio of patient 8 after 3 vaccinations

Fig. 5 Comparison of the IFN+ CD8+ cell ratio between pre – and post – therapy in patients 7 and patient 8

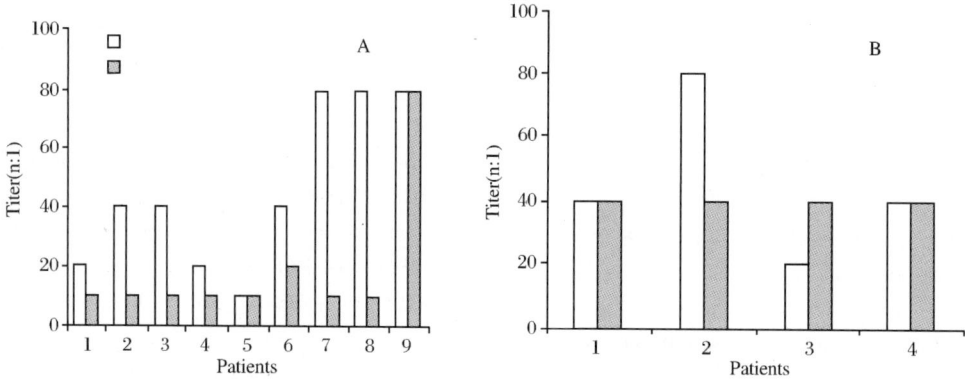

A: The serum of immunized patients were diluted serially, and incubated on the slice with TPA stimulated B95. 8 cell, and then incubated with HRP conjugated anti – IgA antibody to test the titer of IgA/VCA antibody. They were bled in week 0 and week 8 after immunization separately. B: The titer of IgA/VCA antibody in control patients were tested in immuno – enzyme assay as Fig. 6A. They were bled in week 0 and week 8 separately

Fig. 6 Change of the titer of IgA/VCA in patients' serum

antigen – pulsed DCs of different sources has become the focus of interest on DC tumor vaccine[28]. Vaccination therapies with DCs were found to be effective for the treatment of malignant melanoma, renal cell tumor, breast cancer and ovary cancer with encouraging results[24 – 26].

The studies involving the relationship between EBV and NPC and the existence of EBV in tumor tissue constitute the target for gene therapy, in which EBV – LMP2 becomes the most promising target of immunotherapy for NPC. Many workers are pursuing DC vaccine for the treatment of NPC. Lin et al had worked on the Phase I clinical trial with epitope – pulsed DC vaccine in order to evaluate the efficacy and safety of this vaccine[29]. Dendritic cells were generated *in vitro* from blood monocytes of NPC patients, cultured and matured with cytokines, and then pulsed with LMP2 – specific

HLA – restricted peptides. Twelve NPC patients were enrolled in this trial, in which each patient received 4 injections and CTL responses were assayed by intracellular staining of the epitope – specific IFN – γ and cytotoxicity. The results showed that all patients tolerated this treatment very well and the intracellular staining of IFN – γ revealed enhancement of level of LMP2 – specific CTL responses in 6 patients. In addition, 2 patients treated showed partial reduction of tumor growth.

Recombinant adenovirus vector has many merits, including the ability to infect a broad range of mammalian cells, the actions on infection and the expressions of genes in dividing or non – dividing cells. Now, the tranduction of gene into APC by recombinant adenovirus has been regarded as a promising strategy for tumor vaccine. In the present study, although the virus used is the replication – deficient one, we also irradiated this vires with ^{60}Co for safety. It was found that the process of irradiation did not show any change in the surface markers before or after irradiation of DCs (data not shown). So the antigen – presenting function of DCs after irradiation is likely maintained. Our data also indicated that the level of LMP2 – specific CTL responses was increased markedly in 5 of 9 patients. As demonstrated by LDH method, the lysis rates of cytotoxicity in these 5 patients (No. 1, 2, 4, 7 and 8) increased. In addition, the percentage of CD8$^+$ cells producing IFN – γ in response to the stimulation of LMP2 specific peptide in lymphocytes of patients No. 7 and 8 of 4 patients who had assayed with the intracellular staining, also increased significantly. This is in agreement with the result obtained from LDH method. At the same time, the antibody levels of IgA/VCA in these 5 patients were decreased, accounting for a better prognosis for these patients.

The treatment of NPC patients by using the DC vaccine modified by the recombinant adenovirus vector containing EBV – LMP2 gene is a safe procedure, because all the patients can complete the whole course of treatment without any side effects, and this kind of vaccination can induce specific CTL responses in NPC patients. However, there still existed some patients who did not show any significant response; the possible reasons for this might be the ages, general conditions of patients, and the maturation state of DCs which may influence the function of antigen presentation. Several reports suggested that compared with the healthy individuals, the dendritic cells of cancerous patients could not be cultivated to be at the full and functional state of maturation[30,31]. In the present study, there was just one patient who showed a DC maturation rate of less than 40% (data not shown). Hence, whether there is any difference between the NPC patients and the healthy individuals is still unclear. As a whole, multiple immunization may be considered to be a means to improve the efficacy of vaccines, and the exploration of more potent approaches of immunization to get more effective enhancement of specific cytoxicity is desirable in the future.

Acknowledgement

This work was supported by Chinese National High – tech Program (2001AA217091).

[In 《J Microbiol Immunol》 2006, 4 (1): 41 –48]

References

1 Chan AT, Toe PM, Johnson PJ. Nasopharyngeal carcinoma. Ann Oncol, 2002, 13: 1007 – 1025

2 Yu MC. Nasopharyngeal carcinoma: epidemiology and dietary factors. IARC Sci Publ, 1991, 105: 39 – 47

3 Altun M, Fandi A, Dupuis O, Cvitkovic E, Krajina Z, Eschwege F. Undifferentiated nasopharyngeal cancer (UCNT): current diagnostic and therapeutic aspects. Int J Radiat Oncol Biol Phys, 1995, 32: 859 – 877

4 Mould RF, Tai TH. Nasopharyngeal carcinoma: treatments and outcomes in the 20th century. Br J Radiol, 2002, 75: 307 – 339

5 Teo PM, Kwan WH, Lee WY, Leung SF, Johnson PJ. Prognosticators determining survival subsequent to distant metastasis from nasopharyngeal carcinoma. Cancer (Phila), 1996, 77: 2423 – 2431

6 Rickinson AB, Kieff E. Epstein – Barr virus. In: Fields BN, Knipe DM, Howley PM, eds. Fields Virology. Philadelphia: Lippincott – Raven, 1996, 2397 – 2446

7 Niedobitek G. Epstein – Barr virus infection in the pathogenesis of nasopharyngeal carcinoma. Mol Pathol, 2002, 53: 248 – 254

8 Wolf H, Zurhausen H, Becker V. EB viral genomes in epithelial nasopharyngeal carcinoma cells. Nat Rew Biol, 1973, 244: 245 – 247

9 Magrath I. The pathogenesis of Burkitt's lymphoma. Adv Cancer Res, 1990, 55: 133 – 270

10 Deacon EM, Pallesen G, Niedobitek G, Crocker J, Brooks L, Rickinson AB, et al. Epstein – Barr virus and Hodgkin's disease: transcriptional analysis ot virus latency in the malignant cells. J Exp Med, 1993, 177: 339 – 349

11 Shapiro RS, McClain K, Frizzera G, Gaji – Peczalska KJ, Kersey JH, Blazar BR, et al. Epstein – Barr virus associated B cell lymphoproliferative disorders following bone marrow transplantation. Blood, 1988, 71: 1234 – 1243

12 Chang KL, Chen Y – Y, Shibata D, Weiss LM. Description of an *in situ* hybridization methodology for detection of Epstein – Barr virus RNA in paraffinem – bedded tissues, with a survey of normal and neoplastic tissues. Diagn Mol Pathol, 1992, 1: 246 – 255

13 Young LS, Dawson CW, Clark D, Rupani H, Busson P, Tursz T, et al. Epstein – Barr virus gene expression in nasopharyngeal carcinoma. J Gen Virol, 1988, 69 (pt 5): 1051 – 1065

14 Fahraeus R, Fu HL, Emberg I, Finke J, Rowe M, Klein G, et al. Expression of Epstein – Barr virus – encoded proteins in nasopharyngeal carcinoma, Int J Cancer, 1988, 42: 329 – 338

15 Brooks L, Yao QY, Rickinson AB, Young LS. Epstein – Barr virus latent gene transcription in nasopharyngeal carcinoma cells: coexpression of EBNA1, LMP1, and LMP2 transcripts. J Virol, 1992, 66: 2689 – 2697

16 Levitskaya J, Coram M, Levitsky V, Imreh S, Steigerwald – Mullen PM, Klein G, et al. Inhibition of antigen processing by the internal repeat region of the Epstein – Barr virus nuclear antigen – 1. Nature, 1995, 375: 685 – 688

17 Lee SP, Tiemey R J, Thomas WA, Brooks JM, Rickinson AB. Conserved CTL epitopes within EBV latent membrane protein 2: a potential target for CTL – based tumor therapy. J Immunol, 1997, 158 (7): 3325 – 3334

18 Lee SP, Chan AT, Cheung ST, Thomas WA, CroomCarter D, Dawson CW, et al. CTL control of EBV in nasopharyngeal carcinoma: EBV – specific CTL responses in the blood and tumors of NPC patients and the antigen – processing function of the tumor cells. J Immunol, 2000, 165: 573 – 582

19 Khanna R, Busson P, Burrows SR, Raffoux C, Moss D J, Nicholls JM, et al. Molecular characterization of antigen – processing function in nasopharyngeal carcinoma (NPC): evidence for efficient presentation of Epstein – Barr vires cytotoxic T – cell epitopes by NPC cells. Cancer Res,

1998, 58: 310 – 314

20 Whitney BM, Chan AT, Rickinson AB, Lee SP, Lin CK, Johnson PJ. Frequency of Epstein – Barr virus – specific cytotoxic T lymphocytes in the blood of Southern Chinese blood donors and nasopharyngeal carcinoma patients. J Med Viral, 2002, 67: 359 – 363

21 Rooney CM, Smith CA, Ng CY, Loftin S, Li C, Krance RA, et al. Use of gene – modified virus – spefic T lymphocytes to control Epstein – Barr virus related lympholiferation. The Lancet, 1995, 345: 9 – 13

22 Zuo JM, Zhou L, Wang Q, Zeng Y. The *in vitro* and *in vivo* immunogenicity of recombinant adenovirus vaccine containing EBV – latent membrane protein 2. Chinese Journal of Microbiology and Immunology, 2003, 23 (6): 446 – 449

23 Zhang JK. The dendritic cell and tumor immunological therapy. Shantou: The Publishing Company of the University of Shantou, 2001, 65 – 83

24 Ranieri E, Kierstead LS, Zarour H. Dendritic cell/peptide cancer vaccines: clinical responsiveness and epitope spreading. Immunol Invest, 2000, 29 (2): 121 – 125

25 Kugler A, Stuhler G, Walden P, Zoller G, Zobywalski A, Bmssart P, et al. Regression of human metastatic renal cell carcinoma after vaccination with tumor cell – dendritic cell hybrids. Nat – Med, 2000, 6 (3): 332 – 336

26 Bmssart P, Wirths S, Stuhler G, Reichardt VL, Kanz L, Brugger W. Induction of cytotoxic T – lymphocyte responses *in vivo* after vaccinations with peptide – pulsed dendritic cells. Blood, 2000, 96 (9): 3102 – 3108

27 Banchereau J, Steinman RM. Dendritic cells and the control of immunity. Nature (Lend.), 1998, 392: 245 – 252

28 Nestle FO, Banchereau J, Hart D. Dendritic cells: on the move from bench to bedside. Nat Med, 2001, 7: 761 – 765

29 Lin CL, Lo WF, Lee TH, Ren Y, Hwang SL, Cheng YF, et al. Immunization with Epstein – Barr Virus (EBV) peptide – pulsed dendritic cells induces functional CD8$^+$ T – cell immunity and may lead to tumor regression in patients with EBV – positive nasopharyngeal carcinoma. Cancer Res, 2002, 62 (23): 6952 – 6958

30 Katsenelson NS, Shurin GV, Bykovskaia SN, Shogan J, Shurin MR. Human small cell lung carcinoma and carcinoid tumor regulate dendritic cell maturation and function. Mod Pathol, 2001, 14: 40 – 45

31 Inoshima N, Nakanishi Y, Minami T, Izumi M, Takayama K, Yoshino I, et al. The influence of dendritic cell infiltration and vascular endothelial growth factor expression on the prognosis of non – small cell lung cancer. Clin Cancer Res, 2002, 8: 3480 – 3486

98. *In vitro* Anti – Tumor Immune Response Induced by Dendritic Cells Transfected with EBV – LMP2 Recombinant Adenovirus

PAN Ying[1], ZHANG Jin – kun[1], ZHOU Ling[2], ZUO Jian – min[2], ZENG Yi[2]

1. Department of Onco – pathology and the Key Immunopathology Laboratory of Guangdong
Province, Shantou University Medical College;
2. Institute for Viral Disease Control and Prevention, Chinese Center
for Disease Control and Prevention

Summary

Epstein – Barr virus (EBV) – associated nasopharyngeal carcinoma (NPC) is a high – incidence tumor in southern China. Latent membrane proteins 2 (LMP2) is a subdominant antigen of EBV. The present study was to develop a dendritic cells (DCs) – based cancer vaccine (rAd – LMP2 – DC) and to study its biological characteristics and its immune functions. Our results showed that LMP2 gene transfer did not alter the typical morphology of mature DC, and the representative phenotypes of mature DC (CD80, CD83, and CD86) were highly expressed in rAd – LMP2 – DCs. The expression of LMP2 in rAd – LPM2 – DCs was about 84.54%, which suggested efficient gene transfer. Transfected DCs markedly increased antigen – specific T – cell proliferation. The specific cytotoxicity against NPC cell was significantly higher than that in controls (p < 0.05), and enhanced with increased stimulations by transfected DCs. In addition, phenotypic analysis demonstrated that the LMP2 – specific CTLs consisted of both CD4 + and CD8 + T cells. These results showed that development of DC – based vaccine by transfection with malignancy – associated virus antigens could elicit potent CTL response and provide a potential strategy of immunotherapy for EBV – associated NPC.

[**Key words**] Dendritic cell; Nasopharyngeal carcinoma; Epstein – Barr virus; Gene transfer; Cytotoxic T lymphocyte; Cancer vaccine

Introduction

Cytotoxic T lymphocytes (CTLs) recognize peptides derived from the intracellular breakdown of foreign anti – gens and present these peptides at the cell surface as a complex with major histocompatibility complex (MHC) class I molecules. Such CTLs play an important role in controlling virus infection. The virus – induced CTL response tends to focus on a few immunodominant peptide epitopes whose identities are specific for the particular MHC type of the host. This study concerns the CTL response to Epstein – Barr virus (EBV), a herpesvirus commonly associated with nasopharyngeal carcinoma (NPC). Among the EBV – associated NPC patients, the proteins of EBV expressed on tumor cells are very limited, only latent class II EBV anti – gens such as the latent EBV nuclear antigens (EBNA1) and latent membrane proteins (LMP1 and LMP2) can be detected on NPC cells. Many human leukocyte antigen (HLA) class I restricted epitopes of LMP2 have been identified and their sequences are conserved, and LMP2 is thus the most frequently recognized protein by CTLs[1]. LMP2 constitutes potentially the major target antigen for immunotherapy of NPC.

In our study, we sought to develop an efficient protocol to induce a strong LMP2 – specific CTL response against tumor cells of NPC. Dendritic cells (DCs) are highly efficient and specialized antigen – presenting cells (APC) that are the only ones that can stimulate the native T cell and activate antigen – specific CTLs[2,3]. *In vivo*, immature DCs develop from hematopoietic progenitors and are located strategically at body surfaces, where they play a sentinel role in capturing and processing antigens.

Following antigen exposure, DCs migrate to lymphoid organs and acquire potent antigen – presenting function. Mature DCs process antigens efficiently by both MHC class I and II pathways with upregulation of cell surface adhesion molecules such as CD54 (ICAMI) and of costimulatory molecules such as CD80 and CD86. DCs have demonstrated potent anti – tumor properties in a variety of experimental models[4-6].

Recently some reports showed that calcium – signaling agents could induce maturation of DCs derived from peripheral blood monocytes[7,8]. DCs activated with calcium – signaling agents, in the presence of cytokines in serum – free medium, rapidly express mature DC marker, CD83, and high levels of co – stimulatory molecules within 96 h of culture. These activated DCs can efficiently sensitize T cells to recognize tumor cells through tumor antigens expressed by tumor cells. In our study, we prepared DCs by adenoviral transfection with EBV – LPM2 and calcium ionophore treatment. The acquired DC vaccine could stimulate T cells and elicit the potent antigen – specific CTLs activity against NPC cells.

Materials and Methods

1. Nasopharyngeal carcinoma cell culture: Nasopharyngeal carcinoma cell line (CNE – 2) which contains LMP2 gene[9] was obtained from the Institute for Viral Disease Control and Prevention of the Chinese Center for Disease Control and Prevention (Chinese CDC). The CNE – 2 cells were grown in complete RPMI medium 1640 (Gibco, USA) supplemented with 10% heat – inactivated fetal calf serum (FCS, Hyclone), 2 mmol/L L – glutamine, 100 U/ml penicillin, and 100 μg/ mL streptomycin.

2. Preparation of DCs: Human peripheral blood mononuclear cells (PBMCs) were isolated from whole blood of healthy donors by Ficoll – Hypaque (d = 1.077 g/ml) density – gradient centrifugation. Such PBMCs were suspended in RPMI 1640 medium supplemented with 10% heat – in – activated FCS. After incubation for 2 h at 37 ℃ in 5% CO_2, the nonadherent cells were removed. The adherent cells as monocytes were harvested and resuspended in macrophage serum – free medium (Mφ – SFM; Gibco). The monocytes were then plated in a 24 – well tissue – culture plate (Costar, USA) at 2.5×10^6 cells/well supplemented with 50 ng/ml rhGM – CSF (Peprotech,

USA). This combination of Mφ – SFM and rhGM – CSF, which constitutes basal culture medium for all monocytes and DCs in this study, is henceforth referred to in the text simply as SFM/G. The monocytes were cultured for 24 – 48 h at 37 ℃ in 5% CO_2. To obtain mature DCs, the cells were treated with calcium ionophore A23187 (Sigma) at a concentration of 150 ng/ml for additional 48 h. The mature DCs were then collected and were analyzed for DC typical phenotypes by fluorescence – activated cell sorter (FACS) analysis or co – cultured with T cells for sensitization assays.

3. Preparation of adenovirus transfected DCs: Recombinant serotype 5 adenoviruses encoding the LMP2 gene (rAd – LMP2) were obtained from the Institute for Viral Disease Control and Prevention of the Chinese CDC. The virus stocks were proliferated in human embryonic kidney (293) cells in DMEM (Gibco) supplemented with 2% heat – inactivated FCS and purified through cesium chloride (Sigma) gradient ultracentrifugations[10]. Viral particle concentration was determined by UV absorbance at 260 nm[11], and final viral titers were 10^{11} plaque – forming units (pfu).

The monocytes were cultured in SFM/G for 48 h as described previously. The cells were harvested as immature DCs and were resuspended at 1×10^6 cells/200 µl in serum – free medium. The recombinant adenoviruses encoding the LMP2 gene were then added to infect immature DCs at multiplicities of infection (MOI) 200. Infection was allowed to proceed for 2 h at 37 ℃. Then fresh SFM/G was added to bring the cultures to 2 ml per well. One hour after adenoviruses transfection of immature DCs, calcium ionophore A23187 was added at a concentration of 150 ng/ml. Transfected cells were cultured for additional 48 h and the mature rAd – LMP2 – DCs were harvested. To determine the viability of adenoviruses-infected DCs, trypan blue (Sigma) exclusion was used to determine viable cells. The expression of LMP2 protein in rAd – LMP2 – DCs was analyzed by indirect immunofluorescence and FACS assays. In addition, DC phenotypes CD80, CD83, and CD86 were determined as well.

4. Preparation of T lymphocytes: Sterile nylon – wool isolation column (Wako, Japan) was soaked in complete RPM1 1640 medium supplemented with 10% heat – inactivated FCS, 2 mmol/L L – glutamine, 100 U/ml penicillin, and 100 µg/ml streptomycin for 1 h at 37 ℃. Then the non-adherent cells isolated from peripheral blood mononuclear cells described previously were applied on the column and cultured for additional 1 h. T lymphocytes were eluted from the column with 10 ml RPMI 1640/10% FCS. Purity of about 90% was obtained with this method.

5. Flow cytometric analysis of cell populations: DCs were collected and resuspended in cold FACS buffer (phosphate – buffered saline with 0.2% BSA and 0.09% sodium azide). Cells were immunostained with fluorescein isothiocyanate (FITC) conjugated mouse anti – human CDS0, CD83, and CD86 antibodies (eBioscience, USA). Corresponding FITC immunoglobulin G (IgG) isotype control antibody (eBioscience, USA) was used. A total of 1×10^6 cells were incubated overnight at 4℃ with antibodies. The cells were then washed once with FACS buffer, resuspended, and phenotyped on a FACScan (Becton – Dickinson, USA).

An intracellular staining method was used for the detection of LMP2 proteins in rAd – LMP2 – DCs. Mature DCs were fixed in 2% paraformaldehyde. Cell membranes were permeated in 2% Triton X – 100 (Amresco, USA) and then incubated with LMP2 rabbit multiclonal antibody (obtained from Institute for Viral Disease Control and Prevention of Chinese CDC) at 4 ℃ overnight. After washing with PBS twice, the cells were immunostained with FITC – conjugated goat anti – rabbit IgG (Sigma) for 30 min at 37 ℃. The cells were then washed once with FACS buffer, resuspended, and analyzed on a FACScan.

6. Lymphocyte proliferation assays: Lymphocyte proliferation assays were performed by using rAd – LMP2 – DCs, untransfected DCs and CNE – 2 cells as stimulator cells and T lymphocytes as responder cells. Stimulator cells were incubated with Mitomycin C (MMC) at 25 μg/ml at 37 ℃ for 30 min and then washed with PBS twice. T lymphocytes isolated from the peripheral blood mononuclear cells were plated in 96 – well fiat – bottomed culture plate (Costar, USA) at 5×10^5 cells per well. Then stimulators were added and co – cultured with responders at ratios of 1 : 5, 1 : 10, 1 : 20, 1 : 50, 1 : 100, and 1 : 200 for 96 h at 37 ℃ in 5% CO_2. T cells incubated in medium alone served as control. The cells were then incubated with 5 mg/ml metrizamide (MTT; Sigma) 20 μl per well for 4 h. The supernatant was removed and 150 μl dimethyl sulfoxide (DMSO; Amresco, USA) was added to each well and agitated for 10 min to fully dissolve the crystals. Absorbance was measured at 570 nm on automatic ELISA reader (TRITURUS). All determinations were carried out in triplicate and repeated four times. Stimulation index (SI) was calculated as follows: SI = (experimental – blank) / (control – blank).

7. Induction of CTLs by transfected DCs: T cells were harvested by nylonwool separation as described previously. T cells (1×10^6) were co – cultured with rAd – LMP2 – DCs (5×10^4) in a 24 – well tissue culture plate in 1 ml RPMI 1640/20% FCS at 37 ℃ in 5% CO_2. IL – 2 was added at a final concentration of 40 IU/ml to all wells 3 days later and every 2 – 3 days thereafter. Responding T cells were re – stimulated weekly for 2 weeks with transfected DC at a responder T cell – to – stimulator DC ratio of 20 : 1. The CTLs were then collected and used as the effector cells in CTL assays against CNE – 2 cells.

8. Cytotoxicity assays: The target cells were placed in 96 – well tissue culture plates at 1×10^4 cells per well and co – cultured with effector cells (CTLs) at the ratio of 1 : 5, 1 : 10, and 1 : 20 for 48 h at 37 ℃ in 5% CO_2. The cytotoxic activities were determined by MTT assay. Freshly prepared and filtered 20 μl metrizamide (5 mg/ml) was added to each well, and the cells were continuously cultured for 4 h. The supernatant was removed and 150 μl dimethyl sulfoxide was added to each well and agitated for 10 min to fully dissolve the crystals. Absorbance was measured at 570 nm on automatic ELISA reader (TRITURUS). All determinations were carried out in triplicate and repeated four times. Experiments were performed in triplicate. The percentage of specific cytotoxicity was calculated as [(experimental – minimal) / (maximal – minimal)] × 100. Target cells incubated in medium alone or in medium containing 1% Triton X – 100 were used to determine minimal and maximal cytotoxicity, respectively. T cells separated from the peripheral blood mononuclear cells by sterile nylonwool isolation column as described previously and the untransfected DCs were used as controls, respectively.

9. Analysis of LMP2 – specific CTLs populations: T cells were stimulated with rAd – LMP2 – DCs weekly as described previously. After two rounds of stimulations, the induced LMP2 – specific

CTLs were collected on day 14 and then resuspended in cold FACS buffer. Cells were immunostained with FITC/PE/PE – cyanine5 (Cy5) conjugated mouse anti – human CD4/CD8/CD3 antibodies (Jingmei Biotech, China). Corresponding mouse FITC/PE/PE – Cy5 IgG isotype control antibody (Jingmei Biotech, China) was used. A total of 1×10^6 cells were incubated with antibodies for 30 min at 37 ℃. The cells were then washed once with FACS buffer, resuspended, and bidimensional analyzed with a FACScan.

10. Statistical analysis: SPSS11.0 was used for data variation analysis; p values less than 0.05 were considered statistically significant.

Results

1. Morphological features of rAd – LMP2 – DCs

Fully morphologic differentiation of mature DCs activated by calcium ionophore required 72 – 96 h of culture. The rAd – LMP2 – DCs retained typical morphological features of DCs (Fig. 1). While untreated PBMCs maintained their rounded, smooth surface morphology and appeared as dispersed, nonadherent cells in culture, rAd – LMP2 – DCs predominantly gathered in clusters as nonadherent or loosely adherent cells with a larger cell surface and irregular shape.

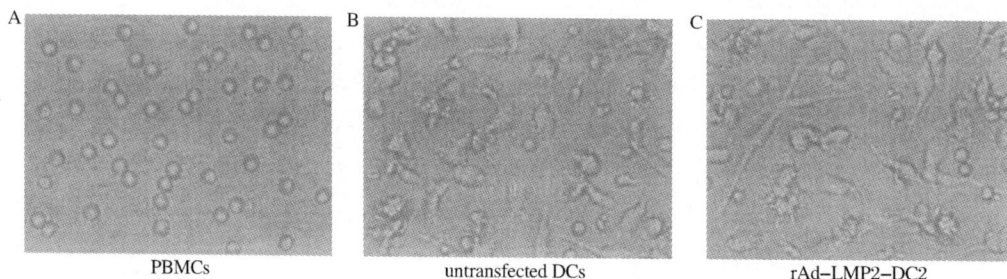

| PBMCs | untransfected DCs | rAd–LMP2–DC2 |

Fig. 1 Morphological characters of PBMCs (A), untransfected DCs (B), and transfected DCs (C). After 96 h culture, PBMCs were grown by suspension. DCs with or without transfection appeared similar morphology with long dendritic projection. Photomicrographs were taken with inverted phase contrast microscope under 200 × magnifications

2. Phenotype of transfected DCs

To determine whether mature DCs transfected with rAd – LMP2 expressed co – stimulatory molecules, mature DCs with or without rAd – LMP2 transfection were analyzed for co – stimulatory molecules (CD80 and CD86) and DC activation marker (CD83). We found these immunophenotypic alterations occurred promptly within the first 20 – 40h of culture with calcium ionophore A23187. Adenoviruses transfection of mature DCs did not result in significant increases or decreases in CD83, CD80, or CD86 expression. Data shown in Fig. 2 are representative of three independent experiments that produced similar results.

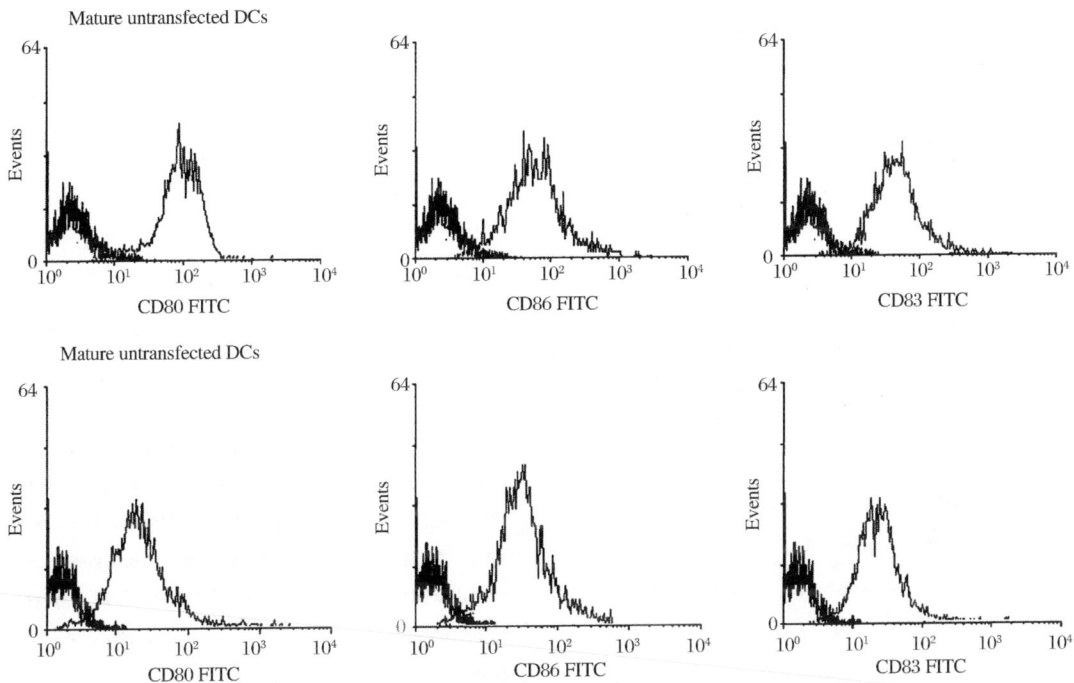

Fig. 2 Phenotype of mature DCs with or without rAd – LMP2 transfection. Cells were incubated with FITC, conjugated mAbs against CD80, CD86, and CD83. The result showed that mature untransfected DCs (top panel) with expressions of CD80, CD83, and CD86 were 86.32%, 85.73% and 86.27%, respectively; mature transfected DCs (bottom panel) were 81.54%, 87.48% and 88.37%, respectively

3. Expression of LMP2 in transfected DCs

Immature DCs were transfected with rAd – LMP2 at MOI 200 for 2 h. The transfected cells were cultured in the presence of calcium ionophore and rhGM – CSF for additional 48 h. When mature rAd – LMP2 – DCs were analyzed by flow cytometry, the percentage of transfected DCs expressing LMP2 was 84.54%, which suggested efficient gene transfer (Fig. 3).

4. Stimulation of T lymphocytes by rAd – LMP2 – DCs

It was found that anti – tumor T cells were generated by a single stimulation with mature DCs transfected with rAd – LMP2. The rAd – LMP2 – DCs were more potent stimulators of T lymphocytes than untransfected DCs ($P < 0.05$) or CNE – 2 cells ($P < 0.01$), respectively. The effect was enhanced with higher ratio of rAd – LMP2 – DCs to T cells (Fig. 4).

5. Cytotoxicity assays

After stimulating twice with rAd – LMP2 – DCs, highly LMP2 – specific anti – tumor CTLs could be induced. The cytotoxic activity was enhanced with increased ratio of effector – to – target cells. MTT assay showed that cytotoxic activity in rAd – LMP2 – DCs group was higher than that in untransfected DCs ($P < 0.05$) and T cell groups ($P < 0.01$), respectively (Fig. 5).

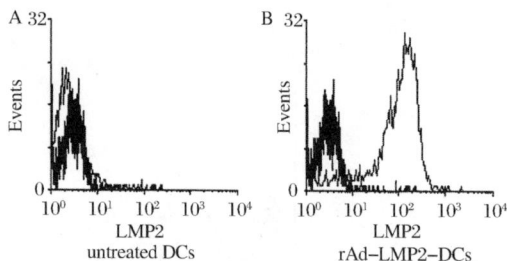

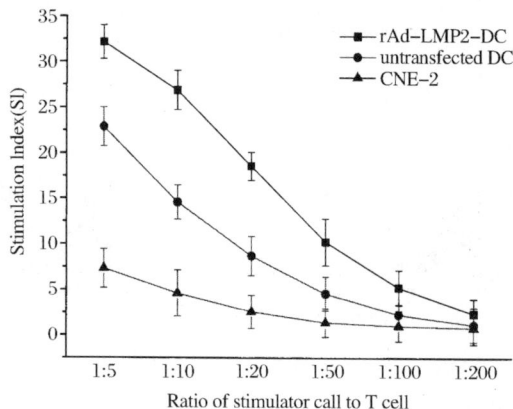

Fig. 3　Expression of LMP2 in untreated and transfected DCs. Flow cytometry indicated expression of LMP2 in rAd – LMP2 – DCs（B）was 84. 54 and 1. 57% in untreated DCs（A）

Fig. 4　T lymphocytes proliferation reaction stimulated by rAd – LMP2 – DCs, untransfected DCs, and CNE – 2. Mature DCs transfected with rAd – LMP2 for 48 h were collected and co – cultured with T cells for 96 h. Specific CTLs were detected by MTT assay. The results are expressed as ($\bar{x} \pm s$) of three replicates. Data indicate that rAd – LMP2 – DCs were potent stimulators of lymphocyte than untransfected DCs（$P < 0.05$）or CNE – 2（$P < 0.01$）, respectively

6. Effect of stimulation on cytotoxicity

T cells were stimulated by rAd – LMP2 – DCs weekly. The induced CTLs were harvested as effector cells. These effector cells were used against target cells in cytotoxicity assay as previously described on day 7, 14, and 21. The results showed that LMP2 – specific cytotoxicity elicited by only a single stimulation of transfected DCs was higher than those by T cell group（$P < 0.01$）and untransfected DC group（$P < 0.05$）. Further more, the cytotoxicity could augment with repeated stimulations. Compared with that on day 7, the specific cytotoxicity was evidently higher on day 14 in all groups, respectively（$P < 0.01$）, but there was no significant difference between those on day 14 and on day 21（$P > 0.05$）in untransfected DC group and rAd – LMP2 – DC group（Fig. 6）. The experiments indicated that two rounds of stimulation were enough to induce potent specific cytotoxicity.

7. Flow cytometric analysis of LMP2 – specific CTLs populations

LMP2 – specific CTLs induced by rAd – LMP2 – DCs on day 14 were collected and analyzed with flow cytometry. We found that the CTLs consisted of CD4 + and CD8 + T cells simultaneously. The component of CD8 + T cell was slightly larger than that of CD4 + T cell（Fig. 7）.

Discussion

Recently, malignancy – associated viruses are used as potential targets for immunotherapeutic vaccines aiming to stimulate T – cell responses against viral antigens expressed in tumor cells[12 – 14]. Here we have shown that the induction of primary antigen – specific CTL responses *in vitro* by human PBMCs – derived DCs adenovirally transfected with LMP2 gene, a subdominant antigen in EBV – associated NPC cells.

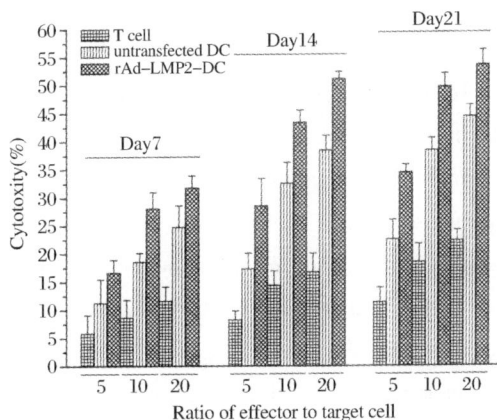

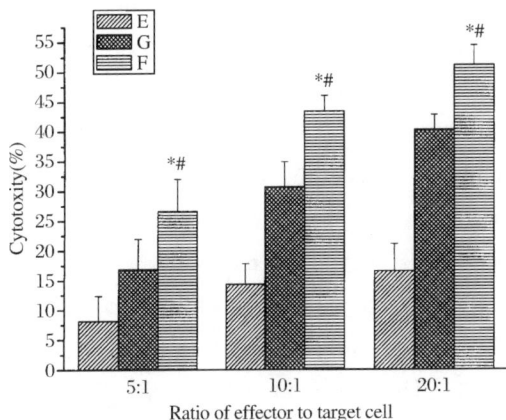

The CNE – 2 cells were placed in 96 – well tissue culture plates at 1 x 10^4 per well and co – cultured with effector cells at the ratio of 1 : 5, 1 : 10, and 1 : 20 for 48 h. Percentage cytotoxicity ($\bar{x} \pm s$) of three replicates) was determined by MTT assay. $*P < 0.01$ vs. T cell group; $\#P < 0.05$ vs. untransfected DC group

Fig. 5 Cytotoxicity of CTLs against nasopharyngeal carcinoma cells

T cells were stimulated by rAd – LMP2 – DCs weekly, and the induced CTLs were harvested as effector cells. These effector cells were used against target cells in cytotoxicity assay on day 7, 14, and 21. Cytotoxicity on day 14 was obviously higher than that on day 7 ($P < 0.01$), but there was no significant difference between those on day 14 and on day 21 ($P > 0.05$)

Fig. 6 Effect of different rounds of stimulation on cytotoxicity

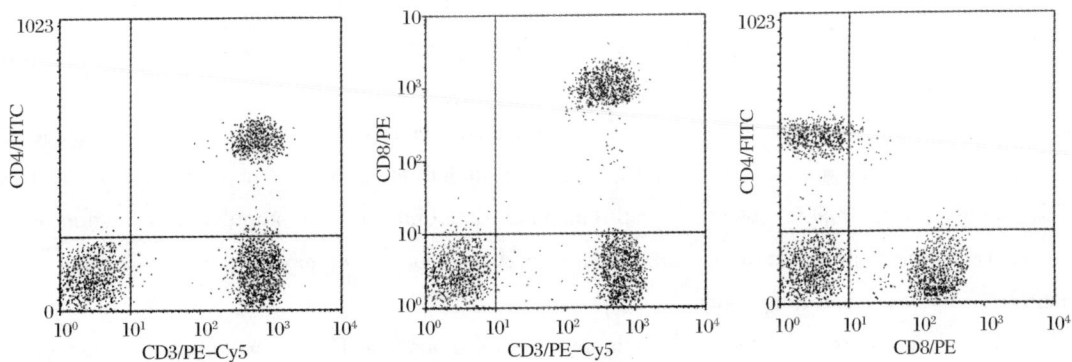

Fig. 7 Flow cytometric analysis of LMP2 – specific CTLs populations. The percentage of CD3$^+$, CD4$^+$, and CD8$^+$ T cells was 61. 73%, 36. 47% and 46. 18%, respectively

DCs are professional APCs that play a critical role in the activation of the immune response to antigen. Mature DCs express high levels of co – stimulatory molecules, necessary components of T cell activation by APCs, which must occur in conjunction with MHC – restricted presentation of the antigen to the T – cell receptor. The co – stimulatory molecules identified on DCs with their respective T – cell receptors are CD54, CD80, CD83, CD86, CD40 and CD40 ligand. In the present study, the co – stimulatory molecules CD80, CD86 and CD83 did not change significantly after rAd – LMP2 modification of DCs compared with untreated DCs, which demonstrated adenovirus transfection had little effect on DC maturation and antigen – presenting function.

Recent laboratory observations indicated that pharmacologic agents that mobilize intracellular calcium can be used to enhance APC functions in human PBMCs[15,16]. The phospholipase C (PLC) – calcium signaling pathway is involved in the maturation of DCs induced by the agonists such as calcium ionophore A23187. A23187 can pump extracellular calcium into the cell. The increased cytoplasmic calcium concentration induced by A23187 could cause calcium – induced calcium release from intracellular stores. On the other hand, increased cytoplasmic calcium levels may induce positive feedback that activates PLC, which causes the second messengers inositol 1, 4, 5 – tri – phosphate (IP$_3$) liberating from the plasma membrane. Thus, calcium release from IP3 – gated stores induces the maturation of DCs. In our study, we tried to use calcium ionophore as the main agent to generate DCs more effectively from human PBMCs and examined the characteristics of DCs including cellular morphology and APC function. Our results showed that fully morphologic differentiation of DCs activated by calcium ionophore required 72 – 96 h of culture, and immunophenotypic alterations occurred promptly within the first 20 – 40 h of culture after CI treatment, including upregulation of CD80 and CD86 expression, and de novo expression of the DC – associated activation marker CD83. Such rapid activation kinetics contrasted to the much slower activation (needed about 9 – 10 days) observed when PBMCs were treated with cytokine combinations such as rhGM – CSF, rhIL – 4, and rhTNF – α[5,6].

Gene transfer is an attractive means to affect the immunostimulatory properties of DCs. We have used a gene – based vaccination strategy by using DCs expressing the tumor antigen to elicit a potent therapeutic anti – tumor immunity. This approach has apparent advantages over protein – or peptide – based immunization[17]. Tumor associated antigen (TAA) gene expression in DCs causes endogenous processing and presentation of multiple and/or undefined antigenic peptides independent of MHC alleles. Furthermore, specific T cell – mediated immunity may be stimulated by vaccine – involved APC without prior knowledge of responder MHC haplotypes or of relevant MHC class I – or class II – restricted peptide epitopes.

Although a variety of vectors are available for gene transfer to DCs, recombinant adenovirus is most efficient. Adenovirus vector is a highly efficient and reproducible method of gene transfer. Indeed, several studies have shown that successful adenoviral gene transfer into human DCs resulted in induction of a T – cell response against tumor[18,19]. Our results showed that the expression of LMP2 in transfected DCs reached a high level of 84. 54% , which indicated efficient gene transfection.

Currently, published reports showed that malignancy – associated virus antigen could induce specific antitumor CTL. EBV is a herpesvirus commonly associated with malignancies such as Hodgkin disease (HD), T – cell lymphoma, and NPC, particularly in immunocompromised hosts. EBV elicits a strong cytotoxic T lymphocyte (CTL) response directed against a broad range of viral antigens that are involved in the control and regulation of latency and in the induction of proliferation and transformation[20]. EBV – associated NPC, a high – incidence tumor in southern China, expresses a limited set of EBV proteins. Only latent class II EBV antigens such as EBNA1 and LMP1, LMP2 can be detected on NPC cells. Among these three antigens, LMP1 is an NPC – associated viral oncogene[21], and EBNA1 is an abundant source of HLA class II – restricted CD4$^+$ T – cell epitopes that contains a Gly – Ala repeat sequence, which can interrupt the presentation of it through HLA class I – restricted subway to T cells[22,23]. LMP2 is a source of subdominant CD8$^+$ T – cell epitopes presented by HLA class I alleles common in the Chinese population[24,25]. In some studies, EBV transformed B lymphoblastoid cell lines (LCLs) have been used to induce EBV – specific CTLs. Adoptive transfer of EBV – specific CTLs has been successfully applied in the treatment of EBV associated post – transplant lymphoproliferative disease[26,27]. Nevertheless, application of this approach to EBV – associated NPC is difficult, because LCLs focus T cell expansion on immunodominant EBV antigens such as the latent EBNA3A, 3B, and 3C that are not expressed in EBV – associated NPC. On the other hand, in adoptive immunotherapy LCLs elicited only EBV – specific memory T cell responses but not native T – cell responses[28]. In our study, we demonstrated that DCs transfected with LMP2 by adenovirus vector were able to stimulate enhanced T – cell proliferation and LMP2 – specific cytotoxic T – cell responses *in vitro*. Analyzing the populations of CTLs elicited by rAd – LMP2 – DCs, we found LMP2 – specific CTLs consisted of both CD4$^+$ and CD8$^+$ T cells simultaneously. Our result was similar to the reports of CTLs induced by DCs transfected with LMP2a in experiment of HD treatment[26]. The specific CTLs lysed carcinoma cells maybe by both MHC class I – and MHC class II – restricted mechanisms. In summary, the results demonstrate that vaccination using DCs simultaneously transfected with malignancy – associated virus antigens can elicit potent CTL response and provide a potential immunotherapy strategy for EBV – associated NPC.

Acknowledgements

This work was supported by the National "863" Project of China (No. 2003AA216071) and the National Natural Science Foundation of China (No. 30270520).

[In 《Biochemical and Biophycial Research Communication》 2006, 347: 551 –557]

References

1 Niedobitek G, Epstein – Barr virus infection in the pathogenesis of nasopharyngeal carcinoma, J Clin Pathol, 2000, 53: 248 –254

2 Steinman R M. The dendritic cell system and its role in immunogenicity, Ann Rev Immunol, 1991, 9: 271 –296

3 Banchereau J, Steinman R. Dendritic cells and the control of immunity, Nature (Lond.), 1998, 392: 245 –252

4 Tomohide T, Andrea G, Paul D R., Walter J S, Interleukin 18 gene transfer expands the repertoire of antitumor Th1 – type immunity elicited by dendritic cell – based vaccines in association with enhanced therapeutic efficacy. Cancer Res, 2002, 62: 5853 –5858

5 Jenne L, Arrighi J F, Jonuleit H, et al. Dendritic cells containing apoptotic melanoma cells prime human CD8$^+$ T cells for efficient tumor cell ly-

sis. Cancer Res, 2000, 60: 4446 – 4452

6　Nouri – Shirazi M, Banchereau J, Bell D, et al. Dendritic cells capture killed tumor cells and present their antigens to elicit tumorspecific immune responses. J Immunol, 2000, 165: 3797 – 3803

7　Bagley K C, Abdelwahab S F, Tuskan R G, Lewis G K. Calcium signaling through phospholipase C activates dendritic cells to mature and is necessary for the activation and maturation of dendritic cells induced by diverse agonists. Clin Diagn Lab Immunol, 2004, 11: 77 – 82

8　Sadovnikova E, Parovichnikova E N, Semikina E L , et al. Adhesion capacity and integrin expression by dendritic – like cells generated from acute myeloid leukemia blasts by calcium ionophore treatment. Exp. Hematol, 2004, 32: 563 – 570

9　Du H J, Zhou L, Zuo J M, et al. A study on killing effect of cytotoxic T cell activated by LMP2 peptides in nasopharyngeal carcinoma cells. J Oncol, 2004, 10: 92 – 94

10　Rosenfeld M A, Yoshimura K, Trapnell B C, et al. *In vivo* transfer of the human cystic fibrosis transmembrane conductance regulator gene to the airway epithelium. Cell, 1992, 68: 143 – 155

11　Mittereder N, March K L, Trapnell B C , Evaluation of the concentration and bioactivity of adenovirus vectors for gene therapy. J Virol, 1996, 70: 7498 – 7509

12　Gottschalk S, Edwards O L, Sili U, et al. Generating CTLs against the subdominant Epstein – Barr virus LMP1 antigen for the adoptive immunotherapy of EBV – associated malignancies. Blood, 2003, 101: 1905 – 1912

13　Wagner H J, Sili U, Gahn B, et al. Expansion of EBV latent membrane protein 2a specific cytotoxic T cells for the adoptive immunotherapy of EBV latency type 2 malignancies: influence of recombinant IL12 and IL15. Cytotherapy, 2003, 5: 231 – 240

14　Santodonato L, Agostino G D, Nisini R, et al. Monocyte – derived dendritic cells generated after a shortterm culture with IFN – alpha and granulocyte – macrophage colony – stimulating factor stimulate a potent Epstein – Barr virus – specific CD8 + T cell response J Immunol, 2003, 170: 5195 – 5202

15　Westers T M, Stam A G , Scheper R J, et al. Rapid generation of antigen – presenting cells from leukaemic blasts in acute myeloid leukaemia, Cancer Immunol. Immunother, 2003, 52: 17 – 27

16　Waclavicek M, Berer A, Oehler L, et al. Calcium ionophore: a single reagent for the differentiation of primary human acute myelogenous leukaemia cells towards dendritic cells. Br J Haematol, 2001, 114: 466 – 473

17　Diebold S S, Cotten M, Koch N, Zenke M. MHC class II presentation of endogenously expressed antigens by transfected dendritic cells. Gene Ther, 2001, 8: 487 – 493

18　Paul L F, Christie L, Jennifer M A , et al. Efficacy of CD40 ligand gene therapy in malignant mesothelioma. Am J Resp Cell Mol Biol, 2003, 29 : 321 – 330

19　Jay M L, Ali M, Reiko Y, et al. Adenovirus vector – mediated overexpression of a truncated form of the p65 nuclear factor κB eDNA in dendritic cells enhances their function resulting in immunemediated suppression of preexisting murine tumors. Clin Cancer Res, 2002, 8: 3561 – 3569

20　Khanna R, Burrows S R. Role of cytotoxic T lymphocytes in Epstein – Barr virus – associated diseases. Annu Rev Microbiol, 2000, 54: 19 – 48

21　Burrows J M, Bromham L, Woolfit M, et al. Selection pressuredriven evolution of the Epstein – Barr virus – encoded oncogene LMP1 in virus isolates from Southeast Asia. J Virol, 2004, 78: 7131 – 7137

22　Munz C, Bickham K L, Subklewe M . Human CD4 (+) T lymphocytes consistently respond to the latent Epstein – Barr virus nuclear antigen EBNA1. J Exp Med, 2000, 191: 1649 – 1660

23   Khanna R, Tellam J, Duraiswamy J, et al. Immunotherapeutic strategies for EBV – associated malignancies. Trends Mol. Med, 2001, 7: 270 – 276

24   Lin C L, Lo W F, Lee T H, et al. Immunization with Epstein – Barr Virus (EBV) peptide – pulsed dendritic cells induces functional CD8[+] T – cell immunity and may lead to tumor regression in patients with EBV – positive nasopharyngeal carcinoma. Cancer Res, 2002, 62: 6952 – 6958

25   Taylor G S, Haigh T A, Gudgeon N H, et al. Dual stimulation of Epstein – Barr Virus (EBV) – specific CD4[+] – and CD8[+] – T – cell responses by a chimeric antigen construct: potential therapeutic vaccine for EBV – positive naso-

pharyngeal carcinoma. J Virol, 2004, 78: 768 – 778

26   Bollard C M, Straathof K C, Huls M H, et al. The generation and characterization of LMP2 – specific CTLs for use as adoptive transfer from patients with relapsed EBV – positive Hodgkin disease. J Immunother, 2004, 27: 317 – 327

27   Chua D, Huang J, Zheng B, et al. Adoptive transfer of autologous Epstein – Barr virus – specific cytotoxic T cells for nasopharyngeal carcinoma. Int J Cancer, 2001, 94: 73 – 80

28   Herr W, Ranieri E, Olson W, et al. Mature dendritic cells pulsed with freeze – thaw cell lysates define an effective in vitro vaccine designed to elicit EBV – specific CD4 (+) and CD8 (+) T lymphocyte responses. Blood, 2000, 96: 1857 – 1864

99.   Synthesis of Stilbene Derivatives with Inhibition of SARS Coronavirus Replication

LI Yue-qing[1], LI Ze-lin[2], ZHAO Wei-jie[1], WEN Rui-xing[2], MENG Qing-wei[1], ZENG Yi[2]

1. State Key Laboratory of Fine Chemicals, Dalian University of Technology;

2. College of Life Science and Bioengineering, Beijing University of Technology

Summary

Stilbene derivatives have wide range of activities. In an effort to find other potential activities of this kind of compounds, 17 derivatives, including resveratrol, were synthesized. Twelve of them were evaluated for their antiviral potential against severe acute respiratory syndrome (SARS) – CoV – induced cytopathicity in Vero E6 cell culture. The result showed that SARS virus was totally inhibited by compounds 17 and 19 ($\leqslant 0.5$ mg $\cdot$ ml^{-1}) and no significant cytotoxic effects were observed in vitro.

[Key words]   Severe acute respiratory syndrome; Coronavirus; Stilbene derivatives

Introduction

Stilbene derivatives are widely distributed in nature, which are thought to be phytoalexins. There is a growing interest in stilbene derivatives because many activities have been observed in some of the naturally occurring as well as some of the synthetic stilbenes. Activities include antimicrobial[1-3], anfioxidant[4,5], antileukemic[6], anti – platelet aggregative[7,8], protein tyrosine kinase inhibitory[9], anti – inflammatory[10,11], anticarcinogenic activity[12,13], anti – HIV[14,15] and anti – herpes simplex virus[16]. In the course of our research for potential activities of stilbene derivatives, we designed different hydroxyl substituted sites against resveratrol and kept the transtructure to simuiate this kind of phytoalexins.

In early 2003, severe acute respiratory syndrome (SARS) broke out in China and other countries. Many scholars and researchers were engaged in the search of anti – SARS agents and vaccines. At the same time, we also used this kind of compounds to the urgent antiviral filtration *in vitro*. The result was exciting that the change of hydroxyl group's site and the introduction of nitrogen atom were beneficial to the anti – SARS activity. Especially the substituted sites of hydroxyl groups in compounds 17 and 19 have been demonstrated to be a key structure element of anti – SARS virus.

No one knows the likelihood of evolution of SARS – CoV in human and animals. Moreover the complete understanding of pathogenesis of SARS remains tentative. In this study, we evaluated such stilbene derivatives for their potential to inhibit SARS – CoV replication for the first time and thought this discovery would be beneficial to the anti – SARS – CoV development.

Chemistry

In order to prepare a variety of stilbene derivatives with hydroxyl groups, we used methoxyl materials as starting materials to avoid the oxidation of hydroxyl compounds. Also we introduced one pyridine ring in place of one of the benzene rings for the purpose of evaluation the activity change. Preparation of aim compounds was accomplished as given in Scheme 1. According to this scheme, 3, 5 – dimethoxybenzyl bromide, 2 – chloromethyl – 3, 4 – dimethoxy pyridine or 2 – chloro – methyl – 3, 5 – dimethyl – 4 – methoxypyridine were heated with triethyl phosphite in the presence of $(n - Bu)_4 NI$ to give the phosphonates (2 – 4) (Michaelis – Arbuzov reaction[17]). In turn, the reaction of aryl aldehyde with the anion of phosphates formed *in situ* with sodium hydride gave (E) – stilbene derivatives (5 – 14) (with almost none of (Z) – isomer) together with water soluble diethyl phosphate (Wittig – Horner reaction[18,19]). But the yield of pyridine containing derivatives is poor. Further methoxyl derivatives were demethylated with BBr_3 in dichloromethane.

Biological Results and Discussion

In this study the antiviral potential of twelve compounds (7, 8, 10, 11, 14 – 21) was evaluated *in vitro* for their inhibitory effect of SARS virus. Cytotoxicity in Vero E6 cells was measured before the antiviral activity. The compounds that showed cytotoxity to Vero E6 were weeded out first. The remainder was studied their inhibitory activity. Only compounds 17 and 19 could inhibit the replication of SARS virus in Vero E6 cells in concentration $\leqslant 0.5$ mg $\cdot$ ml^{-1} (2.05 mmol/L). There was no cytotoxicity to Vero E6 in concentration $\geqslant 2$ mg $\cdot$ ml^{-1} (8.20 mmol/L) (Tab. 3). It showed that both of them could inhibit SARS virus *in vitro*. Compounds 17 and 19 clearly inhibited the cytopathic effect (CPE) induced by infection with SARS – CoV (Fig. 2). As compared with cinanserin which in-

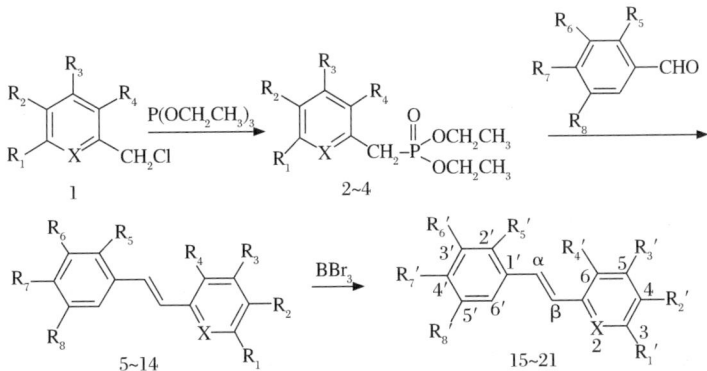

(a) P (OCH$_2$CH$_3$)$_3$, (n – Bu)$_4$NI, 100 – 120 ℃; (b) methoxy benzaldehyde, NaH, THF, room temperature;

(c) BBr$_3$, CH$_2$Cl$_2$, 20 – 40 ℃

Fig. 1   Synthesis of compounds 5 – 21

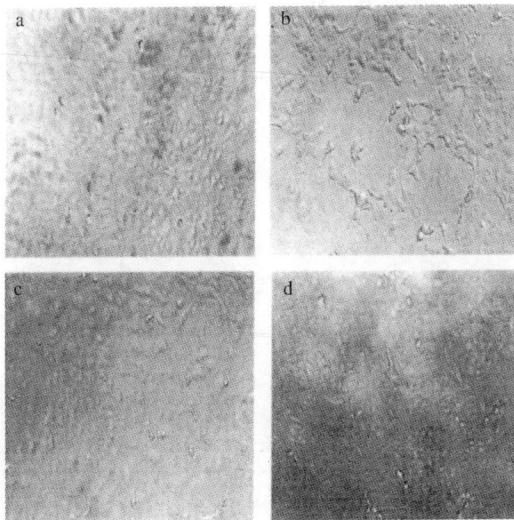

(a) The normal Vero E6 cells were cultivated with Eagle's medium containing 10% fetal calf serum. (b) The Vero E6 cells infected with 100TCID$_{50}$ SARS virus. (c) The infected cells were treated with compound 17 for 72h. (d) The infected cells were treated with compound 19 for 72h

Fig. 2   Cytopathic effect (CPE) of compounds 17 and 19 on replication of SARS – CoV in Vero E6 cells

hibits SARS virus at 66 μmol/L (IC$_9$0) [20], the result in our research seems to be not so good. But the concentration 0.5 mg · ml^{-1} is not the terminal of the inhibition, because our research could not undergo in present situation in our country.

The methoxy stilbene derivatives (7, 8, 10, 11, 14) showed no cytotoxity. Compound 7 and 10, which were methyl ethers of compounds 17 and 19, respectively, could inhibit 50% replication of SARS virus *in Vero* E6 cells in concentration 1 mg · ml^{-1}. Compounds 8, 11 and 14 had no inhibition in the evaluation. Although compounds 17 and 19 had one same part, the compound 21 that had the same part showed no activity. It seemed that the whole molecular structure was necessary to the inhibitory activity and hydroxy group was prior to methoxy group. However, the structure – activity relationship was still unclear in our present study. Maybe the further research will reveal the relationship between the substituent and activity (Tabs. 1 and 2).

Tab. 1 Substituent for compounds 2 – 14

Compounds	X	R_1	R_2	R_3	R_4	R_5	R_6	R_7	R_8
2	C	OCH₃	H	OCH₃	H				
3	N	H	H	OCH₃	OCH₃				
4	N	H	CH₃	OCH₃	CH₃				
5	C	OCH₃	H	OCH₃	H	H	H	OCH₃	H
6	C	OCH₃	H	OCH₃	H	H	OCH₃	H	OCH₃
7	C	OCH₃	H	OCH₃	H	OCH₃	H	H	OCH₃
8	C	OCH₃	H	OCH₃	H	OCH₃	H	OCH₃	H
9	N	H	H	OCH₃	OCH₃	H	OCH₃	H	OCH₃
10	N	H	H	OCH₃	OCH₃	OCH₃	H	H	OCH₃
11	N	H	H	OCH₃	OCH₃	OCH₃	H	OCH₃	H
12	N	H	CH₃	OCH₃	CH₃	H	OCH₃	H	OCH₃
13	N	H	CH₃	OCH₃	CH₃	OCH₃	H	H	OCH₃
14	N	N	CH₃	OCH₃	CH₃	OCH₃	H	OCH₃	H

Tab. 2 Substituent for compounds 15 – 21

Compounds	X	R_1'	R_2'	R_3'	R_4'	R_5'	R_6'	R_7'	R_8'
15	C	OH	H	OH	H	H	H	OH	H
16	C	OH	H	OH	H	H	OH	H	OH
17	C	OH	H	OH	H	OH	H	H	OH
18	N	H	H	OH	OH	H	OH	H	OH
19	N	H	H	OH	OH	OH	H	H	OH
20	N	H	CH₃	OH	CH₃	H	OH	H	OH
21	N	H	CH₃	OH	CH₃	OH	H	H	OH

Tab. 3 Inhibitory effect of compounds 17 and 19 on SARS virus *in vitro*

Compounds	Experiment number	Chemical control	Final concentration of compound (mg ml⁻1)			Viral control (100TCI – D50)
			2	1	0. 5	
17	1	–	–	–	–	+ + +
	2	–	–	–	–	+ + +
	3	–	–	–	–	+ + +
19	1	–	–	–	–	+ + +
	2	–	–	–	–	+ + +
	3	–	–	–	–	+ + +

 – = no cytotoxicity, no CPE; + + + = CPE > 75% cells with CPE

Conclusions

We have synthesized (E) – stilbene derivatives bearing hydroxyl groups and in some of them used pyridine ring in place of one benzene ring. Two of these compounds possessed antiviral activity against SARS in vitro and till now this has never been reported.

Experimental Protocols

1. Synthesis

Melting points were determined in capillary tubes on a Buchi oil bath apparatus and uncorrected. Spectra were obtained as follows: LC/Q – Tof MS, ^{1}H NMR spectra on Varian INOVA 400 MHz spectrometer with TMS as the internal standard, IR spectra on FT – IR Nicolet 20 spectrometer. Thin layer chromatography (TLC) was carried out on Si gel plates (60 F_{254}, Merck).

All reagents were commercially available and used as received.

(1) *Diethyl [3, 5 – dimethoxybenzyl] phosphonate* (2) [21]

Triethyl phosphate (2.7 ml, 16 mmol) was added to the 3, 5 – dimethoxybenzyl bromide (2.3 g, 10 mmol) containing a catalytic amount of tetrabutyl – ammonium iodine, and the mixture was heated at 110 – 130 ℃ for 5 – 6 h. Excess triethyl phosphite was removed by heating at 80 – 90 ℃ under vacuum (< 5 kPa) to yield 2 as a light yellow oil. ^{1}H NMR (CDCl$_3$) δ 1.27 (t, $^3J_{HH}$ = 7.2 Hz, 6H, OCH$_2$CH$_3$), 3.09 (d, $^2J_{HP}$ = 21.6 Hz, 2H, PCH$_2$), 3.78 (s, 6H, OCH$_3$), 4.04 (quint, $^3J_{HH}$ = $^3J_{HP}$ = 7.2 Hz, 4H, OCH$_2$CH$_3$), 6.35 (s, 1H, H$_{para}$ of C$_6$H$_4$), 6.46 (s, 2H, H$_o$rtho of C$_6$H$_4$) ppm; IR (KBr) 2939, 2839, 1685, 1598, 1512, 1461, 1257 (vs, P = O), 1206, 1160 (w, P – O – C), 1029 (vs, P – O), 969, 835 cm^{-1}.

(2) *Diethyl [3, 4 – dimethoxypyridine – 2 – methylene] phosphonate* (3)

The compound was synthesized in the same manner as for 2 as a viscous orange liquid and used directly in the next step. ^{1}H NMR (CDCl$_3$) δ 1.28 (t, $^3J_{HH}$ = 7.2 Hz, 6H, OCH$_2$CH$_3$), 2.23 (d, J = 1.2 Hz, 3H, CH$_3$), 2.32 (s, 3H, CH$_3$), 3.42 (d, $^2J_{HP}$ = 22.0 Hz, 2H, PCH$_2$), 3.75 (s, 3H, OCH$_3$), 4.09 (quint, $^3J_{HH}$ = $^3J_{HP}$ = 7.2 Hz, 4H, OCH$_2$CH$_3$), 8.19 (s, 1H, H of C$_5$NH) ppm; IR (KBr) 2982, 2944, 1583, 1489, 1447, 1425, 1302, 1253 (vs, P = O), 1163 (w, P – O – C), 1053 (vs, P – O), 1027 (vs, P – O), 965, 827, 781 cm$^-$1.

(3) *Diethyl [3, 5 – dimethyl – 4 – methoxypyridine – 2 – methylene] phosphonate* (4)

The compound was synthesized in the same manner as for 2 as a viscous liquid and used directly in the next step. ^{1}H NMR (CDCl$_3$) δ 1.30 (t, $^3J_{HH}$ = 7.2 Hz, 6H, OCH$_2$CH$_3$), 3.48 (d, $^2J_{HP}$ = 22.4 Hz, 2H, PCH$_2$), 3.90 (s, 3H, OCH$_3$), 3.92 (s, 3H, OCH$_3$), 4.13 (quint, $^3J_{HH}$ = $^3J_{HP}$ = 7.2 Hz, 4H, OCH$_2$CH$_3$), 6.76 (dd, $^3J_{HH}$ = 5.6 Hz, $^6J_{HP}$ = 1.6 Hz, 1H, H$_{para}$ of C$_5$NH$_2$), 8.19 (d, $^3J_{HH}$ = 5.6 Hz, 1H, H$_{meta}$ of C$_5$NH$_2$) ppm; IR (KBr) 2983, 2933, 1590, 1566, 1475, 1396, 1253 (vs, P = O), 1164 (w, P – O – C), 1054 (vs, P – O), 1029 (vs, P – O), 965, 873, 850, 807, 777, 753 cm^{-1}.

(4) *(E) – 3, 4', 5 – Trimethoxystilbene* (5)

The product of first step was dissolved in dry THF (20 ml) and stirred at 0 – 5 ℃. Sodium hydride (0.6 g, 25 mmol) added to the well – stirred phosphonate ester solution. After 30 min, the aldehyde (10 mmol) in THF (30 ml) was added dropwise, and the mixture was allowed to stir at room temperature for 8 – 16 h. The mixture was then cooled to 0 ℃, and the excess sodium hydride was quenched with water (10 ml). The reaction mixture was then poured on ice, followed by addition of 1 mol/L HCl to pH 6, and the product was extracted with ethyl acetate (4 × 50 ml). The organic layers were combined and washed with saturated solution of sodium chloride (2 × 30 ml). The ethyl acetate layer was dried over anhydrous sodium sulfate and evaporated. The residue was purified by recrystallization and gave 1. 7 g (65. 9% total yield of the first two steps) of 5 as a slightly yellow crystal, m. p. 56 – 57 ℃ dec. ^{1}H NMR (CDCl$_3$) δ3. 83 (s, 9H, OCH$_3$), 6. 38 (s, 1H, H – 4), 6. 65 (s, 2H, H – 2, 6), 6. 90 (d, J = 8. 8 Hz, 2H, H – 3', 5'), 6. 91 (d, J = 16. 4 Hz, 1H, H – α), 7. 04 (d, J = 16. 4 Hz, 1H, H – β), 7. 44 (d, J = 8. 8 Hz, 2H, H – 2', 6') ppm; IR (KBr) 2936, 2836, 1598, 1512, 1459, 1426, 1253, 1154, 962, 832 cm^{-1}.

(5) (E) – 3, 3', 5, 5' – Tetramethoxystilbene (6)

The compound was synthesized in the same manner as for 5 in 83. 0% yield as a white crystal, m. p. 135 – 136 ℃ dec. ^{1}H NMR (CDCl$_3$) δ 3. 84 (S, 12H, OCH$_3$), 6. 40 (t, J = 2. 0 Hz, 2H, H – 4, 4'), 6. 67 (d, J = 2. 0 Hz, 4H, H – 2, 2', 6, 6'), 7. 01 (s, 2H, H – α, β) ppm; IR (KBr) 2999, 2939, 2839, 1594, 1462, 1428, 1207, 1194, 1151, 1065, 943, 865, 837, 825 cm^{-1}.

(6) (E) – 2', 3, 5, 5' – Tetramethoxystilbene (7)

The compound was synthesized in the same manner as for 5 in 72. 6% yield as a milk white amorphous solid, m. p. 54 – 55 ℃ dec. ^{1}H NMR (CDCl$_3$) δ 3. 81 (s, 3H, OCH$_3$), 3. 83 (s, 6H, OCH$_3$), 3. 84 (s, 3H, OCH$_3$), 6. 39 (t, J = 1. 6 Hz, 1H, H – 4), 6. 69 (d, J = 1. 6 Hz, 2H, H – 2, 6), 6. 79 (dd, ^{3}J = 8. 4 Hz, ^{4}J = 2. 8 Hz, 1H, H – 4'), 6. 83 (d, J = 8. 4 Hz, 1H, H – 3'), 7. 13 (d, J = 2. 8 Hz, 1H, H – 6'), 7. 02, 7. 42 each 1H (d, J = 16. 4 Hz, H – α, β) ppm; IR (KBr) 3005, 2941, 2833, 1593, 1498, 1463, 1240, 1206, 1154, 1063, 1052, 966, 862, 804 cm^{-1}.

(7) (E) – 2', 3, 5, 4' – Tetramethoxystilbene (8)

The compound was synthesized in the same manner as for 5 in 83. 0% yield as a pale yellow crystal, m. p. 81 – 82 ℃ dec. ^{1}H NMR (CDCl$_3$) δ 3. 82 (s, 9H, OCH$_3$), 3. 86 (s, 3H, OCH$_3$), 6. 36 (t, ^{4}J = 2. 0 Hz, 1H, H – 4), 6. 46 (d, ^{4}J = 2. 4 Hz, 1H, H – 3'), 6. 51 (dd, ^{4}J = 2. 4 Hz, ^{3}J = 8. 4 Hz, 1H, H – 5'), 6. 67 (d, ^{4}J = 2. 0 Hz, 2H, H – 2, 6), 7. 49 (d, ^{3}J = 8. 4 Hz, 1H, H – 6'), 7. 36, 6. 94 each 1H (d, ^{3}J = 16. 4 Hz, 2H, H – α, β) ppm; IR (KBr) 3001, 2944, 2840, 1629, 1590, 1504, 1461, 1428, 1295, 1196, 1155, 1062, 1029, 968, 837, 818 cm^{-1}.

(8) (E) – 3, 5, 5, 6 – Tetramethoxystilbene – 2 – nitrogen (9)

The compound was synthesized in the same manner as for 5 in 23. 2% yield as a pale yellow solid, m. p. 88 – 90 ℃ dec. ^{1}H NMR (CDCl$_3$) δ 3. 83 (s, 6H, OCH$_3$), 3. 88 (s, 3H, OCH$_3$), 3. 93 (s, 3H, OCH$_3$), 6. 43 (t, J = 2. 0 Hz, 1H, H – 4'), 6. 78 (d, J = 2. 0 Hz, 2H, H – 2', 6'), 6. 75 (d, J = 5. 2 Hz, 1H, H – 4), 8. 27 (d, J = 5. 2 Hz, 1H, H – 3), 7. 47, 7. 71 each 1H (d, J = 16. 0 Hz, 2H, H – α, β) ppm; IR (KBr) 2927, 2836, 1597,

1476, 1458, 1423, 1280, 1270, 1203, 1158, 1148, 980, 818 cm^{-1}.

(9) (E) $-2'$, 5, 5', 6 $-$ Tetramethoxystilbene $-2-$ nitrogen (10)

The compound was synthesized in the same manner as for 5 in 20.7% yield as a pale yellow crystal, m. p. 100 $-$ 101 ℃ dec. ^{1}H NMR (CDCl$_3$) δ 3.82 (S, 3H, OCH$_3$), 3.85 (s, 3H, OCH$_3$), 3.87 (s, 3H, OCH$_3$), 3.92 (s, 3H, OCH$_3$), 6.73 (d,^{3}J = 5.6 Hz, 1H, H $-$ 4), 7.23 (d,^{4}J = 2.4 Hz, 1H, H $-$ 6'), 6.83 (dd,^{3}J = 8.8 Hz,^{4}J = 2.4 Hz, 1H, H $-$ 4'), 6.85 (d,^{3}J = 8.8 Hz, 1H, H $-$ 3'), 8.28 (d,^{3}J = 5.6 Hz, 1H, H $-$ 3), 8.08, 7.52 each 1H (d,^{3}J = 16.4 Hz, 2H, H $-$ α, β) ppm; IR (KBr) 2995, 2944, 2833, 1629, 1573, 1495, 1286, 1241, 1208, 1069, 994, 978, 854, 817, 802 cm

(10) (E) $-2'$, 5, 4', 6 $-$ Tetramethoxystilbene $-2-$ nitrogen (11)

The compound was synthesized in the same manner as for 5 in 27.5% yield as a white amorphous solid, m. p. 131 $-$ 133 ℃ dec. ^{1}H NMR (CDCl$_3$) δ 3.84 (S, 3H, OCH$_3$), 3.86 (s, 3H, OCH$_3$), 3.87 (s, 3H, OCH$_3$), 3.91 (s, 3H, OCH$_3$), 6.47 (d,^{4}J = 2.0 Hz, 1H, H $-$ 3'), 6.52 (dd,^{3}J = 8.4,^{4}J = 2.0 Hz, 1H, H $-$ 5'), 6.70 (d,^{3}J = 5.2 Hz, 1H, H $-$ 4), 7.61 (d,^{3}J = 8.4 Hz, 1H, H $-$ 6'), 8.27 (d,^{3}J = 5.2 Hz, 1H, H $-$ 3), 7.43, 8.03 each 1H (d,^{3}J = 16.4 Hz, 2H, H $-$ α, β) ppm; IR (KBr) 2961, 2940, 2837, 1624, 1602, 1573, 1504, 1478, 1290, 1278, 1207, 1063, 1031, 998, 939, 821 cm^{-1}.

(11) (E) -4, 6 $-$ Dimethyl $-3'$, 5, 5' $-$ trimethoxystilbene $-2-$ nitrogen (12)

The compound was synthesized in the same manner as for 5 in 34.7% yield as a white needle crystal, m. p. 94 $-$ 95 ℃ dec. ^{1}H NMR (CDCl$_3$): δ 2.27 (s, 3H, CH$_3$), 2.35 (s, 3H, CH$_3$), 3.77 (s, 3H, OCH$_3$), 3.83 (s, 6H, OCH$_3$), 6.42 (t,^{4}J = 2.0 Hz, 1H, H $-$ 4'), 6.75 (d,^{4}J = 2.0 Hz, 2H, H $-$ 2', 6'), 8.26 (s, 1H, H $-$ 3), 7.29, 7.63 each 1H (d,^{3}J = 15.6 Hz, 2H, H $-$ α, β) ppm; IR (KBr) 2959, 2936, 2838, 1632, 1592, 1465, 1266, 1207, 1159, 1071, 994, 964, 819 cm^{-1}.

(12) (E) -4, 6 $-$ Dimethyl $-2'$, 5, 5' $-$ trimethoxystilbene $-2-$ nitrogen (13)

The compound was synthesized in the same manner as for 5 as a golden oil; ^{1}H NMR (CDCl$_3$) δ 2.24 (s, 3H, CH$_3$), 2.33 (s, 3H, CH$_3$), 3.74 (s, 3H, OCH$_3$), 3.79 (s, 3H, OCH$_3$), 3.83 (s, 3H, OCH$_3$), 6.79 (dd,^{3}J = 9.0 Hz,^{4}J = 2.8 Hz, 1H, H $-$ 4'), 6.83 (d,^{3}J = 9.0 Hz, 1H, H $-$ 3'), 7.16 (d,^{4}J = 2.8 Hz, 1H, H $-$ 6'), 8.25 (s, 1H, H $-$ 3), 7.37, 7.95 each 1H (d, J = 16.0 Hz, 2H, H $-$ α, β) ppm; IR (KBr) 2939, 2834, 1629, 1581, 1551, 1496, 1469, 1277, 1220, 1076, 1045, 1027, 978, 880, 853, 803 cm^{-1}.

(13) (E) -4, 6 $-$ Dimethyl $-2'$, 5, 4' $-$ trimethoxystilbene $-2-$ nitrogen (14)

The compound was synthesized in the same manner as for 5 in 34.7% yield as a yellow needle crystal, m. p. 108 $-$ 110.5 ℃ dec. ^{1}H NMR (CDCl$_3$): δ 2.25 (s, 3H, CH$_3$), 2.33 (s, 3H, CH$_3$), 3.75 (s, 3H, OCH$_3$), 3.84 (s, 3H, OCH$_3$), 3.87 (s, 3H, OCH$_3$), 6.51 (dd,^{3}J = 8.4 Hz,^{4}J = 2.0 Hz, 1H, H $-$ 5'), 6.47 (d,^{4}J = 2.0 Hz, 1H, H $-$ 3'), 7.53 (d,^{3}J = 8.4 Hz, 1H, H $-$ 6'), 8.25 (s, 1H, H $-$ 3), 7.30, 7.90 each 1H (d,^{3}J = 15.6 Hz, 2H, H $-$ α, β) ppm; IR (KBr) 2993, 2938, 2837, 1620, 1604, 1580, 1506, 1469, 1296,

1279, 1209, 1104, 1074, 1033, 1004, 986, 943, 818 cm^{-1}.

(14) (E) -3, 4', 5 $-$ *Trihydroxystilbene* (15)

Boron tribromide (3. 4 ml, 36 mmol) in CH$_2$Cl$_2$ (20 ml) was added to a stirred solution of 5 (4 mmol) in CH$_2$Cl$_2$ (50 ml) at -4 to 0 ℃. The mixture was allowed to warm to 25 $-$ 35 ℃, stirred for 3 h, then poured into ice-water, and extracted with ethyl acetate (80 ml × 3). The organic layers were combined and washed with saturated solution of sodium chloride (1 × 30 ml). The ethyl acetate layer was dried over anhydrous sodium sulfate and evaporated the residue was purified by recrystallization and gave 0. 7 $-$0. 9 g (75. 0% $-$90. 2%) of 15 as amorphous solid, m, p, 251 $-$252 ℃ dec[2]. ^{1}H NMR (DMSO $-$d6) δ 6.23 (s, 1H, H $-$4), 6.44 (s, 2H, H $-$2, 6), 6.74 (d,^{3}J = 16. 4 Hz, IH, H $-$α), 6. 90 (d,^{3}J = 16. 4 Hz, 1H, H $-$β), 6.80 (d,^{3}J = 8. 2 Hz, 2H, H $-$3', 5'), 7.31 (d,^{3}J = 8. 2 Hz, 2H, H $-$2', 6') ppm; IR (KBr) 3298, 1606, 1589, 1514, 1445, 1327, 1249, 1154, 1010, 988, 966, 832 cm^{-1}.

(15) (E) -3, 3', 5, 5' $-$ *Tetrahydroxystilbene* (16)

The compound was synthesized in the same manner as for 15 in 96. 0% yield as pale yellow crystal, m. p. > 300 ℃ dec[2]. ^{1}H NMR (DMSO $-$d6) δ 6. 11 (s, 2H, H $-$4, 4'), 6. 36 (d,^{4}J = 1. 6 Hz, 4H, H $-$ 2, 2', 6, 6'), 6. 79 (s, 2H, H $-$ α, β) ppm; IR (KBr) 3538, 3449, 3240, 1602, 1516, 1462, 1387, 1346, 1312, 1259, 1158, 1005, 966, 956, 836 cm^{-1}.

(16) (E) -2', 3, 5', 5 $-$ *Tetrthydroxystilbene* (17)

The compound was synthesized in the same manner as for 15 in 84. 5% yield as pale pink amorphous solid, m. p. 212 $-$ 214 ℃ dec. ^{1}H NMR (DMSO $-$ d$_6$) δ 6. 29 (s, 1H, H $-$ 4), 6. 53 (d,^{4}J = 2. 0 Hz, 2H, H $-$2, 6), 6. 60 (m, 1H, H $-$4), 6. 71 (d, 3J = 8. 4 Hz, 1H, H $-$3), 6. 99 (d,^{4}J =2. 8 Hz, 1H, H $-$6), 7. 34, 6. 87 each 1H (d,^{3}J = 16. 4 Hz, 2H, H $-$α, β) ppm; IR (KBr) 3589, 3359, 1621, 1599, 1507, 1493, 1476, 1351, 1340, 1202, 1163, 1148, 989, 966, 862, 831, 810 cm^{-1}; API $-$ ES: 243 (M $-$ H$^+$), 279 (M + Cl$^-$), 487 (2M $-$ H$^+$), 523 (2M + CI$^-$). Q $-$ TOFMS m/z [M + 1]$^+$ 245. 0811 (calculated for C$_{14}$ H$_{13}$O$_4$, 245. 0814).

(17) (E) -3, 5, 5, 6 $-$ *Tetrhydroxystilbene $-$2 $-$ nitrogen* (18)

The compound was synthesized in the same manner as for 15 in 98. 9% yield as pale yellow amorphous solid, m. p. > 300 ℃, ^{1}H NMR (DMSO $-$ d$_6$) δ 6. 17 (d,^{4}J = 2. 2 Hz, 2H, H $-$2', 6'), 6. 22 (t,^{4}J = 2. 2 Hz, 1H, H $-$4'), 7. 51 (d,^{3}J = 6. 5 Hz, 1H, H'3), 6. 18 (d, 3^J = 6. 5 Hz, 1H, H $-$ 4), 6. 99, 7. 54 each 1H (d,^{3}J = 16. 5 Hz, 2H, H $-$ α, β) ppm; IR (KBr) 3414, 3330, 1620, 1609, 1597, 1555, 1507, 1446, 1356, 1144, 1107, 1063, 1008, 991, 962, 825, 794, 734, 680 cm^{-1}. API $-$ ES: 244 (M $-$ H$^+$), 280 (M + Cl$^-$), 489 (2M $-$ H$^+$), 525 (2M + CI$^-$). Q $-$ TOFMS m/z [M + 1]$^+$ 246. 0767 (calculated for C$_{13}$ H$_{12}$NO$_4$, 246. 0766).

(18) (E) -2', 5, 5', 6 $-$ *Tetrahyroxystilbene $-$2 $-$ nitrogen* (19)

The compound was synthesized in the same manner as for 15 in 99. 7% yield as pale yellow amorphous solid, m. p. > 300 ℃ dec. ^{1}H NMR (DMSO $-$d6) δ 6. 90 (d,^{3}J = 8. 8 Hz, 1H, H $-$3'), 6. 73 (d,^{4}J = 3. 0 Hz, 1H, H $-$6'), 6. 58 (dd,^{4}J = 3. 0 Hz, 3J = 8. 8 Hz, IH, H $-$

4'), 7. 47 (d, ^3J = 6. 7 Hz, 1H, H – 3), 6. 18 (d, ^3J = 6. 7 Hz, 1H, H – 4), 7. 16, 7. 54 each 1H (d, ^3J = 16. 9 Hz, 2H, H – α, β) ppm; IR (KBr) 3246, 3145, 2690, 1639, 1603, 1503, 1414, 1373, 1243, 1211, 1004, 980, 969, 857, 842, 816 cm^{-1}; API – ES: 244 (M – H$^+$), 280 (M + Cl$^-$), 489 (2M – H$^+$), 525 (2M + Cl$^-$). Q – TOFMS m/z [M + 1]$^+$ 246. 0777 (calculated for $C_{13}H_{12}NO_4$, 246. 0766).

(19) (*E*) – 4, 6 – *Dimethyl – 3', 5, 5' – trihydroxystilbene – 2 – nitrogen* (20)

The compound was synthesized in the same manner as for 15 in 94. 0% yield as pale yellow amorphous solid, m. p. 260 ℃ (oxy.) dec. ^{1}H NMR (DMSO – d$_6$) δ 1. 91 (s, 3H, CH$_3$), 2. 06 (s, 3H, CH$_3$), 6. 47 (d, ^4J = 2. 2 Hz, 2H, H – 2', 6'), 6. 23 (t, ^4J = 2. 2 Hz, 1H, H – 4'), 7. 56 (s, 1H, H – 3), 7. 05, 7. 11 each 1H (d, ^3J = 16. 6 Hz, 2H, H – α, β) ppm; IR (KBr) 3239, 1624, 1594, 1478, 1442, 1377, 1343, 1310, 1281, 1157, 1091, 1013, 997, 961, 834 cm^{-1}; API – ES: 292 (M + Cl$^-$), 513 (2M – H$^+$), 549 (2M + Cl$^-$). Q – TOFMS m/z [M + 1]$^+$ 258. 1136 (calculated for $C_{15}H_{16}NO_3$, 258. 1130).

(20) (*E*) – 4, 6 – *Dimethyl – 2', 5, 5' – trihydroxystilbene – 2 – nitrogen* (21)

The compound was synthesized in the same manner as for 15 in 62. 8% yield as pale yellow amorphous solid, m. p. 240 ℃ (oxy.) dec. ^{1}H NMR (DMSO – d$_6$) δ1. 89 (s, 3H, CH$_3$), 2. 03 (s, 3H, CH$_3$), 6. 72 (d, ^3J = 8. 5 Hz, 1H, H – 3'), 6. 95 (d, ^4J = 3. 0 Hz, 1H, H – 6'), 6. 61 (dd, ^4J = 3. 0 Hz, ^3J = 8. 5 Hz, 1H, H – 4'), 7. 47 (s, 1H, H – 3), 7. 14, 7. 39 each 1H (d, ^3J = 16. 6 Hz, 2H, H – α, β) ppm; IR (KBr) 3313, 1641, 1606, 1589, 1537, 1500, 1446, 1379, 1279, 1238, 1173, 1092, 970, 823 cm^{-1}; MS: 258 (M + H), Q – TOFMS m/z [M + 1]$^+$ 258. 1141 (calculated for$C_{15}H_{16}NO_3$, 258. 1130).

2. Assay method

Vero E6 cells were cultured in our laboratory, College of Life Science and Bioengineering, Beijing University of Technology. BJ 9 – 2b SARS virus was provided by Institute of Microbiology and Epidemiology, Academy of Military Medical Sciences. Vero E6 cells (1. 2 × 10^4 cells per well) were inoculated in triplicate in 200 μl of Eagle's medium containing 10% fetal calf serum in a 96 wells plate. After 2 days cultivation, the cells grew to a thin layer, then the cells were infected with 100 TCID$_{50}$ SARS virus in 0. 1 ml medium and incubated at 37 ℃ , 5% CO$_2$, humidified incubator for 2 hours. After a 2 h adsorption, the supernatant was discarded to remove the free virus. Chemicals in a series of 2 × dilation in 0. 2 ml medium were added into the cell wells, the chemical control without virus and the viral control without chemical in 0. 2 ml medium were also added into the cell wells, they were then cultivated in the same incubator for 3 days and the CPE of cells were recorded.

[In 《European Journal of Medicinal Chemistry》 2006, 41: 1084 – 1089]

References

1 Wieslaw P, Bogdan K, Farmaco IL. 1999, 54: 584 – 587

2 Ali M A, Kondo K, Tsuda Y. Chem Pharm Bull. (Tokyo), 1992, 40: 1130 – 1136

3 Boonlaksiri C, Oonanant W, Kongsaeree P, Kittakoop P, Tanticharoen M, Thebtaranonth Y,

Phytochem, 2000, 54: 415 – 417

4　Cai Y J, Fang J G, Ma L P, Yang L, Liu Z L. Biochim Biophys Acta, 2003: 163731 – 38

5　Matsuda H, Morikawa T, Toguchida I, Park J Y, Harima S, Yoshikawa M. Bioorg Med Chem, 2001, 9: 41 – 50

6　Zheng J B, Ramirez V D. Biochem Biophys Res Commun, 1991, 261: 499 – 503

7　Orsini F, Pelizzoni F, Verotta L, Aburjai T. Prod J Nat, 1997, 60: 1082 – 1087

8　Aburjai T A. Phytochem, 2000, 55: 407 – 410

9　Thakkar K, Geahlen R L, Cushman M. Chem J Med, 1993, 36: 2950 – 2955

10　Chen G H, Liu W, Li J X, John W M. Application P C T, 2001, 0142231

11　Chen G H, Liu W, Li J X, John W M, Application P C T, 2002, 02057219

12　Jiang M S, Cai L N, Udeani G O, Slowing K V, Thomas C F, Beecher C W W, Fong H H S, Famsworth N R, Kinghom A D, Mehta R G, Moon R C, Pezzuto J M. Science, 1997, 275: 218 – 220.

13　Yang L M, Lin S J, Hsu F L, Yang T H,

Bioorg Med Chem Lett, 2002, 12: 1013 – 1015

14　Wang X, Heredia A, Song H, Zhang Z, Yu B, Davis C, Redfield R. Sci J Pharm, 2004, 93: 2448 – 2457

15　Likhitwitayawuid K, Sritularak B, Benchanak K, Lipipun V, Mathew J, Schinazi R F. Nat Prod Res, 2005, 19: 177 – 182

16　Docherty J J, Fu M M H, Stiffler B S, Limperos R J, Pokabla C M, DeLucia A L. Antivir Res, 1999, 43: 135 – 145

17　Bhatacharya A K, Thyagarajan G. Chem Rev, 1981, 81: 415 – 430

18　Piechucki C. Synthesis, 1976, 3: 187 – 188

19　Baker R, Sims R J. Synthesis, 1981, 2: 117

20　Chen L, Gui C, Luo X, Yang Q, Giinther S, Scandella E, Drosten C, Bai D, He X, Ludewig B, Chen J, Luo H, Yang Y, Yang Y, Zou J, Thiel V, Chen K, Shen J, Shen X, Jiang H. Virol J, 2005, 11: 7095 – 7103

21　Murias M, Handler N, Erker T, Pleban K, Ecker G, Saiko P, Szekeres T, Jager W. Bioorg Med Chem, 2004, 12: 5571 – 5578.

100.　Immunogenicity of the Recombinant Adenovirus Type 5 Vector with Type 35 Fiber Containing HIV – 1 *gag* Gene

LIU Xin-lei[1,2], YU Shunag-qing[2], FENG Xia[2], WANG Xial-li[1],

LIU Hong-mei[2], ZHANG Xiao-mei[2], LI Hong-xia[2], ZHOU Ling[2], ZENG Yi[1,2]

1. College of Life Science and Bio – engineering, Beijing University of Technology,

2. State Key Laboratory for Infectious Disease Prevention and Control, National Institute for Viral Disease Control and Prevention, Chinese Center for Disease Control and prevention

Summary

The immune efficiency of a recombinant adenovirus type 5 with type 35 fiber containing HIV – 1*gag* gene (rAd5/F35 – mod. *gag*) was investigated in BALB/c mice, in which the rAdS/F35 – mod, *gag* was firstly identified with PCR, then transfected to 293 cells and the *in vitro* expression level of *gag* protein was determined by Western blotting and indirect immuno – fluorescent assay. Mice were immunized with intramuscular injections of rAd5/F35 – mod, *gag*, rAd5 – mod, *gag* or DNA and were boosted after 3 weeks. To test the effect of pre – existing anti – viral immunity on immunization, mice were also injected with Ad5 – GFP vector and then immunized 4 and 7 weeks later with Ad5/F35 – mod. *gag* vector. The P24 – specific IgG antibody in sera of immunized mice was determined by ELISA and the specific cytotoxic T lymphocyte (CTL) response was assayed by intracellular cytokine staining. It was demonstrated that the rAd5/F35 – mod. *gag* vector could express efficiently the HIV *gag* protein in 293 ceils *in vitro* and induce strong HIV – specific immune responses *in vivo*. The strongest CTL and serum IgG response occurred when mice were immunized twice with injection of rAd5/F35 alone, but the anti – AdS antibody after primary, infection with adenovirus could inhibit the specific immune responses induced by rAd5/F35 vector. It is concluded that single immunization with recombinant adenovirus rAd5/F35 – mod, *gag* can induce specific CTL and serum IgG antibody responses in mice, but the immunogenicity of rAd5/F35 is comparably weaker than that of rAd5.

[**Key words**]   HIV – 1; AIDS vaccines; *Adenoviridae*

Introduction

Recombinant adenovirus (rAd) vectors are widely used for both *in vitro* and *in vivo* gene transfer. rAd – based vaccines have a number of advantages over naked DNA vaccines or vaccines based on other viruses such as poxvims or alphavims. These advantages include the ability to prepare high – titer stocks of purified virus easily and the remarkable efficiency of the Ad cell/nucleus entry process leading to high – level transgene expression. Furthermore, it is thought that Ad can provide an adjuvant effect to stimulate of antigen – specific immune responses. It has been shown in diverse *in vivo* models that recombinant adenovirus type 5 (rAd5) has potential as a vehicle to transfer genes for HIV. However, in clinical trials of an early version of the vaccine containing a gene for the HIV *gag* protein, the response to the vaccine was blunted in people even with moderate levels (a titer over 1 : 200) of pre – existing anti – Ad5 antibodies. In the US and Europe about one – third of the population has pre – existing immunity able to significantly reduce vaccine efficacy, and in developing countries that could be as high as 80% of the population[1,2]. The pre – existing antiviral immunity strongly influencing the efficacy of the HIV vaccine is one hurdle of using Ad5 vector. Another hurdle is the hepatocellular tropism of AdS limits the safety of this viral vector[3]. This virus uses the coxsackievirus and adenovirus receptor (CAR) as its primary attachment receptor, which confers tropism for liver parenchymal cells. In response to these two shortcomings, our laboratory has examined the immunogenicity of a replication – defective chimeric Ad5 vector with Ad type 35 fiber (rAd5/F35) (Ad35 virus was classified as subgroup B). The Ad35 fiber showed 25% amino – acid homology with the Ad5 fiber, Cell entry of Ad35 is CAR independent and may involve CD46 receptor, which expresses on most human cells[4]. In the present study, we found that the rAd5/F35 recombinants induced rather strong antigen – specific humoral and cellular immune responses in BALB/c mice.

Material and Methods

1. Recombinant vectors

Recombinant vectors El, E3 – deletion, replication – defective recombinant viruses were constructed by our laboratory. The recombinant virus (Ad5/F35 – mod. *gag*, Ad5 – mod. *gag*) was propagated and purified by vector gene technology company.

2. Expression of rAd5/F35 – mod, *gag* in 293 cells

Human embryonic kidney cell line 293 was maintained in DMEM, supplemented with 10% bovine serum albumin. And 293 cells were infected with the rAd5/F35 – mod, *gag* at the MOI of 10 for 2 h, and then cultured at 37℃ for another two days. Then the cells were collected, the expressed protein *gag* was analysed by Western blotting and immune fluorescence assay (IFA).

3. Animal immunization

Four – to 6 – week old female BALB/c mice were purchased from Institute of Experimental Animal Sciences, Chinese Academy of Medical Sciences. BALB/c mice were divided into several groups with 5 in each randomly. The mice were immunized with intramuscular injection of DNA, of Ad5/F35 – mod. *gag* or Ad5 – mod. *gag* vector and were boosted after 3 weeks. To test the effect of pre – existing antiviral immunity on vaccination, the mice were injected intramuscularly with 8.5×10^{10} viral particles (vp) of Ad5 – GFP vector and then immunized 4 and 7 weeks later with Ad5/F35 – mod. *gag* vector. Inoculation of animals was conducted as scheduled in Tab. 1.

Tab. 1 Immunization schedule of rAd5 and rAd5/F35 vectors

Groups	0 week	4 week	7 week	8 week
1	–	PBS (100μl)	PUS (100 μl)	detection of immune
2	rAd5 – GFP (8.5×10^{10} vp)	rAd5/F35 – *gag* (10^{10} vp)	rAd5/F35 – *gag* (10^{10} vp)	detection of immune
3	–	rAdS – *gag* (3×10^9 vp)	rAd5 – *gag* (3×10^9 vp)	detection of immune
4	–	rAdS/F35 – *gag* (10^{10} vp)	rAdS/F35 – *gag* (10^{10} vp)	detection of immune
5	–	DNA (100 μg)	rAd5 – *gag* (3×10^9 vp)	detection of immune
6	–	DNA (100 μg)	rAd5/F35 – *gag* (10^{10} vp)	detection of immune

4. Intracellular cytokine staining array

Lymphocytes were isolated from the mouse spleen by EZ – Sep^{T}M Mouse 1 × density gradient centrifugation (Dakewe Biotech Company Limited, China). Isolated lymphocytes were cultured (2×10^6 cells/500 μl per well) into 48 – well plates (Costa, Cambridge, MA) in RPMI 1640 containing 10 % fetal calf serum (FCS). The cells suspension were incubated with 6μg/ml of the HIV peptide [P1 (197 – 205): AMQMLKETI; P2 (239 – 247): TISTLQEQI; P3 (291 – 300): EPFRDYVDRF] at 37℃. After 3 h, 2 μg/ml monensin was added. The cells were incubated for another 10 h. The cells were washed with staining buffer (1% bovine serum albumin in PBS), and stained with phycoerythrin (PE) – conjugated anti – mouse CD8 antibody (Ly – 2, BD Pharmingen, USA). The cells were fixed with 4 % formaldehyde/PBS for 15 min at room temperature, permeabilized in 0.3 % saponin (sigma), 1% bovine serum albumin in PBS, and stained with anti-mouse IFN-γ Ab conjugated with fluorescein isothiocyanate (FITC) (BD Pharmingen, USA) at room temperature for 1 h, followed by flow cytometric analysis.

5. Detection of HIV-1-specific and Ad5-specific antibodies

The specific antibody was detected by the enzymelinked immunosorbent assay (ELISA). 96-well microtiter plates were coated with 2 μg/ml of HIV P24 protein prepared by our laboratory or 10^9 vp/ml of Ad5-LMP2 and incubated overnight at 4℃. The well were blocked with PBS containing 5% skimmed milk power for 1 h at 37℃. They were then treated with 100 μl of serially diluted antisera and incubated for an additional 1 h at 37℃. The plates were washed five times with PBS containing 0.5% Tween 20 and incubated for 1 h with a 1 : 20 000 dilution of an affinity purified HRP-1a-beled anti-mouse Ab (Jackson, West Grove, Pennsylvania, US). The plates were then washed five times, developed with tetramethylbenzidine, stopped with 2 mol/L H_2SO_4, and analyzed at 450 nm/630 nm.

Results

1. Detection of the recombinant adenovirus containing the *gag* gene

Supernatant of 293 cells infected with rAd5/F35-mod. *gag* was collected, proteinase K (20 mg/ml) was added into it. The supernatant was incubated in water at 55℃, and then was used as the templet of PCR. The sequences of primers were as follows (the *gag* sequence detection primer): 5'-GTA CCG GCT GAA GCA CAT CGT-3' (sense) and 5'-CAT CAT GAT GCT GGC GGA GTT-3' (antisense); (the fiber sequence detection primer): 5'-TGGGAGGGGGACTTACAGTG-3' (sense) and 5'-GCTTTGCTGCTGGCTACAG-3' (anti-sense). The Fig. 1 and Fig. 2 indicated that the gene *gag* was inserted into rAdS/F35 correctly.

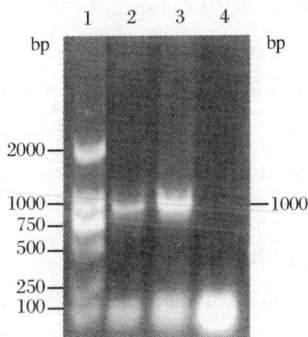

1: DL2000 DNA marker; 2: PCR product of pVR-mod. *gag* as positive control; 3: PCR product of rAd5/F35-mod. *gag*; 4: PCR prodnet of water

Fig. 1   PCR analysis of *gag*

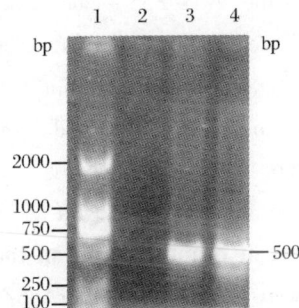

1: DL2000 DNA marker; 2: PCR product of water; 3: PCR product of rAd5/F35-mod, *gag*; 4: PCR product of rAd5/F35-GFP

Fig. 2   PCR analysis of fiber

2. Expression of rAd5/F35 − mod · *gag* in 293 cells

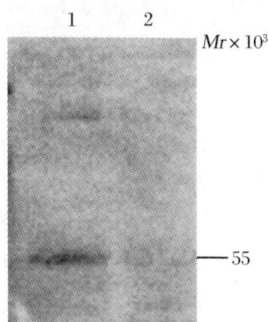

1: 293 cells infected with rAd5/F35 − mod, *gag*; 2: Normal 293 Cells

Fig. 3 Western blotting analysis of expression of *gag* in rAd5/F35 − mod, *gag*

Expression of the *gag* protein in rAd5/F35 − mod. *gag* was analyzed by Western blotting. Extract protein of 293 cells transfected with rAd5/F35 − mod. *gag*. The protein samples were subjected to SDS − PAGE and electroblotted onto nitrocellulose blotting membranes. Blots were blocked with 5% fat free milk in PBS containing 0.05 % Tween 20 and probed with rabbit anti − HIV P24 polyclonal antibody prepared by our laboratory and peroxidase − conjugated goat anti − rabbit immunoglobulin. Proteins were visualized by staining with 3, 3' − diaminobenzidine. The results showed that the modified gene *gag* in rAd5/F35 − mod, *gag* was expressed correctly and efficiently after transfection to 293 cells (Fig. 3).

Expression of the *gag* protein in 293 cells infected with rAd5/F35 − mod, *gag* was analyzed by indirect fluorescence assay. 293 cells were infected with the rAd5/F35 − mod, *gag* at the MOI of 10 for 2 h, and then cultured at 37℃ for another two days. After the 293 cells were collected by centrifugation and resuspended with saline, the 293 cells were incubated with anti − P24 rabbit serum (prepared by our laboratory), and FTTC conjugated anti − rabbit − IgG (1/160, 2 h, Jackson, West Grove, Pennsylvania, US). After cells were washed, they were detected by FIC. It was demonstrated that most 293 cells expressed *gag* when they were infected with rAd5/F35 − mod, *gag* at the MOI of 10 (Fig.4).

3. Immune responses in mice after immunization

The ability of the rAd5/F35 − mod, *gag* to trigger the activation and proliferation of antigen − specific T cells was monitored by the intracellular cytokine staining (ICS). The assay has been widely utilized to distinguish the relative contributions of $CD8^+$ cells to the overall T − cell responses. Immunization with the rAd5/F35 − mod, *gag* alone induced the number of HIV − specific IFN − γ − secreting $CD8^+$ T cells more than other groups, that increased the IFN − γ − secreting $CD8^+$ T cells to 8.828% (Fig. 5). Immunization with the rAd5 − mod, *gag* alone is less than the effect of immunization with the rAd5/F35 − mod, *gag*. Priming with the DNA − *gag* and followed by an rAd5 − mod, *gag* boost increased the IFN − γ − secreting $CD8^+$ T cells to 8.569%, this was significantly more than the effect of boost with rAd5/F35 − mod, *gag*.

4. Effect of immunization on the specific serum IgG antibodies

Mice were vaccinated with rAd5/F35 − mod, *gag* vector to explore the humoral immune response one week after the final immunization. As demonstrated by ELISA the animals immunized with 10^{10} vp of rAd5/F35 − mod, *gag* vector developed a high − tittered anti − P24 antibody (Ab) reponse (Fig. 6).

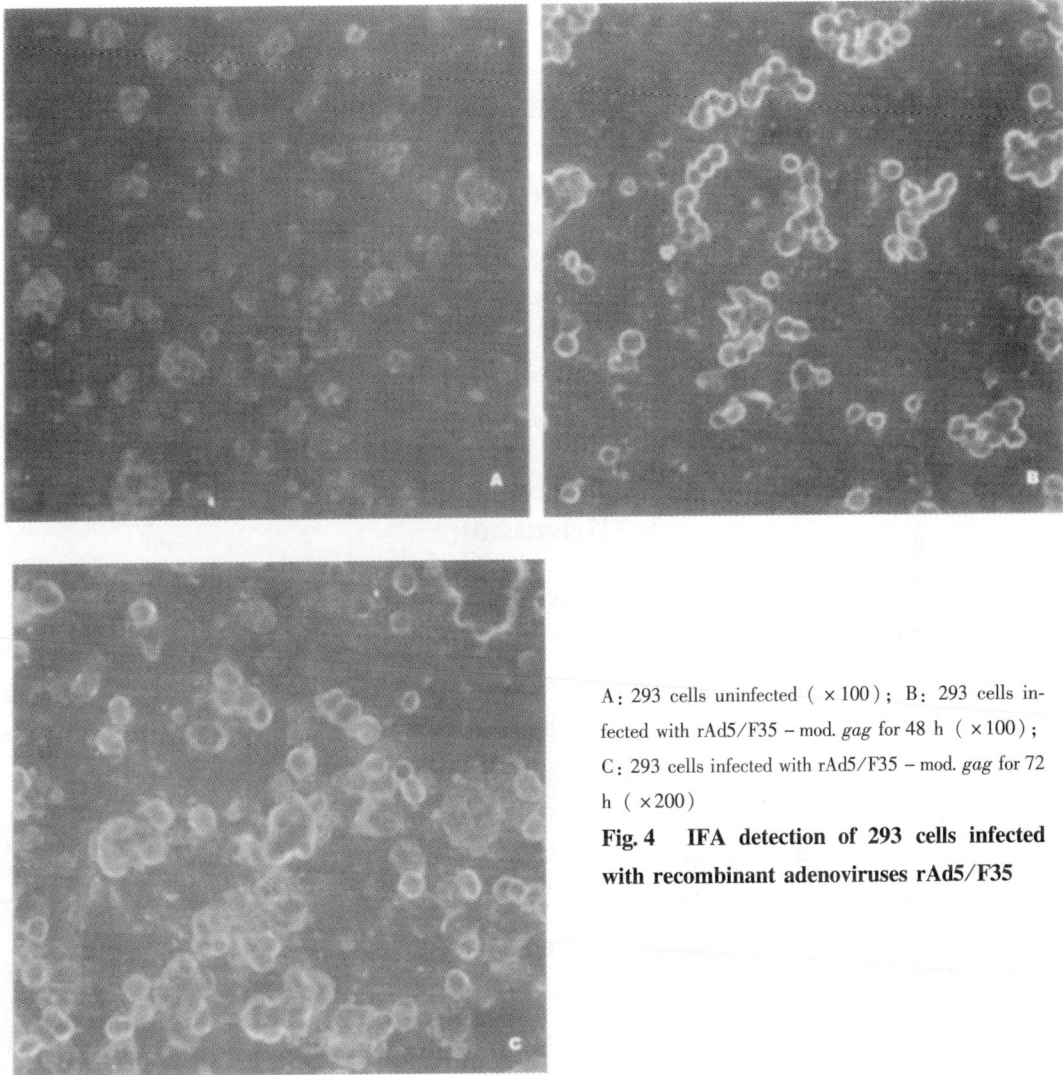

A: 293 cells uninfected (×100); B: 293 cells infected with rAd5/F35 – mod. *gag* for 48 h (×100); C: 293 cells infected with rAd5/F35 – mod. *gag* for 72 h (×200)

Fig. 4   IFA detection of 293 cells infected with recombinant adenoviruses rAd5/F35

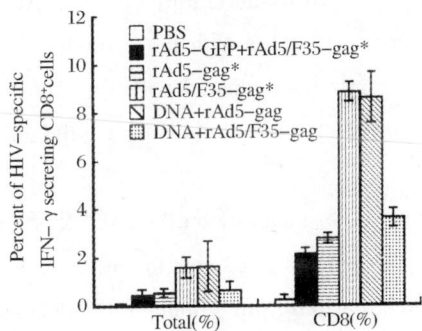

*: Immunization with rAdS/F35 – *gag* or rAd5 – *gag* twice

Fig. 5   HIV – 1 *gag* – specific, IFN – γ – secreting CD8⁺ T cells measured by ICS

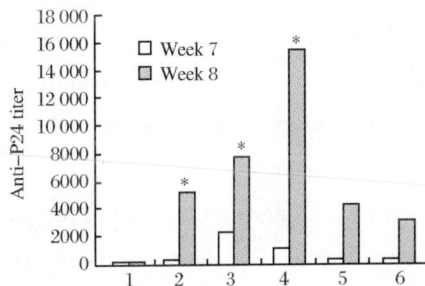

1: PBS; 2: rAd5 – GFP + rAd5/F35 – *gag*; 3: rAd5 – *gag*; 4: rAd5/F35 – *gag*; 5: DNA + rAd5 – *gag*; 6: DNA + rAd5/F35 – *gag*. *: Immunization with rAd5/F35 – *gag* or rAd5 – *gag* twice

Fig. 6   P24 specific antibody in immunized BALB/c mice

5. Effect of immunization on pre – existing immunity

To examine the effect of pre – existing anti – Ad5 immunity on the activity of the rAd5/F35 vector *in vivo*, mice were injected intramuscularly with 8.5×10^9 vp of rAd5 – GFP. After 4 weeks, these animals were immunized with 10^{10} vp of rAd5/F35 – mod. *gag* with high titers of anti – Ad5 Abs (anti – Ad5 neutralizing titer = 1 : 600). The HIV – specific responses were detected by ICS and ELISA after immunization. The magnitude of this response was significantly altered by preimmunization with the rAd5 – GFP (Figs. 5 and 6). Although pre – existing immunity to Ad5 reduced the immune response elicited by both vectors, the rAd5/35 – HIV vector was significantly less susceptible to the pre – existing Ad5 immunity than a comparable rAd5 vector. After injection with rAd5 and rAd5/F35 vector, the anti – rAd5 specific antibody titer was 1 : 600 and less than 1 : 100 respectively.

Discussion

A practically used vaccine for HIV infection should exhibit high immunogenicity, low cost for production and low or even without pathogenicity. Under these requirements, replication – defective Ad5 may be one of the best vectors for HIV vaccine development. However, majority of human population is infected adenovirus type 5[1,2]. In addition, the neutralizing antibodies and the cellular immune responses against the Ad5 fiber capsid protein may reduce the efficacy of the Ad5 vector, when it is used as vaccine in clinical trials[2]. The switching of the serotypes of adenovirus and the substitutive use of animal adenovirus may induce partial bypass of the pre – existing immune responses to Ad5 viruses, but they would bring a few drawbacks, such as lack of knowledge about these viruses, including the tropism to human cells; the potential difficulties in the manufacture and the possibility of *in vitro* recombination with other human viruses leading to the development of unknown diseases.

In this study, a chimeric Ad5 vector with Ad35 fiber related to cell tropism was used, in which the rAd5/F35, similar to Ad5, had a high productive titer in tissue culture cells[5]. Nevertheless, the virus displayed cell tripism to Ad35 fiber protein[6]. Coupled with the evidence that a rAd5/F35 vector transduces human dendritic cells more efficiently as compared with the Ad5 vector[7], these findings suggest that the rAd5/F35 – mod. *gag* vector is a promising candidate for clinical trials in humans.

In the present study, the effect of the pre – existing anti – Ad5 immunity on the rAd5/F35 vector was investigated. The effect of immunization with rAd5/F35 – mod, *gag* twice to native mice appeared to be more prominent than that of immunization of mice that were pre – immune to adenovirus (Figs. 5 and 6). It was well known that introduction or infection with adenoviruses would induce immune responses to hexon[8], penton, and fiber protein antigen. The exchange of fiber can partially reduce inhibition of the pre – existing immunity against the parent adenovirus. Furthermore, the exchanges of other genes, including those for hexon and penton may further reduce inhibition of the pre – existing immunity against Ad5. However, researchers at Merck Co. had encountered two problems with this approach. First, the majority of these chimeric viruses lost the ability to replicate, presumably due to the structural constrains of viral capsid. Second, even though this approach allowed the viruses to overcome anti – Ad5 antibody, yet CTL reactive to other Ad5 proteins blunted the responses[9].

It was demonstrated that high titer of anti – Ad5 antibody was detected after the first immunization with rAd5, but not with rAd5/F35. This antibody could block the activity of rAd5/F35 – mod. *gag*. Also, the effect of immunization with rAd5/F35 – mod, *gag* twice alone was more prominent than that of immunization with rAd5 – mod, *gag* in native mice (Figs. 5 and 6). Therefore, rAd5/F35 – mod. *gag* was less susceptible to the pre – existing immunity of Ad5 than a comparable rAd5 – mod. *gag*. These results were in agreement with those data reported by other studies[10].

To examine the immunogenicity and the protective immunity of rAd5/F35 vector induced, the gene *gag* from HIV – 1 clade B was used in the present study, because this strain of virus had been defined to be conserved in Henan province of China, and from the studies on the immunodominant epitopes, it was found that the *gag* gene products were the most conserved ones among the HIV – 1 subtypes. Moreover, anti – *gag* CTL – response appeared to be inversely correlated with disease progression, suggesting that they should be considered as the first candidate to be chosen for vaccine preparation constructed aiming to induce a broad spectrum of CTL – responses[5]. The *gag* gene was cloned from the peripheral blood mononuclear cells (PBMCs) genome of patients infected with HIV – 1 in Henan province. By analysis and alignment, the common sequence was obtained. To improve the expression level of *gag* protein expressed in mammalian cells, the codons of the consensus *gag* sequence were modified according to mammalian codon usage.

Priming with DNA – *gag* followed by rAd5 – mod. *gag* boosting could induce an increase in thenumbers of the IFN – γ – producing CD8$^+$ T cells, and this was more prominent significantly than the effect of boosting with rAd5/F35 – mod, *gag*, indicating that boosting with rAd5/F35 – mod, *gag* vector elicited slightly lower cellular immune responses and substantially lower humoral immune responses in comparison with that induced by rAd5 – mod. *gag*. Consequently, it appears that rAd5/F35 – mod. *gag* is less immunogenic than rAd5 – mod. *gag*.

All together, it is evident that rAdS/F35 vector can induce strong HIV – specific immune re – sponses in BALB/c mice and efficiently tranduce human dendritic cells *ex vivo*. Based on these ob – servations, rAd5/F35 vectors can be used as avaluable tool for the studies of immunotherapy and-vaccination.

[In 《J Microbiol Immunol》 2006, 4 (4): 306 – 312]

References

1 Vogels R, Zuijdgeest D, Rijnsoever R, Hartkoom E, Damen I, de Bethune MP, et al. Replicationdeficient human adenovirus type 35 vectors for gene transfer and vaccination: efficient human cell infection and bypass of pre – existing adenovirus immunity. J Virol, 2003, 77: 8263 – 8271

2 Barouch DH, Pau MG, Custers JH, Koudstaal W, Kostense S, Havenga M J, et al. Immunogenicity of recombinant adenovirus serotype 35 vaccine in the presence of pre – existing anti – AdS immunity. J Immunol, 2004, 172: 6290 – 6297

3 Thomas CE, Ehrhardt A, Kay MA. Progress and problems with the use of viral vectors for gene therapy. Nat Rev Genet, 2003, 4: 346 – 358

4 Zhang Y, Bergelson JM. Adenovirus receptors. J Virol, 2005, 79: 12125 – 12131

5 Ferrari G, Kostyu DD, Cox J, Dawson DV, Flores J, Weinhold K J, et al. Identification of highly conserved and broadly cross – reactive HIV type1 cytotoxic T lymphocyte epitopes as candidate immunogens for inclusion in *Mycobacterium bovis* BCG – vectored HIV vaccines. AIDS Res Hum Retroviruses, 2000, 16: 1433 – 1443

6 Shayakhmetov DN, Papayannopoulou T, Stamatoy annopoulos G, André Lieber. Efficient gene transfer into human CD34 (+) cells by a retargeted adenovirus vector. J Virol, 2000, 74: 2567 – 2583

7 Ophorst OJ, Kostense S, Goudsmit J, De Swart RL, Verhaagh S, Zakhartehouk A, et al. An adenoviral type 5 vector carrying a type 35 fiber as a vaccine vehicle: DC targeting, cross neutralization, and immunogenieity. Vaccine, 2004, 22: 3035 – 3044

8 Sumida SN, Truitt DM, Lemekert AA, Vogels R, Custers JH, Addo MN, et al. Neutralizing antibodies to adenovirus serotype 5 vaccine vectors are directed primarily against the adenovirus hexon protein. J Immunol, 2005, 174: 7179 – 7185

9 Youil R, Toner TJ, Su Q, Chen M, Tang A, Bett A J, et al. Hexon gene switwh strategy for the generation of chimeric recombinant adenovims. Hum Gene Ther, 2002, 13: 311 – 320

10 Xin KQ, Jounai N , Someya K, Houma K, Nizuguehi H, Naganawa S, et al. Prime – boost vaccination with plasmid DNA and a chimeric adenovirus type 5 vector with type 35 fiber induces protective immunity against HIV. Gene Ther, 2005, 12: 1769 – 1777

101. Trends in Incidence and Mortality of Nasopharyngeal Carcinoma over a 20 – 25 Year Period (1978/1983 – 2002) in Sihui and Cangwu Counties in Southern China

JIA Wei-hua[1], HUANG Qi-hong[2], LIAO Jian[3], YE Wei-min[4], SHUGART YY[5],
LIU Qing[1], CHEN Li-zhen[1], LI Yan-hua[2], LIN Xiao[2], WEN Fa-lin[2],
ADAMI Hans-olov[4], ZENG Yi[6], ZENG Yi-xin[* 1]

1. Departments of Experimental Research, Cancer Center, Sun Yat – sen University,
GuangZhou, State Key Laboratory of Oncology in Southern China, China; 2. Sihui
Cancer Registry, Guangdong, China; 3. Cangwu Cancer Registry, Guangxi, China;
4. Department of Epidemiology and Statistics, Karolinska Institute, Sweden;
5. Department of Epidemiology, Johns Hopkins University, Baltimore, USA;
6. Institute of Virology, Chinese National Center for Disease Control, Beijing, China

Summary

Background: Nasopharyngeal carcinoma (NPC) is a rare malignancy in most parts of the world but is common in southern China. A recent report from the Hong Kong Cancer Registry, a high – risk area for NPC in southern China, showed that incidence rate decreased by 29% for males and by 30% for females from 1980 – 1999, while mortality rate decreased by 43% for males and 50% for females. Changing environmental risk factors and improvements in diagnosis and treatment were speculated to be the major factors contributing to the downward trend of the incidence and mortality rates of NPC. To investigate the secular trends in different Cantonese populations with different socio – economic backgrounds and lifestyles, we report the incidences and mortality rates from two popula- tion – based cancer registries in Sihui and Cangwu counties from 1978 – 2002.

Methods: Incidence and mortality rates were aggregated by 5 – year age groups and 5 calendar years. To adjust for the effect of difference in age composition for different periods, the total and age – specific rates of NPC incidence and mortality rate were adjusted by direct standardization according to the World Standard Population (1960). The Estimated Annual Percentage Change (EAPC) was used as an estimate of the trend.

Results: The incidence rate of NPC has remained stable during the recent two decades in Sihui and in females in Cangwu, with a slight increase observed in males in Cangwu from 17. 81 to 19. 76 per 100 000. The incidence rate in Sihui is 1. 4 – 2. 0 times higher during the corresponding years than in Cangwu, even though the residents of both areas are of Cantonese ethnicity. A progressive decline in mortality rate was observed in females only in Sihui, with an average reduction of 6. 3% ($P = 0.016$) per five – year period.

Conclusion: To summarize, there is great potential to work in the area of NPC prevention and treatment in southern China to decrease NPC risk and improve survival risk rates in order to reduce M: 1 ratios. Future efforts on effective prevention, early detection and treatment strategies were also discussed in this paper. Furthermore, the data quality and completeness also need to be improved.

Background

The epidemiology of nasopharyngeal carcinoma (NPC) shows a uniquely skewed geographic distribution. The disease is rare in most parts of the world, where incidence rates are generally below 1 per 100 000 person – years[1]. However, in southern China and southeastern Asia, the rates can be as high as 20 to 50 per 100 000 person – years [1,2]. The highest risk has been observed in Cantonese of the Guangdong province, thus giving NPC a special name – " Canton tumor". However, no international report has been published on its secular trends in high – risk areas, except for Hong Kong.

Sihui city, located along the Xijiang River in the middle east of Guangdong province, has the highest incidence of NPC in China. In the 1970s, a cancer registry was established in Sihui to report the incidence and mortality of the major cancers, including NPC. Cangwu county is upstream of the Xijiang river in Guangxi province and is on the border between Guangxi and Guangdong. Given the high incidence of NPC in Cangwu, a cancer registry system was also established there in the 1980s.

Cancer Incidence in Five Continents (CI5) is regularly published by the International Agency for Research on Cancer (IARC), providing comparable data on the incidence of cancer in different geographical locations and distinct sub – populations (particularly ethnic) within these locations. However, none of the cancer registries in southern China (except Hong Kong) has taken part in the CI5 project. Until the year 2000, only a few Chinese registries such as Shanghai, Tianjin, Qidong, and Hong Kong have been involved in the project.

Recently, a very informative epidemiological study from the Hong Kong Cancer Registry reported that the age – standardized incidence rate of NPC in Hong Kong steadily decreased between 1980 and 1999; the total decrease amounted to 29% for males and 30% for females over the 20 – year period[3]. This encouraging reduction of NPC incidence in Hong Kong has been attributed to the changing lifestyle among local residents from traditional Chinese to Western diets, followed by rapid economic growth and development, which started in early 1960s. These data suggested that reduction of exposure to environmental risk factors can reduce the incidence of NPC. In fact, a decreasing trend was also observed in the Shanghai female population, with moderate risk during the 23 years from 1972 – 1974 to 1993 – 1994[4]. A recent report on Chinese – Americans living in Los Angeles County and the San Francisco Metropolitan area also suggested that the rates decreased by 37% in men but by just 1% in women from 1992 to 2002, with the overall decline limited primarily to type I tumors. Type I represents well to moderately differentiated squamous cell carcinomas with keratin production, but among Cantonese or Chinese living in Hong Kong, Taiwan and Macao type Ⅲ tumors predominate (undifferentiated carcinoma or lymphoepitheliomas)[5]. In Singapore, NPC incidence rates also started to decrease from the mid – 1990s[6]. A nationwide report for time trends in cancer mortality in China showed that the age – standardized mortality rate of NPC also declined from 1987 to 1999, especially in urban areas[7].

To investigate the secular trends in NPC in different Cantonese populations with different socio – economic backgrounds and lifestyles, we examined the incidence of and mortality from NPC from 1978 to 2002 in Sihui and Cangwu counties to search for possible etiological clues and provide guidelines for primary and secondary prevention in a southern Chinese population.

Methods

1. Cancer registry

Sihui and Cangwu are two counties located in the Guangdong and the Guangxi province, respectively, in southern China. Both counties are located along the Xijiang river that originates from the mountainous region of the Guangxi province and eventually joins the Pearl River and the Southern China Sea (Fig. 1, Map of China with Sihui, Cangwu, and Hong Kong indicated). In the year 2000, there were 393 655 and 649 404 inhabitants in Sihui and Cangwu, respectively. The populations in both counties are of Cantonese origin.

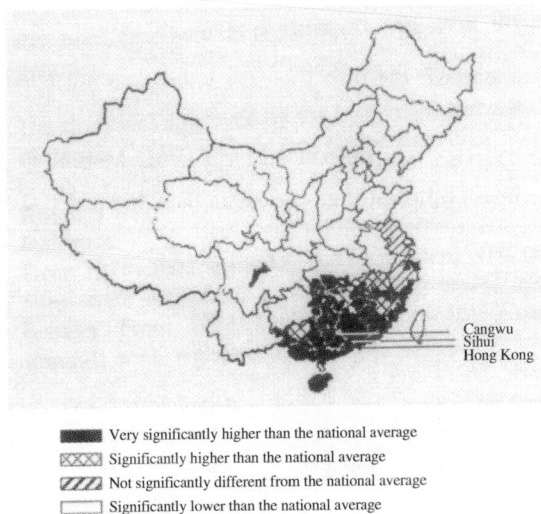

Cangwu
Sihui
Hong Kong

■■■ Very significantly higher than the national average
▨▨▨ Significantly higher than the national average
▨▨▨ Not significantly different from the national average
☐ Significantly lower than the national average

[Atlas of Cancer Mortality in the People's Republic of China (1973 – 1976). Beijing: China Map Press 81 – 2 (1979)]

Fig. 1 Map of China with NPC mortality indicated. The map shows the NPC mortality based on a nationwide investigation carried out from 1973 – 1976. Black areas indicate the regions with the highest mortality from NPC in southern China. Cangwu, Sihui, and Hong Kong are marked

The Sihui and Cangwu Cancer Registries were established in 1977 and 1982, respectively. Since there was no law for compulsory reporting of cancer cases, and people could choose hospitals freely, a Cancer Prevention Network was formed by local governments. The basic unit is the village, in which local general practitioners are responsible for health care of the local residents. They reported incident cancer cases regularly (at least annually) to health officers in the regional hospital of each town. In general, each town consists of 10 to 20 villages, and each county consists of 10 to 20 towns or administrative districts. The cancer registry collected reports from each regional hospital or local clinical oncology departments of the county hospitals.

For each incident cancer case, information including registered identification number (ID), medical ID, China Identity Card Number (unique for each resident), ICD code (9th or 10th version), name, sex, birth date, occupation, ethnicity, resident address, phone number, cancer site, diagnosis basis, and pathological report if available (date of diagnosis, hospital and doctor name for diagnosis) are all registered. To ensure completeness of registration, the list of NPC patients enrolled in surrounding referral hospitals was also collected regularly. Tab. 1 shows the proportion of diagnoses with histological confirmation. Up to 85. 2% – 90. 3% and 95. 5% – 97. 3% of NPC patients were diagnosed by pathology in the Sihui and Cangwu Registry, respectively. The surviving NPC patients were followed up by a Cancer Registry staff member through regular interviews of the patients or their family members once a year by phone call, letter, or home visit until death occurred. They assigned the cause of death by " death certification" which was provided by the local public security bureau and the hospitals where patients were treated and/or died. Death information was also reconfirmed according to annual whole – population and whole – death causes report from local health bureau. For those who died of traffic accident, suicide or other non – NPC related disease, such as heart attack, stroke, the direct causes of death were recorded.

Tab. 1 Proportion (%) of diagnoses type for NPC incident cases, Mortality vs. incidence (M: I) ratios for males and females during different time periods in the Sihui and Cangwu Cancer Registries

Period	Diagnosis type in Sihui			Diagnosis type in Cangwu			M : I Ratio in Sihui		M : I Ratio in Cangwu	
	Histological verification	Clinical deduction	Death certificate	Histological verification	Clinical deduction	Death certificate	Male	Female	Male	Female
1978 – 1982	89. 1	10. 1	0. 80	–	–	–	0. 87	0. 85	–	–
1983 – 1987	89. 6	9. 50	0. 90	95. 5	2. 41	2. 06	0. 71	0. 64	0. 70	0. 67
1988 – 1992	87. 7	10. 0	2. 30	96. 0	1. 65	2. 31	0. 72	0. 67	0. 98	0. 91
1993 – 1997	85. 2	13. 0	1. 80	96. 5	2. 02	1. 44	0. 70	0. 76	0. 95	1. 00
1998 – 2002	90. 3	8. 90	0. 80	97. 3	1. 06	1. 60	0. 71	0. 59	0. 76	0. 65

2. Statistical analysis

The first register for incident cases of NPC in Sihui and Cangwu began in 1977 and 1982, respectively, so we used data from 1978 and 1983 to avoid over – reporting in the first year. Incidence and mortality rates were aggregated by 5 – year age groups and 5 calendar years. To adjust for the effect of difference in age composition for different periods, the total and age – specific rates of NPC incidence and mortality rate were adjusted by direct standardization according to the World Standard Population (1960) [8]. The Estimated Annual Percentage Change (EAPC) [9] was used as an estimate of the trend. Using calendar year as a regression variable, a regression line was fitted to the natural logarithm of the rates, i. e. $y = mx + b$, where $y = \ln$ (rate) and $x =$ calendar year. EAPC was calculated using the equation EAPC $= 100 \times (e^m - 1)$. Testing the hypothesis that the EAPC is equal to zero is equivalent to testing the hypothesis that the slope of the regression line is zero, using the t – distribution of m/SE_m. The number of degrees of freedom equals the number of calendar years minus 2. The standard error of m, i. e. SE_m, is obtained from the fit of the regression line. This calculation assumes that the rates increased/decreased at a constant rate over the entire period.

The study was approved by the human ethical committee of Cancer Center, Sun Yat – Sen University.

Results

From 1978 – 2002, a total of 1710 incident NPC cases in Sihui were registered, including 1159 males and 551 females. From 1983 – 2002, a total of 1297 cases in Cangwu were registered, including 973 males and 324 females. Tab. 2 represents the age – standardized incidence rates for different genders during different periods in the two areas. For example, from 1998 – 2002, the rates were 30. 94 in males and 13. 00 in females in Sihui and 19. 76 in males 7. 33 in females in Cangwu per 100 000.

Tab. 2 Age – standardized incidence and mortality rates of NPC per 100 000 during different time periods and the Estimated Annual Percent Changes (EAPC) in Sihui and Cangwu

	Period					EAPC (p value)
	1978 – 1982	1983 – 1987	1988 – 1992	1993 – 1997	1998 – 2002	
Incidence Rate (number of cases)						
Sihui						
Male	28. 07 (167)	28. 68 (205)	28. 65 (235)	28. 03 (268)	30. 94 (284)	+ 1. 73% (0. 215)
Female	12. 29 (91)	14. 79 (111)	13. 35 (116)	11. 81 (116)	13. 00 (117)	– 1. 12% (0. 739)
Cangwu						
Male		17. 81 (209)	18. 68 (225)	19. 43 (259)	19. 76 (280)	+ 3. 57% (0. 020)
Female		7. 44 (82)	6. 91 (78)	7. 23 (68)	7. 33 (96)	0. 00 (0. 998)
Mortality Rate (number of cases)						
Sihui						
Male	24. 42 (143)	20. 49 (141)	20. 65 (166)	19. 56 (182)	21. 95 (195)	– 2. 56% (0. 413)
Female	10. 47 (80)	9. 43 (73)	8. 95 (76)	9. 01 (89)	7. 73 (70)	– 6. 32% (0. 016)
Cangwu						
Male		12. 53 (137)	18. 38 (213)	18. 41 (227)	15. 00 (201)	+ 5. 56% (0. 622)
Female		4. 97 (54)	6. 27 (71)	7. 26 (87)	4. 73 (61)	– 0. 019% (0. 999)

Tab. 2 and Fig. 2A show that the incidence rates have remained stable in Sihui over 25 years (EAPC 1. 73%, P = 0. 215 vs . EAPC – 1. 12%, P = 0. 739), whereas a slight increase was observed in males in Cangwu from 17. 81 to 19. 76 per 100 000 (EAPC 3. 57%, P =0. 020), but the trend remain stable in females (EAPC 0. 00%, P =0. 998) over 20 years. Moreover, the incidence in Sihui was 1.4 – 2.0 times higher during the same time period (Fig. 2A).

NPC was very rare among populations younger than 30 years, but the rate then rose sharply to reach a peak at age 50 – 59 years in Cangwu. However, in Sihui the peak was advanced and prolonged, beginning at age 40 – 49 years and remaining constant until age 59 years (Tab. 2 and Fig. 2C and 2D).

A. NPC incidence in Sihui and Cangwu by year 1978 – 2002

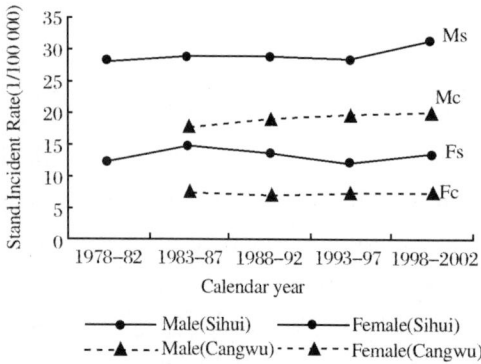

B. NPC mortality in Sihui and Cangwu by year 1978 – 2002

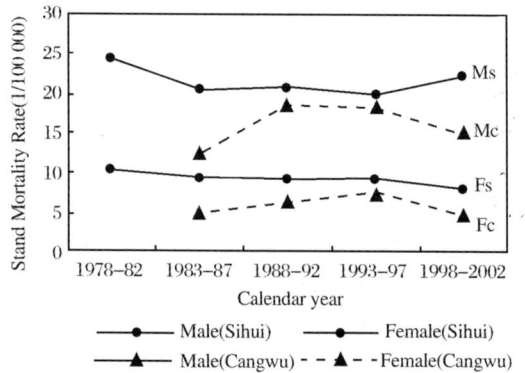

C. NPC incidence in Sihui by age, 1978 – 2002

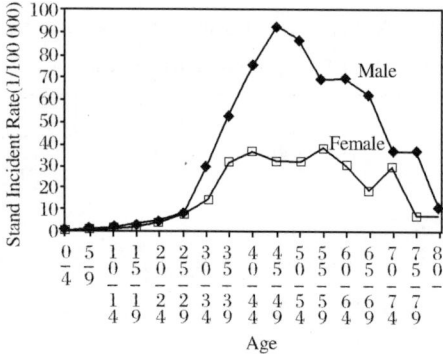

D. NPC incidence in Cangwu by age, 1983 – 2002

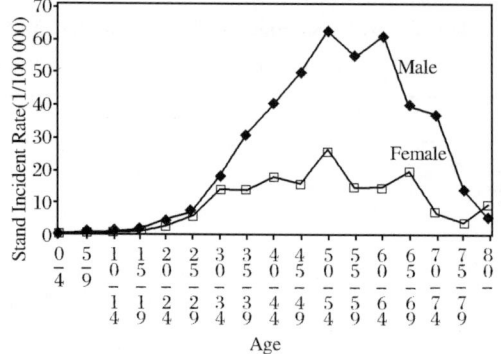

E. NPC mortality in Sihui by age, 1978 – 2002

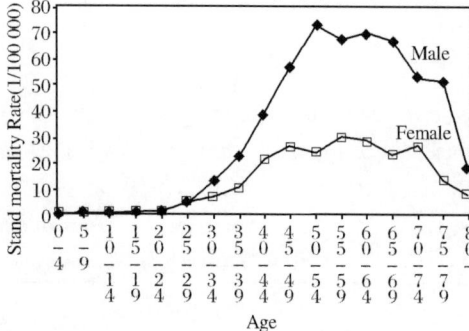

F. NPC mortality in Cangwu by age, 1983 – 2002

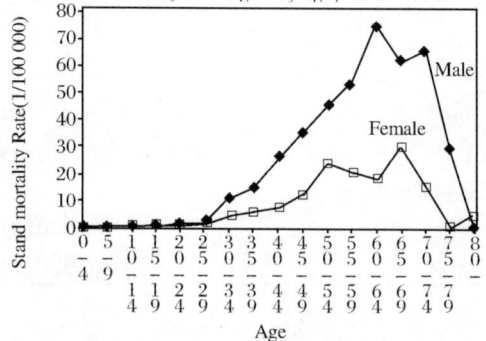

Ms (males in Sihui), Fs (Females in Sihui), Mc (Males in Cangwu), Fc (Females in Cangwu)

Fig. 2 NPC incidence and mortality rates in Sihui (1978 – 2002) and Cangwu (1983 – 2002) by calendar year and age, standardized by world population

Discussion

Commonly suspected risk factors for NPC include Epstein – Barr virus (EBV) infection, environmental factors, and genetic susceptibility. EBV has been consistently identified as an important risk factor, with a dose – response relationship between EBV antibody level and NPC risk[10-13]. Consumption of Cantonese – style salted fish and preserved food, which contain high levels of nitrosamines, especially during weaning period and adolescence, has long been considered an important risk factor for NPC[14-20]. Cigarette smoking, occupational exposures to wood dust, formaldehyde, and chemical fumes, as well as use of Chinese herbs have also been associated with increased NPC risk[21,22]. Tea, especially green tea, has demonstrated to decrease the risk of several cancers. Green tea contains several components including catechins, a category of polyphenols that have chemopreventive properties[23,24]. However, to date, no positive result has been report for the association of green tea with reduced NPC risk. In recent years, a series of studies have also provided evidence that the genetic component is a key etiological factor in the occurrence and development of NPC[25-29].

The most interesting findings from this study is that the incidence rates of NPC have largely remained stable in Sihui and Cangwu in the recent 20 – 25year period (except for a slight increase in males of Gangwu), which differs from a consistent declining trend reported in Hong Kong. As we know, the risk of NPC varies not only in the world, but also in different sub – ethnicities within the Chinese. The IARC reports that the age – standardized incidence rate for the Chinese (adjusted by world population) was 4.46 for males in Shanghai (located in eastern China) and 1.61 in Tianjin (located in northern China) from 1988 – 1992[1]. Before, during, and after Hong Kong was returned to the mainland central government in 1997, many people from northern China moved to Hong Kong, and some people from Hong Kong, who are native Cantonese, moved to other countries. Whether or not the proportion of Cantonese in the total Chinese population has declined in recent years needs to be further investigated, because this could possibly contribute to the decreased rates. More advantage, Sihui and Cangwu have stable population structure, the report seems to reflect NPC secular trend in completely Cantonese.

The 1978 policy of " Reform and Open to Outside of the World" in China Mainland led to changes in both lifestyle and living conditions, which have influenced the cancer profile. For example, the incidence of lung cancer for males in Qidong, China increased from 32.3 to 48.5 per 100 000, while the incidence of stomach cancer decreased from 43.1 to 35.6 in 1983 and 1997, respectively[30]. In southern China, both Sihui and Cangwu are well known for their high incidences of NPC, and therefore preventive measures have been carried out there in the last two decades. The traditional diet has changed progressively, and some identified NPC risk factors in foods such as preserved salted fish are no longer frequently consumed in most households. Previous studies have also suggested that exposure to risk factors such as salted fish in infancy confers a higher NPC risk than in adulthood[22]. Accordingly, those who born after 1978 in the two areas were less than thirty years old during the period of this study and have not reached peak ages for incidence of NPC until now, the significant decline in NPC incidence unable to be observed. Moreover, in Singapore the NPC rates started to show a declined trend 30 years after economical development. The real economic development in China occurred in early 1990s and thus we anticipate the decreasing rate of NPC in China would occur in the next 10 to 20 years, given continued social and economic development and transitions in lifestyle. Surely, there are some possibilities related to registries themselves that could have an influence on the rates. For example, an improved cancer registry system that collecting more cancer cases that could be missed before, and improved diagnostic procedure on NPC that could be misdiagnosed before, etc.

An argument against this could be the fact that environmental risk factors should act as an accumulating effect, and 20 years is long enough to reflect some changes in incidence. Certainly, future studies are required to draw a clearer conclusion.

In our study, the incidence of NPC in Sihui was 1. 4 to 2. 0 times higher in males and females as compared to their counterparts in Cangwu. Cangwu is about 210 kilometer far from Sihui. Historically, it was part of the Guangdong province – Inhabitant in both areas is Cantonese sharing a similar genetic background and speaks the same dialect. Noteworthy, NPC incidences are quite different between the two areas. The reasons why NPC risk is different in two populations have not been thoroughly investigated and few comparative epidemiological study have been carried out until recent years. During recent decades, Sihui still has higher economic development speed than Cangwu, the latter is defined as a county of poverty in China, where inhabitant annual average income just as half as those in Sihui. Social economic level may reflect the variations in related to life style, diet habit, living condition as well as other factors. We also observed a decrease in mortality with an average reduction of 2. 56% and 6. 32% for males and females, respectively, per five – year period from 1978 – 2002 in Sihui, and although it did not reach a significant level in males, we are encouraged by the result. A screening program has been conducted in a high – risk population in Guangdong (including Sihui), and since 1986, 98 180 residents have participated in the program[31]. The study showed that the 5 – year survival rate for NPC patients in this screening project was 79. 87%, significantly higher than the 58. 43% in hospital – based cases during the same period. We would like to note that the Sihui population has improved the importance and awareness of early diagnosis of NPC, – compared with those living in other areas, as a result of the implemented screening program.

In this paper, we analyzed the mortality – incidence ratios (M : I ratio) in two registries (Tab. 1). The M : I ratio, which compares the number of deaths attributed to a specific cancer and the number of incident cases in the same time period, can be interpreted as an indirect indicator of general survival if registration is complete and no marked temporal changes in incidence rates are present[32]. The ratios in Sihui have been stable since 1983 but fluctuated in Cangwu. We noted that mortality rates increased during the middle ten years (1988 – 1997) in Cangwu, which caused an increase in the M : I ratios. After further observation, we found that 51% of the new cases diagnosed from 1991 – 1995 died of the malignancy within one year after their diagnosis, while the 1 – year survival rate was reported to be as high as 98. 2% for new cases who received regular treatment during the same period[33]. Radiotherapy is a key therapy for NPC clinical treatment. We noticed that the price of radiotherapy has doubled in all hospitals in Cangwu since 1991. Some NPC patients might have found the increased costs unaffordable and terminate radiotherapy. On the other hand, this observation may also be due to lack of the completeness of Cangwu cancer registry during that period. In addition, the clinical staging of NPC at initial diagnosis may shed some light on this discrepancy, but we cannot deepen analysis because the records are not completeness in the registries.

Conclusion

NPC is an important health problem in southern China. Effective prevention and treatment strategies need to be further developed, which will greatly benefit residents in high – risk areas. Future efforts could focus on several aspects: changing unhealthy traditional diet habits, developing screening programs or annual physical examinations including EBV serology tests, nasopharyngeal examinations in high – risk populations, improving clinical treatment strategies, making efforts to minimize delayed diagnosis, increasing the accuracy in staging, and promoting treatment strategies and making radiotherapy available for all NPC – patients at a lower cost. Moreover, basic scientific research on genetics, environment and their interaction should be conducted to clarify NPC aetiology in – depth.

Acknowledgements

This work was supported by a grant from the International Collaborative Project of the Ministry of Science and Technology of China (2004DFA05700).

We thank Dr. Juhua Luo for trying analysis using Age – period – cohort model. We also thankWei Ling, Yi – Wei Xu and Qi – Nan Ling for their hard work in Sihui Cancer Registry.

[In 《BMC Concer》 2006, 6: 178]

References

1　Parkin DM, Muir CS. Cancer incidence in five continents Volume 7. Lyon, IARC Sci Publ; Electronic publication, 1997

2　Jeannel D, Bouvier G, Huber A. Nasopharyngeal carcinoma, an epidemiological approach to carcinogenesis. Cancer Surv, 1999, 33: 125 – 155

3　Lee AW, Foo W, Mang O, Sze WM, Chappell R, Lau WH, Ko WM. Changing epidemiology of nasopharyngeal carcinoma in Hong Kong over a 20 – year period (1980 – 99): an encouraging reduction in both incidence and mortality. Int J Cancer, 2003, 20: 680 – 685

4　Jin F, Devesa SS, Chow WH, Zheng W, Ji BT, Fraumeni JF Jr, Gao YT. Cancer incidence trends in urban shanghai, 1972 – 1994: an update, Int J Cancer, 1999, 83: 435 – 440

5　Sun LM, Epplein M, Li CI, Vaughan TL, Weiss NS. Trends in the incidence rates of nasopharyngeal carcinoma among Chinese Americans living in Los Angeles County and the San Francisco metropolitan area, 1992 – 2002. Am J Epidemial, 2005, 162: 1174 – 1178

6　Seow A, Koh WP, Chia KS, Shi LM, Lee HP, Shanmugaramam K. Trends in. Cancer Incidence in Singapore 1968 – 2002. Report No. 6 Singapore Cancer. Registry, 2004

7　Yang L, Parkin DM, Li L, Chen Y. Time trends in cancer mortality in China: 1987 – 1999. Int J Cancer, 2003, 106: 771 – 783

8　Waterhouse J, Muir C, Correa P, Powell J.

Cancer Incidence in Five Continents Lyon, IARC Sci Publ, 1976, 456

9　Kleinbaum DG, Kupper LL, Muller KE. Applied regression analysis and other multivariable methods Boston, PWSKENT Publishing Company, 1988

10　Lin TM, Chen KP, Lin CC, Hsu MM, Tu SM, Chiang TC, Jung PF, Hirayama T. Retrospective study on nasopharyngeal carcinoma. J Natl Cancer Inst, 1973, 51: 1403 – 1408

11　De – The G, Dubouch P, Fontaine C, Wedderburn N, Carter RL, Edwards MB, Cohen B. Natural antibodies to EBV – VCA antigens in common marmosets (Callithrix jacchus) and response after EBV inoculation. Intervirology, 1980, 14: 284 – 291

12　Chen JY, Liu MY, Chen CJ, Hsu MM, Tu SM, Lee HH, Kuo SL, Yang CS. Antibody to Epstein-Barr virus – specific DNase as a marker for the early detection of nasopharyngeal carcinoma. J Med Viral, 1985, 17: 47 – 49

13　Chan CK, Mueller N, Evans A, Harris NL, Comstock GW, Jellum E, Magnus K, Orentreich N, Polk BF, Vogelman J. Epstein – Barr virus antibody patterns preceding the diagnosis of nasopharyngeal carcinoma. Cancer Causes Control. 1991, 2: 125 – 131

14　Ha JH, Huang DP, Fang YY. Salted fish and nasopharyngeal carcinoma in southern Chinese. Lancet, 1978, 2: 626

15　Yu MC, Huang TB, Henderson BE. Diet and nasopharyngeal carcinoma: a case – control study in Guangzhou, China. Int J Cancer, 1989, 43: 1077 – 1082

16　Yu MC, Mo CC, Chang WX, Yeh FS, Henderson BE. Preserved foods and nasopharyngeal carcinoma: a case – control study in Guangxi, China. Cancer Res, 1988, 48: 1954 – 1959

17　Ning JP, Yu MC, Wang QS, Henderson BE. Consumption of salted fish and other risk factors for nasopharyngeal carcinoma (NPC) in Tianjin, a low – risk region for NPC in the People's Republic of China. J Natl Cancer Inst, 1990, 82:

291 – 296

18　Sriamporn S, Vatanasapt V, Pisani P, Yongchaiyudha S, Rungpitarangsri V. Environmental risk factors for nasopharyngeal carcinoma: a case – control study in northeastern Thailand. Cancer Epidemiol Biomarkers Prev, 1992, 1: 345 – 348

19　Lee HP, Gourley L, DufF/SW, Esteve J, Lee J, Day NE. Preserved foods and nasopharyngeal carcinoma: a case – control study among Singapore Chinese. Int J Cancer, 1994, 59: 585 – 590

20　Armstrong RW, Imrey PB, Lye MS, Armstrong MJ, Yu MC, Sani S: Nasopharyngeal carcinoma in Malaysian Chinese: salted fish and other dietary exposures. Int J Cancer, 1998, 77: 228 – 235

21　IARC monograph. Tobacco smoking and tobacco smoke (Group I). Lyon, IARC Sci Publ, 1997

22　Yu MC, Yuan JM: Epidemiology of nasopharyngeal carcinoma. Semin Cancer Biol, 2002, 12: 421 – 429

23　Lee AH, Fraser ML Meng X, Binns CW. Protective effects of green tea against prostate cancer. Expert Rev Anticancer Ther, 2006, 6: 507 – 513

24　Cabrera C, Artacho R, Gimenez R. Beneficial effects of green tea – a review. J Am Coll Nutr, 2006: 79 – 99

25　Zeng YX, Jia WH. Familial nasopharyngeal carcinoma. Semin Cancer Biol, 2002, 12: 443 – 450

26　Jia WH, Collins A, Zeng YX, Feng BJ, Yu xJ, Huang LX, Feng QS, Huang P, Yao MH, 5hugart YY. Complex segregation analysis of nasopharyngeal carcinoma in Guangdong, China: evidence for a multifactorial mode of inheritance. Eur J Hum Genet, 2005, 13: 248 – 252

27　Jia WH, Feng BJ, Xu ZL, Zhang XS, Huang P, Huang LX, Yu XJ, Feng QS, Yao MH, Shugart YY, Zeng YX. Familial risk and clustering of nasopharyngeal carcinoma in Guangdong, China. Cancer, 2004, 101: 363 – 369

28　Grulich AE, McCredie M, Coates M. Cancer incidence in Asian migrants to New South Wales, Australia. Br J Cancer, 1995, 71: 400 – 408

29 Feng BJ, Huang W, Shugart YY, Lee MK, Zhang F, Xia JC, Wang HY, Huang TB, Jian SW, Huang P, Feng QS, Huang IX, Yu XJ, Li D, Chen LZ, Jia WH, Fang Y, Huang HM, Zhu JL, Liu XM, Zhao Y, Liu WQ, Deng MQ, Hu WH, Wu SX, Mo HY, Hang MF, King MC, Chen Z, Zeng YX. Genome – wide scan for familial nasopharyngeal carcinoma reveals evidence of linkage to chromosome 4. Nat Genet, 2002, 31: 395 – 399

30 Website title [http://www – deo. iarc. fr/]

31 Huang TB, Wang HM, Li JL. Screening program for nasopharyngeal carcinoma (NPC), early detection for NPC. Ai Zheng (Chinese), 1997, 16: 245 – 247

32 Parkin DM, Chen VVV, Ferlay J, Galceran J, Storm HH, Whelan SL. Comparability and Quality Control in Cancer Registration. In IARC Technical Report No. 19 Lyon, International Agency for Research on Cancer, 1994

33 Yi – Qin Zhang, Bao – Qing Wei. Improvement on nasopharyngeal carcinoma radiotherapy during 20 years. Chinese Journal of Radiation Ontology (Chinese), 1999, 8: 73 – 76

102. Strategy for AIDS Prevention and Treatment

ZENG Yi

College of Life Science and Bioengineering, Beijing University of Technology

Summary

HIV/AIDS has been circulating in China for over 25 year. While making progress and achievements on HIV/AIDS prevention, there still are great challenge and difficulties such as HIV epidemic controlling and vaccine research.

[**Key words**] HIV/AIDS; Prevention and care; Infection; Vaccine; Drugs

In 1982, HIV in Factor VIII entered China from USA and 4 Chinese Hemophiliac patients were infected with HIV through injection of this factor VIII during 1983 – 1985. In 1985, the first AIDS patient came to China from USA[1]. Afterward HIV came to China and transmitted through drug users, CSWs, and blood plasma donors, then spread all over China. The cumulative estimated members of people living with HIV/AIDS had reached 650 000 in 2005. The cumulative reported cases living with HIV/AIDS were 214 300 by the end of July 2007. These reports indicate that the HIV/AIDS epidemic is spreading in the general population and that the proportion of female HIV cases has increased considerably in recent years. Although the government was slow respond to the epidemic in the late 20th century, it has made a vigorous response in the early 21 st century, a series of important strategies and policies in fighting HIV/AIDS have been adopted by the central government, such as strengthing leadership, surveillance and information systems, clarifying responsibilities, establishing comprehensive HIV/AIDS prevention and control system, providing treatment, care and support for AIDS patients and increasing funding for HIV/AIDS.

A major issue is how to control the HIV/AIDS epidemic. Thailand's great achievement in response to HIV/AIDS is a good example. Since 1990, Thailand has launched a nation – wide HIV education and intervention campaign led by the prime minister along with all levels of the government. They took all kinds of effective measures of education and intervention, especially the 100% Condom Use Program. As a result, STD infection decreased dramatically, and the HIV infection rate also markedly reduced. UNAIDS remarked that Thailand had initiated active educational and intervention programs that prevented several million people from being infected. On the contrary, South Africa government did not take effective measures to response HIV infection; the HIV infection rate is as high as 20% – 25% in South Africa. One study estimated that from 2002 to 2010, without sufficient education and interventions, the global number of HIV infection would raise to 48 million by 2010; but if with effective interventions, 29 million people could be prevented from HIV infection[2].

From this, we can see the importance of HIV education and interventions. Thus, the major strategy is to initiate educational and intervention action widely and profoundly. The point is to undertake these measures among rural areas, migration population and the youth. Education and interventions are long – term works that a casual attitude cannot be adopted.

In terms of HIV vaccine, no successful vaccine has been invented yet after long period of research. Then why we spend enormous efforts and funds on vaccine research?

1. We must realize that, in a microscope, human is the most intelligent creature, and human will continuously deepen their understanding of the rules that how the objective matters develop, and formulate various effective measures to conquer difficulties. Diseases, such as smallpox and poliomyelitis, had once caused great loss to human, but at last, human have defeated, prevented and controlled over them. Things will be the same with HIV/AIDS. In the end, human will win the battle from AIDS.

2. The most sufficient methods to prevent HIV/AIDS at present are undertaking active education and interventions. However, 1/3 of the infections are hard to prevent. Therefore, we need successful vaccine to prevent HIV. Human has invented many effective vaccines, which have played very important roles in fighting against major infectious diseases. In addition, China should take part in this vaccine research actively.

3. Due to the mutation of HIV, the vaccine is still not available; besides, people still know little about the pathogenesis and immuno – response of HIV. For these reasons, we should focus more on fundamental research, which is significant to the vaccine study.

4. The vaccine is not yet available, which brings a major challenge to scientists. Chinese scientists should devote themselves with their wisdom in fighting against HIV/AIDS, and invent HIV vaccine with our own patent, as the contributions to all.

At present, more than 20 drugs have been used to treat HIV/AIDS, which decrease HIV – 1 viral load, or even reduce to undetectable level; increase the number of CD4 cells and improve patients' life quality, meanwhile prolong their life. The decrease of the viral load also reduces the transmission of the virus. However, mutations occur easily, drug resistance happens during the treatment, hence more and chipper new drugs will be needed. China now does not have drugs with our own patent, but the Government has listed HIV/AIDS control and prevention as a major program. As a result, China should strengthen the research of HIV/AIDS drugs, which should include the traditional Chinese medicinal herbs and synthetic drugs. I believe China certainly would succeed in anti-HIV drugs invention.

China's present control and prevention strategies of infectious diseases are in favor of controlling and researching severe infectious diseases now; meanwhile, we should emphasis more on cooperation and com – munications in research. This very issue of *Virologic Sinica* has published HIV/AIDS essays by experts in this field; and I believe enhance academic exchange would benefit our HIV/AIDS prevention and control.

[In 《Virologica Sinica》 2007, 22 (6): 419 – 420]

References

1 Zeng Y, Xu H, Zhang J. Infectious Diseases in China. AIDS in Asia (Lu Y, Essex M. ed.), New York: Kluwer Academic/Plenum Publishers, 2004, 295 – 305

2 Zeng Y. Health Education and Intervention are Major Strategies for AIDS Control. Chinese J Health Education, 2003, 11 (19): 846 – 848

103. Human Papillomavirus and Esophageal Squamous Cell Carcinoma

SHEN Zhong-ying[1], ZENG Yi[2], QING Bo-qiang[3]

1. Medical College of Shantou University; 2. Chinese Academy of Preventive Medicine;
3. Chinese Academy of Medical Sciences

Summary

Esophageal squamous cell carcinoma (ESCC) has a distinct geographic distribution with a high prevalence in certain regions of the world, being the most common carcinoma in China, South Africa and some other areas. The etiology of ESCC remains unclear, but it is most likely multifactorial. The role of human papillomavirus (HPV) in the etiology of ESCC has been much debated over the past 20 years. The aim of this review will demonstrate the relationship between HPV and ESCC *in vivo* and *in vitro*. Detection of HPV relies strictly on molecular analyses of the HPV DNA sequence. In clinically resected esophageal samples HPV was detected frequently through polymerase chain reaction (PCR) and *in situ* hybridization (ISH). The incidence of infection of HPV in ESCC, however, differs markedly depending on the geographic location of the patient population being studied.

· 757 ·

Frequency variation within regions also exists with HPV detection rates varying from 0 to 71%. The highest reported prevalence of HPV occurs in the high ESCC risk regions of China and South Africa. ISH is the manifestation of specific genetic information within a morphological context. It demonstrates that HPV in dot hybridization is found in cancerous, paracancerous and mucosal tissues of the esophagus. Histological evidence of HPV infection is characterized by the presence of koilocytosis' giant and multinucleated cells, dyskeratosis, hyperkeratosis, acanthosis and papillomatosis, accompanying HPV infection in various positive rates.

In vitro, our intention was to explore the role of HPV on esophageal carcinogenesis. At first, recombinant plasmids were constructed by pGEM/HPV18E6E7 plasmid and ligated with a plasmid of associated adenovirus (pAAV3), generating HPV18E6E7 AAV which infected the human embryonic esophageal epithelial cells. The cells were continually cultivated over 120 passages and displayed the multistep process of malignant transformation: including initial (transduction of E6E7 genes of HPV), preimmortalized (by pass senescence), immortalized (activity of telomerase), premalignant (anchorage – independent growth) and malignantly transformed (tumor formation in immunodeficient mice) stages. Our data revealed that the viral oncogene of HPV type 18 was one of the important etiological causes of ESCC. We also explored the synergetic effects of promoters, TPA (12 – O – tetradecanoyl – phorbol – 13 – acetate), sodium butyrate, nitrosamins and radiation, which can all promote the process of esophageal carcinogenesis.

Taken together, HPV may play a role in esophageal carcinogenesis with a more pronounced presence in those regions of the world with a high incidence of the ESCC. HPV cooperates via the synergistic effects of other carcinogens on the carcinogenesis of ESCC.

[**Key words**] Esophageal carcinoma; Human papillomavirus; Epidemiology; Esophageal epithelial cell line; Immortalization; Malignant transformation.

A. Prevalence of Human Papillomavirus in Human Esophageal Squamous Cell Carcinoma

Chinese historical records from 2000 years ago described " dysphagia" syndromes as the obstruction of the esophagus associated with malnutrition and death, just as is the pattern with the current esophageal cancer (EC). EC has long been recognized as a deadly disease. In China, it is the second most common cause of death by cancer in males[1].

EC is not itself a uniform lesion. Squamous cell carcinomas comprise about 90 – 95 percent of EC, with the remaining 5 – 10 percent of primary EC being adenocarcinomas[2], which occur in patients with Barrett's esophagus in the vast majority of cases, or in a columnar epithelium – lined distal esophagus[3]. There are different etiological theories for these two lesions.

Esophageal squamous cell carcinoma (ESCC) has a distinct geographic distribution[4]. There are many areas of high incidence of ESCC in the world. One such area encompasses the southern shore of the Caspian Sea in the west to northeastern Asia in the east and involves the inhabitants of

Turkey, Iran, Central Asia, (which includes Turkenistan, Turkoman and Kazakh tribesmen) Afghanistan, Siberia, Pakistan, Mongolia, China, Korea and Japan[5-7]. Epidemiological data reports that Southern Africa has among the highest incidence in the world[8]. But ESCC rarely occurs in some parts of the world, such as Western Europe or northern Africa. The difference in ESCC incidence can be as high as 600 fold between the most and least affected areas of the same country as in China[9].

ESCC has been a relatively common disease in China. In 1959, epidemiological studies were initiated to survey the incidence and mortality of EC in 181 counties in the three northern Chinese provinces of Henan, Hebei, and Shanxi, covering a population of 50 000 000 people. The entire region had an EC mortality rate of 37.4 per 100 000[10]. In southern China, the Shantou area, containing Nan'ao County in Guangdong Province, is a high-risk area for EC. The annual standardized mortality rate in Nan'ao County for esophageal cancer is $110/10^6$ ($166/10^6$ in men and $69/10^6$ in women)[11] and has persistently remained high from 1973 up to 2005. However, Lufeng County in the same coastal area, about 80 kilometers from Nan'ao, shows a very different mortality rate ($10/10^6$) [12]. " The hypothesis of concentric circles" of EC incidence has suggested that there is an extremely high mortality in the center of the area with mortality rates decreasing with increasing distance from this center. The ratio between the highest mortality rates in the center and the lowest rates in the outermost areas is approximately 100 to 1[10]. In the entire area the male-to-female mortality ratio of patients with ESCC is 2.4 : 1.

The incidence of ESCC shows great geographic variation. The reasons for these major regional variations in the incidence of ESCC are poorly understood, but may be related to differing environmental factors involved in different geographical regions.

The etiology of ESCC remains unclear, but it is most likely multifacial. Epidemiological studies in separate high incidence areas suggest different causative factors. In some studies, a number of different etiological agents have been pinpointed in a single area. Additionally, other studies have raised the possibility of a multiplicity of agents working synergistically to cause malignant changes[12].

Studies of ESCC have suggested that environmental, occupational risk and genetic factors could be important factors in the carcinogenesis of this tumor[13]. Compelling epidemiological and experimental data suggest that some chemical factors (such as nitrosamines, the precursors of mycotoxins, cigarette smoke and excessive alcohol intake) [14,15], nutritional deficiencies (such as a poor, generally monotonous diet lacking in green leafy vegetables, fresh fruits and deficiencies of vitamins A, B, C) [16,17]; the lack of trace elements (molybdenum, manganese, zinc, magnesium, silicon, nickel, iron, bromium, iodine and food grown in mineral deficient soil) [18] and physical factors (such as coarse and hot food and drink) are associated with the development of this malignancy. certain microorganisms and its products, fungi and its production of mycotoxins, bacterium and viruses (such as human papilloma virus, herpes simplex virus, cytomegalovirus, and Epstein-Barr virus) have been shown to impact the esophageal epithelium[19]. Heavy alcohol and cigarette consumption are recognized risk factors in Western Europe and North America[20]. Diets deficient in vitamins and protective antioxidants combined with the intake of carcinogens, like N-nitrosamine, which is present in poorly preserved food, have been implicated in the development of ESCC in Central Asia and China[21]. These factors render the esophageal mucosa more susceptible to injury by carcinogens, but as yet no firm evidence for a role in esophageal carcinogenesis has been provided.

Human papillomaviruse (HPV) play an important role in the development of squamous cell carcinomas in various body sites including the anogenital, upper respiratory, and digestive tracts[22]. Oncogenic types of HPV, most notably HPV16 and HPV18, are recognized as the most significant risk factors of esophageal cancer[23-24]. The association of HPV infection with the development of esophageal carcinoma is currently well established (see below).

1. Molecular epidemiology of human papillomavirus in esophageal squamous cell carcinoma

The role of HPV in the etiology of ESCC has been debated over the past 20 years. Syrjanen first suggested a role of HPV in the etiology of ESCC in 1982, based on the observation of characteristic histological findings suggesting the presence of HPV in benign esophageal epithelia and malignant esophageal tumors[25]. The presence of HPV DNA in ESCC was subsequently confirmed initially by in-situ hybridization[26] and later by the more sensitive method of PCR-Southern hybridization[27]. There is also sero-epidemiological evidence that HPV type 16 infection is a risk factor for ESCC[28]. Many more squamous cell carcinomas have been analysed by Syrjanen from 1982 – 2001: of the 1485 squamous cell carcinomas analysed by *in situ* hybridization, 22.9% were positive for HPV DNA, as were 15.2% of the 2020 cases tested by the polymerase chain reaction (PCR)[29]. Thus, although HPV may play a role in esophageal carcinogenesis, this role may be more pronounced in those regions of the world with a high incidence of the disease, and may be less important in areas with moderate or low risks for esophageal cancer.

HPV types 6, 11, 16, 18, and 31 have all been described in esophageal squamous cell lesions, but the incidence of HPV positivity varies significantly based on the histology of the lesion. HPV 16 and 18 are defined as 'high risk', implying a comparatively high risk for invasive disease. Recent evidence has shown that esophageal infection with HPV, particularly high-risk types 16 and 18, increased esophageal carcinoma morbidity 13-fold[30]. In contrast, other genotypes, such as HPV 6 and 11, are considered as 'low risk' for the development of ESCC.

(1) *Detection of HPV in ESCC samples by PCR*: However, the role of HPV in the causation of ESCC remains controversial. Samples from various world regions have been tested, resulting in the demonstrated presence of HPV DNA in as many as 67% of tumors down to the absence of any detectable HPV DNA. The HPV detection rate in ESCC varies geographically[29,30]. An obvious phenomenon is that the HPV detection rate is absent or significantly lower in areas with a low incidence of esophageal carcinoma[31,32]; but in high incidence areas of the disease in China, the detection rate of HPV is much higher, 43.1% in Linxian[33], 50% in Sichuan[34], 64% in Anyang[35] and 20% –63% in Shantou area[36].

(2) *Detection of HPV in cancerous, paracancerous and normal mucosal tissues by PCR*: In resected samples of ESCC, the samples were subjected to PCR to detect HPV infection using consensus and type-specific primers for HPV type 6, 11, 16, and 18. They revealed different detection

rates of high and low risk HPV types in tissues with different pathological states. The incidence rate of HPV in the consensus primer (HPV type 6, 11, 16 and 18) was 65.5%, 69.1%, and 60% in tissues of cancerous, paracancerous and normal mucosa, respectively[37]. Further analysis of the distribution of HPV types in the three sections of tissues showed that the high – risk HPV types 16 and 18 were found mainly in the cancer cells (43.2%), followed by paracancerous tissues (38.6%), and the lowest in normal mucosa (13.6%). Whereas the low – risk HPV types 6 and 11 were seen mainly in the normal mucosa (52.3%)[37].

HPV infection is not only limited to cancerous and paracancerous tissues, but also to normal epithelial tissues which are HPV positive. High – risk types are predominant in cancerous tissues, indicating that high – risk types may play a more important role in carcinogenesis. Low risk types are more common in healthy tissues, and mixed infections of high and low risk types are also seen[38]. Because low risk types cause proliferation of epithelium, mixed infections of high and low risk types 16, 18, 6, and 11 may have synergic effects in the enhancement of cell proliferation and transformation[39].

(3) *Detection of HPV in the histology of ESCC by ISH*: In our specimens, ISH signals were detected in strongly positive staining in paraffin sections of EC through the use of the HPV – 16/18 probes[40]. We also found all of these three labeling patterns, diffusive, punctate and mixed types in ESCC specimens. Within the sections of ESCC, positive labeling was found only in the nuclei of epithelial tumor cells and not in other cells. The size of these spots was small to medium with ISH. A small proportion of ESCC cells also displayed three to four labeled intranuclear spots corresponding to one to four virus copies. Negative controls gave no dots. This signal type containing integrated HPV is also found in invasive ESCC with a high predominance of the punctate pattern, which showed multiple spots contained in almost all of the nuclei.

ISH also proved a useful factor for differentiating precancerous lesions in dysplasia lesions[41]. In dysplasia II and III with morphological evidence, HPV virus infection took the form of punctate signal (varied – sized dots). In a comparison of both groups of dysplasia, the rate of HPV positivity was similar; dysplasia II had 59.2% and dysplasia III, 59.5%. By ISH determination we found that the specimens for positive HPV 16 sequences in ESCC or in dysplasia lesions did not differ from those containing HPV 18, 33 or 35, when morphological features were reviewed.

HPV types 6, 11, 16, 18, and 31 represent the most common types found in the epithelium of squamous cell hyperplasias, dysplasias, and carcinomas. Xu (2004) reported expression of HPV16 – E6 and E7 oncoprotein in normal mucosa tissue, dysplasia and carcinoma of the esophagus[42]. In this case, the expression rate detailed was 22.2% in simple hyperplasia, 50% in mild dysplasia, 80.6% in moderate dysplasia, 67.8% in severe dysplasia, and 66.7% in SCC. These reports raise speculation regarding the potential for HPV infection in dysplastic and malignant transformation of esophageal squamous lesions.

2. Histology suggesting HPV infection

Syrjanen published his observations on " histological changes identical to those of condylomatous lesions found in esophageal squamous cell carcinoma. " and suggested the possibility that HPV are the etiologic agents of esophageal squamous cell carcinoma in 1982[25]. This hypothesis is also suggested by the animal model of bovine papillomavirus infection, which commonly impacts on the esophageal mucosae, produces papillomas, then undergoes malignant transformation following exposure to carcinogens in ingested bracken ferns[43]. The relationship between HPV and benign proliferation of squamous epithelia has been recognized for a long time. Although a definite correlation between HPV infection and esophageal carcinoma has not been made, it seems reasonable to assume that HPV infection of the esophagus may occur and may, therefore, be responsible for benign proliferations of the squamous epithelia of the esophagus. On the human esophageal mucosae surfaces, these may present as focal epithelial hyperplasias, dysplasia or true papillomas. These data suggest that the majority of esophageal condylomatous lesions may usually be overlooked by endoscopists and pathologists[44].

Histological changes suggesting HPV infection are found adjacent to squamous cell carcinomas or in squamous papillomas of the esophagus. This histological evidence of HPV infection is characterized by the presence of koilocytosis, giant and multinucleated cells, dyskeratosis, hyperkeratosis, acanthosis and papillomatosis[25]. These lesions suggesting HPV infection will appear in 15.4% – 100.0% of the esophageal specimens (Tab. 1).

Tab. 1 Frequency of single histological parameter suggesting HPV infection (%)

Case no.	acanthosis	koilocyte	basal cell Hyperplasia	dyskeratosis	multinucleation hyperkeratosis	Intraepithelial capillary loops	Reference
35	34.6	50.0	/	30.8	15.4	7.7	kim 1992, [45]
41	100	92.7	44.0	52.3	34.1	92.7	Woo 1996, [46]
51	49.0	54.9	15.7	35.3	23.5	60.8	change 1990, [47]

(1) *Papilloma and papillomatous lesions*: Esophageal squamous – cell papilloma is a rare lesion, with 0.01% – 0.04% morbidity as estimated by autopsy data[48]. In Hungary, Szanto reported 155 patients in 59 056 upper digestive panendoscopic examinations with papilloma of the esophagus[49]. To date, 239 esophageal squamous cell papillomas have been analyzed in 29 separate studies using different HPV detection methods, with HPV being detected in 51 (21.3%) cases[50]. In our study, the papillomatous and true papillomas were found on esophageal mucosae adjacent to ESCC mass[51]. (Fig. 1)

(2) *Koilocytotic cells*: The koilocytotic lesions adjacent to ESCC suggest a relationship between HPV infection and esophageal carcinoma[52]. Koilocytotic HPV – type histological changes are often found adjacent to or intermingled with intraepithelial neoplasm and ESCC[45-47]. The cells have clear, vacuolated cytoplasm and pyknotic, often wrinkled, hyperchromatic nuclei and binucleate cells (Fig. 2) with demonstration of the HPV antigens by immunohistochemical and ISH detection.

(3) *Balloon cells and clear-cell acanthosis*: Balloon cells are frequently found adjacent to the carcinoma in areas suggestive of HPV infection. Esophageal balloon cells are characterized by epithelial hyperplasia with swollen clear cells, referred to as clear – cell acanthosis[41]. Such changes

produce whitish patches or whitish mucosa, which appear upon endoscopy. Some authors have considered clear – cell acanthosis as resulting from thermal injury caused by drinking hot beverages. Undoubtedly, clear – cell acanthosis bears some resemblance to HPV – induced epithelial changes observed in the esophageal mucosa[45-47]. Although balloon cells found in clear – cell acanthosis usually lack characteristic nuclear changes as observed in true koilocytes, the possibility is by no means excluded that clear – cell acanthosis could be caused by HPV infection[46] (Fig. 3).

Fig. 1   Scanning electron – microscopic view from papillomatous lesions on the mucosal surface of the esophagus. The multiple spiniform (A) and the flower – bud like (B) lesions were noted with presence of multiple layers of squmous epithelium. Some of the squamae was in the process of exfoliation (SEM, ×7000)

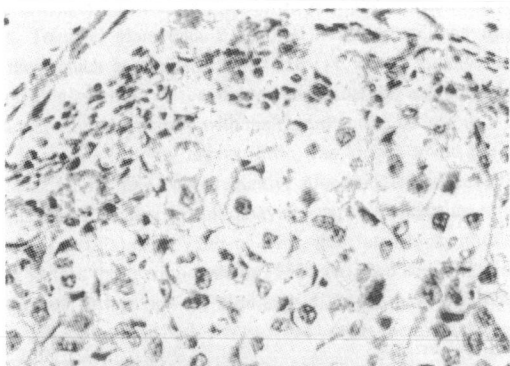

Fig. 2   Koilocytotic cells had clear, vacuolated cytoplasm and pyknotic, often wrinkled, hyperchromatic nuclei and binucleate cells (HE, ×100)

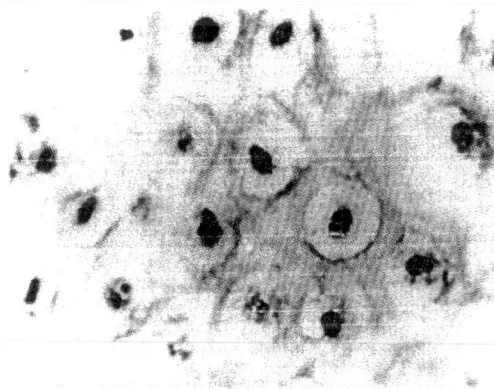

Fig. 3   Clear – cell acanthosis, in the intermediated and superficial cell layers adjacent to the ESCC, subjected to HPV DNA in situ hybridization (ISH, ×200)

In conclusion our results suggest that HPV might play a role in the carcinogenesis of the ESCC.

We still believe that somehow HPV is involved in the pathogenesis of ESCC, either as a promoter in association with other factors or as the prime etiology, due to cases frequently showing histological parameters of HPV infection. We emphasize the significant role of HPV infection in the development of benign and malignant squamous cell lesions of the esophagus, especially ESCC.

3. Analysis of the infection entry pathway of HPV

Papilloma – viruses are highly species – and tissue – specific viruses and are primarily found in higher vertebrates. To date, more than a hundred human papillomaviruses have been identified, some of them, such as the HPV6 and HPV11, cause mainly benign epithelial papillomas or warts on skin and mucosa (condyloma acuminatum) and some of them such as HPV16 and HPV18 are strongly associated with malignant epithelial lesions.

Due to the lack of good infectivity assays, little was known about the initial steps of papillomavirus uptake for a long time. More recently, infectivity assays have been developed by using pseudo virions that consists of papillomavirus capsids enclosing a marker genome termed virus – like particles (VLPs)[53]. VLPs of HPV, generated by synthesis of the capsid proteins L1 and L2 in various expression systems[54] h ave shed some light on the mode of interaction between the cell surface and the viral capsid[55]. It has been demonstrated that the binding capsid on cell surfaces is highly conserved within the animal kingdom and that all cell lines tested were able to bind VLPs with few exceptions[56].

The initial interaction of HPV with cells appears to occur by binding to cell surface receptors. Several viruses, including HPV, use multiple receptors sequentially. One model suggests that cell attachment is initiated by interaction of the viral capsid with a primary receptor of relatively low specificity and proceeds by binding to a secondary receptor resulting in internalization and infection within the cell[57].

Heparan sulfate (HS), a glycosaminoglycan, is widely expressed on cells, extracellular matrices, and basement membranes. The presence of HS on the cell surface is required for infection by HPV – 16 and HPV – 33 pseudo virions. These results indicate that HS can serve as HPV receptors. Indeed, binding to HS on the cell surface is often merely the first step in a series of events between the virus and the cell that is required for virus entry and the initiation of infection[58]. More recently VLPs ofHPV – 11 are shown to bind to heparin and to cell surfaces via HS[59].

Proteoglycans are proteins that contain covalently linked glycosaminoglycans containing the syndecans and glypicans[60]. Syndecans and glypicans are expressed in a cell – tissue – , and development – specific fashion. For papillomaviruses, initiating steps may involve syndecan and glypican and interaction with the minor capsid protein L2 in addition to L1[61,62]. Syndecans and glypicans bind proteins of the extracellular environment via their HS chains, regulating a wide spectrum of biological activities, including cell proliferation and differentiation, morphogenesis, wound repair, and host defense. Under normal conditions, syndecan – 1 expression is modest in basal keratinocytes and induced in suprabasal layers during differentiation[63]. In basal keratinocytes and suprabasal cells exposed by minor trauma or abrasion, overexpression of syndecan – I may occur, and thus may strongly upregulate the ability to attach and internalize papillomaviruses *in vivo*. Syndecan – 1 may serve as a primary attachment receptor for HPV in natural infection.

The integrins are a superfamily of surface proteins present in a wide variety of cells. They are involved in the adhesion of cells to other cells or to specific components of the extracellular matrix. They are heterodimers, containing an α and a β subunit linked noncovalently. The subunits contain extracellular, transmembrane, and intracellular segments. The extracellular segments bound to a variety of ligands α − 6 integrin was proposed as a putative papillomavirus receptor. Alpha − 6 integrin plays an important role as a primary receptor protein in natural HPV infection of keratinocytes. Recently, α − 6 integrin was suggested as the binding receptor for HPV6, HPV 11 or HPV16 VLP binding to cells[64,65].

Beside membrane − associated HS and alpha − 6 integrin, the cultured keratinocytes uniquely secrete a component into the basal extracellular matrix (ECM), which can function to adsorb HPV particles and can then be internalized by adherent cells[66]. This uncharacterized basal ECM adsorption receptor was secreted by normal human epidermal keratinocytes and by the keratinocyte − derived cell lines, but not by non − keratinocyte cell lines. Multiple HPV types bound preferentially to this keratinocyte − specific receptor over the membrane − associated receptor, and binding to the basal ECM adsorption receptor was refractory to inhibition by heparin. Like the membrane − associated receptor, this basal ECM component was functional as an adsorption receptor *in vitro* infection mode. These findings suggest a model for natural infections, in which HPV virions, nonspecifically adsorbed to HS on suprabasal keratinocytes throughout an epithelial wound, might be transferred to mitotically active migrating keratinocytes via an intermediate association with the ECM secreted by these cells as they reestablish the basement membrane.

Infection with native papillomaviruses or VLP uptake into intracellular vesicles has been reported to occur via either clathrin − or caveolas − mediated endocytosis[67]. In dually exposed cells, HPV16 VLPs and BPV virions colocalized intracellularly. BPV VLPs colocalized with AP − 2, a clathrin adapter molecular and a marker of the clathrin − dependent endocytic pathway; and also with transferring receptor, a marker of early endosomes; and Lamp − 2, a marker of late endosomes and lysosomes. Papillomaviruses infect cells via clathrin − dependent receptor − mediated endocytosis.

Treatment of cells with various reagents has disclosed a protein component that is involved in binding to prevent the virus from reaching the target cell.

4. Interaction of HPV with cellular biological behaviors

HPV infected epithelial cells attempt to assemble in terminally differentiating keratinocytes and must first stimulate the cell's cycle progression for the replication of their genome [68]. It is evident that all types of HPV interfere with the proliferative regulation of cells, promoting the induction of cellular DNA synthesis to support viral replication. In the high − risk HPV types, most frequently HPV 16 and HPV 18, these activities can contribute to the development of lesions with the potential for malignant conversion and in the low − risk HPV types, most frequently HPV 6 and HPV 11, can give rise almost exclusively to benign condylomata. The oncogenic activity of the HPV appears to reflect the sum of many parameters, including viral protein function, control of some gene expression and immune response disorder.

Many viral proteins influence the cellular pathways that control cell proliferation and cell death. The oncogenic effect of the high − risk viruses, in particular the importance of the E6 and E7 genes, has been studied extensively using isolated DNA clones to express viral proteins *in vitro* in suitable cell types. The E6 protein may degrade the p53 – encoded protein[69] and the E7 protein may act – on retinoblastoma protein[70]. The degradation and inactivation of the products of the tumor – suppressor genes promote entry into the cell cycle, thus leading to cell proliferation[71]. The viral oncoproteins E6 and E7 are known to help disrupt the cell cycle checkpoint machinery and accelerate chromosomal instability events, which are critical in malignant conversion. The high – risk E6 and E7 oncoproteins tend to show a much more efficient ability to disturb normal cell growth, and these quantitative differences between high – and low – risk viral protein functions are likely to contribute significantly to the differences in malignant potential[72]. Papillomavirus oncogene can regulate apoptosis of epithelial cells[73].

HPV 16 E7 protein could interact with pRb by separating the complex of pRb – E2F – 1, release free E2F – 1, resulting in a disordered cell cycle and cell proliferation[74]. HPV oncoproteins share at least some ability to interfere with important negative growth signals by which host cell proliferation is normally regulated, such as cell cycle regulators (p16, p18, and CDC7) and transcription factors (TAF7L, RFC4, RPA2, and TFDP2) which are generally disordered in HPV positive samples.

Both E6 and E7 of HPV contribute to increasing EGFR levels[75], but with different mechanisms. On the one hand, although E6 can increase EGFR levels, it cannot overcome the resistance of normal esophageal epithelia to excessive EGFR signaling[76]. On the other hand, E7, which alone does not acutely increase EGFR mRNA or protein, allows for EGFR overexpression in esophageal epithelial carcinoma[77].

Normal somatic cells terminate their capacity of replication through a pathway, which is triggered by activation of p53 and/or pRb in response to critically shortened telomere DNA, and results in cellular senescence. High – risk HPVs immortalize keratinocytes by disrupting the Rb/p16 pathway and activating telomerase[78]. While the E7 oncoprotein targets Rb, the E6 oncoprotein induces telomerase activity in human keratinocytes; therefore, E7 may augment the E6 – mediated activation ofhTERT transcription[79].

Survivin has recently been identified as a novel member of the inhibitor of the apoptosis (IAP) gene family[80]. The product of this gene not only suppresses apoptosis but also controls cell division. Survivin is undetectable in most terminally differentiated normal tissues, but is expressed in embryonic and fetal organs and is present in most malignant tumors. HPV – 16 E6, but not E7, was found to significantly transactivate the survivin promoter[81], resulting in the inhibition of apoptosis of cancer cells[82].

The results suggest that in esophageal tissues manifesting high risk HPV infection, changes in hTERT, p16, EGFR, survivin and transcription factors, expression may be interacting with each other, which can influence the progression and carcinogenesis of ESCC.

Despite viral immune evasion, the immune system effectively repels most HPV infections, and is associated with strong localized cell mediated immune responses[83]. HPV has several mechanisms for avoiding the immune system. One way it does this is by down – regulating the expression of interferon genes. In lesions produced by HPV type 16, there is a reduction in the numbers of immune cells, especially Langerhans cells (LC)[84]. HPV – 16 E6 and E7 inhibit macrophage inflammatory protein 3alpha (MIP – 3α) transcription, resulting in suppression of the migration of immature Langerhans precursor – like cells[85]. It has been reported that E6 and E7 of high – risk HPVs interfere with immune mediators in order to suppress the recruitment of immune cells and the antiviral activities of infected cells[86]. The confinement of HPV infection to epithelial puts the epithelial dendritic cell, the LC, in charge of the induction of T cell – dependent immunity. Because HPV – infected keratinocytes cannot reach the regional lymphoid organs, and HPV – infection of LCs does not result in viral gene expression, priming of antiviral T cells exclusively depends on cross – presentation of viral antigens by the LC. Sensitization of the immune system in the regional lymphoid organs elicits systemic anti – HPV immunity as well as intraepithelial immune surveillance by memory – type intraepithelial T cells and locally produced antibodies. These results suggest that one mechanism by which HPV – infected cells suppress the immune response may be through the inhibition of a vital alert signal, thus contributing to the persistence of HPV infection.

B. Immortalization and Malignant Transformation of the Esophageal Epithelial Cells Induced by Human Papillomavirus in the Multistage Process

Many viruses have been shown to have oncogenic potential, such as hepatitis B virus linked to cancer of the liver, Epstein – Barr virus to nasopharyngeal carcinoma, HPV to carcinoma of the cervix and so on. Viral transformation refers to the changes in the function and morphology of a cell. These changes confer properties on the infected cell characteristic of neoplasia, which is always preceded by the immortalized and malignantly transformed stages.

In our survey the majority of esophageal carcinomas in the Shantou area were found to be associated with an infection of HPV[87]. To study the carcinogenesis of HPV *in vitro*, we used E6E7 genes of HPV type18 and transferred it into human embryonic esophageal epithelium and succeeded in establishing a cell line SHEE, which was able to provide powerful model systems for the study of tumorigenesis of human esophageal epithelial cells[88].

Tumorigenesis in humans is a multistep process, as has been reported in lung, gastric, oral, bladder and skin cancers. The carcinogenesis of esophageal epithelial cells induced by HPV is also a multistage process, which goes through the preimmortal, immortal, biphase differential, premalignant, malignant and invasive stages[89]. For neoplastic transformation of normal human cells, they must first be immortalized and then converted into neoplastic cells. It is well known that immortalization is a critical step for the neoplastic transformation of cells[90].

The cell line was examined along with the biological behaviors of carcinogenesis according to cell morphology, cell cycle, apoptosis, cell differentiation, contact inhibition growth, anchorage – dependency, telomere and telomerase activity, molecular and genetic changes, and tumorigenesis.

1. Establishment of SHEE cell line induced by human papillomavirus

（1）*Construction of the recombinant plasmids*: Recombinant plasmids were constructed by standard protocols[91]. pGEM/HPV18 plasmid was prepared by the alkaline lyses method from *E. coli* HMS174. The E6E7 gene of HPV18 was amplified with PCR. PCR product was lengthened in 0. 9 kb and ligated with pGEM – T vector（Promega）, generating pGEM – E6E7, pGEM – E6E7 was digested with SpeI enzyme. The fragment was ligated with pAAV3 vector, generating pAAV – E6E7（Fig. 4）. The adenoviral vectors, which have been well characterized for transferring foreign genes into mammalian cells, are easy to manipulate, and can be grown to high titers. PAAV – E6E7 was cotransfected with plasmid pAd8 into human 293 cell line, a human kidney cell line （HK293）, to produce the infectious virus（HPV18E6E7 – AAV）. The virus was collected from the supernatant of the HK293 culture medium and titers were estimated for these recombinant viruses[92].

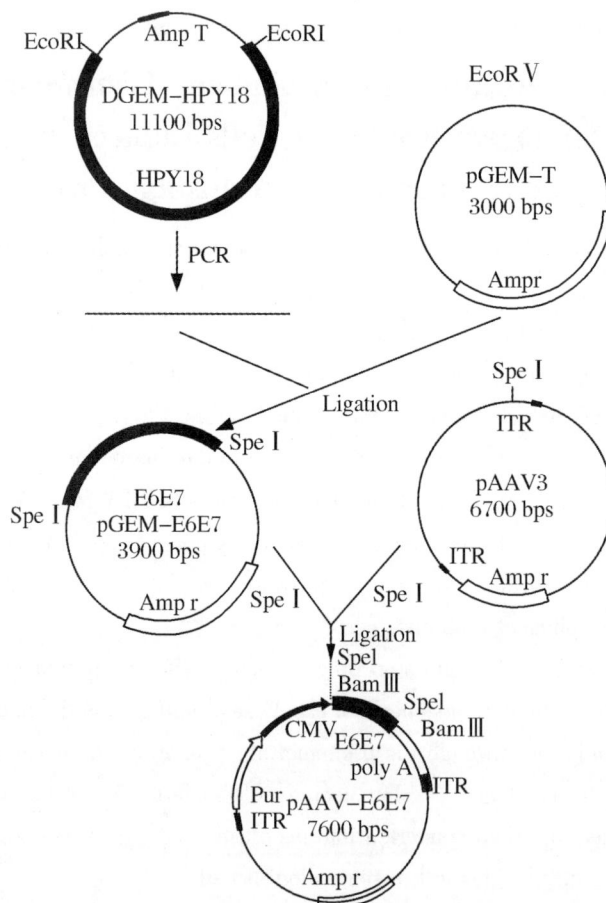

Fig. 4　Construction of pAAV – E6E7. For insertion of E6E7 gene of HPV type 18, pGEM – HPV18 was amplified with PCR and ligated with pGEMT, generating pGEM – E6E7, pGEM – E6E7 was digested with SpeI and ligated with pAAV3（also digested with SpeI）, generating pAAV – E6E7

(2) *Cultivation of human embryonic esophageal mucosa*: The esophagus, obtained from a fetal embryo, which proved to be normal without attachment with any carcinogen, was cut into small pieces and cultivated in medium 199 (Gibco) with 10% fetal bovine serum (FBS) and antibiotics (100 U/ml penicillin, 100 U/ml streptomycin).

(3) *HPV18 E6E7 – AAV infection*: Small pieces of the esophageal tissue became adhered to the wall of the flask. After standing in serum – free medium for two hours, the infectious virus of HPV18E6E7AAV was added and incubated for two hours. Then it was refreshed with a normal culture medium. The cultured human embryonic esophageal epithelial cells grew from the margin of the esophageal tissue and were continually cultured and propagated. This led to the establishment of a new esophageal cell line (SHEE)[93]. In their biological behaviors, the SHEE cells are close to the basal cells of their original fetal esophageal mucosa keeping potency of proliferationand differentiation.

2. Immortalization and malignant transformation of esophageal epithelial cells in the multistage process

The SHEE cells were transferred by gene E6E7 of HPV type 18 in our laboratory and continually cultivated over 100 passages, which had been malignantly transformed. The data collected demonstrate that carcinogenesis of esophageal epithelial cells induced by HPV is a multistage process, which goes through an initial, immortal, biphase differential (premalignant), malignant and invasive stage. The generation of esophageal carcinoma is caused by an accumulation of cellular, genetic and molecular changes. To investigate the multistage process of carcinogenesis further, the progressive alteration of the morphology, telomerase, cytogenesis, oncogenes and tumorigenicity of the process of immortalization and malignant transformation of the SHEE cell was studied.

(1) *The preimmortal stage*[94]: The early SHEE cell, SHEE10 – SHEE16 (10th – 16th passage), grew evenly in a single layer in flasks and appeared typical of squamous epithelium with a multi – angular outline and an oval nucleus. When they came into contact with each other, cell proliferation was arrested, and a significant number of them remained quiescent or died, a phenomenon which is called contact – inhibition growth. SHEE cells contained tonofilaments in their cytoplasm by electron microscopic examination and was shown to be cytokeratin positive by an immunohistochemical procedure. This showed that the cells were squamous epithelium in origin. Subsequently the cell line containing the HPV 18 E6 and E7 genes were demonstrated by fluorescent *in situ* hybridization (FISH) and PCR assay. Cells of continual passage followed telomere length shortened continuously, but without telomerase activation. This was called the preimmortal stage. SHEE cells from the 17th to the 22nd passage exhibited growth retardation. Cells displayed large, flattened terminally – mature senescent cells and abundant apoptotic cells which had small rounded cells with coagulated chromatin in the nucleus and were always detached from the culture flask. Cells in the 17th – 22nd passages tended to become senescent and undergo crises with a decrease of DNA synthesis and an increase of cell apoptosis. To characterize the altered growth kinetic in these passage cultures, DNA analyses by flow cytometry (FCM) in the cell cycle was performed. In cells of initial

passage, SHEE14 cells was demonstrated that 12. 6% of cells were in S phase and 36% were in G_0G_1 phase. When cells entered senescence and crisis stage, SHEE20 cells were accumulated in G_0G_1 and in the sub – GoG_1 phase (total 83. 7%) and in the S phase were only 8. 6%.

(2) *The Immortalized Stage*[95,96] : Then the cells bypassed the senescence checkpoint and commenced proliferating. During the ectopic expression of the RNA subunit of telomerase (hTR and hTERT), the telomerase activity eventually appeared to arrest telomere shortening, and to extend the life span of the cells *in vitro*. The activity of telomerase first appeared in SHEE30 cells, but it was weakly positive, and was positive in SHEE40 cells with its telomere length stabilized. Apparently with telomerase activity, cells overcame senescence and crisis and restored to proliferation again, and entered a proliferate stage. After long – term culture, the phenotype kept the characteristics of primary epithelial cells. In SHEE30, cells underwent crowded growth and connection. Some cells died and detached, retained a network outlook in the culture flask, and some cells revealed proliferation and were weak in contact inhibition growth, and some cells revealed differentiation. The DNA synthesized cells in the SHEE30 cells increased to 31. 6% , indicating an over – growth of the proliferative cells. The data demonstrated that anchorage – dependent growth was decreased in some cells of SHEE30. The cell line then entered the immortal stage. The criteria of immortalization will be as follow. Apart from the activation of telomerase activity, the immortal phenotype of SHEE showed an increase of cell proliferation with a higher DNA synthesis, weakened contact – inhibition growth and anchorage – dependency, maintenance of dose – dependency to EGF, but not tumor formation in immunodeficient mice. This cell line has become immortal and has propagated continuously for more than 50th passage.

(3) *The biphase differential stage*[97,98] : The cultured SHEE cells at its 35th passage, in the early phase of immortalization, grew exuberantly and displayed biphasic morphology, partially, well differentiated and partially undifferentiated basal cells with weakened contact – inhibition growth. The cultured SHEE35 cells were grouped into both the undifferentiated basal epithelium and the differentiated squamous epithelium; the cytokeratin and F – actin were empty in the former and rich in the latter. The contact inhibition and anchorage – dependent growth of these undifferentiated cells were weakened. The modal number of chromosomes in 100 mitotic cells revealed two stem-lines, 56 chromosomes (30%) and 61 chromosomes (24%). By karyotype analysis, SHEE35 were heteroploidy belonging to hyperdiploidy and hypotriploidy. C – myc and p53 genes were upregulated and ras gene was positive. The colony cells (the optimum cells) sought from the soft – agar culture could not grow in SCID mice. The expression of HPV18E6 was positive. The immortalized esophageal epithelium from the 35th to 60th passage displayed differentiated phenotype in biphase. In the 60th passage, based on these findings of morphological phenotype, the double chromosomal modal number of cytogenetics, the upregulation of some oncogenes, and the formation of tumors in one quarter of severely combined immunodeficient (SCID) mice but not nude mice. It is suggested that the SHEE60 cell line is in the precancerous phase.

(4) *The malignantly transformed stage*[99] : Light microscopy showed that the cells in SHEE85

· 770 ·

were crowded together with different sizes and had considerably more giant nucleus cells with multi-nucleoli, which indicated that the cells of SHEE85 were in over proliferation and in poor differentiation. The DNA content of the cells revealed that their distribution in the cell cycle consisted of G_0G_1, 54.1%; S, 21.3%, G_2 M, 22.5%; sub $-G_0G_1$ 2.1%. This meant that an increase in cell proliferation (proliferative index was 43.8%), a decrease in apoptosis and an increase in hyperploidy cells had occurred. In the karyotype of SHEE85 chromosomes, the modal number of chromosomes was 59 – 65. The distribution of chromosomes was diploidy and hypodiploidy, 4.9%; hyperdiploidy, 18.8%; hypotriploidy, 58.8%; hypertriploidy, 17.7%. The chromosomes of SHEE cells were hyperdiploid in the early passages, but became hypotriploid in the 85th passage, showing the property of chromosome instability in SHEE cell lines. Expression of hTERT in SHEE85 cells were determined by RT – PCR. The hTERT expression was positive in the 85th passage. In passage 85 of SHEE the $p53$, $c - myc$, $bcl - 2$ and ras was positive. The $p53$ sequence analysis was aligned to phosphorylation, compared with $p53$ gene from GenBank (gb M22894). Cells were cultivated in soft agar and large colonies formed. When these cells were injected into nude mice, they grew rapidly to form tumors within 20 days. Thus, the cells of SHEE85 were in fully malignant transformation.

(5) *The invasive and metastasis stage*[100,101]: *In vitro* testing of SHEE cell invasion was assessed by using a fresh fetal amnion, which was cultured in 199 medium supplemented with 10 % FBS. In the invasive experiment, on each piece of amnion in one well of a 6 – well plate, 5×10^4 SHEE cells (at 30th or 100th passage) were added, and incubated together for 24 h and 72 h, and then were washed several times with PBS. After inoculation on the amnion, SHEE100 cells were shown to grow in clusters on the amniotic epithelial surface with the formation of pseudopodia that intruded into the gap between intercellular conjunctions. Under microscope SHEE100 cells were observed invading into the amnion stroma. It was not found that the inoculated SHEE35 cells could adhere to or invade into the amnion.

SHEE cell invasion was assessed by transplantation into immunodeficient mice. After being inoculated in the sub – axilla of the SCID mice, SHEE85 cells were noted growing rapidly and forming tumor masses within 20 days. In the subcutaneous tissue of the fight axilla, tumor masses were observed macroscopically forming on the thoracic wall and penetrating into the thoracic cavity. Microscopically, these tumor cells harboring larger nucleus, nucleolus and less cytoplasm could infiltrate and destroy adjacent muscular fibers. Metastasis of tumor cells to axillary lymph nodes occurred after 40 days. By inoculatation into the peritoneal cavity, tumors were observed occurring in the mesentery, pancreas, urinary cyst, sub – diaphragm, etc. with a special propensity for invading the mesentery and pancreas, but there were no signs of hepatic metastases. Histological examination revealed that these tumor cells did not only grow on the surface of organs, but also invaded into adjacent tissues. Once transplanted, these tumors could keep be passed on to other SCID mice. In contrast, SHEE30 cells were not found to form tumors in inoculated tissue by gross and microscopic examination.

This report is based on results from one of these cell lines, from SHEE85 to SHEE100, which acquired the ability to form invasive squamous cell carcinoma in athymic mice during their propagation in culture. SHEE100 cell strains isolated at different passage levels expressed distinct transformed properties, suggesting that progressive change requiring multiple events was involved in the development of the malignant cells.

3. Biological criteria used to monitor the developed process of immortalization and malignant transformation in cultured esophageal epithelial cells

In this section we will try to define the biological criteria of immortalization and malignant transformation. It is considered extremely useful to determine how the process of the cultured esophageal epithelial cells developed and how this related to the variability in cellular and molecular biology between immortalized cells and malignantly transformed cells.

The cell line SHEE was derived from immortalized embryonic esophageal epithelium and was induced by gene E6E7 of HPV 18 in our laboratory after being cultivated and propagated over 100 passages. How to monitor these proceedings from the initial immortal to the malignant stage is an important problem. Changes occurring in SHEE cells from the 10th to the 100th passage, with emphasis on their phenotypes (which included the morphological changes of proliferation, differentiation and apoptosis, cytogenetic changes, telomerase activity, gene – expression profile and tumorigenicity, especially sofi – agar culture and tumor formation in SCID mice) were used to monitor development of carcinogenesis.

(1) *Cellular proliferation, differentiation and cell cycle*: Cellular proliferation, differentiation and apoptosis are fundamental life activities, and are also the growth markers of cells. In cell morphology, the cells in the SHEE11 group (11th passage) were uniform in size and shape, and grew as an even monolayer with characteristics of squamous epithelium (Fig. 5A). SHEE35 cells proliferated and exhibited diphasic differentiation, some cells displaying the undifferentiated basal epithelium and others displaying differentiated squamous epithelium (Fig. 5B). Cells in SHEE65 were crowded together with undifferentiated, and more mitotic nuclei (Fig. 5). Over the SHEE85 cells lost the pattern of maturation and cells displayed growth that overlapped in a differently shaped and poorly differentiated manner. They showed evidence of polymorphism with comparatively giant nucleus cells and multinucleoli (Fig. 5D). According to the DNA content and proliferation index, the cells of SHEE11, SHEE35, SHEE65 and SHEE85 all showed proliferative characteristics. The proliferation index and the distribution of DNA content of SHEE series cells were analyzed by FCM. In the DNA histogram, the proliferative indexes were 21.1%, 32.5%, 33.2% and 57.3% in SHEE11, SHEE35, SHEE65 and SHEE85 respectively (Fig. 6). Accounting DNA < 2n cells, sub – GoG1 cells, the apoptotic indexes were 3.3%, 2.7%, 3.5% and 3.0%, in SHEE11, SHEE35, SHEE65 and SHEE85 respectively. Of all DNA > 4n cells there were SHEE11, 2.5%; SHEE35, 4.7%; SHEE65, 6.1% and SHEE85, 12.2%. This indicated that heteroploid and hyperploid cells increased with the progressive culture of SHEE.

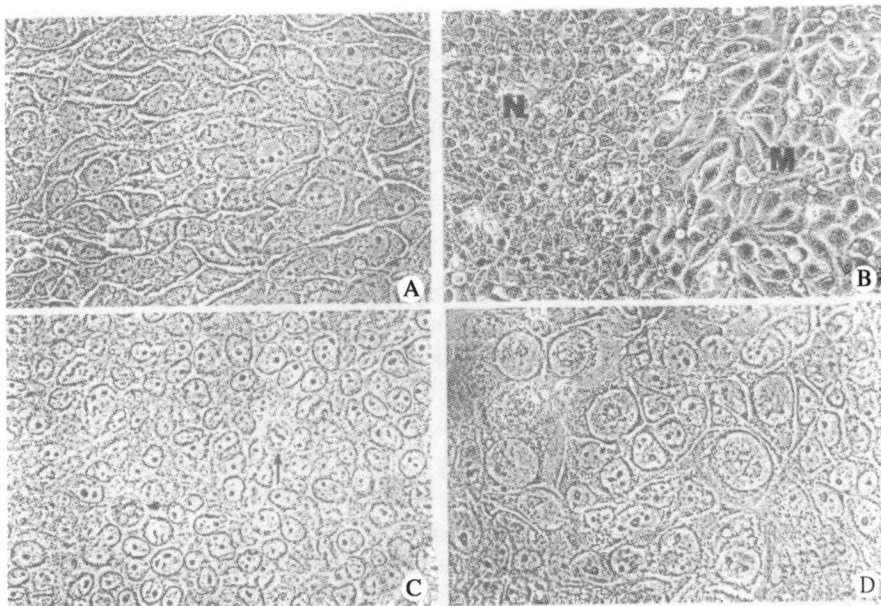

(A), Cells of SHEE11 displayed good differentiation and squarnous epithelium in outlook (×400); (B), Cells of SHEE35 displayed two differentiated directions, partially well differentiated (M) and partial poorly (N) (×200);
(C), Cells of SHEE65 crowded together with more mitosis (arrow) (×400); (D), Cells of SHEE85 displayed poor differentiation with polymorphism and meganuclei (×400)

Fig. 5   Morphology of living cells in the SHEE cell series was observed by phase contrast microscopy

(2) *Telomere and telomerase*: Telomeres are specialized nucleoproteins that play an important role in chromosome structure and function. The telomeric DNA together with its associated proteins protects the chromosomal ends from degradation or aberrant recombination[102]. Because of the mechanism of conventional DNA polymerases, the replication of DNA molecules can be predicted to result in the gradual shortening of the chromosome by the length of a terminal primer at each cell cycle. These data are considered by some to provide direct evidence that telomere shortening controls cellular aging. Aging of normal cells is a result of their limited proliferative capacity. After attaining their finite life span, normal cells cease dividing and senesce. It appears that cells lacking telomerase progressively lose telomeres, resulting in senescence, and it has been suggested that the sequential shortening of telomeric DNA may be an important molecular marker for cell life.

In our data the telomere length of the normal fetal esophageal cell was 30 kb in length, shortened to 17 kb in SHEE11, and 3.5 kb in SHEE35, and then maintained this level continually to SHEE65 and SHEE100. The telomerase activity was absent in normal esophageal epithelium and in the early passages (10th – 25th) of SHEE cells and appeared in SHEE35, SHEE65, SHEE85 and SHEE100[103] (Fig. 7). The activation of telomerase to maintain telomere length was necessary for immortalization and malignancy, but was not sufficient for malignant transformation in the SHEE cell line. Our results indicated that immortalization and malignant transformation of SHEE cells might require activation of telomerase and other genes, which abrogated normal cell proliferation and differ-

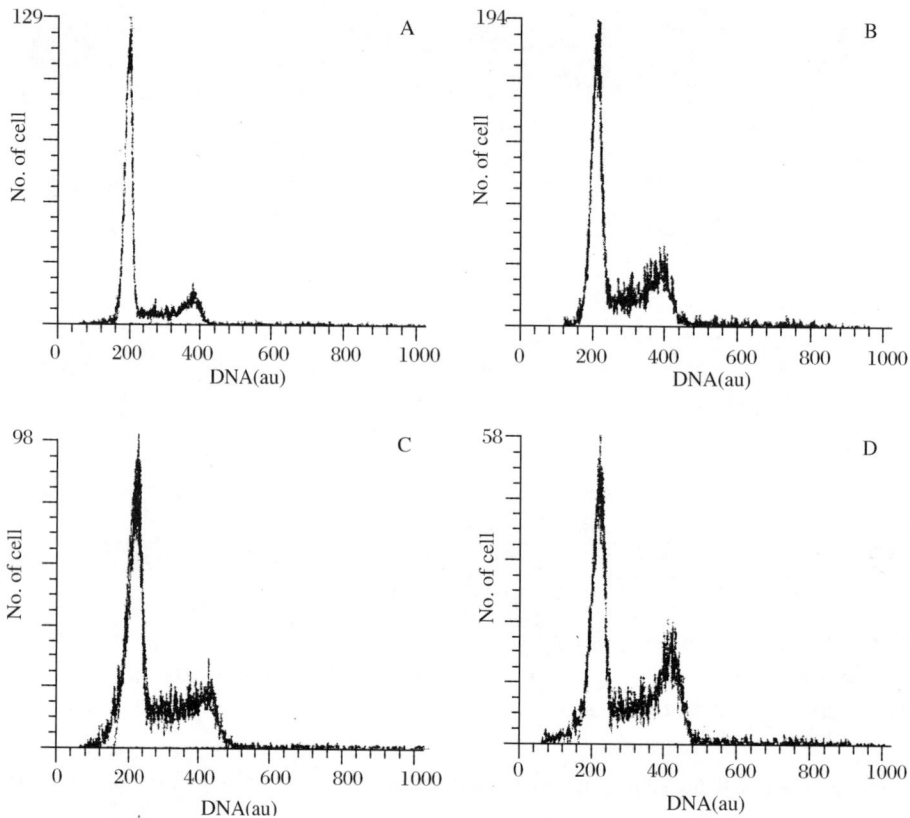

(A). SHEE11; (B), SHEE35; (C), SHEE65; (D), SHEE85. au: arbitrary unit

Fig. 6　DNA histograms of cell cycle

entiation of the cells. For this reason we suggest that telomerase may be a prerequisite and the diagnostic criteria, to immortal cells[104], and that telomerase expression may be a marker of premalignant and malignant lesions[105].

(3) p53, bcl −2, c − Myc and Ras genes: Some genes, such as c − myc, ras, p53 and bcl −2, which we analyzed are related to immortalization and the malignant transformation of SHEE cells[106]. In SHEE10 cells, which were in an early stage of the preimmortal stage, c − myc, ras, p53 and bcl −2 genes were at a low level; when c − myc, ras and p53 (a probable mutation type) were shifted − up from SHEE65 to SHEE85 (Fig. 8), the morphology and tumorigenicity in some of the cells showed the phenotype of malignant transformation.

In our esophageal epithelial cell line SHEE, the tumor suppressor gene p53 was expressed in phosphor − protein (unpublished data). Post − transcriptional modification of p53 by phosphorylation has been proposed to be an important mechanism of p53 stabilization and functional changes[107] by which SHEE cells contributed to proliferation. We have shown p53, c − myc and ras to be up − regulated in SHEE65 and SHEE85, playing a central role in a cell − cell contact, mediated switch mechanism by which cell division versus differentiation is determined[108].

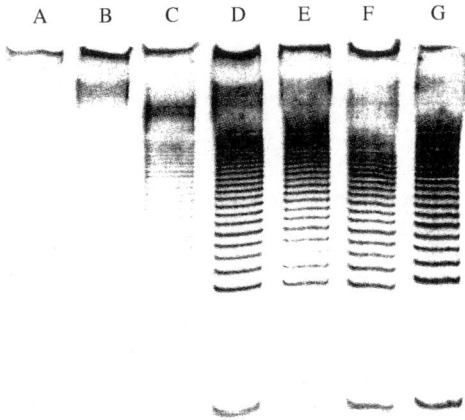

Fig. 7 Detection of telomerase activity using TRAP assay: Normal mucosa epithelium (A) of esophagus and cells of SHEE 11 (B), negative; SHEE 35 (C), weak; SHEE 65 (D), SHEE 85 (E), SHEE 100 (F) and human esophageal squamous cell carcinoma (G), strong positive

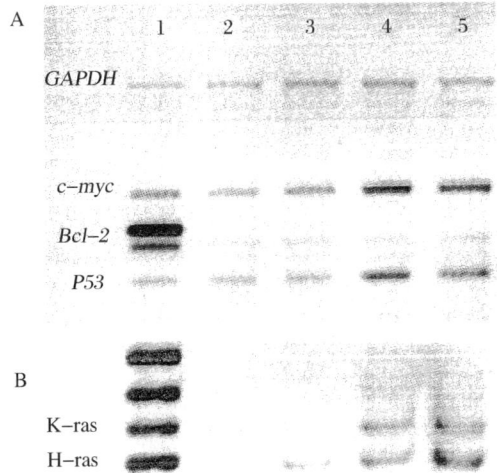

A, p53, bcl − 2, c − myc and GAPDH; B, H − ras and K − ras. Lanes: 1, marker: 2. SHEE10; 3, SHEE35; 4, SHEE65; 5. SHEE85.

Fig. 8 Electrophoretograph of multi − PCR products

The step − wise accumulation of genetic and cytogenetic alterations in SHEE development included viral integration mediated chromosome rearrangements and genetic alterations[109]. Bcl − 2 promoted cell proliferation in concert with p53, c − myc and ras was involved in cell − cycle regulation. When hTERT appeared and cell immortalization occurred in this way, SHEE was responsive to malignant transformation by the expression of H − ras or K − ras oncogene. After infection with HPV, the integration and expression of E6E7 genes induced chromosomal aberrance and inactivation of p53 and pRb, which regulated epithelial cell growth and differentiation. When telomerase activity appeared, the SHEE cells would be immortalized with upregulation of c − myc. Combined with HPV18E6E7, activated ras induced malignant transformation.

(4) Cytogenetic abnormality: Infection of HPV can cause karyotype confusion, such as the breaking of chromosomes, and leading to an abnormal structure and number of chromosomes. It also suggests that immortal esophageal cells induced by HPV 18 E6E7 may affect the changes of chromosomes, and cause instability of genetic characters. The virus genome inserts and integrates with the chromosome of the host, causing the activation and expression of oncogenes.

Chromosome instability, often referred to as karyotypic instability, is one of the major characteristics of SHEE cells. Chromosome instability in most cases reflects the occurrence of defective mitosis, including unequal distribution of chromosomes to daughter cells, which leads to the generation of aneuploid cells. Aneuploidy destabilized the karyotype and thus initiated an autocatalytic karyotype evolution that generated preneoplastic and eventually neoplastic karyotypes. An aneuploid being a discrete chromosome mutation will tend to lead immortal cells to progressively, change toward malignant transformation. The changes of cytogenetics will control the proliferation and differentiation of cells, which will easily undergo malignant transformation by promoters.

In general, the immortalized or transformed cells caused by carcinogens are always accompanied by chromosome abnormality and mutation of the gene. The chromosome's changes are manifested in the structure and the number of chromosomes. All of these changes appear in the procedure from quantitative to qualitative changes.

There were more hyperdiploid and hypotriploid in the chromosomes of four stages of SHEE cell lines. The separate modal number of chromosomes first appeared in SHEE35 and continued to SHEE60. The number of chromosomal sets and the percentage of hyperploid in SHEE35 varied between SHEE10 and SHEE65, and SHEE85 had more hypotriploid cells than the others. All of the above indicated that chromosomes of SHEE series cell lines were unstable and more susceptible to malignant transformation by promoters. The changes of cytogenetics will control the proliferation and differentiation of cells.

The number of chromosomes in SHEE10, SHEE35 SHEE65 and SHEE85 ranged between 32 and 196, these chromosomes were mainly hyperdiploids and hypotriploids, and most hypertriploid cells were found over SHEE65. The modal number of chromosomes at SHEE10 was 51 – 58 with abnormal chromosomes +1, +3, +7, +9, + +17, + +18 (Fig. 9A). SHEE35 and SHEE60 had a bimodal distribution, 55 – 57, 61 – 63 at SHEE35 with abnormal chromosomes, +1, +3, +7, +9, +11, +12, +14, +17, +18, and 58 – 60, 63 – 65 at SHEE60 with abnormal chromosomes, +1, +3, + +7, +9, +11, +12, +13, – 14, +17, +18, +22. The modal number from the 60th to the 85th passage increased slowly[110] (Fig. 9B).

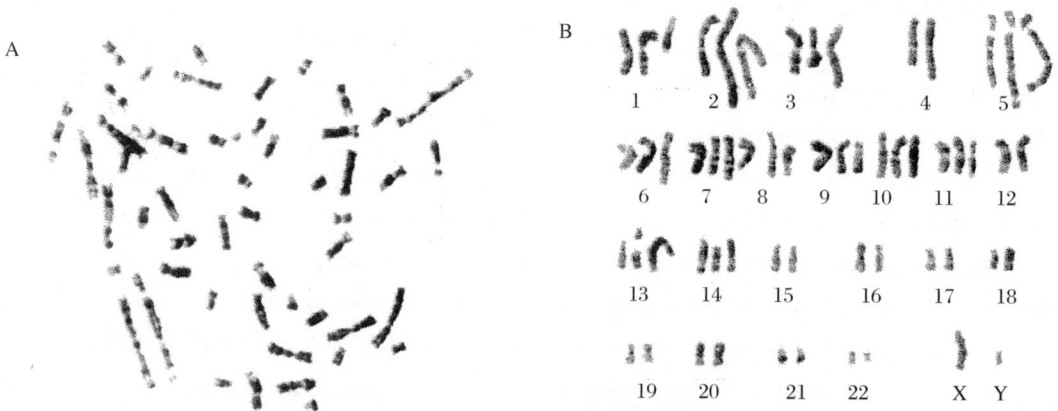

A, Hyperdiploid of SHEE10 (Giemsa, ×1000). B, Karyotype of SHEE 65 with many trisomy, 58, XY, +1, +2, +3, +5, +6, + +7, +9, +10, +11, +13, +14

Fig. 9 Chromosomal analysis

Structural aberration of chromosomes was identified as follows. The composite karyotypes at SHEE10 were del (1) (q12), del (1) (p32); der (4), t (4;?) (q31;?); der (5), t (5;?) (q31;?); der (13), t (13; 21), (p11; q11). At SHEE35 were del (1) (q12), del (1), (p32); der (4), t (4;?) (q31;?); der (13), t (13; 21) (p11; q11). At SHEE61A were del

(1) (p32); der (4), t (4;?) (q31 ;?); der (5), t (5;) (q31;?); der (13), t (13; 13) (p11; q11); der (13), t (13; 14) (p11; q11). In these three cell lines, karyotypic aberrations were mainly partial deletion of chromosomes lq12 and p32, and translocation of 4q, 13p. According to these results, the instability and increased imbalance of chromosomes, and the appearance of pre-neoplastic aneuploidy[111], was probably the genetic background for cell immortalization.

(5) *Tumorigenicity*: Colony formation in soft agar and tumor formation in SCID mice was assessed for tumorigenicity. SHEE11 could not grow on soft agar, but SHEE35, SHEE65 and SHEE85 and SHEE100 could. SHEE35 formed small colonies (less than 20 cells in a colony), and large colonies (more than 50 cells in a colony) on soft agar in SHEE65, SHEE85 and SHEE100, which showed that the characteristics of anchorage – dependent growth lost and tumorigenicity increased.

SCID mice inoculated with cells of SHEE10 and SHEE35 did not form tumors, but one quarter of SCID mice with inoculation of SHEE65 cells formed tumors with a latency period of over 2 months. It was determined that a percentage of cells of SHEE65 manifested malignancy. So we judged that they were at a premalignant stage. SHEE85 and SHEE100 cells were transplanted into nude and SCID mice and developed tumors in all mice with an invasive pattern (Fig. 10A). Tumor cells infiltrated into the muscular layer histopathologically (Fig. 10B) and took metastasis in axillary node (Fig. 10C).

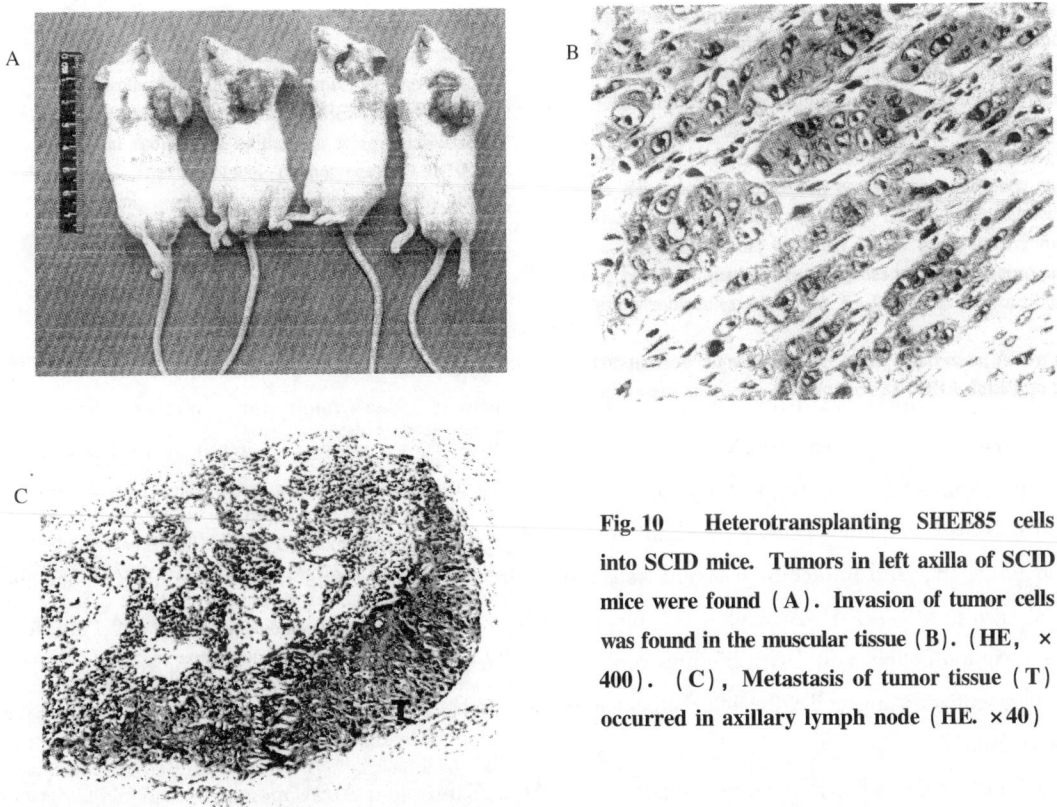

Fig. 10 Heterotransplanting SHEE85 cells into SCID mice. Tumors in left axilla of SCID mice were found (A). Invasion of tumor cells was found in the muscular tissue (B). (HE, × 400). (C), Metastasis of tumor tissue (T) occurred in axillary lymph node (HE. ×40)

The SHEE100 cells inoculated into SCID mice were found to form tumor masses in both visceral organs and soft tissues over a period of 40 days. Pathologically, these tumor cells harboring larger nucleus, nucleolus and less cytoplasm could infiltrate and destroy adjacent tissues. In the *in vitro* study, the inoculated SHEE100 cells were seen to grow in clusters on the amniotic epithelial surface and intrude into the amniotic stroma. In contrast, unrestricted growth and invasive property were not found in the SHEE10 and SHEE35 cells that were used as the control in both *in vivo* and *in vitro* experiments.

(6) *cDNA microarray assay*: Recently global gene – expression profile changes in different cancers demonstrated that the gene expression in cancer cells was complex and the cDNA microarray, a new genomic technique[112], might facilitate considerably the molecular profiling of the ES-CC[113]. The microarray technology is used for analysis of the 588 known gene expression profiles, which concern biological behaviors involving cancer, such as cell – cycle, growth regulators, cytoskeleton; cell apoptosis; oncogenes; DNA damage repair, receptors and signal transduction; cell adhesion; invasion regulators, cell – cell interactions and cytokines.

Analysis of expression profiles of a large number of genes in a cultured esophageal cell line is an essential step toward clarifying the detailed mechanisms of carcinogenesis at the gene level. In our laboratory the immortalized (SHEE30) and malignant transformed (SHEE85) cells of human fetal esophageal epithelium were established by induction of human papillomavirus 18E6E7. The biological phenotypes of SHEE30 and SHEE85, including grade of differentiation, proliferation rate, tumorigenicity, invasive ability and potential for metastasis, were different[95,99]. To investigate gene expression profile changes between SHEE30 and SHEE85, a cDNA microarray was assembled and selected. By using the model of SHEE30 and SHEE85 for gene expression assay, the characters of the samples were (a) pure epithelial cells in cultured medium without mingling with other cells such as connective tissue cell in tissue specimens, and (b) the cells in different stages being heritors originating from the same ancestors of the human fetal esophageal epithelium.

The Atlas [TM] Human cDNA Expression Array (Clontech, CA, USA) was used to study the gene expression profile in immortalized and malignantly transformed esophageal epithelial cells. The Atlas cDNA Expression Array membrane came in duplicate. Each membrane contains cDNAs from 588 known genes and 13 housekeeping genes. These 588 known genes are dotted in duplicate into 6 quadrants broadly according to their function: quadrant A, for oncogenes, tumor suppressor genes and cell cycle regulator genes; quadrant B, for genes responsible for stress response, ion channel and transport, and intracellular signal transduction modulators and effectors; quadrant C, for genes related to apoptosis, DNA synthesis, repair and recombination; quadrant D, for genes coding for transcription factors and DNA binding proteins; quadrant E, for genes coding for receptors, cell surface antigens and cell adhesion molecules; quadrant F, for genes responsible for cell – cell communication.

The patterns of hybridization signals on the Atlas Expression Array membranes in SHEE30 and SHEE85 cells were analyzed. Among the 588 cDNA dotted on the membrane, the intensity of hybridization signals measured with Atlalmag software were different in 21 cDNA, corresponding to 39 undifferential genes, in both membranes. Fifteen up – regulated genes were expressed at a level 2 folds higher in SHEE85 cells than that in SHEE30 epithelial cells and six down – regulated genes cDNA dotted on the array hybridized SHEE85 cells at least two folds lower than SHEE30 cells. Overexpression was found for 15 genes including *CDK*6, *CDKN2D*, *K*18, *K*19, *CASP*9, *RFC*38, *DSH*, *LAMR*1, *cytohesin* – 1, *Ezrin*, *tPA*, *NDKA*, *PuF*, *ras* – *related* C_3 *and JUP*, whereas six genes were down – regulated including *HHR*23*A*, *CAPR*, *BCGF*1, *ET*2, 1*L* – 1 and transmembrane protein sex precursor.

The major difference in gene expression between immortalized (SHEE30) and malignantly transformed (SHEE85) cells was in relation to cell adhesion and invasion, which were expressed at a higher level in malignant transformed cells than immortalized cells. The progressive accumulation of genetic changes in SHEE85 parallels the histopathologic progression from immortalization to malignant change. They may more accurately reflect cell phenotype, which would display in invasive to surround tissues and metastasized to distal organ. Therefore, these data also revealed that such quantitative profiling of gene expression would be important in understanding the relationship between cell phenotype and the expression profile of a specific gene's set. The accumulation of specific gene expression patterns in malignant transformed cells points to the potential utility of gene expression profiling in the identification of malignant markers for molecular diagnosis, which provides a molecular approach to assist in diagnosis. Analyzing variations of gene expression profiles in different stages of carcinogenesis can be useful for diagnosis, therapeutics and prognosis.

4. The promoters effect on the malignant transformation in the initial esophageal epithelial cells

We have shown that the induction of malignant transformation of esophageal epithelial cells (SHEE) by E6E7 genes of HPV type 18 may occur over a long time (four years) after the initiator has been applied. But malignant transformation of SHEE appears to be a rapidly occurring phenomenon requiring minimal doses of promoter agents. In this experiment, a single application of E6E7 genes of HPV as a carcinogen (called an initiator) resulted, however, in the development of tumors when the application was followed by repeated treatment with chemical, irradiation and biological agents (called a promoter). It is of interest to note that morphologic and biochemical changes have been detected in the initiating and promoter stages.

As already stated, a noncarcinogenic agent is able to induce tumors when applied repeatedly several weeks after transferring HPV18E6E7 genes. When applied only a few times, however, the effect of promotion is incomplete, and if application of the promoting agent is permanently discontinued, the promotion phase appears to be reversible. Promoting agents are unable to induce tumors by themselves, such as the TPA and the sodium butyrate given prior to initial cells, which were unable to induce tumors. On the other hand, when given to immortal cells, TPA in two weeks or sodium butyrate in 14 weeks induces tumors.

(1) *Human papillomavirus in synergy with TPA induced the formation of malignant transforma-*

tion of esophageal epithelial cells[114] : In order to investigate the role of promoters in the formation of esophageal carcinoma, cultured cells SHEE were exposed to media with TPA (5ng/ml) for 2 weeks at 10th and 13th passages (TPA groups). Control group (SHEE) was cultured in the same media without TPA. The morphological phenotype of transformation was assessed by microscopy (including light –, electron – and fluorescent – microscope). DNA content and cell cycle were detected by flow cytometry. Morphologically transformed foci were assayed by plating 10^3 cells at passage 10 on 35mm soft – agar dishes (5 dishes for each group). The tumorigenicity was assayed in nude mice, which were injected with 10^6 cells/mouse subcutaneously in 6 animals of each group. HPV18E6E7 gene was detected by PCR and FISH.

TPA group showed that DNA synthesis and the proliferative index increased more than that of control group. The polyploid of DNA in TPA group (5.70%) was more than that in the control group (1.53%). Scoring foci in soft agar dishes, large colonies (as positively transformed foci) with dense, multiplayer cells were found more frequently in the TPA group (4.0%) and less frequently in the control group (0.1%). Tumorigenesis was observed in all six nude mice of the TPA group and non – tumorigenic in the control group. E6E7 gene was found in the nucleus of two cell groups by FISH. The malignant transformation of human embryonic esophageal epithelial cells was induced in vitro by HPV18E6E7 in synergy with TPA.

(2) *Human papillomavirus in synergy with* ^{60}*Co bablt radiation promotes malignant transformation of esophageal epithelial cells*[115] : Cultured esophageal epithelial cells (SHEE) were transfected with HPV18E6E7 AAV and propagated for 13 passages. The transfected cells were then divided into four groups. Three experimental groups were exposed to ^{60}Co radiation (2, 4, 8Gy respectively) once a week for 4 weeks at the 13th passage of the SHEE cells and one control group was cultured in the same medium without radiation. The morphology of the cells was observed under phase – contrast microscope; the DNA content during the cell cycle synthesis was analyzed by flow cytometry; the modal number of chromosomes was analyzed by routine method. The tumorigenicity was assessed by colony formation after cultivating it in soft agar and transplanting the cells into nude mice.

The results revealed that cell necrosis and the apoptosis (crisis) stages appeared after cobalt radiation. At 8 weeks after radiation, cells of the 4Gy group proliferated with a higher proliferative index (34%). Through soft – agar cultivation and injection into nude mice, the radiated SHEE cells (4Gy group) became tumorigenic. In the control group, the proliferative index was 24% and no tumor formation appeared in the nude mice. The modal number of chromosomes was 58 – 62 in the control group and 63 – 65 in the 4Gy group. The data above suggest that the malignant transformation of human embryonic esophageal epithelial cells was induced *in vitro* by HPV18E6E7 in synergy with cobalt radiation and that cobalt radiation was the promoted factor that accelerated the malignant transformation of esophageal epithelial cells.

(3) *The N – Nitrosopiperidine promotes the malignant transformation of esophageal epithelial cells*[116] : Research into the effects promoted by N – nitrosopiperidine on the carcinogenetic process was performed. It was based on the initial passage of human fetal esophageal epithelium induced by

human papillomavirus (HPV) 18E6E7 genes. The esophageal epithelium cell line SHEE was induced by HPV18E6E7. The cells at the 17th passages (SHEE17) were cultured in 50 ml flasks. N – nitrosopiperidine (NPIP) 0, 2, 4, 8 mmol/L was added to the cultured medium of SHEE cells for 3 weeks. The morphology, proliferation and apoptosis of the cells were studied by phase contrast microscopy and flow cytometry. The modal number of chromosomes was analyzed by standard method. Tumorigenicity of the cell was assessed by soft agar colony formation and by transplantation of cells into nude mice. Expression of HPV was detected by Western blot.

When cells were exposed to a high concentration (8 mmol/L) of NPIP, cell death was increased, leaving a few live cells. In normal cultural medium instead of NPIP, the proliferative status of the living cells was restored after 4 weeks and the cells progressed to the proliferation stage with continuous replication and atypical hyperplasia. At the end of the 8th week, the cells appeared with large colonies in soft – agar and tumor formation in transplanted nude mice. When the cells were cultured in 2, 4 mmol/L NPIP the doubling passage was delayed and without tumor formation in nude mice. The modal number of chromosomes was 61 – 65 in 8 mmol/L for the NPIP group. For the control group, the modal number was 56 – 61. Expression of HPV18 appeared in the experimental and control groups. NPIP promotes malignant change of the initial esophageal epithelial cells induced by HPV 18E6E7. HPV 18E6E7 in synergy with NPIP will accelerate malignant transformation in esophageal epithelium.

(4) *Sodium butyrate promotes the malignant transformation of immortalized esophageal epithelial cells*[117,118] : The promoter effects of sodium butyrate, in high or low dosages, on the carcinogenetic process were studied, based on the immortalization of human fetal esophageal epithelial cells induced by human papillomavirus (HPV) 18E6E – genes. The immortalized esophageal epithelium SHEE30 was treated with a high concentration of sodium butyrate (80 mmol/L) and then with a low concentration (5 mmol/L) for 8 weeks respectively. The cells were cultured continuously without sodium butyrate for 14 weeks. The morphology, proliferation and apoptosis of the cells were studied by phase contrast microscopy, immunohistochemistry and flow cytometry. The dead and the viable cells were assayed by fluorescent microscopy with Hoechst 33342 and Propidium Iodide staining. Tumorigenesis of the cells was assessed by soft agar colony formation and by transplantation of cells into nude mice and SCID mice.

When cells were exposed to a high concentration of sodium butyrate, cell death was increased leaving few live cells. When cells were cultured in the medium with a low concentration of sodium butyrate, the first proliferative stage appeared. Removal of the sodium butyrate from the cultural medium caused the cells to enter a crisis stage with a long doubling time resembling senescent cells. After the crisis stage, the cells progressed to the second proliferation stage with continuous replication and atypical hyperplasia. At the end of the second proliferative stage, carcinogenesis of the cells appeared with large colonies in soft – agar and tumor formation in transplanted SCID mice and nude mice. The malignant change of the immortalized epithelium caused by the effects of sodium butyrate is the consequence of a two – stage mortality mechanism: cell death by butyrate cytotoxic activity and cell crisis by abrogation of sodium butyrate. These data reveal that in high dosages, sodium butyrate induces cell death and in low dosages, it induces cell proliferation, which emphasizes the importance of sodium butyrate as a promoter of carcinogenesis.

5. Conclusion

In summary, we have developed a malignant transformed cell model to facilitate the study of carcinogenesis in esophageal carcinoma *in vitro* through immortalization of human fetal esophageal epithelium induced with HPV18E6E7. The multistep process of malignant transformation consists of initial (transduction of E6E7 genes), preimmortalized (cell senescence and crisis), immortalized (activity of telomerase), premalignant (biphase differentiation) and malignant transformation (tumor formation in immunodeficient mice).

Our data indicated that genes E6/E7 of the HPV18 were capable of inducing immortalization in fetal esophageal epithelial cells. The immortal phenotype requires both activation of telomerase and genetic alterations that abrogate normal differentiation and promote cellular proliferation. When the immortalized cell line continues to culture over 85 passages, the cell line will result in malignant transformation through multistage. This cell line can assist us in characterizing the role played by HPV in carcinogenesis.

The mechanisms of the malignant transformation of SHEE may require a synergistic effect generated by the coexistence of these factors, such as promoters: TPA, NPIP, irradiation and sodium butyrate. While HPV genes probably bind to DNA molecules, promoting substances appear to be able to trigger cell proliferation and enable the promotion of malignant changes.

The process of immortalization and malignant change of esophageal epithelium induced by HPV18 E6E7 and promoters was summarized in the Fig. 11.

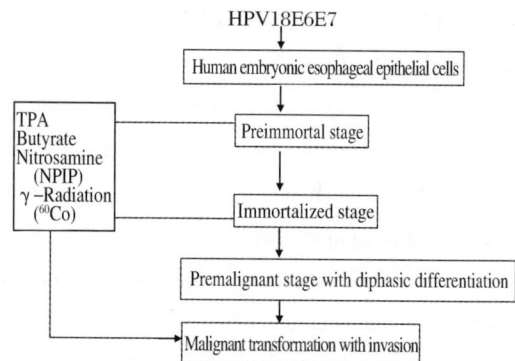

Fig. 11 **The process of immortalization and malignant change of esophageal epithelium is induced by HPV18E6E7 and synergy with promoters**

〔In 《New Research on Esophageal Cancer》 2007, : 95 - 130〕

References

1 Li JY. Epidemiology of esophageal cancer in China. Natl Cancer Inst Monogr, 1982, 62: 113 - 120

2 Pera M, Manterola C, Vidal O, Grande L. Epidemiology of esophageal adenocarcinoma. J Surg Oncol, 2005, 92: 151 - 159

3 DeMeester SR. Adenocarcinoma of the esophagus and cardia: a review of the disease and its treatment. Ann Surg Oncol, 2006, 13: 12 - 30

4 Faivre J, Lepage C, Bouvier AM. Recent data on the epidemiology of esophageal cancer. Gastroenterol Clin Biol, 2005, 29: 534 - 539

5　Badar F, Anwar N, Mahmood S. Geographical variation in the epidemiology of esophageal cancer in Pakistan. Asian Pac J Cancer Prev, 2005, 6: 139 – 142

6　Turkdogan MK, Akman N, Tuncer 1, Uygan I, Kosem M, Ozel S, Kara K, Bozkurt S, Memik F. Epidemiological aspects of endemic upper gastrointestinal cancers in eastern Turkey. Hepatogastroenterology, 2005, 52: 496 – 500

7　Kim WH, Song SI, Kang JK, Park SI, Chai HJ, Moon JG, Hang IK. Detection of human papillomavirus (HPV) DNA in esophageal cancer using polymerase chain reaction (PCR). Korean J Gastroanterol, 1992, 24: 207 – 215

8　Hendricks D, Parker MI. Oesophageal cancer in Africa. IUBMB Life, 2002, 53: 263 – 268

9　Lu JB, Yang WX, Liu JM, Li YS, Qin YM. Trends in morbidity and mortality for oesophageal cancer in Linxian County, 1959 – 1983. Int J Caner, 1985, 36: 643 – 645

10　Li JY, Liu BQ, Li GY, Chen ZJ, Sun XI, Rong SD. Atlas of cancer mortality in the People's Republic of China. An aid for cancer control and research, Int J Epidemiol, 1981, 10: 127 – 133

11　Chen WS, Chen SR, Lin QH, Qiu JW, Xie XK, Fu LN, Li WM, Liao DD, Chen JH, Wu ZR. Epidemiologic investigation of carcinoma of the esophagus and gastric cardia carcinoma in Nanao County, Guangdong province. Zhonghua Zhong Liu Za Zhi, 1986, 8: 265 – 267

12　Lin K, Shen WY, Shen ZY, Wu YN, Lu SS. Dietary exposure and urinary excretion of total N – nitroso compounds, nitrosamino acids and volatile nitrosamine in inhabitants of high – and low – risk areas for esophageal cancer in southern China. Int J Cancer, 2002, 102: 207 – 211

13　Wemli KJ, Fitzgibbons ED, Ray RM, Gao DL, Li W, Seixas NS, Camp JE, Astrakianakis G, Feng Z, Thomas DB, Checkoway H. Occupational risk factors for esophageal and stomach cancers among female textile workers in Shanghai, China. Am J Epidemiol, 2006, 163: 717 – 725

14　Shen ZY, Chen ZP, Lu SX. Investigation on nitrosamines in the diets of the inhabitants of high – risk area for esophageal cancer in the southern china and analysis of the correlation factors (in Chinese). J Hygiene Res, 1987, 26: 266 – 269

15　Garavello W, Negri E, Talamini R, Levi F, Zambon P, Dal Maso L, Bosetti C, Franceschi S, La Vecchia C. Family history of cancer, its combination with smoking and drinking, and risk of squamous cell carcinoma of the esophagus. Cancer Epidemiol Biomarkers Prey, 2005, 14: 1390 – 1393

16　De Stefani E, Boffetta P, Deneo – Pellegrini H, Ronco AL, Correa P, Mendilaharsu M. The role of vegetable and fruit consumption in the aetiology of squamous cell carcinoma of the oesophagus: a case – control study in Uruguay. Int J Cancer, 2005, 116: 130 – 135

17　Siassi F, Ghadirian P. Riboflavin deficiency and esophageal cancer: a case control – household study in the Caspian Littoral of Iran. Cancer Detect Prev, 2005, 29: 464 – 469

18　Shen WY, Shen ZY, Chen MH. A multivariable discriminant analysis on the trace elements content of hair in the districts of high, middle and low incidence of the esophageal cancer. World Elemental Med, 1997, 4: 5 – 8

19　Chang F, Syrjanen S, Wang L, Syrjanen K. Infectious agents in the etiology of esophageal cancer. Gastroenterology, 1992, 103: 1336 – 1348

20　Sakata K, Hoshiyama Y, Morioka S, Hashimoto T, Takeshita T, Tamakoshi A, JACC Study Group. Smoking, alcohol drinking and esophageal cancer: findings from the JACC study. J Epidemiol, 2005, 2: S212 – 219

21　Lin K, Yu SJ, Zhang JJ, Wu YN, Zhang Q, Tan X, Luo J. Study on N – nitroso compound in food and its relevant risk factors for esophageal cancer. Wei Sheng Yan Jiu, 2005, 34: 350 – 352

22　Syrjanen K. HPV infections of the oesophagus. In: Syrjanen K, Syrjanen S. Papillomavirus infections in human pathology. New York: J Wiley

& Sons, 2000, 413 – 428

23 Sur M, Cooper K. The role of the human papilloma virus in esophageal cancer. Pathology, 1998, 30: 348 – 354

24 Franceschi S, Munoz N, Bosch XF, Snijders PJ, Walboomers JM. Human papillomavirus and cancers of the upper aerodigestive tract: a review of epidemiological and experimental evidence. Cancer Epidemiol Biomarkers Prey, 1996, 5: 567 – 575

25 Syrianen KJ. Histological changes identical to those of condylomatous lesions found in esophageal squamous cell carcinomas. Arch Geschwulstforsch, 1982, 52: 283 – 292

26 Chang F, Syrjanen S, Shen Q, Wang L, Syrjanen K. Screening for human papillomavirus infections in esophageal squamous cell carcinomas by in situ hybridization. Cancer, 1993, 72: 2525 – 2530

27 Poljak M, Cerar A, Seme K. Human papillomavirus infection in esophageal carcinomas: a study of 121 lesions using multiple broad – spectrum polymerase chain reactions and literature review. Hum Patholm, 1998, 29: 266 – 271

28 Dillner J, Knekt P, Schiller JT, Hakulinen T. Prospective seroepidemiological evidence that human papillomavirus type 16 infection is a risk factor for oesophageal squamous cell carcinoma. Brit Med J, 1995, 311: 1346

29 Syrjanen KJ. HPV infections and oesophageal cancer. J Clin Pathol, 2002, 55: 721 – 728

30 Chen B, Yin H, Dhurandhar N. Detection of human papillomavirus DNA in esophageal squamous cell carcinomas by the polymerase chain reaction using general consensus primers. Hum Pathol, 1994, 25: 920 – 923

31 Morgan RJ, Perry AC, Newcomb PV, Hardwick RH, Alderson D. Human papillomavirus and oesophageal squamous cell carcinoma in the UK. Eur J Surg Oncol, 1997, 23: 513 – 517

32 Saegusa M, Hashimura M, Takano Y, Ohbu M, Okayasu I. Absence of human papillomavirus genomic sequences detected by the polymerase chain reaction in oesophageal and gastric carcinomas in Japan. Mol Pathol, 1997, 50: 101 – 104

33 Chang F, Syrjanen S, Shen Q, Cintorino M, Santopietro R, Tosi P, Syrjanen K. Human papillomavirus involvement in esophageal carcinogenesis in the high – incidence area of China. A study of 700 cases by screening and type – specific in situ hybridization. Scand J Gastroenterol, 2000, 35: 123 – 130

34 Wang XJ, Wang CJ, Wang XH, Tao DM, Shao HY. Relationship between human papillomavirus infection and the development of esophageal carcinoma. Zhongguo Zhongliu Lingchuang, 1996, 23: 761 – 763

35 Li T, Lu ZM, Chen KN, Guo M, Xing HP, Mei Q, Yang HH, Lechner JF, Ke Y. Human papillomavirus type 16 is an important infectious factor in the high incidence of esophageal cancer in Anyang area of China. Carcinogenesis, 2001, 22: 929, 934

36 Shen ZY. Human papillomavirus and esophageal cancer in Shantou area. In Li CH (ed) Current advances in tumor biology. MMs, Beijing, 1997, 16 – 22

37 Shen ZY, Hu SP, Lu LC, Tang CZ, Kuang ZS, Zhong SP, Zeng Y. Detection of human papillomavirus in esophageal carcinoma. J Med Virol, 2002, 68: 412 – 416

38 de Villiers EM. Human Papillomavirus. Introduction. Semin Cancer Biol, 1999, 9: 377

39 de Villiers EM, Lavergne D, Chang F, Syrjanen K, Tosi P, Cintorino M, Santopietro R, Syrjanen S. An interlaboratory study to determine the presence of human papillomavirus DNA in esophageal carcinoma from China. Int J Cancer, 1999, 81: 225 – 228

40 Lu LZ, Shen ZY, You SJ, Shen J, Cai WJ. Detection of human papillomavirus (HPV) on esophageal carcinoma by PCR and insitu hybridization. Aizheng, 1999, 18: 162 – 164

41 Zhou XB, Quan AP, Zhang W, Lu ZM, Wang QH, Ke Y, Xu NZ. Detection of human papillomavirus in Chinese esophageal squamous cell car-

cinoma and its adjacent normal epithelium. World J Gastroenterol, 2003, 9: 1170 – 1173

42　Xu CL, Qian XL, Zhou XS, Zhao QZ, Li YC. Expression of HPV16 – E6 and E7 oncoproteins in squamous cell carcinoma tissues of esophageal cancer and non – cancer tissues. Ai Zheng, 2004, 23: 165 – 168.

43　Rubio CA. Experimental models. In: Editor De-Meester TR, Cancer of the esophagus, Orlando, Grune&Stratton, 1985, 21

44　Chang F, Syrjanen S, Shen Q, Wang L, Wang D, Syrjanen K. Human papillomavirus (HPV) involvement in esophageal precancerous lesions and squamous cell carcinomas as evidenced by microscopy and different DNA techniques. Scand J Gastroenterol, 1992, 27: 553 – 563

45　Kim WH, Chang JP, Song SY, Chen CY, Lee SI, Kang JK, Park IS, Choi HJ, Kim KH, Kim YS, Park YN, Park HI. Detection of human papillomavirus in paraffin – embedded esophageal cancer tissue. Korean J Gastroenterol, 1992, 24: 427 – 435

46　Woo YJ, Yoon HK. In situ hybridization study on human papillomavirus DNA expression in benign and malignant squamous lesions of the esophagus. J Korean Med Sci, 1996, 11: 467 – 473

47　Chang F, Syrjanen S, Shen Q, Ji HX, Syrjanen K. Human papillomavirus (HPV) DNA in esophageal precancer lesions and squamous cell carcinomas from China. Int J Cancer, 1990, 45: 21 – 25

48　Kao PC, Vecchio JA, Schned LM, Blaszyk H. Esophageal squamous papillomatosis. Eur J Gastroenterol Hepatol, 2005, 17: 1233 – 1237

49　Szanto I, Szentirmay Z, Banai J, Nagy P, Gonda G, Voros A, Kiss J, Bajtai A. Squamous papilloma of the esophagus. Clinical and pathological observations based on 172 papillomas in 155 patients. Orv Hetil, 2005, 146: 547 – 552

50　Syrjanen K, Pyrhonen S, Aukee S, Koskela E. Squamous cell papilloma of the esophagus: a tumour probably caused by human papilloma virus (HPV). Diagn Histopathol, 1982, 5: 291 –

296

51　Shen J, Shen ZY, Zheng RM, Lu LC. The study of micropapillomatosis in adjacent mucosae near esophageal cancinoma. Shijie Huaren Xiaohua Zazhi, 2000, 8: 1289 – 1290

52　Winkler B, Capo V, Reumann W, Ma A, La Porta R, Reilly S, Green PM, Richart RM, Crum CP. Human papillomavirus infection of the esophagus. A clinicopathologic study with demonstration of papillomavirus antigen by the immunoperoxidase technique. Cancer, 1985, 55: 149 – 155

53　Unckell F, Streeck RE, Sapp M. Generation and neutralization of pseudovirions of human papillomavirus type 33. J Virol, 1997, 71: 2934 – 2939

54　Kimbauer R, Booy F, Cheng N, Lowy DR, Schiller JT. Papillomavirus L1 major capsid protein self – assembles into virus – like particles that are highly immunogenic. Proc Natl Acad Sci USA, 1992, 89: 12180 – 12184

55　Kimbauer R, Taub J, Greenstone H, Roden R, Durst M, Gissmann L, Lowy DR, Schiller JT. Efficient self – assembly of human papillomavirus type 16 L1 and L l – L2 into virus – like particles. J Virol, 1993, 67: 6929 – 6936

56　Roden R, Greenstone HL, Kirnbauer R, Booy FP, Jessie J, Lowy DR, Schiller JT. In vitro generation and type – specific neutralization of a human papillomavirus type 16 virion pseudotype. J Virol, 1996, 70: 5875 – 5883

57　Joyce JG, Tung JS, Przysiecki CT, Cook JC, Lehman ED, Sands JA, Jansen KU, Keller PM. The L1 major capsid protein of human papillomavirus type 11 recombinant virus – like particles interacts with heparin and cell – surface glycosaminoglycans on human keratinocytes. J Biol Chem, 1999, 274: 5810 – 5822

58　Shukla D, Spear PG. Herpesviruses and heparan sulfate: an intimate relationship in aid of viral entry. J Clin Invest, 2001, 108: 503 – 510

59　Giroglou T, Florin L, Schafer F, Streeck RE, Sapp M. Human papillomavirus infection requires cell surface heparan sulfate. J Virol, 2001, 75: 1565 – 1570

60 Shafti – Keramat S, Handisurya A, Kriehuber E, Meneguzzi G, Slupetzky K, Kimbauer R. Different heparan sulfate proteoglycans serve as cellular receptors for human papillomaviruses. J Virol, 2003, 77: 13125 – 13135

61 Roden RB, Weissinger EM, Henderson DW, Booy F, Kimbauer R, Mushinski JF, Lowy DR, Schiller JT. Neutralization of bovine papillomavirus by antibodies to L1 and L2 capsid proteins. J Virol, 1994, 68: 7570 – 7574

62 Kawana K, Matsumoto K, Yoshikawa H, Taketani Y, Kawana T, Yoshiike K, Kanda T. A surface immunodeterminant of human papillomavirus type 16 minor capsid protein L2. Virology, 1998, 245: 353 – 359

63 Inki P, Larjava H, Haapasalmi K, Miettinen HM, Grenman R, Jalkanen M. Expression of syndecan – 1 is induced by differentiation and suppressed by malignant transformation of human keratinocytes. Eur J Cell Biol, 1994, 63: 43 – 51

64 Yoon CS, Kim KD, Park SN, Cheong SW. Alpha (6) integrin is the main receptor of human papillomavirus type 16 VLP. Biochem Biophys Res Commun, 2001, 283: 668 – 673

65 Hodivala KJ, Pei XF, Liu QY, Jones PH, Rytina ER, Gilbert C, Singer A, Watt FM. Integrin expression and function in HPV 16 – immortalised human keratinocytes in the presence or absence of v – Ha – ras. Comparison with cervical intraepithelial neoplasia. Oncogene, 1994, 9: 943 – 948

66 Culp TD, Budgeon LR, Christensen ND. Human papillomaviruses bind a basal extracellular matrix component secreted by keratinocytes which is distinct from a membrane – associated receptor. Virology, 2006, 347: 147 – 159

67 Day PM, Lowy DR, Schiller JT. Papillomaviruses infect cells via a clathrin dependent pathway. Virology, 2003, 307: 1 – 11

68 McMurray HR, Nguyen D, Westbrook TF, McAnce DJ. Biology of human papillomaviruses, Int J Exp Pathol, 2001, 82: 15 – 33

69 Demers GW, Halbert CL, Galloway DA. Elevated wild – type p53 protein levels in human epithelial cell lines immortalized by the human papillomavirus type 16 E7 gene. Virology, 1994, 198: 169 – 174

70 Boyer SN, Wazer DE, Band V. E7 protein of human papilloma virus – 16 induces degradation of retinoblastoma protein through the ubiquitin – proteasome pathway. Cancer Res, 1996, 56: 4620 – 4624

71 Zur Hausen H. Papillomaviruses in human cancers. Proc Assoc Am Physicians, 1999, 111: 581 – 587

72 Sashiyama H, Shino Y, Kawamata Y, Tomita Y, Ogawa N, Shimada H, Kobayashi S, Asano T, Ochiai T, Shirasawa H. Immortalization of human esophageal keratinocytes by E6 and E7 of human papillomavirus type 16. Int J Oncol, 2001, 19: 97 – 103

73 Li TT, Zhao LN, Liu ZG, Han Y, Fan DM. Regulation of apoptosis by the papillomavirus E6 oncogene. World J Gastroenterol, 2005, 11 (7): 931 – 937

74 Ueno T, Sasaki K, Yoshida S, Kajitani N, Satsuka A, Nakamura H, Sakai H. Molecular mechanisms of hyperplasia induction by human papillomavirus E7 Oncogene, 2006, 25: 4155 – 4164

75 Akerman GS, Tolleson WH, Brown KL, Zyzak LL, Mouravteva E, Engin TS, Basaraba A, Coker AL, Creek KE, Pirisi L. Human papillomavirus type 16 E6 and E7 cooperate to increase epidermal growth factor receptor (EGFR) mRNA levels, overcoming mechanisms by which excessive EGFR signaling shortens the life span of normal human keratinocytes. Cancer Res, 2001, 61: 3837 – 3843

76 Sunpaweravong P, Sunpaweravong S, Puttawibul P, Mitamun W, Zeng C, Baron AE, Franklin W, Said S, Varella – Garcia M. Epidermal growth factor receptor and cyclin D I are independently amplified and overexpressed in esopha-

geal squamous cell carcinoma. J Cancer Rev Clin Oncol, 2005, 131: 111 – 119

77　Hanawa M, Suzuki S, Dobashi Y, Yamane T, Kono K, Enomoto N, Ooi A. EGFR protein over-expression and gene amplification in squamous cell carcinomas of the esophagus, Int J Cancer, 2006, 118: 1173 – 1180

78　Oh ST, Kyo S, Laimins LA. Telomerase activation by human papillomavirus type 16 E6 protein" induction of human telomerase reverse transcriptase expression through Myc and GC – rich Spl binding sites. J Virol, 2001, 75: 5559 – 5566

79　Veldman T, Horikawa I, Barrett JC, Schlegel R. Transcriptional activation of the telomerase hTERT gene by human papillomavirus type 16 E6 oncoprotein. J Virol, 2001, 75: 4467 – 4472

80　Zhu HX, Wang YH, Zhou CQ, Zhang G, Bai JF, Quan LP, Lin ZH, Xu NZ. Expression of survivin and its significance in esophageal cancer. Zhonghua Zhong Liu Za Zhi, 2005, 27: 22 – 24

81　Borbely AA, Murvai M, Konya J, Beck Z, Gergely L, LiF, Veress G. Effects of human papillomavirus type 16 oncoproteins on survivin gene expression. J Gen Visol, 2006, 87: 287 – 294

82　Kowalczyk AM, Roeder GE, Green K, Stephens DJ, Gaston K. Measuring the induction or inhibition of apoptosis by HPV proteins. Methods Mol Med, 2005, 119: 419 – 432

83　Stanley M. Immune responses to human papillomavirus. Vaccine, 2005, 24 Suppl1: S16 – 22

84　Fausch SC, Fahey LM, Da Silva DM, Kast WM. Human papillomavirus can escape immune recognition through Langerhans cell phosphoinositide 3 – kinase activation. J Immunol, 2005, 174: 7172 – 7178

85　Guess JC, McCance DJ. Decreased migration of Langerhans precursor – like cells in response to human keratinocytes expressing human papillomavirus type 16 E6/E7 is related to reduced macrophage inflammatory protein – 3alpha production. J Virol, 2005, 79: 14852 – 14862

86　Offringa R, de Jong A, Toes RE, van der Burg SH, Melief CJ. Interplay between human papillomaviruses and dendritic cells. Curr Top Microbiol Immunol, 2003, 276: 215 – 240

87　Kuang ZS, Tang CZ, Shen ZY. Detection of human papillomavirus type 16 and 18 in esophageal cancer. Guangdong Yixue, 2000, 21: 305 – 306

88　Shen ZY, Cen S, Shen J, Cai WJ, Xu JJ, Teng ZP, Hu Z, Zeng Y. Study of immortalization and malignant transformation of human embryonic esophageal epithelial cells induced by HPV18E6E7. J Cancer Res Clin Oncol, 2000, 126: 589 – 594

89　Shen ZY, Xu LY, Li EM, Cai WJ, Shen J, Chen MH, Cen S, Tsao SW, Zeng Y. The multistage process of carcinogenesis in human esophageal epithelial cells induced by human papillomavirus. Oncol Rep, 2004, 11: 647 – 654

90　Shen ZY, Chen XH, Shen J, Cai WH, Chen JY, Huang TFH, Zeng Y. Malignant transformation of immortalized human embryonic esophageal epithelial cells induced by human papillomavirus. Bingdu Xuebao, 2000, 16: 97 – 101

91　Zhang YJ, Shen ZY, Guo XC, Zhao J, Zeng Y. Transformation of fetal esophageal epithelial cells by a recombinant adenoassociated virus containing the E6E7 ORFs of human papillomavirus type 16. Bingdu Xuebao, 2003, 19: 1 – 4

92　Cen S, Teng ZP, Zhang Y, Shen ZY, Xu JJ, Du B, Zeng Y. Construction of human papilloma virus type 18 E6E7 genes in adenoassociated virus expression vector and checking its activity for malignant transformation. Zhonghua Shiyan He Linchuang Bingduxue Zazhi, 2003, 17: 5 – 9

93　Shen ZY, Cen S, Xu LY, Cai WJ, Chen MH, Shen J, Zeng Y. E6/E7 genes of human papilloma virus type 18 induced immortalization of human fetal esophageal epithelium. Oncol Rep, 2003, 10: 1431 – 1436

94　Shen ZY, Shen J, Cai WJ, Cen S, Zeng Y. Biological characteristics of human fetal esophageal epithelial cell line immortalized by the E6 and

E7 gene of HPV type 18. Zhonghua Shiyan He Linchuang Bingduxue ZazhL, 1999, 13: 209 – 212

95　Shen ZY, Cen S, Cai WJ, Teng ZP, Shen J, Hu Z, Zeng Y. Immortalization of human fetal esophageal epithelial cells induced by E6 and E7 genes of human papilloma virus 18. Zhonghua Shiyan He Linchuang Bingduxue Zazhi, 1999, 13: 121 – 123

96　Shen ZY, Xu LY, Li EM, Shen J, Zheng RM, Cai WJ, Zeng Y. Immortal phenotype of the esophagealepithelial cells in the process of immortalization, Int J Mol Med, 2002, 10: 641 – 646

97　Shen ZY, Xu LY, Chen MH, Cai WJ, Shen J, Chen JY, Hon CQ, Zeng Y. Biphasic differentiation of immortalized esophageal epitheliums induced by HPV 18 E6E7. Bingdu Xuebao, 2001, 17: 210 – 214

98　Shen ZY, Xu LY, Chen MH, Shen J, Cai WJ, Zeng Y. Progressive transformation of immortalized esophageal epithelial cells. World J Gastroenterol, 2002, 8: 976 – 981

99　Shen ZY, Shen J, Cai WJ, Chen JY, Zeng Y. Malignant transformation of the immortalized esophageal epithelial cells. Chin Y Oncol, 2002, 24: 107 – 109

100　Shen J, Chen MH, Zheng RM, Chen JJ, Shen ZY. Detection of scanning electron microscope on invasion of tumor cells in vitro. J Chin Electr Microscsoc, 2000, 19: 313 – 314

101　Shen ZY, Xu LY, Chen MH, Li EM, Li JT, Wu XY, Zeng Y. Upregulated expression of Ezrin and invasive phenotype in malignantly transformed esophageal epithelial cells. World Y Gastroenterol, 2003, 9: 1182 – 1186

102　Xu LY, Li EM, Shen ZY, Cai WJ, Shen J. A nonradio – labelled method assays to measurethe telomere length of human chromosome. Aibian Qibian Tubian, 2001, 13: 1 – 4

103　Shen ZY, Xu LY, Li C, Cai WJ, Shen J, Chen JY, Zeng Y. A comparative study of telomerase activity and malignant phenotype in multistage carcinogenesis of esophageal epithelial cells induced by human papillomavirus. Int J Mol Med, 2001, 8: 633 – 639

104　Shen ZY, Xu LY, Li EM, Cai WJ, Chen MH, Shen J, Zeng Y. Telomere and telomerase in the initial stage of immortalization of esophageal epithelial cell. Worm Y Gastroenterol, 2002, 8: 357 – 362

105　Xu LY, Shen ZY, Li EM, Cai WJ, Shen J, Li C, Hong CQ, Chen JY, Zeng Y. Telomere length and telomerase activity in immortalized and malignantly transformed human embryonic esophageal epithelial cell lines by E6 and E7 genes of HPV 18 type. Carcinogenesis Teratogenesis and Mutagenesis, 2001, 13: 137 – 140

106　Shen ZY, Xu LY, Chen MH, Cai WJ, Shen J, Ceng JY, Zeng Y. Cytogenetic and molecular genetic changes in malignant transformation of immortalized esophageal epithelial cells. Int J Mol Med, 2003, 12: 219 – 224

107　Lee JM, Shun CT, Wu MT, Chen YY, Yang SY, Hung HI, Chen JS, Hsu HH, Huang PM, Kuo SW, Lee YC. The associations of p53 overexpression with p53 codon 72 genetic polymorphism in esophageal cancer. Mutat Res, 2006, 594: 181 – 188

108　Tsuneoka M, Teye K, Arima N, Soejima M, Otera H, Ohashi K, Koga Y, Fujita H, Shirouzu K, Kimura H, Koda Y. A novel Myc – target gene, mimitin, that is involved in cell proliferation of esophageal squamous cell carcinoma. J Biol Chem, 2005, 280: 19977 – 19985

109　McCabe ML, Dlamini Z. The molecular mechanisms of oesophageal cancer, Int Immunopharmacol, 2005, 5: 1113 – 30

110　Shen ZY, Xu LY, Chen XH, Zeng Y. The genetic events of HPV – immortalized esophageal epithelium cells, Int J Mol Med, 2001, 8: 537 – 542

111　Chen XH, Huang TH, Shen ZY, Huang JM, Shuai ZB. Cytogenetic studies on HPV – 18 –

immortalized cells from esophageal epithelia of human embryo. Acta Anatomica Sinica, 2002, 33: 278 - 282

112 Hu N, Wang C, Hu Y, Yang HH, Giffen C, Tang ZZ, Han XY, Goldstein AM, Emmert Buck MR, Buetow KH, Taylor PR, Lee MP. Genome - wide association study in esophageal cancer using GeneChip mapping 10K array. Cancer Res, 2005, 65: 2542 - 6

113 Kan T, Yamasaki S, Kondo K, Teratani N, Kawabe A, Kaganoi J, Meltzer SJ, Imamura M, Shimada Y. A new specific gene expression in squamous cell carcinoma of the esophagus detected using representational difference analysis and cDNA microarray. Oncology, 2006, 70: 25 - 33

114 Shen ZY, Cai WJ, Shen J, Xu JJ, Cen S, Teng ZP, Hu Z, Zeng Y. Human papilloma virus 18E6E7 in synergy with TPA induced malignant transformation of human embryonic esophageal epithelial cells. Bingdu Xuebao, 1999, 15: 1 - 6

115 Shen ZY, Cen S, Teng ZP, Cai WJ, Shen J, Chen JY, Chen ZJ, Li DR, Zeng Y. Human

papillomavirus in Synergy with ^{60}Co obalt radiation promotes malignant transformation of esophageal epithelial cells. Chin J Virol, 2004, 20: 225 - 229

116 Shen ZY, Teng ZP, Shen J, Cai WJ, Chen MH, Qin S, Chen JY, Zeng Y. The promotive effects of N - nitrosopiperidine on the malighant transformation of the immortalized esophageal epithelium induced by human papillomavirus. Zhonghua Shiyan He Linchuang Bingduxue Zazhi, 2006, 20: 81 - 83

117 Shen ZY, Chen MH, Cai WJ, Shen J, Chen JY, Hong CQ, Zeng Y. The effects of sodium butyrate on proliferation, differentiation and apoptosis in immortalized esophageal epithleial cells. Zhonghua Binglixue Zazhi, 2001, 30: 121 - 124

118 Shen ZY, Shen J, Cai WJ, Wu XY, Zheng Y. The promtor effects of malignant transformation of sodium butyrate on the immortalized esophageal epithelium induced by human papillomavirus. Zhonghuan Binglixue Zazhi, 2002, 31: 327 - 330

104. GSTM1 and GSTT1 Gene Deletions and the Risk for Nasopharyngeal Carcinoma in Han Chinese

GUO Xiu-chan[1,3], O'BRIEN Stephen J[2], ZENG Yi[1], NELSON George W[3], WINKLER Cheryl A[3]

1. State Key Laboratory for Infectious Diseases Prevention and Control, Institute for Viral Disease Control and Prevention, Chinese CDC; 2. Laboratory of Genomic Diversity, Center for Cancer Research, National Cancer Institute, NIH; and 3. Basic Research Program, SAIC - Frederick, Inc. , National Cancer Institute at Frederick, Frederick, Maryland

Summary

Southern China is a major nasopharyngeal carcinoma endemic region. Environmental factors and genetic susceptibility contribute to nasopharyngeal carcinoma development in this area. Polymorphic deletions of GSTM1 and GSTT1 genes involved in the detoxification of potentially carcinogenic agents may be a risk factor for nasopharyngeal carcinoma. To investigate the roles of genetic variations of GSTM1 and GSTT1 in nasopharyngeal carcinoma susceptibility in the Chinese population, we conducted a case – control study of 350 nasopharyngeal carcinoma cases and 622 controls. GSTM1 and GSTT1 deletion variants were genotyped by multiplex PCR assays. Logistic regression analysis was used to estimate odds ratios and 95% confidence intervals (95% CI). No significant association was observed for either GSTM1 – or GSTT1 – null genotype independently in the contribution to nasopharyngeal carcinoma risk. To explore possible joint effects of the GSTM1 – and GSTT1 – null polymorphisms with each other and with other risk factors for nasopharyngeal carcinoma, we examined the association between each combined genotype and the risk for nasopharyngeal carcinoma stratified by gender and EBV replication status. We found that individuals who carried GSTM1/GSTF1 – double null genotype had a higher risk for nasopharyngeal carcinoma in the male population (odds ratio, 1.76; 95% confidence interval, 1.04 – 2.97; $P = 0.03$); however, this was not significant after correction for multiple comparisons. No statistical difference was found between cases and controls in females and the subpopulation positive for immunoglobulin A antibodies to EBV capsid antigen for combined genotypes. Our results suggest that the GSTM1/GSTT1 – double null genotype may be a risk factor for nasopharyngeal carcinoma among males in southern China, but this result warrants confirmation in other studies.

Introduction

Nasopharyngeal carcinoma is a leading cause of cancer deaths in the Cantonese population of southern China and is the 8th cause of cancer mortality overall in China[1, 2]. Nasopharyngeal carcinoma is a fast – growing tumor characterized by a high frequency of nodal and distant metastasis at diagnosis. Nasopharyngeal carcinoma is thought to be caused by the combined effects of EBV, environmental carcinogens, and genetic susceptibility. Case – control studies have indicated a strong role for environmental factors, including traditional southern Chinese foods such as salted fish and other preserved foods containing volatile nitrosamine, which are commonly consumed in high – nasopharyngeal carcinoma incidence areas[3]. An individual's effectiveness in the detoxification of these chemicals is in part ascribed to genetic differences of metabolic activity that may influence susceptibility to malignant disease.

Glutathione S – transferases constitute a superfamily of ubiquitous, multifunctional enzymes, which play a key role in cellular detoxification. Glutathione S – transferases are widely distributed in nature and are found in essentially all eukaryotic species. GSTM1 and GSTT1 are known to be highly polymorphic. This genetic variation may change an individual's susceptibility to carcinogens and toxins, as well as affect the toxicity and efficacy of certain drugs[4]. GSTM1 is located on chromosome 1p13.3 and is a homologous recombination involving left and right 4.2 – kb repeats, resulting in a 16 – kb deletion containing the entire GSTM1 gene. GSTT1 is located at 22q11.2 and, like GSTM1, is a deletion produced by a homologous recombination event involving left and right 403 – bp repeats, resulting in a 54 – kb deletion containing the entire GSTT1 gene[5]. Homozygous deletions of these genes, referred to as GSTM1 null and GSTT1 null, respectively, result in lack of enzyme activity. Null mutations of these genes have been associated with increased risk for a number of cancers in some studies[6-11], but not in others[12-16]. Two studies have reported modest associations between nasopharyngeal carcinoma and GSTM1 and/or GSTT1 deletions; however, these studies were quite small, with less than 100 nasopharyngeal carcinoma cases[17, 18].

Here, we conducted a case – control study with 350 nasopharyngeal carcinoma cases and 622 controls to determine if deletions of the GSTM1 or GSTT1 genes are associated with nasopharyngeal carcinoma risk in southern Chinese. We also tested for the joint effects of these genes with the known nasopharyngeal carcinoma risk factors of chronic EBV replication and sex.

Materials and Methods

1. Patients and Controls: Cases and controls were recruited from an area along the Xijiang River in Guangxi province of southern China from April 2000 to June 2001. Nasopharyngeal carcinoma cases were defined with nasopharyngeal carcinoma by pathologic examination. Controls were the case's spouse or geographically matched residents who were nasopharyngeal carcinoma – flee at the time of study enrollment. Nasopharyngeal carcinoma cases were hospitalized patients at the Wuzhou Red Cross Hospital in Wuzhou City and outpatients at the Cangwu Institute for Nasopharyngeal Carcinoma Control and Prevention in Cangwu County. All participants self – identified as Han Chinese and selfreported 6 or more months of residency in Guangdong or Guangxi Province of China. Immunoglobulin A antibodies to EBV capsid antigen (EBV/IgA/VCA) and immunoglobulin A antibodies to EBV early antigen were confirmed by serologic testing at the time of study enrollment. Blood samples were obtained from 350 nasopharyngeal carcinoma cases (234 males and 116 females) and 622 controls (267 males and 355 females). The mean age was 45 ± 11 and 46 ± 10 years for nasopharyngeal carcinoma cases and controls. Internal review board approval was obtained from all participating institutions, and informed consent was obtained from each study participant.

2. Genomic DNA Extraction: DNA was extracted from whole blood or lymphoblastoid cell lines using a QIAamp DNA blood maxi kit (Qiagen; catalog 51194). More than 80% of the genotypes were determined from DNA directly extracted from whole blood.

(1) *Genotyping* GSTM1 – and GSTT1 – deletion genotypes were determined by a multiplex PCR protocol described by Arand et al[19], and results were recorded as each gene being present or absent because heterozygotes could not be determined. The GSTM1 – or GSTT1 – specific primer set and a primer set for an albumin gene fragment were used for the same amplification reaction. Forty nanograms of target DNA was amplified in total volume of $15\mu l$ PCR mixtures consisting of 10 mmol/L Tris HCl buffer, 50 mol/L KCl, 2. 0 mol/L $MgCl_2$, 0. 2 mol/L deoxynucleotide triphosphate, 3 μg/ml of each GSTM1, 1 μg/ml of each GSTF1 primer, 0. 6 μg/ml of each albumin primer, and 5 units of Taq polymerase in 96 – well plates. Thermal cycling conditions were 94℃ for 2 minutes, followed by 30 cycles at 94℃ for 1 minute, 64℃ for 1 minute, 72℃ for 1 minute, and then a final extension of 72℃ for 5 minutes. The 215 – bp GSTM1 and the 480 – bp GSTT1 fragments were coamplified with the 350 – bp albumin fragments in the same reaction. The albumin fragments served as a

positive control for the success of the amplification reaction. The absence of either *GSTM*1 or *GSTT*1 fragments indicated the corresponding null genotype. PCR products were electrophoresed on 4% agarose gel.

（2）*Statistics* All statistical analyses were carried out using SAS 9.1 software. Present or null gene frequencies were computed and compared between case and controls with the Pearson's χ^2 test or Fisher's exact test. Odd ratios, 95% confidence intervals（95% *CI*）, and *P* values were computed by logistic regression, and all results were adjusted for age. To investigate the influence of sex and EBV/IgA/VCA antibody status, we have also analyzed the associations between *GSTM*1, *GSTT*1 – null genotype, and the occurrence of nasopharyngeal carcinoma in male, female, and EBV/IgA/VCA – positive subpopulations. Because only 16 nasopharyngeal carcinoma cases were EBV/IgA/VCA seronegative, these were not included in the analysis. Joint effects between *GSTM*1 – null and *GSTT*1 – null genotypes and sex or EBV/IgA/VCA antibody status（as a biomarker of EBV replication）were tested. The *P* values presented were shown without adjustment for multiple tests. After adjustment using a Bonferroni correction for 20 independent tests, $P \leqslant 0.0025$ was considered significant.

Results

The deletion polymorphisms for *GSTM*1 and *GSTT*1 were genotyped in 350 nasopharyngeal carcinoma cases and 622 controls. Genotypes were obtained for more than 95% of the participants. Tab. 1 lists the genotype distribution of the *GSTM*1, *GSTT*1, and *GSTM*1/*GSTT*1 in the total cohort, stratified by sex. The *GSTM*1 – and *GSTT*1 – null genotypes were detected in 57.1% and 46.9% of the participants, respectively. *GSTM*1/*GSTT*1 double nulls were detected in 26.7% of the study population. There was no significant difference（P, 0.39 – 0.88）in the *GSTM*1 – null, *GSTT*1 – null, and the *GSTM*1/*GSTT*1 double null genotypes between males and females.

Tab. 1 Distribution of *GSTM*1 – , *GSTT*1 – , and *GSTM*1/*GSTT*1 – null genotypes

Genotype	Male（%）	Female（%）	Cohort（%）
*GSTM*1			
Positive	200（41.6）	199（44.2）	399（42.9）
Null	281（58.4）	251（55.8）	532（57.1）
*GSTT*1			
Positive	260（54.5）	230（51.6）	490（53.1）
Null	217（45.5）	216（48.4）	433（46.9）
*GSTM*1/*GSTT*1			
Positive/positive	108（22.6）	101（22.7）	201（22.7）
Positive/null	91（19.1）	96（21.6）	187（20.3）
Null/positive	152（31.9）	128（28.8）	280（30.4）
Null/null	126（26.4）	120（27.0）	246（26.7）

Tab. 2 presents the distribution of the *GSTM*1 – and *GSTT*1 – null genotypes in cases and controls, and the odds ratios for the association of *GSTM*1, *GSTF*1, and nasopharyngeal carcinoma. No significant difference in the frequencies of *GSTM*1 – and *GSTT*1 – null genotypes was observed between cases and controls. We stratified the analysis by sex and by EBV/IgA/VCA status to test for joint effects. No significant difference in frequencies of the *GSTM*1 – null or *GSTT*1 – null genotypes was found between cases and controls in these different subgroups（Tab. 2）.

To investigate the joint effects of *GSTM*1 – and *GSTT*1 – null genotypes, the association between each combined genotype and the risk for nasopharyngeal carcinoma was tested again, stratifying for sex and EBV/IgA/VCA status (Tab. 3). Using the GSTM1/GSTT1 double positive as a reference, a relationship between risk for nasopharyngeal carcinoma and the double – null genotype was suggested for males (odds ratio, 1.76; 95% *CI*, 1.04 – 2.97; *P* = 0.03); however, this result was not significant after correction for multiple tests.

Tab. 2 Odds ratios for the association of *GSTM*1 – and *GSTT*1 – null genotypes with nasopharyngeal carcinoma risk

Genotype	Cases (%)	Controls (%)	OR* (95% CI)*	P*
All subjects				
*GSTM*1				
Positive	137 (40.2)	262 (44.4)	1.0	
Null	204 (59.8)	328 (55.6)	1.18 (0.90 – 1.55)	0.23
*GSTT*1				
Positive	174 (51.5)	316 (54.0)	1.0	
Null	164 (48.5)	269 (46.0)	1.11 (0.85 – 1.46)	0.44
Male				
*GSTM*1				
Positive	87 (38.2)	113 (44.7)	1.0	
Null	141 (61.8)	140 (55.3)	1.28 (0.89 – 1.85)	0.18
*GSTT*1				
Positive	114 (50.2)	146 (58.4)	1.0	
Null	113 (49.8)	104 (41.6)	1.37 (0.96 – 1.98)	0.09
Female				
*GSTM*1				
Positive	50 (44.2)	149 (44.2)	1.0	
Null	63 (55.8)	188 (55.8)	0.99 (0.64 – 1.52)	0.96
*GSTF*1				
Positive	60 (54.1)	170 (50.7)	1.0	
Null	51 (45.9)	165 (49.3)	0.9 (0.58 – 1.38)	0.62
IgA/VCA +				
*GSTM*1				
Positive	130 (39.9)	120 (44.9)	1.0	
Null	196 (60.1)	147 (55.1)	1.23 (0.89 – 1.71)	0.22
*GSTT*1				
Positive	65 (51.1)	145 (54.5)	1.0	
Null	158 (48.9)	121 (45.5)	1.15 (0.83 – 1.59)	0.41

Abbreviation; OR, odds ratio.

* Adjusted for age but not for multiple comparisons

Tab. 3 Joint effects of GSTM1/GSTT1 and nasopharyngeal carcinoma

GSTM1/GSTT1	Cases (%)	Controls (%)	OR* (95% CI)*	P*
All subjects				
Positive/positive	71 (21.0)	138 (23.6)	1.0	
Positive/null	65 (19.2)	122 (20.9)	1.03 (0.68 – 1.57)	0.90
Null/positive	103 (30.5)	177 (30.3)	1.11 (0.76 – 1.62)	0.58
Null/null	99 (29.3)	147 (25.2)	1.31 (0.89 – 1.92)	0.17
Male				
Positive/positive	44 (19.4)	64 (25.6)	1.0	
Positive/null	43 (18.9)	48 (19.2)	1.27 (0.72 – 2.23)	0.41
Null/positive	70 (30.8)	82 (32.8)	1.20 (0.73 – 1.99)	0.47
Null/null	70 (30.8)	56 (22.4)	1.76 (1.04 – 2.97)	0.03
Female				
Positive/positive	27 (24.3)	74 (22.2)	1.0	
Positive/null	22 (19.8)	74 (22.2)	0.82 (0.43 – 1.56)	0.54
Null/positive	33 (29.7)	95 (28.4)	0.92 (0.51 – 1.68)	0.80
Null/null	29 (26.1)	91 (27.3)	0.88 (0.48 – 1.62)	0.69
IgA/VCA +				
Positive/positive	67 (20.7)	62 (23.4)	1.0	
Positive/null	62 (19.2)	57 (21.5)	1.01 (0.61 – 1.66)	0.98
Null/positive	98 (30.3)	82 (30.9)	1.11 (0.70 – 1.74)	0.66
Null/null	96 (29.7)	64 (24.2)	1.39 (0.87 – 2.22)	0.17

* Adjusted for age but not for multiple comparisons

Discussion

The effects of the *GSTM*1 – and *GSTT*1 – null genotypes were examined for association with nasopharyngeal carcinoma risk. The frequency of the *GSTM*1 – null genotype was 56% in our control population, similar to Europeans (53%) but much higher than in African – Americans (27%; ref. 5). The frequency of the *GSTT*1 – null genotype was 46% in our control population, higher than in Europeans (22%) and in African – Americans (21%; refs. 11, 20). We observed no significant association for *GSTM*1 – or *GSTT*1 – null genotypes either independently or jointly with nasopharyngeal carcinoma. *GSTM*1 – and *GSTT*1 – null genotypes, separately or in combination, do not contribute to overall nasopharyngeal carcinoma risk in this population of females or in persons with EBV reactivation. Only the *GSTM*1/*GSTT*1 – double null genotype combination showed a tendency to increase risk for nasopharyngeal carcinoma (odds ratio, 1.76) but only in males.

Several studies have provided evidence that glutathione S – transferase isoforms exhibiting overlapping substrate specificity with different combinations of various unfavorable deletion genotypes may increase the risk for head and neck cancers[21]. *GSTM*1/*GSTT*1 double deletions have been reported to confer a higher risk for head and neck squamous cell carcinoma[22]. Similar increases in risk for other cancers have been reported for the combined genotypes of *GSTM*1 null and *GSTT*1 null[9, 23]. Although more than 500 studies have examined the association of the null genotypes for GSTM1 and GSTT1 genes with various tumors, very few have investigated the associations between *GSTM*1 – and *GSTT*1 – null genotypes and the risk for nasopharyngeal carcinoma in Chinese populations. One study showed that *GSTM*1 – null genotype was associated with increased risk for nasopharyngeal carcinoma in European – Americans with 83 cases[17]; however, the risk factors in China are likely quite different, making comparisons difficult. A Chinese study enrolling 91 cases showed associations between nasopharyngeal carcinoma and *GSTM*1 – and *GSTT*1 – null genotypes separately and jointly[18]; however, we found no support for this finding in our study that included 350 nasopharyngeal carcinoma cases from southern China. The discrepancies between the two studies may reflect differences in overall study design.

Interestingly, we did observe an increase in risk for males who carried the combination of null genotypes for GSTM1 and GSTT1. A 1996 population – based survey in mainland China reported a prevalence of 63% male and 4% female smokers[24]. We speculate that smoking may be responsible for the increased risk in males because glutathione S – transferase genes have been reported to detoxify nicotine and smoke[13]. Previous studies on the interaction of *GSTM*1 and *GSTT*1 with smoking in tobacco – associated cancer have shown that the deletion of the *GSTM*1 and *GSTT*1 genes may increase cancer risk in smokers[9, 25], whereas another study reported an interaction with tobacco chewing but not with smoking for head and neck squamous cell carcinoma[22]. Further studies on the effects of smoking and *GSTM*1 and *GSTT*1 genotypes will clarify the role of smoking – gene interactions in nasopharyngeal carcinoma.

A limitation of our study is that we did not consider gene copy number in the analysis because heterozygotes cannot be detected by the genotyping assay used. We were assessing the role of homozygosity for the null mutations and comparing the null group to individuals carrying either one or two copies of the gene. It is possible that if the effects were additive or dominant, we may have missed associations. A second limitation is that no smoking exposure data are available for this group of nasopharyngeal carcinoma cases and controls to directly assess the interactions between these genetic factors and smoking.

No previous study has systematically assessed the effects of *GSTM*1/*GSTT*1 – double null genotypes with sex and EBV replication status in nasopharyngeal carcinoma. Studies with detailed data on environmental risk factors for nasopharyngeal carcinoma, such as salted fish and other preserved meat consumption, smoking, and occupational exposures to carcinogens, are needed to fully understand the role of *GSTM*1 and *GSTT*1 gene copy number in nasopharyngeal carcinoma disease.

Acknowledgements

The costs of publication of this article were defrayed in part by the payment of page charges. This article must therefore be hereby marked advertisement in accordance with 18 U. S. C. Section 1734 solely to indicate this fact.

We thank Bailey Kessing for data management and Beth Binns – Roemer, Michael Malasky, and Mary McNally for the excellent technical assistance.

〔In 《Cancer Epidemiol Biomarkers prev》2008, 17 (7): 1760 – 1763〕

References

1   Dai XD, Li LD, Lu FZ, et al. Mortality distribution analysis of nasopharyngeal carcinoma in China from 1990 to 1992. J Pract Oncol, 1998, 12: 81 – 84

2   Li LD, Rao KQ, Zhang SW, Lu FZ, Zou XN. Statistical Analysis of Data from 12 Cancer Registries in China, 1993 – 1997. Bull Chin Cancer, 2002, 11: 497 – 507

3   Chang ET, Adami HO. The enigmatic epidemiology of nasopharyngeal carcinoma. Cancer Epidemiol Biomarkers Prey, 2006, 15: 1765 – 1777

4   Hayes JD, Pulford DJ. The glutathione S – transferase supergene family: regulation of GST and the contribution of the isoenzymes to cancer chemoprotection and drug resistance. Crit Rev Biochem Mol Biol, 1995, 30: 445 – 600

5   Parl FF. Glutathione S – transferase genotypes and cancer risk. Cancer Lett, 2005, 221: 123 – 129

6   Rebbeck TR. Molecular epidemiology of the human glutathione S – transferase genotypes GSTM1 and GSTrl in cancer susceptibility. Cancer Epidemiol Biomarkers Prev, 1997, 6: 733 – 743

7   Helzlsouer KJ, Selmin O, Huang HY, et al. Association between glutathione S – transferase M1, P1, and T1 genetic polymorphisms and development of breast cancer. J Nat Cancer Inst, 1998, 90: 512 – 518

8   Park SK, Yoo KY, Lee SJ, et al. Alcohol consumption, glutathione S – transferase M1 and T1 genetic polymorphisms and breast cancer risk. Pharmacogenetics, 2000, 10: 301 – 309

9   Ates NA, Tamer L, Ates C, et al. Glutathione S – transferase M1, T1, P1 genotypes and risk for development of colorectal cancer. Biochem Genet, 2005, 43: 149 – 163

10 Hohaus S, Massini G, D'Alo F, et al. Association between glutathione S – transferase genotypes and Hodgkin's lymphoma risk and prognosis. Clin Cancer Res, 2003, 9: 3435 – 3440

11 Garcia – Closas M, Malats N, Silverman D, et al. NAT2 slow acetylation, GSTM1 null genotype, and risk of bladder cancer: results from the Spanish Bladder Cancer Study and meta – analyses. Lancet, 2005, 366: 649 – 659

12 Egan KM, Cai Q, Shu XO, et al. Genetic polymorphisms in GSTM1, GSTP1, and GSTT1 and the risk for breast cancer: results from the Shanghai Breast Cancer Study and meta – analysis. Cancer Epidemiol Biomarkers Prey, 2004, 13: 197 – 204

13 Schneider J, Bemges U, Philipp M, Woitowitz HJ. GSTM1, GSTT1, and GSTP1 polymorphism and lung cancer risk in relation to tobacco smoking. Cancer Lett, 2004, 208: 65 – 74

14 Roodi N, Dupont WD, Moore JH, Parl FF. Association of homozygous wild – type glutathione S – transferase M1 genotype with increased breast cancer risk. Cancer Res, 2004, 64: 1233 – 1236

15 Mossner R, Anders N, Konig IR, et al. Variations of the melanocortin – 1 receptor and the glutathione – S transferase T1 and M1 genes in cutaneous malignant melanoma. Arch Dermatol Res, 2007, 298: 371 – 379

16 Colombo J, Rossit AR, Caetano A, Borim AA, Womrath D, Silva AE. GSTT1, GSTM1 and CYP2E1 genetic polymorphisms in gastric cancer

· 795 ·

and chronic gastritis in a Brazilian population. World J Gastroenterol, 2004, 10: 1240 – 1245

17　Nazar – Stewart V, Vaughan TL, Burt RD, Chen C, Berwick M, Swanson GM. Glutathione S – transferase M1 and susceptibility to nasopharyngeal carcinoma. Cancer Epidemiol Biomarkers Prey, 1999, 8: 547 – 551

18　Deng ZL, Wei YP, Ma Y. Frequent genetic deletion of detoxifying enzyme GSTM1 and GSTT1 genes in nasopharyngeal carcinoma patients in Guangxi Province, China. Zhonghua Zhong Liu Za Zhi, 2004, 26: 598 – 600

19　Arand M, Muhlbauer R, Hengstler J, et al. A multiplex polymerase chain reaction protocol for the simultaneous analysis of the glutathione S – transferase GSTM1 and GSTT1 polymorphisms. Anal Biochem, 1996, 236: 184 – 186

20　SNP500Cancer Database cited 2008 Jan 7. Available from: http: // snp500cancer. nci. nih. gov/ snp. cfm? ethnic = true&snp_ id = GSTT1 – 02

21　Hashibe M, Brennan P, Strange RC, et al. Meta –

and pooled analyses of GSTM1, GSTT1, GSTP1, and CYP1A1 genotypes and risk of head and neck cancer. Cancer Epidemiol Biomarkers Prey, 2003, 12: 1509 – 1517

22　Singh M, Shah PP, Singh AP, et al. Association of genetic polymorphisms in glutathione S – transferases and susceptibility to head and neck cancer. Mutat Res, 2008, 638: 184 – 194

23　Saadat I, Saadat M. Glutathione S – transferase M1 and T1 null genotypes and the risk of gastric and colorectal cancers. Cancer Lett, 2001, 169: 21 – 26

24　Yang G, Fan L, Tan J, et al. Smoking in China: findings of the 1996 National Prevalence Survey. JAMA, 1999, 282: 1247 – 1253

25　Kihara M, Kihara M, Kubota A, Furukawa M, Kimura H. GSTM1 gene polymorphism as a possible marker for susceptibility to head and neck cancers among Japanese smokers. Cancer Lett, 1997, 112: 257 – 262

105.　Potent Specific Immune Responses Induced by Prime – Boost – Boost Strategies Based on DNA, Adenovirus, and Sendai Virus Vectors Expressing *gag* Gene of Chinese HIV – 1 Subtype B

YU Shuang – qing[1], FENG Xia[1], SHU Tsugumine[2], MATANO Tetsuro[3], HASEGAWA Mamoru[2], WANG Xiao – li[4], MA Hong – tao[4], LI Hong – xia[1], LI Ze – lin[d], ZENG Yi[1,4]

1. State Key Laboratory for Infectious Disease Prevention and Control, National Institute for Viral Disease Control and Prevention, China CDC; 2. DNAVEC Corporation.
1 – 25 – 11 Kannondai. Tsukuba – shi. Japan; 3. The Institute of Medical Science, The University of Tokyo; 4. Beijing University of Technology

Summary

To study the immune responses elicited by multiple vectors and develop vaccines strategies against prevalent HIV – 1 strains in China, we have examined the potency of vaccine regimens of plasmid DNA, adenovirus, and Sendal virus vectors expressing HIV – 1 *gag* consensus sequence of

HIV – 1 isolates from China for inducing specific immune responses. In BALB/c mice, combination of these vectors induced higher *gag* – specific cellular immune response than any regimen using single vector alone. The prime – boost – boost regimen consisting of the triple heterologous vectors induced *gag* – specific T – cell responses the most efficiently. In rhesus macaques, the prime – boost – boost regimen induced potent *gag* – specific cellular immune responses as well as tong lasting humoral immune response, and each booster resulted in rapid and efficient expansion of *gag* – specific T cells. These results indicate that this prime – boost – boost regimen using triple heterologous vectors is a promising AIDS vaccine candidate for efficiently inducing HIV – 1 – specific cellular and humoral immune responses. Its further studies as a promising scheme for therapeutic and/or prophylactic HIV – 1 vaccines should be grounded.

Introduction

The global HIV epidemic continues to expand, exceeding previous predictions and causing tremendous suffering. At the same time, a rapid increase of HIV infection was also found in China in recent years. The cumulative reported cases living with HIV/AIDS were 214 300 by the end of July 2007[1]. Therefore, developing HIV vaccines targeting the prevalent strains in China is one of the most important tasks of Chinese HIV/AIDS therapy and prevention. Subtype B is found to be prevalent in several epidemic regions where paid blood donors are mainly affected population[2]. HIV – 1 Subtype B isolates in these epidemic regions are relatively conserved, so the prevalent strains isolated from these regions were used as vaccine strain. After primary HIV infection, the initial event is the acute retrovial syndrome which is accompanied by a precipitous decline in CD4 cell counts, high plasma viremia. Then clinical recovery is accompanied by a reduction in plasma viremia, CD4 increase and development of strong immune responses including specific CTL and seroconversion. The median latent period for 50% HIV individuals developing AIDS is about 7 – 8 years. After HAART treatment, many patients can have immune reconstruction. All these data showed that human has strong immune responses to primary HIV infection and last for years. So we try to develop therapeutic vaccine with multiple vectors in combination with HAART treatment for enhancing the host – specific immunity. The study in untreated HIV – 1 infected cohort and HIV – 2 long – term non – progressors indicated that only *gag* – specific responses were associated with lowering viremia and may play an important role in controlling viral load during chronic infection[3-6]. Therefore, we made DNA, adenovirus and Sendai virus vectors expressing *gag* gene of the prevalent Subtype B strains in China as candidate vaccine strategies.

Virus – specific cellular immune responses play an important role in the control of HIV infections[7-9]. DNA vaccines, recombinant viral vector based vaccines, and their combinations are promising AIDS vaccine methods because of their potential for inducing cellular immune responses. DNA vaccine is safe and easy to manufacture, and it is quite effective as a priming or initial immunogen in a bimodal vaccine strategy[10,11]; adenovirus vector's efficiency of inducing target gene specific cellular response has been identified[12]; the Sendai virus vector in this study is a non – transmissible, replication – competent recombinant SeV deleting the envelope Fusion (F) protein gene (SeV/△F). It can efficiently mediate gene transfer in vivo[13,14]. The expression of wild type HIV structural genes by DNA vectors is dependent on the HIV Rev protein and the Rev – responsive element (RRE) in the mammalian cells[15]. In the absence of functional Rev/RRE, mRNAs containing inhibitory sequences (INS) are either retained in the nucleus or degraded rapidly; therefore, little protein can be expressed from these mRNAs. Modification of HIV – 1 sequences, presumably removing the inhibitory sequences, can result in significantly enhanced HIV – 1 protein expression in the absence of Rev/RRE[16,17]. Our unpublished data indicated that recombinant SeV/△F has extraordinary advantages on delivering HIV – 1 wild type *gag* gene expression *in vitro*. Unlike DNA vectors, which require nucleus – to – cytoplasm transport of the transcribed mRNA, the Sendai virus expression system replicates its RNA only in the cytoplasm, thus removing the barrier of nuclear export. In addition, the genome of SeV/△F is still able to replicate in the cytoplasm but incapable of re – infecting neighbouring cells. It is able to further enhance the level of cytoplasmic accumulation of mRNA and the expression of foreign gene[18]. Former studies have demonstrated that DNA – prime/*gag* – expressing Sendai virus vector boost induced highly effective CTLs and resulted in the containment of replication of SHIV89.6PD and SiVmac239[19,20].

Commonly, vaccination protocols require multiple immunizations to achieve robust, protective and sustained immune responses. However, the immunity against viral vectors has limited the repeated use of same candidate vaccine. Most researches have focused on vaccination protocols based on one or two candidates. In this study we addressed the magnitude, sustain and breadth of HIV – 1 *gag* – specific immune responses induced by prime – boost – boost scheme with combination of triple heterologous vectors, i. e. DNA, adenovirus and SeV.

Materials and Methods

1. Amplification and codon – modification of HIV – 1 *gag* gene

Genomic DNA was extracted from whole blood dealt with EDTA anticoagulant uncultured peripheral blood mononuclear cells (PBMCs) of HIV – 1 infected paid blood donors in Henan prevalence region. The *gag* genes were amplified from genomic DNA by nest – PCR using primers G – 1/G – 2 (G – 1: CGA CGC AGG ACT CGG cTr GC, G – 2: CCT GGC TIT AAT TI – r AC) for the first round and primers G – 3/G – 4 (G – 3: GAG ATG GGT GCG AGA GCG TCA, G – 4: GTT GAC AGG TGT AGG TCC TAC) for the second round. Sequence analysis identified that all of them were Subtype B. The consensus sequence based on these prevalent strains was obtained by malignment using Vector NTI software. To improve the expression level of *gag* protein in mammalian cells, the codons of the consensus *gag* sequence were modified according to mammalian codon usage. The codonoptimized *gag* gene was synthesized by Shanghai Sangon Biological Engineering Technology & Service Co., Ltd.

2. Construction of vaccines

The synthetic modified *gag* gene was inserted into pVR[21] under the control of human cytomegalovirus immediate – early promoter to obtain DNA vaccine. The same *gag* gene was constructed to the rAdS. Recombinant adenovirus containing *gag* (rAd5 – *gag*) was generated using Cre/loxP –

based system[22]. In brief, the synthetic *gag* gene was cloned into shuttle vector pDC316 to make pDC316 − *gag* vector; HEK293 cells were co − transfected with the pDC316 − *gag* and the backbone vector pBHGlox △ E1, 3Cre, and then the recombinant adenovirus containing *gag* gene (rAd5 − *gag*) was acquired by Cre − lox recombination in HEK293 cells. Our unpublished data showed that wild type and codons − optimized *gag* gene were equally efficient to mediate *gag* protein expression and immune responses in BALB/c mice using SeV/ △ F vector. Hence one of the wild type *gag* gene was used to construct rSeV vaccine. The identity of amino acid sequence between this wild type *gag* gene and the consensus sequence was 92%. A significant difference between the two sequences was the wild type *gag* had a 10 aa insert mutation at the C − terminal of p17 and a 5 aa insert mutation at the N − terminal of p6. The recombinant Sendai virus containing *gag* (rSeV − *gag*) was constructed and amplified in DNAVEC Corporation as previous described with modification[23]. In brief, (i) EIS sequence and NotI restrictive enzyme sites were adopted to the *gag* gene by PCR and cloned into pBluescript KS (+) named as pBS − HIV*gag*. (ii) pBS − HIV*gag* was digested with NotI and DNA fragment containing *gag* was constructed into pSeV/ △ F that contains SeV full − length cDNA lacking F gene. The generated plasmid was named as pSeV18 + HIV*gag*/ △ E (iii) Recombinant virus containing *gag* gene was recovered by co − tranfection 293T/17 cells with pCAGGS − P (z) / 4C − , pCAGGS − L (TDK), pCAGGS − F5R, pCAGGS − T7 and pSeV18 + HIV*gag*/ △ F. (iv) Recovered virus was cloned and amplified in LLC − MK2/F/Ad cells.

3. Immunization of animals

All animal experiments were reviewed and approved by the Institutional Animal Care and Use Committee at China CDE animal facility. The experiments were performed in accordance with relevant guidelines and regulations. All mice were anesthetized using barbital sodium by i. p. before immunization and sacrifice.

Four − to six − week − old female BALB/c mice ($H − 2^d$) were randomly divided into several groups of 6. To determine the optimal dosage of rAd5 − *gag* and rSeV − *gag*, different amount of the purified vaccines (100 μg of DNA, 5×10^8, 10^7 and 10^6 PFU of rAd5 − *gag*, 4×10^8, 10^7 and 10^6 CIU of rSeV − *gag*, etc.) were administered in 3 weeks interval as describe in results. The immunological response induced by single or prime − boost regimens with two vaccines was compared with optimal dose (100 μg of DNA, 10^7 CU of rSeV − *gag* and 10^7 PFU of rAd5 − *gag*) in 3 weeks interval. To compare the immune responses induced by prime − boost − boost regimens with triple heterologous vectors vaccine, mice were sequentially primed with DNA and boosted with rAd5 − *gag* and rSeV − *gag* in different order at week 0, 3 and 12. In all experiments, the purified vaccines stock was suspended in 100 μl of sterile phosphate − buffered saline and administered in the tibia anterior muscle by intramuscular injection (i. m.).

Twelve adult rhesus macaques were housed at Institute of Experimental Animal Sciences, Chinese Academy of Medical Sciences. All animals were anesthetized using barbital sodium before immunization and bleeding. They were divided into 3 groups of 4. The rhesus macaques (group 2 and 3) were primed with 5 mg DNA in 1 ml sterile PBS at week 0, 2 and boosted with the first viral vector vaccine (10^9 PFU rAdV5 − *gag* or 10^9 CIU rSeV − *gag*) in 1 ml sterile PBS twice at week 6 and

33 by intramuscular injection, then crossboosted with another viral vector vaccine twice at week 45 and 87. The control rhesus macaques (group 1) were injected with 1 ml sterile PBS at all 6 – time points. Immunized animals were bled at indicated time points.

4. Intracellular IFN – γ staining

One week after the last immunization, H – 2^d – restricted *gag* – specific CD8 $^+$ T – cell responses in immunized mice were quantitated using intracellular cytokine staining (ICS). Freshly isolated splenic lymphocytes (2×10^6 cells per well) were suspended in CM (complete culture medium, RPMI – 1640 containing 10% fetal bovine serum and 1% penicillin – streptomycin – L – glutamine) and incubated with H – 2^d – restricted CTL epitope peptides, $gag_{197-205}$ (AMQMLKETI)[24], $gag_{239-247}$ (TTSTLQEQI)[25] and $gag_{291-300}$ (EPFRDYVDRF)[26] for 16 h in incubator at 37 ℃ with 5% CO_2. These three defined epitopes present in both wild type and modified *gag*. The final concentration for each peptide was 10 μg/ml. For the final 12 h, brefeldin A (Sigma, MO, USA) was added at 1 μg/ml. The cells were then washed and resuspended in phosphate – buffered saline –2% FBS. Surface markers were stained with 5 μl of R – PE – conjugated rat anti – mouse CD8a (Ly – 2) monoclonal antibody (BD Pharmingen, CA, USA). The cells were then sequentially fixed with 4% paraformaldehyde, permeabilized with 0.3% saponion for 15 rain, respectively, resuspended in phosphate – buffered saline – 2% FBS, and stained with 1 μl FITC – conjugated rat anti – mouse IFN – γ, monoclonal antibody (BD Pharmingen, CA, USA). Corresponding isotype control was used to define the threshold for positive staining. Samples were analyzed with Coulter EPICS Altra flowmetry using CellQuest software.

5. Elispot assays

gag – specific cellular immune responses in vaccinated rhesus macaques were assessed by IFN – γ, enzyme – linked immunospot (Elispot) assays. 14 – 16aa peptides with 11 aa overlap spanning the entire wild type *gag* obtained from the university of Tokyo were used as stimulator. The peptide pool is homologous with the *gag* expressed in rSeV. The peptide pool was divided into 5 sub – pools (sub – pool AB [aa1 – 105], sub – pool CD [aa95 – 203], sub – pool EF [aa193 – 305], sub – pool GH [aa295 – 408] and sub – pool IJ [aa398 – 516]). The frequency of *gag* – specific IFN – γ, secreting cells was determined using the IFN – γ/Elispot assay kit (U – Cytech, Utrech, The Netherlands) according to the manufacturer's manual. In brief, 2×10^5 fresh PBMCs stimulated in duplicate wells with each of five sub – pools (the final concentration of each peptide was 1 μmol/L) were cultured in an purified anti – monkey IFN – γ, monoclonal antibodies coated 96 – well plate for 20 h at 37 ℃ under 5% CO_2. Cells cultured in 1% DMSO were used to assess the background. After incubation, the spots were developed according to the manufacturer's instructions. Finally, the plates were air – dried and the resulting spots were counted with Immunospot Reader (CTL, Cleveland, OH, USA). Peptide – specific IFN – γ Elispot responses were considered as positive only when the response was fourfold above negative control and the spot – forming cells (SFCs) were >50SFC/10^6 PBMCs.

6. Enzyme – linked immunosorbent assay (ELISA)

Serum antibodies against HIV – 1 *gag* in immunized mice and rhesus macaques were measured by indirect ELISA. In brief, 96 – well microtiter plates were coated with 200 ng/well recombinant HIV – 1 HXB2 *gag* P24 and incubated overnight at 4 ℃. The wells were blocked with PBS containing 5% skimmed milk for 2 h at 37℃, then were washed five times with PBS containing 0.05% Tween – 20. Sera were then added in serial dilutions and incubated for 1 h at 37℃. After washing the plates five times, 100 µl of 1 : 100 000 diluted HRP – conjugated goat anti – mouse or goat anti – monkey IgG antibody (Bethyl, Montgomery, TX, USA) was added, and the plates were incubated for 1 h at 37℃. Samples were developed for 30 rain at 37℃ with TMB (WanTai, Beijing, China), the reactions were stopped with 0.5 mol/L H_2SO_4, and results were analyzed at 450 nm with 630 nm as reference. Endpoint antibody titers were defined as the last reciprocal serial serum dilution at which the absorption at 450 nm was greater than two times the background signal detected.

7. Statistical analysis

Results are given as means with ranges. Statistical analysis was based on Student's test, $P <$ 0.05 was considered as significant.

Results

1. *gag* – specific cellular responses induced by different candidate vaccines in BALB/c mice

To choose an appropriate dosage of rSeV – *gag* for the DNA – prime/SeV – boost scheme and rAd5 – *gag*, mice were immunized with different doses of rAd5 – *gag* twice or 100 µg DNA combined with different doses of rSeV – *gag* in 3 weeks interval. Significant number of HIV – 1 *gag* – specific CD8$^+$ T cells (13.02% ±3.40%, 7.53% ±2.05% and 2.73% ±1.09% in 5 × 10^8, 10^7 and 10^6 PFU group, respectively) was detected in all rAd5 – *gag* immunized groups and it was dose – dependent (Fig. 1B). The frequency of *gag* – specific IFN – γ secreting CD8$^+$ T cells induced by 100 µg DNA combined with 4 × 10^8 and 10^7 ClU rSeV – *gag* (2.41% ± 0.86% and 2.14% ± 0.80%, respectively) was significantly higher than that of 10^6 CIU (0.43 % ± 0.22%) (Fig. 1C). However, there was no significant difference between 10^7 and 10^8 CIU groups. These results are comparative with other studies[27–30], which indicated the good quality of our vaccines.

Then we compared the immune response induced by single or combined constructs with the optimal dose. After single injection of these candidate vaccines, rAd5 – *gag* induced better *gag* – specific CD8$^+$ T cells response than DNA or rSeV – *gag* (Fig. 2A). Similar with the single injection, twice injection of rSeV – *gag* alone elicited moderate *gag* – specific cellular responses (0.96% ± 0.6%), however, the frequency of IFN – γ, secreting CD8$^+$ T cells increased significantly when combined with DNA or rAd5 – *gag* (4.33% ±2.48% and 2.85% ± 0.91%, respectively, Fig. 2B). rAd5 – *gag* alone injected twice induced *gag* – specific cellular responses efficiently (2.73% ± 1.09%), the *gag* – specific cellular responses increased but not statistical significant when combined with DNA – prime (4.18% ±1.29%). In conclusion, rAd5 – *gag* induced better *gag* – specific CD8$^+$ T cells responses than DNA or rSeV – *gag* when vaccines were used alone, however, DNA/rSeV – *gag* induced similar frequency of IFN – γ/secreting CD8$^+$ T cells with DNA/rAd5 –

gag in combined schemes. Priming with rAd5 – *gag* and boosting with rSeV – *gag* induced the highest frequency of *gag* – specific IFN – γ secreting CD8$^+$ T cells (5. 23% $\pm$ 1. 72%). It's interesting that rSeV – prime/rAd5 – boost scheme (2. 85% $\pm$ 0. 91%) did not induce as good response as the rAd5 – prime/rSeV – boost scheme (Fig. 2B). It seems that the order of vaccinations impacted on the effect. Similar trend was reported in the study of prime – boost schemes with rAd5 and poxvirus vectors[31].

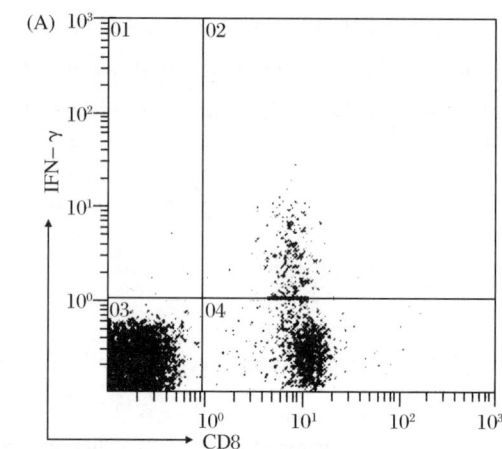

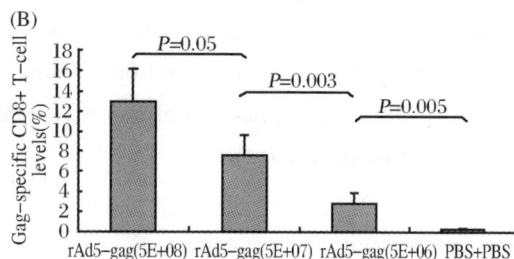

(A) Representative dot plots of IFN – γ induction in CO8$^+$ T cells after *gag* – specific stimulation. (B) BALB/c mice immunized with different doses of fAdS – *gag* twice in 3 weeks interval. (C) BALB/c mice primed with 100 μg DNA and boosted with different doses of rSeV – *gag* in 3 weeks interval

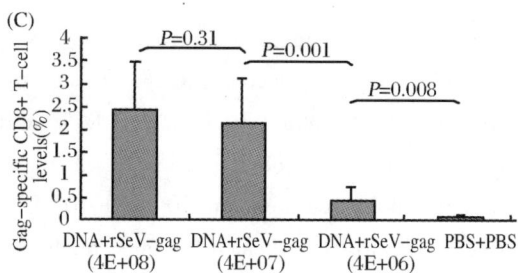

Fig. 1　*gag* – specific T cell immune response induced by different doses of candidate vaccines in BALB/c mice. Mice were immunized with different vaccines twice at 3 weeks interval by intramuscular injection. Immunized mice scarified at week 4. The splenic lymphocytes were stimulated with *gag* – specific dominant CTL peptides and the percentage of *gag* – specific IFN – γ secreting CD8$^+$ T cells was determined by intracellular cytokine staining assay. The results were expressed as the percentage of IFN – γ secreting CD8$^+$ cells in CD8$^+$ cells

To test *gag* – specific cellular responses induced by prime – boost – boost scheme with three heterologous vector vaccines, mice were primed with DNA first and boosted with rAd5 – *gag* and rSeV – *gag* in different order at week 3 and 12. One week after the second heteroiogous vector vaccine boost, the frequency of *gag* – specific CD8$^+$ T cells was 2. 4 to 4. 5 – fold higher than prime – boost groups in DNA/rSeV – *gag*/rAd5 – *gag* group (Fig. 2B, 12. 704% $\pm$ 1. 43%, $P < 0. 01$) and 1. 8 – to 3. 4 – fold in DNA/rAd5 – *gag*/rSeV – *gag* group, respectively (Fig. 2B, 9. 75% $\pm$ 1. 61%, $P < 0. 01$). The results indicated that the second heterologous vector boost improved *gag* – specific responses significantly in mice regardless of the order, but the scheme of DNA/rSeV – *gag*/rAd5 – *gag* was more effective than DNA/rAdS – *gag*/rSeV – *gag* ($P = 0. 01$).

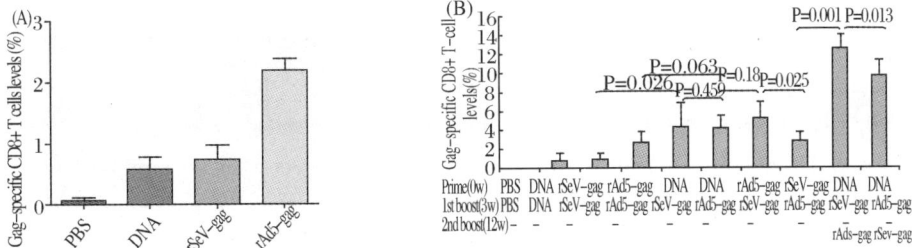

（A）BALB/c mice were immunized with DNA, rSeV – *gag* or rAdS – *gag* once, and the frequency of *gag* – specific IFN – γ secreting CD8 + T cells was tested 3 weeks later. （B）BALB/c mice were injected with candidate vaccines two or three times according the table below the histogram. The frequency of *gag* – specific IFN – γ secreting CD8[+] T cells was determined by intracellular cytokine staining assay 1 week after the last immunization. （ – ）without immunization

Fig. 2　*gag* – specific cellular response induced by different schemes in BALB/c mice. The dose of each vaccine was as follows: 100μg DNA. 4×10^7 CIU rSeV – *gag* and 5×10^7 PFU rAd5 – *gag*

2. *gag* – specific humoral responses were enhanced significantly by the second heterologous vector vaccine boost

Sera of immunized mice were colleted at different time points and P24 specific IgG antibodies were detected by ELISA. The mice primed with DNA and boosted with rSeV – *gag* or rAd5 – *gag* induced high level of anti – P24 antibodies 1 week after boost and persistent at high level at least 8 weeks. After the second boost with another heterologous vector vaccine, the geometric mean anti – P24 IgG titer increased from 1 : 3676 to 1 : 155 210 in DNA/rSeV – *gag*/rAd5 – *gag* group and from 1 : 11 144 to 1 : 67 558 in DNA/rAd5 – *gag*/rSeV – *gag* group （Fig. 3）. These results indicated that the second boost also enhanced the *gag* – specific humoral responses.

Fig. 3　The P24 – specific IgG in serum induced by prime – boost – boost regimens. Mice were primed with DNA and boost with rSeV – *gag* and rAd5 – *gag* sequentially in different order. The up arrows indicated the time points of immunization. Mice were bled at week 0, 4, 8, 12 and 13. P24 – specific IgG in serum were detected by ELISA and quantified by endpoint antibody binding titration

3. Potent and broad *gag* – specific cellular responses were induced by prime – boost – boost regimens in rhesus macaques

To investigate the immunogenicity of these three vaccine candidates in non – human primates and look for better scheme, rhesus macaques were primed with DNA twice and then boosted with each viral vector vaccine in different order twice sequentially. Group 2 was sequentially immunized with DNA/rSeV – *gag*/rAd5 – *gag* and group 3 was with DNA/rAd5 – *gag*/rSeV – *gag*. Specific cellular responses were measured by *gag* – specific IFN – γ Elispot assay using freshly isolated PBMCs at indicated time points （Fig. 4A and B）. *gag* – specific responses were undetectable in all control macaques all the time. In all vaccinated macaques, the *gag* – specific cellular immune re-

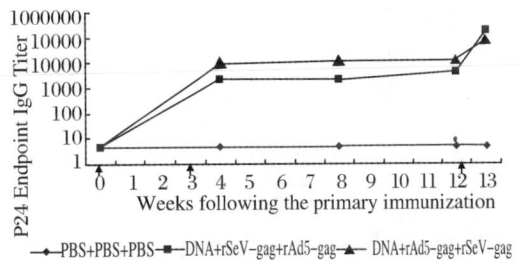

sponses were undetectable before or after two times of DNA – prime (data not show). The response began to increase significantly 4 weeks after the first boost and reached the peak at 6 weeks after the first boost. The average SFCs/million PBMCs were 2450 ± 983 in group 2 and 3110 ± 1461 in group 3 at the first peak response. There was no statistical difference in both groups. Subsequently the specific cellular responses began to drop down. Fourteen weeks after the first boost all macaques maintained at a steady – state level. At that time, half of immunized macaques in each group still persisted at a rather high level (>500 SFCs/million PBMCs). At week 33, all macaques were given the second same vector vaccine boost. The frequency of *gag* – specific cellular immune response increased again 2 weeks later and reached the second peak response in 2 – 6 weeks. At the peak point, the frequency of IFN – γ induction PBMCs increased by 3 – 7 times compared with the steady – state level before the boost in all vaccinated macaques. Two macaques induced higher peak response compared with the first peak (2. 8 – fold in 98R0053 in group 2 and 2. 3 – fold in 00R0227 in group 3).

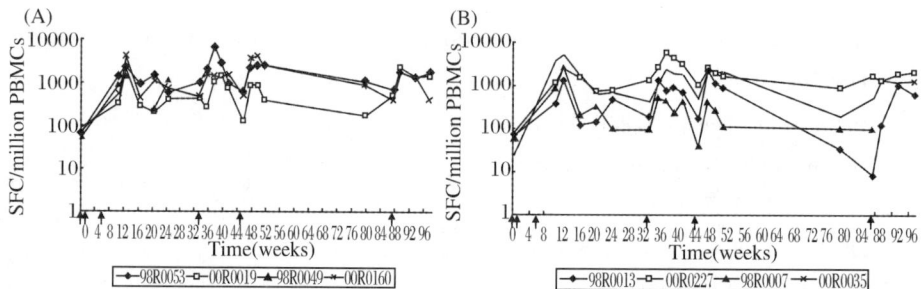

(A) Group 2: macaques were immunized with DNA/rSeV – *gag*/rAd5 – *gag* sequentially. (98R0049 died of unrelated cause at week 33). (B) Group 3: macaques were immunized with DNA/rAdS – *gag*/rSeV – *gag* sequentially (98R0007 died of anaesthesia at week 89)

Fig. 4   The magnitude of the *gag* – specific cellular responses in vaccinated rhesus macaques. Up arrows indicated time – points of immunization (↑ : DNA. ↑ : rSeV – *gag*, ↑ : rAd5 – *gag*). The PBMCs of animals were stimulated by *gag* – derived peptide pools and IFN – γ secreting cells were detected by Elispot at multiple time points. The results were showed as spot – forming cells (SFC) per million PBMCs. The *gag* – specific cellular responses in control group were undetectable all the time (data not shown)

To investigate whether additional immunization with heterologous vector can induce rapid and better response, the third and fourth boost with heterologous vector vaccine was given at week 45 and 87 when the response dropped down to a steady – state level. All vaccinated macaques increased by 4 – 8 times at peak responses compared with the steady – state level before boost. Higher peak response than former was not found in most of immunized macaques except monkey 98R0013 which induced higher peak response (1. 7 – fold, 2358 SFCs/million PBMCs) after the third heterologous vector vaccine boost (rSeV – *gag*) at week 45 compared with the former peak (Fig. 4B, 1413 SFCs/million PBMCs at week 35). No obvious increase of the *gag* – specific responses at steady – state level was observed in both groups.

The repeated boosts resulted in rapid specific cellular response (from 6 weeks after the first boost to 2 weeks after the third and fourth boost in DNA/rSeV – gag/rAd5 – gag group). At the same time, the time maintained at peak level also became longer. After the fourth boost, half of the immunized macaques maintained at peak response more than 8 weeks. The results implied that repeated immunization with heterologous vectors expressing the same antigen could more rapidly expand the specific immune response.

In terms of the breadth of gag – specific responses induced by vaccines, all macaques had the broadest responses at peak responses time points, with specific responses against 4 – 5 subpools. These results indicate that broad responses were induced. The gag – specific responses mainly clustered in the 2 – 3 sub – pools when the specific response decreased and maintained at steady – state level. Since the loss of breath was accompanied with the decrease in magnitude of all specific responses, the loss of breadth was due to the general decrease in magnitude. However, we did not study the breadth of gag – specific responses at single peptide level, we cannot distinguish exactly how many epitopes or which epitopes are dominant. The sub – pools (sub – pool CD, EF and GH) spanned P24 were dominant in all vaccinated macaques (Fig. 5). The result is prospective because P24 is relative conservative and contain many dominant epitopes. One significant difference between the two vaccination regimes is that the responses against sub – pool CD which spanned C – terminal of P17 and N – terminal of P24 were the strongest in animals boosted with rSeV – gag first (group 2), but were very weak in 3 of 4 macaques boosted with rAd5 – gag first (group 3, Fig. 5A). In addition, the dominant responsive sub – pools did not alter all the time in all animals. It may be related to the MHC – 1.

4. Strong and persistent gag – specific humoral responses were induced in rhesus macaques

To determine the gag – specific humoral responses induced by candidate vaccines, the P24 specific IgG in serum were detected by ELISA (Fig. 6). The results showed that good IgG responses against gag were detected and persisted for a long period in macaques in both vaccinated groups. There was no significant difference between the two groups. The results indicated that the prime – boost – boost scheme may also have advantages on inducing persistent humoral responses.

Discussion

For HIV – 1 vaccine development, one important focus is to develop vaccines that induce strong, broad and persistent CD8[+] cytotoxic lymphocytes (CTL). Viral vectors were attractive for potent capacity of inducing trans – gene specific cellular responses. One limitation of viral vector vaccines involves the negative effect of preexisting vector – specific immune responses. Anti – Ad5 immunity has already been shown to suppress the immunogenicity of fad5 vaccines for HIV – 1 in both preclinical studies[32-34] and clinical trials (http: //www. hvtn. org/fgm/1107slides/ McElrath. pdf). A possible obstacle for the application of rSeV vector in humans could be the presence of antibodies against the surface glycoproreins of the human parainfluenza virus type 1, which is known to cross – react with SeV HN – proteins[35]. The anti – vector immune responses can depress

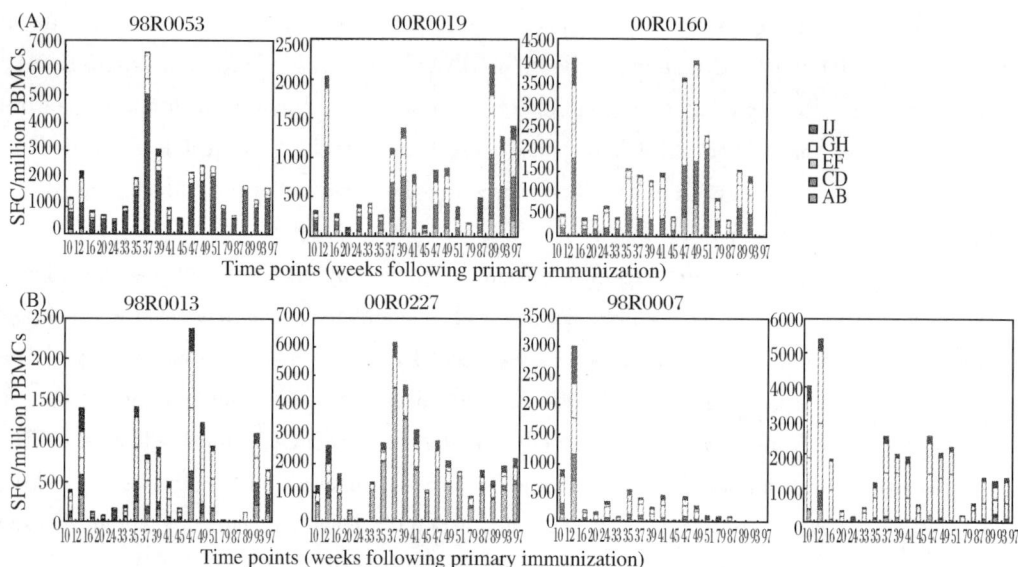

Fig. 5 The magnitude *gag* – specifc cellular responses against each HIV – 1 *gag* peptide sub – pool. Sub – pool AB (aa1 – 105), sub – pool CD (aa95 – 203), sub – pool EF (aa193 – 305). sub – pool GH (aa295 – 408) and sub – pool IJ (aa398 – 516) were used as stimulators, specific IFN – γ production were measured by Elispot. The results were showed as SFC per million PBMCs. The rhesus macaques of 98R0053. 00R0019, and 00R0160 were in group 2; 98R0013, 00R0227, 98R0007 (died of anaesthesia at week 89) and 99R0035 were in group 3

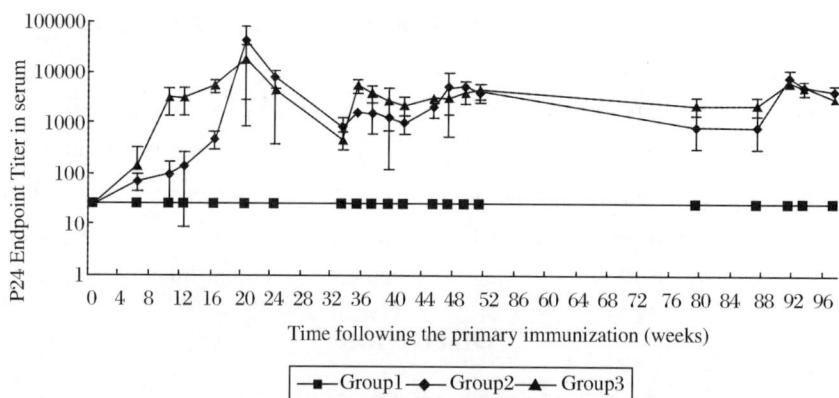

Fig. 6 Time course of anti – p24 antibodies induced by vaccines in rhesus macaques. The P24 – specific IgG in serum were detected by enzyme – linked immunosorbent assay (ELISA) using twofold serially diluted serum and were quantified by endpoint antibody binding titration. The antibody titer was expressed as the mean titer of each group

the re – administration of homologous vector[32,36] , however, multiple immunizations is necessary to induce persistent HlV – specific immune responses especially for therapeutic vaccine. So we tried to use sequential multiple vectors carrying the same antigen. Several candidates used one by one may be a promising means to resolve it. Our results demonstrated that potent *gag* – specific cellular and humoral responses were induced both in mice and macaques by prime – boost – boost scheme.

The immunogenicity of rSeV – *gag* is very moderate and not much better than DNA when used alone according to our results. According to our study, the mice immunized with 10^7 CIU rSeV – *gag* induced high anti – SeV IgG antibody (>1 : 900) 3 weeks later and rapidly increased to higher level after the second immunization at week 3 (the mean titer was 1 : 114 940 at week 4). The high level of antibody persisted for 8 weeks at least and may neutralize the subsequently given rSeV – *gag*. It may be the mostly important reason for the moderate immunogenicity of rSeV – *gag* alone used twice. But the scheme of DNA – prime/rSeV – boost induced similar responses with DNA – prime/rAdS – boost which was significantly better than rSeV – *gag* or DNA alone used twice. The results were consistent with others, which used SIV *gag* as immunogen and found that DNA – prime/ rSeV – boost was better than DNA or rSeV alone[37]. According to Fig. 2B, rSeV – *gag* combined with DNA or rAdS – *gag* both induced very good *gag* – specific cellular immune responses, especially as a booster. So its possible that rSeV is a very good boosting vaccine candidate just as DNA is quite effective as a priming or initial immunogen in a bimodal vaccine strategy.

gag – specific cellular responses were augmented in all vaccinated macaques after each boost (Fig. 4). However, the level of immune response after immunization varied in individual macaques. To look for the possible causes, the IgG against adenovirus or Sendai virus was detected before each immunization. The results showed no significant differences between the good response macaques and low ones (data not shown). The difference seems not because of the level of pre – immunity against vectors.

One purpose of this study was to investigate whether the multiple vector vaccines immunizations can induce higher or faster specific responses. However, the higher peak response than former after heterologous vector vaccine boost was not found in most of immunized macaques. Its different from the results in BALB/c mice. No obvious increase of the *gag* – specific responses at steady – state level was observed in both groups. But one promising result was observed in macaques. The host responded to *gag* faster after repeated boosts in DNA/rSeV – *gag*/rAdS – *gag* group. Since the immune system should respond as soon as possible after HIV – 1entry, the vaccine scheme that can generate fast response should be more effective in theory.

Recent reports have suggested that the *gag* – specific T cell response may play an important role in controlling viral load during chronic infection[4,5,38]. In Lu's study, inactivated whole virus – pulsed DC vaccines suppressed the plasma viral load in chronically HIV – 1 – infected and currently untreated individuals[39]. Several studies indicated that the breadth of cellular responses against *gag*, especially P24, was associated with the virus control[4,5,38,40], however, whether these dominant responses could contribute to the control of HIV – 1 replication and maintain the selective pressure remain to be proven in further studies. Our result that the time maintained at peak level became longer (about 2 months in half vaccinated macaques after 6 immunizations) indicates that the double prime – boost – boost scheme might be promising as a therapeutic vaccine strategy.

One significant difference between the two vaccination regimens is that the responses against sub –

pool CD were the strongest in animals boosted with rSeV – *gag* first (group 2), but were very weak in 3 of 4 macaques boosted with rAd5 – *gag* first (group 3, Fig. 5A). Since one of the major differences of the two *gag* – sequences exists in this segment (wild type *gag* had a 10aa insert mutation at the C – terminal of p17), we speculate the difference in the *gag* – sequence may be one of the reasons for the recognition of different peptide pools. However, even after heterologous boost with rSeV – *gag* later in group 3, which containing wild type *gag* gene, the immune responses against sub – pool CD did not be expanded any more. It has been reported that a narrowing of the TCR repertoire occurs during the secondary responses[41,42]. Whether the dominant responses were determined by the first boost need to be proven in further studies.

In conclusion, we have shown that prime – boost – boost regimens with combination of triple heterologous vectors (DNA, adenovirus and SeV) encoding *gag* gene of the prevalent HIV – 1 strains in China induced broad and sustained high level of specific immune responses. It might be a promising scheme for therapeutic or prophylactic HIV – 1 vaccine against the HIV – 1 epidemic in China.

Acknowledgements

This work was supported by the Chinese National 863 project (grant 2003AA219070), DNAVEC Corporation, NIH CIPRA project and a grant from the Ministry of Education, Culture, Sports, Science, and Technology. We thank Prof. Kongwei at Jilin University in China for his kindly present plasmid vector pVR and Dr. Yongjun Guan at the University of Maryland for assistance with the preparation of the manuscript.

〔In 《Vaccine》 2008, 26: 6124 – 6131〕

References

1　Zeng Y. Strategy for AIDS prevention and treatment. Virol Sin, 2007, 22 (6): 419 – 510

2　Song YH, Xing H, Wang Z, Zhang YZ, Sun F, He X, et al. An analysis of *gp*120 and *gag* gene variation of HIV – 1 strains isolated from Henan and Xinjiang in China. Chin J AIDS STD, 2007, 13 (5): 405

3　Novitsky V, Cao H, Rybak N, Gilbert P, McLane MF, Gaolekwe S, et al. Magnitude and frequency of cytotoxic T – lymphocyte responses: identification of immunodominant regions of human immunodeficiency virus type 1 subtype C. J Virol, 2002, 76 (20): 10155 – 10168

4　Kiepiela P, Ngumbela K, Thobakgale C, Ramduth D, Honeyborne I, Moodley E, et al. CD8 + T – cell responses to different HIV proteins have discordant associations with viral load. Nat Med, 2007, 13 (1): 46 – 53

5　Geldmacher C, Currier JR, Herrmann E, Haule A, Kuta E, McCutchan F, et al. CD8 T – cell recognition of multiple epitopes within specific *gag* regions is associated with maintenance of a low steady – state viremia in human immunodeficiency virus type 1 – seropositive patients. J virol, 2007, 81 (5): 2440 – 2448

6　Leligdowicz A, Yindom LM, Onyango C, Sarge – Njie R, Alabi A, Cotten M, et al. Robust *gag* – specific T cell responses characterize viremia control in HIV – 2 infection, J Clin Invest, 2007, 117 (10): 3067 – 3074

7　Yasutomi Y, Reimann KA, Lord CI, Miller MD, Letvin NL. Simian immunode – ficiency virus – specific CD8 + lymphocyte response in acutely infected rhesus monkeys. J Virol, 1993, 67 (3): 1707 – 1711

8　Borrow P, Lewicki H, Hahn BH, Shaw GM, Oldstone MB. Virus – specific CDS + cytotoxic T – lymphocyte activity associated with control of viremia in primary human immunodeficiency virus type 1 infection. J Virol, 1994, 68 (9): 6103 –6110

9　Altfeld M, Rosenberg ES. The role of CD4⁺ T helper cells in the cytotoxic T lymphocyte response to HIV – 1. Curr Opin Immunol, 2000, 12 (4): 375 – 380

10　Wolff JA. Malone RW. Williams P. Chong W, Acsadi G, Jani A, et al. Direct gene transfer into mouse muscle in vivo. Science, 1990, 247: 1465 – 1468, 4949 Pt 1

11　Prince AM, Whalen R. Brotman B, Successful nucleic acid based immunization of newborn chimpanzees against hepatitis B virus. Vaccine, 1997, 15 (8): 916 –919

12　Letvin NL. Strategies for an HIV vaccine. J Clin Invest, 2002, 110 (1): 15 – 20

13　Shiotani A. Fukumura M, Maeda M, Hou X, Inoue M, Kanamori T, et al. Skeletal muscle regeneration after insulin – like growth factor I gene transfer by recombinant Sendai virus vector. Gene Ther, 2001, 8 (14): 1043 – 1050

14　Ferrari S, Griesenbach U, Shiraki – lida T, Shu T, Hironaka T, Hou X, et al. A defective nontransmissible recombinant Sendai virus mediates efficient gene transfer to airway epithelium in vivo. Gene Ther, 2004, 11 (22): 1659 –1664

15　Felher BK, Hadzopoulou – Cladaras M, Cladaras C, Copeland T, Pavlakis GN. Rev protein of human immunodeficiency virus type 1 affects the stability and transport of the viral mRNA. Proc Natl Acad Sci USA, 1989, 86 (5): 1495 – 1499

16　Andre S, Seed B, Eberle J, Schraut W, Bultmann A, Haas J. Increased immune response elicited by DNA vaccination with a synthetic *gp*120 sequence with optimized codon usage. J Virol, 1998, 72 (2): 1497 –1503

17　Casimiro DR, Tang A, Perry HC, Long RS. Chen M, Heidecker GJ, et al. Vaccine – induced immune responses in rodents and nonhuman primates by use of a humanized human immunodefi-

ciency virus type 1 pol gene. J virol, 2002, 76 (1): 185 – 194

18　Bitzer M, Armeanu S, Lauer UM, Neubert WJ. Sendal virus vectors as an emerging negative – strand RNA viral vector system. J Gene Med, 2003, 5 (7): 543 –553

19　Matano T. Kobayashi M, lgarashi H, Takeda A, Nakamura H. Kano M, et al. Cytotoxic T lymphocyte – based control of simian immunodeficiency virus replication in a preclinical AIDS vaccine trial. J Exp Med, 2004, 199 (12): 1709 – 1718

20　Takeda A, lgarashi H, Nakamura H, Kano M, lida A, Hirata T, et al. Protective efficacy of an AIDS vaccine, a single DNA priming followed by a single booster with a recombinant replication – defective Sendai virus vector, in a macaque AIDS model. J Virol, 2003, 77 (17): 9710 –9715

21　Hong – Xia YX – H MA, Chun – Lai JIANG, Yong – Ge WU, Kong Wei. Construction of eukaryotic expression vector and screening of cell strain for stable expression of HIV – 1 Subtype BJC env. Chin J Biol, 2006, 19 (4): 365 – 367

22　Ng P, Parks RJ, Cummings DT, Evelegh CM, Sankar U, Graham FL. A high – efficiency Cre/ loxP – based system for construction of adenoviral vectors. Hum Gene Ther, 1999, 10 (16): 2667 – 2672

23　Hirata T, lida A, Shiraki – lida T, Kitazato K, Kato A, Nagai Y, et al. An improved method for recovery of F – defective Sendai virus expressing foreign genes from cloned cDNA. J Virol Methods, 2002, 104 (2): 125 – 133

24　Haglund K, Leiner I. Kerksiek K, Buonocore L, Pamer E, Rose JK. High – level primary CDS (+) T – cell response to human immunodeficiency virus type 1 *gag* and env generated by vaccination with recombinant vesicular stomatitis viruses. J Virol, 2002, 76 (6): 2730 – 2738

25　Mata M, Travers PJ, Liu Q, Frankel FR, Paterson Y. The MHC class I – restricted immune response to HIV – *gag* in BALB/c mice selects a single epitope that does not have a predictable MHC – binding motif and binds to Kd through interactions between a glutamine at P3 and pocket D. J Immunol, 1998, 161 (6): 2985 – 2993

26　Billaut – Mulot O, Idziorek T, Loyens M, Capron A, Bahr GM. Modulation of cellular and humoral immune responses to a multiepitopic HIV – 1 DNA vaccine by interleukin – 18 DNA immunization/viral protein boost. Vaccine, 2001, 19 (20 – 22): 2803 – 2811

27　Wu L, Kong WP, Nabel GJ. Enhanced breadth of CD4 T – cell immunity by DNA prime and adenovirus boost immunization to human immunodeficiency virus Env and *gag* immunogens. J Virol, 2005, 79 (13): 8024 – 8031

28　Lin SW, Hensley SE, Tatsis N, Lasaro MO, Ertl HC. Recombinant adeno – associated virus vectors induce functionally impaired transgene product – specific CD8 + T cells in mice. J Clin Invest, 2007, 117 (12): 3958 – 3970

29　Nanda A, Lynch DM, Goudsmit J, Lemckert. AA, Ewald BA, Sumida SM, et al. Immunogenicity of recombinant fiber – chimeric adenovirus serotype 35 vector – based vaccines in mice and rhesus monkeys. J Virol, 2005, 79 (22): 14161 – 14168

30　Tatsis N, Lin SW, Harris – McCoy K, Garber DA, Feinberg MB, Ertl HC. Multiple immunizations with adenovirus and MVA vectors improve CD8 + T cell functionality and mucosal homing. Virology, 2007, 367 (1): 156 – 167

31　Casimiro DR, Bett AJ. Fu TM, Davies ME, Tang A, Wilson KA, et al. Heteroiogous human immunodeficiency virus type 1 priming – boosting immunization strategies involving replication – defective adenovirus and poxvirus vaccine vectors. J virol, 2004, 78 (20): 11434 – 11438

32　Casimiro DR, Chen L, Fu TM, Evans RK, Caulfield MJ, Davies ME, et al. Comparative immunogenicity in rhesus monkeys of DNA plasmid, recombinant vaccinia virus, and replication – defective adenovirus vectors expressing a human immunodeficiency virus type 1 *gag* gene. J virol,

2003, 77 (11): 6305 – 6313

33　Sumida SM, Truitt DM, Lemckert AA, Vogels R, Custers JH, Addo MM. et al. Neutralizing antibodies to adenovirus serotype 5 vaccine vectors are directed primarily against the adenovirus hexon protein. J immunol, 2005, 174 (11): 7179 – 7185

34　Yang ZY, Wyatt LS, Kong WP, Moodie Z, Moss B, Nabel GJ. Overcoming immunity to a viral vaccine by DNA priming before vector boosting. J Virol, 2003, 77 (1): 799 – 803

35　Smith FS, Portner A, Leggiaclro RJ, Turner EV, HurwitzJL. Age – related development of human memory T – helper and B – cell responses toward parainfluenza virus type – 1. Virology, 1994, 205 (2): 453 – 461

36　SHIVER JW. Development of an HIV – 1 vaccine based on replication – defective adenovirus. Keystone Symposium on HIV vaccine development: Progress and Prospects 2004

37　Matano T, Kano M, Nakamura H, Takeda A, Nagai Y. Rapid appearance of secondary immune responses and protection from acute CD4 depletion after a highly pathogenic immunodeficiency virus challenge in macaques vaccinated with a DNA prime/Sendai virus vector boost regimen. J Virol, 2001, 75 (23): 11891 – 11896

38　Geldmacher C, Currier JR, Gerhardt M, Haule A, Maboko L, Birx D, et al. In a mixed subtype epidemic, the HIV – 1 *gag* – specific T – cell response is biased towards the infecting subtype. AIDS, 2007, 21 (2): 135 – 143

39　Lu W, Arraes LC, Ferreira WT, Andrieu JM. Therapeutic dendritic – cell vaccine for chronic HIV – 1 infection. Nat Med, 2004, 10 (12): 1359 – 1365

40　Geldmacher C, Gray C, Nason M, CurrierJR, Haule A, Njovu L, et al. A high viral burden predicts the loss of CD8 T cell responses specific for subdominant *gag* epitopes during chronic HIV infection. J Virol, 2007

41　Busch DH, Pilip I, Pamer EG. Evolution of a complex T cell receptor repertoire during primary

and recall bacterial infection. J Exp Med, 1998,
188 (1): 61 – 70

42 Kedl RM, Kappler Jw, Marrack P. Epitope dom-

inance, competition and T cell affinity maturation.
Curr Opin Immunol, 2003, 15 (1): 120 – 127

106. Prokaryotic Expression and Purification of HIV – 1 *vif* and hAPOBEC3G, Preparation of Polyclonal Antibodies

LI Lan, YANG Yi – shu, LI Ze – lin, ZENG Yi

College of Life Science & Bio – engineering, Beijing University of Technology

Summary

To prepare HIV – 1 *vif* and hAPOBEC3G and to produce their antibodies, the full length gene fragment of HIV – 1 *vif* was amplified by PCR from a plasmid of HIV – 1 NL4. 3 cDNA, and the APOBEC3G gene was obtained by RT – PCR from the total RNA of H9 cells. The resulting DNA construct was cloned into a prokaryotic expression vector (pET – 32a). Recombinant pET – *vif* and pET – APOBEC3G were expressed respectively in *Eserichia coli* BL21 (DE3) as an insoluble protein. The vector also contained a six – histidine tag at the C – terminus for convenient purification and detection. To express and purify the HIV – 1 *vif* and hAPOBEC3G in *E. coli* cells, the accuracy of inserted gene and specificity of proteins were detected by the two enzyme digestion method, SDS – PAGE, and Western blotting. Rabbits were then immunized by *vif* or APOBEC3G protein and serum samples were tested by indirect ELISA to determine the level of antibodies. Immunoenzyme and immunofluorescence assays were performed to identify the specificity of polyclonal antibodies. The titer of the anti – *vif* antibodies was 1 : 204 800, and that of the anti – APOBEC3G antibodies was 1 : 102 400. Thus the antibodies could detect the antigen expression in the cells, demonstrating that fusion proteins with high purity and their corresponding polyclonal antibodies with high titer and specificity were achieved.

〔**Key words**〕 Human immunodeficiency virus type 1 (HIV – 1); Viral infectivity factor; hAPOBEC3G; Protein purification; Polyclonal antibody

Introduction

The viral infectivity factor (*vif*) gene which is located in the genome of human immunodeficiency virus (HIV) was originally discovered in the mid 1980s and was initially called *sor*[10, 3]. The gene product is a 23×10^3 basic protein encoded by all lentiviruses, except for the equine infectious anemia virus, and was later renamed as *vif*. Compared to the *rev* and *gag* sequences, the *vif* sequence is highly conserved across clades[6]. *APOBEC3G*, the apoli – protein B mRNA – editing enzyme – catalytic polypeptide – like – 3G, which is arranged in tandem on chromosome 22, includes eight exons and seven introns. The length of the *APOBEC3G* cDNA is 1155 bp, and encodes 384 amino acid residues[9]. *APOBEC3G* has been identified as a mediator of anti – HIV – 1 activity[1, 2, 7], and its activity has been shown to be suppressed by *vif*[5, 11]. The interaction between HIV – 1 *vif* and APOBEC3G provides a potential novel target for *vif* and its subsequent products in the HIV life cycle. In this study, HIV *vif* and *hAPOBEC3G* were each cloned into a prokaryotic expression vector (pET – 32a) and the resulting recombinant pET – *vif* and pET – APOBEC3G were highly expressed respectively in *E. coli* BL21 (DE3) as an insoluble protein. In addition, rabbits were immunized with the purified protein to produce polycloned antibodies. These products have potential for use in studying the construction of HIV – 1 *vif* and as identifying potential new targets for HIV drugs.

Materials and Methods

1. Reagents

Restriction endonucleases, T4 DNA ligase, Taq DNA polymerase and Agarose Gel DNA Fragment Recovery Kit were obtained form the Takara Company (Dalian, China). Anti – His – Tag monoclonal antibody and goat anti – mouse IgG/HRP were purchased from New England Biolabs. All other chemicals were of analytical reagent grade.

2. Cloning of *vif* by PCR

For PCR amplification of *vif*, a forward primer (5' – tgcaccatggaaaacagatggcaggtgat – 3') and reverse primer (5' – ccgctcgaggggtgtccattcattgtat – 3') were used, which overhang the *Nco* I and *Xho* I sites (underlined) respectively. In a 50μl reaction system, the template used for *vif* open reading frame (ORF) amplification by polymerase chain reaction (PCR) was 1 μl plasmid of HIV – 1 NL4. 3 cDNA. Amplification was accomplished using the following reaction conditions: initial heating at 94℃ for 5 min; then 30 cycles consisting of denaturation at 94℃ for 45 s, annealing at 55℃ for 45 s and elongation at 72℃ for 1 min; final extension at 72℃ for 10 min. The gene products were ligated to the cloning vector pGEM – T. The DNA sequence was determined using an automated sequencer.

3. Cloning of APOBEC3G by RT – PCR

RNA was extracted from H9 cells using TRIzol reagent according to the manufacturer's specifications. APOBEC3G was amplified using these primers: (F): 5' – ccatggctatgaagcctcact – Y, (R): 5' – ctcgagcgttttcctgattctg – 3'. The first step of RT – PCR to synthese cDNA used the reaction condition: 42℃, 30 min. The second step used the reaction condition: initial heating at 94℃ for 5 min; followed by 35 cycles consisting of denaturation at 94℃ for 50 s, annealing at 55℃ for 1 min and elongation at 72℃ for 3 min; and final extension at 72℃ for 10 min. The gene products were ligated to the cloning vector pGEM – T. The DNA sequence was determined using an automated sequencer.

4. Construction of expression vector for *vif* and *APOBEC3G*

The amplified PCR product was digested with *Nco* I and *Xho* I followed by subcloning into the prokaryotic expression vector pET – 32a. This generated the recombinant plasmid pET – *vif* and pET – APOBEC3G respectively, in which the C – terminal domains of *vif* and *APOBEC3G* were fused with a His – tag for improved purification and immune identification. *E. coli* DH5α. competent cells were transformed with the ligation reaction, and the DNA was subsequently purified and checked by 1% agarose – gel electro – phoresis after digesting with *Nco* I and *Xho* I. The samples were then sequenced to confirm the correct insert by the Shanghai Sangon Biology Technology Limited Company (sequencing primer is T7 termi – nator primer). The results were compared with the expected sequence using the vector NTI 8. 0 software.

5. Recombinant pET – *vif* and pET – APOBEC3G expression *E. coli*

E. coli BL21 (DE3) competent cells were transformed with pET – *vif* and pET – APOBEC3G respectively, and the transformant cells were cultivated at 37℃ with shaking at 190 r/min in 5ml of LB medium supplemented with 100 μg of ampicillin and 34 μg of chloramphenicol per ml. Cultures were induced for expression at an A_{600} of 0. 6 – 0. 8 with 1 mmol/LIPTG, and continued growth with shaking for 4 h. Cells were then harvested by centrifugation at 12 000 r/min for 1 min. Total cellular pellets were analysed by 12% SDS – PAGE and Western blotting.

6. Purification of recombinant *vif* and APOBEC3G

Identified pET – *vif*/BL21 (DE3) and p ETAPOBEC3G/BL21 (DE3) cells were cultivated overnight. 8 ml Cultures were added to 800 mL 2 × YT medium supplemented with 100 μg of ampicillin and 34μg of chloramphenicol per mL, and were cultivated at 37℃ with shaking at 190 r/min. Cultures were induced for expression at an A_{600} of 0. 6 – 0. 8 with 1 mmol/L IPTG, and continued growth with shaking for 4 h. Cells were harvested by centrifugation at 5000 r/rain for 20 min at 4℃. Cell pellets were thoroughly suspended in 10 ml of lysis buffer (10 mmol/L Tris – HCl, 100 mmol/L NaH_2PO_4, pH 8. 0) per gram of cell paste. The cells were then sonicated at 50% amplitude on ice for 300 s (30s on/30s out) followed by centrifugation at 8000 r/min for 30 min at 4℃. The pellet washing was repeated for a total of three washes and the final pellet was resuspended in 10 mL extraction buffer (10 mmol/L Tris – HCl, 100mmol/L NaH_2PO_4, and 7. 5 mol/L guanidine – HCl, pH 8. 0) per gram of cells, followed by incubation ovemight at 15℃ with gentle mixing. Following the step, the sample was centrifuged at 12 000 g for 30 min at 4℃. 2 ml supernatant was then filtered through a 0. 45 μm syringe filter and applied to an equilibrated gel filtration chromatography column (superdex 75 200) using a 2 ml Superloop at 1 ml/min on the FPLC, and eluted by buffer B (100 mmol/L NaH_2PO_4, 10 mmol/L Tris, 8 mol/L Urea, pH 6. 2). The purified productions were analyzed by 12% SDS – PAGE and Western blotting.

7. Western blotting analysis for recombinant protein

For Western blotting, samples were separated using 12% SDS – PAGE and then transferred to pre – cut nitrocellulose membrane. The membranes were blocked for 1 h at ambient temperature in phosphate – buffered saline (PBS) with 5% nonfat dried milk, and then incubated for 1 h with rat anti – His monoclonal antibody (1 : 200). The membranes were rinsed thrice with PBST (1 × PBS,

0.05% Tween 20, pH 7.5), followed by incubation with goat anti – rat IgG antibodies conjugated to horseradish peroxidase (HRP) (1 : 400) for 1 h at ambient temperature. The *vif* and APOBEC3G samples were visualized by incubation with DAB (0.06 mg/mL) and 0.1% H_2O_2.

8. Protein concentration detection

Concentration of protein was detected by the Bradford method.

9. Preparation polyclonal antibodies

Polyclonal anti – recombinant protein antibodies were raised in two rabbits. Rabbit sera were taken prior to immunization as negative control. Rabbits were immunized on days 1 and 28 with 500μg recombinant protein in their legs. Each immunization was done with 1 ml 500 μg/ml recombinant protein, which was diluted to 2 ml with Freund's complete adjuvants. Serum samples were collected two weeks niter the last injection and stored at – 70℃. Polyclonal antibodies in sera were tested by indirect ELISA.

10. Indirect ELISA

Microtiter plates were coated overnight at 4℃ with 60 ng *vif* and 15 ng APOBEC3G per well in coating buffer (carbonate – buffer, pH 9.6). Unbound antigen was removed by washing with PBST solution (1 × PBS, 0.2% Tween 20). 2% bovine serum albumin (BSA) wan added to block the coated wells for 2 h at 37℃. Serum samples diluted 200 – 204 800 fold were added to the coated wells and incubated for 90 min at 37℃ in a humidified atmosphere. Wells were washed with PBST solution and incubated with 100 μl of HRP – conjugated goat anti – rabbit immunoglobulins. After 60 min of incubation at 37℃, wells were washed and the antigen – antibody complexes were detected by the addition of 100 μl of A and B solution. After 15 min of incubation at room temperature, the enzymatic reaction was stopped by adding 50 μl of 2 mol/L sulfuric acid to each well. Optical densities (A) were measured at 450/630 nm. The cutoff value was set two standard deviations above the average of negative control sample which was taken before the immunization.

11. Construction of pEGFP – *vif* and transient cell transfection

Expression vectors pEGFP – *vif* expressing HIV – 1 *vif*. The *vif* gene sequence was amplified by PCR from the HIV plasmid NL4.3 cDNA. For DNA transfection, 6 – well plates were seeded with Hela – CD4 – LTR – β – gal – CXCR4 cells at a density of 1×10^5 cells/well in 2 ml of DMEM cell culture medium and maintained at 37℃ and 5% CO_2. The following day, the optimal amount of pEGFP – *vif* for transfection was determined to be 6 μg. Lipofectamine 2000 (In vitrogen, Carlsbad, California) was used essentially according to the manufacturer's instructions, at 3 μl Lipofectamine in FBS – and antibiotic – free DMEM. At 6 h post – transfection, the medium of the cells was replaced with fresh DMEM containing 10% FBS.

12. Cell preparation

Hela – CD4 – LTR – β – gal – CXCR4 cells transfected by pEGFP – *vif*, H9 cells and MT4 cells were fixed in 4% formalin for 10 min at room temperature.

13. Immunoenzyme and immunofluorescence assay

Cells were incubated with polyclonal antibodies for 1 hour at 37℃. After washing three times with PBS, cells were incubated with HRP – conjugated or TRITC – conjugated goat anti – rabbit antibodies for 45 minutes at 37℃. Immunostaining was revealed by exposure to fresh DAB solution for 5 min, and rinsed with distilled water. The result of immunofluoresence was observed directly by fluorescent microscope.

Results

1. Identification of the pET – *vif* and pET – APOBEC3G

Two fragments were obtained from the recombinant plasmid of pET – *vif* by digesting with *Nco* I and *Xho* I, and the bulks of two fragments were about 576 bp and 5.9 kb respectively (Fig. 1). By the same method, digested pET – APOBEC3G had two fragments with 1152 bp and 5.9 kb (Fig. 2). Plasmid pET – *vif* and pET – APOBEC3G were sequenced by the Shanghai Sangon Biology Technology Limited Company. Using Vector NTI AlignX software, the pET – *vif* sequencing result was compared with the HIV – 1 NL4.3 *vif* sequence (GenBank No.: AF070521). This indicated that the homologous fragment of 576 bp had high similarity to the expected sequence and the recombinant plasmid pET – *vif* was correctly constructed. The pET – APOBEC3G sequencing result was compared with the *APOBEC3G* sequence (GenBank No.: NM – 021822) and it was similarly demonstrated that the homologous fragment of 1152 bp had high similarity and the recombinant pET – APOBEC3G gene was correctly constructed.

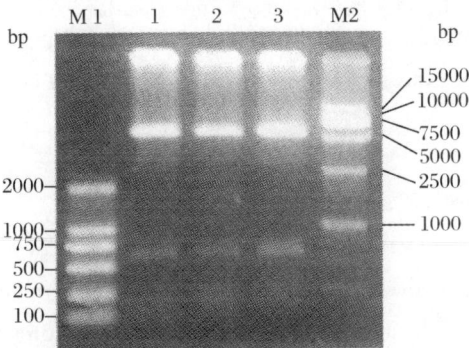

1 – 3, pET – *vif*; M1/M2, DNA ladders

Fig. 1   Analysis of recombinant plasmid pET – *vif* by *Nco* I & *Xho* I digestion

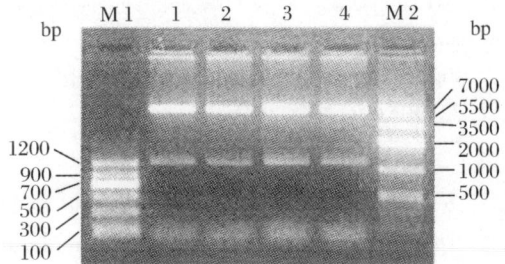

1 – 4, pET – APOBEC3G; M1/M2, DNA ladders

Fig. 2   Analysis of recombinant plasmid pET – APOBEC3G by *Nco* I & *Xho* I digestion

2. Expression of *vif* and APOBEC3G fusion protein

Extracts of *E. coli* BL21 (DE3) cells transformed with the recombinant plasmid and induced by IPTG treatment were analysed by polyacrylamide gel electrophoresis followed by staining with Coomassie Brilliant Blue and Western blotting. Fig. 3 shows strong expression of the *vif* – His fusion protein of about 45×10^3 (lane 4, arrowed), which was consistent with the expected molecular mass of the desired protein. In the lane containing the blank control, the band of the *vif* fusion protein was not detected (lane 3). There were faint expression before induction (lane 1), and after induction a band of about 22×10^3 was visible which was consistent with the enhanced solubility of the protein

· 815 ·

(lane 2, arrowed). Its sequence was then introduced immediately upstream of the *vif* sequence in the vector. The clearly visible band of APOBEC3G fusion protein was present in the total cell extracts of *E. coli* BL21 (DE3) cells transformed with pET – APOBEC3G after induction (lane 6, arrowed) but not the cell extracts before induction (lane 5). The relative amount of *vif* was estimated by densitometry of the stained gel, and was found to be about 17.7% of the total protein; similarly the relative amount of APOBEC3G was 22.3%.

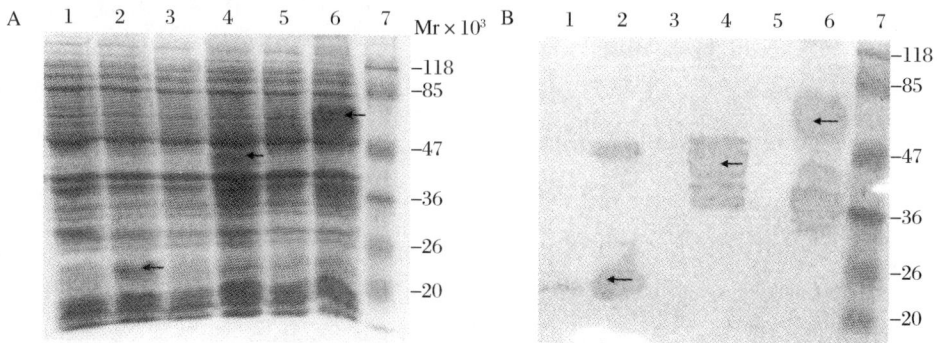

A: 1, pET – 32a total cell extract (uninduced); 2, pET – 32a total cell extract (induced); 3, pET – *vif* total cell extract (uninduced); 4, pET – *vif* total cell extract (induced); 5, pET – APOBEC3G total cell extract (uninduced); 6, pET – APOBEC3G total cell extract (induced). B: 1 – 6, Western blotting of A1 – 6 respectively; 7: Protein marker

Fig. 3 SDS – PAGE and Western blotting analysis of expressed protein by pET – *vif* and pET – APOBEC3G

3. Purification and identification of recombinant *vif* and APOBEC3G

Total cell content after sonication, including both the supernatant fluid (soluble – protein fraction) and the pellet (inclusion – body fraction), were analysed by SDS – PAGE and Western blotting. Most of the expressed *vif* protein accumulated in inclusion bodies (data not shown). After the inclusion bodies were washed and dissolved, the supernant was passed through gel filtration chromatography and eluted by buffer B. The purified proteins were analysed by 12% SDS – PAGE and Western blotting. Figure 4A shows the *vif* fusion protein was -45×10^3 (lane 2) and APOBEC3G was -68×10^3 (lane 7). These results were consistent with the expected molecular mass of the each protein. Furthermore, Westem blotting analysis of these fractions using an anti – His monoclonal antibody confirms this result (arrowed if Fig. 4B). The concentration of purified *vif* protein was 0.45 mg/ml, and that of APOBEC3G was 0.87 mg/ml.

4. Immunogenicity evaluation of recombinant *vif* and APOBEC3G

After repeated inoculation, the sera of rabbits were collected and antibodies to anti – *vif* and anti – APOBEC3G measured by indirect ELISA. To confirm appropriate quantities of the antigen, serial dilutions of recombinant protein were made down to 60 ng/well for the *vif* protein and 15 ng/well for the APOBEC3G. The antibody titer was calculated from the mean of four wells. The sera before immunization were used as negative control. The highest antibodies level was elicited after the immu-

A: 1, pET – *vif* total cell extract (induced); 2, Purified fusion protein of *vif*; 3, pET – 32a total cell extract (induced); 4, Purified protein of enhancing solubility; 5, Protein marker; 6, pET – APOBEC3G total cell extract (induced); 7, Purified fusion protein of APOBEC3G. B: 1, Protein marker; 2, Purified fusion protein of *vif*; 3, Purified protein of enhancing solubility; 4, Purified fusion protein of APOBEC3G

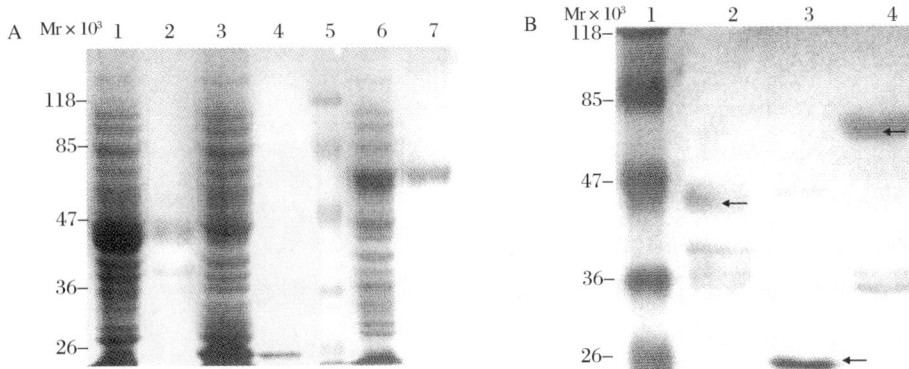

Fig. 4   SDS – PAGE and Western blotting analysis of purified fusion protein

nization. The titer of antibodies to anti – *vif* reached 1 : 204 800, and to anti – APOBEC3G reached 1 : 102 400 (Fig. 5). In this study, the cutoff value of *vif* was 0. 038, and that of APOBEC3G was 0. 134.

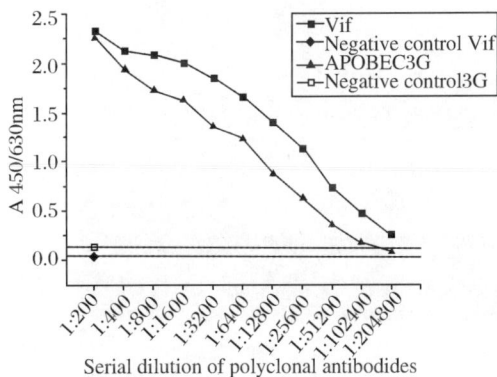

Fig. 5   Titer of polycloned antibodies detected by indirect ELISA

1, pEGFP – *vif*; M1/M2, DNA ladders

Fig. 6   Analysis of recombinant plasmid pEGFP – *vif* by *Xho* I & *BamH* I digestion

5. Identification of the pEGFP – *vif*

The recombinant plasmid of pEGFP – *vif* was generated by cloning the corresponding *vif* coding sequence obtained from the HIV – 1 NL4. 3 into the pEGFP – C3 vector, pEGFP – *vif* was digested with *Xho* I and *BamH* I. Two fragments were obtained at about 576 bp and 4. 7 kb respectively (Fig. 6). pEGFP – *vif* was then sequenced, and the result was compared with the submitted HIV – 1 NL4. 3 *vif* sequence (GenBank No. : AF070521). The result demonstrated that the homologous fragment of 576 bp maintained high similarity.

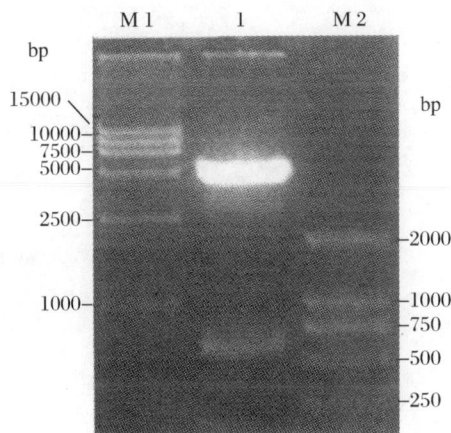

6. Specificity evaluation of anti – *vif* polyclonal antibodies

Hela – CD4 – LTR – β – gal – CXCR4 cells were transected with pEGEP – *vif*, which expresses HIV – 1 *vif* with a GFP tag, and the cells transfected with pEGFP – C3 only express green fluorescent protein. Observed by fluorescent microscope, Hela – CD4 – LTR – β – gal – CXCR4 cells with green signal (Fig. 7A) demonstrated pEGFP – *vif* had been transfected into cells, and HIV – 1 *vif* and been expressed successfully. Figure 7B showed green fluorescent protein had been expressed, as the negative control. After immunofuorescence staining, the cells which expressed *vif* showed saffron yellow fluorescence (Fig. 7C), and there was no change in cells which expressed green fluorescent protein only (Fig. 7D).

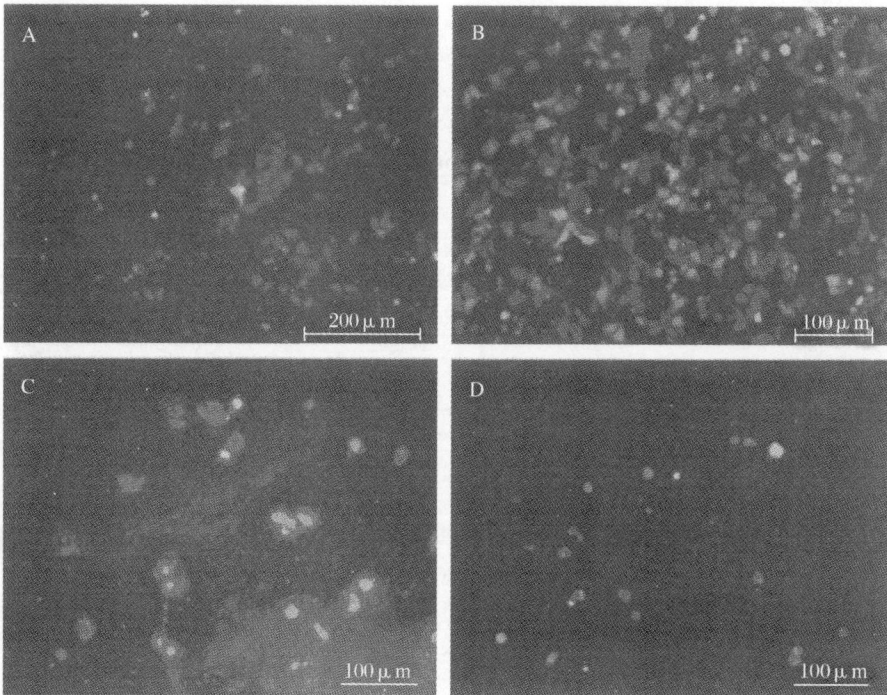

A: Hela – CD4 – LTR – β – gal – CXCR4 cells transfected with pEGFP – *vif*. B: Hela – CD4 – LTR – β – gaI – CXCR4 ceils transfected with pEGFP – C3. C: Cells expressing *vif* indicating fluorescent staining. D. Cells expressing GFP indicating fluorescent staining

Fig. 7   Micrographs of immunofluorescence assay

7. Specificity evaluation of anti – APOBEC3G polyclonal antibodies

Human cells can be divided into two categories based on whether they are able to support the replication of *vif* – deficient HIV – 1 (permissive cells) or unable to do so (non – permissive cells)[8]. H9 cells (non – permissive T cells) expressed endogenous hA3G, whereas MT4 cells (permissive cells) failed to express hA3G. Immunoenzyme assay studies of H9 and MT4 cells (Fig. 8) revealed that H9 cells were positive for anti – APOBEC3G polyclonal antibodies, and MT4 cells were negative.

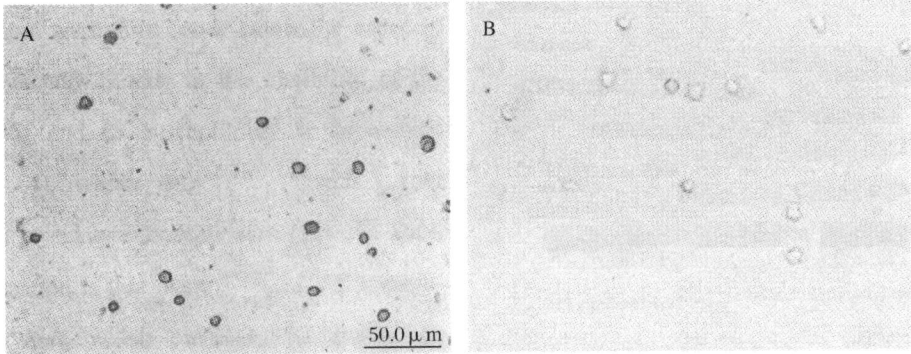

A: Stained H9 cells are clearly visible. B: No staining of MT4 cells is observed

Fig. 8   Results of immunoenzyme assay

Discussion

The viral infectivity of HIV is modulated by *vif*. HIV – 1 *vif* is a 192 – aminoacid protein conserved in all known lentiviruses with the exception of equine infectious anemia virus. Moreover, since *vif* is essential for viral infection, its molecular characterization should provide information on potential antiretroviral strategies. The purpose of this study was to obtain the *vif* protein and confirm its immunogenicity and antigenicity.

The T7 – RNA – polymerase – promoter expression systemis a tightly controlled bacterial expression system. The 22×10^3 protein located before the *vif* could enhance expression and solubility. In our experiments, v/f or APOBEC3G DNA was inserted into the T7 – promoter – controlled bacterial expression vector pET – 32a and the resulting *vif* or APOBEC3G was successfully expressed at high levels within a few hours of IPTG inductino. Total cell content after sonication, including both the supernatant fluid (soluble – protein fraction) and the pellet (inclusionbody fraction), were analysed by SDS – PAGE and Western blotting. Most of the expressed *vif* or APOBEC3G protein was accumulated in inclusion bodies. The recombinant protein could not be detected in the soluble fraction. The expressed recombinant protein in the inclusionbody contained histidine residues as a C – terminal marker sequence to facilitate purification and detection, which were beneficial in achieving high purity protein SDS – PAGE and Western blotting indicated that no band wsa present priot to IPTG induction. After induction, the strong expression of a *vif* – His fusion protein of about 45×10^3 was detected corresponding to the expected molecular mass of the desired protein. However, there were faint bands below the expected band and this may be due to the instability of the expressed *vif* and its susceptibility to proteolytic degradation[4], which is consistent with earlier observations in eukaryotic expression[12]. An alternative explanation is that overexpressed *vif* accumulated in *E. coli*, which promoted its degradtion. Nevertheless, it has been demonstrated a homogenous *vif* protein with high purity was obtained using this technique.

The purified protein was used to immune the rabbits, and achived polyclonal antibodies which had high titer and excellent specificity. The titer of anti − *vif* antibodies in srea of rabbits after immunization with this recombinant protein reached up to 1 ∶ 204 800, and that of anti − APOBEC3G antibodies reached to 1 ∶ 102 400. These results ensure the immunogenicity and antigenicity of the purified recombinant proteins.

Immunoenzyme staining demonstrated that the anti − APOBEC3G polyclonal antibodies could detect APO − BEC3G in H9 cells, but not in MT4 cells which did not express APOBEC3G. Anti − *vif* polyclonal anti − bodies could detct HIV − 1 *vif* in Hela − CD4 − LTR − β − gaI − CXCR4 cells by immunofluorescence assay. The present study demonstrates the specificity of antibodies.

In summary, the results described above show that we have been able to obtain large amounts of purified protein using the pET − *vif* or pET − APOBEC3G construct and can prepare high effectivity antibodies. The producion of *vif* will be a subject of further structure study and a target of an anti − HIV drug.

〔In 《Virological Sinica》 2008, 23 (3): 173 − 182〕

References

1 Doehle B P, Schafer A, Cullen B R. Human APOBEC3B is a potent inhibiror of HIV − 1 infectivity and is reexistant to HIV − 1 *vif*. Virology, 2005, 339 (2): 281 − 288

2 Dussart S, Douaisi M, Courcoul M, *et al.* APOBEC3G ubiquitination by Nedd4 − 1 favors its packaging into HIV − 1 particles. J Mole Biol, 2005, 345 (3): 547 − 558.

3 Fisher A G, Ensoli B, Ivanoff L, *et al.* The sor gene of HIV − 1 is required for efficient virus transmissino in vitro. Science, 1987, 237: 888 − 893

4 Fujita M, Akari H, Sakurai A, *et al.* Expression of HIV − 1 accessory protein *vif* is controlled uniquely to be low and optimal by proteasome degradation. Microbes Infect, 2004, 6: 791 − 798

5 Gabuzda D H, Lawrence K, Langhoff E, *el al.* Role of *vif* in replication of human immunodeficiency virus type 1 in CD4 + T lymphocytes. J Virol, 1992, 66: 6489 − 6495

6 Lee S K, Dykxboorn D M, Kumar P, *el al.* Lentiviral delivery of short hairpin RNAs protects CD4 T cells from multiple clades and primary isolates of HIV. Blood, 2005, 106 (3): 818 −

826

7 Luo K, Liu B, Xiao Z, *et al.* Amino − terminal region of the human immunodeficiency virus type I nucleocapsid is required for human APOBEC3G packaging. J Virol, 2004, 78 (21): 11841 − 11852

8 Rose K M, Marin M, Kozak S L, *et al.* The Viral infectivity factou (*vif*) of HIV − 1 unveiled. Trends Mol Med, 2004, 10 (6): 291 − 297

9 Sheehy A M, Gaddis N C, Choi J D, *et al.* Isolation of a human gene that inhibits HIV − 1 infection and is suppressed by the viral *vif* protein. Nature, 2002, 418: 646 − 650

10 Strebel K, Daugherty D, CIouse K, *et al.* The HIV 'A' (sor) gene product is essential for vius infectivity. Nature, 1987, 328: 728 − 730

11 Von S U, Song J, Aiken C, *et al.* *Vif* is crucial for human immunodeficiency virus type I proviral DNA synthesis in infected cells. J Virol, 1993, 67: 4945 − 4955

12 Wang H, Sakurai A, Khamsri B, *et al.* Unique characteristics of HIV − 1 *vif* expression. Microbes Infect, 2005, 7 (3): 385 − 390

107. Immunogenicity of DNA and Recombinant Sendai Virus Vaccines Expressing the HIV – 1 *gag* Gene

FENG Xia[1], YU Shuang – qing[1], SHU Tsugumine[2], MATANO Tetsuro[3], HASEGAWA Mamoru[2], WANG Xiao – li[4], MA Hong – tao[4], LI Hong – xia[1], ZENG Yi[1,4]

1. State Key Laboratory for Infectious Disease Prevention and Control, National Institute for Viral Disease Control and Prevention, Chinese Center for Disease Control and Prevention; 2. DNAVEC Corporation, Japan; 3. Department of Microbiology, Graduate School of Medicine, University of Tokyo, Japan; 4. College of Life Science and Bioengineering, Beijing University of Technology

Summary

Combinations of DNA and recombinant – viral – vector based vaccines are promising AIDS vaccine methods because of their potential for inducing cellular immune responses. It was found that *gag* – specific cytotoxic lymphocyte (CTL) responses were associated with lowering viremia in an untreated HIV – 1 infected cohort. The main objectives of our studies were the construction of DNA and recombinant Sendai virus vector (rSeV) vaccines containing a *gag* gene from the prevalent Thailand subtype B strain in China and trying to use these vaccines for therapeutic and prophylactic vaccines. The candidate plasmid DNA vaccine pcDNA3.1 (+) – *gag* and recombinant Sendai virus vaccine (rSeV – *gag*) were constructed separately. It was verified by Western blotting analysis that both DNA and rSeV – *gag* vaccines expressed the HIV – 1 *gag* protein correctly and efficiently. Balb/c mice were immunized with these two vaccines in different administration schemes. HIV – 1 *gag* – specific CTL responses and antibody levels were detected by intracellular cytokine staining assay and enzyme – linked immunosorbant assay (ELISA) respectively. Combined vaccines in a DNA prime/rSeV – *gag* boost vaccination regimen induced the strongest and most long – lasting *gag* – specific CTL and antibody responses. It maintained relatively high levels even 9 weeks post immunization. This data indicated that the prime – boost regimen with DNA and rSeV – *gag* vaccines may offer promising HIV vaccine regimens.

[**Key words**]　HIV – 1 vaccines; *gag* gene; DNA vector; Sendai virus

Introduction

The global HIV epidemic continues to expand, exceeding previous predictions and causing tremendous suffering. It has been the leading cause of death in Africa. At the same time, a rapid increase of HIV infection has also been found in China in recent years. Therefore, developing HIV vaccines targeting the prevalent strains in China is one of the most important tasks for Chinese HIV/AIDS control and prevention.

Virus – specific cellular immune responses play an important role in the control of HIV infections[1, 6, 23]. Recently some studies indicated that only *gag* – specific responses were associated with lowering viremia in an untreated HIV – 1 infected cohort and HIV – 2 long – term nonprogressors, while Env – and Nef – specific responses are positively correlated with a high viral load[4, 7, 13, 15, 16]. Thus *gag* would be the preferred antigen in HIV candidate vaccines. DNA vaccines, recombinant – viral – vector based vaccines, and their combinations are promising AIDS vaccine methods because of their potential for inducing cellular immune responses. DNA vaccine is safe and easy to manufacture, and it is quite effective as a priming or initial immunogen in a bimodal vaccine strategy[17, 22]; Some of the extraordinary features of the SeV vector are the remarkably brief contact time that is necessary for cellular uptake, a strong but adjustable expression of foreign genes, and an exclusively cytoplasmic replication cycle without any risk of chromosomal integration[5]. An extremely efficient antigen expression system in mammalian cell cultures using recombinant SeV were established in the late 1990's[8, 12, 24]. Using this system Kano *et al.* demonstrated the excellent protective efficacy of DNA priming followed by simian immunodeficiency virus (SIV) *gag* – expressing Sendai virus boosting against a pathogenic simian – human immunodeficiency virus (SHIV89. 6PD) infection in macaques[10]. Takeda *et al.* obtained similar results by using a DNA prime/replication defective SIV *gag* – expressing SeV boost system[19]. In the present study a HIV – 1 *gag* gene was used to construct DNA and replication defective rSeV vaccines. High – frequency CTL responses may be elicited by combining these two vaccines.

Although the discontinuation of the Merck rAd5 Phase Ⅱ proof – of – concept STEP study is undoubtedly a significant setback for the field of HIV vaccine development[18], it does not indicate the failure of the T – cell vaccine concept. In their study a trivalent recombinant adenovirus type 5 vaccine expressing *gag*, Pol, and Nef respectively was tested. The issues of lack of vaccine efficacy and differential infection rates between vaccine and placebo groups with previous Ad5 immunity are complex ones and require careful consideration. Selection of T – cell vaccine antigens for chronic persistent viral infections has been largely empirical. Though immunization with more than one immunogen (co – immunization) is an efficient regimen to induce immunity to multiple antigens, Toapanta *et al* found that when HIV – 1 Env (*gp*120) and HIV – 1 *gag* (p55) DNA plasmids were co – inoculated, there was a reduction in the immune responses elicited to HIV – 1 *gag* (p55). This anti – HIV – 1 *gag* immune interference was specific to coimmunizations with HIV – 1 Env (*gp*120) and may involve a yet undefined immunological mechanism[20]. Recently Kiepiela et al performed a comprehensive analysis of the 160 dominant CD8 [+] T cell responses in 578 untreated HIV – infected individuals from KwaZulu – Natal, South Africa. They found only *gag* – specific responses were associated with lowering viremia[13]. So in our study *gag* – expressing vaccines would be investigated first, and only if excellent immune efficacy could be induced would other antigens be included in a further study.

Most worldwide vaccine developments have been focused on inducing protection against prevalent strains in North America and Europe. Developing HIV vaccines targeting the prevalent strains in China is of great importance Thailand subtype B is found to be prevalent in several epidemic regions where paid blood donors are the principally affected population. The HIV – 1 Subtype B in these epidemic regions is relatively well conserved, so the prevalent strains isolated from these regions was used as the vaccine strain. The construction of DNA and rSeV vector vaccines containing the *gag* gene of prevalent Thailand B strains in China was therefore the main objectives of our studies.

Materials and Methods

1. Construction of DNA and rSeV vaccines

The eodon – modified consensus *gag* gene from HIV – 1 prevalent strains in Henan province was used to construct a DNA vaccine. Firstly, the *gag* genes were amplified by nest – PCR using specific primers from the DNA from the peripheral blood mononuclear cells (PBMC) of HIV – 1 infected patients in Henan province. Twelve *gag* sequences were obtained from 20 patients and 6 of them had complete open reading frames (ORF). The consensus *gag* gene was generated by sequence malignment of the 6 *gag* genes. The consensus *gag* sequence had different amino acids in 6 positions compared with reference sequences of subtype B (B. FR. HXB2 – LAI – IIIB – BRU and B. TH. BK132). 95 % and 96 % of amino acid sequence homology were found with B. FR. HXB2 – LAI-IIIB – BRU and B. TH. BK132 respectively. To increase the expression level of *gag* protein in DNA vector the eodons of the consensus *gag* gene were modified according to mammalian condon usage. The codonmodified *gag* gene was synthesized and inserted into plasmid pUC57 by the Shanghai Sangon Biological Engineering Technology & Service Co. , Ltd (Shanghai, China). *Kpn* I and *Xho* I restriction enzyme sites were inserted into the *gag* gene by PCR and cloned into pcDNA3. 1 (+) which was named peDNA3. 1 (+) – *gag*. The pcDNA3. 1 (+) – *gag* was standardized at 1 mg/ml in endotoxin – free Buffer TE.

The wild – type *gag* gene was used for construction of the recombinant Sendai virus containing *gag* (rSeV – *gag*). The rSeV – *gag* was constructed and amplified by the DNAVEC Corporation (Tsukuba, Japan) as follows: (i) EIS sequence and *Not* I restriction enzyme sites were incorporated into the *gag* gene by PCR, cloned into pBluescript KS (+) and was named pBS – HIV*gag*. (ii) pBS – HIV*gag* was digested with *Not* I and the DNA fragment containing *gag* was inserted into pSeV/△F (a viral genomic RNA – encoding plasmid) that contains SeV full – length eDNA lacking the F gene. The generated plasmid was named pSeV18 + HIV*gag*/△F. (iii) Recombinant virus containing the *gag* gene was recovered by co – transfeetion of 293T/17 cells with pCAGGS – P (z) /4C – (P protein – expressing plasmid), pCAGGS – NP (NP protein – expressing plasmid), pCAGGS – L (TDK) (L protein – expressing plasmid), pCAGGS – F5R (modified – F protein – expressing plasmid), pCAGGS – T7 (T7 RNA polymerase – expressing plasmid) and pSeV18 + HIV*gag*/△F. (iv) Recovered virus was cloned and amplified in LLC – MK2/F/A cells expressing the F protein. Virus yield is expressed in cell infectious units (CIU). The recombinant Sendai virus was titrated as follows: LLC – MK2 cells were seeded into a 6 – well plate at a cell density of 2×10^5 cells/2ml/well and incubated at37℃, 5% CO_2 for 72 h until 100% confluence. 10 – fold dilutions of the virus stock were prepared in PBS containing 1% BSA. The confluent monolayer was washed once with PBS and then was inoculated in duplicate with 0. 1 ml of the virus samples/well. The cells were incubated for 1 h at 37℃, 5% CO_2. and the plate was tilted every 15 min. Virus inoculum was removed, the cells were washed with PBS and supplemented with 2 ml of DMEM and incubated at 37℃, 5% CO_2 for 48 h. The cells were washed with PBS and fixed with methanol for 10 min at 4℃. After removing methanol the plate was dried for 10 min at room temperature. 0.5 ml of 1 : 500 diluted polyclonal rabbit – anti SeV antibodies (prepared by the DNAVEC Corporation) was added into each well. The plate was incubated for 45 min at 37℃ then was washed twice with PBS. 0.5 ml of 1 : 200 diluted goat anti – rabbit immunoglobulin G (Zhongshan Goldenbridge Biotech – nology Co., LTD, Beijing, China) was added into each well. The plate was incubated for 45 min at 37℃ then was washed twice with PBS. The number of positive cells was counted under the fluorescence microscope. The virus titer was calculated by the formula: mean numbers of positive cells in dupli-cate wells × dilution multiple × 10 (CIU/ml). The titer of generated rSeV – *gag* was 1.8 × 10^{10} CIU/ml.

2. Identification of *gag* gene in rSeV – *gag* by RT – PCR

LLC – MK2 cells were seeded into a 6 – well plate at a cell density of 2 × 10^5 cells/2ml/well and incubated at 37℃, 5% CO_2 for 72 h to 100% confluence. The confluent monolayer was washed with PBS, then was inoculated with 0.1 ml of the virus samples/well (MOI of 5, to count the cell number per well, the cells in one well was trypsinized and counted under a microscope). The cells were incubated for 1 h at 37℃, 5% CO_2. and the plate was tilted every 15 min. Virus inoculum was removed, the monolayer was washed with PBS and supplemented with 2ml of DMEM and incu-bated at 37℃, 5 % CO_2 for 48 h. The virus which did not contain the foreign gene (rSeV – control) was used as the control. 48 h later, total RNA was isolated from infected cells using Trizol reagent (Promega, Madison, USA). One – Step RT – PCR was used to test the presence of the *gag* gene in the recombinant virus (wt269F: 5' – GGCAAGCAGGGAACTAGAAC – 3', wt269R: 5' AGAACCG-GTCTACATAGTCTC – 3').

3. Identification of *gag* expression in DNA and rSeV – *gag* vaccines

LLC – MK2 cells were maintained in DMEM and supplemented with 10% fetal bovine serum (FBS). Cells were transfected with pcDNA3.1 (+) – *gag* or infected with rSeV – *gag* at a MOI of 5. Proteins of transfected or infected LLC – MK2 cells were extracted 48 h later using TRIzol rea-gent. The protein samples were subjected to SDS – PAGE and electroblotted onto nitrocellulose blot-ting membranes. Blots were blocked with 5% fat free milk in phosphate buffered saline (PBS) con-taining 0.05% Tween 20 and probed with mouse anti – HIV – 1 P24 monoclonal antibody (NIH AIDS Research and Reference Reagent program, Germantown, USA) and peroxidase – conjugated goat anti – mouse immunoglobulin G (Zhongshan Goldenbridge Biotechnology Co., LTD, Beijing, China). Proteins were visualized by staining with 3, 3' – Diamino – benzidine.

4. Animals and immunization

Four to six – week – old Balb/c female mice were purchased from Institute of Experimental Ani-mal Sciences, Chinese Academy of Medical Science (Beijing, China). To compare the immunoge-nicity of the DNA and rSeV – *gag* vaccines applied in single or combined vaccines, mice were inoc-ulated with these vaccines intramuscularly either in single or combined modality at week 0 and week

3. The immunization dose for the DNA vaccine was $100\mu g$ per animal and for the rSeV vaccine was 1.8×10^7 CIU per animal. Balb/c mice were randomly divided into four groups of 16. Inoculation was conducted according to the schedule in Tab. 1. The splenocytes and sera of immunized mice were collected at 1, 5 and 9 weeks post immunization and the cellular and humoral immune responses were analyzed.

Tab. 1 Immunization schedule of DNA and rSeV vaccines

Groups	0w	3w	1w pi (post immunization)	5w pi	9w pi
1	PBS	PBS	6 sacrificed	5 sacrificed	5 sacrificed
2	DNA	DNA	6 sacrificed	5 sacrificed	5 sacrificed
3	rSeV – *gag*	rSeV – *gag*	6 sacrificed	5 sacrificed	5 sacrificed
4	DNA	rSeV – *gag*	6 sacrificed	5 sacrificed	5 sacrificed

5. Intracellular IFN – γ staining

Freshly isolated splenic lymphocytes (2×10^6 cells) were suspended in 10% FBS RPMI 1640 medium and incubated with $H - 2^d$ – restricted CTL epitope peptides, $gag_{197-205}$ (AMQMLKETI), $gag_{239-247}$ (TTSTLQEQI) and $gag_{291-300}$ (EPFRDYVDRF) for antigen – specific stimulation or without peptides for mock stimulation. The three peptides were pooled together and the final concentration for each peptide was $10\mu g/ml$. Cells were cultured for 16h at 37℃. For the final 12 h, brefeldin A (Sigma, Saint Louis, USA) was added at $5\mu g/ml$. After stimulation, the cells were stained (60 min, 25℃) for surface markers with 5 μl of R – PE – conjugated rat anti – mouse CD8a (Ly – 2) monoclonal antibody (BD Pharmingen, San Diego, USA). Then the cells were sequentially fixed with 4% paraformal – dehyde and permeabilized with 0.3 % saponion for 15 min respectively, and stained for 60 min with 1 μl of FITC – conjugated rat anti – mouse IFN – γ monoclonal antibody (BD Pharmingen, San Diego, USA). Stained samples were collected by a Coulter EPICS Altra flow cytometer (Beckman, Fullerton, USA) and analyzed using the CellQuest software package. Gating was performed on mononuclear cells and then on CD8$^+$ subpopulations. 50 000 of CD8$^+$ T cells were collected in total. From the ratio of CD8$^+$ IFN – γ$^+$ cells to CD8$^+$ cells, the frequency of CD8$^+$ IFN – γ$^+$ cells in the total CD8$^+$ T cells was calculated. *gag* – specific T – cell frequencies were calculated by subtracting the CD8$^+$ IFN – γ$^+$ – cell frequencies after mock stimulation from those after *gag* – specific peptides stimulation.

6. Detection of HIV – 1 *gag* – specific antibodies in immunized mice

Specific antibodies were detected by the enzymelinked immunosorbant assay (ELISA). 96 – well microtiter plates were coated with 200ng/well of HIV P24 protein prepared by our laboratory and incubated overnight at 4℃. The wells were blocked with PBS containing 5% fat free milk for 1h at 37℃. They were then treated with 100 μl of serially diluted mice sera and incubated for an additional 1 h at 37℃. The plates were washed five times with PBS containing 0.05% Tween – 20 and incubated for 1h with 1 : 20 000 diluted goat anti – mouse IgG/HRP. The plates were then washed five times, developed with tetramethylbenzidine, stopped with 2 mol/L H_2SO_4, and analyzed at $A_{450nm/630nm}$.

Results

1. Restriction endoenzyme analysis of pcDNA3. 1 (+) −*gag*

Restriction endoenzyme analysis was used to identify the correct insertion of the *gag* gene into the pcDNA3. 1 (+) vector. 5. 4 kb and 1. 5 kb of fragments were obtained when pcDNA3. 1 (+) −*gag* was digested with *Kpn* I/*Xho* I. 4. 3 kb, 1. 6 kb, 0. 5 kb, 0. 3 kb and 0. 2 kb of fragments were obtained after digestion with *Pst* I. *Kpn* I/*Sal* I digestion gave 2. 3 kb, 2. 2 kb, 1. 5 kb and 0. 9 kb bands (Fig. 1). The results were consistent with expectations and indicated that the *gag* gene was correctly inserted into the pcDNA3. 1 (+) vector.

2. *gag* gene was inserted into rSeV − *gag* correctly

RT − PCR was preformed using the RNA from LLC − MK2 cells infected with rSeV − control and rSeV − *gag*. A specific fragment of 0. 8 kb was amplified in LLC − MK2 cells infected with rSeV − *gag* while no band was found in the rSeV − control infected cells (Fig. 2). It indicated that *gag* gene was inserted into genome of rSeV vector.

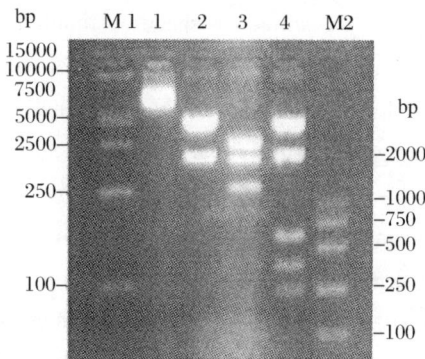

MI, DL15000 DNA marker; 1, pcDNA3. 1 (+) −*gag*; 2, pcDNA3. 1 (+) −*gag*/*Kpn* I +*Xho* I ; 3, pcDNA3. 1 (+) − *gag*/*Kpn* I + *Sal* I; 4, pcDNA3. 1 (+) −*gag*/*Pst* I; M2, DL2000 DNA marker

Fig. 1   Restriction endoenzyme analysis of pcD-NA3. 1 (+) −*gag*

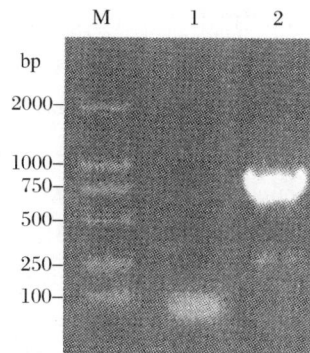

M, DL2000 DNA marker; 1, RT − PCR products of RNA extracted from rSeV − control infected LLC − MK2 cells; 2, RT − PCR products of RNA extracted from rSeV − *gag* infected LLC − MK2 cells

Fig. 2   RT − PCR detection of *gag* gene in recombinant Sendai viruses

3. DNA and rSeV − *gag* vaccines express HIV − 1 *gag* protein efficiently

Expression of the *gag* Protein in DNA and rSeV vaccines was analyzed by Western blotting. The modified *gag* gene in plasmid pcDNA3. 1 (+) −*gag* was expressed correctly and efficiently after transfecting to LLC − MK2 cells. As shown in Fig. 3 , a specific band of 55 $\times 10^3$ could be detected in pcDNA3. 1, (+) −*gag* transfected cells. The recombinant virus rSeV − *gag* gave a similar level of expression of *gag* protein, while no bands could be seen in mock cells.

4. Comparison of *gag* – specific CTL responses in mice

To find a better immunization scheme for these two vaccines, immunogenicty with single or combined modality was compared. As shown in Fig. 4, two individual immunizations with the DNA or rSeV – *gag* vaccine alone could induce low levels of *gag* – specific CTL responses, while DNA vaccine priming and rSeV – *gag* boosting elicited high frequency of and long – lasting CTL responses targeting *gag*. One week post immunization, the percentage of IFN – γ secreting CD8$^+$ T cells in total CD8$^+$ T cells reached 13. 1% ± 2. 2% in the combined immunization group. Although the *gag* – specific CD8$^+$ T cells level declined

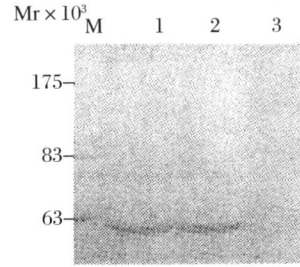

M, Prestained protein marker; 1, LLC – MK2 cells transfected with pcDNA3. 1 (+) – *gag*; 2, LLC – MK2 cells infected with rSeV – *gag*; 3, Mock LLC – MK2 cells

Fig. 3   Western blotting analysis of expression of *gag* gene in DNA and rSeV – *gag* vaccine

with time, 8. 1% ±5. 7% and 2. 5% ±1. 3% *gag* – specific CD8$^+$ T cells were detected at 5 and 9 weeks post immunization respectively in this group. Only 1. 3% ±0. 3% and 1. 9% ±0. 8% *gag* – specific CD8$^+$ T cells could be induced at 1 week post immunization in the DNA group and rSeV – *gag* group. They decreased to much lower levels at 5 and 9 weeks post immunization in these two groups.

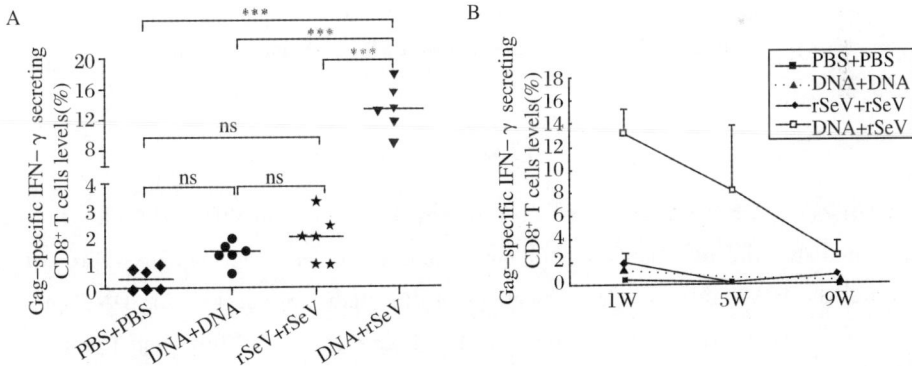

A: *gag* – specific CD8$^+$ T cells level in mice at 1 week post immunization. Newman – Keuls Multiple Comparison Test was used for comparison between two groups, ns: $P > 0.05$; $* * *$: $P < 0.001$. B: Comparison of *gag* – specific CD8$^+$ T cells levels in mice at different time point post immunization

Fig. 4   Comparison of *gag* – specific CD8$^+$ T cells levels in mice immunized with different vaccines. Numbers of *gag* – specific CD8$^+$ T cells are shown as percentages of IFN – γ secreting CD8$^+$ T cells in the total CD8$^+$ T cells

5. Comparison of HIV – 1 *gag* – specific antibodies levels in immunized mice

Sera of immunized mice were collected at different time points and P24 – specific antibodies were detected by ELISA. As shown in Fig. 5, the mice immunized with DNA and rSeV – *gag* vaccines induced high level of anti – P24 antibodies at 1 week post immunization. The anti – P24 antibodies reached peak levels at 5 weeks post immunization (1 : 6400) and then decreased but still kept high levels compared with other groups. Two Immunizations with DNA or rSeV – *gag* vaccine did not induce P24 – specific antibodies at 1 week post immunization. In mice immunized with DNA vaccine twice, P24 – specific antibodies were detected at 5 and 9 weeks post immunization, while in mice immunized with rSeV – *gag* vaccine twice anti – P24 antibodies was undetectable even at 5 and 9 weeks post immunization.

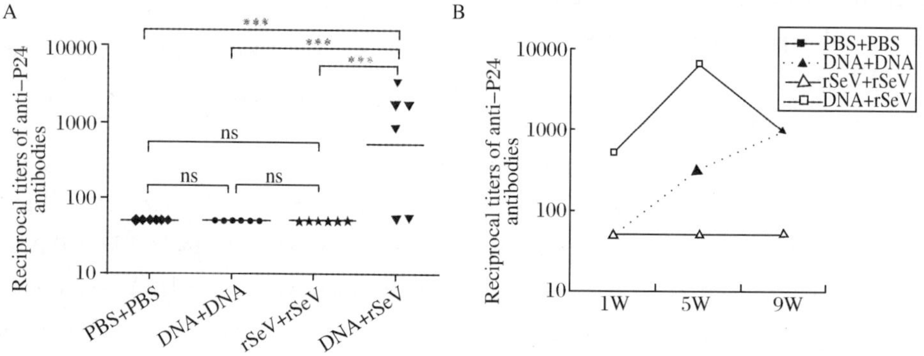

A: Anti – P24 antibodies levels in mice at one week post immunization. Newman – Keuls Multiple Comparison Test was used for comparison between two groups, ns: $P > 0.05$; * * * : $P < 0.001$. B: Comparison of anti – P24 antibodies levels in mice at different time point post immunization

Fig. 5 P24 – specific antibodies levels in immunized mice

Discussion

The prevalence of HIV/AIDS poses a severe threat to human health worldwide. Because drugs that could eliminate HIV infection are still not available, an effective vaccine represents the best hope to curtail the HIV epidemic. The objectives of this study are constructing DNA and rSeV vaccines based on prevalent Thailand subtype B HIV – 1 *gag* genes in China and trying to use these vaccines for therapeutic and prophylactic vaccines.

A number of studies have suggested that plasmid DNA is quite effective as a priming or initial immunogen in a bimodal vaccine strategy[2]. Expression of HIV structural proteins in plasmids DNA has been hampered by the fact that their expression is dependent on the HIV Rev protein and the Rev – responsive element. These proteins could be expressed efficiently only in the presence of *rev* gene in the plasmid DNA, otherwise their expression level is very low. Changes in the codon usage of HIV structural proteins to those employed by highly expressed human codons resulted in increased Rev – independent expression[3, 14, 25]. In order to increase the expression level of *gag* in DNA vaccine, the codons of the consensus Thailand B *gag* sequence from Henan were modified according to mammalian codon usage. And it was verified that expression level of *gag* was improved largely by codon – modification (data not shown). The codon modified consensus *gag* gene was used to construct the DNA vaccine.

The SeV vector emerged as a member of a new class of viral vectors with the development of reverse genetics technology. SeV is an enveloped virus with a negative – sense RNA genome. It causes fatal pneumonia in mice, its natural host, but is thought to be nonpathogenic in primates, including humans[11, 21]. Some of the extraordinary features of SeV make it particularly suitable for expressing HIV structural proteins. Because its replication cycle is exclusively in cytoplasm, the HIV gene expressed in this vector does not contain a nuclear phase, therefore Rev is not necessary and the *gag* gene in this vector could be expressed efficiently without codon modification. Because of this, the wild type Thailand B *gag* gene from Henan province was used to construct a rSev vaccine. It was verified that *gag* protein could be expressed well in this vector. The SeV vector used in this study was developed by DNAVEC Corporation. They had confirmed that SIV *gag* – expressing rSeV vaccine combined with DNA vaccine could induce high frequency of *gag* – specific CTL responses and exerted excellent protective efficacy in macaque models[10, 19]. Based on these results we tried to develop vaccines expressing HIV – 1 *gag* derived from Chinese prevalent strains using the same rSeV vectors. In the studies performed by the DNAVEC Corporation rSeV the vaccines were introduced by intranasal route. In the present study intramuscular immunization was performed for both DNA and rSeV vaccines, and high level of *gag* – specific CTL responses and antibodies were elicited in DNA prime/rSeV boost group.

There are many studies in recent years which indicate that the combinations of DNA and recombinant – viral – vector based vaccines are promising AIDS vaccine methods because of their potential for inducing cellular immune responses[2, 9]. However, in late 2007 the failure of Merck rAd5 Phase II proof – of – concept STEP study began to influence thinking on the design of T – cell vaccines. The candidate may have failed for several reasons. It may not have contained the right mix of HIV components. Alternatively, the vector (Adenovirus type 5) may not be robust enough. Scientists at Merck are looking hard at these issues and collaborating with others to help decipher the data. Given that T cells do not work until infection has occurred, prevention of infection is very likely too high a bar for a T cell vaccine. The more realistic goal for such a vaccine would be to significantly reduce the viral load of individuals who have been infected with HIV. So in our design the vaccines would be first used for therapeutic immunization to control HIV replication in the future clinical trials. If this works then prophylactic immunization could be studied further.

In the present study we compared the immunogenicity of DNA and rSeV – *gag* vaccines applied in single or combined vaccines. Our results demonstrated that combined vaccines in the DNA prime/ rSeV – *gag* boost vaccination regimen induced the strongest and most long – lasting *gag* – specific CTL and antibody responses compared with single vaccines. The CTL responses were induced earlier compared with humoral immune responses. It reached a peak at one week post immunization and then decreased with time. Even at 9 weeks post immunization comparative *gag* – specific CD8 $^+$ T cells could be detected in this group. The *gag* – specific antibody peak occurred at 5 weeks post immunization in this group. DNA vaccine or rSeV – *gag* vaccine alone elicited a low frequency of HIV –

specific CTL responses. *gag* – specific antibodies were not detected in mice immunized with rSeV – *gag* vaccine alone, while in mice immunized with the DNA vaccine antibody responses could be induced and the antibody peak occurred at 9 weeks post immuniztion. As we expected the DNA vaccine showed low immunogenicity when used alone, but why when the rSeV vaccine was used alone could it not induce a high level of humoral response? We speculate that anti – SeV antibodies were elicited after the first immunization. These antibodies would interfere with the effect of the second immunization. We performed another test to verify this hypothesis. Mice were immunized with 1.8×10^7 CIU rSeV vaccine, 3 weeks later anti – SeV antibodies were detected, the titer was 1 : 919. This indicated that though not high, anti – rSeV antibodies were induced and may disturb the further use of this vaccine. Another possible reason is the antigen presentation pathway of SeV is different from other vectors; further studies are needed to demonstrate the mechanism.

In conclusion our data demonstrated that DNA prime/rSeV – *gag* boost vaccination regimen could induce high level of cellular and humoral immune responses in mice.

Acknowledgements

This work was supported by the National 863 project (grant 2003AA219070) and DNAVEC Corporation. We thank the excellent staffin Beijing University of Technology that conducted the flow cytometry analyses.

［In 《Virol Sin》 2008, 23 (4): 295 –304］

References

1　Altfeld M, Rosenberg E S. The role of CD4⁺ T helper cells in the cytotoxic T lymphocyte response to HIV – 1. Curr Opin Immunol, 2000, 12 (4): 375 – 380

2　Amara R R, Villinger F, Altman J D, et al. Control of a mucosal challenge and prevention of AIDS by a multi – protein DNA/MVA vaccine. Science, 2001, 292 (5514): 69 – 74

3　Andre S, Seed B, Eberle J, et al. Increased immune response elicited by DNA vaccination with a synthetic gpl20 sequence with optimized codon usage. J Virol, 1998, 72 (2): 1497 – 1503

4　Betts M R, Ambrozak D R, Douek D C, et al. Analysis of total human immunodeficiency virus (HIV) – specific CD4 (+) and CD8 (+) T – cell responses: relationship to viral load in untreated HIV infection. J Virol, 2001, 75 (24): 11983 – 1991

5　Bitzer M, Armeanu S, Lauer U M, et al. Sendai virus vectors as an emerging negative – strand RNA viral vector system. J Gene Med, 2003, 5

(7): 543 – 553

6　Borrow P, Lewicki H, Hahn B H, et al. Virus – specific CD8⁺ cytotoxic T – lymphocyte activity associated with control of viremia in primary human immunodeficiency virus type 1 infection. J Virol, 1994, 68 (9): 6103 – 6110

7　Geldmaeher C, Currier J R, Herrmann E, et al. CD8 T – cell recognition of multiple epitopes within specific gag regions is associated with maintenance of a low steady – state viremia in human immunodeficiency virus type 1 – seropositive patients. J Virol, 2007, 81 (5): 2440 – 2448

8　Hasan M K, Kato A, Shioda T, et al. Creation of an infectious recombinant Sendai virus expressing the firefly luciferase gene from the 3' proximal first locus. J Gen Virol, 1997, 78: 2813 – 2820

9　Im E J, Nkolola J P, di Gloria K, et al. Induction of long – lasting multispecific CD8⁺ T cells by a four – component DNA – MVA/HIVA – RENTA candidate HIV – 1 vaccine in rhesus macaques. Eur J Immunol, 2006, 36 (10): 2574 –

2584

10　Kano M, Matano T, Nakamura H, et al. Elicitation of protective immunity against simian immunodeficiency virus infection by a recombinant Sendai virus expressing the *gag* protein. AIDS, 2000, 14 (9): 1281 – 1282

11　Kano M, Matano T, Kato A, et al. Primary replication of a recombinant Sendai virus vector in macaques. J Gen Virol, 2002, 83 (Pt6): 1377 – 1386

12　Kato A, Sakai Y, Shioda T, et al. Initiation of Sendai virus multiplication from transfected cDNA or RNA with negative or positive sense, Genes Cells, 1996, 1: 569 – 579

13　Kiepiela P, Nqumbela K, Thobakqale C, et al. CD8 + T – cell responses to different HIV proteins have discordant associations with viral load. Nat Med, 2007, 13 (1): 46 – 53

14　Kong W, Tian C, Liu B, et al. Stable expression of primary human immunodeficiency virus type 1 structural gene products by use of a noncytopathic sindbis virus vector. J Virol, 2002, 76 (22): 11434 – 11439.

15　Leligdowiez A, Yindom L M, Onyango C, et al. Robust *gag* – specific T cell responses characterize viremia control in HIV – 2 infection. J Clin Invest, 2007, 117 (10): 3067 – 3074.

16　Novitsky V, Cao H, Rybak N, et al. Magnitude and frequency of cytotoxic T – lymphocyte responses: identification of immunodominant regions of human immunodeficiency virus type 1 subtype C. J Virol, 2002, 76 (20): 10155 – 10168

17　Prince A M, Whalen R, Brotman B. Successful nucleic acid based immunization of newborn chimpanzees against hepatitis B virus. Vaccine, 1997, 15 (8): 916 – 919

18　Schoenly K A, Weiner D B. Human immunodeficiency virus type 1 Vaccine Development: recent Advances in the CTL Platform "Spotty Business". J Virol, 2008, 82 (7): 3166 – 3180.

19　Takeda A, Igarashi H, Nakamura H, et al. Protective efficacy of an AIDS vaccine, a single DNA priming followed by a single booster with a recombinant replication – defective Sendal virus vector, in a macaque AIDS model. J Virol, 2003, 77 (17): 9710 – 9715

20　Toapanta F R, Craigo J K, Montelaro R C, et al. Reduction of anti – HIV – 1 *gag* immune responses during co – immunization: immune interference by the HIV – 1 envelope. Cnrr HIV Res, 2007, 25 (2): 199 – 209

21　Urwitz J L, Soike K F, Sangster M Y, et al. Intranasal Sendai virus vaccine protects African green monkeys from infection with human parainfluenza virus – type one. Vaccine, 1997, 15 (5): 533 – 540

22　Wolff J A, Malone R W, Williams P, et al. Direct gene transfer into mouse muscle *in vivo*. Science, 247 (4949 Pt1): 1465 – 1468

23　Yasutomi Y, Reimann K A, Lord C I, et al. Simian immunodeficiency virus – specific CD8 + lymphocyte response in acutely infected rhesus monkeys. J Virol, 1993, 67 (3): 1707 – 1711

24　Yu D, Shioda T, Kato A, et al. Sendai virus – based expression of HIV – 1 *gp*120: reinforcement by the V (–) version. Genes Cells, 1997, 2: 457 – 466

25　Zur Megede J, Chen M C, Doe B, et al. Increased expression and immunogenicity of sequence – modified human immunodeficiency virus type 1 *gag* gene. J Virol, 2002, 74 (6): 2628 – 2635

108.　Evaluation of Nonviral Risk Factors for Nasopharyngeal Carcinoma in A High – Risk Population of Southern China

GUO Xiu – chan[1,2], JOHNSON Randall C[1], DENG Hong[3], LIAO Jian[4], GUAN Li[1],
NELSON George W[1], TANG Ming – zhong[3], ZHENG Yu – ming[3], de – The Guy[5],
O'BRIEN Stephen J[6], WINKLER Cheryl A[1], ZENG Yi[2]

1. Laboratory of Genomic Diversity, SAIC – Frederick, Inc. , NCI – Frederick, Frederick, MD;
2. State Key Laboratory for Infectious Diseases Prevention and Control, Institute for Viral
Disease Control and Prevention, Chinese CDC; 3. Cancer Center, Wuzhou Red Cross Hospital;
4. Cangwu Institute for Nasopharyngeal Carcinoma Control and Prevention, Wuzhou;
5. Institut Pasteur, Paris, France; 6 Laboratory of Genomic Diversity, Center for Cancer
Research, National Cancer Institute, National Institutes of Health, Frederick. MD

Summary

To understand the role of environmental and genetic influences on nasopharyngeal carcinoma (NPC) in populations at high risk of NPC, we have performed a case – control study in Guangxi Province of Southern China in 2004 – 2005. NPC cases (n = 1049) were compared with 785 NPC – free matched controls who were seropositive for IgA antibodies (IgA) to Epstein – Barr virus (EBV) capsid antigen (VCA) – a predictive marker for NPC in Chinese populations. A questionnaire was used to capture exposure and NPC family history data. Risk factors associated with NPC in a multi-variant analysis model were the following: (i) a first, second or third degree relative with NPC [attributable risk (AR) = 6%, odds ratio (OR) = 3.1, 95% confidence interval (CI) = 2.0 – 4.9, P < 0.001]; (ii) consumption of salted fish 3 or more than 3 times per month (AR = 3%, OR = 1.9, 95% CI = 1.1 – 3.5, P = 0.035); (iii) exposure to domestic wood cooking fires for more than 10 years (AR = 69%, OR = 5.8, 95% CI = 2.5 – 13.6, P < 0.001); and (iv) exposure to occupational solvents for 10 or less years (AR = 4%, OR = 2.6, 95% CI = 1.4 – 4.8, P = 0.002). Consumption of preserved meats or a history of tobacco smoking were not associated with NPC (P > 0.05). We also assessed the contribution of EBV/IgA/VCA antibody serostatus to NPC risk – 32.2 % of NPC can be explained by IgA + status. However, family history and environmental risk factors cumulatively explained only 2.7% of NPC development in NPC high risk population. These findings should have important public health implications for NPC risk reduction in endemic regions.

[**Key words**] Nasopharyngeal carcinoma; Risk factor; Epidemiology; Southern China; Epstein – Barr Virus

Introduction

Nasopharyngeal carcinoma (NPC) is rare in most regions of the world; however, it is a common cancer in Southern China, especially in persons of Cantonese origin. The incidence rate of NPC for males in the southern Chinese provinces of Guangdong and Guangxi is more than 20 per 100 000 person – years and up to 25 – 40 per 100 000 person – years in some areas bordering the Xijiang River and Pearl River drainages in these 2 provinces[1,2]. In Southeast Asia, the incidence of NPC seems to vary by degree of racial and cultural Cantonese admixture[3]. In Hong Kong, a region with a high immigrant Cantonese population, a recent 20 – year longitudinal study has shown that the NPC incidence for males is more than 20 per 100 000 person – years[3]. Intermediate incidence levels were observed in the Thai, Macaonese and Malay indigenous populations, who have a history of intermarriage with Cantonese. In northern China, such as Beijing and Tianjing, the incidence of NPC in males is less 2 per 100 000 person – years[3]. The incidence of NPC begins to increase after the age of 30; 93% of NPC patients are over 30 years old at diagnosis and the NPC incidence peaks at 45 – 55 years of age. NPC is the 8th leading cause of cancer death in China and a leading cause of cancer death in the Guangdong and Guangxi population[5,6].

It has been well established that reactivation of Epstein – Barr virus (EBV) in the epithelial mucosal lining of the nasopharynx is strongly associated with NPC[7]. The presence of IgA antibodies to EBV capsid antigens (EBV/IgA/VCA) may be a biomarker for EBV reactivation and serves as a predictive marker for NPC in Chinese populations[8,9]. A 10 – year longitudinal study in Guangxi province indicated that NPC incidence was 467, 2 and 28 per 100 000 people – years for EBV/IgA/VCA positive (IgA +), EBV/IgA/VCA negative (IgA –) participants, and the general population, respectively[10]. The same study showed that although 93% of NPC patients came from the 5% of the population that were IgA + population, less than 5% of IgA + individuals will ever develop NPC. Consequently, the population at highest risk of NPC is the IgA + population in Southern China. The incidence of NPC in the IgA – population in Southern China is similar to the general population in northern China[3].

Familial aggregation of NPC cases has been observed in both high and low risk populations in different geographic regions[11,13]. The nonviral environmental exposures most consistently associated with NPC are traditional southern Chinese consumption of salted fish and other preserved meats containing volatile nitrosamine[14-19]. Although cigarette smoking is a known risk factor for cancers of various organ systems, previous studies on the association between smoking and NPC in NPC endemic areas have not been consistent. Several case control studies reported that heavy smoking increased NPC risk by 2 – to 4 – fold[20-24]; however, other studies found no association[14,25-27]. Discrepant results were also reported for domestic wood cooking fire exposures and occupational exposures to solvents with some studies but not other reporting associations with NPC[28-33].

Most previous epidemiological studies to identify nonviral environmental risk factors for NPC have not considered EBV/IgA/VCA serum status, although IgA + status appears to confer the highest known risk of NPC. To systemically evaluate non – EBV environmental risk factors in the development of NPC in a high – risk population, we conducted a case – control study with 1049 NPC cases and 785 IgA + controls in the Guangxi province of Southern China. In this study, we examined familial recurrence, ethnicity and environmental factors, including dietary, smoking, household and occupational exposures in NPC cases and IgA + controls to evaluate nonviral risk factors in NPC development in a high – risk population in Southern China.

Material and Methods

1. Cohort design

Subjects residing in the catchment area along the Xijiang River on the border of the Guangxi and Guangdong provinces of Southern China were invited to enroll in the study from November. 2004 to October, 2005. Cases were incident or prevalent, biopsy – confirmed NPC cases (n = 1049). The incident NPC were cases who were diagnosed with NPC during the enrollment period and referred to our study. The prevalent cases were the cases who were diagnosed as NPC between January 2001 and October 2004. In our cohort, 81.6% of cases were prevalent cases and 18.4% cases were incident cases, Treatment infonnation was available for 95% cases. Of these, 99.4% received radiotherapy, 0.4% received chemotherapy, 5.2% received both radiotherapy and chemotherapy and 0.2% rejected any therapy. Treatment plans were not affected by enrollment in this study. Controls (n = 785) were EBV/IgA/VCA positive (IgA +), NPC free at the time of study enrollment and matched to NPC cases on age, sex and district/township of residence. Controls were determined to be NPC free by physical examination of the nasopharyngeal cavity by oncologists using indirect naso – endoscope. NPC cases were hospitalized patients at Wuzhou Red Cross Hospital in Wuzhou City and outpatients at Cangwu Institute for NPC Control and Prevention in Cangwu County. Controls were identilied from records of EBV/IgA/VCA screening that occurred during 2001 to 2003 and those who were EBV/IgA/VCA positive were asked to participate in the study. All participants self – identilied as Han Chinese, self reported 6 or more months of residency in Guangdong or Guangxi, and self – reported Guangdong or Guangxi Provincial ancestry for either maternal or paternal ancestry tor at least 3 generations. Persons of minority ethnicity or who had blood relatives enrolled in the study were excluded. Institutional review board approval was obtained from all participating institutes, and informed consent was obtained from all study participants or their respective legal guardians.

2. Questionnaire

Trained local health workers administered a questionnaire to each participant at enrollment. The questionnaire captured family history of NPC, parental ancestry for 3 generations, age, sex, dietary and smoking habits, household exposures to wood fires, and occupational exposures to solvents. Responses were recorded by double – entry and verification of all data was performed to avoid data entry errors. Probands were asked if there was a family history of NPC in first (children, siblings or parents), second (aunts or uncles, nieces or nephews and grandparents) or third degree relatives (first cousins). Information on salty fish and preserved meat captured frequency of consumption per month (⩾3 times/month, <3 times/month). Questions on cigarette smoking included current and past smoking habits and number of cigarettes smoked per day. Questions on household and occupational exposures captured data on domestic exposure to wood fires for cooking and occupational exposures to solvents (e. g. , formaldehyde, acetone, toluene or xylene), and duration of exposure (> 10 years or ⩽ 10 years). The response rate for the questions on family history of NPC, salted – lish, preserved meat, smoking, wood fire and solvent were 99.9% , 96.4% , 96% , 98.9% , 98.6% and 98.1%.

Laboratory and clinical data. Date of NPC diagnoses, clinical stage at presentation, tumor histological results, therapeutic treatment (radiation, and/or chemotherapy) and response to treatment were obtained from NPC cases clinical records. Serological testing for IgA antibodies to EBV capsid antigens (EBV/IgA/VCA) and lgA antibodies to EBV early antigen (EBV/IgA/EA) was perfonned on a serum sampled from peripheral blood samples obtained by venipuncture at study enrollment. All serological testing was done by immunoenzymatic assay at the Wuzhou Red Cross Hospital, Wuzhou City. B95 – 8 cells were used for the detection of VCA – IgA antibodies, and Raji cells were used for EA – IgA test. Sera at 1 : 10 to 1 : 640 (1 : 5 to 1 : 160 for EA – IgA) dilutions were added to cells in separate wells and the slides were incubated at 37℃ for 30 min at a humid atmosphere and washed 3 times with PBS. Horseradish peroxidase labeled antihuman IgA antibody at the appropriate dilution was added to the slides, and the slides were further incubated at 37℃ for 30 min, washed 3 times with PBS and immersed into diaminobenzene solution with H_2O_2 for 10 min. Positive and negative controls were included in each experiment[34]. The cutoff value is 1 : 10 and 1 : 5 for IgA/VCA and IgA/EA, respectively.

3. Biological blood sample collection

Peripheral blood samples were collected from all cases and controis, processed and stored as described previously[35].

4. Statistical analysis

EBV/IgA/EA antibody presence and EBV/IgA/VCA titer were compared using Pearson's conelation coefficient. The association of individual risk factors with NPC was analyzed using multiple logistic regressions, with the reference group being individuals with no reported exposure. A corresponding multivariate analysis, including measures for all risk factors, was also explored. For each explanatory variable, we conditioned the multivariate analysis on NPC family history, dietary salted fish, dietary preserved meat, smoking, domestic exposure to wood fire and occupational exposure to solvents. Reported measures of association are odds ratios (OR), 95% confidence intervals (CIs) and p values. Because of small sample sizes in some of the risk groups, boot strapped CI estimates, using 10 000 repetitions, are reported for odds ratios comparing the joint effects of risk faclors. The attributable risk (AR) and explained fraction (EF) were calculated using the method described by Nelson and O'Brien[36]. We used the IgA + proportion (5%) in the general population from a 10 – year follow – up study to estimate prevalence in the general population and to calculate AR and EF[10]. Because our disease is rare we can substitute the OR for the relative risk when calculating the AR for NPC family history, salty fish, exposure to wood fire anti solvent. The fonnula is as follows: AR = prevalence of exposure in NPC × (OR – 1) /OR. EBV/IgA/VCA antibody negative NPC cases (n = 39) were not included in the analysis in the Tasb. 2 – 4. All statistical analyses were carried out using R version 2. 6. 2 (R Foundation for Statistical Computing, Vienna, Austria) or SAS 9. 1 software.

Results

The mean age of 1. 049 cases at NPC diagnosis was 45 ± 11 (range, 8 – 77) years and of 785 IgA + controls at enrolhnent was 47 ± 12 (range, 20 – 84) years (P > 0. 05); the male to female ratio was 2. 6 for both cases and Controls. The majority of the NPC cases (n = 1010, 96. 3%)

were positive for EBV/IgA/VCA antibodies and EBV/IgA/EA (n = 752, 72%). The mean titers for the NPC cases were 1 : 92 and 1 : 23 for EBV/IgA/VCA and EBV/IgA/EA antibodies, respectively. There was a strong positive correlation between the presence of EBV/IgA/EA antibodies and titer of IgA/VCA antibodies (r^2 = 0.81. P < 0.0001) (Fig. 1). The mean EBV/IgA/VCA titer was lower in IgA + controls (1 : 15) compared with titers in NPC cases (P < 0.0001). Fig. 2 presents the EBV/IgA/VCA titer in NPC cases and IgA + controls. Only 5% of EBV/IgA/VCA positive control subjects were EBV/IgA/EA positive.

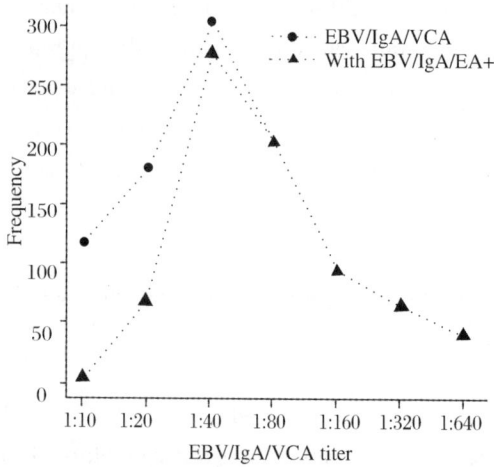

[Color figure can be viewed in the online issue, which is available at www. interscience. wiley. com.]

Fig. 1 EBV/IgA/VCA antibody titer distribution and conelation with EBV/IgA/EA antibody positive status for NPC patients

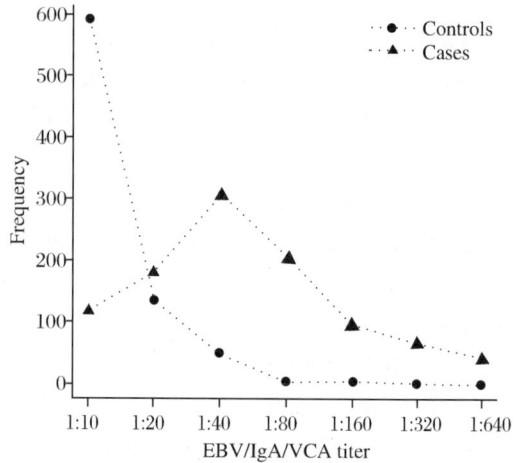

[Color figure can be viewed in the online issue, which is available at www. interscience. wiley. com.]

Fig. 2 EBV/IgA/VCA antibody titer distribution in NPC cases and IgA + controls

Tumor histological types were available for 1038 out of 1049 NPC cases (99%). Using the World Health Organization (WHO) classification for NPC (1991) criteria,[37] 14.9% of NPC patients had keratinizing squamous cell carcinoma (KSCC) and 85.1% NPC cases had non–keratinizing carcinoma (NKC). Among the 39 EBV/IgA/VCA negative NPC cases, 28.2% had KSCC and 71.8% had NKC. The clinical stage at diagnosis was available for 1043 (99.4%) cases, 38.9% of whom were in early stage (Stage I and II) and 61.1% were late stage (Stage III and IV) at presentation. Among EBV/IgA/VCA negative NPC cases, 53.8% were early stage and 46.2% were late stage at presentation. For those with keratinizing squamous cell carcinoma, 50.3% were early stage and 49.7% were late stage. EBV/IgA/VCA negative patients have a higher rate of keratinizing squamous cell carcinoma (P = 0.02) and lower rate of diagnosis at late stage (P = 0.055). Patients with non–keratinizing carcinoma had a higher rate of late stage NPC at diagnosis (P = 0.002).

1. Risk associated with a family history of NPC

Tab. 1 lists the characteristics of NPC cases who had first–, second– or third–degree relatives with a history of NPC (familial NPC), and NPC cases who reported having no NPC–affected relatives (nonfamilial NPC). There were no significant differences between familial NPC and nonfamilial NPC on gender, age of onset, histological types, clinical stage, EBV/IgA/VCA and EBV/IgA/EA antibody status. Among NPC cases, 104 of 1049 cases (9.9%) reported having a first −, second − or third − degree blood relative with NPC. More NPC cases (9.9%) than controls (3.7%) reported having one or more first, second or third degree relatives with NPC. Comparing NPC cases with IgA + and IgA + controls, we found individuals with a first, second or third − degree relative with NPC were 3 − fold more likely to develop NPC ($P < 0.001$), after adjusting for all other risk factors, i. e. , salted fish, preserved meat, smoking, wood fire and solvents (see below).

2. Risk associated with environmental factors

Tab. 2 lists the information and distribution of NPC family history, salty fish and preserved meat consumption, smoking habits, and wood fire and solvent exposures in NPC cases and IgA + controls, plus the association between these factors and NPC risk in simple and multivariant models, adjusted for all environmental exposures. In a simple analysis, consumption of salty fish and preserved meats, domestic exposure to wood cooking fires, and occupational exposure to solvents were NPC risk factors ($OR = 1.58 − 3.53$; $P < 0.002$). After adjusting for all risk factors in a multivariant analysis, only consuming salty fish 3 or more than 3 times per month at the time of the interview ($OR = 1.9$, 95% $CI = 1.1 − 3.5$), exposure to wood cooking fires for more than

Tab. 1 Clinical characteristics for familial and nonfamilial NPC cases

Item	Familial NPC ($n = 104$)	Nonfamilial NPC ($n = 945$)	p value
M/F ratio	3.3 : 1 (80/24)	2.5 : 1 (674/271)	0.25
Age[1]	43.7 ± 11.3	45.2 ± 10.8	0.18
EBV/IgA/VCA[2]	96.2%	96.3%	0.79
EBV/IgA/EA[3]	77.9%	71.2%	0.17
Histology.			0.46
KSCC[4]	11.8%	15.3%	
NKC[5]	88.2%	84.7%	
Stage			0.53
Early	35.60%	39.3%	
Late	64.40%	60.7%	

Notes: 1. $\bar{x} \pm s$; − 2. IgA antibody to EBV viral capsid antigen, titer ≥ 1 : 10; − 3. IgA antibody to EBV early antigen, titer ≥ 1 : 5; − 4. Keratinizing squamous cell carcinoma; − 5. Non − keratinizing carcinoma

10 years ($OR = 5.8$, 95% $CI = 2.5 − 13.6$) and exposure to solvents for 10 or less 10 years ($OR = 2.6$, 95% $CI = 1.4 − 4.8$) remained significant NPC risk factors. The exposure rates in cases and controls for occupational solvents, salty fish and domestic exposures to wood cooking fires were 6%, 30% and 98%, respectively. Half of cases (51.7%) and controls (50.8%) reported a history of smoking, but there were no significant associations between smoking and NPC.

To investigate the possible joint effects of family history of NPC, salty fish, wood fire and solvent exposure, we examined the association between 1 or 2 of these risk factors and NPC. Because the most frequent risk exposure was to wood cooking lires, we used wood fire exposure as the referent group in the analysis. Tab. 3 presents the results of single and pairwise combinations of exposure factors and family history on NPC risk. Individuals with no exposures to these risk factors were significantly less likely to develop NPC ($OR = 0.2$, 95% $CI = 0 − 0.7$). On the other hand, individuals reporting any 2 combination of risk factors were at greater risk ($OR = 1.5 − 2.6$). We also analyzed the risk for individuals having none or having ≥ 2 risk factors compared with the reference group who have only 1 risk factor (Tab. 4). Consistent with the results presented in Tab. 3, there was increasing NPC risk for individuals with multiple environmental risk factors ($OR = 2.47$, 95% $CI = 1.87 - 3.38$).

Tab. 2 Odds ratios for the association of NPC family history, dietary salt fish preserved Meat, smoking, wood fire and solvent to risk of NPC

Item	Cases (%) NPC with IgA +	Controls (%) IgA +	Simple model OR	95% CI	p – value	Multivariant model[1] OR	95% CI	p – value	AR²(%)	EF³ (%)
Relative with NPC										
1st degree	75 (7.4)	23 (2.9)	2.65	1.64 – 4.27	<0.001	2.81	1.69 – 4.67	<0.001	5	
1st, 2nd and 3rd degree	100 (9.9)	29 (3.7)	2.86	1.87 – 4.37	<0.001	3.09	1.97 – 4.86	<0.001	6	0.6
Salty fish										0.5
Never	633 (65.1)	573 (75.6)								
<3 times/month	274 (28.2)	156 (20.6)	1.59	1.27 – 1.99	<0.001	1.23	0.83 – 1.81	0.304		
≥3 times/month	65 (6.7)	29 (3.8)	2.03	1.29 – 3.19	0.002	1.9	1.05 – 3.47	0.035	3	
Preserved meat										
Never	691 (71.2)	612 (81.2)								
<3 times/month	238 (24.5)	119 (15.8)	1.77	1.39 – 2.26	<0.001	1.29	0.84 – 1.96	0.24		
≥3 times/month	41 (4.2)	23 (3.1)	1.58	0.94 – 2.66	0.086	1.03	0.51 – 2.05	0.94		
SmokingNever	488 (48.3)	384 (49.2)								
Smoker	522 (51.7)	396 (50.8)	1.04	0.86 – 1.25	0.701	0.92	0.76 – 1.12	0.405		
<10 cig/day⁴	33 (3.3)	32 (4.1)	0.81	0.49 – 1.34	0.417	0.7	0.41 – 1.19	0.186		
10 – 20 cig./day	385 (38.5)	307 (39.6)	0.99	0.81 – 1.21	0.897	0.89	0.72 – 1.10	0.272		
>20 cig./day	94 (9.4)	53 (6.8)	1.4	0.97 – 2.01	0.071	1.23	0.84 – 1.81	0.286		
Wood fire										0.9
Never	9 (0.9)	23 (2.9)								
Exposure	999 (99.1)	761 (97.1)	3.35	1.54 – 7.30	0.002	5.04	2.18 – 11.65	<0.001	70	
≤ 10 years	22 (2.2)	56 (7.2)	1	0.40 – 2.51	0.993	1.3	0.49 – 3.46	0.601		
> 10 years	963 (96.9)	697 (89.8)	3.53	1.62 – 7.68	0.001	5.82	2.50 – 13.57	<0.001	69	
Solvent										0.7
Never	914 (91.7)	739 (96.1)								
Exposure	83 (8.3)	30 (3.9)	2.24	1.46 – 3.44	<0.001	2.17	1.32 – 3.58	0.002	5	
≤10 years	62 (6.2)	17 (2.2)	2.95	1.71 – 5.09	<0.001	2.61	1.43 – 4.78	0.002	4	
>10 years	20 (2.0)	8 (1.0)	2.02	0.88 – 4.62	0.095	1.56	0.66 – 3.70	0.315		
Sum of factor EFs										2.7

Note: 1 Adjusting for all environmental exposures; –2 Attributable risk; –3 Explained fraction; –4 Cigarettes

Tab. 3 Joint effects of NPC family history, dietary salty fish, wood fire and solvent exposure with the risk of NPC

Risk factors	No. cases	No. controls	OR	95% CI
None	3	12	0.22	0 – 0.67
Solvent	1	3	0.29	0 – lnf¹
Wood fire²	750	659	1.0	
NPC family history	5	3	1.46	0.29 – lnf
Salty fish/Wood fire	47	25	1.65	1.02 – 2.82
Wood fire/Solvent	56	21	2.34	1.44 – 4.22
Family/Wood fire	73	25	2.57	1.66 – 4.33

Note: 1 Infinity; –2 Wood fire was used as a reference

Tab. 4 The association of having no or 2 or more risk factors¹ with NPC

No. of risk factors	No. of cases	No. of controls	OR	95% CI
0	3	12	0.22	0 – 0.65
1²	756	665	1.0	
≥2	205	73	2.47	1.87 – 3.38

Note: 1 NPC family history, dietary salty fish, domestic wood fire or occupational solvent exposures; –2 Having 1 risk factor was used as a reference group

To measure the impact of multiple environmental factors on NPC, the AR was determined.

The AR for NPC to occupational exposure to solvents, NPC family history, salty fish consumption and domestic exposure to wood cooking fires were 5%, 6%, 3% and 70%, respectively (Tab. 2). We also determined the risk attributable to EBV replication status using an NPC prevalence rate (59/100 000, IgA + proportion (5%) in the general population, IgA + rate (95%) in NPC cases and the incident of NPC (4.6 %) in IgA + population in the geographic catchment area of this study[10]. The AR of NPC due to IgA + status was 93% and explained fraction (EF) was 32.2%. In contrast, the family history and environmental risk factors cumulatively explained only 2.7% of NPC.

Because 15% of the NPC cases were KSCC, we sought to determine whether the risk factors were significantly different between KSCC and NKC. We found a significant association between preserved meat and NPC – KSCC but not with NPC – NKC; other factors were not significantly different between the 2 forms of NPC.

Discussion

This study enrolled participants from a catchment area of 6 million people in Southern China reporting one of the world's highest NPC prevalence rates. The peak age of NPC at diagnosis was 35 to 55 years old with 96% of NPC cases seropositive for EBV/IgA/VCA antibodies in agreement with previous sampling[36-38]. Significant risk predictors for NPC include: (i) first, second or third degree relative with NPC; (ii) diet that included consumption of salty fish 3 or more times a month; (iii) exposure to home wood cooking fires; and (iv) occupational exposures to solvents.

In the catchment area over 95% of NPC cases were EBV/IgA/VCA antibody positive compared with only 5% of the general population. Our study is unique in that we compared NPC cases who were EBV/IgA/VCA antibody positive to controls who were also EBV/IgA/VCA antibody positive to determine familial and environmental factors associated with NPC in the setting of active EBV replication. To our knowledge, this is the largest investigation to date to systematically evalute the role of NPC family history, salty fish, preserved meat, domestic and occupational exposures on risk of NPC in NPC high – risk population. EBV is widely believed to be a necessary (but not sufficient) cause of NPC, EBV/IgA/VCA antibodies have been used as a prospective marker of NPC in Southern China for over 2 decades,[39-42] However, as we show in this study, EBV/IgA/VCA antibody status explains only 32% portion of the risk. Our findings indicated that EBV/IgA/VCA titer may offer a quantitative indicator for NPC risk as titers are significantly higher in NPC cases than in EBV/IgA/VCA controls ($P < 0.0001$, Fig. 2).

Consistent with many other studies in Southeast Asia that report that 6% – 8% of NPC have a first degree relative with NPC in NPC cases[14,21,43], we also found that 7% of NPC cases had a first degree relative with NPC. In the absence of twin studies, it is not possible to determine if the recurrence rate in families is due to shared genetic or shared environmental factors or both. There were no significant differences on sex, age onset, tumor histological types and clinical stages between familial and nonfamilial NPC, confirming previous reports[13,43] Several researchers have reported the excess risk was frona 6 – to 19 – fold among individuals with a first – degree relatives of NPC[41,43,44], much more than we estimate in this study possibly because those studies were not controlling for EBV/IgA/VCA antibody status. It is also possible that recall bias may at least partially explain discrepancies among the studies. A recent prospective study in Singapore reported a 2 – to 3 – fold increased risk of NPC among first – degree relative with NPC[20] , similar to our results.

Salted fish consumption has been consistently reported to be moderate to strong risk predictor of NPC in Southern China in a series of small studies enrolling fewer than 300 cases from 1983 to 1998[14–16,19,27,45,46] . A large study (935 NPC subjects) in Shanghai, China also reported that dietary, salted fish was a risk factor for NPC in Shanghai, China[17]. NPC risk is also moderately associated with other preserved foods[3]. Along with economic development after 1978, there has been a gradual and current change in eating habits and life style in China. Our data show that less than one – third of population were eating salted fish or preserved meat at enrolhnent and heavy consumption of these food products is less than 4% in the control group. This study provides evidence that only regular consumption (≥3 times/month) of salted fish is a significant NPC risk factor. The consumption of preserved meat was not significantly associated with risk of NPC after adjusting for all other factors in our study.

Although occupational exposures to solvents are uncommon in the general population, the use of wood as a domestic cooking fuel is very frequent in the countryside of Southern China. About 98% of the study participants reported exposure to wood cooking fireone of the strongest risk predictors for NPC in this study, confirming an earlier report from the same region[46]. Those exposed to wood fire for more than 10 years had a nearly 6 – fold excess risk of NPC.

Formaldehyde is known to cause tumors by experimental observation in rodents,[47,48] but epidemiologic evidence in human is limited, especially in the NPC endemic region. A meta – analysis of available studies suggested that formaldehyde exposure was associated with a 2 – fold increase in risk of NPC[49]. In our study population, only 6% of the population reported unspecified, occupational solvent exposure. This study confirms earlier studies that individuals seropositive for EBV/VCA/IgA are at a 2 – fold greater risk of developing NPC if they are exposed to solvents in the workplace[23,32]. In Tab. 2 and 2 earlier studies[23,32], an association is apparent for short (≤10 years) , but not long – term exposure to solvents. The lack of association between NPC and long – term exposure (>10 years) may reflect a frailty bias; e. g. , long – term exposure to solvents to may lead to mortality due to other causes.

The populations in Southern China are exposed to multiple environmental risk factors. Our results show that a combination of risk factors increased the risk of NPC in the 5% of the population that are also EBV/IgA/VCA seropositive (Tab. 3 and 4). Our results indicate that the AR of wood fire exposure for NPC is 70% ; however, the AR measures how much of one disease state can be attributed to a given factor, but not how much of the absence of the disease state can be attributed to the absence of the factor. This study shows that in high risk population who were EBV/IgA/VCA positive, only 2. 7% of NPC can be explained by family history with NPC consumption of salted fish, exposure to domestic wood fires, and exposure to occupational solvents. Clearly, there are additional causal influences yet to be discovered, including genetic factors, which determine susceptibility to NPC in the EBV/IgA/VCA positive population. These results may provide the basis for a quantitative risk assessment instrument to screen the EBV/IgA/VCA seropositive population in endemic areas for NPC risk.

We did not find that cigarette smoking was associated with NPC risk. This result is consistent with some previous studies reporting that smoking is not a risk factor for NPC in endemic areas[14,25,46,50]. A population – based case – control study in the United States reported an association between cigarette smoking and keratinizing squamous cell carcinoma (KSCC) of NPC, but there was no evidence that non – keratinizing carcinoma (NKC) were associated with cigarette smoking[24]. In our cases, 15% of NPC patients were KSCC. We analysis the association stratified by the histologic type and observed no associations between smoking, intensity of smoking and KSCC or NKC in either the unadjusted or adjusted analysis (data not shown).

A limitation of our study may have been recall bias – particularly for childhood exposures to wood fires or dietary fish. Family members may also not have precise knowledge of family members with NPC. During the face to face interview, the interviewers found most volunteers were not confident in recalling the events that occurred in childhood or early adulthood, such as childhood con – sumption salty fish, preserved meat or age smoking was initiated. We did not include questions, which may not be accurate due to obvious recall bias. Although lower social class or educational level may be considered as a risk factor of NPC, it is difficult to define socioeconomic status during a period of rapid economic development in China during past 30 years. For remediation, all the samples we collected came from same geographic region and the cases and controls were matched by district. Although 81.6% of NPC cases were prevalent, 97% of the prevalent cases developed NPC within 3 – 4 years of study. It is therefore unlikely that entry recall bias for recent events is different between seroprevalent and seroincident NPC cases. The positive associations between dietary salty fish and wood – cooking fires and NPC may be due to other socioeconomic factors tracked by these exposures.

In summary, we observed a consistent association between NPC family history, consumption of salted fish, exposure to domestic wood fire and occupational exposures to solvents with NPC risk in Southern China. Familial aggregation among family members suggests that, in addition to viral and environmental factors, predisposing genetic factors may be important. An understanding of the interactions among genetic, viral and environmental factors will lead to a better understanding of the carcinogenic pathways leading to NPC and provide insights into improving therapeutic options. In addition, this study indicates that public health policy aimed at reducing exposures to nitrosamines found in preserved foods, wood cooking fires and solvents will reduce incidence of NPC.

Acknowledgements

The authors thank Drs. James Lautenberger for invaluable statistical advice, Mr. Bailey Kessing for efficient data management, and Ms. Jan Martenson for her organizational expertise and her assistance in developing the questionnaire.

[In 《Int J Cancer》 2009, 124: 2942 – 2947]

References

1 Wu J J, Guo H, Su R. Analysis and forecast of incidence and mortality of nasopharynx cancer by time series in Zhongshan city. Chin J Hospital Stat, 2001, 8: 16 – 19

2　Jia WH, Huang QH, Liao J, Ye W, Shugart YY, Liu Q, Chen LZ, Li YH, Lin X, Wen FL, Adami HO, Zeng Y, et al. Trends in incidence and mortality of nasopharyngeal carcinoma over a 20 – 25 year period (1978/1983 – 2002) in Sihui and Cangwu counties in southern China. BMC Cancer, 2006, 6: 178

3　Chang ET. Adami HO. The enigmatic epidemiology of nasopharyngeal carcinoma. Cancer Epidemiol Biomarkers Prev, 2006, 15: 1765 – 1777

4　Lee AW, Foo W, Mang O, Sze WM, Chappell R, Lau WH, Ko WM. Changing epidemiology of nasopharyngeal carcinoma in Hong Kong over a 20 – year period 11980 – 99): an encouraging reduction in both incidence and mortality. Int J Cancer, 2003, 103: 680 – 685

5　Li LD, Lu FZ, Zhang SW, Mu R. Sun XD, Huangpu XM, Sun J, Zhou YS, Ouyang NH, Rao KQ, Chen YD. Sun AM, et al. Trends and recent prediction in mortality of cancer during 20 years in China. Chin J Oncol, 1997, 19: 3 – 9

6　Dai XD, Li LD, Lu FZ, Zhang SW, Sun J, Lin Y J, Sun XW, Shi YP. Trends and recent prediction in mortality of nasopharyngeal carcinoma during 20 years in China. J Pract Oncol, 1999, 13: 1 – 5

7　Hildesheim A, Levine PH. Etiology of nasopharyngeal carcinoma: a review. Epidemiol Rev, 1993, 15: 466 – 485

8　Henle G, Henle W. Epstein – Barr virus – specific IgA serum antibodies as an outstanding feature of nasopharyngeal carcinoma. Int J Cancer, 1976, 17: 1 – 7

9　Raab – Traub N. Epstein – Barr virus in the pathogenesis of NPC. Semin Cancer Biol, 2002, 12: 431 – 441

10　Zeng Y, Deng H. A 10 – year prospective study on nasopharyngeal carcinoma in Wuzhou City and Cangwu County, Guangxi, China. In: Tursz T, Pagano JS, Ablashi DV, De The G, Lenoir G, Pearson GR, eds. The Epstein – Barr virus and associated diseases. France: Colloque INSERM/John Libbey Eurotext, 1993, 735 – 741

11　Brown TM, Heath CW, Lang RM, Lee SK, Whalley BW. Nasopharyngeal cancer in Bermuda. Cancer, 1976, 37: 1464 – 1468

12　Coffin CM. Rich SS. Dehner LP. Familial aggregation of nasopharyngeal carcinoma and other malignancies. A clinicopathologic description. Cancer, 1991, 68: 1323 – 1328

13　Loh KS, Goh BC, Lu J, Hsieh WS, Tan L. Familial nasopharyngeal carcinoma in a cohort of 200 patients. Arch Otolaryngol Head Neck Surg, 2006, 132: 82 – 85

14　Yu MC, Ho JH. Lai SH, Henderson BE. Cantonese – style salted fish as a cause of nasopharyngeal carcinoma: report of a case – control study in Hong Kong. Cancer Res, 1986, 46: 956 – 961

15　Yu MC, Huang TB, Henderson BE. Diet and nasopharyngeal carcinoma: a case – control study in Guangzhou, China. Int J Cancer, 1989, 43: 1077 – 1082

16　Annstrong RW, hnrey PB, Lye MS, Armstrong MJ, Yu MC, Sani S. Nasopharyngeal carcinoma in Malaysian Chinese: salted fish and other dietary exposures, lnt J Cancer, 1998, 77: 228 – 235

17　Yuan JM, Wang XL. Xiang YB, Gao YT, Ross RK. Yu MC. Preserved foods in relation to risk of nasopharyngeal carcinoma in Shanghai, China. Int J Cancer, 2000, 85: 358 – 363

18　Zou J, Sun Q, Akiba S, Yuan Y, Zha Y, Tao Z, Wei L, Sugahara T. A case – control study of nasopharyngeal carcinoma in the high background radiation areas of Yangjiang, China. J Radiat Res, 2000, 41 Suppl: 53 – 62

19　Yu MC, Mo CC, Chong WX, Yeh FS. Henderson BE. Preserved foods and nasopharyngeal carcinoma: a case – control study in Guangxi, China. Cancer Res, 1988, 48: 1954 – 1959

20　Friborg JT, Yuan JM, Wang R, Koh WP, Lee HP, Yu MC. A prospective study of tobacco and alcohol use as risk factors for pharyngeal carcinomas in Singapore Chinese. Cancer, 2007, 109: 1183 – 1191

21　Yu MC, Garabrant DH, Huang TB, Henderson BE. Occupational and other non – dietary risk factors for nasopharyngeal carcinoma in Guang-zhou, China. Int J Cancer, 1990, 45: 1033 – 1039

22 Cheng YJ, Hildesheim A, Hsu MM, Chen IH, Brinton LA, Levine PH, Chen CJ, Yang CS. Cigarette smoking, alcohol consumption and risk of nasopharyngeal carcinoma in Taiwan. Cancer Causes Control, 1999, 10: 201 – 207

23 West S, Hildesheim A, DosemecJ M. Non – viral risk factors for naso – pharyngeal carcinoma in the Philippines: results from a case – control study. Int J Cancer, 1993, 55: 722 – 727

24 Vaughan TL, Shapiro JA, Burt RD, Swanson GM, Berwick M, Lynch CF, Lyon JL. Nasopha-ryngeal cancer in a low – risk population: defining risk factors by histological type. Cancer Epidemi-ol Biomarkers Prey, 1996, 5: 587 – 593

25 Sriampom S, Vatanasapt V, Pisani P, Yongchaiyudha S, Rungpitarmgsri V. Environmental risk factors for nasopharyngeal carcinoma: a case – control study in northeastern Thailand. Cancer Epidemiol Biomark-ers Prev, 1992, 1: 345 – 348

26 Zheng X, Yan L, Nilsson B, Eklund G, Drettner B. Epstein – Barr virus infection, salted fish and nasopharyngeal carcinoma. A case – control study in southern China. Acta Oncol, 1994, 33: 867 – 872

27 Ning JP, Yu MC, Wang QS, Henderson BE. Consumption of salted fish and other risk factors fo," nasopharyngeal carcinoma (NPC) in Tian-jin, a low – risk region for NPC in the People's Republic of China. J Natl Cancer Inst, 1990, 82: 291 – 296

28 Vaughan TL, Stewart PA, Teschke K, Lynch CF, Swanson GM, Lyon JL, Berwick M. Occu-pational exposure to formaldehyde and wood tust and nasopharyngeal carcinoma. Occup Envir Med, 2000, 57: 376 – 384

29 Armstrong RW, Imrey PB, Lye MS, Armstrong M J, Yu MC, Sani S. Nasopharyngeal carcinoma in Malaysian Chinese: occupational exposures to particles, formaldehyde and heat. Int J Epidemi-ol, 2000, 29: 991 – 998

30 Marsh GM, Youk AO, Buchanich JM, Cassidy LD, Lucas LJ, Esmen NA, Gathuru IM. Pharyn-geal cancer mortality among chemical plant work-ers exposed to formaldehyde. Toxicol lnd Health, 2002, 18: 257 – 268

31 Marsh GM, Youk AO. Reevahmtion of mortality risks from nasopharyngeal cancer in the formalde-hyde cohort study of the National Cancer Institu-te. Regui Toxicol Pharmacol, 2005, 42: 275 – 283

32 Hildesheim A, Dosemeci M, Chan CC, Chen CJ, Cheng Y J, Hsu MM, Chen IH, Mittl BF, Sun B, Levine PH, Chen JY, Brinton LA, et al. Oc-cupational exposure to wood, formaldehyde, and solvents and risk of nasopharyngeal carcinoma. Cancer Epidemiol Biomarkers Prev, 2001, 10: 1145 – 1153

33 Li W, Ray RM, Gao DL, Fitzgibbons ED, Seixas NS, Camp JE, Wernli KJ, Astrakianakis G, Feng Z, Thomas DB. Checkoway H. Occupation-al risk factors for nasopharyngeal cancer among female textile workers in Shanghai, China. Occup Envir Med, 2006, 63: 39 – 44

34 Deng H, Zeng Y, Lei Y, Zhao Z, Wang P, Li B. Pi Z. Tan B, Zheng Y, Pan W. Serological survey of nasopharyngeal carcinoma in 21 cities of south China. Chin Med J (Engl), 1995, 108: 300 – 303

35 Guo XC, Scott K, Liu Y, Dean M, David V, Nelson GW, Johnson RC, Dilks HH, Lauten-berger J, Kessing B. Martenson J, Guan L, et al. Genetic factors leading to chronic Epstein – Barr virus infection and nasopharyngeal carcinoma in Southeast China: study design, methods and feasibility. Hum Genomics, 2006, 2: 365 – 375

36 Nelson GW, O'Brien SJ. Using mutual informa-tion to measure the impact of multiple genetic fac-tors on AIDS. J Acquit Immune Defic Syndr, 2006, 42: 347 – 354

37 Ou SH, Zell JA. Ziogas A, Anton – Culver H. Epidemiology of nasopharyngeal carcinoma in the United States: improved survival of Chinese pa-tients within the keratinizing squamous cell carci-

noma histology. Ann Oncol, 2007, 18: 29 - 35

38 Guo X, O'Brien SJ, Zeng Y, Nelson GW, Winkler CA. GSTM1 and GSTT1 gene deletions and the risk for nasopharyngeal carcinoma in han chinese. Cancer Epidemiol Biomarkers Prey, 2008, 17: 1760 - 1763

39 Zeng Y, Pi GH, Deng H, Zhang JM, Wang PC, Wolf H, De The G. Epstein - Barr virus seroepidemiology in China. AIDS Res, 1986, 2 (Suppl1): S7 - S15

40 Zeng Y, Liu Y, Liu C, Chen S, Wei J, Zhu J. Zai H. Application of immunoenzymic method and immunoautoradiographic method for the mass survey of nasopharyngeal carcinoma. Chin J Oncol, 1979, 1: 2 - 7

41 Zeng Y, Zhang LG, Li HY, Jan MG, Zhang Q. Wu YC, Wang YS, Su GR. Serological mass storey for early detection of nasopharyngeal Carcinoma in Wuzhou City, China. Int J Cancer, 1982, 29: 139 - 141

42 Zeng Y, Zhang LG, Wu YC, ttuang YS, Huang NQ, Li JY, Wang YB, Jiang MK, Fang Z, Meng NN. Prospective studies on nasopharyngeal carcinoma in Epstein - Barr virus IgA/VCA antibody - positive persons in Wuzhou City, China. lnt J Cancer, 1985, 36: 545 - 547

43 Ung A, Chen CJ, Levine PH. Cheng YJ. Brinton LA, Chen IH, Gold - stein AM, Hsu MM, Chhabra SK, Chen JY. Apple RJ, Yang CS, et al. Familial and sporadic cases of nasopharyngeal carcinoma in Taiwan. Anticancer Res, 1999, 19: 661 - 665

44 Chen CJ, Liang KY, Chang YS. Wang YF. Hsieh T, Hsu MM, Chen JY, Liu MY. Multiple risk factors of nasopharyngeal carcinoma: Epstein - Barr virus, malarial infection, cigarette smoking and familial tendency. Anticancer Res, 1990, 10: 547 - 553

45 Armstrong RW, Armstrong M J, Yu MC, Henderson BE. Salted fish and inhalants as risk factors for nasopharyngeal carcinoma in Malaysian Chinese. Cancer Res, 1983, 43: 2967 - 2970

46 Zheng YM, Tuppin P, Hubert A, Jeannel D, Pan YJ, Zeng Y, De The G. Environmental and dietary risk factors for nasopharyngeal carcinoma: a case - control study in Zangwu County, Gumlgxi, China. Br J Cancer, 1994, 69: 508 - 514

47 Swenberg JA, Kerns WD, Mitchell R1, Gralla E J, Pavkov KL. Induction of squamous cell carcinomas of the rat nasal cavity by inhalation exposure to formaldehyde vapor. Cancer Res, 1980, 40: 3398 - 3402

48 Albert RE, Sellakumar AR, Laskin S, Kuschner M, Nelson N, Snyder CA. Gaseous formaldehyde and hydrogen chloride induction of nasal Cancer in the rat. J Natl Cancer lnst, 1982, 68: 597 - 603

49 Partanen T. Formaldehyde exposure and respiratory cancer - a metaanalysis of the epidemiologic evidence. Scand J Work Environ Health, 1993, 19: 8 - 15

50 Lee HP, Gourley L, Duffy SW, Esteve J, Lee J, Day NE. Preserved foods and nasopharyngeal carcinoma: a case - control study among Singapore Chinese. Int J Cancer, 1994, 59: 585 - 590

109. Analysis of Epstein – Barr Viral DNA Load, EBV – LMP2 Specific Cytotoxic T – lymphocytes and Levels of CD4+ CD25+ T Cells in Patients with Nasopharyngeal Carcinomas Positive for IgA Antibody to EBV Viral Capsid Antigen

MO Wu – ning[1], TANG An – zhou[2], ZHOU Ling[3], HUANG Guang – wu[2], WANG Zhan[3], ZENG Yi[3]

Laboratory Department 1, Department of Otolaryngology & Head and Neck Surgery 2, First Affiliated Hospital, Guangxi Medical University; State Key Laboratory for Infectious Disease Prevention and Control, Institute for Viral Disease Control and Prevention Chinese Center for Disease Control and Prevention[3]

Summary

Background Epstein – Barr virus (EBV) is a herpesvirus commonly associated with several malignant diseases including nasopharyngeal carcinoma (NPC), which is a common cancer in Southeastern Asia. Previous studies showed that plasma levels of EBV – DNA might be a sensitive and reliable biomarker for the diagnosis, staging and evaluating of therapy for NPC. There are a few analyses of the levels of EBV – latent membrane protein 2 (LMP2), specific cytotoxic T – lymphocytes (CTLs) in patients with NPC. This study was conducted to investigate the levels of EBV – LMP2 – specific CTLs, EBV – DNA load and the level of CD4+ CD25+ T cells in such patients.

Methods From February 2006 to April 2006, 62 patients with NPC, 40 healthy virus carriers positive for EBV viral capsid antigen (EBV – IgA – VCA) and 40 controls were enrolled in the study. We used a highly sensitive ELISPOT assay, real – time polymerase chain reaction (PCR) and flow cytometry to measure the EBV – LMP2 – specific CTL response, the EBV DNA load and the level of CD4+ CD25+ T cells, respectively.

Results The EBV – LMP2 – specific CTL responses of the samples from the control, healthy virus carriers and patients with NPC were significantly different from the LMP2 epitopes, with the control and healthy virus carrier samples displaying a stronger response in three cases. There were significant differences in EBV DNA load in serum between NPC and the healthy groups; patients with NPC at stages III or IV had significantly higher viral loads compared with those at stages I or II. A significantly higher percentage of CD4+ CD25+ T lymphocytes were detected in the patients, compared with healthy virus carriers and healthy controls. Moreover, patients with advanced stages of NPC (III and IV) had significantly higher percentages than the patients with early stages (I and II).

Conclusions Patients with NPC are frequently unable to establish or maintain sufficient immunosurveillance to control proliferating B cells harboring EBV and to destroy the tumor cells that express immunodominant LMP2 proteins. Controlling the activity of CD4+ CD25+ T cells and elevating

CD8$^+$ cells specific for LMP2 epitopes could be an effective immunotherapy for patients with NPC.

[**Key words**] Nasopharyngeal carcinoma; Cell immunity; Epstein – Barr virus; Latent membrane protein 2; Cytotoxic T lymphocyte

Introduction

Epstein – Barr virus (EBV), a widespread gamma herpesvirus, can persist as a latent chronic infection in memory B lymphocytes[1] mainly controlled by a specific CD8$^+$ cytoxic T – lymphocyte (CTL) – mediated immune response[2]. This response could play a major role in the pathogenesis of EBV – related diseases, including lymphoproliferative disorders associated with immunocompromise, Burkitt lymphoma, Hodgkin disease, undifferentiated nasopharyngeal carcinoma (NPC) and various T – cell lymphomas. Thus, the levels of EBV – specific CTLs in patients with NPC might decrease before the tumor appears. Furthermore, elevated EBV DNA loads can also be found in the plasma of patients with NPC. These findings suggest that loss of the EBV – specific cellular immune control might favor EBV reactivation and NPC progression. The aim of this study was to analyze EBV DNA load, EBV – latent membrane protein 2 (LMP2) – specific CTL and the level of CD4$^+$ CD25$^+$ T cells in controls, healthy virus carriers and patients with NPC. EBV – specific CTL levels were assessed by a highly sensitive ELISPOT assay, using EBV – LMP2 antigens adapted to the HLA type I context of the Chinese population. EBV reactivation was evaluated by assessing EBV DNA load in serum, with real – time quantitative polymerase chain reaction (PCR).

Methods

1. Patients

Sixty – two patients with NPC (mean age 43 years, range 29 – 60 years) were identified at the Department of Pathology, First Affiliated Hospital, Guangxi Medical University, China. The NPC diagnosis was based on pathology assessment of paraffin wax – embedded tumor biopsy specimens, TNM staging for tumor size (T), lymph node involvement (N) and metastasis (M) was done using the 1997 criteria of the Union International Contre le Cancer[3] for all patients, using clinical measurements and computed tomography scans as part of the routine patient workup. Approval of the local medical ethics committee was obtained. Forty healthy virus carriers of EBV viral capsid antigen (EBV – IgA – VCA) and 40 controls (normal individuals) were matched to cases by gender and age in five – year age groups.

2. EBV DNA extraction

Serum was prepared within 4 hours for collection and stored frozen until analysis. Total DNA was extracted from 0.2 ml of serum using the Total Nucleic Acid kit (DaAn Gene Company Ltd. of Sun Yat – sen University, China) according to the manufacturer's instructions. Nucleic acid was eluted into 20 μl of the buffer provided in the extraction kit.

3. EBV DNA load measurement

The EBV DNA load in serum was determined by a real – time PCR that targets a highly conserved region of the BamH1W sequence of EBV. The primers used in this assay were: forward primer W – 44F 5' – AGTCTCTGCCTCCAGGCA – 3' and reverse primer W119R 5' – ACGAGGGGCCT-

GTCCTGTCCACCG – 3'). The fluorigenic internal hybridization probe sequence was 5' – （FAM） – CACTGTGTAAAGTAAAGTCCAGCCTCC – （TAMRA） – 3'.

Tenfold dilutions of spectrophotometrically quantified plasmid DNA containing the BamH1W target sequence were used to create a standard curve. Each run included positive and negative controls extracted in parallel with experimental samples. Reaction conditions were as follows: a volume of 45 µl with 10 mmol/L of each primer, 50 mmol/L KCl, 1. 5 mmol/L MgCl$_2$, 200 µmol of each deoxynucleoside triphosphate, 10 mmol/L Tris （pH 8. 5） and 5 U of Taq polymerase. Amplification was done for 10 cycles of 1 minute at 93℃ and 45 seconds at 55℃. The first denaturation step and the last elongation step were extended for 2 minutes. Data were collected and analyzed using Detection System software （Bio – Rad iCycler, Bio – Rad Laboratories, Inc. , Hercules, CA, USA）.

The EBV load was expressed as the number of copies of the EBV genome per milliliter of serum. Samples with detectable EBV. genomes were defined as positive. The clinical status of these samples was blinded during all laboratory procedures.

4. Peptides

The peptides used in this study, their corresponding EBV – LMP2 protein and their HLA restriction type are listed in Tab. 1 using EBV – LMP2 antigens adapted to the HLA I context of the China population （Beijing Scilight Biotechnology Ltd. Co. , Beijing, China）. The peptides were dissolved in dimethylsulfoxide （DMSO; Sigma – Aldrich, St Louis, MO, USA） at 5 mg/ml.

5. Interferon gamma （IFN – γ） ELISPOT assay

Peripheral blood mononuclear cells （PBMCs） were isolated from venous blood by Ficoll – Hypaque density gradient centrifugation, washed twice in RPMI – 1640 medium and cryopreserved for future ELISPOT analyses. To determine the frequencies of CTLs specific for the EBV – LMP2 peptide among the PBMCs, an IFN – γELISPOT assay was used to determine the number of cells secreting IFN – γ. This assay was carried out according to the manufacturer's instructions

Tab. 1　Epitope peptides used

Designation	Sequence	Protein	Amino acid residues	HLA restriction
LLS	LLSAWILTA	LMP2	447 – 455	A * 0203
LLW	LLWTLVVL	LMP2	329 – 337	A * 0201
LTA	LTAGFLIFL	LMP2	453 – 461	A * 0206
SSC	SSCSSCPLSK	LMP2	340 – 349	A * 1101
TYG	TYGPVFMSL	LMP2	419 – 427	A * 2402
IED	IEDPPFNSL	LMP2	200 – 208	B * 4011

（U – CyTech biotech, Utrecht, the Netherlands）. PBMCs were thawed and cultured in complete medium （RPMI 1640, 2 mmol/L glutamine, 10 mmol/L HEPES, 50 mg/ml gentamicin） overnight before being placed in 96 – well plates coated with anti – IFN – γ monoclonal antibodies （mAbs）. After overnight incubation at 4℃, the wells were washed six times with RPMI 1640 medium （Life Technologies, Gaithersberg, MD, USA） and then blocked with 1% bovine serum albumin （BSA） in phosphate – buffered saline （PBS） for 1 hour at 37℃. PBMCs （viability 80% – 95%） were added to duplicate wells at concentrations of 5×10^5 cells in 100 ml complete medium. Peptides, diluted to 1 mg/ml in complete medium, were added to the appropriate wells to give a final concentration of 10 µg/ml of each peptide. Negative control wells received 10 µl of complete medium containing 0.1% v/v DMSO, while positive control wells received 10 μl PHA (Life Technologies). The plates were incubated undisturbed overnight at 37℃ in 5% CO_2. The next day, the wells were washed once with ice – cold deionized water and washed 10 times with PBS with 0.05% Tween 20 (Sigma – Aldrich). After the last wash, 50 μl of the biotinylated anti – IFN – γ mAb (biotinylated detector antibody) was added to each well. The plates were incubated at 37℃ for 1 hour, after which they were washed 5 times with PBS – Tween; 50 ml of diluted φ – labeled anti – biotin antibodies (GABA) was added to each well and the plates were incubated at 37℃ for 1 hour. After washing five times with PBS – Tween, 30 μl Activator I / II (freshly prepared) was added to each well. The plates were incubated at room temperature until spots had appeared in the control wells, at which time the plates were washed thoroughly with tap water. Then the plates were air – dried. The number of spots per well was counted using a stereomicroscope. Only well – defined spots consisting of a dark center surrounded by a lighter " halo" were counted. The result was judged according to the following criteria. Where the negative control had 0 to 5 spots, a positively reactive sample was indicated if (assay spot count) – (negative control count) ≥6; where the negative control had ≥ 6 spots, a positively reactive sample was indicated if (assay spot count) ≥2 × (negative control count).

6. Antibodies

The following antibodies were used: fluorescein isothiocyanate (FITC) – conjugated anti – CD4, phycoerythrin (PE) – conjugated anti – CD25, PE – conjugated anti – CD8, CyChrome – anti – CD3 from BD Biosciences PharMingen (San Diego, CA, USA).

7. Fluorescence – activated cell sorting (FACS) analysis

Cells (1×10^5 – 5×10^5) were stained with a fluorochrome – conjugated antibody for 30 minutes on ice in the dark. The cells were then washed in cold FACS buffer and fixed with 1% formaldehyde in PBS for 15 minutes on ice in the dark. Flow cytometry of samples was performed on FAC-Scan machines (Coulter EPICS XL, Beckman Coulter Corp, USA) using Cell Quest software for acquisition and analysis of data. Lymphocytes were gated by plotting forward scatter versus side scatter. Appropriate isotype controls were performed for each experiment.

8. Statistical analysis

Associations between clinical status and qualitative (positive vs undetectable) or quantitative (expressed in copies/ml or 1×10^6 PBMCs or as the mean ± standard deviation) measures were assessed by one way analysis of variance (ANOVA), chi – squared test or Spearman's rank correlation test as appropriate. Patients with advanced and early stage NPCs were compared using the Mann – Whitney rank – sum test or unpaired sample Student's t tests. $P < 0.05$ was considered statistically significant. All analyses were performed using SPSS software (version 13.0, SPSS Inc, Chicago, IL, USA).

Results

1. EBV loads and EBV – specific immune responses against EBV – LMP2 antigens

EBV loads in patients with NPC, healthy virus carriers and healthy controls are shown in Tab. 2 and Fig. 1. The median viral load and positive rates in healthy controls, healthy virus carriers and patients with NPC were 0 copies/ml (range 0 – 0) , 0 copies/ml (range 0 – 49) and 1062 copies/ml (range 0 – 7 320 000) , respectively and positive rates were 0% , 5% and 87% , respectively. As expected, there were significant differences EBV DNA load in serum between disease and non – disease groups with a P < 0. 05 by the Mann – Whitney rank – sum test.

The results of the EBV – specific immune responses against EBV – LMP2 antigens are summarized in Tab. 2 and Fig. 1. EBV – specific CTL – mediated immune responses could be found in all groups, albeit often at low levels. The maximum frequency was found in the control samples. A positive specific cellular immune response against EBV – LMP2 epitopes was evidenced in most of the healthy controls (32/40, 80%) ; whereas a T – cell – mediated immune response against EBV – 1MP2 epitopes was evidenced in a few patients with NPC (13/62, 21%). The PBMC samples of most of the patients with NPC (49/62, 79%) did not demonstrate a response, with healthy virus carriers having EBV specific CTL immune responses (20 of 40, 50%). There were significant differences between the three groups (P < 0. 05).

Tab. 2　EBV DNA load and EBV – specific immune response in three groups

Groups	n	EBV – DNA load			EBV – specific immune response	
		Positive (n (%))	Median	Range of positives (copies/ml)	Positive (n (%))	Range of spots (1 ×10^6 PBMCs)
Patients with NPC	62	54 (87)	1062	0 – 7 320 000	13 (21)	0 – 94
Healthy virus carriers	40	2 (5)	0	049	20 (50)	0 – 92
Healthy controls	40	0	–	–	32 (80)	0 – 166

2. Relation of TNM staging with EBV DNA load and EBV – specific immune responses against EBV – LMP2 antigens in patients with NPC

The distribution of clinical stages in all 62 patients with NPC is shown in Tab. 3. The median viral loads and positive rates in early and advanced stage patients were 108 copies/ml and 1319 copies/ml respectively and the positive rates were 60% and 100% respectively. Patients with stage Ⅲ or Ⅳ disease had significantly higher viral load compared with patients with stage Ⅰ or Ⅱ disease (P < 0. 05, Mann – Whitney rank – sum test). However, there was no significant difference between patients with stages Ⅲ or Ⅳ. Positive specific cellular immune response against EBV – LMP2 epitopes in early and advanced stage patients were 30% and 16. 7% respectively. Patients with advanced stages had lower cellular immune responses compared with patients with early stages of the disease. However, there was no significant difference between these two groups.

3. Levels of CD4$^+$CD25$^+$T cells and T lymphocyte subsets

The results are summarized in Tab. 4. The peripheral blood CD4$^+$CD25$^+$T cell count of the NPC group was significantly higher than the level of the healthy virus carriers and healthy controls (F = 8. 852, P = 0. 000). However, there was no significant difference between healthy virus carriers and healthy controls. The CD3$^+$ and CD4$^+$CD3$^+$T cell levels of the NPC group were significantly lower than those of the healthy virus carriers and healthy controls (F = 16. 911, P = 0. 000 and

$F = 14.634$, $P = 0.000$, respectively). However, there was also no significant difference between healthy virus carriers and healthy controls. No significantly different percentage of $CD8^+ CD3^+$ T lymphocytes was detected among the three groups.

4. Relation of TNM staging with $CD4^+ CD25^+$ T cells and T lymphocyte subsets in patients with NPC

The percentages of $CD4^+ CD25^+$, $CD3^+$ and $CD4^+ CD3^+$ T cells of advanced stage patients varied significantly ($F = 3.914$, $P = 0.013$; $F = 7.220$, $P = 0.000$

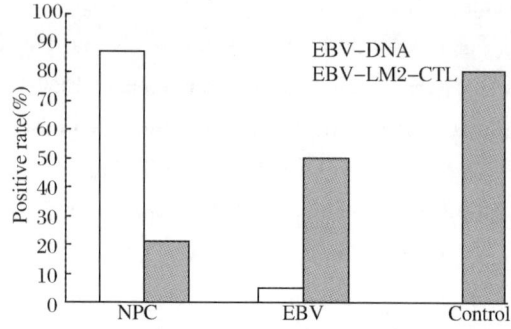

Fig. 1 **The positive rate of EBV DNA loads and EBV – specific immune responses in patients with NPC, healthy virus carriers and healthy controls**

and $F = 16.101$, $P = 0.000$, respectively), and were significantly different between patients with early and advanced stage tumors (all $P = 0.000$, t test).

Tab. 3 **EBV DNA load and EBV – specific immune response by clinical status**

Stages	n	EBV – DNA load			EBV – specific immune response Groups	
		Positive (n (%))	Median	Range of positives (copies/ml)	Positive (n (%))	Range of spots (1×10^6 PBMCs)
I	2	1 (50)	25	0 – 50	1 (50)	0 – 38
II	18	11 (61)	108	0 – 5540	5 (28)	0 – 94
III	21	21 (100)	1318	264 – 8000	4 (19)	0 – 52
IV	21	21 (100 ~ , 4)	1320	278 – 7320000	3 (14)	0 – 34

Tab. 4 **Levels of $CD4^+ CD25^+$ T cells and T lymphocyte subsets in three groups ($(\bar{x} \pm s)$ %)**

Groups	n	$CD4^+ CD25^+$	$CD3^+$	$CD4^+ CD3^+$	$CD8^+ CD3^+$
Patients with NPC	62	3.01 ± 1.15	59.07 ± 12.18	30.44 ± 8.42	21.40 ± 6.87
Healthy virus carriers	40	2.35 ± 1.06	68.30 ± 8.84	36.64 ± 5.40	22.43 ± 7.62
Healthy controls	40	2.17 ± 0.95	69.23 ± 6.24	36.80 ± 5.24	23.63 ± 5.81

Tab. 5 **Relation of TNM stage and levels of $CD4^+ CD25^+$ T cells and T lymphocyte subsets in patients with NPC [$(\bar{x} \pm s)$ %]**

Stages	n	$CD4^+ CD25^+$	$CD3^+$	$CD4^+ CD3^+$	$CD8^+ CD3^+$
I	2	2.45 ± 0.81	65.75 ± 2.19	37.35 ± 7.28	24.10 ± 10.04
II	18	2.34 ± 0.72	68.43 ± 8.62	38.69 ± 7.44	21.97 ± 5.55
III	21	3.20 ± 0.68	54.34 ± 15.45	27.19 ± 6.92	20.50 ± 7.61
IV	21	3.45 ± 1.31	55.12 ± 5.24	25.97 ± 4.52	21.56 ± 7.29

Discussion

Epstein – Barr virus (EBV) is a ubiquitous gamma herpesvirus that can establish both latent and lytic infections. Most primary infections with EBV occur during childhood and often do not produce obvious symptoms, but the EBV can stay in the body for the subject's whole life. Once the EBV is activated, it can act as the nosogenesis of many diseases including malignant diseases. NPC is a common cancer in Southeastern Asia, especially in southern China, with a yearly incidence rate between 10 and 50 per 100 000. The definite function of EBV in the etiology of NPC is not clear, but several antigens to EBV, such as EBNA1, LMP1 and LMP2 can be detected in NPC cells and tumor biopsies. EBNA1 contains a glycine – alanine repeat (GAr) domain that prevents degradation by the proteasome complex[4] and protected from processing and presentation to HLA class I – restricted T cells. Many HLA – I restricted epitopes of LMP2 have been identified that are conserved among different populations and LMP2 can be found to persist in NPC cells. Therefore, attention has been focused on responses to the LMP2 proteins. Previous studies *in vitro* using the regression of EBV transformation assay, which measures T cell – mediated regression and death of EBV – infected B cells, have shown that virus – specific T cell immunity is significantly reduced in patients with NPC compared with that in healthy virus carriers[5]. Clinical studies have already demonstrated that infusions of autologous EBV – specific CTLs expanded *in vitro* can safely and effectively control EBV – positive posttransplant lymphoproliferative disease in bone marrow transplant recipients[6] and patients with advanced NPC[7,8].

Recent studies suggest that the plasma EBV – DNA level might be a sensitive and reliable biomarker for diagnosis, staging and evaluating therapeutic effects at a molecular level/clinical practice[9 – 12], but other results indicate that the circulating EBV DNA load is independent of serological parameters and does not reflect intact tumor cells. Thus, the primary diagnostic value of the EBV DNA load for the detection of NPC is limited[13,14].

Immunoregulatory $CD4^+CD25^+$ cells have been studied extensively. They suppress autoimmune T – cell responses and maintain peripheral tolerance[15,16]. They might also simultaneously prevent the host from mounting an immune response to EBV or tumor antigens. There have been reports that the peripheral blood of patients with breast and pancreatic cancers[17], hepatocellular carcinomas[18], colorectal carcinomas[19], lung cancer[20], and NPC[21] have higher $CD4^+CD25^+$ cell counts than healthy individuals. Consistent with this concept, experimental depletion of $CD4^+CD25^+$ cells in mice with tumors improved the immune – mediated tumor clearance[22] and enhanced the response to immune – based therapy[23].

In the present study, we used a highly sensitive ELISPOT assay to measure the EBV – LMP2 – specific CTL responses in peripheral blood samples from 40 healthy controls, 40 healthy virus carriers and 62 patients newly diagnosed with NPC. The EBV DNA load in serum and the level of $CD4^+CD25^+$ T cells in the peripheral blood were measured by real – time PCR and FACS analysis respectively in the same individuals. We found that the EBV – LMP2 – specific CTL responses of all three groups were statistically different for the LMP2 epitopes. The EBV – LMP2 – specific CTL response detection rate in controls (80%) was significantly higher than that in healthy virus carriers (50%) and in patients with NPC (21%; $P < 0.05$); with control samples detecting positive rate and level displaying a stronger response in three cases. The results indicate that the levels of EBV – specific CTLs in patients with NPC might decrease before the tumor appears. There were significant differences EBV DNA loads in serum between the NPC and non – disease groups. Patients with stage Ⅲ or Ⅳ disease had significantly higher viral loads than patients with stages Ⅰ or Ⅱ disease. The results support the idea that the serum EBV – DNA level is an indicator for the staging and prognosis of NPC. The positive rate of cellular immune responses to EBV – LMP2 in patients with NPC is lower, together with a higher detection rate of serum EBV DNA load. This suggests that patients with NPC frequently are unable to establish or maintain sufficient immuno – surveillance to destroy the tumor cells expressing immunodominant LMP2 proteins.

We also analyzed $CD4^+CD25^+$ T cell levels by FACS in the three groups. A significantly higher percentage of $CD4^+CD25^+$ T lymphocytes were detected in the patients compared with healthy virus carriers and healthy controls ($P < 0.05$). Patients with advanced stages of NPC had significantly higher percentage than did patients in early stages ($P < 0.05$). The percentages of $CD3^+$ and $CD3^+CD4^+$ T cells were decreased significantly in the patients with NPC. The level of $CD4^+CD25^+$ T cells in NPC being higher with cellular immune responses lower shows that $CD4^+CD25^+$ T cells might suppress cellular immune responses in NPC.

Our result indicates that plasma EBV – DNA level could be a biomarker for TNM staging and prognosis of patients with NPC. LMP2 is potentially a major target antigen for developing immunotherapy. Controlling the activity of $CD4^+CD25^+$ T cells and improving the EBV – LMP2 – specific CTL immune response could be a strategy for immunotherapy in such patients.

〔In 《Chin Med J》 2009, 122 (10): 1173 – 1178〕

References

1 Miyashita EM, Yang B, Babcock G J, Thorley – Lawson DA. Identification of the site of Epstein – Barr virus persistence *in vivo* as a resting B cell. J Virol, 1997, 71: 4882 – 4891

2 Lee SP, Chan AT, Cheung ST, Thomas WA, CroomCarter D, Dawson CW, et al. CTL control of EBV in nasopharyngeal carcinoma (NPC): EBV – specific CTL responses in the blood and tumors of NPC patients and the antigen – processing function of the tumor cells. J Immunol, 2000, 165: 573 – 582

3 Sobin LH, Fleming ID. TNM Classification of Malignant Tumors, fifth edition (1997). Union Internationale Contre le Cancer and the American Joint Committee on Cancer. Cancer, 1997, 80: 1803 – 1804

4 Levitskaya J, Shapiro A, Leonchiks A, Ciechanover A, Masucci MG. Inhibition of ubiquitin/ proteasome – dependent protein degradation by the Gly – Ala repeat domain of the Epstein – Barr virus nuclear antigen 1. Proc Natl Acad Sci U S A, 1997, 94: 12616 – 12621

5 Moss D J, Chan SH, Burrows SR, Chew TS, Kane RG, Staples JA, et al. Epstein – Barr virus T cell response in nasopharyngeal carcinoma patients. Int J Cancer, 1983, 32: 301 – 305

6 Rooney CM, Smith CA, Ng CY, Loftin SK, Sixbey JW, Gan YJ, et al. Infusion of cytotoxic T cells for the prevention and treatment of Epstein – Barr virus – induced lymphoma in allogeneic transplant recipients. Blood, 1998, 92: 1549 – 1555

7 Straathof KC, Bollard CM, Popat U, Huls MH, Lopez T, Morriss MC, et al. Treatment of nasopharyngeal carcinoma with Epstein – Barr virus – specific T lymphocytes. Blood, 2005, 105: 1898 –

1904

8　Masmoudi A, Toumi N, Khanfir A, Kallel – Slimi L, Daoud J, Karray H, et al. Epstein – Barr virus – targeted immunotherapy for nasopharyngeal carcinoma. Cancer Treat Rev, 2007, 33: 499 – 505

9　Yang X, Goldstein AM, Chen CJ. Distribution of Epstein – Barr viral load in serum of individuals from nasopharyngeal carcinoma high – risk families in Taiwan. Int J Cancer, 2006, 118: 780 – 784

10　Shao JY, Zhang Y, Li YH, Gao HY, Feng HX, Wu QL, et al. Comparison of Epstein – Barr virus DNA level in plasma, peripheral blood cell and tumor tissue in nasopharyngeal carcinoma. Anti-cancer Res, 2004, 24: 4059 – 4066

11　Fan H, Nicholls J, Chua D, Chan KH, Sham J, Lee S, et al. Laboratory markers of tumor burden in nasopharyngeal carcinoma: a comparison of viral load and serologic tests for Epstein – Barr virus. Int J Cancer, 2004, 112: 1036 – 1041

12　Zhang Y, Gao HY, Feng HX, Deng L, Huang MY, Hu B, et al. Quantitative analysis of Epstein – Barr virus DNA in plasma and peripheral blood cells in patients with nasopharyngeal Carcinoma. Natl Med J Chin (Chin), 2004, 84: 982 – 986

13　Stevens S J, Verkuijlen SA, Hariwiyanto B, Harijadi, Fachiroh J, Paramita DK, et al. Diagnostic value of measuring Epstein – Barr virus (EBV) DNA load and carcinoma – specific viral mRNA in relation to anti – EBV immunoglobulin A (IgA) and IgG antibody levels in blood of naso-pharyngeal carcinoma patients from Indonesia. J Clin Microbiol, 2005, 43: 3066 – 3073

14　Kalpoe JS, Dekker PB, van Krieken JH, de Jong RJ, Kroes AC. Role of Epstein – Barr virus DNA measurement in plasma in the clinical manage-ment of nasopharyngeal carcinoma in a low risk area. J Clin Pathol, 2006, 59: 537 – 541

15　Baecher – Allan C, Viglietta V, Hailer DA. Human CD4 + CD25 + regulatory T cells. Semin Immunol, 2004, 16: 89 – 98

16　Shevach EM. CD4 + CD25 + suppressor T cells: more questions than answers. Nat Rev Immunol, 2002, 2: 389 – 400

17　Liyanage UK, Moore TT, Joo HG, Tanaka Y, Herrmann V, Doherty G, et al. Prevalence of regulatory T cells is increased in peripheral blood and tumor microenvironment of patients with pan-creas or breast adenocarcinoma. J Immunol, 2002, 169: 2756 – 2761

18　Ormandy LA, Hillemann T, Wedemeyer H, Manns MP, Greten TF, Korangy F. Increased populations of regulatory T cells in peripheral blood of patients with hepatocellular carcinoma. Cancer Res, 2005, 65: 2457 – 2464

19　Somasundaram R, Jacob L, Swoboda R, Caputo L, Song, Basak S, et al. Inhibition of cytolytic T lymphocyte proliferation by autologous CD4 + / CD25 + regulatory T cells in a colorectal carcino-ma patient is mediated by transforming growth fac-tor – beta. Cancer Res, 2002, 62: 5267 – 5272

20　Wolf AM, Wolf D, Steurer M, Gastl G, Gunsili-us E, Grubeck – Loebenstein B. Increase of regu-latory T cells in the peripheral blood of cancer pa-tients. Clin Cancer Res, 2003, 9: 606 – 612

21　Li J, Zeng XH, Mo HY, Rol6n U, Gao YF, Zhang XS, et al. Masucci MG. Functional inacti-vation of EBV – specific T – lymphocytes in naso-pharyngeal carcinoma: implications for tumor im-munotherapy. PLoS ONE, 2007, 2: e1122

22　Shimizu J, Yamazaki S, Sakaguchi S. Induction of tumor immunity by removing CD25 + CD4 + T cells: a common basis between tumor immunity and autoimmunity. J Immunol, 1999, 163: 5211 – 5218

23　Steitz J, Bruck J, Lenz J, Knop J, Tuting T. De-pletion of CD25 (+) CD4 (+) T cells and treatment with tyrosinase – related protein 2 – transduced dendritic cells enhance the interferon alpha – induced, CD8 (+) T – cell – dependent immune defense of B16 melanoma. Cancer Res, 2001, 61 : 8643 – 8646

110.　Haplotype – dependent HLA Susceptibility to Nasopharyngeal Carcinoma in A Southern Chinese Population

TANG M[1,2], ZNEG Y[3], POISSON A[4], MARTI D[4], GUAN L[5], ZHENG Y[2],
DENG H[2], LIAO J[6], GUO X[3,5], SUN S[7], NELSON G[5], de The G[8],
WINKLER CA[5], O'Brien SJ[1], CARRINGTON M[4], GAO X[4]

1. Laboratory of Genomic Diversity National Cancer Institute – Frederick, Frederick, MD, USA;
2. Cancer Center, Wuzhou Red Cross Hospital, Guangxi, China; 3. State Key Laboratory for Infectious
Diseases Prevention and Control, Institute for Viral Disease Control and Prevention, Chinese
Center for Disease Control and Prevention, Beijing, China; 4. Cancer and Inflammation Program,
Laboratory of Experimental Immunology, National Cancer Institute – Frederick, Frederick,
MD, USA; 5. Labolatory of Genomic Diversity SAIC – Frederick, Inc. , National Cancer
Institute – Frederick, Frederick, MD, USA; 6. Department of Epidemiology, Cangwu Institute
for Nasopharyngeal Carcinoma Control and Prevention Guangxi, China; 7. Conservation
Biology Building College of Life Sciences Peking University, Beijing, China,
8. Oncogenic Virus Epidemiology and Pathophysiology, Institute Pasteur, Paris, France

Summary

We have conducted a comprehensive case – control study of a nasopharyngeal carcinoma (NPC) population cohort from Guangxi Province of Southern China, a region with one of the highest NPC incidences on record. A total of 1 407 individuals including NPC patients, healthy controls, and their adult children were examined for the human leukocyte antigen (HLA) association, which is so far the largest NPC cohort reported for such studies. Stratified analysis performed in this study clearly demonstrated that while NPC protection is associated with independent HLA alleles, most NPC susceptibility is strictly associated with HLA haplotypes. Our study also detected for the first time that A* 0206, a unique A2 subtype to South and Southeast Asia is also associated with a high risk for NPC. HLA – A* 0206, HLA – B* 3802 alleles plus the A* 0207 – B* 4601 and A* 3303 – B* 5801 haplotypes conferred high risk for NPC showing a combined odds ratio (OR) of 2. 6 (P < 0. 0001). HLA alleles that associate with low risk for NPC include HLA – A* 1101. B* 27. and B* 55 with a combined OR of 0. 42 (P < 0. 0001). The overall high frequency of NPC – susceptible HLA factors in the Guangxi Population is likely to have contributed to the high – NPC incidence in this region.

〔**Key words**〕　HLA; Nasopharyngeal carcinoma; Haplotype; Stratified analysis

Introduction

Nasopharyngeal carcinoma (NPC) is an epithelial malignancy caused by a combination of Epstein – Barr virus (EBV) infection and environmental factors[1-3]. Genetic predisposition also has a function in NPC susceptibility causing the observed familial aggregation[4] and distinct racial and geographical distribution of the disease incidence[5-8]. Southern China and Southeast Asia have a disproportionally high incidence of NPC. In these regions, NPC is one of the most common cancers,[9,10] but in Caucasians, it is a rare disease. [11]

Among the host genetic markers that have been associated with NPC, the class I human leukocyte antigen (HLA) genes have shown a strong and consistent association with disease risk. As it was first reported by Simons et al. [12] in a Singaporean Chinese cohort in the mid 1970s, the HLA association with NPC has been widely detected in different racial groups even though the exact HLA factors (alleles and haplotypes) associated with the disease sometimes vary among racial groups due to population – dependent HLA distributions. Populations ot Southern China and Southeast Asia share high – NPC incidence as well as common HLA characteristics. Mainland Chinese[9,13], Taiwanese[14] and Singaporean Chinese[15], NPC cohorts also show similar HLA associations, where HLA – A*11 and B*13 seem protective against NPC, and A*02 (A*0207), A*33, B*46, and B*58 associate with susceptibility to this disease (Fig. 1). Other HLA loci including HLA – C and the class Ⅱ loci, DR and DQ, have not shown independent associations with NPC even though certain class Ⅱ alleles might be part of the extended HLA haplotypes associated with the disease as described by Hildesheim et al. [14]

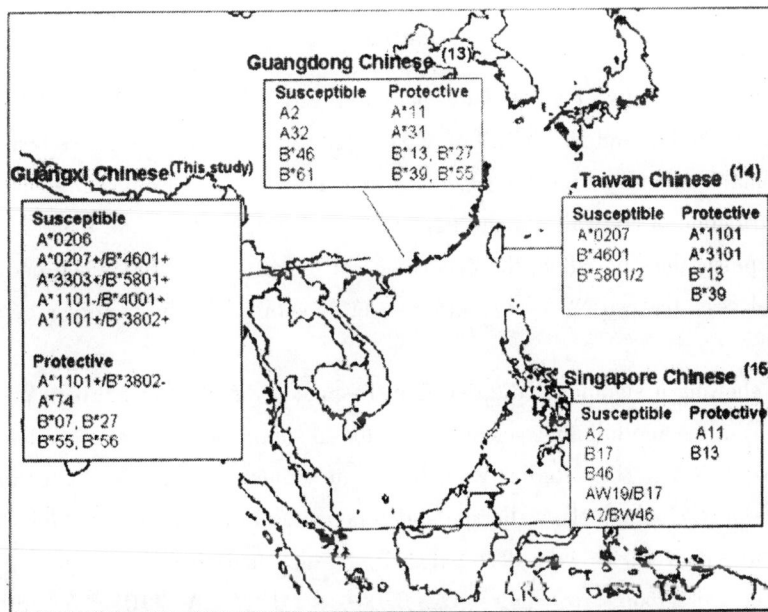

Fig. 1　HLA associations with NPC in Chinese population.

Association of HLA polymorphism with human disease may indicate direct involvement of the HLA molecule in the disease pathogenesis. To date, however, there has been little *in vitro* or *in vivo* evidence that the NPC – associated HLA alleles affect differentially NPC – related EBV replica – tion on pathogenesis. The role of HLA – A and – B alleles or haplotypes that include combinations of these alleles in NPC pathogenesis remain unknown. It is possible that the associated HLA class I alleles are simply marking by linkage disequilibrium (LD) the true NPC – causing gene (s). Large – scale cohort studies exploring combinations of associated HLA alleles may provide useful insights into the influence of HLA on NPC.

An early indicator of NPC development is the occurrence of immunoglobulin (Ig) A antibodies to EBV capsid antigens (EBV – IgA/VCA) [16-18]. Even though > 95% of adults in the general population of all ethnic groups are healthy carriers of EBV, < 2.5% are EBV – IgA/VCA antibody positive. In comparison, > 95% of all NPC patients are EBV – IgA/VCA antibody positive. [10] If HLA diversity is indeed directly responsible for the individually varied NPC risks, it is plausible that the development of the EBV – IgA/VCA antibodies in EBV – positive individuals may also be affected by the HLA polymorphism.

Here, we have conducted a case – control study of an NPC cohort recruited from Guangxi Province In Southern China where the NPC incidence is as high as 25 – 50 cases per 10 000 individuals. DNA – based high – resolution HLA typing was performed on a total of 1407 individuals including NPC cases, matched controls, and offspring of the study subjects. This large study has allowed a comprehensive stratification of NPC – associated HLA factors and provided novel insights into the nature of the HLA association with the disease.

Results

HLA typing was informative for 356 NPC patients, 287 NPC free EBV – IgA/VCA antibody positive healthy individuals, and 342 NPC free EBV – IgA/VCA antibody negative healthy individuals. Comparative analyses between the two healthy groups failed to detect any significant deviation in the frequency distribution of HLA alleles and haplotypes (Supplementary Tab. 1 and 2), indicating that HLA polymorphism does not affect the occurrence of the EBV – IgA/VCA antibody. Therefore, in subsequent analyses, the two NPC – free groups were combined as the control group ($n = 629$) for NPC cases.

HLA alleles showing a significant difference in frequency distribution between cases and controls are listed in Tab. 1, and full analyses are presented in supplementary Tab. 3 and 4. For the reason of most of former NPC, HLA studies were based on serology typing, for better understanding the HLA influence in NPC, both HLA allotype and genotype were present in this table. For the HLA – A locus, 31 four – digit alleles were detected, 12 of which had an allele frequency > 1% in either the case or control group. Five alleles, A* 0206, A* 0207, A* 1101, A* 3303, and A* 7401/7402, showed a significant difference in frequency distribution between the case and control groups. After correction, however, only two alleles, A* 1101 and A* 3303, remained significant. Of the two detected A* 11 alleles, A* 1101 showed a reduced presence in patients compared with controls ($P < 0.0001$), whereas the less common A* 1102 showed no difference between these groups, but due to its low frequency, it need to be confirmed in even large study cohort. Four common A* 02 subtypes, A* 0201, A* 0203, A* 0206, and A* 0207, were detected. Two of alleles,

A * 0206 and A * 0207, showed an elevated frequency in patients. A * 3303 also associated with increased risk of NPC ($P = 0.0004$).

Fifty – five HLA – B alleles were detected, but only 14 had a frequency $> 1\%$. Among the 14 B alleles, seven showed a significantly different distribution between cases and controls, including B * 1301, B * 2704, B * 3802, B * 4001, B * 5502, B * 5601, and B * 5801. After P – value correction, however, only the decrease of B * 5502 in the patient group remained significant (P corrected = 0.003). Three B * 27subtypes, B * 2704, B * 2705, and B * 2706 were detected only in the control group with a combined frequency of 1.59% ($P = 0.0014$). B * 3802, B * 4001, and B * 5801 were each observed more frequently among the case group but the significance disappeared after correction.

Tab. 1 Gene frequencies (%) of the two and four – digit HLA – A, – B, and – C alleles detected in NPC patients ($n = 356$) and controls ($n = 629$)

	Allele	Frequency		P – value	$P_c$value *
		NPC patients	Controls		
HLA – A	02	38.62 (275)	32.75 (412)	0.01	NS
	0206	5.2 (37)	2.31 (29)	0.0006	NS
	0207	14.89 (106)	11.29 (143)	0.02	NS
	11	23.74 (169)	32.35 (407)	0.0001	0.003
	1101	19.52 (139)	29.09 (366)	<0.0001	0.0003
	33	20.51 (146)	14.31 (180)	0.0004	0.02
	3303	20.51 (146)	14.31 (180)	0.0004	0.04
	74	0.14 (1)	1.11 (14)	0.02	NS
HLA – B	07	0.0 (0)	1.03 (13)	0.01	NS
	13	8.43 (60)	13.04 (164)	0.002	NS
	1301	7.72 (55)	12.08 (152)	0.002	NS
	27	0.0 (0)	1.59 (20)	0.001	NS
	2704	0.0 (0)	1.27 (16)	0.005	NS
	38	12.78 (91)	9.78 (123)	0.04	NS
	3802	12.64 (90)	9.30 (117)	0.02	NS
	39	0.28 (2)	1.19 (15)	0.04	NS
	40	16.29 (116)	12.32 (155)	0.01	NS
	4001	14.75 (105)	11.05 (139)	0.02	NS
	55	0.56 (4)	4.21 (53)	<0.0001	0.0002
	5502	0.56 (4)	3.66 (46)	<0.0001	0.003
	56	0.84 (6)	2.23 (28)	0.02	NS
	5601	0.28 (2)	1.67 (21)	0.006	NS
	58	19.10 (136)	14.07 (177)	0.003	NS
	5801	19.10 (136)	14.07 (177)	0.003	NS
HLA – C	0302	18.12 (129)	13.43 (169)	0.005	NS
	0403	0.98 (7)	2.23 (28)	0.05	NS
	12	1.12 (8)	4.29 (54)	0.0001	0.006
	1202	0.56 (4)	2.23 (28)	0.005	NS
	1203	0.56 (4)	2.07 (26)	0.01	NS
	15	1.12 (8)	2.62 (33)	0.03	NS

Abbreviations: HLA, human leukocyte antigen; NPC, nasopharyngeal carcinoma

* 'NS' = $P > 0.05$.

Twenty – four alleles were detected at the HLA – C locus, 13 of which had a frequency >1%. Cw*0403, Cw*1202, and Cw*1203 were observed at a lower frequency in cases and Cw*0302 was the only allele showing an elevated frequency in cases. After correction, however, none of the four – digit HLA – C alleles remained significant.

HLA typing was informative for 422 children of the study subjects, which enabled us to directly determine HLA haplotypes in 179 patients and 379 controls. For the remaining 177 patients and 250 controls, HLA haplotypes were assigned by population – based estimation methods. HLA – A/B haplotype frequencies were calculated on the basis of both methods of haplotype assignments. Stratification analyses were performed to determine tne nature ot the observed HLA association (Tab. 2 and 3). Tab. 2 compares the frequency distribution of all HLA – A/B (where the HLA – A allele showed individual association with NPC) and – B/A (where the HLA – B allele showed individual association with NPC) haplotypes related to individual NPC associated alleles between cases and controls to determine whether the observed allele association with NPC might be due to particular haplotypes or to an individual allele. Four of the five potentially NPC – related HLA – A alleles, A*0206, A*0207, A*1101, and A*3303, were each observed on HLA – A/B haplotypes with frequencies of >1% and were therefore included in Tab. 2A. Both the dominant A*0206 – B*1502 haplotype (P = 0.0278) and all other A*0206 – associated haplotypes combined (P = 0.0071) showed elevated frequencies in the patient group. A*0207 is predominantly associated with B*4601, and this was the only A*0207 haplotype associating with susceptibility to NPC (P = 0.0095). Similarly, A*3303 – B*5801 was the only A*3303 haplotype associating with NPC susceptibility (P = 0.0003). The protective A*1101 allele was found on seven HLA – A/B haplotypes with frequencies >1% and these haplotypes showed inconsistent associations with NPC. A*1101 – B*1301 (P = 0.0029), A*1101 – B*4601 (P = 0.0198), and A*1101 – B* others (P = 0.0009) showed reduced frequencies in the NPC group, whereas A*1101 – B*3802 haplotype (P = 0.0228) showed an elevated frequency in the patient group.

Five of the seven NPC – associated HLA – B alleles composed haplotypes with frequencies >1%. The B*5801 – A*3303 haplotype showed an elevated frequency in NPC patients, whereas the other B*5801 haplotypes did not. The other NPC – associated B alleles were all found on at least two common HLA – B/A haplotypes. Both B*13011 and B*3802 were significantly more common in the control group only when A*1101 was also present on these haplotypes (P = 0.0029 and P = 0.0228, respectively). B*4001 was associated with four haplotypes with a frequency >1%. All haplotypes containing this allele were observed more frequently in the patients (B*4001 – A*0203 and B*4001 – A*1102 reaching significance; P = 0.0167 and P = 0.0138, respectively), except for the A*1101 – associated B*4001 haplotype, which was more common in the controls albeit not significantly.

Tab. 3 stratifies six common NPC – associated HLA – A/B allelic combinations to examine independent as well as collective effects of HLA alleles on NPC risk analyzed in Tab. 2. For each tested HLA – A/B phenotypic combination, the study subjects were divided into three groups: those having both of the relevant HLA – A and HLA – B alleles and those that had one, but not the other.

Tab. 2 HLA – A/B and B/A haplotype frequencies （%）

in NPC patients （n =356） and controls （n =629）ᵃ

Haplotype		Frequency		P – value *	Allele/haplotype effectᵇ
		NPC patients	Controls		
（A）					
HLA – A/B					
0206	1502	1. 26	0. 4	0. 03	Allele A * 0206 （S）
	Others	3. 93	1. 91	0. 007	
0207	4601	12. 22	8. 59	0. 01	Haplotype A * 0207 – B * 4601 （S）
	Others	2. 67	2. 7	NS	
1101	1301	3. 79	7. 07	0. 003	Allele A * 1101 （P）
	1502	4. 21	5. 72	NS	
	3802	1. 97	0. 79	0. 02	Allele offset
	4001	2. 53	3. 74	NS	
	4601	1. 12	2. 7	0. 02	Allele A * 1101 （P）
	5101	1. 4	1. 43	NS	
	5401	1. 4	1. 03	NS	
	Others	3. 09	6. 6	0. 0009	
3303	5801	17. 13	11. 29	0. 0003	Haplotype A * 3303 – B * 5801 （S）
	Others	3. 37	3. 02	NS	
（B）					
HLA – B – A					
1301	1101	3. 79	7. 07	0. 003	Allele A * 1101 （P）
	2402	1. 83	2. 85	NS	
	Others	2. 11	2. 15	NS	
3802	0203	9. 41	7. 07	NS	
	1101	1. 97	0. 79	0. 03	Allele offset
	Others	1. 26	1. 43	NS	
4001	0203	3. 23	1. 59	0. 02	Allele B * 4001 （S）
	1101	2. 53	3. 74	NS	
	1102	1. 54	0. 48	0. 01	Allele B * 4001 （S）
	2402	4. 07	3. 34	NS	
	Others	3. 37	1. 91	NS	
5502	0203	0. 42	2. 07	0. 004	Allele B * 5502 （P）
	others	0. 14	1. 59	0. 003	
5801	3303	17. 13	11. 29	0. 0003	Haplotype B * 5801 – A * 3303 （S）
	others	1. 97	2. 78	NS	

Notes: HLA, human leukocyte antigen; NPC, nasopharyngeal carcinoma.

ᵃHLA – A/B haplotypes （Tab. 2A） are those containing HLA – A alleles that showed significant association with NPC in the analysis of individual alleles and HLA – B/A haplotypes （Tab. 2B） are those containing HLA – B alleles that showed significant association with NPC. Only the haplotypes with at least one allele showing significantly different gene frequencies between patients and controls are shown. Lowfrequency haplotypes （ <1. 0% in both groups） associated with a particular A or B allele are combined as 'others'. A * 7400, which associated with protection （Tab. 1）, was not found on any haplotype with a frequency of >1%, and thus is not listed. This was also the case for B * 2704 and B * 5601

ᵇ'P' = Protective effect, 'S' = Susceptible effect

* 'NS' = P >0. 05

The protective A * 1101 and the susceptible B * 4001 allele combination is included first in Tab. 3 to show the collective effect of these two offsetting NPC – associated alleles. The protective B * 5502 is in LD with A * 0203, and stratification for this allele combination showed that the protective effect of B * 5502 could be detected in the presence ［odds ratio （OR） = 0. 18, P = 0. 0018］ or absence （OR = 0. 11, P = 0. 0088） of A * 0203. A * 1101, which is commonly linked to the NPC – related B alleles B * 1301 and B * 4001, is the most common protective allele in the Guangxi NPC cohort. Stratified analysis showed that the A * 1101 protection was stronger when B * 1301 was also present （OR = 0. 5, P = 0. 0010; without B * 1301; OR = 0. 73, P = 0. 0280）. However, B * 1301 had no effect ($OR = 0.9$, $P > 0.05$) in the absence of A*1101. Remarkably, B*4001 actually associated strongly with risk of developing NPC if A*1101 was missing ($OR = 2.56$, $P < 0.0001$), but in the presence of A*1101, the susceptibility effect of B*4001 was completely abrogated ($OR = 0.66$, $P > 0.05$). The susceptibiliy effect of A*0206 seems to be largely independent of HLA – B, though the A*0206 and B*1502 combination was somewhat stronger ($OR = 3.63$, $P = 0.0065$) than that in the presence of other HLA – B alleles ($OR = 1.99$, $P = 0.0184$). Both A*0207 (Tab. 1) and B*4601 [14,19] were thought to be independent high – risk factors for NPC in Southern Chinese and Southeast Asia. Then, two alleles are in strong LD in the Guangxi cohort, as in most Southern Chinese populations, and the A*0207 – B*4601 haplotype associated with susceptibility (Tab. 3). However, A*0207 showed no effect in the absence of B*4601, whereas B*4601 had a weak protective effect in the absence of A*0207 ($OR = 0.61$, $P = 0.0325$). Finally, A*3303 and B*5801 forms the most common A – B haplotype in the Guangxi cohort and both alleles were associated with an elevated NPC risk (Tab. 1). Stratification of the A*3303 and B*5801 combination showed that the NPC effect was more strongly associated with the presence of both alleles ($OR = 1.79$, $P = 0.0001$) as one allele without the other had no effect.

Tab. 3   Effect of allelic combinations of NPC – associated HLA alleles on the disease development

Allele combination[a] A/B	NPC patients ($n = 356$) % (n)	Controls ($n = 629$) % (n)	OR (95% CI)	P – value*	Allele/haplotype effect
Protective					
1101/4001					Allele A*1101
– / +	19.38 (69)	8.59 (54)	2.56 (1.76 – 3.72)	< 0.0001	
+ / –	28.37 (101)	38.95 (254)	0.62 (0.47 – 0.82)	0.0008	
+ / +	8.15 (29)	11.92 (75)	0.66 (0.42 – 1.02)	NS	
0203/5502					Allele B*5502
– / +	0.28 (1)	2.54 (16)	0.11 (0.02 – 0.57)	0.009	
+ / –	27.53 (98)	23.37 (147)	1.25 (0.93 – 1.68)	NS	
+ / +	0.84 (3)	4.45 (28)	0.18 (0.06 – 0.53)	0.002	
1101/1301					Allele A*1101
– / +	5.9 (21)	6.52 (41)	0.9 (0.52 – 1.55)	NS	
+ / –	27.25 (97)	34.02 (214)	0.73 (0.55 – 0.97)	0.03	
+ / +	9.27 (33)	16.85 (106)	0.5 (0.33 – 0.76)	0.001	
Susceptible					
0206/1502					Allele A*0206
– / +	13.2 (47)	15.74 (99)	0.81 (0.56 – 1.18)	NS	
+ / –	7.02 (25)	3.66 (23)	1.99 (1.12 – 3.53)	0.02	
+ / +	3.37 (12)	0.95 (6)	3.62 (1.43 – 9.15)	0.007	
0207/4601					Haplotype A*0207 – B*4601
– / +	7.87 (28)	12.24 (77)	0.61 (0.39 – 0.96)	0.03[h]	
+ / –	3.93 (14)	4.93 (31)	0.79 (0.4 – 1.50)	NS	
+ / +	23.31 (83)	16.38 (103)	1.55 (1.12 – 2.14)	0.008	
3303/5801					Haplotype A*3303 – B*5801
– / +	3.37 (12)	4.45 (28)	0.75 (0.28 – 1.49)	NS	
+ / –	5.06 (18)	5.41 (34)	0.93 (0.52 – 1.68)	NS	
+ / +	32.87 (117)	21.46 (135)	1.79 (1.34 – 2.39)	0.0001	

Notes: CI. confidence interval; HLA, human leukocyte antigen; NPC, nasopharyngeal carcinoma; OR, odds ratio.
[a] ' + ' = positive for the allele; ' – ' = negative for the allele. [b] The A0207 – /B4601 + signal ($P = 0.0325$) is a weak 'protective' confirmed, the opposite of the strong susceptible haplotype A0207/B4601
* 'NS' = $P > 0.05$

Tab. 4 summarizes the HLA alleles and the allele combinations associated with the risk of developing NPC. HLA protection against NPC mainly involves independent allele influence except that the A*1101protection may be overridden by the presence of the high – risk B*3802. The presence of one or more of the six protective factors has a frequency of 58. 98% within the control group and 34. 83% within the NPC patients. Together, these protective factors delivered a combined OR of 0. 37 (P <0. 0001). NPC susceptibility, on the other hand, associates most strongly with certain allele combinations, as opposed to single alleles, with the exception of the A*0206 effect. Ind ividual positive for any of the five high – risk factors accounted for 50. 08 and 73. 03% of the controls and cases, respectively. These high – risk factors showed individual ORs between 2. 05 and 3. 56 and a collective OR of 2. 70 (P <0. 0001).

Tab. 4 Protective and susceptible HLA alleles, haplotypes, and genotypes for the development of NPC

HLA factor	NPC cases (n =356) % (n)	Controls (n =629) % (n)	OR (95% CI)	P – value
Protective				
A*1101 +/B*3802	32. 58 (116)	49. 76 (313)	0. 49 (0. 37 –0. 64)	<0. 0001
A*74	0. 28 (1)	2. 07 (13)	0. 13 (0. 05 –0. 32)	0. 03
B*07	0. 0 (0)	2. 07 (13)	0. 07 (0. 01 –0. 56)	0. 01
B*27	0. 0 (0)	3. 18 (20)	0. 04 (0. 01 –0. 29)	0. 001
B*55	1. 12 (4)	7. 95 (50)	0. 13 (0. 05 –0. 32)	<0. 0001
B*56	1. 69 (6)	4. 45 (28)	0. 37 (0. 1 6 –0. 87)	0. 02
Combined	34. 83 (124)	58. 98 (371)	0. 37 (0. 28 –0. 49)	<0. 0001
Susceptible				
A*0206	10. 39 (37)	4. 61 (29)	2. 40 (1. 47 –3. 92)	0. 0005
A*0207 +/B*4601 +	23. 03 (82)	16. 38 (103)	1. 53 (1. 11 –2. 11)	0. 01
A*3303 +/B*5801 +	32. 87 (117)	21. 14 (133)	1. 83 (1. 37 –2. 44)	<0. 0001
A*1101 –/B*4001 +	23. 03 (82)	14. 15 (89)	1. 82 (1. 30 –2. 53)	0. 0004
A*1101 +/B*3802 +	3. 93 (14)	1. 59 (10)	2. 53 (1. 14 –5. 61)	0. 02
Combined	73. 03 (260)	50. 08 (315)	2. 70 (2. 05 –3. 56)	<0. 0001

Notes: *CI*, confidence interval; HLA, human leukocyte antigen; NPC, nasopharyngeal carcinoma; *OR*, odds ratio

On tne basis of analyses shown, HLA haplotypes were classified as NPC protective (P), susceptible (S), and neutral (N), generating six genotypes. Fig. 2 shows the ORs of the six genotypes for NPC development.

Interestingly, neither the protective nor the susceptible genotypes seemed dominant over the other. The exact values are presented in Supplementary Tab. 5. The ORs of the remaining genotypes were distributed in a manner that was expected given that the protective and susceptible genotypes were defined in this same cohort. The validity of this scheme will be particularly interesting to test in an independent cohort.

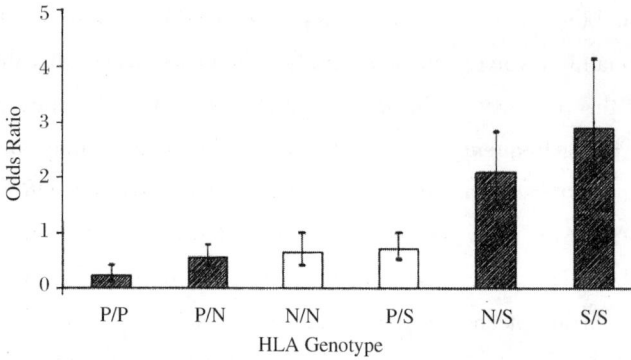

P: protective haplotype; S: susceptible haplotype; N: neutral haplotype.
Columns with a pattern indicate *ORs* with *P* values <0.05

Fig. 2 *ORs* of HLA genotypes for NPC development

The influence of HLA genotypes on NPC risk was analyzed by comparing all individual genotypes in HLA – A and – B loci separately and the compound genotypes of HLA – A and B. Two HLA – A, six HLA – B genotypes and six HLA – A – B compound genotypes were significantly associated with either elevated NPC risk or protection(Supplementary Tab. 6). These genotypes all involve at least one allele showing a higher risk or protection for the disease.

Discussion

Southern China has a disproportionally high incidence of NPC compared with other parts of the world. The study population from Wuzhou City of Guangxi Province in Southern China holds, perhaps, the highest recorded NPC incidence. [9,10] Apart from unique environmental factors (mainly traditional diet), our data support a role for genetic predisposition contributing to the high – disease incidence. The results from this study confirm and extend previously reported HLA, and NPC associations in Southern Chinese populations, [12 – 14,19] such as the protective effect associated with HLA – A * 1101 and susceptibility effects associated with B * 0207, A * 3303, and B * 5801. This study provides further insights into the nature of the HLA association with NPC, such as the observation that the dominant A * ll protection can be entirely attributed to the major subtype A * 1101. The other A * 11 subtype detected in this population, A * 1102, did not show any protective effect even though the two A * 11 subtypes differ by only a single amino acid in position 19 outside of the peptide – binding groove.

Our stratified analyses demonstrated that a proportion of the HLA – associated NPC susceptibility is likely to be haplotype – dependent in this cohort (Tab. 3). In earlier cohort studies from Southern China[13] and Taiwan, [14] the A * 0207 – B * 4601 allele combination was consistently associated with a high – NPC risk, but there are contradictory reports on whether the two alleles have independent effects. In our study cohort, two major high – risk HLA – A/B allele combinations, A * 0207 – B * 4601 and A * 3303 – B * 5801, were identified. Patients with either of these two allele combinations accounted for half of the NPC cohort and susceptibility conferred by these HLA – A/B combinations was strictly dependent on the presence of both alleles. The alleles comprising both pairs are in strong LD, forming the two most common HLA – A/B haplotypes in the Guangxi population.

Haplotype – dependent disease associations may indicate one of the two possibilities: (1) the two alleles on the haplotype are behaving in an epistatic manner to reach functional synergy or (2) the disease – associated HLA haplotype is tracking an unidentified disease locus present on that specific haplotype. An example of the former is that the structural stability of the HLA – DQ $\alpha\beta$ dimer is affected by the type of DQA and DQB alleles carried on HLA haplotypes, which in turn influences the risk of insulin – dependent diabetes mellitus (IDDM)[20]. Functional synergy between HLA – A and – B molecule and any potential clinical relevance such synergy may have, however, are yet to be established. In terms of the alternative HLA marker model, there are no data that unequivocally support or refute the hypothesis in NPC studies, despite efforts to identify non – HLA disease genes in the extended MHC region[21,22]. Still, direct involvement of HLA molecules in NPC pathogenesis, including that involving the causative EBV infection, also lacks functional data support. Interpretations of the present results tend to support the locus tracking hypothesis in that there is evidence for (1) a lack of HLA association with the occurrence of IgA antibodies to EBV IgA/VCA, a well – known precursor of NPC, (2) haplotype – dependent but allele – independent disease susceptibility, and (3) the lack of association of A*1102 despite its identical peptide – binding structure shared with the well – documented protective A*1101 allele[14]. Furthermore, distinct population – specific HLA alleles associate with susceptibility to NPC, which is unexpected if HLA is directly involved in disease pathogenesis.

In contrast to the haplotype – dependent but allele – independent susceptibility exhlited by the two major high – risk haplotypes A*0207 – B*4601 and A*3303 – B*5801, the dominant protective effect conferred by A*1101 did not show a clear haplotypic dependence. In fact, most A*1101 – linked A/B haplotypes showed a trend of underrepresentation in the case group except the one linked with the susceptible B*3802. A*1101 seems to protect against other viral infections as well. For example, A*1101 may confer protection against AIDS through restriction of immunodominant HIV peptide,[23,24] but the structural and functional basis for its direct involvement in NPC pathogenesis has not been thoroughly explored. The detected allele and haplotype associations with NPC seem to be independent of HLA genotypes as genotypes showing significant ORs are essentially those with least two copies of NPC – associated alleles or haplotypes. Also, no particular genotypes showed stronger association than individual alleles or haplotypes.

Another noteworthy HLA association observed in the Guangxi NPC cohort is an opportune 'complete' protection of B*27 against disease development. B*27 was detected with a typical frequency in the control group (1.59%), but was completely absent in the patient group (OR = 0.04, P = 0.001). Earlier cohort studies from Southern China and Taiwan have also detected B*27 as a low – risk allele though the protection was never 'complete'[14,19,25]. B*27 is well known for its high – risk association with the inflammatory autoimmune disease ankylosing spondylitis and it confers protection against AIDS progression apparently through its peptide – binding properties for immunodominant HIV epitopes.[26] Perhaps B*27 is also involved in EBV – related pathogenesis, impacting the risk of NPC development. Given the low frequency of B*27 in the study population, however, the absence of B*27 in the patient group should be confirmed in larger replication cohort studies.

The overall HLA profile of a population is likely to influence the incidence of HLA – associated diseases. An example is the correlation between the rates of IDDM incidence and the frequency distribution of the IDDM associated HLA – DQ and DR alleles in world populations[27] The NPC incidence in a population may also correlate positively witn the frequencies of NPC – susceptible HLA alleles and haplotypes of the population. The two dominant susceptible HLA haplotypes, A*0207 – B*4601 and A*3303 – B*5801, are the two most common HLA – A/B haplotypes in Guangxi and together they account for 20% of the total haplotype frequency of the study Wuznou population, which is one of the highest in Far East Asian populations. Overall, the susceptible HLA factors were found in about half of the individuals in the Wuzhou population, which is also one of the highest. Therefore, the unusually high presence of NPC – associated HLA alleles and haplotypes may partly explain the disproportionally high – NPC incidence in Southern China and Southeast Asia.

Materials and Methods

Study cohorts

NPC cases and controls were re-cruited from Wuzhou City and Cang-wu County of Guangxi Province. [10] All study subjects were of Han ethnic origin. Informed consent was obtained from all study participants. An effort was made to enroll triads consisting of a proband (either NPC patient or NPC – free but EBV – IgA/VCA anti-body positive), an unaffected spouse and an adult child. The case group included 356 unrelated patients with biopsy – confirmed NPC. The mean age was

Tab. 5 Gender and age of NPC patients and EBV – IgA/VCA positive and negative controls

	NPC patients	Controls	
		EBV – IgA/ VCA positive	EBV – IgA/ VCA negative
	N = 356	N = 287	N = 342
Male/female	237/119	141/146	129/213
(% male)	(67%)	(49%)	(37%)
Age	50. 09 ± 10. 83	45. 69 ± 8. 90	46. 87 ± 10. 24
EBV – IgA/VCA	95. 50%	100%	0%

Abbreviations: EBV, Epstein – Barr virus; NPC, nasopharyngeal carcinoma.

50. 1 years (range 19 – 80), 95. 5% of them being EBV – IgA/VCA antibody positive. Two groups of control were the case's spouse or geographically matched residents who were NPC free at the time of study enrollment. An antibody to EBV capsid antigen (EBV – IgA/VCA) were confirmed by serologic testing at the time of study enrollment. One group was positive (n = 287) and the other negative (n = 342) for the EBV – IgA/VCA antibody. The mean age was 45. 7 and 46. 9, respectively, for the antibody positive and negative groups. The controls were matched to the cases by age, ethnicity, and geographic residence (Tab. 5). In addition, 422 adult children of the study subjects in the case and control groups were recruited to allow elucidation of HLA haplotypes, but they were excluded in all other analyses.

Detection of EBV – IgA/VCA antibody

The presence of the EBV – IgA/VCA antibody was detected using the immunoperoxidase assay as described earlier. [28] EBV positive B95 – 8 cells fixed on slides were incubated with multiple dilutions of the testing serum followed by incubation with antihuman IgA horseradish peroxidase and staining with diaminobenzidine. Testing sera with a staining titer of 1: 10 or higher dilution were considered positive for the EBV – IgA/VCA antibody.

HLA typing

HLA class I alleles were characterized using a PCR – SSOP (sequence – specific oligonucleotide probe) typing protocol developed by the 13th International histocompatibility Workshop. [29]

Briefly, the gene fragment spanning exon 2, intron 2, and exon 3 was amplified using locusspecific primers for HLA – A, – B, and – C separately. The PCR products were immobilized on nylon membranes and hybridized with a panel of P^{32} – labeled oligonucleotide (19 mers) matching all known sequence variations of the HLA genes. Typing results were interpreted by SSOP hybridization patterns based on sequences of known HLA alleles. Typing ambiguities were resolved by sequencing exons 2 and 3 completely. For sequencing analysis, the PCR product of HLA – A, – B, or – C was used as the template for the sequencing reaction. For each of the HLA genes, two sequencing reactions (one for exon 2 and one for exon 3) were performed using exon – specific sequencing primers. The sequencing analysis was per formed using the ABI Big Dye Terminator Cycle Sequencing Kit and ABl3730 × 1 DNA analyzer (Applied Biosystems, Foster City, CA, USA). HLA alleles were assigned on the basis of the sequence database of known alleles with the help of the AS-SIGN software developed by Conexio Genomics (Conexio Genomics, Perth City, WA, Australia). Ambiguous heterozygous genotypes were resolved by additional PCR and sequencing procedures using allele – specific PCR primers to selectively amplify only one of the two alleles.

Statistical analyses

HLA allele frequencies were calculated based on observed genotypes, and HLA – A and – B haplotype frequencies were estimated using one of the two methods: (1) unambiguous assignment based on familial segregation for the cases and controls using recruited spouse and child genotype data or (2) indirect assignment based on maximum likelihood estimation for the study subjects without recruited family members. For the latter, the haplotypic analysis was performed using the BLOCK-HEAD genetic analysis software developed by George Nelson in the Laboratory of Genomic Diversity, National Cancer Institute.

The effect of HLA alleles on the development of NPC and EBV – IgA/VCA antibody was evaluated by computing *ORs* and 95% confidence intervals as well as exact *P* – values using the FREQ procedure of the SAS 9. 1 software (The SAS Institute, NC, USA). A correction of 0. 5 was applied on every cell of the 2 × 2 table that contains a zero. The analyses were performed at four – digit and two – digit resolution levels separately. *P* – value was calculated by χ^2 – test for each allele and was corrected by multiplying the number of all detected alleles. Significance was considered at P < 0. 05. Both uncorrected and corrected *P* – values were presented in our tables. Stratified analyses were applied to evaluate the effect of haplotypes of HLA A and – B that contained alleles showing individual significant associations with NPC risk.

Acknowledgements: This project has been funded in whole or in part with federal funds from the National Cancer Institute, National Institutes of Health, under Contract No. HHSN261200800001E. The content of this publication does not necessarily reflect the views or policies of the Department of Health and Human Services, nor does mention of trade names, commercial products, or organizations imply endorsement by the US Government. This Research was supported in part by the Intramural Research Program of the NIH, National Cancer Institute, Center for Cancer Research.

〔In 《Genes and immunity》 2010, 1 – 9〕

References

1 Brown TM, Heath CW, Lang RM, et al. Nasopharyngeal cancer in Bermuda. Cancer, 1976, 37: 1464 – 1468

2 Yu MC, Garabrant DH, Huang TB, et al. Occupational and other non – dietary risk factors for nasopharyngeal carcinoma in Guangzhou, China. Int J Cancer, 1990, 45: 1033 – 1039

3 Guo X, Johnson RC, Deng H, et al. Evaluation of non – viral risk factors for nasopharyngeal carcinoma in a high – risk population of Southern China. Int J Cancer, 2009, 124: 2942 – 2947

4 Coffin CM, Rich SS, Dehner LP. Familial aggregation of nasopharyngeal carcinoma and other malignancies. A clinicopathologic description. Cancer, 1991, 68: 1323 – 1328

5 Jeannel D, Hubert A, de Vlathaire F, et al. Diet, living conditions and nasopharyngeal carcinoma in Tunisia – a case – control study. Int I Cancer, 1990; 46: 421 – 425

6 Laramore GE, Clubb B, Quick C, el al. Nasopharyngeal carcinoma in Saudi Arabia: a retro spective study of 166 cases treated with curative intent. Int J Radiat Oncol Biof PhtIs, 1988, 15: 1119 – 1127

7 Johansen LV, Mestre M, Overgaard J. Carcinoma of the nasopharynx: analysis of treatment results in 167 consecutively admitted patients. Head Neck, 1992, 14: 200 – 207

8 Lee AW, Foo W, Mang O. Changing epidemiology of nasopharyngeal carcinoma in Hong Kong over a 20 – year period (1980 – 99): an encouraging reduction in both incidence and mortality. Int Cancer, 2003, 103: 680 – 685

9 Lu SJ, Day NE, Degos L, et al. Linkage of a nasopharyngeal carcinoma susceptibility locus to the HLA region. Nature, 1990, 346: 470 – 471

10 Guo XC, Scott K, Liu Y, et al. Genetic factors leading to chronic Epstein – Barr virus infection and nasopnaryngeal carcinoma in South East China: study design, methods and feasibility. Hum Genomics, 2006, 2: 365 – 375

11 Burt RD, Vaughan TL, McKnight B, et al. Associations between human leukocyte antigen type and nasopharyngeal carcinoma in Caucasians in the United States. Cancer Epidemiof Biomarkers Prev, 1996, 5: 879 – 887

12 Simons MJ, Wee GB, Chan SH, et al. Immunogenetic aspects of nasopharyngeal carcinoma (NPC) III. HL – a type as a genetic marker of NPC premsposition to test the hypothesis that Epstein – Barr virus is an etiological factor in NPC. IARC Sci Publ, 1975, 249 – 258

13 Hu SP, Day NE, Li DR, et al. Further evidence for an HLA – related recessive mutation in nasopnaryngeal carcinoma among the Chinese. BR J Cancer, 2005, 92: 967 – 970

14 Hildesheim A, Apple RJ, Chen CJ, et al. Association Of HLA class I and II alleles and extended haplotypes with nasopharyngeal carcinoma in Taiwan. J Natl Cancer Inst, 2002, 94: 1780 – 1789

15 Chan SH, Day NE, Kunaratnam N, et al. HLA and nasopharyngeal carcinoma in Chinese – a further study. Int J Cancer, 1983, 32: 171 – 176

16 Henle G, Henle W. Epstein – Barr virus – specific IgA serum antibodies as an outstanding feature of nasopharyngeal carcinoma. Int J Cancer, 1976, 17: 1 – 7

17 Zeng Y, Zhang LG, Li HY, et al. Serological mass survey for early detection of nasopharyngeal carcinoma in Wuzhou City, China. Int J Cancer, 1982, 29: 139 – 141

18 Zeng Y, Zhang LG, Wu YC, et al. Prospective studies on nasopharyngeal carcinoma in Epstein – Barr virus IgA/VCA antibody – positive persons in Wuzhou City, China. Int J Cancer, 1985, 36: 545 – 547

19 Wu SB, Hwang SJ, Chang AS, et al. Human leukocyte antigen (HLA) frequency among patients with nasopharyngeal carcinoma in Taiwan. Anticancer Res, 1989, 9: 1649 – 1653

20 Kwok WW, Schwarz D. Nepom BS, et al. HLA

- DQ molecules form alpha - beta heterodimers of mixed allotype. J Immunol, 1988, 141: 3123 -3127

21 Lu CC, Chen Jc, Tsai ST, et al. Nasopharyngeal carcinoma - susceptibility locus is localized to a 132 kb segment containing HLA - A using high - resolution microsatellite mapping. Int J Cancer, 2005, 115: 742 -746

22 Ooi EE, Ren EC, Chan SH. Association between microsatellites within the human MHC and nasopharyngeal carcinoma. Int J Cancer, 1997, 74: 229 -232

23 Beyrer C, Artenstein AW. Rugpao S, et al. Epidemiologic and biologic characterization of a cohort of human immunodeficiency virus type 1 highly exposed, persistently seronegative female sex workers in northern Thailand. Chiang Mai HEPS Working Group. J Infect Dis, 1999, 179: 59 - 67

24 Fukada K, Tomiyama H, Chujoh Y, et al. HLA - A * 1101 - restricted cytotoxic T lymphocyte recognition for a novel epitope derived from the HIV - 1 Env protein. Aids, 1999, 13: 2597 - 2599

25 Yu MC, Huang TB, Henderson BE. Diet and nasopharyngeal carcinoma: a case - control study in Guangzhou, China. Int J Cancer, 1989, 43: 1077 -1082

26 Gao X, Bashirova A, Iversen AK, et al. AIDS restriction HLA allotypes target distinct intervals of HIV - 1 pathogenesis. Nat Med, 2005, 11: 1290 -1292

27 Bao MZ, Wang JX, Dorman JS, et al. HLA - DQ beta non ASP - 57 allele and incidence of diabetes in China and the USA. Lancet 1989, 2: 497 -498

28 Cevenini R, Donati M, Rumpianesi F, et al. An immunoperoxidase assay for the detection of specific IgA antibody in Epstein - Barr vieua infections. J Clin Pathol, 1984, 37: 440 -443

29 Gao X, Nelson GW, Karacki P, et al. Effect of a single amino acid change in MHC class I molecules on the rate of progression to AIDS. N Engl J Med, 2001, 344: 1668 -1675

111. Clinical Laboratories on A Chip for Human Immunodeficiency Virus Assay

GUO Wen – penh[1], MA Xue – mei[1], ZENG Yi[1.2]

1. College of Life Science and Bioengineering, Beijing University of Technology, Beijing, P. R. China; 2. Chinese Center for Disease Control and Prevention, Beijing, P. R. China

Summary

Automated chip – based technologies for clinical diagnosis may have great facilities in the area of life science and medicine. This paper presents the lab – on – a – chip design for the assay of HIV, which includes the sample preparation, reaction, and signal amplification module. A laser induced fluorescence system is also designed fot real – time monitor of the signals.

[**Key words**] Clinical diagnosis; HIV; lab – on – a – chip; PCR

Introduction

Miniaturization and automation have revolutionized the world of microelectronics. Much smaller and more powerful notebook, palm – tops, and game consoles have replaced large room – sized computers of the 1950s and 60s. The advent of lab – on – a – chip technologies that enable the laboratoty ptocedutres and instrumectation miniarurized will undergo a similarly substantial change in years to come[1-2].

The development of lab – on – a – chip devices for biochemical analysis has rapidly evolved too many applications such as clinical analysis (blood gas analysis, glucose/lactate analysis, etc.), DNA analysis (including – mucleic acid sequence analysis), ptoteomics analysis (proteins and peptides), combinatorial synthesis/analysis, immunoassays, toxicity monitoring, and even forensic analysis[3-9]. But, a significant application area for this kind of technologies is still in clinical diagnosis, which wok for the eatly, rapid, safety, sensitive and specificity detection of the disease states as a vital goal.

Micro – sevice has some merits, such as safety, small sample and reagent volumes, speed of reaction, portability, and low cost, which will greatly meet the needs of clinical laboratories. Especially for malignant infectious diseases such as HIV (human immunodeficiency virus), which have become more seriously across the globe. The total number of people living with the HIV rose in 2004 to reach its highest level ever: an estimated 39. 4 million [35. 9 million – 44. 3 million] people are living with the virus[10].

In this paper, we present a plastic integrated microfluidic chip design for the possible application in HIV assay.

Methodology

Current methods available for diagnosis of HIV infection have improved the care of many patients at risk for HIV infection or currently infected with the virus[11]. All HIV diagnostic tests are based on detection of one or more of the molecules that make up an HIV virus particle or detection of the antibodies that human hosts make against HIV particles.

Traditional methods to assess the accuracy of clinical diagnostic data express the results using indices such as sensitivity, specificity and predictive valves as gold standard. The sensitivity is the probability of achieving a correct (i. e, positives) result for individuals who truly have the disease; the specificity is the probability of a correct (i. e. negative) result for individuals who truly do not have the disease; and the positive and negative predictive values give the probabilities of having the disease conditional on having received a positive or negative diagnostic test result, respectively. In modern clinical diagnosis, the indices are also very important especially in microsystems, which reduce not only the dosage of the solution but also the molecules in the sample.

1) Overview of HIV diagnostic tests; HIV is essentially a capsule composed of proteins, glycoproteins (proteins covalently attached to different sugar molecules), fat molecules and the ribonucleic acids (RNA) that carry the genetic material of HIV. Each HIV virus particle is encapsulated in a viral envelope that is derived from the fatty cell membrane of the human host's cells. Seventy – two glycoprotein projections composed of two different glycoproteins called *gp*120 and gp41 extend from the interior of the virus particle through the viral envelope to the exterior of the particle. Within the viral envelope lies the nucleocapsid or the viral core. The outer layer of this core is composed of a protein called p17 while the inner core is made up of p24 proteins. The core of the viral particle holds the HIV genome, which is made up of two single – stranded RNA molecules. These molecules are associated with several other proteins needed for HIV ptopagation and survival, including a reverse transcriptase, which converts the HIV RNA genome to DNA, and an integrase, which integrates HIV DNA into the host cell's genome. These proteins and glycoproteins are often referred to as antigens because they can evoke an immune response from the infected host[12].

HIV diagnostic tests function cither by detecting host antibodies made against different HIV proteins or by directly detecting the whole virus itself or components of the virus (such as the HIV antigen or HIV RNA)[13]. The goal of most HIV diagnostic tests is to detect HIV infection as early as possible, thereby decreasing the length of the diagnostic window[14]. Certain HIV diagnostic tests, however, aim to distinguish between recent and longstanding HIV infections. This is done primarily for epidemiological reasons, in order to estimate the incidence of HIV (the frequency at which new HIV infections arise) within a given population[15,16].

The sensitivity for detecting HIV in HIV infected people has been increasing systematically by the use of sophisticated molecular techniques. Presectly, the assay that attains maximal sensitivity and dynamic range of viral copy number is nucleic acid amplification of HIV - 1 RNA in plasma[17]. In addition to mucleic acid testing for the determination of HIV - 1 viremia, several serologic assays to detect the HIV core antigen have been developed. The HIV core antigen is detectable in the blood of infectied people during the acute phase of infection and late in the disease. Enzyme - linked immunosorbent assay (ELISA) methods for quantification if HIV core antigen have been relatively insensitive. The development of high sensitive detection method of HIV core antigen is considered a simple and inexpensive alternative to HIV - 1 RNA testing for monitoring treatment and protecting the blood supply.

2) Lab - on - chip design for HIV assay: In the view of clinical worker, the key step in the design of a biomedical device is to fully understand the principle of biological or medical process, which have great influence on the success of the designs, Accordingly, a good researcher should first know what's the real needs in the area, and then apply his ideas to make the design simple, ease of use, and of course inexpensive.

As regards ourselves, we have done HIV research for many years in our lab, leading by Prifessor Zeng Yi, the academician of Chinese Academy of Science. The chip - based designs for HIV assay inevitably become our chiefly attention.

Begin with the serological samples, the whole system integrate on chip ELISA and laser induced fluorescence detection module, which is fully automated and can give the result without any artificial interference. The sketch map is shown in Fig. 1 (brief).

Results and Discussion

A. Experimental results

HIV core Antigen is detectable in serum or plasma during the acute phase of primary HIV infection and during the late symptomatic stages of infection.

In our previous work, we have developed two monoclonal antibodies against HIV core antigen, and reached high sensitivity and specificity by double antibody - sandwich ELISA. Through modifications, the sensivity can be increase greatly, and the process can be easily achieved on chip.

B. The micro - fluidic Chip Design

The main frame of our clinical laboratories on a chip detection system is shown in Fig. 3. The total system mainly contains three modules, micro - fluidic chip, sensor and actuator executors, and a CCD collection system. The home - made controller precisely manipulates the total system.

The choice of material, which influence greatly the success of the experiments, is very important for micro - fluidic chip design and fabrication. At present polycarbonate (PS), polydimethylsiloxane (PDMS) and polymethylmethacrylate (PMMA) are most commonly used. Although those offer certain attractive chatacteristics tor biochips, we have chosen the cyclic olefin copolymers (COC), for it has significantly better material ptoperties than others, which is highly UV - transparent, has low water absorption, and is chemically resistant to hydrolysis, acids and organic polar solvents that make it suitable for lab – on – a – chip applications. The properties is shown in Tab. 1 and Fig. 4.

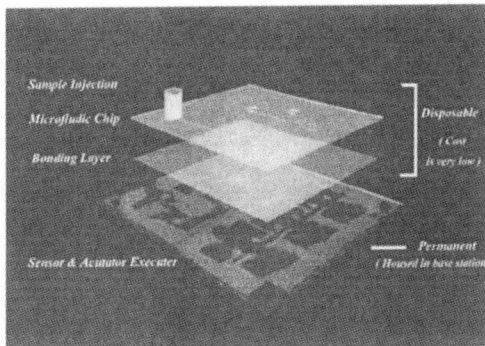

Fig. 2   The subject framework of the chip-based system for HIV assay

Tab. 1   Special properties between COC and other plastics

Chemical Resistance	The Type of Plastics		
	COC	PC	PMMA
Acids	Yes	Yes	Yes
Alkali	Yes	No	Yes
Hydroxide	Yes	Medium	No
Ketone	Yes	No	No
Ester	Yes	No	No
Chloric	No	No	No

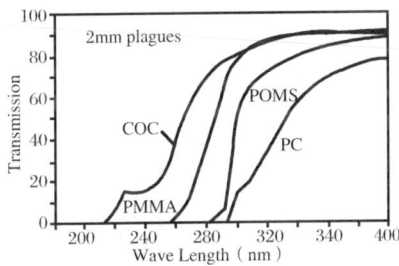

Fig. 3   The properties of UV light transmission

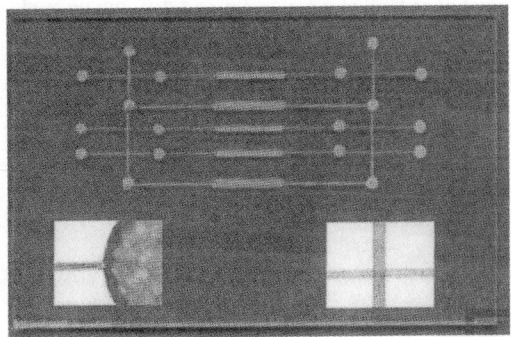

Fig. 4   PCR part of the chip – based clinical diagnosis system

C. Polymerase Chain Reaction Controlling system

The polymerase chain reaction (PCR) is a well described method for the selective identical replication of DNA molecules. By an enzymatic in – vitro amplification cycling the concentration of a DNA species is nearly doubled in a cycle, stepping through three different temperatures (93℃ denaturing, 55℃ annealing and 72℃ entending). In this way, the DNA molecules can be multiplied more than million – fold by 20 to 30 cycles of temperature[18].

Since the first report of specific DNA amplification using the PCR in 1985, the number of different applications has grown steadily in the room – based or chip – based laboratories. For the importance role of PCR in has played in the detection processing, the different designing experiments have been made.

The chip disigned for PCR is shown in Fig. 5 (brief). The PCR chamber volume is 1 μl. The depth of the chamber was 50 μm. The square dimensions of the chambers have been chosen for the easier optical detection where laser beams shine into the chamber and the fluorescent signals is detected and reported in real time.

The sensors and acutators are implimented below the chip at proper places where are chosen based on the ANSYS themal simulation[19]. Validation of the design is performed by setting the heater and sensor line thermal conductivity equal to that of the membrane. The Fig. 5 (brief) shows the heating fields the chamber.

Conclusion

HIV diagnostic tests continue to be developed and improved with the hope that widely available, simple and highly sensitive assays will continue to contribute to early clinical diagnosis of HIV infection. Early detection will ensure the safety of patient dependent on blood thansfusions. But most importantly, it will facilitate early counseling and treatment of HIV – infected patients and possibly help to prevent the spread of this debilitating virus.

The disign is engaged to satisfy the clinical demands, and help to effectively control the spreading of malignant infectious disease again.

Once this system is successful developed, it will greatly improve the clinical diagnosetics of HIV, and may have positive effect on other malignant infectious disease. Point – of – care diagnosis, personal medicine will be speeded.

Acknowledgements

The authors would like to appreciate financial support from the Beijing Science Nova Program, under the contract number 2004B05.

〔In 《Proceedings of the 2005 IEEE Engineering in Medicine and Biology 27th Annual Conference Shanghai, China, September 1 – 4》: 2005, 1274 – 1277〕

References

1　Weigl B H, Hedine K. Lab – on – a – chip – based separation and detection technology for clinical diagnostics, American Clinic Lab. 2002, 49 (10): 8 – 13

2　Manz A, Graber N, Widmer H M. Miniaturized total chemical analysis system: a movel concept for chemical sensing. Sensors and Actuators, B1, 1990, 14 (1): 244 – 248

3　Vespoorte E. Mecrofluidic chips for clinical and forcnsic analysis. Electrophoresis, 2002, 23: 677 – 712

4　Ehrnstrom R. Miniaturization and integration: Challenges and breakthroughs in microfluidics. Lab Chip, 2002, 2: 26N – 30N

5　Service R. Lab. on a chip : Coming soon: The pocket DNA sequencer. Science, 1998, 282 (5338): 399 – 401

6　Vinet F, Chaton P, Fouillet Y. Microarrays and microfluidic devices: Miniaturized systems for bilological analysis. Microelectron Eng, 2002, 61 – 62: 41 – 47

7　Malins C, Niggemann M, MacCraith B D. Mulri – analyte lptical chemical sensor employing a plastic substrate. Measure Sci Tech, 2000, 11 (8): 1105 – 1110

8　Bergstrom P, Wise K, Patel S, Schwank J. A micromachined surface work – function gas sensor for low – pressure oxygen detection. Sens Acruators B, 1997, 42 (3): 195 – 204

9　Cao L, Mantell S, Polla D. Design and simulation of an implantable medical drug delivery system using microelectromechanical systens technology, Sens Acrtuators, 2001, 94: 117 – 125

10　Joint United Nations Programme on HIV/AIDS (UNAIDS). global Summary of the HIV/AIDS Epidemic, December 2004. Available: http: //

www. unaids. org/wad2004/report_ pdf. html

11 Bartelett J G, Serologic tests for the diagnosis of HIV Infection, Official Reprint from up to date Available: http: //www. up to date. com

12 Goldsby R A, Kindt T J, Osborne B A. AIDS and other immunodeficiencies. In Kuby Immunology, 4th ed. New York: W. H. Freeman and Company, 2000: 467 – 96

13 Gurtler L. Difficulties and strategies of HIV diagnosis, Lancet, 1996, 348: 176 – 179

14 Ly T D, Laperche S, Courouce A M. Early detection of human immundeficiency virus infection using third and fourth generation screening assays. Eur J Clin Microbiol Infect Dis, 2001, 20: 104 – 10

15 Parekh B S, Kennedy M S, Dobbs T. Quantitative detection of incteasing HIV type 1antibodies after seroconversion: a simple assay for detecting recent HIV infection and estimating incedence. AIDS Res

Hum Retovituses, 2002, 18: 295 – 307

16 Janssen R S, Satten G A, Stramer S L. New testing strategy to detect early HIV – 1 infection for use iv incidence cstenates and for clinical prevention purposes. JAMA, 1998, 280: 42 – 48

17 Barletta J M, Edelman D C, Constantine N T. Lowering the detection limits of HIV – 1viral load using real – time immuno – PCR for HIV – 1 p24 antigen. Am J Clin Pathol, 2004, 122: 20 – 27

18 Erlich H A. PCR Technology: Principles and Applications for DNA amplification. Stockton Press, 1989

19 Ankur J, Kevin N, Angie M C, Jianh L N, Kenneth G. Design fabrication and thermal characterization of a MEMS device for control of nerve cell growth. in Proc of International Mechanical Engineering Congress & Expositon IMECE'03 washington. D. C. 1 – 7

112. Design, Synthesis abd Cu^{2+} Recognition of β-Diketoacid and Quinoxalone Derivatives Bearing Caffeoyl or Galloyl Moieties Linked by Arylamide as Potential HIV Integrase Inhibitors

XU Yi – sheng[1], ZENG Cheng – chu[1], LI Xue – mei[1], ZHONG Ru – gang[1], ZENG Yi[2]

1. College of Life Science & Bio – Engineering, Beijing University of Technology, Beijing 100022, China

2. Chinese Center for Disease Control and Prevention, Beijing, China

Summary

An efficient procedure for the synthesis of caffeoyl and galloyl – containing β – diketoacid derivatives linked by arylamide was reported by, in the key step, dissolving the corresponding phenyl methyl ketone in THF/DME in the presence of NaOMe as base and dimethyl oxalate as oxalylation reagent, and then separating the sodium ketoenolate ester. The resulting β – diketoacids underwent further condensation reaction with o – phenylenediamine to generate quinoxalone derivatives in good yield, rather than 2 – benzimidazol. The preliminary ion binding properties of quinoxalone derivatives were also investigated. UV – Vis spectra showed that these compounds could selectively recognize Cu^{2+} ion in ethanol and form a 1 : 2 complex.

[**Key words**] Diketoacid; Quinoxalone derivative; HIV integrase inhibitor; Cu^{2+} recognition

Introduction

At present, there is a great and urgent need for the development of new anti – HIV drugs that act specifically at different steps of the viral infection and replication cycle, due to the increased emergency of drug – resistant strains, limited bioavailability and high toxicity of known reverse transcriptase (RT) and protease (PR) inhibitors. The HIV integrase (IN) is an example of such a specific target and because no cellular homologue has been found in humans, HIV IN is thus being considered as a promising target for the development ot selective anti – AIDS drugs. [1]

Among numerous classes of molecules as potential HIV integrase inhibitors, [2-4] natural and synthetic polyhydroxylated aromatics, namely caffeic acid phenethyl ester (CAPE) [5], dicaffeoylquinic acid (DCQA, dicaffeoyltartaric acid (DCTA), [6-7] L – chicoric acid, [8] and (3R, 4R, 5S) –bis (3, 4, 5 – trihydroxy – benzoyloxy) – 5 – hydroxymethyldihydrofuran – 2 – one (1), [9] have demonstrated potential activity against HIV IN and can inhibit HIV replication with moderate anti – HIV activity in cell culture. Common structural features of these reported analogues are caffeic acid ester or galloic acid ester separated by aliphatic, alicyclic or aromatic linker (Fig. 1).

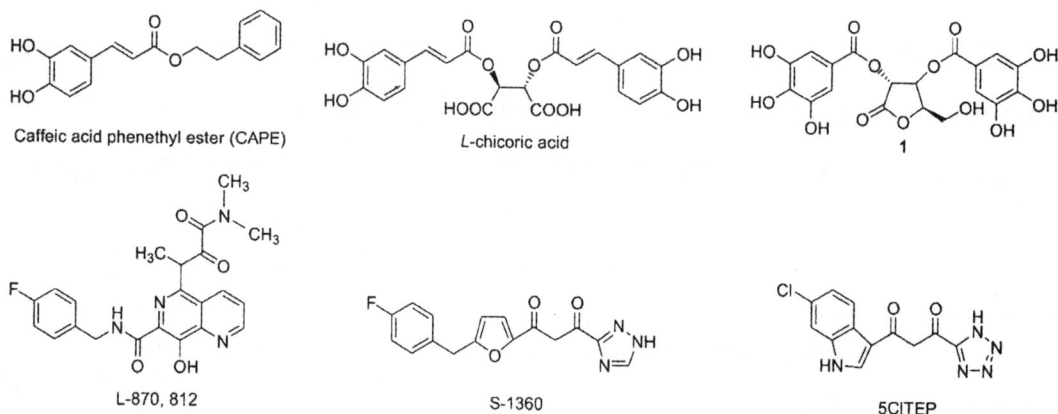

Fig. 1 **Several typical polyhydroxylated aromatics and aryl diketoacid derivatives as HIV integrase inhibitors**

In addition to polyhydroxylated aromatics, aryldiketoacid – containing molecules (their usual names are ADKs and aroylpyruvic acids but their structure correspond to 4 – aryl – 4 – oxo – 2 – hydroxy – 2 – butenoic acids) have been developed as the most promising drug candidates. These molecules are selective for HIV integrase and have been demonstrated to attord preferential inhibition of ST versus 3' – P reactions. 5 CITEP, [10-11] which belongs to the aryldiketoacid – containing compounds since the tetrazole group is a well – known bioisostere of a carboxylic acid, cocrystallized with the enzyme providing the first X – ray crystal structure of an inhibitor in complex with the HIV

−1 IN. Later, S − 1360[12−13] from Shionogi Company and L − 870, 812[14] from Merck Company are the only two currently available integrase inhibitors under clinical studies (Fig. 1). All these compounds are composed of a central aryl ring and a 4 − oxo − 2 − hydroxy − 2 − butenoic acids side chain, and both of them are absolutely required and essential for the inhibitory activity.

Based on the HIV integrase inhibitory activity of caffeic acid esters or galloic acid esters and the aryldiketoacid molecules, as the first attempt, we have designed and synthesized a series of novel compounds bearing both diketoacid and caffeoyl or galloyl entities in one molecule linked by arylamide. The presence of the biologically labile 1, 3 − diketoacid moiety within the molecules is also a concern for development of these compounds as chemotherapeutic agents.[15] Replacement of the essential pharmacophore led to the synthesis of corresponding quinoxalone derivatives, hypothesizing a similar coplanar arrangement of the central phenyl ring with 3 − [2 − oxo − 2 − ethylidene] − 3, 4 − dihydro − 1H − quinoxalin − 2 − one moieties since experimental results show that the 1.3 − diketoacid moiety enolizes at the α − position, and the resultant conjugated Z − 4 − oxo − 2 − hydroxy − 2 − butenoic acid side chain is coplanar with the central benzene ring (Fig. 2). Herein we report the efficient sythesis of such novel diketoacid and quinoxalone derivatives bear-

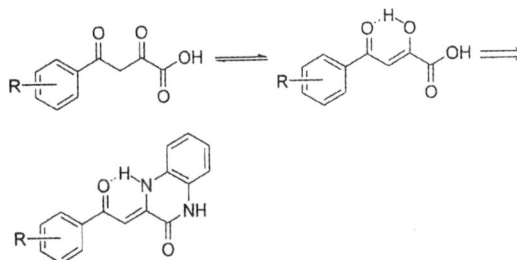

Fig. 2   Structure of diketoacid and designed quinoxalone molecules

ing caffeoyl or galloyl entities linked by amide, as well as their divalent metal ions binding properties.

Results and Discussion

Synthesis of caffeoyl − or galloyl − containing phenyl diketoacids 5 and 9

The general synthesis of diketoacid derivatives bearing caffeoyl or galloyl moiety linked by arylamide is described in Scheme 1. Treatment of *meta* − or *para* − aminoacetophenone 1 and dimethylcaffeoyl chloride or trimethylgalloyl chloride, which were prepared by the direct reaction of the corresponding dimethylcaffeoylic acid or trimethylgallic acid with excessive thinoyl chloride gave caffeoylamide − or galloylamide − substituted acetophenones 2 and 6. The resultant acetophenones underwent. Claisen condensation with di − methyl oxalate in the presence of NaOMe as condensing reagent and subsequent hydrolysis of the methyl ester 3 and 7 with 1 mol/L aqueous sodium hydroxide resulted in the formation of phenyl 1, 3 − diketoacid derivatives 4 and 8. Removal of methyl groups under excessive BBr₃ afforded the desired caffeoyl − or galloyl − containing phenyl diketoacids 5 and 9.

The oxalylation of dimethylcaffeoylamide − or trimethylgalloylamide − substituted acetophenone 2 and 6 was the key step to afford the desired phenyl diketoacid derivatives, which, in general, was achieved by the oxalylation of the corresponding aryl methyl ketones in the presence of base followed by either alkaline or acidic hydrolysis. However, due to the structures of the aryl methyl ketones, mainly their solubility, there was absent of a general procedure to obtain the corresponding diketoacid ester. Many articles provided methods of oxalyation of aryl methyl ketones to generate the diketoacid ester utilizing diethyl or dimethyl oxalate in the presence of NaH (benzene, toluene or DMF), NaOEt or NaOMe[16-18] (in the corresponding alcohol). The yields of reactions were quoted in a wide range of 40%—95% depending on the structure of the aryl methyl ketone and the reaction conditions (for example: base, solvent, reaction time and temperature). Attempts to carry out our reactions according to these procedures gave inconsistent results with much lower yields. Very recently, Cotelle et al. [19] rovided a rapid alternative method for the synthesis of phenyl diketoacid esters in which the phenyl methyl ketones were dissolved in diethyl ether and resulting sodium ketoenolate esters were first separated by simple filtering followed by acidification to afford the diketo acid ester. However, in the present case, the dimethylcaffeoylamide – or trimethylgalloylamide – substituted acetophenones 2 or 6 could not dissolve in diethyl ester and thus no products were detected.

In our previous work on the reaction of sulfonamide – containing phenyl methyl ketone with dimethyl oxalate, we found by trial and error that, in the presence of new – prepared. NaOMe as base, THF/DME ($V:V=1:1$) as solvent,[20] the reaction can proceed smoothly and efficiently to afford the desired diketo acid ester. Therefore, the syntheses of dimethylcaffeoyl – or trimethylgalloyl – containing phenyl diketoacids were also performed on the same conditions. Considering the tedious extractive work – up and column chromatographic isolation in literature methods, we simplified this procedure by separating the corresponding sodium ketoenolate esters from simple filtering and washing and then neutralizing to afford diketoacid ester derivatives in good total yleld (80% – 90%). To the best of our knowledge, most of the literature – reported dimethyl oxalate/NaOCH$_3$ method is suitable for substrates possessing halide, alkyl, nitro or ether on the arylring, and it did not work well for the amino – substituted analogues. For example, 4 – (4 – N, N – dibenzylaminophenyl) – 4 – oxo – 2 – nydroxy – 2 – butenoic acid methyl ester was obtained in a poor 14% yield, what was more, no product was obtained from benzyloxycarbonylaminophenyl methyl ketone. [21] Therefore, this alternative method provided an efficient procedure for an expanded range of aryl methyl ketone including amine – substitution on the central aromatic ring.

It is noteworthy that although the structures of dimethylcaffeoyl – or trimethylgalloyl – containing phenyl methyl ketones 2a and 2b or 6a and 6b are very similar, differing only in the position of the substituent, the reaction time is significantly different. 2b and 6b took longer time comparing to 2a and 6a, respectively, still due to the slightly different solubility in the same mixture of solution. Fortunately, it could be overcome by warming the reaction mixture.

To remove the methyl groups, boron tribromide was employed due to its high selectivity and mild reaction conditions. [22-25] Boron tribromide is a Lewis acid and will form Lewis acid – base complex with the basic group possessing lone – pair electron within the substrate molecule and thus consume more BBr$_3$. Then 4 and 8, instead of 3 and 7 were chosen to react with BBr$_3$ to obtain polyhydroxylated diketoacid derivatives 5 and 9, although the quantity of BBr$_3$ still needs an excess of 10—12 equiv.

The spectroscopic data were in agreement with the structure of products. It should be pointed out that diketoacid esters 3, 7 and acids 4, 8 are present as keto – enolic tautomer in solution. In the

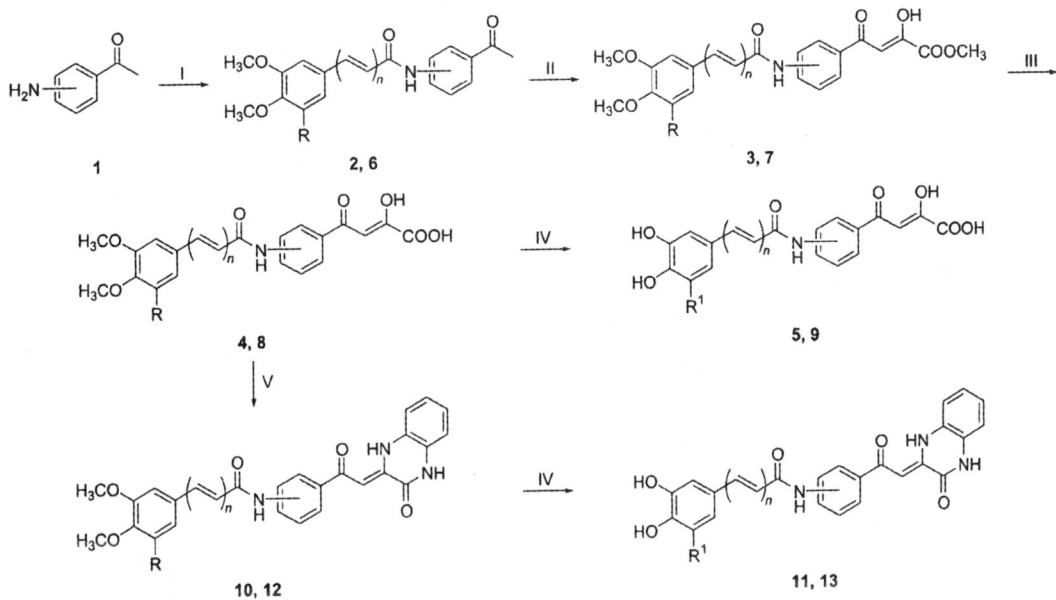

$n = 1$, R = H, R^1 = H, caffeoyl derivatives, 2a—2b, 3a—3b, 4a—4b, 5a—5b, 10a—10b, 11a—11b, $n = 0$, R = OCH$_3$, R^1 = OH, galloyl derivatives, 6a—6b, 7a—7b, 8a—8b, 9a—9b, 12a—12b, 13a—13b $a = meta$, $b = para$

Reagents and conditions: (I) **dimethylcaffeoyl chloride or trimethylgalloyl chloride, CH$_2$Cl$_2$, pyridine, r. t. , overnight; (II) NaOMe, dimethyl oxalate, THF/DME, r. t. , overnight, then H$_2$O, filter, HCl to pH 3—4; (III) 1 mol/L NaOH, CH$_3$OH/THF ($V : V = 1 : 1$), Then 2 mol/L HCl to pH = 2—3; (IV) CH$_2$Cl$_2$, BBr$_3$, r. t. ; (V) o – phenylenediamine, H$_2$O, 120 ℃, 3 h**

^{13}C NMR spectra of these compounds, there was a signal at about δ 98 which attributed to the CH enolic carbon whereas the resonance of the corresponding CH$_2$ carbon of the keto form was predicted to reside at about δ 56—60 upfield. The proton signal of the corresponding CH was weak and sometimes disappeared in the ^{1}H NMR spectra of these compounds.

Synthesis of caffeoyl – or galloyl – containing quinoxalinone derivatives 11 and 13

With 4 and 8 in hand, we then moved to synthesize the corresponding quinoxalone derivatives. Heating the mixture of 4 or 8 and phenylenediamine in water over refluxing for 2—4 h,[26] after cooling and filtering, the quinoxalone derivatives 10 or 12 were obtained in almost quantitative yield and the subsequent deprotection of methyl groups generated the desired 11 and 13, respectively. Regarding the formation of compounds analogous to 10—13 from diketoacid, to the best of our knowledge, most of the known examples are confirmed to reactions of 3 – polyfluoroacyl – or 3 – pepntafluorinated benzoylpyruvates with o – phenylenediamine.[27] For the reaction of nonfluorinated benzoylpyruvic acid, such as 4 – phenyl – 2, 4 – dioxobutanoic acid and hydroxy – substituted 4 – phenyl – 2. 4 – dioxobutanoic acid with o—phenylenediamine hydrochloride, very recently, it was reported to generate2 – benzimidazole derivatives, instead of quinoxalone derivatives.[28] The synthesis of such quinoxalone derivatives proceeds easily and conveniently, therefore we provide here an efficient procedure to afford quinoxalone derivatives from non – fluorinated benzoylpyruvic acid.

The structures of all quinoxalone derivatives 10, 11, 12 and 13 were confirmed by ^{1}H NMR, ^{13}C NMR, IR and HRMS (ESI) spectra. In the ^{1}H NMR spectra of these compounds, two sharp singlets in the range of δ 13.6—12.0 and one rather broad singlet around δ 11.0 were observed. They disappeared after D$_2$O exchange and therefore were attributed to the two protons of N—H of the quinoxalone ring and the arylamide proton, respectively. The deshielding of the NH proton on quinoxalone ring demonstrated that these compounds were in the Z – form due to intramolecular hydrogen bonding with the benzoyl oxygen atom. The signal of the proton of the ethylidene CH afforded a singlet at about δ 6.8 and its corresponding carbon was at about δ89 in ^{13}C NMR spectrum. The signals of the four protons on the quinoxalone ring centered at about δ 7.1 and 7.2. In the ^{13}C NMR spectra of these compounds, one C = O peak at about δ 185 and one around δ 160 was corresponded to signals of benzoyl and amide fragments, respectively. The IR spectra of these compounds further support their postulated structure. There were several weak to medium bands near 3000 cm^{-1} corresponding to O—H, N—H and C—H bonds and a strong band at 1670—680 cm^{-1} due to the C = O bond of the quinoxaline moiety. The additional strong bands between 1555 and 1621 cm^{-1} were the characteristics of enaminones.[29]

Inspired by the encouraging result, we explored the possible mechanism of the coupling reaction. Based on the fact that the reaction of methyl benzoylpyruvate with aniline affords the corresponding enamine at the C – 2 carbon atom, we speculated that the amino group of the dinucleophilic phenylenediamine might react first with the C – 2 electrophilic center to generate enamine. Then the attack of the second amino group proceeded at the C – 1 center to form six – membered heterocyclic quinoxalone due to thermodynamic factors.

Preliminary UV – Vis studies

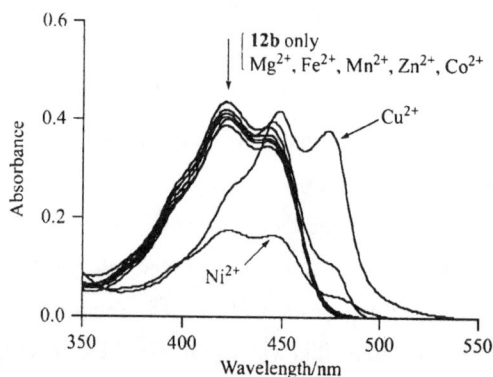

Fig. 3　UV – Vis absorption spectra of free 1 2b and 12b in the presence of metal ions in ethanol. [12b] = 1.2 × 10^{-5} mol · dm^{-3}, [M^{2+}] = 5 × 10^{-4} mol · dm^{-3}

All samples (5, 9, 10—13) were sent to test their anti – HIV activity. Unfortunately. due to their lower solubility in water, such experiments were unable to carry out. Considering that these compounds were designed to coordinate with divalent metal ions in the HIV integrase active domain. UV – Vis spectroscopy in ethanol was utilized as an alternative method to assay their possible binding properties with metal ions. Therefore, taking quinoxalone derivatives as examples, the preliminary ion binding properties of 10, 11, 1 2 and 13 were investigated. Fig. 3 shows the representative chromogenic behaviors of 12b in ethanol ([12b] = 1.2 × 10^{-5} mol · dm^{-3}, [M^{2+}] = 5 × 10^{-4} mol · dm^{-3}). The UV – Vis spectra of free compound 12b contain two significant maxima in the region of 421 and 443 nm about 22 nm apart in tne visible range. Upon addition of about 40 equiv of metal ions of Co^{2+}, Fe^{2+}, Mg^{2+}, Ni^{2+},

Mn^{2+} and Zn^{2+} (in perchlorate), the absorption behaviors of these compounds were not significantly affected. However the addition of Cu^{2+} gives rise to a red shift of 53 nm, inducing the absorption wavelength from around 421 to 474 nm. At the same time, the maxima in region of 443 nm have no significant change. So it is apparent that these compounds have a high selectivity for the recognition of Cu^{2+} in ethanol.

The absorption and ion binding properties of other compounds 10—13 are very similar to compound 12b. Their absorption wavelength and the wavelength changes in the presence of Cu^{2+} are summarized in Tab. 1. It should be pointed out that p – substituted quinoxalone derivatives (10b, 11b, 12b and 13b) showed longer wavelength, as well as larger peak shift ($\Delta\lambda_{max}$) on addition of Cu^{2+} in ethanol compared with the corresponding m – substituted 10a, 11a, 12a and 13a, respectlvely. The results of the UV – Vis spectrum are in accordance with the optimized structure of these compounds by the *ab initio* calculation. The *para* – substituted quinoxalone derivatives have coplanary structure. However, for the corresponding *meta* – substituted derivatives, the plane formed by quinoxalone ring derivated from the benzene ring and therefore less degree of conjugation formed.

Fig. 4 shows the changes in the absorption spectra of 12a in ethanol solution upon addition of Cu^{2+} and Ni^{2+}. As the titration continued, the main absorption of Cu^{2+} at 416 nm of 12a decreased. and a new band centered at 465 nm arose. but the absorption at 441 nm has no distinct change. The absorbance behavior of 12a did not greatly respond to Ni^{2+} from the titration curve compared with Cu^{2+}. The presence of a distinct isosbestic point at 450 nm with the two cations demonstrates an existence of equilibrium and nonlinear least – square curve fitting (Correlation coefficient: $R_{Cu} = 0.9989$, $R_{Ni} = 0.9983$) at 420 nm clearly indicates that the spectra changes can be ascribed to the formation of a 1:2 complex of Cu^{2+} or Ni^{2+} with 12a. The binding constant (K_S) was determined to be 2.42×10^{10} and 3.96×10^9 (mol $\cdot$ L^{-1})$^{-2}$ at 25 ℃ in ethanol, respectively. [30,31] Fig. 5 shows the chelating mode of Cu^{2+} with 12b due to replacement of the hydrogen atom of quinoxalone ring by Cu^{2+}. [26]

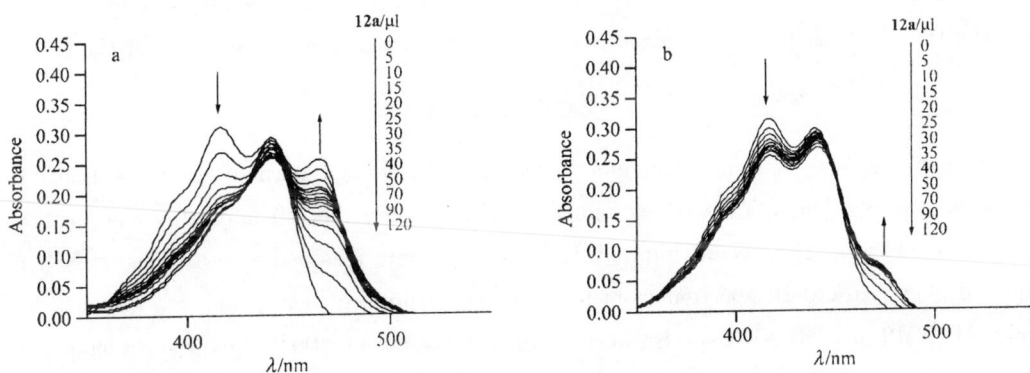

Fig. 4 Changes in UV – Vis absorption of 12a at 25℃ in ethanol on addition of 5—120 μl of (a) [Cu^{2+}] $= 1.0 \times 10^{-3}$ mol $\cdot$ dm^{-3} and (b) [Ni^{2+}] $= 1.0 \times 10^{-3}$ mol $\cdot$ dm^{-3}. [12a] $= 1.2 \times 10^{-5}$ mol $\cdot$ dm^{-3}.

Tab. 1　Absorption peak (λ_{max}) change for compounds 10—13 in the presence of Cu^{2+}

Compound	λ_{max}^{a}/nm	λ_{max}^{b}/nm	$\Delta\lambda_{max}$/nm
10a	417	465	48
10b	423	476	53
11a	416	465	49
11b	423	474	51
12a	416	465	49
12b	421	474	53
13a	416	465	49
13b	421	472	51

Notes: ^{a} Absorption wavelength(λ_{max}) of free compounds.
^{b} Absorption wavelength(λ_{max}) of compounds on addition of Cu^{2+}

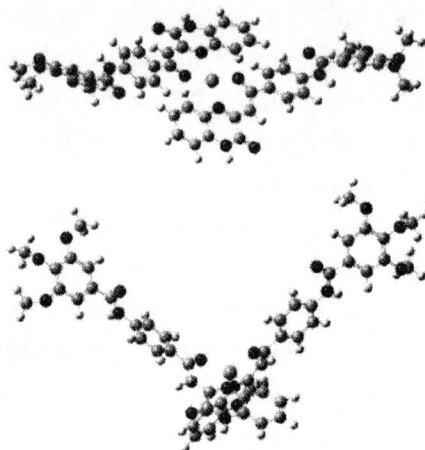

Fig. 5　The chelating mode of Cu^{2+} with 12b in different angle of view

Conclusion

In conclusion, we have designed and synthesized a series of galloyl and caffeoyl substituted phenyl diketoacid and the corresponding quinoxalone derivatives as potential HIV integrase inhibitors. In the key step of oxalylation reaction, the corresponding phenyl methyl ketone and dimethyl oxalate were dissolved in THF/DME in the presence of NaOMe as base. The resulting sodium keto enolate ester was first separated by filtering, thus avoided tedious workup and column chromatographic isolation. Instead of 2 – benzimidazol, the resulting β – diketoacid underwent further condensation reaction with phenylenediamine to generate quinoxalone derivatives in good yield, which offered an efficient approach to afford quinoxalone derivatives from diketoacid. Due to the lower water solubility of all target compounds, anti – HIV assay was unable to carry out. UV – Vis spectroscopy in organic solvent was utilized as an alternative method to assay their possible binding properties with metal ions. The UV – Vis spectra of quinoxalone compounds demonstrated a good selectivity for Cu^{2+}.

Experimental

All solvents were of commercial quality and were dried and purified by conventional methods. Melting points (m. p.) were determined on XT4A electrothermal apparatus equipped with a microscope and were uncorrected. Infrared (IR) spectra were recorded as thin films on KBr plates with a Bruker vertex70 IR spectrophotometer and were expressed in v (cm^{-1}). Unless otherwise stated, ^{1}H NMR and ^{13}C NMR spectra were obtained using an AV 400M Bruker spectrometer in solvent (CDCl$_3$ or DMSO – d_6) with TMS as internal reference. MS spectra (ESI) were recorded on a Bruker esquire 6000 mass spectrometer. HRMS spectra were recorded in the negative ion mode using APEX Ⅱ. FT – ICRMS of Bruker Dalaonics Inc. UV – Vis spectra were measured on a Bruker U – 3010 spectrophotometer. Molecular model simulation was optimized with Gaussian A03 suite of program on the basis set of B3LYP/6 – 31g (d) of density functional theory (DFT) on the IBM/

RS6000 workstation.

General procedure for preparations of 2 and 6

To a solution of amino acetophenone 1 (9 mmol) and dry pyridine (9 mmol) in CH_2Cl_2 (20 ml) was added dropwise dimethylacffeoyl chloride or trimethylgalloyl chloride in CH_2Cl_2, which was prepared by the direct reaction of the corresponding dimethylcaffeoylic acid or trimethylgalloyl with excessive thinoyl chloride. The reaction mixture was stirred overnight at room temperature and then removed CH_2Cl_2, extracted with EtOAc, and washed with water. The residue was recrystalized form EtOAc/petroleum ether to afford 2 or 6 as white solid.

3 - (3, 4 - Dimethoxystyrylcarbonylamino) - 1 - aceto - phenone (2a): Yield 66%; m. p. 154—155 ℃; ^{1}H NMR (400 MHz, CDCl$_3$) δ: 2.65 (s, 3H), 3.92 (s, 3H), 3.94 (s, 3H), 6.56 (d, $J=15.6$Hz, 1H), 6.89 (d, $J=8.4$ Hz, 1H), 7.08 (s, 1H), 7.14 (d, $J=8.0$ Hz, 1H), 7.48 (t, $J=8.0$ Hz, 1H), 7.72 (d, 1H), 7.75 (d, $J=15.6$ Hz, 1H), 7.91 (s, 1H), 8.06 (d, $J=7.56$ Hz, 1H), 8.18 (s, 1H); MS (ESI) m/z: 348.0522 $[M+Na]^+$.

4 - (3, 4 - Dimethoxystyrylcarbonylamino) - 1 - acetophenone (2b): Yield 65%: m. P. 199—201℃; ^{1}H NMR (400 MHz, CDCl$_3$): δ: 2.58 (s, 3H), 3.91 (s, 3H), 3.94 (s, 3H), 6.45 (d, $J=15.6$ Hz, 1H), 6.86 (d, $J=8.3$ Hz, 1H), 7.04 (s, 1H), 7.12 (d, $J=8.3$ Hz, 1H), 7.71 (d, $J=8.4$ Hz, 2H), 7.74 (d, $J=15.6$ Hz, 1H), 7.95 (d, $J=8.4$ Hz, 2H); MS (ESI) m/z: 348.0318 $[M+Na]^+$.

3 - (3, 4, 5 - Trimethoxyphenylcarbonylamino) - 1 - acetophenone (6a): Yield 74%: m. p. 141—142 ℃; ^{1}H NMR (400 MHz, CDCl$_3$) δ: 2.64 (s, 3H), 3.93 (s, 3H), 3.95 (s, 6H), 7.12 (s, 2H), 7.50 (t, $J=8.0$ Hz, 1H), 7.74 (d, $J=8.0$ Hz, 1H), 8.07 (s, 1H), 8.08 (d, $J=8.0$ Hz, 1H), 8.13 (s, 1H); MS (ESI) m/z: 352.0460. $[M+Na]^+$.

4 - (3, 4, 5 - trimethoxyphenylcarbonylamino) - 1 - acetophenone (6b): Yield 76%: m. P. 190—192 ℃; ^{1}H NMR (400 MHz, DMSO $-d_6$) δ: 2.54 (s, 3H), 3.73 (s, 3H), 3.86 (s, 6H), 7.28 (s, 2H), 7.89 (d, $J=8.8$ Hz, 2H), 7.98 (d, $J=8.8$ Hz, 2H), 10.41 (s, 1H); MS (ESI) m/z: 352.0615 $[M+Na]^+$.

General procedure for preparations of 3 and 7

To a solution of NaOMe (40 mmol) in dry THF was added dropwise a mixture of the appropriate phenyl methyl ketone 2 or 6 (10 mmol) and dimethyl oxalate (20 mmol) in dry THF/DME (V : V=1:1). The mixture was stirred overnight and was titrated by methanol. The resulting orange - yellow solid was filtered off, washed with methanol and diethyl ether and dried. Sodium ketoenolate ester was obtained. The sodium ketoenolate ester was dissolved in water at room temperature for 1 h. Then the solution was acidified by adding 1 mol/L aqueous HCl to pH 3—4 and kept at 4 ℃ for 2 h. The light yellow precipitate was filtered off, washed with water and dried.

4 - (3 - (3, 4 - Dimethoxystyrylcarbonylaminophenyl) - 1 - yl) - 4 - oxo - 2 - hydroxy - 2 - butenoic acid methyl ester (3a): Yield 83%: m. P. 167—169 ℃; ^{1}H NMR (400 MHz, DM-SO $-d_6$) δ: 3.78 (s, 3H), 3.80 (s, 3H), 3.85 (s, 3H), 6.68 (d, $J=15.6$ Hz,

1H), 7.0 (d, $J=8.0$Hz, 1H), 7.05 (s, 1 H), 7.19 (d, $J=8.2$ Hz, 1 H), 7.21 (s, 1H), 7.53 (t, $J=8-0$ Hz, 1H), 7155 (d, $J=15.6$ Hz, 1H), 7.76 (d, $J=8.0$ Hz, 1H), 7.99 (d, $J=8.0$Hz, 1H), 8.39 (s, 1H) 10.40 (s, 1H); ^{13}C NMR (100 MHz, DMSO $-d_6$) δ: 190.1, 168.6, 164.3, 162.1, 150.7, 149.1, 141.0, 140.3, 135.1, 129.9, 127.4, 124.5, 122.8, 122.1, 119.5, 117.9, 111.9, 110.2, 98.2, 55.7, 55.5, 53.3: IR (KBr) v: 3391, 2956, 1715, 1680, 1632, 1260 cm^{-1}: MS (ESI) m/z: 434.0872 [M + Na]$^+$.

4 - (4 - (3 -4 - Dimethoxystyrylcarbonylaminophenyl) -1 -yl) -4 - oxo -2 - hydroxy -2 - butenoic acid methyl ester (3b): Yield 81%: m. P. 209—212 ℃; ^{1}H NMR (300 MHz, DMSO - d_6) δ: 3.79 (s, 3H), 3.81 (s, 3H), 3.85 (s, 3H), 6.72 (d, $J=15.6$ Hz, 1H), 7.02 (d, $J=8.1$ Hz, 1H), 7.11 (s, 1H), 7.21 (d, $J=8.1$ Hz, 1H), 7.23 (s, 1H), 7.58 (d, $J=15.6$ Hz, 1H), 7.88 (d, $J=8.7$ Hz, 2H,), 8.09 (d, $J=8.7$ Hz, 2H), 10.59 (s, 1H); ^{13}C NMR (100 MHz, DMSO - d_6) δ: 190.1, 168.3, 164.9, 162.7, 151.2, 149.4, 145.4, 142.0, 130.1, 129.1, 127.7, 122.6, 119.7, 119.2, 112.3, 110.6, 98.3, 56.1, 55.9, 53.5; IR (KBr) v: 3368, 2955, 1722, 1687, 1626, 1598, 1512, 1282, 1261, 1151, 1l41 cm^{-1}: MS (ESI) m/z; 434.1087 [M + Na]$^+$.

4 - (3 - (3, 4, 5 - Trimethoxyphenylcarbonylaminophenyl) - 1 - yl) - 4 - oxo -2 - hydroxy -2 - butenoic acid methyl ester (7a): Yield 82%: m. p. 106—108 ℃; ^{1}H NMR (400 MHz, DMSO - d_6) δ: 3.73 (s, 3H), 3.86 (s, 3H), 3.87 (s, 6H), 7.08 (s, 1 H), 7.30 (s, 2H), 7.57 (t, $J=7.6$ Hz, 1H), 7.81 (d, $J=7.6$ Hz, 1H), 8.12 (d, $J=8.0$ Hz, 1H), 8.41 (s, 1H), 10.37 (s, 1H); ^{13}C NMR (100 MHz, DMSO - d_6) δ: 190.4, 169.1, 165.6, 162.5, 153.1, 141.0, 140.4, 135.3, 130.1, 126.3, 123.6, 119.8, 105.9, 105.9, 98.6, 60.6, 56.6, 53.6: IR (KBr) v: 3261, 2946, 2838, 1728, 1642, 1585, 1533, 1129 cm^{-1}: MS (ESI) m/z: 438.1092 [M + Na]$^+$.

4 - (4 - (3, 4, 5 - Trimethoxyphenylcarbonylaminophenyl) - 1 - yl) - 4 - oxo -2 - hydroxy -2 - butenoic acid methyl ester (7b): Yield 84%: m. P. 174—175 ℃; ^{1}H NMR (400 MHz, DMSO - d_6) δ: 3.92 (s, 3H), 3.94 (s, 6H), 3.95 (s, 3H), 7.08 (s, 1H), 7.08 (s, 2H), 7.94 (s, 1H), 7.81 (d, $J=8.8$ Hz, 2H), 8.05 (d, $J=8.8$ Hz, 2H), 15.36 (s, 1H); ^{13}C NMR (100 MHz, DMSO - d_6) δ: 190.1, 168.4, 165.9, 162.7, 153.1, 145.2, 141.2, 129.8, 120.4, 106.1, 98.4, 60.6, 56.7, 53.5; IR (KBr) v: 3370, 2944, 2836, 1731, 1624, 1521, 1266, 1128 cm^{-1}; MS (ESI) m/z: 438.0885 [M + Na]$^+$.

General procedure for preparations of 4 and 8

To a suspension of appropriated phenyl methyl diketoacid ester (5 mmol) in THF/CH$_3$OH ($V:V=1:1$) was added dropwise 1 mol/L aqueous NaOH (4 equiv.). The resulting clear solution was stirred at room temperature for about 0.5 h. After dilution with water, the mixture was acidified with 1 mol/L HCl to pH 1—2. The light yellow or orange precipitate that formed was filtered off, washed with water and recrystallized from EtOAc.

4 - (3 - (3, 4 - Dimethoxystyrylcarbonylaminophenyl) -1 - yl) -4 - oxo -2 - hydroxy -

2 − butenoic acid (4a): Yield 95%: m. P. 118—120 ℃; ^{1}H NMR (300 MHz, DMSO − d_6) δ: 3. 79 (s, 3H), 3. 81 (s, 3H), 6. 68 (d, J = 15.6Hz, 1H), 7. 01 (d, J = 8.4 Hz, 1H), 7. 04 (s, 1H), 7. 18 (d, J = 8.2 Hz, 1H), 7. 21 (s, 1H), 7. 52 (t, J = 8.0 Hz, 1H), 7. 55 (d, J = 15.6 Hz, 1H), 7. 74 (d, J = 7.8 Hz, 1H), 7. 99 (d, J = 8.1 Hz, 1H), 8. 40 (s, 1H), 10. 40 (s, 1H); ^{13}C NMR (100 MHz, DMSO − d_6) δ: 55. 6, 55. 8, 98. 1, 110. 3, 111. 9, 118. 1, 119. 6, 122. 1, 122. 8, 124. 4, 127. 5, 129. 9, 135. 4, 140. 4, 141. 1, 149. 2, 150. 7, 163. 3, 164. 4, 169. 9, 190. 4: IR (KBr) v: 3308, 2985, 1720, 1708, 1660, 1599, 1548, 1510, 1021 cm^{-1}: MS (ESI) m/z: 420. 1160 [M + Na]$^+$.

4 − (4 − (3, 4 − Dimethoxystyrylcarbonylaminophenyl) −1 − yl) −4 − oxo − 2 − hydroxy − 2 − butenoic acid (4b): Yield 95%: m. P. 228—230 ℃; ^{1}H NMR (400 MHz, DMSO − d_6) δ: 3. 78 (s, 3H), 3. 80 (s, 3H), 6. 71 (d, J = 15.6 Hz, 1H), 7. 01 (d, J = 8.0 Hz, 1H), 7. 02, (s, 1H), 7. 20 (d, J = 11.2 Hz, 1H), 7. 57 (d, J = 15.6 Hz, 1H), 7. 85 (d, J = 8.4 Hz, 2H), 8. 04 (d, J = 8.4 Hz, 2H), 10. 54 (s, 1H); ^{13}C NMR (100MHz, DMSO − d_6) δ: 190. 0, 168. 2, 164. 9, 163. 8, 151. 1, 149. 4, 145. 2, 141. 9, 129. 9, 129. 4, 127. 7, 122. 6, 119. 7, 119. 2, 112. 3, 110. 6, 98. 0, 56. 1, 55. 9; IR (KBr) v: 3344, 2918, 1743, 1672, 1597, 1511, 1251 cm^{-1}; MS (ESI) m/z: 420. 0533 [M + Na]$^+$.

4 − (3 − (3, 4, 5 − Trimethoxyphenylcarbonylaminophen − y1) −1 − y1) −4 − oxo − 2 − hydroxy − 2 − butenoic acid (8a): Yield 90%: m. P. 195—197 ℃; ^{1}H NMR (400 MHz, DMSO − d_6) δ: 3. 72 (s, 3H), 3. 86 (s, 6H), 7. 06 (s, 1H), 7. 30 (s, 2H), 7. 55 (t, J = 8.0 Hz, 1H), 7. 79 (d, J = 8.0 Hz, 1H), 8. 11 (d, J = 8.0 Hz, 1H), 8. 41 (s, 1H), 10. 35 (s, 1H); ^{13}C NMR (100 MHz, DMSO − d_6) δ: 190. 6, 170. 4, 165. 6, 163. 6, 153. 1, 141. 0, 140. 3, 135. 5, 130. 07, 126. 1, 123. 5, 119. 8, 105. 9, 98. 3, 60. 6, 56. 6, 56. 6; IR (KBr) v: 3434, 3336, 2939, 1739, 1628, 1585, 1536, 1121 cm^{-1}; MS (ESI) m/z: 424. 0323 [M + Na]$^+$.

4 − (4 − (3, 4, 5 − Trimethoxyphenylcarbonylaminophen − y1) −1 − y1) −4 − oxo − 2 − hydroxy − 2 − butenoic acid (8b): Yield 88%: m. P. 225—227 ℃; ^{1}H NMR (400 MHz, DMSO − d_6) δ: 3. 72 (s, 3H), 3. 86 (s, 6H), 7. 08 (s, 1H), 7. 27 (s, 2H), 7. 98 (d, J = 9.2 Hz, 2H), 8. 10 (d, J = 9.2 Hz, 2H), 10. 48 (s, 1H); ^{13}C NMR (100 MHz, DMSO − d_6) δ: 190. 2, 169. 8, 165. 8, 163. 7, 153. 1, 145. 0, 141. 2, 130. 2, 129. 7, 120. 4, 106. 0, 98. 1, 60. 6, 56. 6, 56. 6; IR (KBr) v: 3435, 3338, 2939, 1738, 1643, 1620, 1585, 1527, 1243, 1122 cm^{-1}: MS (ESI) m/z: 424. 0473 [M + Na]$^+$.

General procedure for preparations of 5, 9, 11 and 13

BBr$_3$ (1 mol/L in CH$_2$Cl$_2$, 20 mmol for 4, 11 and 24 mmol for 8, 13) was added dropwise to a solution of 4, 8, 10 or 12 (2 mmol) in CH$_2$Cl$_2$ (30 ml), the color of the mixture immediately change from yellow to dark red. The mixture was stirred for 3 h, and then water was added dropwise and continuously stirred overnight. The solid was filtered off and washed with water till the filtrate is not red. The compound was usually recrystallized from acetone or ethanol/water.

4 − (3 − (3, 4 − Dihydroxystyrylcarbonylaminophenyl) −1 − y1) −4 − oxo − 2 − hydroxy − 2 − butenoi − c acid (5a): Yield 92%: m. p. 196—198 ℃; ^{1}H NMR (300 MHz, DMSO − d_6)

δ: 6.57 (d, $J=15.6$ Hz, 1H), 6.82 (d, $J=8.0$ Hz, 1 H), 6.97 (d, $J=8.0$ Hz, 1H), 7.07 (s, 1H), 7.48 (d, $J=15.6$ Hz, 1H、), 7.54 (t, $J=8.0$ Hz, 1H), 7.74 (d, $J=7.2$ Hz, 1H), 7.97 (d, $J=8.0$ Hz, 1H), 8.40 (s, 1H), 10.37 (s, 1H); ^{13}C NMR (100 MHz, DMSO$-d_6$) δ: 192.8, 171.7, 167.3, 165.8, 150.5, 148.2, 144.3, 142.8, 138, 132.6, 128.9, 127.2, 125.4, 123.9, 120.6, 120.6, 118.6, 116.9, 100.8; IR (KBr) v: 3400, 1731, 1609, 1555, 1287 cm^{-1}; MS (ESI) m/z: 370.0340 [M+Na]$^+$.

4 - (4 - (3, 4 - Dihydroxystyrylcarbonylaminophenyl) -1-y1) -4-oxo-2-hydroxy-2-butenoic acid (5b): Yield 91%: m. P. 245—248 ℃; ^{1}H NMR (DMSO$-d_6$) δ: 6.55 (d, $J=15.6$ Hz, 1H), 6.77 (d, $J=8.4$ Hz, 1H), 6.94 (dd, $J=8.4$, $J=1.6$ Hz, 1 H), 7.02 (d, $J=1.6$ Hz, 1H), 7.44 (d, $J=15.6$ Hz, 1H), 7.84 (d, $J=8.4$ Hz, 2H), 8.03 (d, $J=7.6$ Hz, 2H、), 10.54 (s, 0.2H); ^{13}C NMR (100 MHz, DMSO$-d_6$) δ: 192.4, 171.8, 167.4, 165.9, 150.7, 148.3, 147.5, 144.9, 132.3, 131.8, 128.8, 124.1, 121.5, 120.5, 118.6, 116.9, 100.6; IR (KBr) v: 3429, 1721, 1598, 1524, 1249, 1179 cm^{-1}; MS (ESI) m/z: 370.0178 [M+H]$^+$.

4 - (3 - (3, 4, 5 - Trihydroxyphenylcarbonylaminophen-y1) -1-y1) -4-oxo-2-hydroxy-2-butenoic acid (9a): Yield 92%: m. p. 205—207 ℃; ^{1}H NMR (400 MHz, DMSO$-d_6$) δ: 6.98 (s, 2H), 7.04 (s, 1H), 7.50 (t, $J=9.3$ Hz, 1H、), 7.73 (d, $J=7.5$ Hz, 1H), 8.12 (d, $J=7.8$ Hz, 1H), 8.45 (s, 1H), 8.87 (br, 0.87H), 9.19 (br, 1.8H), 10.14 (s, 1H); ^{13}C NMR (100 MHz, DMSO$-d_6$) δ: 190.7, 170.2, 166.3, 163.6, 145.9, 140.5, 137.5, 135.4, 129.9, 125.8, 124.9, 123.0, 119.5, 107.6, 98.2; IR (KBr) v: 3421, 1733, 1625, 1538, 1264 cm^{-1}; MS (ESI) m/z: 381.9888 [M+Na]$^+$.

4 - (4 - (3, 4, 5 - Trihydroxyphenylcarbonylaminophen-y1) -1-y1) -4-oxo-2-hydroxy-2-butenoic acid (9b): Yield 92%, m. p. 237—239 ℃; ^{1}H NMR (400 Hz, DMSO$-d_6$) δ: 6.95 (s, 2H、), 7.92 (d, $J=8.8$ Hz, 2H), 8.03 (d, $J=8.8$ Hz, 1 H); ^{13}C NMR (100 MHz, DMSO$-d_6$) δ: 194.5, 173.0, 170.9, 168.0, 150.2, 149.7, 142.0, 134.1, 133.9, 129.2, 124.4, 112.2, 102.5; IR (KBr) v: 3408, 1716, 1597, 1517, 1327, 1182 cm^{-1}: MS (ESI) m/z: 382.0007 [M+Na]$^+$.

(Z) -3- [2-Oxo-2- (3- (3, 4-dihydroxystyrylcarbonylaminphenyl) -1-y1) -ethylidene] -3, 4-dihydro-1H-quinoxalin-2-one (11a): Yield 90%: m. P. > 300℃; ^{1}H NMR (DMSO$-d_6$) δ: 6.54 (d, $J=15.6$ Hz, 1H), 6.78 (d, $J=8.1$ Hz, 1H), 6.81 (s, 1 H), 6.92 (d, $J=8.1$ Hz, 1H), 7.02 (s, 1H), 7.11—7.14 (m, 3H), 7.44 (d, J=15.6 Hz, 1H), 7.47 (t, J=7.2 Hz, 2H), 7.51—7.52 (m, 1H), 7.65 (d, J=7.6 Hz, 1H), 7.92 (d, J=8.0 Hz, 1H), 8.35 (s, 1H), 10.28 (s, 1H、), 12.04 (s, 1H), 13.63 (s, 1H、); ^{13}C NMR (100 MHz, DMSO$-$d$_6$) δ: 188.6, 164.7, 156.2, 148.3, 146.1, 141.6, 140.4, 139.6, 129.7, 127.2, 126.6, 124.57, 124.49, 124.16, 122.7, 122.1, 121.4, 118.6, 118.1, 117.1, 116.3, 115.8, 114.5, 89.5; IR (KBr) v: 3206, 2918, 2849, 1652, 1605, 1541, 1265 cm^{-1}; MS (ESI) m/z: 464.1518 [$M+Na$]$^+$.

(Z) -3- [2-oxo-2- (4- (3, 4-dihydroxystyrylcarbonylaminophenyl) -1-y1) -

ethylidene] − 3, 4 − *dihydro* − 1H − *quinoxalin* − 2 − *one* (11b): Yield 91%; m. P. > 300 ℃; ^1H NMR (400 MHz, DMSO − d$_6$) δ: 6.56 (d, J = 15.6 Hz, 1H), 6.77 (d, J = 8.0 Hz, 1H), 6.79 (s, 1H), 6.91 (d, J = 8.0 Hz, 1H), 7.02 (s, 1H), 7.09—7.11 (m, 3H), 7.43 (d, J = 15.6 Hz, 1H), 7.45—7.47 (m, 1H) 7.82 (d, J = 8.5 Hz, 2H), 7.97 (d, J = 8.5 Hz, 2H), 10.36 (s, 1H), 11.96 (s, 1H), 13.63 (s, 1H); ^{13}C NMR (100 MHz, DMSO − d$_6$) δ: 187.9, 164.7, 156.1, 148.2, 145.9, 145.5, 143.3, 141.8. 133.5, 128.6, 126.9, 126.4, 124.6, 124.1, 124.0, 121.4. 118.9, 118.4, 116.7, 116.2, 115.7, 114.4, 89.4; IR (KBr) v: 3338, 1679, 1590, 1588, 1514, 1245 cm^{-1}; MS (ESI) m/z: 464.1173 [M + Na]$^+$; HRMS m/z [M—H]$^-$ calcd for $C_{25}H_{18}N_3O_5$ 440.1252; found 440.1253.

(Z) − 3 − [2 − Oxo − 2 − (3 − (3, 4, 5 − *trihydroxyphenylcarbonylaminophenyl*) − 1 − y1) − *ethylidene*] − 3, 4 − *dihydro* − 1H − *quinoxalin* − 2 − *one* (13a): Yield 88%; m. P. > 300 ℃; ^1H NMR (DMSO − d$_6$) δ. 82 (s, 1H), 6.98 (s, 2H), 7.12—7.13 (m, 3H), 7.46 (t, J = 7.8 Hz, 1H), 7.51 (s, 1H), 7.65 (d, J = 7.6 Hz, 1H), 8.01 (d, J = 8.0 Hz, 1H), 8.42 (s, 1H), 8.83 (s, 1H), 9.16 (s, 2H), 10.10 (s, 1H), 12.04 (s, 1H), 13.64 (s, 1H); ^{13}C NMR (100 MHz, DMSO − d$_6$) δ: 188.8, 166.2, 156.2, 145.9, 140.5, 139.4, 137.4, 129.3, 127.1, 125.2, 124.6, 124.4, 124.1, 123.7, 122.0, 119.1, 117.0, 115.8, 107.7, 89.6; IR (KBr) v: 3421, 1698, 1683, 1607, cm^{-1}; MS (ESI) m/z: 454.0752 [M + Na]$^+$.; HRMS m/z [M − H]$^-$ calcd for $C_{23}H_{16}N_3O_6$ 430.1044; found 430.1043.

(Z) − 3 − [2 − Oxo − 2 − (4 − (3, 4, 5 − *trihydroxyphenylcarbonylaminophenyl*) − 1 − y1) − *ethylidene*] — 3, 4 − *dihydro* − 1H − *quinoxalin* − 2 − *one* (13b): Yield 93%; m. P. > 300 ℃; ^1H NMR (400 MHz, DMSO − d_6) δ: 6.79 (s, 1H), 6.96 (s, 2H), 7.11—7.45 (m, 4H), 7.89 (d, J = 8.7 Hz, 2H), 7.94 (d, J = 8.8 Hz, 2H); ^{13}C NMR (100 MHz, DMSO − d_6) δ: 187.8, 165.9, 155.9, 145.6, 145.2, 143.0, 137.2, 133.3, 128.0, 128.0, 126.6, 124.7, 124.3, 123.8, 119.6, 116.4, 115.4, 107.4, 89.2; IR (KBr) v: 3341, 1667, 1595, 1592, 1499, 1208 cm^{-1}: MS (ESI) m/z: 454.1274 [M + Na]$^+$.

General procedure for preparations of 10 and 12

A mixture of appropriate diketo acid 4 or 8 (3 mmol) o − phenylenediamine (3.6 mmol) and 30 ml H$_2$O was refluxed under stirring for 2—4 h. A greenish or yellowish sticky solid appeared during the reaction time. The reaction mixture was cooled to room temperature and collected by filtration, washed with water and then with ether to remove the residual diamine. The compound was usually recrystallized from ethanol or purified by flash column chromatography (eluent: Methanol/ CHCl$_3$, V : V = 1 : 10). All the resulting compounds have fluorescent color from the TLC with 365 nm UV light.

(Z) − 3 − [2 − Oxo − 2 − (3 − (3, 4 − dimethoxystyrylcarbonylaminophenyl) − 1 − y1) − ethylidene] − 3, 4 − *dihydro* − 1H − *quinoxalin* − 2 − *one* (10a): Yield 94%; m. P. > 300 ℃; ^1H NMR (400 MHz, DMSO − d_6) δ: 3.80 (s, 3H), 3.83 (s, 3H), 6.71 (d, J = 15.6 Hz, 1H), 6.81 (s, 1H), 7.01 (d, J = 8.0 Hz, 1H), 7.13—7.16 (m, 3H), 7.21 (d, J =

10 Hz, 1H), 7.23 (s, 1H), 7.47—7.49 (m, 2H), 7.57 (d, $J = 15.6$ Hz, 1H), 7.67 (d, $J = 7.6$ Hz, 1H), 7.89 (d, $J = 7.6$ Hz, 1H), 8.38 (s, 1H); ^{13}C NMR (100 MHz, DMSO $- d_6$) δ: 188.8, 164.6, 156.2, 150.9, 149.4, 145.7, 141.2, 140.1, 139.7, 129.7, 127.8, 126.9, 124.5, 124.4, 124.3, 122.8, 122.4, 122.3, 119.9, 118.1, 116.8, 115.8, 112.2, 110.6, 89.6, 56.0, 55.9; IR (KBr) v: 3317, 3016, 1671, 1596, 1533, 1514, 1360, 1260 cm^{-1}; MS (ESI) m/z: 492.1268 [M + Na]$^+$.

(Z) −3− [2−Oxo−2−(4−(3, 4−dimethoxystyrylcarbonylaminophenyl) −1−yl) −ethylidene] −3, 4−dihydro−1H−quinoxalin−2−one (10b): Yield 93%; m. p. >300 ℃; ^{1}H NMR (400 MHz, DMSO $- d_6$) δ: 3186 (s, 3H), 3.88 (s, 3H), 6.78 (d, $J = 15.6$ Hz, 1H), 6.87 (s, 1H), 7.07 (d, $J = 8.4$ Hz, 1H), 7.16—7.18 (m, 3H), 7.26 (d, $J = 10$ Hz, 1H), 7.28 (s, 1 H), 7.52—7.53 (m, 1H), 7.62 (d, $J = 15.6$ Hz, 1 H), 7.90 (d, $J = 8.0$ Hz, 2H), 8.04 (d, $J = 8.0$ Hz, 2H), 10.48 (s, 1H), 12.04 (s, 1H), 13.71 (s, 1 H); ^{13}C NMR (100 MHz, DMSO $- d_6$) δ: 183.8, 160.4, 152.1, 146.8, 145.2, 141.4, 139.1, 137.3, 129.5, 124.5, 123.6, 122.8, 120.5, 120.0, 119.9, 118.2, 115.7, 114.9, 112.6, 111.6, 108.0, 106.3, 85.3, 51.7, 51.8; IR (KBr) v: 3236, 1675, 1598, 1513, 1257 cm^{-1}: MS (ESI) m/z: 470.1540 [M + Na]$^+$.

(Z) −3− [2−Oxo−2 (3−(3, 4, 5−trimethoxyphenylcarbonylaminophenyl) −1−yl) −ethylidene] −3, 4−dihydro−1H−quinoxalin−2−one (12a): Yield 97%; m. p. 308—310 ℃; ^{1}H NMR (300 MHz, DMSO $- d_6$) δ: 3.73 (s, 3H), 3.88 (s, 6H), 6.84 (s, 1H), 7.13—7.15 (m, 3H), 7.31 (s, 2H), 7.51 (m, 2H), 7.72 (d, $J = 7.2$ Hz, 1 H), 8.02 (d, $J = 7.2$ Hz, 1H), 8.41 (s, 1H), 10.33 (s, 1H), 12.07 (s, 1H), 13.66 (s, 1H); ^{13}C NMR (100 MHz, DMSO $- d_6$) δ: 188.1, 165.1, 155.7, 152.6, 145.6, 140.5, 139.5, 139.1, 129.8, 129.1, 126.7, 124.1, 124.1, 123.7, 122.2, 119.1, 116.6, 115.4, 105.4, 89.1, 79.1, 60.1, 56.1; IR (KBr) v: 3243, 3010, 1677, 1640, 1610, 1505, 1341, 1134 cm^{-1}; MS (ESI) m/z: 474.1004 [M + H]$^+$.

(Z) −3− [2−Oxo−2−(4−(3, 4, 5−trimethoxyphenylcarbonylaminophenyl) −1−yl) −ethylidene] −3, 4−dihydro−1H−quinoxalin−2−one (12b): Yield 98%: m. p. > 300 ℃; ^{1}H NMR (300 MHz, DMSO $- d_6$) δ: 3.73 (s, 3H), 3.87 (s, 6H), 6.83 (s, 1H), 7.11—7.12 (m, 3H), 7.29 (s, 2H), 7.46—7.47 (m, 1 H), 7.91 (d, $J = 9.0$ Hz, 2H), 8.01 (d, $J = 9.0$ Hz, 2H), 10.37 (s, 1H), 11.99 (s, 1H), 13.64 (s, 1H、); ^{13}C NMR (100 MHz, DMSO $- d_6$) δ: 188.0, 165.6, 156.3, 153.1, 145.7, 143.0, 141.0, 134.1, 130.2, 128.4, 127.0, 124.6, 124.2, 124.1, 120.4, 116.8, 115.8, 105.9, 89.5, 60.6, 56.6; IR (KBr) v: 3331, 3003, 2939, 1686, 1593, 1552, 1502, 1124 cm^{-1}; MS (ESI) m/z: 474.1632 [M + H]$^+$.

[In 《Chinese journal of chemistry》 2006, 24: 1086 – 1094]

References

1   De clercq E. Rev Med Virol, 2000, 10: 255

2   Neamati N. Expert Opin Invest Drugs,2001,10:281

3   Neamati N. Expert Opin Ther Pat,2002,12:709

4   Dayam R. , Neamati N. Curr Pharm Des, 2003, 9: 1789

5   Mazumder A. , Wang S, Neamari N, et al. J Med Chem, 1996, 39: 2472

6   Robinson W E Jr, Cordeiro M, Abdel – Malek, S, et al. Mol phar macol 1996, 50: 846

7   Kai Z, Mara L C. , Jocelyn A, et al. J Virol, 1999, 73: 3309

8   Robinson W E Jr, Reinecke M G, Abdel – Malek S. Prow Natl Acad Sci U S A, 1996, 93: 6326

9   Hwang D J, Kim S N, Choi J H, et al. Bioorg Med Chem, 2001, 9: 1429

10 Herr R J. Bioorg Med Chem, 2002, 10: 3379

11 Goldgur Y, Craigie R. , Cohen G H, et al. Proc Natl Acaa Sci U S A, 1999, 23, 13040

12 Yoshiinaga T, Fujishita A S T, Fujiwara T. 9th Conference on Retroviru and Opportunistic Infections, Seaattle, 2002

13 Billich A. Curr Opin Investig Drugs, 2003, 4: 206

14 Hazuda D J, Steve D Y, Guare J P, et al. Science, 2004, 305: 528

15 (a)Grobler J A,Stillmock K,Hu B,et al. Proc Natl Acad Sci U S A,2002,99:6661

    (b)Hazida D J,Felock P,Witmar M,et al. Science, 2000,287:646

16 Drysdale M J, Hind S L, Jansen M, et al. J Med Chem, 2000, 43: 123

17 Wai J S, Egbertson M S, Payne L S, et al. J Med Chem, 2000, 43: 4923

18 Pais G C, Zhang X, Marchand C, et al, Jr J Med Chem, 2002, 45: 3184

19 Cedric M, Fabrice B, Philippe C. Tetrahedron, 2004, 60: 6479

20 Xu Y S, Zeng C C, Li X M, et al, Acta crystallogr Sect E, 2005, E61: 03802

21 Jiang X H, Song L D, Long Y Q J Org Chem, 2003, 68: 7555

22 Rice K C J. Med Chem, 1977, 20: 164

23 Dupont R,Cotelle P. Tertrahedron,2001,57:5585

24 Dupont R, Cotelle P. Tetrahedron Lett, 2001, 42: 597

25 Li W Z. ,Yun L,Qin B Y. Chin J Med Chem. 1998, 8:141

26 (a)Viktor I S,Yanina V B,Kappe C O,et al. Heterocycles,200,52:1411

    (b)Markees D G. J Heterocycl Chem,1989,26:29

27 (a)Elena V P,Olga P K,Elena A G,et al. J Fluorine Chem. 2003,121:201

    (b)Saloutin V I,Perevalov S G,J Fluorine Chem, 1999,96:87

28 Ferro S, Rao A, Zappala M, et al. Heterocycles, 2004, 63: 2727

29 Greenhill J V. Chem Soc Rev, 1997, 6: 280

30 Liu Z Q, Shi M, Li F Y, et al, Org Lett, 2005, 7: 5481

31 Meng L Z, Mei G X, He Y B, et al, Acta Chim Sinica, 2005, 63: 416（in Chinese）